AF856217

Thinking Mathematically

Second Custom Edition for Columbia College Chicago

Robert F. Blitzer

| PEARSON COLLECTIONS |

PEARSON

Cover Art: Courtesy of Photodisc/Getty Images

Pearson Learning Solutions, 501 Boylston Street, Suite 900, Boston, MA 02116

A Pearson Education Company
www.pearsoned.com

PEARSON

ISBN 10: 1323560912

ISBN 13: 9781323560914

Printed in the USA

20 2019

Table of Contents

Contents

About the Author

Bob Blitzer is a native of Manhattan and received a Bachelor of Arts degree with dual majors in mathematics and psychology (minor: English literature) from the City College of New York. His unusual combination of academic interests led him toward a Master of Arts in mathematics from the University of Miami and a doctorate in behavioral sciences from Nova University. Bob's love for teaching mathematics was nourished for nearly 30 years at Miami Dade College, where he received numerous teaching awards, including Innovator of the Year from the League for Innovations in the Community College and an endowed chair based on excellence in the classroom. In addition to *Thinking Mathematically*, Bob has written textbooks covering introductory algebra, intermediate algebra, college algebra, algebra and trigonometry, precalculus, trigonometry, and liberal arts mathematics for high school students, all published by Pearson. When not secluded in his Northern California writer's cabin, Bob can be found hiking the beaches and trails of Point Reyes National Seashore, and tending to the chores required by his beloved entourage of horses, chickens, and irritable roosters.

Bob and his buddy Casper Cockatoo

Preface

Thinking Mathematically, Sixth Edition provides a general survey of mathematical topics that are useful in our contemporary world. My primary purpose in writing the book was to show students how mathematics can be applied to their lives in interesting, enjoyable, and meaningful ways. The book's variety of topics and flexibility of sequence make it appropriate for a one- or two-term course in liberal arts mathematics, quantitative reasoning, finite mathematics, as well as for courses specifically designed to meet state-mandated requirements in mathematics.

I wrote the book to help diverse students, with different backgrounds and career plans, to succeed. **Thinking Mathematically**, **Sixth Edition**, has four major goals:

1. To help students acquire knowledge of fundamental mathematics.
2. To show students how mathematics can solve authentic problems that apply to their lives.
3. To enable students to understand and reason with quantitative issues and mathematical ideas they are likely to encounter in college, career, and life.
4. To enable students to develop problem-solving skills, while fostering critical thinking, within an interesting setting.

One major obstacle in the way of achieving these goals is the fact that very few students actually read their textbook. This has been a regular source of frustration for me and my colleagues in the classroom. Anecdotal evidence gathered over years highlights two basic reasons why students do not take advantage of their textbook:

"I'll never use this information."
"I can't follow the explanations."

I've written every page of the Sixth Edition with the intent of eliminating these two objections. The ideas and tools I've used to do so are described for the student in "A Brief Guide to Getting the Most from This Book," which appears inside the front cover.

What's New in the Sixth Edition?

- **New Applications and Real-World Data.** I'm on a constant search for real-world data that can be used to illustrate unique mathematical applications. I researched hundreds of books, magazines, newspapers, almanacs, and online sites to prepare the Sixth Edition. This edition contains 366 worked-out examples and application exercises based on new data sets.
- **Concept and Vocabulary Checks.** The Sixth Edition contains 653 new short-answer exercises, mainly fill-in-the-blank and true/false items, that assess students' understanding of the definitions and concepts presented in each section. The Concept and Vocabulary Checks appear as separate features preceding the Exercise Sets.
- **Great Question!** This feature takes the content of each Study Tip in the Fifth Edition and presents it in the context of a student question. Answers to questions offer suggestions for problem solving, point out common errors to avoid, and provide informal hints and suggestions. 'Great Question!' should draw students' attention and curiosity more than the 'Study Tips.' As a secondary benefit, this new feature should help students not to feel anxious or threatened when asking questions in class. The feature is extended to the learning objectives at the beginning of each section, which are now framed in the context of a student question: What am I supposed to learn?
- **New Blitzer Bonuses.** The Sixth Edition contains a variety of new but optional enrichment essays. There are more new Blitzer Bonuses in this edition than in any previous revision of *Thinking Mathematically*. These include "Are You Smart Enough to Work at Google?" (Section 1.1), "Science and Math Tattoos" (Section 2.1), "NUMB3RS: Solving Crime with Mathematics" (Section 5.3), "Using Algebra to Measure Blood-Alcohol Concentration" (Section 6.1), "Testing Your Financial Literacy" (Section 8.1), "The Bottom Line on Investments" (Section 8.5), "Financing Your Car" (Section 8.6), "Reducing Rental Costs" (Section 8.7), "College Students and Credit Cards" (Section 8.8), "Big Fears and Their Odds" (Section 11.6), "Using Means to Compare How the U.S. Measures Up" (Section 12.2), "The 2012 Presidential Election" (Section 13.4), and "A Family Tree: The Sopranos" (Section 14.4).
- **Brief Reviews.** The book's Brief Review boxes summarize mathematical skills that students should have learned previously, but which many students still need to review. This feature appears whenever a particular skill is first needed and eliminates the need to reteach that skill.
- **Sample Homework Assignments.** Within each Exercise Set, I have chosen odd-numbered problems from the Practice Exercises and the Application Exercises that can serve as sample homework assignments. These are indicated by a red underline in the Annotated Instructor's Edition. Based on the goals and objectives of your course, you may wish to enrich each sample homework assignment with additional exercises from the other categories in the Exercise Set.

- **Learning Guide.** This study aid is organized by objective and provides support for note-taking, practice, and video review. The Learning Guide is available as PDFs and customizable Word files in MyMathLab. They can also be packaged with the textbook and MyMathLab access code.
- **Thinking Mathematically with Integrated Review.** For courses where students do require more extensive prerequisite review, we have created a version of the *Thinking Mathematically* MyMathLab course called *Thinking Mathematically with Integrated Review* that includes just-in-time review of select topics where appropriate. Students are asked to check their skills with an assignment at the start of each chapter to assess their understanding of requisite, developmental material. For those students who do require further review, resources include the eText, short objective-based videos, Integrated Review Worksheets, and Integrated Review Homework to help provide students with a solid foundation on the review topics needed for their *Thinking Mathematically* course.

What Content and Organizational Changes Have Been Made to the Sixth Edition?

- **Section 3.3 (Truth Tables for Negation, Conjunction, and Disjunction)** opens with a new application on the distribution of looks for U.S. men and women. The application reappears in an example on determining the truth value of a compound statement.
- **Section 4.2 (Number Bases in Positional Systems)** contains a new discussion (within the context of the Great Question! feature) on the use of octal and hexadecimal systems by computer programmers.
- **Section 5.5 (Real Numbers and Their Properties; Clock Addition)** integrates material from Chapter 13 (Mathematical Systems) of the Fifth Edition by applying properties of real numbers to clock arithmetic and discussing related symmetries.
- **Section 6.3 (Applications of Linear Equations)** has a new example on the starting salaries for college graduates with undergraduate degrees. The example from the Fifth Edition on comparing long-distance telephone plans has been replaced by an example on choosing between texting plans.
- **Chapter 8 (Personal Finance)** has been renamed and expanded to include relevant information, both mathematical and non-mathematical, to help students manage their finances. The Sixth Edition contains separate sections on income tax, cars, home ownership, and credit cards.
- **Section 8.2 (Income Tax)** expands the discussion of income tax from the Fifth Edition, giving a broader understanding of the terms and complexities involved in calculating taxes. Included in this new section are discussions of Social Security and Medicare (FICA), as well as an example related to taxes for working students.
- **Section 8.6 (Cars)** is a new section that uses the mathematics of financing a car to develop the loan payment formula for fixed installment loans. The section includes new objectives on the pros and cons of leasing versus buying a car, understanding the different kinds of car insurance, comparing monthly payments on new and used cars, and solving problems related to owning and operating a car.
- **Section 8.7 (The Cost of Home Ownership)** is a new section that applies the loan payment formula that was developed for cars in Section 8.6 to mortgages. The section includes new objectives on solving problems involving what one can afford to spend for a mortgage, and understanding the pros and cons of renting versus buying.
- **Section 8.8 (Credit Cards)** is a new section devoted entirely to credit cards. Objectives unique to the Sixth Edition include understanding the pros and cons of using credit cards, understanding the difference between credit cards and debit cards, knowing what is contained in a credit report, and understanding credit scores as measures of creditworthiness.
- **Section 9.2 (Measuring Area and Volume)** and **Section 9.3 (Measuring Weight and Temperature)** incorporate new discussions on measuring dosages of medication, including dosages based on weight in the metric system.
- **Section 12.4 (The Normal Distribution)** uses the activities U.S. adults say they dread to illustrate a poll's margin of error.

What Familiar Features Have Been Retained in the Sixth Edition?

- **Chapter-Opening and Section-Opening Scenarios.** Every chapter and every section open with a scenario presenting a unique application of mathematics in students' lives outside the classroom. These scenarios are revisited in the course of the chapter or section in an example, discussion, or exercise. The often humorous tone of these openers is intended to help fearful and reluctant students overcome their negative perceptions about math. A new feature in the Sixth Edition, "Here's Where You'll Find These Applications," is included with each chapter opener.
- **Detailed Worked-Out Examples.** Each example is titled, making the purpose of the example clear. Examples are clearly written and provide students with detailed step-by-step solutions. No steps are omitted and each step is thoroughly explained to the right of the mathematics.
- **Explanatory Voice Balloons.** Voice Balloons are used in a variety of ways to demystify mathematics. They translate mathematical language into everyday English, help clarify problem-solving procedures, present alternative ways of understanding concepts, and connect problem solving to concepts students have already learned.

- **Check Point Examples.** Each example is followed by a similar matched problem, called a Check Point, offering students the opportunity to test for conceptual understanding by working a similar exercise. The answers to the Check Points are provided in the answer section in the back of the book. Worked-out video solutions are in the MyMathLab courses or on YouTube.
- **Extensive and Varied Exercise Sets.** An abundant collection of exercises is included in an Exercise Set at the end of each section. Exercises are organized within seven category types: Practice Exercises, Practice Plus Exercises, Application Exercises, Writing in Mathematics, Critical Thinking Exercises, Technology Exercises, and Group Exercises.
- **Practice Plus Problems.** This category of exercises contains practice problems that often require students to combine several skills or concepts, providing instructors the option of creating assignments that take Practice Exercises to a more challenging level.
- **Section Objectives (What Am I Supposed to Learn?).** Learning objectives are clearly stated at the beginning of each section. These objectives help students recognize and focus on the section's most important ideas. The objectives are restated in the margin at their point of use.
- **Chapter Summaries.** Each chapter contains a review chart that summarizes the definitions and concepts in every section of the chapter. Examples that illustrate these key concepts are also referenced in the chart.
- **End-of-Chapter Materials.** A comprehensive collection of review exercises for each of the chapter's sections follows the Summary. This is followed by a Chapter Test that enables students to test their understanding of the material covered in the chapter. Worked-out video solutions are available for every Chapter Test Prep problem in the MyMathLab course or on YouTube.

I hope that my love for learning, as well as my respect for the diversity of students I have taught and learned from over the years, is apparent throughout this new edition. By connecting mathematics to the whole spectrum of learning, it is my intent to show students that their world is profoundly mathematical, and indeed, π is in the sky.

Robert Blitzer

Dynamic Resources

MyMathLab (access code required)

MyMathLab from Pearson is the world's leading online resource in mathematics, integrating interactive homework, assessment, and media in a flexible, easy to use format. It provides **engaging experiences** that personalize, stimulate, and measure learning for each student. And, it comes from an **experienced partner** with educational expertise and an eye on the future.

To learn more about how MyMathLab combines proven learning applications with powerful assessment, visit **www.mymathlab.com** or contact your Pearson representative.

Blitzer's MyMathLab course provides access to innovative and engaging study solutions to increase student success.

Ready to Go Course

This new MyMathLab course option provides students with all the same great MyMathLab features, but makes it easier for instructors to get started, with premade and pre-assigned assignments.

Integrated Review Course Solution

For courses where students require more extensive remediation, we have created the Integrated Review solution that includes just-in-time review of select topics where appropriate. For students who require this review, resources include the eText, short objective-based videos, Integrated Review Worksheets, and Integrated Review Homework to help provide students with a solid foundation for success in their *Thinking Mathematically* course.

Getting Ready

Getting Ready exercises are now available for online review in the Standard and Ready to Go MyMathLab courses. These skill review quizzes test on prerequisite knowledge, allowing students to refresh forgotten concepts.

MathTalk Videos

These fun, application-based videos connect the math in Blitzer to the real world. These instructional videos have a sense of humor while demonstrating how everyday life is full of math applications.

Concept and Vocabulary Check

New Concept and Vocabulary Check exercises provide a quick check of understanding of concepts. Assignable in MyMathLab, these also test for reading comprehension before moving onto the homework exercise sets.

Check Point Videos

These videos show instructors working out every Check Point problem in the text to ensure understanding. New to the Sixth Edition MyMathLab course are assignable Check Point exercises that correspond with each video, ensuring that students watched the video and understood the concepts presented.

Chapter Test Prep Videos

Students can watch instructors work through step-by-step solutions to all the Chapter Test Prep exercises from the textbook. These are available in MyMathLab and on YouTube.

Student Success Module

A new Student Success Module supports students toward continued success in college. This module provides tutorials and guidance on topics such as transition to college, online learning, time management, and more. Additionally, there is content to help students develop professional skills such as resume development and interview preparation.

Instructor Resources

Annotated Instructor's Edition

ISBN-13: 978-0321-91487-3
ISBN-10: 0-321-91487-2

Additional resources can be downloaded from www.pearsonhighered.com:

TestGen

Powerpoint Lecture Slides

Instructor's Solutions Manual

Instructor's Testing Manual

Student Resources

Learning Guide – NEW!

This study aid is organized by objective and provides support for note-taking, practice, and video review. The Learning Guide is available as PDFs and customizable Word files in MyMathLab. It can also be packaged with the textbook and MyMathLab access code.

Student's Solutions Manual

This manual contains fully worked solutions to odd-numbered exercises and all Check Points.

To the Student

The bar graph shows some of the qualities that students say make a great teacher. It was my goal to incorporate each of these qualities throughout the pages of this book to help you gain control over the part of your life that involves numbers and mathematical ideas.

Explains Things Clearly

I understand that your primary purpose in reading *Thinking Mathematically* is to acquire a solid understanding of the required topics in your liberal arts math course. In order to achieve this goal, I've carefully explained each topic. Important definitions and procedures are set off in boxes, and worked-out examples that present solutions in a step-by-step manner appear in every section. Each example is followed by a similar matched problem, called a Check Point, for you to try so that you can actively participate in the learning process as you read the book. (Answers to all Check Points appear in the back of the book and video solutions are in MyMathLab.)

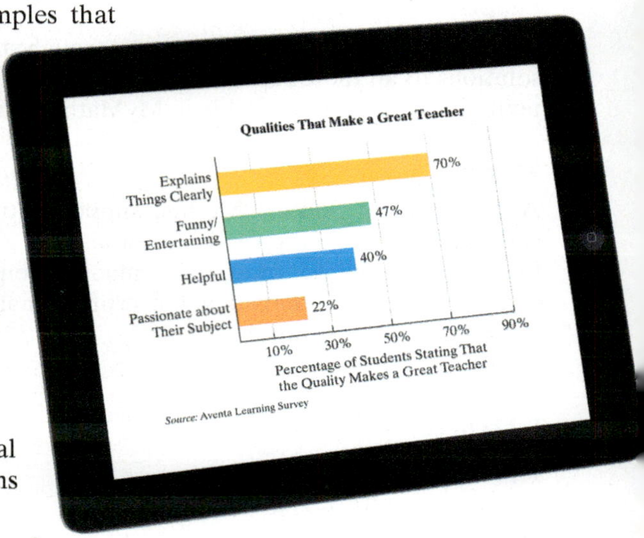

Funny & Entertaining

Who says that a math textbook can't be entertaining? From our quirky cover to the photos in the chapter and section openers, prepare to expect the unexpected. I hope some of the book's enrichment essays, called Blitzer Bonuses, will put a smile on your face from time to time.

Helpful

I designed the book's features to help you acquire knowledge of fundamental mathematics, as well as to show you how math can solve authentic problems that apply to your life. These helpful features include

- **Explanatory Voice Balloons:** Voice balloons are used in a variety of ways to make math less intimidating. They translate mathematical language into everyday English, help clarify problem-solving procedures, present alternative ways of understanding concepts, and connect new concepts to concepts you have already learned.
- **Great Question!:** The book's Great Question! boxes are based on questions students ask in class. The answers to these questions give suggestions for problem solving, point out common errors to avoid, and provide informal hints and suggestions.
- **Chapter Summaries:** Each chapter contains a review chart that summarizes the definitions and concepts in every section of the chapter. Examples from the chapter that illustrate these key concepts are also referenced in the chart. Review these summaries and you'll know the most important material in the chapter!

Passionate about the Subject

I passionately believe that no other discipline comes close to math in offering a more extensive set of tools for application and development of your mind. I wrote the book in Point Reyes National Seashore, 40 miles north of San Francisco. The park consists of 75,000 acres with miles of pristine surf-washed beaches, forested ridges, and bays bordered by white cliffs. It was my hope to convey the beauty and excitement of mathematics using nature's unspoiled beauty as a source of inspiration and creativity. Enjoy the pages that follow as you empower yourself with the mathematics needed to succeed in college, your career, and in your life.

Regards,

Bob

Robert Blitzer

Acknowledgments

An enormous benefit of authoring a successful textbook is the broad-based feedback I receive from students, dedicated users, and reviewers. Every change to this edition is the result of their thoughtful comments and suggestions. I would like to express my appreciation to all the reviewers, whose collective insights form the backbone of this revision. In particular, I would like to thank the following people for reviewing *Thinking Mathematically* for this Sixth Edition.

Deana Alexander, *Indiana University—Purdue University*
Nina Bohrod, *Anoka-Ramsey Community College*
Kim Caldwell, *Volunteer State Community College*
Elizabeth T. Dameron, *Tallahassee Community College*
Cornell Grant, *Georgia Piedmont Technical College*
Elizabeth Kiedaisch, *College of DuPage*
Susan Knights, *College of Western Idaho*
Dennine LaRue, *Farmont State University*
Carla A. Monticelli, *Camden County College*
Cindy Vanderlaan, *Indiana Purdue University—Fort Wayne*
Alexandra Verkhovtseva, *Anoka-Ramsey Community College*

Each reviewer from every edition has contributed to the success of this book and I would like to also continue to offer my thanks to them.

David Allen, *Iona College*; Carl P. Anthony, *Holy Family University*; Laurel Berry, *Bryant and Stratton College*; Kris Bowers, *Florida State University*; Gerard Buskes, *University of Mississippi*; Fred Butler, *West Virginia University*; Jimmy Chang, *St. Petersburg College*; Jerry Chen, *Suffolk County Community College*; Ivette Chuca, *El Paso Community College*; David Cochener, *Austin Peay State University*; Stephanie Costa, *Rhode Island College*; Tristen Denley, *University of Mississippi*; Suzanne Feldberg, *Nassau Community College*; Margaret Finster, *Erie Community College*; Maryanne Frabotta, *Community Campus of Beaver County*; Lyn Geisler III, *Randolph-Macon College*; Patricia G. Granfield, *George Mason University*; Dale Grussing, *Miami Dade College*; Cindy Gubitose, *Southern Connecticut State University*; Virginia Harder, *College at Oneonta*; Joseph Lloyd Harris, *Gulf Coast Community College*; Julia Hassett, *Oakton Community College*; Sonja Hensler, *St. Petersburg College*; James Henson, *Edinboro University of Pennsylvania*; Larry Hoehn, *Austin Peay State University*; Diane R. Hollister, *Reading Area Community College*; Kalynda Holton, *Tallahassee Community College*; Alec Ingraham, *New Hampshire College*; Linda Kuroski, *Erie Community College—City Campus*; Jamie Langille, *University of Nevada, Las Vegas*; Veronique Lanuqueitte, *St. Petersburg College*; Julia Ledet, *Louisiana State University*; Mitzi Logan, *Pitt Community College*; Dmitri Logvnenko, *Phoenix College*; Linda Lohman, *Jefferson Community College*; Richard J. Marchand, *Slippery Rock University*; Mike Marcozzi, *University of Nevada, Las Vegas*; Diana Martelly, *Miami Dade College*; Jim Matovina, *Community College of Southern Nevada*; Erik Matsuoka, *Leeward Community College*; Marcel Maupin, *Oklahoma State University*; Carrie McCammon, *Ivy Tech Community College*; Diana McCammon, *Delgado Community College*; Mex McKinley, *Florida Keys Community College*; Taranna Amani Miller, *Indian River State College*; Paul Mosbos, *State University of New York—Cortland*; Tammy Muhs, *University of Central Florida*; Cornelius Nelan, *Quinnipiac University*; Lawrence S. Orilia, *Nassau Community College*; Richard F. Patterson, *University of North Florida*; Frank Pecchioni, *Jefferson Community College*; Stan Perrine, *Charleston Southern University*; Anthony Pettofrezzo, *University of Central Florida*; Val Pinciu, *Southern Connecticut State University*; Evelyn Pupplo-Cody, *Marshall University*; Virginia S. Powell, *University of Louisiana at Monroe*; Kim Query, *Lindenwood College*; Anne Quinn, *Edinboro University of Pennsylvania*; Bill Quinn, *Frederick Community College*; Sharonda Ragland, *ECPI College of Technology*; Shawn Robinson, *Valencia Community College*; Gary Russell, *Brevard Community College*; Mary Lee Seitz, *Erie Community College*; Laurie A. Stahl, *State University of New York—Fredonia*; Abolhassan Taghavy, *Richard J. Daley College & Chicago State University*; Diane Tandy, *New Hampshire Technical Institute*; Ann Thrower, *Kilgore College*; Mike Tomme, *Community College of Southern Nevada*; Sherry Tornwall, *University of Florida*; Linda Tully, *University of Pittsburgh at Johnstown*; Christopher Scott Vaughen, *Miami Dade College*; Bill Vaughters, *Valencia Community College*; Karen Villareal, *University of New Orleans*; Don Warren, *Edison Community College*; Shirley Wilson, *North Central College*; James Wooland, *Florida State University*; Clifton E. Webb, *Virginia Union University*; Cindy Zarske, *Fullerton College*; Marilyn Zopp, *McHenry County College*

Additional acknowledgments are extended to Brad Davis, for preparing the answer section and annotated answers, serving as accuracy checker, and co-writing the new Learning Guide; Diane Hollister for co-writing the new Learning Guide; Dan Miller and Kelly Barber, for preparing the solutions manuals; Tamsen Herrick for accuracy checking the solutions manuals; the cMPreparé formatting team for the book's brilliant paging; Brian Morris and Kevin Morris at Scientific Illustrators, for superbly illustrating the book; and Rebecca Dunn, project manager, and Kathleen Manley, production editor, whose collective talents kept every aspect of this complex project moving through its many stages, and Kerri Consalvo who helped keep everything organized.

I would like to thank my editor at Pearson, Marnie Greenhut, and editorial assistant, Chris Tominich, who guided and coordinated the book from manuscript through production. Thanks to Beth Paquin for the great cover and interior design. Finally, thanks to marketing manager Alicia Frankel and marketing assistant Brooke Smith for your innovative marketing efforts, and to the entire Pearson sales force, for your confidence and enthusiasm about the book.

Robert Blitzer

1

Problem Solving and Critical Thinking

HOW WOULD YOUR LIFESTYLE CHANGE IF A GALLON OF GAS COST $9.15? OR IF THE PRICE OF A staple such as milk was $15? That's how much those products would cost if their prices had increased at the same rate college tuition has increased since 1980.

TUITION AND FEES AT FOUR-YEAR COLLEGES

	School Year Ending 2000	School Year Ending 2010
Public	$3362	$7020
Private	$15,518	$26,273

Source: The College Board

If these trends continue, what can we expect for the rest of this decade and beyond? We can answer this question by using estimation techniques that allow us to represent the data mathematically. With such representations, called *mathematical models*, we can gain insights and predict what might occur in the future on a variety of issues, ranging from college costs to global warming.

Here's where you'll find these applications:

Mathematical models involving college costs are developed in Example 8 and Check Point 8 of Section 1.2. In Exercises 51 and 52 in Exercise Set 1.2, you will approach our climate crisis mathematically by developing models for data related to global warming.

1.1

WHAT AM I SUPPOSED TO LEARN?

After you have read this section, you should be able to:

1 Understand and use inductive reasoning.

2 Understand and use deductive reasoning.

Inductive and Deductive Reasoning

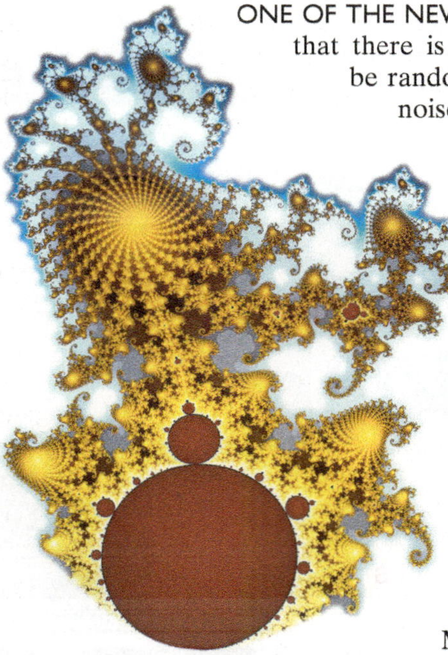

A magnification of the Mandelbrot set
Richard F. Voss

ONE OF THE NEWER FRONTIERS OF MATHEMATICS SUGGESTS that there is an underlying order in things that appear to be random, such as the hiss and crackle of background noises as you tune a radio. Irregularities in the heartbeat, some of them severe enough to cause a heart attack, or irregularities in our sleeping patterns, such as insomnia, are examples of chaotic behavior. Chaos in the mathematical sense does not mean a complete lack of form or arrangement. In mathematics, chaos is used to describe something that appears to be random but is not actually random. The patterns of chaos appear in images like the one shown on the left, called the Mandelbrot set. Magnified portions of this image yield repetitions of the original structure, as well as new and unexpected patterns. The Mandelbrot set transforms the hidden structure of chaotic events into a source of wonder and inspiration.

Many people associate mathematics with tedious computation, meaningless algebraic procedures, and intimidating sets of equations. The truth is that mathematics is the most powerful means we have of exploring our world and describing how it works. The word *mathematics* comes from the Greek word *mathematikos*, which means "inclined to learn." To be mathematical literally means to be inquisitive, open-minded, and interested in a lifetime of pursuing knowledge!

Mathematics and Your Life

A major goal of this book is to show you how mathematics can be applied to your life in interesting, enjoyable, and meaningful ways. The ability to think mathematically and reason with quantitative issues will help you so that you can:

- order and arrange your world by using sets to sort and classify information (Chapter 2, Set Theory)
- use logic to evaluate the arguments of others and become a more effective advocate for your own beliefs (Chapter 3, Logic)
- understand the relationship between cutting-edge technology and ancient systems of number representation (Chapter 4, Number Representation and Calculation)
- put the numbers you encounter in the news, from contemplating the national debt to grasping just how colossal $1 trillion actually is, into perspective (Chapter 5, Number Theory and the Real Number System)
- use mathematical models to gain insights into a variety of issues, including the positive benefits that humor and laughter can have on your life (Chapter 6, Algebra: Equations and Inequalities)

- use basic ideas about savings, loans, and investments to achieve your financial goals (Chapter 8, Personal Finance)
- use geometry to study the shape of your world, enhancing your appreciation of nature's patterns and beauty (Chapter 10, Geometry)
- develop an understanding of the fundamentals of statistics and how these numbers are used to make decisions (Chapter 12, Statistics)
- understand the mathematical paradoxes of voting in a democracy, increasing your ability to function as a more fully aware citizen (Chapter 13, Voting and Apportionment)
- use graph theory to examine how mathematics is used to solve problems in the business world (Chapter 14, Graph Theory)

"It is better to take what may seem to be too much math rather than too little. Career plans change, and one of the biggest roadblocks in undertaking new educational or training goals is poor preparation in mathematics. Furthermore, not only do people qualify for more jobs with more math, they are also better able to perform their jobs."
—Occupational Outlook Quarterly

Mathematics and Your Career

Generally speaking, the income of an occupation is related to the amount of education required. This, in turn, is usually related to the skill level required in language and mathematics. With our increasing reliance on technology, the more mathematics you know, the more career choices you will have.

Mathematics and Your World

Mathematics is a science that helps us recognize, classify, and explore the hidden patterns of our universe. Focusing on areas as different as planetary motion, animal markings, shapes of viruses, aerodynamics of figure skaters, and the very origin of the universe, mathematics is the most powerful tool available for revealing the underlying structure of our world. Within the last 30 years, mathematicians have even found order in chaotic events such as the uncontrolled storm of noise in the nerve cells of the brain during an epileptic seizure.

Understand and use inductive reasoning.

Inductive Reasoning

Mathematics involves the study of patterns. In everyday life, we frequently rely on patterns and routines to draw conclusions. Here is an example:

> The last six times I went to the beach, the traffic was light on Wednesdays and heavy on Sundays. My conclusion is that weekdays have lighter traffic than weekends.

This type of reasoning process is referred to as *inductive reasoning*, or *induction*.

INDUCTIVE REASONING

Inductive reasoning is the process of arriving at a general conclusion based on observations of specific examples.

Although inductive reasoning is a powerful method of drawing conclusions, we can never be absolutely certain that these conclusions are true. For this reason, the conclusions are called **conjectures**, **hypotheses**, or educated guesses. A strong inductive argument does not guarantee the truth of the conclusion, but rather provides strong support for the conclusion. If there is just one case for which the conjecture does not hold, then the conjecture is false. Such a case is called a **counterexample**.

EXAMPLE 1 *Finding a Counterexample*

The ten symbols that we use to write numbers, namely 0, 1, 2, 3, 4, 5, 6, 7, 8, and 9, are called **digits**. In each example shown below, the sum of two two-digit numbers is a three-digit number.

$$\begin{array}{r} 47 \\ +73 \\ \hline 120 \end{array} \qquad \begin{array}{r} 56 \\ +46 \\ \hline 102 \end{array}$$

Two-digit numbers

Three-digit sums

Is the sum of two two-digit numbers always a three-digit number? Find a counterexample to show that the statement

The sum of two two-digit numbers is a three-digit number

is false.

SOLUTION

There are many counterexamples, but we need to find only one. Here is an example that makes the statement false:

$$\begin{array}{r} 56 \\ +43 \\ \hline 99 \end{array}$$

Two-digit numbers

This is a two-digit sum, not a three-digit sum.

This example is a counterexample that shows the statement

The sum of two two-digit numbers is a three-digit number

is false.

CHECK POINT 1 Find a counterexample to show that the statement

The product of two two-digit numbers is a three-digit number

is false.

GREAT QUESTION!

Why is it so important to work each of the book's Check Points?

You learn best by doing. Do not simply look at the worked examples and conclude that you know how to solve them. To be sure you understand the worked examples, try each Check Point. Check your answer in the answer section before continuing your reading. Expect to read this book with pencil and paper handy to work the Check Points.

Here are two examples of inductive reasoning:

- **Strong Inductive Argument** In a random sample of 380,000 freshmen at 722 four-year colleges, 25% said they frequently came to class without completing readings or assignments (*Source*: National Survey of Student Engagement). We can conclude that there is a 95% probability that between 24.84% and 25.15% of all college freshmen frequently come to class unprepared.

In Chapter 12, you will learn how observations from a randomly selected group, one in which each member of the population has an equal chance of being selected, can provide probabilities of what is true about an entire population.

- **Weak Inductive Argument** Neither my dad nor my boyfriend has ever cried in front of me. Therefore, men have difficulty expressing their feelings.

When generalizing from observations about your own circumstances and experiences, avoid jumping to hasty conclusions based on a few observations. Psychologists theorize that we do this - that is, place everyone in a neat category - to feel more secure about ourselves and our relationships to others.

Inductive reasoning is extremely important to mathematicians. Discovery in mathematics often begins with an examination of individual cases to reveal patterns about numbers.

EXAMPLE 2 *Using Inductive Reasoning*

Identify a pattern in each list of numbers. Then use this pattern to find the next number.

a. 3, 12, 21, 30, 39, _____

b. 3, 12, 48, 192, 768, _____

c. 3, 4, 6, 9, 13, 18, _____

d. 3, 6, 18, 36, 108, 216, _____

"For thousands of years, people have loved numbers and found patterns and structures among them. The allure of numbers is not limited to or driven by a desire to change the world in a practical way. When we observe how numbers are connected to one another, we are seeing the inner workings of a fundamental concept."

—Edward B. Burger and Michael Starbird, *Coincidences, Chaos, and All That Math Jazz*, W.W. Norton and Company, 2005

SOLUTION

a. Because 3, 12, 21, 30, 39, _____ is increasing relatively slowly, let's use addition as the basis for our individual observations.

3, 12, 21, 30, 39, _____

$3 + 9 = 12$ $12 + 9 = 21$ $21 + 9 = 30$ $30 + 9 = 39$

Generalizing from these observations, we conclude that each number after the first is obtained by adding 9 to the previous number. Using this pattern, the next number is $39 + 9$, or 48.

b. Because 3, 12, 48, 192, 768, _____ is increasing relatively rapidly, let's use multiplication as the basis for our individual observations.

3, 12, 48, 192, 768, _____

$3 \times 4 = 12$ $12 \times 4 = 48$ $48 \times 4 = 192$ $192 \times 4 = 768$

Generalizing from these observations, we conclude that each number after the first is obtained by multiplying the previous number by 4. Using this pattern, the next number is 768×4, or 3072.

c. Because 3, 4, 6, 9, 13, 18, _____ is increasing relatively slowly, let's use addition as the basis for our individual observations.

3, 4, 6, 9, 13, 18, _____

$3 + 1 = 4$ $4 + 2 = 6$ $6 + 3 = 9$ $9 + 4 = 13$ $13 + 5 = 18$

Generalizing from these observations, we conclude that each number after the first is obtained by adding a counting number to the previous number. The additions begin with 1 and continue through each successive counting number. Using this pattern, the next number is $18 + 6$, or 24.

d. Because 3, 6, 18, 36, 108, 216, _____ is increasing relatively rapidly, let's use multiplication as the basis for our individual observations.

3, 6, 18, 36, 108, 216, _____

$3 \times 2 = 6$ $6 \times 3 = 18$ $18 \times 2 = 36$ $36 \times 3 = 108$ $108 \times 2 = 216$

Generalizing from these observations, we conclude that each number after the first is obtained by multiplying the previous number by 2 or by 3. The multiplications begin with 2 and then alternate, multiplying by 2, then 3, then 2, then 3, and so on. Using this pattern, the next number is 216×3, or 648.

CHECK POINT 2 Identify a pattern in each list of numbers. Then use this pattern to find the next number.

a. 3, 9, 15, 21, 27, _____

b. 2, 10, 50, 250, _____

c. 3, 6, 18, 72, 144, 432, 1728, _____

d. 1, 9, 17, 3, 11, 19, 5, 13, 21, _____

In our next example, the patterns are a bit more complex than the additions and multiplications we encountered in Example 2.

EXAMPLE 3 *Using Inductive Reasoning*

Identify a pattern in each list of numbers. Then use this pattern to find the next number.

a. 1, 1, 2, 3, 5, 8, 13, 21, _____ **b.** 23, 54, 95, 146, 117, 98, _____

SOLUTION

As this tree branches, the number of branches forms the Fibonacci sequence.

a. We begin with 1, 1, 2, 3, 5, 8, 13, 21. Starting with the third number in the list, let's form our observations by comparing each number with the two numbers that immediately precede it.

1, 1, 2, 3, 5, 8, 13, 21, _____

2	3	5	8	13	21
preceded by 1 and 1: $1 + 1 = 2$	preceded by 1 and 2: $1 + 2 = 3$	preceded by 2 and 3: $2 + 3 = 5$	preceded by 3 and 5: $3 + 5 = 8$	preceded by 5 and 8: $5 + 8 = 13$	preceded by 8 and 13: $8 + 13 = 21$

Generalizing from these observations, we conclude that the first two numbers are 1. Each number thereafter is the sum of the two preceding numbers. Using this pattern, the next number is $13 + 21$, or 34. (The numbers 1, 1, 2, 3, 5, 8, 13, 21, and 34 are the first nine terms of the *Fibonacci sequence*, discussed in Chapter 5, Section 5.7.)

b. Now, we consider 23, 54, 95, 146, 117, 98. Let's use the digits that form each number as the basis for our individual observations. Focus on the sum of the digits, as well as the final digit increased by 1.

$2 + 3 = 5$ $5 + 4 = 9$ $9 + 5 = 14$ $1 + 4 + 6 = 11$ $1 + 1 + 7 = 9$

23, 54, 95, 146, 117, 98, _____

$3 + 1 = 4$ $4 + 1 = 5$ $5 + 1 = 6$ $6 + 1 = 7$ $7 + 1 = 8$

Generalizing from these observations, we conclude that for each number after the first, we obtain the first digit or the first two digits by adding the digits of the previous number. We obtain the last digit by adding 1 to the final digit of the preceding number. Applying this pattern to find the number that follows 98, the first two digits are $9 + 8$, or 17. The last digit is $8 + 1$, or 9. Thus, the next number in the list is 179.

GREAT QUESTION!

Can a list of numbers have more than one pattern?

Yes. Consider the illusion in **Figure 1.1**. This ambiguous figure contains two patterns, where it is not clear which pattern should predominate. Do you see a wine goblet or two faces looking at each other? Like this ambiguous figure, some lists of numbers can display more than one pattern, particularly if only a few numbers are given. Inductive reasoning can result in more than one probable next number in a list.

FIGURE 1.1

Example: 1, 2, 4, _____

Pattern: Each number after the first is obtained by multiplying the previous number by 2. The missing number is 4×2, or 8.

Pattern: Each number after the first is obtained by adding successive counting numbers, starting with 1, to the previous number. The second number is $1 + 1$, or 2. The third number is $2 + 2$, or 4. The missing number is $4 + 3$, or 7.

Inductive reasoning can also result in different patterns that produce the same probable next number in a list.

Example: 1, 4, 9, 16, 25, _____

Pattern: Start by adding 3 to the first number. Then add successive odd numbers, 5, 7, 9, and so on. The missing number is $25 + 11$, or 36.

Pattern: Each number is obtained by squaring its position in the list: The first number is $1^2 = 1 \times 1 = 1$, the second number is $2^2 = 2 \times 2 = 4$, the third number is $3^2 = 3 \times 3 = 9$, and so on. The missing sixth number is $6^2 = 6 \times 6$, or 36.

The numbers that we found in Examples 2 and 3 are probable numbers. Perhaps you found patterns other than the ones we pointed out that might have resulted in different answers.

CHECK POINT 3 Identify a pattern in each list of numbers. Then use this pattern to find the next number.

a. 1, 3, 4, 7, 11, 18, 29, 47, ____

b. 2, 3, 5, 9, 17, 33, 65, 129, ____

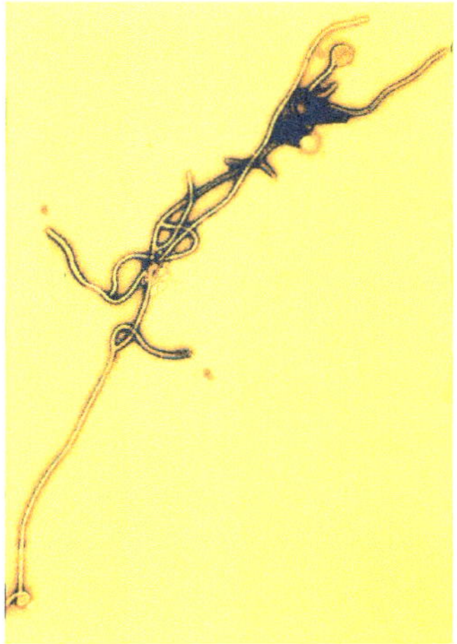

This electron microscope photograph shows the knotty shape of the Ebola virus.

Mathematics is more than recognizing number patterns. It is about the patterns that arise in the world around us. For example, by describing patterns formed by various kinds of knots, mathematicians are helping scientists investigate the knotty shapes and patterns of viruses. One of the weapons used against viruses is based on recognizing visual patterns in the possible ways that knots can be tied.

Our next example deals with recognizing visual patterns.

EXAMPLE 4 *Finding the Next Figure in a Visual Sequence*

Describe two patterns in this sequence of figures. Use the patterns to draw the next figure in the sequence.

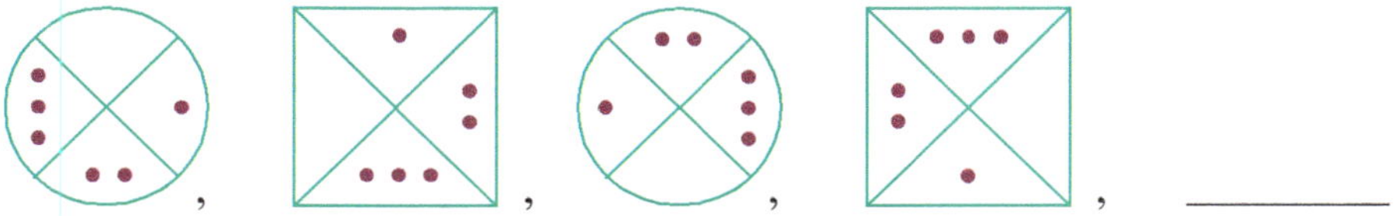

SOLUTION

The more obvious pattern is that the figures alternate between circles and squares. We conclude that the next figure will be a circle. We can identify the second pattern in the four regions containing no dots, one dot, two dots, and three dots. The dots are placed in order (no dots, one dot, two dots, three dots) in a clockwise direction. However, the entire pattern of the dots rotates counterclockwise as we follow the figures from left to right. This means that the next figure should be a circle with a single dot in the right-hand region, two dots in the bottom region, three dots in the left-hand region, and no dots in the top region.

The missing figure in the visual sequence on the previous page, a circle with a single dot in the right-hand region, two dots in the bottom region, three dots in the left-hand region, and no dots in the top region, is drawn in **Figure 1.2.**

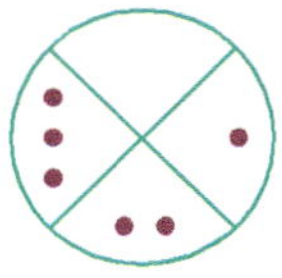

FIGURE 1.2

CHECK POINT 4 Describe two patterns in this sequence of figures. Use the patterns to draw the next figure in the sequence.

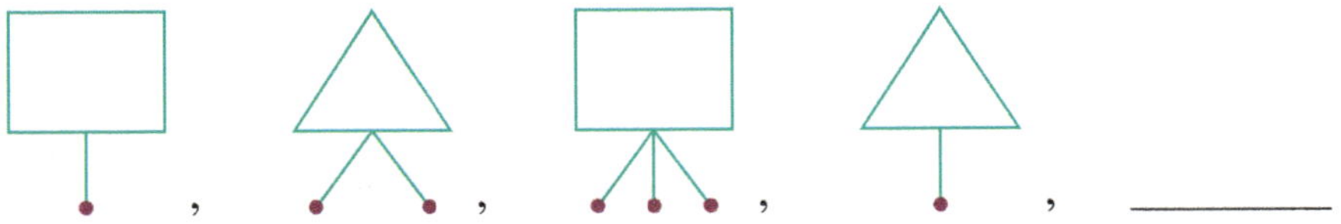

Blitzer Bonus

Are You Smart Enough to Work at Google?

In *Are You Smart Enough to Work at Google?* (Little, Brown, and Company, 2012), author William Poundstone guides readers through the surprising solutions to challenging job-interview questions. The book covers the importance of creative thinking in inductive reasoning, estimation, and problem solving. Best of all, Poundstone explains the answers.

Whether you're preparing for a job interview or simply want to increase your critical thinking skills, we highly recommend tackling the puzzles in *Are You Smart Enough to Work at Google?* Here is a sample of two of the book's problems that involve inductive reasoning. We've provided hints to help you recognize the pattern in each sequence. The answers appear in the answer section.

1. Determine the next entry in the sequence.
   SSS, SCC, C, SC, ___?___
   Hint: Think of the capital letters in the English alphabet. A is made up of three straight lines. B consists of one straight line and two curved lines. C is made up of one curved line.
2. Determine the next line in this sequence of digits.

1

1 1 — The first row contains one 1.

2 1 — The second row contains two 1s.

1 2 1 1 — The third row contains one 2 and one 1.

1 1 1 2 2 1

? ? ? ? ? ?

2 Understand and use deductive reasoning.

Deductive Reasoning

We use inductive reasoning in everyday life. Many of the conjectures that come from this kind of thinking seem highly likely, although we can never be absolutely certain that they are true. Another method of reasoning, called *deductive reasoning*, or *deduction*, can be used to prove that some conjectures are true.

DEDUCTIVE REASONING

Deductive reasoning is the process of proving a specific conclusion from one or more general statements. A conclusion that is proved to be true by deductive reasoning is called a **theorem**.

Deductive reasoning allows us to draw a specific conclusion from one or more general statements. On the next page are two examples of deductive reasoning. Notice that in both everyday situations, the general statement from which the conclusion is drawn is implied rather than directly stated.

Everyday Situation	Deductive Reasoning
One player to another in a Scrabble game: "You have to remove those five letters. You can't use TEXAS as a word."	• All proper names are prohibited in Scrabble. (general statement) TEXAS is a proper name. Therefore, TEXAS is prohibited in Scrabble. (conclusion)
Advice to college freshmen on choosing classes: "Never sign up for a 7 A.M. class. Yes, you did it in high school, but Mom was always there to keep waking you up, and if by some miracle you do make it to an early class, you will sleep through the lecture when you get there." (*Source: How to Survive Your Freshman Year*, Hundreds of Heads Books, 2004)	• All people need to sleep at 7 A.M. (general statement) You sign up for a class at 7 A.M. Therefore, you'll sleep through the lecture or not even make it to class. (conclusion) In Chapter 3, you'll learn how to prove this conclusion from the general statement in the first line. But is the general statement really true? Can we make assumptions about the sleeping patterns of all people, or are we using deductive reasoning to reinforce an untrue reality assumption?

Our next example illustrates the difference between inductive and deductive reasoning. The first part of the example involves reasoning that moves from specific examples to a general statement, illustrating inductive reasoning. The second part of the example begins with the general case rather than specific examples and illustrates deductive reasoning. To begin the general case, we use a letter to represent any one of various numbers. A letter used to represent any number in a collection of numbers is called a **variable**.

A BRIEF REVIEW

In case you have forgotten some basic terms of arithmetic, the following list should be helpful.

Sum: the result of addition

Difference: the result of subtraction

Product: the result of multiplication

Quotient: the result of division

EXAMPLE 5 *Using Inductive and Deductive Reasoning*

Consider the following procedure:

Select a number. Multiply the number by 6. Add 8 to the product. Divide this sum by 2. Subtract 4 from the quotient.

a. Repeat this procedure for at least four different numbers. Write a conjecture that relates the result of this process to the original number selected.

b. Use the variable n to represent the original number and use deductive reasoning to prove the conjecture in part (a).

SOLUTION

a. First, let us pick our starting numbers. We will use 4, 7, 11, and 100, but we could pick any four numbers. Next we will apply the procedure given in this example to 4, 7, 11, and 100, four individual cases, in **Table 1.1**.

TABLE 1.1 Applying a Procedure to Four Individual Cases

Select a number.	4	7	11	100
Multiply the number by 6.	$4 \times 6 = 24$	$7 \times 6 = 42$	$11 \times 6 = 66$	$100 \times 6 = 600$
Add 8 to the product.	$24 + 8 = 32$	$42 + 8 = 50$	$66 + 8 = 74$	$600 + 8 = 608$
Divide this sum by 2.	$\frac{32}{2} = 16$	$\frac{50}{2} = 25$	$\frac{74}{2} = 37$	$\frac{608}{2} = 304$
Subtract 4 from the quotient.	$16 - 4 = 12$	$25 - 4 = 21$	$37 - 4 = 33$	$304 - 4 = 300$

Because we are asked to write a conjecture that relates the result of this process to the original number selected, let us focus on the result of each case.

Original number selected	4	7	11	100
Result of the process	12	21	33	300

Do you see a pattern? Our conjecture is that the result of the process is three times the original number selected. We have used inductive reasoning.

b. Now we begin with the general case rather than specific examples. We use the variable n to represent any number.

Select a number.	n
Multiply the number by 6.	$6n$ (This means 6 times n.)
Add 8 to the product.	$6n + 8$
Divide this sum by 2.	$\frac{6n+8}{2} = \frac{6n}{2} + \frac{8}{2} = 3n + 4$
Subtract 4 from the quotient.	$3n + 4 - 4 = 3n$

Using the variable n to represent any number, the result is $3n$, or three times the number n. This proves that the result of the procedure is three times the original number selected for any number. We have used deductive reasoning. Observe how algebraic notation gives us variables to represent the general case.

CHECK POINT 5 Consider the following procedure:

Select a number. Multiply the number by 4. Add 6 to the product. Divide this sum by 2. Subtract 3 from the quotient.

a. Repeat this procedure for at least four different numbers. Write a conjecture that relates the result of this process to the original number selected.

b. Use the variable n to represent the original number and use deductive reasoning to prove the conjecture in part (a).

Concept and Vocabulary Check

GREAT QUESTION!

What am I supposed to do with the exercises in the Concept and Vocabulary Check?

An important component of thinking mathematically involves knowing the special language and notation used in mathematics. The exercises in the Concept and Vocabulary Check, mainly fill-in-the-blank and true/false items, test your understanding of the definitions and concepts presented in each section. **Work all of the exercises in the Concept and Vocabulary Check** regardless of which exercises your professor assigns in the Exercise Set that follows.

Fill in each blank so that the resulting statement is true.

1. The statement $3 + 3 = 6$ serves as a/an __________ to the conjecture that the sum of two odd numbers is an odd number.

2. Arriving at a specific conclusion from one or more general statements is called _______ reasoning.

3. Arriving at a general conclusion based on observations of specific examples is called _______ reasoning.

4. True or False: A theorem cannot have counterexamples. ____

Exercise Set 1.1

GREAT QUESTION!

Any way that I can perk up my brain before working the book's Exercise Sets?

Researchers say the mind can be strengthened, just like your muscles, with regular training and rigorous practice. Think of the book's Exercise Sets as brain calisthenics. If you're feeling a bit sluggish before any of your mental workouts, try this warmup:

> In the list below, say the color the word is printed in, not the word itself. Once you can do this in 15 seconds without an error, the warmup is over and it's time to move on to the assigned exercises.
>
> **Blue Yellow Red Green Yellow Green Blue Red Yellow Red**

Practice Exercises

In Exercises 1–8, find a counterexample to show that each of the statements is false.

1. No U.S. president has been younger than 65 at the time of his inauguration.
2. No singers appear in movies.
3. If a number is multiplied by itself, the result is even.
4. The sum of two three-digit numbers is a four-digit number.
5. Adding the same number to both the numerator and the denominator (top and bottom) of a fraction does not change the fraction's value.
6. If the difference between two numbers is odd, then the two numbers are both odd.
7. If a number is added to itself, the sum is greater than the original number.
8. If 1 is divided by a number, the quotient is less than that number.

In Exercises 9–38, identify a pattern in each list of numbers. Then use this pattern to find the next number. (More than one pattern might exist, so it is possible that there is more than one correct answer.)

9. 8, 12, 16, 20, 24, ____
10. 19, 24, 29, 34, 39, ____
11. 37, 32, 27, 22, 17, ____
12. 33, 29, 25, 21, 17, ____
13. 3, 9, 27, 81, 243, ____
14. 2, 8, 32, 128, 512, ____
15. 1, 2, 4, 8, 16, ____
16. 1, 5, 25, 125, ____
17. 1, 4, 1, 8, 1, 16, 1, ____
18. 1, 4, 1, 7, 1, 10, 1, ____
19. 4, 2, 0, −2, −4, ____
20. 6, 3, 0, −3, −6, ____
21. $\frac{1}{2}, \frac{1}{6}, \frac{1}{10}, \frac{1}{14}, \frac{1}{18}$, ____
22. $1, \frac{1}{2}, \frac{1}{3}, \frac{1}{4}, \frac{1}{5}$, ____
23. $1, \frac{1}{3}, \frac{1}{9}, \frac{1}{27}$, ____
24. $1, \frac{1}{2}, \frac{1}{4}, \frac{1}{8}$, ____
25. 3, 7, 12, 18, 25, 33, ____
26. 2, 5, 9, 14, 20, 27, ____
27. 3, 6, 11, 18, 27, 38, ____
28. 2, 5, 10, 17, 26, 37, ____
29. 3, 7, 10, 17, 27, 44, ____
30. 2, 5, 7, 12, 19, 31, ____
31. 2, 7, 12, 5, 10, 15, 8, 13, ____
32. 3, 9, 15, 5, 11, 17, 7, 13, ____
33. 3, 6, 5, 10, 9, 18, 17, 34, ____
34. 2, 6, 5, 15, 14, 42, 41, 123, ____
35. 64, −16, 4, −1, ____
36. 125, −25, 5, −1, ____
37. (6, 2), (0, −4), $(7\frac{1}{2}, 3\frac{1}{2})$, (2, −2), (3, ____)
38. $(\frac{2}{3}, \frac{4}{9})$, $(\frac{1}{5}, \frac{1}{25})$, (7, 49), $(-\frac{5}{6}, \frac{25}{36})$, $(-\frac{4}{7},$ ____)

In Exercises 39–42, identify a pattern in each sequence of figures. Then use the pattern to find the next figure in the sequence.

39. 

40. 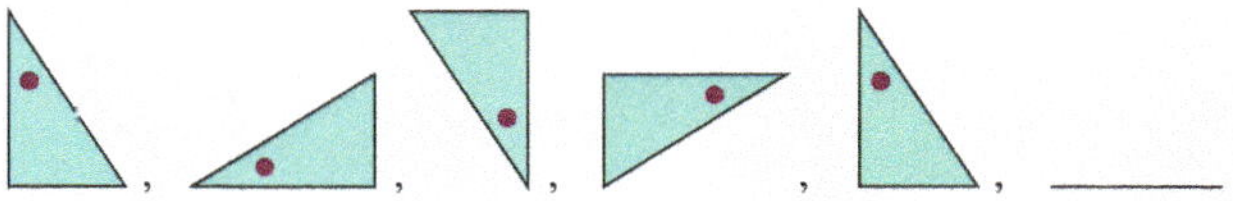

41. 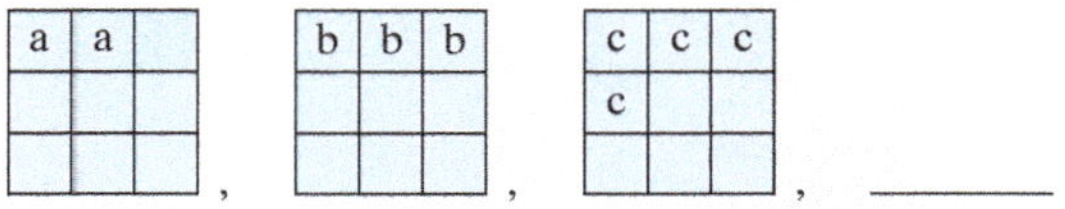

42. 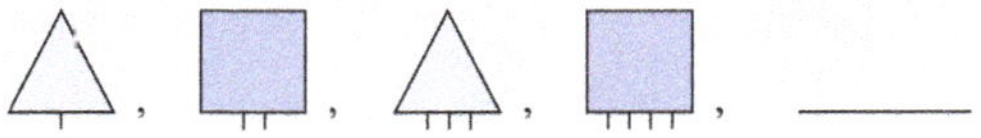

Exercises 43–46 describe procedures that are to be applied to numbers. In each exercise,

a. *Repeat the procedure for four numbers of your choice. Write a conjecture that relates the result of the process to the original number selected.*

b. *Use the variable n to represent the original number and use deductive reasoning to prove the conjecture in part (a).*

43. Select a number. Multiply the number by 4. Add 8 to the product. Divide this sum by 2. Subtract 4 from the quotient.
44. Select a number. Multiply the number by 3. Add 6 to the product. Divide this sum by 3. Subtract the original selected number from the quotient.
45. Select a number. Add 5. Double the result. Subtract 4. Divide by 2. Subtract the original selected number.
46. Select a number. Add 3. Double the result. Add 4. Divide by 2. Subtract the original selected number.

In Exercises 47–52, use inductive reasoning to predict the next line in each sequence of computations. Then use a calculator or perform the arithmetic by hand to determine whether your conjecture is correct.

47.
$$1 + 2 = \frac{2 \times 3}{2}$$
$$1 + 2 + 3 = \frac{3 \times 4}{2}$$
$$1 + 2 + 3 + 4 = \frac{4 \times 5}{2}$$
$$1 + 2 + 3 + 4 + 5 = \frac{5 \times 6}{2}$$

48.
$$3 + 6 = \frac{6 \times 3}{2}$$
$$3 + 6 + 9 = \frac{9 \times 4}{2}$$
$$3 + 6 + 9 + 12 = \frac{12 \times 5}{2}$$
$$3 + 6 + 9 + 12 + 15 = \frac{15 \times 6}{2}$$

49.
$$1 + 3 = 2 \times 2$$
$$1 + 3 + 5 = 3 \times 3$$
$$1 + 3 + 5 + 7 = 4 \times 4$$
$$1 + 3 + 5 + 7 + 9 = 5 \times 5$$

50.
$$\frac{1}{1 \times 2} + \frac{1}{2 \times 3} = \frac{2}{3}$$
$$\frac{1}{1 \times 2} + \frac{1}{2 \times 3} + \frac{1}{3 \times 4} = \frac{3}{4}$$
$$\frac{1}{1 \times 2} + \frac{1}{2 \times 3} + \frac{1}{3 \times 4} + \frac{1}{4 \times 5} = \frac{4}{5}$$

51.
$$9 \times 9 + 7 = 88$$
$$98 \times 9 + 6 = 888$$
$$987 \times 9 + 5 = 8888$$
$$9876 \times 9 + 4 = 88{,}888$$

52.
$$1 \times 9 - 1 = 8$$
$$21 \times 9 - 1 = 188$$
$$321 \times 9 - 1 = 2888$$
$$4321 \times 9 - 1 = 38{,}888$$

Practice Plus

In Exercises 53–54, use inductive reasoning to predict the next line in each sequence of computations. Then use a calculator or perform the arithmetic by hand to determine whether your conjecture is correct.

53.
$$33 \times 3367 = 111{,}111$$
$$66 \times 3367 = 222{,}222$$
$$99 \times 3367 = 333{,}333$$
$$132 \times 3367 = 444{,}444$$

54.
$$1 \times 8 + 1 = 9$$
$$12 \times 8 + 2 = 98$$
$$123 \times 8 + 3 = 987$$
$$1234 \times 8 + 4 = 9876$$
$$12{,}345 \times 8 + 5 = 98{,}765$$

55. Study the pattern in these examples:

$$a^2 \# a^4 = a^{10} \quad a^3 \# a^2 = a^7 \quad a^5 \# a^3 = a^{11}.$$

Select the equation that describes the pattern.

a. $a^x \# a^y = a^{2x+y}$

b. $a^x \# a^y = a^{x+2y}$

c. $a^x \# a^y = a^{x+y+4}$

d. $a^x \# a^y = a^{xy+2}$

56. Study the pattern in these examples:

$$a^5 * a^3 * a^2 = a^5 \quad a^3 * a^7 * a^2 = a^6 \quad a^2 * a^4 * a^8 = a^7.$$

Select the equation that describes the pattern.

a. $a^x * a^y * a^z = a^{x+y+z}$

b. $a^x * a^y * a^z = a^{\frac{xyz}{2}}$

c. $a^x * a^y * a^z = a^{\frac{x+y+z}{2}}$

d. $a^x * a^y * a^z = a^{\frac{xy}{2}+z}$

Application Exercises

In Exercises 57–60, identify the reasoning process, induction or deduction, in each example. Explain your answer.

57. It can be shown that

$$1 + 2 + 3 + \cdots + n = \frac{n(n+1)}{2}.$$

I can use this formula to conclude that the sum of the first one hundred counting numbers, $1 + 2 + 3 + \cdots + 100$, is

$$\frac{100(100+1)}{2} = \frac{100(101)}{2} = 50(101), \text{ or } 5050.$$

58. An HMO does a follow-up study on 200 randomly selected patients given a flu shot. None of these people became seriously ill with the flu. The study concludes that all HMO patients be urged to get a flu shot in order to prevent a serious case of the flu.

59. The data in the graph are from a random sample of 1200 full-time four-year undergraduate college students on 100 U.S. campuses.

The Greatest Problems on Campus

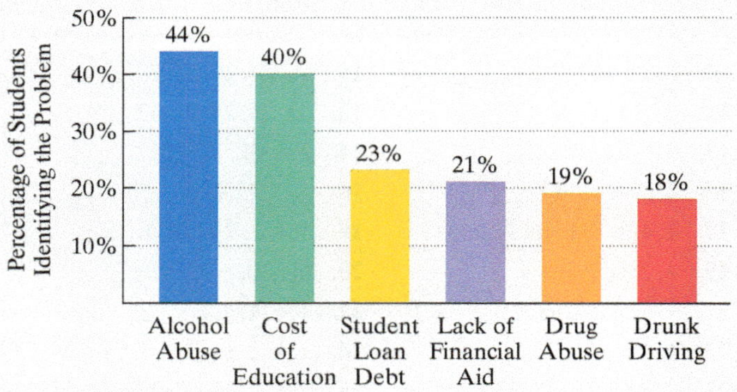

Source: Student Monitor LLC

We can conclude that there is a high probability that approximately 44% of all full-time four-year college students in the United States believe that alcohol abuse is the greatest problem on campus.

60. The course policy states that work turned in late will be marked down a grade. I turned in my report a day late, so it was marked down from B to C.

61. The ancient Greeks studied **figurate numbers**, so named because of their representations as geometric arrangements of points.

Triangular Numbers

1 3 6 10 15 21

Square Numbers

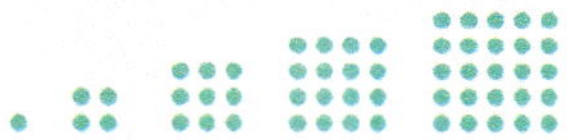

1 4 9 16 25

Pentagonal Numbers

1 5 12 22

a. Use inductive reasoning to write the five triangular numbers that follow 21.

b. Use inductive reasoning to write the five square numbers that follow 25.

c. Use inductive reasoning to write the five pentagonal numbers that follow 22.

d. Use inductive reasoning to complete this statement: If a triangular number is multiplied by 8 and then 1 is added to the product, a _____ number is obtained.

62. The triangular arrangement of numbers shown below is known as **Pascal's triangle**, credited to French mathematician Blaise Pascal (1623–1662). Use inductive reasoning to find the six numbers designated by question marks.

1
1 1
1 2 1
1 3 3 1
1 4 6 4 1
? ? ? ? ? ?

Writing in Mathematics

Writing about mathematics will help you to learn mathematics. For all writing exercises in this book, use complete sentences to respond to the questions. Some writing exercises can be answered in a sentence; others require a paragraph or two. You can decide how much you need to write as long as your writing clearly and directly answers the question in the exercise. Standard references such as a dictionary and a thesaurus may be helpful.

63. The word *induce* comes from a Latin term meaning *to lead.* Explain what leading has to do with inductive reasoning.

64. Describe what is meant by deductive reasoning. Give an example.

65. Give an example of a decision that you made recently in which the method of reasoning you used to reach the decision was induction. Describe your reasoning process.

Critical Thinking Exercises

Make Sense? *In Exercises 66–69, determine whether each statement makes sense or does not make sense, and explain your reasoning.*

66. I use deductive reasoning to draw conclusions that are not certain, but likely.

67. Additional information may strengthen or weaken the probability of my inductive arguments.

68. I used the data shown in the bar graph, which summarizes a random sample of 752 college seniors, to conclude with certainty that 51% of all graduating college females expect to earn $30,000 or less after graduation.

First-Year Income Expectations of Graduating College Seniors

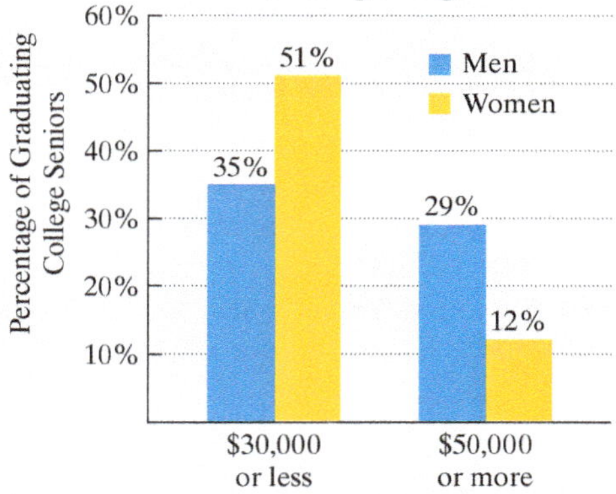

Source: Duquesne University Senior Economic Expectation research survey

69. I used the data shown in the bar graph for Exercise 68, which summarizes a random sample of 752 college seniors, to conclude inductively that a greater percentage of male graduates expect higher first-year income than female graduates.

70. If $(6-2)^2 = 36 - 24 + 4$ and $(8-5)^2 = 64 - 80 + 25$, use inductive reasoning to write a compatible expression for $(11-7)^2$.

71. The rectangle shows an array of nine numbers represented by combinations of the variables a, b, and c.

$a + b$	$a - b - c$	$a + c$
$a - b + c$	a	$a + b - c$
$a - c$	$a + b + c$	$a - b$

a. Determine the nine numbers in the array for $a = 10$, $b = 6$, and $c = 1$. What do you observe about the sum of the numbers in all rows, all columns, and the two diagonals?

b. Repeat part (a) for $a = 12$, $b = 5$, and $c = 2$.

c. Repeat part (a) for values of a, b, and c of your choice.

d. Use the results of parts (a) through (c) to make an inductive conjecture about the rectangular array of nine numbers represented by a, b, and c.

e. Use deductive reasoning to prove your conjecture in part (d).

72. Write a list of numbers that has two patterns so that the next number in the list can be 15 or 20.

73. a. Repeat the following procedure with at least five people. Write a conjecture that relates the result of the procedure to each person's birthday.

Take the number of the month of your birthday (January = 1, February = 2, . . . , December = 12), multiply by 5, add 6, multiply this sum by 4, add 9, multiply this new sum by 5, and add the number of the day on which you were born. Finally, subtract 165.

b. Let M represent the month number and let D represent the day number of any person's birthday. Use deductive reasoning to prove your conjecture in part (a).

Technology Exercises

74. **a.** Use a calculator to find 6×6, 66×66, 666×666, and 6666×6666.

b. Describe a pattern in the numbers being multiplied and the resulting products.

c. Use the pattern to write the next two multiplications and their products. Then use your calculator to verify these results.

d. Is this process an example of inductive or deductive reasoning? Explain your answer.

75. **a.** Use a calculator to find 3367×3, 3367×6, 3367×9, and 3367×12.

b. Describe a pattern in the numbers being multiplied and the resulting products.

c. Use the pattern to write the next two multiplications and their products. Then use your calculator to verify these results.

d. Is this process an example of inductive or deductive reasoning? Explain your answer.

Group Exercise

76. Stereotyping refers to classifying people, places, or things according to common traits. Prejudices and stereotypes can function as assumptions in our thinking, appearing in inductive and deductive reasoning. For example, it is not difficult to find inductive reasoning that results in generalizations such as these, as well as deductive reasoning in which these stereotypes serve as assumptions:

School has nothing to do with life.

Intellectuals are nerds.

People on welfare are lazy.

Each group member should find one example of inductive reasoning and one example of deductive reasoning in which stereotyping occurs. Upon returning to the group, present each example and then describe how the stereotyping results in faulty conjectures or prejudging situations and people.

1.2 Estimation, Graphs, and Mathematical Models

WHAT AM I SUPPOSED TO LEARN?

After you have read this section, you should be able to:

1 Use estimation techniques to arrive at an approximate answer to a problem.

2 Apply estimation techniques to information given by graphs.

3 Develop mathematical models that estimate relationships between variables.

IF PRESENT TRENDS CONTINUE, IS IT POSSIBLE THAT our descendants could live to be 200 years of age? To answer this question, we need to examine data for life expectancy and develop estimation techniques for representing the data mathematically. In this section, you will learn estimation methods that will enable you to obtain mathematical representations of data displayed by graphs, using these representations to predict what might occur in the future.

Estimation

1 Use estimation techniques to arrive at an approximate answer to a problem.

Estimation is the process of arriving at an approximate answer to a question. For example, companies estimate the amount of their products consumers are likely to use, and economists estimate financial trends. If you are about to cross a street, you may estimate the speed of oncoming cars so that you know whether or not to wait before crossing. Rounding numbers is also an estimation method. You might round a number without even being aware that you are doing so. You may say that you are 20 years old, rather than 20 years 5 months, or that you will be home in about a half-hour, rather than 25 minutes.

You will find estimation to be equally valuable in your work for this class. Making mistakes with a calculator or a computer is easy. Estimation can tell us whether the answer displayed for a computation makes sense.

In this section, we demonstrate several estimation methods. In the second part of the section, we apply these techniques to information given by graphs.

Performing computations using rounded numbers is one way to check whether an answer displayed by a calculator or a computer is reasonable. Rounding whole numbers depends on knowing the place values of the digits. (The digits that we use in base ten are 0, 1, 2, 3, 4, 5, 6, 7, 8, and 9. In Chapter 4, we'll explore other ways of representing numbers, including bases other than ten.) The place that a digit occupies in a number tells us its value in that number. Here is an example using world population at 7:49 A.M. Eastern Time on January 10, 2013.

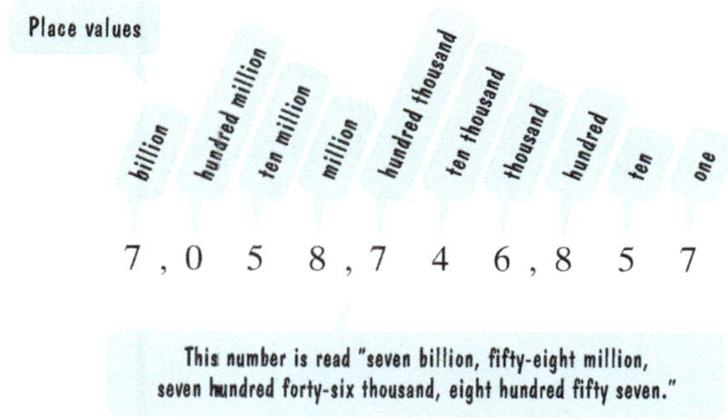

ROUNDING WHOLE NUMBERS

1. Look at the digit to the right of the digit where rounding is to occur.

2. a. If the digit to the right is 5 or greater, add 1 to the digit to be rounded. Replace all digits to the right with zeros.

b. If the digit to the right is less than 5, do not change the digit to be rounded. Replace all digits to the right with zeros.

The symbol $\approx$ means *is approximately equal to*. We will use this symbol when rounding numbers.

EXAMPLE 1 *Rounding a Whole Number*

Round world population (7,058,746,857) as follows:

a. to the nearest hundred million

b. to the nearest million

c. to the nearest hundred thousand.

SOLUTION

a. 7,058,746,857 $\approx$ 7,100,000,000

Hundred millions digit, where rounding is to occur | Digit to the right is 5. | Add 1 to the digit to be rounded. | Replace all digits to the right with zeros.

World population to the nearest hundred million is seven billion, one-hundred million.

b. 7,058,746,857 $\approx$ 7,059,000,000

Millions digit, where rounding is to occur | Digit to the right is greater than 5. | Add 1 to the digit to be rounded. | Replace all digits to the right with zeros.

World population to the nearest million is seven billion, fifty-nine million.

c. 7,058,746,857 $\approx$ 7,058,700,000

Hundred thousands digit, where rounding is to occur | Digit to the right is less than 5. | Do not change the digit to be rounded. | Replace all digits to the right with zeros.

World population to the nearest hundred thousand is seven billion, fifty-eight million, seven-hundred thousand.

CHECK POINT 1 Round world population (7,058,746,857) as follows:

a. to the nearest billion

b. to the nearest ten thousand.

Rounding can also be applied to decimal notation, used to denote a part of a whole. Once again, the place that a digit occupies tells us its value. Here's an example using the first seven digits of the number π (pi). (We'll have more to say about π, whose digits extend endlessly with no repeating pattern, in Chapter 5.)

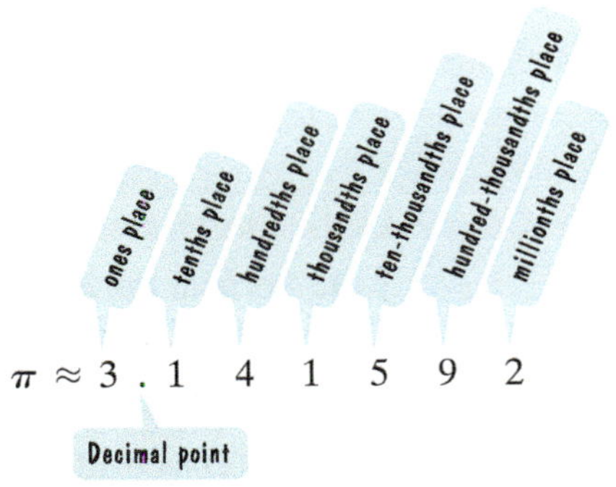

We round the decimal part of a decimal number in nearly the same way that we round whole numbers. The only difference is that we drop the digits to the right of the rounding place rather than replacing these digits with zeros.

EXAMPLE 2 *Rounding the Decimal Part of a Number*

Round 3.141592, the first seven digits of π, as follows:

a. to the nearest hundredth

b. to the nearest thousandth.

SOLUTION

a. $3.141592 \approx 3.14$

Hundredths digit, where rounding is to occur | Digit to the right is less than 5. | Do not change the digit to be rounded. | Drop all digits to the right.

The number π to the nearest hundredth is three and fourteen hundredths.

b. $3.141592 \approx 3.142$

Thousandths digit, where rounding is to occur | Digit to the right is 5. | Add 1 to the digit to be rounded. | Drop all digits to the right.

The number π to the nearest thousandth is three and one hundred forty-two thousandths.

CHECK POINT 2 Round 3.141592, the first seven digits of π, as follows:

a. to the nearest tenth

b. to the nearest ten-thousandth.

Blitzer Bonus

Estimating Support for a Cause

Police often need to estimate the size of a crowd at a political demonstration. One way to do this is to select a reasonably sized rectangle within the crowd and estimate (or count) the number of people within the rectangle. Police then estimate the number of such rectangles it would take to completely fill the area occupied by the crowd. The police estimate is obtained by multiplying the number of such rectangles by the number of demonstrators in the representative rectangle. The organizers of the demonstration might give a larger estimate than the police to emphasize the strength of their support.

EXAMPLE 3 *Estimation by Rounding*

You purchased bread for \$2.59, detergent for \$2.17, a sandwich for \$3.65, an apple for \$0.47, and coffee for \$5.79. The total bill was given as \$18.67. Is this amount reasonable?

SOLUTION

If you are in the habit of carrying a calculator to the store, you can answer the question by finding the exact cost of the purchase. However, estimation can be used to determine if the bill is reasonable even if you do not have a calculator. We will round the cost of each item to the nearest dollar.

Round to the nearest dollar. — Use digits in the tenths place to do the rounding.

		Round to the nearest dollar
Bread	\$2.59	≈ \$3.00
Detergent	\$2.17	≈ \$2.00
Sandwich	\$3.65	≈ \$4.00
Apple	\$0.47	≈ \$0.00
Coffee	\$5.79	≈ \$6.00
		\$15.00

The total bill that you were given, \$18.67, seems a bit high compared to the \$15.00 estimate. You should check the bill before paying it. Adding the prices of all five items gives the true total bill of \$14.67.

CHECK POINT 3 You and a friend ate lunch at Ye Olde Cafe. The check for the meal showed soup for \$3.40, tomato juice for \$2.25, a roast beef sandwich for \$5.60, a chicken salad sandwich for \$5.40, two coffees totaling \$3.40, apple pie for \$2.85, and chocolate cake for \$3.95.

a. Round the cost of each item to the nearest dollar and obtain an estimate for the food bill.

b. The total bill before tax was given as \$29.85. Is this amount reasonable?

EXAMPLE 4 *Estimation by Rounding*

A carpenter who works full time earns \$28 per hour.

a. Estimate the carpenter's weekly salary.

b. Estimate the carpenter's annual salary.

SOLUTION

a. In order to simplify the calculation, we can round the hourly rate of \$28 to \$30. Be sure to write out the units for each number in the calculation. The work week is 40 hours per week, and the rounded salary is \$30 per hour. We express this as

$$\frac{40 \text{ hours}}{\text{week}} \quad \text{and} \quad \frac{\$30}{\text{hour}}.$$

The word *per* is represented by the division bar. We multiply these two numbers to estimate the carpenter's weekly salary. We cancel out units that are identical if they are above and below the division bar.

$$\frac{40 \text{ hours}}{\text{week}} \times \frac{\$30}{\text{hour}} = \frac{\$1200}{\text{week}}$$

Thus, the carpenter earns approximately \$1200 per week, written $\approx \$1200$.

b. For the estimate of annual salary, we may round 52 weeks to 50 weeks. The annual salary is approximately the product of \$1200 per week and 50 weeks per year:

$$\frac{\$1200}{\text{week}} \times \frac{50 \text{ weeks}}{\text{year}} = \frac{\$60{,}000}{\text{year}}.$$

Thus, the carpenter earns approximately \$60,000 per year, or \$60,000 annually, written $\approx \$60{,}000$.

GREAT QUESTION!

Is it OK to cancel identical units if one unit is singular and the other is plural?

Yes. It does not matter whether a unit is singular, such as *week*, or plural, such as *weeks*. *Week* and *weeks* are identical units and can be canceled out, as shown on the right.

CHECK POINT 4 A landscape architect who works full time earns \$52 per hour.

a. Estimate the landscape architect's weekly salary.

b. Estimate the landscape architect's annual salary.

2 Apply estimation techniques to information given by graphs.

Estimation with Graphs

Magazines, newspapers, and websites often display information using circle, bar, and line graphs. The following examples illustrate how rounding and other estimation techniques can be applied to data displayed in each of these types of graphs.

Circle graphs, also called **pie charts**, show how a whole quantity is divided into parts. Circle graphs are divided into pieces, called **sectors**. **Figure 1.3** shows a circle graph that indicates how Americans disagree as to when "old age" begins.

Americans' Definition of Old Age

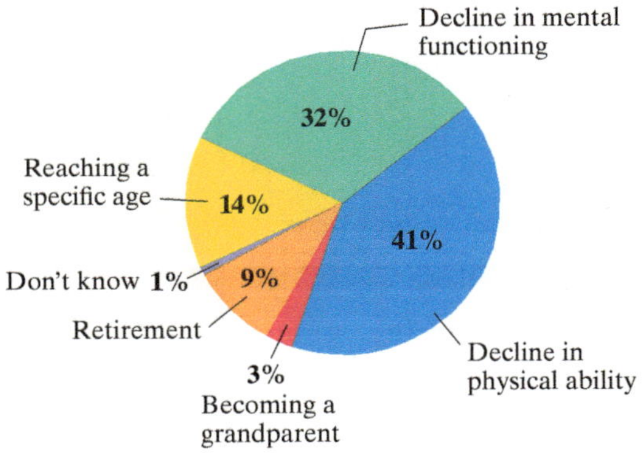

FIGURE 1.3
Source: American Demographics

A BRIEF REVIEW *Percents*

- **Percents** are the result of expressing numbers as part of 100. The word *percent* means *per hundred*. For example, the circle graph in **Figure 1.3** shows that 41% of Americans define old age by a decline in physical ability. Thus, 41 out of every 100 Americans define old age in this manner: $41\% = \frac{41}{100}$.
- To convert a number from percent form to decimal form, move the decimal point two places to the left and drop the percent sign. Example:

$$41\% = 41.\% = 0.41\%$$

Thus, $41\% = 0.41$.

- Many applications involving percent are based on the following formula:

A is *P* percent of B.

$$A = P \cdot B.$$

Note that the word *of* implies multiplication.

In our next example, we will use the information in the circle graph on page 18 to estimate a quantity. Although different rounding results in different estimates, the whole idea behind the rounding process is to make calculations simple.

EXAMPLE 5 *Applying Estimation Techniques to a Circle Graph*

According to the U.S. Census Bureau, in 2012, there were 229,937,770 Americans 20 years and older. Assuming the circle graph in **Figure 1.3** is representative of this age group,

a. Use the appropriate information displayed by the graph to determine a calculation that shows the number of Americans 20 years and older who define old age by a decline in physical ability.

b. Use rounding to find a reasonable estimate for this calculation.

SOLUTION

a. The circle graph in **Figure 1.3** indicates that 41% of Americans define old age by a decline in physical ability. Among the 229,937,770 Americans 20 years and older, the number who define old age in this manner is determined by finding 41% of 229,937,770.

The number of Americans 20 and older who define old age by a decline in physical ability | is | 41% | of | the number of Americans 20 and older.

$$= 0.41 \times 229{,}937{,}770$$

b. We can use rounding to obtain a reasonable estimate of $0.41 \times 229{,}937{,}770$.

Round to the nearest ten million.

$$0.41 \times 229{,}937{,}770 \approx 0.4 \times 230{,}000{,}000 = 92{,}000{,}000$$

Round to the nearest tenth.

$$\begin{array}{r} 230{,}000{,}000 \\ \times \quad 0.4 \\ \hline 92{,}000{,}000.0 \end{array}$$

Our answer indicates that approximately 92,000,000 (92 million) Americans 20 years and older define old age by a decline in physical ability.

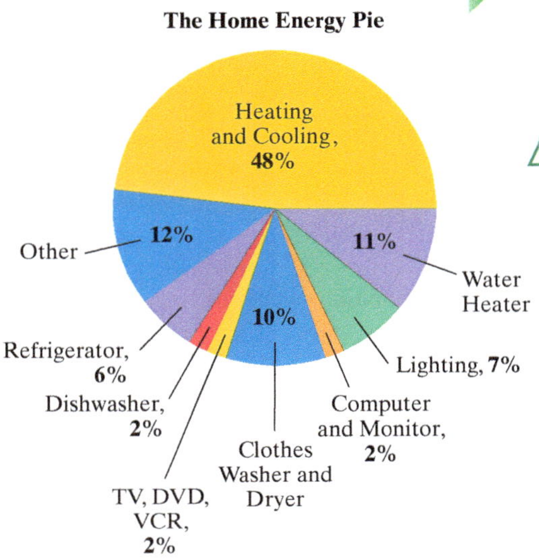

FIGURE 1.4

Source: Natural Home and Garden

CHECK POINT 5 Being aware of which appliances and activities in your home use the most energy can help you make sound decisions that allow you to decrease energy consumption and increase savings. The circle graph in **Figure 1.4** shows how energy consumption is distributed throughout a typical home.

Suppose that last year your family spent $2148.72 on natural gas and electricity. Assuming the circle graph in **Figure 1.4** is representative of your family's energy consumption,

a. Use the appropriate information displayed by the graph to determine a calculation that shows the amount your family spent on heating and cooling for the year.

b. Use rounding to find a reasonable estimate for this calculation.

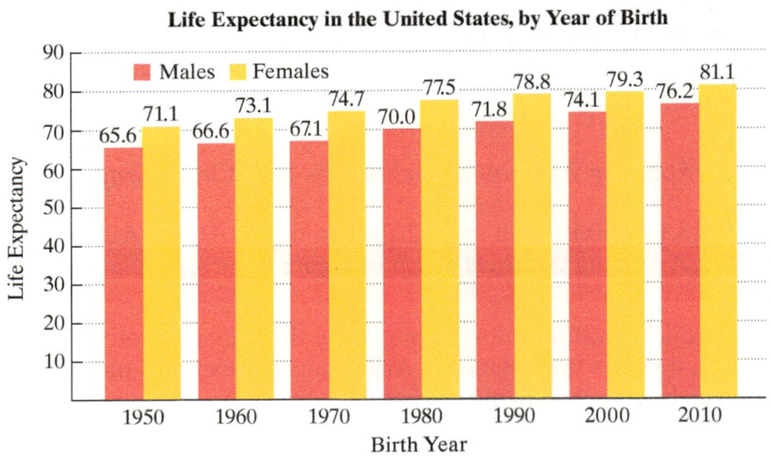

FIGURE 1.5 *Source:* National Center for Health Statistics

Bar graphs are convenient for comparing some measurable attribute of various items. The bars may be either horizontal or vertical, and their heights or lengths are used to show the amounts of different items. **Figure 1.5** is an example of a typical bar graph. The graph shows life expectancy for American men and American women born in various years from 1950 through 2010.

EXAMPLE 6 *Applying Estimation and Inductive Reasoning to Data in a Bar Graph*

Use the data for men in **Figure 1.5** to estimate each of the following:

a. a man's increased life expectancy, rounded to the nearest hundredth of a year, for each subsequent birth year

b. the life expectancy of a man born in 2020.

SOLUTION

a. One way to estimate increased life expectancy for each subsequent birth year is to generalize from the information given for 1950 (male life expectancy: 65.6 years) and for 2010 (male life expectancy: 76.2 years). The average yearly increase in life expectancy is the change in life expectancy from 1950 to 2010 divided by the change in time from 1950 to 2010.

$$\text{Yearly increase in life expectancy} \text{ is approximately } \frac{\text{change in life expectancy from 1950 to 2010}}{\text{change in time from 1950 to 2010}}.$$

$$\approx \frac{76.2 - 65.6}{2010 - 1950}$$

life expectancy in 2010 minus life expectancy in 1950

Change in time is 2010 − 1950, or 60 years.

$$\approx 0.18$$

Use a calculator. See the Technology box below.

For each subsequent birth year, a man's life expectancy is increasing by approximately 0.18 year.

TECHNOLOGY

Here is the calculator keystroke sequence needed to perform the computation in Example 6(a).

[(] 76.2 [−] 65.6 [)] [÷] [(] 2010 [−] 1950 [)]

Press [=] on a scientific calculator or [ENTER] on a graphing calculator to display the answer. As specified, we round to the nearest hundredth.

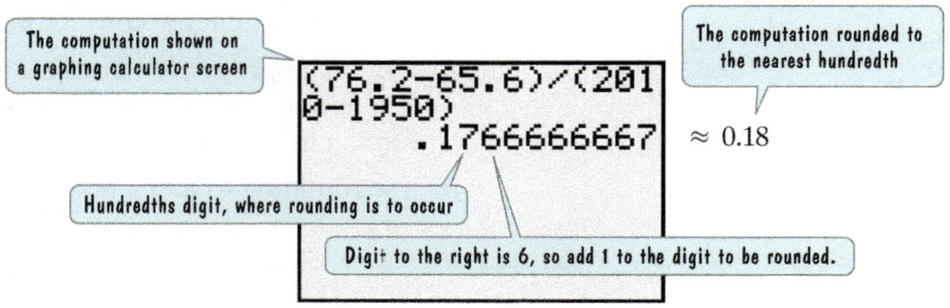

b. We can use our computation in part (a) to estimate the life expectancy of an American man born in 2020. The bar graph indicates that men born in 1950 had a life expectancy of 65.6 years. The year 2020 is 70 years after 1950, and life expectancy is increasing by approximately 0.18 year for each subsequent birth year.

Life expectancy for a man born in 2020	is approximately	life expectancy for a man born in 1950	plus	yearly increase in life expectancy	times	the number of years from 1950 to 2020.
	$\approx$	65.6	$+$	0.18	$\times$	70

$$= 65.6 + 12.6 = 78.2$$

An American man born in 2020 will have a life expectancy of approximately 78.2 years.

CHECK POINT 6 Use the data for women in **Figure 1.5** to estimate each of the following:

a. a woman's increased life expectancy, rounded to the nearest hundredth of a year, for each subsequent birth year

b. the life expectancy, to the nearest tenth of a year, of a woman born in 2050.

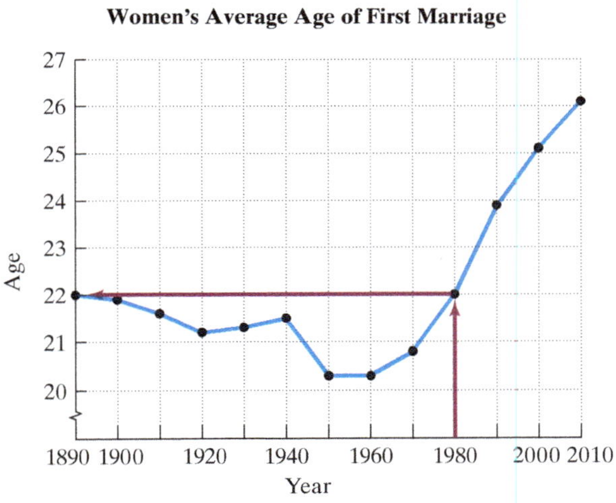

FIGURE 1.6
Source: U.S. Census Bureau

Line graphs are often used to illustrate trends over time. Some measure of time, such as months or years, frequently appears on the horizontal axis. Amounts are generally listed on the vertical axis. Points are drawn to represent the given information. The graph is formed by connecting the points with line segments.

Figure 1.6 is an example of a typical line graph. The graph shows the average age at which women in the United States married for the first time from 1890 through 2010. The years are listed on the horizontal axis, and the ages are listed on the vertical axis. The symbol ⟡ on the vertical axis shows that there is a break in values between 0 and 20. Thus, the first tick mark on the vertical axis represents an average age of 20.

Figure 1.6 shows how to find the average age at which women married for the first time in 1980.

Step 1 Locate 1980 on the horizontal axis.

Step 2 Locate the point on the line graph above 1980.

Step 3 Read across to the corresponding age on the vertical axis.

The age is 22. Thus, in 1980, women in the United States married for the first time at an average age of 22.

EXAMPLE 7 *Using a Line Graph*

The line graph in **Figure 1.7** shows the percentage of U.S. college students who smoked cigarettes from 1982 through 2010.

a. Find an estimate for the percentage of college students who smoked cigarettes in 2010.

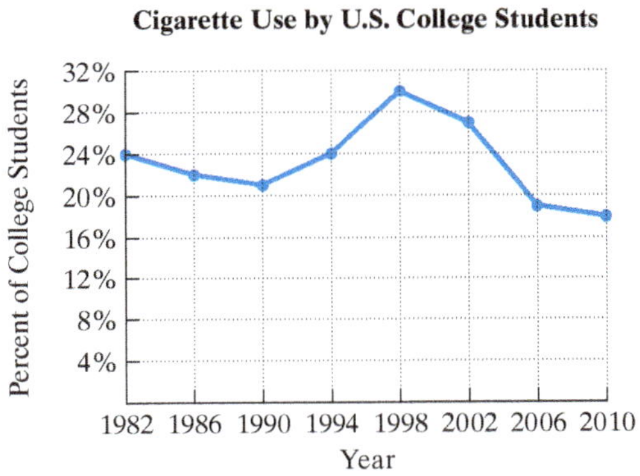

FIGURE 1.7
Source: Rebecca Donatelle, *Health The Basics*, 10th Edition, Pearson, 2013.

b. In which four-year period did the percentage of college students who smoked cigarettes decrease at the greatest rate?

c. In which year did 30% of college students smoke cigarettes?

SOLUTION

a. Estimating the Percentage Smoking Cigarettes in 2010

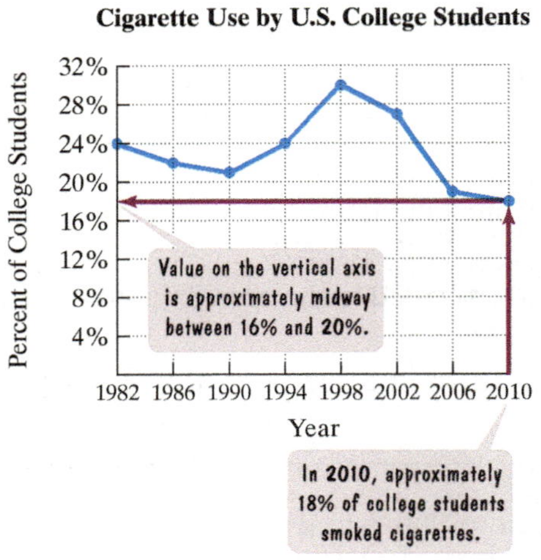

b. Identifying the Period of the Greatest Rate of Decreasing Cigarette Smoking

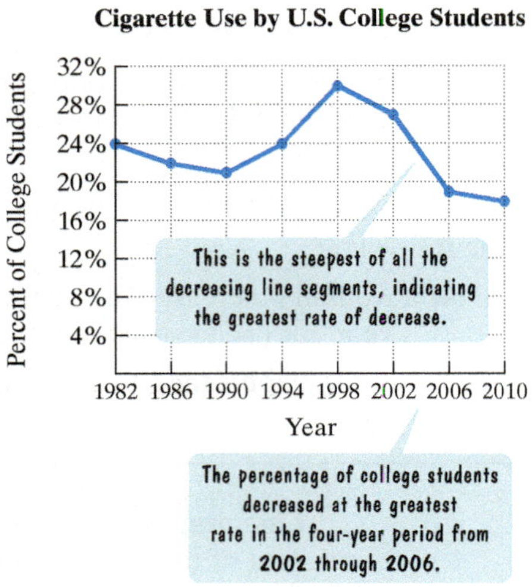

c. Identifying the Year when 30% of College Students Smoked Cigarettes

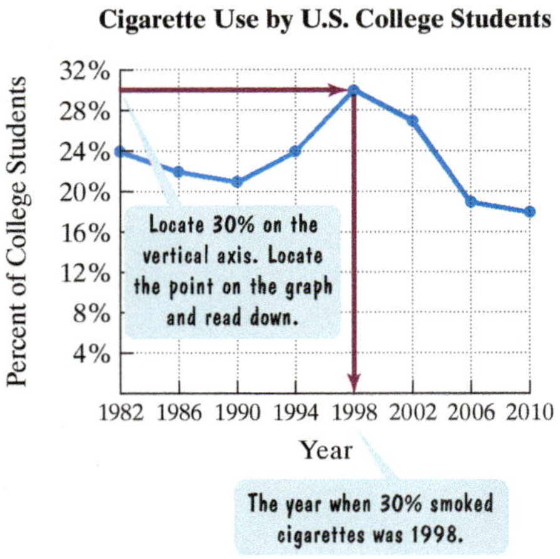

CHECK POINT 7 Use the line graph in **Figure 1.7** at the bottom of the previous page to solve this exercise.

a. Find an estimate for the percentage of college students who smoked cigarettes in 1986.

b. In which four-year period did the percentage of college students who smoked cigarettes increase at the greatest rate?

c. In which years labeled on the horizontal axis did 24% of college students smoke cigarettes?

3 Develop mathematical models that estimate relationships between variables.

Mathematical Models

We have seen that American men born in 1950 have a life expectancy of 65.6 years, increasing by approximately 0.18 year for each subsequent birth year. We can use variables to express the life expectancy, E, for American men born x years after 1950.

Life expectancy for American men	is	life expectancy for a man born in 1950	plus	yearly increase in life expectancy	times the number of birth years after 1950.
E	$=$	65.6	$+$	0.18	x

A **formula** is a statement of equality that uses letters to express a relationship between two or more variables. Thus, $E = 65.6 + 0.18x$ is a formula describing life expectancy, E, for American men born x years after 1950. Be aware that this formula provides *estimates* of life expectancy, as shown in **Table 1.2**.

TABLE 1.2 Comparing Given Data with Estimates Determined by a Formula

Birth Year	Life Expectancy: Given Data	Life Expectancy: Formula Estimate $E = 65.6 + 0.18x$
1950	65.6	$E = 65.6 + 0.18(0) = 65.6 + 0 = 65.6$
1960	66.6	$E = 65.6 + 0.18(10) = 65.6 + 1.8 = 67.4$
1970	67.1	$E = 65.6 + 0.18(20) = 65.6 + 3.6 = 69.2$
1980	70.0	$E = 65.6 + 0.18(30) = 65.6 + 5.4 = 71.0$
1990	71.8	$E = 65.6 + 0.18(40) = 65.6 + 7.2 = 72.8$
2000	74.1	$E = 65.6 + 0.18(50) = 65.6 + 9.0 = 74.6$
2010	76.2	$E = 65.6 + 0.18(60) = 65.6 + 10.8 = 76.4$

In each row, we substitute the number of years after 1950 for x. The better estimates occur in 1950, 2000, and 2010.

The process of finding formulas to describe real-world phenomena is called **mathematical modeling**. Such formulas, together with the meaning assigned to the variables, are called **mathematical models**. We often say that these formulas model, or describe, the relationships among the variables.

EXAMPLE 8 *Modeling the Cost of Attending a Public College*

The bar graph in **Figure 1.8** shows the average cost of tuition and fees for public four-year colleges, adjusted for inflation.

a. Estimate the yearly increase in tuition and fees. Round to the nearest dollar.

b. Write a mathematical model that estimates the average cost of tuition and fees, T, at public four-year colleges for the school year ending x years after 2000.

c. Use the mathematical model from part (b) to project the average cost of tuition and fees at public four-year colleges for the school year ending in 2016.

Average Cost of Tuition and Fees at Public Four-Year U.S. Colleges

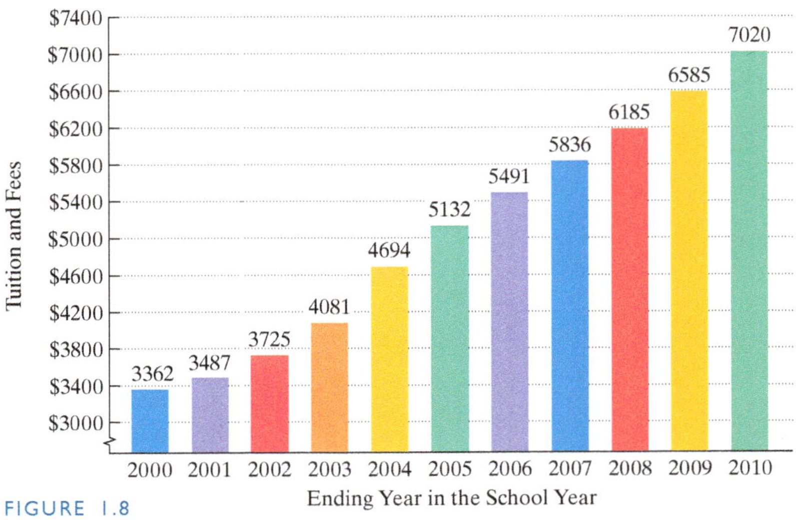

FIGURE 1.8
Source: The College Board

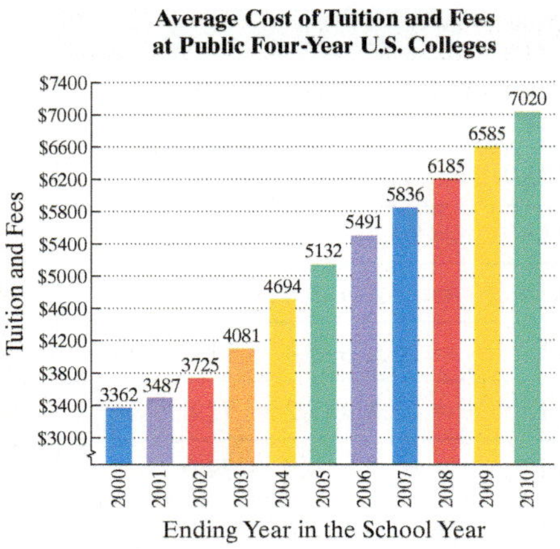

FIGURE 1.8 (repeated)
Source: The College Board

SOLUTION

a. We can use the data in **Figure 1.8** from 2000 and 2010 to estimate the yearly increase in tuition and fees.

Yearly increase in tuition and fees | is approximately | $\dfrac{\text{change in tuition and fees from 2000 to 2010}}{\text{change in time from 2000 to 2010}}$.

$$\approx \frac{7020 - 3362}{2010 - 2000}$$

$$= \frac{3658}{10} = 365.8 \approx 366$$

Each year the average cost of tuition and fees for public four-year colleges is increasing by approximately \$366.

b. Now we can use variables to obtain a mathematical model that estimates the average cost of tuition and fees, T, for the school year ending x years after 2000.

The average cost of tuition and fees | is | tuition and fees in 2000 | plus | yearly increase in tuition and fees | times the number of years after 2000.

$$T \quad = \quad 3362 \quad + \quad 366x$$

The mathematical model $T = 3362 + 366x$ estimates the average cost of tuition and fees, T, at public four-year colleges for the school year ending x years after 2000.

c. Now let's use the mathematical model to project the average cost of tuition and fees for the school year ending in 2016. Because 2016 is 16 years after 2000, we substitute 16 for x.

$T = 3362 + 366x$ — This is the mathematical model from part (b).

$T = 3362 + 366(16)$ — Substitute 16 for x.

$= 3362 + 5856$ — Multiply: $366(16) = 5856$.

$= 9218$ — Add. On a calculator, enter 3362 [+] 366 [×] 16 and press [=] or [ENTER].

Our model projects that the average cost of tuition and fees at public four-year colleges for the school year ending in 2016 will be \$9218.

CHECK POINT 8 The bar graph in **Figure 1.9** on the next page shows the average cost of tuition and fees for private four-year colleges, adjusted for inflation.

a. Estimate the yearly increase in tuition and fees. Round to the nearest dollar.

b. Write a mathematical model that estimates the average cost of tuition and fees, T, at private four-year colleges for the school year ending x years after 2000.

c. Use the mathematical model from part (b) to project the average cost of tuition and fees at private four-year colleges for the school year ending in 2014.

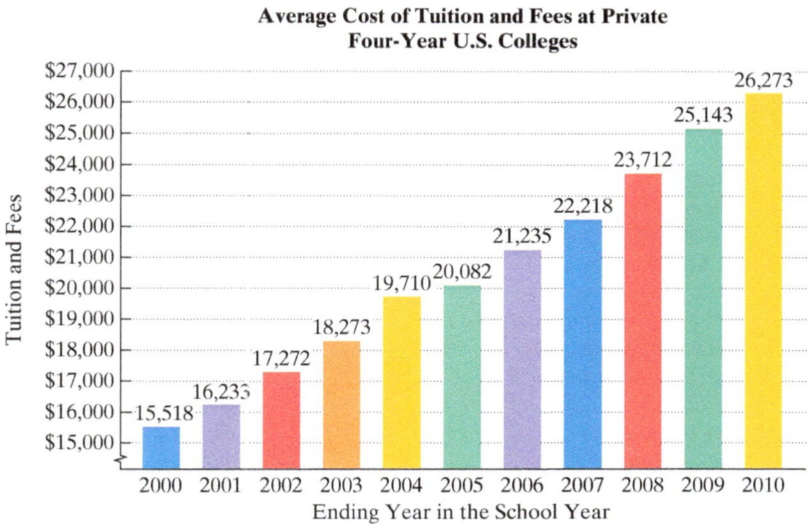

FIGURE 1.9
Source: The College Board

Sometimes a mathematical model gives an estimate that is not a good approximation or is extended to include values of the variable that do not make sense. In these cases, we say that **model breakdown** has occurred. Models that accurately describe data for the past 10 years might not serve as reliable predictions for what can reasonably be expected to occur in the future. Model breakdown can occur when formulas are extended too far into the future.

Concept and Vocabulary Check

Fill in each blank so that the resulting statement is true.

1. The process of arriving at an approximate answer to a computation such as 0.79×403 is called ________.
2. A graph that shows how a whole quantity is divided into parts is called a/an ________.
3. A formula that approximates real-world phenomena is called a/an ______________.
4. True or False: Decimal numbers are rounded by using the digit to the right of the digit where rounding is to occur. _____
5. True or False: Line graphs are often used to illustrate trends over time. _____
6. True or False: Mathematical modeling results in formulas that give exact values of real-world phenomena over time. _____

Exercise Set 1.2

Practice Exercises

The bar graph gives the 2011 populations of the ten most populous states in the United States. Use the appropriate information displayed by the graph to solve Exercises 1–2.

1. Round the population of New York to the nearest **a.** hundred, **b.** thousand, **c.** ten thousand, **d.** hundred thousand, **e.** million, **f.** ten million.
2. Select any state other than New York. For the state selected, round the population to the nearest **a.** hundred, **b.** thousand, **c.** ten thousand, **d.** hundred thousand, **e.** million, **f.** ten million.

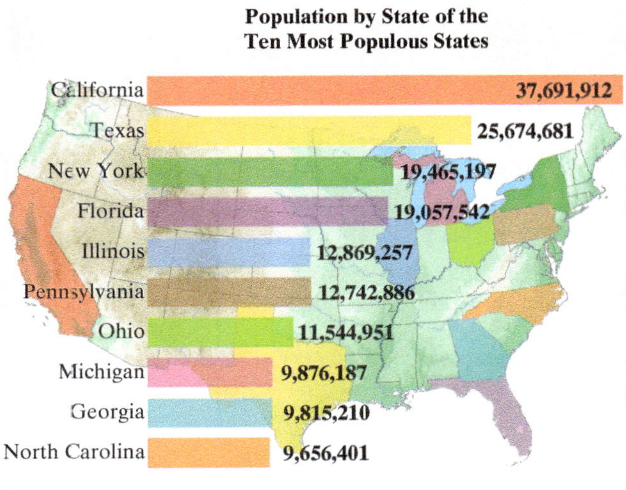

Source: U.S. Census Bureau

Pi goes on and on and on ...
And e is just as cursed.
I wonder: Which is larger
When their digits are reversed?

Martin Gardner

Although most people are familiar with π, the number e is more significant in mathematics, showing up in problems involving population growth and compound interest, and at the heart of the statistical bell curve. One way to think of e is the dollar amount you would have in a savings account at the end of the year if you invested \$1 at the beginning of the year and the bank paid an annual interest rate of 100% compounded continuously (compounding interest every trillionth of a second, every quadrillionth of a second, etc.). Although continuous compounding sounds terrific, at the end of the year your \$1 would have grown to a mere \$e, or \$2.72, rounded to the nearest cent. Here is a better approximation for e.

$$e \approx 2.718281828459045$$

In Exercises 3–8, use this approximation to round e as specified.

3. to the nearest thousandth

4. to the nearest ten-thousandth

5. to the nearest hundred-thousandth

6. to the nearest millionth

7. to nine decimal places

8. to ten decimal places

In Exercises 9–34, because different rounding results in different estimates, there is not one single, correct answer to each exercise.

In Exercises 9–22, obtain an estimate for each computation by rounding the numbers so that the resulting arithmetic can easily be performed by hand or in your head. Then use a calculator to perform the computation. How reasonable is your estimate when compared to the actual answer?

9. $359 + 596$

10. $248 + 797$

11. $8.93 + 1.04 + 19.26$

12. $7.92 + 3.06 + 24.36$

13. $32.15 - 11.239$

14. $46.13 - 15.237$

15. 39.67×5.5

16. 78.92×6.5

17. 0.79×414

18. 0.67×211

19. $47.83 \div 2.9$

20. $54.63 \div 4.7$

21. 32% of 187,253

22. 42% of 291,506

In Exercises 23–34, determine each estimate without using a calculator. Then use a calculator to perform the computation necessary to obtain an exact answer. How reasonable is your estimate when compared to the actual answer?

23. Estimate the total cost of six grocery items if their prices are \$3.47, \$5.89, \$19.98, \$2.03, \$11.85, and \$0.23.

24. Estimate the total cost of six grocery items if their prices are \$4.23, \$7.79, \$28.97, \$4.06, \$13.43, and \$0.74.

25. A full-time employee who works 40 hours per week earns \$19.50 per hour. Estimate that person's annual income.

26. A full-time employee who works 40 hours per week earns \$29.85 per hour. Estimate that person's annual income.

27. You lease a car at \$605 per month for 3 years. Estimate the total cost of the lease.

28. You lease a car at \$415 per month for 4 years. Estimate the total cost of the lease.

29. A raise of \$310,000 is evenly distributed among 294 professors. Estimate the amount each professor receives.

30. A raise of \$310,000 is evenly distributed among 196 professors. Estimate the amount each professor receives.

31. If a person who works 40 hours per week earns \$61,500 per year, estimate that person's hourly wage.

32. If a person who works 40 hours per week earns \$38,950 per year, estimate that person's hourly wage.

33. The average life expectancy in Canada is 80.1 years. Estimate the country's life expectancy in hours.

34. The average life expectancy in Mozambique is 40.3 years. Estimate the country's life expectancy in hours.

Practice Plus

In Exercises 35–36, obtain an estimate for each computation without using a calculator. Then use a calculator to perform the computation. How reasonable is your estimate when compared to the actual answer?

35. $\dfrac{0.19996 \times 107}{0.509}$

36. $\dfrac{0.47996 \times 88}{0.249}$

37. Ten people ordered calculators. The least expensive was \$19.95 and the most expensive was \$39.95. Half ordered a \$29.95 calculator. Select the best estimate of the amount spent on calculators.

a. \$240 **b.** \$310 **c.** \$345 **d.** \$355

38. Ten people ordered calculators. The least expensive was \$4.95 and the most expensive was \$12.95. Half ordered a \$6.95 calculator. Select the best estimate of the amount spent on calculators.

a. \$160 **b.** \$105 **c.** \$75 **d.** \$55

39. Traveling at an average rate of between 60 and 70 miles per hour for 3 to 4 hours, select the best estimate for the distance traveled.

a. 90 miles **b.** 190 miles
c. 225 miles **d.** 275 miles

40. Traveling at an average rate of between 40 and 50 miles per hour for 3 to 4 hours, select the best estimate for the distance traveled.

a. 120 miles **b.** 160 miles
c. 195 miles **d.** 210 miles

41. Imagine that you counted 60 numbers per minute and continued to count nonstop until you reached 10,000. Determine a reasonable estimate of the number of hours it would take you to complete the counting.

42. Imagine that you counted 60 numbers per minute and continued to count nonstop until you reached one million. Determine a reasonable estimate of the number of days it would take you to complete the counting.

Application Exercises

The circle graph shows the most important problems for the 16,503,611 high school teenagers in the United States. Use this information to solve Exercises 43–44.

Most Important Problems for High School Teenagers

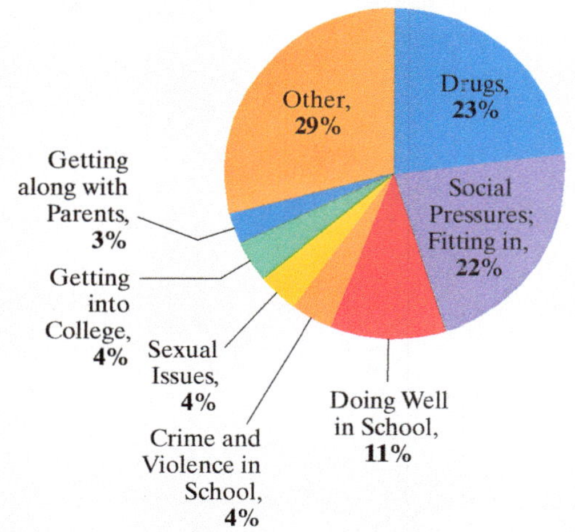

Source: Columbia University

43. Without using a calculator, estimate the number of high school teenagers for whom doing well in school is the most important problem.

44. Without using a calculator, estimate the number of high school teenagers for whom social pressures and fitting in is the most important problem.

An online test of English spelling looked at how well people spelled difficult words. The bar graph shows how many people per 100 spelled each word correctly. Use this information to solve Exercises 45–46.

Number of People per 100 Spelling Various Words Correctly

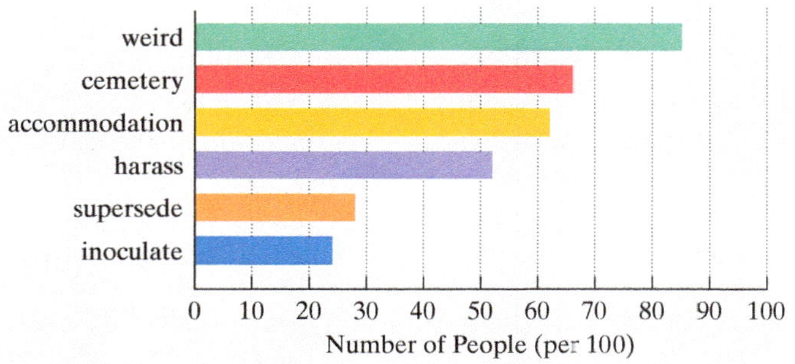

Source: Vivian Cook, *Accomodating Brocolli in the Cemetary or Why Can't Anybody Spell?*, Simon and Schuster, 2004

45. a. Estimate the number of people per 100 who spelled *weird* correctly.

b. In a group consisting of 8729 randomly selected people, estimate how many more people can correctly spell *weird* than *inoculate*.

46. a. Estimate the number of people per 100 who spelled *cemetery* correctly.

b. In a group consisting of 7219 randomly selected people, estimate how many more people can correctly spell *cemetery* than *supersede*.

The percentage of U.S. college freshmen claiming no religious affiliation has risen in recent decades. The bar graph shows the percentage of first-year college students claiming no religious affiliation for four selected years from 1980 through 2010. Use this information to solve Exercises 47–48.

Percentage of First-Year U.S. College Students Claiming No Religious Affiliation

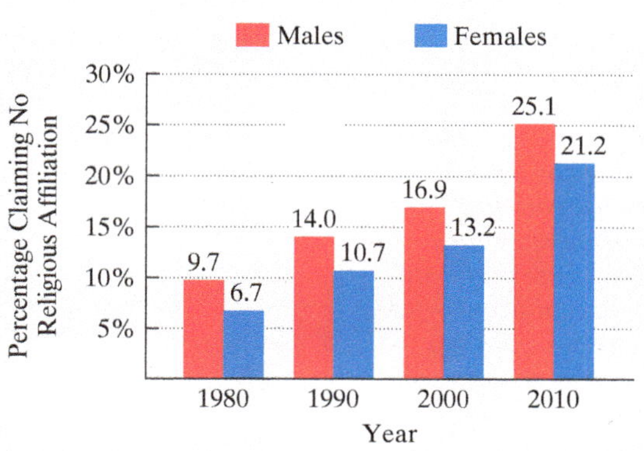

Source: John Macionis, *Sociology*, 14th Edition, Pearson, 2012.

47. a. Estimate the average yearly increase in the percentage of first-year college males claiming no religious affiliation. Round the percentage to the nearest tenth.

b. Estimate the percentage of first-year college males who will claim no religious affiliation in 2020.

48. a. Estimate the average yearly increase in the percentage of first-year college females claiming no religious affiliation. Round the percentage to the nearest tenth.

b. Estimate the percentage of first-year college females who will claim no religious affiliation in 2020.

According to a Gallup poll, in 2012 almost a third of Americans named Iran their country's greatest enemy. That number matched public sentiment in the United States toward Saddam Hussein's Iraq at the time of the 2003 invasion. The line graphs show the percentage of Americans who considered either Iraq or Iran their country's greatest enemy from 2001 through 2012. Use this information to solve Exercises 49–50.

Countries Americans Consider Their Greatest Enemy

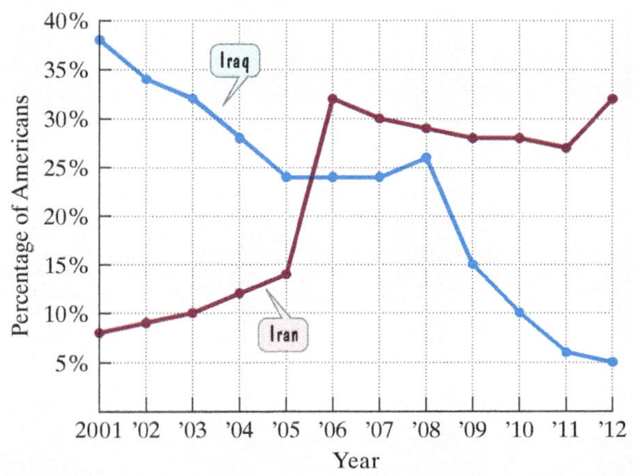

Source: Gallup

(In Exercises 49–50, refer to the graph at the bottom of the previous page.)

49. a. Find an estimate for the percentage of Americans who considered Iraq their country's greatest enemy in 2001.

b. Between which two years did the percentage of Americans who considered Iraq their country's greatest enemy decrease at the greatest rate?

c. In which year did 32% of Americans consider Iraq their country's greatest enemy?

50. a. Find an estimate for the percentage of Americans who considered Iran their country's greatest enemy in 2001.

b. Between which two years did the percentage of Americans who considered Iran their country's greatest enemy increase at the greatest rate?

c. In which year did 12% of Americans consider Iran their country's greatest enemy?

There is a strong scientific consensus that human activities are changing the Earth's climate. Scientists now believe that there is a striking correlation between atmospheric carbon dioxide concentration and global temperature. As both of these variables increase at significant rates, there are warnings of a planetary emergency that threatens to condemn coming generations to a catastrophically diminished future. The bar graphs give the average atmospheric concentration of carbon dioxide and the average global temperature for seven selected years. Use this information to solve Exercises 51–52.

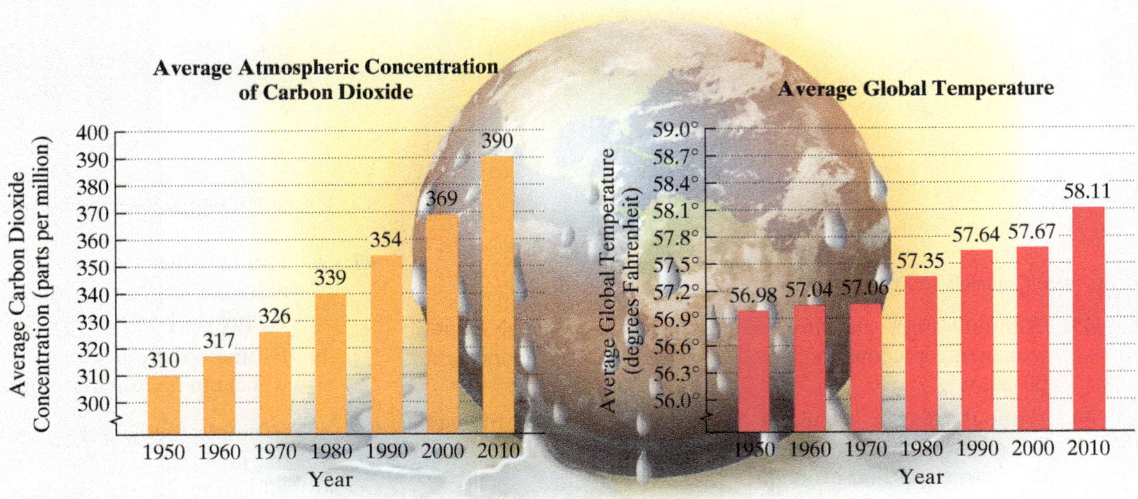

Source: National Oceanic and Atmospheric Administration

51. a. Estimate the yearly increase in the average atmospheric concentration of carbon dioxide. Express the answer in parts per million, rounded to the nearest hundredth.

b. Write a mathematical model that estimates the average atmospheric concentration of carbon dioxide, C, in parts per million, x years after 1950.

c. If the trend shown by the data continues, use your mathematical model from part (b) to project the average atmospheric concentration of carbon dioxide in 2050.

52. a. Estimate the yearly increase in the average global temperature, rounded to the nearest hundredth of a degree.

b. Write a mathematical model that estimates the average global temperature, T, in degrees Fahrenheit, x years after 1950.

c. If the trend shown by the data continues, use your mathematical model from part (b) to project the average global temperature in 2050.

Writing in Mathematics

53. What is estimation? When is it helpful to use estimation?

54. Explain how to round 218,543 to the nearest thousand and to the nearest hundred thousand.

55. Explain how to round 14.26841 to the nearest hundredth and to the nearest thousandth.

56. What does the $\approx$ symbol mean?

57. In this era of calculators and computers, why is there a need to develop estimation skills?

58. Describe a circle graph.

59. Describe a bar graph.

60. Describe a line graph.

61. What does it mean when we say that a formula models real-world phenomena?

62. In 1970, 5% of lawyers in the United States were women. By 2012, 32% of lawyers were women, indicating an increase of approximately 0.64% per year. (*Source*: Bureau of Labor Statistics) Describe how to use this information to write a mathematical model that estimates the percentage of women lawyers, W, x years after 1970.

63. Explain how to use the mathematical model from Exercise 62 to predict the percentage of U.S. lawyers who will be women in 2020.

64. Describe one way in which you use estimation in a nonacademic area of your life.

65. A forecaster at the National Hurricane Center needs to estimate the time until a hurricane with high probability of striking South Florida will hit Miami. Is it better to overestimate or underestimate? Explain your answer.

Critical Thinking Exercises

Make Sense? *In Exercises 66–69, determine whether each statement makes sense or does not make sense, and explain your reasoning.*

66. When buying several items at the market, I use estimation before going to the cashier to be sure I have enough money to pay for the purchase.

67. It's not necessary to use estimation skills when using my calculator.

68. Being able to compute an exact answer requires a different ability than estimating the reasonableness of the answer.

69. My mathematical model estimates the data for the past 10 years extremely well, so it will serve as an accurate prediction for what will occur in 2050.

70. Take a moment to read the verse preceding Exercises 3–8 that mentions the numbers π and e, whose decimal representations continue infinitely with no repeating patterns. The verse was written by the American mathematician (and accomplished amateur magician!) Martin Gardner (1914–2010), author of more than 60 books and best known for his "Mathematical Games" column, which ran in *Scientific American* for 25 years. Explain the humor in Gardner's question.

In Exercises 71–74, match the story with the correct graph. The graphs are labeled (a), (b), (c), and (d).

71. As the blizzard got worse, the snow fell harder and harder.

72. The snow fell more and more softly.

73. It snowed hard, but then it stopped. After a short time, the snow started falling softly.

74. It snowed softly, and then it stopped. After a short time, the snow started falling hard.

a.

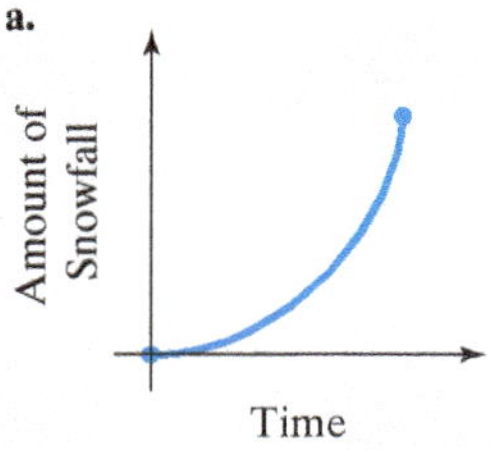

b.

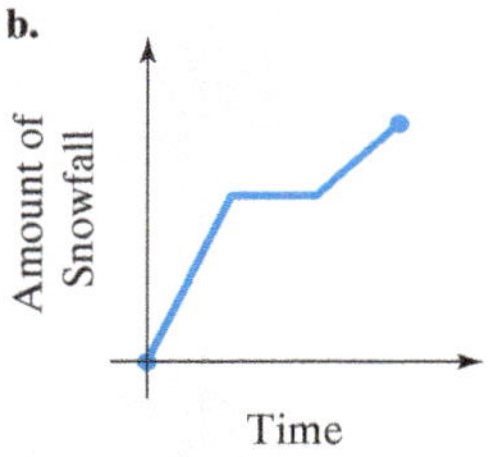

c.

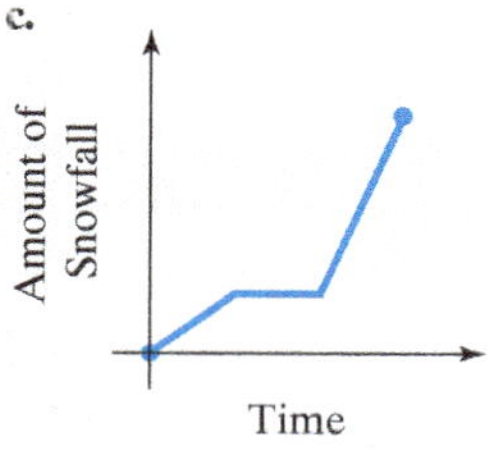

d.

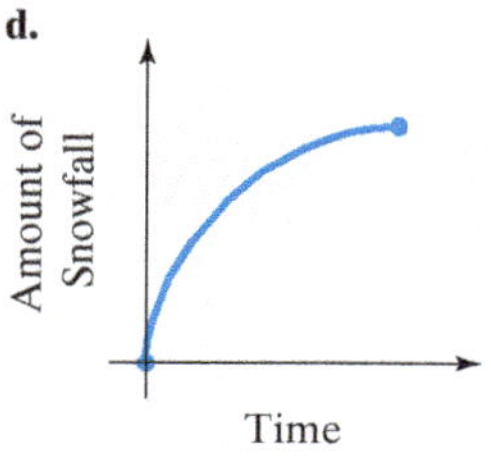

75. American children ages 2 to 17 spend 19 hours 40 minutes per week watching television. (*Source*: TV-Turnoff Network) From ages 2 through 17, inclusive, estimate the number of days an American child spends watching television. How many years, to the nearest tenth of a year, is that?

76. If you spend $1000 each day, estimate how long it will take to spend a billion dollars.

Group Exercises

77. Group members should devise an estimation process that can be used to answer each of the following questions. Use input from all group members to describe the best estimation process possible.

- **a.** Is it possible to walk from San Francisco to New York in a year?
- **b.** How much money is spent on ice cream in the United States each year?

78. Group members should begin by consulting an almanac, newspaper, magazine, or the Internet to find two graphs that show "intriguing" data changing from year to year. In one graph, the data values should be increasing relatively steadily. In the second graph, the data values should be decreasing relatively steadily. For each graph selected, write a mathematical model that estimates the changing variable x years after the graph's starting date. Then use each mathematical model to make predictions about what might occur in the future. Are there circumstances that might affect the accuracy of the prediction? List some of these circumstances.

1.3 Problem Solving

WHAT AM I SUPPOSED TO LEARN?

After you have read this section, you should be able to:

1 Solve problems using the organization of the four-step problem-solving process.

CRITICAL THINKING AND PROBLEM SOLVING are essential skills in both school and work. A model for problem solving was established by the charismatic teacher and mathematician George Polya (1887–1985) in *How to Solve It* (Princeton University Press, Princeton, NJ, 1957). This book, first published in 1945, has sold more than one million copies and is available in 17 languages. Using a four-step procedure for problem solving, Polya's book demonstrates how to think clearly in any field.

1 Solve problems using the organization of the four-step problem-solving process.

"If you don't know where you're going, you'll probably end up some place else."
—Yogi Berra

POLYA'S FOUR STEPS IN PROBLEM SOLVING

Step 1 **Understand the problem.** Read the problem several times. The first reading can serve as an overview. In the second reading, write down what information is given and determine exactly what it is that the problem requires you to find.

Step 2 **Devise a plan.** The plan for solving the problem might involve one or more of these suggested problem-solving strategies:

- Use inductive reasoning to look for a pattern.
- Make a systematic list or a table.
- Use estimation to make an educated guess at the solution. Check the guess against the problem's conditions and work backward to eventually determine the solution.
- Try expressing the problem more simply and solve a similar simpler problem.
- Use trial and error.
- List the given information in a chart or table.
- Try making a sketch or a diagram to illustrate the problem.
- Relate the problem to a similar problem that you have seen before. Try applying the procedures used to solve the similar problem to the new one.
- Look for a "catch" if the answer seems too obvious. Perhaps the problem involves some sort of trick question deliberately intended to lead the problem solver in the wrong direction.
- Use the given information to eliminate possibilities.
- Use common sense.

Step 3 **Carry out the plan and solve the problem.**

Step 4 **Look back and check the answer.** The answer should satisfy the conditions of the problem. The answer should make sense and be reasonable. If this is not the case, recheck the method and any calculations. Perhaps there is an alternate way to arrive at a correct solution.

GREAT QUESTION!

Should I memorize Polya's four steps in problem solving?

Not necessarily. Think of Polya's four steps as guidelines that will help you organize the process of problem solving, rather than a list of rigid rules that need to be memorized. You may be able to solve certain problems without thinking about or using every step in the four-step process.

The very first step in problem solving involves evaluating the given information in a deliberate manner. Is there enough given to solve the problem? Is the information relevant to the problem's solution, or are some facts not necessary to arrive at a solution?

EXAMPLE 1 *Finding What Is Missing*

Which necessary piece of information is missing and prevents you from solving the following problem?

> A man purchased five shirts, each at the same discount price. How much did he pay for them?

SOLUTION

Step 1 **Understand the problem.** Here's what is given:

Number of shirts purchased: 5.

We must find how much the man paid for the five shirts.

Step 2 **Devise a plan.** The amount that the man paid for the five shirts is the number of shirts, 5, times the cost of each shirt. The discount price of each shirt is not given. This missing piece of information makes it impossible to solve the problem.

CHECK POINT 1 Which necessary piece of information is missing and prevents you from solving the following problem?

The bill for your meal totaled $20.36, including the tax. How much change should you receive from the cashier?

EXAMPLE 2 *Finding What Is Unnecessary*

In the following problem, one more piece of information is given than is necessary for solving the problem. Identify this unnecessary piece of information. Then solve the problem.

A roll of E-Z Wipe paper towels contains 100 sheets and costs $1.38. A comparable brand, Kwik-Clean, contains five dozen sheets per roll and costs $1.23. If you need three rolls of paper towels, which brand is the better value?

SOLUTION

Step 1 **Understand the problem.** Here's what is given:

E-Z Wipe: 100 sheets per roll; $1.38

Kwik-Clean: 5 dozen sheets per roll; $1.23

Needed: 3 rolls.

We must determine which brand offers the better value.

Step 2 **Devise a plan.** The brand with the better value is the one that has the lower price per sheet. Thus, we can compare the two brands by finding the cost for one sheet of E-Z Wipe and one sheet of Kwik-Clean. The price per sheet, or the *unit price*, is the price of a roll divided by the number of sheets in the roll. The fact that three rolls are required is not relevant to the problem. This unnecessary piece of information is not needed to find which brand is the better value.

Step 3 **Carry out the plan and solve the problem.**

E-Z Wipe:

$$\text{price per sheet} = \frac{\text{price of a roll}}{\text{number of sheets per roll}} = \frac{\$1.38}{100\text{ sheets}} = \$0.0138 \approx \$0.01$$

Kwik-Clean:

$$\text{price per sheet} = \frac{\text{price of a roll}}{\text{number of sheets per roll}} = \frac{\$1.23}{60\text{ sheets}} = \$0.0205 \approx \$0.02$$

5 dozen = 5 × 12, or 60 sheets

By comparing unit prices, we see that E-Z Wipe, at approximately $0.01 per sheet, is the better value.

Step 4 Look back and check the answer. We can double-check the arithmetic in each of our unit-price computations. We can also see if these unit prices satisfy the problem's conditions. The product of each brand's price per sheet and the number of sheets per roll should result in the given price for a roll.

E-Z Wipe: Check \$0.0138 Kwik-Clean: Check \$0.0205

$$\$0.0138 \times 100 = \$1.38 \qquad \$0.0205 \times 60 = \$1.23$$

These are the given prices for a roll of each respective brand.

The unit prices satisfy the problem's conditions.

A generalization of our work in Example 2 allows you to compare different brands and make a choice among various products of different sizes. When shopping at the supermarket, a useful number to keep in mind is a product's *unit price*. The **unit price** is the total price divided by the total units. Among comparable brands, the best value is the product with the lowest unit price, assuming that the units are kept uniform.

The word *per* is used to state unit prices. For example, if a 12-ounce box of cereal sells for \$3.00, its unit price is determined as follows:

$$\text{Unit price} = \frac{\text{total price}}{\text{total units}} = \frac{\$3.00}{12 \text{ ounces}} = \$0.25 \text{ per ounce.}$$

Blitzer Bonus

Unit Prices and Sneaky Pricejacks

In *200% of Nothing* (John Wiley & Sons, 1993), author A. K. Dewdney writes, "It must be something of a corporate dream come true when a company charges more for a product and no one notices." He gives two examples of "sneaky pricejacks," both easily detected using unit prices. The manufacturers of Mennen Speed Stick deodorant increased the size of the package that held the stick, left the price the same, and reduced the amount of actual deodorant in the stick from 2.5 ounces to 2.25. Fabergé's Brut left the price and size of its cologne jar the same, but reduced its contents from 5 ounces to 4. Surprisingly, the new jar read, "Now, more Brut!" *Consumer Reports* contacted Fabergé to see how this could be possible. Their response: The new jar contained "more fragrance." *Consumer Reports* moaned, "Et tu Brut?"

CHECK POINT 2 Solve the following problem. If the problem contains information that is not relevant to its solution, identify this unnecessary piece of information.

A manufacturer packages its apple juice in bottles and boxes. A 128-ounce bottle costs \$5.39, and a 9-pack of 6.75-ounce boxes costs \$3.15. Which packaging option is the better value?

EXAMPLE 3 *Applying the Four-Step Procedure*

By paying \$100 cash up front and the balance at \$20 a week, how long will it take to pay for a bicycle costing \$680?

SOLUTION

Step 1 Understand the problem. Here's what is given:

Cost of the bicycle: \$680

Amount paid in cash: \$100

Weekly payments: \$20.

If necessary, consult a dictionary to look up any unfamiliar words. The word *balance* means the amount still to be paid. We must find the balance to determine the number of weeks required to pay off the bicycle.

Step 2 Devise a plan. Subtract the amount paid in cash from the cost of the bicycle. This results in the amount still to be paid. Because weekly payments are \$20, divide the amount still to be paid by 20. This will give the number of weeks required to pay for the bicycle.

GREAT QUESTION!

Is there a strategy I can use to determine whether I understand a problem?

An effective way to see if you understand a problem is to restate the problem in your own words.

Step 3 **Carry out the plan and solve the problem.** Begin by finding the balance, the amount still to be paid for the bicycle.

$680	cost of the bicycle
−$100	amount paid in cash
$580	amount still to be paid

Now divide the $580 balance by $20, the payment per week. The result of the division is the number of weeks needed to pay off the bicycle.

$$\frac{\$580}{\frac{\$20}{\text{week}}} = \$580 \times \frac{\text{week}}{\$20} = \frac{580 \text{ weeks}}{20} = 29 \text{ weeks}$$

It will take 29 weeks to pay for the bicycle.

Step 4 **Look back and check the answer.** We can certainly double-check the arithmetic either by hand or with a calculator. We can also see if the answer, 29 weeks to pay for the bicycle, satisfies the condition that the bicycle costs $680.

This is the answer we are checking.

$20	weekly payment	$580	total of weekly payments
× 29	number of weeks	+$100	amount paid in cash
$580	total of weekly payments	$680	cost of bicycle

The answer of 29 weeks satisfies the condition that the cost of the bicycle is $680.

CHECK POINT 3 By paying $350 cash up front and the balance at $45 per month, how long will it take to pay for a computer costing $980?

Making lists is a useful strategy in problem solving.

EXAMPLE 4 *Solving a Problem by Making a List*

Suppose you are an engineer programming the automatic gate for a 50-cent toll. The gate should accept exact change only. It should not accept pennies. How many coin combinations must you program the gate to accept?

SOLUTION

Step 1 **Understand the problem.** The total change must always be 50 cents. One possible coin combination is two quarters. Another is five dimes. We need to count all such combinations.

Step 2 **Devise a plan.** Make a list of all possible coin combinations. Begin with the coins of larger value and work toward the coins of smaller value.

Step 3 **Carry out the plan and solve the problem.** First we must find all of the coins that are not pennies but can combine to form 50 cents. This includes half-dollars, quarters, dimes, and nickels. Now we can set up a table. We will use these coins as table headings.

Half-Dollars	Quarters	Dimes	Nickels

Blitzer Bonus

Trick Questions

Think about the following questions carefully before answering because each contains some sort of trick or catch.

Sample: Do they have a fourth of July in England?

Answer: Of course, they do. However, there is no national holiday on that date!

See if you can answer the questions that follow without developing mental whiplash. The answers appear in the answer section.

1. A farmer had 17 sheep. All but 12 died. How many sheep does the farmer have left?
2. Some months have 30 days. Some have 31. How many months have 28 days?
3. A doctor had a brother, but this brother had no brothers. What was the relationship between doctor and brother?
4. If you had only one match and entered a log cabin in which there was a candle, a fireplace, and a woodburning stove, which should you light first?

Half-Dollars	Quarters	Dimes	Nickels

Each row in the table (repeated above) will represent one possible combination for exact change. We start with the largest coin, the half-dollar. Only one half-dollar is needed to make exact change. No other coins are needed. Thus, we put a 1 in the half-dollars column and 0s in the other columns to represent the first possible combination.

Half-Dollars	Quarters	Dimes	Nickels
1	0	0	0

Likewise, two quarters are also exact change for 50 cents. We put a 0 in the half-dollars column, a 2 in the quarters column, and 0s in the columns for dimes and nickels.

Half-Dollars	Quarters	Dimes	Nickels
1	0	0	0
0	2	0	0

In this manner, we can find all possible combinations for exact change for the 50-cent toll. These combinations are shown in **Table 1.3**.

TABLE 1.3 Exact Change for 50 Cents: No Pennies

Half-Dollars	Quarters	Dimes	Nickels
1	0	0	0
0	2	0	0
0	1	2	1
0	1	1	3
0	1	0	5
0	0	5	0
0	0	4	2
0	0	3	4
0	0	2	6
0	0	1	8
0	0	0	10

Count the coin combinations shown in **Table 1.3**. How many coin combinations must the gate accept? You must program the gate to accept 11 coin combinations.

Step 4 **Look back and check the answer.** Double-check **Table 1.3** to make sure that no possible combinations have been omitted and that the total in each row is 50 cents. Double-check your count of the number of combinations.

CHECK POINT 4 Suppose you are an engineer programming the automatic gate for a 30-cent toll. The gate should accept exact change only. It should not accept pennies. How many coin combinations must you program the gate to accept?

Sketches and diagrams are sometimes useful in problem solving.

EXAMPLE 5 *Solving a Problem by Using a Diagram*

Four runners are in a one-mile race: Maria, Aretha, Thelma, and Debbie. Points are awarded only to the women finishing first or second. The first-place winner gets more points than the second-place winner. How many different arrangements of first- and second-place winners are possible?

SOLUTION

Step 1 **Understand the problem.** Three possibilities for first and second position are

Maria-Aretha
Maria-Thelma
Aretha-Maria.

Notice that Maria finishing first and Aretha finishing second is a different outcome than Aretha finishing first and Maria finishing second. Order makes a difference because the first-place winner gets more points than the second-place winner. We must count all possibilities for first and second position.

Step 2 **Devise a plan.** If Maria finishes first, then each of the other three runners could finish second:

First place	Second place	Possibilities for first and second place
Maria	Aretha	Maria-Aretha
	Thelma	Maria-Thelma
	Debbie	Maria-Debbie

Similarly, we can list each woman as the possible first-place runner. Then we will list the other three women as possible second-place runners. Next we will determine the possibilities for first and second place. This diagram will show how the runners can finish first or second.

Step 3 **Carry out the plan and solve the problem.** Now we complete the diagram started in step 2. The diagram is shown in **Figure 1.10**.

First place	Second place	Possibilities for first and second place
Maria	Aretha	Maria-Aretha
	Thelma	Maria-Thelma
	Debbie	Maria-Debbie
Aretha	Maria	Aretha-Maria
	Thelma	Aretha-Thelma
	Debbie	Aretha-Debbie
Thelma	Maria	Thelma-Maria
	Aretha	Thelma-Aretha
	Debbie	Thelma-Debbie
Debbie	Maria	Debbie-Maria
	Aretha	Debbie-Aretha
	Thelma	Debbie-Thelma

Because of the way **Figure 1.10** branches from first to second place, it is called a **tree diagram.** We will be using tree diagrams in Chapter 11 as a problem-solving tool in the study of uncertainty and probability.

FIGURE 1.10 Possible ways for four runners to finish first and second

Count the number of possibilities shown under the third column, "Possibilities for first and second place." Can you see that there are 12 possibilities? Therefore, 12 different arrangements of first- and second-place winners are possible.

Step 4 **Look back and check the answer.** Check the diagram in **Figure 1.10** to make sure that no possible first- and second-place outcomes have been left out. Double-check your count for the winning pairs of runners.

CHECK POINT 5 Your "lecture wardrobe" is rather limited—just two pairs of jeans to choose from (one blue, one black) and three T-shirts to choose from (one beige, one yellow, and one blue). How many different outfits can you form?

In Chapter 14, we will be studying diagrams, called *graphs*, that provide structures for describing relationships. In Example 6, we use such a diagram to illustrate the relationship between cities and one-way airfares between them.

EXAMPLE 6 *Using a Reasonable Option to Solve a Problem with More Than One Solution*

A sales director who lives in city A is required to fly to regional offices in cities B, C, D, and E. Other than starting and ending the trip in city A, there are no restrictions as to the order in which the other four cities are visited.

The one-way fares between each of the cities are given in **Table 1.4**. A diagram that illustrates this information is shown in **Figure 1.11**.

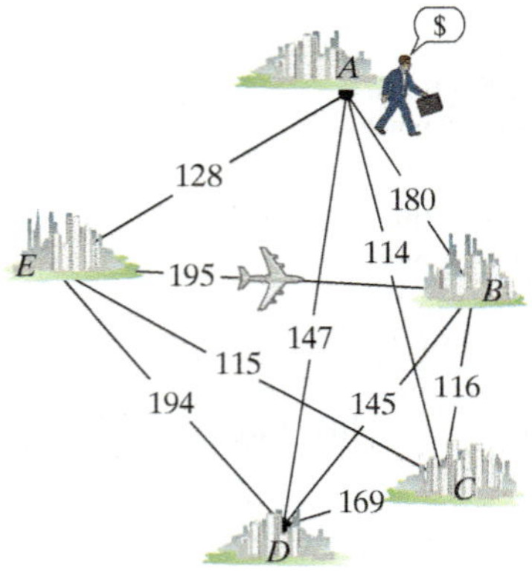

FIGURE 1.11

TABLE 1.4 One-Way Airfares

	A	B	C	D	E
A	*	$180	$114	$147	$128
B	$180	*	$116	$145	$195
C	$114	$116	*	$169	$115
D	$147	$145	$169	*	$194
E	$128	$195	$115	$194	*

Give the sales director an order for visiting cities B, C, D, and E once, returning home to city A, for less than $750.

SOLUTION

Step 1 Understand the problem. There are many ways to visit cities B, C, D, and E once, and return home to A. One route is

$$A, E, D, C, B, A.$$

Fly from A to E to D to C to B and then back to A.

The cost of this trip involves the sum of five costs, shown in both **Table 1.4** and **Figure 1.11**:

$$\$128 + \$194 + \$169 + \$116 + \$180 = \$787.$$

We must find a route that costs less than $750.

Step 2 Devise a plan. The sales director starts at city A. From there, fly to the city to which the airfare is cheapest. Then from there fly to the next city to which the airfare is cheapest, and so on. From the last of the cities, fly home to city A. Compute the cost of this trip to see if it is less than $750. If it is not, use trial and error to find other possible routes and select an order (if there is one) whose cost is less than $750.

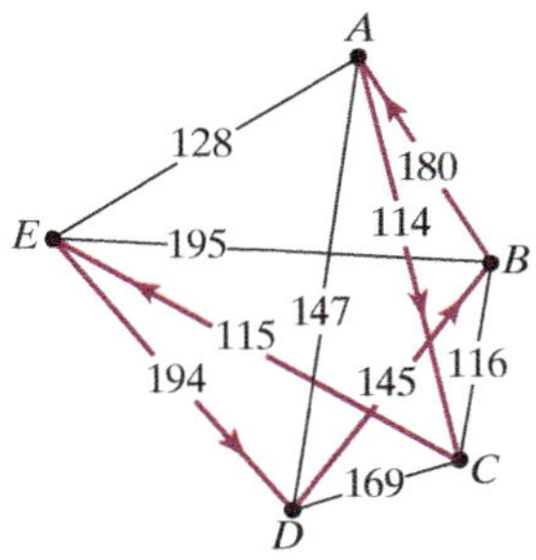

FIGURE 1.12

Step 3 **Carry out the plan and solve the problem.** See **Figure 1.12.** The route is indicated using red lines with arrows.

- Start at A.
- Choose the line segment with the smallest number: 114. Fly from A to C. (cost: \$114)
- From C, choose the line segment with the smallest number that does not lead to A: 115. Fly from C to E. (cost: \$115)
- From E, choose the line segment with the smallest number that does not lead to a city already visited: 194. Fly from E to D. (cost: \$194)
- From D, there is little choice but to fly to B, the only city not yet visited. (cost: \$145)
- From B, return home to A. (cost: \$180)

The route that we are considering is

$$A, C, E, D, B, A.$$

Let's see if the cost is less than \$750. The cost is

$$\$114 + \$115 + \$194 + \$145 + \$180 = \$748.$$

Because the cost is less than \$750, the sales director can follow the order A, C, E, D, B, A.

Step 4 **Look back and check the answer.** Use **Table 1.4** on the previous page or **Figure 1.12** to verify that the five numbers used in the sum shown above are correct. Use estimation to verify that \$748 is a reasonable cost for the trip.

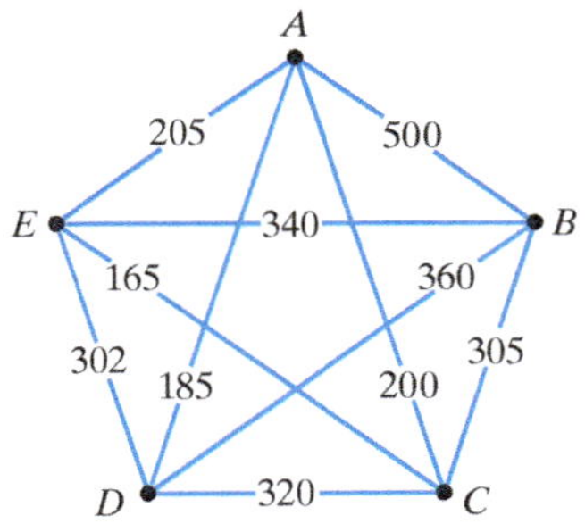

FIGURE 1.13

CHECK POINT 6 As in Example 6, a sales director who lives in city A is required to fly to regional offices in cities B, C, D, and E. The diagram in **Figure 1.13** shows the one-way airfares between any two cities. Give the sales director an order for visiting cities B, C, D, and E once, returning home to city A, for less than \$1460.

Concept and Vocabulary Check

Fill in each blank so that the resulting statement is true.

1. The first step in problem solving is to read the problem several times in order to __________ the problem.
2. The second step in problem solving is to __________ for solving the problem.
3. True or False: Polya's four steps in problem solving make it possible to obtain answers to problems even if necessary pieces of information are missing. _____
4. True or False: When making a choice between various sizes of a product, the best value is the size with the lowest price. _____

Exercise Set 1.3

Everyone can become a better, more confident problem solver. As in learning any other skill, learning problem solving requires hard work and patience. Work as many problems as possible in this Exercise Set. You may feel confused once in a while, but do not be discouraged. Thinking about a particular problem and trying different methods can eventually lead to new insights. Be sure to check over each answer carefully!

Practice and Application Exercises

In Exercises 1–4, what necessary piece of information is missing that prevents solving the problem?

1. If a student saves \$35 per week, how long will it take to save enough money to buy a computer?
2. If a steak sells for \$8.15, what is the cost per pound?

3. If it takes you 4 minutes to read a page in a book, how many words can you read in one minute?

4. By paying $1500 cash and the balance in equal monthly payments, how many months would it take to pay for a car costing $12,495?

In Exercises 5–8, one more piece of information is given than is necessary for solving the problem. Identify this unnecessary piece of information. Then solve the problem.

5. A salesperson receives a weekly salary of $350. In addition, $15 is paid for every item sold in excess of 200 items. How much extra is received from the sale of 212 items?

6. You have $250 to spend and you need to purchase four new tires. If each tire weighs 21 pounds and costs $42 plus $2.50 tax, how much money will you have left after buying the tires?

7. A parking garage charges $2.50 for the first hour and $0.50 for each additional hour. If a customer gave the parking attendant $20.00 for parking from 10 A.M. to 3 P.M., how much did the garage charge?

8. An architect is designing a house. The scale on the plan is 1 inch = 6 feet. If the house is to have a length of 90 feet and a width of 30 feet, how long will the line representing the house's length be on the blueprint?

Use Polya's four-step method in problem solving to solve Exercises 9–40.

9. a. Which is the better value: a 15.3-ounce box of cereal for $3.37 or a 24-ounce box of cereal for $4.59?

b. The supermarket displays the unit price for the 15.3-ounce box in terms of cost per ounce, but displays the unit price for the 24-ounce box in terms of cost per pound. What are the unit prices, to the nearest cent, given by the supermarket?

c. Based on your work in parts (a) and (b), does the better value always have the lower displayed unit price? Explain your answer.

10. a. Which is the better value: a 12-ounce jar of honey for $2.25 or an 18-ounce jar of honey for $3.24?

b. The supermarket displays the unit price for the 12-ounce jar in terms of cost per ounce, but displays the unit price for the 18-ounce jar in terms of cost per quart. Assuming 32 ounces in a quart, what are the unit prices, to the nearest cent, given by the supermarket?

c. Based on your work in parts (a) and (b), does the better value always have the lower displayed unit price? Explain your answer.

11. One person earns $48,000 per year. Another earns $3750 per month. How much more does the first person earn in a year than the second?

12. At the beginning of a year, the odometer on a car read 25,124 miles. At the end of the year, it read 37,364 miles. If the car averaged 24 miles per gallon, how many gallons of gasoline did it use during the year?

13. A television sells for $750. Instead of paying the total amount at the time of the purchase, the same television can be bought by paying $100 down and $50 a month for 14 months. How much is saved by paying the total amount at the time of the purchase?

14. In a basketball game, the Bulldogs scored 34 field goals, each counting 2 points, and 13 foul goals, each counting 1 point. The Panthers scored 38 field goals and 8 foul goals. Which team won? By how many points did it win?

15. Calculators were purchased at $65 per dozen and sold at $20 for three calculators. Find the profit on six dozen calculators.

16. Pens are bought at $0.95 per dozen and sold in groups of four for $2.25. Find the profit on 15 dozen pens.

17. Each day a small business owner sells 200 pizza slices at $1.50 per slice and 85 sandwiches at $2.50 each. If business expenses come to $60 per day, what is the owner's profit for a 10-day period?

18. A college tutoring center pays math tutors $8.15 per hour. Tutors earn an additional $2.20 per hour for each hour over 40 hours per week. A math tutor worked 42 hours one week and 45 hours the second week. How much did the tutor earn in this two-week period?

19. A car rents for $220 per week plus $0.25 per mile. Find the rental cost for a two-week trip of 500 miles for a group of three people.

20. A college graduate receives a salary of $2750 a month for her first job. During the year she plans to spend $4800 for rent, $8200 for food, $3750 for clothing, $4250 for household expenses, and $3000 for other expenses. With the money that is left, she expects to buy as many shares of stock at $375 per share as possible. How many shares will she be able to buy?

21. Charlene decided to ride her bike from her home to visit her friend Danny. Three miles away from home, her bike got a flat tire and she had to walk the remaining two miles to Danny's home. She could not repair the tire and had to walk all the way back home. How many more miles did Charlene walk than she rode?

22. A store received 200 containers of juice to be sold by April 1. Each container cost the store $0.75 and sold for $1.25. The store signed a contract with the manufacturer in which the manufacturer agreed to a $0.50 refund for every container not sold by April 1. If 150 containers were sold by April 1, how much profit did the store make?

23. A storeowner ordered 25 calculators that cost $30 each. The storeowner can sell each calculator for $35. The storeowner sold 22 calculators to customers. He had to return 3 calculators and pay a $2 charge for each returned calculator. Find the storeowner's profit.

24. New York City and Washington, D.C. are about 240 miles apart. A car leaves New York City at noon traveling directly south toward Washington, D.C. at 55 miles per hour. At the same time and along the same route, a second car leaves Washington, D.C. bound for New York City traveling directly north at 45 miles per hour. How far has each car traveled when the drivers meet for lunch at 2:24 P.M.?

25. An automobile purchased for $23,000 is worth $2700 after 7 years. Assuming that the car's value depreciated steadily from year to year, what was it worth at the end of the third year?

26. An automobile purchased for $34,800 is worth $8550 after 7 years. Assuming that the car's value depreciated steadily from year to year, what was it worth at the end of the third year?

27. A vending machine accepts nickels, dimes, and quarters. Exact change is needed to make a purchase. How many ways can a person with five nickels, three dimes, and two quarters make a 45-cent purchase from the machine?

28. How many ways can you make change for a quarter using only pennies, nickels, and dimes?

29. The members of the Student Activity Council on your campus are meeting to select two speakers for a month-long event celebrating artists and entertainers. The choices are Johnny Depp, Jamie Foxx, Jon Stewart, and Hilary Swank. How many different ways can the two speakers be selected?

30. The members of the Student Activity Council on your campus are meeting to select two speakers for a month-long event exploring why some people are most likely to succeed. The choices are Bill Clinton, Sean Combs, Donald Trump, Oprah Winfrey, and Tiger Woods. How many different ways can the two speakers be selected?

31. If you spend \$4.79, in how many ways can you receive change from a five-dollar bill?

32. If you spend \$9.74, in how many ways can you receive change from a ten-dollar bill?

33. You throw three darts at the board shown. Each dart hits the board and scores a 1, 5, or 10. How many different total scores can you make?

34. Suppose that you throw four darts at the board shown. With these four darts, there are 16 ways to hit four different numbers whose sum is 100. Describe one way you can hit four different numbers on the board that total 100.

35. Five housemates (A, B, C, D, and E) agreed to share the expenses of a party equally. If A spent \$42, B spent \$10, C spent \$26, D spent \$32, and E spent \$30, who owes money after the party and how much do they owe? To whom is money owed, and how much should they receive? In order to resolve these discrepancies, who should pay how much to whom?

36. Six houses are spaced equally around a circular road. If it takes 10 minutes to walk from the first house to the third house, how long would it take to walk all the way around the road?

37. If a test has four true/false questions, in how many ways can there be three answers that are false and one answer that is true?

38. There are five people in a room. Each person shakes the hand of every other person exactly once. How many handshakes are exchanged?

39. Five runners, Andy, Beth, Caleb, Darnell, and Ella, are in a one-mile race. Andy finished the race 7 seconds before Caleb. Caleb finished the race 2 seconds before Beth. Beth finished the race 6 seconds after Darnell. Ella finished the race 8 seconds after Darnell. In which order did the runners finish the race?

40. Eight teams are competing in a volleyball tournament. Any team that loses a game is eliminated from the tournament. How many games must be played to determine the tournament winner?

In Exercises 41–42, you have three errands to run around town, although in no particular order. You plan to start and end at home. You must go to the bank, the post office, and the dry cleaners. Distances, in miles, between any two of these locations are given in the diagram.

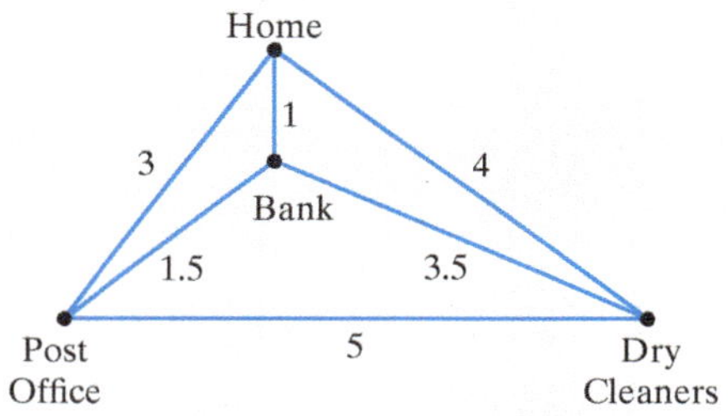

41. Determine a route whose distance is less than 12 miles for running the errands and returning home.

42. Determine a route whose distance exceeds 12 miles for running the errands and returning home.

43. The map shows five western states. Trace a route on the map that crosses each common state border exactly once.

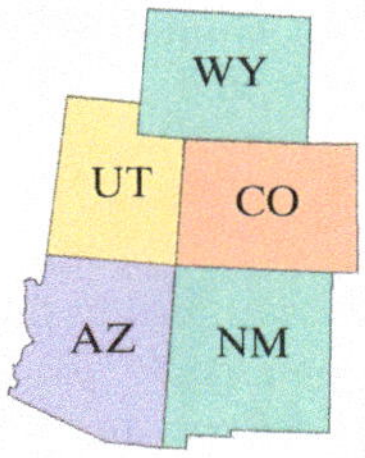

44. The layout of a city with land masses and bridges is shown. Trace a route that shows people how to walk through the city so as to cross each bridge exactly once.

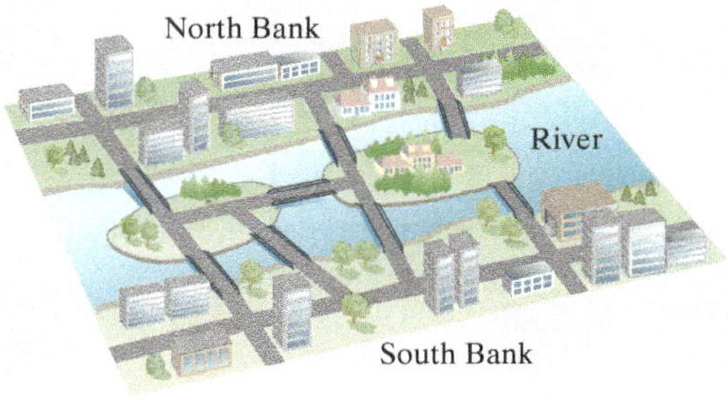

45. Jose, Bob, and Tony are college students living in adjacent dorm rooms. Bob lives in the middle dorm room. Their majors are business, psychology, and biology, although not necessarily in that order. The business major frequently uses the new computer in Bob's dorm room when Bob is in class. The psychology major and Jose both have 8 A.M. classes, and the psychology major knocks on Jose's wall to make sure he is awake. Determine Bob's major.

46. The figure represents a map of 13 countries. If countries that share a common border cannot be the same color, what is the minimum number of colors needed to color the map?

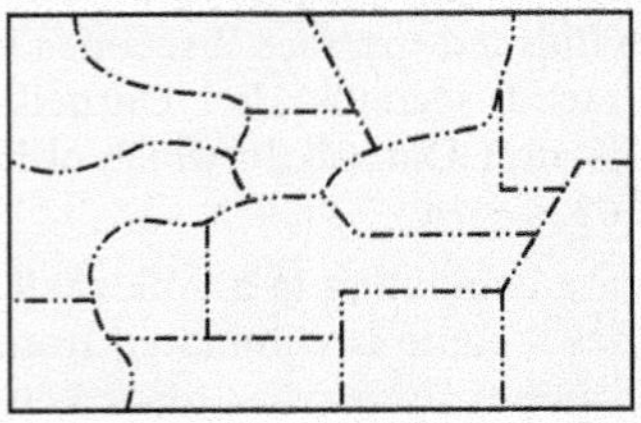

The sudoku (pronounced: sue-DOE-koo) craze, a number puzzle popular in Japan, hit the United States in 2005. A sudoku ("single number") puzzle consists of a 9-by-9 grid of 81 boxes subdivided into nine 3-by-3 squares. Some of the square boxes contain numbers. Here is an example:

2						6	7	
4	5							1
	9		7					
			3					
		4				2	8	
	1			2	4			
	8	3	9		7			
6	2			5		4	1	
							9	

The objective is to fill in the remaining squares so that every row, every column, and every 3-by-3 square contains each of the digits from 1 through 9 exactly once. (You can work this puzzle in Exercise 66, perhaps consulting one of the dozens of sudoku books in which the numerals 1 through 9 have created a cottage industry for publishers. There's even a Sudoku for Dummies.*)*

Trying to slot numbers into small checkerboard grids is not unique to sudoku. In Exercises 47–50, we explore some of the intricate patterns in other arrays of numbers, including magic squares. A **magic square** *is a square array of numbers arranged so that the numbers in all rows, all columns, and the two diagonals have the same sum. Here is an example of a magic square in which the sum of the numbers in each row, each column, and each diagonal is 15:*

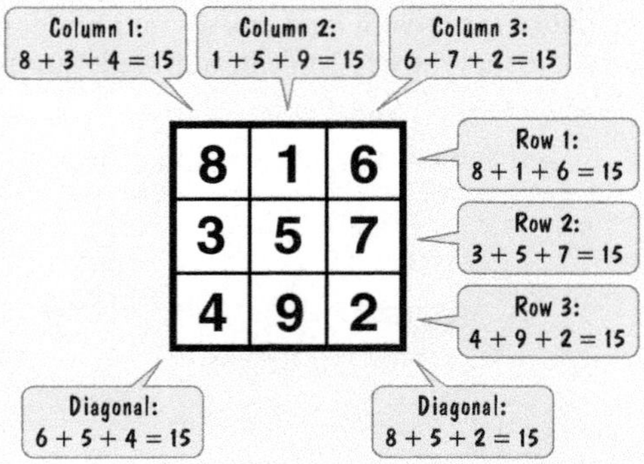

Exercises 47–48 are based on magic squares. (Be sure you have read the preceding discussion.)

47. a. Use the properties of a magic square to fill in the missing numbers.

5		18
	15	
		25

b. Show that the number of letters in the word for each number in the square in part (a) generates another magic square.

48. a. Use the properties of a magic square to fill in the missing numbers.

96		37	45
	43		
		25	57
23		82	78

b. Show that if you reverse the digits for each number in the square in part (a), another magic square is generated. (*Source* for the *alphamagic square* in Exercise 47 and the *mirrormagic square* in Exercise 48: Clifford A. Pickover, *A Passion for Mathematics*, John Wiley & Sons, Inc., 2005)

49. As in sudoku, fill in the missing numbers in the 3-by-3 square so that it contains each of the digits from 1 through 9 exactly once. Furthermore, in this *antimagic square*, the rows, the columns, and the two diagonals must have *different sums*.

9		7
	1	
3		5

50. The missing numbers in the 4-by-4 array are one-digit numbers. The sums for each row, each column, and one diagonal are listed in the voice balloons outside the array. Find the missing numbers.

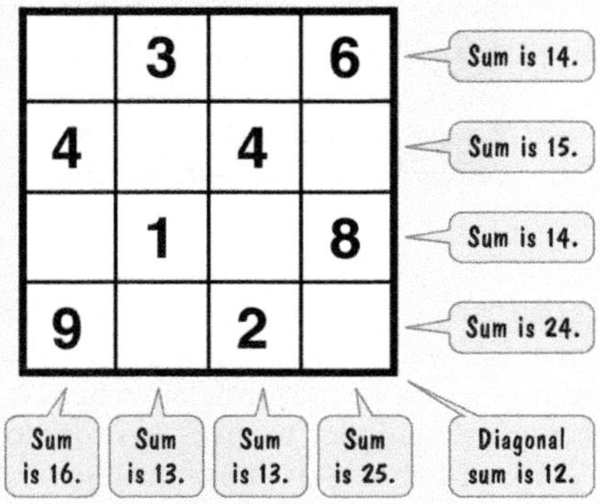

51. Some numbers in the printing of a division problem have become illegible. They are designated below by *. Fill in the blanks.

```
      1**
  **)4***
     28
     *56
     ***
      ***
      ***
        0
```

Writing in Mathematics

In Exercises 52–54, explain the plan needed to solve the problem.

52. If you know how much was paid for several pounds of steak, find the cost of one pound.

53. If you know a person's age, find the year in which that person was born.

54. If you know how much you earn each hour, find your yearly income.

55. Write your own problem that can be solved using the four-step procedure. Then use the four steps to solve the problem.

Critical Thinking Exercises

Make Sense? *In Exercises 56–59, determine whether each statement makes sense or does not make sense, and explain your reasoning.*

56. Polya's four steps in problem solving make it possible for me to solve any mathematical problem easily and quickly.

57. I used Polya's four steps in problem solving to deal with a personal problem in need of a creative solution.

58. I find it helpful to begin the problem-solving process by restating the problem in my own words.

59. When I get bogged down with a problem, there's no limit to the amount of time I should spend trying to solve it.

60. Gym lockers are to be numbered from 1 through 99 using metal numbers to be nailed onto each locker. How many 7s are needed?

61. You are on vacation in an isolated town. Everyone in the town was born there and has never left. You develop a toothache and check out the two dentists in town. One dentist has gorgeous teeth and one has teeth that show the effects of poor dental work. Which dentist should you choose and why?

62. India Jones is standing on a large rock in the middle of a square pool filled with hungry, man-eating piranhas. The edge of the pool is 20 feet away from the rock. India's mom wants to rescue her son, but she is standing on the edge of the pool with only two planks, each $19\frac{1}{2}$ feet long. How can India be rescued using the two planks?

63. One person tells the truth on Monday, Tuesday, Wednesday, and Thursday, but lies on all other days. A second person lies on Tuesday, Wednesday, and Thursday, but tells the truth on all other days. If both people state "I lied yesterday," then what day of the week is it today?

64. (This logic problem dates back to the eighth century.) A farmer needs to take his goat, wolf, and cabbage across a stream. His boat can hold him and one other passenger (the goat, wolf, or cabbage). If he takes the wolf with him, the goat will eat the cabbage. If he takes the cabbage, the wolf will eat the goat. Only when the farmer is present are the cabbage and goat safe from their respective predators. How does the farmer get everything across the stream?

65. As in sudoku, fill in the missing numbers along the sides of the triangle so that it contains each of the digits from 1 through 9 exactly once. Furthermore, each side of the triangle should contain four digits whose sum is 17.

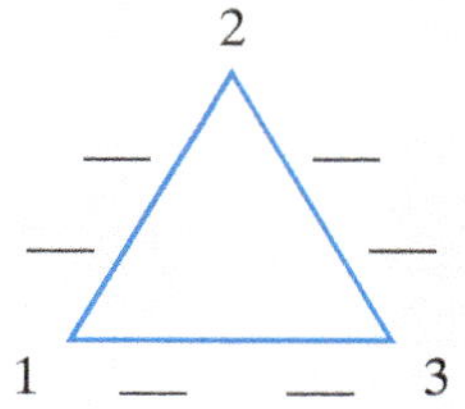

66. Solve the sudoku puzzle in the middle of the left column on page 40.

67. A version of this problem, called the *missing dollar problem*, first appeared in 1933. Three people eat at a restaurant and receive a total bill for \$30. They divide the amount equally and pay \$10 each. The waiter gives the bill and the \$30 to the manager, who realizes there is an error: The correct charge should be only \$25. The manager gives the waiter five \$1 bills to return to the customers, with the restaurant's apologies. However, the waiter is dishonest, keeping \$2 and giving back only \$3 to the customers. In conclusion, each of the three customers has paid \$9 and the waiter has stolen \$2, giving a total of \$29. However, the original bill was \$30. Where has the missing dollar gone?

68. A firefighter spraying water on a fire stood on the middle rung of a ladder. When the smoke became less thick, the firefighter moved up 4 rungs. However it got too hot, so the firefighter backed down 6 rungs. Later, the firefighter went up 7 rungs and stayed until the fire was out. Then, the firefighter climbed the remaining 4 rungs and entered the building. How many rungs does the ladder have?

69. The Republic of Margaritaville is composed of four states: A, B, C, and D. According to the country's constitution, the congress will have 30 seats, divided among the four states according to their respective populations. The table shows each state's population.

POPULATION OF MARGARITAVILLE BY STATE

State	A	B	C	D	Total
Population (in thousands)	275	383	465	767	1890

Allocate the 30 congressional seats among the four states in a fair manner.

Group Exercises

Exercises 70–74 describe problems that have many plans for finding an answer. Group members should describe how the four steps in problem solving can be applied to find a solution. It is not necessary to actually solve each problem. Your professor will let the group know if the four steps should be described verbally by a group spokesperson or in essay form.

70. How much will it cost to install bicycle racks on campus to encourage students to use bikes, rather than cars, to get to campus?

71. How many new counselors are needed on campus to prevent students from waiting in long lines for academic advising?

72. By how much would taxes in your state have to be increased to cut tuition at community colleges and state universities in half?

73. Is your local electric company overcharging its customers?

74. Should solar heating be required for all new construction in your community?

75. Group members should describe a problem in need of a solution. Then, as in Exercises 70–74, describe how the four steps in problem solving can be applied to find a solution.

Chapter Summary, Review, and Test

SUMMARY – DEFINITIONS AND CONCEPTS	EXAMPLES
1.1 Inductive and Deductive Reasoning	
a. Inductive reasoning is the process of arriving at a general conclusion based on observations of specific examples. The conclusion is called a conjecture or a hypothesis. A case for which a conjecture is false is called a counterexample.	Ex. 1, p. 4; Ex. 2, p. 5; Ex. 3, p. 6; Ex. 4, p. 7
b. Deductive reasoning is the process of proving a specific conclusion from one or more general statements. The statement that is proved is called a theorem.	Ex. 5, p. 9
1.2 Estimation, Graphs, and Mathematical Models	
a. The procedure for rounding whole numbers is given in the box on page 15. The symbol $\approx$ means *is approximately equal to*.	Ex. 1, p. 15
b. Decimal parts of numbers are rounded in nearly the same way as whole numbers. However, digits to the right of the rounding place are dropped.	Ex. 2, p. 16
c. Estimation is the process of arriving at an approximate answer to a question. Computations can be estimated by using rounding that results in simplified arithmetic.	Ex. 3, p. 17; Ex. 4, p. 17
d. Estimation is useful when interpreting information given by circle, bar, or line graphs.	Ex. 5, p. 19; Ex. 6, p. 20; Ex. 7, p. 21
e. The process of finding formulas to describe real-world phenomena is called mathematical modeling. Such formulas, together with the meaning assigned to the variables, are called mathematical models.	Ex. 8, p. 23
1.3 Problem Solving	
Polya's Four Steps in Problem Solving **1.** Understand the problem. **2.** Devise a plan. **3.** Carry out the plan and solve the problem. **4.** Look back and check the answer.	Ex. 1, p. 30; Ex. 2, p. 31; Ex. 3, p. 32; Ex. 4, p. 33; Ex. 5, p. 35; Ex. 6, p. 36

Review Exercises

1.1

1. Which reasoning process is shown in the following example? Explain your answer.

All books by Stephen King have made the best-seller list. *Carrie* is a novel by Stephen King. Therefore, *Carrie* was on the best-seller list.

2. Which reasoning process is shown in the following example? Explain your answer.

All books by Stephen King have made the best-seller list. Therefore, it is highly probable that the novel King is currently working on will make the best-seller list.

In Exercises 3–10, identify a pattern in each list of numbers. Then use this pattern to find the next number.

3. 4, 9, 14, 19, ____

4. 7, 14, 28, 56, ____

5. 1, 3, 6, 10, 15, ____

6. $\frac{3}{4}, \frac{3}{5}, \frac{1}{2}, \frac{3}{7}$, ____

7. 40, −20, 10, −5, ____

8. 40, −20, −80, −140, ____

9. 2, 2, 4, 6, 10, 16, 26, ____

10. 2, 6, 12, 36, 72, 216, ____

11. Identify a pattern in the following sequence of figures. Then use the pattern to find the next figure in the sequence.

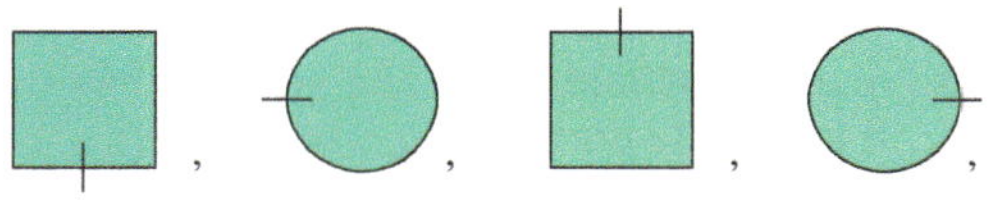

____

In Exercises 12–13, use inductive reasoning to predict the next line in each sequence of computations. Then perform the arithmetic to determine whether your conjecture is correct.

12.

$$2 = 4 - 2$$
$$2 + 4 = 8 - 2$$
$$2 + 4 + 8 = 16 - 2$$
$$2 + 4 + 8 + 16 = 32 - 2$$

13.

$$111 \div 3 = 37$$
$$222 \div 6 = 37$$
$$333 \div 9 = 37$$

14. Consider the following procedure:

Select a number. Double the number. Add 4 to the product. Divide the sum by 2. Subtract 2 from the quotient.

a. Repeat the procedure for four numbers of your choice. Write a conjecture that relates the result of the process to the original number selected.

b. Represent the original number by the variable n and use deductive reasoning to prove the conjecture in part (a).

1.2

15. The number 923,187,456 is called a *pandigital square* because it uses all the digits from 1 to 9 once each and is the square of a number:

$$30{,}384^2 = 30{,}384 \times 30{,}384 = 923{,}187{,}456.$$

(*Source:* David Wells, *The Penguin Dictionary of Curious and Interesting Numbers*)

Round the pandigital square 923,187,456 to the nearest

a. hundred.

b. thousand.

c. hundred thousand.

d. million.

e. hundred million.

16. A magnified view of the boundary of this black "buglike" shape, called the Mandelbrot set, was illustrated in the Section 1.1 opener on page 2.

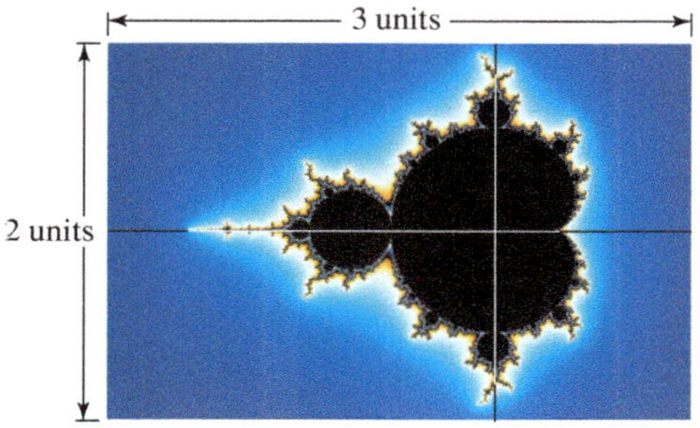

The area of the yellow rectangular region is the product of its length, 3 units, and its width, 2 units, or 6 square units. It is conjectured that the area of the black buglike region representing the Mandelbrot set is

$$\sqrt{6\pi - 1} - e \approx 1.5065916514855 \text{ square units.}$$

(*Source:* Robert P. Munafo, *Mandelbrot Set Glossary and Encyclopedia*)

Round the area of the Mandelbrot set to

a. the nearest tenth.

b. the nearest hundredth.

c. the nearest thousandth.

d. seven decimal places.

In Exercises 17–20, obtain an estimate for each computation by rounding the numbers so that the resulting arithmetic can easily be performed by hand or in your head. Then use a calculator to perform the computation. How reasonable is your estimate when compared to the actual answer?

17. $1.57 + 4.36 + 9.78$

18. 8.83×49

19. $19.894 \div 4.179$

20. 62.3% of 3847.6

In Exercises 21–24, determine each estimate without using a calculator. Then use a calculator to perform the computation necessary to obtain an exact answer. How reasonable is your estimate when compared to the actual answer?

21. Estimate the total cost of six grocery items if their prices are \$8.47, \$0.89, \$2.79, \$0.14, \$1.19, and \$4.76.

22. Estimate the salary of a worker who works for 78 hours at \$6.85 per hour.

23. At a yard sale, a person bought 21 books at \$0.85 each, two chairs for \$11.95 each, and a ceramic plate for \$14.65. Estimate the total amount spent.

24. The circle graph shows how the 17,487,475 students enrolled in U.S. colleges and universities in 2012 funded college costs. Estimate the number of students who covered these costs through grants and scholarships.

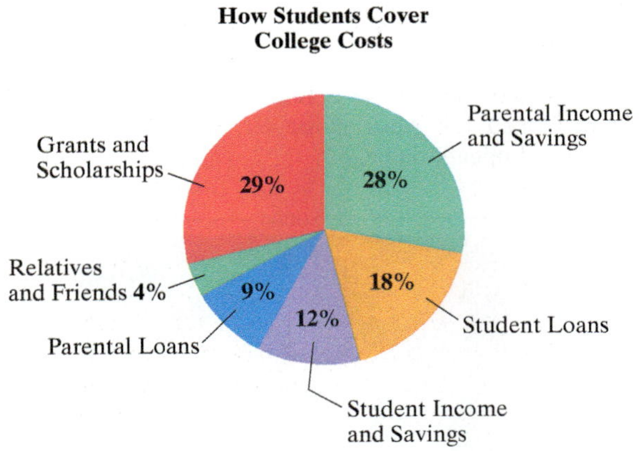

Source: Time, October 29, 2012

25. A small private school employs 10 teachers with salaries ranging from $817 to $992 per week. Which of the following is the best estimate of the monthly payroll for the teachers?

a. $30,000 **b.** $36,000
c. $42,000 **d.** $50,000

26. Select the best estimate for the number of seconds in a day.

a. 1500 **b.** 15,000
c. 86,000 **d.** 100,000

27. Imagine the entire global population as a village of precisely 200 people. The bar graph shows some numeric observations based on this scenario.

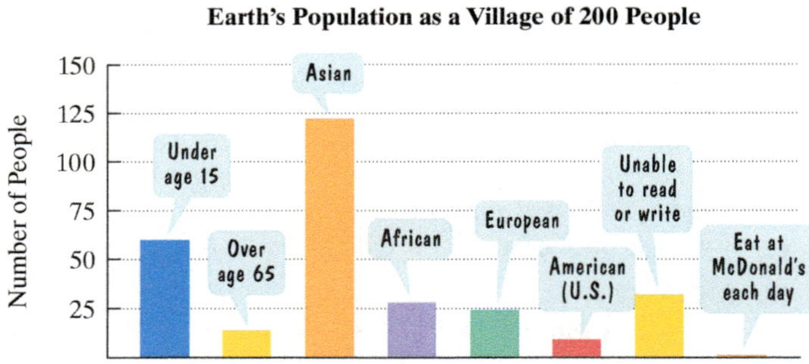

Source: Gary Rimmer, *Number Freaking*, The Disinformation Company Ltd.

a. Which group in the village has a population that exceeds 100? Estimate this group's population.

b. World population is approximately 33 million times the population of the village of 200 people. Use this observation to estimate the number of people in the world, in millions, unable to read or write.

28. The bar graph shows the percentage of people 25 years of age and older who were college graduates in the United States for seven selected years.

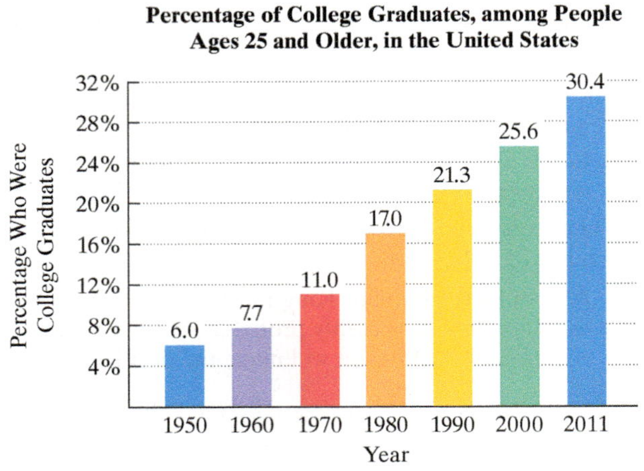

Source: U.S. Census Bureau

a. Estimate the average yearly increase in the percentage of college graduates.

b. If the trend shown by the graph continues, estimate the percentage of people 25 years of age and older who will be college graduates in 2020.

29. During a diagnostic evaluation, a 33-year-old woman experienced a panic attack a few minutes after she had been asked to relax her whole body. The graph shows the rapid increase in heart rate during the panic attack.

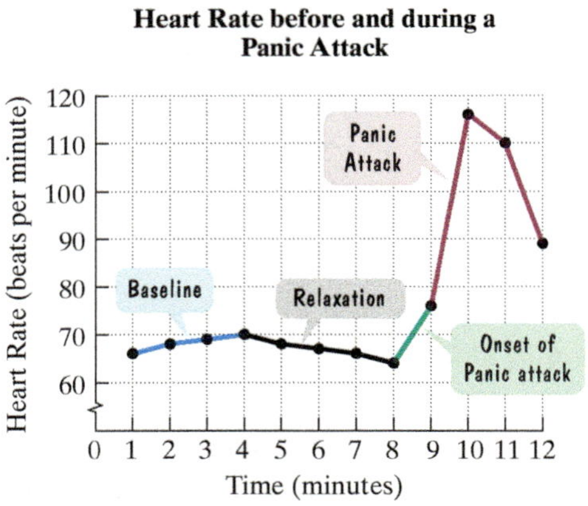

Source: Davis and Palladino, *Psychology*, Fifth Edition, Prentice Hall, 2007.

a. Use the graph to estimate the woman's maximum heart rate during the first 12 minutes of the diagnostic evaluation. After how many minutes did this occur?

b. Use the graph to estimate the woman's minimum heart rate during the first 12 minutes of the diagnostic evaluation. After how many minutes did this occur?

c. During which time period did the woman's heart rate increase at the greatest rate?

d. After how many minutes was the woman's heart rate approximately 75 beats per minute?

30. The bar graph shows the population of the United States, in millions, for five selected years.

Population of the United States

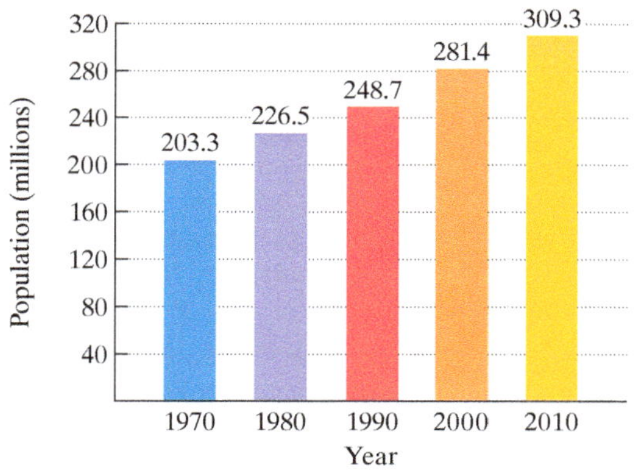

Source: U.S. Census Bureau

a. Estimate the yearly increase in the U.S. population. Express the answer in millions and do not round.

b. Write a mathematical model that estimates the U.S. population, p, in millions, x years after 1970.

c. Use the mathematical model from part (b) to project the U.S. population, in millions, in 2020.

1.3

31. What necessary piece of information is missing that prevents solving the following problem?

If 3 milligrams of a medicine is given for every 20 pounds of body weight, how many milligrams should be given to a 6-year-old child?

32. In the following problem, there is one more piece of information given than is necessary for solving the problem. Identify this unnecessary piece of information. Then solve the problem.

A taxicab charges $3.00 for the first mile and $0.50 for each additional half-mile. After a 6-mile trip, a customer handed the taxi driver a $20 bill. Find the cost of the trip.

Use the four-step method in problem solving to solve Exercises 33–38.

33. If there are seven frankfurters in one pound, how many pounds would you buy for a picnic to supply 28 people with two frankfurters each?

34. A car rents for $175 per week plus $0.30 per mile. Find the rental cost for a three-week trip of 1200 miles.

35. You are choosing between two plans at a discount warehouse. Plan A offers an annual membership fee of $100 and you pay 80% of the manufacturer's recommended list price. Plan B offers an annual membership fee of $40 and you pay 90% of the manufacturer's recommended list price. If you anticipate purchasing $1500 of merchandise in a year, which plan offers the better deal? By how much?

36. Miami is on Eastern Standard Time and San Francisco is on Pacific Standard Time, three hours earlier than Eastern Standard Time. A flight leaves Miami at 10 A.M. Eastern Standard Time, stops for 45 minutes in Houston, Texas, and arrives in San Francisco at 1:30 P.M. Pacific time. What is the actual flying time from Miami to San Francisco?

37. An automobile purchased for $37,000 is worth $2600 after eight years. Assuming that the value decreased steadily each year, what was the car worth at the end of the fifth year?

38. Suppose you are an engineer programming the automatic gate for a 35-cent toll. The gate is programmed for exact change only and will not accept pennies. How many coin combinations must you program the gate to accept?

Chapter 1 Test

1. Which reasoning process is shown in the following example?

The course policy states that if you turn in at least 80% of the homework, your lowest exam grade will be dropped. I turned in 90% of the homework, so my lowest grade will be dropped.

2. Which reasoning process is shown in the following example?

We examine the fingerprints of 1000 people. No two individuals in this group of people have identical fingerprints. We conclude that for all people, no two people have identical fingerprints.

In Exercises 3–6, find the next number, computation, or figure, as appropriate.

3. 0, 5, 10, 15, ____

4. $\frac{1}{6}, \frac{1}{12}, \frac{1}{24}, \frac{1}{48}$, ____

5.
$3367 \times 3 = 10{,}101$
$3367 \times 6 = 20{,}202$
$3367 \times 9 = 30{,}303$
$3367 \times 12 = 40{,}404$ ____________

6. 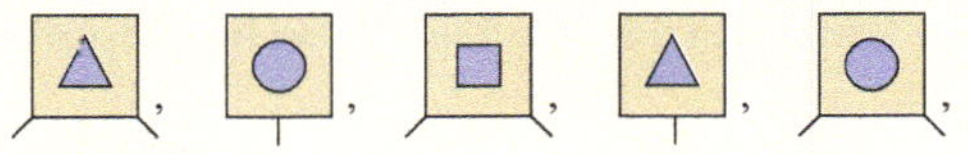 ____

7. Consider the following procedure:
Select a number. Multiply the number by 4. Add 8 to the product. Divide the sum by 2. Subtract 4 from the quotient.

a. Repeat this procedure for three numbers of your choice. Write a conjecture that relates the result of the process to the original number selected.

b. Represent the original number by the variable n and use deductive reasoning to prove the conjecture in part (a).

8. Round 3,279,425 to the nearest hundred thousand.

9. Round 706.3849 to the nearest hundredth.

In Exercises 10–13, determine each estimate without using a calculator. Different rounding results in different estimates, so there is not one single correct answer to each exercise. Use rounding to make the resulting calculations simple.

10. For a spring break vacation, a student needs to spend $47.00 for gas, $311.00 for food, and $405.00 for a hotel room. If the student takes $681.79 from savings, estimate how much more money is needed for the vacation.

11. The cost for opening a restaurant is $485,000. If 19 people decide to share equally in the business, estimate the amount each must contribute.

12. Find an estimate of 0.48992×121.976.

13. The graph shows the composition of a typical American community's trash.

Types of Trash in an American Community by Percentage of Total Weight

Source: U.S. Environmental Protection Agency

Across the United States, people generate approximately 512 billion pounds of trash per year. Estimate the number of pounds of trash in the form of plastic.

14. If the odometer of a car reads 71,911.5 miles and it averaged 28.9 miles per gallon, select the best estimate for the number of gallons of gasoline used.

a. 2400 **b.** 3200 **c.** 4000 **d.** 4800 **e.** 5600

15. The stated intent of the 1994 "don't ask, don't tell" policy was to reduce the number of discharges of gay men and lesbians from the military. Nearly 14,000 active-duty gay servicemembers were dismissed under the policy, which officially ended in 2011, after 18 years. The line graph shows the number of discharges under "don't ask, don't tell" from 1994 through 2010.

Number of Active-Duty Gay Servicemembers Discharged from the Military for Homosexuality

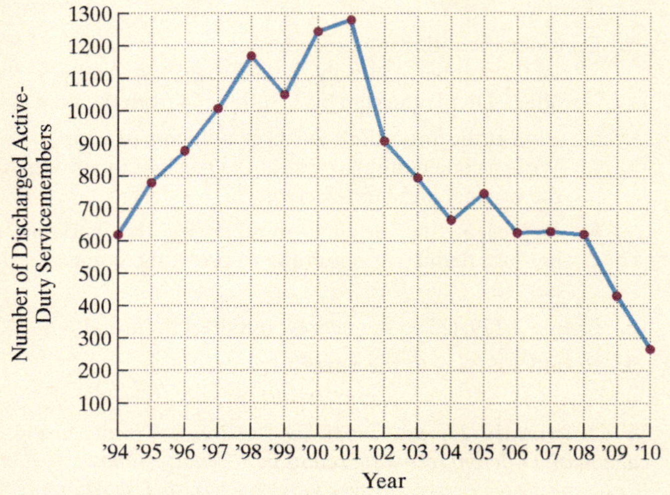

Source: General Accountability Office

a. For the period shown, in which year did the number of discharges reach a maximum? Find a reasonable estimate of the number of discharges for that year.

b. For the period shown, in which year did the number of discharges reach a minimum? Find a reasonable estimate of the number of discharges for that year.

c. In which one-year period did the number of discharges decrease at the greatest rate?

d. In which year were approximately 1000 gay servicemembers discharged under the "don't ask, don't tell" policy?

16. The bar graph shows a dramatic change in the high school grades of students who had just entered college between 1968 and 2010.

Average High School Grades of U.S. Students Entering College

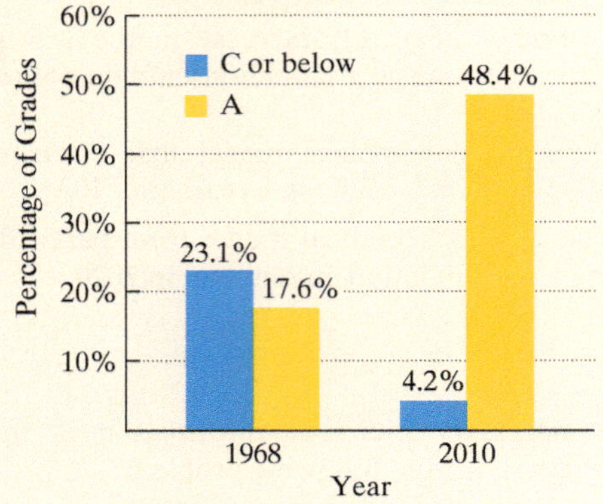

Source: John Macionis, *Sociology*, 14th Edition, Pearson, 2012.

a. Estimate the average yearly increase in the percentage of high school grades of A. Round to the nearest tenth of a percent.

b. Write a mathematical model that estimates the percentage of high school grades of A, p, x years after 1968.

c. If the trend shown by the graph continues, use your mathematical model from part (b) to project the percentage of high school grades of A in 2020.

17. The cost of renting a boat from Estes Rental is $9 per 15 minutes. The cost from Ship and Shore Rental is $20 per half-hour. If you plan to rent the boat for three hours, which business offers the better deal and by how much?

18. A bus operates between Miami International Airport and Miami Beach, 10 miles away. It makes 20 round trips per day carrying 32 passengers per trip. If the fare each way is $11.00, how much money is taken in from one day's operation?

19. By paying $50 cash up front and the balance at $35 a week, how long will it take to pay for a computer costing $960?

20. In 2000, the population of Greece was 10,600,000, with projections of a population decrease of 28,000 people per year. In the same year, the population of Belgium was 10,200,000, with projections of a population decrease of 12,000 people per year. (*Source:* United Nations)

According to these projections, which country will have the greater population in 2035 and by how many more people?

5

Number Theory and the Real Number System

SURFING THE WEB, YOU HEAR POLITICIANS DISCUSSING the problem of the national debt that exceeds $15 trillion. They state that the interest on the debt equals government spending on veterans, homeland security, education, and transportation combined. They make it seem like the national debt is a real problem, but later you realize that you don't really know what a number like 15 trillion means. If the national debt were evenly divided among all citizens of the country, how much would every man, woman, and child have to pay? Is economic doomsday about to arrive?

Here's where you'll find this application:

Literacy with numbers, called numeracy, is a prerequisite for functioning in a meaningful way personally, professionally, and as a citizen. In this chapter, our focus is on understanding numbers, their properties, and their applications.

- The problem of placing a national debt that exceeds $15 trillion in perspective appears as Example 9 in Section 5.6.
- Confronting a national debt in excess of $15 trillion starts with grasping just how colossal $1 trillion actually is. The Blitzer Bonus on page 319 should help provide insight into this mind-boggling number.

5.1 Number Theory: Prime and Composite Numbers

WHAT AM I SUPPOSED TO LEARN?

After you have read this section, you should be able to:

1. Determine divisibility.
2. Write the prime factorization of a composite number.
3. Find the greatest common divisor of two numbers.
4. Solve problems using the greatest common divisor.
5. Find the least common multiple of two numbers.
6. Solve problems using the least common multiple.

Number Theory and Divisibility

YOU ARE ORGANIZING AN intramural league at your school. You need to divide 40 men and 24 women into all-male and all-female teams so that each team has the same number of people. The men's teams should have the same number of players as the women's teams. What is the largest number of people that can be placed on a team?

This problem can be solved using a branch of mathematics called **number theory**. Number theory is primarily concerned with the properties of numbers used for counting, namely 1, 2, 3, 4, 5, and so on. The set of counting numbers is also called the set of **natural numbers**. As we saw in Chapter 2, we represent this set by the letter **N**.

THE SET OF NATURAL NUMBERS

$$\mathbf{N} = \{1, 2, 3, 4, 5, 6, 7, 8, 9, 10, 11, \dots\}$$

We can solve the intramural league problem. However, to do so we must understand the concept of divisibility. For example, there are a number of different ways to divide the 24 women into teams, including

1 team with all 24 women:	$1 \times 24 = 24$
2 teams with 12 women per team:	$2 \times 12 = 24$
3 teams with 8 women per team:	$3 \times 8 = 24$
4 teams with 6 women per team:	$4 \times 6 = 24$
6 teams with 4 women per team:	$6 \times 4 = 24$
8 teams with 3 women per team:	$8 \times 3 = 24$
12 teams with 2 women per team:	$12 \times 2 = 24$
24 teams with 1 woman per team:	$24 \times 1 = 24.$

The natural numbers that are multiplied together resulting in a product of 24 are called *factors* of 24. Any natural number can be expressed as a product of two or more natural numbers. The natural numbers that are multiplied are called the **factors** of the product. Notice that a natural number may have many factors.

$2 \times 12 = 24$ (Factors of 24) $\quad 3 \times 8 = 24$ (Factors of 24) $\quad 6 \times 4 = 24$ (Factors of 24)

The numbers 1, 2, 3, 4, 6, 8, 12, and 24 are all factors of 24. Each of these numbers divides 24 without a remainder.

In general, let a and b represent natural numbers. We say that a is **divisible** by b if the operation of dividing a by b leaves a remainder of 0.

A natural number is divisible by all of its factors. Thus, 24 is divisible by 1, 2, 3, 4, 6, 8, 12, and 24. Using the factor 8, we can express this divisibility in a number of ways:

24 is **divisible** by 8.

8 is a **divisor** of 24.

8 **divides** 24.

Mathematicians use a special notation to indicate divisibility.

GREAT QUESTION!

What's the difference between a factor and a divisor?

There is no difference. The words *factor* and *divisor* mean the same thing. Thus, 8 is a factor and a divisor of 24.

GREAT QUESTION!

What's the difference between $b|a$ and b/a?

It's easy to confuse these notations. The symbol $b|a$ means b divides a. The symbol b/a means b divided by a (that is, $b \div a$, the quotient of b and a). For example, $5|35$ means 5 divides 35, whereas $5/35$ means 5 divided by 35, which is equivalent to the fraction $\frac{1}{7}$.

DIVISIBILITY

If a and b are natural numbers, a is **divisible** by b if the operation of dividing a by b leaves a remainder of 0. This is the same as saying that b is a **divisor** of a, or b **divides** a. All three statements are symbolized by writing

$$b|a.$$

Using this new notation, we can write

$$12|24.$$

Twelve divides 24 because 24 divided by 12 leaves a remainder of 0. By contrast, 13 does not divide 24 because 24 divided by 13 does not leave a remainder of 0. The notation

$$13\nmid 24$$

means that 13 does not divide 24.

Determine divisibility.

Table 5.1 shows some common rules for divisibility. Divisibility rules for 7 and 11 are difficult to remember and are not included in the table.

TABLE 5.1 Rules of Divisibility

Divisible By	Test	Example
2	The last digit is 0, 2, 4, 6, or 8.	5,892,796 is divisible by 2 because the last digit is 6.
3	The sum of the digits is divisible by 3.	52,341 is divisible by 3 because the sum of the digits is $5 + 2 + 3 + 4 + 1 = 15$, and 15 is divisible by 3.
4	The last two digits form a number divisible by 4.	3,947,136 is divisible by 4 because 36 is divisible by 4.
5	The number ends in 0 or 5.	28,160 and 72,805 end in 0 and 5, respectively. Both are divisible by 5.
6	The number is divisible by both 2 and 3. (In other words, the number is even and the sum of its digits is divisible by 3.)	954 is divisible by 2 because it ends in 4. 954 is also divisible by 3 because the digit sum is 18, which is divisible by 3. Because 954 is divisible by both 2 and 3, it is divisible by 6.
8	The last three digits form a number that is divisible by 8.	593,777,832 is divisible by 8 because 832 is divisible by 8.
9	The sum of the digits is divisible by 9.	5346 is divisible by 9 because the sum of the digits, 18, is divisible by 9.
10	The last digit is 0.	998,746,250 is divisible by 10 because the number ends in 0.
12	The number is divisible by both 3 and 4. (In other words, the sum of the digits is divisible by 3 and the last two digits form a number divisible by 4.)	614,608,176 is divisible by 3 because the digit sum is 39, which is divisible by 3. It is also divisible by 4 because the last two digits form 76, which is divisible by 4. Because 614,608,176 is divisible by both 3 and 4, it is divisible by 12.

TECHNOLOGY

Calculators and Divisibility

You can use a calculator to verify each result in Example 1. Consider part (a):

$$4 \mid 3{,}754{,}086.$$

Divide 3,754,086 by 4 using the following keystrokes:

3754086 [÷] 4.

Press [=] or [ENTER]. The number displayed is 938521.5. This is not a natural number. The 0.5 shows that the division leaves a nonzero remainder. Thus, 4 does not divide 3,754,086. The given statement is false.

Now consider part (b):

$$9 \nmid 4{,}119{,}706{,}413.$$

Use your calculator to divide the number on the right by 9:

4119706413 [÷] 9.

Press [=] or [ENTER]. The display is 457745157. This is a natural number. The remainder of the division is 0, so 9 does divide 4,119,706,413. The given statement is false.

GREAT QUESTION!

Can a prime number be even?

The number 2 is the only even prime number. Every other even number has at least three factors: 1, 2, and the number itself.

EXAMPLE 1 *Using the Rules of Divisibility*

Which one of the following statements is true?

a. $4 \mid 3{,}754{,}086$ **b.** $9 \nmid 4{,}119{,}706{,}413$ **c.** $8 \mid 677{,}840$

SOLUTION

a. $4 \mid 3{,}754{,}086$ states that 4 divides 3,754,086. **Table 5.1** indicates that for 4 to divide a number, the last two digits must form a number that is divisible by 4. Because 86 is not divisible by 4, the given statement is false.

b. $9 \nmid 4{,}119{,}706{,}413$ states that 9 does *not* divide 4,119,706,413. Based on **Table 5.1**, if the sum of the digits is divisible by 9, then 9 does indeed divide this number. The sum of the digits is $4 + 1 + 1 + 9 + 7 + 0 + 6 + 4 + 1 + 3 = 36$, which is divisible by 9. Because 4,119,706,413 is divisible by 9, the given statement is false.

c. $8 \mid 677{,}840$ states that 8 divides 677,840. **Table 5.1** indicates that for 8 to divide a number, the last three digits must form a number that is divisible by 8. Because 840 is divisible by 8, then 8 divides 677,840, and the given statement is true.

The statement given in part (c) is the only true statement.

CHECK POINT 1 Which one of the following statements is true?

a. $8 \mid 48{,}324$ **b.** $6 \mid 48{,}324$ **c.** $4 \nmid 48{,}324$

Prime Factorization

2 Write the prime factorization of a composite number.

By developing some other ideas of number theory, we will be able to solve the intramural league problem. We begin with the definition of a prime number.

PRIME NUMBERS

A **prime number** is a natural number greater than 1 that has only itself and 1 as factors.

Using this definition, we see that the number 7 is a prime number because it has only 1 and 7 as factors. Said in another way, 7 is prime because it is divisible by only 1 and 7. The first ten prime numbers are 2, 3, 5, 7, 11, 13, 17, 19, 23, and 29. Each number in this list has exactly two divisors, itself and 1. By contrast, 9 is not a prime number; in addition to being divisible by 1 and 9, it is also divisible by 3. The number 9 is an example of a *composite number*.

COMPOSITE NUMBERS

A **composite number** is a natural number greater than 1 that is divisible by a number other than itself and 1.

Using this definition, the first ten composite numbers are 4, 6, 8, 9, 10, 12, 14, 15, 16, and 18. Each number in this list has at least three distinct divisors.

By the definitions above, both prime numbers and composite numbers must be natural numbers *greater than* 1, so **the natural number 1 is neither prime nor composite.**

Every composite number can be expressed as the product of prime numbers. For example, the composite number 45 can be expressed as

$$45 = 3 \times 3 \times 5.$$

Note that 3 and 5 are prime numbers. Expressing a composite number as the product of prime numbers is called **prime factorization**. The prime factorization of 45 is $3 \times 3 \times 5$. The order in which we write these factors does not matter. This means that

$$45 = 3 \times 3 \times 5$$
$$\text{or } 45 = 5 \times 3 \times 3$$
$$\text{or } 45 = 3 \times 5 \times 3.$$

In Chapter 1, we defined a **theorem** as a statement that can be proved using deductive reasoning. The ancient Greeks proved that if the order of the factors is disregarded, there is only one prime factorization possible for any given composite number. This statement is called the **Fundamental Theorem of Arithmetic**.

THE FUNDAMENTAL THEOREM OF ARITHMETIC

Every composite number can be expressed as a product of prime numbers in one and only one way (if the order of the factors is disregarded).

One method used to find the prime factorization of a composite number is called a **factor tree**. To use this method, begin by selecting any two numbers, other than 1, whose product is the number to be factored. One or both of the factors may not be prime numbers. Continue to factor composite numbers. Stop when all numbers are prime.

EXAMPLE 2 *Prime Factorization Using a Factor Tree*

Find the prime factorization of 700.

SOLUTION

Start with any two numbers, other than 1, whose product is 700, such as 7 and 100. This forms the first branch of the tree. Continue factoring the composite number or numbers that result (in this case 100), branching until each branch ends with a prime number.

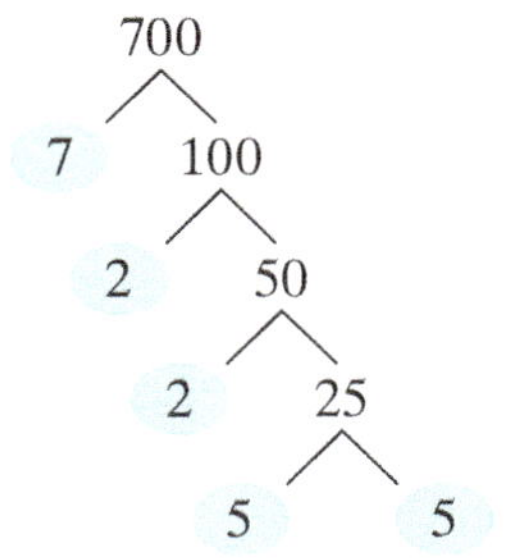

The prime factors are shown on light blue ovals. Thus, the prime factorization of 700 is

$$700 = 7 \times 2 \times 2 \times 5 \times 5.$$

We can use exponents to show the repeated prime factors:

$$700 = 7 \times 2^2 \times 5^2.$$

Using a dot to indicate multiplication and arranging the factors from least to greatest, we can write

$$700 = 2^2 \cdot 5^2 \cdot 7.$$

GREAT QUESTION!

In Example 2, do I have to start the factor tree for 700 with 7 · 100?

No. It does not matter how you begin a factor tree. For example, in Example 2 you can factor 700 by starting with 5 and 140. $(5 \times 140 = 700)$

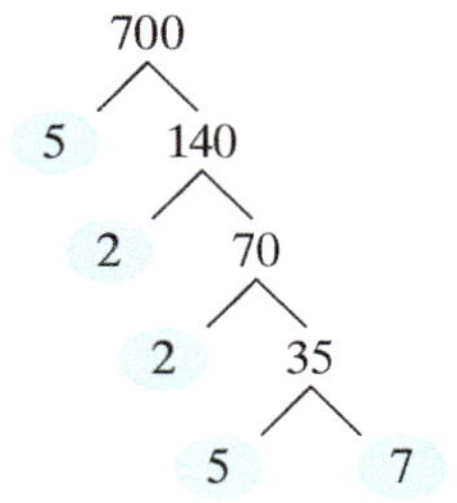

The prime factorization of 700 is

$$700 = 5 \times 2 \times 2 \times 5 \times 7$$
$$= 2^2 \times 5^2 \times 7.$$

This is the same prime factorization we obtained in Example 2.

CHECK POINT 2 Find the prime factorization of 120.

3 Find the greatest common divisor of two numbers.

Greatest Common Divisor

The greatest common divisor of two or more natural numbers is the largest number that is a divisor (or factor) of all the numbers. For example, 8 is the greatest common divisor of 32 and 40 because it is the largest natural number that divides both 32 and 40. Some pairs of numbers have 1 as their greatest common divisor. Such number pairs are said to be **relatively prime**. For example, the greatest common divisor of 5 and 26 is 1. Thus, 5 and 26 are relatively prime.

The greatest common divisor can be found using prime factorizations.

Blitzer Bonus

Simple Questions with No Answers

In number theory, a good problem is one that can be stated quite simply, but whose solution turns out to be particularly difficult, if not impossible.

In 1742, the mathematician Christian Goldbach (1690–1764) wrote a letter to Leonhard Euler (1707–1783) in which he proposed, without a proof, that every even number greater than 2 is the sum of two primes. For example,

Even number | Sum of two primes

$$4 = 2 + 2$$
$$6 = 3 + 3$$
$$8 = 3 + 5$$
$$10 = 5 + 5$$
$$12 = 7 + 5$$

and so on.

Two and a half centuries later, it is still not known if this conjecture is true or false. Inductively, it appears to be true; computer searches have written even numbers as large as 400 trillion as the sum of two primes. Deductively, no mathematician has been able to prove that the conjecture is true. Even a reward of one million dollars for a proof offered by the publishing house Farber and Farber in 2000 to help publicize the novel *Uncle Petros and Goldbach's Conjecture* went unclaimed.

FINDING THE GREATEST COMMON DIVISOR USING PRIME FACTORIZATIONS

To find the greatest common divisor of two or more numbers,

1. Write the prime factorization of each number.
2. Select each prime factor with the smallest exponent that is common to each of the prime factorizations.
3. Form the product of the numbers from step 2. The greatest common divisor is the product of these factors.

EXAMPLE 3 *Finding the Greatest Common Divisor*

Find the greatest common divisor of 216 and 234.

SOLUTION

Step 1 Write the prime factorization of each number. Begin by writing the prime factorizations of 216 and 234.

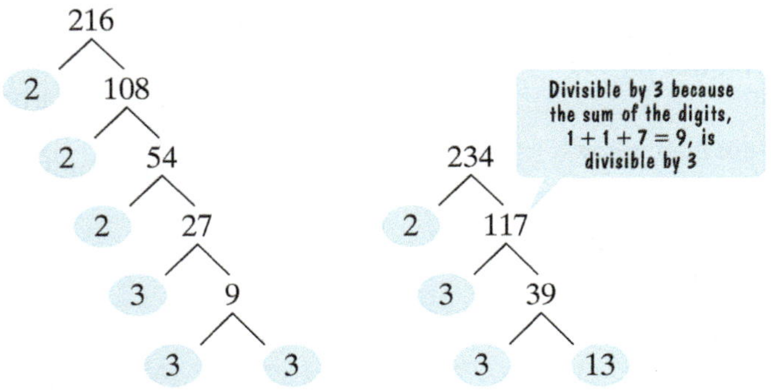

The factor tree on the left indicates that

$$216 = 2^3 \times 3^3.$$

The factor tree on the right indicates that

$$234 = 2 \times 3^2 \times 13.$$

Step 2 Select each prime factor with the smaller exponent that is common to each of the prime factorizations. Look at the factorizations of 216 and 234 from step 1. Can you see that 2 is a prime number common to the factorizations of 216 and 234? Likewise, 3 is also a prime number common to the two factorizations. By contrast, 13 is a prime number that is not common to both factorizations.

$$216 = 2^3 \times 3^3$$
$$234 = 2 \times 3^2 \times 13$$

2 is a prime number common to both factorizations. | 3 is a prime number common to both factorizations.

Now we need to use these prime factorizations to determine which exponent is appropriate for 2 and which exponent is appropriate for 3. The appropriate exponent is the smaller exponent associated with the prime number in the factorizations. The exponents associated with 2 in the factorizations are 1 and 3, so we select 1. Therefore, one factor for the greatest common divisor is 2^1, or 2. The exponents associated with 3 in the factorizations are 2 and 3, so we select 2. Therefore, another factor for the greatest common divisor is 3^2.

$$216 = 2^3 \times 3^3$$

The smaller exponent on 2 is 1.   The smaller exponent on 3 is 2.

$$234 = 2^1 \times 3^2 \times 13$$

Step 3 Form the product of the numbers from step 2. The greatest common divisor is the product of these factors.

$$\text{Greatest common divisor} = 2 \times 3^2 = 2 \times 9 = 18$$

The greatest common divisor of 216 and 234 is 18.

CHECK POINT 3 Find the greatest common divisor of 225 and 825.

4 Solve problems using the greatest common divisor.

EXAMPLE 4 *Solving a Problem Using the Greatest Common Divisor*

For an intramural league, you need to divide 40 men and 24 women into all-male and all-female teams so that each team has the same number of people. What is the largest number of people that can be placed on a team?

SOLUTION

Because 40 men are to be divided into teams, the number of men on each team must be a divisor of 40. Because 24 women are to be divided into teams, the number of women placed on a team must be a divisor of 24. Although the teams are all-male and all-female, the same number of people must be placed on each team. The largest number of people that can be placed on a team is the largest number that will divide into 40 and 24 without a remainder. This is the greatest common divisor of 40 and 24.

To find the greatest common divisor of 40 and 24, begin with their prime factorizations.

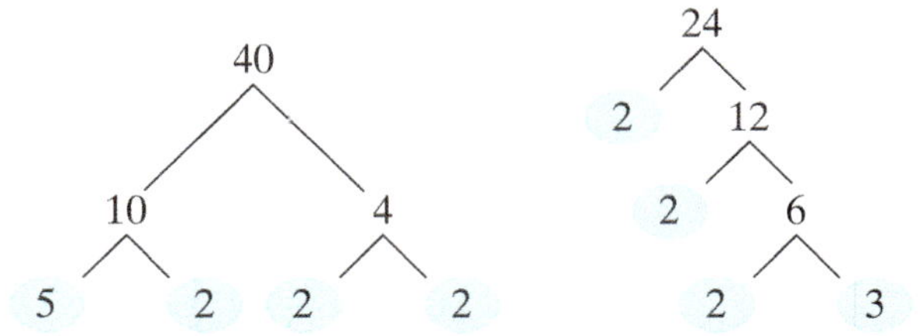

The factor trees indicate that

$$40 = 2^3 \times 5 \quad \text{and} \quad 24 = 2^3 \times 3.$$

We see that 2 is a prime number common to both factorizations. The exponents associated with 2 in the factorizations are 3 and 3, so we select 3.

$$\text{Greatest common divisor} = 2^3 = 2 \times 2 \times 2 = 8$$

The largest number of people that can be placed on a team is 8. Thus, the 40 men can form five teams with 8 men per team. The 24 women can form three teams with 8 women per team.

Blitzer Bonus

GIMPS

A prime number of the form $2^p - 1$, where p is prime, is called a **Mersenne prime**, named for the seventeenth-century monk Marin Mersenne (1588–1648), who stated that $2^p - 1$ is prime for $p = 2, 3, 5, 7, 13, 17, 19, 31, 67, 127$, and 257. Without calculators and computers, it is not known how Mersenne arrived at these assertions, although it is now known that $2^p - 1$ is *not* prime for $p = 67$ and 257; it *is* prime for $p = 61, 89$, and 107.

In 1995, the American computer scientist George Woltman began the *Great Internet Mersenne Prime Search (GIMPS)*. By pooling the combined efforts of thousands of people interested in finding new Mersenne primes, GIMPS's participants have yielded several important results, including the record prime $2^{57,885,161} - 1$, a number with 17,425,170 digits that was discovered in 2013 by Curtis Cooper, a professor of computer science at the University of Central Missouri.

CHECK POINT 4 A choral director needs to divide 192 men and 288 women into all-male and all-female singing groups so that each group has the same number of people. What is the largest number of people that can be placed in each singing group?

5 Find the least common multiple of two numbers.

Least Common Multiple

The **least common multiple** of two or more natural numbers is the smallest natural number that is divisible by all of the numbers. One way to find the least common multiple is to make a list of the numbers that are divisible by each number. This list represents the **multiples** of each number. For example, if we wish to find the least common multiple of 15 and 20, we can list the sets of multiples of 15 and multiples of 20.

Numbers Divisible by 15: Multiples of 15:	{15, 30, 45, 60, 75, 90, 105, 120, . . .}
Numbers Divisible by 20: Multiples of 20:	{20, 40, 60, 80, 100, 120, 140, 160, . . .}

Two common multiples of 15 and 20 are 60 and 120. The least common multiple is 60. Equivalently, 60 is the smallest number that is divisible by both 15 and 20.

Sometimes a partial list of the multiples for each of two numbers does not reveal the smallest number that is divisible by both given numbers. A more efficient method for finding the least common multiple is to use prime factorizations.

FINDING THE LEAST COMMON MULTIPLE USING PRIME FACTORIZATIONS

To find the least common multiple of two or more numbers,

1. Write the prime factorization of each number.
2. Select every prime factor that occurs, raised to the greatest power to which it occurs, in these factorizations.
3. Form the product of the numbers from step 2. The least common multiple is the product of these factors.

EXAMPLE 5 *Finding the Least Common Multiple*

Find the least common multiple of 144 and 300.

SOLUTION

Step 1 Write the prime factorization of each number. Write the prime factorizations of 144 and 300.

$$144 = 2^4 \times 3^2$$
$$300 = 2^2 \times 3 \times 5^2$$

Step 2 Select every prime factor that occurs, raised to the greater power to which it occurs, in these factorizations. The prime factors that occur are 2, 3, and 5. The greater exponent that appears on 2 is 4, so we select 2^4. The greater exponent that appears on 3 is 2, so we select 3^2. The only exponent that occurs on 5 is 2, so we select 5^2. Thus, we have selected 2^4, 3^2, and 5^2.

Step 3 Form the product of the numbers from step 2. The least common multiple is the product of these factors.

$$\text{Least common multiple} = 2^4 \times 3^2 \times 5^2 = 16 \times 9 \times 25 = 3600$$

The least common multiple of 144 and 300 is 3600. The smallest natural number divisible by both 144 and 300 is 3600.

CHECK POINT 5 Find the least common multiple of 18 and 30.

Blitzer Bonus

Palindromic Primes

May a moody baby doom a yam? Leaving aside the answer to this question, what makes the sentence interesting is that it reads the same from left to right and from right to left! Such a sentence is called a **palindrome**. Some prime numbers are also palindromic. For example, the prime number 11 reads the same forward and backward, although a more provocative example containing all ten digits is

1,023,456,987,896,543,201.

In the following pyramid of palindromic primes, each number is obtained by adding two digits to the beginning and end of the previous prime.

2
30203
133020331
1713302033171
12171330203317121
151217133020331712151
1815121713302033171215181
16181512171330203317121518161

A huge palindromic prime was discovered in 2003 by David Broadhurst, a retired electrical engineer. The number contains 30,803 digits (and, yes, 30,803 is also a palindromic prime!).

Source: Clifford A. Pickover, *A Passion for Mathematics,* John Wiley and Sons, Inc., 2005.

6 Solve problems using the least common multiple.

EXAMPLE 6 *Solving a Problem Using the Least Common Multiple*

A movie theater runs its films continuously. One movie runs for 80 minutes and a second runs for 120 minutes. Both movies begin at 4:00 P.M. When will the movies begin again at the same time?

SOLUTION

The shorter movie lasts 80 minutes, or 1 hour, 20 minutes. It begins at 4:00, so it will be shown again at 5:20. The longer movie lasts 120 minutes, or 2 hours. It begins at 4:00, so it will be shown again at 6:00. We are asked to find when the movies will begin again at the same time. Therefore, we are looking for the least common multiple of 80 and 120. Find the least common multiple and then add this number of minutes to 4:00 P.M.

Begin with the prime factorizations of 80 and 120:

$$80 = 2^4 \times 5$$
$$120 = 2^3 \times 3 \times 5.$$

Now select each prime factor, with the greater exponent from each factorization.

$$\text{Least common multiple} = 2^4 \times 3 \times 5 = 16 \times 3 \times 5 = 240$$

Therefore, it will take 240 minutes, or 4 hours, for the movies to begin again at the same time. By adding 4 hours to 4:00 P.M., they will start together again at 8:00 P.M.

GREAT QUESTION!

Can I solve Example 6 by making a partial list of starting times for each movie?

Yes. Here's how it's done:

Shorter Movie (Runs 1 hour, 20 minutes):

4:00, 5:20, 6:40, 8:00, . . .

Longer Movie (Runs 2 hours):

4:00, 6:00, 8:00, . . .

The list reveals that both movies start together again at 8:00 P.M.

CHECK POINT 6 A movie theater runs two documentary films continuously. One documentary runs for 40 minutes and a second documentary runs for 60 minutes. Both movies begin at 3:00 P.M. When will the movies begin again at the same time?

Concept and Vocabulary Check

Fill in each blank so that the resulting statement is true.

1. A natural number greater than 1 that has only itself and 1 as factors is called a/an ________ number.
2. A natural number greater than 1 that is divisible by a number other than itself and 1 is called a/an ________ number.
3. The largest number that is a factor of two or more natural numbers is called their ________________.
4. The smallest number that is divisible by two or more natural numbers is called their ________________.

In Exercises 5–8, determine whether each statement is true or false. If the statement is false, make the necessary change(s) to produce a true statement.

5. The notation $b|a$ means that b is divisible by a. ________
6. $b \nmid a$ means that b does not divide a. ______
7. The words *factor* and *divisor* have opposite meanings. ______
8. A number can only be divisible by exactly one number. ________

Exercise Set 5.1

Practice Exercises

Use rules of divisibility to determine whether each number given in Exercises 1–10 is divisible by

a. 2 **b.** 3 **c.** 4 **d.** 5 **e.** 6
f. 8 **g.** 9 **h.** 10 **i.** 12.

1. 6944 **2.** 7245 **3.** 21,408 **4.** 25,025
5. 26,428 **6.** 89,001 **7.** 374,832 **8.** 347,712
9. 6,126,120 **10.** 5,941,221

*In Exercises 11–24, use a calculator to determine whether each statement is true or false. If the statement is true, explain why this is so using one of the rules of divisibility in **Table 5.1** on page 249.*

11. $3|5958$ **12.** $3|8142$ **13.** $4|10{,}612$
14. $4|15{,}984$ **15.** $5|38{,}814$ **16.** $5|48{,}659$
17. $6|104{,}538$ **18.** $6|163{,}944$ **19.** $8|20{,}104$
20. $8|28{,}096$ **21.** $9|11{,}378$ **22.** $9|23{,}772$
23. $12|517{,}872$ **24.** $12|785{,}172$

In Exercises 25–44, find the prime factorization of each composite number.

25. 75 **26.** 45 **27.** 56
28. 48 **29.** 105 **30.** 180
31. 500 **32.** 360 **33.** 663
34. 510 **35.** 885 **36.** 999
37. 1440 **38.** 1280 **39.** 1996
40. 1575 **41.** 3675 **42.** 8316
43. 85,800 **44.** 30,600

In Exercises 45–56, find the greatest common divisor of the numbers.

45. 42 and 56 **46.** 25 and 70 **47.** 16 and 42
48. 66 and 90 **49.** 60 and 108 **50.** 96 and 212
51. 72 and 120 **52.** 220 and 400 **53.** 342 and 380
54. 224 and 430 **55.** 240 and 285 **56.** 150 and 480

In Exercises 57–68, find the least common multiple of the numbers.

57. 42 and 56 **58.** 25 and 70 **59.** 16 and 42
60. 66 and 90 **61.** 60 and 108 **62.** 96 and 212
63. 72 and 120 **64.** 220 and 400 **65.** 342 and 380
66. 224 and 430 **67.** 240 and 285 **68.** 150 and 480

Practice Plus

In Exercises 69–74, determine all values of d that make each statement true.

69. $9|12{,}34d$ **70.** $9|23{,}42d$ **71.** $8|76{,}523{,}45d$
72. $8|88{,}888{,}82d$ **73.** $4|963{,}23d$ **74.** $4|752{,}67d$

*A **perfect number** is a natural number that is equal to the sum of its factors, excluding the number itself. In Exercises 75–78, determine whether or not each number is perfect.*

75. 28 **76.** 6 **77.** 20 **78.** 50

*A prime number is an **emirp** ("prime" spelled backward) if it becomes a different prime number when its digits are reversed. In Exercises 79–82, determine whether or not each prime number is an emirp.*

79. 41 **80.** 43 **81.** 107 **82.** 113

*A prime number p such that $2p + 1$ is also a prime number is called a **Germain prime**, named after the German mathematician Sophie Germain (1776–1831), who made major contributions to number theory. In Exercises 83–86, determine whether or not each prime number is a Germain prime.*

83. 13 **84.** 11 **85.** 241 **86.** 97

87. Find the product of the greatest common divisor of 24 and 27 and the least common multiple of 24 and 27. Compare this result to the product of 24 and 27. Write a conjecture based on your observation.

88. Find the product of the greatest common divisor of 48 and 72 and the least common multiple of 48 and 72. Compare this result to the product of 48 and 72. Write a conjecture based on your observation.

Application Exercises

89. In Carl Sagan's novel *Contact,* Ellie Arroway, the book's heroine, has been working at SETI, the Search for Extraterrestrial Intelligence, listening to the crackle of the cosmos. One night, as the radio telescopes are turned toward Vega, they suddenly pick up strange pulses through the background noise. Two pulses are followed by a pause, then three pulses, five, seven,

$$11, \quad 13, \quad 17, \quad 19, \quad 23, \quad 29, \quad 31, \ldots$$

continuing through 97. Then it starts all over again. Ellie is convinced that only intelligent life could generate the structure in the sequence of pulses. "It's hard to imagine some radiating plasma sending out a regular set of mathematical signals like this." What is it about the structure of the pulses that the book's heroine recognizes as the sign of intelligent life? Asked in another way, what is significant about the numbers of pulses?

90. There are two species of insects, *Magicicada septendecim* and *Magicicada tredecim,* that live in the same environment. They have a life cycle of exactly 17 and 13 years, respectively. For all but their last year, they remain in the ground feeding on the sap of tree roots. Then, in their last year, they emerge en masse from the ground as fully formed cricketlike insects, taking over the forest in a single night. They chirp loudly, mate, eat, lay eggs, then die six weeks later.

(*Source:* Marcus du Sautoy, *The Music of the Primes,* HarperCollins, 2003)

a. Suppose that the two species have life cycles that are not prime, say 18 and 12 years, respectively. List the set of multiples of 18 that are less than or equal to 216. List the set of multiples of 12 that are less than or equal to 216. Over a 216-year period, how many times will the two species emerge in the same year and compete to share the forest?

b. Recall that both species have evolved prime-number life cycles, 17 and 13 years, respectively. Find the least common multiple of 17 and 13. How often will the two species have to share the forest?

c. Compare your answers to parts (a) and (b). What explanation can you offer for each species having a prime number of years as the length of its life cycle?

91. A relief worker needs to divide 300 bottles of water and 144 cans of food into groups that each contain the same number of items. Also, each group must have the same type of item (bottled water or canned food). What is the largest number of relief supplies that can be put in each group?

92. A choral director needs to divide 180 men and 144 women into all-male and all-female singing groups so that each group has the same number of people. What is the largest number of people that can be placed in each singing group?

93. You have in front of you 310 five-dollar bills and 460 ten-dollar bills. Your problem: Place the five-dollar bills and the ten-dollar bills in stacks so that each stack has the same number of bills, and each stack contains only one kind of bill (five-dollar or ten-dollar). What is the largest number of bills that you can place in each stack?

94. Harley collects sports cards. He has 360 football cards and 432 baseball cards. Harley plans to arrange his cards in stacks so that each stack has the same number of cards. Also, each stack must have the same type of card (football or baseball). Every card in Harley's collection is to be placed in one of the stacks. What is the largest number of cards that can be placed in each stack?

95. You and your brother both work the 4:00 P.M. to midnight shift. You have every sixth night off. Your brother has every tenth night off. Both of you were off on June 1. Your brother would like to see a movie with you. When will the two of you have the same night off again?

96. A movie theater runs its films continuously. One movie is a short documentary that runs for 40 minutes. The other movie is a full-length feature that runs for 100 minutes. Each film is shown in a separate theater. Both movies begin at noon. When will the movies begin again at the same time?

97. Two people are jogging around a circular track in the same direction. One person can run completely around the track in 15 minutes. The second person takes 18 minutes. If they both start running in the same place at the same time, how long will it take them to be together at this place if they continue to run?

98. Two people are in a bicycle race around a circular track. One rider can race completely around the track in 40 seconds. The other rider takes 45 seconds. If they both begin the race at a designated starting point, how long will it take them to be together at this starting point again if they continue to race around the track?

Writing in Mathematics

99. If a is a factor of c, what does this mean?

100. How do you know that 45 is divisible by 5?

101. What does "a is divisible by b" mean?

102. Describe the difference between a prime number and a composite number.

103. What does the Fundamental Theorem of Arithmetic state?

104. What is the greatest common divisor of two or more natural numbers?

105. Describe how to find the greatest common divisor of two numbers.

106. What is the least common multiple of two or more natural numbers?

107. Describe how to find the least common multiple of two natural numbers.

108. The process of finding the greatest common divisor of two natural numbers is similar to finding the least common multiple of the numbers. Describe how the two processes differ.

109. What does the Blitzer Bonus on page 252 have to do with Gödel's discovery about mathematics and logic, described on page 177?

Critical Thinking Exercises

Make Sense? *In Exercises 110–113, determine whether each statement makes sense or does not make sense, and explain your reasoning.*

110. I'm working with a prime number that intrigues me because it has three natural number factors.

111. When I find the greatest common factor, I select common prime factors with the greatest exponent and when I find the least common multiple, I select common prime factors with the smallest exponent.

112. I need to separate 70 men and 175 women into all-male or all-female teams with the same number of people on each team. By finding the least common multiple of 70 and 175, I can determine the largest number of people that can be placed on a team.

113. (If you have not yet done so, read the Blitzer Bonus "GIMPS" on page 253.) I can find a prime number larger than the record prime $2^{57,885,161} - 1$ by simply writing $2^{57,885,162} - 1$.

114. Write a four-digit natural number that is divisible by 4 and not by 8.

115. Find the greatest common divisor and the least common multiple of $2^{17} \cdot 3^{25} \cdot 5^{31}$ and $2^{14} \cdot 3^{37} \cdot 5^{30}$. Express answers in the same form as the numbers given.

116. A middle-aged man observed that his present age was a prime number. He also noticed that the number of years in which his age would again be prime was equal to the number of years ago in which his age was prime. How old is the man?

117. A movie theater runs its films continuously. One movie runs for 85 minutes and a second runs for 100 minutes. The theater has a 15-minute intermission after each movie, at which point the movie is shown again. If both movies start at noon, when will the two movies start again at the same time?

118. The difference between consecutive prime numbers is always an even number, except for two particular prime numbers. What are those numbers?

Technology Exercises

*Use the divisibility rules listed in **Table 5.1** on page 249 to answer the questions in Exercises 119–121. Then, using a calculator, perform the actual division to determine whether your answer is correct.*

119. Is 67,234,096 divisible by 4?

120. Is 12,541,750 divisible by 3?

121. Is 48,201,651 divisible by 9?

Group Exercises

The following topics from number theory are appropriate for either individual or group research projects. A report should be given to the class on the researched topic. Useful references include liberal arts mathematics textbooks, books about numbers and number theory, books whose purpose is to excite the reader about mathematics, history of mathematics books, encyclopedias, and the Internet.

122. Euclid and Number Theory

123. An Unsolved Problem from Number Theory

124. Perfect Numbers

125. Deficient and Abundant Numbers

126. Formulas That Yield Primes

127. The Sieve of Eratosthenes

5.2 The Integers; Order of Operations

WHAT AM I SUPPOSED TO LEARN?

After you have read this section, you should be able to:

1. Define the integers.
2. Graph integers on a number line.
3. Use the symbols $<$ and $>$.
4. Find the absolute value of an integer.
5. Perform operations with integers.
6. Use the order of operations agreement.

CAN YOU CHEAT DEATH? LIFE expectancy for the average American man is 75.2 years; for a woman, it's 80.4. But what's in your hands if you want to eke out a few more birthday candles? In this section, we use operations on a set of numbers called the *integers* to indicate factors within your control that can stretch your probable life span. Start by flossing. (See Example 5 on page 264.)

Mirror II (1963), George Tooker. Addison Gallery, Phillips Academy, MA. © George Tooker.

Defining the Integers

In Section 5.1, we applied some ideas of number theory to the set of natural, or counting, numbers:

$$\text{Natural numbers} = \{1, 2, 3, 4, 5, \ldots\}.$$

1 Define the integers.

When we combine the number 0 with the natural numbers, we obtain the set of **whole numbers**:

$$\text{Whole numbers} = \{0, 1, 2, 3, 4, 5, \ldots\}.$$

The whole numbers do not allow us to describe certain everyday situations. For example, if the balance in your checking account is \$30 and you write a check for \$35, your checking account is overdrawn by \$5. We can write this as −5, read *negative* 5. The set consisting of the natural numbers, 0, and the negatives of the natural numbers is called the set of **integers**.

$$\text{Integers} = \{\underbrace{\ldots, -4, -3, -2, -1}_{\text{Negative integers}}, 0, \underbrace{1, 2, 3, 4, \ldots}_{\text{Positive integers}}\}$$

Notice that the term *positive integers* is another name for the natural numbers. The positive integers can be written in two ways:

1. Use a "+" sign. For example, +4 is "positive four."
2. Do not write any sign. For example, 4 is assumed to be "positive four."

The Number Line; The Symbols < and >

The **number line** is a graph we use to visualize the set of integers, as well as sets of other numbers. The number line is shown in **Figure 5.1**.

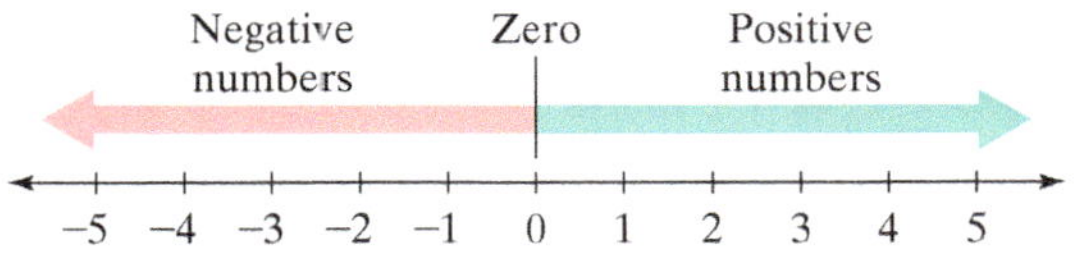

FIGURE 5.1 The number line

The number line extends indefinitely in both directions, shown by the arrows on the left and the right. Zero separates the positive numbers from the negative numbers on the number line. The positive integers are located to the right of 0 and the negative integers are located to the left of 0. **Zero is neither positive nor negative.** For every positive integer on a number line, there is a corresponding negative integer on the opposite side of 0.

2 Graph integers on a number line.

Integers are graphed on a number line by placing a dot at the correct location for each number.

EXAMPLE 1 *Graphing Integers on a Number Line*

Graph: **a.** −3 **b.** 4 **c.** 0.

SOLUTION

Place a dot at the correct location for each integer.

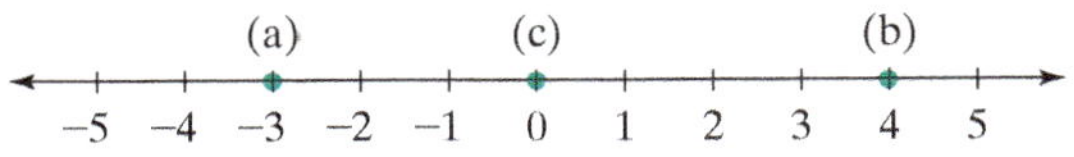

CHECK POINT 1 Graph:

a. −4 **b.** 0 **c.** 3.

3 Use the symbols $<$ and $>$.

We will use the following symbols for comparing two integers:

$<$ means "is less than."

$>$ means "is greater than."

On the number line, the integers increase from left to right. The *lesser* of two integers is the one farther to the *left* on a number line. The *greater* of two integers is the one farther to the *right* on a number line.

Look at the number line in **Figure 5.2**. The integers -4 and -1 are graphed.

FIGURE 5.2

Observe that -4 is to the left of -1 on the number line. This means that -4 is less than -1.

$$-4 < -1$$

−4 is less than −1 because −4 is to the **left** of −1 on the number line.

In **Figure 5.2**, we can also observe that -1 is to the right of -4 on the number line. This means that -1 is greater than -4.

$$-1 > -4$$

−1 is greater than −4 because −1 is to the **right** of −4 on the number line.

The symbols $<$ and $>$ are called **inequality symbols**. These symbols always point to the lesser of the two integers when the inequality statement is true.

−4 is less than −1. $\quad -4 < -1 \quad$ The symbol points to -4, the lesser number.

−1 is greater than −4. $\quad -1 > -4 \quad$ The symbol still points to -4, the lesser number.

EXAMPLE 2 *Using the Symbols $<$ and $>$*

Insert either $<$ or $>$ in the shaded area between the integers to make each statement true:

a. -4 ■ 3 **b.** -1 ■ -5 **c.** -5 ■ -2 **d.** 0 ■ -3.

SOLUTION

The solution is illustrated by the number line in **Figure 5.3**.

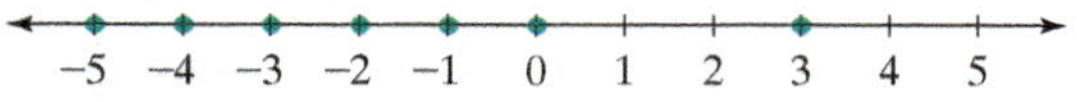

FIGURE 5.3

a. $-4 < 3$ (negative 4 is less than 3) because -4 is to the left of 3 on the number line.

b. $-1 > -5$ (negative 1 is greater than negative 5) because -1 is to the right of -5 on the number line.

c. $-5 < -2$ (negative 5 is less than negative 2) because -5 is to the left of -2 on the number line.

d. $0 > -3$ (zero is greater than negative 3) because 0 is to the right of -3 on the number line.

CHECK POINT 2 Insert either $<$ or $>$ in the shaded area between the integers to make each statement true:

a. 6 ■ -7 **b.** -8 ■ -1 **c.** -25 ■ -2 **d.** -14 ■ 0.

GREAT QUESTION!

Other than using a number line, is there another way to remember that -1 is greater than -5?

Yes. Think of negative integers as amounts of money that you *owe*. It's better to owe less, so

$$-1 > -5.$$

The symbols $<$ and $>$ may be combined with an equal sign, as shown in the following table:

This inequality is true if either the $<$ part or the $=$ part is true.

This inequality is true if either the $>$ part or the $=$ part is true.

Symbols	Meaning	Examples	Explanation
$a \leq b$	a is less than or equal to b.	$2 \leq 9$ $9 \leq 9$	Because $2 < 9$ Because $9 = 9$
$b \geq a$	b is greater than or equal to a.	$9 \geq 2$ $2 \geq 2$	Because $9 > 2$ Because $2 = 2$

4 Find the absolute value of an integer.

Absolute Value

Absolute value describes distance from 0 on a number line. If a represents an integer, the symbol $|a|$ represents its absolute value, read "the absolute value of a." For example,

$$|-5| = 5.$$

The absolute value of -5 is 5 because -5 is 5 units from 0 on a number line.

ABSOLUTE VALUE

The **absolute value** of an integer a, denoted by $|a|$, is the distance from 0 to a on the number line. Because absolute value describes a distance, it is never negative.

EXAMPLE 3 *Finding Absolute Value*

Find the absolute value:

a. $|-3|$ **b.** $|5|$ **c.** $|0|$.

SOLUTION

The solution is illustrated in **Figure 5.4**.

a. $|-3| = 3$ *The absolute value of -3 is 3 because -3 is 3 units from 0.*

b. $|5| = 5$ *5 is 5 units from 0.*

c. $|0| = 0$ *0 is 0 units from itself.*

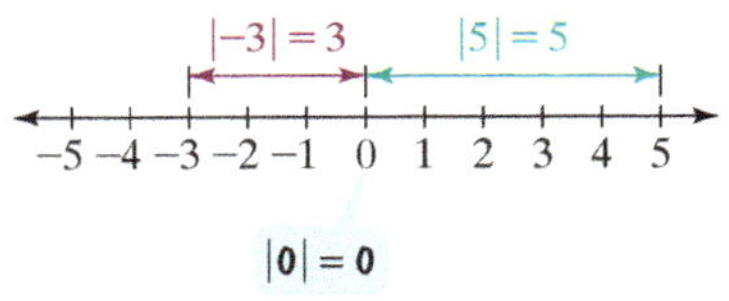

FIGURE 5.4 Absolute value describes distance from 0 on a number line.

Example 3 illustrates that the absolute value of a positive integer or 0 is the number itself. The absolute value of a negative integer, such as -3, is the number without the negative sign. Zero is the only real number whose absolute value is 0: $|0| = 0$. **The absolute value of any integer other than 0 is always positive.**

GREAT QUESTION!

What's the difference between $|-3|$ and $-|3|$?

They're easy to confuse.

$|-3| = 3$ — -3 is 3 units from 0.

$-|3| = -3$ — The negative is not inside the absolute value bars and is not affected by the absolute value.

CHECK POINT 3 Find the absolute value:

a. $|-8|$ **b.** $|6|$ **c.** $-|8|$.

5 Perform operations with integers.

Addition of Integers

It has not been a good day! First, you lost your wallet with \$50 in it. Then, you borrowed \$10 to get through the day, which you somehow misplaced. Your loss of \$50 followed by a loss of \$10 is an overall loss of \$60. This can be written

$$-50 + (-10) = -60.$$

The result of adding two or more numbers is called the **sum** of the numbers. The sum of -50 and -10 is -60.

You can think of gains and losses of money to find sums. For example, to find $17 + (-13)$, think of a gain of \$17 followed by a loss of \$13. There is an overall gain of \$4. Thus, $17 + (-13) = 4$. In the same way, to find $-17 + 13$, think of a loss of \$17 followed by a gain of \$13. There is an overall loss of \$4, so $-17 + 13 = -4$.

Using gains and losses, we can develop the following rules for adding integers:

RULES FOR ADDITION OF INTEGERS

Rule	Examples
If the integers have the same sign, **1.** Add their absolute values. **2.** The sign of the sum is the same as the sign of the two numbers.	$-11 + (-15) = -26$ Add absolute values: $11 + 15 = 26$. Use the common sign.
If the integers have different signs, **1.** Subtract the smaller absolute value from the larger absolute value. **2.** The sign of the sum is the same as the sign of the number with the larger absolute value.	$-13 + 4 = -9$ Subtract absolute values: $13 - 4 = 9$. Use the sign of the number with the greater absolute value. $13 + (-6) = 7$ Subtract absolute values: $13 - 6 = 7$. Use the sign of the number with the greater absolute value.

TECHNOLOGY

Calculators and Adding Integers

You can use a calculator to add integers. Here are the keystrokes for finding $-11 + (-15)$:

Scientific Calculator

11 [+/−] [+] 15 [+/−] [=]

Graphing Calculator

[(−)] 11 [+] [(−)] 15 [ENTER]

Here are the keystrokes for finding $-13 + 4$:

Scientific Calculator

13 [+/−] [+] 4 [=]

Graphing Calculator

[(−)] 13 [+] 4 [ENTER]

GREAT QUESTION!

Other than gains and losses of money, is there another good analogy for adding integers?

Yes. Think of temperatures above and below zero on a thermometer. Picture the thermometer as a number line standing straight up. For example,

$-11 + (-15) = -26$ — If it's 11 below zero and the temperature falls 15 degrees, it will then be 26 below zero.

$-13 + 4 = -9$ — If it's 13 below zero and the temperature rises 4 degrees, the new temperature will be 9 below zero.

$13 + (-6) = 7.$ — If it's 13 above zero and the temperature falls 6 degrees, it will then be 7 above zero.

Using the analogies of gains and losses of money or temperatures can make the formal rules for addition of integers easy to use.

Can you guess what number is displayed if you use a calculator to find a sum such as $18 + (-18)$? If you gain 18 and then lose 18, there is neither an overall gain nor loss. Thus,

$$18 + (-18) = 0.$$

We call 18 and −18 **additive inverses**. Additive inverses have the same absolute value, but lie on opposite sides of zero on the number line. Thus, −7 is the additive inverse of 7, and 5 is the additive inverse of −5. In general, the sum of any integer and its additive inverse is 0:

$$a + (-a) = 0.$$

Subtraction of Integers

Suppose that a computer that normally sells for \$1500 has a price reduction of \$600. The computer's reduced price, \$900, can be expressed in two ways:

$$1500 - 600 = 900 \quad \text{or} \quad 1500 + (-600) = 900.$$

This means that

$$1500 - 600 = 1500 + (-600).$$

To subtract 600 from 1500, we add 1500 and the additive inverse of 600. Generalizing from this situation, we define subtraction as follows:

DEFINITION OF SUBTRACTION

For all integers a and b,

$$a - b = a + (-b).$$

In words, to subtract b from a, add the additive inverse of b to a. The result of subtraction is called the **difference**.

TECHNOLOGY

Calculators and Subtracting Integers

You can use a calculator to subtract integers. Here are the keystrokes for finding 17 − (−11):

Scientific Calculator

17 [−] 11 [+/−] [=]

Graphing Calculator

17 [−] [(−)] 11 [ENTER]

Here are the keystrokes for finding −18 − (−5):

Scientific Calculator

18 [+/−] [−] 5 [+/−] [=]

Graphing Calculator

[(−)] 18 [−] [(−)] 5 [ENTER]

Don't confuse the subtraction key on a graphing calculator, [−], with the sign change or additive inverse key, [(−)]. What happens if you do?

EXAMPLE 4 *Subtracting Integers*

Subtract:

a. 17 − (−11) **b.** −18 − (−5) **c.** −18 − 5.

SOLUTION

a. $17 - (-11) = 17 + 11 = 28$

Change the subtraction to addition. Replace −11 with its additive inverse.

b. $-18 - (-5) = -18 + 5 = -13$

Change the subtraction to addition. Replace −5 with its additive inverse.

c. $-18 - 5 = -18 + (-5) = -23$

Change the subtraction to addition. Replace 5 with its additive inverse.

CHECK POINT 4 Subtract:

a. 30 − (−7) **b.** −14 − (−10) **c.** −14 − 10.

GREAT QUESTION!

Is there a practical way to think about what it means to subtract a negative integer?

Yes. Think of taking away a debt. Let's apply this analogy to 17 − (−11). Your checking account balance is \$17 after an erroneous \$11 charge was made against your account. When you bring this error to the bank's attention, they will take away the \$11 debit and your balance will go up to \$28:

$$17 - (-11) = 28.$$

Subtraction is used to solve problems in which the word *difference* appears. The difference between integers a and b is expressed as $a - b$.

EXAMPLE 5 *An Application of Subtraction Using the Word* Difference

Life expectancy for the average American man is 75.2 years; for a woman, it's 80.4 years. The number line in **Figure 5.5**, with points representing eight integers, indicates factors, many within our control, that can stretch or shrink one's probable life span.

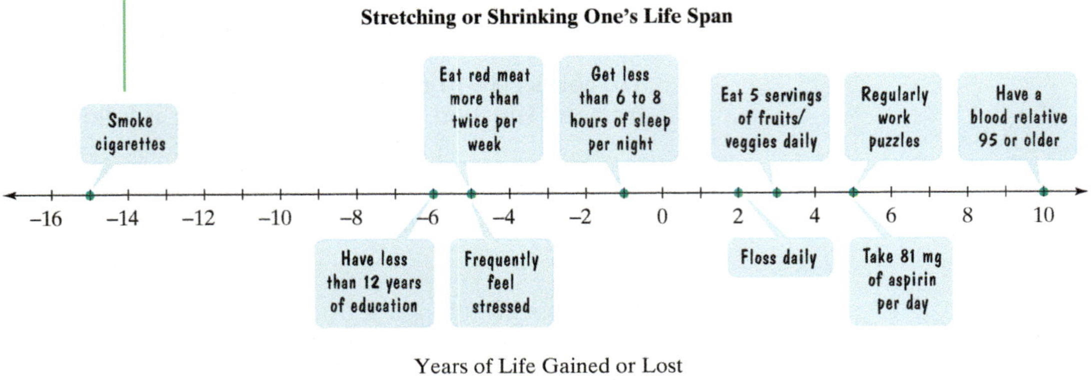

FIGURE 5.5
Source: Newsweek

a. What is the difference in the life span between a person who regularly works puzzles and a person who eats red meat more than twice per week?

b. What is the difference in the life span between a person with less than 12 years of education and a person who smokes cigarettes?

SOLUTION

a. We begin with the difference in the life span between a person who regularly works puzzles and a person who eats red meat more than twice per week. Refer to **Figure 5.5** to determine years of life gained or lost.

The difference	is	the change in life span for a person who regularly works puzzles	minus	the change in life span for a person who eats red meat more than twice per week.
	=	5	−	(−5)

$= 5 + 5 = 10$

The difference in the life span is 10 years.

b. Now we consider the difference in the life span between a person with less than 12 years of education and a person who smokes cigarettes.

The difference	is	the change in life span for a person with less than 12 years of education	minus	the change in life span for a person who smokes cigarettes.
	=	−6	−	(−15)

$= -6 + 15 = 9$

The difference in the life span is 9 years.

CHECK POINT 5 Use the number line in **Figure 5.5** to answer the following questions:

a. What is the difference in the life span between a person who eats five servings of fruits/veggies daily and a person who frequently feels stressed?

b. What is the difference in the life span between a person who gets less than 6 to 8 hours of sleep per night and a person who smokes cigarettes?

Multiplication of Integers

The result of multiplying two or more numbers is called the **product** of the numbers. You can think of multiplication as repeated addition or subtraction that starts at 0. For example,

$$3(-4) = 0 + (-4) + (-4) + (-4) = -12$$

The numbers have different signs and the product is negative.

and

$$(-3)(-4) = 0 - (-4) - (-4) - (-4) = 0 + 4 + 4 + 4 = 12.$$

The numbers have the same sign and the product is positive.

These observations give us the following rules for multiplying integers:

RULES FOR MULTIPLYING INTEGERS

Rule	Examples
1. The product of two integers with different signs is found by multiplying their absolute values. The product is negative.	• $7(-5) = -35$
2. The product of two integers with the same sign is found by multiplying their absolute values. The product is positive.	• $(-6)(-11) = 66$
3. The product of 0 and any integer is 0: $a \cdot 0 = 0$ and $0 \cdot a = 0$.	• $-17(0) = 0$
4. If no number is 0, a product with an odd number of negative factors is found by multiplying absolute values. The product is negative.	• $-2(-3)(-5) = -30$ Three (odd) negative factors
5. If no number is 0, a product with an even number of negative factors is found by multiplying absolute values. The product is positive.	• $-2(3)(-5) = 30$ Two (even) negative factors

Exponential Notation

Because exponents indicate repeated multiplication, rules for multiplying real numbers can be used to evaluate exponential expressions.

EXAMPLE 6 *Evaluating Exponential Expressions*

Evaluate:

a. $(-6)^2$ **b.** -6^2 **c.** $(-5)^3$ **d.** $(-2)^4$.

SOLUTION

a. $(-6)^2 = (-6)(-6) = 36$

Base is −6. Same signs give positive product.

b. $-6^2 = -(6 \cdot 6) = -36$

Base is 6. The negative is not inside parentheses and is not taken to the second power.

c. $(-5)^3 = (-5)(-5)(-5) = -125$

An odd number of negative factors gives a negative product.

d. $(-2)^4 = (-2)(-2)(-2)(-2) = 16$

An even number of negative factors gives a positive product.

CHECK POINT 6 Evaluate:

a. $(-5)^2$ **b.** -5^2 **c.** $(-4)^3$ **d.** $(-3)^4$.

Blitzer Bonus

Integers, Karma, and Exponents

On Friday the 13th, are you a bit more careful crossing the street even if you don't consider yourself superstitious? Numerology, the belief that certain integers have greater significance and can be lucky or unlucky, is widespread in many cultures.

Integer	Connotation	Culture	Origin	Example
4	Negative	Chinese	The word for the number 4 sounds like the word for death.	Many buildings in China have floor-numbering systems that skip 40–49.
7	Positive	United States	In dice games, this prime number is the most frequently rolled number with two dice.	There was a spike in the number of couples getting married on 7/7/07.
8	Positive	Chinese	It's considered a sign of prosperity.	The Beijing Olympics began at 8 P.M. on 8/8/08.
13	Negative	Various	Various reasons, including the number of people at the Last Supper	Many buildings around the world do not label any floor "13."
18	Positive	Jewish	The Hebrew letters spelling *chai*, or living, are the 8th and 10th in the alphabet, adding up to 18	Monetary gifts for celebrations are often given in multiples of 18.
666	Negative	Christian	The New Testament's Book of Revelation identifies 666 as the "number of the beast," which some say refers to Satan.	In 2008, Reeves, Louisiana, eliminated 666 as the prefix of its phone numbers.

Source: The New York Times

Although your author is not a numerologist, he is intrigued by curious exponential representations for 666:

$$666 = 6 + 6 + 6 + 6^3 + 6^3 + 6^3$$
$$666 = 1^3 + 2^3 + 3^3 + 4^3 + 5^3 + 6^3 + 5^3 + 4^3 + 3^3 + 2^3 + 1^3$$
$$666 = 2^2 + 3^2 + 5^2 + 7^2 + 11^2 + 13^2 + 17^2$$

Sum of the squares of the first seven prime numbers

$$666 = 1^6 - 2^6 + 3^6$$

The number 666 is even interesting in Roman numerals:

$$666 = \text{DCLXVI}.$$

Contains all Roman numerals from D(500) to I (1) in decending order

Division of Integers

The result of dividing the integer a by the nonzero integer b is called the **quotient** of the numbers. We can write this quotient as $a \div b$ or $\frac{a}{b}$.

A relationship exists between multiplication and division. For example,

$$\frac{-12}{4} = -3 \text{ means that } 4(-3) = -12.$$

$$\frac{-12}{-4} = 3 \text{ means that } -4(3) = -12.$$

Because of the relationship between multiplication and division, the rules for obtaining the sign of a quotient are the same as those for obtaining the sign of a product.

TECHNOLOGY

Multiplying and Dividing on a Calculator

Example: $(-173)(-256)$

Scientific Calculator

173 [+/−] [×] 256 [+/−] [=]

Graphing Calculator

[(−)] 173 [×] [(−)] 256 [ENTER]

The number 44288 should be displayed.

Division is performed in the same manner, using [÷] instead of [×]. What happens when you divide by 0? Try entering

8 [÷] 0

and pressing [=] or [ENTER].

RULES FOR DIVIDING INTEGERS

Rule	Examples
1. The quotient of two integers with different signs is found by dividing their absolute values. The quotient is negative.	• $\frac{80}{-4} = -20$ • $\frac{-15}{5} = -3$
2. The quotient of two integers with the same sign is found by dividing their absolute values. The quotient is positive.	• $\frac{27}{9} = 3$ • $\frac{-45}{-3} = 15$
3. Zero divided by any nonzero integer is zero.	• $\frac{0}{-5} = 0$ (because $-5(0) = 0$)
4. Division by 0 is undefined.	• $\frac{-8}{0}$ is undefined (because 0 cannot be multiplied by an integer to obtain -8).

6 Use the order of operations agreement.

Order of Operations

Suppose that you want to find the value of $3 + 7 \cdot 5$. Which procedure shown below is correct?

$$3 + 7 \cdot 5 = 3 + 35 = 38 \quad \text{or} \quad 3 + 7 \cdot 5 = 10 \cdot 5 = 50$$

If you know the answer, you probably know certain rules, called the **order of operations**, that make sure there is only one correct answer. One of these rules states that if a problem contains no parentheses, perform multiplication before addition. Thus, the procedure on the left is correct because the multiplication of 7 and 5 is done first. Then the addition is performed. The correct answer is 38.

Here are the rules for determining the order in which operations should be performed:

GREAT QUESTION!

How can I remember the order of operations?

This sentence may help: Please excuse my dear Aunt Sally.

Please	Parentheses
Excuse	Exponents
My Dear	Multiplication Division
Aunt Sally	Addition Subtraction

ORDER OF OPERATIONS

1. Perform all operations within grouping symbols.
2. Evaluate all exponential expressions.
3. Do all multiplications and divisions in the order in which they occur, working from left to right.
4. Finally, do all additions and subtractions in the order in which they occur, working from left to right.

In the third step in the order of operations, be sure to do all multiplications and divisions *as they occur* from left to right. For example,

$$8 \div 4 \cdot 2 = 2 \cdot 2 = 4$$ Do the division first because it occurs first.

$$8 \cdot 4 \div 2 = 32 \div 2 = 16.$$ Do the multiplication first because it occurs first.

EXAMPLE 7 *Using the Order of Operations*

Simplify: $6^2 - 24 \div 2^2 \cdot 3 + 1$.

SOLUTION

There are no grouping symbols. Thus, we begin by evaluating exponential expressions. Then we multiply or divide. Finally, we add or subtract.

$6^2 - 24 \div 2^2 \cdot 3 + 1$

$= 36 - 24 \div 4 \cdot 3 + 1$ Evaluate exponential expressions: $6^2 = 6 \cdot 6 = 36$ and $2^2 = 2 \cdot 2 = 4$.

$= 36 - 6 \cdot 3 + 1$ Perform the multiplications and divisions from left to right. Start with $24 \div 4 = 6$.

$= 36 - 18 + 1$ Now do the multiplication: $6 \cdot 3 = 18$.

$= 18 + 1$ Finally, perform the additions and subtractions from left to right. Subtract: $36 - 18 = 18$.

$= 19$ $18 + 1 = 19$.

CHECK POINT 7 Simplify: $7^2 - 48 \div 4^2 \cdot 5 + 2$.

EXAMPLE 8 *Using the Order of Operations*

Simplify: $(-6)^2 - (5 - 7)^2(-3)$.

SOLUTION

Because grouping symbols appear, we perform the operation within parentheses first.

$(-6)^2 - (5 - 7)^2(-3)$

$= (-6)^2 - (-2)^2(-3)$ Work inside parentheses first: $5 - 7 = 5 + (-7) = -2$.

$= 36 - 4(-3)$ Evaluate exponential expressions: $(-6)^2 = (-6)(-6) = 36$ and $(-2)^2 = (-2)(-2) = 4$.

$= 36 - (-12)$ Multiply: $4(-3) = -12$.

$= 48$ Subtract: $36 - (-12) = 36 + 12 = 48$.

CHECK POINT 8 Simplify: $(-8)^2 - (10 - 13)^2(-2)$.

Concept and Vocabulary Check

Fill in each blank so that the resulting statement is true.

1. The integers are defined by the set ______________.
2. If $a < b$, then a is located to the ________ of b on a number line.
3. On a number line, the absolute value of a, denoted $|a|$, represents ______________.
4. Two integers that have the same absolute value, but lie on opposite sides of zero on a number line, are called ____________.

In Exercises 5–8, determine whether each statement is true or false. If the statement is false, make the necessary change(s) to produce a true statement.

5. The sum of a positive integer and a negative integer is always a positive integer. _______
6. The difference between 0 and a negative integer is always a positive integer. _______
7. The product of a positive integer and a negative integer is never a positive integer. _______
8. The quotient of 0 and a negative integer is undefined. _______

Exercise Set 5.2

Practice Exercises

In Exercises 1–4, start by drawing a number line that shows integers from −5 to 5. Then graph each of the following integers on your number line.

1. 3 **2.** 5 **3.** −4 **4.** −2

In Exercises 5–12, insert either < or > in the shaded area between the integers to make the statement true.

5. −2 ■ 7 **6.** −1 ■ 13 **7.** −13 ■ −2
8. −1 ■ −13 **9.** 8 ■ −50 **10.** 7 ■ −9
11. −100 ■ 0 **12.** 0 ■ −300

In Exercises 13–18, find the absolute value.

13. $|-14|$ **14.** $|-16|$ **15.** $|14|$
16. $|16|$ **17.** $|-300{,}000|$ **18.** $|-1{,}000{,}000|$

In Exercises 19–30, find each sum.

19. $-7 + (-5)$ **20.** $-3 + (-4)$
21. $12 + (-8)$ **22.** $13 + (-5)$
23. $6 + (-9)$ **24.** $3 + (-11)$
25. $-9 + (+4)$ **26.** $-7 + (+3)$
27. $-9 + (-9)$ **28.** $-13 + (-13)$
29. $9 + (-9)$ **30.** $13 + (-13)$

In Exercises 31–42, perform the indicated subtraction.

31. $13 - 8$ **32.** $14 - 3$
33. $8 - 15$ **34.** $9 - 20$
35. $4 - (-10)$ **36.** $3 - (-17)$
37. $-6 - (-17)$ **38.** $-4 - (-19)$
39. $-12 - (-3)$ **40.** $-19 - (-2)$
41. $-11 - 17$ **42.** $-19 - 21$

In Exercises 43–52, find each product.

43. $6(-9)$ **44.** $5(-7)$
45. $(-7)(-3)$ **46.** $(-8)(-5)$
47. $(-2)(6)$ **48.** $(-3)(10)$
49. $(-13)(-1)$ **50.** $(-17)(-1)$
51. $0(-5)$ **52.** $0(-8)$

In Exercises 53–66, evaluate each exponential expression.

53. 5^2 **54.** 6^2
55. $(-5)^2$ **56.** $(-6)^2$
57. 4^3 **58.** 2^3
59. $(-5)^3$ **60.** $(-4)^3$
61. $(-5)^4$ **62.** $(-4)^4$
63. -3^4 **64.** -1^4
65. $(-3)^4$ **66.** $(-1)^4$

In Exercises 67–80, find each quotient, or, if applicable, state that the expression is undefined.

67. $\frac{-12}{4}$ **68.** $\frac{-40}{5}$
69. $\frac{21}{-3}$ **70.** $\frac{60}{-6}$
71. $\frac{-90}{-3}$ **72.** $\frac{-66}{-6}$
73. $\frac{0}{-7}$ **74.** $\frac{0}{-8}$
75. $\frac{-7}{0}$ **76.** $\frac{0}{0}$
77. $(-480) \div 24$ **78.** $(-300) \div 12$
79. $(465) \div (-15)$ **80.** $(-594) \div (-18)$

In Exercises 81–100, use the order of operations to find the value of each expression.

81. $7 + 6 \cdot 3$ **82.** $-5 + (-3) \cdot 8$
83. $(-5) - 6(-3)$ **84.** $-8(-3) - 5(-6)$
85. $6 - 4(-3) - 5$ **86.** $3 - 7(-1) - 6$
87. $3 - 5(-4 - 2)$ **88.** $3 - 9(-1 - 6)$
89. $(2 - 6)(-3 - 5)$ **90.** $9 - 5(6 - 4) - 10$
91. $3(-2)^2 - 4(-3)^2$ **92.** $5(-3)^2 - 2(-2)^3$
93. $(2 - 6)^2 - (3 - 7)^2$
94. $(4 - 6)^2 - (5 - 9)^3$
95. $6(3 - 5)^3 - 2(1 - 3)^3$
96. $-3(-6 + 8)^3 - 5(-3 + 5)^3$
97. $8^2 - 16 \div 2^2 \cdot 4 - 3$
98. $10^2 - 100 \div 5^2 \cdot 2 - (-3)$
99. $24 \div [3^2 \div (8 - 5)] - (-6)$
100. $30 \div [5^2 \div (7 - 12)] - (-9)$

Practice Plus

In Exercises 101–110, use the order of operations to find the value of each expression.

101. $8 - 3[-2(2 - 5) - 4(8 - 6)]$

102. $8 - 3[-2(5 - 7) - 5(4 - 2)]$

103. $-2^2 + 4[16 \div (3 - 5)]$

104. $-3^2 + 2[20 \div (7 - 11)]$

105. $4|10 - (8 - 20)|$

106. $-5|7 - (20 - 8)|$

107. $[-5^2 + (6 - 8)^3 - (-4)] - [|-2|^3 + 1 - 3^2]$

108. $[-4^2 + (7 - 10)^3 - (-27)] - [|-2|^5 + 1 - 5^2]$

109. $\dfrac{12 \div 3 \cdot 5|2^2 + 3^2|}{7 + 3 - 6^2}$

110. $\dfrac{-3 \cdot 5^2 + 89}{(5 - 6)^2 - 2|3 - 7|}$

In Exercises 111–114, express each sentence as a single numerical expression. Then use the order of operations to simplify the expression.

111. Cube -2. Subtract this exponential expression from -10.

112. Cube -5. Subtract this exponential expression from -100.

113. Subtract 10 from 7. Multiply this difference by 2. Square this product.

114. Subtract 11 from 9. Multiply this difference by 2. Raise this product to the fourth power.

Application Exercises

115. The peak of Mount McKinley, the highest point in the United States, is 20,320 feet above sea level. Death Valley, the lowest point in the United States, is 282 feet below sea level. What is the difference in elevation between the peak of Mount McKinley and Death Valley?

116. The peak of Mount Kilimanjaro, the highest point in Africa, is 19,321 feet above sea level. Qattara Depression, Egypt, the lowest point in Africa, is 436 feet below sea level. What is the difference in elevation between the peak of Mount Kilimanjaro and the Qattara Depression?

In Exercises 117–126, we return to the number line that shows factors that can stretch or shrink one's probable life span.

Stretching or Shrinking One's Life Span

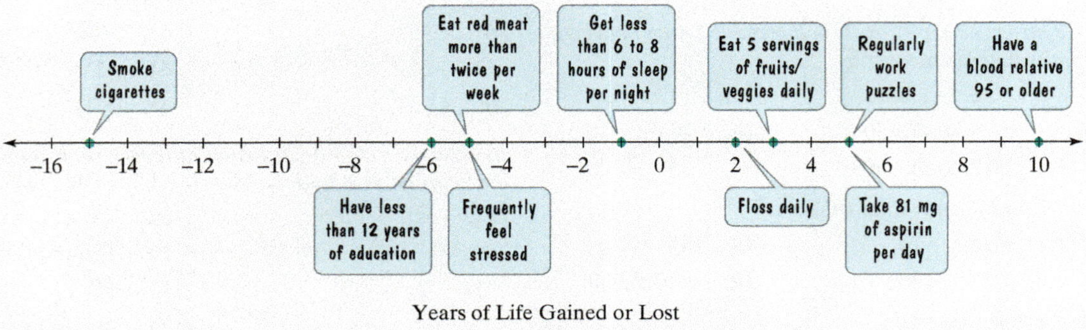

Source: Newsweek

117. If you have a blood relative 95 or older and you smoke cigarettes, do you stretch or shrink your life span? By how many years?

118. If you floss daily and eat red meat more than twice per week, do you stretch or shrink your life span? By how many years?

119. If you frequently feel stressed and have less than 12 years of education, do you stretch or shrink your life span? By how many years?

120. If you get less than 6 to 8 hours of sleep per night and smoke cigarettes, do you stretch or shrink your life span? By how many years?

121. What happens to the life span for a person who takes 81 mg of aspirin per day and eats red meat more than twice per week?

122. What happens to the life span for a person who regularly works puzzles and a person who frequently feels stressed?

123. What is the difference in the life span between a person who has a blood relative 95 or older and a person who smokes cigarettes?

124. What is the difference in the life span between a person who has a blood relative 95 or older and a person who has less than 12 years of education?

125. What is the difference in the life span between a person who frequently feels stressed and a person who has less than 12 years of education?

126. What is the difference in the life span between a person who gets less than 6 to 8 hours of sleep per night and a person who frequently feels stressed?

The accompanying bar graph shows the amount of money, in billions of dollars, collected and spent by the U.S. government in selected years from 2001 through 2011. Use the information from the graph to solve Exercises 127–130. Express answers in billions of dollars.

Money Collected and Spent by the United States Government

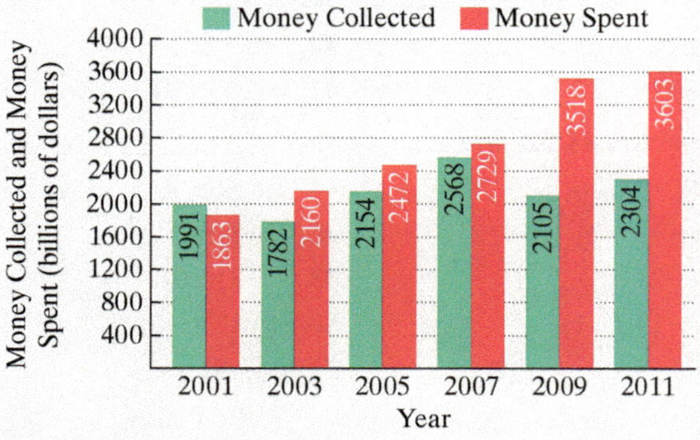

Source: Budget of the U.S. Government

127. a. In 2001, what was the difference between the amount of money collected and the amount spent? Was there a budget surplus or deficit in 2001?

b. In 2011, what was the difference between the amount of money collected and the amount spent? Was there a budget surplus or deficit in 2011?

c. What is the difference between the 2001 surplus and the 2011 deficit?

128. a. In 2001, what was the difference between the amount of money collected and the amount spent? Was there a budget surplus or deficit in 2001?

b. In 2009, what was the difference between the amount of money collected and the amount spent? Was there a budget surplus or deficit in 2009?

c. What is the difference between the 2001 surplus and the 2009 deficit?

129. What is the difference between the 2007 deficit and the 2011 deficit?

130. What is the difference between the 2007 deficit and the 2009 deficit?

The way that we perceive the temperature on a cold day depends on both air temperature and wind speed. The windchill is what the air temperature would have to be with no wind to achieve the same chilling effect on the skin. In 2002, the National Weather Service issued new windchill temperatures, shown in the table below. Use the information from the table to solve Exercises 131–134.

New Windchill Temperature Index

Wind Speed (miles per hour)	Air Temperature (F)											
	30	25	20	15	10	5	0	−5	−10	−15	−20	−25
5	25	19	13	7	1	−5	−11	−16	−22	−28	−34	−40
10	21	15	9	3	−4	−10	−16	−22	−28	−35	−41	−47
15	19	13	6	0	−7	−13	−19	−26	−32	−39	−45	−51
20	17	11	4	−2	−9	−15	−22	−29	−35	−42	−48	−55
25	16	9	3	−4	−11	−17	−24	−31	−37	−44	−51	−58
30	15	8	1	−5	−12	−19	−26	−33	−39	−46	−53	−60
35	14	7	0	−7	−14	−21	−27	−34	−41	−48	−55	−62
40	13	6	−1	−8	−15	−22	−29	−36	−43	−50	−57	−64
45	12	5	−2	−9	−16	−23	−30	−37	−44	−51	−58	−65
50	12	4	−3	−10	−17	−24	−31	−38	−45	−52	−60	−67
55	11	4	−3	−11	−18	−25	−32	−39	−46	−54	−61	−68
60	10	3	−4	−11	−19	−26	−33	−40	−48	−55	−62	−69

Frostbite occurs in 15 minutes or less.

Source: National Weather Service

131. What is the difference between how cold the temperature feels with winds at 10 miles per hour and 25 miles per hour when the air temperature is 15°F?

132. What is the difference between how cold the temperature feels with winds at 5 miles per hour and 30 miles per hour when the air temperature is 10°F?

133. What is the difference in the windchill temperature between an air temperature of 5°F with winds at 50 miles per hour and an air temperature of −10°F with winds at 5 miles per hour?

134. What is the difference in the windchill temperature between an air temperature of 5°F with winds at 55 miles per hour and an air temperature of −5°F with winds at 10 miles per hour?

Writing in Mathematics

135. How does the set of integers differ from the set of whole numbers?

136. Explain how to graph an integer on a number line.

137. If you are given two integers, explain how to determine which one is smaller.

138. Explain how to add integers.

139. Explain how to subtract integers.

140. Explain how to multiply integers.

141. Explain how to divide integers.

142. Describe what it means to raise a number to a power. In your description, include a discussion of the difference between -5^2 and $(-5)^2$.

143. Why is $\frac{0}{4}$ equal to 0, but $\frac{4}{0}$ undefined?

Critical Thinking Exercises

Make Sense? *In Exercises 144–147, determine whether each statement makes sense or does not make sense, and explain your reasoning.*

144. Without adding integers, I can see that the sum of −227 and 319 is greater than the sum of 227 and −319.

145. I found the variation in U.S. temperature by subtracting the record low temperature, a negative integer, from the record high temperature, a positive integer.

146. I've noticed that the sign rules for dividing integers are slightly different than the sign rules for multiplying integers.

147. The rules for the order of operations avoid the confusion of obtaining different results when I simplify the same expression.

In Exercises 148–149, insert one pair of parentheses to make each calculation correct.

148. $8 - 2 \cdot 3 - 4 = 10$

149. $8 - 2 \cdot 3 - 4 = 14$

Technology Exercises

Scientific calculators that have parentheses keys allow for the entry and computation of relatively complicated expressions in a single step. For example, the expression $15 + (10 - 7)^2$ *can be evaluated by entering the following keystrokes:*

15 [+] [(] 10 [−] 7 [)] [y^x] 2 [=].

Find the value of each expression in Exercises 150–152 in a single step on your scientific calculator.

150. $8 - 2 \cdot 3 - 9$

151. $(8 - 2) \cdot (3 - 9)$

152. $5^3 + 4 \cdot 9 - (8 + 9 \div 3)$

5.3

WHAT AM I SUPPOSED TO LEARN?

After you have read this section, you should be able to:

1. Define the rational numbers.
2. Reduce rational numbers.
3. Convert between mixed numbers and improper fractions.
4. Express rational numbers as decimals.
5. Express decimals in the form $\frac{a}{b}$.
6. Multiply and divide rational numbers.
7. Add and subtract rational numbers.
8. Use the order of operations agreement with rational numbers.
9. Apply the density property of rational numbers.
10. Solve problems involving rational numbers.

The Rational Numbers

YOU ARE MAKING EIGHT DOZEN CHOCOLATE chip cookies for a large neighborhood block party. The recipe lists the ingredients needed to prepare five dozen cookies, such as $\frac{3}{4}$ cup sugar. How do you adjust the amount of sugar, as well as the amounts of each of the other ingredients, given in the recipe?

Adapting a recipe to suit a different number of portions usually involves working with numbers that are not integers. For example, the number describing the amount of sugar, $\frac{3}{4}$ (cup), is not an integer, although it consists of the quotient of two integers, 3 and 4. Before returning to the problem of changing the size of a recipe, we study a new set of numbers consisting of the quotients of integers.

1 Define the rational numbers.

Defining the Rational Numbers

If two integers are added, subtracted, or multiplied, the result is always another integer. This, however, is not always the case with division. For example, 10 divided by 5 is the integer 2. By contrast, 5 divided by 10 is $\frac{1}{2}$, and $\frac{1}{2}$ is not an integer. To permit divisions such as $\frac{5}{10}$, we enlarge the set of integers, calling the new collection the *rational numbers*. The set of **rational numbers** consists of all the numbers that can be expressed as a quotient of two integers, with the denominator not 0.

THE RATIONAL NUMBERS

The set of **rational numbers** is the set of all numbers which can be expressed in the form $\frac{a}{b}$, where a and b are integers and b is not equal to 0. The integer a is called the **numerator**, and the integer b is called the **denominator**.

GREAT QUESTION!

Is the rational number $\frac{-3}{4}$ the same as $-\frac{3}{4}$?

We know that the quotient of two numbers with different signs is a negative number. Thus,

$$\frac{-3}{4} = -\frac{3}{4} \quad \text{and} \quad \frac{3}{-4} = -\frac{3}{4}.$$

The following numbers are examples of rational numbers:

$$\frac{1}{2}, \frac{-3}{4}, 5, 0.$$

The integer 5 is a rational number because it can be expressed as the quotient of integers: $5 = \frac{5}{1}$. Similarly, 0 can be written as $\frac{0}{1}$.

In general, every integer a is a rational number because it can be expressed in the form $\frac{a}{1}$.

2 Reduce rational numbers.

Reducing Rational Numbers

A rational number is **reduced to its lowest terms**, or **simplified**, when the numerator and denominator have no common divisors other than 1. Reducing rational numbers to lowest terms is done using the **Fundamental Principle of Rational Numbers**.

THE FUNDAMENTAL PRINCIPLE OF RATIONAL NUMBERS

If $\frac{a}{b}$ is a rational number and c is any number other than 0,

$$\frac{a \cdot c}{b \cdot c} = \frac{a}{b}.$$

The rational numbers $\frac{a}{b}$ and $\frac{a \cdot c}{b \cdot c}$ are called **equivalent fractions**.

When using the Fundamental Principle to reduce a rational number, the simplification can be done in one step by finding the greatest common divisor of the numerator and the denominator, and using it for *c*. Thus, **to reduce a rational number to its lowest terms, divide both the numerator and the denominator by their greatest common divisor.**

For example, consider the rational number $\frac{12}{100}$. The greatest common divisor of 12 and 100 is 4. We reduce to lowest terms as follows:

$$\frac{12}{100} = \frac{3 \cdot \cancel{4}}{25 \cdot \cancel{4}} = \frac{3}{25} \quad \text{or} \quad \frac{12}{100} = \frac{12 \div 4}{100 \div 4} = \frac{3}{25}.$$

EXAMPLE 1 *Reducing a Rational Number*

Reduce $\frac{130}{455}$ to lowest terms.

SOLUTION

Begin by finding the greatest common divisor of 130 and 455.

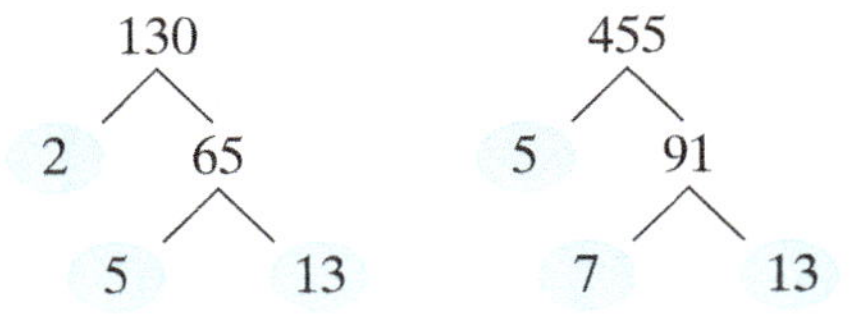

Thus, $130 = 2 \cdot 5 \cdot 13$ and $455 = 5 \cdot 7 \cdot 13$. The greatest common divisor is $5 \cdot 13$, or 65. Divide the numerator and the denominator of the given rational number by $5 \cdot 13$ or by 65.

$$\frac{130}{455} = \frac{2 \cdot \cancel{5} \cdot \cancel{13}}{\cancel{5} \cdot 7 \cdot \cancel{13}} = \frac{2}{7} \quad \text{or} \quad \frac{130}{455} = \frac{130 \div 65}{455 \div 65} = \frac{2}{7}$$

There are no common divisors of 2 and 7 other than 1. Thus, the rational number $\frac{2}{7}$ is in its lowest terms.

CHECK POINT 1 Reduce $\frac{72}{90}$ to lowest terms.

3 Convert between mixed numbers and improper fractions.

Mixed Numbers and Improper Fractions

A **mixed number** consists of the sum of an integer and a rational number, expressed without the use of an addition sign. Here is an example of a mixed number:

$$3\frac{4}{5}.$$

The integer is 3 and the rational number is $\frac{4}{5}$. $3\frac{4}{5}$ means $3 + \frac{4}{5}$.

An **improper fraction** is a rational number whose numerator is greater than its denominator. An example of an improper fraction is $\frac{19}{5}$.

The mixed number $3\frac{4}{5}$ can be converted to the improper fraction $\frac{19}{5}$ using the following procedure:

GREAT QUESTION!

How do I read the mixed number $3\frac{4}{5}$?

It's read "three and four-fifths."

CONVERTING A POSITIVE MIXED NUMBER TO AN IMPROPER FRACTION

1. Multiply the denominator of the rational number by the integer and add the numerator to this product.
2. Place the sum in step 1 over the denominator in the mixed number.

EXAMPLE 2 Converting from a Mixed Number to an Improper Fraction

Convert $3\frac{4}{5}$ to an improper fraction.

SOLUTION

$$3\frac{4}{5} = \frac{5 \cdot 3 + 4}{5}$$

Multiply the denominator by the integer and add the numerator.

Place the sum over the mixed number's denominator.

$$= \frac{15 + 4}{5} = \frac{19}{5}$$

CHECK POINT 2 Convert $2\frac{5}{8}$ to an improper fraction.

When converting a negative mixed number to an improper fraction, copy the negative sign and then follow the previous procedure. For example,

$$-2\frac{3}{4} = -\frac{4 \cdot 2 + 3}{4} = -\frac{8 + 3}{4} = -\frac{11}{4}.$$

Copy the negative sign from step to step and convert $2\frac{3}{4}$ to an improper fraction.

GREAT QUESTION!

Does $-2\frac{3}{4}$ mean that I need to add $\frac{3}{4}$ to -2?

No. $-2\frac{3}{4}$ means

$$-(2\tfrac{3}{4}) \quad \text{or} \quad -(2 + \tfrac{3}{4}).$$

$-2\frac{3}{4}$ does not mean

$$-2 + \tfrac{3}{4}.$$

A positive improper fraction can be converted to a mixed number using the following procedure:

CONVERTING A POSITIVE IMPROPER FRACTION TO A MIXED NUMBER

1. Divide the denominator into the numerator. Record the quotient and the remainder.
2. Write the mixed number using the following form:

$$\text{quotient}\ \frac{\text{remainder}}{\text{original denominator}}.$$

integer part (quotient); rational number part (fraction)

EXAMPLE 3 Converting from an Improper Fraction to a Mixed Number

Convert $\frac{42}{5}$ to a mixed number.

SOLUTION

Step 1 Divide the denominator into the numerator.

$$\begin{array}{r} 8 \\ 5\overline{)42} \\ \underline{40} \\ 2 \end{array}$$

8: quotient; 2: remainder

Step 2 Write the mixed number using quotient $\dfrac{\textbf{remainder}}{\textbf{original denominator}}$. Thus,

$$\frac{42}{5} = 8\frac{2}{5}.$$

8: quotient; 2: remainder; 5: original denominator

GREAT QUESTION!

When should I use mixed numbers and when are improper fractions preferable?

In applied problems, answers are usually expressed as mixed numbers, which many people find more meaningful than improper fractions. However, improper fractions are often easier to work with when performing operations with fractions.

CHECK POINT 3 Convert $\frac{5}{3}$ to a mixed number.

When converting a negative improper fraction to a mixed number, copy the negative sign and then follow the previous procedure. For example,

$$-\frac{29}{8} = -3\frac{5}{8}.$$

Convert $\frac{29}{8}$ to a mixed number.

$$\begin{array}{r} 3 \leftarrow \text{quotient} \\ 8\overline{)29} \\ \underline{24} \\ 5 \leftarrow \text{remainder} \end{array}$$

Copy the negative sign.

4 Express rational numbers as decimals.

Tens	Ones	Tenths	Hundredths	Thousandths	Ten-Thousandths	Hundred-Thousandths
10	1	$\frac{1}{10}$	$\frac{1}{100}$	$\frac{1}{1000}$	$\frac{1}{10,000}$	$\frac{1}{100,000}$

decimal point

Rational Numbers and Decimals

We have seen that a rational number is the quotient of integers. Rational numbers can also be expressed as decimals. As shown in the place-value chart in the margin, it is convenient to represent rational numbers with denominators of 10, 100, 1000, and so on as decimals. For example,

$$\frac{7}{10} = 0.7, \quad \frac{3}{100} = 0.03, \quad \text{and} \quad \frac{8}{1000} = 0.008.$$

Any rational number $\frac{a}{b}$ can be expressed as a decimal by dividing the denominator, b, into the numerator, a.

EXAMPLE 4 *Expressing Rational Numbers as Decimals*

Express each rational number as a decimal:

a. $\frac{5}{8}$ **b.** $\frac{7}{11}$.

SOLUTION

In each case, divide the denominator into the numerator.

a.

$$\begin{array}{r} 0.625 \\ 8\overline{)5.000} \\ \underline{4\,8} \\ 20 \\ \underline{16} \\ 40 \\ \underline{40} \\ 0 \end{array} \qquad \frac{5}{8} = 0.625$$

b.

$$\begin{array}{r} 0.6363\ldots \\ 11\overline{)7.0000\ldots} \\ \underline{6\,6} \\ 40 \\ \underline{33} \\ 70 \\ \underline{66} \\ 40 \\ \underline{33} \\ 70 \\ \vdots \end{array} \qquad \frac{7}{11} = 0.6363\ldots$$

In Example 4, the decimal for $\frac{5}{8}$, namely 0.625, stops and is called a **terminating decimal**. Other examples of terminating decimals are

$$\frac{1}{4} = 0.25, \quad \frac{2}{5} = 0.4, \quad \text{and} \quad \frac{7}{8} = 0.875.$$

By contrast, the division process for $\frac{7}{11}$ results in 0.6363. . . , with the digits 63 repeating over and over indefinitely. To indicate this, write a bar over the digits that repeat. Thus,

$$\frac{7}{11} = 0.\overline{63}.$$

The decimal for $\frac{7}{11}$, $0.\overline{63}$, is called a **repeating decimal**. Other examples of repeating decimals are

$$\frac{1}{3} = 0.333\ldots = 0.\overline{3} \quad \text{and} \quad \frac{2}{3} = 0.666\ldots = 0.\overline{6}.$$

RATIONAL NUMBERS AND DECIMALS

Any rational number can be expressed as a decimal. The resulting decimal will either terminate (stop), or it will have a digit that repeats or a block of digits that repeats.

CHECK POINT 4 Express each rational number as a decimal:

a. $\frac{3}{8}$ **b.** $\frac{5}{11}$.

5 Express decimals in the form $\frac{a}{b}$.

Reversing Directions: Expressing Decimals as Quotients of Two Integers

Terminating decimals can be expressed with denominators of 10, 100, 1000, 10,000, and so on. Use the place-value chart shown in the margin. The digits to the right of the decimal point are the numerator of the rational number. To find the denominator, observe the last digit to the right of the decimal point. The place-value of this digit will indicate the denominator.

Tens	Ones	Tenths	Hundredths	Thousandths	Ten-Thousandths	Hundred-Thousandths
10	1	$\frac{1}{10}$	$\frac{1}{100}$	$\frac{1}{1000}$	$\frac{1}{10,000}$	$\frac{1}{100,000}$

decimal point

EXAMPLE 5 *Expressing Terminating Decimals in $\frac{a}{b}$ Form*

Express each terminating decimal as a quotient of integers:

a. 0.7 **b.** 0.49 **c.** 0.048.

SOLUTION

a. $0.7 = \frac{7}{10}$ because the 7 is in the tenths position.

b. $0.49 = \frac{49}{100}$ because the last digit on the right, 9, is in the hundredths position.

c. $0.048 = \frac{48}{1000}$ because the digit on the right, 8, is in the thousandths position. Reducing to lowest terms, $\frac{48}{1000} = \frac{48 \div 8}{1000 \div 8} = \frac{6}{125}$.

CHECK POINT 5 Express each terminating decimal as a quotient of integers, reduced to lowest terms:

a. 0.9 **b.** 0.86 **c.** 0.053.

A BRIEF REVIEW ***Solving One-Step Equations***

- Solving an equation involves determining all values that result in a true statement when substituted into the equation. Such values are solutions of the equation.

  Example

  The solution of $x - 4 = 10$ is 14 because $14 - 4 = 10$ is a true statement.

- Two basic rules can be used to solve equations:
  1. We can add or subtract the same quantity on both sides of an equation.
  2. We can multiply or divide both sides of an equation by the same quantity, as long as we do not multiply or divide by zero.

Examples of Equations That Can Be Solved in One Step

Equation	How to Solve	Solving the Equation	The Equation's Solution
$x - 4 = 10$	Add 4 to both sides.	$x - 4 + 4 = 10 + 4$ $x = 14$	14
$y + 12 = 17$	Subtract 12 from both sides.	$y + 12 - 12 = 17 - 12$ $y = 5$	5
$99n = 53$	Divide both sides by 99.	$\frac{99n}{99} = \frac{53}{99}$ $n = \frac{53}{99}$	$\frac{53}{99}$
$\frac{z}{5} = 9$	Multiply both sides by 5.	$5 \cdot \frac{z}{5} = 5 \cdot 9$ $z = 45$	45

Equations whose solutions require more than one step are discussed in Chapter 6.

Why have we provided this brief review of equations that can be solved in one step? If you are given a rational number as a repeating decimal, there is a technique for expressing the number as a quotient of integers that requires solving a one-step equation. We begin by illustrating the technique with an example. Then we will summarize the steps in the procedure and apply them to another example.

EXAMPLE 6 Expressing a Repeating Decimal in $\frac{a}{b}$ Form

Express $0.\overline{6}$ as a quotient of integers.

SOLUTION

Step 1 Let n equal the repeating decimal. Let $n = 0.\overline{6}$, so that $n = 0.66666\ldots.$

Step 2 If there is one repeating digit, multiply both sides of the equation in step 1 by 10.

$n = 0.66666\ldots$ This is the equation from step 1.

$10n = 10(0.66666\ldots)$ Multiply both sides by 10.

$10n = 6.66666\ldots$ Multiplying by 10 moves the decimal point one place to the right.

Step 3 Subtract the equation in step 1 from the equation in step 2. Be sure to line up the decimal points before subtracting.

Remember from algebra that n means $1n$. Thus, $10n - 1n = 9n$.

$$\begin{aligned} 10n &= 6.66666\ldots \quad \text{This is the equation from step 2.} \\ -\ \ n &= 0.66666\ldots \quad \text{This is the equation from step 1.} \\ \hline 9n &= 6 \end{aligned}$$

Step 4 Divide both sides of the equation in step 3 by the number in front of n and solve for n. We solve $9n = 6$ for n by dividing both sides by 9.

$9n = 6$ This is the equation from step 3.

$\frac{9n}{9} = \frac{6}{9}$ Divide both sides by 9.

$n = \frac{6}{9} = \frac{2}{3}$ Reduce $\frac{6}{9}$ to lowest terms: $\frac{6}{9} = \frac{2 \cdot \cancel{3}}{3 \cdot \cancel{3}} = \frac{2}{3}$.

We began the solution process with $n = 0.\overline{6}$, and now we have $n = \frac{2}{3}$. Therefore,

$$0.\overline{6} = \frac{2}{3}.$$

Here are the steps for expressing a repeating decimal as a quotient of integers. Assume that the repeating digit or digits begin directly to the right of the decimal point.

EXPRESSING A REPEATING DECIMAL AS A QUOTIENT OF INTEGERS

Step 1 Let n equal the repeating decimal.

Step 2 Multiply both sides of the equation in step 1 by 10 if one digit repeats, by 100 if two digits repeat, by 1000 if three digits repeat, and so on.

Step 3 Subtract the equation in step 1 from the equation in step 2.

Step 4 Divide both sides of the equation in step 3 by the number in front of n and solve for n.

CHECK POINT 6 Express $0.\overline{2}$ as a quotient of integers.

EXAMPLE 7 *Expressing a Repeating Decimal in $\frac{a}{b}$ Form*

Express $0.\overline{53}$ as a quotient of integers.

SOLUTION

Step 1 Let n equal the repeating decimal. Let $n = 0.\overline{53}$, so that $n = 0.535353\ldots$.

Step 2 If there are two repeating digits, multiply both sides of the equation in step 1 by 100.

$$n = 0.535353\ldots \quad \text{This is the equation from step 1.}$$

$$100n = 100(0.535353\ldots) \quad \text{Multiply both sides by 100.}$$

$$100n = 53.535353\ldots \quad \text{Multiplying by 100 moves the decimal point two places to the right.}$$

Step 3 Subtract the equation in step 1 from the equation in step 2.

$$\begin{aligned} 100n &= 53.535353\ldots && \text{This is the equation from step 2.} \\ -\ n &= \ \ 0.535353\ldots && \text{This is the equation from step 1.} \\ \hline 99n &= 53 \end{aligned}$$

Step 4 Divide both sides of the equation in step 3 by the number in front of n and solve for n. We solve $99n = 53$ for n by dividing both sides by 99.

$$99n = 53 \quad \text{This is the equation from step 3.}$$

$$\frac{99n}{99} = \frac{53}{99} \quad \text{Divide both sides by 99.}$$

$$n = \frac{53}{99}$$

Because n equals $0.\overline{53}$ and n equals $\frac{53}{99}$,

$$0.\overline{53} = \frac{53}{99}.$$

CHECK POINT 7 Express $0.\overline{79}$ as a quotient of integers.

6 Multiply and divide rational numbers.

Multiplying and Dividing Rational Numbers

The product of two rational numbers is found as follows:

MULTIPLYING RATIONAL NUMBERS

The product of two rational numbers is the product of their numerators divided by the product of their denominators.

If $\frac{a}{b}$ and $\frac{c}{d}$ are rational numbers, then $\frac{a}{b} \cdot \frac{c}{d} = \frac{a \cdot c}{b \cdot d}$.

EXAMPLE 8 *Multiplying Rational Numbers*

Multiply. If possible, reduce the product to its lowest terms:

a. $\frac{3}{8} \cdot \frac{5}{11}$ **b.** $\left(-\frac{2}{3}\right)\left(-\frac{9}{4}\right)$ **c.** $\left(3\frac{2}{3}\right)\left(1\frac{1}{4}\right)$.

SOLUTION

a. $\frac{3}{8} \cdot \frac{5}{11} = \frac{3 \cdot 5}{8 \cdot 11} = \frac{15}{88}$

b. $\left(-\frac{2}{3}\right)\left(-\frac{9}{4}\right) = \frac{(-2)(-9)}{3 \cdot 4} = \frac{18}{12} = \frac{3 \cdot \not{6}}{2 \cdot \not{6}} = \frac{3}{2}$ or $1\frac{1}{2}$

c. $\left(3\frac{2}{3}\right)\left(1\frac{1}{4}\right) = \frac{11}{3} \cdot \frac{5}{4} = \frac{11 \cdot 5}{3 \cdot 4} = \frac{55}{12}$ or $4\frac{7}{12}$

CHECK POINT 8 Multiply. If possible, reduce the product to its lowest terms:

a. $\frac{4}{11} \cdot \frac{2}{3}$ **b.** $\left(-\frac{3}{7}\right)\left(-\frac{14}{4}\right)$ **c.** $\left(3\frac{2}{5}\right)\left(1\frac{1}{2}\right)$.

GREAT QUESTION!

Is it OK if I divide by common factors before I multiply?

Yes. You can divide numerators and denominators by common factors *before* performing multiplication. Then multiply the remaining factors in the numerators and multiply the remaining factors in the denominators. For example,

$$\frac{7}{15} \cdot \frac{20}{21} = \frac{\overset{1}{\not{7}}}{\underset{3}{\not{15}}} \cdot \frac{\overset{4}{\not{20}}}{\underset{3}{\not{21}}} = \frac{1 \cdot 4}{3 \cdot 3} = \frac{4}{9}.$$

Two numbers whose product is 1 are called **reciprocals**, or **multiplicative inverses**, of each other. Thus, the reciprocal of 2 is $\frac{1}{2}$ and the reciprocal of $\frac{1}{2}$ is 2 because $2 \cdot \frac{1}{2} = 1$. In general, if $\frac{c}{d}$ is a nonzero rational number, its reciprocal is $\frac{d}{c}$ because $\frac{c}{d} \cdot \frac{d}{c} = 1$.

Reciprocals are used to find the quotient of two rational numbers.

DIVIDING RATIONAL NUMBERS

The quotient of two rational numbers is the product of the first number and the reciprocal of the second number.

If $\frac{a}{b}$ and $\frac{c}{d}$ are rational numbers and $\frac{c}{d}$ is not 0, then $\frac{a}{b} \div \frac{c}{d} = \frac{a}{b} \cdot \frac{d}{c} = \frac{a \cdot d}{b \cdot c}$.

EXAMPLE 9 *Dividing Rational Numbers*

Divide. If possible, reduce the quotient to its lowest terms:

a. $\frac{4}{5} \div \frac{1}{10}$ **b.** $-\frac{3}{5} \div \frac{7}{11}$ **c.** $4\frac{3}{4} \div 1\frac{1}{2}$.

SOLUTION

a. $\frac{4}{5} \div \frac{1}{10} = \frac{4}{5} \cdot \frac{10}{1} = \frac{4 \cdot 10}{5 \cdot 1} = \frac{40}{5} = 8$

b. $-\frac{3}{5} \div \frac{7}{11} = -\frac{3}{5} \cdot \frac{11}{7} = \frac{-3(11)}{5 \cdot 7} = -\frac{33}{35}$

c. $4\frac{3}{4} \div 1\frac{1}{2} = \frac{19}{4} \div \frac{3}{2} = \frac{19}{4} \cdot \frac{2}{3} = \frac{19 \cdot 2}{4 \cdot 3} = \frac{38}{12} = \frac{19 \cdot 2}{6 \cdot 2} = \frac{19}{6}$ or $3\frac{1}{6}$

CHECK POINT 9 Divide. If possible, reduce the quotient to its lowest terms:

a. $\frac{9}{11} \div \frac{5}{4}$ **b.** $-\frac{8}{15} \div \frac{2}{5}$ **c.** $3\frac{3}{8} \div 2\frac{1}{4}$.

7 Add and subtract rational numbers.

Adding and Subtracting Rational Numbers

Rational numbers with identical denominators are added and subtracted using the following rules:

ADDING AND SUBTRACTING RATIONAL NUMBERS WITH IDENTICAL DENOMINATORS

The sum or difference of two rational numbers with identical denominators is the sum or difference of their numerators over the common denominator.

If $\frac{a}{b}$ and $\frac{c}{b}$ are rational numbers, then $\frac{a}{b} + \frac{c}{b} = \frac{a+c}{b}$ and $\frac{a}{b} - \frac{c}{b} = \frac{a-c}{b}$.

EXAMPLE 10 *Adding and Subtracting Rational Numbers with Identical Denominators*

Perform the indicated operations:

a. $\frac{3}{7} + \frac{2}{7}$ **b.** $\frac{11}{12} - \frac{5}{12}$ **c.** $-5\frac{1}{4} - \left(-2\frac{3}{4}\right)$.

SOLUTION

a. $\frac{3}{7} + \frac{2}{7} = \frac{3+2}{7} = \frac{5}{7}$

b. $\frac{11}{12} - \frac{5}{12} = \frac{11-5}{12} = \frac{6}{12} = \frac{1 \cdot 6}{2 \cdot 6} = \frac{1}{2}$

c. $-5\frac{1}{4} - \left(-2\frac{3}{4}\right) = -\frac{21}{4} - \left(-\frac{11}{4}\right) = -\frac{21}{4} + \frac{11}{4} = \frac{-21+11}{4} = \frac{-10}{4} = -\frac{5}{2}$ or $-2\frac{1}{2}$

CHECK POINT 10 Perform the indicated operations:

a. $\frac{5}{12} + \frac{3}{12}$ **b.** $\frac{7}{4} - \frac{1}{4}$ **c.** $-3\frac{3}{8} - \left(-1\frac{1}{8}\right)$.

If the rational numbers to be added or subtracted have different denominators, we use the least common multiple of their denominators to rewrite the rational numbers. The least common multiple of the denominators is called the **least common denominator**.

Rewriting rational numbers with a least common denominator is done using the Fundamental Principle of Rational Numbers, discussed at the beginning of this section. Recall that if $\frac{a}{b}$ is a rational number and c is a nonzero number, then

$$\frac{a}{b} = \frac{a}{b} \cdot \frac{c}{c} = \frac{a \cdot c}{b \cdot c}.$$

Multiplying the numerator and the denominator of a rational number by the same nonzero number is equivalent to multiplying by 1, resulting in an equivalent fraction.

EXAMPLE 11 *Adding Rational Numbers with Unlike Denominators*

Find the sum: $\frac{3}{4} + \frac{1}{6}$.

SOLUTION

The smallest number divisible by both 4 and 6 is 12. Therefore, 12 is the least common multiple of 4 and 6, and will serve as the least common denominator. To obtain a denominator of 12, multiply the denominator and the numerator of the first rational number, $\frac{3}{4}$, by 3. To obtain a denominator of 12, multiply the denominator and the numerator of the second rational number, $\frac{1}{6}$, by 2.

$$\frac{3}{4}+\frac{1}{6}=\frac{3}{4}\cdot\frac{3}{3}+\frac{1}{6}\cdot\frac{2}{2}$$

Rewrite each rational number as an equivalent fraction with a denominator of 12; $\frac{3}{3} = 1$ and $\frac{2}{2} = 1$, and multiplying by 1 does not change a number's value.

$$=\frac{9}{12}+\frac{2}{12}$$

Multiply.

$$=\frac{11}{12}$$

Add numerators and put this sum over the least common denominator.

CHECK POINT 11 Find the sum: $\frac{1}{5}+\frac{3}{4}$.

If the least common denominator cannot be found by inspection, use prime factorizations of the denominators and the method for finding their least common multiple, discussed in Section 5.1.

EXAMPLE 12 *Subtracting Rational Numbers with Unlike Denominators*

Perform the indicated operation: $\frac{1}{15}-\frac{7}{24}$.

SOLUTION

We need to first find the least common denominator, which is the least common multiple of 15 and 24. What is the smallest number divisible by both 15 and 24? The answer is not obvious, so we begin with the prime factorization of each number.

$$15 = 5\cdot 3$$
$$24 = 8\cdot 3 = 2^3\cdot 3$$

The different factors are 5, 3, and 2. Using the greater number of times each factor appears in either factorization, we find that the least common multiple is $5\cdot 3\cdot 2^3 = 5\cdot 3\cdot 8 = 120$. We will now express each rational number with a denominator of 120, which is the least common denominator. For the first rational number, $\frac{1}{15}$, 120 divided by 15 is 8. Thus, we will multiply the numerator and the denominator by 8. For the second rational number, $\frac{7}{24}$, 120 divided by 24 is 5. Thus, we will multiply the numerator and the denominator by 5.

$$\frac{1}{15}-\frac{7}{24}=\frac{1}{15}\cdot\frac{8}{8}-\frac{7}{24}\cdot\frac{5}{5}$$

Rewrite each rational number as an equivalent fraction with a denominator of 120.

$$=\frac{8}{120}-\frac{35}{120}$$

Multiply.

$$=\frac{8-35}{120}$$

Subtract the numerators and put this difference over the least common denominator.

$$=\frac{-27}{120}$$

Perform the subtraction.

$$=\frac{-9\cdot\cancel{3}}{40\cdot\cancel{3}}$$

Reduce to lowest terms.

$$=-\frac{9}{40}$$

CHECK POINT 12 Perform the indicated operation: $\frac{3}{10}-\frac{7}{12}$.

TECHNOLOGY

Here is a possible keystroke sequence on a graphing calculator for the subtraction problem in Example 12:

1 [÷] 15 [−] 7 [÷] 24

[► Frac] [ENTER].

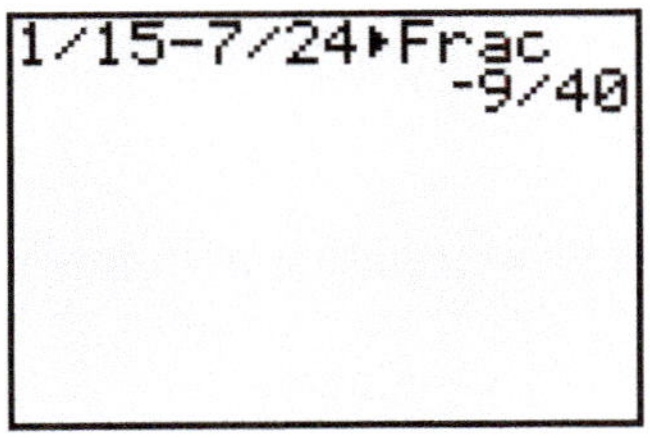

The calculator display reads −9/40, serving as a check for our answer in Example 12.

8 Use the order of operations agreement with rational numbers.

Order of Operations with Rational Numbers

In the previous section, we presented rules for determining the order in which operations should be performed: operations in grouping symbols; exponential expressions; multiplication/division (left to right); addition/subtraction (left to right). In our next example, we apply the order of operations to an expression with rational numbers.

EXAMPLE 13 *Using the Order of Operations*

Simplify: $\left(\frac{1}{2}\right)^3 - \left(\frac{1}{2} - \frac{3}{4}\right)^2(-4)$.

SOLUTION

Because grouping symbols appear, we perform the operation within parentheses first.

$$\left(\frac{1}{2}\right)^3 - \left(\frac{1}{2} - \frac{3}{4}\right)^2(-4)$$

$$= \left(\frac{1}{2}\right)^3 - \left(-\frac{1}{4}\right)^2(-4)$$ Work inside parentheses first: $\frac{1}{2} - \frac{3}{4} = \frac{2}{4} - \frac{3}{4} = \frac{2}{4} + \left(-\frac{3}{4}\right) = -\frac{1}{4}$.

$$= \frac{1}{8} - \frac{1}{16}(-4)$$ Evaluate exponential expressions: $\left(\frac{1}{2}\right)^3 = \frac{1}{2}\cdot\frac{1}{2}\cdot\frac{1}{2} = \frac{1}{8}$ and $\left(-\frac{1}{4}\right)^2 = \left(-\frac{1}{4}\right)\left(-\frac{1}{4}\right) = \frac{1}{16}$.

$$= \frac{1}{8} - \left(-\frac{1}{4}\right)$$ Multiply: $\frac{1}{16}\cdot\left(\frac{-4}{1}\right) = -\frac{4}{16} = -\frac{1}{4}$.

$$= \frac{3}{8}$$ Subtract: $\frac{1}{8} - \left(-\frac{1}{4}\right) = \frac{1}{8} + \frac{1}{4} = \frac{1}{8} + \frac{2}{8} = \frac{3}{8}$.

CHECK POINT 13 Simplify: $\left(-\frac{1}{2}\right)^2 - \left(\frac{7}{10} - \frac{8}{15}\right)^2(-18)$.

9 Apply the density property of rational numbers.

Density of Rational Numbers

It is always possible to find a rational number between any two distinct rational numbers. Mathematicians express this idea by saying that the set of rational numbers is **dense**.

DENSITY OF THE RATIONAL NUMBERS

If r and t represent rational numbers, with $r < t$, then there is a rational number s such that s is between r and t:

$$r < s < t.$$

One way to find a rational number between two given rational numbers is to find the rational number halfway between them. Add the given rational numbers and divide their sum by 2, thereby finding the average of the numbers.

EXAMPLE 14 *Illustrating the Density Property*

Find the rational number halfway between $\frac{1}{2}$ and $\frac{3}{4}$.

GREAT QUESTION!

Why does $\frac{1}{2} < \frac{5}{8} < \frac{3}{4}$ seem meaningless to me?

The inequality $\frac{1}{2} < \frac{5}{8} < \frac{3}{4}$ is more obvious if all denominators are changed to 8:

$$\frac{4}{8} < \frac{5}{8} < \frac{6}{8}.$$

SOLUTION

First, add $\frac{1}{2}$ and $\frac{3}{4}$.

$$\frac{1}{2} + \frac{3}{4} = \frac{2}{4} + \frac{3}{4} = \frac{5}{4}$$

Next, divide this sum by 2.

$$\frac{5}{4} \div \frac{2}{1} = \frac{5}{4} \cdot \frac{1}{2} = \frac{5}{8}$$

The number $\frac{5}{8}$ is halfway between $\frac{1}{2}$ and $\frac{3}{4}$. Thus,

$$\frac{1}{2} < \frac{5}{8} < \frac{3}{4}.$$

We can repeat the procedure of Example 14 and find the rational number halfway between $\frac{1}{2}$ and $\frac{5}{8}$. Repeated application of this procedure implies the following surprising result:

Between any two distinct rational numbers are *infinitely many* rational numbers.

CHECK POINT 14 Find the rational number halfway between $\frac{1}{3}$ and $\frac{1}{2}$.

10 Solve problems involving rational numbers.

Problem Solving with Rational Numbers

A common application of rational numbers involves preparing food for a different number of servings than what the recipe gives. The amount of each ingredient can be found as follows:

$$\text{Amount of ingredient needed} = \frac{\text{desired serving size}}{\text{recipe serving size}} \times \text{ingredient amount in the recipe.}$$

EXAMPLE 15 *Adjusting the Size of a Recipe*

A chocolate-chip cookie recipe for five dozen cookies requires $\frac{3}{4}$ cup sugar. If you want to make eight dozen cookies, how much sugar is needed?

SOLUTION

Amount of sugar needed

$$= \frac{\text{desired serving size}}{\text{recipe serving size}} \times \text{sugar amount in recipe}$$

$$= \frac{8 \text{ dozen}}{5 \text{ dozen}} \times \frac{3}{4} \text{ cup}$$

The amount of sugar needed, in cups, is determined by multiplying the rational numbers:

$$\frac{8}{5} \times \frac{3}{4} = \frac{8 \cdot 3}{5 \cdot 4} = \frac{24}{20} = \frac{6 \cdot \cancel{4}}{5 \cdot \cancel{4}} = 1\frac{1}{5}.$$

Thus, $1\frac{1}{5}$ cups of sugar is needed. (Depending on the measuring cup you are using, you may need to round the sugar amount to $1\frac{1}{4}$ cups.)

CHECK POINT 15 A chocolate-chip cookie recipe for five dozen cookies requires two eggs. If you want to make seven dozen cookies, exactly how many eggs are needed? Now round your answer to a realistic number that does not involve a fractional part of an egg.

Blitzer Bonus

NUMB3RS: Solving Crime with Mathematics

NUMB3RS was a prime-time TV crime series. The show's hero, Charlie Eppes, is a brilliant mathematician who uses his powerful skills to help the FBI identify and catch criminals. The episodes are entertaining and the basic premise shows how math is a powerful weapon in the never-ending fight against crime. *NUMB3RS* is significant because it was the first popular weekly drama that revolved around mathematics. A team of mathematician advisors ensured that the equations seen in the scripts were real and relevant to the episodes. The mathematical content of the show included many topics from this book, ranging from prime numbers, probability theory, and basic geometry.

Episodes of *NUMB3RS* begin with a spoken tribute about the importance of mathematics:

"We all use math everywhere. To tell time, to predict the weather, to handle money ... Math is more than formulas and equations. Math is more than numbers. It is logic. It is rationality. It is using your mind to solve the biggest mysteries we know."

Concept and Vocabulary Check

Fill in each blank so that the resulting statement is true.

1. The set of ________ numbers is the set of all numbers which can be expressed in the form $\frac{a}{b}$, where a and b are ________ and b is not equal to ________.
2. The number $\frac{17}{5}$ is an example of a/an ________ fraction because ________________.
3. Numbers in the form $\frac{a}{b}$ (see Exercise 1) can be expressed as decimals. The decimals either ____________ or ____________.
4. The quotient of two fractions is the product of the first number and the ________________ of the second number.

In Exercises 5–8, determine whether each statement is true or false. If the statement is false, make the necessary change(s) to produce a true statement.

5. $\frac{1}{2} + \frac{1}{5} = \frac{2}{7}$ ______
6. $\frac{1}{2} \div 4 = 2$ ______
7. Every fraction has infinitely many equivalent fractions. ______
8. $\frac{3+7}{30} = \frac{\cancel{3}^{1}+7}{\cancel{30}_{10}} = \frac{8}{10} = \frac{4}{5}$ ______

Exercise Set 5.3

Practice Exercises

In Exercises 1–12, reduce each rational number to its lowest terms.

1. $\frac{10}{15}$
2. $\frac{18}{45}$
3. $\frac{15}{18}$
4. $\frac{16}{64}$
5. $\frac{24}{42}$
6. $\frac{32}{80}$
7. $\frac{60}{108}$
8. $\frac{112}{128}$
9. $\frac{342}{380}$
10. $\frac{210}{252}$
11. $\frac{308}{418}$
12. $\frac{144}{300}$

In Exercises 13–18, convert each mixed number to an improper fraction.

13. $2\frac{3}{8}$
14. $2\frac{7}{9}$
15. $-7\frac{3}{5}$
16. $-6\frac{2}{5}$
17. $12\frac{7}{16}$
18. $11\frac{5}{16}$

In Exercises 19–24, convert each improper fraction to a mixed number.

19. $\frac{23}{5}$
20. $\frac{47}{8}$
21. $-\frac{76}{9}$
22. $-\frac{59}{9}$
23. $\frac{711}{20}$
24. $\frac{788}{25}$

In Exercises 25–36, express each rational number as a decimal.

25. $\frac{3}{4}$
26. $\frac{3}{5}$
27. $\frac{7}{20}$
28. $\frac{3}{20}$
29. $\frac{7}{8}$
30. $\frac{5}{16}$
31. $\frac{9}{11}$
32. $\frac{3}{11}$
33. $\frac{22}{7}$
34. $\frac{20}{3}$
35. $\frac{2}{7}$
36. $\frac{5}{7}$

In Exercises 37–48, express each terminating decimal as a quotient of integers. If possible, reduce to lowest terms.

37. 0.3
38. 0.9
39. 0.4
40. 0.6
41. 0.39
42. 0.59
43. 0.82
44. 0.64
45. 0.725
46. 0.625
47. 0.5399
48. 0.7006

In Exercises 49–56, express each repeating decimal as a quotient of integers. If possible, reduce to lowest terms.

49. $0.\overline{7}$ **50.** $0.\overline{1}$ **51.** $0.\overline{9}$ **52.** $0.\overline{3}$

53. $0.\overline{36}$ **54.** $0.\overline{81}$ **55.** $0.\overline{257}$ **56.** $0.\overline{529}$

In Exercises 57–104, perform the indicated operations. If possible, reduce the answer to its lowest terms.

57. $\frac{3}{8}\cdot\frac{7}{11}$ **58.** $\frac{5}{8}\cdot\frac{3}{11}$ **59.** $\left(-\frac{1}{10}\right)\left(\frac{7}{12}\right)$

60. $\left(-\frac{1}{8}\right)\left(\frac{5}{9}\right)$ **61.** $\left(-\frac{2}{3}\right)\left(-\frac{9}{4}\right)$ **62.** $\left(-\frac{5}{4}\right)\left(-\frac{6}{7}\right)$

63. $\left(3\frac{3}{4}\right)\left(1\frac{3}{5}\right)$ **64.** $\left(2\frac{4}{5}\right)\left(1\frac{1}{4}\right)$ **65.** $\frac{5}{4}\div\frac{3}{8}$

66. $\frac{5}{8}\div\frac{4}{3}$ **67.** $-\frac{7}{8}\div\frac{15}{16}$ **68.** $-\frac{13}{20}\div\frac{4}{5}$

69. $6\frac{3}{5}\div 1\frac{1}{10}$ **70.** $1\frac{3}{4}\div 2\frac{5}{8}$ **71.** $\frac{2}{11}+\frac{3}{11}$

72. $\frac{5}{13}+\frac{2}{13}$ **73.** $\frac{5}{6}-\frac{1}{6}$ **74.** $\frac{7}{12}-\frac{5}{12}$

75. $\frac{7}{12}-\left(-\frac{1}{12}\right)$ **76.** $\frac{5}{16}-\left(-\frac{5}{16}\right)$ **77.** $\frac{1}{2}+\frac{1}{5}$

78. $\frac{1}{3}+\frac{1}{5}$ **79.** $\frac{3}{4}+\frac{3}{20}$ **80.** $\frac{2}{5}+\frac{2}{15}$

81. $\frac{5}{24}+\frac{7}{30}$ **82.** $\frac{7}{108}+\frac{55}{144}$ **83.** $\frac{13}{18}-\frac{2}{9}$

84. $\frac{13}{15}-\frac{2}{45}$ **85.** $\frac{4}{3}-\frac{3}{4}$ **86.** $\frac{3}{2}-\frac{2}{3}$

87. $\frac{1}{15}-\frac{27}{50}$ **88.** $\frac{4}{15}-\frac{1}{6}$ **89.** $2\frac{2}{3}+1\frac{3}{4}$

90. $2\frac{1}{8}+3\frac{3}{4}$ **91.** $3\frac{2}{3}-2\frac{1}{2}$

92. $3\frac{3}{4}-2\frac{1}{3}$ **93.** $-5\frac{2}{3}+3\frac{1}{6}$

94. $-2\frac{1}{2}+1\frac{3}{4}$ **95.** $-1\frac{4}{7}-\left(-2\frac{5}{14}\right)$

96. $-1\frac{4}{9}-\left(-2\frac{5}{18}\right)$ **97.** $\left(\frac{1}{2}-\frac{1}{3}\right)\div\frac{5}{8}$

98. $\left(\frac{1}{2}+\frac{1}{4}\right)\div\left(\frac{1}{2}+\frac{1}{3}\right)$ **99.** $-\frac{9}{4}\left(\frac{1}{2}\right)+\frac{3}{4}\div\frac{5}{6}$

100. $\left[-\frac{4}{7}-\left(-\frac{2}{5}\right)\right]\left[-\frac{3}{8}+\left(-\frac{1}{9}\right)\right]$

101. $\dfrac{\frac{7}{9}-3}{\frac{5}{6}}\div\dfrac{3}{2}+\dfrac{3}{4}$ **102.** $\dfrac{\frac{17}{25}}{\frac{3}{5}-4}\div\dfrac{1}{5}+\dfrac{1}{2}$

103. $\frac{1}{4}-6(2+8)\div\left(-\frac{1}{3}\right)\left(-\frac{1}{9}\right)$

104. $\frac{3}{4}-4(2+7)\div\left(-\frac{1}{2}\right)\left(-\frac{1}{6}\right)$

In Exercises 105–110, find the rational number halfway between the two numbers in each pair.

105. $\frac{1}{4}$ and $\frac{1}{3}$ **106.** $\frac{2}{3}$ and $\frac{5}{6}$ **107.** $\frac{1}{2}$ and $\frac{2}{3}$

108. $\frac{3}{5}$ and $\frac{2}{3}$ **109.** $-\frac{2}{3}$ and $-\frac{5}{6}$ **110.** -4 and $-\frac{7}{2}$

Different operations with the same rational numbers usually result in different answers. Exercises 111–112 illustrate some curious exceptions.

111. Show that $\frac{13}{4}+\frac{13}{9}$ and $\frac{13}{4}\times\frac{13}{9}$ give the same answer.

112. Show that $\frac{169}{30}+\frac{13}{15}$ and $\frac{169}{30}\div\frac{13}{15}$ give the same answer.

Practice Plus

In Exercises 113–116, perform the indicated operations. Leave denominators in prime factorization form.

113. $\dfrac{5}{2^2\cdot 3^2}-\dfrac{1}{2\cdot 3^2}$ **114.** $\dfrac{7}{3^2\cdot 5^2}-\dfrac{1}{3\cdot 5^3}$

115. $\dfrac{1}{2^4\cdot 5^3\cdot 7}+\dfrac{1}{2\cdot 5^4}-\dfrac{1}{2^3\cdot 5^2}$

116. $\dfrac{1}{2^3\cdot 17^8}+\dfrac{1}{2\cdot 17^9}-\dfrac{1}{2^2\cdot 3\cdot 17^8}$

In Exercises 117–120, express each rational number as a decimal. Then insert either < or > in the shaded area between the rational numbers to make the statement true.

117. $\dfrac{6}{11}$ ■ $\dfrac{7}{12}$ **118.** $\dfrac{29}{36}$ ■ $\dfrac{28}{35}$

119. $-\dfrac{5}{6}$ ■ $-\dfrac{8}{9}$ **120.** $-\dfrac{1}{125}$ ■ $-\dfrac{3}{500}$

Application Exercises

As a younger teen, how did you deal with stress? A study of teens under the age of 18 revealed the answers to this question. The circle graphs show the breakdown of the number of men and women who responded to stress in four different ways. Use this information to solve Exercises 121–122.

How Teens Deal with Stress

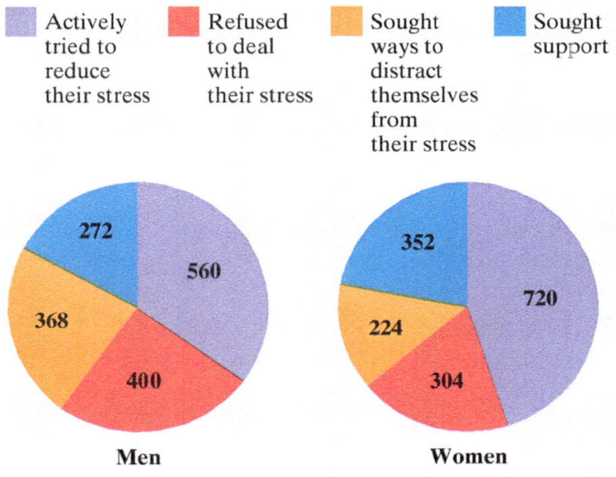

Source: Johns Hopkins Bloomberg School of Public Health

121. a. What fractional part of the men actively tried to reduce their stress? Reduce this fraction to its lowest terms.

b. Express the rational number in part (a) as a decimal. What percentage of teen males actively tried to reduce their stress?

c. What fractional part of the women actively tried to reduce their stress? Reduce this fraction to its lowest terms.

d. Express the rational number in part (c) as a decimal. What percentage of teen females actively tried to reduce their stress?

e. What is the difference between the percentage of teen females and teen males who actively tried to reduce their stress?

122. a. What fractional part of the men refused to deal with their stress? Reduce this fraction to its lowest terms.

b. Express the rational number in part (a) as a decimal. What percentage of teen males refused to deal with their stress?

c. What fractional part of the women refused to deal with their stress? Reduce this fraction to its lowest terms.

d. Express the rational number in part (c) as a decimal. What percentage of teen females refused to deal with their stress?

e. What is the difference between the percentage of teen males and teen females who refused to deal with their stress?

Use the following list of ingredients for chocolate brownies to solve Exercises 123–128.

Ingredients for 16 Brownies

$\frac{2}{3}$ *cup butter, 5 ounces unsweetened chocolate,* $1\frac{1}{2}$ *cups sugar, 2 teaspoons vanilla, 2 eggs, 1 cup flour*

123. How much of each ingredient is needed to make 8 brownies?

124. How much of each ingredient is needed to make 12 brownies?

125. How much of each ingredient is needed to make 20 brownies?

126. How much of each ingredient is needed to make 24 brownies?

127. With only one cup of butter, what is the greatest number of brownies that you can make? (Ignore part of a brownie.)

128. With only one cup of sugar, what is the greatest number of brownies that you can make? (Ignore part of a brownie.)

A mix for eight servings of instant potatoes requires $2\frac{2}{3}$ *cups of water. Use this information to solve Exercises 129–130.*

129. If you want to make 11 servings, how much water is needed?

130. If you want to make six servings, how much water is needed?

The sounds created by plucked or bowed strings of equal diameter and tension produce various notes depending on the lengths of the strings. If a string is half as long as another, its note will be an octave higher than the longer string. Using a length of 1 unit to represent middle C, the diagram shows different fractions of the length of this unit string needed to produce the notes D, E, F, G, A, B, and c one octave higher than middle C.

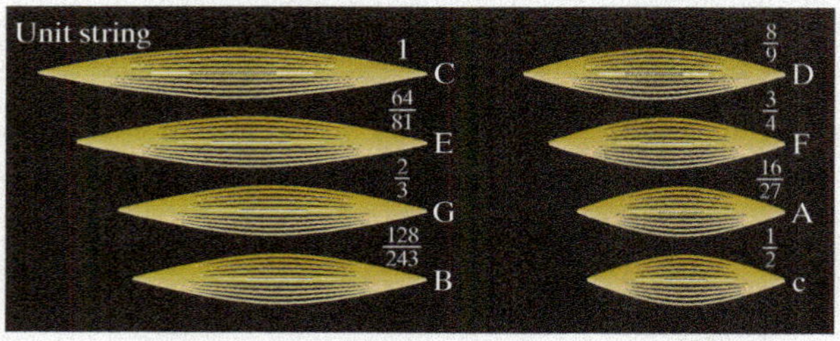

For many of the strings, the length is $\frac{8}{9}$ *of the length of the previous string. For example, the A string is* $\frac{8}{9}$ *of the length of the G string:* $\frac{8}{9} \cdot \frac{2}{3} = \frac{16}{27}$*. Use this information to solve Exercises 131–132.*

131. a. Which strings from D through c are $\frac{8}{9}$ of the length of the preceding string?

b. How is your answer to part (a) shown on this one-octave span of the piano keyboard?

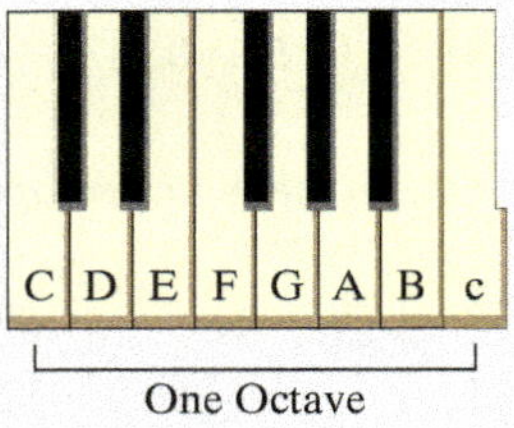

132. a. Which strings from D through c are not $\frac{8}{9}$ of the length of the preceding string?

b. How is your answer to part (a) shown on the one-octave span on the piano keyboard in Exercise 131(b)?

133. A board $7\frac{1}{2}$ inches long is cut from a board that is 2 feet long. If the width of the saw cut is $\frac{1}{16}$ inch, what is the length of the remaining piece?

134. A board that is $7\frac{1}{4}$ inches long is cut from a board that is 3 feet long. If the width of the saw cut is $\frac{1}{16}$ inch, what is the length of the remaining piece?

135. A franchise is owned by three people. The first owns $\frac{5}{12}$ of the business and the second owns $\frac{1}{4}$ of the business. What fractional part of the business is owned by the third person?

136. At a workshop on enhancing creativity, $\frac{1}{4}$ of the participants are musicians, $\frac{2}{5}$ are artists, $\frac{1}{10}$ are actors, and the remaining participants are writers. What fraction of the people attending the workshop are writers?

137. If you walk $\frac{3}{4}$ mile and then jog $\frac{2}{5}$ mile, what is the total distance covered? How much farther did you walk than jog?

138. Some companies pay people extra when they work more than a regular 40-hour work week. The overtime pay is often $1\frac{1}{2}$ times the regular hourly rate. This is called time and a half. A summer job for students pays $12 an hour and offers time and a half for the hours worked over 40. If a student works 46 hours during one week, what is the student's total pay before taxes?

139. A will states that $\frac{3}{5}$ of the estate is to be divided among relatives. Of the remaining estate, $\frac{1}{4}$ goes to charity. What fraction of the estate goes to charity?

140. The legend of a map indicates that 1 inch = 16 miles. If the distance on the map between two cities is $2\frac{3}{8}$ inches, how far apart are the cities?

Writing in Mathematics

141. What is a rational number?

142. Explain how to reduce a rational number to its lowest terms.

143. Explain how to convert from a mixed number to an improper fraction. Use $7\frac{2}{3}$ as an example.

144. Explain how to convert from an improper fraction to a mixed number. Use $\frac{47}{5}$ as an example.

145. Explain how to write a rational number as a decimal.

146. Explain how to write 0.083 as a quotient of integers.

147. Explain how to write $0.\overline{9}$ as a quotient of integers.

148. Explain how to multiply rational numbers. Use $\frac{5}{6} \cdot \frac{1}{2}$ as an example.

149. Explain how to divide rational numbers. Use $\frac{5}{6} \div \frac{1}{2}$ as an example.

150. Explain how to add rational numbers with different denominators. Use $\frac{5}{6} + \frac{1}{2}$ as an example.

151. What does it mean when we say that the set of rational numbers is dense?

Critical Thinking Exercises

Make Sense? *In Exercises 152–155, determine whether each statement makes sense or does not make sense, and explain your reasoning.*

152. I saved money by buying a computer for $\frac{3}{2}$ of its original price.

153. I find it easier to multiply $\frac{1}{5}$ and $\frac{3}{4}$ than to add them.

154. My calculator shows the decimal form for the rational number $\frac{3}{11}$ as 0.2727273, so $\frac{3}{11} = 0.2727273$.

155. The value of $\frac{|3-7|-2^3}{(-2)(-3)}$ is the rational number that results when $\frac{1}{3}$ is subtracted from $-\frac{1}{3}$.

156. Shown below is a short excerpt from "The Star-Spangled Banner." The time is $\frac{3}{4}$, which means that each measure must contain notes that add up to $\frac{3}{4}$. The values of the different notes tell musicians how long to hold each note.

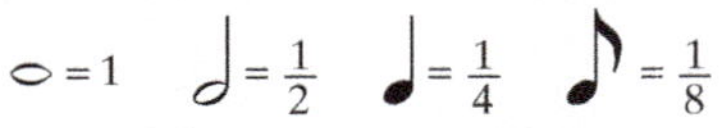

Use vertical lines to divide this line of "The Star-Spangled Banner" into measures.

157. Use inductive reasoning to predict the addition problem and the sum that will appear in the fourth row. Then perform the arithmetic to verify your conjecture.

$$\frac{1}{1\cdot 2} + \frac{1}{2\cdot 3} = \frac{2}{3}$$

$$\frac{1}{1\cdot 2} + \frac{1}{2\cdot 3} + \frac{1}{3\cdot 4} = \frac{3}{4}$$

$$\frac{1}{1\cdot 2} + \frac{1}{2\cdot 3} + \frac{1}{3\cdot 4} + \frac{1}{4\cdot 5} = \frac{4}{5}$$

Technology Exercises

158. Use a calculator to express the following rational numbers as decimals.

a. $\frac{197}{800}$ **b.** $\frac{4539}{3125}$ **c.** $\frac{7}{6250}$

159. Some calculators have a fraction feature. This feature allows you to perform operations with fractions and displays the answer as a fraction reduced to its lowest terms. If your calculator has this feature, use it to verify any five of the answers that you obtained in Exercises 57–104.

Group Exercise

160. Each member of the group should present an application of rational numbers. The application can be based on research or on how the group member uses rational numbers in his or her life. If you are not sure where to begin, ask yourself how your life would be different if fractions and decimals were concepts unknown to our civilization.

5.4 The Irrational Numbers

WHAT AM I SUPPOSED TO LEARN?

After you have read this section, you should be able to:

1. Define the irrational numbers.
2. Simplify square roots.
3. Perform operations with square roots.
4. Rationalize denominators.

Shown here is Renaissance artist Raphael Sanzio's (1483–1520) image of Pythagoras from *The School of Athens* mural. Detail of left side. Stanza della Segnatura, Vatican Palace, Vatican State. Scala/Art Resource, NY.

FOR THE FOLLOWERS OF THE GREEK MATHEMATICIAN PYTHAGORAS IN THE sixth century B.C., numbers took on a life-and-death importance. The "Pythagorean Brotherhood" was a secret group whose members were convinced that properties of whole numbers were the key to understanding the universe. Members of the Brotherhood (which admitted women) thought that all numbers that were not whole numbers could be represented as the ratio of whole numbers. A crisis occurred for the Pythagoreans when they discovered the existence of a number that was not rational. Because the Pythagoreans viewed numbers with reverence and awe, the punishment for speaking about this number was death. However, a member of the Brotherhood revealed the secret of the number's existence. When he later died in a shipwreck, his death was viewed as punishment from the gods.

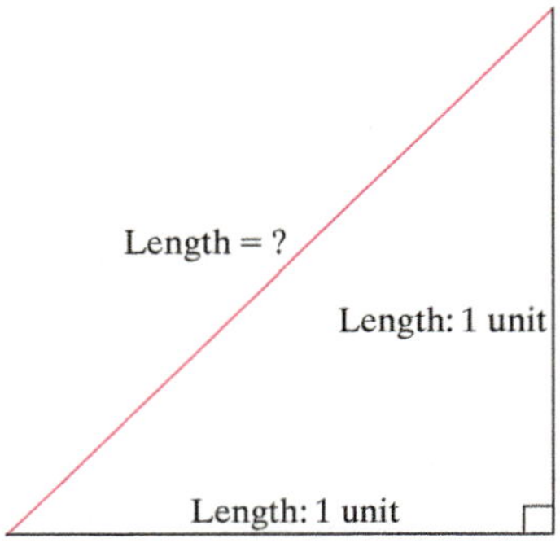

FIGURE 5.6

The triangle in **Figure 5.6** led the Pythagoreans to the discovery of a number that could not be expressed as the quotient of integers. Based on their understanding of the relationship among the sides of this triangle, they knew that the length of the side shown in red had to be a number that, when squared, is equal to 2. The Pythagoreans discovered that this number seemed to be close to the rational numbers

$$\frac{14}{10}, \frac{141}{100}, \frac{1414}{1000}, \frac{14{,}142}{10{,}000}, \text{ and so on.}$$

However, they were shocked to find that there is no quotient of integers whose square is equal to 2.

The positive number whose square is equal to 2 is written $\sqrt{2}$. We read this "the square root of 2," or "radical 2." The symbol $\sqrt{\ }$ is called the **radical sign**. The number under the radical sign, in this case 2, is called the **radicand**. The entire symbol $\sqrt{2}$ is called a **radical**.

Using deductive reasoning, mathematicians have proved that $\sqrt{2}$ cannot be represented as a quotient of integers. This means that there is no terminating or repeating decimal that can be multiplied by itself to give 2. We can, however, give a decimal approximation for $\sqrt{2}$. We use the symbol $\approx$, which means "is approximately equal to." Thus,

$$\sqrt{2} \approx 1.414214.$$

We can verify that this is only an approximation by multiplying 1.414214 by itself. The product is not exactly 2:

$$1.414214 \times 1.414214 = 2.000001237796.$$

Define the irrational numbers.

A number like $\sqrt{2}$, whose decimal representation does not come to an end and does not have a block of repeating digits, is an example of an **irrational number**.

THE IRRATIONAL NUMBERS

The set of **irrational numbers** is the set of numbers whose decimal representations are neither terminating nor repeating.

Perhaps the best known of all the irrational numbers is π (pi). This irrational number represents the distance around a circle (its circumference) divided by the diameter of the circle. In the *Star Trek* episode "Wolf in the Fold," Spock foils an evil computer by telling it to "compute the last digit in the value of π." Because π is an irrational number, there is no last digit in its decimal representation:

$$\pi = 3.14159265358979323846264338327 95. . . .$$

The nature of the irrational number π has fascinated mathematicians for centuries. Amateur and professional mathematicians have taken up the challenge of calculating π to more and more decimal places. Although such an exercise may seem pointless, it serves as the ultimate stress test for new high-speed computers and also as a test for the long-standing, but still unproven, conjecture that the distribution of digits in π is completely random.

Blitzer Bonus

The Best and Worst of π

In 2011, Japanese mathematician Shigeru Kondo calculated π to ten trillion decimal places. The calculations used 8900 hours, or nearly 371 days, of computer time.

The most inaccurate version of π came from the 1897 General Assembly of Indiana. Bill No. 246 stated that "π was by law 4."

TECHNOLOGY

You can obtain decimal approximations for irrational numbers using a calculator. For example, to approximate $\sqrt{2}$, use the following keystrokes:

Scientific Calculator	Graphing Calculator
2 [√] or 2 [2ND/INV] [x^2]	[√] 2 [ENTER] or [2ND/INV] [x^2] 2 [ENTER]

Some graphing calculators show an open parenthesis after displaying √. In this case, enter a closed parenthesis, [ )], after 2.

The display may read 1.41421356237, although your calculator may show more or fewer digits. Between which two integers would you graph $\sqrt{2}$ on a number line?

Square Roots

The United Nations Building in New York was designed to represent its mission of promoting world harmony. Viewed from the front, the building looks like three rectangles stacked upon each other. In each rectangle, the width divided by the height is $\sqrt{5} + 1$ to 2, approximately 1.618 to 1. The ancient Greeks believed that such a rectangle, called a **golden rectangle**, was the most pleasing of all rectangles. The comparison 1.618 to 1 is approximate because $\sqrt{5}$ is an irrational number.

The U.N. building is designed with three golden rectangles.

The **principal square root** of a nonnegative number n, written $\sqrt{n}$, is the nonnegative number that when multiplied by itself gives n. Thus,

$$\sqrt{36} = 6 \text{ because } 6 \cdot 6 = 36$$

and

$$\sqrt{81} = 9 \text{ because } 9 \cdot 9 = 81.$$

Notice that both $\sqrt{36}$ and $\sqrt{81}$ are rational numbers because 6 and 9 are terminating decimals. Thus, **not all square roots are irrational.**

Numbers such as 36 and 81 are called *perfect squares*. A **perfect square** is a number that is the square of a whole number. The first few perfect squares are as follows.

$\mathbf{0} = 0^2$	$\mathbf{16} = 4^2$	$\mathbf{64} = 8^2$	$\mathbf{144} = 12^2$
$\mathbf{1} = 1^2$	$\mathbf{25} = 5^2$	$\mathbf{81} = 9^2$	$\mathbf{169} = 13^2$
$\mathbf{4} = 2^2$	$\mathbf{36} = 6^2$	$\mathbf{100} = 10^2$	$\mathbf{196} = 14^2$
$\mathbf{9} = 3^2$	$\mathbf{49} = 7^2$	$\mathbf{121} = 11^2$	$\mathbf{225} = 15^2$

The principal square root of a perfect square is a whole number. For example,

$$\sqrt{0} = 0,\ \sqrt{1} = 1,\ \sqrt{4} = 2,\ \sqrt{9} = 3,\ \sqrt{16} = 4,\ \sqrt{25} = 5,\ \sqrt{36} = 6,$$

and so on.

2 Simplify square roots.

Simplifying Square Roots

A rule for simplifying square roots can be generalized by comparing $\sqrt{25 \cdot 4}$ and $\sqrt{25} \cdot \sqrt{4}$. Notice that

$$\sqrt{25 \cdot 4} = \sqrt{100} = 10 \quad \text{and} \quad \sqrt{25} \cdot \sqrt{4} = 5 \cdot 2 = 10.$$

Because we obtain 10 in both situations, the original radicals must be equal. That is,

$$\sqrt{25 \cdot 4} = \sqrt{25} \cdot \sqrt{4}.$$

This result is a particular case of the **product rule for square roots** that can be generalized as follows:

THE PRODUCT RULE FOR SQUARE ROOTS

If a and b represent nonnegative numbers, then

$$\sqrt{ab} = \sqrt{a} \cdot \sqrt{b} \quad \text{and} \quad \sqrt{a} \cdot \sqrt{b} = \sqrt{ab}.$$

The square root of a product is the product of the square roots.

GREAT QUESTION!

Is the square root of a sum the sum of the square roots?

No. There are no addition or subtraction rules for square roots:

$$\sqrt{a+b} \neq \sqrt{a} + \sqrt{b}$$
$$\sqrt{a-b} \neq \sqrt{a} - \sqrt{b}.$$

For example, if $a = 9$ and $b = 16$,

$$\sqrt{9+16} = \sqrt{25} = 5$$

and

$$\sqrt{9} + \sqrt{16} = 3 + 4 = 7.$$

Thus,

$$\sqrt{9+16} \neq \sqrt{9} + \sqrt{16}.$$

Example 1 shows how the product rule is used to remove from the square root any perfect squares that occur as factors.

EXAMPLE 1 Simplifying Square Roots

Simplify, if possible:

a. $\sqrt{75}$ **b.** $\sqrt{500}$ **c.** $\sqrt{17}$.

SOLUTION

a. $\sqrt{75} = \sqrt{25 \cdot 3}$ 25 is the greatest perfect square that is a factor of 75.

$= \sqrt{25} \cdot \sqrt{3}$ $\sqrt{ab} = \sqrt{a} \cdot \sqrt{b}$

$= 5\sqrt{3}$ Write $\sqrt{25}$ as 5.

b. $\sqrt{500} = \sqrt{100 \cdot 5}$ 100 is the greatest perfect square factor of 500.

$= \sqrt{100} \cdot \sqrt{5}$ $\sqrt{ab} = \sqrt{a} \cdot \sqrt{b}$

$= 10\sqrt{5}$ Write $\sqrt{100}$ as 10.

c. Because 17 has no perfect square factors (other than 1), $\sqrt{17}$ cannot be simplified.

CHECK POINT 1 Simplify, if possible:

a. $\sqrt{12}$ **b.** $\sqrt{60}$ **c.** $\sqrt{55}$.

3 Perform operations with square roots.

Multiplying Square Roots

If a and b are nonnegative, then we can use the product rule

$$\sqrt{a} \cdot \sqrt{b} = \sqrt{a \cdot b}$$

to multiply square roots. The product of the square roots is the square root of the product. Once the square roots are multiplied, simplify the square root of the product when possible.

EXAMPLE 2 *Multiplying Square Roots*

Multiply:

a. $\sqrt{2}\cdot\sqrt{5}$ **b.** $\sqrt{7}\cdot\sqrt{7}$ **c.** $\sqrt{6}\cdot\sqrt{12}$.

SOLUTION

a. $\sqrt{2}\cdot\sqrt{5} = \sqrt{2\cdot 5} = \sqrt{10}$

b. $\sqrt{7}\cdot\sqrt{7} = \sqrt{7\cdot 7} = \sqrt{49} = 7$

c. $\sqrt{6}\cdot\sqrt{12} = \sqrt{6\cdot 12} = \sqrt{72} = \sqrt{36\cdot 2} = \sqrt{36}\cdot\sqrt{2} = 6\sqrt{2}$

It is possible to multiply irrational numbers and obtain a rational number for the product.

CHECK POINT 2 Multiply:

a. $\sqrt{3}\cdot\sqrt{10}$ **b.** $\sqrt{10}\cdot\sqrt{10}$ **c.** $\sqrt{6}\cdot\sqrt{2}$.

Dividing Square Roots

Another property for square roots involves division.

THE QUOTIENT RULE FOR SQUARE ROOTS

If a and b represent nonnegative numbers and $b \neq 0$, then

$$\frac{\sqrt{a}}{\sqrt{b}} = \sqrt{\frac{a}{b}} \quad \text{and} \quad \sqrt{\frac{a}{b}} = \frac{\sqrt{a}}{\sqrt{b}}.$$

The quotient of two square roots is the square root of the quotient.

Once the square roots are divided, simplify the square root of the quotient when possible.

EXAMPLE 3 *Dividing Square Roots*

Find the quotient:

a. $\frac{\sqrt{75}}{\sqrt{3}}$ **b.** $\frac{\sqrt{90}}{\sqrt{2}}$.

SOLUTION

a. $\frac{\sqrt{75}}{\sqrt{3}} = \sqrt{\frac{75}{3}} = \sqrt{25} = 5$

b. $\frac{\sqrt{90}}{\sqrt{2}} = \sqrt{\frac{90}{2}} = \sqrt{45} = \sqrt{9\cdot 5} = \sqrt{9}\cdot\sqrt{5} = 3\sqrt{5}$

CHECK POINT 3 Find the quotient:

a. $\frac{\sqrt{80}}{\sqrt{5}}$ **b.** $\frac{\sqrt{48}}{\sqrt{6}}$.

Adding and Subtracting Square Roots

The number that multiplies a square root is called the square root's **coefficient**. For example, in $3\sqrt{5}$, 3 is the coefficient of the square root.

Square roots with the same radicand can be added or subtracted by adding or subtracting their coefficients:

$$a\sqrt{c} + b\sqrt{c} = (a + b)\sqrt{c} \qquad a\sqrt{c} - b\sqrt{c} = (a - b)\sqrt{c}.$$

Sum of coefficients times the common square root

Difference of coefficients times the common square root

EXAMPLE 4 *Adding and Subtracting Square Roots*

Add or subtract as indicated:

a. $7\sqrt{2} + 5\sqrt{2}$ **b.** $2\sqrt{5} - 6\sqrt{5}$ **c.** $3\sqrt{7} + 9\sqrt{7} - \sqrt{7}$.

SOLUTION

a. $7\sqrt{2} + 5\sqrt{2} = (7 + 5)\sqrt{2}$
$= 12\sqrt{2}$

b. $2\sqrt{5} - 6\sqrt{5} = (2 - 6)\sqrt{5}$
$= -4\sqrt{5}$

c. $3\sqrt{7} + 9\sqrt{7} - \sqrt{7} = 3\sqrt{7} + 9\sqrt{7} - 1\sqrt{7}$ Write $\sqrt{7}$ as $1\sqrt{7}$.
$= (3 + 9 - 1)\sqrt{7}$
$= 11\sqrt{7}$

CHECK POINT 4 Add or subtract as indicated:

a. $8\sqrt{3} + 10\sqrt{3}$

b. $4\sqrt{13} - 9\sqrt{13}$

c. $7\sqrt{10} + 2\sqrt{10} - \sqrt{10}$.

GREAT QUESTION!

Can I combine $\sqrt{2} + \sqrt{7}$?

No. Sums or differences of square roots that cannot be simplified and that do not contain a common radicand cannot be combined into one term by adding or subtracting coefficients. Some examples:

- $5\sqrt{3} + 3\sqrt{5}$ cannot be combined by adding coefficients. The square roots, $\sqrt{3}$ and $\sqrt{5}$, are different.
- $28 + 7\sqrt{3}$, or $28\sqrt{1} + 7\sqrt{3}$, cannot be combined by adding coefficients. The square roots, $\sqrt{1}$ and $\sqrt{3}$, are different.

In some situations, it is possible to add and subtract square roots that do not contain a common square root by first simplifying.

EXAMPLE 5 *Adding and Subtracting Square Roots by First Simplifying*

Add or subtract as indicated:

a. $\sqrt{2} + \sqrt{8}$ **b.** $4\sqrt{50} - 6\sqrt{32}$.

SOLUTION

a. $\sqrt{2} + \sqrt{8}$

$= \sqrt{2} + \sqrt{4 \cdot 2}$ Split 8 into two factors such that one factor is a perfect square.

$= 1\sqrt{2} + 2\sqrt{2}$ $\sqrt{4 \cdot 2} = \sqrt{4} \cdot \sqrt{2} = 2\sqrt{2}$

$= (1 + 2)\sqrt{2}$ Add coefficients and retain the common square root.

$= 3\sqrt{2}$ Simplify.

b. $4\sqrt{50} - 6\sqrt{32}$

$= 4\sqrt{25 \cdot 2} - 6\sqrt{16 \cdot 2}$ — 25 is the greatest perfect square factor of 50 and 16 is the greatest perfect square factor of 32.

$= 4 \cdot 5\sqrt{2} - 6 \cdot 4\sqrt{2}$ — $\sqrt{25 \cdot 2} = \sqrt{25}\sqrt{2} = 5\sqrt{2}$ and $\sqrt{16 \cdot 2} = \sqrt{16}\sqrt{2} = 4\sqrt{2}$

$= 20\sqrt{2} - 24\sqrt{2}$ — Multiply.

$= (20 - 24)\sqrt{2}$ — Subtract coefficients and retain the common square root.

$= -4\sqrt{2}$ — Simplify.

CHECK POINT 5 Add or subtract as indicated:

a. $\sqrt{3} + \sqrt{12}$ **b.** $4\sqrt{8} - 7\sqrt{18}$.

4 Rationalize denominators.

Rationalizing Denominators

The calculator screen in **Figure 5.7** shows approximate values for $\frac{1}{\sqrt{3}}$ and $\frac{\sqrt{3}}{3}$. The two approximations are the same. This is not a coincidence:

FIGURE 5.7 The calculator screen shows approximate values for $\frac{1}{\sqrt{3}}$ and $\frac{\sqrt{3}}{3}$.

$$\frac{1}{\sqrt{3}} = \frac{1}{\sqrt{3}} \cdot \boxed{\frac{\sqrt{3}}{\sqrt{3}}} = \frac{\sqrt{3}}{\sqrt{9}} = \frac{\sqrt{3}}{3}$$

Any number divided by itself is 1. Multiplication by 1 does not change the value of $\frac{1}{\sqrt{3}}$.

This process involves rewriting a radical expression as an equivalent expression in which the denominator no longer contains any radicals. The process is called **rationalizing the denominator**. If the denominator contains the square root of a natural number that is not a perfect square, **multiply the numerator and the denominator by the smallest number that produces the square root of a perfect square in the denominator**.

GREAT QUESTION!

What exactly does rationalizing a denominator do to an irrational number in the denominator?

Rationalizing a numerical denominator makes that denominator a rational number.

EXAMPLE 6 *Rationalizing Denominators*

Rationalize the denominator:

a. $\frac{15}{\sqrt{6}}$ **b.** $\sqrt{\frac{3}{5}}$ **c.** $\frac{12}{\sqrt{8}}$.

SOLUTION

a. If we multiply the numerator and the denominator of $\frac{15}{\sqrt{6}}$ by $\sqrt{6}$, the denominator becomes $\sqrt{6} \cdot \sqrt{6} = \sqrt{36} = 6$. Therefore, we multiply by 1, choosing $\frac{\sqrt{6}}{\sqrt{6}}$ for 1.

$$\frac{15}{\sqrt{6}} = \frac{15}{\sqrt{6}} \cdot \frac{\sqrt{6}}{\sqrt{6}} = \frac{15\sqrt{6}}{\sqrt{36}} = \frac{15\sqrt{6}}{6} = \frac{5\sqrt{6}}{2}$$

Multiply by 1.

Simplify: $\frac{15}{6} = \frac{5 \cdot 3}{2 \cdot 3} = \frac{5}{2}$.

b. $$\sqrt{\frac{3}{5}} = \frac{\sqrt{3}}{\sqrt{5}} = \frac{\sqrt{3}}{\sqrt{5}} \cdot \frac{\sqrt{5}}{\sqrt{5}} = \frac{\sqrt{15}}{\sqrt{25}} = \frac{\sqrt{15}}{5}$$

Multiply by 1.

c. The *smallest* number that will produce a perfect square in the denominator of $\frac{12}{\sqrt{8}}$ is $\sqrt{2}$, because $\sqrt{8} \cdot \sqrt{2} = \sqrt{16} = 4$. We multiply by 1, choosing $\frac{\sqrt{2}}{\sqrt{2}}$ for 1.

$$\frac{12}{\sqrt{8}} = \frac{12}{\sqrt{8}} \cdot \frac{\sqrt{2}}{\sqrt{2}} = \frac{12\sqrt{2}}{\sqrt{16}} = \frac{12\sqrt{2}}{4} = 3\sqrt{2}$$

GREAT QUESTION!

Can I rationalize the denominator of $\frac{12}{\sqrt{8}}$ by multiplying by $\frac{\sqrt{8}}{\sqrt{8}}$?

Yes. However, it takes more work to simplify the result.

CHECK POINT 6 Rationalize the denominator:

a. $\frac{25}{\sqrt{10}}$ **b.** $\sqrt{\frac{2}{7}}$ **c.** $\frac{5}{\sqrt{18}}$.

Blitzer Bonus

Golden Rectangles

The early Greeks believed that the most pleasing of all rectangles were **golden rectangles**, whose ratio of width to height is

$$\frac{w}{h} = \frac{\sqrt{5} + 1}{2}.$$

The Parthenon at Athens fits into a golden rectangle once the triangular pediment is reconstructed.

Irrational Numbers and Other Kinds of Roots

Irrational numbers appear in the form of roots other than square roots. The symbol $\sqrt[3]{\ }$ represents the **cube root** of a number. For example,

$$\sqrt[3]{8} = 2 \text{ because } 2 \cdot 2 \cdot 2 = 8 \quad \text{and} \quad \sqrt[3]{64} = 4 \text{ because } 4 \cdot 4 \cdot 4 = 64.$$

Although these cube roots are rational numbers, most cube roots are not. For example,

$$\sqrt[3]{217} \approx 6.0092 \text{ because } (6.0092)^3 \approx 216.995, \text{ not exactly } 217.$$

There is no end to the kinds of roots for numbers. For example, $\sqrt[4]{\ }$ represents the **fourth root** of a number. Thus, $\sqrt[4]{81} = 3$ because $3 \cdot 3 \cdot 3 \cdot 3 = 81$. Although the fourth root of 81 is rational, most fourth roots, fifth roots, and so on tend to be irrational.

Blitzer Bonus

A Radical Idea: Time Is Relative

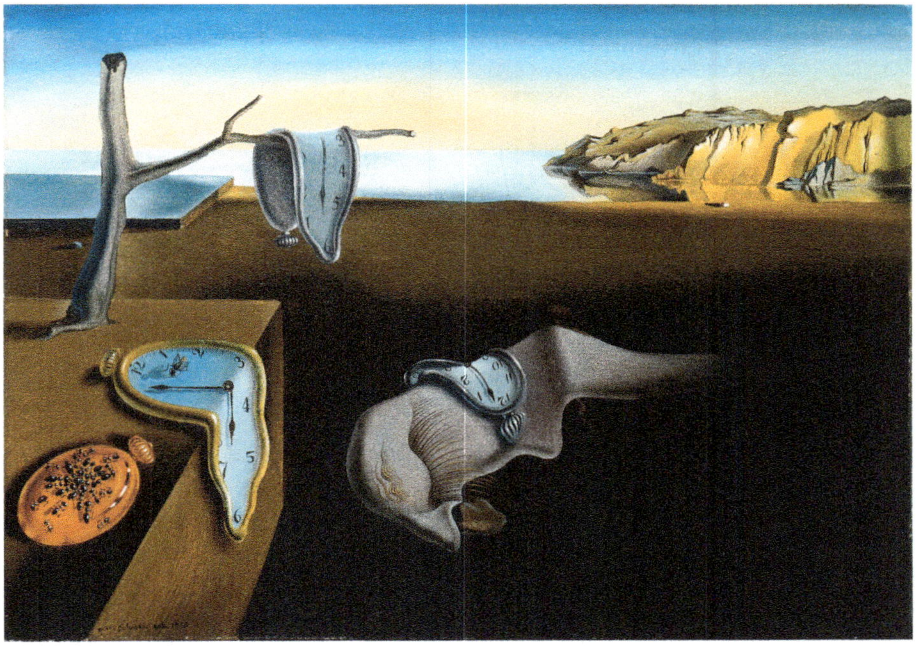

The Persistence of Memory (1931), Salvador Dali.

What does travel in space have to do with square roots? Imagine that in the future we will be able to travel at velocities approaching the speed of light (approximately 186,000 miles per second). According to Einstein's theory of special relativity, time would pass more quickly on Earth than it would in the moving spaceship. The special-relativity equation

$$R_a = R_f\sqrt{1-\left(\frac{v}{c}\right)^2}$$

gives the aging rate of an astronaut, R_a, relative to the aging rate of a friend, R_f, on Earth. In this formula, v is the astronaut's speed and c is the speed of light. As the astronaut's speed approaches the speed of light, we can substitute c for v.

$$R_a = R_f\sqrt{1-\left(\frac{v}{c}\right)^2}$$ Einstein's equation gives the aging rate of an astronaut, R_a, relative to the aging rate of a friend, R_f, on Earth.

$$R_a = R_f\sqrt{1-\left(\frac{c}{c}\right)^2}$$ The velocity, v, is approaching the speed of light, c, so let $v = c$.

$$= R_f\sqrt{1-1}$$ $\left(\frac{c}{c}\right)^2 = 1^2 = 1 \cdot 1 = 1$

$$= R_f\sqrt{0}$$ Simplify the radicand: $1 - 1 = 0$.

$$= R_f \cdot 0$$ $\sqrt{0} = 0$

$$= 0$$ Multiply: $R_f \cdot 0 = 0$.

Close to the speed of light, the astronaut's aging rate, R_a, relative to that of a friend, R_f, on Earth is nearly 0. What does this mean? As we age here on Earth, the space traveler would barely get older. The space traveler would return to an unknown futuristic world in which friends and loved ones would be long gone.

Concept and Vocabulary Check

Fill in each blank so that the resulting statement is true.

1. The set of irrational numbers is the set of numbers whose decimal representations are neither ________ nor ________.
2. The irrational number ________ represents the circumference of a circle divided by the diameter of the circle.
3. The square root of n, represented by ____, is the nonnegative number that when multiplied by itself gives ____.
4. $\sqrt{49 \cdot 6} = \sqrt{\_} \cdot \sqrt{\_} =$ ____
5. The number that multiplies a square root is called the square root's ________.
6. $8\sqrt{3} + 10\sqrt{3} = (\_ + \_)\sqrt{3} =$ ____
7. $\sqrt{50} + \sqrt{32} = \sqrt{25 \cdot 2} + \sqrt{16 \cdot 2}$
   $= \sqrt{25} \cdot \sqrt{2} + \sqrt{16} \cdot \sqrt{2} = \_\sqrt{2} + \_\sqrt{2} =$ ____
8. The process of rewriting a radical expression as an equivalent expression in which the denominator no longer contains any radicals is called ____________________.

9. The number $\sqrt{\frac{2}{7}}$ can be rewritten without a radical in the denominator by multiplying the numerator and denominator by ________.

10. The number $\frac{5}{\sqrt{12}}$ can be rewritten without a radical in the denominator by multiplying the numerator and denominator by ________, which is the smallest number that will produce a perfect square in the denominator.

Exercise Set 5.4

Practice Exercises

Evaluate each expression in Exercises 1–10.

1. $\sqrt{9}$ **2.** $\sqrt{16}$ **3.** $\sqrt{25}$ **4.** $\sqrt{49}$

5. $\sqrt{64}$ **6.** $\sqrt{100}$ **7.** $\sqrt{121}$ **8.** $\sqrt{144}$

9. $\sqrt{169}$ **10.** $\sqrt{225}$

In Exercises 11–16, use a calculator with a square root key to find a decimal approximation for each square root. Round the number displayed to the nearest **a.** *tenth,* **b.** *hundredth,* **c.** *thousandth.*

11. $\sqrt{173}$ **12.** $\sqrt{3176}$ **13.** $\sqrt{17{,}761}$

14. $\sqrt{779{,}264}$ **15.** $\sqrt{\pi}$ **16.** $\sqrt{2\pi}$

In Exercises 17–24, simplify the square root.

17. $\sqrt{20}$ **18.** $\sqrt{50}$ **19.** $\sqrt{80}$

20. $\sqrt{12}$ **21.** $\sqrt{250}$ **22.** $\sqrt{192}$

23. $7\sqrt{28}$ **24.** $3\sqrt{52}$

In Exercises 25–56, perform the indicated operation. Simplify the answer when possible.

25. $\sqrt{7}\cdot\sqrt{6}$ **26.** $\sqrt{19}\cdot\sqrt{3}$

27. $\sqrt{6}\cdot\sqrt{6}$ **28.** $\sqrt{5}\cdot\sqrt{5}$

29. $\sqrt{3}\cdot\sqrt{6}$ **30.** $\sqrt{12}\cdot\sqrt{2}$

31. $\sqrt{2}\cdot\sqrt{26}$ **32.** $\sqrt{5}\cdot\sqrt{50}$

33. $\frac{\sqrt{54}}{\sqrt{6}}$ **34.** $\frac{\sqrt{75}}{\sqrt{3}}$

35. $\frac{\sqrt{90}}{\sqrt{2}}$ **36.** $\frac{\sqrt{60}}{\sqrt{3}}$

37. $\frac{-\sqrt{96}}{\sqrt{2}}$ **38.** $\frac{-\sqrt{150}}{\sqrt{3}}$

39. $7\sqrt{3}+6\sqrt{3}$ **40.** $8\sqrt{5}+11\sqrt{5}$

41. $4\sqrt{13}-6\sqrt{13}$ **42.** $6\sqrt{17}-8\sqrt{17}$

43. $\sqrt{5}+\sqrt{5}$ **44.** $\sqrt{3}+\sqrt{3}$

45. $4\sqrt{2}-5\sqrt{2}+8\sqrt{2}$

46. $6\sqrt{3}+8\sqrt{3}-16\sqrt{3}$

47. $\sqrt{5}+\sqrt{20}$ **48.** $\sqrt{3}+\sqrt{27}$

49. $\sqrt{50}-\sqrt{18}$ **50.** $\sqrt{63}-\sqrt{28}$

51. $3\sqrt{18}+5\sqrt{50}$ **52.** $4\sqrt{12}+2\sqrt{75}$

53. $\frac{1}{4}\sqrt{12}-\frac{1}{2}\sqrt{48}$ **54.** $\frac{1}{5}\sqrt{300}-\frac{2}{3}\sqrt{27}$

55. $3\sqrt{75}+2\sqrt{12}-2\sqrt{48}$

56. $2\sqrt{72}+3\sqrt{50}-\sqrt{128}$

In Exercises 57–66, rationalize the denominator.

57. $\frac{5}{\sqrt{3}}$ **58.** $\frac{12}{\sqrt{5}}$ **59.** $\frac{21}{\sqrt{7}}$

60. $\frac{30}{\sqrt{5}}$ **61.** $\frac{12}{\sqrt{30}}$ **62.** $\frac{15}{\sqrt{50}}$

63. $\frac{15}{\sqrt{12}}$ **64.** $\frac{13}{\sqrt{40}}$ **65.** $\sqrt{\frac{2}{5}}$

66. $\sqrt{\frac{5}{7}}$

Practice Plus

In Exercises 67–74, perform the indicated operations. Simplify the answer when possible.

67. $3\sqrt{8}-\sqrt{32}+3\sqrt{72}-\sqrt{75}$

68. $3\sqrt{54}-2\sqrt{24}-\sqrt{96}+4\sqrt{63}$

69. $3\sqrt{7}-5\sqrt{14}\cdot\sqrt{2}$

70. $4\sqrt{2}-8\sqrt{10}\cdot\sqrt{5}$

71. $\frac{\sqrt{32}}{5}+\frac{\sqrt{18}}{7}$ **72.** $\frac{\sqrt{27}}{2}+\frac{\sqrt{75}}{7}$

73. $\frac{\sqrt{2}}{\sqrt{3}}+\frac{\sqrt{3}}{\sqrt{2}}$ **74.** $\frac{\sqrt{2}}{\sqrt{7}}+\frac{\sqrt{7}}{\sqrt{2}}$

Application Exercises

The formula

$$d=\sqrt{\frac{3h}{2}}$$

models the distance, d, in miles, that a person h feet high can see to the horizon. Use this formula to solve Exercises 75–76.

75. The pool deck on a cruise ship is 72 feet above the water. How far can passengers on the pool deck see? Write the answer in simplified radical form. Then use the simplified radical form and a calculator to express the answer to the nearest tenth of a mile.

76. The captain of a cruise ship is on the star deck, which is 120 feet above the water. How far can the captain see? Write the answer in simplified radical form. Then use the simplified radical form and a calculator to express the answer to the nearest tenth of a mile.

Police use the formula $v=2\sqrt{5L}$ to estimate the speed of a car, v, in miles per hour, based on the length, L, in feet, of its skid marks upon sudden braking on a dry asphalt road. Use the formula to solve Exercises 77–78.

77. A motorist is involved in an accident. A police officer measures the car's skid marks to be 245 feet long. Estimate the speed at which the motorist was traveling before braking. If the posted speed limit is 50 miles per hour and the motorist tells the officer he was not speeding, should the officer believe him? Explain.

78. A motorist is involved in an accident. A police officer measures the car's skid marks to be 45 feet long. Estimate the speed at which the motorist was traveling before braking. If the posted speed limit is 35 miles per hour and the motorist tells the officer she was not speeding, should the officer believe her? Explain.

79. The graph shows the median heights for boys of various ages in the United States from birth through 60 months, or five years old.

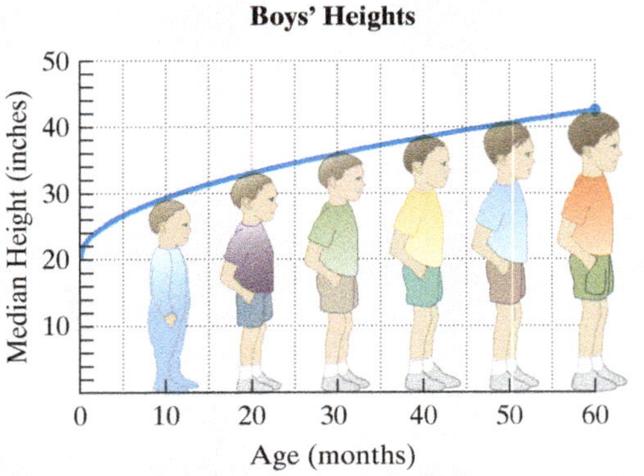

Source: Laura Walther Nathanson, *The Portable Pediatrician for Parents*

a. Use the graph to estimate the median height, to the nearest inch, of boys who are 50 months old.

b. The formula $h = 2.9\sqrt{x} + 20.1$ models the median height, h, in inches, of boys who are x months of age. According to the formula, what is the median height of boys who are 50 months old? Use a calculator and round to the nearest tenth of an inch. How well does your estimate from part (a) describe the median height obtained from the formula?

80. The graph shows the median heights for girls of various ages in the United States from birth through 60 months, or five years old.

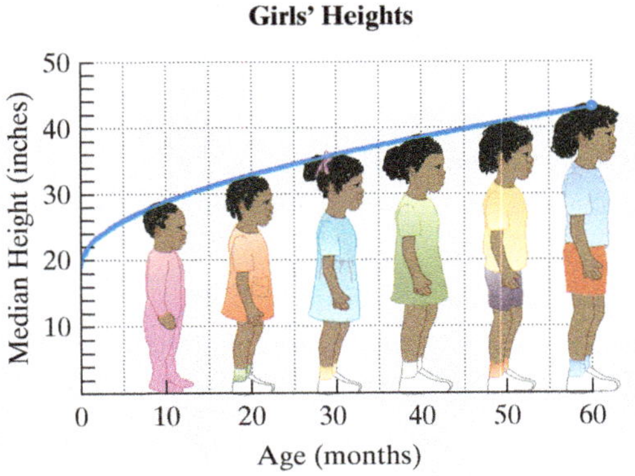

Source: Laura Walther Nathanson, *The Portable Pediatrician for Parents*

a. Use the graph to estimate the median height, to the nearest inch, of girls who are 50 months old.

b. The formula $h = 3.1\sqrt{x} + 19$ models the median height, h, in inches, of girls who are x months of age. According to the formula, what is the median height of girls who are 50 months old? Use a calculator and round to the nearest tenth of an inch. How well does your estimate from part (a) describe the median height obtained from the formula?

Autism is a neurological disorder that impedes language and derails social and emotional development. New findings suggest that the condition is not a sudden calamity that strikes children at the age of 2 or 3, but a developmental problem linked to abnormally rapid brain growth during infancy. The graphs show that the heads of severely autistic children start out smaller than average and then go through a period of explosive growth. Exercises 81–82 involve mathematical models for the data shown by the graphs.

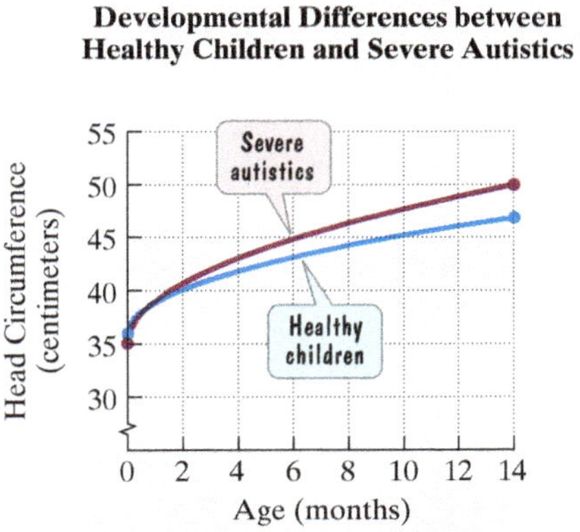

Source: *The Journal of the American Medical Association*

81. The data for one of the two groups shown by the graphs can be modeled by

$$H = 2.9\sqrt{x} + 36,$$

where H is the head circumference, in centimeters, at age x months, $0 \le x \le 14$.

a. According to the model, what is the head circumference at birth?

b. According to the model, what is the head circumference at 9 months?

c. According to the model, what is the head circumference at 14 months? Use a calculator and round to the nearest tenth of a centimeter.

d. Use the values that you obtained in parts (a) through (c) and the graphs shown above to determine if the given model describes healthy children or severe autistics.

82. The data for one of the two groups shown by the graphs can be modeled by

$$H = 4\sqrt{x} + 35,$$

where H is the head circumference, in centimeters, at age x months, $0 \le x \le 14$.

a. According to the model, what is the head circumference at birth?

b. According to the model, what is the head circumference at 9 months?

c. According to the model, what is the head circumference at 14 months? Use a calculator and round to the nearest centimeter.

d. Use the values that you obtained in parts (a) through (c) and the graphs shown above to determine if the given model describes healthy children or severe autistics.

83. The popular comic strip *FoxTrot* follows the off-the-wall lives of the Fox family. Youngest son Jason is forever obsessed by his love of math. In the math-themed strip shown below, Jason shares his opinion in a coded message about the mathematical abilities of his sister Paige.

Foxtrot © 2000 Bill Amend. Reprinted with permission of Universal Uclick. All rights reserved.

Solve problems A through Z in the left panel. Then decode Jason Fox's message involving his opinion about the mathematical abilities of his sister Paige shown on the first line.

Hints: Here is the solution for problem C and partial solutions for problems Q and U.

These are from trigonometry.

This is from calculus.

$$C = \sin\frac{\pi}{2} = \sin 90° = 1$$

$$Q = \int_0^2 9x^2 dx = 3x^3 \Big|_0^2 = 3 \cdot 2^3 - 3 \cdot 0^3 = \_\_\_\_$$

$$U = -3\cos\pi = -3\cos 180° = -3(-1) = \_\_\_\_$$

Note: The comic strip *FoxTrot* is now printed in more than one thousand newspapers. What made cartoonist Bill Amend, a college physics major, put math in the comic? "I always try to use math in the strip to make the joke accessible to anyone," he said. "But if you understand math, hopefully you'll like it that much more!" We highly recommend the math humor in Amend's *FoxTrot* collection *Math, Science, and Unix Underpants* (Andrews McMeel Publishing, 2009).

The Blitzer Bonus on page 295 gives Einstein's special-relativity equation

$$R_a = R_f\sqrt{1 - \left(\frac{v}{c}\right)^2}$$

for the aging rate of an astronaut, R_a, relative to the aging rate of a friend on Earth, R_f, where v is the astronaut's speed and c is the speed of light. Take a few minutes to read the essay and then solve Exercises 84–85.

84. You are moving at 80% of the speed of light. Substitute $0.8c$ in the equation shown above. What is your aging rate relative to a friend on Earth? If 100 weeks have passed for your friend, how long were you gone?

85. You are moving at 90% of the speed of light. Substitute $0.9c$ in the equation shown above. What is your aging rate, correct to two decimal places, relative to a friend on Earth? If 100 weeks have passed for your friend, how long, to the nearest week, were you gone?

Writing in Mathematics

86. Describe the difference between a rational number and an irrational number.

87. Describe what is wrong with this statement: $\pi = \frac{22}{7}$.

88. Using $\sqrt{50}$, explain how to simplify a square root.

89. Describe how to multiply square roots.

90. Explain how to add square roots with the same radicand.

91. Explain how to add $\sqrt{3} + \sqrt{12}$.

92. Describe what it means to rationalize a denominator. Use $\frac{2}{\sqrt{5}}$ in your explanation.

93. Read the Blitzer Bonus on page 295. The future is now: You have the opportunity to explore the cosmos in a starship traveling near the speed of light. The experience will enable you to understand the mysteries of the universe in deeply personal ways, transporting you to unimagined levels of knowing and being. The down side: You return from your two-year journey to a futuristic world in which friends and loved ones are long gone. Do you explore space or stay here on Earth? What are the reasons for your choice?

Critical Thinking Exercises

Make Sense? *In Exercises 94–97, determine whether each statement makes sense or does not make sense, and explain your reasoning.*

94. The humor in this cartoon is based on the fact that the football will never be hiked.

Foxtrot © 2000 Bill Amend. Reprinted with permission of Universal Uclick.

95. The humor in this cartoon is based on the fact that the enemy will never be charged.

Hagar ©1995 by King Features Syndicate, Inc. World rights reserved.

96. I rationalized a numerical denominator and the simplified denominator still contained an irrational number.

97. I simplified $\sqrt{20}$ and $\sqrt{75}$, and then I was able to perform the addition $2\sqrt{20} + 4\sqrt{75}$ by combining the sum into one square root.

In Exercises 98–101, determine whether each statement is true or false. If the statement is false, make the necessary change(s) to produce a true statement.

98. The product of any two irrational numbers is always an irrational number.

99. $\sqrt{9} + \sqrt{16} = \sqrt{25}$

100. $\sqrt{\sqrt{16}} = 2$

101. $\dfrac{\sqrt{64}}{2} = \sqrt{32}$

In Exercises 102–104, insert either < or > in the shaded area between the numbers to make each statement true.

102. $\sqrt{2}$ ■ 1.5

103. $-\pi$ ■ -3.5

104. $-\dfrac{3.14}{2}$ ■ $-\dfrac{\pi}{2}$

105. How does doubling a number affect its square root?

106. Between which two consecutive integers is $-\sqrt{47}$?

107. Simplify: $\sqrt{2} + \sqrt{\dfrac{1}{2}}$.

108. Create a counterexample to show that the following statement is false: The difference between two irrational numbers is always an irrational number.

Group Exercises

The following topics related to irrational numbers are appropriate for either individual or group research projects. A report should be given to the class on the researched topic.

109. A History of How Irrational Numbers Developed

110. Pi: Its History, Applications, and Curiosities

111. Proving That $\sqrt{2}$ Is Irrational

112. Imaginary Numbers: Their History, Applications, and Curiosities

113. The Golden Rectangle in Art and Architecture

5.5 Real Numbers and Their Properties; Clock Addition

WHAT AM I SUPPOSED TO LEARN?

After you have read this section, you should be able to:

1. Recognize subsets of the real numbers.
2. Recognize properties of real numbers.
3. Apply properties of real numbers to clock addition.

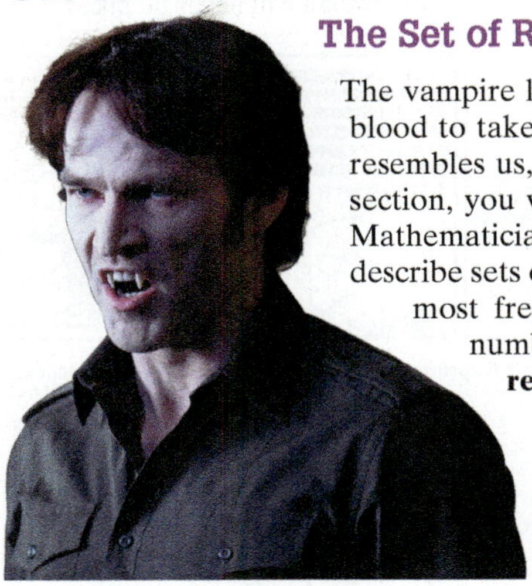

The Set of Real Numbers

The vampire legend is death as seducer; he/she sucks our blood to take us to a perverse immortality. The vampire resembles us, but appears hidden among mortals. In this section, you will find vampires in the world of numbers. Mathematicians even use the labels *vampire* and *weird* to describe sets of numbers. However, the label that appears most frequently is *real*. The union of the rational numbers and the irrational numbers is the set of **real numbers**.

1 Recognize subsets of the real numbers.

The sets that make up the real numbers are summarized in **Table 5.2**. We refer to these sets as **subsets** of the real numbers, meaning that all elements in each subset are also elements in the set of real numbers.

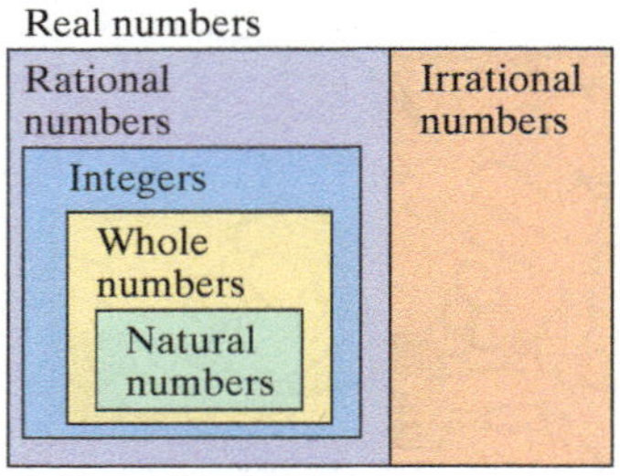

This diagram shows that every real number is rational or irrational.

TABLE 5.2 Important Subsets of the Real Numbers

Name	Description	Examples
Natural numbers	$\{1, 2, 3, 4, 5, \ldots\}$ These are the numbers that we use for counting.	$2, 3, 5, 17$
Whole numbers	$\{0, 1, 2, 3, 4, 5, \ldots\}$ The set of whole numbers includes 0 and the natural numbers.	$0, 2, 3, 5, 17$
Integers	$\{\ldots, -5, -4, -3, -2, -1, 0, 1, 2, 3, 4, 5, \ldots\}$ The set of integers includes the whole numbers and the negatives of the natural numbers.	$-17, -5, -3, -2, 0, 2, 3, 5, 17$
Rational numbers	$\left\{\frac{a}{b} \,\middle\vert\, a \text{ and } b \text{ are integers and } b \neq 0\right\}$ The set of rational numbers is the set of all numbers that can be expressed as a quotient of two integers, with the denominator not 0. Rational numbers can be expressed as terminating or repeating decimals.	$-17 = \frac{-17}{1}, -5 = \frac{-5}{1}, -3,$ $-2, 0, 2, 3, 5, 17,$ $\frac{2}{5} = 0.4,$ $\frac{-2}{3} = -0.6666\ldots = -0.\overline{6}$
Irrational numbers	The set of irrational numbers is the set of all numbers whose decimal representations are neither terminating nor repeating. Irrational numbers cannot be expressed as a quotient of integers.	$\sqrt{2} \approx 1.414214$ $-\sqrt{3} \approx -1.73205$ $\pi \approx 3.142$ $-\frac{\pi}{2} \approx -1.571$

EXAMPLE 1 *Classifying Real Numbers*

Consider the following set of numbers:

$$\left\{-7, -\frac{3}{4}, 0, 0.\overline{6}, \sqrt{5}, \pi, 7.3, \sqrt{81}\right\}.$$

List the numbers in the set that are

a. natural numbers. **b.** whole numbers. **c.** integers.
d. rational numbers. **e.** irrational numbers. **f.** real numbers.

Blitzer Bonus

Weird Numbers

Mathematicians use the label **weird** to describe a number if

1. The sum of its factors, excluding the number itself, is greater than the number.
2. No partial collection of the factors adds up to the number.

The number 70 is weird. Its factors other than itself are 1, 2, 5, 7, 10, 14, and 35. The sum of these factors is 74, which is greater than 70. Two or more numbers in the list of factors cannot be added to obtain 70.

Weird numbers are rare. Below 10,000, the weird numbers are 70, 836, 4030, 5830, 7192, 7912, and 9272. It is not known whether an odd weird number exists.

SOLUTION

a. Natural numbers: The natural numbers are the numbers used for counting. The only natural number in the set is $\sqrt{81}$ because $\sqrt{81} = 9$. (9 multiplied by itself, or 9^2, is 81.)

b. Whole numbers: The whole numbers consist of the natural numbers and 0. The elements of the set that are whole numbers are 0 and $\sqrt{81}$.

c. Integers: The integers consist of the natural numbers, 0, and the negatives of the natural numbers. The elements of the set that are integers are $\sqrt{81}$, 0, and -7.

d. Rational numbers: All numbers in the set that can be expressed as the quotient of integers are rational numbers. These include $-7\left(-7 = \frac{-7}{1}\right), -\frac{3}{4}, 0\left(0 = \frac{0}{1}\right)$, and $\sqrt{81}\left(\sqrt{81} = \frac{9}{1}\right)$. Furthermore, all numbers in the set that are terminating or repeating decimals are also rational numbers. These include $0.\overline{6}$ and 7.3.

e. Irrational numbers: The irrational numbers in the set are $\sqrt{5}\,(\sqrt{5} \approx 2.236)$ and $\pi\,(\pi \approx 3.14)$. Both $\sqrt{5}$ and π are only approximately equal to 2.236 and 3.14, respectively. In decimal form, $\sqrt{5}$ and π neither terminate nor have blocks of repeating digits.

f. Real numbers: All the numbers in the given set are real numbers.

CHECK POINT 1 Consider the following set of numbers:

$$\left\{-9, -1.3, 0, 0.\overline{3}, \frac{\pi}{2}, \sqrt{9}, \sqrt{10}\right\}.$$

List the numbers in the set that are

a. natural numbers. **b.** whole numbers.
c. integers. **d.** rational numbers.
e. irrational numbers. **f.** real numbers.

Blitzer Bonus

Vampire Numbers

Like legendary vampires that lie concealed among humans, vampire numbers lie hidden within the set of real numbers, mostly undetected. By definition, vampire numbers have an even number of digits. Furthermore, they are the product of two numbers whose digits all survive, in scrambled form, in the vampire. For example, 1260, 1435, and 2187 are vampire numbers.

$$21 \times 60 = 1260 \qquad 35 \times 41 = 1435 \qquad 27 \times 81 = 2187$$

The digits 2, 1, 6, and 0 lie scrambled in the vampire number.

The digits 3, 5, 4, and 1 lurk within the vampire number.

The digits 2, 7, 8, and 1 survive in the vampire number.

As the real numbers grow increasingly larger, is it necessary to pull out a wooden stake with greater frequency? How often can you expect to find vampires hidden among the giants? And is it possible to find a weird vampire?

On the right of the equal sign is a 40-digit vampire number that was discovered using a Pascal program on a personal computer:

$$98{,}765{,}432{,}198{,}765{,}432{,}198 \times 98{,}765{,}432{,}198{,}830{,}604{,}534 = 9{,}754{,}610{,}597{,}415{,}368{,}368{,}844{,}499{,}268{,}390{,}128{,}385{,}732.$$

Source: Clifford Pickover, *Wonders of Numbers,* Oxford University Press, 2001.

2 Recognize properties of real numbers.

Properties of the Real Numbers

When you use your calculator to add two real numbers, you can enter them in either order. The fact that two real numbers can be added in either order is called the **commutative property of addition**. You probably use this property, as well as other properties of the real numbers listed in **Table 5.3**, without giving it much thought. The properties of the real numbers are especially useful in algebra, as we shall see in Chapter 6.

Blitzer Bonus

The Associative Property and the English Language

In the English language, phrases can take on different meanings depending on the way the words are associated with commas.

Here are three examples.

- Woman, without her man, is nothing.
  Woman, without her, man is nothing.
- In the parade will be several hundred students carrying flags, and many teachers.
  In the parade will be several hundred students, carrying flags and many teachers.
- Population of Amsterdam broken down by age and sex
  Population of Amsterdam, broken down by age and sex

TABLE 5.3 Properties of the Real Numbers

Name	Meaning	Examples
Closure Property of Addition	The sum of any two real numbers is a real number.	$4\sqrt{2}$ is a real number and $5\sqrt{2}$ is a real number, so $4\sqrt{2} + 5\sqrt{2}$, or $9\sqrt{2}$, is a real number.
Closure Property of Multiplication	The product of any two real numbers is a real number.	10 is a real number and $\frac{1}{2}$ is a real number, so $10 \cdot \frac{1}{2}$, or 5, is a real number.
Commutative Property of Addition	Changing order when adding does not affect the sum. $a + b = b + a$	• $13 + 7 = 7 + 13$ • $\sqrt{2} + \sqrt{5} = \sqrt{5} + \sqrt{2}$
Commutative Property of Multiplication	Changing order when multiplying does not affect the product. $ab = ba$	• $13 \cdot 7 = 7 \cdot 13$ • $\sqrt{2} \cdot \sqrt{5} = \sqrt{5} \cdot \sqrt{2}$
Associative Property of Addition	Changing grouping when adding does not affect the sum. $(a + b) + c = a + (b + c)$	$(7 + 2) + 5 = 7 + (2 + 5)$ $9 + 5 = 7 + 7$ $14 = 14$
Associative Property of Multiplication	Changing grouping when multiplying does not affect the product. $(ab)c = a(bc)$	$(7 \cdot 2) \cdot 5 = 7 \cdot (2 \cdot 5)$ $14 \cdot 5 = 7 \cdot 10$ $70 = 70$
Distributive Property of Multiplication over Addition	Multiplication distributes over addition. $a \cdot (b + c) = a \cdot b + a \cdot c$	$7(4 + \sqrt{3}) = 7 \cdot 4 + 7 \cdot \sqrt{3}$ $= 28 + 7\sqrt{3}$
Identity Property of Addition	Zero can be deleted from a sum. $a + 0 = a$ $0 + a = a$	• $\sqrt{3} + 0 = \sqrt{3}$ • $0 + \pi = \pi$
Identity Property of Multiplication	One can be deleted from a product. $a \cdot 1 = a$ $1 \cdot a = a$	• $\sqrt{3} \cdot 1 = \sqrt{3}$ • $1 \cdot \pi = \pi$
Inverse Property of Addition	The sum of a real number and its additive inverse gives 0, the additive identity. $a + (-a) = 0$ $(-a) + a = 0$	• $\sqrt{3} + (-\sqrt{3}) = 0$ • $-\pi + \pi = 0$
Inverse Property of Multiplication	The product of a nonzero real number and its multiplicative inverse gives 1, the multiplicative identity. $a \cdot \frac{1}{a} = 1, a \neq 0$ $\frac{1}{a} \cdot a = 1, a \neq 0$	• $\sqrt{3} \cdot \frac{1}{\sqrt{3}} = 1$ • $\frac{1}{\pi} \cdot \pi = 1$

GREAT QUESTION!

Is there an easy way to distinguish between the commutative and associative properties?

Commutative: Changes *order.*
Associative: Changes *grouping.*

EXAMPLE 2 *Identifying Properties of the Real Numbers*

Name the property illustrated:

a. $\sqrt{3} \cdot 7 = 7 \cdot \sqrt{3}$

b. $(4 + 7) + 6 = 4 + (7 + 6)$

c. $2(3 + \sqrt{5}) = 6 + 2\sqrt{5}$

d. $\sqrt{2} + (\sqrt{3} + \sqrt{7}) = \sqrt{2} + (\sqrt{7} + \sqrt{3})$

e. $17 + (-17) = 0$

f. $\sqrt{2} \cdot 1 = \sqrt{2}.$

SOLUTION

a. $\sqrt{3} \cdot 7 = 7 \cdot \sqrt{3}$ Commutative property of multiplication

b. $(4 + 7) + 6 = 4 + (7 + 6)$ Associative property of addition

c. $2(3 + \sqrt{5}) = 6 + 2\sqrt{5}$ Distributive property of multiplication over addition

d. $\sqrt{2} + (\sqrt{3} + \sqrt{7}) = \sqrt{2} + (\sqrt{7} + \sqrt{3})$ The only change between the left and the right sides is in the order that $\sqrt{3}$ and $\sqrt{7}$ are added. The order is changed from $\sqrt{3} + \sqrt{7}$ to $\sqrt{7} + \sqrt{3}$ using the commutative property of addition.

e. $17 + (-17) = 0$ Inverse property of addition

f. $\sqrt{2} \cdot 1 = \sqrt{2}$ Identity property of multiplication

CHECK POINT 2 Name the property illustrated:

a. $(4 \cdot 7) \cdot 3 = 4 \cdot (7 \cdot 3)$

b. $3(\sqrt{5} + 4) = 3(4 + \sqrt{5})$

c. $3(\sqrt{5} + 4) = 3\sqrt{5} + 12$

d. $2(\sqrt{3} + \sqrt{7}) = (\sqrt{3} + \sqrt{7})2$

e. $1 + 0 = 1$

f. $-4\left(-\frac{1}{4}\right) = 1.$

Although the entire set of real numbers is closed with respect to addition and multiplication, some of the subsets of the real numbers do not satisfy the closure property for a given operation. If an operation on a set results in just one number that is not in that set, then the set is not closed for that operation.

EXAMPLE 3 *Verifying Closure*

a. Are the integers closed with respect to multiplication?

b. Are the irrational numbers closed with respect to multiplication?

c. Are the natural numbers closed with respect to division?

SOLUTION

a. Consider some examples of the multiplication of integers:

$$3 \cdot 2 = 6 \qquad 3(-2) = -6 \qquad -3(-2) = 6 \qquad -3 \cdot 0 = 0.$$

The product of any two integers is always a positive integer, a negative integer, or zero, which is an integer. Thus, the integers are closed under the operation of multiplication.

b. If we multiply two irrational numbers, must the product always be an irrational number? The answer is no. Here is an example:

$$\sqrt{7} \cdot \sqrt{7} = \sqrt{49} = 7.$$

Both irrational | Not an irrational number

This means that the irrational numbers are not closed under the operation of multiplication.

c. If we divide any two natural numbers, must the quotient always be a natural number? The answer is no. Here is an example:

$$4 \div 8 = \frac{1}{2}.$$

Both natural numbers | Not a natural number

Thus, the natural numbers are not closed under the operation of division.

CHECK POINT 3

a. Are the natural numbers closed with respect to multiplication?

b. Are the integers closed with respect to division?

The commutative property involves a change in order with no change in the final result. However, changing the order in which we subtract and divide real numbers can produce different answers. For example,

$$7 - 4 \neq 4 - 7 \quad \text{and} \quad 6 \div 2 \neq 2 \div 6.$$

Because the real numbers are not commutative with respect to subtraction and division, it is important that you enter numbers in the correct order when using a calculator to perform these operations.

The associative property does not hold for the operations of subtraction and division. The examples below show that if we change groupings when subtracting or dividing three numbers, the answer may change.

$$(6 - 1) - 3 \neq 6 - (1 - 3) \qquad (8 \div 4) \div 2 \neq 8 \div (4 \div 2)$$
$$5 - 3 \neq 6 - (-2) \qquad 2 \div 2 \neq 8 \div 2$$
$$2 \neq 8 \qquad 1 \neq 4$$

Blitzer Bonus

Beyond the Real Numbers

Only real numbers greater than or equal to zero have real number square roots. The square root of -1, $\sqrt{-1}$, is not a real number. This is because there is no real number that can be multiplied by itself that results in -1. Multiplying any real number by itself can never give a negative product. In the sixteenth century, mathematician Girolamo Cardano (1501–1576) wrote that square roots of negative numbers would cause "mental tortures." In spite of these "tortures," mathematicians invented a new number, called i, to represent $\sqrt{-1}$. The number i is not a real number; it is called an **imaginary number**. Thus, $\sqrt{9} = 3$, $-\sqrt{9} = -3$, but $\sqrt{-9}$ is not a real number. However, $\sqrt{-9}$ is an imaginary number, represented by $3i$. The adjective *real* as a way of describing what we now call the real numbers was first used by the French mathematician and philosopher René Descartes (1596–1650) in response to the concept of imaginary numbers.

3 Apply properties of real numbers to clock addition.

Properties of the Real Numbers and Clock Arithmetic

FIGURE 5.8

Mathematics is about the patterns that arise in the world about us. Mathematicians look at a snowflake in terms of its underlying structure. Notice that if the snowflake in **Figure 5.8** is rotated by any multiple of 60° ($\frac{1}{6}$ of a rotation), it will always look about the same. A **symmetry** of an object is a motion that moves the object back onto itself. In a symmetry, you cannot tell, at the end of the motion, that the object has been moved.

The snowflake in **Figure 5.8** has **sixfold rotational symmetry**. After six 60° turns, the snowflake is back to its original position. If it takes m equal turns to restore an object to its original position and each of these turns is a figure that is identical to the original figure, the object has ***m*-fold rotational symmetry**.

The sixfold rotational symmetry of the snowflake in **Figure 5.8** can be studied using the set $\{0, 1, 2, 3, 4, 5\}$ and an operation called *clock addition*. The "6-hour clock" in **Figure 5.9** exhibits the sixfold rotational symmetry of the snowflake.

FIGURE 5.9 Sixfold rotational symmetry in the face of a clock and a flower

Using **Figure 5.9**, we can define **clock addition** as follows: Add by moving the hour hand in a clockwise direction. The symbol $\oplus$ is used to designated clock addition. **Figure 5.10** illustrates that

$$2 \oplus 3 = 5, \quad 4 \oplus 5 = 3, \quad \text{and} \quad 3 \oplus 4 = 1.$$

FIGURE 5.10 Addition in a 6-hour clock system

TABLE 5.4 6-Hour Clock Addition

$\oplus$	**0**	**1**	**2**	**3**	**4**	**5**
0	0	1	2	3	4	5
1	1	2	3	4	5	0
2	2	3	4	5	0	1
3	3	4	5	0	1	2
4	4	5	0	1	2	3
5	5	0	1	2	3	4

Table 5.4 is the addition table for clock addition in a 6-hour clock system.

TABLE 5.4 (repeated)

⊕	0	1	2	3	4	5
0	0	1	2	3	4	5
1	1	2	3	4	5	0
2	2	3	4	5	0	1
3	3	4	5	0	1	2
4	4	5	0	1	2	3
5	5	0	1	2	3	4

EXAMPLE 4 *Properties of the Real Numbers Applied to the 6-Hour Clock System*

Table 5.4, the table for clock addition in a 6-hour clock system, is repeated in the margin.

a. How can you tell that the set $\{0, 1, 2, 3, 4, 5\}$ is closed under the operation of clock addition?

b. Verify the associative property:

$$(2 \oplus 3) \oplus 4 = 2 \oplus (3 \oplus 4).$$

c. What is the identity element in the 6-hour clock system?

d. Find the inverse of each element in the 6-hour clock system.

e. Verify two cases of the commutative property:

$$4 \oplus 3 = 3 \oplus 4 \quad \text{and} \quad 5 \oplus 4 = 4 \oplus 5.$$

SOLUTION

a. The Closure Property. The set $\{0, 1, 2, 3, 4, 5\}$ is closed under the operation of clock addition because the entries in the body of **Table 5.4** are all elements of the set.

b. The Associative Property. We were asked to verify one case of the associative property.

$$\begin{aligned}(2 \oplus 3) \oplus 4 &= 2 \oplus (3 \oplus 4)\\ 5 \oplus 4 &= 2 \oplus 1\\ 3 &= 3\end{aligned}$$

Locate 2 on the left and 3 on the top of **Table 5.4**. Intersecting lines show $2 \oplus 3 = 5$.

Locate 3 on the left and 4 on the top of **Table 5.4**. Intersecting lines show $3 \oplus 4 = 1$.

c. The Identity Property. Look for the element in **Table 5.4** that does not change anything when used in clock addition. **Table 5.4** shows that the column under 0 is the same as the column with boldface numbers on the left. Thus, $0 \oplus 0 = 0$, $1 \oplus 0 = 1$, $2 \oplus 0 = 2$, $3 \oplus 0 = 3$, $4 \oplus 0 = 4$, and $5 \oplus 0 = 5$. The table also shows that the row next to 0 is the same as the row with boldface numbers on top. Thus, $0 \oplus 0 = 0$, $0 \oplus 1 = 1$, $0 \oplus 2 = 2$, up through $0 \oplus 5 = 5$. Each element of the set does not change when we perform clock addition with 0. Thus, 0 is the identity element. The identity property is satisfied because 0 is contained in the given set.

d. The Inverse Property. When an element is added to its inverse, the result is the identity element. Because the identity element is 0, we can find the inverse of each element in $\{0, 1, 2, 3, 4, 5\}$ by answering the question: What must be added to each element to obtain 0?

$$\text{element} + ? = 0$$

Figure 5.11 illustrates how we answer the question. If each element in the set has an inverse, then a 0 will appear in every row and column of the table. This is, indeed, the case. Use the 0 in each row. Because each element in $\{0, 1, 2, 3, 4, 5\}$ has an inverse within the set, the inverse property is satisfied.

⊕	0	1	2	3	4	5	
0	0	1	2	3	4	5	$0 \oplus 0 = 0$: The inverse of 0 is 0.
1	1	2	3	4	5	0	$1 \oplus 5 = 0$: The inverse of 1 is 5.
2	2	3	4	5	0	1	$2 \oplus 4 = 0$: The inverse of 2 is 4.
3	3	4	5	0	1	2	$3 \oplus 3 = 0$: The inverse of 3 is 3.
4	4	5	0	1	2	3	$4 \oplus 2 = 0$: The inverse of 4 is 2.
5	5	0	1	2	3	4	$5 \oplus 1 = 0$: The inverse of 5 is 1.

FIGURE 5.11

e. The Commutative Property. We were asked to verify two cases of the commutative property.

$$4 \oplus 3 = 3 \oplus 4$$
$$1 = 1$$

Locate 4 on the left and 3 on the top of **Table 5.4.** Intersecting lines show $4 \oplus 3 = 1$.

Locate 3 on the left and 4 on the top of **Table 5.4.** Intersecting lines show $3 \oplus 4 = 1$.

$$5 \oplus 4 = 4 \oplus 5$$
$$3 = 3$$

Locate 5 on the left and 4 on the top of **Table 5.4.** Intersecting lines show $5 \oplus 4 = 3$.

Locate 4 on the left and 5 on the top of **Table 5.4.** Intersecting lines show $4 \oplus 5 = 3$.

Figure 5.12 illustrates four types of rotational symmetry.

Fourfold rotational symmetry — Fivefold rotational symmetry — Eightfold rotational symmetry — 18–fold rotational symmetry

FIGURE 5.12 Types of rotational symmetry

The fourfold rotational symmetry shown on the left in **Figure 5.12** can be explored using the 4-hour clock in **Figure 5.13** and **Table 5.5**, the table for clock addition in the 4-hour clock system.

FIGURE 5.13 A 4-hour clock

TABLE 5.5 4-Hour Clock Addition

$\oplus$	**0**	**1**	**2**	**3**
0	0	1	2	3
1	1	2	3	0
2	2	3	0	1
3	3	0	1	2

CHECK POINT 4 Use **Table 5.5** which shows clock addition in the 4-hour clock system to solve this exercise.

a. How can you tell that the set $\{0, 1, 2, 3\}$ is closed under the operation of clock addition?

b. Verify the associative property: $(2 \oplus 2) \oplus 3 = 2 \oplus (2 \oplus 3)$.

c. What is the identity element in the 4-hour clock system?

d. Find the inverse of each element in the 4-hour clock system.

e. Verify two cases of the commutative property: $1 \oplus 3 = 3 \oplus 1$ and $3 \oplus 2 = 2 \oplus 3$

Concept and Vocabulary Check

Fill in each blank so that the resulting statement is true.

1. Every real number is either ________ or ________.
2. The ________ property of addition states that the sum of any two real numbers is a real number.
3. If a and b are real numbers, the commutative property of multiplication states that ________.
4. If a, b, and c are real numbers, the associative property of addition states that ____________________.
5. If a, b, and c are real numbers, the distributive property states that ______________.
6. The ________ property of addition states that zero can be deleted from a sum.
7. The ________ property of multiplication states that ________ can be deleted from a product.
8. The product of a nonzero real number and its ________________________ gives 1, the ______________.
9. Shown in the figure is a 5-hour clock. Clock addition is performed by moving the hour hand in a clockwise direction. Thus,

$1 \oplus 4 =$ ___

$3 \oplus 3 =$ ___

and $4 \oplus 2 =$ ___.

10. True or False: The 5-hour clock in Exercise 9 could be used to describe the rotational symmetry of this flower. _______

Exercise Set 5.5

Practice Exercises

In Exercises 1–4, list all numbers from the given set that are

a. *natural numbers.* **b.** *whole numbers.*
c. *integers.* **d.** *rational numbers.*
e. *irrational numbers.* **f.** *real numbers.*

1. $\{-9, -\frac{4}{5}, 0, 0.25, \sqrt{3}, 9.2, \sqrt{100}\}$
2. $\{-7, -0.\overline{6}, 0, \sqrt{49}, \sqrt{50}\}$
3. $\{-11, -\frac{5}{6}, 0, 0.75, \sqrt{5}, \pi, \sqrt{64}\}$
4. $\{-5, -0.\overline{3}, 0, \sqrt{2}, \sqrt{4}\}$
5. Give an example of a whole number that is not a natural number.
6. Give an example of an integer that is not a whole number.
7. Give an example of a rational number that is not an integer.
8. Give an example of a rational number that is not a natural number.
9. Give an example of a number that is an integer, a whole number, and a natural number.
10. Give an example of a number that is a rational number, an integer, and a real number.
11. Give an example of a number that is an irrational number and a real number.
12. Give an example of a number that is a real number, but not an irrational number.

Complete each statement in Exercises 13–15 to illustrate the commutative property.

13. $3 + (4 + 5) = 3 + (5 +$ ___$)$
14. $\sqrt{5} \cdot 4 = 4 \cdot$ ___
15. $9 \cdot (6 + 2) = 9 \cdot (2 +$ ___$)$

Complete each statement in Exercises 16–17 to illustrate the associative property.

16. $(3 + 7) + 9 =$ ___ $+ (7 +$ ___$)$
17. $(4 \cdot 5) \cdot 3 =$ ___ $\cdot (5 \cdot$ ___$)$

Complete each statement in Exercises 18–20 to illustrate the distributive property.

18. $3 \cdot (6 + 4) = 3 \cdot 6 + 3 \cdot$ ___
19. ___ $\cdot (4 + 5) = 7 \cdot 4 + 7 \cdot 5$
20. $2 \cdot ($___ $+ 3) = 2 \cdot 7 + 2 \cdot 3$

Use the distributive property to simplify the radical expressions in Exercises 21–28.

21. $5(6 + \sqrt{2})$
22. $4(3 + \sqrt{5})$
23. $\sqrt{7}(3 + \sqrt{2})$
24. $\sqrt{6}(7 + \sqrt{5})$
25. $\sqrt{3}(5 + \sqrt{3})$
26. $\sqrt{7}(9 + \sqrt{7})$
27. $\sqrt{6}(\sqrt{2} + \sqrt{6})$
28. $\sqrt{10}(\sqrt{2} + \sqrt{10})$

In Exercises 29–44, state the name of the property illustrated.

29. $6 + (-4) = (-4) + 6$
30. $11 \cdot (7 + 4) = 11 \cdot 7 + 11 \cdot 4$
31. $6 + (2 + 7) = (6 + 2) + 7$
32. $6 \cdot (2 \cdot 3) = 6 \cdot (3 \cdot 2)$
33. $(2 + 3) + (4 + 5) = (4 + 5) + (2 + 3)$
34. $7 \cdot (11 \cdot 8) = (11 \cdot 8) \cdot 7$
35. $2(-8 + 6) = -16 + 12$

36. $-8(3 + 11) = -24 + (-88)$

37. $(2\sqrt{3}) \cdot \sqrt{5} = 2(\sqrt{3} \cdot \sqrt{5})$

38. $\sqrt{2}\pi = \pi\sqrt{2}$

39. $\sqrt{17} \cdot 1 = \sqrt{17}$

40. $\sqrt{17} + 0 = \sqrt{17}$

41. $\sqrt{17} + (-\sqrt{17}) = 0$

42. $\sqrt{17} \cdot \frac{1}{\sqrt{17}} = 1$

43. $\frac{1}{\sqrt{2} + \sqrt{7}}(\sqrt{2} + \sqrt{7}) = 1$

44. $(\sqrt{2} + \sqrt{7}) + -(\sqrt{2} + \sqrt{7}) = 0$

In Exercises 45–49, use two numbers to show that

45. the natural numbers are not closed with respect to subtraction.

46. the natural numbers are not closed with respect to division.

47. the integers are not closed with respect to division.

48. the irrational numbers are not closed with respect to subtraction.

49. the irrational numbers are not closed with respect to multiplication.

50. Shown in the figure is a 7-hour clock and the table for clock addition in the 7-hour clock system.

$\oplus$	0	1	2	3	4	5	6
0	0	1	2	3	4	5	6
1	1	2	3	4	5	6	0
2	2	3	4	5	6	0	1
3	3	4	5	6	0	1	2
4	4	5	6	0	1	2	3
5	5	6	0	1	2	3	4
6	6	0	1	2	3	4	5

a. How can you tell that the set $\{0, 1, 2, 3, 4, 5, 6\}$ is closed under the operation of clock addition?

b. Verify the associative property:
$(3 \oplus 5) \oplus 6 = 3 \oplus (5 \oplus 6)$.

c. What is the identity element in the 7-hour clock system?

d. Find the inverse of each element in the 7-hour clock system.

e. Verify two cases of the commutative property:
$4 \oplus 5 = 5 \oplus 4$ and $6 \oplus 1 = 1 \oplus 6$.

51. Shown in the figure is an 8-hour clock and the table for clock addition in the 8-hour clock system.

$\oplus$	0	1	2	3	4	5	6	7
0	0	1	2	3	4	5	6	7
1	1	2	3	4	5	6	7	0
2	2	3	4	5	6	7	0	1
3	3	4	5	6	7	0	1	2
4	4	5	6	7	0	1	2	3
5	5	6	7	0	1	2	3	4
6	6	7	0	1	2	3	4	5
7	7	0	1	2	3	4	5	6

a. How can you tell that the set $\{0, 1, 2, 3, 4, 5, 6, 7\}$ is closed under the operation of clock addition?

b. Verify the associative property:
$(4 \oplus 6) \oplus 7 = 4 \oplus (6 \oplus 7)$.

c. What is the identity element in the 8-hour clock system?

d. Find the inverse of each element in the 8-hour clock system.

e. Verify two cases of the commutative property:
$5 \oplus 6 = 6 \oplus 5$ and $4 \oplus 7 = 7 \oplus 4$.

Practice Plus

In Exercises 52–55, determine whether each statement is true or false. Do not use a calculator.

52. $468(787 + 289) = 787 + 289(468)$

53. $468(787 + 289) = 787(468) + 289(468)$

54. $58 \cdot 9 + 32 \cdot 9 = (58 + 32) \cdot 9$

55. $58 \cdot 9 \cdot 32 \cdot 9 = (58 \cdot 32) \cdot 9$

In Exercises 56–57, name the property used to go from step to step each time that (why?) occurs.

56. $7 + 2(x + 9)$
$= 7 + (2x + 18)$ (why?)
$= 7 + (18 + 2x)$ (why?)
$= (7 + 18) + 2x$ (why?)
$= 25 + 2x$
$= 2x + 25$ (why?)

57. $5(x + 4) + 3x$
$= (5x + 20) + 3x$ (why?)
$= (20 + 5x) + 3x$ (why?)
$= 20 + (5x + 3x)$ (why?)
$= 20 + (5 + 3)x$ (why?)
$= 20 + 8x$
$= 8x + 20$ (why?)

The tables show the operations $\square$ and $\triangle$ on the set {a, b, c, d, e}. Use these tables to solve Exercises 58–65.

$\square$	*a*	*b*	*c*	*d*	*e*
a	*a*	*b*	*c*	*d*	*e*
b	*b*	*c*	*d*	*e*	*a*
c	*c*	*d*	*e*	*a*	*b*
d	*d*	*e*	*a*	*b*	*c*
e	*e*	*a*	*b*	*c*	*d*

$\triangle$	*a*	*b*	*c*	*d*	*e*
a	*a*	*a*	*a*	*a*	*a*
b	*a*	*b*	*c*	*d*	*e*
c	*a*	*c*	*e*	*b*	*d*
d	*a*	*d*	*b*	*e*	*c*
e	*a*	*e*	*d*	*c*	*b*

58. a. Show that $e \triangle (c \square d) = (e \triangle c) \square (e \triangle d)$.

b. What property of the real numbers is illustrated in part (a)?

59. a. Show that $c \triangle (d \square e) = (c \triangle d) \square (c \triangle e)$.

b. What property of the real numbers is illustrated in part (a)?

60. Find $c \triangle [c \square (c \triangle c)]$.

61. Find $d \triangle [d \square (d \triangle d)]$.

In Exercises 62–65, replace x with a, b, c, d, or e to form a true statement.

62. $x \square d = e$

63. $x \square d = a$

64. $x \triangle (e \square c) = d$

65. $x \triangle (e \square d) = b$

66. If $\begin{bmatrix} a & b \\ c & d \end{bmatrix} \times \begin{bmatrix} e & f \\ g & h \end{bmatrix} = \begin{bmatrix} ae + bg & af + bh \\ ce + dg & cf + dh \end{bmatrix}$, find

a. $\begin{bmatrix} 2 & 3 \\ 4 & 7 \end{bmatrix} \times \begin{bmatrix} 0 & 1 \\ 5 & 6 \end{bmatrix}$

b. $\begin{bmatrix} 0 & 1 \\ 5 & 6 \end{bmatrix} \times \begin{bmatrix} 2 & 3 \\ 4 & 7 \end{bmatrix}$

c. Draw a conclusion about one of the properties discussed in this section in terms of these arrays of numbers under multiplication.

Application Exercises

In Exercises 67–70, use the definition of vampire numbers from the Blitzer Bonus on page 301 to determine which products are vampires.

67. $15 \times 93 = 1395$

68. $80 \times 86 = 6880$

69. $20 \times 51 = 1020$

70. $146 \times 938 = 136{,}948$

A **narcissistic number** *is an n-digit number equal to the sum of each of its digits raised to the nth power. Here's an example:*

$$153 = 1^3 + 5^3 + 3^3.$$

Three digits, so exponents are 3

In Exercises 71–74, determine which real numbers are narcissistic.

71. 370

72. 371

73. 372

74. 9474

75. The algebraic expressions

$$\frac{D(A + 1)}{24} \quad \text{and} \quad \frac{DA + D}{24}$$

describe the drug dosage for children between the ages of 2 and 13. In each algebraic expression, D stands for an adult dose and A represents the child's age.

a. Name the property that explains why these expressions are equal for all values of D and A.

b. If an adult dose of ibuprofen is 200 milligrams, what is the proper dose for a 12-year-old child? Use both forms of the algebraic expressions to answer the question. Which form is easier to use?

76. Closure illustrates that a characteristic of a set is not necessarily a characteristic of all of its subsets. The real numbers are closed with respect to multiplication, but the irrational numbers, a subset of the real numbers, are not. Give an example of a set that is not mathematical that has a particular characteristic, but which has a subset without this characteristic.

Name the kind of rotational symmetry shown in Exercises 77–78.

77.

Native American design

78.

Mercedes Benz symbol

Writing in Mathematics

79. What does it mean when we say that the rational numbers are a subset of the real numbers?

80. What does it mean if we say that a set is closed under a given operation?

81. State the commutative property of addition and give an example.

82. State the commutative property of multiplication and give an example.

83. State the associative property of addition and give an example.

84. State the associative property of multiplication and give an example.

85. State the distributive property of multiplication over addition and give an example.

86. Does $7 \cdot (4 \cdot 3) = 7 \cdot (3 \cdot 4)$ illustrate the commutative property or the associative property? Explain your answer.

87. Explain how to use the 8-hour clock shown in Exercise 51 to find $6 \oplus 5$.

Critical Thinking Exercises

Make Sense? *In Exercises 88–91, determine whether each statement makes sense or does not make sense, and explain your reasoning.*

88. The humor in this cartoon is based on the fact that "rational" and "real" have different meanings in mathematics and in everyday speech.

89. The number of pages in this book is a real number.

90. The book that I'm reading on the history of π appropriately contains an irrational number of pages.

91. Although the integers are closed under the operation of addition, I was able to find a subset that is not closed under this operation.

In Exercises 92–99, determine whether each statement is true or false. If the statement is false, make the necessary change(s) to produce a true statement.

92. Every rational number is an integer.

93. Some whole numbers are not integers.

94. Some rational numbers are not positive.

95. Irrational numbers cannot be negative.

96. Subtraction is a commutative operation.

97. $(24 \div 6) \div 2 = 24 \div (6 \div 2)$

98. $7 \cdot a + 3 \cdot a = a \cdot (7 + 3)$

99. $2 \cdot a + 5 = 5 \cdot a + 2$

5.6 Exponents and Scientific Notation

WHAT AM I SUPPOSED TO LEARN?

After you have read this section, you should be able to:

1. Use properties of exponents.
2. Convert from scientific notation to decimal notation.
3. Convert from decimal notation to scientific notation.
4. Perform computations using scientific notation.
5. Solve applied problems using scientific notation.

Bigger than the biggest thing ever and then some. Much bigger than that in fact, really amazingly immense, a totally stunning size, real 'wow, that's big', time...Gigantic multiplied by colossal multiplied by staggeringly huge is the sort of concept we're trying to get across here.

—DOUGLAS ADAMS, *The Restaurant at the End of the Universe*

Although Adams's description may not quite apply to this $15.2 trillion national debt, exponents can be used to explore the meaning of this "staggeringly huge" number. In this section, you will learn to use exponents to provide a way of putting large and small numbers in perspective.

Properties of Exponents

1 Use properties of exponents.

We have seen that exponents are used to indicate repeated multiplication. Now consider the multiplication of two exponential expressions, such as $b^4 \cdot b^3$. We are multiplying 4 factors of b and 3 factors of b. We have a total of 7 factors of b:

$$b^4 \cdot b^3 = \underbrace{(b \cdot b \cdot b \cdot b)}_{\text{4 factors of } b}\underbrace{(b \cdot b \cdot b)}_{\text{3 factors of } b} = b^7.$$

Total: 7 factors of b

The product is exactly the same if we add the exponents:

$$b^4 \cdot b^3 = b^{4+3} = b^7.$$

Properties of exponents allow us to perform operations with exponential expressions without having to write out long strings of factors. Three such properties are given in **Table 5.6**.

TABLE 5.6 Properties of Exponents

Property	Meaning	Examples
The Product Rule $b^m \cdot b^n = b^{m+n}$	When multiplying exponential expressions with the same base, add the exponents. Use this sum as the exponent of the common base.	$9^6 \cdot 9^{12} = 9^{6+12} = 9^{18}$
The Power Rule $(b^m)^n = b^{m \cdot n}$	When an exponential expression is raised to a power, multiply the exponents. Place the product of the exponents on the base and remove the parentheses.	$(3^4)^5 = 3^{4 \cdot 5} = 3^{20}$ $(5^3)^8 = 5^{3 \cdot 8} = 5^{24}$
The Quotient Rule $\frac{b^m}{b^n} = b^{m-n}$	When dividing exponential expressions with the same base, subtract the exponent in the denominator from the exponent in the numerator. Use this difference as the exponent of the common base.	$\frac{5^{12}}{5^4} = 5^{12-4} = 5^8$ $\frac{9^{40}}{9^5} = 9^{40-5} = 9^{35}$

The third property in **Table 5.6**, $\frac{b^m}{b^n} = b^{m-n}$, called the quotient rule, can lead to a zero exponent when subtracting exponents. Here is an example:

$$\frac{4^3}{4^3} = 4^{3-3} = 4^0.$$

We can see what this zero exponent means by evaluating 4^3 in the numerator and the denominator:

$$\frac{4^3}{4^3} = \frac{4 \cdot 4 \cdot 4}{4 \cdot 4 \cdot 4} = \frac{64}{64} = 1.$$

This means that 4^0 must equal 1. This example illustrates the zero exponent rule.

THE ZERO EXPONENT RULE

If b is any real number other than 0,

$$b^0 = 1.$$

EXAMPLE 1 *Using the Zero Exponent Rule*

Use the zero exponent rule to simplify:

a. 7^0 **b.** π^0 **c.** $(-5)^0$ **d.** -5^0.

SOLUTION

a. $7^0 = 1$ **b.** $\pi^0 = 1$ **c.** $(-5)^0 = 1$ **d.** $-5^0 = -1$

Only 5 is raised to the 0 power.

CHECK POINT 1 Use the zero exponent rule to simplify:

a. 19^0 **b.** $(3\pi)^0$ **c.** $(-14)^0$ **d.** -14^0.

The quotient rule can result in a negative exponent. Consider, for example, $4^3 \div 4^5$:

$$\frac{4^3}{4^5} = 4^{3-5} = 4^{-2}.$$

We can see what this negative exponent means by evaluating the numerator and the denominator:

$$\frac{4^3}{4^5} = \frac{\cancel{4} \cdot \cancel{4} \cdot \cancel{4}}{\cancel{4} \cdot \cancel{4} \cdot \cancel{4} \cdot 4 \cdot 4} = \frac{1}{4^2}.$$

Notice that $\frac{4^3}{4^5}$ equals both 4^{-2} and $\frac{1}{4^2}$. This means that 4^{-2} must equal $\frac{1}{4^2}$. This example is a particular case of the negative exponent rule.

GREAT QUESTION!

What's the difference between $\frac{4^3}{4^5}$ and $\frac{4^5}{4^3}$?

$\frac{4^3}{4^5}$ and $\frac{4^5}{4^3}$ represent different numbers:

$$\frac{4^3}{4^5} = 4^{3-5} = 4^{-2} = \frac{1}{4^2} = \frac{1}{16}$$

$$\frac{4^5}{4^3} = 4^{5-3} = 4^2 = 16.$$

THE NEGATIVE EXPONENT RULE

If b is any real number other than 0 and m is a natural number,

$$b^{-m} = \frac{1}{b^m}.$$

EXAMPLE 2 *Using the Negative Exponent Rule*

Use the negative exponent rule to simplify:

a. 8^{-2} **b.** 5^{-3} **c.** 7^{-1}.

SOLUTION

a. $8^{-2} = \frac{1}{8^2} = \frac{1}{8 \cdot 8} = \frac{1}{64}$ **b.** $5^{-3} = \frac{1}{5^3} = \frac{1}{5 \cdot 5 \cdot 5} = \frac{1}{125}$

c. $7^{-1} = \frac{1}{7^1} = \frac{1}{7}$

CHECK POINT 2 Use the negative exponent rule to simplify:

a. 9^{-2} **b.** 6^{-3} **c.** 12^{-1}.

Powers of Ten

Exponents and their properties allow us to represent and compute with numbers that are large or small. For example, one billion, or 1,000,000,000 can be written as 10^9. In terms of exponents, 10^9 might not look very large, but consider this: If you can count to 200 in one minute and decide to count for 12 hours a day at this rate, it would take you in the region of 19 years, 9 days, 5 hours, and 20 minutes to count to 10^9!

Powers of ten follow two basic rules:

1. **A positive exponent tells how many 0s follow the 1.** For example, 10^9 (one billion) is a 1 followed by nine zeros: 1,000,000,000. A googol, 10^{100}, is a 1 followed by one hundred zeros. (A googol far exceeds the number of protons, neutrons, and electrons in the universe.) A googol is a veritable pipsqueak compared to the googolplex, 10 raised to the googol power, or $10^{10^{100}}$; that's a 1 followed by a googol zeros. (If each zero in a googolplex were no larger than a grain of sand, there would not be enough room in the universe to represent the number.)
2. **A negative exponent tells how many places there are to the right of the decimal point.** For example, 10^{-9} (one billionth) has nine places to the right of the decimal point. The nine places include eight 0s and the 1.

$$10^{-9} = 0.\underbrace{000000001}_{\text{nine places}}$$

Blitzer Bonus

Earthquakes and Powers of Ten

The earthquake that ripped through northern California on October 17, 1989, measured 7.1 on the Richter scale, killed more than 60 people, and injured more than 2400. Shown here is San Francisco's Marina district, where shock waves tossed houses off their foundations and into the street.

The Richter scale is misleading because it is not actually a 1 to 8, but rather a 1 to 10 million scale. Each level indicates a tenfold increase in magnitude from the previous level, making a 7.0 earthquake a million times greater than a 1.0 quake.

The following is a translation of the Richter scale:

Richter Number (R)	**Magnitude** (10^{R-1})
1	$10^{1-1} = 10^0 = 1$
2	$10^{2-1} = 10^1 = 10$
3	$10^{3-1} = 10^2 = 100$
4	$10^{4-1} = 10^3 = 1000$
5	$10^{5-1} = 10^4 = 10{,}000$
6	$10^{6-1} = 10^5 = 100{,}000$
7	$10^{7-1} = 10^6 = 1{,}000{,}000$
8	$10^{8-1} = 10^7 = 10{,}000{,}000$

TABLE 5.7 Names of Large Numbers	
10^2	hundred
10^3	thousand
10^6	million
10^9	billion
10^{12}	trillion
10^{15}	quadrillion
10^{18}	quintillion
10^{21}	sextillion
10^{24}	septillion
10^{27}	octillion
10^{30}	nonillion
10^{100}	googol
10^{googol}	googolplex

Scientific Notation

As of December 2011, the national debt of the United States was about \$15.2 trillion. This is the amount of money the government has had to borrow over the years, mostly by selling bonds, because it has spent more than it has collected in taxes. A stack of \$1 bills equaling the national debt would measure more than 950,000 miles. That's more than two round trips from Earth to the moon. Because a trillion is 10^{12} (see **Table 5.7**), the national debt can be expressed as

$$15.2 \times 10^{12}.$$

Because $15.2 = 1.52 \times 10$, the national debt can be expressed as

$$15.2 \times 10^{12} = (1.52 \times 10) \times 10^{12} = 1.52 \times (10 \times 10^{12})$$
$$= 1.52 \times 10^{1+12} = 1.52 \times 10^{13}$$

The number 1.52×10^{13} is written in a form called *scientific notation.*

SCIENTIFIC NOTATION

A positive number is written in **scientific notation** when it is expressed in the form

$$a \times 10^n,$$

where a is a number greater than or equal to 1 and less than 10 ($1 \le a < 10$) and n is an integer.

It is customary to use the multiplication symbol, ×, rather than a dot, when writing a number in scientific notation.

Here are three examples of numbers in scientific notation:

- The universe is 1.375×10^{10} years old.
- In 2010, humankind generated 1.2 zettabytes, or 1.2×10^{21} bytes, of digital information. (A byte consists of eight binary digits, or bits, 0 or 1.)
- The length of the AIDS virus is 1.1×10^{-4} millimeter.

2 Convert from scientific notation to decimal notation.

We can use n, the exponent on the 10 in $a \times 10^n$, to change a number in scientific notation to decimal notation. If n is **positive**, move the decimal point in a to the **right** n places. If n is **negative**, move the decimal point in a to the **left** $|n|$ places.

EXAMPLE 3 *Converting from Scientific to Decimal Notation*

Write each number in decimal notation:

a. 1.375×10^{10} **b.** 1.1×10^{-4}.

SOLUTION

In each case, we use the exponent on the 10 to move the decimal point. In part (a), the exponent is positive, so we move the decimal point to the right. In part (b), the exponent is negative, so we move the decimal point to the left.

a. $1.375 \times 10^{10} = 13{,}750{,}000{,}000$

$n = 10$

Move the decimal point 10 places to the right.

b. $1.1 \times 10^{-4} = 0.00011$

$n = -4$

Move the decimal point $|-4|$ places, or 4 places, to the left.

CHECK POINT 3 Write each number in decimal notation:

a. 7.4×10^9 **b.** 3.017×10^{-6}.

3 Convert from decimal notation to scientific notation.

To convert a positive number from decimal notation to scientific notation, we reverse the procedure of Example 3.

CONVERTING FROM DECIMAL TO SCIENTIFIC NOTATION

Write the number in the form $a \times 10^n$.

- Determine a, the numerical factor. Move the decimal point in the given number to obtain a number greater than or equal to 1 and less than 10.
- Determine n, the exponent on 10^n. The absolute value of n is the number of places the decimal point was moved. The exponent n is positive if the given number is greater than 10 and negative if the given number is between 0 and 1.

EXAMPLE 4 *Converting from Decimal Notation to Scientific Notation*

Write each number in scientific notation:

a. 4,600,000 **b.** 0.000023.

SOLUTION

a. $4{,}600{,}000 = 4.6 \times 10^6$

This number is greater than 10, so n is positive in $a \times 10^n$.

Move the decimal point in 4,600,000 to get $1 \le a < 10$.

The decimal point moved 6 places from 4,600,000 to 4.6.

b. $0.000023 = 2.3 \times 10^{-5}$

This number is less than 1, so n is negative in $a \times 10^n$.

Move the decimal point in 0.000023 to get $1 \le a < 10$.

The decimal point moved 5 places from 0.000023 to 2.3.

CHECK POINT 4 Write each number in scientific notation:

a. 7,410,000,000 **b.** 0.000000092.

TECHNOLOGY

You can use your calculator's [EE] (enter exponent) or [EXP] key to convert from decimal to scientific notation. Here is how it's done for 0.000023:

Many Scientific Calculators

Keystrokes	Display
.000023 [EE] [=]	2.3 − 05

Many Graphing Calculators

Use the mode setting for scientific notation.

Keystrokes	Display
.000023 [ENTER]	2.3 E −5

EXAMPLE 5 *Expressing the U.S. Population in Scientific Notation*

As of December 2011, the population of the United States was approximately 312 million. Express the population in scientific notation.

SOLUTION

Because a million is 10^6, the 2011 population can be expressed as

$$312 \times 10^6.$$

This factor is not between 1 and 10, so the number is not in scientific notation.

The voice balloon indicates that we need to convert 312 to scientific notation.

$$312 \times 10^6 = (3.12 \times 10^2) \times 10^6 = 3.12 \times 10^{2+6} = 3.12 \times 10^8$$

$312 = 3.12 \times 10^2$

In scientific notation, the population is 3.12×10^8.

GREAT QUESTION!

I read that the U.S. population exceeds $\frac{3}{10}$ of a billion. Yet you described it as 312 million. Which description is correct?

Both descriptions are correct. We can use exponential properties to express 312 million in billions.

$$312 \text{ million} = 312 \times 10^6 = (0.312 \times 10^3) \times 10^6 = 0.312 \times 10^{3+6} = 0.312 \times 10^9$$

Because 10^9 is a billion, U.S. population exceeds $\frac{3}{10}$ of a billion.

CHECK POINT 5 Express 410×10^7 in scientitic notation.

4 Perform computations using scientific notation.

Computations with Scientific Notation

We use the product rule for exponents to multiply numbers in scientific notation:

$$(a \times 10^n) \times (b \times 10^m) = (a \times b) \times 10^{n+m}.$$

Add the exponents on 10 and multiply the other parts of the numbers separately.

TECHNOLOGY

$(3.4 \times 10^9)(2 \times 10^{-5})$ on a Calculator:

Many Scientific Calculators

3.4 [EE] 9 [×] 2 [EE] 5 [+/−] [=]

Display: 6.8 04

Many Graphing Calculators

3.4 [EE] 9 [×] 2 [EE] [(−)] 5 [ENTER]

Display: 6.8 E 4

EXAMPLE 6 *Multiplying Numbers in Scientific Notation*

Multiply: $(3.4 \times 10^9)(2 \times 10^{-5})$. Write the product in decimal notation.

SOLUTION

$(3.4 \times 10^9)(2 \times 10^{-5}) = (3.4 \times 2) \times (10^9 \times 10^{-5})$ Regroup factors.

$= 6.8 \times 10^{9+(-5)}$ Add the exponents on 10 and multiply the other parts.

$= 6.8 \times 10^4$ Simplify.

$= 68{,}000$ Write the product in decimal notation.

CHECK POINT 6 Multiply: $(1.3 \times 10^7)(4 \times 10^{-2})$. Write the product in decimal notation.

We use the quotient rule for exponents to divide numbers in scientific notation:

$$\frac{a \times 10^n}{b \times 10^m} = \left(\frac{a}{b}\right) \times 10^{n-m}.$$

Subtract the exponents on 10 and divide the other parts of the numbers separately.

EXAMPLE 7 *Dividing Numbers in Scientific Notation*

Divide: $\frac{8.4 \times 10^{-7}}{4 \times 10^{-4}}$. Write the quotient in decimal notation.

SOLUTION

$$\frac{8.4 \times 10^{-7}}{4 \times 10^{-4}} = \left(\frac{8.4}{4}\right) \times \left(\frac{10^{-7}}{10^{-4}}\right)$$ Regroup factors.

$$= 2.1 \times 10^{-7-(-4)}$$ Subtract the exponents on 10 and divide the other parts.

$$= 2.1 \times 10^{-3}$$ Simplify: $-7 - (-4) = -7 + 4 = -3$.

$$= 0.0021$$ Write the quotient in decimal notation.

CHECK POINT 7 Divide: $\frac{6.9 \times 10^{-8}}{3 \times 10^{-2}}$. Write the quotient in decimal notation.

Multiplication and division involving very large or very small numbers can be performed by first converting each number to scientific notation.

EXAMPLE 8 *Using Scientific Notation to Multiply*

Multiply: $0.00064 \times 9{,}400{,}000{,}000$. Express the product in **a.** scientific notation and **b.** decimal notation.

SOLUTION

a. $0.00064 \times 9{,}400{,}000{,}000$

$$= 6.4 \times 10^{-4} \times 9.4 \times 10^{9}$$ Write each number in scientific notation.

$$= (6.4 \times 9.4) \times (10^{-4} \times 10^{9})$$ Regroup factors.

$$= 60.16 \times 10^{-4+9}$$ Add the exponents on 10 and multiply the other parts.

$$= 60.16 \times 10^{5}$$ Simplify.

$$= (6.016 \times 10) \times 10^{5}$$ Express 60.16 in scientific notation.

$$= 6.016 \times 10^{6}$$ Add exponents on 10: $10^1 \times 10^5 = 10^{1+5} = 10^6$.

b. The answer in decimal notation is obtained by moving the decimal point in 6.016 six places to the right. The product is 6,016,000.

CHECK POINT 8 Multiply: $0.0036 \times 5{,}200{,}000$. Express the product in **a.** scientific notation and **b.** decimal notation.

5 Solve applied problems using scientific notation.

Applications: Putting Numbers in Perspective

Due to tax cuts and spending increases, the United States began accumulating large deficits in the 1980s. To finance the deficit, the government had borrowed \$15.2 trillion as of December 2011. The graph in **Figure 5.14** on the next page shows the national debt increasing over time.

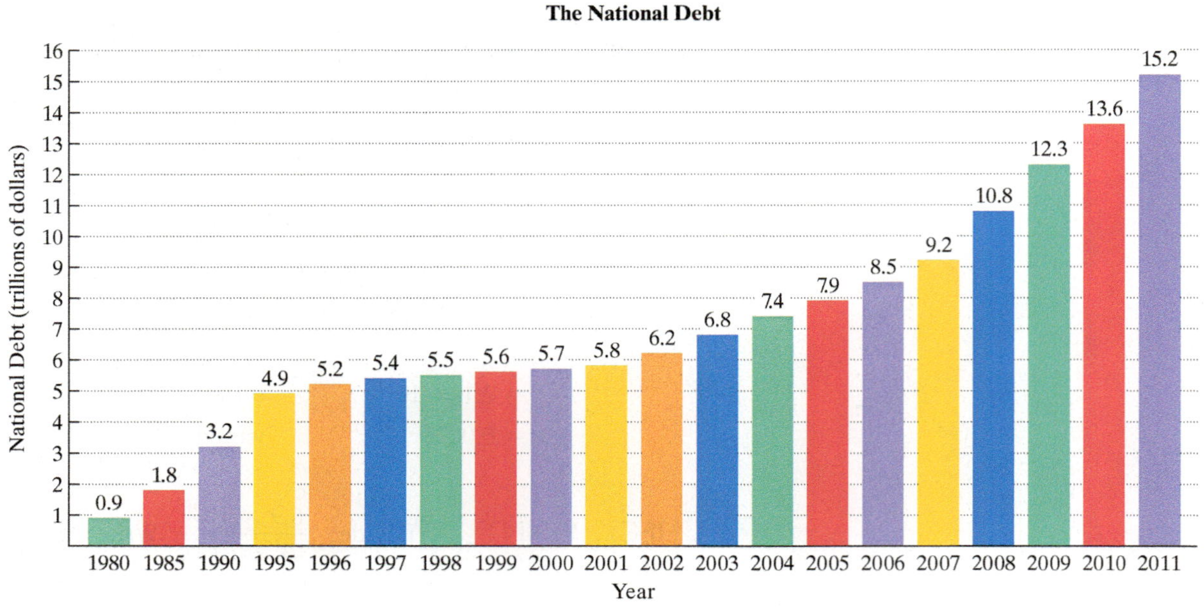

FIGURE 5.14
Source: Office of Management and Budget

Example 9 shows how we can use scientific notation to comprehend the meaning of a number such as 15.2 trillion.

EXAMPLE 9 *The National Debt*

As of December 2011, the national debt was \$15.2 trillion, or 15.2×10^{12} dollars. At that time, the U.S. population was approximately 312,000,000 (312 million), or 3.12×10^8. If the national debt was evenly divided among every individual in the United States, how much would each citizen have to pay?

SOLUTION

The amount each citizen must pay is the total debt, 15.2×10^{12} dollars, divided by the number of citizens, 3.12×10^8.

$$\frac{15.2 \times 10^{12}}{3.12 \times 10^8} = \left(\frac{15.2}{3.12}\right) \times \left(\frac{10^{12}}{10^8}\right)$$
$$\approx 4.87 \times 10^{12-8}$$
$$= 4.87 \times 10^4$$
$$= 48{,}700$$

Every U.S. citizen would have to pay approximately \$48,700 to the federal government to pay off the national debt.

TECHNOLOGY

Here is the keystroke sequence for solving Example 9 using a calculator:

15.2 [EE] 12 [÷] 3.12 [EE] 8.

The quotient is displayed by pressing [=] on a scientific calculator or [ENTER] on a graphing calculator. The answer can be displayed in scientific or decimal notation. Consult your manual.

If a number is written in scientific notation, $a \times 10^n$, the digits in a are called **significant digits**.

National Debt: 15.2×10^{12} U.S. Population: 3.12×10^8

Three significant digits Three significant digits

Because these were the given numbers in Example 9, we rounded the answer, 4.87×10^4, to three significant digits. When multiplying or dividing in scientific notation where rounding is necessary and rounding instructions are not given, **round the scientific notation answer to the least number of significant digits found in any of the given numbers.**

CHECK POINT 9 As of December 2011, the United States had spent \$2.6 trillion for the wars in Iraq and Afghanistan. (Source: costsofwar.org) At that time, the U.S. population was approximately 312 million (3.12×10^8). If the cost of these wars was evenly divided among every individual in the United States, how much would each citizen have to pay?

Blitzer Bonus

Seven Ways to Spend \$1 Trillion

Confronting a national debt of \$15.2 trillion starts with grasping just how colossal \$1 trillion (1×10^{12}) actually is. To help you wrap your head around this mind-boggling number and to put the national debt in further perspective, consider what \$1 trillion will buy:

- 40,816,326 new cars based on an average sticker price of \$24,500 each
- 5,574,136 homes based on the national median price of \$179,400 for existing single-family homes
- one year's salary for 14.7 million teachers based on the average teacher salary of \$68,000 in California
- the annual salaries of all 535 members of Congress for the next 10,742 years based on current salaries of \$174,000 per year
- the salary of basketball superstar LeBron James for 50,000 years based on an annual salary of \$20 million
- annual base pay for 59.5 million U.S. privates (that's 100 times the total number of active-duty soldiers in the Army) based on basic pay of \$16,794 per year
- salaries to hire all 2.8 million residents of the state of Kansas in full-time minimum-wage jobs for the next 23 years based on the federal minimum wage of \$7.25 per hour

Source: Kiplinger.com

Concept and Vocabulary Check

Fill in each blank so that the resulting statement is true.

1. When multiplying exponential expressions with the same base, _________ the exponents.

2. When an exponential expression is raised to a power, _________ the exponents.

3. When dividing exponential expressions with the same base, _________ the exponents.

4. Any nonzero real number raised to the zero power is equal to _________.

5. A positive number is written in scientific notation when the first factor is _________ and the second factor is _________.

In Exercises 6–10, determine whether each statement is true or false. If the statement is false, make the necessary change(s) to produce a true statement.

6. $2^3 \cdot 2^5 = 4^8$ ______

7. $\frac{10^8}{10^4} = 10^2$ ______

8. $5^{-2} = -5^2$ ______

9. A trillion is one followed by 12 zeros. ______

10. According to *Mother Jones* magazine, sending all U.S. high school graduates to private colleges would cost \$347 billion. Because a billion is 10^9, the cost in scientific notation is 347×10^9 dollars. ______

Exercise Set 5.6

Practice Exercises

In Exercises 1–12, use properties of exponents to simplify each expression. First express the answer in exponential form. Then evaluate the expression.

1. $2^2 \cdot 2^3$

2. $3^3 \cdot 3^2$

3. $4 \cdot 4^2$

4. $5 \cdot 5^2$

5. $(2^2)^3$

6. $(3^3)^2$

7. $(1^4)^5$

8. $(1^3)^7$

9. $\frac{4^7}{4^5}$

10. $\frac{6^7}{6^5}$

11. $\frac{2^8}{2^4}$

12. $\frac{3^8}{3^4}$

In Exercises 13–24, use the zero and negative exponent rules to simplify each expression.

13. 3^0 **14.** 9^0 **15.** $(-3)^0$ **16.** $(-9)^0$

17. -3^0 **18.** -9^0 **19.** 2^{-2} **20.** 3^{-2}

21. 4^{-3} **22.** 2^{-3} **23.** 2^{-5} **24.** 2^{-6}

In Exercises 25–30, use properties of exponents to simplify each expression. First express the answer in exponential form. Then evaluate the expression.

25. $3^4 \cdot 3^{-2}$ **26.** $2^5 \cdot 2^{-2}$ **27.** $3^{-3} \cdot 3$

28. $2^{-3} \cdot 2$ **29.** $\frac{2^3}{2^7}$ **30.** $\frac{3^4}{3^7}$

In Exercises 31–42, use properties of exponents to simplify each expression. Express answers in exponential form with positive exponents only. Assume that any variables in denominators are not equal to zero.

31. $(x^5 \cdot x^3)^{-2}$ **32.** $(x^2 \cdot x^4)^{-3}$ **33.** $\frac{(x^3)^4}{(x^2)^7}$

34. $\frac{(x^2)^5}{(x^3)^4}$ **35.** $\left(\frac{x^5}{x^2}\right)^{-4}$ **36.** $\left(\frac{x^7}{x^2}\right)^{-3}$

37. $\frac{2x^5 \cdot 3x}{15x^6}$ **38.** $\frac{4x^7 \cdot 5x}{10x^8}$

39. $(-2x^3y^{-4})(3x^{-1}y)$ **40.** $(-5x^4y^{-3})(4x^{-1}y)$

41. $\frac{30x^2y^5}{-6x^8y^{-3}}$ **42.** $\frac{24x^2y^{13}}{-8x^5y^{-2}}$

In Exercises 43–58, express each number in decimal notation.

43. 2.7×10^2 **44.** 4.7×10^3

45. 9.12×10^5 **46.** 8.14×10^4

47. 8×10^7 **48.** 7×10^6

49. 1×10^5 **50.** 1×10^8

51. 7.9×10^{-1} **52.** 8.6×10^{-1}

53. 2.15×10^{-2} **54.** 3.14×10^{-2}

55. 7.86×10^{-4} **56.** 4.63×10^{-5}

57. 3.18×10^{-6} **58.** 5.84×10^{-7}

In Exercises 59–78, express each number in scientific notation.

59. 370 **60.** 530

61. 3600 **62.** 2700

63. 32,000 **64.** 64,000

65. 220,000,000 **66.** 370,000,000,000

67. 0.027 **68.** 0.014

69. 0.0037 **70.** 0.00083

71. 0.00000293 **72.** 0.000000647

73. 820×10^5 **74.** 630×10^8

75. 0.41×10^6 **76.** 0.57×10^9

77. 2100×10^{-9} **78.** $97{,}000 \times 10^{-11}$

In Exercises 79–92, perform the indicated operation and express each answer in decimal notation.

79. $(2 \times 10^3)(3 \times 10^2)$ **80.** $(5 \times 10^2)(4 \times 10^4)$

81. $(2 \times 10^9)(3 \times 10^{-5})$ **82.** $(4 \times 10^8)(2 \times 10^{-4})$

83. $(4.1 \times 10^2)(3 \times 10^{-4})$ **84.** $(1.2 \times 10^3)(2 \times 10^{-5})$

85. $\frac{12 \times 10^6}{4 \times 10^2}$ **86.** $\frac{20 \times 10^{20}}{10 \times 10^{15}}$

87. $\frac{15 \times 10^4}{5 \times 10^{-2}}$ **88.** $\frac{18 \times 10^2}{9 \times 10^{-3}}$

89. $\frac{6 \times 10^3}{2 \times 10^5}$ **90.** $\frac{8 \times 10^4}{2 \times 10^7}$

91. $\frac{6.3 \times 10^{-6}}{3 \times 10^{-3}}$ **92.** $\frac{9.6 \times 10^{-7}}{3 \times 10^{-3}}$

In Exercises 93–102, perform the indicated operation by first expressing each number in scientific notation. Write the answer in scientific notation.

93. (82,000,000)(3,000,000,000)

94. (94,000,000)(6,000,000,000)

95. (0.0005)(6,000,000) **96.** (0.000015)(0.004)

97. $\frac{9{,}500{,}000}{500}$ **98.** $\frac{30{,}000}{0.0005}$

99. $\frac{0.00008}{200}$ **100.** $\frac{0.0018}{0.0000006}$

101. $\frac{480{,}000{,}000{,}000}{0.00012}$ **102.** $\frac{0.000000096}{16{,}000}$

Practice Plus

In Exercises 103–106, perform the indicated operations. Express each answer as a fraction reduced to its lowest terms.

103. $\frac{2^4}{2^5} + \frac{3^3}{3^5}$ **104.** $\frac{3^5}{3^6} + \frac{2^3}{2^6}$

105. $\frac{2^6}{2^4} - \frac{5^4}{5^6}$ **106.** $\frac{5^6}{5^4} - \frac{2^4}{2^6}$

In Exercises 107–110, perform the indicated computations. Express answers in scientific notation.

107. $(5 \times 10^3)(1.2 \times 10^{-4}) \div (2.4 \times 10^2)$

108. $(2 \times 10^2)(2.6 \times 10^{-3}) \div (4 \times 10^3)$

109. $\frac{(1.6 \times 10^4)(7.2 \times 10^{-3})}{(3.6 \times 10^8)(4 \times 10^{-3})}$

110. $\frac{(1.2 \times 10^6)(8.7 \times 10^{-2})}{(2.9 \times 10^6)(3 \times 10^{-3})}$

Application Exercises

The bar graph shows the total amount Americans paid in federal taxes, in trillions of dollars, and the U.S. population, in millions, from 2007 through 2010. Exercises 111–112 are based on the numbers displayed by the graph.

Federal Taxes and the United States Population

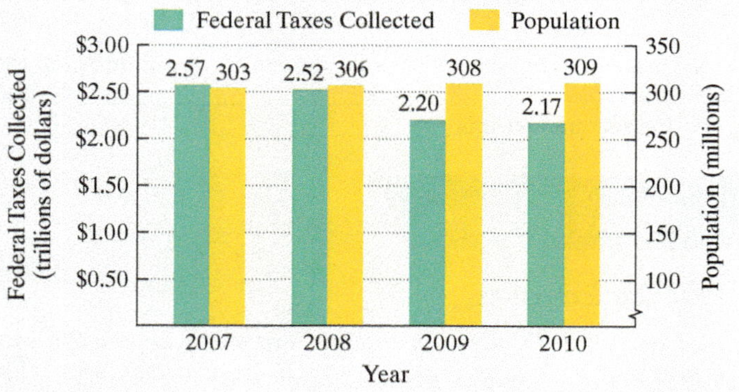

Sources: Internal Revenue Service and U.S. Census Bureau

111. **a.** In 2010, the United States government collected \$2.17 trillion in taxes. Express this number in scientific notation.

b. In 2010, the population of the United States was approximately 309 million. Express this number in scientific notation.

c. Use your scientific notation answers from parts (a) and (b) to answer this question: If the total 2010 tax collections were evenly divided among all Americans, how much would each citizen pay? Express the answer in scientific and decimal notations.

112. **a.** In 2009, the United States government collected \$2.20 trillion in taxes. Express this number in scientific notation.

b. In 2009, the population of the United States was approximately 308 million. Express this number in scientific notation.

c. Use your scientific notation answers from parts (a) and (b) to answer this question: If the total 2009 tax collections were evenly divided among all Americans, how much would each citizen pay? Express the answer in scientific and decimal notations.

The bar graph quantifies our love for movies by showing the number of tickets sold, in millions, and the average price per ticket for five selected years. Exercises 113–114 are based on the numbers displayed by the graph. Do not round the answers.

United States Film Admissions and Admission Charges

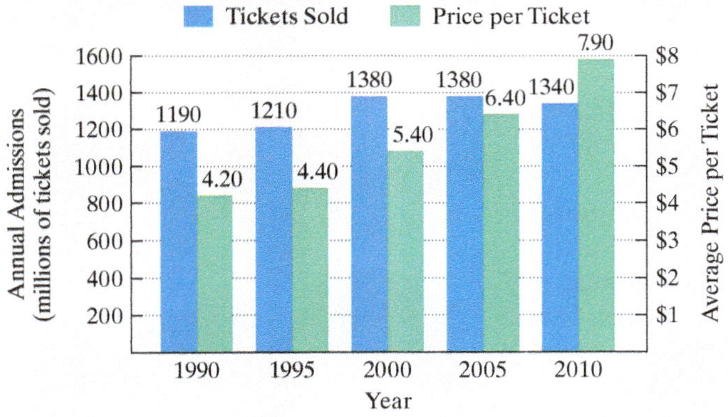

Source: Motion Picture Association of America

113. Use scientific notation to compute the amount of money that the motion picture industry made from box-office receipts in 2010. Express the answer in scientific notation.

114. Use scientific notation to compute the amount of money that the motion picture industry made from box-office receipts in 2005. Express the answer in scientific notation.

115. The mass of one oxygen molecule is 5.3×10^{-23} gram. Find the mass of 20,000 molecules of oxygen. Express the answer in scientific notation.

116. The mass of one hydrogen atom is 1.67×10^{-24} gram. Find the mass of 80,000 hydrogen atoms. Express the answer in scientific notation.

117. There are approximately 3.2×10^7 seconds in a year. According to the United States Department of Agriculture, Americans consume 127 chickens per second. How many chickens are eaten per year in the United States? Express the answer in scientific notation.

118. Convert 365 days (one year) to hours, to minutes, and, finally, to seconds, to determine how many seconds there are in a year. Express the answer in scientific notation.

Writing in Mathematics

119. Explain the product rule for exponents. Use $2^3 \cdot 2^5$ in your explanation.

120. Explain the power rule for exponents. Use $(3^2)^4$ in your explanation.

121. Explain the quotient rule for exponents. Use $\frac{5^8}{5^2}$ in your explanation.

122. Explain the zero exponent rule and give an example.

123. Explain the negative exponent rule and give an example.

124. How do you know if a number is written in scientific notation?

125. Explain how to convert from scientific to decimal notation and give an example.

126. Explain how to convert from decimal to scientific notation and give an example.

127. Suppose you are looking at a number in scientific notation. Describe the size of the number you are looking at if the exponent on ten is **a.** positive, **b.** negative, **c.** zero.

128. Describe one advantage of expressing a number in scientific notation over decimal notation.

Critical Thinking Exercises

Make Sense? *In Exercises 129–132, determine whether each statement makes sense or does not make sense, and explain your reasoning.*

129. If 5^{-2} is raised to the third power, the result is a number between 0 and 1.

130. The expression $\frac{a^n}{b^0}$ is undefined because division by 0 is undefined.

131. For a recent year, total tax collections in the United States were $\$2.02 \times 10^7$.

132. I just finished reading a book that contained approximately 1.04×10^5 words.

In Exercises 133–140, determine whether each statement is true or false. If the statement is false, make the necessary change(s) to produce a true statement.

133. $4^{-2} < 4^{-3}$

134. $5^{-2} > 2^{-5}$

135. $(-2)^4 = 2^{-4}$

136. $5^2 \cdot 5^{-2} > 2^5 \cdot 2^{-5}$

137. $534.7 = 5.347 \times 10^3$

138. $\frac{8 \times 10^{30}}{4 \times 10^{-5}} = 2 \times 10^{25}$

139. $(7 \times 10^5) + (2 \times 10^{-3}) = 9 \times 10^2$

140. $(4 \times 10^3) + (3 \times 10^2) = 43 \times 10^2$

141. Give an example of a number for which there is no advantage to using scientific notation instead of decimal notation. Explain why this is the case.

142. The mad Dr. Frankenstein has gathered enough bits and pieces (so to speak) for $2^{-1} + 2^{-2}$ of his creature-to-be. Write a fraction that represents the amount of his creature that must still be obtained.

Technology Exercises

143. Use a calculator in a fraction mode to check your answers in Exercises 19–24.

144. Use a calculator to check any three of your answers in Exercises 43–58.

145. Use a calculator to check any three of your answers in Exercises 59–78.

146. Use a calculator with an EE or EXP key to check any four of your computations in Exercises 79–102. Display the result of the computation in scientific notation and in decimal notation.

Group Exercises

147. Putting Numbers into Perspective. A large number can be put into perspective by comparing it with another number. For example, we put the \$15.2 trillion national debt (Example 9) and the \$2.17 trillion the government collected in taxes (Exercise 111) into perspective by comparing these numbers to the number of U.S. citizens.

For this project, each group member should consult an almanac, a newspaper, or the Internet to find a number greater than one million. Explain to other members of the group the context in which the large number is used. Express the number in scientific notation. Then put the number into perspective by comparing it with another number.

148. Refer to the Blitzer Bonus on page 319. Group members should use scientific notation to verify any three of the bulleted items on ways to spend \$1 trillion.

5.7 Arithmetic and Geometric Sequences

WHAT AM I SUPPOSED TO LEARN?

After you have read this section, you should be able to:

1. Write terms of an arithmetic sequence.
2. Use the formula for the general term of an arithmetic sequence.
3. Write terms of a geometric sequence.
4. Use the formula for the general term of a geometric sequence.

Sequences

Many creations in nature involve intricate mathematical designs, including a variety of spirals. For example, the arrangement of the individual florets in the head of a sunflower forms spirals. In some species, there are 21 spirals in the clockwise direction and 34 in the counterclockwise direction. The precise numbers depend on the species of sunflower: 21 and 34, or 34 and 55, or 55 and 89, or even 89 and 144.

This observation becomes even more interesting when we consider a sequence of numbers investigated by Leonardo of Pisa, also known as Fibonacci, an Italian mathematician of the thirteenth century. The **Fibonacci sequence** of numbers is an infinite sequence that begins as follows:

$$1, 1, 2, 3, 5, 8, 13, 21, 34, 55, 89, 144, 233, \ldots.$$

The first two terms are 1. Every term thereafter is the sum of the two preceding terms. For example, the third term, 2, is the sum of the first and second terms: $1 + 1 = 2$. The fourth term, 3, is the sum of the second and third terms: $1 + 2 = 3$, and so on. Did you know that the number of spirals in a daisy or a sunflower, 21 and 34, are two Fibonacci numbers? The number of spirals in a pinecone, 8 and 13, and a pineapple, 8 and 13, are also Fibonacci numbers.

We can think of a **sequence** as a list of numbers that are related to each other by a rule. The numbers in a sequence are called its **terms**. The letter a with a subscript is used to represent the terms of a sequence. Thus, a_1 represents the first term of the sequence, a_2 represents the second term, a_3 the third term, and so on. This notation is shown for the first six terms of the Fibonacci sequence:

1,	1,	2,	3,	5,	8.
$a_1 = 1$	$a_2 = 1$	$a_3 = 2$	$a_4 = 3$	$a_5 = 5$	$a_6 = 8$

Blitzer Bonus

Fibonacci Numbers on the Piano Keyboard

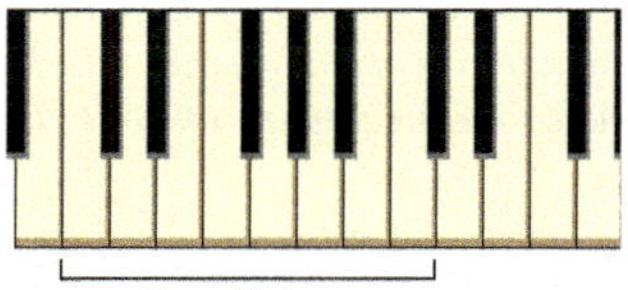

One Octave

Numbers in the Fibonacci sequence can be found in an octave on the piano keyboard. The octave contains 2 black keys in one cluster, 3 black keys in another cluster, a total of 5 black keys, 8 white keys, and a total of 13 keys altogether. The numbers 2, 3, 5, 8, and 13 are the third through seventh terms of the Fibonacci sequence.

Arithmetic Sequences

The bar graph in **Figure 5.15** is based on a mathematical model that shows how much Americans spent on their pets, to the nearest billion dollars, each year from 2001 through 2010.

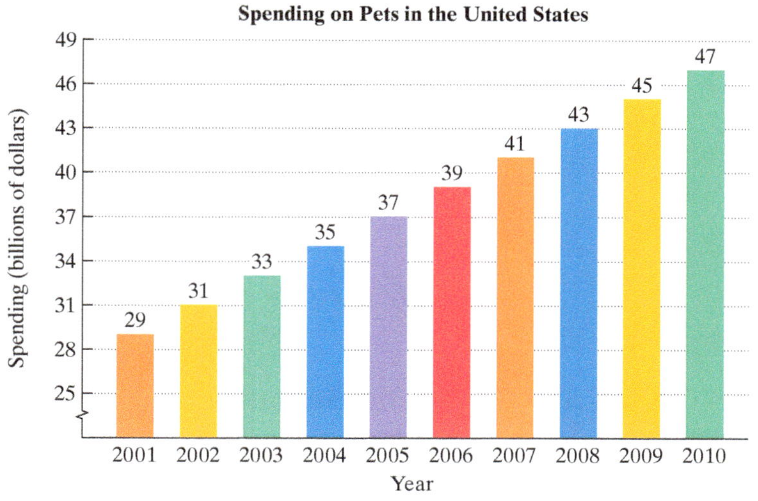

FIGURE 5.15
Source: American Pet Products Manufacturers Association

The graph illustrates that each year spending increased by $2 billion. The sequence of annual spending

$$29, 31, 33, 35, 37, 39, 41, \ldots$$

shows that each term after the first, 29, differs from the preceding term by a constant amount, namely 2. This sequence is an example of an *arithmetic sequence.*

DEFINITION OF AN ARITHMETIC SEQUENCE

An **arithmetic sequence** is a sequence in which each term after the first differs from the preceding term by a constant amount. The difference between consecutive terms is called the **common difference** of the sequence.

The common difference, d, is found by subtracting any term from the term that directly follows it. In the following examples, the common difference is found by subtracting the first term from the second term: $a_2 - a_1$.

Arithmetic Sequence	Common Difference
$29, 31, 33, 35, 37, \ldots$	$d = 31 - 29 = 2$
$-5, -2, 1, 4, 7, \ldots$	$d = -2 - (-5) = -2 + 5 = 3$
$8, 3, -2, -7, -12, \ldots$	$d = 3 - 8 = -5$

If the first term of an arithmetic sequence is a_1, each term after the first is obtained by adding d, the common difference, to the previous term.

1 Write terms of an arithmetic sequence.

EXAMPLE 1 *Writing the Terms of an Arithmetic Sequence*

Write the first six terms of the arithmetic sequence with first term 6 and common difference 4.

SOLUTION

The first term is 6. The second term is $6 + 4$, or 10. The third term is $10 + 4$, or 14, and so on. The first six terms are

$$6, 10, 14, 18, 22, \text{ and } 26.$$

CHECK POINT 1 Write the first six terms of the arithmetic sequence with first term 100 and common difference 20.

EXAMPLE 2 *Writing the Terms of an Arithmetic Sequence*

Write the first six terms of the arithmetic sequence with $a_1 = 5$ and $d = -2$.

SOLUTION

The first term, a_1, is 5. The common difference, d, is -2. To find the second term, we add -2 to 5, giving 3. For the next term, we add -2 to 3, and so on. The first six terms are

$$5, 3, 1, -1, -3, \text{ and } -5.$$

CHECK POINT 2 Write the first six terms of the arithmetic sequence with $a_1 = 8$ and $d = -3$.

2 Use the formula for the general term of an arithmetic sequence.

The General Term of an Arithmetic Sequence

Consider an arithmetic sequence whose first term is a_1 and whose common difference is d. We are looking for a formula for the general term, a_n. Let's begin by writing the first six terms. The first term is a_1. The second term is $a_1 + d$. The third term is $a_1 + d + d$, or $a_1 + 2d$. Thus, we start with a_1 and add d to each successive term. The first six terms are

$$a_1, \quad a_1 + d, \quad a_1 + 2d, \quad a_1 + 3d, \quad a_1 + 4d, \quad a_1 + 5d.$$

a_1, first term; a_2, second term; a_3, third term; a_4, fourth term; a_5, fifth term; a_6, sixth term

Applying inductive reasoning to the pattern of the terms results in the following formula for the general term, or the nth term, of an arithmetic sequence:

GENERAL TERM OF AN ARITHMETIC SEQUENCE

The nth term (the general term) of an arithmetic sequence with first term a_1 and common difference d is

$$a_n = a_1 + (n - 1)d.$$

EXAMPLE 3 *Using the Formula for the General Term of an Arithmetic Sequence*

Find the eighth term of the arithmetic sequence whose first term is 4 and whose common difference is -7.

SOLUTION

To find the eighth term, a_8, we replace n in the formula with 8, a_1 with 4, and d with -7.

$$a_n = a_1 + (n - 1)d$$
$$a_8 = 4 + (8 - 1)(-7) = 4 + 7(-7) = 4 + (-49) = -45$$

The eighth term is -45. We can check this result by writing the first eight terms of the sequence:

$$4, -3, -10, -17, -24, -31, -38, -45.$$

CHECK POINT 3 Find the ninth term of the arithmetic sequence whose first term is 6 and whose common difference is -5.

In Chapter 1, we saw that the process of finding formulas to describe real-world phenomena is called mathematical modeling. Such formulas, together with the meaning assigned to the variables, are called mathematical models. Example 4 illustrates how the formula for the general term of an arithmetic sequence can be used to develop a mathematical model.

EXAMPLE 4 *Using an Arithmetic Sequence to Model Changes in the U.S. Population*

The graph in **Figure 5.16** shows the percentage of the U.S. population by race/ethnicity for 2010, with projections by the U.S. Census Bureau for 2050.

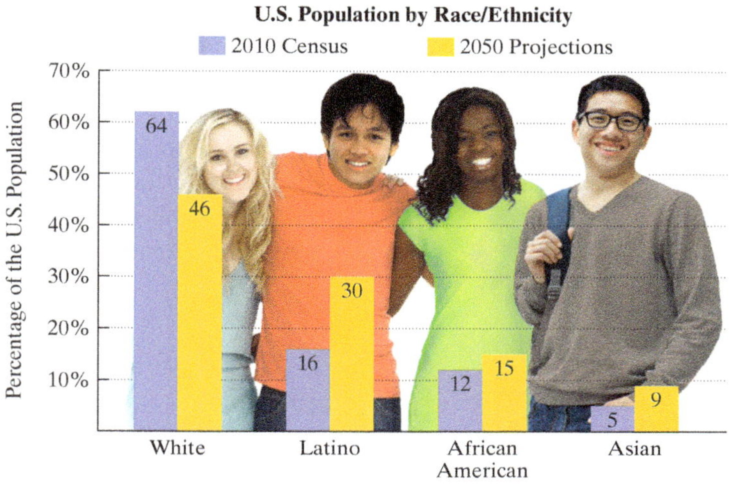

FIGURE 5.16
Source: U.S. Census Bureau

The data show that in 2010, 64% of the U.S. population was white. On average, this is projected to decrease by approximately 0.45% per year.

a. Write a formula for the nth term of the arithmetic sequence that describes the percentage of the U.S. population that will be white n years after 2009.

b. What percentage of the U.S. population is projected to be white in 2030?

SOLUTION

a. With a yearly decrease of 0.45%, we can express the percentage of the white population by the following arithmetic sequence:

$$64, \quad 64 - 0.45 = 63.55, \quad 63.55 - 0.45 = 63.10, \quad \ldots .$$

a_1: percentage of whites in the population in 2010, 1 year after 2009

a_2: percentage of whites in the population in 2011, 2 years after 2009

a_3: percentage of whites in the population in 2012, 3 years after 2009

In this sequence, 64, 63.55, 63.10, . . . , the first term, a_1, represents the percentage of the population that was white in 2010. Each subsequent year this amount decreases by 0.45%, so $d = -0.45$. We use the formula for the general term of an arithmetic sequence to write the nth term of the sequence that describes the percentage of whites in the population n years after 2009.

$a_n = a_1 + (n-1)d$ — This is the formula for the general term of an arithmetic sequence.

$a_n = 64 + (n-1)(-0.45)$ — $a_1 = 64$ and $d = -0.45$.

$a_n = 64 - 0.45n + 0.45$ — Distribute -0.45 to each term in parentheses.

$a_n = -0.45n + 64.45$ — Simplify.

Thus, the percentage of the U.S. population that will be white n years after 2009 can be described by

$$a_n = -0.45n + 64.45.$$

b. Now we need to project the percentage of the population that will be white in 2030. The year 2030 is 21 years after 2009. Thus, $n = 21$. We substitute 21 for n in $a_n = -0.45n + 64.45$.

$$a_{21} = -0.45(21) + 64.45 = 55$$

The 21st term of the sequence is 55. Thus, 55% of the U.S. population is projected to be white in 2030.

CHECK POINT 4 The data in **Figure 5.16** on the previous page show that in 2010, 16% of the U.S. population was Latino. On average, this is projected to increase by approximately 0.35% per year.

a. Write a formula for the nth term of the arithmetic sequence that describes the percentage of the U.S. population that will be Latino n years after 2009.

b. What percentage of the U.S. population is projected to be Latino in 2030?

Geometric Sequences

Figure 5.17 shows a sequence in which the number of squares is increasing. From left to right, the number of squares is 1, 5, 25, 125, and 625. In this sequence, each term after the first, 1, is obtained by multiplying the preceding term by a constant amount, namely 5. This sequence of increasing numbers of squares is an example of a *geometric sequence.*

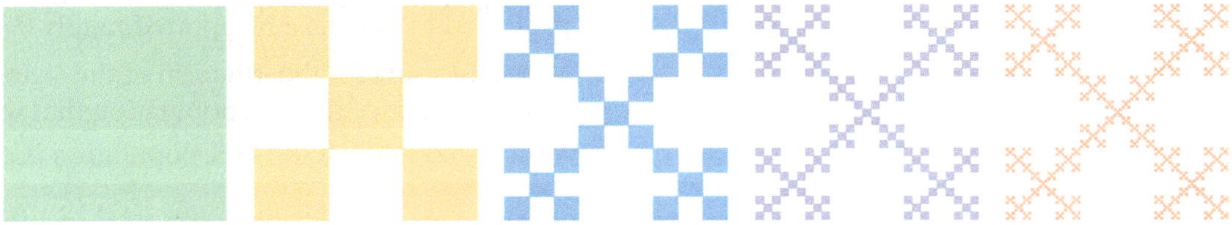

FIGURE 5.17 A geometric sequence of squares

DEFINITION OF A GEOMETRIC SEQUENCE

A **geometric sequence** is a sequence in which each term after the first is obtained by multiplying the preceding term by a fixed nonzero constant. The amount by which we multiply each time is called the **common ratio** of the sequence.

The common ratio, r, is found by dividing any term after the first term by the term that directly precedes it. In the examples below, the common ratio is found by dividing the second term by the first term: $\frac{a_2}{a_1}$.

Geometric sequence	Common ratio
1, 5, 25, 125, 625, . . .	$r = \frac{5}{1} = 5$
4, 8, 16, 32, 64, . . .	$r = \frac{8}{4} = 2$
6, −12, 24, −48, 96, . . .	$r = \frac{-12}{6} = -2$
9, −3, 1, $-\frac{1}{3}$, $\frac{1}{9}$, . . .	$r = \frac{-3}{9} = -\frac{1}{3}$

GREAT QUESTION!

What happens to the terms of a geometric sequence when the common ratio is negative?

When the common ratio of a geometric sequence is negative, the signs of the terms alternate.

3 Write terms of a geometric sequence.

How do we write out the terms of a geometric sequence when the first term and the common ratio are known? We multiply the first term by the common ratio to get the second term, multiply the second term by the common ratio to get the third term, and so on.

EXAMPLE 5 *Writing the Terms of a Geometric Sequence*

Write the first six terms of the geometric sequence with first term 6 and common ratio $\frac{1}{3}$.

SOLUTION

The first term is 6. The second term is $6 \cdot \frac{1}{3}$, or 2. The third term is $2 \cdot \frac{1}{3}$, or $\frac{2}{3}$. The fourth term is $\frac{2}{3} \cdot \frac{1}{3}$, or $\frac{2}{9}$, and so on. The first six terms are

$$6, 2, \frac{2}{3}, \frac{2}{9}, \frac{2}{27}, \text{ and } \frac{2}{81}.$$

CHECK POINT 5 Write the first six terms of the geometric sequence with first term 12 and common ratio $-\frac{1}{2}$.

4 Use the formula for the general term of a geometric sequence.

The General Term of a Geometric Sequence

Consider a geometric sequence whose first term is a_1 and whose common ratio is r. We are looking for a formula for the general term, a_n. Let's begin by writing the first six terms. The first term is a_1. The second term is $a_1 r$. The third term is $a_1 r \cdot r$, or $a_1 r^2$. The fourth term is $a_1 r^2 \cdot r$, or $a_1 r^3$, and so on. Starting with a_1 and multiplying each successive term by r, the first six terms are

$$a_1, \quad a_1 r, \quad a_1 r^2, \quad a_1 r^3, \quad a_1 r^4, \quad a_1 r^5.$$

a_1, first term; a_2, second term; a_3, third term; a_4, fourth term; a_5, fifth term; a_6, sixth term

Applying inductive reasoning to the pattern of the terms results in the following formula for the general term, or the nth term, of a geometric sequence:

GENERAL TERM OF A GEOMETRIC SEQUENCE

The nth term (the general term) of a geometric sequence with first term a_1 and common ratio r is

$$a_n = a_1 r^{n-1}.$$

EXAMPLE 6 *Using the Formula for the General Term of a Geometric Sequence*

Find the eighth term of the geometric sequence whose first term is -4 and whose common ratio is -2.

SOLUTION

To find the eighth term, a_8, we replace n in the formula with 8, a_1 with -4, and r with -2.

$$a_n = a_1 r^{n-1}$$
$$a_8 = -4(-2)^{8-1} = -4(-2)^7 = -4(-128) = 512$$

The eighth term is 512. We can check this result by writing the first eight terms of the sequence: $-4, 8, -16, 32, -64, 128, -256, 512$.

GREAT QUESTION!

When using $a_1 r^{n-1}$ to find the nth term of a geometric sequence, what should I do first?

Be careful with the order of operations when evaluating

$$a_1 r^{n-1}.$$

First, subtract 1 in the exponent and then raise r to that power. Finally, multiply the result by a_1.

CHECK POINT 6 Find the seventh term of the geometric sequence whose first term is 5 and whose common ratio is -3.

EXAMPLE 7 *Geometric Population Growth*

The table shows the population of the United States in 2000 and 2010, with estimates given by the Census Bureau for 2001 through 2009.

Year	**2000**	**2001**	**2002**	**2003**	**2004**	**2005**	**2006**	**2007**	**2008**	**2009**	**2010**
Population (millions)	281.4	284.0	286.6	289.3	292.0	294.7	297.4	300.2	303.0	305.8	308.7

a. Show that the population is increasing geometrically.

b. Write the general term for the geometric sequence modeling the population of the United States, in millions, n years after 1999.

c. Project the U.S. population, in millions, for the year 2020.

SOLUTION

a. First, we use the sequence of population growth, 281.4, 284.0, 286.6, 289.3, and so on, to divide the population for each year by the population in the preceding year.

$$\frac{284.0}{281.4} \approx 1.009, \quad \frac{286.6}{284.0} \approx 1.009, \quad \frac{289.3}{286.6} \approx 1.009$$

Continuing in this manner, we will keep getting approximately 1.009. This means that the population is increasing geometrically with $r \approx 1.009$. The population of the United States in any year shown in the sequence is approximately 1.009 times the population the year before.

b. The sequence of the U.S. population growth is

$$281.4, 284.0, 286.6, 289.3, 292.0, 294.7, \ldots.$$

Because the population is increasing geometrically, we can find the general term of this sequence using

$$a_n = a_1 r^{n-1}.$$

In this sequence, $a_1 = 281.4$ and [from part (a)] $r \approx 1.009$. We substitute these values into the formula for the general term. This gives the general term for the geometric sequence modeling the U.S. population, in millions, n years after 1999.

$$a_n = 281.4(1.009)^{n-1}$$

c. We can use the formula for the general term, a_n, in part (b) to project the U.S. population for the year 2020. The year 2020 is 21 years after 1999—that is, $2020 - 1999 = 21$. Thus, $n = 21$. We substitute 21 for n in $a_n = 281.4(1.009)^{n-1}$.

$$a_{21} = 281.4(1.009)^{21-1} = 281.4(1.009)^{20} \approx 336.6$$

The model projects that the United States will have a population of approximately 336.6 million in the year 2020.

CHECK POINT 7 Write the general term for the geometric sequence

$$3, 6, 12, 24, 48, \ldots.$$

Then use the formula for the general term to find the eighth term.

Blitzer Bonus

Geometric Population Growth

Economist Thomas Malthus (1766–1834) predicted that population growth would increase as a geometric sequence and food production would increase as an arithmetic sequence. He concluded that eventually population would exceed food production. If two sequences, one geometric and one arithmetic, are increasing, the geometric sequence will eventually overtake the arithmetic sequence, regardless of any head start that the arithmetic sequence might initially have.

Concept and Vocabulary Check

Fill in each blank so that the resulting statement is true.

1. A sequence in which each term after the first differs from the preceding term by a constant amount is called a/an ________ sequence. The difference between consecutive terms is called the ______________ of the sequence.
2. The nth term of the sequence described in Exercise 1 is given by the formula ________________, where a_1 is _________ and d is ________________.
3. A sequence in which each term after the first is obtained by multiplying the preceding term by a fixed nonzero number is called a/an ________ sequence. The amount by which we multiply each time is called the ___________ of the sequence.
4. The nth term of the sequence described in Exercise 3 is given by the formula _________, where a_1 is _________ and r is ______________.

Exercise Set 5.7

Practice Exercises

In Exercises 1–20, write the first six terms of the arithmetic sequence with the first term, a_1, and common difference, d.

1. $a_1 = 8, d = 2$
2. $a_1 = 5, d = 3$
3. $a_1 = 200, d = 20$
4. $a_1 = 300, d = 50$
5. $a_1 = -7, d = 4$
6. $a_1 = -8, d = 5$
7. $a_1 = -400, d = 300$
8. $a_1 = -500, d = 400$
9. $a_1 = 7, d = -3$
10. $a_1 = 9, d = -5$
11. $a_1 = 200, d = -60$
12. $a_1 = 300, d = -90$
13. $a_1 = \frac{5}{2}, d = \frac{1}{2}$
14. $a_1 = \frac{3}{4}, d = \frac{1}{4}$
15. $a_1 = \frac{3}{2}, d = \frac{1}{4}$
16. $a_1 = \frac{3}{2}, d = -\frac{1}{4}$
17. $a_1 = 4.25, d = 0.3$
18. $a_1 = 6.3, d = 0.25$
19. $a_1 = 4.5, d = -0.75$
20. $a_1 = 3.5, d = -1.75$

In Exercises 21–40, find the indicated term for the arithmetic sequence with first term, a_1, and common difference, d.

21. Find a_6, when $a_1 = 13, d = 4$.
22. Find a_{16}, when $a_1 = 9, d = 2$.
23. Find a_{50}, when $a_1 = 7, d = 5$.
24. Find a_{60}, when $a_1 = 8, d = 6$.
25. Find a_9, when $a_1 = -5, d = 9$.
26. Find a_{10}, when $a_1 = -8, d = 10$.
27. Find a_{200}, when $a_1 = -40, d = 5$.
28. Find a_{150}, when $a_1 = -60, d = 5$.
29. Find a_{10}, when $a_1 = 8, d = -10$.
30. Find a_{11}, when $a_1 = 10, d = -6$.
31. Find a_{60}, when $a_1 = 35, d = -3$.
32. Find a_{70}, when $a_1 = -32, d = 4$.
33. Find a_{12}, when $a_1 = 12, d = -5$.
34. Find a_{20}, when $a_1 = -20, d = -4$.
35. Find a_{90}, when $a_1 = -70, d = -2$.
36. Find a_{80}, when $a_1 = 106, d = -12$.
37. Find a_{12}, when $a_1 = 6, d = \frac{1}{2}$.
38. Find a_{14}, when $a_1 = 8, d = \frac{1}{4}$.
39. Find a_{50}, when $a_1 = 14, d = -0.25$.
40. Find a_{110}, when $a_1 = -12, d = -0.5$.

In Exercises 41–48, write a formula for the general term (the nth term) of each arithmetic sequence. Then use the formula for a_n to find a_{20}, the 20th term of the sequence.

41. 1, 5, 9, 13, . . .
42. 2, 7, 12, 17, . . .
43. 7, 3, −1, −5, . . .
44. 6, 1, −4, −9, . . .
45. $a_1 = 9, d = 2$
46. $a_1 = 6, d = 3$
47. $a_1 = -20, d = -4$
48. $a_1 = -70, d = -5$

In Exercises 49–70, write the first six terms of the geometric sequence with the first term, a_1, and common ratio, r.

49. $a_1 = 4, r = 2$
50. $a_1 = 2, r = 3$
51. $a_1 = 1000, r = 1$
52. $a_1 = 5000, r = 1$
53. $a_1 = 3, r = -2$
54. $a_1 = 2, r = -3$
55. $a_1 = 10, r = -4$
56. $a_1 = 20, r = -4$
57. $a_1 = 2000, r = -1$
58. $a_1 = 3000, r = -1$
59. $a_1 = -2, r = -3$
60. $a_1 = -4, r = -2$
61. $a_1 = -6, r = -5$
62. $a_1 = -8, r = -5$
63. $a_1 = \frac{1}{4}, r = 2$
64. $a_1 = \frac{1}{2}, r = 2$
65. $a_1 = \frac{1}{4}, r = \frac{1}{2}$
66. $a_1 = \frac{1}{5}, r = \frac{1}{2}$
67. $a_1 = -\frac{1}{16}, r = -4$
68. $a_1 = -\frac{1}{8}, r = -2$
69. $a_1 = 2, r = 0.1$
70. $a_1 = -1000, r = 0.1$

In Exercises 71–90, find the indicated term for the geometric sequence with first term, a_1, and common ratio, r.

71. Find a_7, when $a_1 = 4, r = 2$.
72. Find a_5, when $a_1 = 4, r = 3$.
73. Find a_{20}, when $a_1 = 2, r = 3$.
74. Find a_{20}, when $a_1 = 2, r = 2$.
75. Find a_{100}, when $a_1 = 50, r = 1$.
76. Find a_{200}, when $a_1 = 60, r = 1$.
77. Find a_7, when $a_1 = 5, r = -2$.
78. Find a_4, when $a_1 = 4, r = -3$.
79. Find a_{30}, when $a_1 = 2, r = -1$.
80. Find a_{40}, when $a_1 = 6, r = -1$.
81. Find a_6, when $a_1 = -2, r = -3$.
82. Find a_5, when $a_1 = -5, r = -2$.
83. Find a_8, when $a_1 = 6, r = \frac{1}{2}$.
84. Find a_8, when $a_1 = 12, r = \frac{1}{2}$.

85. Find a_6, when $a_1 = 18$, $r = -\frac{1}{3}$.

86. Find a_4, when $a_1 = 9$, $r = -\frac{1}{3}$.

87. Find a_{40}, when $a_1 = 1000$, $r = -\frac{1}{2}$.

88. Find a_{30}, when $a_1 = 8000$, $r = -\frac{1}{2}$.

89. Find a_8, when $a_1 = 1{,}000{,}000$, $r = 0.1$.

90. Find a_8, when $a_1 = 40{,}000$, $r = 0.1$.

In Exercises 91–98, write a formula for the general term (the nth term) of each geometric sequence. Then use the formula for a_n to find a_7, the seventh term of the sequence.

91. 3, 12, 48, 192, . . .

92. 3, 15, 75, 375, . . .

93. $18, 6, 2, \frac{2}{3}, \ldots$

94. $12, 6, 3, \frac{3}{2}, \ldots$

95. 1.5, −3, 6, −12, . . .

96. $5, -1, \frac{1}{5}, -\frac{1}{25}, \ldots$

97. 0.0004, −0.004, 0.04, −0.4, . . .

98. 0.0007, −0.007, 0.07, −0.7, . . .

Determine whether each sequence in Exercises 99–114 is arithmetic or geometric. Then find the next two terms.

99. 2, 6, 10, 14, . . .

100. 3, 8, 13, 18, . . .

101. 5, 15, 45, 135, . . .

102. 15, 30, 60, 120, . . .

103. −7, −2, 3, 8, . . .

104. −9, −5, −1, 3, . . .

105. $3, \frac{3}{2}, \frac{3}{4}, \frac{3}{8}, \ldots$

106. $6, 3, \frac{3}{2}, \frac{3}{4}, \ldots$

107. $\frac{1}{2}, 1, \frac{3}{2}, 2, \ldots$

108. $\frac{2}{3}, 1, \frac{4}{3}, \frac{5}{3}, \ldots$

109. 7, −7, 7, −7, . . .

110. 6, −6, 6, −6, . . .

111. 7, −7, −21, −35, . . .

112. 6, −6, −18, −30, . . .

113. $\sqrt{5}, 5, 5\sqrt{5}, 25, \ldots$

114. $\sqrt{3}, 3, 3\sqrt{3}, 9, \ldots$

Practice Plus

The sum, S_n, of the first n terms of an arithmetic sequence is given by

$$S_n = \frac{n}{2}(a_1 + a_n),$$

in which a_1 is the first term and a_n is the nth term. The sum, S_n, of the first n terms of a geometric sequence is given by

$$S_n = \frac{a_1(1 - r^n)}{1 - r},$$

in which a_1 is the first term and r is the common ratio ($r \neq 1$). In Exercises 115–122, determine whether each sequence is arithmetic or geometric. Then use the appropriate formula to find S_{10}, the sum of the first ten terms.

115. 4, 10, 16, 22, . . .

116. 7, 19, 31, 43, . . .

117. 2, 6, 18, 54, . . .

118. 3, 6, 12, 24, . . .

119. 3, −6, 12, −24, . . .

120. 4, −12, 36, −108, . . .

121. −10, −6, −2, 2, . . .

122. −15, −9, −3, 3, . . .

123. Use the appropriate formula shown above to find $1 + 2 + 3 + 4 + \cdots + 100$, the sum of the first 100 natural numbers.

124. Use the appropriate formula shown above to find $2 + 4 + 6 + 8 + \cdots + 200$, the sum of the first 100 positive even integers.

Application Exercises

The bar graph shows changes in the percentage of college graduates for Americans ages 25 and older from 1990 to 2010. Exercises 125–126 involve developing arithmetic sequences that model the data.

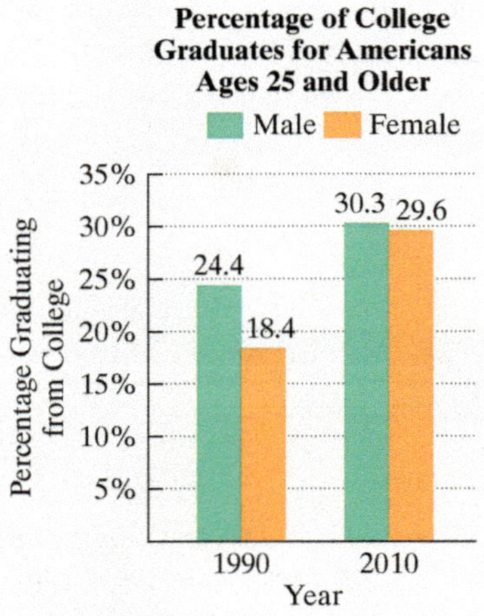

Source: U.S. Census Bureau

125. In 1990, 18.4% of American women ages 25 and older had graduated from college. On average, this percentage has increased by approximately 0.6 each year.

a. Write a formula for the *n*th term of the arithmetic sequence that models the percentage of American women ages 25 and older who had graduated from college *n* years after 1989.

b. Use the model from part (a) to project the percentage of American women ages 25 and older who will be college graduates by 2019.

126. In 1990, 24.4% of American men ages 25 and older had graduated from college. On average, this percentage has increased by approximately 0.3 each year.

a. Write a formula for the *n*th term of the arithmetic sequence that models the percentage of American men ages 25 and older who had graduated from college *n* years after 1989.

b. Use the model from part (a) to project the percentage of American men ages 25 and older who will be college graduates by 2019.

127. Company A pays $24,000 yearly with raises of $1600 per year. Company B pays $28,000 yearly with raises of $1000 per year. Which company will pay more in year 10? How much more?

128. Company A pays $23,000 yearly with raises of $1200 per year. Company B pays $26,000 yearly with raises of $800 per year. Which company will pay more in year 10? How much more?

In Exercises 129–130, suppose you save $1 the first day of a month, $2 the second day, $4 the third day, and so on. That is, each day you save twice as much as you did the day before.

129. What will you put aside for savings on the fifteenth day of the month?

130. What will you put aside for savings on the thirtieth day of the month?

131. A professional baseball player signs a contract with a beginning salary of $3,000,000 for the first year with an annual increase of 4% per year beginning in the second year. That is, beginning in year 2, the athlete's salary will be 1.04 times what it was in the previous year. What is the athlete's salary for year 7 of the contract? Round to the nearest dollar.

132. You are offered a job that pays $30,000 for the first year with an annual increase of 5% per year beginning in the second year. That is, beginning in year 2, your salary will be 1.05 times what it was in the previous year. What can you expect to earn in your sixth year on the job? Round to the nearest dollar.

In Exercises 133–134, you will develop geometric sequences that model the population growth for California and Texas, the two most populated U.S. states.

133. The table shows the population of California for 2000 and 2010, with estimates given by the U.S. Census Bureau for 2001 through 2009.

Year	2000	2001	2002	2003	2004	2005
Population in millions	33.87	34.21	34.55	34.90	35.25	35.60

Year	2006	2007	2008	2009	2010
Population in millions	36.00	36.36	36.72	37.09	37.25

a. Divide the population for each year by the population in the preceding year. Round to two decimal places and show that California has a population increase that is approximately geometric.

b. Write the general term of the geometric sequence modeling California's population, in millions, n years after 1999.

c. Use your model from part (b) to project California's population, in millions, for the year 2020. Round to two decimal places.

134. The table shows the population of Texas for 2000 and 2010, with estimates given by the U.S. Census Bureau for 2001 through 2009.

Year	2000	2001	2002	2003	2004	2005
Population in millions	20.85	21.27	21.70	22.13	22.57	23.02

Year	2006	2007	2008	2009	2010
Population in millions	23.48	23.95	24.43	24.92	25.15

a. Divide the population for each year by the population in the preceding year. Round to two decimal places and show that Texas has a population increase that is approximately geometric.

b. Write the general term of the geometric sequence modeling Texas's population, in millions, n years after 1999.

c. Use your model from part (b) to project Texas's population, in millions, for the year 2020. Round to two decimal places.

Writing in Mathematics

135. What is a sequence? Give an example with your description.

136. What is an arithmetic sequence? Give an example with your description.

137. What is the common difference in an arithmetic sequence?

138. What is a geometric sequence? Give an example with your description.

139. What is the common ratio in a geometric sequence?

140. If you are given a sequence that is arithmetic or geometric, how can you determine which type of sequence it is?

141. For the first 30 days of a flu outbreak, the number of students on your campus who become ill is increasing. Which is worse: The number of students with the flu is increasing arithmetically or is increasing geometrically? Explain your answer.

Critical Thinking Exercises

Make Sense? *In Exercises 142–145, determine whether each statement makes sense or does not make sense, and explain your reasoning.*

142. Now that I've studied sequences, I realize that the joke in the accompanying cartoon is based on the fact that you can't have a negative number of sheep.

WHEN MATHEMATICIANS CAN'T SLEEP

143. The sequence for the number of seats per row in our movie theater as the rows move toward the back is arithmetic with $d = 1$ so people don't block the view of those in the row behind them.

144. There's no end to the number of geometric sequences that I can generate whose first term is 5 if I pick nonzero numbers r and multiply 5 by each value of r repeatedly.

145. I've noticed that the big difference between arithmetic and geometric sequences is that arithmetic sequences are based on addition and geometric sequences are based on multiplication.

In Exercises 146–153, determine whether each statement is true or false. If the statement is false, make the necessary change(s) to produce a true statement.

146. The common difference for the arithmetic sequence given by $1, -1, -3, -5, \ldots$ is 2.

147. The sequence $1, 4, 8, 13, 19, 26, \ldots$ is an arithmetic sequence.

148. The nth term of an arithmetic sequence whose first term is a_1 and whose common difference is d is $a_n = a_1 + nd$.

149. If the first term of an arithmetic sequence is 5 and the third term is -3, then the fourth term is -7.

150. The sequence 2, 6, 24, 120, . . . is an example of a geometric sequence.

151. Adjacent terms in a geometric sequence have a common difference.

152. A sequence that is not arithmetic must be geometric.

153. If a sequence is geometric, we can write as many terms as we want by repeatedly multiplying by the common ratio.

154. A person is investigating two employment opportunities. They both have a beginning salary of \$20,000 per year. Company A offers an increase of \$1000 per year. Company B offers 5% more than during the preceding year. Which company will pay more in the sixth year?

Group Exercise

155. Enough curiosities involving the Fibonacci sequence exist to warrant a flourishing Fibonacci Association. It publishes a quarterly journal. Do some research on the Fibonacci sequence by consulting the research department of your library or the Internet, and find one property that interests you. After doing this research, get together with your group to share these intriguing properties.

Chapter Summary, Review, and Test

SUMMARY – DEFINITIONS AND CONCEPTS | EXAMPLES

5.1 Number Theory: Prime and Composite Numbers

a. The set of natural numbers is $\{1, 2, 3, 4, 5, \ldots\}$. $b \mid a$ (b divides a: a is divisible by b) for natural numbers a and b if the operation of dividing a by b leaves a remainder of 0. Rules of divisibility are given in Table 5.1 on page 249.	Ex. 1, p. 250
b. A prime number is a natural number greater than 1 that has only itself and 1 as factors. A composite number is a natural number greater than 1 that is divisible by a number other than itself and 1. The Fundamental Theorem of Arithmetic: Every composite number can be expressed as a product of prime numbers in one and only one way (if the order of the factors is disregarded).	Ex. 2, p. 251
c. The greatest common divisor of two or more natural numbers is the largest number that is a divisor (or factor) of all the numbers. The procedure for finding the greatest common divisor is given in the box on page 252.	Ex. 3, p. 252; Ex. 4, p. 253
d. The least common multiple of two or more natural numbers is the smallest natural number that is divisible by all of the numbers. The procedure for finding the least common multiple is given in the box on page 254.	Ex. 5, p. 254; Ex. 6, p. 255

5.2 The Integers; Order of Operations

a. The set of whole numbers is $\{0, 1, 2, 3, 4, 5, \ldots\}$. The set of integers is $\{\ldots, -3, -2, -1, 0, 1, 2, 3, \ldots\}$. Integers are graphed on a number line by placing a dot at the correct location for each number.	Ex. 1, p. 259
b. $a < b$ (a is less than b) means a is to the left of b on a number line. $a > b$ (a is greater than b) means a is to the right of b on a number line.	Ex. 2, p. 260
c. $\lvert a \rvert$, the absolute value of a, is the distance of a from 0 on a number line. The absolute value of a positive number is the number itself. The absolute value of 0 is 0: $\lvert 0 \rvert = 0$. The absolute value of a negative number is the number without the negative sign. For example, $\lvert -8 \rvert = 8$.	Ex. 3, p. 261
d. Rules for performing operations with integers are given in the boxes on pages 262, 265, and 267.	Ex. 4, p. 263; Ex. 5, p. 264; Ex. 6, p. 265
e. Order of Operations **1.** Perform all operations within grouping symbols. **2.** Evaluate all exponential expressions. **3.** Do all multiplications and divisions from left to right. **4.** Do all additions and subtractions from left to right.	Ex. 7, p. 268; Ex. 8, p. 268

5.3 The Rational Numbers

a.	The set of rational numbers is the set of all numbers which can be expressed in the form $\frac{a}{b}$, where a and b are integers and b is not equal to 0.	
b.	A rational number is reduced to its lowest terms, or simplified, by dividing both the numerator and the denominator by their greatest common divisor.	Ex. 1, p. 273
c.	A mixed number consists of the sum of an integer and a rational number, expressed without the use of an addition sign. An improper fraction is a rational number whose numerator is greater than its denominator. Procedures for converting between these forms are given in the boxes on pages 273 and 274.	Ex. 2, p. 274; Ex. 3, p. 274
d.	Any rational number can be expressed as a decimal. The resulting decimal will either terminate (stop), or it will have a digit that repeats or a block of digits that repeats. The rational number $\frac{a}{b}$ is expressed as a decimal by dividing b into a.	Ex. 4, p. 275
e.	To express a terminating decimal as a quotient of integers, the digits to the right of the decimal point are the numerator. The place-value of the last digit to the right of the decimal point determines the denominator.	Ex. 5, p. 276
f.	To express a repeating decimal as a quotient of integers, use the boxed procedure on page 278.	Ex. 6, p. 277; Ex. 7, p. 278
g.	The product of two rational numbers is the product of their numerators divided by the product of their denominators.	Ex. 8, p. 279
h.	Two numbers whose product is 1 are called reciprocals, or multiplicative inverses, of each other. The quotient of two rational numbers is the product of the first number and the reciprocal of the second number.	Ex. 9, p. 279
i.	The sum or difference of two rational numbers with identical denominators is the sum or difference of their numerators over the common denominator.	Ex. 10, p. 280
j.	Add or subtract rational numbers with unlike denominators by first expressing each rational number with the least common denominator and then following item (i) above.	Ex. 11, p. 280; Ex. 12, p. 281
k.	The order of operations can be applied to an expression with rational numbers.	Ex. 13, p. 282
l.	**Density of the Rational Numbers** Given any two distinct rational numbers, there is always a rational number between them. To find the rational number halfway between two rational numbers, add the rational numbers and divide their sum by 2.	Ex. 14, p. 282

5.4 The Irrational Numbers

a.	The set of irrational numbers is the set of numbers whose decimal representations are neither terminating nor repeating. Examples of irrational numbers are $\sqrt{2} \approx 1.414$ and $\pi \approx 3.142$.	
b.	Simplifying square roots: Use the product rule, $\sqrt{ab} = \sqrt{a} \cdot \sqrt{b}$, to remove from the square root any perfect squares that occur as factors.	Ex. 1, p. 290
c.	Multiplying square roots: $\sqrt{a} \cdot \sqrt{b} = \sqrt{ab}$. The product of square roots is the square root of the product.	Ex. 2, p. 291
d.	Dividing square roots: $\frac{\sqrt{a}}{\sqrt{b}} = \sqrt{\frac{a}{b}}$. The quotient of square roots is the square root of the quotient.	Ex. 3, p. 291
e.	Adding and subtracting square roots: If the radicals have the same radicand, add or subtract their coefficients. The answer is the sum or difference of the coefficients times the common square root. Addition or subtraction is sometimes possible by first simplifying the square roots.	Ex. 4, p. 292; Ex. 5, p. 292
f.	Rationalizing denominators: Multiply the numerator and the denominator by the smallest number that produces a perfect square radicand in the denominator.	Ex. 6, p. 293

5.5 Real Numbers and Their Properties; Clock Addition

a. The set of real numbers is obtained by combining the rational numbers with the irrational numbers. The important subsets of the real numbers are summarized in Table 5.2 on page 300. A diagram representing the relationships among the subsets of the real numbers is given to the left of Table 5.2. Ex. 1, p. 300

b. Properties of real numbers, including closure properties ($a + b$ and ab are real numbers), commutative properties ($a + b = b + a; ab = ba$), associative properties $[(a + b) + c = a + (b + c);\ (ab)c = a(bc)]$, the distributive property $[a(b + c) = ab + ac]$, identity properties ($a + 0 = a; 0 + a = a;\ a \cdot 1 = a; 1 \cdot a = a$), and inverse properties $[a + (-a) = 0; (-a) + a = 0; a, \frac{1}{a} = 1, a \neq 0; \frac{1}{a} \cdot a = 1, a \neq 0]$ are summarized in Table 5.3 on page 302. Ex. 2, p. 303; Ex. 3, p. 303

c. Clock addition is defined by moving a clock's hour hand in a clockwise direction. Tables for clock addition show that the operation satisfies closure, associative, identity, inverse, and commutative properties. Clock addition can be used to explore various kinds of rotational symmetry. Ex. 4, p. 306

5.6 Exponents and Scientific Notation

a. Properties of Exponents Table 5.6, p. 311; Ex. 1, p. 312; Ex. 2, p. 313

- Product rule: $b^m \cdot b^n = b^{m+n}$
- Zero exponent rule: $b^0 = 1, b \neq 0$
- Power rule: $(b^m)^n = b^{m \cdot n}$
- Negative exponent rule: $b^{-m} = \frac{1}{b^m}, b \neq 0$
- Quotient rule: $\frac{b^m}{b^n} = b^{m-n}, b \neq 0$

b. A positive number in scientific notation is expressed as $a \times 10^n$, where $1 \leq a < 10$ and n is an integer.

c. Changing from Scientific to Decimal Notation: If n is positive, move the decimal point in a to the right n places. If n is negative, move the decimal point in a to the left $|n|$ places. Ex. 3, p. 314

d. Changing from Decimal to Scientific Notation: Move the decimal point in the given number to obtain a, where $1 \leq a < 10$. The number of places the decimal point moves gives the absolute value of n in $a \times 10^n$; n is positive if the number is greater than 10 and negative if the number is less than 1. Ex. 4, p. 315

e. The product and quotient rules for exponents are used to multiply and divide numbers in scientific notation. If a number is written in scientific notation, $a \times 10^n$, the digits in a are called significant digits. If rounding is necessary, round the scientific notation answer to the least number of significant digits found in any of the given numbers. Ex. 6, p. 316; Ex. 7, p. 317; Ex. 8, p. 317; Ex. 9, p. 318

5.7 Arithmetic and Geometric Sequences

a. In an arithmetic sequence, each term after the first differs from the preceding term by a constant, the common difference. Subtract any term from the term that directly follows it to find the common difference. Ex. 1, p. 323; Ex. 2, p. 324

b. The general term, or the nth term, of an arithmetic sequence is

$$a_n = a_1 + (n - 1)d,$$

where a_1 is the first term and d is the common difference. Ex. 3, p. 324; Ex. 4, p. 325

c. In a geometric sequence, each term after the first is obtained by multiplying the preceding term by a nonzero constant, the common ratio. Divide any term after the first by the term that directly precedes it to find the common ratio. Ex. 5, p. 327

d. The general term, or the nth term, of a geometric sequence is

$$a_n = a_1 r^{n-1},$$

where a_1 is the first term and r is the common ratio. Ex. 6, p. 327; Ex. 7, p. 328

Review Exercises

5.1

In Exercises 1 and 2, determine whether the number is divisible by each of the following numbers: 2, 3, 4, 5, 6, 8, 9, 10, and 12. If you are using a calculator, explain the divisibility shown by your calculator using one of the rules of divisibility.

1. 238,632 **2.** 421,153,470

In Exercises 3–5, find the prime factorization of each composite number.

3. 705 **4.** 960 **5.** 6825

In Exercises 6–8, find the greatest common divisor and the least common multiple of the numbers.

6. 30 and 48

7. 36 and 150

8. 216 and 254

9. For an intramural league, you need to divide 24 men and 60 women into all-male and all-female teams so that each team has the same number of people. What is the largest number of people that can be placed on a team?

10. The media center at a college runs videotapes of two lectures continuously. One videotape runs for 42 minutes and a second runs for 56 minutes. Both videotapes begin at 9:00 A.M. When will the videos of the two lectures begin again at the same time?

5.2

In Exercises 11–12, insert either $<$ or $>$ in the shaded area between the integers to make the statement true.

11. $-93 \; \blacksquare \; 17$ **12.** $-2 \; \blacksquare \; -200$

In Exercises 13–15, find the absolute value.

13. $|-860|$ **14.** $|53|$ **15.** $|0|$

Perform the indicated operations in Exercises 16–28.

16. $8 + (-11)$ **17.** $-6 + (-5)$

18. $-7 - 8$ **19.** $-7 - (-8)$

20. $(-9)(-11)$ **21.** $5(-3)$

22. $\frac{-36}{-4}$ **23.** $\frac{20}{-5}$

24. $-40 \div 5 \cdot 2$ **25.** $-6 + (-2) \cdot 5$

26. $6 - 4(-3 + 2)$ **27.** $28 \div (2 - 4^2)$

28. $36 - 24 \div 4 \cdot 3 - 1$

29. For the year 2015, the Congressional Budget Office projects a budget deficit of $-\$57$ billion. For the same year, the Brookings Institution forecasts a budget deficit of $-\$715$ billion. What is the difference between the CBO projection and the Brookings projection?

5.3

In Exercises 30–32, reduce each rational number to its lowest terms.

30. $\frac{40}{75}$ **31.** $\frac{36}{150}$ **32.** $\frac{165}{180}$

In Exercises 33–34, convert each mixed number to an improper fraction.

33. $5\frac{9}{11}$ **34.** $-3\frac{2}{7}$

In Exercises 35–36, convert each improper fraction to a mixed number.

35. $\frac{27}{5}$ **36.** $-\frac{17}{9}$

In Exercises 37–40, express each rational number as a decimal.

37. $\frac{4}{5}$ **38.** $\frac{3}{7}$ **39.** $\frac{5}{8}$ **40.** $\frac{9}{16}$

In Exercises 41–44, express each terminating decimal as a quotient of integers in lowest terms.

41. 0.6 **42.** 0.68 **43.** 0.588 **44.** 0.0084

In Exercises 45–47, express each repeating decimal as a quotient of integers in lowest terms.

45. $0.\overline{5}$ **46.** $0.\overline{34}$ **47.** $0.\overline{113}$

In Exercises 48–58, perform the indicated operations. Where possible, reduce the answer to lowest terms.

48. $\frac{3}{5} \cdot \frac{7}{10}$ **49.** $\left(3\frac{1}{3}\right)\left(1\frac{3}{4}\right)$ **50.** $\frac{4}{5} \div \frac{3}{10}$

51. $-1\frac{2}{3} \div 6\frac{2}{3}$ **52.** $\frac{2}{9} + \frac{4}{9}$ **53.** $\frac{7}{9} + \frac{5}{12}$

54. $\frac{3}{4} - \frac{2}{15}$ **55.** $\frac{1}{3} + \frac{1}{2} \cdot \frac{4}{5}$ **56.** $\frac{3}{8}\left(\frac{1}{2} + \frac{1}{3}\right)$

57. $\frac{1}{2} - \frac{2}{3} \div \frac{5}{9} + \frac{3}{10}$

58. $\left(\frac{1}{2} + \frac{1}{3}\right) \div \left(\frac{1}{4} - \frac{3}{8}\right)$

In Exercises 59–60, find the rational number halfway between the two numbers in each pair.

59. $\frac{1}{7}$ and $\frac{1}{8}$ **60.** $\frac{3}{4}$ and $\frac{3}{5}$

61. A recipe for coq au vin is meant for six people and requires $4\frac{1}{2}$ pounds of chicken. If you want to serve 15 people, how much chicken is needed?

62. The gas tank of a car is filled to its capacity. The first day, $\frac{1}{4}$ of the tank's gas is used for travel. The second day, $\frac{1}{3}$ of the tank's original amount of gas is used for travel. What fraction of the tank is filled with gas at the end of the second day?

5.4

In Exercises 63–66, simplify the square root.

63. $\sqrt{28}$ **64.** $\sqrt{72}$

65. $\sqrt{150}$ **66.** $\sqrt{300}$

In Exercises 67–75, perform the indicated operation. Simplify the answer when possible.

67. $\sqrt{6} \cdot \sqrt{8}$ **68.** $\sqrt{10} \cdot \sqrt{5}$

69. $\frac{\sqrt{24}}{\sqrt{2}}$ **70.** $\frac{\sqrt{27}}{\sqrt{3}}$

71. $\sqrt{5} + 4\sqrt{5}$ **72.** $7\sqrt{11} - 13\sqrt{11}$

73. $\sqrt{50} + \sqrt{8}$ **74.** $\sqrt{3} - 6\sqrt{27}$

75. $2\sqrt{18} + 3\sqrt{8}$

In Exercises 76–77, rationalize the denominator.

76. $\frac{30}{\sqrt{5}}$ **77.** $\sqrt{\frac{2}{3}}$

78. Paleontologists use the mathematical model $W = 4\sqrt{2x}$ to estimate the walking speed of a dinosaur, W, in feet per second, where x is the length, in feet, of the dinosaur's leg. What is the walking speed of a dinosaur whose leg length is 6 feet? Express the answer in simplified radical form. Then use your calculator to estimate the walking speed to the nearest tenth of a foot per second.

5.5

79. Consider the set

$$\left\{-17, -\frac{9}{13}, 0, 0.75, \sqrt{2}, \pi, \sqrt{81}\right\}.$$

List all numbers from the set that are **a.** natural numbers, **b.** whole numbers, **c.** integers, **d.** rational numbers, **e.** irrational numbers, **f.** real numbers.

80. Give an example of an integer that is not a natural number.

81. Give an example of a rational number that is not an integer.

82. Give an example of a real number that is not a rational number.

In Exercises 83–90, state the name of the property illustrated.

83. $3 + 17 = 17 + 3$

84. $(6 \cdot 3) \cdot 9 = 6 \cdot (3 \cdot 9)$

85. $\sqrt{3}(\sqrt{5} + \sqrt{3}) = \sqrt{15} + 3$

86. $(6 \cdot 9) \cdot 2 = 2 \cdot (6 \cdot 9)$

87. $\sqrt{3}(\sqrt{5} + \sqrt{3}) = (\sqrt{5} + \sqrt{3})\sqrt{3}$

88. $(3 \cdot 7) + (4 \cdot 7) = (4 \cdot 7) + (3 \cdot 7)$

89. $-3\left(-\frac{1}{3}\right) = 1$

90. $\sqrt{7} \cdot 1 = \sqrt{7}$

In Exercises 91–92, give an example to show that

91. The natural numbers are not closed with respect to division.

92. The whole numbers are not closed with respect to subtraction.

93. Shown in the figure is a 5-hour clock and the table for clock addition in the 5-hour system.

⊕	0	1	2	3	4
0	0	1	2	3	4
1	1	2	3	4	0
2	2	3	4	0	1
3	3	4	0	1	2
4	4	0	1	2	3

a. How can you tell that the set $\{0, 1, 2, 3, 4\}$ is closed under the operation of clock addition?

b. Verify the associative property:

$(4 \oplus 2) \oplus 3 = 4 \oplus (2 \oplus 3)$

c. What is the identity element in the 5-hour clock system?

d. Find the inverse of each element in the 5-hour clock system.

e. Verify two cases of the commutative property:

$3 \oplus 4 = 4 \oplus 3$ and $3 \oplus 2 = 2 \oplus 3$

5.6

In Exercises 94–104, evaluate each expression.

94. $6 \cdot 6^2$ **95.** $2^3 \cdot 2^3$ **96.** $(2^2)^2$

97. $(3^3)^2$ **98.** $\frac{5^6}{5^4}$ **99.** 7^0

100. $(-7)^0$ **101.** 6^{-3} **102.** 2^{-4}

103. $\frac{7^4}{7^6}$ **104.** $3^5 \cdot 3^{-2}$

In Exercises 105–108, express each number in decimal notation.

105. 4.6×10^2 **106.** 3.74×10^4

107. 2.55×10^{-3} **108.** 7.45×10^{-5}

In Exercises 109–114, express each number in scientific notation.

109. 7520 **110.** 3,590,000

111. 0.00725 **112.** 0.000000409

113. 420×10^{11} **114.** 0.97×10^{-4}

In Exercises 115–118, perform the indicated operation and express each answer in decimal notation.

115. $(3 \times 10^7)(1.3 \times 10^{-5})$ **116.** $(5 \times 10^3)(2.3 \times 10^2)$

117. $\frac{6.9 \times 10^3}{3 \times 10^5}$ **118.** $\frac{2.4 \times 10^{-4}}{6 \times 10^{-6}}$

In Exercises 119–122, perform the indicated operation by first expressing each number in scientific notation. Write the answer in scientific notation.

119. (60,000)(540,000)

120. (91,000)(0.0004)

121. $\frac{8{,}400{,}000}{4000}$ **122.** $\frac{0.000003}{0.00000006}$

In 2011, the United States government spent more than it had collected in taxes, resulting in a budget deficit of \$1.3 trillion. In Exercises 123–125, you will use scientific notation to put a number like 1.3 trillion in perspective. Use 10^{12} for 1 trillion.

123. Express 1.3 trillion in scientific notation.

124. There are approximately 32,000,000 seconds in a year. Express this number in scientific notation.

125. Use your scientific notation answers from Exercises 123 and 124 to answer this question: How many years is 1.3 trillion seconds? (*Note*: 1.3 trillion seconds would take us back in time to a period when Neanderthals were using stones to make tools.)

126. The human body contains approximately 3.2×10^4 microliters of blood for every pound of body weight. Each microliter of blood contains approximately 5×10^6 red blood cells. Express in scientific notation the approximate number of red blood cells in the body of a 180-pound person.

5.7

In Exercises 127–129, write the first six terms of the arithmetic sequence with the first term, a_1, and common difference, d.

127. $a_1 = 7, d = 4$

128. $a_1 = -4, d = -5$

129. $a_1 = \frac{3}{2}, d = -\frac{1}{2}$

In Exercises 130–132, find the indicated term for the arithmetic sequence with first term, a_1, and common difference, d.

130. Find a_6, when $a_1 = 5, d = 3$.

131. Find a_{12}, when $a_1 = -8, d = -2$.

132. Find a_{14}, when $a_1 = 14, d = -4$.

In Exercises 133–134, write a formula for the general term (the nth term) of each arithmetic sequence. Then use the formula for a_n to find a_{20}, the 20th term of the sequence.

133. $-7, -3, 1, 5, \ldots$

134. $a_1 = 200, d = -20$

In Exercises 135–137, write the first six terms of the geometric sequence with the first term, a_1, and common ratio, r.

135. $a_1 = 3, r = 2$

136. $a_1 = \frac{1}{2}, r = \frac{1}{2}$

137. $a_1 = 16, r = -\frac{1}{2}$

In Exercises 138–140, find the indicated term for the geometric sequence with first term, a_1, and common ratio, r.

138. Find a_4, when $a_1 = 2, r = 3$.

139. Find a_6, when $a_1 = 16, r = \frac{1}{2}$.

140. Find a_5, when $a_1 = -3, r = 2$.

In Exercises 141–142, write a formula for the general term (the nth term) of each geometric sequence. Then use the formula for a_n to find a_8, the eighth term of the sequence.

141. $1, 2, 4, 8, \ldots$ **142.** $100, 10, 1, \frac{1}{10}, \ldots$

Determine whether each sequence in Exercises 143–146 is arithmetic or geometric. Then find the next two terms.

143. $4, 9, 14, 19, \ldots$ **144.** $2, 6, 18, 54, \ldots$

145. $1, \frac{1}{4}, \frac{1}{16}, \frac{1}{64}, \ldots$ **146.** $0, -7, -14, -21, \ldots$

147. The bar graph shows the number of hours per week devoted to housework by wives and husbands in 1965 and 2010.

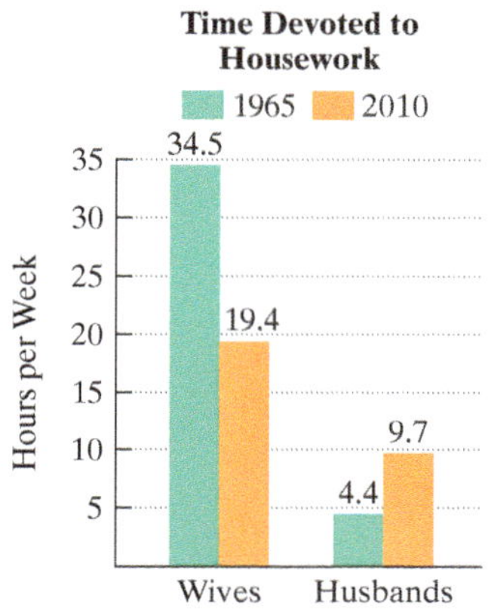

Source: James Henslin, *Sociology*, Eleventh Edition, Pearson, 2012.

In 1965, wives averaged 34.5 hours per week doing housework. On average, this has decreased by approximately 0.3 hour per year since then.

a. Write a formula for the *n*th term of the arithmetic sequence that describes the number of hours per week devoted to housework by wives *n* years after 1964.

b. Use the model to project the number of hours per week wives will devote to housework in 2020.

148. The table shows the population of Florida for 2000 and 2010, with estimates given by the U.S. Census Bureau for 2001 through 2009.

Year	2000	2001	2002	2003	2004	2005
Population in millions	15.98	16.24	16.50	16.76	17.03	17.30

Year	2006	2007	2008	2009	2010
Population in millions	17.58	17.86	18.15	18.44	18.80

a. Divide the population for each year by the population in the preceding year. Round to two decimal places and show that Florida has a population increase that is approximately geometric.

b. Write the general term of the geometric sequence modeling Florida's population, in millions, *n* years after 1999.

c. Use your the model from part (b) to project Florida's population, in millions, for the year 2030. Round to two decimal places.

Chapter 5 Test

1. Which of the numbers 2, 3, 4, 5, 6, 8, 9, 10, and 12 divide 391,248?

2. Find the prime factorization of 252.

3. Find the greatest common divisor and the least common multiple of 48 and 72.

Perform the indicated operations in Exercises 4–6.

4. $-6 - (5 - 12)$ **5.** $(-3)(-4) \div (7 - 10)$

6. $(6 - 8)^2(5 - 7)^3$ **7.** Express $\frac{7}{12}$ as a decimal.

8. Express $0.\overline{64}$ as a quotient of integers in lowest terms.

In Exercises 9–11, perform the indicated operations. Where possible, reduce the answer to its lowest terms.

9. $\left(-\frac{3}{7}\right) \div \left(-2\frac{1}{7}\right)$ **10.** $\frac{19}{24} - \frac{7}{40}$ **11.** $\frac{1}{2} - 8\left(\frac{1}{4} + 1\right)$

12. Find the rational number halfway between $\frac{1}{2}$ and $\frac{2}{3}$.

13. Multiply and simplify: $\sqrt{10} \cdot \sqrt{5}$.

14. Add: $\sqrt{50} + \sqrt{32}$.

15. Rationalize the denominator: $\frac{6}{\sqrt{2}}$.

16. List all the rational numbers in this set:

$$\left\{-7, -\frac{4}{5}, 0, 0.25, \sqrt{3}, \sqrt{4}, \frac{22}{7}, \pi\right\}.$$

In Exercises 17–18, state the name of the property illustrated.

17. $3(2 + 5) = 3(5 + 2)$

18. $6(7 + 4) = 6 \cdot 7 + 6 \cdot 4$

In Exercises 19–21, evaluate each expression.

19. $3^3 \cdot 3^2$ **20.** $\frac{4^6}{4^3}$ **21.** 8^{-2}

22. Multiply and express the answer in decimal notation.

$$(3 \times 10^8)(2.5 \times 10^{-5})$$

23. Divide by first expressing each number in scientific notation. Write the answer in scientific notation.

$$\frac{49{,}000}{0.007}$$

In Exercises 24–26 use 10^6 *for one million and* 10^9 *for one billion to rewrite the number in each statement in scientific notation.*

24. The 2009 economic stimulus package allocated \$53.6 billion for grants to states for education.

25. The population of the United States at the time the economic stimulus package was voted into law was approximately 307 million.

26. Use your scientific notation answers from Exercises 24 and 25 to answer this question:

If the cost for grants to states for education was evenly divided among every individual in the United States, how much would each citizen have to pay?

27. Write the first six terms of the arithmetic sequence with first term, a_1, and common difference, d.

$$a_1 = 1, d = -5$$

28. Find a_9, the ninth term of the arithmetic sequence, with the first term, a_1, and common difference, d.

$$a_1 = -2, d = 3$$

29. Write the first six terms of the geometric sequence with first term, a_1, and common ratio, r.

$$a_1 = 16, r = \frac{1}{2}$$

30. Find a_7, the seventh term of the geometric sequence, with the first term, a_1, and common ratio, r.

$$a_1 = 5, r = 2$$

6

Algebra: Equations and Inequalities

THE BELIEF THAT HUMOR AND LAUGHTER CAN HAVE POSITIVE EFFECTS ON OUR LIVES IS NOT NEW. THE BIBLE TELLS US, "A MERRY HEART DOETH good like a medicine, but a broken spirit drieth the bones" (Proverbs 17:22).

Some random humor factoids: • The average adult laughs 15 times each day (Newhouse News Service). • Forty-six percent of people who are telling a joke laugh more than the people they are telling it to (*U.S. News and World Report*). • Eighty percent of adult laughter does not occur in response to jokes or funny situations (*Independent*). • Algebra can be used to model the influence that humor plays in our responses to negative life events (Bob Blitzer, *Thinking Mathematically*).

That last tidbit that your author threw into the list is true. Based on our sense of humor, there is actually a formula that predicts how we will respond to difficult life events. Formulas can be used to explain what is happening in the present and to make predictions about what might occur in the future. In this chapter, you will learn to use formulas and mathematical models in new ways that will help you to recognize patterns, logic, and order in a world that can appear chaotic to the untrained eye.

Here's where you'll find this application:

Humor opens Section 6.2, and the advantage of having a sense of humor becomes laughingly evident in the models in Example 6 on page 356.

6.1

WHAT AM I SUPPOSED TO LEARN?

After you have read this section, you should be able to:

1 Evaluate algebraic expressions.
2 Use mathematical models.
3 Understand the vocabulary of algebraic expressions.
4 Simplify algebraic expressions.

Algebraic Expressions and Formulas

YOU ARE THINKING ABOUT BUYING A high-definition television. How much distance should you allow between you and the TV for pixels to be undetectable and the image to appear smooth?

Algebraic Expressions

Let's see what the distance between you and your TV has to do with algebra. The biggest difference between arithmetic and algebra is the use of *variables* in algebra. A **variable** is a letter that represents a variety of different numbers. For example, we can let x represent the diagonal length, in inches, of a high-definition television. The industry rule for most of the current HDTVs on the market is to multiply this diagonal length by 2.5 to get the distance, in inches, at which a person with perfect vision can see a smooth image. This can be written $2.5 \cdot x$, but it is usually expressed as $2.5x$. Placing a number and a letter next to one another indicates multiplication.

Notice that $2.5x$ combines the number 2.5 and the variable x using the operation of multiplication. A combination of variables and numbers using the operations of addition, subtraction, multiplication, or division, as well as powers or roots, is called an **algebraic expression**. Here are some examples of algebraic expressions:

$$x + 2.5 \qquad x - 2.5 \qquad 2.5x \qquad \frac{x}{2.5} \qquad 3x + 5 \qquad \sqrt{x} + 7.$$

1 Evaluate algebraic expressions.

Evaluating Algebraic Expressions

Evaluating an algebraic expression means finding the value of the expression for a given value of the variable. For example, we can evaluate $2.5x$ (the ideal distance between you and your x-inch TV) for $x = 50$. We substitute 50 for x. We obtain $2.5 \cdot 50$, or 125. This means that if the diagonal length of your TV is 50 inches, your distance from the screen should be 125 inches. Because 12 inches = 1 foot, this distance is $\frac{125}{12}$ feet, or approximately 10.4 feet.

Many algebraic expressions contain more than one operation. Evaluating an algebraic expression correctly involves carefully applying the order of operations agreement that we studied in Chapter 5.

THE ORDER OF OPERATIONS AGREEMENT

1. Perform operations within the innermost parentheses and work outward. If the algebraic expression involves a fraction, treat the numerator and the denominator as if they were each enclosed in parentheses.
2. Evaluate all exponential expressions.
3. Perform multiplications and divisions as they occur, working from left to right.
4. Perform additions and subtractions as they occur, working from left to right.

EXAMPLE 1 Evaluating an Algebraic Expression

Evaluate $7 + 5(x - 4)^3$ for $x = 6$.

SOLUTION

$$\begin{aligned} 7 + 5(x - 4)^3 &= 7 + 5(6 - 4)^3 && \text{Replace } x \text{ with 6.} \\ &= 7 + 5(2)^3 && \text{First work inside parentheses: } 6 - 4 = 2. \\ &= 7 + 5(8) && \text{Evaluate the exponential expression: } 2^3 = 2 \cdot 2 \cdot 2 = 8. \\ &= 7 + 40 && \text{Multiply: } 5(8) = 40. \\ &= 47 && \text{Add.} \end{aligned}$$

CHECK POINT 1 Evaluate $8 + 6(x - 3)^2$ for $x = 13$.

GREAT QUESTION!

Is there a difference between evaluating x^2 for $x = -6$ and evaluating $-x^2$ for $x = 6$?

Yes. Notice the difference between these evaluations:

- x^2 for $x = -6$
  $x^2 = (-6)^2 = (-6)(-6) = 36$
- $-x^2$ for $x = 6$
  $-x^2 = -6^2 = -6 \cdot 6 = -36$

The negative is not inside parentheses and is not taken to the second power.

Work carefully when evaluating algebraic expressions with exponents and negatives.

EXAMPLE 2 Evaluating an Algebraic Expression

Evaluate $x^2 + 5x - 3$ for $x = -6$.

SOLUTION

We substitute -6 for each of the two occurrences of x. Then we use the order of operations to evaluate the algebraic expression.

$$\begin{aligned} &x^2 + 5x - 3 && \text{This is the given algebraic expression.} \\ &= (-6)^2 + 5(-6) - 3 && \text{Substitute } -6 \text{ for each } x. \\ &= 36 + 5(-6) - 3 && \text{Evaluate the exponential expression: } (-6)^2 = (-6)(-6) = 36. \\ &= 36 + (-30) - 3 && \text{Multiply: } 5(-6) = -30. \\ &= 6 - 3 && \text{Add and subtract from left to right. First add: } 36 + (-30) = 6. \\ &= 3 && \text{Subtract.} \end{aligned}$$

CHECK POINT 2 Evaluate $x^2 + 4x - 7$ for $x = -5$.

EXAMPLE 3 Evaluating an Algebraic Expression

Evaluate $-2x^2 + 5xy - y^3$ for $x = 4$ and $y = -2$.

SOLUTION

We substitute 4 for each x and -2 for each y. Then we use the order of operations to evaluate the algebraic expression.

$$\begin{aligned} &-2x^2 + 5xy - y^3 && \text{This is the given algebraic expression.} \\ &= -2 \cdot 4^2 + 5 \cdot 4(-2) - (-2)^3 && \text{Substitute 4 for } x \text{ and } -2 \text{ for } y. \\ &= -2 \cdot 16 + 5 \cdot 4(-2) - (-8) && \text{Evaluate the exponential expressions: } 4^2 = 4 \cdot 4 = 16 \text{ and } (-2)^3 = (-2)(-2)(-2) = -8. \\ &= -32 + (-40) - (-8) && \text{Multiply: } -2 \cdot 16 = -32 \text{ and } 5(4)(-2) = 20(-2) = -40. \\ &= -72 - (-8) && \text{Add and subtract from left to right. First add: } -32 + (-40) = -72. \\ &= -64 && \text{Subtract: } -72 - (-8) = -72 + 8 = -64. \end{aligned}$$

CHECK POINT 3 Evaluate $-3x^2 + 4xy - y^3$ for $x = 5$ and $y = -1$.

2 Use mathematical models.

Formulas and Mathematical Models

An **equation** is formed when an equal sign is placed between two algebraic expressions. One aim of algebra is to provide a compact, symbolic description of the world. These descriptions involve the use of *formulas*. A **formula** is an equation that uses variables to express a relationship between two or more quantities.

Here are two examples of formulas related to heart rate and exercise.

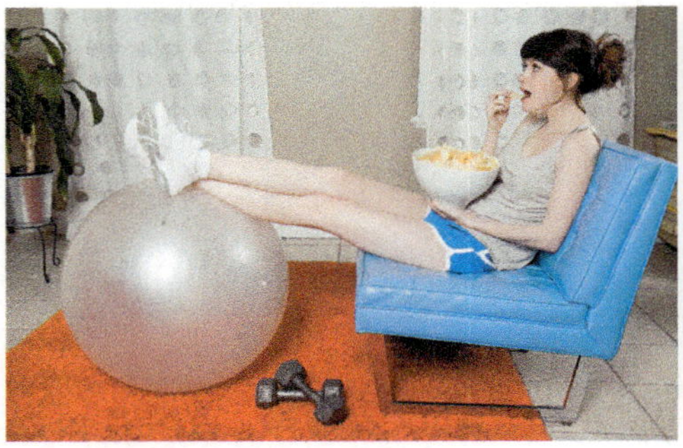

Couch-Potato Exercise

$$H = \frac{1}{5}(220 - a)$$

Heart rate, in beats per minute, | is | $\frac{1}{5}$ of | the difference between 220 and your age.

Working It

$$H = \frac{9}{10}(220 - a)$$

Heart rate, in beats per minute, | is | $\frac{9}{10}$ of | the difference between 220 and your age.

These important definitions are repeated from earlier chapters in case your course did not cover this material.

The process of finding formulas to describe real-world phenomena is called **mathematical modeling**. Such formulas, together with the meaning assigned to the variables, are called **mathematical models**. We often say that these formulas model, or describe, the relationships among the variables.

EXAMPLE 4 *Modeling Caloric Needs*

The bar graph in **Figure 6.1** shows the estimated number of calories per day needed to maintain energy balance for various gender and age groups for moderately active lifestyles. (Moderately active means a lifestyle that includes physical activity equivalent to walking 1.5 to 3 miles per day at 3 to 4 miles per hour, in addition to the light physical activity associated with typical day-to-day life.)

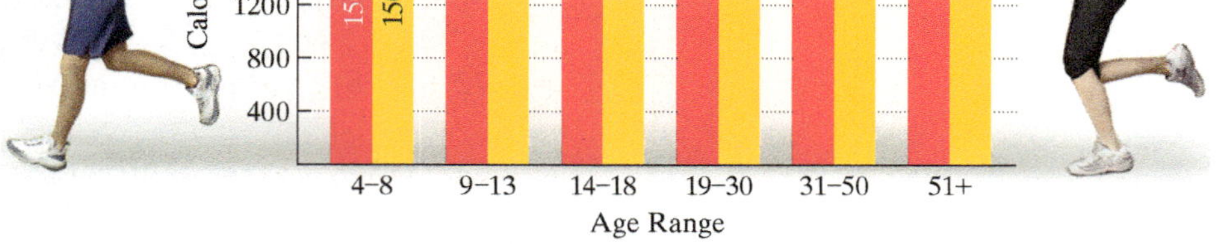

FIGURE 6.1
Source: USDA

The mathematical model

$$W = -66x^2 + 526x + 1030$$

describes the number of calories needed per day, W, by women in age group x with moderately active lifestyles. According to the model, how many calories per day are needed by women between the ages of 19 and 30, inclusive, with this lifestyle? Does this underestimate or overestimate the number shown by the graph in **Figure 6.1**? By how much?

SOLUTION

Because the 19–30 age range is designated as group 4, we substitute 4 for x in the given model. Then we use the order of operations to find W, the number of calories needed per day by women between the ages of 19 and 30.

$W = -66x^2 + 526x + 1030$	This is the given mathematical model.
$W = -66 \cdot 4^2 + 526 \cdot 4 + 1030$	Replace each occurrence of x with 4.
$W = -66 \cdot 16 + 526 \cdot 4 + 1030$	Evaluate the exponential expression: $4^2 = 4 \cdot 4 = 16$.
$W = -1056 + 2104 + 1030$	Multiply from left to right: $-66 \cdot 16 = -1056$ and $526 \cdot 4 = 2104$.
$W = 2078$	Add.

The formula indicates that women in the 19–30 age range with moderately active lifestyles need 2078 calories per day. **Figure 6.1** indicates that 2100 calories are needed. Thus, the mathematical model underestimates caloric needs by $2100 - 2078$ calories, or by 22 calories per day.

CHECK POINT 4 The mathematical model

$$M = -120x^2 + 998x + 590$$

describes the number of calories needed per day, M, by men in age group x with moderately active lifestyles. According to the model, how many calories per day are needed by men between the ages of 19 and 30, inclusive, with this lifestyle? Does this underestimate or overestimate the number shown by the graph in **Figure 6.1**? By how much?

3 Understand the vocabulary of algebraic expressions.

The Vocabulary of Algebraic Expressions

We have seen that an algebraic expression combines numbers and variables. Here is another example of an algebraic expression:

$$7x - 9y - 3.$$

The **terms** of an algebraic expression are those parts that are separated by addition. For example, we can rewrite $7x - 9y - 3$ as

$$7x + (-9y) + (-3).$$

This expression contains three terms, namely $7x$, $-9y$, and -3.

The numerical part of a term is called its **coefficient**. In the term $7x$, the 7 is the coefficient. In the term $-9y$, the -9 is the coefficient.

Coefficients of 1 and -1 are not written. Thus, the coefficient of x, meaning $1x$, is 1. Similarly, the coefficient of $-y$, meaning $-1y$, is -1.

A term that consists of just a number is called a **numerical term** or a **constant**. The numerical term of $7x - 9y - 3$ is -3.

The parts of each term that are multiplied are called the **factors** of the term. The factors of the term $7x$ are 7 and x.

Like terms are terms that have the same variable factors. For example, $3x$ and $7x$ are like terms.

4 Simplify algebraic expressions.

Simplifying Algebraic Expressions

The properties of real numbers that we discussed in Chapter 5 can be applied to algebraic expressions.

PROPERTIES OF REAL NUMBERS

Property	Example
Commutative Property of Addition $a + b = b + a$	$13x^2 + 7x = 7x + 13x^2$
Commutative Property of Multiplication $ab = ba$	$x \cdot 6 = 6x$
Associative Property of Addition $(a + b) + c = a + (b + c)$	$3 + (8 + x) = (3 + 8) + x = 11 + x$
Associative Property of Multiplication $(ab)c = a(bc)$	$-2(3x) = (-2 \cdot 3)x = -6x$
Distributive Property	
$a(b + c) = ab + ac$	$5(3x + 7) = 5 \cdot 3x + 5 \cdot 7 = 15x + 35$
$a(b - c) = ab - ac$	$4(2x - 5) = 4 \cdot 2x - 4 \cdot 5 = 8x - 20$

GREAT QUESTION!

Do I have to use the distributive property to combine like terms? Can't I just do it in my head?

Yes, you can combine like terms mentally. Add or subtract the coefficients of the terms. Use this result as the coefficient of the terms' variable factor(s).

The distributive property in the form

$$ba + ca = (b + c)a$$

enables us to add or subtract like terms. For example,

$$3x + 7x = (3 + 7)x = 10x$$
$$7y^2 - y^2 = 7y^2 - 1y^2 = (7 - 1)y^2 = 6y^2.$$

This process is called **combining like terms.**

An algebraic expression is **simplified** when parentheses have been removed and like terms have been combined.

EXAMPLE 5 Simplifying an Algebraic Expression

Simplify: $5(3x - 7) - 6x$.

SOLUTION

$5(3x - 7) - 6x$

$= 5 \cdot 3x - 5 \cdot 7 - 6x$ Use the distributive property to remove the parentheses.

$= 15x - 35 - 6x$ Multiply.

$= (15x - 6x) - 35$ Group like terms.

$= 9x - 35$ Combine like terms: $15x - 6x = (15 - 6)x = 9x$.

CHECK POINT 5 Simplify: $7(2x - 3) - 11x$.

EXAMPLE 6 *Simplifying an Algebraic Expression*

Simplify: $6(2x^2 + 4x) + 10(4x^2 + 3x)$.

SOLUTION

$6(2x^2 + 4x) + 10(4x^2 + 3x)$

$= 6 \cdot 2x^2 + 6 \cdot 4x + 10 \cdot 4x^2 + 10 \cdot 3x$ — Use the distributive property to remove the parentheses.

$= 12x^2 + 24x + 40x^2 + 30x$ — Multiply.

$= (12x^2 + 40x^2) + (24x + 30x)$ — Group like terms.

$= 52x^2 + 54x$ — Combine like terms: $12x^2 + 40x^2 = (12 + 40)x^2 = 52x^2$ and $24x + 30x = (24 + 30)x = 54x$.

$52x^2$ and $54x$ are not like terms. They contain different variable factors, x^2 and x, and cannot be combined.

CHECK POINT 6 Simplify: $7(4x^2 - 3x) + 2(5x^2 + x)$.

It is not uncommon to see algebraic expressions with parentheses preceded by a negative sign or subtraction. An expression of the form $-(a + b)$ can be simplified as follows:

$$-(a + b) = -1(a + b) = (-1)a + (-1)b = -a + (-b) = -a - b.$$

Do you see a fast way to obtain the simplified expression on the right? **If a negative sign or a subtraction symbol appears outside parentheses, drop the parentheses and change the sign of every term within the parentheses.** For example,

$$-(3x^2 - 7x - 4) = -3x^2 + 7x + 4.$$

EXAMPLE 7 *Simplifying an Algebraic Expression*

Simplify: $8x + 2[5 - (x - 3)]$.

SOLUTION

$8x + 2[5 - (x - 3)]$

$= 8x + 2[5 - x + 3]$ — Drop parentheses and change the sign of each term in parentheses: $-(x - 3) = -x + 3$.

$= 8x + 2[8 - x]$ — Simplify inside brackets: $5 + 3 = 8$.

$= 8x + 16 - 2x$ — Apply the distributive property: $2[8 - x] = 2 \cdot 8 - 2x = 16 - 2x$.

$= (8x - 2x) + 16$ — Group like terms.

$= 6x + 16$ — Combine like terms: $8x - 2x = (8 - 2)x = 6x$.

CHECK POINT 7 Simplify: $6x + 4[7 - (x - 2)]$.

Blitzer Bonus

Using Algebra to Measure Blood-Alcohol Concentration

The amount of alcohol in a person's blood is known as blood-alcohol concentration (BAC), measured in grams of alcohol per deciliter of blood. A BAC of 0.08, meaning 0.08%, indicates that a person has 8 parts alcohol per 10,000 parts blood. In every state in the United States, it is illegal to drive with a BAC of 0.08 or higher.

How Do I Measure My Blood-Alcohol Concentration?
Here's a formula that models BAC for a person who weighs w pounds and who has n drinks* per hour.

$$\text{BAC} = \frac{600n}{w(0.6n + 169)}$$

Blood-alcohol concentration

Body weight, in pounds

Number of drinks consumed in an hour

* A drink can be a 12-ounce can of beer, a 5-ounce glass of wine, or a 1.5-ounce shot of liquor. Each contains approximately 14 grams, or $\frac{1}{2}$ ounce, of alcohol.

Blood-alcohol concentration can be used to quantify the meaning of "tipsy."

BAC	Effects on Behavior
0.05	Feeling of well-being; mild release of inhibitions; absence of observable effects
0.08	Feeling of relaxation; mild sedation; exaggeration of emotions and behavior; slight impairment of motor skills; increase in reaction time
0.12	Muscle control and speech impaired; difficulty performing motor skills; uncoordinated behavior
0.15	Euphoria; major impairment of physical and mental functions; irresponsible behavior; some difficulty standing, walking, and talking
0.35	Surgical anesthesia; lethal dosage for a small percentage of people
0.40	Lethal dosage for 50% of people; severe circulatory and respiratory depression; alcohol poisoning/overdose

Source: National Clearinghouse for Alcohol and Drug Information

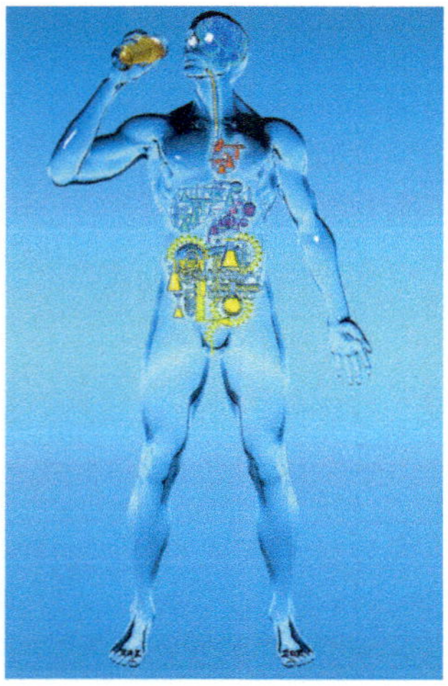

Keeping in mind the meaning of "tipsy," we can use our model to compare blood-alcohol concentrations of a 120-pound person and a 200-pound person for various numbers of drinks. We determined each BAC using a calculator, rounding to three decimal places.

Blood-Alcohol Concentrations of a 120-Pound Person

$$\text{BAC} = \frac{600n}{120(0.6n + 169)}$$

n (number of drinks per hour)	1	2	3	4	5	6	7	8	9	10
BAC (blood-alcohol concentration)	0.029	0.059	0.088	0.117	0.145	0.174	0.202	0.230	0.258	0.286

Illegal to drive (3–10 drinks)

Blood-Alcohol Concentrations of a 200-Pound Person

$$\text{BAC} = \frac{600n}{200(0.6n + 169)}$$

n (number of drinks per hour)	1	2	3	4	5	6	7	8	9	10
BAC (blood-alcohol concentration)	0.018	0.035	0.053	0.070	0.087	0.104	0.121	0.138	0.155	0.171

Illegal to drive (5–10 drinks)

Like all mathematical models, the formula for BAC gives approximate rather than exact values. There are other variables that influence blood-alcohol concentration that are not contained in the model. These include the rate at which an individual's body processes alcohol, how quickly one drinks, sex, age, physical condition, and the amount of food eaten prior to drinking.

Concept and Vocabulary Check

Fill in each blank so that the resulting statement is true.

1. Finding the value of an algebraic expression for a given value of the variable is called ________ the expression.
2. When an equal sign is placed between two algebraic expressions, an ________ is formed.
3. The parts of an algebraic expression that are separated by addition are called the ________ of the expression.
4. In the algebraic expression $7x$, 7 is called the ________ because it is the numerical part.
5. In the algebraic expression $7x$, 7 and x are called ________ because they are multiplied together.
6. The algebraic expressions $3x$ and $7x$ are called ________ because they contain the same variable to the same power.

Exercise Set 6.1

Practice Exercises

In Exercises 1–34, evaluate the algebraic expression for the given value or values of the variables.

1. $5x + 7;\quad x = 4$
2. $9x + 6;\quad x = 5$
3. $-7x - 5;\quad x = -4$
4. $-6x - 13;\quad x = -3$
5. $x^2 + 4;\quad x = 5$
6. $x^2 + 9;\quad x = 3$
7. $x^2 - 6;\quad x = -2$
8. $x^2 - 11;\quad x = -3$
9. $-x^2 + 4;\quad x = 5$
10. $-x^2 + 9;\quad x = 3$
11. $-x^2 - 6;\quad x = -2$
12. $-x^2 - 11;\quad x = -3$
13. $x^2 + 4x;\quad x = 10$
14. $x^2 + 6x;\quad x = 9$
15. $8x^2 + 17;\quad x = 5$
16. $7x^2 + 25;\quad x = 3$
17. $x^2 - 5x;\quad x = -11$
18. $x^2 - 8x;\quad x = -5$
19. $x^2 + 5x - 6;\quad x = 4$
20. $x^2 + 7x - 4;\quad x = 6$
21. $4 + 5(x - 7)^3;\quad x = 9$
22. $6 + 5(x - 6)^3;\quad x = 8$
23. $x^2 - 3(x - y);\quad x = 2, y = 8$
24. $x^2 - 4(x - y);\quad x = 3, y = 8$
25. $2x^2 - 5x - 6;\quad x = -3$
26. $3x^2 - 4x - 9;\quad x = -5$
27. $-5x^2 - 4x - 11;\quad x = -1$
28. $-6x^2 - 11x - 17;\quad x = -2$
29. $3x^2 + 2xy + 5y^2;\quad x = 2, y = 3$
30. $4x^2 + 3xy + 2y^2;\quad x = 3, y = 2$
31. $-x^2 - 4xy + 3y^3;\quad x = -1, y = -2$
32. $-x^2 - 3xy + 4y^3;\quad x = -3, y = -1$
33. $\dfrac{2x + 3y}{x + 1};\quad x = -2, y = 4$
34. $\dfrac{2x + y}{xy - 2x};\quad x = -2, y = 4$

The formula

$$C = \frac{5}{9}(F - 32)$$

expresses the relationship between Fahrenheit temperature, F, and Celsius temperature, C. In Exercises 35–36, use the formula to convert the given Fahrenheit temperature to its equivalent temperature on the Celsius scale.

35. 50°F
36. 86°F

A football was kicked vertically upward from a height of 4 feet with an initial speed of 60 feet per second. The formula

$$h = 4 + 60t - 16t^2$$

describes the ball's height above the ground, h, in feet, t seconds after it was kicked. Use this formula to solve Exercises 37–38.

37. What was the ball's height 2 seconds after it was kicked?
38. What was the ball's height 3 seconds after it was kicked?

In Exercises 39–40, name the property used to go from step to step each time that "(why?)" occurs.

39. $7 + 2(x + 9)$
$= 7 + (2x + 18)$ (why?)
$= 7 + (18 + 2x)$ (why?)
$= (7 + 18) + 2x$ (why?)
$= 25 + 2x$
$= 2x + 25$ (why?)

40. $5(x + 4) + 3x$
$= (5x + 20) + 3x$ (why?)
$= (20 + 5x) + 3x$ (why?)
$= 20 + (5x + 3x)$ (why?)
$= 20 + (5 + 3)x$ (why?)
$= 20 + 8x$
$= 8x + 20$ (why?)

In Exercises 41–62, simplify each algebraic expression.

41. $7x + 10x$
42. $5x + 13x$
43. $5x^2 - 8x^2$
44. $7x^2 - 10x^2$
45. $3(x + 5)$
46. $4(x + 6)$
47. $4(2x - 3)$
48. $3(4x - 5)$
49. $5(3x + 4) - 4$
50. $2(5x + 4) - 3$
51. $5(3x - 2) + 12x$
52. $2(5x - 1) + 14x$
53. $7(3y - 5) + 2(4y + 3)$
54. $4(2y - 6) + 3(5y + 10)$
55. $5(3y - 2) - (7y + 2)$
56. $4(5y - 3) - (6y + 3)$
57. $3(-4x^2 + 5x) - (5x - 4x^2)$
58. $2(-5x^2 + 3x) - (3x - 5x^2)$
59. $7 - 4[3 - (4y - 5)]$

60. $6 - 5[8 - (2y - 4)]$

61. $8x - 3[5 - (7 - 6x)]$

62. $7x - 4[6 - (8 - 5x)]$

Practice Plus

In Exercises 63–66, simplify each algebraic expression.

63. $18x^2 + 4 - [6(x^2 - 2) + 5]$

64. $14x^2 + 5 - [7(x^2 - 2) + 4]$

65. $2(3x^2 - 5) - [4(2x^2 - 1) + 3]$

66. $4(6x^2 - 3) - [2(5x^2 - 1) + 1]$

Application Exercises

The maximum heart rate, in beats per minute, that you should achieve during exercise is 220 minus your age:

$$220 - a.$$

This algebraic expression gives maximum heart rate in terms of age, a.

The bar graph shows the target heart rate ranges for four types of exercise goals. The lower and upper limits of these ranges are fractions of the maximum heart rate, 220 – a. Exercises 67–68 are based on the information in the graph.

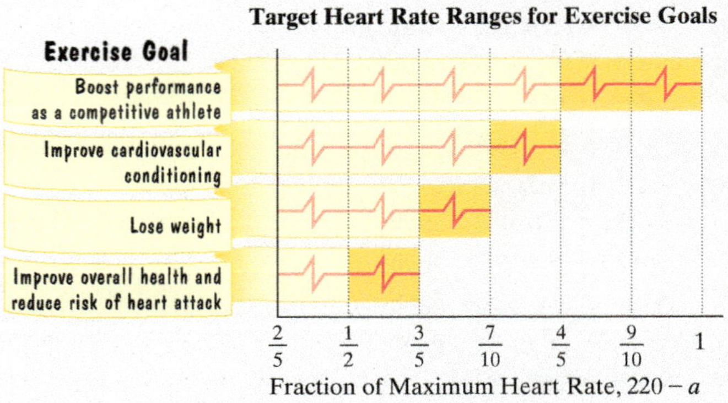

67. If your exercise goal is to improve cardiovascular conditioning, the graph shows the following range for target heart rate, H, in beats per minute:

Lower limit of range: $H = \frac{7}{10}(220 - a)$

Upper limit of range: $H = \frac{4}{5}(220 - a)$.

a. What is the lower limit of the heart rate range, in beats per minute, for a 20-year-old with this exercise goal?

b. What is the upper limit of the heart rate range, in beats per minute, for a 20-year-old with this exercise goal?

68. If your exercise goal is to improve overall health, the graph in the previous column shows the following range for target heart rate, H, in beats per minute:

Lower limit of range: $H = \frac{1}{2}(220 - a)$

Upper limit of range: $H = \frac{3}{5}(220 - a)$.

a. What is the lower limit of the heart rate range, in beats per minute, for a 30-year-old with this exercise goal?

b. What is the upper limit of the heart rate range, in beats per minute, for a 30-year-old with this exercise goal?

The bar graph shows the estimated number of calories per day needed to maintain energy balance for various gender and age groups for sedentary lifestyles. (Sedentary means a lifestyle that includes only the light physical activity associated with typical day-to-day life.)

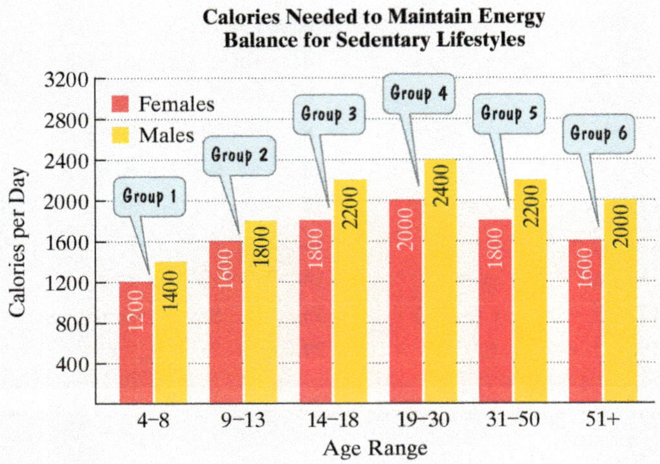

Source: USDA

Use the appropriate information displayed by the graph to solve Exercises 69–70.

69. The mathematical model

$$F = -82x^2 + 654x + 620$$

describes the number of calories needed per day, F, by females in age group x with sedentary lifestyles. According to the model, how many calories per day are needed by females between the ages of 19 and 30, inclusive, with this lifestyle? Does this underestimate or overestimate the number shown by the graph? By how much?

70. The mathematical model

$$M = -96x^2 + 802x + 660$$

describes the number of calories needed per day, M, by males in age group x with sedentary lifestyles. According to the model, how many calories per day are needed by males between the ages of 19 and 30, inclusive, with this lifestyle? Does this underestimate or overestimate the number shown by the graph? By how much?

Salary after College. *In 2010, MonsterCollege surveyed 1250 U.S. college students expecting to graduate in the next several years. Respondents were asked the following question:*

What do you think your starting salary will be at your first job after college?

The line graph shows the percentage of college students who anticipated various starting salaries.

Anticipated Starting Salary at First Job after College

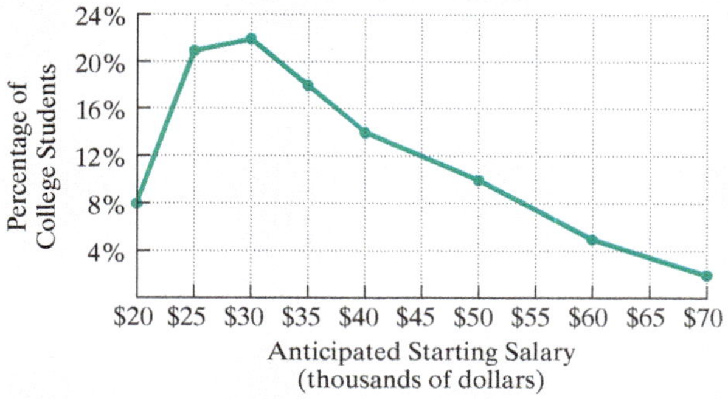

Source: MonsterCollege™

The mathematical model

$$p = -0.01s^2 + 0.8s + 3.7$$

describes the percentage of college students, p, who anticipated a starting salary, s, in thousands of dollars. Use this information to solve Exercises 71–72.

71. **a.** Use the line graph to estimate the percentage of students who anticipated a starting salary of $30 thousand.

b. Use the formula to find the percentage of students who anticipated a starting salary of $30 thousand. How does this compare with your estimate in part (a)?

72. **a.** Use the line graph to estimate the percentage of students who anticipated a starting salary of $40 thousand.

b. Use the formula to find the percentage of students who anticipated a starting salary of $40 thousand. How does this compare with your estimate in part (a)?

73. Read the Blitzer Bonus on page 346. Use the formula

$$\text{BAC} = \frac{600n}{w(0.6n + 169)}$$

and replace w with your body weight. Using this formula and a calculator, compute your BAC for integers from $n = 1$ to $n = 10$. Round to three decimal places. According to this model, how many drinks can you consume in an hour without exceeding the legal measure of drunk driving?

Writing in Mathematics

74. What is an algebraic expression? Provide an example with your description.

75. What does it mean to evaluate an algebraic expression? Provide an example with your description.

76. What is a term? Provide an example with your description.

77. What are like terms? Provide an example with your description.

78. Explain how to add like terms. Give an example.

79. What does it mean to simplify an algebraic expression?

80. An algebra student incorrectly used the distributive property and wrote $3(5x + 7) = 15x + 7$. If you were that student's teacher, what would you say to help the student avoid this kind of error?

Critical Thinking Exercises

Make Sense? *In Exercises 81–84, determine whether each statement makes sense or does not make sense, and explain your reasoning.*

81. I did not use the distributive property to simplify $3(2x + 5x)$.

82. The terms $13x^2$ and $10x$ both contain the variable x, so I can combine them to obtain $23x^3$.

83. Regardless of what real numbers I substitute for x and y, I will always obtain zero when evaluating $2x^2y - 2yx^2$.

84. A model that describes the number of lobbyists x years after 2000 cannot be used to estimate the number in 2000.

In Exercises 85–92, determine whether each statement is true or false. If the statement is false, make the necessary change(s) to produce a true statement.

85. The term x has no coefficient.

86. $5 + 3(x - 4) = 8(x - 4) = 8x - 32$

87. $-x - x = -x + (-x) = 0$

88. $x - 0.02(x + 200) = 0.98x - 4$

89. $3 + 7x = 10x$

90. $b \cdot b = 2b$

91. $(3y - 4) - (8y - 1) = -5y - 3$

92. $-4y + 4 = -4(y + 4)$

93. A business that manufactures small alarm clocks has weekly fixed costs of $5000. The average cost per clock for the business to manufacture x clocks is described by

$$\frac{0.5x + 5000}{x}.$$

a. Find the average cost when $x = 100, 1000,$ and $10{,}000$.

b. Like all other businesses, the alarm clock manufacturer must make a profit. To do this, each clock must be sold for at least 50¢ more than what it costs to manufacture. Due to competition from a larger company, the clocks can be sold for $1.50 each and no more. Our small manufacturer can only produce 2000 clocks weekly. Does this business have much of a future? Explain.

6.2

WHAT AM I SUPPOSED TO LEARN?

After you have read this section, you should be able to:

1. Solve linear equations.
2. Solve linear equations containing fractions.
3. Solve proportions.
4. Solve problems using proportions.
5. Identify equations with no solution or infinitely many solutions.

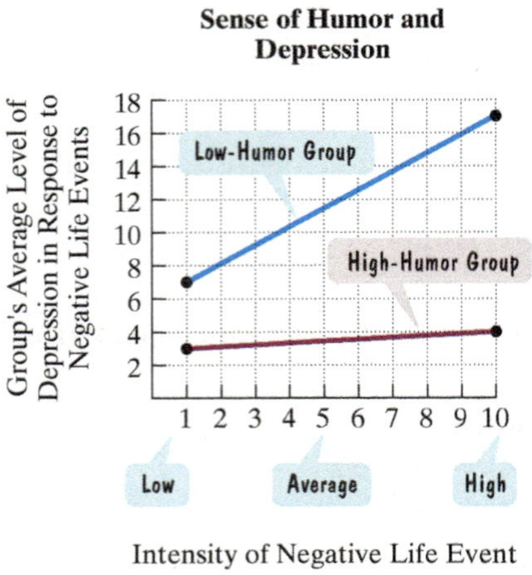

FIGURE 6.2

Source: Steven Davis and Joseph Palladino, *Psychology*, 5th Edition. Prentice Hall, 2007.

Linear Equations in One Variable and Proportions

THE BELIEF THAT HUMOR AND LAUGHTER can have positive benefits on our lives is not new. The graphs in **Figure 6.2** indicate that persons with a low sense of humor have higher levels of depression in response to negative life events than those with a high sense of humor. These graphs can be modeled by the following formulas:

Low-Humor Group: $$D = \frac{10}{9}x + \frac{53}{9}$$

High-Humor Group: $$D = \frac{1}{9}x + \frac{26}{9}.$$

In each formula, x represents the intensity of a negative life event (from 1, low, to 10, high) and D is the level of depression in response to that event.

Suppose that the low-humor group averages a level of depression of 10 in response to a negative life event. We can determine the intensity of that event by substituting 10 for D in the low-humor model, $D = \frac{10}{9}x + \frac{53}{9}$:

$$10 = \frac{10}{9}x + \frac{53}{9}.$$

The two sides of an equation can be reversed. So, we can also express this equation as

$$\frac{10}{9}x + \frac{53}{9} = 10.$$

Notice that the highest exponent on the variable is 1. Such an equation is called a *linear equation in one variable*. In this section, we will study how to solve such equations. We return to the models for sense of humor and depression later in the section.

1 Solve linear equations.

Solving Linear Equations in One Variable

We begin with the general definition of a linear equation in one variable.

DEFINITION OF A LINEAR EQUATION

A **linear equation in one variable x** is an equation that can be written in the form

$$ax + b = 0,$$

where a and b are real numbers, and $a \neq 0$.

An example of a linear equation in one variable is

$$4x + 12 = 0.$$

Solving an equation in x involves determining all values of x that result in a true statement when substituted into the equation. Such values are **solutions**, or **roots**, of the equation. For example, substitute -3 for x in $4x + 12 = 0$. We obtain

$$4(-3) + 12 = 0, \quad \text{or} \quad -12 + 12 = 0.$$

This simplifies to the true statement $0 = 0$. Thus, -3 is a solution of the equation $4x + 12 = 0$. We also say that -3 **satisfies** the equation $4x + 12 = 0$, because when we substitute -3 for x, a true statement results. The set of all such solutions is called the equation's **solution set**. For example, the solution set of the equation $4x + 12 = 0$ is $\{-3\}$.

Two or more equations that have the same solution set are called **equivalent equations**. For example, the equations

$$4x + 12 = 0 \quad \text{and} \quad 4x = -12 \quad \text{and} \quad x = -3$$

are equivalent equations because the solution set for each is $\{-3\}$. To solve a linear equation in x, we transform the equation into an equivalent equation one or more times. Our final equivalent equation should be of the form

$$x = \text{a number}.$$

The solution set of this equation is the set consisting of the number.

To generate equivalent equations, we will use the following properties:

THE ADDITION AND MULTIPLICATION PROPERTIES OF EQUALITY

The Addition Property of Equality

The same real number or algebraic expression may be added to both sides of an equation without changing the equation's solution set.

$$a = b \text{ and } a + c = b + c \text{ are equivalent equations.}$$

The Multiplication Property of Equality

The same nonzero real number may multiply both sides of an equation without changing the equation's solution set.

$$a = b \text{ and } ac = bc \text{ are equivalent equations as long as } c \neq 0.$$

Because subtraction is defined in terms of addition, the addition property also lets us subtract the same number from both sides of an equation without changing the equation's solution set. Similarly, because division is defined in terms of multiplication, the multiplication property of equality can be used to divide both sides of an equation by the same nonzero number to obtain an equivalent equation.

Table 6.1 illustrates how these properties are used to isolate x to obtain an equation of the form x = a number.

TABLE 6.1 Using Properties of Equality to Solve Equations

These equations are solved using the Addition Property of Equality.

These equations are solved using the Multiplication Property of Equality.

Equation	How to Isolate x	Solving the Equation	The Equation's Solution Set
$x - 3 = 8$	Add 3 to both sides.	$x - 3 + 3 = 8 + 3$ $x = 11$	{11}
$x + 7 = -15$	Subtract 7 from both sides.	$x + 7 - 7 = -15 - 7$ $x = -22$	{–22}
$6x = 30$	Divide both sides by 6 (or multiply both sides by $\frac{1}{6}$).	$\frac{6x}{6} = \frac{30}{6}$ $x = 5$	{5}
$\frac{x}{5} = 9$	Multiply both sides by 5.	$5 \cdot \frac{x}{5} = 5 \cdot 9$ $x = 45$	{45}

EXAMPLE 1 *Using Properties of Equality to Solve an Equation*

Solve and check: $2x + 3 = 17$.

SOLUTION

Our goal is to obtain an equivalent equation with x isolated on one side and a number on the other side.

$2x + 3 = 17$	This is the given equation.
$2x + 3 - 3 = 17 - 3$	Subtract 3 from both sides.
$2x = 14$	Simplify.
$\frac{2x}{2} = \frac{14}{2}$	Divide both sides by 2.
$x = 7$	Simplify: $\frac{2x}{2} = 1x = x$ and $\frac{14}{2} = 7$.

Now we check the proposed solution, 7, by replacing x with 7 in the original equation.

$2x + 3 = 17$	This is the original equation.
$2 \cdot 7 + 3 \stackrel{?}{=} 17$	Substitute 7 for x. The question mark indicates that we do not yet know if the two sides are equal.
$14 + 3 \stackrel{?}{=} 17$	Multiply: $2 \cdot 7 = 14$.
This statement is true. $17 = 17$	Add: $14 + 3 = 17$.

Because the check results in a true statement, we conclude that the solution set of the given equation is $\{7\}$.

CHECK POINT 1 Solve and check: $4x + 5 = 29$.

Here is a step-by-step procedure for solving a linear equation in one variable. Not all of these steps are necessary to solve every equation.

SOLVING A LINEAR EQUATION

1. Simplify the algebraic expression on each side by removing grouping symbols and combining like terms.
2. Collect all the variable terms on one side and all the constants, or numerical terms, on the other side.
3. Isolate the variable and solve.
4. Check the proposed solution in the original equation.

EXAMPLE 2 *Solving a Linear Equation*

Solve and check: $2(x - 4) - 5x = -5$.

SOLUTION

Step 1 Simplify the algebraic expression on each side.

$2(x - 4) - 5x = -5$	This is the given equation.
$2x - 8 - 5x = -5$	Use the distributive property.
$-3x - 8 = -5$	Combine like terms: $2x - 5x = -3x$.

Step 2 Collect variable terms on one side and constants on the other side. The only variable term in $-3x - 8 = -5$ is $-3x$, and $-3x$ is already on the left side. We will collect constants on the right side by adding 8 to both sides.

$$-3x - 8 + 8 = -5 + 8 \quad \text{Add 8 to both sides.}$$
$$-3x = 3 \quad \text{Simplify.}$$

Step 3 Isolate the variable and solve. We isolate the variable, x, by dividing both sides of $-3x = 3$ by -3.

$$\frac{-3x}{-3} = \frac{3}{-3} \quad \text{Divide both sides by } -3.$$
$$x = -1 \quad \text{Simplify: } \frac{-3x}{-3} = 1x = x \text{ and } \frac{3}{-3} = -1.$$

Step 4 Check the proposed solution in the original equation. Substitute -1 for x in the original equation.

$$2(x - 4) - 5x = -5 \quad \text{This is the original equation.}$$
$$2(-1 - 4) - 5(-1) \stackrel{?}{=} -5 \quad \text{Substitute } -1 \text{ for } x.$$
$$2(-5) - 5(-1) \stackrel{?}{=} -5 \quad \text{Simplify inside parentheses: } -1 - 4 = -1 + (-4) = -5.$$
$$-10 - (-5) \stackrel{?}{=} -5 \quad \text{Multiply: } 2(-5) = -10 \text{ and } 5(-1) = -5.$$
$$-5 = -5 \quad -10 - (-5) = -10 + 5 = -5$$

This statement is true.

Because the check results in a true statement, we conclude that the solution set of the given equation is $\{-1\}$.

CHECK POINT 2 Solve and check: $6(x - 3) - 10x = -10$.

GREAT QUESTION!

What are the differences between what I'm supposed to do with algebraic expressions and algebraic equations?

We simplify algebraic expressions. We solve algebraic equations. Although basic rules of algebra are used in both procedures, notice the differences between the procedures:

Simplifying an Algebraic Expression

Simplify: $3(x - 7) - (5x - 11)$.

This is not an equation. There is no equal sign.

Solution
$$\begin{aligned} &3(x - 7) - (5x - 11) \\ &= 3x - 21 - 5x + 11 \\ &= (3x - 5x) + (-21 + 11) \\ &= -2x + (-10) \\ &= -2x - 10 \end{aligned}$$

Stop! Further simplification is not possible. Avoid the common error of setting $-2x - 10$ equal to 0.

Solving an Algebraic Equation

Solve: $3(x - 7) - (5x - 11) = 14$.

This is an equation. There is an equal sign.

Solution
$$\begin{aligned} 3(x - 7) - (5x - 11) &= 14 \\ 3x - 21 - 5x + 11 &= 14 \\ -2x - 10 &= 14 \\ -2x - 10 + 10 &= 14 + 10 \quad \text{Add 10 to both sides.} \\ -2x &= 24 \\ \frac{-2x}{-2} &= \frac{24}{-2} \quad \text{Divide both sides by } -2. \\ x &= -12 \end{aligned}$$

The solution set is $\{-12\}$.

EXAMPLE 3 *Solving a Linear Equation*

Solve and check: $5x - 12 = 8x + 24$.

SOLUTION

Step 1 Simplify the algebraic expression on each side. Neither side contains grouping symbols or like terms that can be combined. Therefore, we can skip this step.

Step 2 Collect variable terms on one side and constants on the other side. One way to do this is to collect variable terms on the left and constants on the right. This is accomplished by subtracting $8x$ from both sides and adding 12 to both sides.

$$5x - 12 = 8x + 24$$ This is the given equation.

$$5x - 12 - 8x = 8x + 24 - 8x$$ Subtract $8x$ from both sides.

$$-3x - 12 = 24$$ Simplify: $5x - 8x = -3x$.

$$-3x - 12 + 12 = 24 + 12$$ Add 12 to both sides and collect constants on the right side.

$$-3x = 36$$ Simplify.

Step 3 Isolate the variable and solve. We isolate the variable, x, by dividing both sides of $-3x = 36$ by -3.

$$\frac{-3x}{-3} = \frac{36}{-3}$$ Divide both sides by -3.

$$x = -12$$ Simplify.

Step 4 Check the proposed solution in the original equation. Substitute -12 for x in the original equation.

$$5x - 12 = 8x + 24$$ This is the original equation.

$$5(-12) - 12 \stackrel{?}{=} 8(-12) + 24$$ Substitute -12 for x.

$$-60 - 12 \stackrel{?}{=} -96 + 24$$ Multiply: $5(-12) = -60$ and $8(-12) = -96$.

$$-72 = -72$$ Add: $-60 + (-12) = -72$ and $-96 + 24 = -72$.

This statement is true.

Because the check results in a true statement, we conclude that the solution set of the given equation is $\{-12\}$.

GREAT QUESTION!

Do I have to solve $5x - 12 = 8x + 24$ by collecting variable terms on the left and numbers on the right?

No. If you prefer, you can solve the equation by collecting variable terms on the right and numbers on the left. To collect variable terms on the right, subtract $5x$ from both sides:

$$5x - 12 - 5x = 8x + 24 - 5x$$
$$-12 = 3x + 24.$$

To collect numbers on the left, subtract 24 from both sides:

$$-12 - 24 = 3x + 24 - 24$$
$$-36 = 3x.$$

Now isolate x by dividing both sides by 3:

$$\frac{-36}{3} = \frac{3x}{3}$$
$$-12 = x.$$

This is the same solution that we obtained in Example 3.

CHECK POINT 3 Solve the equation: $2x + 9 = 8x - 3$.

EXAMPLE 4 *Solving a Linear Equation*

Solve and check: $2(x - 3) - 17 = 13 - 3(x + 2)$.

SOLUTION

Step 1 Simplify the algebraic expression on each side.

Do not begin with $13 - 3$. Multiplication (the distributive property) is applied before subtraction.

$$2(x - 3) - 17 = 13 - 3(x + 2)$$ This is the given equation.

$$2x - 6 - 17 = 13 - 3x - 6$$ Use the distributive property.

$$2x - 23 = -3x + 7$$ Combine like terms.

Step 2 Collect variable terms on one side and constants on the other side. We will collect variable terms of $2x - 23 = -3x + 7$ on the left by adding $3x$ to both sides. We will collect the numbers on the right by adding 23 to both sides.

$$2x - 23 + 3x = -3x + 7 + 3x \quad \text{Add } 3x \text{ to both sides.}$$
$$5x - 23 = 7 \quad \text{Simplify: } 2x + 3x = 5x.$$
$$5x - 23 + 23 = 7 + 23 \quad \text{Add 23 to both sides.}$$
$$5x = 30 \quad \text{Simplify.}$$

Step 3 Isolate the variable and solve. We isolate the variable, x, by dividing both sides of $5x = 30$ by 5.

$$\frac{5x}{5} = \frac{30}{5} \quad \text{Divide both sides by 5.}$$
$$x = 6 \quad \text{Simplify.}$$

Step 4 Check the proposed solution in the original equation. Substitute 6 for x in the original equation.

$$2(x - 3) - 17 = 13 - 3(x + 2) \quad \text{This is the original equation.}$$
$$2(6 - 3) - 17 \stackrel{?}{=} 13 - 3(6 + 2) \quad \text{Substitute 6 for } x.$$
$$2(3) - 17 \stackrel{?}{=} 13 - 3(8) \quad \text{Simplify inside parentheses.}$$
$$6 - 17 \stackrel{?}{=} 13 - 24 \quad \text{Multiply.}$$
$$-11 = -11 \quad \text{Subtract.}$$

The true statement $-11 = -11$ verifies that the solution set is $\{6\}$.

CHECK POINT 4 Solve and check: $4(2x + 1) = 29 + 3(2x - 5)$.

2 Solve linear equations containing fractions.

Linear Equations with Fractions

Equations are easier to solve when they do not contain fractions. How do we remove fractions from an equation? We begin by multiplying both sides of the equation by the least common denominator of any fractions in the equation. The least common denominator is the smallest number that all denominators will divide into. Multiplying every term on both sides of the equation by the least common denominator will eliminate the fractions in the equation. Example 5 shows how we "clear an equation of fractions."

EXAMPLE 5 *Solving a Linear Equation Involving Fractions*

Solve and check: $\frac{3x}{2} = \frac{8x}{5} - 4$.

SOLUTION

The denominators are 2 and 5. The smallest number that is divisible by 2 and 5 is 10. We begin by multiplying both sides of the equation by 10, the least common denominator.

$$\frac{3x}{2} = \frac{8x}{5} - 4 \quad \text{This is the given equation.}$$
$$10 \cdot \frac{3x}{2} = 10\left(\frac{8x}{5} - 4\right) \quad \text{Multiply both sides by 10.}$$
$$10 \cdot \frac{3x}{2} = 10 \cdot \frac{8x}{5} - 10 \cdot 4 \quad \text{Use the distributive property. Be sure to multiply all terms by 10.}$$

$$\overset{5}{\cancel{10}} \cdot \frac{3x}{\underset{1}{\cancel{2}}} = \overset{2}{\cancel{10}} \cdot \frac{8x}{\underset{1}{\cancel{5}}} - 40$$ Divide out common factors in the multiplications.

$$15x = 16x - 40$$ Complete the multiplications. The fractions are now cleared.

At this point, we have an equation similar to those we have previously solved. Collect the variable terms on one side and the constants on the other side.

$$15x - 16x = 16x - 40 - 16x$$ Subtract 16x from both sides to get the variable terms on the left.

$$-x = -40$$ Simplify.

We're not finished. A negative sign should not precede x.

Isolate x by multiplying or dividing both sides of this equation by -1.

$$\frac{-x}{-1} = \frac{-40}{-1}$$ Divide both sides by -1.

$$x = 40$$ Simplify.

Check the proposed solution. Substitute 40 for x in the original equation. You should obtain $60 = 60$. This true statement verifies that the solution set is $\{40\}$.

CHECK POINT 5 Solve and check: $\frac{2x}{3} = 7 - \frac{x}{2}$.

EXAMPLE 6 *An Application: Responding to Negative Life Events*

In the section opener, we introduced line graphs, repeated in **Figure 6.2**, indicating that persons with a low sense of humor have higher levels of depression in response to negative life events than those with a high sense of humor. These graphs can be modeled by the following formulas:

Low-Humor Group $$D = \frac{10}{9}x + \frac{53}{9}$$

High-Humor Group $$D = \frac{1}{9}x + \frac{26}{9}.$$

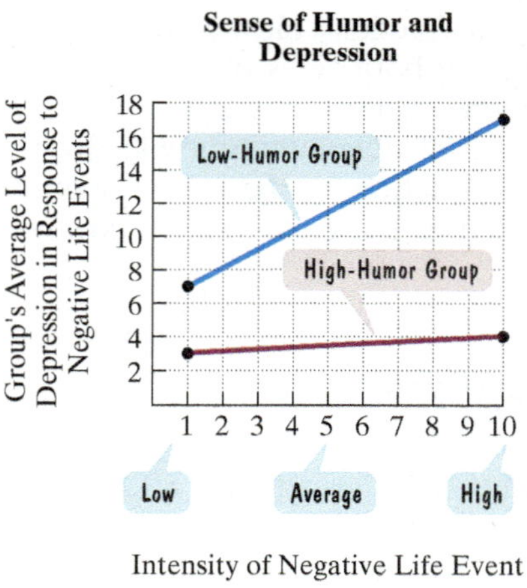

FIGURE 6.2 (repeated)

In each formula, x represents the intensity of a negative life event (from 1, low, to 10, high) and D is the average level of depression in response to that event. If the high-humor group averages a level of depression of 3.5, or $\frac{7}{2}$, in response to a negative life event, what is the intensity of that event? How is the solution shown on the red line graph in **Figure 6.2**?

SOLUTION

We are interested in the intensity of a negative life event with an average level of depression of $\frac{7}{2}$ for the high-humor group. We substitute $\frac{7}{2}$ for D in the high-humor model and solve for x, the intensity of the negative life event.

$$D = \frac{1}{9}x + \frac{26}{9}$$ This is the given formula for the high-humor group.

$$\frac{7}{2} = \frac{1}{9}x + \frac{26}{9}$$ Replace D with $\frac{7}{2}$.

$$18 \cdot \frac{7}{2} = 18\left(\frac{1}{9}x + \frac{26}{9}\right)$$ Multiply both sides by 18, the least common denominator.

$$18 \cdot \frac{7}{2} = 18 \cdot \frac{1}{9}x + 18 \cdot \frac{26}{9}$$ Use the distributive property.

$$\overset{9}{\cancel{18}} \cdot \frac{7}{\underset{1}{\cancel{2}}} = \overset{2}{\cancel{18}} \cdot \frac{1}{\underset{1}{\cancel{9}}}x + \overset{2}{\cancel{18}} \cdot \frac{26}{\underset{1}{\cancel{9}}}$$ Divide out common factors in the multiplications.

$$63 = 2x + 52$$ Complete the multiplications. The fractions are now cleared.

$$63 - 52 = 2x + 52 - 52$$ Subtract 52 from both sides to get constants on the left.

$$11 = 2x$$ Simplify.

$$\frac{11}{2} = \frac{2x}{2}$$ Divide both sides by 2.

$$\frac{11}{2} = x$$ Simplify.

The formula indicates that if the high-humor group averages a level of depression of 3.5 in response to a negative life event, the intensity of that event is $\frac{11}{2}$, or 5.5. This is illustrated on the line graph for the high-humor group in **Figure 6.3**.

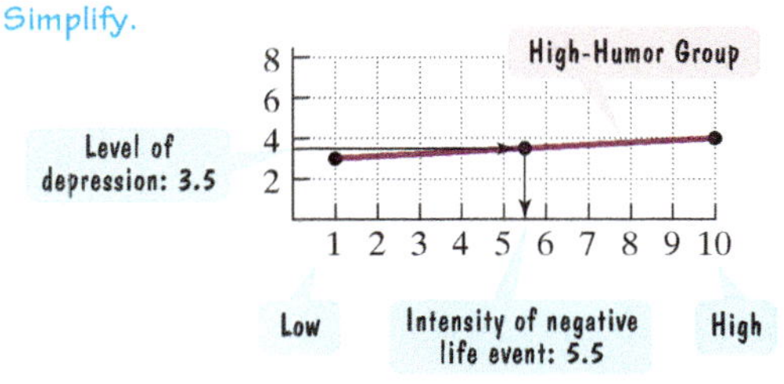

FIGURE 6.3

CHECK POINT 6 Use the model for the low-humor group given in Example 6 on the previous page to solve this problem. If the low-humor group averages a level of depression of 10 in response to a negative life event, what is the intensity of that event? How is the solution shown on the blue line graph in **Figure 6.2**?

3 Solve proportions.

Proportions

A **ratio** compares quantities by division. For example, a group contains 60 women and 30 men. The ratio of women to men is $\frac{60}{30}$. We can express this ratio as a fraction reduced to lowest terms:

$$\frac{60}{30} = \frac{2 \cdot \cancel{30}}{1 \cdot \cancel{30}} = \frac{2}{1}.$$

This ratio can be expressed as 2:1, or 2 to 1.

A **proportion** is a statement that says that two ratios are equal. If the ratios are $\frac{a}{b}$ and $\frac{c}{d}$, then the proportion is

$$\frac{a}{b} = \frac{c}{d}.$$

We can clear this equation of fractions by multiplying both sides by bd, the least common denominator:

$$\frac{a}{b} = \frac{c}{d}$$ This is the given proportion.

$$bd \cdot \frac{a}{b} = bd \cdot \frac{c}{d}$$ Multiply both sides by bd ($b \neq 0$ and $d \neq 0$). Then simplify. On the left, $\frac{\cancel{b}d}{1} \cdot \frac{a}{\cancel{b}} = da = ad$. On the right, $\frac{b\cancel{d}}{1} \cdot \frac{c}{\cancel{d}} = bc$.

$$ad = bc.$$

We see that the following principle is true for any proportion:

THE CROSS-PRODUCTS PRINCIPLE FOR PROPORTIONS

If $\frac{a}{b} = \frac{c}{d}$, then $ad = bc$. ($b \neq 0$ and $d \neq 0$)

The cross products ad and bc are equal.

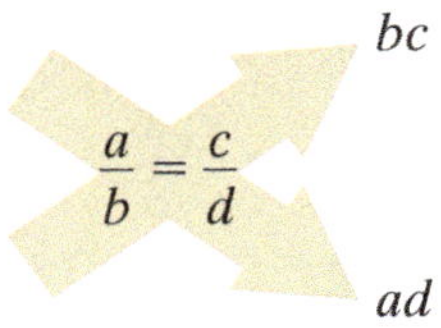

The cross-products principle: $ad = bc$

For example, since $\frac{2}{3} = \frac{6}{9}$, we see that $2 \cdot 9 = 3 \cdot 6$, or $18 = 18$. We can also use $\frac{2}{3} = \frac{6}{9}$ and conclude that $3 \cdot 6 = 2 \cdot 9$. When using the cross-products principle, it does not matter on which side of the equation each product is placed.

If three of the numbers in a proportion are known, the value of the missing quantity can be found by using the cross-products principle. This idea is illustrated in Example 7(a).

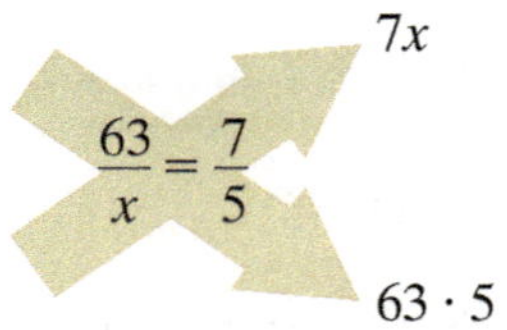

Cross products

EXAMPLE 7 *Solving Proportions*

Solve each proportion and check:

a. $\dfrac{63}{x} = \dfrac{7}{5}$ **b.** $\dfrac{20}{x - 10} = \dfrac{30}{x}$.

SOLUTION

a.

$\dfrac{63}{x} = \dfrac{7}{5}$	This is the given proportion.
$63 \cdot 5 = 7x$	Apply the cross-products principle.
$315 = 7x$	Simplify.
$\dfrac{315}{7} = \dfrac{7x}{7}$	Divide both sides by 7.
$45 = x$	Simplify.

The solution set is $\{45\}$.

Check

$\dfrac{63}{45} \stackrel{?}{=} \dfrac{7}{5}$	Substitute 45 for x in $\dfrac{63}{x} = \dfrac{7}{5}$.
$\dfrac{7 \cdot \cancel{9}}{5 \cdot \cancel{9}} \stackrel{?}{=} \dfrac{7}{5}$	Reduce $\dfrac{63}{45}$ to lowest terms.
$\dfrac{7}{5} = \dfrac{7}{5}$	This true statement verifies that the solution set is $\{45\}$.

b.

$\dfrac{20}{x - 10} = \dfrac{30}{x}$	This is the given proportion.
$20x = 30(x - 10)$	Apply the cross-products principle.
$20x = 30x - 30 \cdot 10$	Use the distributive property.
$20x = 30x - 300$	Simplify.
$20x - 30x = 30x - 300 - 30x$	Subtract $30x$ from both sides.
$-10x = -300$	Simplify.
$\dfrac{-10x}{-10} = \dfrac{-300}{-10}$	Divide both sides by -10.
$x = 30$	Simplify.

The solution set is $\{30\}$.

Check

$\dfrac{20}{30 - 10} \stackrel{?}{=} \dfrac{30}{30}$	Substitute 30 for x in $\dfrac{20}{x - 10} = \dfrac{30}{x}$.
$\dfrac{20}{20} \stackrel{?}{=} \dfrac{30}{30}$	Subtract: $30 - 10 = 20$.
$1 = 1$	This true statement verifies that the solution is 30.

CHECK POINT 7 Solve each proportion and check:

a. $\frac{10}{x} = \frac{2}{3}$ **b.** $\frac{22}{60 - x} = \frac{2}{x}$.

4 Solve problems using proportions.

Applications of Proportions

We now turn to practical application problems that can be solved using proportions. Here is a procedure for solving these problems:

SOLVING APPLIED PROBLEMS USING PROPORTIONS

1. Read the problem and represent the unknown quantity by x (or any letter).
2. Set up a proportion by listing the given ratio on one side and the ratio with the unknown quantity on the other side. Each respective quantity should occupy the same corresponding position on each side of the proportion.
3. Drop units and apply the cross-products principle.
4. Solve for x and answer the question.

EXAMPLE 8 *Applying Proportions: Calculating Taxes*

The property tax on a house with an assessed value of \$480,000 is \$5760. Determine the property tax on a house with an assessed value of \$600,000, assuming the same tax rate.

SOLUTION

Step 1 Represent the unknown by x. Let x = the tax on the \$600,000 house.

Step 2 Set up a proportion. We will set up a proportion comparing taxes to assessed value.

$\frac{\text{Tax on \$480,000 house}}{\text{Assessed value (\$480,000)}}$ equals $\frac{\text{Tax on \$600,000 house}}{\text{Assessed value (\$600,000)}}$

Given ratio $\left\{ \frac{\$5760}{\$480,000} \right.$ $=$ $\frac{\$x}{\$600,000}$ ← Unknown ($\$x$); ← Given quantity ($\$600,000$)

Step 3 Drop the units and apply the cross-products principle. We drop the dollar signs and begin to solve for x.

$$\frac{5760}{480,000} = \frac{x}{600,000} \quad \text{This is the proportion that models the problem's conditions.}$$

$$480,000x = (5760)(600,000) \quad \text{Apply the cross-products principle.}$$

$$480,000x = 3,456,000,000 \quad \text{Multiply.}$$

Step 4 Solve for x and answer the question.

$$\frac{480,000x}{480,000} = \frac{3,456,000,000}{480,000} \quad \text{Divide both sides by 480,000.}$$

$$x = 7200 \quad \text{Simplify.}$$

The property tax on the \$600,000 house is \$7200.

GREAT QUESTION!

Are there other proportions that I can use in step 2 to model the problem's conditions?

Yes. Here are three other correct proportions you can use:

- $\frac{\$480,000 \text{ value}}{\$5760 \text{ tax}} = \frac{\$600,000 \text{ value}}{\$x \text{ tax}}$
- $\frac{\$480,000 \text{ value}}{\$600,000 \text{ value}} = \frac{\$5760 \text{ tax}}{\$x \text{ tax}}$
- $\frac{\$600,000 \text{ value}}{\$480,000 \text{ value}} = \frac{\$x \text{ tax}}{\$5760 \text{ tax}}$

Each proportion gives the same cross product obtained in step 3.

CHECK POINT 8 The property tax on a house with an assessed value of \$250,000 is \$3500. Determine the property tax on a house with an assessed value of \$420,000, assuming the same tax rate.

EXAMPLE 9 *Applying Proportions: Estimating Wildlife Population*

Wildlife biologists catch, tag, and then release 135 deer back into a wildlife refuge. Two weeks later they observe a sample of 140 deer, 30 of which are tagged. Assuming the ratio of tagged deer in the sample holds for all deer in the refuge, approximately how many deer are in the refuge?

SOLUTION

Step 1 Represent the unknown by *x*. Let $x =$ the total number of deer in the refuge.

Step 2 Set up a proportion.

Original number of tagged deer / Total number of deer (Unknown → Total number of deer) equals Number of tagged deer in the observed sample / Total number of deer in the observed sample (Known ratio)

$$\frac{135}{x} = \frac{30}{140}$$

Steps 3 and 4 Apply the cross-products principle, solve, and answer the question.

$$\frac{135}{x} = \frac{30}{140}$$ This is the proportion that models the problem's conditions.

$$(135)(140) = 30x$$ Apply the cross-products principle.

$$18{,}900 = 30x$$ Multiply.

$$\frac{18{,}900}{30} = \frac{30x}{30}$$ Divide both sides by 30.

$$630 = x$$ Simplify.

There are approximately 630 deer in the refuge.

CHECK POINT 9 Wildlife biologists catch, tag, and then release 120 deer back into a wildlife refuge. Two weeks later they observe a sample of 150 deer, 25 of which are tagged. Assuming the ratio of tagged deer in the sample holds for all deer in the refuge, approximately how many deer are in the refuge?

5 Identify equations with no solution or infinitely many solutions.

Equations with No Solution or Infinitely Many Solutions

Thus far, each equation or proportion that we have solved has had a single solution. However, some equations are not true for even one real number. By contrast, other equations are true for all real numbers.

If you attempt to solve an equation with no solution, you will eliminate the variable and obtain a false statement, such as $2 = 5$. If you attempt to solve an equation that is true for every real number, you will eliminate the variable and obtain a true statement, such as $4 = 4$.

EXAMPLE 10 *Attempting to Solve an Equation with No Solution*

Solve: $2x + 6 = 2(x + 4)$.

SOLUTION

$$2x + 6 = 2(x + 4)$$ This is the given equation.

$$2x + 6 = 2x + 8$$ Use the distributive property.

$$2x + 6 - 2x = 2x + 8 - 2x$$ Subtract 2x from both sides.

$$6 = 8$$ Simplify.

Keep reading. 6 = 8 is not the solution.

The original equation, $2x + 6 = 2(x + 4)$, is equivalent to the statement $6 = 8$, which is false for every value of x. The equation has no solution. The solution set is $\varnothing$, the empty set.

CHECK POINT 10 Solve: $3x + 7 = 3(x + 1)$.

EXAMPLE 11 *Solving an Equation for Which Every Real Number Is a Solution*

Solve: $4x + 6 = 6(x + 1) - 2x$.

SOLUTION

$4x + 6 = 6(x + 1) - 2x$ — This is the given equation.

$4x + 6 = 6x + 6 - 2x$ — Apply the distributive property on the right side.

$4x + 6 = 4x + 6$ — Combine like terms on the right side: $6x - 2x = 4x$.

Can you see that the equation $4x + 6 = 4x + 6$ is true for every value of x? Let's continue solving the equation by subtracting $4x$ from both sides.

$$4x + 6 - 4x = 4x + 6 - 4x$$

$$6 = 6$$

Keep reading. $6 = 6$ is not the solution.

The original equation is equivalent to the statement $6 = 6$, which is true for every value of x. Thus, the solution set consists of the set of all real numbers, expressed in set-builder notation as $\{x | x \text{ is a real number}\}$. Try substituting any real number of your choice for x in the original equation. You will obtain a true statement.

GREAT QUESTION!

Do I have to use sets to write the solution of an equation?

Because of the fundamental role that sets play in mathematics, it's a good idea to use set notation to express an equation's solution. If an equation has no solution, its solution set is $\varnothing$, the empty set. If an equation with variable x is true for every real number, its solution set is $\{x | x \text{ is a real number}\}$.

CHECK POINT 11 Solve: $7x + 9 = 9(x + 1) - 2x$.

Concept and Vocabulary Check

Fill in each blank so that the resulting statement is true.

1. An equation in the form $ax + b = 0$, $a \neq 0$, such as $3x + 17 = 0$, is called a/an ________ equation in one variable.
2. Two or more equations that have the same solution set are called ________ equations.
3. The addition property of equality states that if $a = b$, then $a + c =$ ________.
4. The multiplication property of equality states that if $a = b$ and $c \neq 0$, then $ac =$ ________.
5. The first step in solving $7 + 3(x - 2) = 2x + 10$ is to ____________.
6. The algebraic expression $7(x - 4) + 2x$ can be ________, whereas the algebraic equation $7(x - 4) + 2x = 35$ can be ________.
7. The equation

$$\frac{x}{4} = 2 + \frac{x}{3}$$

can be cleared of fractions by multiplying both sides by the ______________ of $\frac{x}{4}$ and $\frac{x}{3}$, which is ________.

8. A statement that two ratios are equal is called a/an ________.
9. The cross-products principle states that if $\frac{a}{b} = \frac{c}{d}$ ($b \neq 0$ and $d \neq 0$), then ________.

10. In solving an equation, if you eliminate the variable and obtain a statement such as $2 = 3$, the equation has ________ solution. The solution set can be expressed using the symbol ________.

11. In solving an equation with variable x, if you eliminate the variable and obtain a statement such as $6 = 6$, the equation is ________ for every value of x. The solution set can be expressed in set-builder notation as ________________.

In Exercises 12–15, determine whether each statement is true or false. If the statement is false, make the necessary change(s) to produce a true statement.

12. The equation $2x + 5 = 0$ is equivalent to $2x = 5$. _____

13. The equation $x + \frac{1}{3} = \frac{1}{2}$ is equivalent to $x + 2 = 3$. _____

14. The equation $3x = 2x$ has no solution. _____

15. The equation $3(x + 4) = 3(4 + x)$ has precisely one solution. _____

Exercise Set 6.2

Practice Exercises

In Exercises 1–58, solve and check each equation.

1. $x - 7 = 3$
2. $x - 3 = -17$
3. $x + 5 = -12$
4. $x + 12 = -14$
5. $\frac{x}{3} = 4$
6. $\frac{x}{5} = 3$
7. $5x = 45$
8. $6x = 18$
9. $8x = -24$
10. $5x = -25$
11. $-8x = 2$
12. $-6x = 3$
13. $5x + 3 = 18$
14. $3x + 8 = 50$
15. $6x - 3 = 63$
16. $5x - 8 = 72$
17. $4x - 14 = -82$
18. $9x - 14 = -77$
19. $14 - 5x = -41$
20. $25 - 6x = -83$
21. $9(5x - 2) = 45$
22. $10(3x + 2) = 70$
23. $5x - (2x - 10) = 35$
24. $11x - (6x - 5) = 40$
25. $3x + 5 = 2x + 13$
26. $2x - 7 = 6 + x$
27. $8x - 2 = 7x - 5$
28. $13x + 14 = -5 + 12x$
29. $7x + 4 = x + 16$
30. $8x + 1 = x + 43$
31. $8y - 3 = 11y + 9$
32. $5y - 2 = 9y + 2$
33. $2(4 - 3x) = 2(2x + 5)$
34. $3(5 - x) = 4(2x + 1)$
35. $8(y + 2) = 2(3y + 4)$
36. $3(3y - 1) = 4(3 + 3y)$
37. $3(x + 1) = 7(x - 2) - 3$
38. $5x - 4(x + 9) = 2x - 3$
39. $5(2x - 8) - 2 = 5(x - 3) + 3$
40. $7(3x - 2) + 5 = 6(2x - 1) + 24$
41. $6 = -4(1 - x) + 3(x + 1)$
42. $100 = -(x - 1) + 4(x - 6)$
43. $10(z + 4) - 4(z - 2) = 3(z - 1) + 2(z - 3)$
44. $-2(z - 4) - (3z - 2) = -2 - (6z - 2)$
45. $\frac{2x}{3} - 5 = 7$
46. $\frac{3x}{4} - 9 = -6$
47. $\frac{x}{3} + \frac{x}{2} = \frac{5}{6}$
48. $\frac{x}{4} - \frac{x}{5} = 1$
49. $20 - \frac{z}{3} = \frac{z}{2}$
50. $\frac{z}{5} - \frac{1}{2} = \frac{z}{6}$
51. $\frac{y}{3} + \frac{2}{5} = \frac{y}{5} - \frac{2}{5}$
52. $\frac{y}{12} + \frac{1}{6} = \frac{y}{2} - \frac{1}{4}$
53. $\frac{3x}{4} - 3 = \frac{x}{2} + 2$
54. $\frac{3x}{5} - \frac{2}{5} = \frac{x}{3} + \frac{2}{5}$
55. $\frac{3x}{5} - x = \frac{x}{10} - \frac{5}{2}$
56. $2x - \frac{2x}{7} = \frac{x}{2} + \frac{17}{2}$
57. $\frac{x - 3}{5} - 1 = \frac{x - 5}{4}$
58. $\frac{x - 2}{3} - 4 = \frac{x + 1}{4}$

In Exercises 59–72, solve each proportion and check.

59. $\frac{24}{x} = \frac{12}{7}$
60. $\frac{56}{x} = \frac{8}{7}$
61. $\frac{x}{6} = \frac{18}{4}$
62. $\frac{x}{32} = \frac{3}{24}$
63. $\frac{-3}{8} = \frac{x}{40}$
64. $\frac{-3}{8} = \frac{6}{x}$
65. $\frac{x}{12} = -\frac{3}{4}$
66. $\frac{x}{64} = -\frac{9}{16}$
67. $\frac{x - 2}{12} = \frac{8}{3}$
68. $\frac{x - 4}{10} = \frac{3}{5}$
69. $\frac{x}{7} = \frac{x + 14}{5}$
70. $\frac{x}{5} = \frac{x - 3}{2}$
71. $\frac{y + 10}{10} = \frac{y - 2}{4}$
72. $\frac{2}{y - 5} = \frac{3}{y + 6}$

In Exercises 73–92, solve each equation. Use set notation to express solution sets for equations with no solution or equations that are true for all real numbers.

73. $3x - 7 = 3(x + 1)$
74. $2(x - 5) = 2x + 10$
75. $2(x + 4) = 4x + 5 - 2x + 3$
76. $3(x - 1) = 8x + 6 - 5x - 9$
77. $7 + 2(3x - 5) = 8 - 3(2x + 1)$
78. $2 + 3(2x - 7) = 9 - 4(3x + 1)$
79. $4x + 1 - 5x = 5 - (x + 4)$
80. $5x - 5 = 3x - 7 + 2(x + 1)$
81. $4(x + 2) + 1 = 7x - 3(x - 2)$
82. $5x - 3(x + 1) = 2(x + 3) - 5$
83. $3 - x = 2x + 3$
84. $5 - x = 4x + 5$
85. $\frac{x}{3} + 2 = \frac{x}{3}$
86. $\frac{x}{4} + 3 = \frac{x}{4}$

87. $\frac{x}{3} = \frac{x}{2}$

88. $\frac{x}{4} = \frac{x}{3}$

89. $\frac{x-2}{5} = \frac{3}{10}$

90. $\frac{x+4}{8} = \frac{3}{16}$

91. $\frac{x}{2} - \frac{x}{4} + 4 = x + 4$

92. $\frac{x}{2} + \frac{2x}{3} + 3 = x + 3$

Practice Plus

93. Evaluate $x^2 - x$ for the value of x satisfying $4(x-2) + 2 = 4x - 2(2-x)$.

94. Evaluate $x^2 - x$ for the value of x satisfying $2(x-6) = 3x + 2(2x-1)$.

95. Evaluate $x^2 - (xy - y)$ for x satisfying $\frac{x}{5} - 2 = \frac{x}{3}$ and y satisfying $-2y - 10 = 5y + 18$.

96. Evaluate $x^2 - (xy - y)$ for x satisfying $\frac{3x}{2} + \frac{3x}{4} = \frac{x}{4} - 4$ and y satisfying $5 - y = 7(y+4) + 1$.

In Exercises 97–104, solve each equation.

97. $[(3+6)^2 \div 3] \cdot 4 = -54x$

98. $2^3 - [4(5-3)^3] = -8x$

99. $5 - 12x = 8 - 7x - [6 \div 3(2 + 5^3) + 5x]$

100. $2(5x + 58) = 10x + 4(21 \div 3.5 - 11)$

101. $0.7x + 0.4(20) = 0.5(x + 20)$

102. $0.5(x + 2) = 0.1 + 3(0.1x + 0.3)$

103. $4x + 13 - \{2x - [4(x-3) - 5]\} = 2(x - 6)$

104. $-2\{7 - [4 - 2(1-x) + 3]\} = 10 - [4x - 2(x-3)]$

Application Exercises

The latest guidelines, which apply to both men and women, give healthy weight ranges, rather than specific weights, for your height. The further you are above the upper limit of your range, the greater are the risks of developing weight-related health problems. The bar graph shows these ranges for various heights for people between the ages of 19 and 34, inclusive.

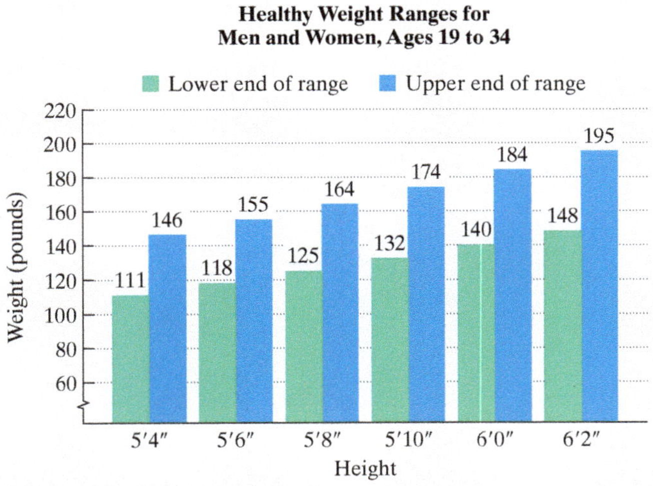

Source: U.S. Department of Health and Human Services

The mathematical model

$$\frac{W}{2} - 3H = 53$$

describes a weight, W, in pounds, that lies within the healthy weight range for a person whose height is H inches over 5 feet. Use this information to solve Exercises 105–106.

105. Use the formula to find a healthy weight for a person whose height is 5'6". (*Hint:* $H = 6$ because this person's height is 6 inches over 5 feet.) How many pounds is this healthy weight below the upper end of the range shown by the bar graph at the bottom of the previous column?

106. Use the formula to find a healthy weight for a person whose height is 6'0". (*Hint:* $H = 12$ because this person's height is 12 inches over 5 feet.) How many pounds is this healthy weight below the upper end of the range shown by the bar graph at the bottom of the previous column?

In the years after warning labels were put on cigarette packs, the number of smokers dropped from approximately two in five adults to one in five. The bar graph shows the percentage of American adults who smoked cigarettes for selected years from 1970 through 2010.

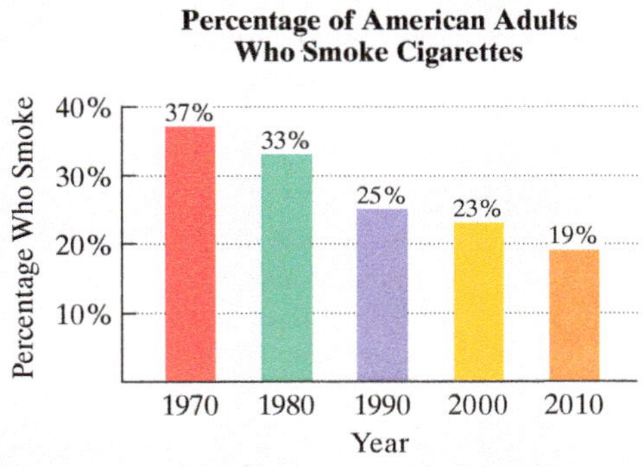

Source: Centers for Disease Control and Prevention

The mathematical model

$$p + \frac{x}{2} = 37$$

describes the percentage of Americans who smoked cigarettes, p, x years after 1970. Use this model to solve Exercises 107–108.

107. **a.** Does the mathematical model underestimate or overestimate the percentage of American adults who smoked cigarettes in 2010? By how much?

b. Use the mathematical model to project the year when only 7% of American adults will smoke cigarettes.

108. **a.** Does the mathematical model underestimate or overestimate the percentage of American adults who smoked cigarettes in 2000? By how much?

b. Use the mathematical model to project the year when only 2% of American adults will smoke cigarettes.

109. The volume of blood in a person's body is proportional to body weight. A person who weighs 160 pounds has approximately 5 quarts of blood. Estimate the number of quarts of blood in a person who weighs 200 pounds.

110. The number of gallons of water used when taking a shower is proportional to the time in the shower. A shower lasting 5 minutes uses 30 gallons of water. How much water is used in a shower lasting 11 minutes?

111. An alligator's tail length is proportional to its body length. An alligator with a body length of 4 feet has a tail length of 3.6 feet. What is the tail length of an alligator whose body length is 6 feet?

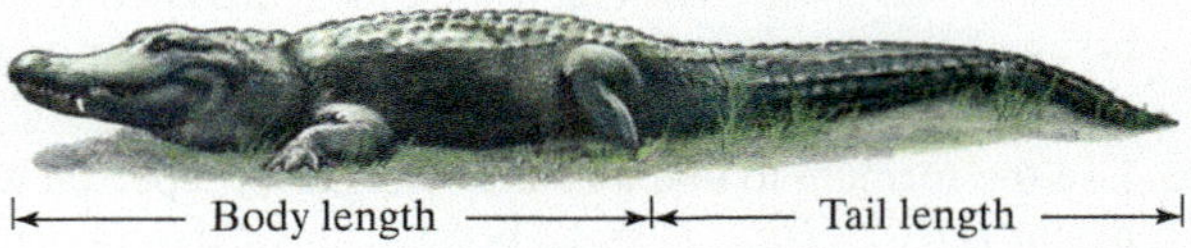

112. An object's weight on the moon is proportional to its weight on Earth. Neil Armstrong, the first person to step on the moon on July 20, 1969, weighed 360 pounds on Earth (with all of his equipment on) and 60 pounds on the moon. What is the moon weight of a person who weighs 186 pounds on Earth?

113. St. Paul Island in Alaska has 12 fur seal rookeries (breeding places). In 1961, to estimate the fur seal pup population in the Gorbath rookery, 4963 fur seal pups were tagged in early August. In late August, a sample of 900 pups was observed and 218 of these were found to have been previously tagged. Estimate the total number of fur seal pups in this rookery.

114. To estimate the number of bass in a lake, wildlife biologists tagged 50 bass and released them in the lake. Later they netted 108 bass and found that 27 of them were tagged. Approximately how many bass are in the lake?

Writing in Mathematics

115. What is the solution set of an equation?

116. State the addition property of equality and give an example.

117. State the multiplication property of equality and give an example.

118. What is a proportion? Give an example with your description.

119. Explain how to solve a proportion. Illustrate your explanation with an example.

120. How do you know whether an equation has one solution, no solution, or infinitely many solutions?

121. What is the difference between solving an equation such as $2(x - 4) + 5x = 34$ and simplifying an algebraic expression such as $2(x - 4) + 5x$? If there is a difference, which topic should be taught first? Why?

122. Suppose that you solve $\frac{x}{5} - \frac{x}{2} = 1$ by multiplying both sides by 20, rather than the least common denominator of $\frac{x}{5}$ and $\frac{x}{2}$ (namely, 10). Describe what happens. If you get the correct solution, why do you think we clear the equation of fractions by multiplying by the *least* common denominator?

123. Suppose you are an algebra teacher grading the following solution on an examination:

$$\begin{aligned} \text{Solve: } & -3(x - 6) = 2 - x. \\ \text{Solution: } & -3x - 18 = 2 - x \\ & -2x - 18 = 2 \\ & -2x = -16 \\ & x = 8. \end{aligned}$$

You should note that 8 checks, and the solution set is $\{8\}$. The student who worked the problem therefore wants full credit. Can you find any errors in the solution? If full credit is 10 points, how many points should you give the student? Justify your position.

124. Although the formulas in Example 6 on page 356 are correct, some people object to representing the variables with numbers, such as a 1-to-10 scale for the intensity of a negative life event. What might be their objection to quantifying the variables in this situation?

Critical Thinking Exercises

Make Sense? *In Exercises 125–128, determine whether each statement makes sense or does not make sense, and explain your reasoning.*

125. Although I can solve $3x + \frac{1}{5} = \frac{1}{4}$ by first subtracting $\frac{1}{5}$ from both sides, I find it easier to begin by multiplying both sides by 20, the least common denominator.

126. Because I know how to clear an equation of fractions, I decided to clear the equation $0.5x + 8.3 = 12.4$ of decimals by multiplying both sides by 10.

127. The number 3 satisfies the equation $7x + 9 = 9(x + 1) - 2x$, so $\{3\}$ is the equation's solution set.

128. I can solve $\frac{x}{9} = \frac{4}{6}$ by using the cross-products principle or by multiplying both sides by 18, the least common denominator.

129. Write three equations whose solution set is $\{5\}$.

130. If x represents a number, write an English sentence about the number that results in an equation with no solution.

131. A woman's height, h, is related to the length of the femur, f (the bone from the knee to the hip socket), by the formula $f = 0.432h - 10.44$. Both h and f are measured in inches. A partial skeleton is found of a woman in which the femur is 16 inches long. Police find the skeleton in an area where a woman slightly over 5 feet tall has been missing for over a year. Can the partial skeleton be that of the missing woman? Explain.

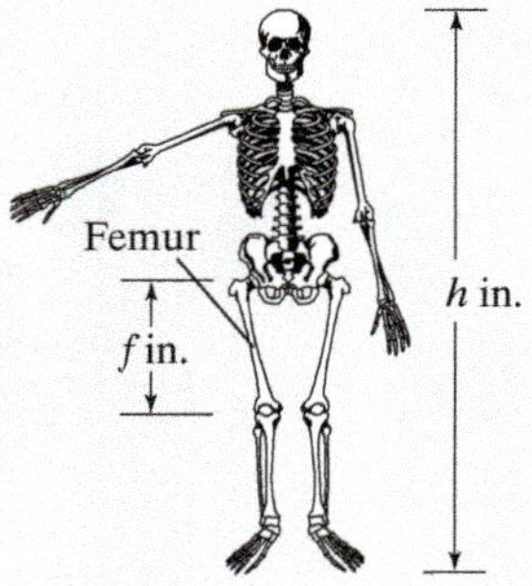

6.3 Applications of Linear Equations

WHAT AM I SUPPOSED TO LEARN?

After you have read this section, you should be able to:

1 Use linear equations to solve problems.

2 Solve a formula for a variable.

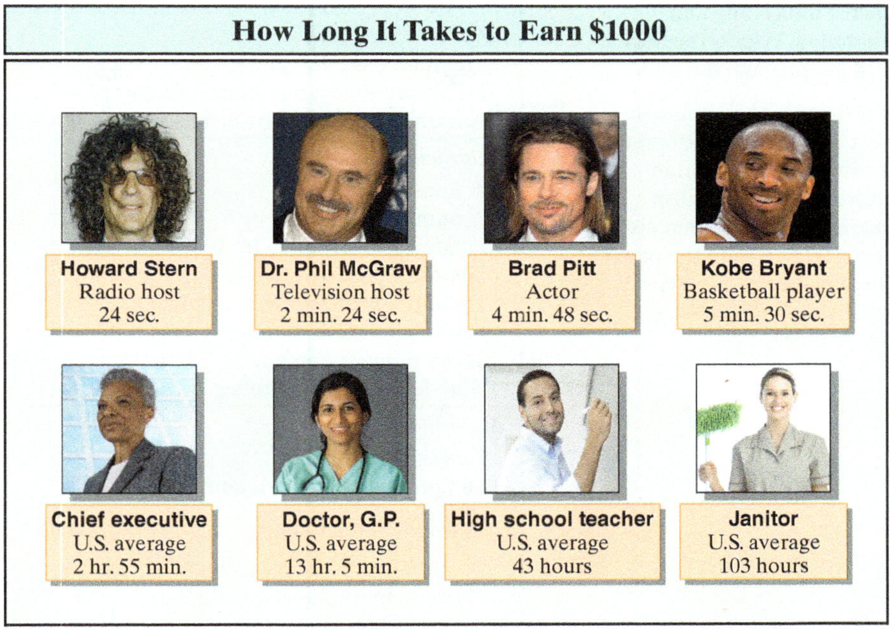

Source: Time

In this section, you'll see examples and exercises focused on how much money Americans earn. These situations illustrate a step-by-step strategy for solving problems. As you become familiar with this strategy, you will learn to solve a wide variety of problems.

1 Use linear equations to solve problems.

Problem Solving with Linear Equations

We have seen that a model is a mathematical representation of a real-world situation. In this section, we will be solving problems that are presented in English. This means that we must obtain models by translating from the ordinary language of English into the language of algebraic equations. To translate, however, we must understand the English prose and be familiar with the forms of algebraic language. Following are some general steps we will follow in solving word problems.

STRATEGY FOR SOLVING WORD PROBLEMS

Step 1 Read the problem carefully several times until you can state in your own words what is given and what the problem is looking for. Let x (or any variable) represent one of the unknown quantities in the problem.

Step 2 If necessary, write expressions for any other unknown quantities in the problem in terms of x.

Step 3 Write an equation in x that models the verbal conditions of the problem.

Step 4 Solve the equation and answer the problem's question.

Step 5 Check the solution *in the original wording* of the problem, not in the equation obtained from the words.

The most difficult step in this process is step 3 because it involves translating verbal conditions into an algebraic equation. Translations of some commonly used English phrases are listed in **Table 6.2** on the next page. We choose to use x to represent the variable, but we can use any letter.

GREAT QUESTION!

Table 6.2 looks long and intimidating. What's the best way to get through the table?

Cover the right column with a sheet of paper and attempt to formulate the algebraic expression for the English phrase in the left column on your own. Then slide the paper down and check your answer. Work through the entire table in this manner.

TABLE 6.2 Algebraic Translations of English Phrases

English Phrase	Algebraic Expression
Addition	
The sum of a number and 7	$x + 7$
Five more than a number; a number plus 5	$x + 5$
A number increased by 6; 6 added to a number	$x + 6$
Subtraction	
A number minus 4	$x - 4$
A number decreased by 5	$x - 5$
A number subtracted from 8	$8 - x$
The difference between a number and 6	$x - 6$
The difference between 6 and a number	$6 - x$
Seven less than a number	$x - 7$
Seven minus a number	$7 - x$
Nine fewer than a number	$x - 9$
Multiplication	
Five times a number	$5x$
The product of 3 and a number	$3x$
Two-thirds of a number (used with fractions)	$\frac{2}{3}x$
Seventy-five percent of a number (used with decimals)	$0.75x$
Thirteen multiplied by a number	$13x$
A number multiplied by 13	$13x$
Twice a number	$2x$
Division	
A number divided by 3	$\frac{x}{3}$
The quotient of 7 and a number	$\frac{7}{x}$
The quotient of a number and 7	$\frac{x}{7}$
The reciprocal of a number	$\frac{1}{x}$
More than one operation	
The sum of twice a number and 7	$2x + 7$
Twice the sum of a number and 7	$2(x + 7)$
Three times the sum of 1 and twice a number	$3(1 + 2x)$
Nine subtracted from 8 times a number	$8x - 9$
Twenty-five percent of the sum of 3 times a number and 14	$0.25(3x + 14)$
Seven times a number, increased by 24	$7x + 24$
Seven times the sum of a number and 24	$7(x + 24)$

EXAMPLE 1 *Starting Salaries for College Graduates with Undergraduate Degrees*

The bar graph in **Figure 6.4** shows the ten most popular college majors with median, or middlemost, starting salaries for recent college graduates.

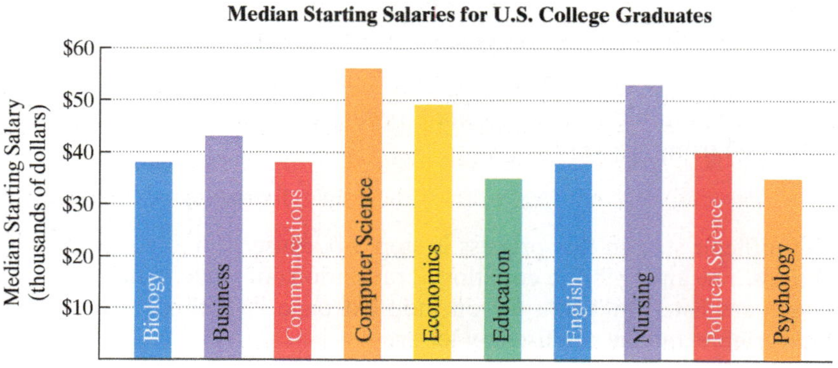

FIGURE 6.4
Source: PayScale (2010 data)

The median starting salary of a business major exceeds that of a psychology major by \$8 thousand. The median starting salary of an English major exceeds that of a psychology major by \$3 thousand. Combined, their median starting salaries are \$116 thousand. Determine the median starting salaries of psychology majors, business majors, and English majors with bachelor's degrees.

SOLUTION

Step 1 Let x represent one of the unknown quantities. We know something about the median starting salaries of business majors and English majors: Business majors earn \$8 thousand more than psychology majors and English majors earn \$3 thousand more than psychology majors. We will let

x = the median starting salary, in thousands of dollars, of psychology majors.

Step 2 Represent other unknown quantities in terms of x. Because business majors earn \$8 thousand more than psychology majors, let

Salary exceeds a psychology major, x, by \$8 thousand.

$x + 8$ = the median starting salary, in thousands of dollars, of business majors.

Because English majors earn \$3 thousand more than psychology majors, let

Salary exceeds a psychology major, x, by \$3 thousand.

$x + 3$ = the median starting salary, in thousands of dollars, of English majors.

Step 3 Write an equation in x that models the conditions. Combined, the median starting salaries for psychology, business, and English majors are \$116 thousand.

The median starting salary for psychology majors	plus	the median starting salary for business majors	plus	the median starting salary for English majors	is	\$116 thousand.
x	$+$	$(x + 8)$	$+$	$(x + 3)$	$=$	116

Step 4 Solve the equation and answer the question.

$$x + (x + 8) + (x + 3) = 116$$ This is the equation that models the problem's conditions.

$$3x + 11 = 116$$ Remove parentheses, regroup, and combine like terms.

$$3x = 105$$ Subtract 11 from both sides.

$$x = 35$$ Divide both sides by 3.

Thus,

starting salary of psychology majors $= x = 35$

starting salary of business majors $= x + 8 = 35 + 8 = 43$

starting salary of English majors $= x + 3 = 35 + 3 = 38.$

The median starting salary of psychology majors is \$35 thousand, the median starting salary of business majors is \$43 thousand, and the median starting salary of English majors is \$38 thousand.

Step 5 Check the proposed solution in the original wording of the problem. The problem states that combined, the median starting salaries are \$116 thousand. Using the median salaries we determined in step 4, the sum is

\$35 thousand + \$43 thousand + \$38 thousand,

or \$116 thousand, which verifies the problem's conditions.

GREAT QUESTION!

Example 1 involves using the word *exceeds* to represent two of the unknown quantities. Can you help me to write algebraic expressions for quantities described using *exceeds*?

Modeling with the word *exceeds* can be a bit tricky. It's helpful to identify the smaller quantity. Then add to this quantity to represent the larger quantity. For example, suppose that Tim's height exceeds Tom's height by a inches. Tom is the shorter person. If Tom's height is represented by x, then Tim's height is represented by $x + a$.

CHECK POINT 1 Three of the bars in **Figure 6.4** on page 366 represent median starting salaries of education, computer science, and economics majors. The median starting salary of a computer science major exceeds that of an education major by \$21 thousand. The median starting salary of an economics major exceeds that of an education major by \$14 thousand. Combined, their median starting salaries are \$140 thousand. Determine the median starting salaries of education majors, computer science majors, and economics majors with bachelor's degrees.

EXAMPLE 2 *Modeling Attitudes of College Freshmen*

Your author teaching math in 1969

Researchers have surveyed college freshmen every year since 1969. **Figure 6.5** shows that attitudes about some life goals have changed dramatically over the years. In particular, the freshman class of 2010 was more interested in making money than the freshmen of 1969 had been. In 1969, 42% of first-year college students considered "being well-off financially" essential or very important. For the period from 1969 through 2010, this percentage increased by approximately 0.9 each year. If this trend continues, by which year will all college freshmen consider "being well-off financially" essential or very important?

Life Objectives of College Freshmen, 1969–2010

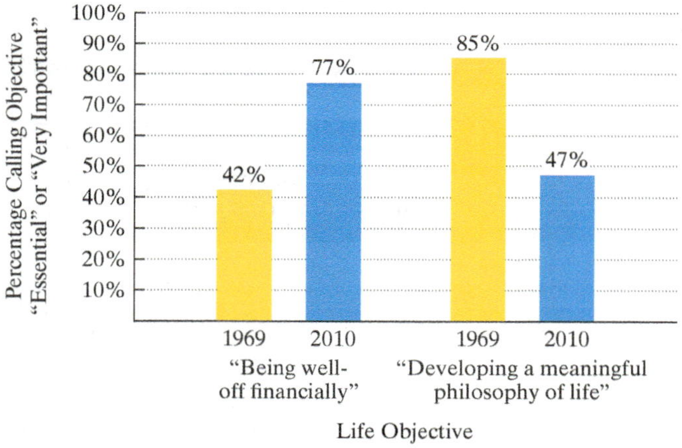

FIGURE 6.5
Source: Higher Education Research Institute

SOLUTION

Step 1 Let x represent one of the unknown quantities. We are interested in the year when all college freshmen, or 100% of the freshmen, will consider this life objective essential or very important. Let

x = the number of years after 1969 when all freshmen will consider "being well-off financially" essential or very important.

Step 2 Represent other unknown quantities in terms of x. There are no other unknown quantities to find, so we can skip this step.

Step 3 Write an equation in x that models the conditions.

The 1969 percentage	increased by	0.9 each year for x years	equals	100% of the freshmen.
42	+	$0.9x$	=	100

Step 4 Solve the equation and answer the question.

$$42 + 0.9x = 100$$ This is the equation that models the problem's conditions.

$$42 - 42 + 0.9x = 100 - 42$$ Subtract 42 from both sides.

$$0.9x = 58$$ Simplify.

$$\frac{0.9x}{0.9} = \frac{58}{0.9}$$ Divide both sides by 0.9.

$$x = 64.\overline{4} \approx 64$$ Simplify and round to the nearest whole number.

Using current trends, by approximately 64 years after 1969, or in 2033, all freshmen will consider "being well-off financially" essential or very important.

Step 5 Check the proposed solution in the original wording of the problem. The problem states that all freshmen (100%, represented by 100 using the model) will consider the objective essential or very important. Does this approximately occur if we increase the 1969 percentage, 42%, by 0.9 each year for 64 years, our proposed solution?

$$42 + 0.9(64) = 42 + 57.6 = 99.6 \approx 100$$

This verifies that using trends shown in **Figure 6.5**, all first-year college students will consider the objective essential or very important approximately 64 years after 1969.

A BRIEF REVIEW *Clearing an Equation of Decimals*

- You can clear an equation of decimals by multiplying each side by a power of 10. The exponent on 10 will be equal to the greatest number of digits to the right of any decimal point in the equation.
- Multiplying a decimal number by 10^n has the effect of moving the decimal point n places to the right.

Example

$$42 + 0.9x = 100$$

The greatest number of digits to the right of any decimal point in the equation is 1. Multiply each side by 10^1, or 10.

$$\begin{aligned} 10(42 + 0.9x) &= 10(100) \\ 10(42) + 10(0.9x) &= 10(100) \\ 420 + 9x &= 100 \\ 420 - 420 + 9x &= 1000 - 420 \\ 9x &= 580 \\ \frac{9x}{9} &= \frac{580}{9} \\ x &= 64.\overline{4} \approx 64 \end{aligned}$$

It is not a requirement to clear decimals before solving an equation. Compare this solution to the one in step 4 of Example 2. Which method do you prefer?

CHECK POINT 2 **Figure 6.5** shows that the freshman class of 2010 was less interested in developing a philosophy of life than the freshmen of 1969 had been. In 1969, 85% of the freshmen considered this objective essential or very important. Since then, this percentage has decreased by approximately 0.9 each year. If this trend continues, by which year will only 25% of college freshmen consider "developing a meaningful philosophy of life" essential or very important?

GREAT QUESTION!

Why are algebraic word problems important?

There is great value in reasoning through the steps for solving a word problem. This value comes from the problem-solving skills that you will attain and is often more important than the specific problem or its solution.

EXAMPLE 3 *Selecting a Monthly Text Message Plan*

You are choosing between two texting plans. Plan A has a monthly fee of \$20.00 with a charge of \$0.05 per text. Plan B has a monthly fee of \$5.00 with a charge of \$0.10 per text. Both plans include photo and video texts. For how many text messages will the costs for the two plans be the same?

SOLUTION

Step 1 Let x represent one of the unknown quantities. Let

x = the number of text messages for which the two plans cost the same.

Step 2 Represent other unknown quantities in terms of x. There are no other unknown quantities, so we can skip this step.

Step 3 Write an equation in x that models the conditions. The monthly cost for plan A is the monthly fee, \$20.00, plus the per-text charge, \$0.05, times the number of text messages, x. The monthly cost for plan B is the monthly fee, \$5.00, plus the per-text charge, \$0.10, times the number of text messages, x.

The monthly cost for plan A | must equal | the monthly cost for plan B.

$$20 + 0.05x = 5 + 0.10x$$

Step 4 Solve the equation and answer the question.

$$20 + 0.05x = 5 + 0.10x$$ This is the equation that models the problem's conditions.

$$20 = 5 + 0.05x$$ Subtract $0.05x$ from both sides.

$$15 = 0.05x$$ Subtract 5 from both sides.

$$\frac{15}{0.05} = \frac{0.05x}{0.05}$$ Divide both sides by 0.05.

$$300 = x$$ Simplify.

Because x represents the number of text messages for which the two plans cost the same, the costs will be the same for 300 texts per month.

Step 5 Check the proposed solution in the original wording of the problem. The problem states that the costs for the two plans should be the same. Let's see if they are with 300 text messages:

$$\text{Cost for plan A} = \$20 + \$0.05(300) = \$20 + \$15 = \$35$$

Monthly fee | Per-text charge

$$\text{Cost for plan B} = \$5 + \$0.10(300) = \$5 + \$30 = \$35.$$

With 300 text messages, both plans cost \$35 for the month. Thus, the proposed solution, 300 text messages, satisfies the problem's conditions.

CHECK POINT 3 You are choosing between two texting plans. Plan A has a monthly fee of \$15.00 with a charge of \$0.08 per text. Plan B has a monthly fee of \$3.00 with a charge of \$0.12 per text. For how many text messages will the costs for the two plans be the same?

EXAMPLE 4 *A Price Reduction on a Digital Camera*

Your local computer store is having a terrific sale on digital cameras. After a 40% price reduction, you purchase a digital camera for \$276. What was the camera's price before the reduction?

SOLUTION

Step 1 Let x represent one of the unknown quantities. We will let

x = the original price of the digital camera prior to the reduction.

Step 2 Represent other unknown quantities in terms of x. There are no other unknown quantities to find, so we can skip this step.

Step 3 Write an equation in x that models the conditions. The camera's original price minus the 40% reduction is the reduced price, \$276.

Original price	minus	the reduction (40% of the original price)	is	the reduced price, \$276.
x	$-$	$0.4x$	$=$	276

Step 4 Solve the equation and answer the question.

$$x - 0.4x = 276 \quad \text{This is the equation that models the problem's conditions.}$$

$$0.6x = 276 \quad \text{Combine like terms: } x - 0.4x = 1x - 0.4x = 0.6x.$$

$$\frac{0.6x}{0.6} = \frac{276}{0.6} \quad \text{Divide both sides by 0.6.}$$

$$x = 460 \quad \text{Simplify: } 0.6\overline{)276.0} = 460.$$

The digital camera's price before the reduction was \$460.

Step 5 Check the proposed solution in the original wording of the problem. The price before the reduction, \$460, minus the 40% reduction should equal the reduced price given in the original wording, \$276:

$$460 - 40\% \text{ of } 460 = 460 - 0.4(460) = 460 - 184 = 276.$$

This verifies that the digital camera's price before the reduction was \$460.

GREAT QUESTION!

Why is the 40% reduction written as $0.4x$ in Example 4?

- 40% is written 0.40 or 0.4.
- "Of" represents multiplication, so 40% of the original price is $0.4x$.

Notice that the original price, x, reduced by 40% is $x - 0.4x$ and *not* $x - 0.4$.

CHECK POINT 4 After a 30% price reduction, you purchase a new computer for \$840. What was the computer's price before the reduction?

2 Solve a formula for a variable.

Solving a Formula for One of Its Variables

We know that solving an equation is the process of finding the number (or numbers) that make the equation a true statement. All of the equations we have solved contained only one letter, x.

By contrast, formulas contain two or more letters, representing two or more variables. An example is the formula for the perimeter of a rectangle:

$$P = 2l + 2w. \quad \text{A rectangle's perimeter is the sum of twice its length and twice its width.}$$

We say that this formula is solved for the variable P because P is alone on one side of the equation and the other side does not contain a P.

Solving a formula for a variable means rewriting the formula so that the variable is isolated on one side of the equation. It does not mean obtaining a numerical value for that variable.

To solve a formula for one of its variables, treat that variable as if it were the only variable in the equation. Think of the other variables as if they were numbers. Isolate all terms with the specified variable on one side of the equation and all terms without the specified variable on the other side. Then divide both sides by the same nonzero quantity to get the specified variable alone. The next two examples show how to do this.

EXAMPLE 5 *Solving a Formula for a Variable*

Solve the formula $P = 2l + 2w$ for l.

SOLUTION

First, isolate $2l$ on the right by subtracting $2w$ from both sides. Then solve for l by dividing both sides by 2.

We need to isolate l.

$P = 2l + 2w$	This is the given formula.
$P - 2w = 2l + 2w - 2w$	Isolate $2l$ by subtracting $2w$ from both sides.
$P - 2w = 2l$	Simplify.
$\frac{P - 2w}{2} = \frac{2l}{2}$	Solve for l by dividing both sides by 2.
$\frac{P - 2w}{2} = l$	Simplify.

Equivalently, $l = \frac{P - 2w}{2}$.

CHECK POINT 5 Solve the formula $P = 2l + 2w$ for w.

EXAMPLE 6 *Solving a Formula for a Variable*

The total price of an article purchased on a monthly deferred payment plan is described by the following formula:

$$T = D + pm.$$

In this formula, T is the total price, D is the down payment, p is the monthly payment, and m is the number of months one pays. Solve the formula for p.

SOLUTION

First, isolate pm on the right by subtracting D from both sides. Then, isolate p from pm by dividing both sides of the formula by m.

We need to isolate p.

$T = D + pm$	This is the given formula. We want p alone.
$T - D = D - D + pm$	Isolate pm by subtracting D from both sides.
$T - D = pm$	Simplify.
$\frac{T - D}{m} = \frac{pm}{m}$	Now isolate p by dividing both sides by m.
$\frac{T - D}{m} = p$	Simplify: $\frac{pm}{m} = \frac{p\cancel{m}}{\cancel{m}} = \frac{p}{1} = p$.

CHECK POINT 6 Solve the formula $T = D + pm$ for m.

Concept and Vocabulary Check

Fill in each blank so that the resulting statement is true.

1. According to the U.S. Office of Management and Budget, the 2011 budget for defense exceeded the budget for education by \$658.6 billion. If x represents the budget for education, in billions of dollars, the budget for defense can be represented by ________.
2. In 2000, 31% of U.S. adults viewed a college education as essential for success. For the period from 2000 through 2010, this percentage increased by approximately 2.4 each year. The percentage of U.S. adults who viewed a college education as essential for success x years after 2000 can be represented by ________.
3. A text message plan costs \$4.00 per month plus \$0.15 per text. The monthly cost for x text messages can be represented by ________.
4. I purchased a computer after a 15% price reduction. If x represents the computer's original price, the reduced price can be represented by ____________.
5. Solving a formula for a variable means rewriting the formula so that the variable is ______________.
6. In order to solve $y = mx + b$ for x, we first ________ and then ________.

Exercise Set 6.3

Practice Exercises

Use the five-step strategy for solving word problems to find the number or numbers described in Exercises 1–10.

1. When five times a number is decreased by 4, the result is 26. What is the number?
2. When two times a number is decreased by 3, the result is 11. What is the number?
3. When a number is decreased by 20% of itself, the result is 20. What is the number?
4. When a number is decreased by 30% of itself, the result is 28. What is the number?
5. When 60% of a number is added to the number, the result is 192. What is the number?
6. When 80% of a number is added to the number, the result is 252. What is the number?
7. 70% of what number is 224?
8. 70% of what number is 252?
9. One number exceeds another by 26. The sum of the numbers is 64. What are the numbers?
10. One number exceeds another by 24. The sum of the numbers is 58. What are the numbers?

Practice Plus

In Exercises 11–18, write each English phrase as an algebraic expression. Then simplify the expression. Let x represent the number.

11. A number decreased by the sum of the number and four
12. A number decreased by the difference between eight and the number
13. Six times the product of negative five and a number
14. Ten times the product of negative four and a number
15. The difference between the product of five and a number and twice the number
16. The difference between the product of six and a number and negative two times the number
17. The difference between eight times a number and six more than three times the number
18. Eight decreased by three times the sum of a number and six

Application Exercises

How will you spend your average life expectancy of 78 years? The bar graph shows the average number of years you will devote to each of your most time-consuming activities. Exercises 19–20 are based on the data displayed by the graph.

How You Will Spend Your Average Life Expectancy of 78 Years

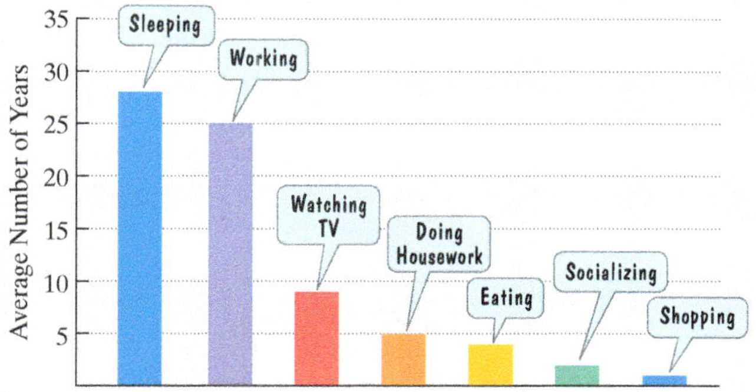

Source: U.S. Bureau of Labor Statistics

19. According to the U.S. Bureau of Labor Statistics, you will devote 37 years to sleeping and watching TV. The number of years sleeping will exceed the number of years watching TV by 19. Over your lifetime, how many years will you spend on each of these activities?
20. According to the U.S. Bureau of Labor Statistics, you will devote 32 years to sleeping and eating. The number of years sleeping will exceed the number of years eating by 24. Over your lifetime, how many years will you spend on each of these activities?

The bar graph shows average yearly earnings in the United States for people with a college education, by final degree earned. Exercises 21–22 are based on the data displayed by the graph.

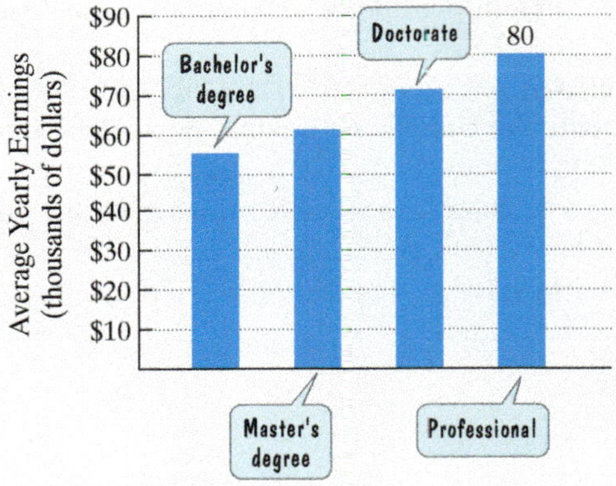

Source: U.S. Census Bureau

21. The average yearly salary of an American whose final degree is a master's is \$49 thousand less than twice that of an American whose final degree is a bachelor's. Combined, two people with each of these educational attainments earn \$116 thousand. Find the average yearly salary of Americans with each of these final degrees.

22. The average yearly salary of an American whose final degree is a doctorate is \$39 thousand less than twice that of an American whose final degree is a bachelor's. Combined, two people with each of these educational attainments earn \$126 thousand. Find the average yearly salary of Americans with each of these final degrees.

Even as Americans increasingly view a college education as essential for success, many believe that a college education is becoming less available to qualified students. Exercises 23–24 are based on the data displayed by the graph.

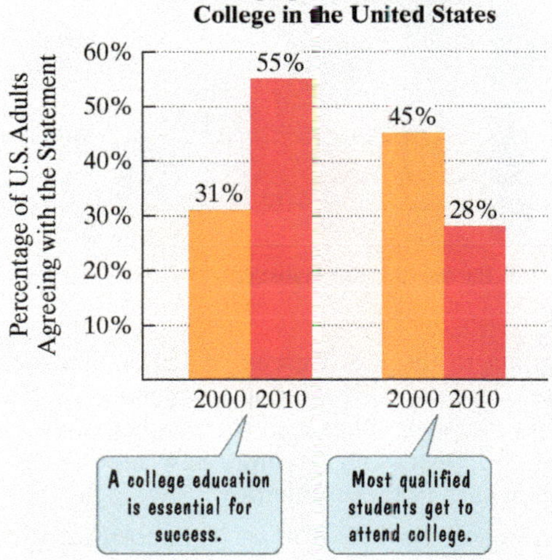

Source: Public Agenda

23. In 2000, 31% of U.S. adults viewed a college education as essential for success. For the period 2000 through 2010, the percentage viewing a college education as essential for success increased on average by approximately 2.4 each year. If this trend continues, by which year will 67% of all American adults view a college education as essential for success?

24. The data displayed by the graph at the bottom of the previous column indicate that in 2000, 45% of U.S. adults believed most qualified students get to attend college. For the period from 2000 through 2010, the percentage who believed that a college education is available to most qualified students decreased by approximately 1.7 each year. If this trend continues, by which year will only 11% of all American adults believe that most qualified students get to attend college?

On average, every minute of every day, 158 babies are born. The bar graph represents the results of a single day of births, deaths, and population increase worldwide. Exercises 25–26 are based on the information displayed by the graph.

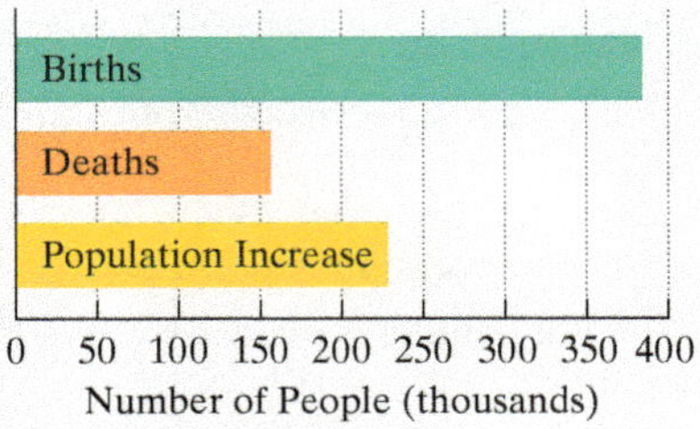

Source: James Henslin, *Sociology*, Eleventh Edition, Pearson, 2012.

25. Each day, the number of births in the world is 84 thousand less than three times the number of deaths.

a. If the population increase in a single day is 228 thousand, determine the number of births and deaths per day.

b. If the population increase in a single day is 228 thousand, by how many millions of people does the worldwide population increase each year? Round to the nearest million.

c. Based on your answer to part (b), approximately how many years does it take for the population of the world to increase by an amount greater than the entire U.S. population (315 million)?

26. Each day, the number of births in the world exceeds twice the number of deaths by 72 thousand.

a. If the population increase in a single day is 228 thousand, determine the number of births and deaths per day.

b. If the population increase in a single day is 228 thousand, by how many millions of people does the worldwide population increase each year? Round to the nearest million.

c. Based on your answer to part (b), approximately how many years does it take for the population of the world to increase by an amount greater than the entire U.S. population (315 million)?

27. A new car worth \$24,000 is depreciating in value by \$3000 per year. After how many years will the car's value be \$9000?

28. A new car worth \$45,000 is depreciating in value by \$5000 per year. After how many years will the car's value be \$10,000?

29. You are choosing between two health clubs. Club A offers membership for a fee of \$40 plus a monthly fee of \$25. Club B offers membership for a fee of \$15 plus a monthly fee of \$30. After how many months will the total cost at each health club be the same? What will be the total cost for each club?

30. You need to rent a rug cleaner. Company A will rent the machine you need for \$22 plus \$6 per hour. Company B will rent the same machine for \$28 plus \$4 per hour. After how many hours of use will the total amount spent at each company be the same? What will be the total amount spent at each company?

31. The bus fare in a city is \$1.25. People who use the bus have the option of purchasing a monthly discount pass for \$15.00. With the discount pass, the fare is reduced to \$0.75. Determine the number of times in a month the bus must be used so that the total monthly cost without the discount pass is the same as the total monthly cost with the discount pass.

32. A discount pass for a bridge costs \$30.00 per month. The toll for the bridge is normally \$5.00, but it is reduced to \$3.50 for people who have purchased the discount pass. Determine the number of times in a month the bridge must be crossed so that the total monthly cost without the discount pass is the same as the total monthly cost with the discount pass.

33. You are choosing between two plans at a discount warehouse. Plan A offers an annual membership fee of \$100 and you pay 80% of the manufacturer's recommended list price. Plan B offers an annual membership fee of \$40 and you pay 90% of the manufacturer's recommended list price. How many dollars of merchandise would you have to purchase in a year to pay the same amount under both plans? What will be the cost for each plan?

34. You are choosing between two plans at a discount warehouse. Plan A offers an annual membership fee of \$300 and you pay 70% of the manufacturer's recommended list price. Plan B offers an annual membership fee of \$40 and you pay 90% of the manufacturer's recommended list price. How many dollars of merchandise would you have to purchase in a year to pay the same amount under both plans? What will be the cost for each plan?

35. In 2010, there were 13,300 students at college A, with a projected enrollment increase of 1000 students per year. In the same year, there were 26,800 students at college B, with a projected enrollment decline of 500 students per year. According to these projections, when will the colleges have the same enrollment? What will be the enrollment in each college at that time?

36. In 2000, the population of Greece was 10,600,000, with projections of a population decrease of 28,000 people per year. In the same year, the population of Belgium was 10,200,000, with projections of a population decrease of 12,000 people per year. (*Source:* United Nations) According to these projections, when will the two countries have the same population? What will be the population at that time?

37. After a 20% reduction, you purchase a television for \$336. What was the television's price before the reduction?

38. After a 30% reduction, you purchase a dictionary for \$30.80. What was the dictionary's price before the reduction?

39. Including 8% sales tax, an inn charges \$162 per night. Find the inn's nightly cost before the tax is added.

40. Including 5% sales tax, an inn charges \$252 per night. Find the inn's nightly cost before the tax is added.

Exercises 41–42 involve markup, the amount added to the dealer's cost of an item to arrive at the selling price of that item.

41. The selling price of a refrigerator is \$584. If the markup is 25% of the dealer's cost, what is the dealer's cost of the refrigerator?

42. The selling price of a scientific calculator is \$15. If the markup is 25% of the dealer's cost, what is the dealer's cost of the calculator?

In Exercises 43–60, solve each formula for the specified variable. Do you recognize the formula? If so, what does it describe?

43. $A = LW$ for L

44. $D = RT$ for R

45. $A = \frac{1}{2}bh$ for b

46. $V = \frac{1}{3}Bh$ for B

47. $I = Prt$ for P

48. $C = 2\pi r$ for r

49. $E = mc^2$ for m

50. $V = \pi r^2 h$ for h

51. $y = mx + b$ for m

52. $P = C + MC$ for M

53. $A = \frac{1}{2}h(a + b)$ for a

54. $A = \frac{1}{2}h(a + b)$ for b

55. $S = P + Prt$ for r

56. $S = P + Prt$ for t

57. $Ax + By = C$ for x

58. $Ax + By = C$ for y

59. $a_n = a_1 + (n - 1)d$ for n

60. $a_n = a_1 + (n - 1)d$ for d

Writing in Mathematics

61. In your own words, describe a step-by-step approach for solving algebraic word problems.

62. Write an original word problem that can be solved using a linear equation. Then solve the problem.

63. Explain what it means to solve a formula for a variable.

64. Did you have difficulties solving some of the problems that were assigned in this Exercise Set? Discuss what you did if this happened to you. Did your course of action enhance your ability to solve algebraic word problems?

Critical Thinking Exercises

Make Sense? *In Exercises 65–68, determine whether each statement makes sense or does not make sense, and explain your reasoning.*

65. By modeling attitudes of college freshmen from 1969 through 2010, I can make precise predictions about the attitudes of the freshman class of 2020.

66. I find the hardest part in solving a word problem is writing the equation that models the verbal conditions.

67. I solved a formula for one of its variables, so now I have a numerical value for that variable.

68. After a 35% reduction, a computer's price is \$780, so I determined the original price, x, by solving $x - 0.35 = 780$.

69. The price of a dress is reduced by 40%. When the dress still does not sell, it is reduced by 40% of the reduced price. If the price of the dress after both reductions is \$72, what was the original price?

70. In a film, the actor Charles Coburn plays an elderly "uncle" character criticized for marrying a woman when he is 3 times her age. He wittily replies, "Ah, but in 20 years time I shall only be twice her age." How old is the "uncle" and the woman?

71. Suppose that we agree to pay you 8¢ for every problem in this chapter that you solve correctly and fine you 5¢ for every problem done incorrectly. If at the end of 26 problems we do not owe each other any money, how many problems did you solve correctly?

72. It was wartime when the Ricardos found out Mrs. Ricardo was pregnant. Ricky Ricardo was drafted and made out a will, deciding that \$14,000 in a savings account was to be divided between his wife and his child-to-be. Rather strangely, and certainly with gender bias, Ricky stipulated that if the child were a boy, he would get twice the amount of the mother's portion. If it were a girl, the mother would get twice the amount the girl was to receive. We'll never know what Ricky was thinking of, for (as fate would have it) he did not return from the war. Mrs. Ricardo gave birth to twins—a boy and a girl. How was the money divided?

73. A thief steals a number of rare plants from a nursery. On the way out, the thief meets three security guards, one after another. To each security guard, the thief is forced to give one-half the plants that he still has, plus 2 more. Finally, the thief leaves the nursery with 1 lone palm. How many plants were originally stolen?

In Exercises 74–75, solve each proportion for x.

74. $\dfrac{x + a}{a} = \dfrac{b + c}{c}$

75. $\dfrac{ax - b}{b} = \dfrac{c - d}{d}$

Group Exercise

76. One of the best ways to learn how to *solve* a word problem in algebra is to *design* word problems of your own. Creating a word problem makes you very aware of precisely how much information is needed to solve the problem. You must also focus on the best way to present information to a reader and on how much information to give. As you write your problem, you gain skills that will help you solve problems created by others.

The group should design five different word problems that can be solved using linear equations. All of the problems should be on different topics. For example, the group should not have more than one problem on price reduction. The group should turn in both the problems and their algebraic solutions.

6.4 Linear Inequalities in One Variable

WHAT AM I SUPPOSED TO LEARN?

After you have read this section, you should be able to:

1 Graph subsets of real numbers on a number line.

2 Solve linear inequalities.

3 Solve applied problems using linear inequalities.

RENT-A-HEAP, A CAR RENTAL company, charges \$125 per week plus \$0.20 per mile to rent one of their cars. Suppose you are limited by how much money you can spend for the week: You can spend at most \$335. If we let x represent the number of miles you drive the heap in a week, we can write an inequality that models the given conditions:

The weekly charge of \$125 | plus | the charge of \$0.20 per mile for x miles | must be less than or equal to | \$335.

$$125 + 0.20x \leq 335.$$

Notice that the highest exponent on the variable in $125 + 0.20x \leq 335$ is 1. Such an inequality is called a *linear inequality in one variable*. The symbol between the two sides of an inequality can be $\leq$ (is less than or equal to), $<$ (is less than), $\geq$ (is greater than or equal to), or $>$ (is greater than).

In this section, we will study how to solve linear inequalities such as $125 + 0.20x \leq 335$. **Solving an inequality** is the process of finding the set of numbers that makes the inequality a true statement. These numbers are called the **solutions** of the inequality and we say that they **satisfy** the inequality. The set of all solutions is called the **solution set** of the inequality. We begin by discussing how to represent these solution sets, which are subsets of real numbers, on a number line.

Graph subsets of real numbers on a number line.

Graphing Subsets of Real Numbers on a Number Line

Table 6.3 shows how to represent various subsets of real numbers on a number line. Open dots (∘) indicate that a number is not included in a set. Closed dots (•) indicate that a number is included in a set.

TABLE 6.3 Graphs of Subsets of Real Numbers

Let a and b be real numbers such that $a < b$.

Set-Builder Notation		Graph
$\{x \mid x < a\}$	x is a real number less than a.	
$\{x \mid x \leq a\}$	x is a real number less than or equal to a.	
$\{x \mid x > b\}$	x is a real number greater than b.	
$\{x \mid x \geq b\}$	x is a real number greater than or equal to b.	
$\{x \mid a < x < b\}$	x is a real number greater than a and less than b.	
$\{x \mid a \leq x \leq b\}$	x is a real number greater than or equal to a and less than or equal to b.	
$\{x \mid a \leq x < b\}$	x is a real number greater than or equal to a and less than b.	
$\{x \mid a < x \leq b\}$	x is a real number greater than a and less than or equal to b.	

EXAMPLE 1 *Graphing Subsets of Real Numbers*

Graph each set:

a. $\{x|x < 3\}$ **b.** $\{x|x \geq -1\}$ **c.** $\{x|-1 < x \leq 3\}$.

SOLUTION

a. $\{x \mid x < 3\}$ — x is a real number less than 3.

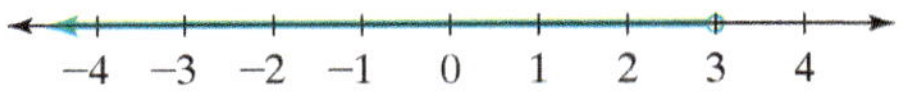

b. $\{x \mid x \geq -1\}$ — x is a real number greater than or equal to -1.

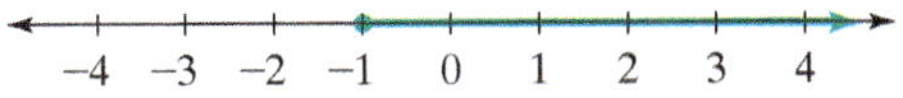

c. $\{x \mid -1 < x \leq 3\}$ — x is a real number greater than -1 and less than or equal to 3.

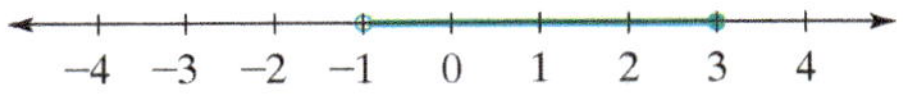

CHECK POINT 1 Graph each set:

a. $\{x|x < 4\}$ **b.** $\{x|x \geq -2\}$ **c.** $\{x|-4 \leq x < 1\}$.

2 Solve linear inequalities.

Solving Linear Inequalities in One Variable

We know that a linear equation in x can be expressed as $ax + b = 0$. A **linear inequality in x** can be written in one of the following forms:

$$ax + b < 0, \quad ax + b \leq 0, \quad ax + b > 0, \quad ax + b \geq 0.$$

In each form, $a \neq 0$.

Back to our question that opened this section: How many miles can you drive your Rent-a-Heap car if you can spend at most \$335? We answer the question by solving

$$0.20x + 125 \leq 335$$

for x. The solution procedure is nearly identical to that for solving

$$0.20x + 125 = 335.$$

Our goal is to get x by itself on the left side. We do this by subtracting 125 from both sides to isolate $0.20x$:

$$0.20x + 125 \leq 335 \quad \text{This is the given inequality.}$$

$$0.20x + 125 - 125 \leq 335 - 125 \quad \text{Subtract 125 from both sides.}$$

$$0.20x \leq 210. \quad \text{Simplify.}$$

Finally, we isolate x from $0.20x$ by dividing both sides of the inequality by 0.20:

$$\frac{0.20x}{0.20} \leq \frac{210}{0.20} \quad \text{Divide both sides by 0.20.}$$

$$x \leq 1050. \quad \text{Simplify.}$$

With at most \$335 per week to spend, you can travel at most 1050 miles.

We started with the inequality $0.20x + 125 \leq 335$ and obtained the inequality $x \leq 1050$ in the final step. Both of these inequalities have the same solution set, namely $\{x|x \leq 1050\}$. Inequalities such as these, with the same solution set, are said to be **equivalent**.

We isolated x from $0.20x$ by dividing both sides of $0.20x \leq 210$ by 0.20, a positive number. Let's see what happens if we divide both sides of an inequality by a negative number. Consider the inequality $10 < 14$. Divide 10 and 14 by -2:

$$\frac{10}{-2} = -5 \quad \text{and} \quad \frac{14}{-2} = -7.$$

Because -5 lies to the right of -7 on the number line, -5 is greater than -7:

$$-5 > -7.$$

Notice that the direction of the inequality symbol is reversed:

$$10 < 14$$
$$\updownarrow$$
$$-5 > -7.$$

Dividing by -2 changes the direction of the inequality symbol.

In general, **when we multiply or divide both sides of an inequality by a negative number, the direction of the inequality symbol is reversed.** When we reverse the direction of the inequality symbol, we say that we change the *sense* of the inequality.

GREAT QUESTION!

What are some common English phrases and sentences that I can model with linear inequalities?

English phrases such as "at least" and "at most" can be represented by inequalities.

English Sentence	Inequality
x is at least 5.	$x \geq 5$
x is at most 5.	$x \leq 5$
x is no more than 5.	$x \leq 5$
x is no less than 5.	$x \geq 5$

We can summarize our discussion with the following statement:

SOLVING LINEAR INEQUALITIES

The procedure for solving linear inequalities is the same as the procedure for solving linear equations, with one important exception: When multiplying or dividing both sides of the inequality by a negative number, reverse the direction of the inequality symbol, changing the sense of the inequality.

EXAMPLE 2 *Solving a Linear Inequality*

Solve and graph the solution set: $4x - 7 \geq 5$.

SOLUTION

Our goal is to get x by itself on the left side. We do this by first getting $4x$ by itself, adding 7 to both sides.

$$4x - 7 \geq 5 \quad \text{This is the given inequality.}$$
$$4x - 7 + 7 \geq 5 + 7 \quad \text{Add 7 to both sides.}$$
$$4x \geq 12 \quad \text{Simplify.}$$

Next, we isolate x from $4x$ by dividing both sides by 4. The inequality symbol stays the same because we are dividing by a positive number.

$$\frac{4x}{4} \geq \frac{12}{4} \quad \text{Divide both sides by 4.}$$
$$x \geq 3 \quad \text{Simplify.}$$

The solution set consists of all real numbers that are greater than or equal to 3, expressed in set-builder notation as $\{x | x \geq 3\}$. The graph of the solution set is shown as follows:

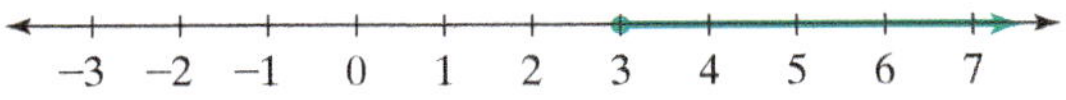

We cannot check all members of an inequality's solution set, but we can take a few values to get an indication of whether or not it is correct. In Example 2, we found that the solution set of $4x - 7 \geq 5$ is $\{x | x \geq 3\}$. Show that 3 and 4 satisfy the inequality, whereas 2 does not.

CHECK POINT 2 Solve and graph the solution set: $5x - 3 \leq 17$.

EXAMPLE 3 *Solving Linear Inequalities*

Solve and graph the solution set:

a. $\frac{1}{3}x < 5$ **b.** $-3x < 21$.

SOLUTION

In each case, our goal is to isolate x. In the first inequality, this is accomplished by multiplying both sides by 3. In the second inequality, we can do this by dividing both sides by -3.

a. $\frac{1}{3}x < 5$ This is the given inequality.

$3 \cdot \frac{1}{3}x < 3 \cdot 5$ Isolate x by multiplying by 3 on both sides. The symbol $<$ stays the same because we are multiplying both sides by a positive number.

$x < 15$ Simplify.

The solution set is $\{x \mid x < 15\}$. The graph of the solution set is shown as follows:

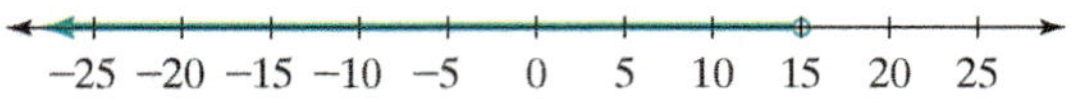

b. $-3x < 21$ This is the given inequality.

$\frac{-3x}{-3} > \frac{21}{-3}$ Isolate x by dividing by -3 on both sides. The symbol $<$ must be reversed because we are dividing both sides by a negative number.

$x > -7$ Simplify.

The solution set is $\{x \mid x > -7\}$. The graph of the solution set is shown as follows:

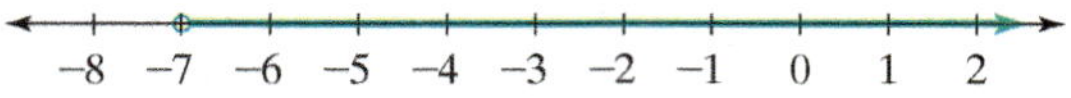

CHECK POINT 3 Solve and graph the solution set:

a. $\frac{1}{4}x < 2$ **b.** $-6x < 18$.

EXAMPLE 4 *Solving a Linear Inequality*

Solve and graph the solution set: $6x - 12 > 8x + 2$.

SOLUTION

We will get x by itself on the left side. We begin by subtracting $8x$ from both sides so that the variable term appears on the left.

$6x - 12 > 8x + 2$ This is the given inequality.

$6x - 8x - 12 > 8x - 8x + 2$ Subtract $8x$ on both sides with the goal of isolating x on the left.

$-2x - 12 > 2$ Simplify.

Next, we get $-2x$ by itself, adding 12 to both sides.

$-2x - 12 + 12 > 2 + 12$ Add 12 to both sides.

$-2x > 14$ Simplify.

In order to solve $-2x > 14$, we isolate x from $-2x$ by dividing both sides by -2. The direction of the inequality symbol must be reversed because we are dividing by a negative number.

$\frac{-2x}{-2} < \frac{14}{-2}$ Divide both sides by -2 and change the sense of the inequality.

$x < -7$ Simplify.

The solution set is $\{x \mid x < -7\}$. The graph of the solution set is shown as follows:

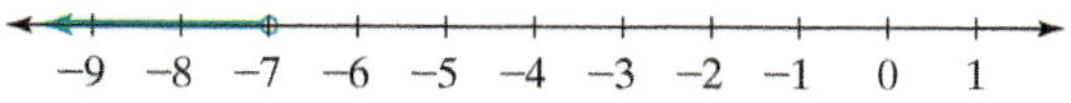

CHECK POINT 4 Solve and graph the solution set: $7x - 3 > 13x + 33$.

EXAMPLE 5 *Solving a Linear Inequality*

Solve and graph the solution set:

$$2(x - 3) + 5x \le 8(x - 1).$$

SOLUTION

Begin by simplifying the algebraic expression on each side.

$$2(x - 3) + 5x \le 8(x - 1) \quad \text{This is the given inequality.}$$

$$2x - 6 + 5x \le 8x - 8 \quad \text{Use the distributive property.}$$

$$7x - 6 \le 8x - 8 \quad \text{Add like terms on the left: } 2x + 5x = 7x.$$

We will get x by itself on the left side. Subtract $8x$ from both sides.

$$7x - 8x - 6 \le 8x - 8x - 8$$

$$-x - 6 \le -8$$

Next, we get $-x$ by itself, adding 6 to both sides.

$$-x - 6 + 6 \le -8 + 6$$

$$-x \le -2$$

To isolate x, we must eliminate the negative sign in front of the x. Because $-x$ means $-1x$, we can do this by dividing both sides of the inequality by -1. This reverses the direction of the inequality symbol.

$$\frac{-x}{-1} \ge \frac{-2}{-1} \quad \text{Divide both sides by } -1 \text{ and change the sense of the inequality.}$$

$$x \ge 2 \quad \text{Simplify.}$$

The solution set is $\{x \mid x \ge 2\}$. The graph of the solution set is shown as follows:

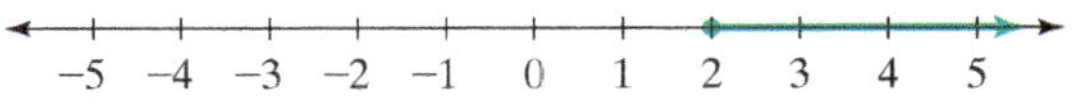

GREAT QUESTION!

Do I have to solve $7x - 6 \le 8x - 8$ by isolating the variable on the left?

No. You can solve

$$7x - 6 \le 8x - 8$$

by isolating x on the right side. Subtract $7x$ from both sides and add 8 to both sides:

$$7x - 6 - 7x \le 8x - 8 - 7x$$

$$-6 \le x - 8$$

$$-6 + 8 \le x - 8 + 8$$

$$2 \le x.$$

This last inequality means the same thing as

$$x \ge 2.$$

Solution sets, in this case $\{x \mid x \ge 2\}$, are expressed with the variable on the left and the constant on the right.

CHECK POINT 5 Solve and graph the solution set:

$$2(x - 3) - 1 \le 3(x + 2) - 14.$$

In our next example, the inequality has three parts:

$$-3 < 2x + 1 \le 3.$$

$2x + 1$ is greater than -3 and less than or equal to 3.

By performing the same operation on all three parts of the inequality, our goal is to **isolate x in the middle**.

EXAMPLE 6 *Solving a Three-Part Inequality*

Solve and graph the solution set:

$$-3 < 2x + 1 \le 3.$$

SOLUTION

We would like to isolate x in the middle. We can do this by first subtracting 1 from all three parts of the inequality. Then we isolate x from $2x$ by dividing all three parts of the inequality by 2.

$$-3 < 2x + 1 \le 3 \qquad \text{This is the given inequality.}$$

$$-3 - 1 < 2x + 1 - 1 \le 3 - 1 \qquad \text{Subtract 1 from all three parts.}$$

$$-4 < 2x \le 2 \qquad \text{Simplify.}$$

$$\frac{-4}{2} < \frac{2x}{2} \le \frac{2}{2} \qquad \text{Divide each part by 2.}$$

$$-2 < x \le 1 \qquad \text{Simplify.}$$

The solution set consists of all real numbers greater than -2 and less than or equal to 1, represented by $\{x \mid -2 < x \le 1\}$. The graph is shown as follows:

CHECK POINT 6 Solve and graph the solution set on a number line: $1 \le 2x + 3 < 11$.

As you know, different professors may use different grading systems to determine your final course grade. Some professors require a final examination; others do not. In our next example, a final exam is required *and* it counts as two grades.

3 Solve applied problems using linear inequalities.

EXAMPLE 7 *An Application: Final Course Grade*

To earn an A in a course, you must have a final average of at least 90%. On the first four examinations, you have grades of 86%, 88%, 92%, and 84%. If the final examination counts as two grades, what must you get on the final to earn an A in the course?

SOLUTION

We will use our five-step strategy for solving algebraic word problems.

Steps 1 and 2 Represent unknown quantities in terms of *x*. Let

$$x = \text{your grade on the final examination.}$$

Step 3 Write an inequality in *x* that models the conditions. The average of the six grades is found by adding the grades and dividing the sum by 6.

$$\text{Average} = \frac{86 + 88 + 92 + 84 + x + x}{6}$$

Because the final counts as two grades, the x (your grade on the final examination) is added twice. This is also why the sum is divided by 6.

To get an A, your average must be at least 90. This means that your average must be greater than or equal to 90.

Your average	must be greater than or equal to	90.
$\dfrac{86+88+92+84+x+x}{6}$	$\geq$	90

Step 4 Solve the inequality and answer the problem's question.

$$\frac{86+88+92+84+x+x}{6} \geq 90 \quad \text{This is the inequality that models the given conditions.}$$

$$\frac{350+2x}{6} \geq 90 \quad \text{Combine like terms in the numerator.}$$

$$6\left(\frac{350+2x}{6}\right) \geq 6(90) \quad \text{Multiply both sides by 6, clearing the fraction.}$$

$$350+2x \geq 540 \quad \text{Multiply.}$$

$$350+2x-350 \geq 540-350 \quad \text{Subtract 350 from both sides.}$$

$$2x \geq 190 \quad \text{Simplify.}$$

$$\frac{2x}{2} \geq \frac{190}{2} \quad \text{Divide both sides by 2.}$$

$$x \geq 95 \quad \text{Simplify.}$$

You must get at least 95% on the final examination to earn an A in the course.

Step 5 Check. We can perform a partial check by computing the average with any grade that is at least 95. We will use 96. If you get 96% on the final examination, your average is

$$\frac{86+88+92+84+96+96}{6} = \frac{542}{6} = 90\frac{1}{3}.$$

Because $90\frac{1}{3} > 90$, you earn an A in the course.

CHECK POINT 7 To earn a B in a course, you must have a final average of at least 80%. On the first three examinations, you have grades of 82%, 74%, and 78%. If the final examination counts as two grades, what must you get on the final to earn a B in the course?

Concept and Vocabulary Check

Fill in each blank so that the resulting statement is true.

1. On a number line, an open dot indicates that a number ________ in a solution set, and a closed dot indicates that a number ________ in a solution set.
2. If an inequality's solution set consists of all real numbers, x, that are less than a, the solution set is represented in set-builder notation as ________.
3. If an inequality's solution set consists of all real numbers, x, that are greater than a and less than or equal to b, the solution set is represented in set-builder notation as ________.
4. When both sides of an inequality are multiplied or divided by a/an ________ number, the direction of the inequality symbol is reversed.

In Exercises 5–8, determine whether each statement is true or false. If the statement is false, make the necessary change(s) to produce a true statement.

5. The inequality $x - 3 > 0$ is equivalent to $x < 3$. ____
6. The statement "x is at most 5" is written $x < 5$. ____
7. The inequality $-4x < -20$ is equivalent to $x > -5$. ____
8. The statement "the sum of x and 6% of x is at least 80" is modeled by $x + 0.06x \geq 80$. ____

Exercise Set 6.4

Practice Exercises

In Exercises 1–12, graph each set of real numbers on a number line.

1. $\{x \mid x > 6\}$

2. $\{x \mid x > -2\}$

3. $\{x \mid x < -4\}$

4. $\{x \mid x < 0\}$

5. $\{x \mid x \geq -3\}$

6. $\{x \mid x \geq -5\}$

7. $\{x \mid x \leq 4\}$

8. $\{x \mid x \leq 7\}$

9. $\{x \mid -2 < x \leq 5\}$

10. $\{x \mid -3 \leq x < 7\}$

11. $\{x \mid -1 < x < 4\}$

12. $\{x \mid -7 \leq x \leq 0\}$

In Exercises 13–66, solve each inequality and graph the solution set on a number line.

13. $x - 3 > 2$

14. $x + 1 < 5$

15. $x + 4 \leq 9$

16. $x - 5 \geq 1$

17. $x - 3 < 0$

18. $x + 4 \geq 0$

19. $4x < 20$

20. $6x \geq 18$

21. $3x \geq -15$

22. $7x < -21$

23. $2x - 3 > 7$

24. $3x + 2 \leq 14$

25. $3x + 3 < 18$

26. $8x - 4 > 12$

27. $\frac{1}{2}x < 4$

28. $\frac{1}{2}x > 3$

29. $\frac{x}{3} > -2$

30. $\frac{x}{4} < -1$

31. $-3x < 15$

32. $-7x > 21$

33. $-3x \geq -15$

34. $-7x \leq -21$

35. $3x + 4 \leq 2x + 7$

36. $2x + 9 \leq x + 2$

37. $5x - 9 < 4x + 7$

38. $3x - 8 < 2x + 11$

39. $-2x - 3 < 3$

40. $14 - 3x > 5$

41. $3 - 7x \leq 17$

42. $5 - 3x \geq 20$

43. $-x < 4$

44. $-x > -3$

45. $5 - x \leq 1$

46. $3 - x \geq -3$

47. $2x - 5 > -x + 6$

48. $6x - 2 \geq 4x + 6$

49. $2x - 5 < 5x - 11$

50. $4x - 7 > 9x - 2$

51. $3(x + 1) - 5 < 2x + 1$

52. $4(x + 1) + 2 \geq 3x + 6$

53. $8x + 3 > 3(2x + 1) - x + 5$

54. $7 - 2(x - 4) < 5(1 - 2x)$

55. $\frac{x}{4} - \frac{3}{2} \leq \frac{x}{2} + 1$

56. $\frac{3x}{10} + 1 \geq \frac{1}{5} - \frac{x}{10}$

57. $1 - \frac{x}{2} > 4$

58. $7 - \frac{4}{5}x < \frac{3}{5}$

59. $6 < x + 3 < 8$

60. $7 < x + 5 < 11$

61. $-3 \leq x - 2 < 1$

62. $-6 < x - 4 \leq 1$

63. $-11 < 2x - 1 \leq -5$

64. $3 \leq 4x - 3 < 19$

65. $-3 \leq \frac{2}{3}x - 5 < -1$

66. $-6 \leq \frac{1}{2}x - 4 < -3$

Practice Plus

In Exercises 67–70, write an inequality with x isolated on the left side that is equivalent to the given inequality.

67. $Ax + By > C$; Assume $A > 0$.

68. $Ax + By \leq C$; Assume $A > 0$.

69. $Ax + By > C$; Assume $A < 0$.

70. $Ax + By \leq C$; Assume $A < 0$.

In Exercises 71–76, use set-builder notation to find all real numbers satisfying the given conditions.

71. A number increased by 5 is at least two times the number.

72. A number increased by 12 is at least four times the number.

73. Twice the sum of four and a number is at most 36.

74. Three times the sum of five and a number is at most 48.

75. If the quotient of three times a number and five is increased by four, the result is no more than 34.

76. If the quotient of three times a number and four is decreased by three, the result is no less than 9.

Application Exercises

The graphs show that the three components of love, namely passion, intimacy, and commitment, progress differently over time. Passion peaks early in a relationship and then declines. By contrast, intimacy and commitment build gradually. Use the graphs to solve Exercises 77–84. Assume that x represents years in a relationship.

The Course of Love Over Time

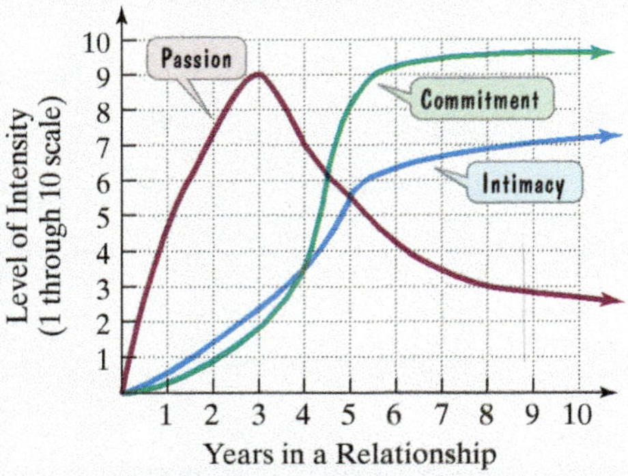

Source: R. J. Sternberg, A Triangular Theory of Love, *Psychological Review*, 93, 119–135.

77. Use set-builder notation to write an inequality that expresses for which years in a relationship intimacy is greater than commitment.

78. Use set-builder notation to write an inequality that expresses for which years in a relationship passion is greater than or equal to intimacy.

79. What is the relationship between passion and intimacy for $\{x \mid 5 \leq x < 7\}$?

80. What is the relationship between intimacy and commitment for $\{x \mid 4 \leq x < 7\}$?

81. What is the relationship between passion and commitment for $\{x|6 < x < 8\}$?

82. What is the relationship between passion and commitment for $\{x|7 < x < 9\}$?

83. What is the maximum level of intensity for passion? After how many years in a relationship does this occur?

84. After approximately how many years do levels of intensity for commitment exceed the maximum level of intensity for passion?

In more U.S. marriages, spouses have different faiths. The bar graph shows the percentage of households with an interfaith marriage in 1988 and 2008. Also shown is the percentage of households in which a person of faith is married to someone with no religion.

Percentage of U.S. Households in Which Married Couples Do Not Share the Same Faith

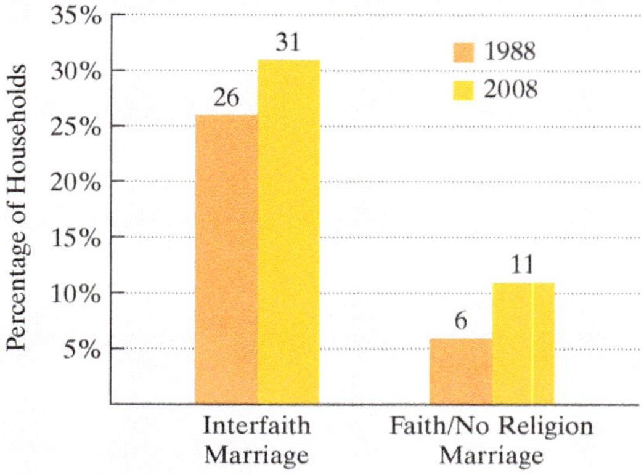

Source: General Social Survey, University of Chicago

The formula

$$I = \frac{1}{4}x + 26$$

models the percentage of U.S. households with an interfaith marriage, I, x years after 1988. The formula

$$N = \frac{1}{4}x + 6$$

models the percentage of U.S. households in which a person of faith is married to someone with no religion, N, x years after 1988. Use these models to solve Exercises 85–86.

85. a. In which years will more than 33% of U.S. households have an interfaith marriage?

b. In which years will more than 14% of U.S. households have a person of faith married to someone with no religion?

c. Based on your answers to parts (a) and (b), in which years will more than 33% of households have an interfaith marriage and more than 14% have a faith/no religion marriage?

86. a. In which years will more than 34% of U.S. households have an interfaith marriage?

b. In which years will more than 15% of U.S. households have a person of faith married to someone with no religion?

c. Based on your answers to parts (a) and (b), in which years will more than 34% of households have an interfaith marriage and more than 15% have a faith/no religion marriage?

87. On two examinations, you have grades of 86 and 88. There is an optional final examination, which counts as one grade. You decide to take the final in order to get a course grade of A, meaning a final average of at least 90.

a. What must you get on the final to earn an A in the course?

b. By taking the final, if you do poorly, you might risk the B that you have in the course based on the first two exam grades. If your final average is less than 80, you will lose your B in the course. Describe the grades on the final that will cause this to happen.

88. On three examinations, you have grades of 88, 78, and 86. There is still a final examination, which counts as one grade.

a. In order to get an A, your average must be at least 90. If you get 100 on the final, compute your average and determine if an A in the course is possible.

b. To earn a B in the course, you must have a final average of at least 80. What must you get on the final to earn a B in the course?

89. A car can be rented from Continental Rental for \$80 per week plus 25 cents for each mile driven. How many miles can you travel if you can spend at most \$400 for the week?

90. A car can be rented from Basic Rental for \$60 per week plus 50 cents for each mile driven. How many miles can you travel if you can spend at most \$600 for the week?

91. An elevator at a construction site has a maximum capacity of 3000 pounds. If the elevator operator weighs 245 pounds and each cement bag weighs 95 pounds, up to how many bags of cement can be safely lifted on the elevator in one trip?

92. An elevator at a construction site has a maximum capacity of 2800 pounds. If the elevator operator weighs 265 pounds and each cement bag weighs 65 pounds, up to how many bags of cement can be safely lifted on the elevator in one trip?

93. A basic cellphone plan costs \$20 per month for 60 calling minutes. Additional time costs \$0.40 per minute. The formula

$$C = 20 + 0.40(x - 60)$$

gives the monthly cost for this plan, C, for x calling minutes, where $x > 60$. How many calling minutes are possible for a monthly cost of at least \$28 and at most \$40?

94. The formula for converting Fahrenheit temperature, F, to Celsius temperature, C, is

$$C = \frac{5}{9}(F - 32).$$

If Celsius temperature ranges from 15° to 35°, inclusive, what is the range for the Fahrenheit temperature?

Writing in Mathematics

95. When graphing the solutions of an inequality, what is the difference between an open dot and a closed dot?

96. When solving an inequality, when is it necessary to change the direction of the inequality symbol? Give an example.

97. Describe ways in which solving a linear inequality is similar to solving a linear equation.

98. Describe ways in which solving a linear inequality is different than solving a linear equation.

Critical Thinking Exercises

Make Sense? *In Exercises 99–102, determine whether each statement makes sense or does not make sense, and explain your reasoning.*

99. I can check inequalities by substituting 0 for the variable: When 0 belongs to the solution set, I should obtain a true statement, and when 0 does not belong to the solution set, I should obtain a false statement.

100. In an inequality such as $5x + 4 < 8x - 5$, I can avoid division by a negative number depending on which side I collect the variable terms and on which side I collect the constant terms.

101. I solved $-2x + 5 \geq 13$ and concluded that -4 is the greatest integer in the solution set.

102. I began the solution of $5 - 3(x + 2) > 10x$ by simplifying the left side, obtaining $2x + 4 > 10x$.

103. A car can be rented from Basic Rental for \$260 per week with no extra charge for mileage. Continental charges \$80 per week plus 25 cents for each mile driven to rent the same car. How many miles must be driven in a week to make the rental cost for Basic Rental a better deal than Continental's?

104. A company manufactures and sells personalized stationery. The weekly fixed cost is \$3000 and it cost \$3.00 to produce each package of stationery. The selling price is \$5.50 per package. How many packages of stationery must be produced and sold each week for the company to generate a profit?

6.5 Quadratic Equations

WHAT AM I SUPPOSED TO LEARN?

After you have read this section, you should be able to:

1. Multiply binomials using the FOIL method.
2. Factor trinomials.
3. Solve quadratic equations by factoring.
4. Solve quadratic equations using the quadratic formula.
5. Solve problems modeled by quadratic equations.

I'm very well acquainted, too, with matters mathematical,
I understand equations, both simple and quadratical.
About binomial theorem I'm teeming with a lot of news,
With many cheerful facts about the square of the hypotenuse.

—Gilbert and Sullivan, *The Pirates of Penzance*

EQUATIONS QUADRATICAL? CHEERFUL NEWS ABOUT THE SQUARE OF THE hypotenuse? You've come to the right place. In this section, we study two methods for solving *quadratic equations*, equations in which the highest exponent on the variable is 2. (Yes, it's *quadratic* and not *quadratical*, despite the latter's rhyme with mathematical.) In Chapter 10 (Section 10.2), we look at an application of quadratic equations, introducing (cheerfully, of course) the Pythagorean Theorem and the square of the hypotenuse.

1 Multiply binomials using the FOIL method.

Multiplying Two Binomials Using the FOIL Method

Before we learn about the first method for solving quadratic equations, factoring, we need to consider the FOIL method for multiplying two binomials. A **binomial** is a simplified algebraic expression that contains two terms in which each exponent that appears on a variable is a whole number.

Examples of Binomials

$$x + 3, \quad x + 4, \quad 3x + 4, \quad 5x - 3$$

Two binomials can be quickly multiplied by using the FOIL method, in which **F** represents the product of the **first** terms in each binomial, **O** represents the product of the **outside** terms, **I** represents the product of the two **inside** terms, and **L** represents the product of the **last,** or second, terms in each binomial.

USING THE FOIL METHOD TO MULTIPLY BINOMIALS

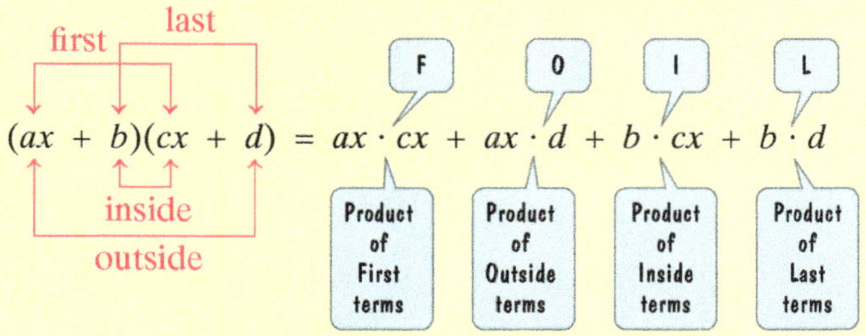

$$(ax + b)(cx + d) = ax \cdot cx + ax \cdot d + b \cdot cx + b \cdot d$$

Once you have multiplied first, outside, inside, and last terms, combine all like terms.

EXAMPLE 1 *Using the FOIL Method*

Multiply: $(x + 3)(x + 4)$.

SOLUTION

F : First terms	$= x \cdot x = x^2$	$(x + 3)(x + 4)$
O: Outside terms	$= x \cdot 4 = 4x$	$(x + 3)(x + 4)$
I : Inside terms	$= 3 \cdot x = 3x$	$(x + 3)(x + 4)$
L : Last terms	$= 3 \cdot 4 = 12$	$(x + 3)(x + 4)$

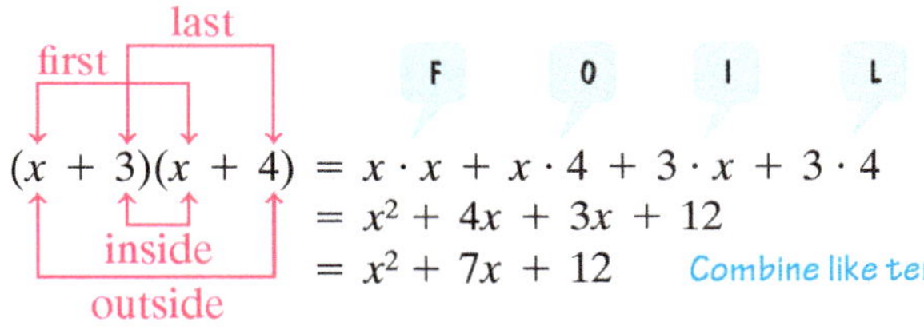

$$\begin{aligned}(x + 3)(x + 4) &= x \cdot x + x \cdot 4 + 3 \cdot x + 3 \cdot 4\\ &= x^2 + 4x + 3x + 12\\ &= x^2 + 7x + 12\end{aligned}$$ Combine like terms.

CHECK POINT 1 Multiply: $(x + 5)(x + 6)$.

EXAMPLE 2 *Using the FOIL Method*

Multiply: $(3x + 4)(5x - 3)$.

SOLUTION

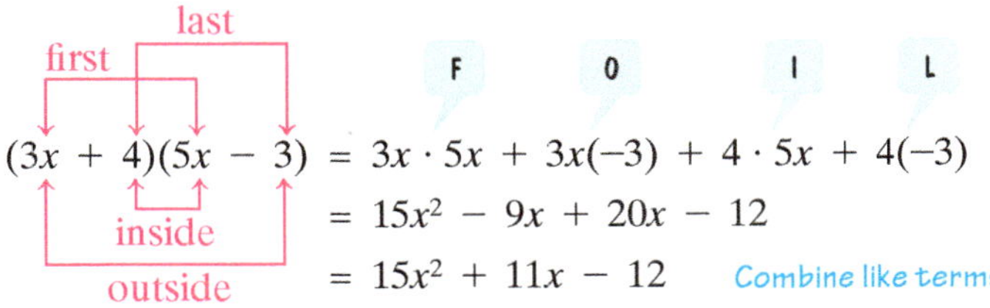

$$\begin{aligned}(3x + 4)(5x - 3) &= 3x \cdot 5x + 3x(-3) + 4 \cdot 5x + 4(-3)\\ &= 15x^2 - 9x + 20x - 12\\ &= 15x^2 + 11x - 12\end{aligned}$$ Combine like terms.

CHECK POINT 2 Multiply: $(7x + 5)(4x - 3)$.

2 Factor trinomials.

Factoring a Trinomial Where the Coefficient of the Squared Term Is 1

The algebraic expression $x^2 + 7x + 12$ is called a trinomial. A **trinomial** is a simplified algebraic expression that contains three terms in which all variables have whole number exponents.

We can use the FOIL method to multiply two binomials to obtain the trinomial $x^2 + 7x + 12$:

Factored Form		F	O	I	L		Trinomial Form
$(x + 3)(x + 4)$	=	x^2	$+ 4x$	$+ 3x$	$+ 12$	=	$x^2 + 7x + 12$

Because the product of $x + 3$ and $x + 4$ is $x^2 + 7x + 12$, we call $x + 3$ and $x + 4$ the **factors** of $x^2 + 7x + 12$. **Factoring** an algebraic expression containing the sum or difference of terms means finding an equivalent expression that is a product. Thus, to factor $x^2 + 7x + 12$, we write

$$x^2 + 7x + 12 = (x + 3)(x + 4).$$

We can make several important observations about the factors on the right side.

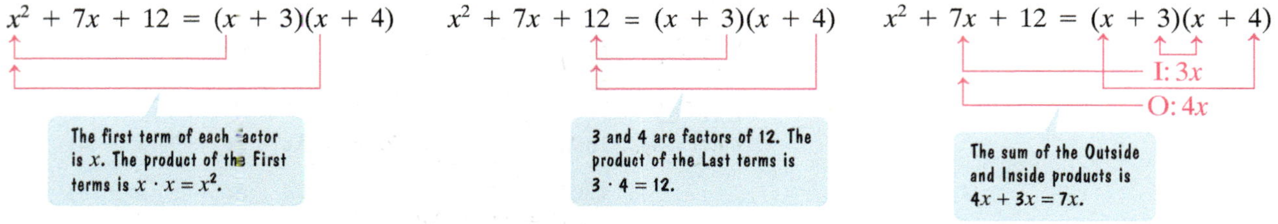

These observations provide us with a procedure for factoring $x^2 + bx + c$.

A STRATEGY FOR FACTORING $x^2 + bx + c$

1. Enter x as the first term of each factor.

$$(x \quad\;)(x \quad\;) = x^2 + bx + c$$

2. List pairs of factors of the constant c.
3. Try various combinations of these factors as the second term in each set of parentheses. Select the combination in which the sum of the Outside and Inside products is equal to bx.

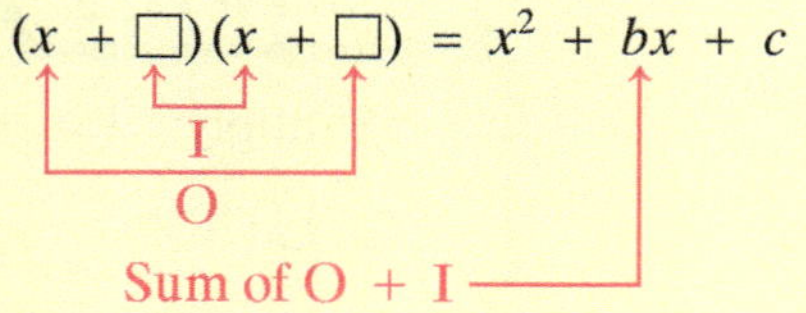

4. Check your work by multiplying the factors using the FOIL method. You should obtain the original trinomial.

If none of the possible combinations yield an Outside product and an Inside product whose sum is equal to bx, the trinomial cannot be factored using integers and is called **prime**.

EXAMPLE 3 *Factoring a Trinomial in $x^2 + bx + c$ Form*

Factor: $x^2 + 6x + 8$.

SOLUTION

Step 1 Enter x as the first term of each factor.

$$x^2 + 6x + 8 = (x \quad)(x \quad)$$

To find the second term of each factor, we must find two integers whose product is 8 and whose sum is 6.

Step 2 List all pairs of factors of the constant, 8.

Factors of 8	8, 1	4, 2	−8, −1	−4, −2

Step 3 Try various combinations of these factors. The correct factorization of $x^2 + 6x + 8$ is the one in which the sum of the Outside and Inside products is equal to $6x$. Here is a list of the possible factorizations:

Possible Factorizations of $x^2 + 6x + 8$	Sum of Outside and Inside Products (Should Equal $6x$)
$(x + 8)(x + 1)$	$x + 8x = 9x$
$(x + 4)(x + 2)$	$2x + 4x = 6x$
$(x - 8)(x - 1)$	$-x - 8x = -9x$
$(x - 4)(x - 2)$	$-2x - 4x = -6x$

This is the required middle term.

Thus, $x^2 + 6x + 8 = (x + 4)(x + 2)$.

Step 4 Check this result by multiplying the right side using the FOIL method. You should obtain the original trinomial. Because of the commutative property, the factorization can also be expressed as

$$x^2 + 6x + 8 = (x + 2)(x + 4).$$

GREAT QUESTION!

Is there a way to eliminate some of the combinations of factors for $x^2 + bx + c$ when c is positive?

Yes. To factor $x^2 + bx + c$ when c is positive, find two numbers with the same sign as the middle term.

$$x^2 + 6x + 8 = (x + 2)(x + 4)$$

Same signs

$$x^2 - 5x + 6 = (x - 3)(x - 2)$$

Same signs

Using this observation, it is not necessary to list the last two factorizations in step 3 on the right.

CHECK POINT 3 Factor: $x^2 + 5x + 6$.

EXAMPLE 4 *Factoring a Trinomial in $x^2 + bx + c$ Form*

Factor: $x^2 + 2x - 35$.

SOLUTION

Step 1 Enter x as the first term of each factor.

$$x^2 + 2x - 35 = (x \quad)(x \quad)$$

To find the second term of each factor, we must find two integers whose product is −35 and whose sum is 2.

Step 2 List pairs of factors of the constant, −35.

Factors of −35	35, −1	−35, 1	−7, 5	7, −5

Step 3 Try various combinations of these factors. The correct factorization of $x^2 + 2x - 35$ is the one in which the sum of the Outside and Inside products is equal to $2x$. Here is a list of the possible factorizations:

Possible Factorizations of $x^2 + 2x - 35$	Sum of Outside and Inside Products (Should Equal $2x$)
$(x - 1)(x + 35)$	$35x - x = 34x$
$(x + 1)(x - 35)$	$-35x + x = -34x$
$(x - 7)(x + 5)$	$5x - 7x = -2x$
$(x + 7)(x - 5)$	$-5x + 7x = 2x$

This is the required middle term.

Thus, $x^2 + 2x - 35 = (x + 7)(x - 5)$ or $(x - 5)(x + 7)$.

Step 4 Verify the factorization using the FOIL method.

F O I L

$$(x + 7)(x - 5) = x^2 - 5x + 7x - 35 = x^2 + 2x - 35$$

Because the product of the factors is the original trinomial, the factorization is correct.

GREAT QUESTION!

Is there a way to eliminate some of the combinations of factors for $x^2 + bx + c$ when c is negative?

Yes. To factor $x^2 + bx + c$ when c is negative, find two numbers with opposite signs whose sum is the coefficient of the middle term.

$x^2 + 2x - 35 = (x + 7)(x - 5)$

Negative   Opposite signs

CHECK POINT 4 Factor: $x^2 + 3x - 10$.

Factoring a Trinomial Where the Coefficient of the Squared Term Is Not 1

How do we factor a trinomial such as $3x^2 - 20x + 28$? Notice that the coefficient of the squared term is 3. We must find two binomials whose product is $3x^2 - 20x + 28$. The product of the First terms must be $3x^2$:

$$(3x \quad)(x \quad).$$

From this point on, the factoring strategy is exactly the same as the one we use to factor trinomials for which the coefficient of the squared term is 1.

EXAMPLE 5 *Factoring a Trinomial*

Factor: $3x^2 - 20x + 28$.

SOLUTION

Step 1 Find two First terms whose product is $3x^2$.

$$3x^2 - 20x + 28 = (3x \quad)(x \quad)$$

Step 2 List all pairs of factors of the constant, 28. The number 28 has pairs of factors that are either both positive or both negative. Because the middle term, $-20x$, is negative, both factors must be negative. The negative factorizations of 28 are $-1(-28)$, $-2(-14)$, and $-4(-7)$.

GREAT QUESTION!

When factoring trinomials, must I list every possible factorization before getting the correct one?

With practice, you will find that it is not necessary to list every possible factorization of the trinomial. As you practice factoring, you will be able to narrow down the list of possible factors to just a few. When it comes to factoring, practice makes perfect.

Step 3 Try various combinations of these factors. The correct factorization of $3x^2 - 20x + 28$ is the one in which the sum of the Outside and Inside products is equal to $-20x$. Here is a list of the possible factorizations:

Possible Factorizations of $3x^2 - 20x + 28$	Sum of Outside and Inside Products (Should Equal $-20x$)
$(3x - 1)(x - 28)$	$-84x - x = -85x$
$(3x - 28)(x - 1)$	$-3x - 28x = -31x$
$(3x - 2)(x - 14)$	$-42x - 2x = -44x$
$(3x - 14)(x - 2)$	$-6x - 14x = -20x$
$(3x - 4)(x - 7)$	$-21x - 4x = -25x$
$(3x - 7)(x - 4)$	$-12x - 7x = -19x$

This is the required middle term.

Thus,

$$3x^2 - 20x + 28 = (3x - 14)(x - 2) \quad \text{or} \quad (x - 2)(3x - 14).$$

Step 4 Verify the factorization using the FOIL method.

$$\begin{aligned}(3x - 14)(x - 2) &= \overset{\text{F}}{3x \cdot x} + \overset{\text{O}}{3x(-2)} + \overset{\text{I}}{(-14) \cdot x} + \overset{\text{L}}{(-14)(-2)} \\ &= 3x^2 - 6x - 14x + 28 \\ &= 3x^2 - 20x + 28\end{aligned}$$

Because this is the trinomial we started with, the factorization is correct.

CHECK POINT 5 Factor: $5x^2 - 14x + 8$.

EXAMPLE 6 *Factoring a Trinomial*

Factor: $8y^2 - 10y - 3$.

SOLUTION

Step 1 Find two first terms whose product is $8y^2$.

$$8y^2 - 10y - 3 \stackrel{?}{=} (8y \quad)(y \quad)$$
$$8y^2 - 10y - 3 \stackrel{?}{=} (4y \quad)(2y \quad)$$

Step 2 List all pairs of factors of the constant, -3. The possible factorizations are $1(-3)$ and $-1(3)$.

Step 3 Try various combinations of these factors. The correct factorization of $8y^2 - 10y - 3$ is the one in which the sum of the Outside and Inside products is equal to $-10y$. Here is a list of the possible factorizations:

Possible Factorizations of $8y^2 - 10y - 3$	Sum of Outside and Inside Products (Should Equal $-10y$)
$(8y + 1)(y - 3)$	$-24y + y = -23y$
$(8y - 3)(y + 1)$	$8y - 3y = 5y$
$(8y - 1)(y + 3)$	$24y - y = 23y$
$(8y + 3)(y - 1)$	$-8y + 3y = -5y$
$(4y + 1)(2y - 3)$	$-12y + 2y = -10y$
$(4y - 3)(2y + 1)$	$4y - 6y = -2y$
$(4y - 1)(2y + 3)$	$12y - 2y = 10y$
$(4y + 3)(2y - 1)$	$-4y + 6y = 2y$

These four factorizations are $(8y \quad)(y \quad)$ with $1(-3)$ and $-1(3)$ as factorizations of -3.

These four factorizations are $(4y \quad)(2y \quad)$ with $1(-3)$ and $-1(3)$ as factorizations of -3.

This is the required middle term.

Thus, $8y^2 - 10y - 3 = (4y + 1)(2y - 3)$.

By the commutative property,

$$8y^2 - 10y - 3 = (4y + 1)(2y - 3) \quad \text{or} \quad (2y - 3)(4y + 1).$$

Show that either of these factorizations is correct by multiplying the factors using the FOIL method. You should obtain the original trinomial.

CHECK POINT 6 Factor: $6y^2 + 19y - 7$.

3 Solve quadratic equations by factoring.

Solving Quadratic Equations by Factoring

We have seen that in a linear equation, the highest exponent on the variable is 1. We now define a quadratic equation, in which the greatest exponent on the variable is 2.

DEFINITION OF A QUADRATIC EQUATION

A **quadratic equation** in x is an equation that can be written in the form

$$ax^2 + bx + c = 0,$$

where a, b, and c are real numbers, with $a \neq 0$.

Here is an example of a quadratic equation:

$$x^2 - 7x + 10 = 0.$$

$a = 1$ $b = -7$ $c = 10$

Notice that we can factor the left side of this equation.

$$x^2 - 7x + 10 = 0$$
$$(x - 5)(x - 2) = 0$$

If a quadratic equation has zero on one side and a factored trinomial on the other side, it can be solved using the **zero-product principle**:

THE ZERO-PRODUCT PRINCIPLE

If the product of two factors is zero, then one (or both) of the factors must have a value of zero.

If $AB = 0$, then $A = 0$ or $B = 0$.

EXAMPLE 7 *Solving a Quadratic Equation Using the Zero-Product Principle*

Solve: $(x - 5)(x - 2) = 0$.

SOLUTION

The product $(x - 5)(x - 2)$ is equal to zero. By the zero-product principle, the only way that this product can be zero is if at least one of the factors is zero. We set each individual factor equal to zero and solve each resulting equation for x.

$$(x - 5)(x - 2) = 0$$
$$x - 5 = 0 \quad \text{or} \quad x - 2 = 0$$
$$x = 5 \qquad\qquad x = 2$$

Check the proposed solutions by substituting each one separately for x in the original equation.

Check 5:	**Check 2:**
$(x - 5)(x - 2) = 0$	$(x - 5)(x - 2) = 0$
$(5 - 5)(5 - 2) \stackrel{?}{=} 0$	$(2 - 5)(2 - 2) \stackrel{?}{=} 0$
$0(3) \stackrel{?}{=} 0$	$-3(0) \stackrel{?}{=} 0$
$0 = 0$, true	$0 = 0$, true

The resulting true statements indicate that the solutions are 5 and 2. The solution set is $\{2, 5\}$.

CHECK POINT 7 Solve: $(x + 6)(x - 3) = 0$.

SOLVING A QUADRATIC EQUATION BY FACTORING

1. If necessary, rewrite the equation in the form $ax^2 + bx + c = 0$, moving all terms to one side, thereby obtaining zero on the other side.
2. Factor.
3. Apply the zero-product principle, setting each factor equal to zero.
4. Solve the equations in step 3.
5. Check the solutions in the original equation.

GREAT QUESTION!

Can all quadratic equations be solved by factoring?

No. The method on the right does not apply if $ax^2 + bx + c$ is not factorable, or prime.

EXAMPLE 8 *Solving a Quadratic Equation by Factoring*

Solve: $x^2 - 2x = 35$.

SOLUTION

Step 1 Move all terms to one side and obtain zero on the other side. Subtract 35 from both sides and write the equation in $ax^2 + bx + c = 0$ form.

$$x^2 - 2x = 35$$
$$x^2 - 2x - 35 = 35 - 35$$
$$x^2 - 2x - 35 = 0$$

Step 2 Factor.

$$(x - 7)(x + 5) = 0$$

Steps 3 and 4 Set each factor equal to zero and solve each resulting equation.

$$x - 7 = 0 \quad \text{or} \quad x + 5 = 0$$
$$x = 7 \qquad\qquad x = -5$$

Step 5 Check the solutions in the original equation.

Check 7:	**Check −5:**
$x^2 - 2x = 35$	$x^2 - 2x = 35$
$7^2 - 2 \cdot 7 \stackrel{?}{=} 35$	$(-5)^2 - 2(-5) \stackrel{?}{=} 35$
$49 - 14 \stackrel{?}{=} 35$	$25 + 10 \stackrel{?}{=} 35$
$35 = 35$, true	$35 = 35$, true

The resulting true statements indicate that the solutions are 7 and -5. The solution set is $\{-5, 7\}$.

CHECK POINT 8 Solve: $x^2 - 6x = 16$.

EXAMPLE 9 Solving a Quadratic Equation by Factoring

Solve: $5x^2 - 33x + 40 = 0$.

SOLUTION

All terms are already on the left and zero is on the other side. Thus, we can factor the trinomial on the left side. $5x^2 - 33x + 40$ factors as $(5x - 8)(x - 5)$.

$$5x^2 - 33x + 40 = 0 \quad \text{This is the given quadratic equation.}$$

$$(5x - 8)(x - 5) = 0 \quad \text{Factor.}$$

$$5x - 8 = 0 \quad \text{or} \quad x - 5 = 0 \quad \text{Set each factor equal to zero.}$$

$$5x = 8 \qquad x = 5 \quad \text{Solve the resulting equations.}$$

$$x = \frac{8}{5}$$

Check these values in the original equation to confirm that the solution set is $\{\frac{8}{5}, 5\}$.

CHECK POINT 9 Solve: $2x^2 + 7x - 4 = 0$.

4 Solve quadratic equations using the quadratic formula.

Solving Quadratic Equations Using the Quadratic Formula

The solutions of a quadratic equation cannot always be found by factoring. Some trinomials are difficult to factor, and others cannot be factored (that is, they are prime). However, there is a formula that can be used to solve all quadratic equations, whether or not they contain factorable trinomials. The formula is called the *quadratic formula.*

GREAT QUESTION!

Is it ok if I write

$$x = -b \pm \frac{\sqrt{b^2 - 4ac}}{2a}?$$

No. The entire numerator of the quadratic formula must be divided by $2a$. Always write the fraction bar all the way across the numerator.

$$x = \frac{-b \pm \sqrt{b^2 - 4ac}}{2a}$$

THE QUADRATIC FORMULA

The solutions of a quadratic equation in the form $ax^2 + bx + c = 0$, with $a \neq 0$, are given by the **quadratic formula**

$$x = \frac{-b \pm \sqrt{b^2 - 4ac}}{2a}.$$

x equals negative b plus or minus the square root of $b^2 - 4ac$, all divided by $2a$.

To use the quadratic formula, be sure that the quadratic equation is expressed with all terms on one side and zero on the other side. It may be necessary to begin by rewriting the equation in this form. Then determine the numerical values for a (the coefficient of the x^2-term), b (the coefficient of the x-term), and c (the constant term). Substitute the values of a, b, and c into the quadratic formula and evaluate the expression. The $\pm$ sign indicates that there are two solutions of the equation.

EXAMPLE 10 Solving a Quadratic Equation Using the Quadratic Formula

Solve using the quadratic formula: $2x^2 + 9x - 5 = 0$.

SOLUTION

The given equation is in the desired form, with all terms on one side and zero on the other side. Begin by identifying the values for a, b, and c.

$$2x^2 + 9x - 5 = 0$$

$a = 2 \quad b = 9 \quad c = -5$

Substituting these values into the quadratic formula and simplifying gives the equation's solutions.

$$x = \frac{-b \pm \sqrt{b^2 - 4ac}}{2a}$$ Use the quadratic formula.

$$x = \frac{-9 \pm \sqrt{9^2 - 4(2)(-5)}}{2(2)}$$ Substitute the values for a, b, and c: $a = 2, b = 9$, and $c = -5$.

$$= \frac{-9 \pm \sqrt{81 + 40}}{4}$$ $9^2 - 4(2)(-5) = 81 - (-40) = 81 + 40$

$$= \frac{-9 \pm \sqrt{121}}{4}$$ Add under the radical sign.

$$= \frac{-9 \pm 11}{4}$$ $\sqrt{121} = 11$

Now we will evaluate this expression in two different ways to obtain the two solutions. On the left, we will *add* 11 to −9. On the right, we will *subtract* 11 from −9.

$$x = \frac{-9 + 11}{4} \quad \text{or} \quad x = \frac{-9 - 11}{4}$$

$$= \frac{2}{4} = \frac{1}{2} \qquad\qquad = \frac{-20}{4} = -5$$

The solution set is $\left\{-5, \frac{1}{2}\right\}$.

CHECK POINT 10 Solve using the quadratic formula: $8x^2 + 2x - 1 = 0$.

GREAT QUESTION!

What's the bottom line on whether I can use factoring to solve a quadratic equation?

Compute $b^2 - 4ac$, which appears under the radical sign in the quadratic formula. If $b^2 - 4ac$ is a perfect square, such as 4, 25, or 121, then the equation can be solved by factoring.

The quadratic equation in Example 10 has rational solutions, namely −5 and $\frac{1}{2}$. The equation can also be solved by factoring. Take a few minutes to do this now and convince yourself that you will arrive at the same two solutions.

Any quadratic equation that has rational solutions can be solved by factoring or using the quadratic formula. However, quadratic equations with irrational solutions cannot be solved by factoring. These equations can be readily solved using the quadratic formula.

EXAMPLE 11 *Solving a Quadratic Equation Using the Quadratic Formula*

Solve using the quadratic formula: $2x^2 = 4x + 1$.

SOLUTION

The quadratic equation must have zero on one side to identify the values for a, b, and c. To move all terms to one side and obtain zero on the right, we subtract $4x + 1$ from both sides. Then we can identify the values for a, b, and c.

$$2x^2 = 4x + 1$$ This is the given equation.

$$2x^2 - 4x - 1 = 0$$ Subtract $4x + 1$ from both sides.

$a = 2$ $b = -4$ $c = -1$

Substituting these values, $a = 2, b = -4$, and $c = -1$, into the quadratic formula and simplifying gives the equation's solutions.

$$x = \frac{-b \pm \sqrt{b^2 - 4ac}}{2a} \qquad \text{Use the quadratic formula.}$$

$$x = \frac{-(-4) \pm \sqrt{(-4)^2 - 4(2)(-1)}}{2(2)} \qquad \text{Substitute the values for } a, b, \text{ and } c\text{: } a = 2, b = -4, \text{ and } c = -1.$$

$$= \frac{4 \pm \sqrt{16 - (-8)}}{4} \qquad -(-4) = 4, (-4)^2 = (-4)(-4) = 16, \text{ and } 4(2)(-1) = -8$$

$$= \frac{4 \pm \sqrt{24}}{4} \qquad 16 - (-8) = 16 + 8 = 24$$

The solutions are $\frac{4 + \sqrt{24}}{4}$ and $\frac{4 - \sqrt{24}}{4}$. These solutions are irrational numbers. You can use a calculator to obtain a decimal approximation for each solution. However, in situations such as this that do not involve applications, it is better to leave the irrational solutions in radical form as exact answers. In some cases, we can simplify this radical form. Using methods for simplifying square roots discussed in Section 5.4, we can simplify $\sqrt{24}$:

$$\sqrt{24} = \sqrt{4 \cdot 6} = \sqrt{4}\sqrt{6} = 2\sqrt{6}.$$

Now we can use this result to simplify the two solutions. First, use the distributive property to factor out 2 from both terms in the numerator. Then, divide the numerator and the denominator by 2.

$$x = \frac{4 \pm \sqrt{24}}{4} = \frac{4 \pm 2\sqrt{6}}{4} = \frac{\overset{1}{\cancel{2}}(2 \pm \sqrt{6})}{\underset{2}{\cancel{4}}} = \frac{2 \pm \sqrt{6}}{2}$$

In simplified radical form, the equation's solution set is

$$\left\{\frac{2 + \sqrt{6}}{2}, \frac{2 - \sqrt{6}}{2}\right\}.$$

TECHNOLOGY

Using a Calculator to Approximate $\frac{4 + \sqrt{24}}{4}$:

Many Scientific Calculators

(4 + 24 √) ÷ 4 =

Many Graphing Calculators

(4 + √ 24) ÷ 4 ENTER

Correct to the nearest tenth,

$$\frac{4 + \sqrt{24}}{4} \approx 2.2.$$

GREAT QUESTION!

The simplification of the irrational solutions in Example 11 was kind of tricky. Any suggestions to guide the process?

Many students use the quadratic formula correctly until the last step, where they make an error in simplifying the solutions. Be sure to factor the numerator before dividing the numerator and the denominator by the greatest common factor.

$$\frac{4 \pm 2\sqrt{6}}{4} = \frac{2(2 \pm \sqrt{6})}{4} = \frac{\overset{1}{\cancel{2}}(2 \pm \sqrt{6})}{\underset{2}{\cancel{4}}} = \frac{2 \pm \sqrt{6}}{2}$$

You cannot divide just one term in the numerator and the denominator by their greatest common factor.

Incorrect!

$$\frac{\overset{1}{\cancel{4}} \pm 2\sqrt{6}}{\underset{1}{\cancel{4}}} = 1 \pm 2\sqrt{6} \qquad \frac{4 \pm \overset{1}{\cancel{2}}\sqrt{6}}{\underset{2}{\cancel{4}}} = \frac{4 \pm \sqrt{6}}{2}$$

Examples 10 and 11 illustrate that the solutions of quadratic equations can be rational or irrational numbers. In Example 10, the expression under the square root was 121, a perfect square ($\sqrt{121} = 11$), and we obtained rational solutions. In Example 11, this expression was 24, which is not a perfect square (although we simplified $\sqrt{24}$ to $2\sqrt{6}$), and we obtained irrational solutions. If the expression under the square root simplifies to a negative number, then the quadratic equation has **no real solution**. The solution set consists of *imaginary numbers,* discussed in the Blitzer Bonus on page 304.

CHECK POINT 11 Solve using the quadratic formula: $2x^2 = 6x - 1$.

5 Solve problems modeled by quadratic equations.

Applications

EXAMPLE 12 *Blood Pressure and Age*

The graphs in **Figure 6.6** illustrate that a person's normal systolic blood pressure, measured in millimeters of mercury (mm Hg), depends on his or her age. The formula

$$P = 0.006A^2 - 0.02A + 120$$

models a man's normal systolic pressure, P, at age A.

a. Find the age, to the nearest year, of a man whose normal systolic blood pressure is 125 mm Hg.

b. Use the graphs in **Figure 6.6** to describe the differences between the normal systolic blood pressures of men and women as they age.

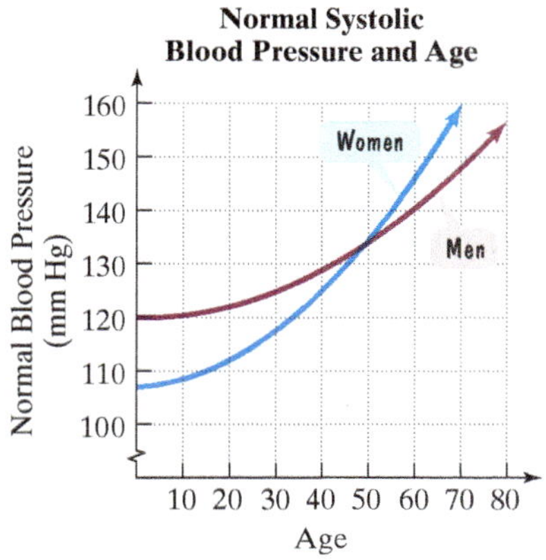

FIGURE 6.6

SOLUTION

a. We are interested in the age of a man with a normal systolic blood pressure of 125 millimeters of mercury. Thus, we substitute 125 for P in the given formula for men. Then we solve for A, the man's age.

$$P = 0.006A^2 - 0.02A + 120 \quad \text{This is the given formula for men.}$$

$$125 = 0.006A^2 - 0.02A + 120 \quad \text{Substitute 125 for } P.$$

$$0 = 0.006A^2 - 0.02A - 5 \quad \text{Subtract 125 from both sides and obtain zero on one side.}$$

$a = 0.006$ $b = -0.02$ $c = -5$

Because the trinomial on the right side of the equation is prime, we solve using the quadratic formula.

Notice that the variable is A, rather than the usual x.

$$A = \frac{-b \pm \sqrt{b^2 - 4ac}}{2a} \quad \text{Use the quadratic formula.}$$

$$= \frac{-(-0.02) \pm \sqrt{(-0.02)^2 - 4(0.006)(-5)}}{2(0.006)} \quad \text{Substitute the values for } a, b, \text{ and } c\text{: } a = 0.006, b = -0.02, \text{ and } c = -5.$$

$$= \frac{0.02 \pm \sqrt{0.1204}}{0.012} \quad \text{Use a calculator to simplify the expression under the square root.}$$

$$\approx \frac{0.02 \pm 0.347}{0.012} \quad \text{Use a calculator: } \sqrt{0.1204} \approx 0.347.$$

$$A \approx \frac{0.02 + 0.347}{0.012} \quad \text{or} \quad A \approx \frac{0.02 - 0.347}{0.012}$$

$$A \approx 31 \qquad A \approx -27 \quad \text{Use a calculator and round to the nearest integer.}$$

Reject this solution. Age cannot be negative.

TECHNOLOGY

On most calculators, here is how to approximate

$$\frac{0.02 + \sqrt{0.1204}}{0.012}.$$

Many Scientific Calculators

[( ] .02 [+] .1204 [√] [ )] [÷] .012 [=]

Many Graphing Calculators

[( ] .02 [+] [√] .1204 [ )] [÷] .012 [ENTER]

If your calculator displays an open parenthesis after √, you'll need to enter another closed parenthesis here.

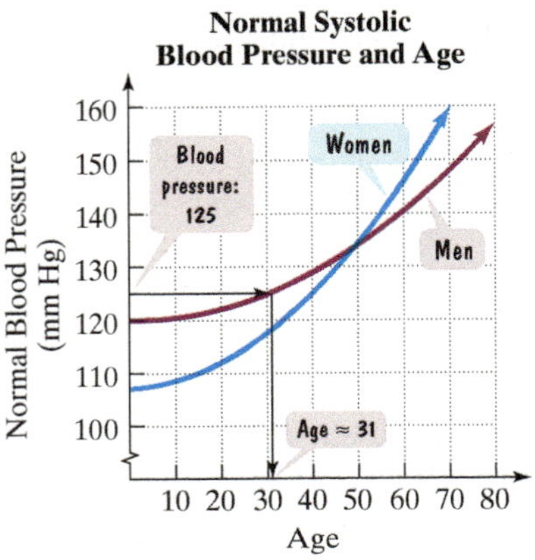

FIGURE 6.7

The positive solution, $A \approx 31$, indicates that 31 is the approximate age of a man whose normal systolic blood pressure is 125 mm Hg. This is illustrated by the black lines with the arrows on the red graph representing men in **Figure 6.7**.

b. Take a second look at the graphs in **Figure 6.6** or **Figure 6.7**. Before approximately age 50, the blue graph representing women's normal systolic blood pressure lies below the red graph representing men's normal systolic blood pressure. Thus, up to age 50, women's normal systolic blood pressure is lower than men's, although it is increasing at a faster rate. After age 50, women's normal systolic blood pressure is higher than men's.

CHECK POINT 12 The formula $P = 0.01A^2 + 0.05A + 107$ models a woman's normal systolic blood pressure, P, at age A. Use this formula to find the age, to the nearest year, of a woman whose normal systolic blood pressure is 115 mm Hg. Use the blue graph in **Figure 6.6** on the previous page to verify your solution.

Blitzer Bonus

Art, Nature, and Quadratic Equations

A **golden rectangle** can be a rectangle of any size, but its long side must be Φ times as long as its short side, where $\Phi \approx 1.6$. Artists often use golden rectangles in their work because they are considered to be more visually pleasing than other rectangles.

In *The Bathers at Asnières*, by the French impressionist Georges Seurat (1859–1891), the artist positions parts of the painting as though they were inside golden rectangles.

Bathers at Asnières (1884), Georges Seurat. Oil on canvas, 201 cm × 300 cm (79 in × 118 in). National Gallery, London/Art Resource, New York.

If a golden rectangle is divided into a square and a rectangle, as in **Figure 6.8(a)**, the smaller rectangle is a golden rectangle. If the smaller golden rectangle is divided again, the same is true of the yet smaller rectangle, and so on. The process of repeatedly dividing each golden rectangle in this manner is illustrated in **Figure 6.8(b)**. We've also created a spiral by connecting the opposite corners of all the squares with a smooth curve. This spiral matches the spiral shape of the chambered nautilus shell shown in **Figure 6.8(c)**. The shell spirals out at an ever-increasing rate that is governed by this geometry.

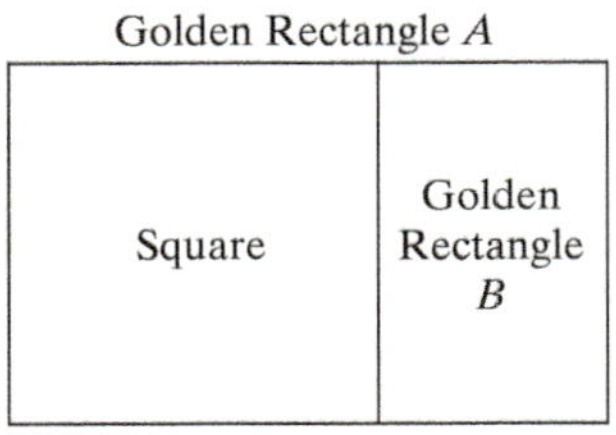

FIGURE 6.8(a)

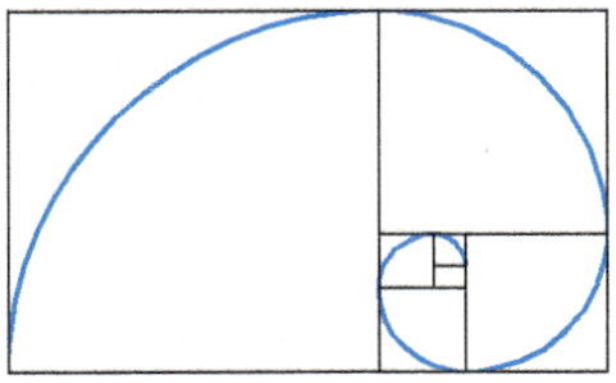

FIGURE 6.8(b)

FIGURE 6.8(c)

In the Exercise Set that follows, you will use the golden rectangles in **Figure 6.8(a)** to obtain an exact value for Φ, the ratio of the long side to the short side in a golden rectangle of any size. Your model will involve a quadratic equation that can be solved by the quadratic formula. (See Exercise 87.)

Concept and Vocabulary Check

Fill in each blank so that the resulting statement is true.

1. For $(x + 5)(2x + 3)$, the product of the first terms is ________, the product of the outside terms is ________, the product of the inside terms is ________, and the product of the last terms is ________.

2. $x^2 + 13x + 30 = (x + 3)(x$ ____$)$

3. $x^2 - 9x + 18 = (x - 3)(x$ ____$)$

4. $x^2 - x - 30 = (x - 6)(x$ ____$)$

5. $x^2 - 5x - 14 = (x + 2)(x$ ____$)$

6. $8x^2 - 10x - 3 = (4x + 1)(2x$ ____$)$

7. $12x^2 - x - 20 = (4x + 5)(3x$ ____$)$

8. $2x^2 - 5x + 3 = (x - 1)($______$)$

9. $6x^2 + 17x + 12 = (2x + 3)($______$)$

10. An equation that can be written in the form $ax^2 + bx + c = 0$, $a \neq 0$, is called a/an ________ equation.

11. The zero-product principle states that if $AB = 0$, then ____________.

12. The equation $5x^2 + x = 18$ can be written in the form $ax^2 + bx + c = 0$ by __________ on both sides.

13. The solutions of $ax^2 + bx + c = 0$, $a \neq 0$, are given by ________________, called the ______________

In Exercises 14–18, determine whether each statement is true or false. If the statement is false, make the necessary change(s) to produce a true statement.

14. One factor of $x^2 + x + 20$ is $x + 5$. ______

15. If $(x + 3)(x - 4) = 2$, then $(x + 3) = 0$ or $(x - 4) = 0$. ____

16. In using the quadratic formula to solve the quadratic equation $5x^2 = 2x - 7$, we have $a = 5$, $b = 2$, and $c = -7$. ______

17. The quadratic formula can be expressed as

$$x = -b \pm \frac{\sqrt{b^2 - 4ac}}{2a}.$$ ______

18. The solutions $\frac{4 \pm \sqrt{3}}{2}$ can be simplified to $2 \pm \sqrt{3}$. ______

Exercise Set 6.5

Practice Exercises

Use FOIL to find the products in Exercises 1–8.

1. $(x + 3)(x + 5)$

2. $(x + 7)(x + 2)$

3. $(x - 5)(x + 3)$

4. $(x - 1)(x + 2)$

5. $(2x - 1)(x + 2)$

6. $(2x - 5)(x + 3)$

7. $(3x - 7)(4x - 5)$

8. $(2x - 9)(7x - 4)$

Factor the trinomials in Exercises 9–32, or state that the trinomial is prime. Check your factorization using FOIL multiplication.

9. $x^2 + 5x + 6$

10. $x^2 + 8x + 15$

11. $x^2 - 2x - 15$

12. $x^2 - 4x - 5$

13. $x^2 - 8x + 15$

14. $x^2 - 14x + 45$

15. $x^2 - 9x - 36$

16. $x^2 - x - 90$

17. $x^2 - 8x + 32$

18. $x^2 - 9x + 81$

19. $x^2 + 17x + 16$

20. $x^2 - 7x - 44$

21. $2x^2 + 7x + 3$

22. $3x^2 + 7x + 2$

23. $2x^2 - 17x + 30$

24. $5x^2 - 13x + 6$

25. $3x^2 - x - 2$

26. $2x^2 + 5x - 3$

27. $3x^2 - 25x - 28$

28. $3x^2 - 2x - 5$

29. $6x^2 - 11x + 4$

30. $6x^2 - 17x + 12$

31. $4x^2 + 16x + 15$

32. $8x^2 + 33x + 4$

In Exercises 33–36, solve each equation using the zero-product principle.

33. $(x - 8)(x + 3) = 0$

34. $(x + 11)(x - 5) = 0$

35. $(4x + 5)(x - 2) = 0$

36. $(x + 9)(3x - 1) = 0$

Solve the quadratic equations in Exercises 37–52 by factoring.

37. $x^2 + 8x + 15 = 0$

38. $x^2 + 5x + 6 = 0$

39. $x^2 - 2x - 15 = 0$

40. $x^2 + x - 42 = 0$

41. $x^2 - 4x = 21$

42. $x^2 + 7x = 18$

43. $x^2 + 9x = -8$

44. $x^2 - 11x = -10$

45. $x^2 - 12x = -36$

46. $x^2 - 14x = -49$

47. $2x^2 = 7x + 4$

48. $3x^2 = x + 4$

49. $5x^2 + x = 18$

50. $3x^2 - 4x = 15$

51. $x(6x + 23) + 7 = 0$

52. $x(6x + 13) + 6 = 0$

Solve the equations in Exercises 53–72 using the quadratic formula.

53. $x^2 + 8x + 15 = 0$

54. $x^2 + 8x + 12 = 0$

55. $x^2 + 5x + 3 = 0$

56. $x^2 + 5x + 2 = 0$

57. $x^2 + 4x = 6$

58. $x^2 + 2x = 4$

59. $x^2 + 4x - 7 = 0$

60. $x^2 + 4x + 1 = 0$

61. $x^2 - 3x = 18$

62. $x^2 - 3x = 10$

63. $6x^2 - 5x - 6 = 0$

64. $9x^2 - 12x - 5 = 0$

65. $x^2 - 2x - 10 = 0$

66. $x^2 + 6x - 10 = 0$

67. $x^2 - x = 14$

68. $x^2 - 5x = 10$

69. $6x^2 + 6x + 1 = 0$

70. $3x^2 = 5x - 1$

71. $4x^2 = 12x - 9$

72. $9x^2 + 6x + 1 = 0$

Practice Plus

In Exercises 73–80, solve each equation by the method of your choice.

73. $\frac{3x^2}{4} - \frac{5x}{2} - 2 = 0$

74. $\frac{x^2}{3} - \frac{x}{2} - \frac{3}{2} = 0$

75. $(x - 1)(3x + 2) = -7(x - 1)$

76. $x(x + 1) = 4 - (x + 2)(x + 2)$

77. $(2x - 6)(x + 2) = 5(x - 1) - 12$

78. $7x(x - 2) = 3 - 2(x + 4)$

79. $2x^2 - 9x - 3 = 9 - 9x$

80. $3x^2 - 6x - 3 = 12 - 6x$

81. When the sum of 6 and twice a positive number is subtracted from the square of the number, 0 results. Find the number.

82. When the sum of 1 and twice a negative number is subtracted from twice the square of the number, 0 results. Find the number.

Application Exercises

The formula

$$N = \frac{t^2 - t}{2}$$

describes the number of football games, N, that must be played in a league with t teams if each team is to play every other team once. Use this information to solve Exercises 83–84.

83. If a league has 36 games scheduled, how many teams belong to the league, assuming that each team plays every other team once?

84. If a league has 45 games scheduled, how many teams belong to the league, assuming that each team plays every other team once?

A substantial percentage of the United States population is foreign-born. The bar graph shows the percentage of foreign-born Americans for selected years from 1920 through 2010.

Percentage of the United States Population That Was Foreign-Born, 1920–2010

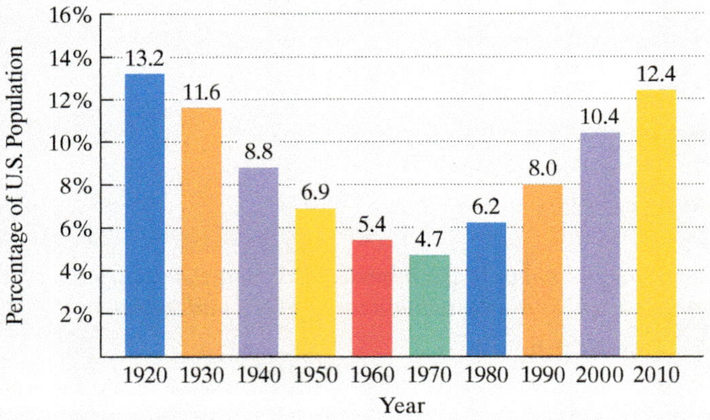

Source: U.S. Census Bureau

The percentage, p, of the United States population that was foreign-born x years after 1920 can be modeled by the formula

$$p = 0.004x^2 - 0.36x + 14.$$

Use the formula to solve Exercises 85–86.

85. a. According to the model, what percentage of the U.S. population was foreign-born in 2000? Does the model underestimate or overestimate the actual number displayed by the bar graph? By how much?

b. If trends shown by the model continue, in which year will 18% of the U.S. population be foreign-born?

86. a. According to the model, what percentage of the U.S. population was foreign-born in 1990? Does the model underestimate or overestimate the actual number displayed by the bar graph on the previous page? By how much?

b. If trends shown by the model continue, in which year will 23% of the U.S. population be foreign-born? Round to the nearest year.

87. If you have not yet done so, read the Blitzer Bonus on page 398. In this exercise, you will use the golden rectangles shown to obtain an exact value for Φ, the ratio of the long side to the short side in a golden rectangle of any size.

Golden Rectangle A

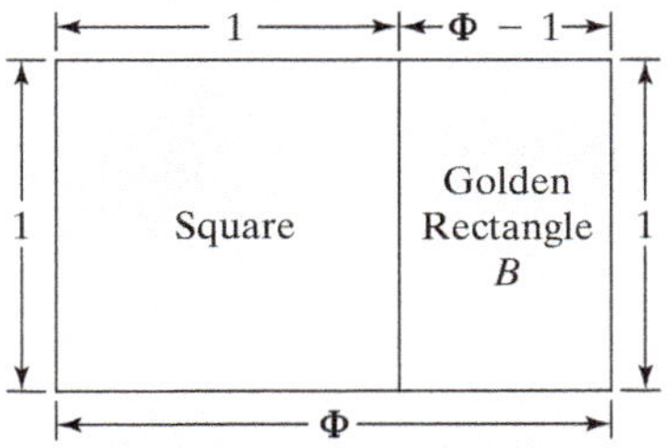

a. The golden ratio in rectangle A, or the ratio of the long side to the short side, can be modeled by $\frac{\Phi}{1}$. Write a fractional expression that models the golden ratio in rectangle B.

b. Set the expression for the golden ratio in rectangle A equal to the expression for the golden ratio in rectangle B. Solve the resulting proportion using the quadratic formula. Express Φ as an exact value in simplified radical form.

c. Use your solution from part (b) to complete this statement: The ratio of the long side to the short side in a golden rectangle of any size is _____ to 1.

Writing in Mathematics

88. Explain how to multiply two binomials using the FOIL method. Give an example with your explanation.

89. Explain how to factor $x^2 - 5x + 6$.

90. Explain how to solve a quadratic equation by factoring. Use the equation $x^2 + 6x + 8 = 0$ in your explanation.

91. Explain how to solve a quadratic equation using the quadratic formula. Use the equation $x^2 + 6x + 8 = 0$ in your explanation.

92. Describe the trend shown by the data for the percentage of foreign-born Americans in the graph for Exercises 85–86. Do you believe that this trend is likely to continue or might something occur that would make it impossible to extend the model into the future? Explain your answer.

Critical Thinking Exercises

Make Sense? *In Exercises 93–96, determine whether each statement makes sense or does not make sense, and explain your reasoning.*

93. I began factoring $x^2 - 17x + 72$ by finding all number pairs with a sum of -17.

94. It's easy to factor $x^2 + x + 1$ because of the relatively small numbers for the constant term and the coefficient of x.

95. The fastest way for me to solve $x^2 - x - 2 = 0$ is to use the quadratic formula.

96. I simplified $\frac{3 + 2\sqrt{3}}{2}$ to $3 + \sqrt{3}$ because 2 is a factor of $2\sqrt{3}$.

97. The radicand of the quadratic formula, $b^2 - 4ac$, can be used to determine whether $ax^2 + bx + c = 0$ has solutions that are rational, irrational, or not real numbers. Explain how this works. Is it possible to determine the kinds of answers that one will obtain to a quadratic equation without actually solving the equation? Explain.

In Exercises 98–99, find all positive integers b so that the trinomial can be factored.

98. $x^2 + bx + 15$

99. $x^2 + 4x + b$

100. Factor: $x^{2n} + 20x^n + 99$.

101. Solve: $x^2 + 2\sqrt{3}x - 9 = 0$.

Chapter Summary, Review, and Test

SUMMARY – DEFINITIONS AND CONCEPTS

6.1 Algebraic Expressions and Formulas

SUMMARY – DEFINITIONS AND CONCEPTS	EXAMPLES
a. An algebraic expression combines variables and numbers using addition, subtraction, multiplication, division, powers, or roots.	
b. Evaluating an algebraic expression means finding its value for a given value of the variable or for given values of the variables. Once these values are substituted, follow the order of operations agreement in the box on page 340.	Ex. 1, p. 341; Ex. 2, p. 341; Ex. 3, p. 341
c. An equation is a statement that two expressions are equal. Formulas are equations that express relationships among two or more variables. Mathematical modeling is the process of finding formulas to describe real-world phenomena. Such formulas, together with the meaning assigned to the variables, are called mathematical models. The formulas are said to model, or describe, the relationships among the variables.	Ex. 4, p. 342

d. Terms of an algebraic expression are separated by addition. Like terms have the same variables with the same exponents on the variables. To add or subtract like terms, add or subtract the coefficients and copy the common variable.	
e. An algebraic expression is simplified when parentheses have been removed (using the distributive property) and like terms have been combined.	Ex. 5, p. 344; Ex. 6, p. 345; Ex. 7, p. 345

6.2 Linear Equations in One Variable and Proportions

a. A linear equation in one variable can be written in the form $ax + b = 0$, where a and b are real numbers, and $a \neq 0$.	
b. Solving a linear equation is the process of finding the set of numbers that makes the equation a true statement. These numbers are the solutions. The set of all such solutions is the solution set.	
c. Equivalent equations have the same solution set. Properties for generating equivalent equations are given in the box on page 351.	Ex. 1, p. 352
d. A step-by-step procedure for solving a linear equation is given in the box on page 352.	Ex. 2, p. 352; Ex. 3, p. 354; Ex. 4, p. 354
e. If an equation contains fractions, begin by multiplying both sides of the equation by the least common denominator of the fractions in the equation, thereby clearing fractions.	Ex. 5, p. 355; Ex. 6, p. 356
f. The ratio of a to b is written $\frac{a}{b}$, or $a : b$.	
g. A proportion is a statement in the form $\frac{a}{b} = \frac{c}{d}$.	
h. The cross-products principle states that if $\frac{a}{b} = \frac{c}{d}$, then $ad = bc$.	Ex. 7, p. 358
i. A step-by-step procedure for solving applied problems using proportions is given in the box on page 359.	Ex. 8, p. 359; Ex. 9, p. 360
j. If a false statement (such as $-6 = 7$) is obtained in solving an equation, the equation has no solution. The solution set is $\varnothing$, the empty set.	Ex. 10, p. 360
k. If a true statement (such as $-6 = -6$) is obtained in solving an equation, the equation has infinitely many solutions. The solution set is the set of all real numbers, written $\{x \mid x \text{ is a real number}\}$.	Ex. 11, p. 361

6.3 Applications of Linear Equations

a. Algebraic translations of English phrases are given in Table 6.2 on page 366.	
b. A step-by-step strategy for solving word problems using linear equations is given in the box on page 365.	Ex. 1, p. 366; Ex. 2, p. 368; Ex. 3, p. 370; Ex. 4, p. 371
c. Solving a formula for a variable means rewriting the formula so that the variable is isolated on one side of the equation.	Ex. 5, p. 372; Ex. 6, p. 372

6.4 Linear Inequalities in One Variable

A procedure for solving a linear inequality is given in the box on page 379. Remember to reverse the direction of the inequality symbol when multiplying or dividing both sides of an inequality by a negative number, thereby changing the sense of the inequality.	Ex. 2, p. 379; Ex. 3, p. 379; Ex. 4, p. 380; Ex. 5, p. 381; Ex. 6, p. 382

6.5 Quadratic Equations

a. A quadratic equation can be written in the form $ax^2 + bx + c = 0, a \neq 0$.	
b. Some quadratic equations can be solved using factoring and the zero-product principle. A step-by-step procedure is given in the box on page 393.	Ex. 8, p. 393; Ex. 9, p. 394
c. All quadratic equations in the form $ax^2 + bx + c = 0$ can be solved using the quadratic formula: $x = \frac{-b \pm \sqrt{b^2 - 4ac}}{2a}$.	Ex. 10, p. 394; Ex. 11, p. 395; Ex. 12, p. 397

Review Exercises

6.1

In Exercises 1–3, evaluate the algebraic expression for the given value of the variable.

1. $6x + 9; x = 4$
2. $7x^2 + 4x - 5; x = -2$
3. $6 + 2(x - 8)^3; x = 5$
4. The diversity index, from 0 (no diversity) to 100, measures the chance that two randomly selected people are a different race or ethnicity. The diversity index in the United States varies widely from region to region, from as high as 81 in Hawaii to as low as 11 in Vermont. The bar graph shows the national diversity index for the United States for four years in the period from 1980 through 2010.

Chance That Two Randomly Selected Americans Are a Different Race or Ethnicity

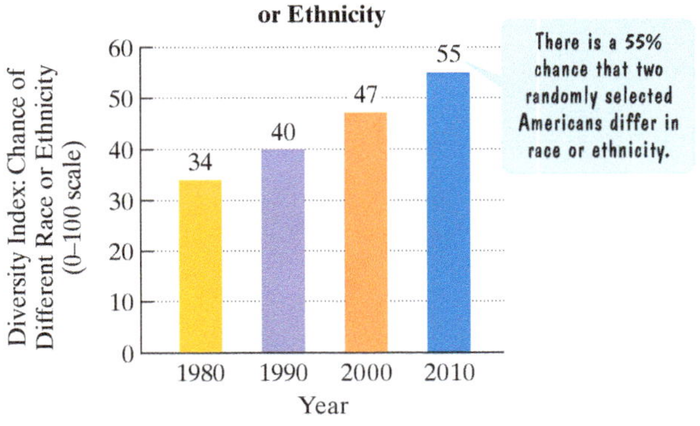

Source: USA Today

The data in the graph can be modeled by the formula

$$D = 0.005x^2 + 0.55x + 34,$$

where D is the national diversity index in the United States x years after 1980. According to the formula, what was the U.S. diversity index in 2010? How does this compare with the index displayed by the bar graph?

In Exercises 5–7, simplify each algebraic expression.

5. $5(2x - 3) + 7x$
6. $3(4y - 5) - (7y - 2)$
7. $2(x^2 + 5x) + 3(4x^2 - 3x)$

6.2

In Exercises 8–14, solve each equation.

8. $4x + 9 = 33$
9. $5x - 3 = x + 5$
10. $3(x + 4) = 5x - 12$
11. $2(x - 2) + 3(x + 5) = 2x - 2$
12. $\frac{2x}{3} = \frac{x}{6} + 1$
13. $7x + 5 = 5(x + 3) + 2x$
14. $7x + 13 = 2(2x - 5) + 3x + 23$

In Exercises 15–18, solve each proportion.

15. $\frac{3}{x} = \frac{15}{25}$
16. $\frac{-7}{5} = \frac{91}{x}$
17. $\frac{x + 2}{3} = \frac{4}{5}$
18. $\frac{5}{x + 7} = \frac{3}{x + 3}$
19. If a school board determines that there should be 3 teachers for every 50 students, how many teachers are needed for an enrollment of 5400 students?
20. To determine the number of trout in a lake, a conservationist catches 112 trout, tags them, and returns them to the lake. Later, 82 trout are caught, and 32 of them are found to be tagged. How many trout are in the lake?
21. The line graph shows the cost of inflation. What cost \$10,000 in 1982 would cost the amount shown by the graph in subsequent years.

The Cost of Inflation

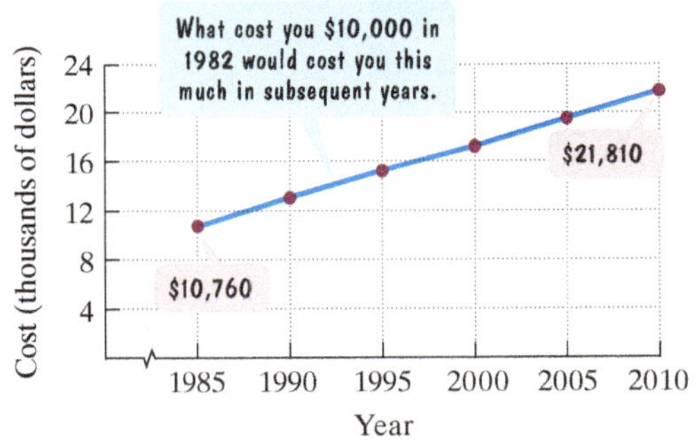

Source: U.S. Bureau of Labor Statistics

Here are two mathematical models for the data shown by the graph. In each formula, C represents the cost x years after 1985 of what cost \$10,000 in 1982.

Model 1 $C = 438x + 10{,}800$

Model 2 $C = 0.3x^2 + 430x + 10{,}824$

a. Use the graph to estimate the cost in 1995, to the nearest thousand dollars, of what cost \$10,000 in 1982.

b. Use model 1 to determine the cost in 1995. How well does this describe your estimate from part (a)?

c. Use model 2 to determine the cost in 1995. How well does this describe your estimate from part (a)?

d. Use model 1 to determine in which year the cost will be \$28,320 for what cost \$10,000 in 1982.

6.3

22. Destined for Gory. As sequels to horror films increase, so does the body count. Wes Craven's slasher *Scream* series adheres to that axiom.

Whether it's knife to the back, knife to the gut, or knife to the head, the body count in *Scream 2* exceeds the departed in *Scream* by 2. Appropriately, the number of characters killed off in *Scream 3* exceeds the departed in *Scream* by 3. The total body count in the four *Scream* films shown in the graphic is 33. Find the body count in *Scream*, *Scream 2*, and *Scream 3*.

Body Count in Wes Craven's *Scream* Series

23. The bar graph shows the average price of a movie ticket for selected years from 1980 through 2010. The graph indicates that in 1980, the average movie ticket price was $2.69. For the period from 1980 through 2010, the price increased by approximately $0.15 per year. If this trend continues, by which year will the average price of a movie ticket be $8.69?

1980
Ordinary People
Ticket Price $2.69

Average Price of a U.S. Movie Ticket

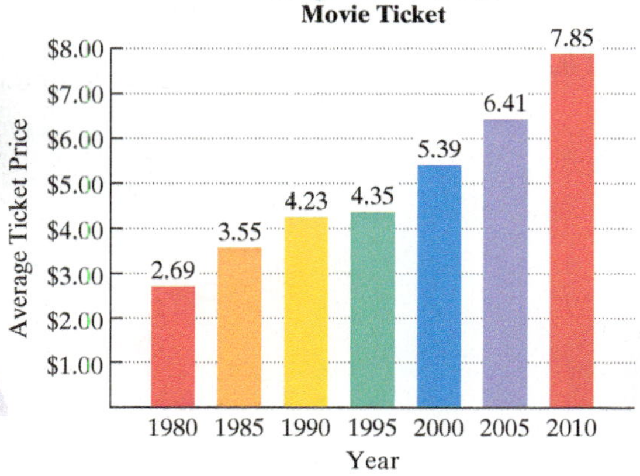

2010
Alice in Wonderland
Ticket Price $7.85

Sources: Motion Picture Association of America, National Association of Theater Owners (NATO), and Bureau of Labor Statistics (BLS)

24. You are choosing between two texting plans. One plan has a monthly fee of $15 with a charge of $0.05 per text. The other plan has a monthly fee of $5 with a charge of $0.07 per text. For how many text messages will the costs for the two plans be the same?

25. After a 20% price reduction, a cordless phone sold for $48. What was the phone's price before the reduction?

26. A salesperson earns $300 per week plus 5% commission on sales. How much must be sold to earn $800 in a week?

In Exercises 27–30, solve each formula for the specified variable.

27. $Ax - By = C$ for x

28. $A = \frac{1}{2}bh$ for h

29. $A = \frac{B + C}{2}$ for B

30. $vt + gt^2 = s$ for g

6.4

In Exercises 31–37, solve each inequality and graph the solution set on a number line.

31. $2x - 5 < 3$

32. $\frac{x}{2} > -4$

33. $3 - 5x \leq 18$

34. $4x + 6 < 5x$

35. $6x - 10 \geq 2(x + 3)$

36. $4x + 3(2x - 7) \leq x - 3$

37. $-1 < 4x + 2 \leq 6$

38. To pass a course, a student must have an average on three examinations of at least 60. If a student scores 42 and 74 on the first two tests, what must be earned on the third test to pass the course?

6.5

Use FOIL to find the products in Exercises 39–40.

39. $(x + 9)(x - 5)$

40. $(4x - 7)(3x + 2)$

Factor the trinomials in Exercises 41–46, or state that the trinomial is prime.

41. $x^2 - x - 12$

42. $x^2 - 8x + 15$

43. $x^2 + 2x + 3$

44. $3x^2 - 17x + 10$

45. $6x^2 - 11x - 10$

46. $3x^2 - 6x - 5$

Solve the quadratic equations in Exercises 47–50 by factoring.

47. $x^2 + 5x - 14 = 0$

48. $x^2 - 4x = 32$

49. $2x^2 + 15x - 8 = 0$

50. $3x^2 = -21x - 30$

Solve the quadratic equations in Exercises 51–54 using the quadratic formula.

51. $x^2 - 4x + 3 = 0$

52. $x^2 - 5x = 4$

53. $2x^2 + 5x - 3 = 0$

54. $3x^2 - 6x = 5$

55. As gas prices surge, more Americans are cycling as a way to save money, stay fit, or both. In 2010, Boston installed 20 miles of bike lanes and New York City added more than 50 miles. The bar graph shows the number of bicycle-friendly U.S. communities, as designated by the League of American Bicyclists, for selected years from 2003 through 2011.

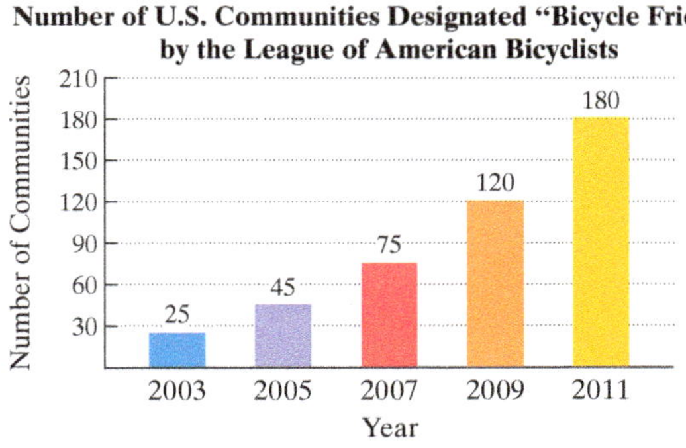

Source: League of American Bicyclists

The formula

$$B = 1.7x^2 + 6x + 26$$

models the number of bicycle-friendly communities, B, x years after 2003.

a. Use the formula to find the number of bicycle-friendly communities in 2011. Round to the nearest whole number. Does this rounded value underestimate or overestimate the number shown by the graph? By how much?

b. Use the formula to determine the year in which 826 U.S. communities will be bicycle friendly.

Chapter 6 Test

1. Evaluate $x^3 - 4(x - 1)^2$ when $x = -2$.

2. Simplify: $5(3x - 2) - (x - 6)$.

In Exercises 3–6, solve each equation.

3. $12x + 4 = 7x - 21$

4. $3(2x - 4) = 9 - 3(x + 1)$

5. $3(x - 4) + x = 2(6 + 2x)$

6. $\frac{x}{5} - 2 = \frac{x}{3}$

7. Solve for y: $By - Ax = A$.

8. The bar graph in the next column shows the percentage of American adults reporting personal gun ownership for selected years from 1980 through 2010.

Here are two mathematical models for the data shown by the graph. In each formula, p represents the percentage of American adults who reported personal gun ownership x years after 1980.

Model 1 $p = -0.3x + 30$

Model 2 $p = -0.003x^2 - 0.22x + 30$

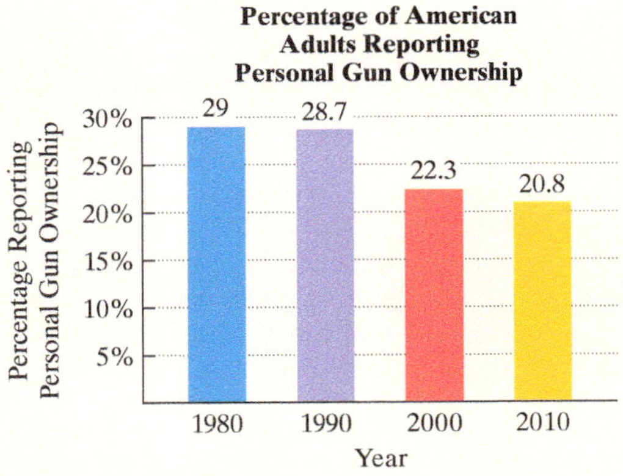

Source: General Social Survey

a. According to model 1, what percentage of American adults reported personal gun ownership in 2010? Does this underestimate or overestimate the percentage shown by the graph? By how much?

b. According to model 2, what percentage of American adults reported personal gun ownership in 2010? Does this underestimate or overestimate the percentage shown by the graph? By how much?

c. If trends shown by the data continue, use model 1 to determine in which year 17.7% of American adults will report personal gun ownership.

In Exercises 9–10, solve each proportion.

9. $\frac{5}{8} = \frac{x}{12}$

10. $\frac{x+5}{8} = \frac{x+2}{5}$

11. Park rangers catch, tag, and release 200 tule elk back into a wildlife refuge. Two weeks later they observe a sample of 150 elk, of which 5 are tagged. Assuming that the ratio of tagged elk in the sample holds for all elk in the refuge, how many elk are there in the park?

12. What's the last word in capital punishment? An analysis of the final statements of all men and women Texas has executed since the Supreme Court reinstated the death penalty in 1976 revealed that "love" is by far the most frequently uttered word. The bar graph shows the number of times various words were used in final statements by Texas death-row inmates.

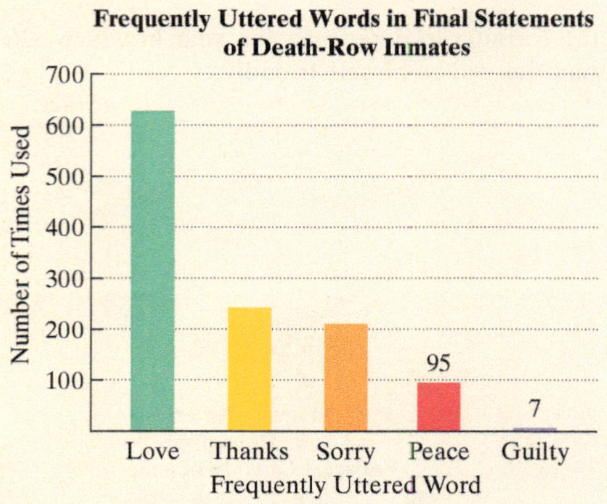

Source: Texas Department of Criminal Justice

The number of times "love" was used exceeded the number of times "sorry" was used by 419. The number of utterances of "thanks" exceeded the number of utterances of "sorry" by 32. Combined, these three words were used 1084 times. Determine the number of times each of these words was used in final statements by Texas inmates.

13. You bought a new car for \$50,750. Its value is decreasing by \$5500 per year. After how many years will its value be \$12,250?

14. You are choosing between two texting plans. Plan A charges \$25 per month for unlimited texting. Plan B has a monthly fee of \$13 with a charge of S0.06 per text. For how many text messages, will the costs for the two plans be the same?

15. After a 60% reduction, a jacket sold for \$20. What was the jacket's price before the reduction?

In Exercises 16–18, solve each inequality and graph the solution set on a number line.

16. $6 - 9x \geq 33$

17. $4x - 2 > 2(x + 6)$

18. $-3 \leq 2x + 1 < 6$

19. A student has grades on three examinations of 76, 80, and 72. What must the student earn on a fourth examination in order to have an average of at least 80?

20. Use FOIL to find this product: $(2x - 5)(3x + 4)$.

21. Factor: $2x^2 - 9x + 10$.

22. Solve by factoring: $x^2 + 5x = 36$.

23. Solve using the quadratic formula: $2x^2 + 4x = -1$.

The graphs show the amount being paid in Social Security benefits and the amount going into the system. All data are expressed in billions of dollars. Amounts from 2012 through 2024 are projections.

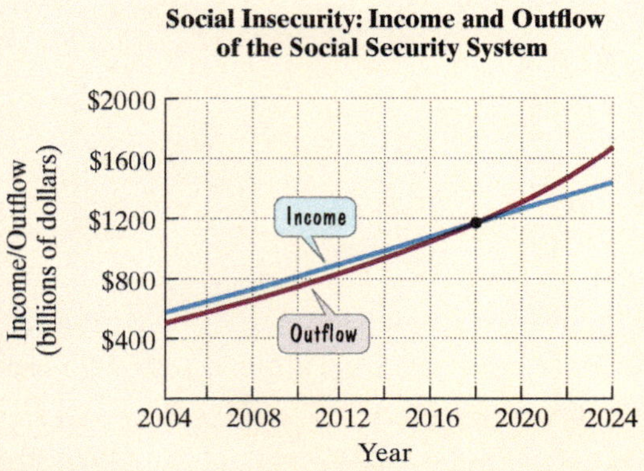

Source: 2004 Social Security Trustees Report

Exercises 24–26 are based on the data shown by the graphs.

24. In 2004, the system's income was \$575 billion, projected to increase at an average rate of \$43 billion per year. In which year will the system's income be \$1177 billion?

25. The data for the system's outflow can be modeled by the formula

$$B = 0.07x^2 + 47.4x + 500,$$

where B represents the amount paid in benefits, in billions of dollars, x years after 2004. According to this model, when will the amount paid in benefits be \$1177 billion? Round to the nearest year.

26. How well do your answers to Exercises 24 and 25 model the data shown by the graphs?

7

Algebra: Graphs, Functions, and Linear Systems

TELEVISION, MOVIES, AND MAGAZINES PLACE GREAT EMPHASIS ON PHYSICAL BEAUTY. OUR CULTURE emphasizes physical appearance to such an extent that it is a central factor in the perception and judgment of others. The modern emphasis on thinness as the ideal body shape has been suggested as a major cause of eating disorders among adolescent women.

Cultural values of physical attractiveness change over time. During the 1950s, actress Jayne Mansfield embodied the postwar ideal: curvy, buxom, and big-hipped. Men, too, have been caught up in changes of how they "ought" to look. The 1960s' ideal was the soft and scrawny hippie. Today's ideal man is tough and muscular.

Given the importance of culture in setting standards of attractiveness, how can you establish a healthy weight range for your age and height? In this chapter, we will use systems of inequalities to explore these skin-deep issues.

Here's where you'll find these applications:

You'll find a weight that fits you using the models (mathematical, not fashion) in Example 4 of Section 7.4 and Exercises 45–48 in Exercise Set 7.4. Exercises 51–52 use graphs and a formula for body-mass index to indicate whether you are obese, overweight, borderline overweight, normal weight, or underweight.

7.1

WHAT AM I SUPPOSED TO LEARN?

After you have read this section, you should be able to:

1 Plot points in the rectangular coordinate system.
2 Graph equations in the rectangular coordinate system.
3 Use function notation.
4 Graph functions.
5 Use the vertical line test.
6 Obtain information about a function from its graph.

Graphing and Functions

THE BEGINNING OF THE SEVENTEENTH CENTURY was a time of innovative ideas and enormous intellectual progress in Europe. English theatergoers enjoyed a succession of exciting new plays by Shakespeare. William Harvey proposed the radical notion that the heart was a pump for blood rather than the center of emotion. Galileo, with his new-fangled invention called the telescope, supported the theory of Polish astronomer Copernicus that the sun, not the Earth, was the center of the solar system. Monteverdi was writing the world's first grand operas. French mathematicians Pascal and Fermat invented a new field of mathematics called probability theory.

Into this arena of intellectual electricity stepped French aristocrat René Descartes (1596–1650). Descartes (pronounced "day cart"), propelled by the creativity surrounding him, developed a new branch of mathematics that brought together algebra and geometry in a unified way—a way that visualized numbers as points on a graph, equations as geometric figures, and geometric figures as equations. This new branch of mathematics, called *analytic geometry,* established Descartes as one of the founders of modern thought and among the most original mathematicians and philosophers of any age. We begin this section by looking at Descartes's deceptively simple idea, called the **rectangular coordinate system** or (in his honor) the **Cartesian coordinate system**.

1 Plot points in the rectangular coordinate system.

Points and Ordered Pairs

Descartes used two number lines that intersect at right angles at their zero points, as shown in **Figure 7.1**. The horizontal number line is the ***x*-axis**. The vertical number line is the ***y*-axis**. The point of intersection of these axes is their zero points, called the **origin**. Positive numbers are shown to the right and above the origin. Negative numbers are shown to the left and below the origin. The axes divide the plane into four quarters, called **quadrants**. The points located on the axes are not in any quadrant.

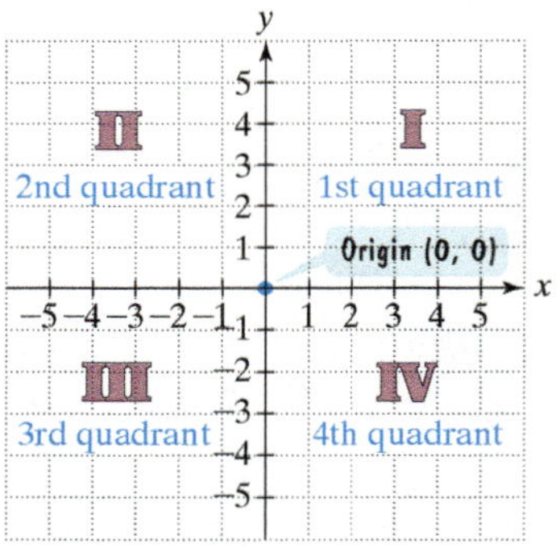

FIGURE 7.1 The rectangular coordinate system

Each point in the rectangular coordinate system corresponds to an **ordered pair** of real numbers, (x, y). Examples of such pairs are $(-5, 3)$ and $(3, -5)$. The first number in each pair, called the ***x*-coordinate**, denotes the distance and direction from the origin along the x-axis. The second number in each pair, called the ***y*-coordinate**, denotes vertical distance and direction along a line parallel to the y-axis or along the y-axis itself.

Figure 7.2 shows how we **plot**, or locate, the points corresponding to the ordered pairs $(-5, 3)$ and $(3, -5)$. We plot $(-5, 3)$ by going 5 units from 0 to the left along the x-axis. Then we go 3 units up parallel to the y-axis. We plot $(3, -5)$ by going 3 units from 0 to the right along the x-axis and 5 units down parallel to the y-axis. The phrase "the points corresponding to the ordered pairs $(-5, 3)$ and $(3, -5)$" is often abbreviated as "the points $(-5, 3)$ and $(3, -5)$."

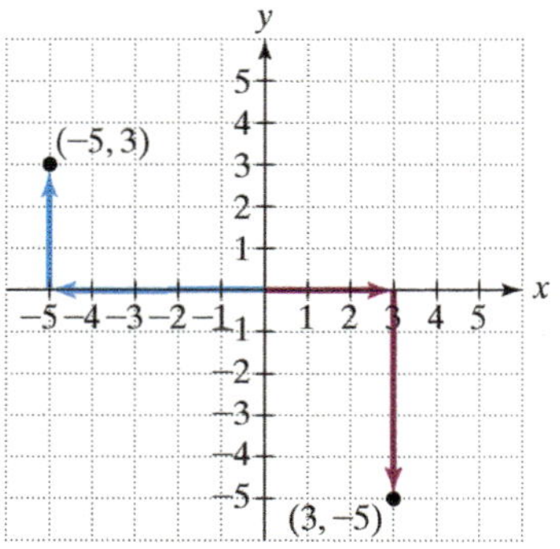

FIGURE 7.2 Plotting $(-5, 3)$ and $(3, -5)$

GREAT QUESTION!

What's the significance of the word *ordered* when describing a pair of real numbers?

The phrase *ordered pair* is used because order is important. The order in which coordinates appear makes a difference in a point's location. This is illustrated in **Figure 7.2**.

EXAMPLE 1 *Plotting Points in the Rectangular Coordinate System*

Plot the points: $A(-3, 5)$, $B(2, -4)$, $C(5, 0)$, $D(-5, -3)$, $E(0, 4)$, and $F(0, 0)$.

SOLUTION

See **Figure 7.3**. We move from the origin and plot the points in the following way:

$A(-3, 5)$:	3 units left, 5 units up
$B(2, -4)$:	2 units right, 4 units down
$C(5, 0)$:	5 units right, 0 units up or down
$D(-5, -3)$:	5 units left, 3 units down
$E(0, 4)$:	0 units right or left, 4 units up
$F(0, 0)$:	0 units right or left, 0 units up or down

Notice that the origin is represented by (0, 0).

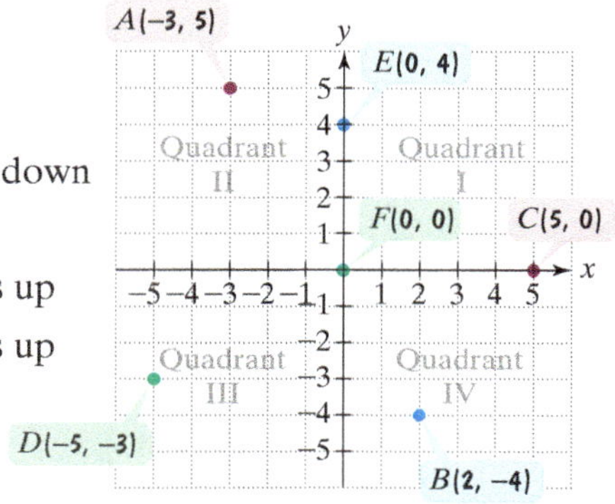

FIGURE 7.3 Plotting points

CHECK POINT 1 Plot the points: $A(-2, 4)$, $B(4, -2)$, $C(-3, 0)$, and $D(0, -3)$.

2 Graph equations in the rectangular coordinate system.

Graphs of Equations

A relationship between two quantities can sometimes be expressed as an **equation in two variables**, such as

$$y = 4 - x^2.$$

A **solution of an equation in two variables**, x and y, is an ordered pair of real numbers with the following property: When the x-coordinate is substituted for x and the y-coordinate is substituted for y in the equation, we obtain a true statement. For example, consider the equation $y = 4 - x^2$ and the ordered pair $(3, -5)$. When 3 is substituted for x and -5 is substituted for y, we obtain the statement $-5 = 4 - 3^2$, or $-5 = 4 - 9$, or $-5 = -5$. Because this statement is true, the ordered pair $(3, -5)$ is a solution of the equation $y = 4 - x^2$. We also say that $(3, -5)$ **satisfies** the equation.

We can generate as many ordered-pair solutions as desired to $y = 4 - x^2$ by substituting numbers for x and then finding the corresponding values for y. For example, suppose we let $x = 3$:

Start with x. Compute y. Form the ordered pair (x, y).

x	$y = 4 - x^2$	Ordered Pair (x, y)
3	$y = 4 - 3^2 = 4 - 9 = -5$	$(3, -5)$

Let $x = 3$. $(3, -5)$ is a solution of $y = 4 - x^2$.

The **graph of an equation in two variables** is the set of all points whose coordinates satisfy the equation. One method for graphing such equations is the **point-plotting method**. First, we find several ordered pairs that are solutions of the equation. Next, we plot these ordered pairs as points in the rectangular coordinate system. Finally, we connect the points with a smooth curve or line. This often gives us a picture of all ordered pairs that satisfy the equation.

EXAMPLE 2 Graphing an Equation Using the Point-Plotting Method

Graph $y = 4 - x^2$. Select integers for x, starting with -3 and ending with 3.

SOLUTION

For each value of x, we find the corresponding value for y.

Start with x. Compute y. Form the ordered pair (x, y).

We selected integers from -3 to 3, inclusive, to include three negative numbers, 0, and three positive numbers. We also wanted to keep the resulting computations for y relatively simple.

x	$y = 4 - x^2$	Ordered Pair (x, y)
-3	$y = 4 - (-3)^2 = 4 - 9 = -5$	$(-3, -5)$
-2	$y = 4 - (-2)^2 = 4 - 4 = 0$	$(-2, 0)$
-1	$y = 4 - (-1)^2 = 4 - 1 = 3$	$(-1, 3)$
0	$y = 4 - 0^2 = 4 - 0 = 4$	$(0, 4)$
1	$y = 4 - 1^2 = 4 - 1 = 3$	$(1, 3)$
2	$y = 4 - 2^2 = 4 - 4 = 0$	$(2, 0)$
3	$y = 4 - 3^2 = 4 - 9 = -5$	$(3, -5)$

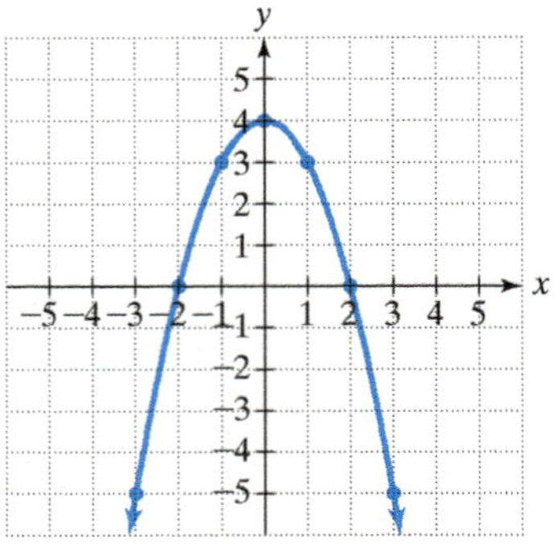

FIGURE 7.4 The graph of $y = 4 - x^2$

Now we plot the seven points and join them with a smooth curve, as shown in **Figure 7.4**. The graph of $y = 4 - x^2$ is a curve where the part of the graph to the right of the y-axis is a reflection of the part to the left of it and vice versa. The arrows on the left and the right of the curve indicate that it extends indefinitely in both directions.

CHECK POINT 2 Graph $y = 4 - x$. Select integers for x, starting with -3 and ending with 3.

Part of the beauty of the rectangular coordinate system is that it allows us to "see" formulas and visualize the solution to a problem. This idea is demonstrated in Example 3.

EXAMPLE 3 An Application Using Graphs of Equations

The toll to a bridge costs \$2.50. Commuters who use the bridge frequently have the option of purchasing a monthly discount pass for \$21.00. With the discount pass, the toll is reduced to \$1.00. The monthly cost, y, of using the bridge x times can be described by the following formulas:

Without the discount pass:

$y = 2.5x$ — The monthly cost, y, is \$2.50 times the number of times, x, that the bridge is used.

With the discount pass:

$y = 21 + 1 \cdot x$
$y = 21 + x.$ — The monthly cost, y, is \$21 for the discount pass plus \$1 times the number of times, x, that the bridge is used.

a. Let $x = 0, 2, 4, 10, 12, 14,$ and 16. Make a table of values for each equation showing seven solutions for the equation.

b. Graph the equations in the same rectangular coordinate system.

c. What are the coordinates of the intersection point for the two graphs? Interpret the coordinates in practical terms.

SOLUTION

a. Tables of values showing seven solutions for each equation follow.

WITHOUT THE DISCOUNT PASS

x	$y = 2.5x$	(x, y)
0	$y = 2.5(0) = 0$	(0, 0)
2	$y = 2.5(2) = 5$	(2, 5)
4	$y = 2.5(4) = 10$	(4, 10)
10	$y = 2.5(10) = 25$	(10, 25)
12	$y = 2.5(12) = 30$	(12, 30)
14	$y = 2.5(14) = 35$	(14, 35)
16	$y = 2.5(16) = 40$	(16, 40)

WITH THE DISCOUNT PASS

x	$y = 21 + x$	(x, y)
0	$y = 21 + 0 = 21$	(0, 21)
2	$y = 21 + 2 = 23$	(2, 23)
4	$y = 21 + 4 = 25$	(4, 25)
10	$y = 21 + 10 = 31$	(10, 31)
12	$y = 21 + 12 = 33$	(12, 33)
14	$y = 21 + 14 = 35$	(14, 35)
16	$y = 21 + 16 = 37$	(16, 37)

b. Now we are ready to graph the two equations. Because the x- and y-coordinates are nonnegative, it is only necessary to use the origin, the positive portions of the x- and y-axes, and the first quadrant of the rectangular coordinate system. The x-coordinates begin at 0 and end at 16. We will let each tick mark on the x-axis represent two units. However, the y-coordinates begin at 0 and get as large as 40 in the formula that describes the monthly cost without the coupon book. So that our y-axis does not get too long, we will let each tick mark on the y-axis represent five units. Using this setup and the two tables of values, we construct the graphs of $y = 2.5x$ and $y = 21 + x$, shown in **Figure 7.5**.

c. The graphs intersect at (14, 35). This means that if the bridge is used 14 times in a month, the total monthly cost without the discount pass is the same as the total monthly cost with the discount pass, namely $35.

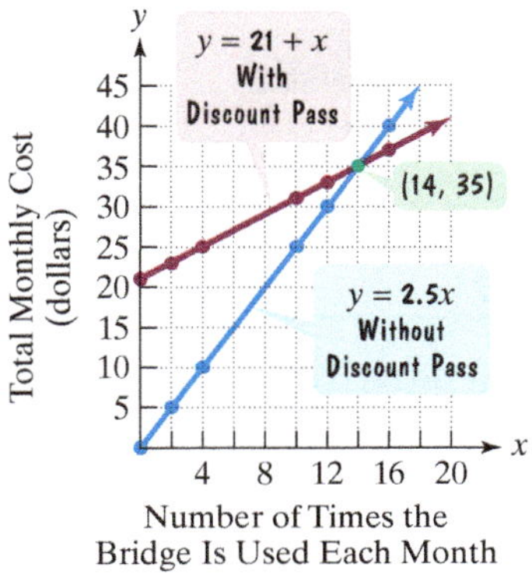

FIGURE 7.5 Options for a toll

In **Figure 7.5**, look at the two graphs to the right of the intersection point (14, 35). The red graph of $y = 21 + x$ lies below the blue graph of $y = 2.5x$. This means that if the bridge is used more than 14 times in a month ($x > 14$), the (red) monthly cost, y, with the discount pass is less than the (blue) monthly cost, y, without the discount pass.

CHECK POINT 3 The toll to a bridge costs $2. If you use the bridge x times in a month, the monthly cost, y, is $y = 2x$. With a $10 discount pass, the toll is reduced to $1. The monthly cost, y, of using the bridge x times in a month with the discount pass is $y = 10 + x$.

a. Let $x = 0, 2, 4, 6, 8, 10$, and 12. Make tables of values showing seven solutions of $y = 2x$ and seven solutions of $y = 10 + x$.

b. Graph the equations in the same rectangular coordinate system.

c. What are the coordinates of the intersection point for the two graphs? Interpret the coordinates in practical terms.

3 Use function notation.

Functions

Reconsider one of the equations from Example 3, $y = 2.5x$. Recall that this equation describes the monthly cost, y, of using the bridge x times, with a toll cost of $2.50 each time the bridge is used. The monthly cost, y, depends on the number of times the bridge is used, x. For each value of x, there is one and only one value of y. If an equation in two variables (x and y) yields precisely one value of y for each value of x, we say that y is a **function** of x.

The notation $y = f(x)$ indicates that the variable y is a function of x. The notation $f(x)$ is read "f of x."

For example, the formula for the cost of the bridge

$$y = 2.5x$$

can be expressed in function notation as

$$f(x) = 2.5x.$$

We read this as "f of x is equal to $2.5x$." If, say, x equals 10 (meaning that the bridge is used 10 times), we can find the corresponding value of y (monthly cost) using the equation $f(x) = 2.5x$.

$$f(x) = 2.5x$$
$$f(10) = 2.5(10) \quad \text{To find } f(10)\text{, read "}f\text{ of 10," replace } x \text{ with 10.}$$
$$= 25$$

Because $f(10) = 25$ (f of 10 equals 25), this means that if the bridge is used 10 times in a month, the total monthly cost is \$25.

Table 7.1 compares our previous notation with the new notation of functions.

TABLE 7.1 Function Notation

"y Equals" Notation	"$f(x)$ Equals" Notation
$y = 2.5x$	$f(x) = 2.5x$
If $x = 10$, $y = 2.5(10) = 25.$	$f(10) = 2.5(10) = 25$ f of 10 equals 25.

GREAT QUESTION!

Doesn't $f(x)$ indicate that I need to multiply f and x?

The notation $f(x)$ does *not* mean "f times x." The notation describes the "output" for the function f when the "input" is x. Think of $f(x)$ as another name for y.

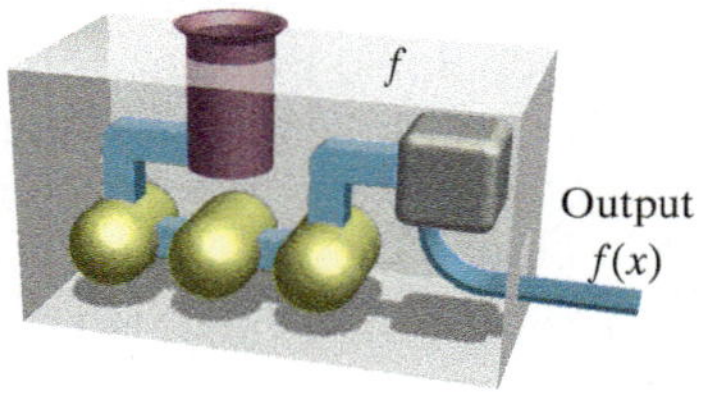

In our next example, we will apply function notation to three different functions. It would be awkward to call all three functions f. We will call the first function f, the second function g, and the third function h. These are the letters most frequently used to name functions.

EXAMPLE 4 *Using Function Notation*

Find each of the following:

a. $f(4)$ for $f(x) = 2x + 3$ **b.** $g(-2)$ for $g(x) = 2x^2 - 1$

c. $h(-5)$ for $h(r) = r^3 - 2r^2 + 5$.

SOLUTION

a.
$$f(x) = 2x + 3 \quad \text{This is the given function.}$$
$$f(4) = 2 \cdot 4 + 3 \quad \text{To find } f \text{ of 4, replace } x \text{ with 4.}$$
$$= 8 + 3 \quad \text{Multiply: } 2 \cdot 4 = 8.$$
$$f(4) = 11 \quad \text{Add.}$$

f of 4 is 11.

b.
$$g(x) = 2x^2 - 1 \quad \text{This is the given function.}$$
$$g(-2) = 2(-2)^2 - 1 \quad \text{To find } g \text{ of } -2\text{, replace } x \text{ with } -2.$$
$$= 2(4) - 1 \quad \text{Evaluate the exponential expression: } (-2)^2 = 4.$$
$$= 8 - 1 \quad \text{Multiply: } 2(4) = 8.$$
$$g(-2) = 7 \quad \text{Subtract.}$$

g of -2 is 7.

c.
$$h(r) = r^3 - 2r^2 + 5 \quad \text{The function's name is } h \text{ and } r \text{ represents the function's input.}$$
$$h(-5) = (-5)^3 - 2(-5)^2 + 5 \quad \text{To find } h \text{ of } -5\text{, replace each occurrence of } r \text{ with } -5.$$
$$= -125 - 2(25) + 5 \quad \text{Evaluate exponential expressions: } (-5)^3 = -125 \text{ and } (-5)^2 = 25.$$
$$= -125 - 50 + 5 \quad \text{Multiply: } 2(25) = 50.$$
$$h(-5) = -170 \quad -125 - 50 = -175 \text{ and } -175 + 5 = -170.$$

h of -5 is -170.

CHECK POINT 4 Find each of the following:

a. $f(6)$ for $f(x) = 4x + 5$

b. $g(-5)$ for $g(x) = 3x^2 - 10$

c. $h(-4)$ for $h(r) = r^2 - 7r + 2$.

EXAMPLE 5 *An Application Involving Function Notation*

Tailgaters beware: If your car is going 35 miles per hour on dry pavement, your required stopping distance is 160 feet, or the width of a football field. At 65 miles per hour, the distance required is 410 feet, or approximately the length of one and one-tenth football fields. **Figure 7.6** shows stopping distances for cars at various speeds on dry roads and on wet roads. **Figure 7.7** uses a line graph to represent stopping distances at various speeds on dry roads.

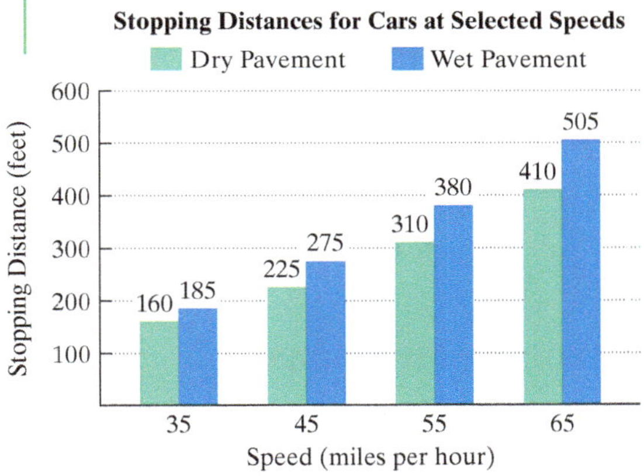

FIGURE 7.6

Source: National Highway Traffic Safety Administration

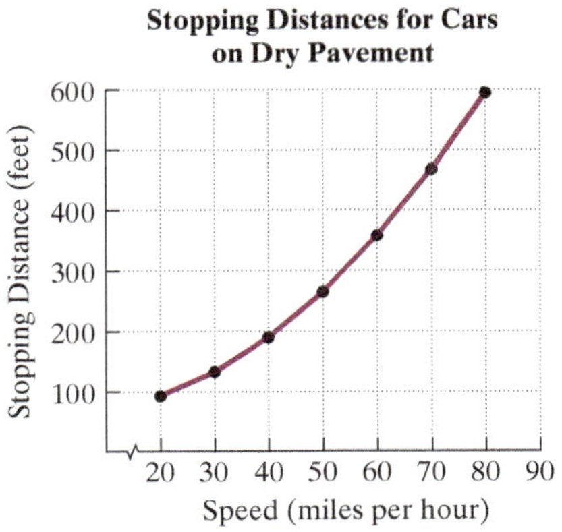

FIGURE 7.7

a. Use the line graph in **Figure 7.7** to estimate a car's required stopping distance at 60 miles per hour on dry pavement. Round to the nearest 10 feet.

b. The function

$$f(x) = 0.0875x^2 - 0.4x + 66.6$$

models a car's required stopping distance, $f(x)$, in feet, on dry pavement traveling at x miles per hour. Use this function to find the required stopping distance at 60 miles per hour. Round to the nearest foot.

SOLUTION

a. The required stopping distance at 60 miles per hour is estimated using the point shown in **Figure 7.8**. The second coordinate of this point extends slightly more than midway between 300 and 400 on the vertical axis. Thus, 360 is a reasonable estimate. We conclude that at 60 miles per hour on dry pavement, the required stopping distance is approximately 360 feet.

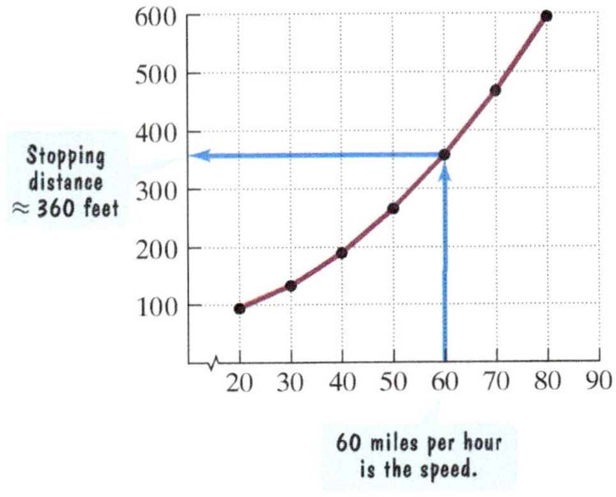

FIGURE 7.8

TECHNOLOGY

On most calculators, here is how to find

$0.0875(60)^2 - 0.4(60) + 66.6.$

Many Scientific Calculators

.0875 [×] 60 [x^2] [−]

.4 [×] 60 [+] 66.6 [=]

Many Graphing Calculators

.0875 [×] 60 [∧] 2 [−]

.4 [×] 60 [+] 66.6 [ENTER]

b. Now we use the given function to determine the required stopping distance at 60 miles per hour. We need to find $f(60)$. The arithmetic gets somewhat "messy," so it is probably a good idea to use a calculator.

$$f(x) = 0.0875x^2 - 0.4x + 66.6$$ This function models stopping distance, $f(x)$, at x miles per hour.

$$f(60) = 0.0875(60)^2 - 0.4(60) + 66.6$$ Replace each x with 60.

$$= 0.0875(3600) - 0.4(60) + 66.6$$ Use the order of operations, first evaluating the exponential expression.

$$= 315 - 24 + 66.6$$ Perform the multiplications.

$$= 357.6$$ Subtract and add as indicated.

$$\approx 358$$ As directed, we've rounded to the nearest foot.

We see that $f(60) \approx 358$—that is, f of 60 is approximately 358. The model indicates that the required stopping distance on dry pavement at 60 miles per hour is approximately 358 feet.

CHECK POINT 5 **a.** Use the line graph in **Figure 7.7** on the previous page to estimate a car's required stopping distance at 40 miles per hour on dry pavement. Round to the nearest ten feet.

b. Use the function in Example 5(b), $f(x) = 0.0875x^2 - 0.4x + 66.6$, to find the required stopping distance at 40 miles per hour. Round to the nearest foot.

4 Graph functions.

Graphing Functions

The **graph of a function** is the graph of its ordered pairs. In our next example, we will graph two functions.

EXAMPLE 6 *Graphing Functions*

Graph the functions $f(x) = 2x$ and $g(x) = 2x + 4$ in the same rectangular coordinate system. Select integers for x from −2 to 2, inclusive.

SOLUTION

For each function, we use the suggested values for x to create a table of some of the coordinates. These tables are shown below. Then, we plot the five points in each table and connect them, as shown in **Figure 7.9**. The graph of each function is a straight line. Do you see a relationship between the two graphs? The graph of g is the graph of f shifted vertically up 4 units.

x	$f(x) = 2x$	(x, y) or $(x, f(x))$
−2	$f(-2) = 2(-2) = -4$	(−2, −4)
−1	$f(-1) = 2(-1) = -2$	(−1, −2)
0	$f(0) = 2 \cdot 0 = 0$	(0, 0)
1	$f(1) = 2 \cdot 1 = 2$	(1, 2)
2	$f(2) = 2 \cdot 2 = 4$	(2, 4)

Choose x. Compute $f(x)$ by evaluating f at x. Form the ordered pair.

x	$g(x) = 2x + 4$	(x, y) or $(x, g(x))$
−2	$g(-2) = 2(-2) + 4 = 0$	(−2, 0)
−1	$g(-1) = 2(-1) + 4 = 2$	(−1, 2)
0	$g(0) = 2 \cdot 0 + 4 = 4$	(0, 4)
1	$g(1) = 2 \cdot 1 + 4 = 6$	(1, 6)
2	$g(2) = 2 \cdot 2 + 4 = 8$	(2, 8)

Choose x. Compute $g(x)$ by evaluating g at x. Form the ordered pair.

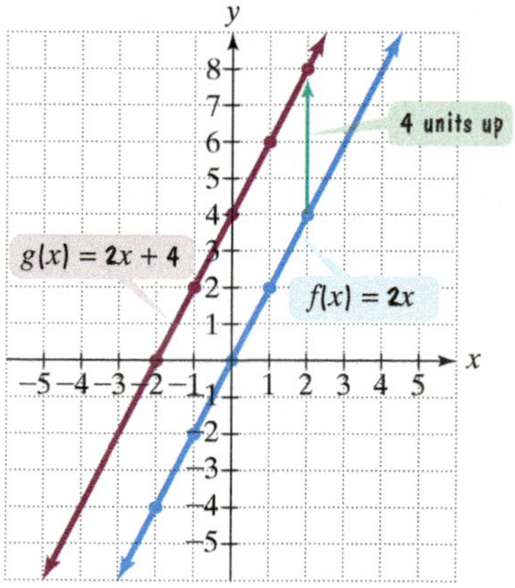

FIGURE 7.9

CHECK POINT 6 Graph the functions $f(x) = 2x$ and $g(x) = 2x - 3$ in the same rectangular coordinate system. Select integers for x from −2 to 2, inclusive. How is the graph of g related to the graph of f?

5 Use the vertical line test.

TECHNOLOGY

A graphing calculator is a powerful tool that quickly generates the graph of an equation in two variables. Here is the graph of $y = 4 - x^2$ that we drew by hand in **Figure 7.4** on page 410.

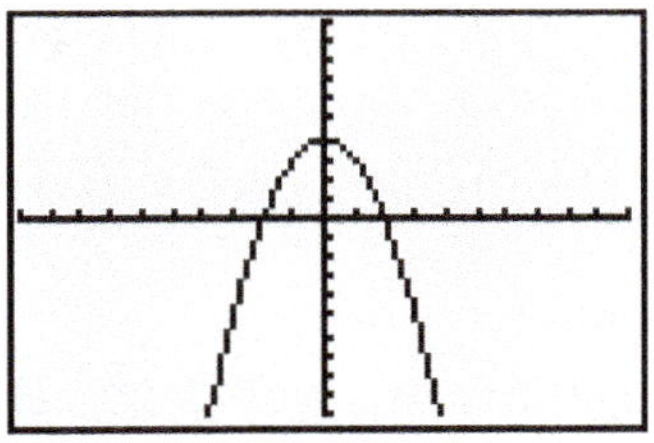

What differences do you notice between this graph and the graph we drew by hand? This graph seems a bit "jittery." Arrows do not appear on the left and right ends of the graph. Furthermore, numbers are not given along the axes. For the graph shown above, the x-axis extends from -10 to 10 and the y-axis also extends from -10 to 10. The distance represented by each consecutive tick mark is one unit. We say that the **viewing window** is $[-10, 10, 1]$ by $[-10, 10, 1]$.

To graph an equation in x and y using a graphing calculator, enter the equation, which must be solved for y, and specify the size of the viewing window. The size of the viewing window sets minimum and maximum values for both the x- and y-axes. Enter these values, as well as the values between consecutive tick marks, on the respective axes. The $[-10, 10, 1]$ by $[-10, 10, 1]$ viewing rectangle used above is called the **standard viewing window**.

The Vertical Line Test

Not every graph in the rectangular coordinate system is the graph of a function. The definition of a function specifies that no value of x can be paired with two or more different values of y. Consequently, if a graph contains two or more different points with the same first coordinate, the graph cannot represent a function. This is illustrated in **Figure 7.10**. Observe that points sharing a common first coordinate are vertically above or below each other.

This observation is the basis of a useful test for determining whether a graph defines y as a function of x. The test is called the **vertical line test**.

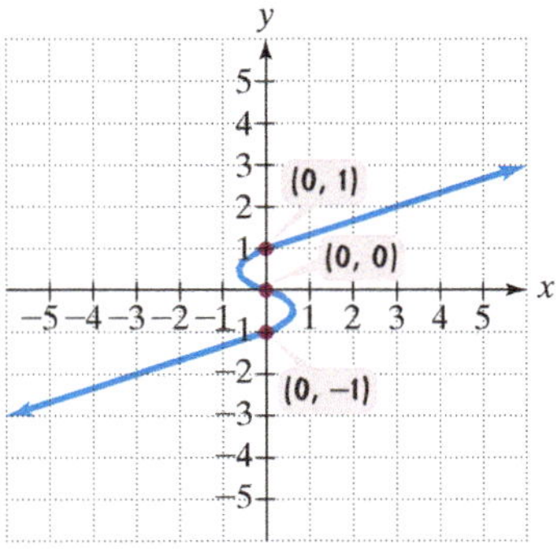

FIGURE 7.10 y is not a function of x because 0 is paired with three values of y, namely, 1, 0, and -1.

THE VERTICAL LINE TEST FOR FUNCTIONS

If any vertical line intersects a graph in more than one point, the graph does not define y as a function of x.

EXAMPLE 7 *Using the Vertical Line Test*

Use the vertical line test to identify graphs in which y is a function of x.

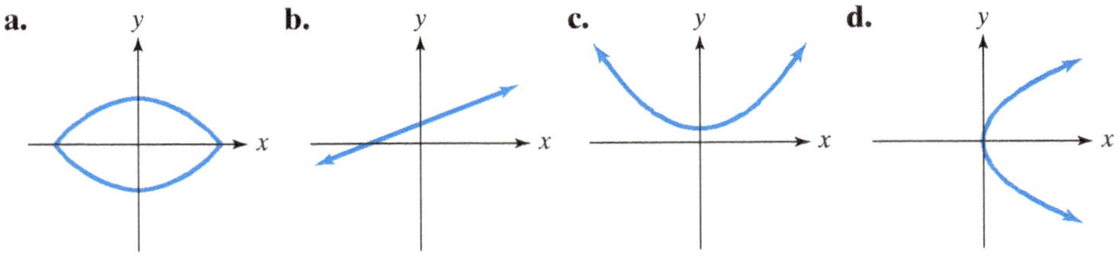

SOLUTION

y is a function of x for the graphs in (b) and (c).

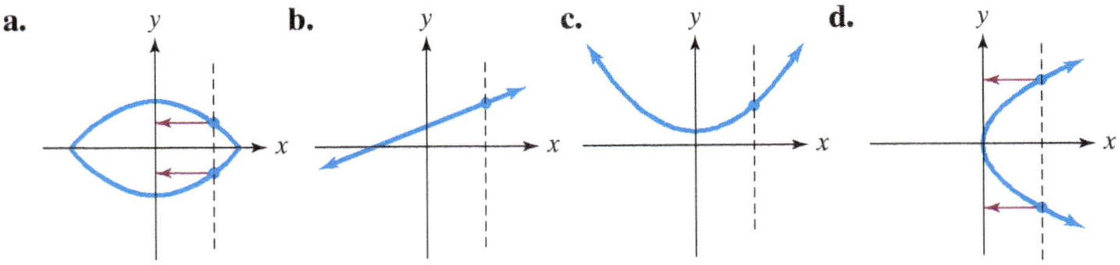

a. y **is not a function** of x. Two values of y correspond to one x-value.

b. y **is a function** of x.

c. y **is a function** of x.

d. y **is not a function** of x. Two values of y correspond to one x-value.

CHECK POINT 7 Use the vertical line test to identify graphs in which y is a function of x.

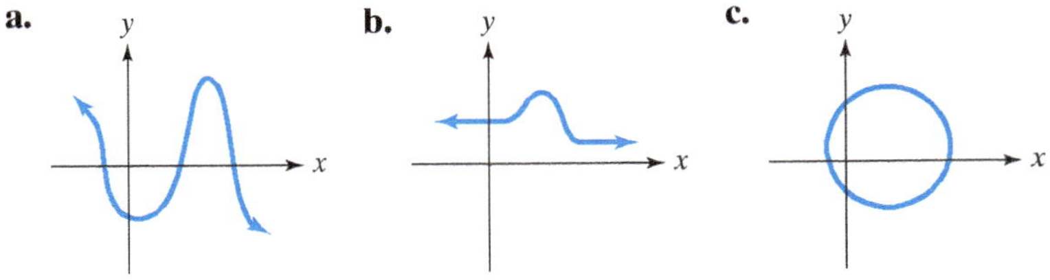

6 Obtain information about a function from its graph.

Obtaining Information from Graphs

Example 8 illustrates how to obtain information about a function from its graph.

EXAMPLE 8 *Analyzing the Graph of a Function*

Too late for that flu shot now! It's only 8 A.M. and you're feeling lousy. Fascinated by the way that algebra models the world (your author is projecting a bit here), you construct a graph showing your body temperature from 8 A.M. through 3 P.M. You decide to let x represent the number of hours after 8 A.M. and y represent your body temperature at time x. The graph is shown in **Figure 7.11**. The symbol ⸗ on the y-axis shows that there is a break in values between 0 and 98. Thus, the first tick mark on the y-axis represents a temperature of 98°F.

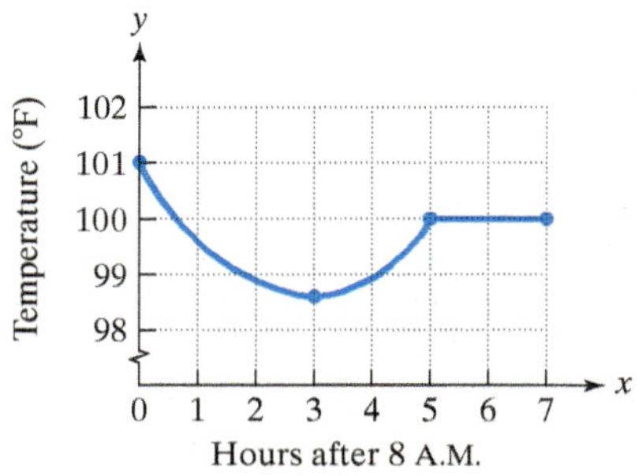

FIGURE 7.11 Body temperature from 8 A.M. through 3 P.M.

a. What is your temperature at 8 A.M.?

b. During which period of time is your temperature decreasing?

c. Estimate your minimum temperature during the time period shown. How many hours after 8 A.M. does this occur? At what time does this occur?

d. During which period of time is your temperature increasing?

e. Part of the graph is shown as a horizontal line segment. What does this mean about your temperature and when does this occur?

f. Explain why the graph defines y as a function of x.

SOLUTION

a. Because x is the number of hours after 8 A.M., your temperature at 8 A.M. corresponds to $x = 0$. Locate 0 on the horizontal axis and look at the point on the graph above 0. **Figure 7.12** shows that your temperature at 8 A.M. is 101°F.

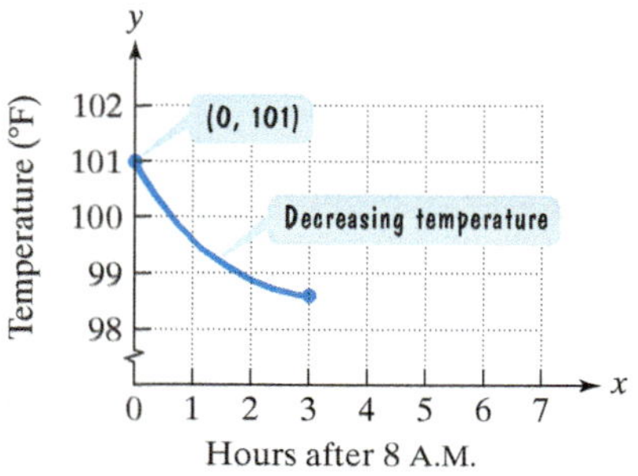

FIGURE 7.12

b. Your temperature is decreasing when the graph falls from left to right. This occurs between $x = 0$ and $x = 3$, also shown in **Figure 7.12**. Because x represents the number of hours after 8 A.M., your temperature is decreasing between 8 A.M. and 11 A.M.

c. Your minimum temperature can be found by locating the lowest point on the graph. This point lies above 3 on the horizontal axis, shown in **Figure 7.13**. The y-coordinate of this point falls more than midway between 98 and 99, at approximately 98.6. The lowest point on the graph, (3, 98.6), shows that your minimum temperature, 98.6°F, occurs 3 hours after 8 A.M., at 11 A.M.

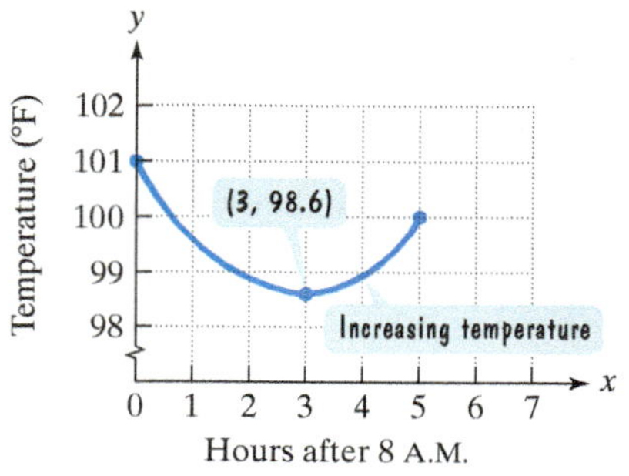

FIGURE 7.13

d. Your temperature is increasing when the graph rises from left to right. This occurs between $x = 3$ and $x = 5$, shown in **Figure 7.13**. Because x represents the number of hours after 8 A.M., your temperature is increasing between 11 A.M. and 1 P.M.

e. The horizontal line segment shown in **Figure 7.14** indicates that your temperature is neither increasing nor decreasing. Your temperature remains the same, 100°F, between $x = 5$ and $x = 7$. Thus, your temperature is at a constant 100°F between 1 P.M. and 3 P.M.

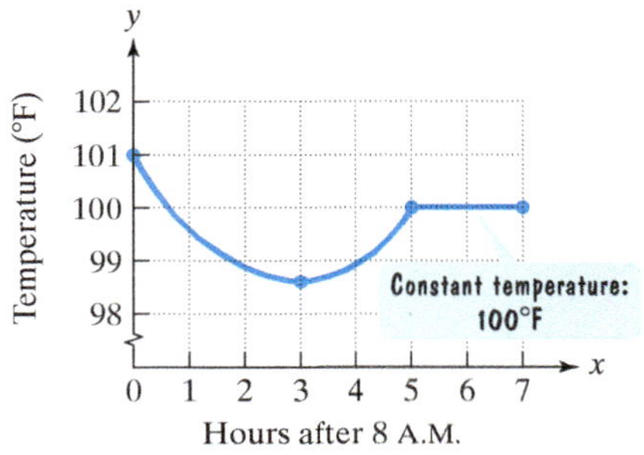

FIGURE 7.14

f. The complete graph of your body temperature from 8 A.M. through 3 P.M. is shown in **Figure 7.14**. No vertical line can be drawn that intersects this blue graph more than once. By the vertical line test, the graph defines y as a function of x. In practical terms, this means that your body temperature is a function of time. Each hour (or fraction of an hour) after 8 A.M., represented by x, yields precisely one body temperature, represented by y.

CHECK POINT 8 When a person receives a drug injected into a muscle, the concentration of the drug in the body, measured in milligrams per 100 milliliters, depends on the time elapsed after the injection, measured in hours. **Figure 7.15** shows the graph of drug concentration over time, where x represents hours after the injection and y represents the drug concentration at time x.

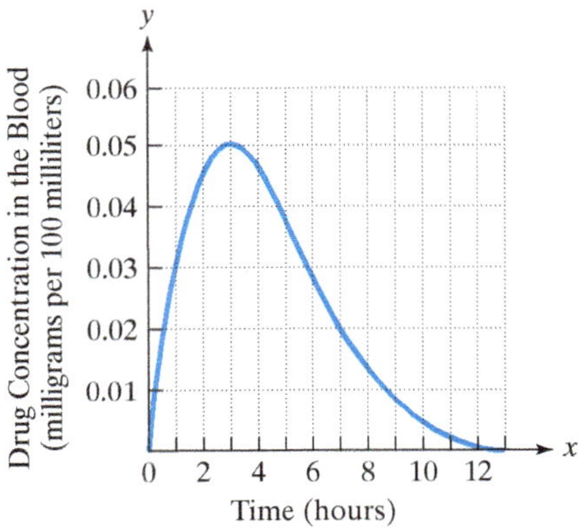

FIGURE 7.15

a. During which period of time is the drug concentration increasing?

b. During which period of time is the drug concentration decreasing?

c. What is the drug's maximum concentration and when does this occur?

d. What happens by the end of 13 hours?

e. Explain why the graph defines y as a function of x.

Concept and Vocabulary Check

Fill in each blank so that the resulting statement is true.

1. In the rectangular coordinate system, the horizontal number line is called the ________.
2. In the rectangular coordinate system, the vertical number line is called the ________.
3. In the rectangular coordinate system, the point of intersection of the horizontal axis and the vertical axis is called the________.
4. The axes of the rectangular coordinate system divide the plane into regions, called ________. There are ________ of these regions.
5. The first number in an ordered pair such as (3, 8) is called the ________. The second number in such an ordered pair is called the ________.
6. The ordered pair (1, 3) is a/an ________ of the equation $y = 5x - 2$ because when 1 is substituted for x and 3 is substituted for y, we obtain a true statement. We also say that (1, 3) ________ the equation.
7. If an equation in two variables (x and y) yields precisely one value of ________ for each value of ________, we say that y is a/an ________ of x.
8. If $f(x) = 3x + 5$, we can find $f(6)$ by replacing ________ with ________.
9. If any vertical line intersects a graph ________, the graph does not define y as a/an ________ of x.

Exercise Set 7.1

Practice Exercises

In Exercises 1–20, plot the given point in a rectangular coordinate system.

1. $(1, 4)$
2. $(2, 5)$
3. $(-2, 3)$
4. $(-1, 4)$
5. $(-3, -5)$
6. $(-4, -2)$
7. $(4, -1)$
8. $(3, -2)$
9. $(-4, 0)$
10. $(-5, 0)$
11. $(0, -3)$
12. $(0, -4)$
13. $(0, 0)$
14. $\left(-3, -1\frac{1}{2}\right)$
15. $\left(-2, -3\frac{1}{2}\right)$
16. $(-5, -2.5)$
17. $(3.5, 4.5)$
18. $(2.5, 3.5)$
19. $(1.25, -3.25)$
20. $(2.25, -4.25)$

Graph each equation in Exercises 21–32. Select integers for x from −3 to 3, inclusive.

21. $y = x^2 - 2$
22. $y = x^2 + 2$
23. $y = x - 2$
24. $y = x + 2$
25. $y = 2x + 1$
26. $y = 2x - 4$
27. $y = -\frac{1}{2}x$
28. $y = -\frac{1}{2}x + 2$
29. $y = x^3$
30. $y = x^3 - 1$
31. $y = |x| + 1$
32. $y = |x| - 1$

In Exercises 33–46, evaluate each function at the given value of the variable.

33. $f(x) = x - 4$ **a.** $f(8)$ **b.** $f(1)$
34. $f(x) = x - 6$ **a.** $f(9)$ **b.** $f(2)$
35. $f(x) = 3x - 2$ **a.** $f(7)$ **b.** $f(0)$
36. $f(x) = 4x - 3$ **a.** $f(7)$ **b.** $f(0)$
37. $g(x) = x^2 + 1$ **a.** $g(2)$ **b.** $g(-2)$
38. $g(x) = x^2 + 4$ **a.** $g(3)$ **b.** $g(-3)$
39. $g(x) = -x^2 + 2$ **a.** $g(4)$ **b.** $g(-3)$
40. $g(x) = -x^2 + 1$ **a.** $g(5)$ **b.** $g(-4)$
41. $h(r) = 3r^2 + 5$ **a.** $h(4)$ **b.** $h(-1)$
42. $h(r) = 2r^2 - 4$ **a.** $h(5)$ **b.** $h(-1)$
43. $f(x) = 2x^2 + 3x - 1$ **a.** $f(3)$ **b.** $f(-4)$
44. $f(x) = 3x^2 + 4x - 2$ **a.** $f(2)$ **b.** $f(-1)$
45. $f(x) = \dfrac{x}{|x|}$ **a.** $f(6)$ **b.** $f(-6)$
46. $f(x) = \dfrac{|x|}{x}$ **a.** $f(5)$ **b.** $f(-5)$

In Exercises 47–54, evaluate $f(x)$ for the given values of x. Then use the ordered pairs $(x, f(x))$ from your table to graph the function.

47. $f(x) = x^2 - 1$

x	$f(x) = x^2 - 1$
−2	
−1	
0	
1	
2	

48. $f(x) = x^2 + 1$

x	$f(x) = x^2 + 1$
−2	
−1	
0	
1	
2	

49. $f(x) = x - 1$

x	$f(x) = x - 1$
−2	
−1	
0	
1	
2	

50. $f(x) = x + 1$

x	$f(x) = x + 1$
−2	
−1	
0	
1	
2	

51. $f(x) = (x - 2)^2$

x	$f(x) = (x - 2)^2$
0	
1	
2	
3	
4	

52. $f(x) = (x + 1)^2$

x	$f(x) = (x + 1)^2$
−3	
−2	
−1	
0	
1	

53. $f(x) = x^3 + 1$

x	$f(x) = x^3 + 1$
−3	
−2	
−1	
0	
1	

54. $f(x) = (x + 1)^3$

x	$f(x) = (x + 1)^3$
−3	
−2	
−1	
0	
1	

For Exercises 55–62, use the vertical line test to identify graphs in which y is a function of x.

55.

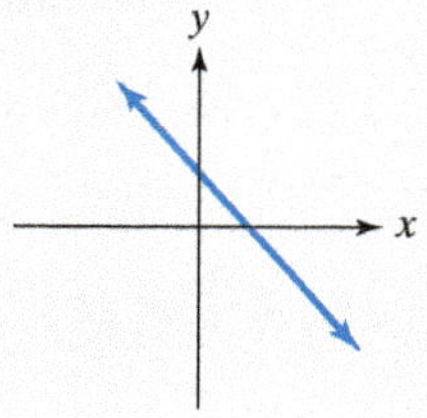

56.

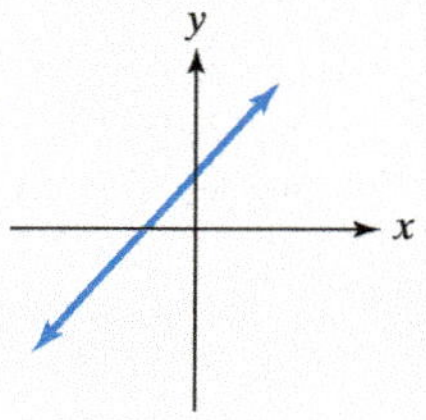

57.

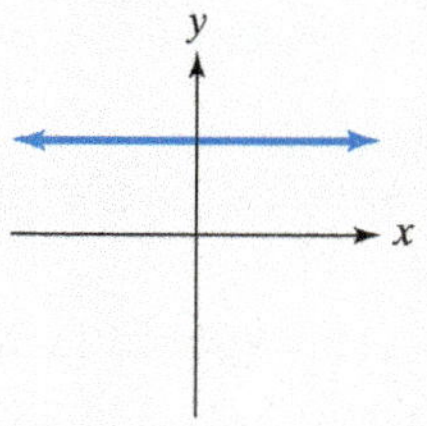

58.

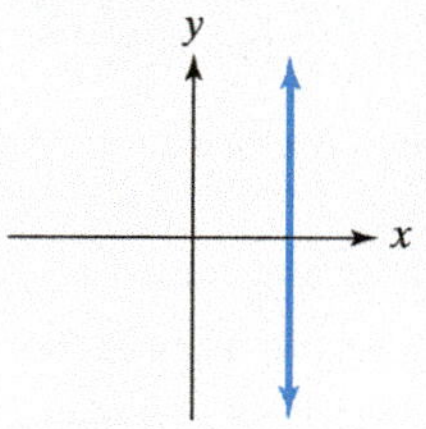

59.

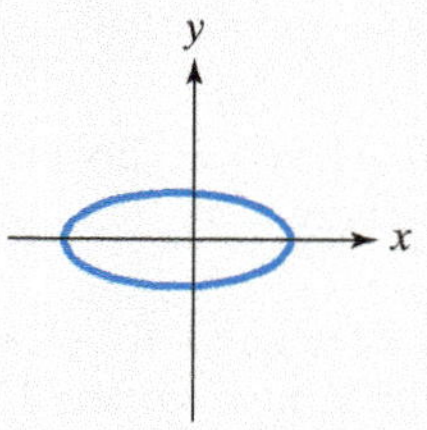

60.

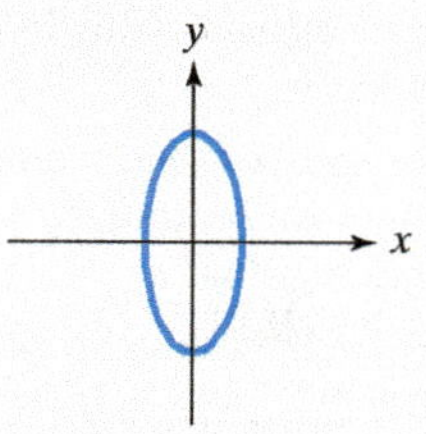

61.

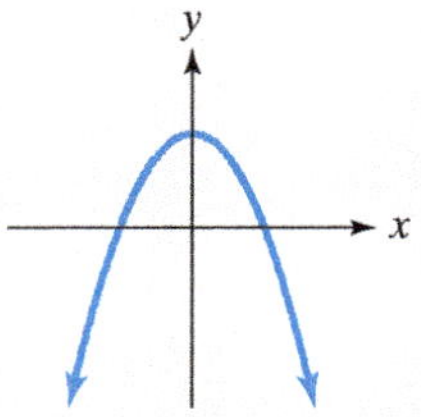

62.

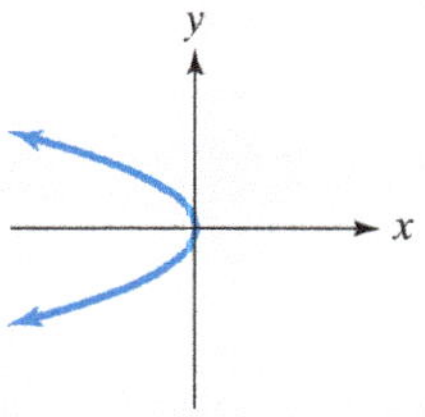

Practice Plus

In Exercises 63–64, let $f(x) = x^2 - x + 4$ *and* $g(x) = 3x - 5$.

63. Find $g(1)$ and $f(g(1))$.

64. Find $g(-1)$ and $f(g(-1))$.

In Exercises 65–66, let f and g be defined by the following table:

x	$f(x)$	$g(x)$
−2	6	0
−1	3	4
0	−1	1
1	−4	−3
2	0	−6

65. Find $\sqrt{f(-1) - f(0)} - [g(2)]^2 + f(-2) \div g(2) \cdot g(-1)$.

66. Find $|f(1) - f(0)| - [g(1)]^2 + g(1) \div f(-1) \cdot g(2)$.

In Exercises 67–70, write each English sentence as an equation in two variables. Then graph the equation.

67. The y-value is four more than twice the x-value.

68. The y-value is the difference between four and twice the x-value.

69. The y-value is three decreased by the square of the x-value.

70. The y-value is two more than the square of the x-value.

Application Exercises

A football is thrown by a quarterback to a receiver. The points in the figure show the height of the football, in feet, above the ground in terms of its distance, in yards, from the quarterback. Use this information to solve Exercises 71–76.

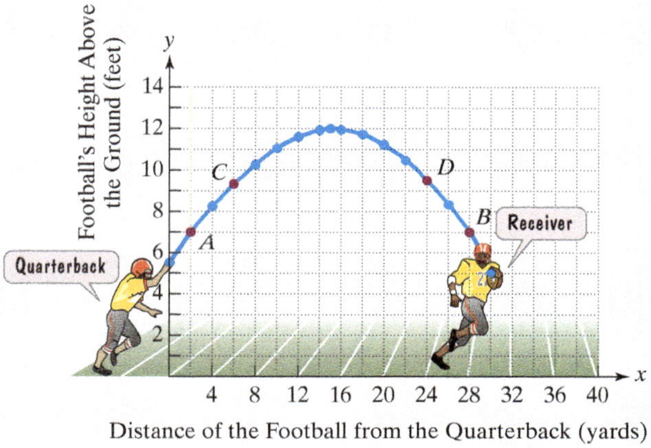

71. Find the coordinates of point A. Then interpret the coordinates in terms of the information given.

72. Find the coordinates of point B. Then interpret the coordinates in terms of the information given.

73. Estimate the coordinates of point C.

74. Estimate the coordinates of point D.

75. What is the football's maximum height? What is its distance from the quarterback when it reaches its maximum height?

76. What is the football's height when it is caught by the receiver? What is the receiver's distance from the quarterback when he catches the football?

The wage gap is used to compare the status of women's earnings relative to men's. The wage gap is expressed as a percent and is calculated by dividing the median, or middlemost, annual earnings for women by the median annual earnings for men. The bar graph shows the wage gap for selected years from 1980 through 2010.

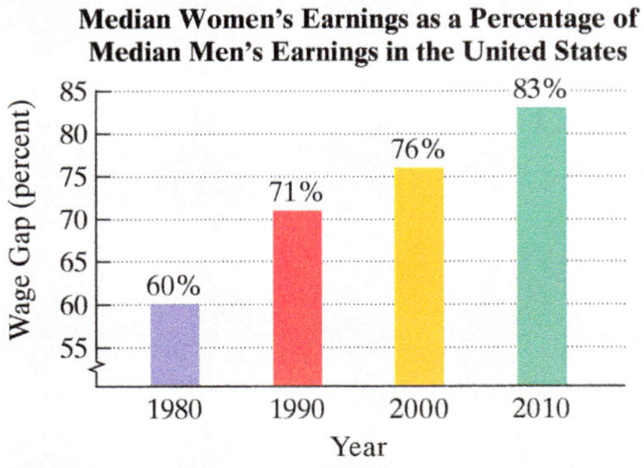

Source: Bureau of Labor Statistics

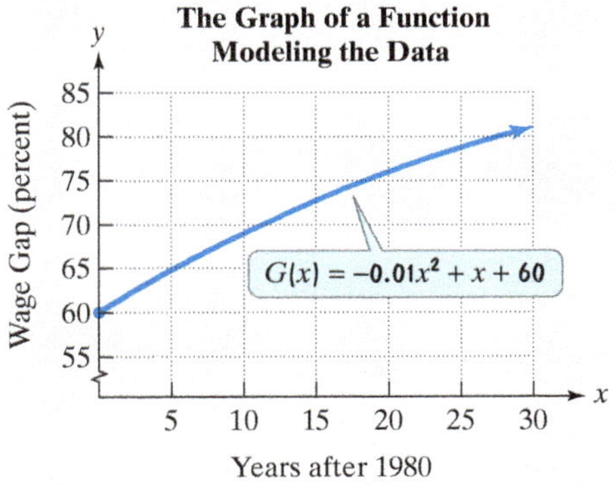

The function $G(x) = -0.01x^2 + x + 60$ *models the wage gap, as a percent, x years after 1980. The graph of function G is shown to the right of the actual data. Use this information to solve Exercises 77–78.*

77. **a.** Find and interpret $G(30)$. Identify this information as a point on the graph of the function.

b. Does $G(30)$ overestimate or underestimate the actual data shown by the bar graph? By how much?

78. **a.** Find and interpret $G(10)$. Identify this information as a point on the graph of the function.

b. Does $G(10)$ overestimate or underestimate the actual data shown by the bar graph? By how much?

The function $f(x) = 0.4x^2 - 36x + 1000$ *models the number of accidents,* $f(x)$*, per 50 million miles driven as a function of a driver's age,* x*, in years, where* x *includes drivers from ages 16 through 74, inclusive. The graph of f is shown. Use the equation for f to solve Exercises 79–82.*

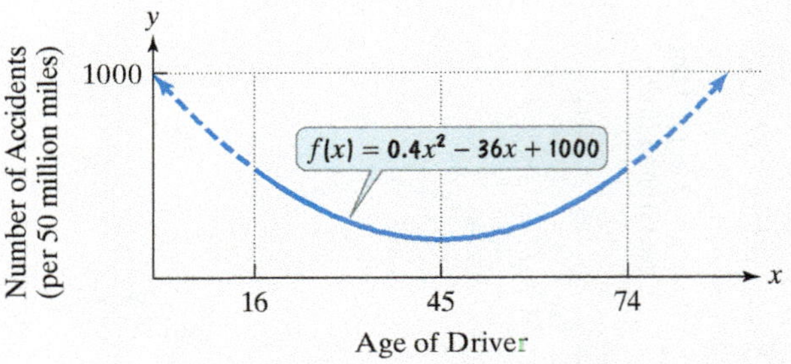

79. Find and interpret $f(20)$. Identify this information as a point on the graph of f.

80. Find and interpret $f(50)$. Identify this information as a point on the graph of f.

81. For what value of x does the graph reach its lowest point? Use the equation for f to find the minimum value of y. Describe the practical significance of this minimum value.

82. Use the graph to identify two different ages for which drivers have the same number of accidents. Use the equation for f to find the number of accidents for drivers at each of these ages.

Writing in Mathematics

83. What is the rectangular coordinate system?

84. Explain how to plot a point in the rectangular coordinate system. Give an example with your explanation.

85. Explain why $(5, -2)$ and $(-2, 5)$ do not represent the same ordered pair.

86. Explain how to graph an equation in the rectangular coordinate system.

87. What is a function?

88. Explain how the vertical line test is used to determine whether a graph represents a function.

Critical Thinking Exercises

Make Sense? *In Exercises 89–92, determine whether each statement makes sense or does not make sense, and explain your reasoning.*

89. My body temperature is a function of the time of day.

90. Using $f(x) = 3x + 2$, I found $f(50)$ by applying the distributive property to $(3x + 2)50$.

91. I knew how to use point plotting to graph the equation $y = x^2 - 1$, so there was really nothing new to learn when I used the same technique to graph the function $f(x) = x^2 - 1$.

92. The graph of my function revealed aspects of its behavior that were not obvious by just looking at its equation.

In Exercises 93–96, use the graphs of f and g to find each number.

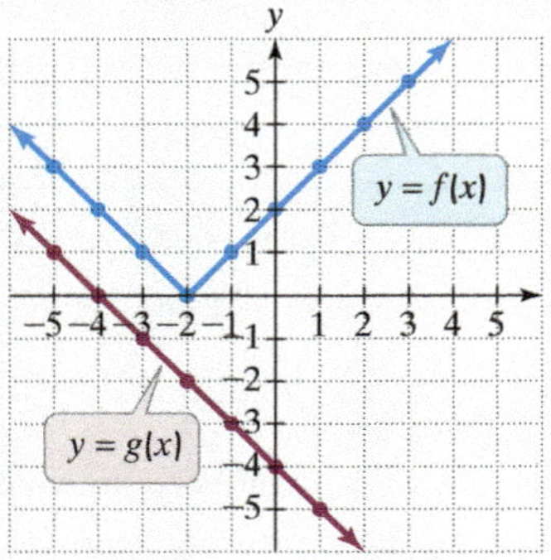

93. $f(-1) + g(-1)$

94. $f(1) + g(1)$

95. $f(g(-1))$

96. $f(g(1))$

Technology Exercise

97. Use a graphing calculator to verify the graphs that you drew by hand in Exercises 47–54.

7.2 Linear Functions and Their Graphs

WHAT AM I SUPPOSED TO LEARN?

After you have read this section, you should be able to:

1. Use intercepts to graph a linear equation.
2. Calculate slope.
3. Use the slope and y-intercept to graph a line.
4. Graph horizontal or vertical lines.
5. Interpret slope as rate of change.
6. Use slope and y-intercept to model data.

IT'S HARD TO BELIEVE THAT THIS gas-guzzler, with its huge fins and overstated design, was available in 1957 for approximately \$1800. Sadly, its elegance quickly faded, depreciating by \$300 per year, often sold for scrap just six years after its glorious emergence from the dealer's showroom.

From these casual observations, we can obtain a mathematical model and its graph. The model is

$$y = -300x + 1800.$$

The car is depreciating by \$300 per year for x years.

The new car is worth \$1800.

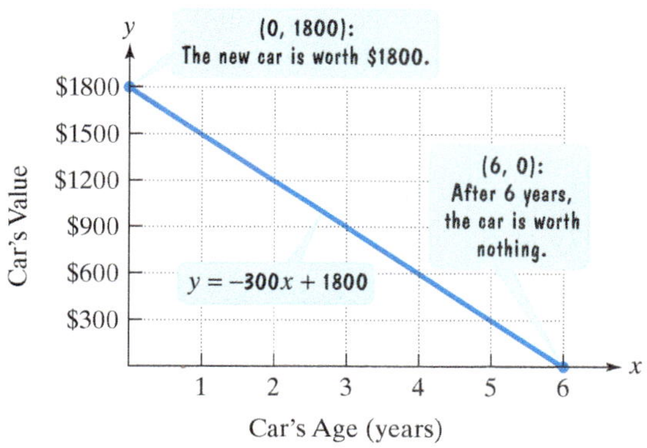

FIGURE 7.16

In this model, y is the car's value after x years. **Figure 7.16** shows the equation's graph. Using function notation, we can rewrite the equation as

$$f(x) = -300x + 1800.$$

A function such as this, whose graph is a straight line, is called a **linear function**. In this section, we will study linear functions and their graphs.

Use intercepts to graph a linear equation.

Graphing Using Intercepts

There is another way that we can write the equation

$$y = -300x + 1800.$$

We will collect the x- and y-terms on the left side. This is done by adding $300x$ to both sides:

$$300x + y = 1800.$$

All equations of the form $Ax + By = C$ are straight lines when graphed, as long as A and B are not both zero. Such equations are called **linear equations in two variables**. We can quickly obtain the graph for equations in this form when none of A, B, or C is zero by finding the points where the graph intersects the x-axis and the y-axis. The x-coordinate of the point where the graph intersects the x-axis is called the ***x*-intercept**. The y-coordinate of the point where the graph intersects the y-axis is called the ***y*-intercept**.

The graph of $300x + y = 1800$ in **Figure 7.16** intersects the x-axis at $(6, 0)$, so the x-intercept is 6. The graph intersects the y-axis at $(0, 1800)$, so the y-intercept is 1800.

LOCATING INTERCEPTS

To locate the x-intercept, set $y = 0$ and solve the equation for x.

To locate the y-intercept, set $x = 0$ and solve the equation for y.

An equation of the form $Ax + By = C$ as described above can be graphed by finding the x- and y-intercepts, plotting the intercepts, and drawing a straight line through these points. When graphing using intercepts, it is a good idea to use a third point, a checkpoint, before drawing the line. A checkpoint can be obtained by selecting a value for x, other than 0 or the x-intercept, and finding the corresponding value for y. The checkpoint should lie on the same line as the x- and y-intercepts. If it does not, recheck your work and find the error.

EXAMPLE 1 *Using Intercepts to Graph a Linear Equation*

Graph: $3x + 2y = 6$.

SOLUTION

Note that $3x + 2y = 6$ is of the form $Ax + By = C$.

$$3x + 2y = 6$$

$A = 3$ $B = 2$ $C = 6$

In this case, none of A, B, or C is zero.

Find the *x*-intercept by letting $y = 0$ and solving for x.	**Find the *y*-intercept by letting $x = 0$ and solving for y.**
$3x + 2y = 6$	$3x + 2y = 6$
$3x + 2 \cdot 0 = 6$	$3 \cdot 0 + 2y = 6$
$3x = 6$	$2y = 6$
$x = 2$	$y = 3$

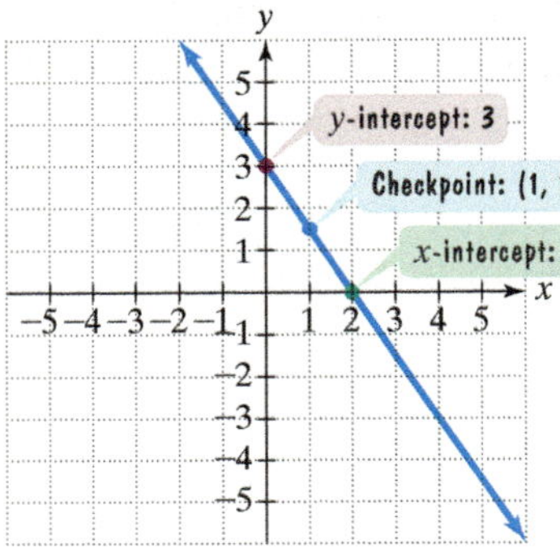

FIGURE 7.17 The graph of $3x + 2y = 6$

The x-intercept is 2, so the line passes through the point $(2, 0)$. The y-intercept is 3, so the line passes through the point $(0, 3)$.

For our checkpoint, we choose a value for x other than 0 or the x-intercept, 2. We will let $x = 1$ and find the corresponding value for y.

$$\begin{aligned} 3x + 2y &= 6 && \text{This is the given equation.} \\ 3\cdot 1 + 2y &= 6 && \text{Substitute 1 for } x. \\ 3 + 2y &= 6 && \text{Simplify.} \\ 2y &= 3 && \text{Subtract 3 from both sides.} \\ y &= \tfrac{3}{2} && \text{Divide both sides by 2.} \end{aligned}$$

The checkpoint is the ordered pair $\left(1, \frac{3}{2}\right)$, or $(1, 1.5)$.

The three points in **Figure 7.17** lie along the same line. Drawing a line through the three points results in the graph of $3x + 2y = 6$. The arrowheads at the ends of the line show that the line continues indefinitely in both directions.

CHECK POINT 1 Graph: $2x + 3y = 6$.

2 Calculate slope.

Slope

Mathematicians have developed a useful measure of the steepness of a line, called the *slope* of the line. Slope compares the vertical change (the **rise**) to the horizontal change (the **run**) when moving from one fixed point to another along the line. To calculate the slope of a line, we use a ratio that compares the change in y (the rise) to the change in x (the run).

DEFINITION OF SLOPE

The **slope** of the line through the distinct points (x_1, y_1) and (x_2, y_2) is

$$\frac{\text{Change in } y}{\text{Change in } x} = \frac{\text{Rise}}{\text{Run}} = \frac{y_2 - y_1}{x_2 - x_1}$$

Rise: Vertical change
Run: Horizontal change

where $x_2 - x_1 \neq 0$.

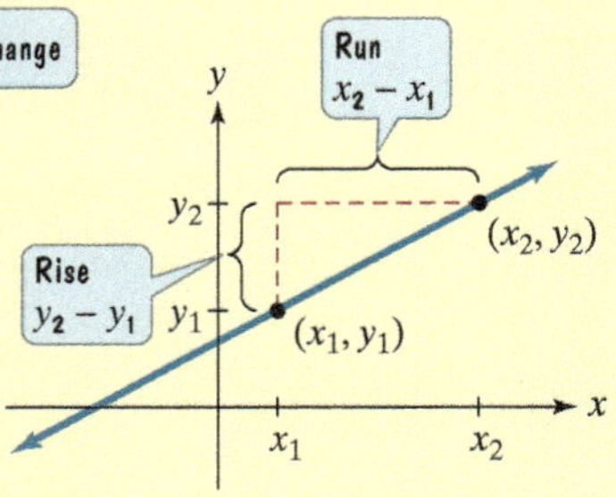

It is common notation to let the letter m represent the slope of a line. The letter m is used because it is the first letter of the French verb *monter,* meaning "to rise," or "to ascend."

EXAMPLE 2 *Using the Definition of Slope*

Find the slope of the line passing through each pair of points:

a. $(-3, -1)$ and $(-2, 4)$ **b.** $(-3, 4)$ and $(2, -2)$.

SOLUTION

a. Let $(x_1, y_1) = (-3, -1)$ and $(x_2, y_2) = (-2, 4)$. We obtain the slope as follows:

$$m = \frac{\text{Change in } y}{\text{Change in } x} = \frac{y_2 - y_1}{x_2 - x_1} = \frac{4 - (-1)}{-2 - (-3)} = \frac{5}{1} = 5.$$

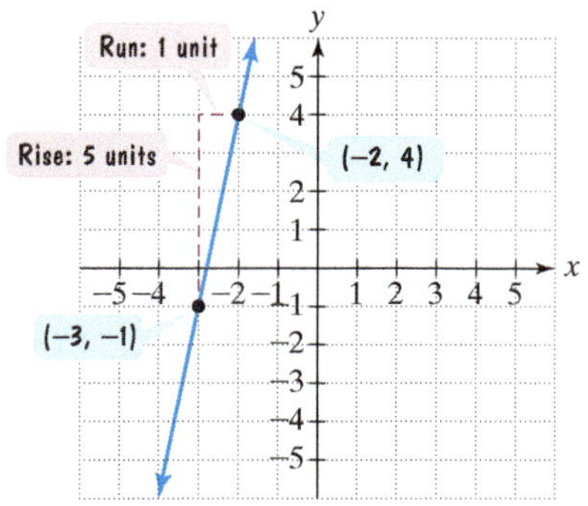

FIGURE 7.18 Visualizing a slope of 5

The situation is illustrated in **Figure 7.18**. The slope of the line is 5, indicating that there is a vertical change, a rise, of 5 units for each horizontal change, a run, of 1 unit. The slope is positive and the line rises from left to right.

GREAT QUESTION!

When using the definition of slope, how do I know which point to call (x_1, y_1) and which point to call (x_2, y_2)?

When computing slope, it makes no difference which point you call (x_1, y_1) and which point you call (x_2, y_2). If we let $(x_1, y_1) = (-2, 4)$ and $(x_2, y_2) = (-3, -1)$, the slope is still 5:

$$m = \frac{\text{Change in } y}{\text{Change in } x} = \frac{y_2 - y_1}{x_2 - x_1} = \frac{-1 - 4}{-3 - (-2)} = \frac{-5}{-1} = 5.$$

However, you should not subtract in one order in the numerator $(y_2 - y_1)$ and then in the opposite order in the denominator $(x_1 - x_2)$. The slope is *not* −5:

$$\frac{-1 - 4}{-2 - (-3)} = \frac{-5}{1} = -5. \quad \text{incorrect}$$

b. We can let $(x_1, y_1) = (-3, 4)$ and $(x_2, y_2) = (2, -2)$. The slope of the line shown in **Figure 7.19** is computed as follows:

$$m = \frac{\text{Change in } y}{\text{Change in } x} = \frac{y_2 - y_1}{x_2 - x_1} = \frac{-2 - 4}{2 - (-3)} = \frac{-6}{5} = -\frac{6}{5}.$$

The slope of the line is $-\frac{6}{5}$. For every vertical change of −6 units (6 units down), there is a corresponding horizontal change of 5 units. The slope is negative and the line falls from left to right.

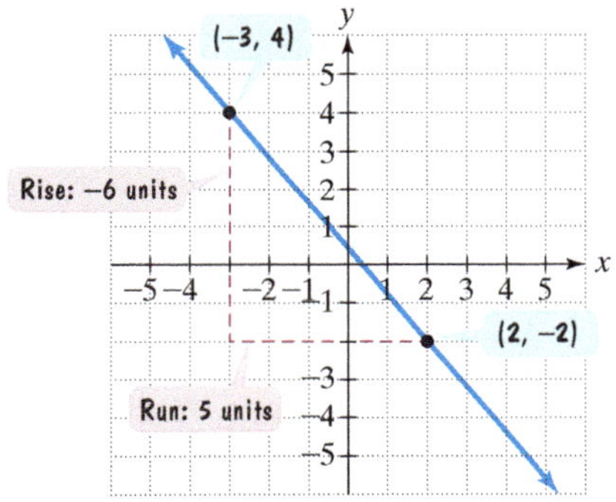

FIGURE 7.19 Visualizing a slope of $-\frac{6}{5}$

CHECK POINT 2 Find the slope of the line passing through each pair of points:

a. $(-3, 4)$ and $(-4, -2)$

b. $(4, -2)$ and $(-1, 5)$.

Example 2 illustrates that a line with a positive slope is rising from left to right and a line with a negative slope is falling from left to right. By contrast, a horizontal line neither rises nor falls and has a slope of zero. A vertical line has no horizontal change, so $x_2 - x_1 = 0$ in the formula for slope. Because we cannot divide by zero, the slope of a vertical line is undefined. This discussion is summarized in **Table 7.2**.

TABLE 7.2 Possibilities for a Line's Slope

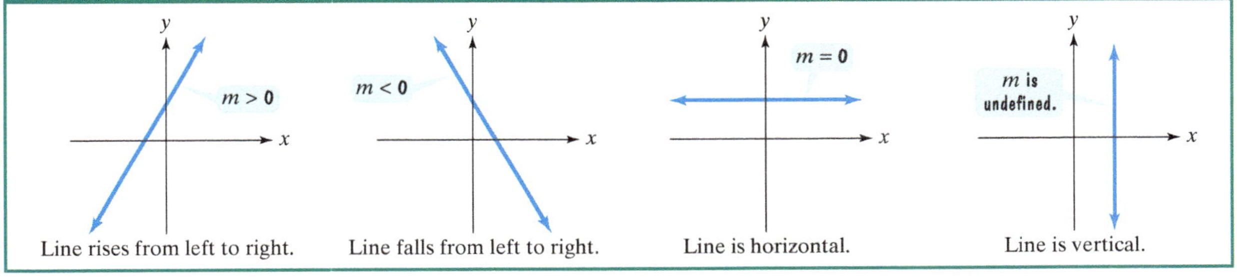

3 Use the slope and y-intercept to graph a line.

The Slope-Intercept Form of the Equation of a Line

We can use the definition of slope to write the equation of any nonvertical line with slope m and y-intercept b. Because the y-intercept is b, the point $(0, b)$ lies on the line. Now, let (x, y) represent any other point on the line, shown in **Figure 7.20**. Keep in mind that the point (x, y) is arbitrary and is not in one fixed position. By contrast, the point $(0, b)$ is fixed.

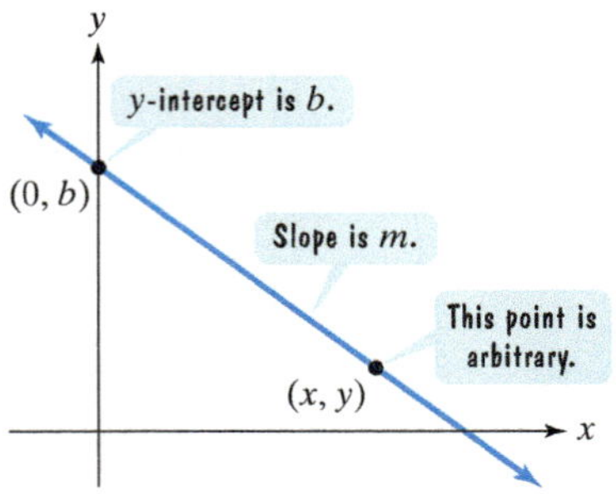

FIGURE 7.20 A line with slope m and y-intercept b

Regardless of where the point (x, y) is located, the steepness of the line in **Figure 7.20** remains the same. Thus, the ratio for slope stays a constant m. This means that for all points along the line

$$m = \frac{\text{Change in } y}{\text{Change in } x} = \frac{y - b}{x - 0} = \frac{y - b}{x}.$$

We can clear the fraction by multiplying both sides by x, the denominator. Note that x is not zero since (x, y) is distinct from $(0, b)$, the only point on the line with first coordinate 0.

$$m = \frac{y - b}{x} \quad \text{This is the slope of the line in Figure 7.20.}$$

$$mx = \frac{y - b}{x} \cdot x \quad \text{Multiply both sides by } x.$$

$$mx = y - b \quad \text{Simplify: } \frac{y - b}{x} \cdot x = y - b.$$

$$mx + b = y - b + b \quad \text{Add } b \text{ to both sides and solve for } y.$$

$$mx + b = y \quad \text{Simplify.}$$

Now, if we reverse the two sides, we obtain the *slope-intercept form* of the equation of a line.

SLOPE-INTERCEPT FORM OF THE EQUATION OF A LINE

The **slope-intercept form of the equation** of a nonvertical line with slope m and y-intercept b is

$$y = mx + b.$$

The slope-intercept form of a line's equation, $y = mx + b$, can be expressed in function notation by replacing y with $f(x)$:

$$f(x) = mx + b.$$

We have seen that functions in the form $f(x) = mx + b$ are called **linear functions**. Thus, in the equation of a linear function, the x-coefficient is the line's slope and the constant term is the y-intercept. Here are two examples:

$$y = 2x - 4 \qquad f(x) = \frac{1}{2}x + 2.$$

The slope is 2. The y-intercept is -4. The slope is $\frac{1}{2}$. The y-intercept is 2.

If a linear function's equation is in slope-intercept form, we can use the y-intercept and the slope to obtain its graph.

GREAT QUESTION!

If the slope is an integer, such as 2, why should I express it as $\frac{2}{1}$ for graphing purposes?

Writing the slope, m, as a fraction allows you to identify the rise (the fraction's numerator) and the run (the fraction's denominator).

GRAPHING $y = mx + b$ USING THE SLOPE AND y-INTERCEPT

1. Plot the point containing the y-intercept on the y-axis. This is the point $(0, b)$.
2. Obtain a second point using the slope, m. Write m as a fraction, and use rise over run, starting at the point containing the y-intercept, to plot the second point.
3. Use a straightedge to draw a line through the two points. Draw arrowheads at the ends of the line to show that the line continues indefinitely in both directions.

EXAMPLE 3 *Graphing by Using the Slope and y-Intercept*

Graph the linear function $y = \frac{2}{3}x + 2$ by using the slope and y-intercept.

SOLUTION

The equation of the linear function is in the form $y = mx + b$. We can find the slope, m, by identifying the coefficient of x. We can find the y-intercept, b, by identifying the constant term.

$$y = \frac{2}{3}x + 2$$

The slope is $\frac{2}{3}$. The y-intercept is 2.

Now that we have identified the slope and the y-intercept, we use the three-step procedure to graph the equation.

Step 1 Plot the point containing the y-intercept on the y-axis. The y-intercept is 2. We plot $(0, 2)$, shown in **Figure 7.21**.

Step 2 Obtain a second point using the slope, m. Write m as a fraction, and use rise over run, starting at the point containing the y-intercept, to plot the second point. The slope, $\frac{2}{3}$, is already written as a fraction:

$$m = \frac{2}{3} = \frac{\text{Rise}}{\text{Run}}.$$

We plot the second point on the line by starting at $(0, 2)$, the first point. Based on the slope, we move 2 units *up* (the rise) and 3 units to the *right* (the run). This puts us at a second point on the line, $(3, 4)$, shown in **Figure 7.21**.

Step 3 Use a straightedge to draw a line through the two points. The graph of $y = \frac{2}{3}x + 2$ is shown in **Figure 7.21**.

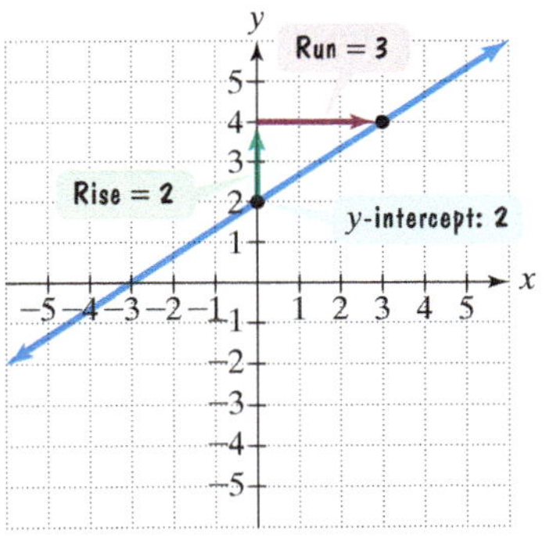

FIGURE 7.21 The graph of $y = \frac{2}{3}x + 2$

CHECK POINT 3 Graph the linear function $y = \frac{3}{5}x + 1$ by using the slope and y-intercept.

Earlier in this section, we considered linear functions of the form $Ax + By = C$. We used x- and y-intercepts, as well as checkpoints, to graph these functions. It is also possible to obtain the graphs by using the slope and y-intercept. To do this, begin by solving $Ax + By = C$ for y. This will put the equation in slope-intercept form. Then use the three-step procedure to graph the equation. This is illustrated in Example 4.

EXAMPLE 4 *Graphing by Using the Slope and y-Intercept*

Graph the linear function $2x + 5y = 0$ by using the slope and y-intercept.

SOLUTION

We put the equation in slope-intercept form by solving for y.

$$2x + 5y = 0 \quad \text{This is the given equation.}$$

$$2x - 2x + 5y = 0 - 2x \quad \text{Subtract } 2x \text{ from both sides.}$$

$$5y = -2x + 0 \quad \text{Simplify.}$$

$$\frac{5y}{5} = \frac{-2x + 0}{5} \quad \text{Divide both sides by 5.}$$

$$y = \frac{-2x}{5} + \frac{0}{5} \quad \text{Divide each term in the numerator by 5.}$$

$$y = -\frac{2}{5}x + 0 \quad \text{Simplify. Equivalently, } f(x) = -\tfrac{2}{5}x + 0.$$

Now that the equation is in slope-intercept form, we can use the slope and y-intercept to obtain its graph. Examine the slope-intercept form:

$$y = -\frac{2}{5}x + 0.$$

slope: $-\frac{2}{5}$ y-intercept: 0

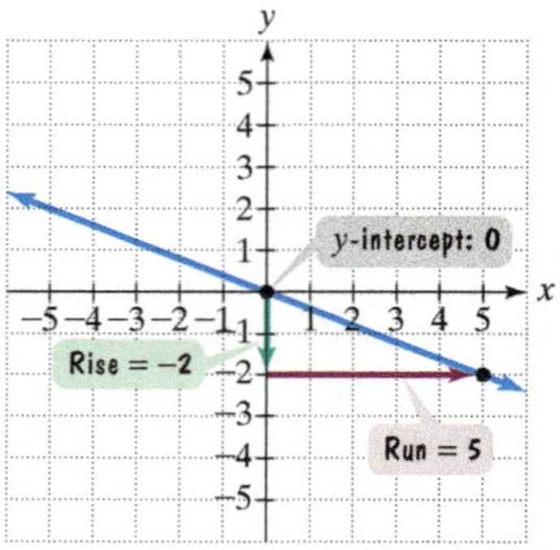

FIGURE 7.22 The graph of $2x + 5y = 0$, or $y = -\frac{2}{5}x$

Note that the slope is $-\frac{2}{5}$ and the y-intercept is 0. Use the y-intercept to plot $(0, 0)$ on the y-axis. Then locate a second point by using the slope.

$$m = -\frac{2}{5} = \frac{-2}{5} = \frac{\text{Rise}}{\text{Run}}$$

Because the rise is -2 and the run is 5, move *down* 2 units and to the *right* 5 units, starting at the point $(0, 0)$. This puts us at a second point on the line, $(5, -2)$. The graph of $2x + 5y = 0$ is the line drawn through these points, shown in **Figure 7.22**.

The equation $2x + 5y = 0$ in Example 4 is of the form $Ax + By = C$ with $C = 0$. If you try graphing $2x + 5y = 0$ by using intercepts, you will find that the x-intercept is 0 and the y-intercept is 0. This means that the graph passes through the origin. A second point must be found to graph the line. In Example 4, the line's slope gave us the second point.

CHECK POINT 4 Graph the linear function $3x + 4y = 0$ by using the slope and y-intercept.

4 Graph horizontal or vertical lines.

Equations of Horizontal and Vertical Lines

If a line is horizontal, its slope is zero: $m = 0$. Thus, the equation $y = mx + b$ becomes $y = b$, where b is the y-intercept. All horizontal lines have equations of the form $y = b$.

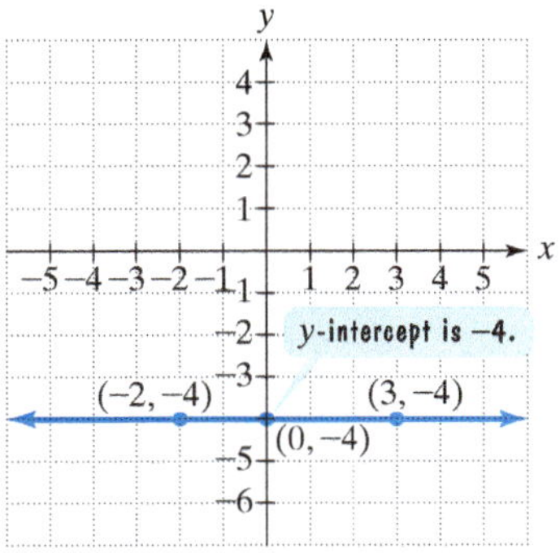

FIGURE 7.23 The graph of $y = -4$ or $f(x) = -4$

EXAMPLE 5 *Graphing a Horizontal Line*

Graph $y = -4$ in the rectangular coordinate system.

SOLUTION

All ordered pairs that are solutions of $y = -4$ have a value of y that is always -4. Any value can be used for x. (Think of $y = -4$ as $0x + 1y = -4$.) In the table at the right, we have selected three of the possible values for x: -2, 0, and 3. The table shows that three ordered pairs that are solutions of $y = -4$ are $(-2, -4)$, $(0, -4)$, and $(3, -4)$. Drawing a line that passes through the three points gives the horizontal line shown in **Figure 7.23**.

x	$y = -4$	(x, y)
−2	−4	(−2, −4)
0	−4	(0, −4)
3	−4	(3, −4)

For all choices of x, y is a constant -4.

CHECK POINT 5 Graph $y = 3$ in the rectangular coordinate system.

Next, let's see what we can discover about the graph of an equation of the form $x = a$ by looking at an example.

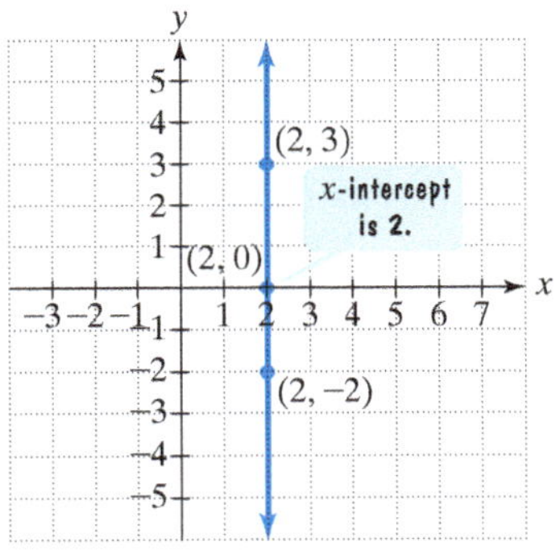

FIGURE 7.24 The graph of $x = 2$

EXAMPLE 6 *Graphing a Vertical Line*

Graph $x = 2$ in the rectangular coordinate system.

SOLUTION

All ordered pairs that are solutions of $x = 2$ have a value of x that is always 2. Any value can be used for y. (Think of $x = 2$ as $1x + 0y = 2$.) In the table at the right, we have selected three of the possible values for y: -2, 0, and 3. The table shows that three ordered pairs that are solutions of $x = 2$ are $(2, -2)$, $(2, 0)$, and $(2, 3)$. Drawing a line that passes through the three points gives the vertical line shown in **Figure 7.24**.

For all choices of y,

$x = 2$	y	(x, y)
2	−2	(2, −2)
2	0	(2, 0)
2	3	(2, 3)

x is always 2.

Does a vertical line represent the graph of a linear function? No. Look at the graph of $x = 2$ in **Figure 7.24**. A vertical line drawn through $(2, 0)$ intersects the graph infinitely many times. This shows that infinitely many outputs are associated with the input 2. **No vertical line represents a linear function.** All other lines are graphs of functions.

HORIZONTAL AND VERTICAL LINES

The graph of $y = b$ or $f(x) = b$ is a horizontal line. The y-intercept is b.

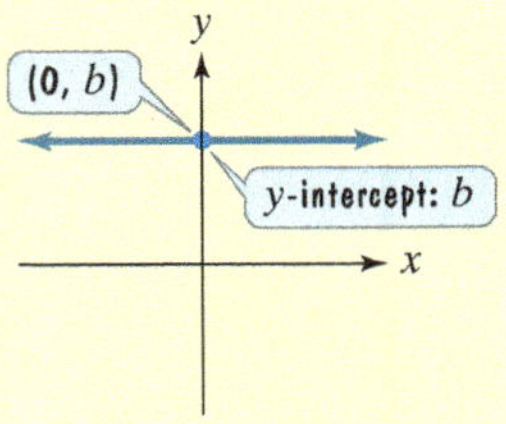

The graph of $x = a$ is a vertical line. The x-intercept is a.

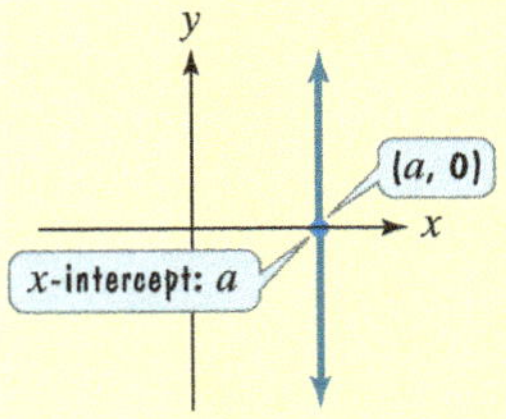

CHECK POINT 6 Graph $x = -2$ in the rectangular coordinate system.

5 Interpret slope as rate of change.

Slope as Rate of Change

Slope is defined as the ratio of a change in y to a corresponding change in x. Our next example shows how slope can be interpreted as a **rate of change** in an applied situation.

EXAMPLE 7 *Slope as a Rate of Change*

The line graphs in **Figure 7.25** show the percentage of American men and women ages 20 to 24 who were married from 1970 through 2010. Find the slope of the line segment representing women. Describe what the slope represents.

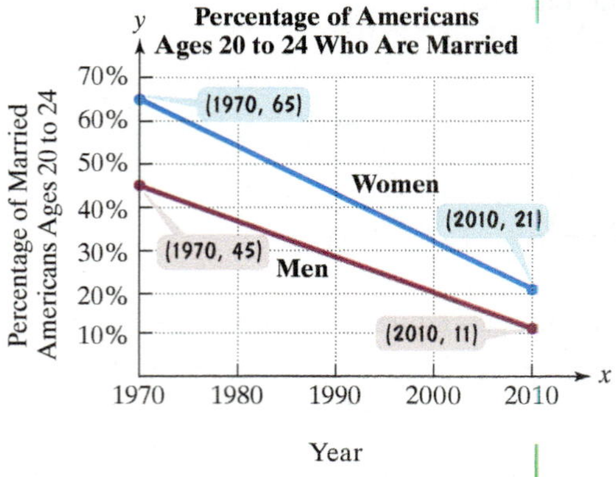

FIGURE 7.25
Source: U.S. Census Bureau

SOLUTION

We let x represent a year and y the percentage of married women ages 20–24 in that year. The two points shown on the line segment for women have the following coordinates:

$$(1970, 65) \quad \text{and} \quad (2010, 21).$$

In 1970, 65% of American women ages 20 to 24 were married.

In 2010, 21% of American women ages 20 to 24 were married.

Now we compute the slope.

$$m = \frac{\text{Change in } y}{\text{Change in } x} = \frac{21 - 65}{2010 - 1970}$$

$$= \frac{-44}{40} = -1.1$$

The unit in the numerator is the *percentage of married women ages 20 to 24.*

The unit in the denominator is *year.*

The slope indicates that for the period from 1970 through 2010, the percentage of married women ages 20 to 24 decreased by 1.1 per year. The rate of change is -1.1% per year.

CHECK POINT 7 Find the slope of the line segment representing men in **Figure 7.25**. Use your answer to complete this statement:

For the period from 1970 through 2010, the percentage of married men ages 20 to 24 decreased by ________ per year. The rate of change is ________ per ________.

6 Use slope and y-intercept to model data.

Modeling Data with the Slope-Intercept Form of the Equation of a Line

The slope-intercept form for equations of lines is useful for obtaining mathematical models for data that fall on or near a line. For example, the bar graph in **Figure 7.26(a)** at the top of the next page shows the percentage of the U.S. population who had graduated from high school and from college in 1960 and 2010. The data are displayed as points in a rectangular coordinate system in **Figure 7.26(b)**.

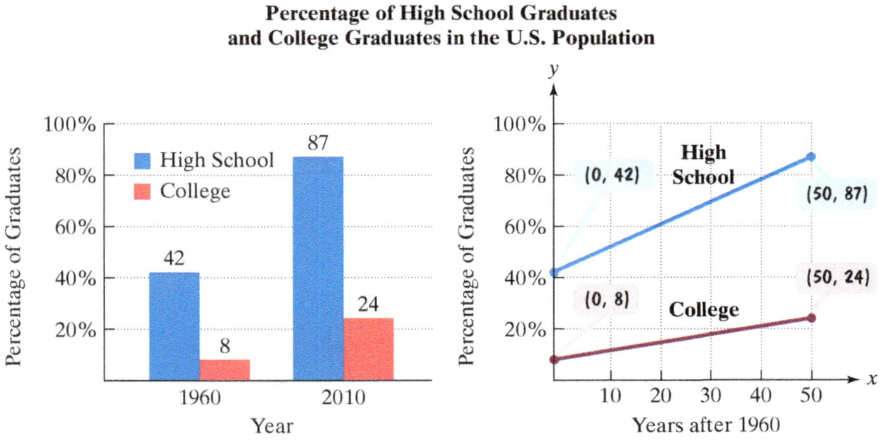

FIGURE 7.26(a) FIGURE 7.26(b)

Source: James M. Henslin, *Essentials of Sociology*, Ninth Edition, Pearson, 2011.

Example 8 illustrates how we can use the equation $y = mx + b$ to obtain a model for the data and make predictions about what might occur in the future.

EXAMPLE 8 *Modeling with the Slope-Intercept Form of the Equation*

a. Use the two points for high school in **Figure 7.26(b)** to find a function in the form $H(x) = mx + b$ that models the percentage of high school graduates in the U.S. population, $H(x)$, x years after 1960.

b. Use the model to project the percentage of high school graduates in 2020.

SOLUTION

a. We will use the line segment for high school using the points $(0, 42)$ and $(50, 87)$ to obtain a model. We need values for m, the slope, and b, the y-intercept.

$$y = mx + b$$

$$m = \frac{\text{Change in } y}{\text{Change in } x} = \frac{87 - 42}{50 - 0} = 0.9$$

The point $(0, 42)$ lies on the line segment, so the y-intercept is 42: $b = 42$.

The percentage of the U.S. population who graduated from high school, $H(x)$, x years after 1960 can be modeled by the linear function

$$H(x) = 0.9x + 42.$$

The slope, 0.9, indicates an increase in the percentage of high school graduates of 0.9% per year from 1960 through 2010.

b. Now let's use this model to project the percentage of high school graduates in 2020. Because 2020 is 60 years after 1960, substitute 60 for x in $H(x) = 0.9x + 42$ and evaluate the function at 60.

$$H(60) = 0.9(60) + 42 = 54 + 42 = 96$$

Our model projects that 96% of the U.S. population will have graduated from high school in 2020.

CHECK POINT 8

a. Use the two points for college in **Figure 7.26(b)** on the previous page to find a function in the form $C(x) = mx + b$ that models the percentage of college graduates in the U.S. population, $C(x)$, x years after 1960.

b. Use the model to project the percentage of college graduates in 2020.

Concept and Vocabulary Check

Fill in each blank so that the resulting statement is true.

1. The x-coordinate of a point where a graph crosses the x-axis is called a/an ____________.

2. The point $(0, 3)$ lies along a line, so 3 is a/an ____________ of that line.

3. The slope of the line through the distinct points (x_1, y_1) and (x_2, y_2) is ____________.

4. The slope-intercept form of the equation of a line is ________, where m represents the ________ and b represents the ________.

5. The slope of the linear function whose equation is $f(x) = -4x + 3$ is ________ and the y-intercept of its graph is ________.

6. In order to graph the line whose equation is $y = \frac{2}{5}x + 3$, begin by plotting the point ________. From this point, we move ________ units up (the rise) and ________ units to the right (the run).

7. The graph of the equation $y = 3$ is a/an ________ line.

8. The graph of the equation $x = -2$ is a/an ________ line.

Exercise Set 7.2

Practice Exercises

In Exercises 1–8, use the x- and y-intercepts to graph each linear equation.

1. $x - y = 3$
2. $x + y = 4$
3. $3x - 4y = 12$
4. $2x - 5y = 10$
5. $2x + y = 6$
6. $x + 3y = 6$
7. $5x = 3y - 15$
8. $3x = 2y + 6$

In Exercises 9–20, calculate the slope of the line passing through the given points. If the slope is undefined, so state. Then indicate whether the line rises, falls, is horizontal, or is vertical.

9. $(2, 6)$ and $(3, 5)$
10. $(4, 2)$ and $(3, 4)$
11. $(-2, 1)$ and $(2, 2)$
12. $(-1, 3)$ and $(2, 4)$
13. $(-2, 4)$ and $(-1, -1)$
14. $(6, -4)$ and $(4, -2)$
15. $(5, 3)$ and $(5, -2)$
16. $(3, -4)$ and $(3, 5)$
17. $(2, 0)$ and $(0, 8)$
18. $(3, 0)$ and $(0, -9)$
19. $(5, 1)$ and $(-2, 1)$
20. $(-2, 3)$ and $(1, 3)$

In Exercises 21–32, graph each linear function using the slope and y-intercept.

21. $y = 2x + 3$
22. $y = 2x + 1$
23. $y = -2x + 4$
24. $y = -2x + 3$
25. $y = \frac{1}{2}x + 3$
26. $y = \frac{1}{2}x + 2$
27. $f(x) = \frac{2}{3}x - 4$
28. $f(x) = \frac{3}{4}x - 5$
29. $y = -\frac{3}{4}x + 4$
30. $y = -\frac{2}{3}x + 5$
31. $f(x) = -\frac{5}{3}x$
32. $f(x) = -\frac{4}{3}x$

In Exercises 33–40,

a. *Put the equation in slope-intercept form by solving for y.*
b. *Identify the slope and the y-intercept.*
c. *Use the slope and y-intercept to graph the line.*

33. $3x + y = 0$
34. $2x + y = 0$
35. $3y = 4x$
36. $4y = 5x$
37. $2x + y = 3$
38. $3x + y = 4$
39. $7x + 2y = 14$
40. $5x + 3y = 15$

In Exercises 41–48, graph each horizontal or vertical line.

41. $y = 4$
42. $y = 2$
43. $y = -2$
44. $y = -3$
45. $x = 2$
46. $x = 4$
47. $x + 1 = 0$
48. $x + 5 = 0$

Practice Plus

In Exercises 49–52, find the slope of the line passing through each pair of points or state that the slope is undefined. Assume that all variables represent positive real numbers. Then indicate whether the line through the points rises, falls, is horizontal, or is vertical.

49. $(0, a)$ and $(b, 0)$
50. $(-a, 0)$ and $(0, -b)$
51. (a, b) and $(a, b + c)$
52. $(a - b, c)$ and $(a, a + c)$

In Exercises 53–54, find the slope and y-intercept of each line whose equation is given. Assume that $B \neq 0$.

53. $Ax + By = C$
54. $Ax = By - C$

In Exercises 55–56, find the value of y if the line through the two given points is to have the indicated slope.

55. $(3, y)$ and $(1, 4)$, $m = -3$

56. $(-2, y)$ and $(4, -4)$, $m = \frac{1}{3}$

Use the figure to make the lists in Exercises 57–58.

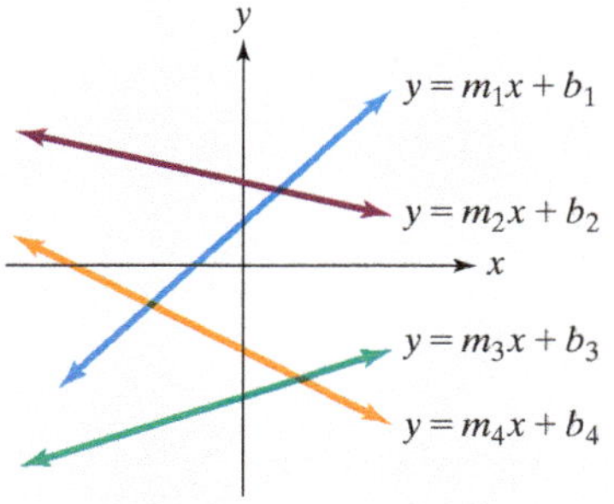

57. List the slopes m_1, m_2, m_3, and m_4 in order of decreasing size.

58. List the y-intercepts b_1, b_2, b_3, and b_4 in order of decreasing size.

Application Exercises

59. Older, Calmer. As we age, daily stress and worry decrease and happiness increases, according to an analysis of 340,847 U.S. adults, ages 18–85, in the journal *Proceedings of the National Academy of Sciences*. The graphs show a portion of the research.

Percentage of Americans Reporting "a Lot" of Stress, by Age

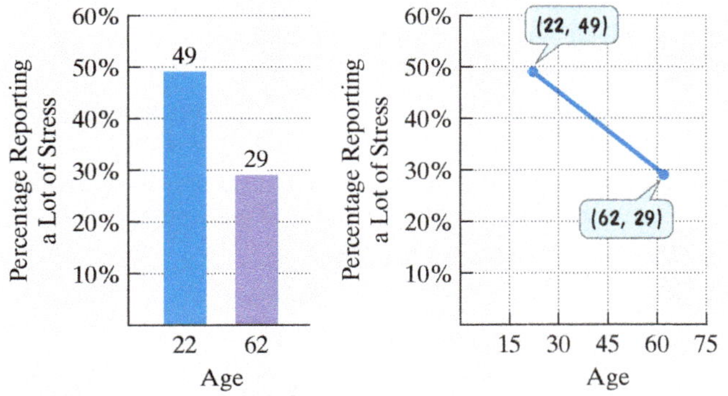

Source: National Academy of Sciences

a. Find the slope of the line passing through the two points shown by the voice balloons. Express the slope as a decimal.

b. Use your answer from part (a) to complete the statement:

For each year of aging, the percentage of Americans reporting "a lot" of stress decreases by ________%. The rate of change is ________% per __________.

60. Exercise is useful not only in preventing depression, but also as a treatment. The graphs at the top of the next column show the percentage of patients with depression in remission when exercise (brisk walking) was used as a treatment. (The control group that engaged in no exercise had 11% of the patients in remission.)

Exercise and Percentage of Patients with Depression in Remission

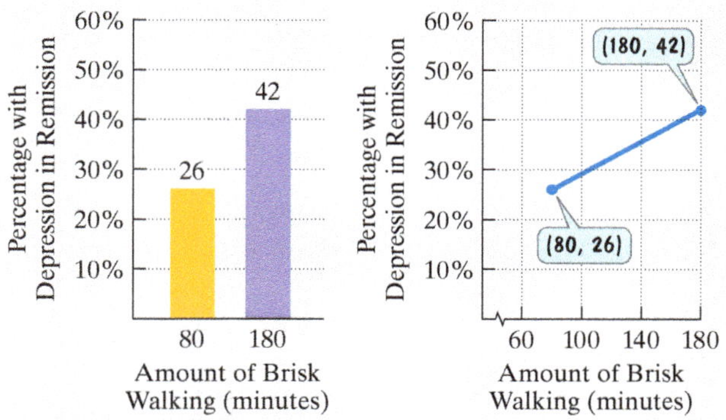

Source: Newsweek, March 26, 2007

a. Find the slope of the line passing through the two points shown by the voice balloons. Express the slope as a decimal.

b. Use your answer from part (a) to complete this statement:

For each minute of brisk walking, the percentage of patients with depression in remission increased by ________%. The rate of change is ________% per ________________.

Grade Inflation in U.S. High Schools. *In recent decades, high school teachers have given higher and higher grades to students. The bar graph shows the percentage of grades of A and C for high school students entering college.*

High School Grades of Students Entering College

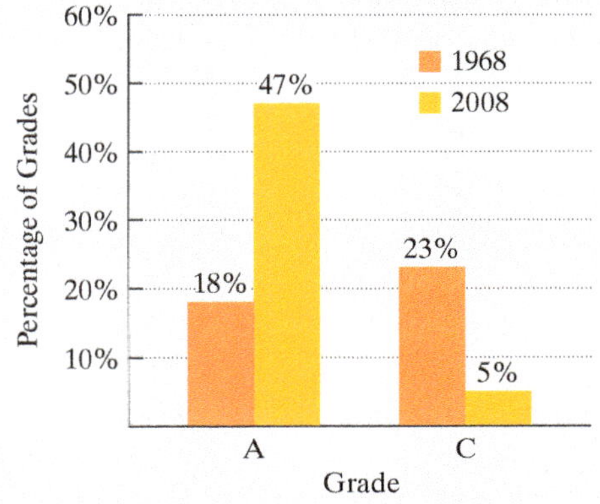

Source: UCLA Higher Education Research Institute

In Exercises 61–62, find a linear function in slope-intercept form that models the given description. Each function should model the percentage of the particular high school grade, P(x), of students entering college x years after 1968.

61. In 1968, 18% of the grades for students entering college were A (A+, A, or A−). This has increased at an average rate of approximately 0.725% per year since then.

62. In 1968, 23% of the grades for students entering college were C (C+, C, or C−). This has decreased at an average rate of approximately 0.45% per year since then.

Big (Lack of) Men on Campus. *The bar graph shows the number of bachelor's degrees, in thousands, awarded to men and women in the United States for four selected years from 1980 to 2010. The trend indicated by the graphs is among the hottest topics of debate among college-admissions officers. Some private liberal arts colleges have quietly begun special efforts to recruit men—including admissions preferences for them.*

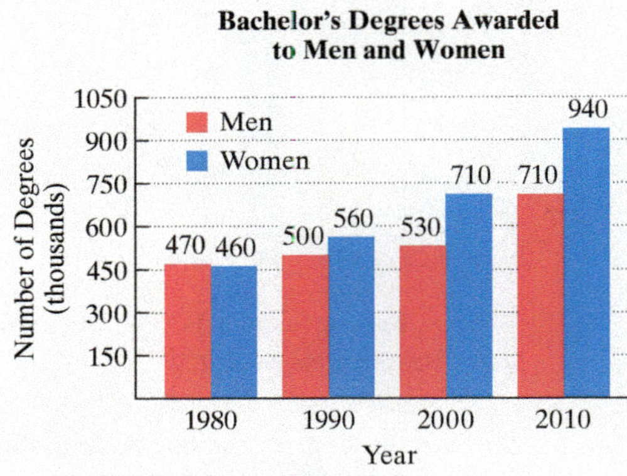

Source: Department of Education

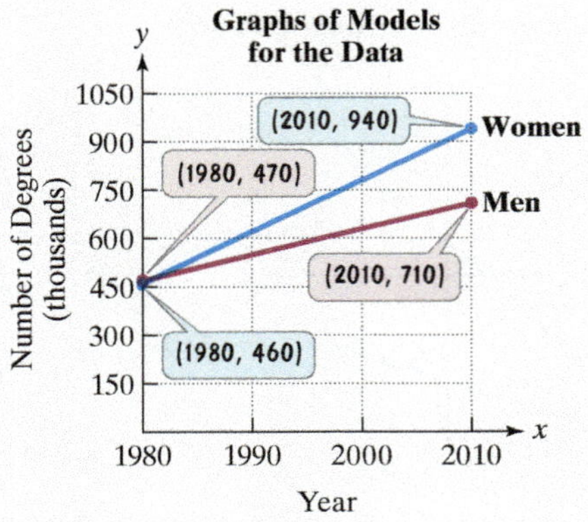

Exercises 63–64 involve the graphs of models for the data shown in the rectangular coordinate system.

63. a. Use the two points for women shown by the blue voice balloons to find a function in the form $W(x) = mx + b$ that models the number of bachelor's degrees, $W(x)$, in thousands, awarded to women x years after 1980.

b. Use the model from part (a) to project the number of bachelor's degrees that will be awarded to women in 2020.

64. a. Use the two points for men shown by the red voice balloons to find a function in the form $M(x) = mx + b$ that models the number of bachelor's degrees, $M(x)$, in thousands, awarded to men x years after 1980.

b. Use the model from part (a) to project the number of bachelor's degrees that will be awarded to men in 2020.

Writing in Mathematics

65. Describe how to find the x-intercept of a linear equation.

66. Describe how to find the y-intercept of a linear equation.

67. What is the slope of a line?

68. Describe how to calculate the slope of a line passing through two points.

69. Describe how to graph a line using the slope and y-intercept. Provide an original example with your description.

70. What does it mean if the slope of a line is 0?

71. What does it mean if the slope of a line is undefined?

72. What is the least number of points needed to graph a line? How many should actually be used? Explain.

73. Explain why the y-values can be any number for the equation $x = 5$. How is this shown in the graph of the equation?

Critical Thinking Exercises

Make Sense? *In Exercises 74–77, determine whether each statement makes sense or does not make sense, and explain your reasoning.*

74. When finding the slope of the line passing through $(-1, 5)$ and $(2, -3)$, I must let (x_1, y_1) be $(-1, 5)$ and (x_2, y_2) be $(2, -3)$.

75. A linear function that models tuition and fees at public four-year colleges from 2000 through 2010 has negative slope.

76. Because the variable m does not appear in $Ax + By = C$, equations in this form make it impossible to determine the line's slope.

77. If I drive m miles in a year, the function $c(x) = 0.61m + 3500$ models the annual cost, $c(x)$, in dollars, of operating my car, so the function shows that with no driving at all, the cost is \$3500, and the rate of increase in this cost is \$0.61 for each mile that I drive.

In Exercises 78–81, determine whether each statement is true or false. If the statement is false, make the necessary change(s) to produce a true statement.

78. The equation $y = mx + b$ shows that no line can have a y-intercept that is numerically equal to its slope.

79. Every line in the rectangular coordinate system has an equation that can be expressed in slope-intercept form.

80. The line $3x + 2y = 5$ has slope $-\frac{3}{2}$.

81. The line $2y = 3x + 7$ has a y-intercept of 7.

82. The relationship between Celsius temperature, C, and Fahrenheit temperature, F, can be described by a linear equation in the form $F = mC + b$. The graph of this equation contains the point $(0, 32)$: Water freezes at 0°C or at 32°F. The line also contains the point $(100, 212)$: Water boils at 100°C or at 212°F. Write the linear equation expressing Fahrenheit temperature in terms of Celsius temperature.

Technology Exercises

83. Use a graphing utility to verify any three of your hand-drawn graphs in Exercises 21–32.

84. Use a graphing utility to verify any three of your hand-drawn graphs in Exercises 33–40. Solve the equation for y before entering it.

7.3

Systems of Linear Equations in Two Variables

WHAT AM I SUPPOSED TO LEARN?

After you have read this section, you should be able to:

1. Decide whether an ordered pair is a solution of a linear system.
2. Solve linear systems by graphing.
3. Solve linear systems by substitution.
4. Solve linear systems by addition.
5. Identify systems that do not have exactly one ordered-pair solution.
6. Solve problems using systems of linear equations.

RESEARCHERS IDENTIFIED COLLEGE STUDENTS WHO generally were procrastinators or nonprocrastinators. The students were asked to report throughout the semester how many symptoms of physical illness they had experienced. **Figure 7.27** shows that by late in the semester, all students experienced increases in symptoms. Early in the semester, procrastinators reported fewer symptoms, but late in the semester, as work came due, they reported more symptoms than their nonprocrastinating peers.

The data in **Figure 7.27** can be analyzed using a pair of linear models in two variables. The figure shows that by week 6, both groups reported the same number of symptoms of illness, an average of approximately 3.5 symptoms per group. In this section, you will learn two algebraic methods, called *substitution* and *addition,* that will reinforce this graphic observation, verifying (6, 3.5) as the point of intersection.

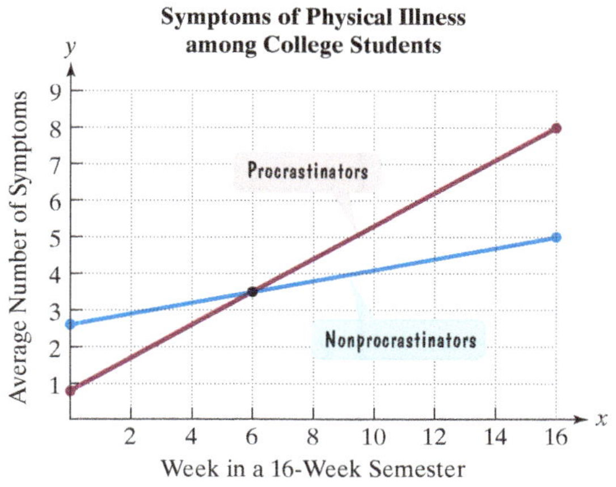

FIGURE 7.27
Source: Richard Gerrig, *Psychology and Life,* 20th Edition, Pearson, 2013.

1 Decide whether an ordered pair is a solution of a linear system.

Systems of Linear Equations and Their Solutions

We have seen that all equations in the form $Ax + By = C$, A and B not both zero, are straight lines when graphed. Two such equations are called a **system of linear equations** or a **linear system**. A **solution to a system of linear equations in two variables** is an ordered pair that satisfies both equations in the system. For example, (3, 4) satisfies the system

$$\begin{cases} x + y = 7 & \text{(3 + 4 is, indeed, 7.)} \\ x - y = -1. & \text{(3 − 4 is, indeed, −1.)} \end{cases}$$

Thus, (3, 4) satisfies both equations and is a solution of the system. The solution can be described by saying that $x = 3$ and $y = 4$. The solution can also be described using set notation. The solution set of the system is $\{(3, 4)\}$—that is, the set consisting of the ordered pair (3, 4).

A system of linear equations can have exactly one solution, no solution, or infinitely many solutions. We begin with systems having exactly one solution.

EXAMPLE 1 *Determining Whether an Ordered Pair Is a Solution of a System*

Determine whether (1, 2) is a solution of the system:

$$\begin{cases} 2x - 3y = -4 \\ 2x + y = 4. \end{cases}$$

SOLUTION

Because 1 is the x-coordinate and 2 is the y-coordinate of $(1, 2)$, we replace x with 1 and y with 2.

$$\begin{aligned} 2x - 3y &= -4 \\ 2(1) - 3(2) &\stackrel{?}{=} -4 \\ 2 - 6 &\stackrel{?}{=} -4 \\ -4 &= -4, \quad \text{true} \end{aligned} \qquad\qquad \begin{aligned} 2x + y &= 4 \\ 2(1) + 2 &\stackrel{?}{=} 4 \\ 2 + 2 &\stackrel{?}{=} 4 \\ 4 &= 4, \quad \text{true} \end{aligned}$$

The pair $(1, 2)$ satisfies both equations: It makes each equation true. Thus, the pair is a solution of the system.

CHECK POINT 1 Determine whether $(-4, 3)$ is a solution of the system:

$$\begin{cases} x + 2y = 2 \\ x - 2y = 6. \end{cases}$$

2 Solve linear systems by graphing.

Solving Linear Systems by Graphing

The solution to a system of linear equations can be found by graphing both of the equations in the same rectangular coordinate system. For a system with one solution, **the coordinates of the point of intersection of the lines is the system's solution.**

EXAMPLE 2 *Solving a Linear System by Graphing*

Solve by graphing:

$$\begin{cases} x + 2y = 2 \\ x - 2y = 6. \end{cases}$$

SOLUTION

We find the solution by graphing both $x + 2y = 2$ and $x - 2y = 6$ in the same rectangular coordinate system. We will use intercepts to graph each equation.

$$\mathbf{x + 2y = 2}$$

x-intercept: Set $y = 0$.

$$\begin{aligned} x + 2 \cdot 0 &= 2 \\ x &= 2 \end{aligned}$$

The line passes through $(2, 0)$.

y-intercept: Set $x = 0$.

$$\begin{aligned} 0 + 2y &= 2 \\ 2y &= 2 \\ y &= 1 \end{aligned}$$

The line passes through $(0, 1)$.

We graph $x + 2y = 2$ as a blue line in **Figure 7.28**.

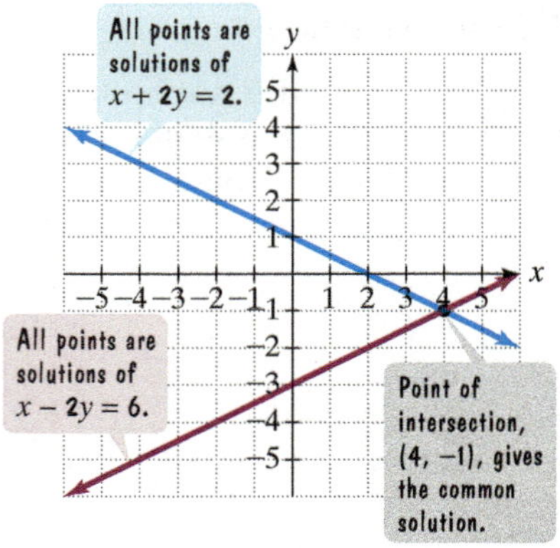

FIGURE 7.28 Visualizing a system's solution

$$\mathbf{x - 2y = 6}$$

x-intercept: Set $y = 0$.

$$\begin{aligned} x - 2 \cdot 0 &= 6 \\ x &= 6 \end{aligned}$$

The line passes through $(6, 0)$.

y-intercept: Set $x = 0$.

$$\begin{aligned} 0 - 2y &= 6 \\ -2y &= 6 \\ y &= -3 \end{aligned}$$

The line passes through $(0, -3)$.

We graph $x - 2y = 6$ as a red line in **Figure 7.28**.

The system is graphed in **Figure 7.28**. To ensure that the graph is accurate, check the coordinates of the intersection point, $(4, -1)$, in both equations.

GREAT QUESTION!

Can I use a rough sketch on scratch paper to solve a linear system by graphing?

No. When solving linear systems by graphing, neatly drawn graphs are essential for determining points of intersection.

- Use rectangular coordinate graph paper.
- Use a ruler or straightedge.
- Use a pencil with a sharp point.

We replace x with 4 and y with -1.

$$\begin{aligned} x + 2y &= 2 \\ 4 + 2(-1) &\stackrel{?}{=} 2 \\ 4 + (-2) &\stackrel{?}{=} 2 \\ 2 &= 2, \quad \text{true} \end{aligned} \qquad \begin{aligned} x - 2y &= 6 \\ 4 - 2(-1) &\stackrel{?}{=} 6 \\ 4 - (-2) &\stackrel{?}{=} 6 \\ 4 + 2 &\stackrel{?}{=} 6 \\ 6 &= 6, \quad \text{true} \end{aligned}$$

The pair $(4, -1)$ satisfies both equations—that is, it makes each equation true. This verifies that the system's solution set is $\{(4, -1)\}$.

CHECK POINT 2 Solve by graphing:

$$\begin{cases} 2x + 3y = 6 \\ 2x + y = -2. \end{cases}$$

3 Solve linear systems by substitution.

Solving Linear Systems by the Substitution Method

Finding the solution to a linear system by graphing equations may not be easy to do. For example, a solution of $\left(-\frac{2}{3}, \frac{157}{29}\right)$ would be difficult to "see" as an intersection point on a graph.

Let's consider a method that does not depend on finding a system's solution visually: the substitution method. This method involves converting the system to one equation in one variable by an appropriate substitution.

GREAT QUESTION!

In the first step of the substitution method, how do I know which variable to isolate and in which equation?

You can choose both the variable and the equation. If possible, solve for a variable whose coefficient is 1 or -1 to avoid working with fractions.

SOLVING LINEAR SYSTEMS BY SUBSTITUTION

1. Solve either of the equations for one variable in terms of the other. (If one of the equations is already in this form, you can skip this step.)
2. Substitute the expression found in step 1 into the *other* equation. This will result in an equation in one variable.
3. Solve the equation containing one variable.
4. Back-substitute the value found in step 3 into the equation from step 1. Simplify and find the value of the remaining variable.
5. Check the proposed solution in both of the system's given equations.

EXAMPLE 3 *Solving a System by Substitution*

Solve by the substitution method:

$$\begin{cases} y = -x - 1 \\ 4x - 3y = 24. \end{cases}$$

SOLUTION

Step 1 Solve either of the equations for one variable in terms of the other. This step has already been done for us. The first equation, $y = -x - 1$, is solved for y in terms of x.

Step 2 Substitute the expression from step 1 into the other equation. We substitute the expression $-x - 1$ for y into the other equation:

$$y = \boxed{-x - 1} \qquad 4x - 3\boxed{y} = 24 \qquad \text{Substitute } -x - 1 \text{ for } y.$$

This gives us an equation in one variable, namely

$$4x - 3(-x - 1) = 24.$$

The variable y has been eliminated.

Step 3 Solve the resulting equation containing one variable.

$4x - 3(-x - 1) = 24$	This is the equation containing one variable.
$4x + 3x + 3 = 24$	Apply the distributive property.
$7x + 3 = 24$	Combine like terms.
$7x = 21$	Subtract 3 from both sides.
$x = 3$	Divide both sides by 7.

Step 4 Back-substitute the obtained value into the equation from step 1. We now know that the x-coordinate of the solution is 3. To find the y-coordinate, we back-substitute the x-value into the equation from step 1.

$y = -x - 1$	This is the equation from step 1.
Substitute 3 for x.	
$y = -3 - 1$	
$y = -4$	Simplify.

With $x = 3$ and $y = -4$, the proposed solution is $(3, -4)$.

Step 5 Check. Check the proposed solution, $(3, -4)$, in both of the system's given equations. Replace x with 3 and y with -4.

$$y = -x - 1$$
$$-4 \stackrel{?}{=} -3 - 1$$
$$-4 = -4, \quad \text{true}$$

$$4x - 3y = 24$$
$$4(3) - 3(-4) \stackrel{?}{=} 24$$
$$12 + 12 \stackrel{?}{=} 24$$
$$24 = 24, \quad \text{true}$$

The pair $(3, -4)$ satisfies both equations. The system's solution set is $\{(3, -4)\}$.

TECHNOLOGY

A graphing calculator can be used to solve the system in Example 3. Graph each equation and use the intersection feature. The calculator displays the solution $(3, -4)$, as

$x = 3, y = -4$.

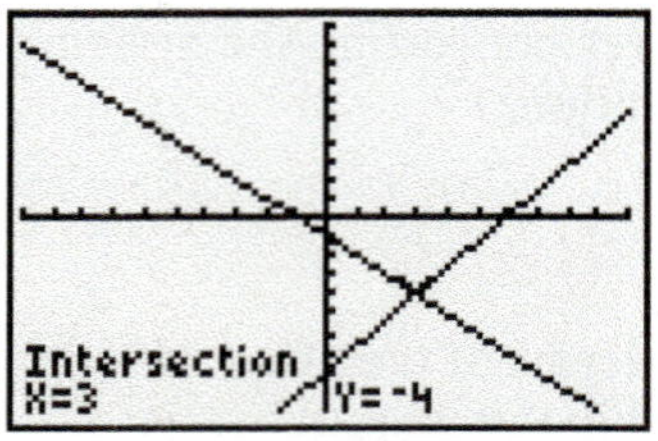

[−10, 10, 1] by [−10, 10, 1]

CHECK POINT 3 Solve by the substitution method:

$$\begin{cases} y = 3x - 7 \\ 5x - 2y = 8. \end{cases}$$

EXAMPLE 4 *Solving a System by Substitution*

Solve by the substitution method:

$$\begin{cases} 5x - 4y = 9 \\ x - 2y = -3. \end{cases}$$

SOLUTION

Step 1 Solve either of the equations for one variable in terms of the other. We begin by isolating one of the variables in either of the equations. By solving for x in the second equation, which has a coefficient of 1, we can avoid fractions.

$x - 2y = -3$	This is the second equation in the given system.
$x = 2y - 3$	Solve for x by adding $2y$ to both sides.

Step 2 Substitute the expression from step 1 into the other equation. We substitute $2y - 3$ for x in the first equation.

$$x = \boxed{2y - 3} \qquad 5\boxed{x} - 4y = 9$$

Substituting $2y - 3$ for x in $5x - 4y = 9$ gives us an equation in one variable, namely

$$5(2y - 3) - 4y = 9.$$

The variable x has been eliminated.

Step 3 Solve the resulting equation containing one variable.

$$5(2y - 3) - 4y = 9 \quad \text{This is the equation containing one variable.}$$
$$10y - 15 - 4y = 9 \quad \text{Apply the distributive property.}$$
$$6y - 15 = 9 \quad \text{Combine like terms: } 10y - 4y = 6y.$$
$$6y = 24 \quad \text{Add 15 to both sides.}$$
$$y = 4 \quad \text{Divide both sides by 6.}$$

Step 4 Back-substitute the obtained value into the equation from step 1. Now that we have the y-coordinate of the solution, we back-substitute 4 for y in the equation $x = 2y - 3$.

$$x = 2y - 3 \quad \text{Use the equation obtained in step 1.}$$
$$x = 2(4) - 3 \quad \text{Substitute 4 for } y.$$
$$x = 8 - 3 \quad \text{Multiply.}$$
$$x = 5 \quad \text{Subtract.}$$

With $x = 5$ and $y = 4$, the proposed solution is $(5, 4)$.

Step 5 Check. Take a moment to show that (5, 4) satisfies both given equations, $5x - 4y = 9$ and $x - 2y = -3$. The solution set is $\{(5, 4)\}$.

GREAT QUESTION!

If my solution satisfies one of the equations in the system, do I have to check the solution in the other equation?

Yes. Get into the habit of checking ordered-pair solutions in both equations of the system.

CHECK POINT 4 Solve by the substitution method:

$$\begin{cases} 3x + 2y = -1 \\ x - y = 3. \end{cases}$$

4 Solve linear systems by addition.

Solving Linear Systems by the Addition Method

The substitution method is most useful if one of the given equations has an isolated variable. A third, and frequently the easiest, method for solving a linear system is the addition method. Like the substitution method, the addition method involves eliminating a variable and ultimately solving an equation containing only one variable. However, this time we eliminate a variable by adding the equations.

For example, consider the following system of linear equations:

$$\begin{cases} 3x - 4y = 11 \\ -3x + 2y = -7. \end{cases}$$

When we add these two equations, the x-terms are eliminated. This occurs because the coefficients of the x-terms, 3 and -3, are opposites (additive inverses) of each other:

$$\begin{cases} 3x - 4y = 11 \\ -3x + 2y = -7 \end{cases}$$
$$\text{Add: } -2y = 4 \quad \text{The sum is an equation in one variable.}$$
$$y = -2 \quad \text{Divide both sides by } -2 \text{ and solve for } y.$$

Now we can back-substitute -2 for y into one of the original equations to find x. It does not matter which equation we use: We will obtain the same value for x in either case. If we use either equation, we can show that $x = 1$ and the solution $(1, -2)$ satisfies both equations in the system.

When we use the addition method, we want to obtain two equations whose sum is an equation containing only one variable. The key step is to **obtain, for one of the variables, coefficients that differ only in sign**. To do this, we may need to multiply one or both equations by some nonzero number so that the coefficients of one of the variables, x or y, become opposites. Then when the two equations are added, this variable is eliminated.

GREAT QUESTION!

Isn't the addition method also called the elimination method?

Although the addition method is also known as the elimination method, variables are eliminated when using both the substitution and addition methods. The name *addition method* specifically tells us that the elimination of a variable is accomplished by adding two equations.

SOLVING LINEAR SYSTEMS BY ADDITION

1. If necessary, rewrite both equations in the form $Ax + By = C$.
2. If necessary, multiply either equation or both equations by appropriate nonzero numbers so that the sum of the x-coefficients or the sum of the y-coefficients is 0.
3. Add the equations in step 2. The sum is an equation in one variable.
4. Solve the equation in one variable.
5. Back-substitute the value obtained in step 4 into either of the given equations and solve for the other variable.
6. Check the solution in both of the original equations.

EXAMPLE 5 *Solving a System by the Addition Method*

Solve by the addition method:

$$\begin{cases} 3x + 2y = 48 \\ 9x - 8y = -24. \end{cases}$$

SOLUTION

Step 1 Rewrite both equations in the form $Ax + By = C$. Both equations are already in this form. Variable terms appear on the left and constants appear on the right.

Step 2 If necessary, multiply either equation or both equations by appropriate numbers so that the sum of the x-coefficients or the sum of the y-coefficients is 0. We can eliminate x or y. Let's eliminate x. Consider the terms in x in each equation, that is, $3x$ and $9x$. To eliminate x, we can multiply each term of the first equation by -3 and then add the equations.

$$\begin{cases} 3x + 2y = 48 \\ 9x - 8y = -24 \end{cases} \xrightarrow[\text{No change}]{\text{Multiply by } -3.} \begin{cases} -9x - 6y = -144 \\ 9x - 8y = -24 \end{cases}$$

Step 3 Add the equations. Add: $-14y = -168$

Step 4 Solve the equation in one variable. We solve $-14y = -168$ by dividing both sides by -14.

$$\frac{-14y}{-14} = \frac{-168}{-14} \quad \text{Divide both sides by } -14.$$

$$y = 12 \quad \text{Simplify.}$$

Step 5 Back-substitute and find the value for the other variable. We can back-substitute 12 for y into either one of the given equations. We'll use the first one.

$$3x + 2y = 48 \quad \text{This is the first equation in the given system.}$$
$$3x + 2(12) = 48 \quad \text{Substitute 12 for } y.$$
$$3x + 24 = 48 \quad \text{Multiply.}$$
$$3x = 24 \quad \text{Subtract 24 from both sides.}$$
$$x = 8 \quad \text{Divide both sides by 3.}$$

We found that $y = 12$ and $x = 8$. The proposed solution is $(8, 12)$.

Step 6 Check. Take a few minutes to show that (8, 12) satisfies both of the original equations in the system. The solution set is $\{(8, 12)\}$.

CHECK POINT 5 Solve by the addition method:

$$\begin{cases} 4x + 5y = 3 \\ 2x - 3y = 7. \end{cases}$$

EXAMPLE 6 *Solving a System by the Addition Method*

Solve by the addition method:

$$\begin{cases} 7x = 5 - 2y \\ 3y = 16 - 2x. \end{cases}$$

SOLUTION

Step 1 Rewrite both equations in the form $Ax + By = C$. We first arrange the system so that variable terms appear on the left and constants appear on the right. We obtain

$$\begin{cases} 7x + 2y = 5 & \text{Add } 2y \text{ to both sides of the first equation.} \\ 2x - 3y = 16 & \text{Add } 2x \text{ to both sides of the second equation.} \end{cases}$$

Step 2 If necessary, multiply either equation or both equations by appropriate numbers so that the sum of the x-coefficients or the sum of the y-coefficients is 0. We can eliminate x or y. Let's eliminate y by multiplying the first equation by 3 and the second equation by -2.

$$\begin{cases} 7x + 2y = 5 \\ 2x + 3y = 16 \end{cases} \quad \begin{array}{c} \xrightarrow{\text{Multiply by 3.}} \\ \xrightarrow{\text{Multiply by } -2.} \end{array} \quad \begin{cases} 21x + 6y = 15 \\ -4x - 6y = -32 \end{cases}$$

Step 3 Add the equations. Add:

$$\begin{aligned} 17x + 0y &= -17 \\ 17x &= -17 \end{aligned}$$

Step 4 Solve the equation in one variable. We solve $17x = -17$ by dividing both sides by 17.

$$\begin{aligned} \frac{17x}{17} &= \frac{-17}{17} && \text{Divide both sides by 17.} \\ x &= -1 && \text{Simplify.} \end{aligned}$$

Step 5 Back-substitute and find the value for the other variable. We can back-substitute -1 for x into either one of the given equations. We'll use the second one.

$$\begin{aligned} 3y &= 16 - 2x && \text{This is the second equation in the given system.} \\ 3y &= 16 - 2(-1) && \text{Substitute } -1 \text{ for } x. \\ 3y &= 16 + 2 && \text{Multiply.} \\ 3y &= 18 && \text{Add.} \\ y &= 6 && \text{Divide both sides by 3.} \end{aligned}$$

With $x = -1$ and $y = 6$, the proposed solution is $(-1, 6)$.

Step 6 Check. Take a moment to show that $(-1, 6)$ satisfies both given equations. The solution is $(-1, 6)$ and the solution set is $\{(-1, 6)\}$.

CHECK POINT 6 Solve by the addition method:

$$\begin{cases} 3x = 2 - 4y \\ 5y = -1 - 2x. \end{cases}$$

5 Identify systems that do not have exactly one ordered-pair solution.

Linear Systems Having No Solution or Infinitely Many Solutions

We have seen that a system of linear equations in two variables represents a pair of lines. The lines either intersect at one point, are parallel, or are identical. Thus, there are three possibilities for the number of solutions to a system of two linear equations.

THE NUMBER OF SOLUTIONS TO A SYSTEM OF TWO LINEAR EQUATIONS

The number of solutions to a system of two linear equations in two variables is given by one of the following. (See **Figure 7.29**.)

Number of Solutions	What This Means Graphically
Exactly one ordered-pair solution	The two lines intersect at one point.
No solution	The two lines are parallel.
Infinitely many solutions	The two lines are identical.

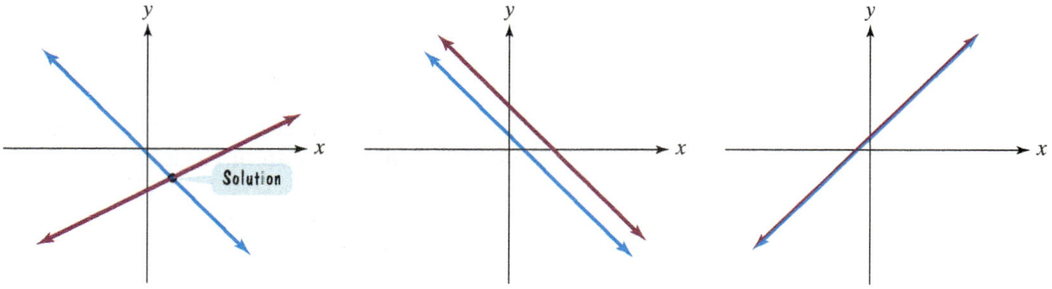

Exactly one solution

No solution (parallel lines)

Infinitely many solutions (lines coincide)

FIGURE 7.29 Possible graphs for a system of two linear equations in two variables

EXAMPLE 7 *A System with No Solution*

Solve the system:

$$\begin{cases} 4x + 6y = 12 \\ 6x + 9y = 12. \end{cases}$$

SOLUTION

Because no variable is isolated, we will use the addition method. To obtain coefficients of x that differ only in sign, we multiply the first equation by 3 and the second equation by -2.

$$\begin{cases} 4x + 6y = 12 \\ 6x + 9y = 12 \end{cases} \quad \begin{array}{c} \xrightarrow{\text{Multiply by 3.}} \\ \xrightarrow{\text{Multiply by } -2.} \end{array} \quad \begin{cases} 12x + 18y = 36 \\ -12x - 18y = -24 \end{cases}$$

$$\text{Add:} \qquad 0 = 12$$

There are no values of x and y for which $0 = 12$. No values of x and y satisfy $0x + 0y = 12$.

The false statement $0 = 12$ indicates that the system has no solution. The solution set is the empty set, $\varnothing$.

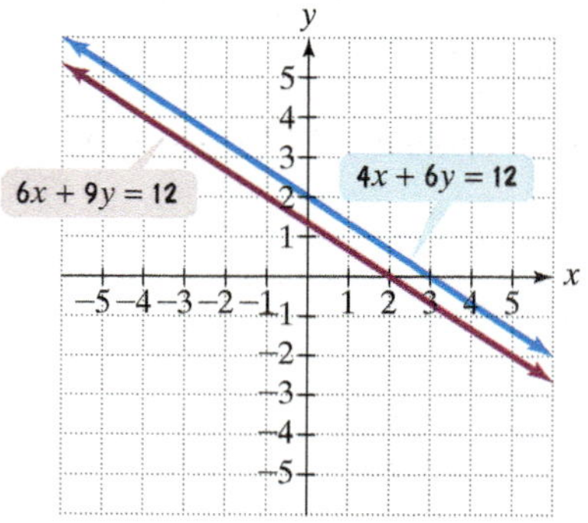

FIGURE 7.30 The graph of a system with no solution

The lines corresponding to the two equations in Example 7 are shown in **Figure 7.30**. The lines are parallel and have no point of intersection.

CHECK POINT 7 Solve the system:

$$\begin{cases} x + 2y = 4 \\ 3x + 6y = 13. \end{cases}$$

EXAMPLE 8 A System with Infinitely Many Solutions

Solve the system:

$$\begin{cases} y = 3x - 2 \\ 15x - 5y = 10. \end{cases}$$

SOLUTION

Because the variable y is isolated in $y = 3x - 2$, the first equation, we can use the substitution method. We substitute the expression for y into the second equation.

$y = \boxed{3x - 2}$ $\quad 15x - 5\boxed{y} = 10$ Substitute $3x - 2$ for y.

$15x - 5(3x - 2) = 10$ The substitution results in an equation in one variable.

$15x - 15x + 10 = 10$ Apply the distributive property.

$10 = 10$ Simplify.

This statement is true for all values of x and y.

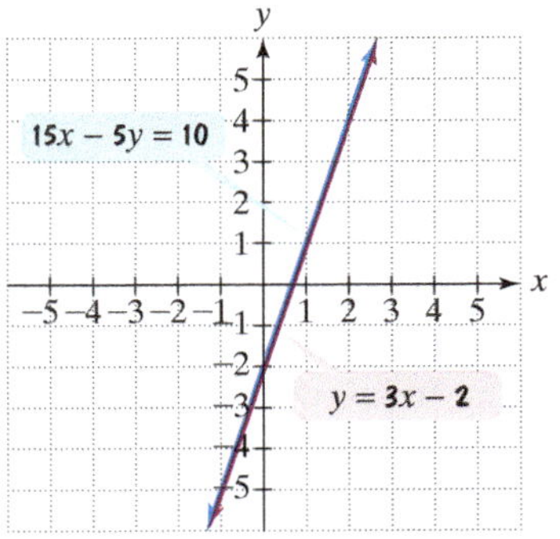

FIGURE 7.31 The graph of a system with infinitely many solutions

In our final step, both variables have been eliminated and the resulting statement, $10 = 10$, is true. This true statement indicates that the system has infinitely many solutions. The solution set consists of all points (x, y) lying on either of the coinciding lines, $y = 3x - 2$ or $15x - 5y = 10$, as shown in **Figure 7.31**.

We express the solution set for the system in one of two equivalent ways:

$$\{(x, y) \mid y = 3x - 2\} \quad \text{or} \quad \{(x, y) \mid 15x - 5y = 10\}.$$

The set of all ordered pairs (x, y) such that $y = 3x - 2$

The set of all ordered pairs (x, y) such that $15x - 5y = 10$

GREAT QUESTION!

The system in Example 8 has infinitely many solutions. Does that mean that any ordered pair of numbers is a solution?

No. Although the system in Example 8 has infinitely many solutions, this does not mean that any ordered pair of numbers you can form will be a solution. The ordered pair (x, y) must satisfy one of the system's equations, $y = 3x - 2$ or $15x - 5y = 10$, and there are infinitely many such ordered pairs. Because the graphs are coinciding lines, the ordered pairs that are solutions of one of the equations are also solutions of the other equation.

CHECK POINT 8 Solve the system:

$$\begin{cases} y = 4x - 4 \\ 8x - 2y = 8. \end{cases}$$

LINEAR SYSTEMS HAVING NO SOLUTION OR INFINITELY MANY SOLUTIONS

If both variables are eliminated when solving a system of linear equations by substitution or addition, one of the following applies:

1. There is no solution if the resulting statement is false.
2. There are infinitely many solutions if the resulting statement is true.

6 Solve problems using systems of linear equations.

Modeling with Systems of Equations: Making Money (and Losing It)

What does every entrepreneur, from a kid selling lemonade to Mark Zuckerberg, want to do? Generate profit, of course. The profit made is the money taken in, or the revenue, minus the money spent, or the cost.

REVENUE AND COST FUNCTIONS

A company produces and sells x units of a product. Its **revenue** is the money generated by selling x units of the product. Its **cost** is the cost of producing x units of the product.

Revenue Function

$$R(x) = (\text{price per unit sold})x$$

Cost Function

$$C(x) = \text{fixed cost} + (\text{cost per unit produced})x$$

The point of intersection of the graphs of the revenue and cost functions is called the **break-even point**. The x-coordinate of the point reveals the number of units that a company must produce and sell so that money coming in, the revenue, is equal to money going out, the cost. The y-coordinate of the break-even point gives the amount of money coming in and going out. Example 9 illustrates the use of the substitution method in determining a company's break-even point.

EXAMPLE 9 *Finding a Break-Even Point*

Technology is now promising to bring light, fast, and beautiful wheelchairs to millions of disabled people. A company is planning to manufacture these radically different wheelchairs. Fixed cost will be \$500,000 and it will cost \$400 to produce each wheelchair. Each wheelchair will be sold for \$600.

a. Write the cost function, C, of producing x wheelchairs.

b. Write the revenue function, R, from the sale of x wheelchairs.

c. Determine the break-even point. Describe what this means.

SOLUTION

a. The cost function is the sum of the fixed cost and variable cost.

Fixed cost of \$500,000 — plus — Variable cost: \$400 for each chair produced

$$C(x) = 500{,}000 + 400x$$

b. The revenue function is the money generated from the sale of x wheelchairs. We are given that each wheelchair will be sold for \$600.

Revenue per chair, \$600, times — the number of chairs sold

$$R(x) = 600x$$

c. The break-even point occurs where the graphs of C and R intersect. Thus, we find this point by solving the system

$$\begin{cases} C(x) = 500{,}000 + 400x \\ R(x) = 600x \end{cases} \quad \text{or} \quad \begin{cases} y = 500{,}000 + 400x \\ y = 600x. \end{cases}$$

Using substitution, we can substitute $600x$ for y in the first equation:

$600x = 500{,}000 + 400x$ — Substitute $600x$ for y in $y = 500{,}000 + 400x$.

$200x = 500{,}000$ — Subtract $400x$ from both sides.

$x = 2500$ — Divide both sides by 200.

Back-substituting 2500 for x in either of the system's equations (or functions), $C(x) = 500{,}000 + 400x$ or $R(x) = 600x$, we obtain

$$R(2500) = 600(2500) = 1{,}500{,}000.$$

We used $R(x) = 600x$.

The break-even point is (2500, 1,500,000). This means that the company will break even if it produces and sells 2500 wheelchairs. At this level, the money coming in is equal to the money going out: $1,500,000.

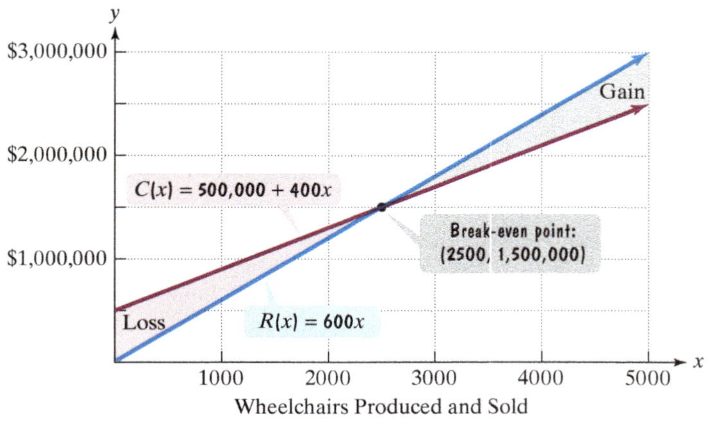

FIGURE 7.32

Figure 7.32 shows the graphs of the revenue and cost functions for the wheelchair business. Similar graphs and models apply no matter how small or large a business venture may be.

The intersection point confirms that the company breaks even by producing and selling 2500 wheelchairs. Can you see what happens for $x < 2500$? The red cost graph lies above the blue revenue graph. The cost is greater than the revenue and the business is losing money. Thus, if they sell fewer than 2500 wheelchairs, the result is a *loss*. By contrast, look at what happens for $x > 2500$. The blue revenue graph lies above the red cost graph. The revenue is greater than the cost and the business is making money. Thus, if they sell more than 2500 wheelchairs, the result is a *gain*.

CHECK POINT 9 A company that manufactures running shoes has a fixed cost of $300,000. Additionally, it costs $30 to produce each pair of shoes. They are sold at $80 per pair.

a. Write the cost function, C, of producing x pairs of running shoes.

b. Write the revenue function, R, from the sale of x pairs of running shoes.

c. Determine the break-even point. Describe what this means.

The profit generated by a business is the money taken in (its revenue) minus the money spent (its cost). Thus, once a business has modeled its revenue and cost with a system of equations, it can determine its *profit function*, $P(x)$.

THE PROFIT FUNCTION

The profit, $P(x)$, generated after producing and selling x units of a product is given by the **profit function**

$$P(x) = R(x) - C(x),$$

where R and C are the revenue and cost functions, respectively.

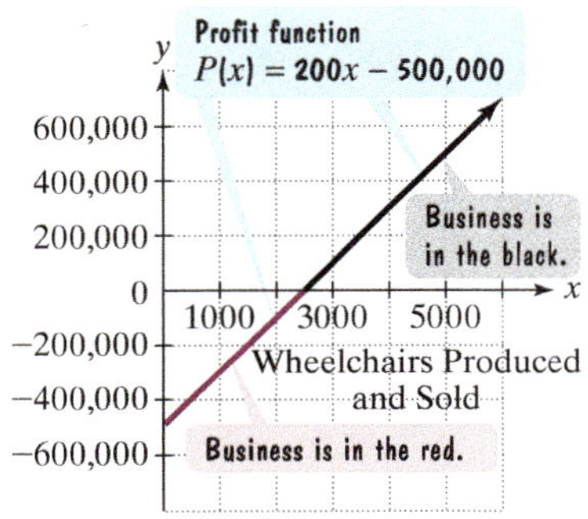

FIGURE 7.33

The profit function for the wheelchair business in Example 9 is

$$\begin{aligned} P(x) &= R(x) - C(x) \\ &= 600x - (500{,}000 + 400x) \\ &= 200x - 500{,}000. \end{aligned}$$

The graph of this profit function is shown in **Figure 7.33**. The red portion lies below the x-axis and shows a loss when fewer than 2500 wheelchairs are sold. The business is "in the red." The black portion lies above the x-axis and shows a gain when more than 2500 wheelchairs are sold. The wheelchair business is "in the black."

Concept and Vocabulary Check

Fill in each blank so that the resulting statement is true.

1. A solution to a system of linear equations in two variables is an ordered pair that ______________.
2. When solving a system of linear equations by graphing, the system's solution is determined by locating ______________.
3. When solving
$$\begin{cases} 3x - 2y = 5 \\ y = 3x - 3 \end{cases}$$
by the substitution method, we obtain $x = \frac{1}{3}$, so the solution set is ________.
4. When solving
$$\begin{cases} 2x + 10y = 9 \\ 8x + 5y = 7 \end{cases}$$
by the addition method, we can eliminate y by multiplying the second equation by ________ and then adding the equations.
5. When solving
$$\begin{cases} 4x - 3y = 15 \\ 3x - 2y = 10 \end{cases}$$
by the addition method, we can eliminate y by multiplying the first equation by 2 and the second equation by ________ and then adding the equations.
6. When solving
$$\begin{cases} 12x - 21y = 24 \\ 4x - 7y = 7 \end{cases}$$
by the addition method, we obtain $0 = 3$, so the solution set is ________. If you attempt to solve such a system by graphing, you will obtain two lines that are ________.
7. When solving
$$\begin{cases} x = 3y + 2 \\ 5x - 15y = 10 \end{cases}$$
by the substitution method, we obtain $10 = 10$, so the solution set is ______________. If you attempt to solve such a system by graphing, you will obtain two lines that ______________.
8. A company's ________ function is the money generated by selling x units of its product. The difference between this function and the company's cost function is called its ________ function.
9. A company has a graph that shows the money it generates by selling x units of its product. It also has a graph that shows its cost of producing x units of its product. The point of intersection of these graphs is called the company's ______________.

Exercise Set 7.3

Practice Exercises

In Exercises 1–4, determine whether the given ordered pair is a solution of the system.

1. $(2, 3)$
$$\begin{cases} x + 3y = 11 \\ x - 5y = -13 \end{cases}$$
2. $(-3, 5)$
$$\begin{cases} 9x + 7y = 8 \\ 8x - 9y = -69 \end{cases}$$
3. $(2, 5)$
$$\begin{cases} 2x + 3y = 17 \\ x + 4y = 16 \end{cases}$$
4. $(8, 5)$
$$\begin{cases} 5x - 4y = 20 \\ 3y = 2x + 1 \end{cases}$$

In Exercises 5–12, solve each system by graphing. Check the coordinates of the intersection point in both equations.

5. $\begin{cases} x + y = 6 \\ x - y = 2 \end{cases}$
6. $\begin{cases} x + y = 2 \\ x - y = 4 \end{cases}$
7. $\begin{cases} 2x - 3y = 6 \\ 4x + 3y = 12 \end{cases}$
8. $\begin{cases} 4x + y = 4 \\ 3x - y = 3 \end{cases}$
9. $\begin{cases} y = x + 5 \\ y = -x + 3 \end{cases}$
10. $\begin{cases} y = x + 1 \\ y = 3x - 1 \end{cases}$
11. $\begin{cases} y = -x - 1 \\ 4x - 3y = 24 \end{cases}$
12. $\begin{cases} y = 3x - 4 \\ 2x + y = 1 \end{cases}$

In Exercises 13–24, solve each system by the substitution method. Be sure to check all proposed solutions.

13. $\begin{cases} x + y = 4 \\ y = 3x \end{cases}$
14. $\begin{cases} x + y = 6 \\ y = 2x \end{cases}$
15. $\begin{cases} x + 3y = 8 \\ y = 2x - 9 \end{cases}$
16. $\begin{cases} 2x - 3y = -13 \\ y = 2x + 7 \end{cases}$
17. $\begin{cases} x + 3y = 5 \\ 4x + 5y = 13 \end{cases}$
18. $\begin{cases} y = 2x + 7 \\ 2x - y = -5 \end{cases}$
19. $\begin{cases} 2x - y = -5 \\ x + 5y = 14 \end{cases}$
20. $\begin{cases} 2x + 3y = 11 \\ x - 4y = 0 \end{cases}$
21. $\begin{cases} 2x - y = 3 \\ 5x - 2y = 10 \end{cases}$
22. $\begin{cases} -x + 3y = 10 \\ 2x + 8y = -6 \end{cases}$
23. $\begin{cases} x + 8y = 6 \\ 2x + 4y = -3 \end{cases}$
24. $\begin{cases} -4x + y = -11 \\ 2x - 3y = 5 \end{cases}$

In Exercises 25–36, solve each system by the addition method. Be sure to check all proposed solutions.

25. $\begin{cases} x + y = 1 \\ x - y = 3 \end{cases}$
26. $\begin{cases} x + y = 6 \\ x - y = -2 \end{cases}$
27. $\begin{cases} 2x + 3y = 6 \\ 2x - 3y = 6 \end{cases}$
28. $\begin{cases} 3x + 2y = 14 \\ 3x - 2y = 10 \end{cases}$
29. $\begin{cases} x + 2y = 2 \\ -4x + 3y = 25 \end{cases}$
30. $\begin{cases} 2x - 7y = 2 \\ 3x + y = -20 \end{cases}$
31. $\begin{cases} 4x + 3y = 15 \\ 2x - 5y = 1 \end{cases}$
32. $\begin{cases} 3x - 7y = 13 \\ 6x + 5y = 7 \end{cases}$
33. $\begin{cases} 3x - 4y = 11 \\ 2x + 3y = -4 \end{cases}$
34. $\begin{cases} 2x + 3y = -16 \\ 5x - 10y = 30 \end{cases}$
35. $\begin{cases} 2x = 3y - 4 \\ -6x + 12y = 6 \end{cases}$
36. $\begin{cases} 5x = 4y - 8 \\ 3x + 7y = 14 \end{cases}$

In Exercises 37–44, solve by the method of your choice. Identify systems with no solution and systems with infinitely many solutions, using set notation to express their solution sets.

37. $\begin{cases} x = 9 - 2y \\ x + 2y = 13 \end{cases}$

38. $\begin{cases} 6x + 2y = 7 \\ y = 2 - 3x \end{cases}$

39. $\begin{cases} y = 3x - 5 \\ 21x - 35 = 7y \end{cases}$

40. $\begin{cases} 9x - 3y = 12 \\ y = 3x - 4 \end{cases}$

41. $\begin{cases} 3x - 2y = -5 \\ 4x + y = 8 \end{cases}$

42. $\begin{cases} 2x + 5y = -4 \\ 3x - y = 11 \end{cases}$

43. $\begin{cases} x + 3y = 2 \\ 3x + 9y = 6 \end{cases}$

44. $\begin{cases} 4x - 2y = 2 \\ 2x - y = 1 \end{cases}$

Practice Plus

Use the graphs of the linear functions to solve Exercises 45–46.

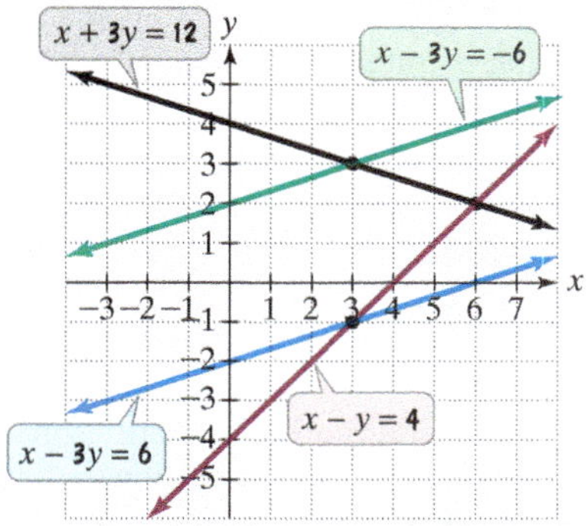

45. Write the linear system whose solution set is $\{(6, 2)\}$. Express each equation in the system in slope-intercept form.

46. Write the linear system whose solution set is $\varnothing$. Express each equation in the system in slope-intercept form.

In Exercises 47–48, solve each system for x and y, expressing either value in terms of a or b, if necessary. Assume that $a \neq 0$ and $b \neq 0$.

47. $\begin{cases} 5ax + 4y = 17 \\ ax + 7y = 22 \end{cases}$

48. $\begin{cases} 4ax + by = 3 \\ 6ax + 5by = 8 \end{cases}$

49. For the linear function $f(x) = mx + b, f(-2) = 11$ and $f(3) = -9$. Find m and b.

50. For the linear function $f(x) = mx + b, f(-3) = 23$ and $f(2) = -7$. Find m and b.

Application Exercises

The figure shows the graphs of the cost and revenue functions for a company that manufactures and sells small radios. Use the information in the figure to solve Exercises 51–56.

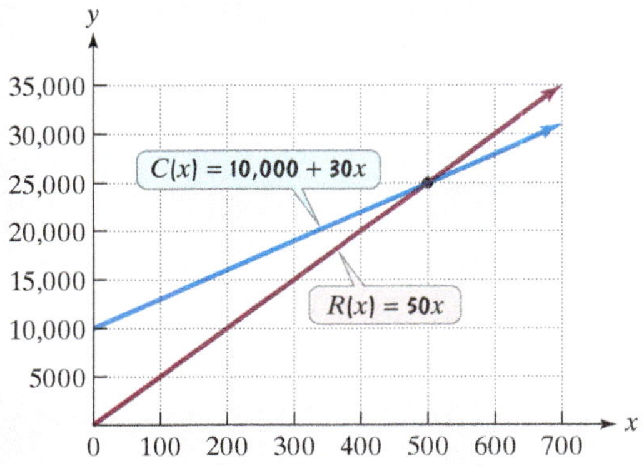

51. How many radios must be produced and sold for the company to break even?

52. More than how many radios must be produced and sold for the company to have a profit?

53. Use the formulas shown in the voice balloons to find $R(200) - C(200)$. Describe what this means for the company.

54. Use the formulas shown in the voice balloons to find $R(300) - C(300)$. Describe what this means for the company.

55. **a.** Use the formulas shown in the voice balloons to write the company's profit function, P, from producing and selling x radios.

b. Find the company's profit if 10,000 radios are produced and sold.

56. **a.** Use the formulas shown in the voice balloons to write the company's profit function, P, from producing and selling x radios.

b. Find the company's profit if 20,000 radios are produced and sold.

Exercises 57–60 describe a number of business ventures. For each exercise,

a. *Write the cost function, C.*

b. *Write the revenue function, R.*

c. *Determine the break-even point. Describe what this means.*

57. A company that manufactures small canoes has a fixed cost of \$18,000. It costs \$20 to produce each canoe. The selling price is \$80 per canoe. (In solving this exercise, let x represent the number of canoes produced and sold.)

58. A company that manufactures bicycles has a fixed cost of \$100,000. It costs \$100 to produce each bicycle. The selling price is \$300 per bike. (In solving this exercise, let x represent the number of bicycles produced and sold.)

59. You invest in a new play. The cost includes an overhead of \$30,000, plus production costs of \$2500 per performance. A sold-out performance brings in \$3125. (In solving this exercise, let x represent the number of sold-out performances.)

60. You invested \$30,000 and started a business writing greeting cards. Supplies cost 2 cents per card and you are selling each card for 50 cents. (In solving this exercise, let x represent the number of cards produced and sold.)

An important application of systems of equations arises in connection with supply and demand. As the price of a product increases, the demand for that product decreases. However, at higher prices, suppliers are willing to produce greater quantities of the product. The price at which supply and demand are equal is called the **equilibrium price**. *The quantity supplied and demanded at that price is called the* **equilibrium quantity**. *Exercises 61–62 involve supply and demand.*

61. The following models describe wages for low-skilled labor.

Demand Model	**Supply Model**
$p = -0.325x + 5.8$	$p = 0.375x + 3$

Demand Model: p = Price of labor (per hour); x = Millions of workers employers will hire

Supply Model: p = Price of labor (per hour); x = Millions of available workers

Source: O'Sullivan and Sheffrin, *Economics*, Prentice Hall, 2007.

a. Solve the system and find the equilibrium number of workers, in millions, and the equilibrium hourly wage.

b. Use your answer from part (a) to complete this statement: If workers are paid ____ per hour, there will be ____ million available workers and ____ million workers will be hired.

c. In 2007, the federal minimum wage was set at \$5.15 per hour. Substitute 5.15 for p in the demand model, $p = -0.325x + 5.8$, and determine the millions of workers employers will hire at this price.

d. At a minimum wage of \$5.15 per hour, use the supply model, $p = 0.375x + 3$, to determine the millions of available workers. Round to one decimal place.

e. At a minimum wage of \$5.15 per hour, use your answers from parts (c) and (d) to determine how many more people are looking for work than employers are willing to hire.

62. The following models describe demand and supply for three-bedroom rental apartments.

Demand Model	**Supply Model**
$p = -50x + 2000$	$p = 50x$
p: Monthly rental price	p: Monthly rental price
x: Number of apartments demanded, in thousands	x: Number of apartments supplied, in thousands

a. Solve the system and find the equilibrium quantity and the equilibrium price.

b. Use your answer from part (a) to complete this statement:

When rents are ______ per month, consumers will demand ______ apartments and suppliers will offer ______ apartments for rent.

63. We opened this section with a study showing that late in the semester, procrastinating students reported more symptoms of physical illness than their nonprocrastinating peers.

a. At the beginning of the semester, procrastinators reported an average of 0.8 symptoms, increasing at a rate of 0.45 symptoms per week. Write a function that models the average number of symptoms, y, after x weeks.

b. At the beginning of the semester, nonprocrastinators reported an average of 2.6 symptoms, increasing at a rate of 0.15 symptoms per week. Write a function that models the average number of symptoms, y, after x weeks.

c. By which week in the semester did both groups report the same number of symptoms of physical illness? For that week, how many symptoms were reported by each group? How is this shown in **Figure 7.27** on page 433?

64. a. In 1960, 5% of U.S. adults lived alone, increasing at a rate of 0.2% per year. Write a function that models the percentage of U.S. adults living alone, y, x years after 1960.

b. In 1960, 47% of U.S. adults were married, living with kids, decreasing at a rate of 0.4% per year. Write a function that models the percentage of married U.S. adults living with kids, y, x years after 1960.

c. Use the models from parts (a) and (b) to project the year in which the percentage of adults living alone will be the same as the percentage of married adults living with kids. What percentage of U.S. adults will belong to each group during that year?

Writing in Mathematics

65. What is a system of linear equations? Provide an example with your description.

66. What is the solution to a system of linear equations?

67. Explain how to solve a system of equations using graphing.

68. Explain how to solve a system of equations using the substitution method. Use $y = 3 - 3x$ and $3x + 4y = 6$ to illustrate your explanation.

69. Explain how to solve a system of equations using the addition method. Use $3x + 5y = -2$ and $2x + 3y = 0$ to illustrate your explanation.

70. What is the disadvantage to solving a system of equations using the graphing method?

71. When is it easier to use the addition method rather than the substitution method to solve a system of equations?

72. When using the addition or substitution method, how can you tell whether a system of linear equations has infinitely many solutions? What is the relationship between the graphs of the two equations?

73. When using the addition or substitution method, how can you tell whether a system of linear equations has no solution? What is the relationship between the graphs of the two equations?

74. Describe the break-even point for a business.

Critical Thinking Exercises

Make Sense? *In Exercises 75–78, determine whether each statement makes sense or does not make sense, and explain your reasoning.*

75. Even if a linear system has a solution set involving fractions, such as $\{(\frac{8}{11}, \frac{43}{11})\}$, I can use graphs to determine if the solution set is reasonable.

76. Each equation in a system of linear equations has infinitely many ordered-pair solutions.

77. Every system of linear equations has infinitely many ordered-pair solutions.

78. I find it easiest to use the addition method when one of the equations has a variable on one side by itself.

79. Write a system of equations having $\{(-2, 7)\}$ as a solution set. (More than one system is possible.)

80. One apartment is directly above a second apartment. The resident living downstairs calls his neighbor living above him and states, "If one of you is willing to come downstairs, we'll have the same number of people in both apartments." The upstairs resident responds, "We're all too tired to move. Why don't one of you come up here? Then we will have twice as many people up here as you've got down there." How many people are in each apartment?

81. A set of identical twins can only be distinguished by the characteristic that one always tells the truth and the other always lies. One twin tells you of a lucky number pair: "When I multiply my first lucky number by 3 and my second lucky number by 6, the addition of the resulting numbers produces a sum of 12. When I add my first lucky number and twice my second lucky number, the sum is 5." Which twin is talking?

7.4

WHAT AM I SUPPOSED TO LEARN?

After you have read this section, you should be able to:

1 Graph a linear inequality in two variables.
2 Use mathematical models involving linear inequalities.
3 Graph a system of linear inequalities.

Linear Inequalities in Two Variables

WE OPENED THE CHAPTER NOTING THAT THE modern emphasis on thinness as the ideal body shape has been suggested as a major cause of eating disorders. In this section (Example 4), as well as in the Exercise Set (Exercises 45–48), we use systems of linear inequalities in two variables that will enable you to establish a healthy weight range for your height and age.

Linear Inequalities in Two Variables and Their Solutions

We have seen that equations in the form $Ax + By = C$, where A and B are not both zero, are straight lines when graphed. If we change the symbol $=$ to $>$, $<$, $\geq$, or $\leq$, we obtain a **linear inequality in two variables**. Some examples of linear inequalities in two variables are $x + y > 2$, $3x - 5y \leq 15$, and $2x - y < 4$.

A **solution of an inequality in two variables**, x and y, is an ordered pair of real numbers with the following property: When the x-coordinate is substituted for x and the y-coordinate is substituted for y in the inequality, we obtain a true statement. For example, $(3, 2)$ is a solution of the inequality $x + y > 1$. When 3 is substituted for x and 2 is substituted for y, we obtain the true statement $3 + 2 > 1$, or $5 > 1$. Because there are infinitely many pairs of numbers that have a sum greater than 1, the inequality $x + y > 1$ has infinitely many solutions. Each ordered-pair solution is said to **satisfy** the inequality. Thus, $(3, 2)$ satisfies the inequality $x + y > 1$.

1 Graph a linear inequality in two variables.

The Graph of a Linear Inequality in Two Variables

We know that the graph of an equation in two variables is the set of all points whose coordinates satisfy the equation. Similarly, the **graph of an inequality in two variables** is the set of all points whose coordinates satisfy the inequality.

Let's use **Figure 7.34** to get an idea of what the graph of a linear inequality in two variables looks like. Part of the figure shows the graph of the linear equation $x + y = 2$. The line divides the points in the rectangular coordinate system into three sets. First, there is the set of points along the line satisfying $x + y = 2$. Next, there is the set of points in the green region above the line. Points in the green region satisfy the linear inequality $x + y > 2$. Finally, there is the set of points in the purple region below the line. Points in the purple region satisfy the linear inequality $x + y < 2$.

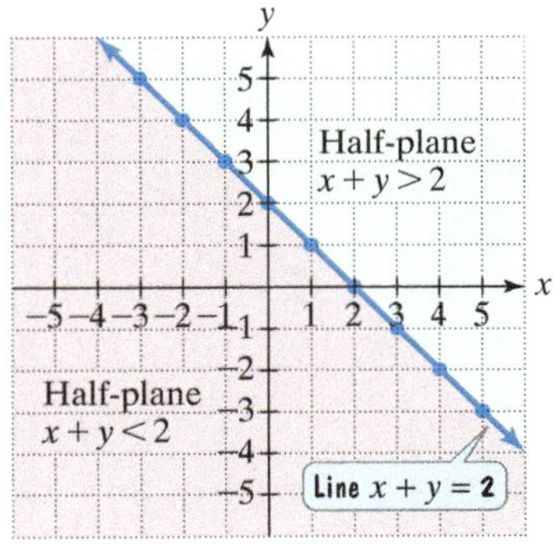

FIGURE 7.34

A **half-plane** is the set of all the points on one side of a line. In **Figure 7.34**, the green region is a half-plane. The purple region is also a half-plane. A half-plane is the graph of a linear inequality that involves $>$ or $<$. The graph of an inequality that involves $\geq$ or $\leq$ is a half-plane and a line. A solid line is used to show that a line is part of a graph. A dashed line is used to show that a line is not part of a graph.

GRAPHING A LINEAR INEQUALITY IN TWO VARIABLES

1. Replace the inequality symbol with an equal sign and graph the corresponding linear equation. Draw a solid line if the original inequality contains a $\leq$ or $\geq$ symbol. Draw a dashed line if the original inequality contains a $<$ or $>$ symbol.
2. Choose a test point from one of the half-planes. (Do not choose a point on the line.) Substitute the coordinates of the test point into the inequality.
3. If a true statement results, shade the half-plane containing this test point. If a false statement results, shade the half-plane not containing this test point.

EXAMPLE 1 *Graphing a Linear Inequality in Two Variables*

Graph: $3x - 5y \geq 15$.

SOLUTION

Step 1 Replace the inequality symbol by = and graph the linear equation. We need to graph $3x - 5y = 15$. We can use intercepts to graph this line.

We set $y = 0$ to find the x-intercept.	We set $x = 0$ to find the y-intercept.
$3x - 5y = 15$	$3x - 5y = 15$
$3x - 5 \cdot 0 = 15$	$3 \cdot 0 - 5y = 15$
$3x = 15$	$-5y = 15$
$x = 5$	$y = -3$

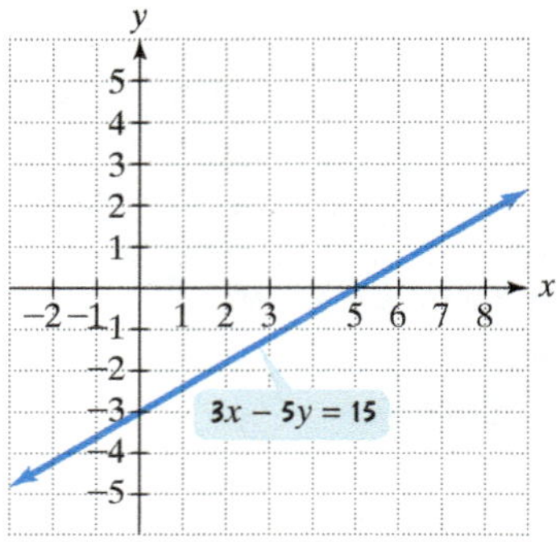

FIGURE 7.35 Preparing to graph $3x - 5y \geq 15$

The x-intercept is 5, so the line passes through $(5, 0)$. The y-intercept is -3, so the line passes through $(0, -3)$. Using the intercepts, the line is shown in **Figure 7.35** as a solid line. The line is solid because the inequality $3x - 5y \geq 15$ contains a $\geq$ symbol, in which equality is included.

Step 2 Choose a test point from one of the half-planes and not from the line. Substitute its coordinates into the inequality. The line $3x - 5y = 15$ divides the plane into three parts—the line itself and two half-planes. The points in one half-plane satisfy $3x - 5y > 15$. The points in the other half-plane satisfy $3x - 5y < 15$. We need to find which half-plane belongs to the solution of $3x - 5y \geq 15$. To do so, we test a point from either half-plane. The origin, $(0, 0)$, is the easiest point to test.

$3x - 5y \geq 15$ — This is the given inequality.

$3 \cdot 0 - 5 \cdot 0 \overset{?}{\geq} 15$ — Test (0, 0) by substituting 0 for x and 0 for y.

$0 - 0 \overset{?}{\geq} 15$ — Multiply.

$0 \geq 15$ — This statement is false.

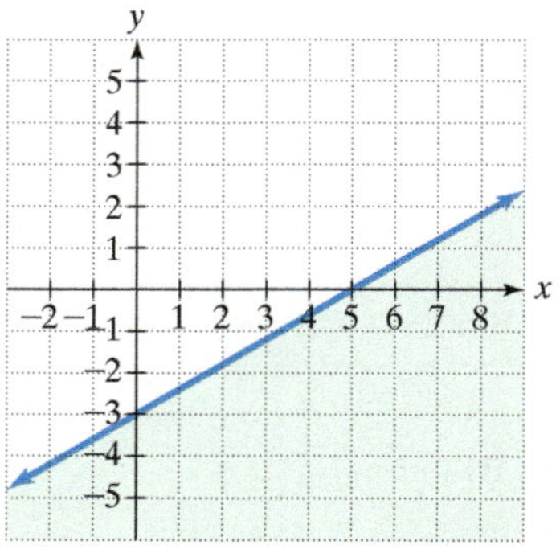

FIGURE 7.36 The graph of $3x - 5y \geq 15$

Step 3 If a false statement results, shade the half-plane not containing the test point. Because 0 is not greater than or equal to 15, the test point, $(0, 0)$, is not part of the solution set. Thus, the half-plane below the solid line $3x - 5y = 15$ is part of the solution set. The solution set is the line and the half-plane that does not contain the point $(0, 0)$, indicated by shading this half-plane. The graph is shown using green shading and a blue line in **Figure 7.36**.

CHECK POINT 1 Graph: $2x - 4y \geq 8$.

When graphing a linear inequality, test a point that lies in one of the half-planes and *not on the line separating the half-planes*. The test point $(0, 0)$ is convenient because it is easy to calculate when 0 is substituted for each variable. However, if $(0, 0)$ lies on the dividing line and not in a half-plane, a different test point must be selected.

EXAMPLE 2 *Graphing a Linear Inequality in Two Variables*

Graph: $y > -\frac{2}{3}x$.

SOLUTION

Step 1 Replace the inequality symbol by = and graph the linear equation. Because we are interested in graphing $y > -\frac{2}{3}x$, we begin by graphing $y = -\frac{2}{3}x$. We can use the slope and the y-intercept to graph this linear function.

$$y = -\frac{2}{3}x + 0$$

Slope $= \frac{-2}{3} = \frac{\text{rise}}{\text{run}}$ y-intercept $= 0$

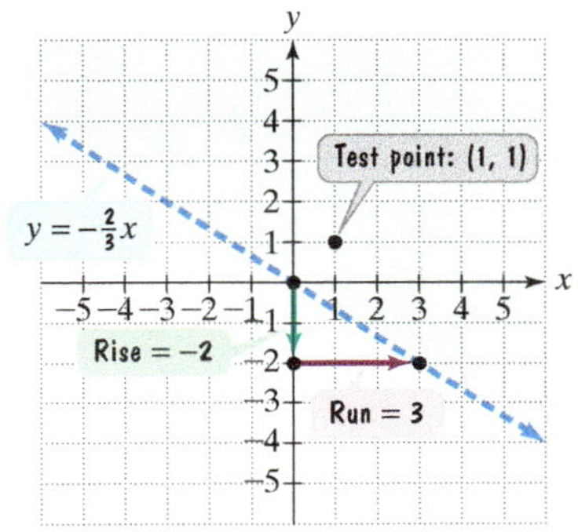

FIGURE 7.37 The graph of $y > -\frac{2}{3}x$

The y-intercept is 0, so the line passes through $(0,0)$. Using the y-intercept and the slope, the line is shown in **Figure 7.37** as a dashed line. The line is dashed because the inequality $y > -\frac{2}{3}x$ contains a $>$ symbol, in which equality is not included.

Step 2 Choose a test point from one of the half-planes and not from the line. Substitute its coordinates into the inequality. We cannot use $(0, 0)$ as a test point because it lies on the line and not in a half-plane. Let's use $(1, 1)$, which lies in the half-plane above the line.

$$y > -\frac{2}{3}x \quad \text{This is the given inequality.}$$

$$1 \overset{?}{>} -\frac{2}{3}\cdot 1 \quad \text{Test } (1, 1) \text{ by substituting 1 for } x \text{ and 1 for } y.$$

$$1 > -\frac{2}{3} \quad \text{This statement is true.}$$

Step 3 If a true statement results, shade the half-plane containing the test point. Because 1 is greater than $-\frac{2}{3}$, the test point, $(1, 1)$, is part of the solution set. All the points on the same side of the line $y = -\frac{2}{3}x$ as the point $(1, 1)$ are members of the solution set. The solution set is the half-plane that contains the point $(1, 1)$, indicated by shading this half-plane. The graph is shown using green shading and a dashed blue line in **Figure 7.37**.

CHECK POINT 2 Graph: $y > -\frac{3}{4}x$.

Graphing Linear Inequalities without Using Test Points

You can graph inequalities in the form $y > mx + b$ or $y < mx + b$ without using test points. The inequality symbol indicates which half-plane to shade.

- If $y > mx + b$, shade the half-plane above the line $y = mx + b$.
- If $y < mx + b$, shade the half-plane below the line $y = mx + b$.

Observe how this is illustrated in **Figure 7.37**. The graph of $y > -\frac{2}{3}x$ is the half-plane above the line $y = -\frac{2}{3}x$.

It is also not necessary to use test points when graphing inequalities involving half-planes on one side of a vertical or a horizontal line.

For the Vertical Line $x = a$:

- If $x > a$, shade the half-plane to the right of $x = a$.
- If $x < a$, shade the half-plane to the left of $x = a$.

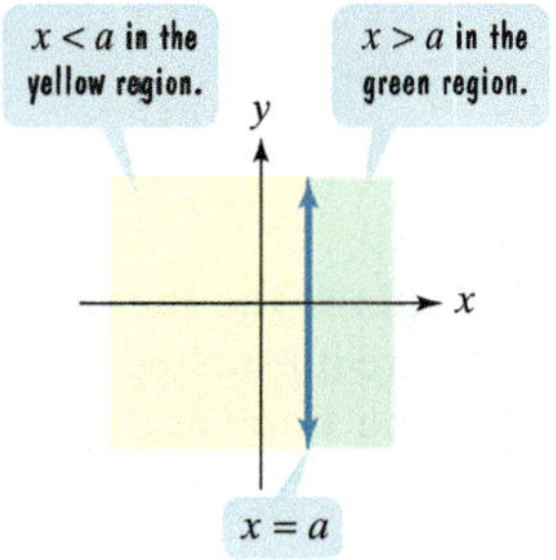

For the Horizontal Line $y = b$:

- If $y > b$, shade the half-plane above $y = b$.
- If $y < b$, shade the half-plane below $y = b$.

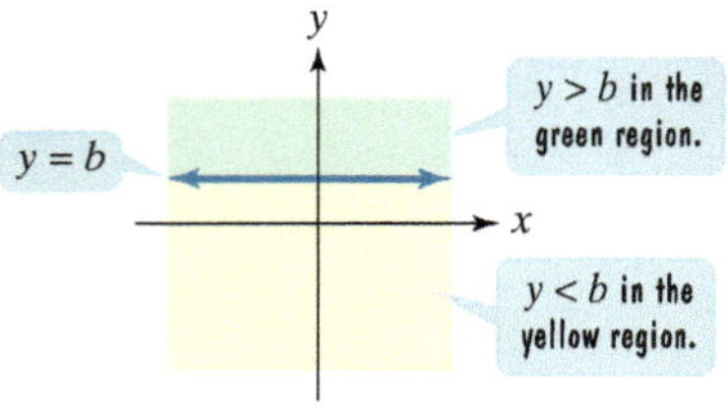

> **GREAT QUESTION!**
>
> **When is it important to use test points to graph linear inequalities?**
>
> Continue using test points to graph inequalities in the form $Ax + By > C$ or $Ax + By < C$. The graph of $Ax + By > C$ can lie above or below the line given by $Ax + By = C$, depending on the values of A and B. The same comment applies to the graph of $Ax + By < C$.

EXAMPLE 3 *Graphing Inequalities without Using Test Points*

Graph each inequality in a rectangular coordinate system:

a. $y \leq -3$ **b.** $x > 2$.

SOLUTION

a. $y \leq -3$

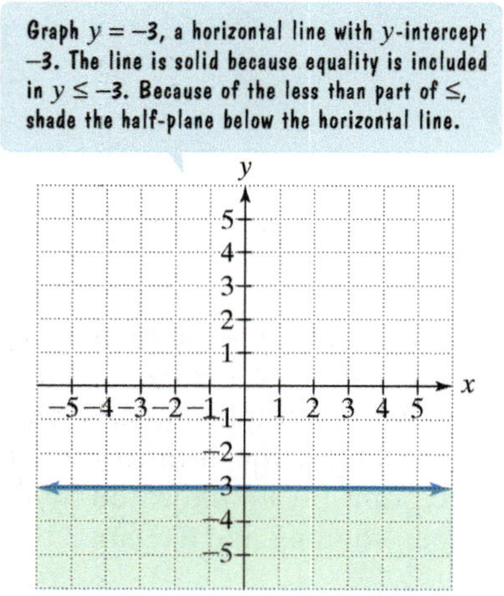

b. $x > 2$

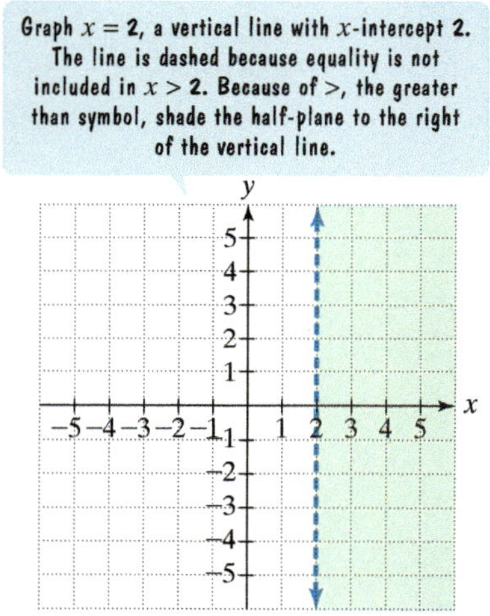

CHECK POINT 3 Graph each inequality in a rectangular coordinate system:

a. $y > 1$ **b.** $x \leq -2$.

2 Use mathematical models involving linear inequalities.

Modeling with Systems of Linear Inequalities

Just as two or more linear equations make up a system of linear equations, two or more linear inequalities make up a **system of linear inequalities**. A **solution of a system of linear inequalities** in two variables is an ordered pair that satisfies each inequality in the system.

EXAMPLE 4 *Does Your Weight Fit You?*

The latest guidelines, which apply to both men and women, give healthy weight ranges, rather than specific weights, for your height. **Figure 7.38** shows the healthy weight region for various heights for people between the ages of 19 and 34, inclusive.

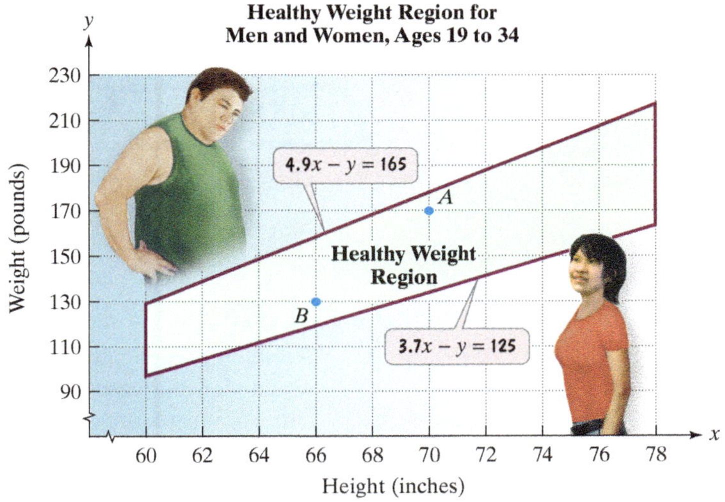

FIGURE 7.38
Source: U.S. Department of Health and Human Services

If x represents height, in inches, and y represents weight, in pounds, the healthy weight region in **Figure 7.38** can be modeled by the following system of linear inequalities:

$$\begin{cases} 4.9x - y \geq 165 \\ 3.7x - y \leq 125. \end{cases}$$

Show that point A in **Figure 7.38** is a solution of the system of inequalities that describes healthy weight.

SOLUTION

Point A has coordinates (70, 170). This means that if a person is 70 inches tall, or 5 feet 10 inches, and weighs 170 pounds, then that person's weight is within the healthy weight region. We can show that (70, 170) satisfies the system of inequalities by substituting 70 for x and 170 for y in each inequality in the system.

$$\begin{aligned} 4.9x - y &\geq 165 \\ 4.9(70) - 170 &\geq 165 \\ 343 - 170 &\geq 165 \\ 173 &\geq 165, \quad \text{true} \end{aligned}$$

$$\begin{aligned} 3.7x - y &\leq 125 \\ 3.7(70) - 170 &\leq 125 \\ 259 - 170 &\leq 125 \\ 89 &\leq 125, \quad \text{true} \end{aligned}$$

The coordinates (70, 170) make each inequality true. Thus, (70, 170) satisfies the system for the healthy weight region and is a solution of the system.

CHECK POINT 4 Show that point B in **Figure 7.38** is a solution of the system of inequalities that describes healthy weight.

3 Graph a system of linear inequalities.

Graphing Systems of Linear Inequalities

The **solution set of a system of linear inequalities in two variables** is the set of all ordered pairs that satisfy each inequality in the system. Thus, to graph a system of inequalities in two variables, begin by graphing each individual inequality in the same rectangular coordinate system. Then find the region, if there is one, that is common to every graph in the system. This region of intersection gives a picture of the system's solution set.

EXAMPLE 5 *Graphing a System of Linear Inequalities*

Graph the solution set of the system:

$$\begin{cases} x - y < 1 \\ 2x + 3y \geq 12. \end{cases}$$

SOLUTION

Replacing each inequality symbol in $x - y < 1$ and $2x + 3y \geq 12$ with an equal sign indicates that we need to graph $x - y = 1$ and $2x + 3y = 12$. We can use intercepts to graph these lines.

$x - y = 1$		$2x + 3y = 12$
x-intercept: $x - 0 = 1$; $x = 1$. The line passes through (1, 0).	Set $y = 0$ in each equation.	x-intercept: $2x + 3 \cdot 0 = 12$; $2x = 12$; $x = 6$. The line passes through (6, 0).
y-intercept: $0 - y = 1$; $-y = 1$; $y = -1$. The line passes through (0, −1).	Set $x = 0$ in each equation.	y-intercept: $2 \cdot 0 + 3y = 12$; $3y = 12$; $y = 4$. The line passes through (0, 4).

Now we are ready to graph the solution set of the system of linear inequalities.

Graph $x - y < 1$. The blue line, $x - y = 1$, is dashed: Equality is not included in $x - y < 1$. Because (0, 0) makes the inequality true ($0 - 0 < 1$, or $0 < 1$, is true), shade the half-plane containing (0, 0) in yellow.

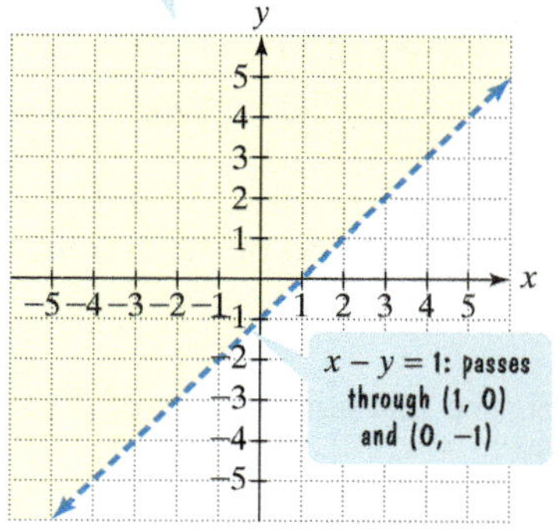

The graph of $x - y < 1$

Add the graph of $2x + 3y \geq 12$. The red line, $2x + 3y = 12$, is solid: Equality is included in $2x + 3y \geq 12$. Because (0, 0) makes the inequality false ($2 \cdot 0 + 3 \cdot 0 \geq 12$, or $0 \geq 12$, is false), shade the half-plane not containing (0, 0) using green vertical shading.

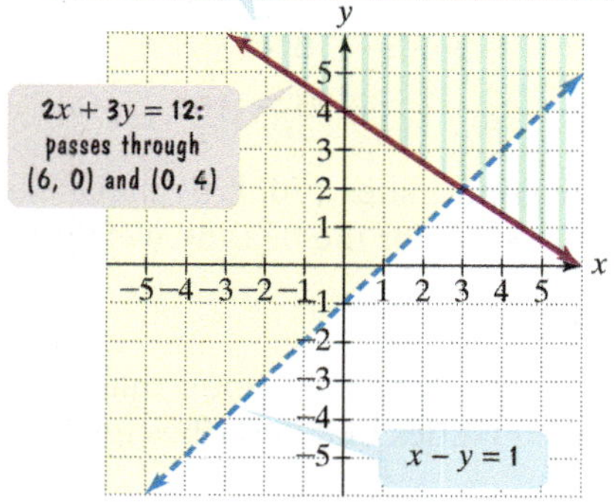

Adding the graph of $2x + 3y \geq 12$

The solution set of the system is graphed as the intersection (the overlap) of the two half-planes. This is the region in which the yellow shading and the green vertical shading overlap.

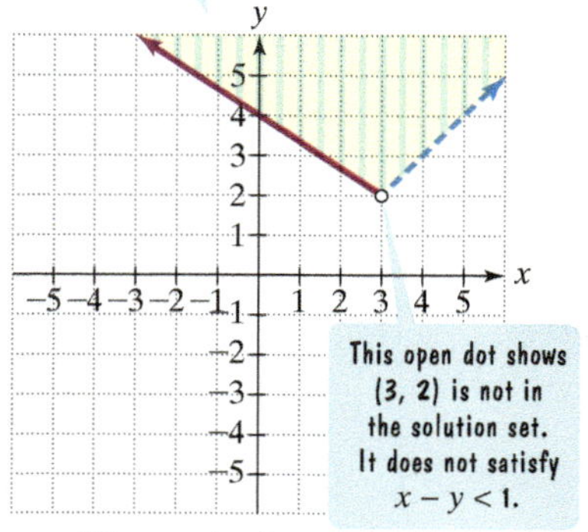

The graph of $x - y < 1$ and $2x + 3y \geq 12$

CHECK POINT 5 Graph the solution set of the system:

$$\begin{cases} x + 2y > 4 \\ 2x - 3y \le -6. \end{cases}$$

EXAMPLE 6 *Graphing a System of Linear Inequalities*

Graph the solution set of the system:

$$\begin{cases} x \le 4 \\ y > -2. \end{cases}$$

SOLUTION

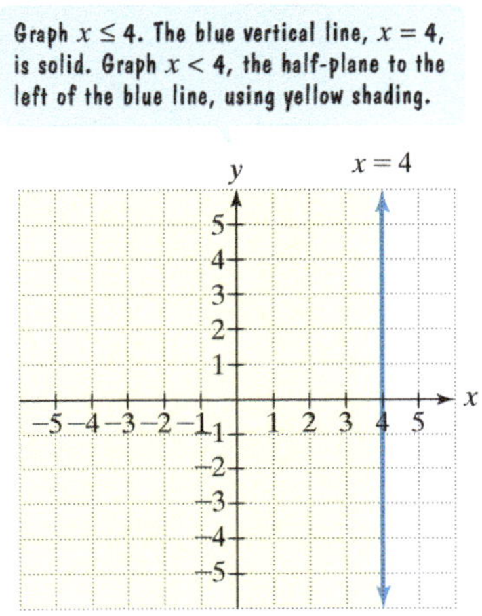

The graph of $x \le 4$

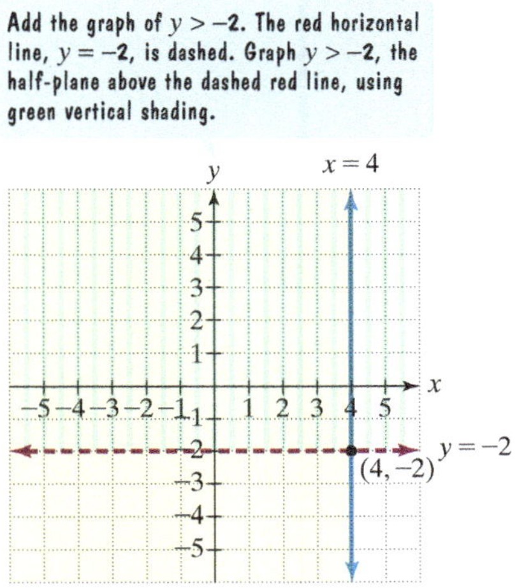

Adding the graph of $y > -2$

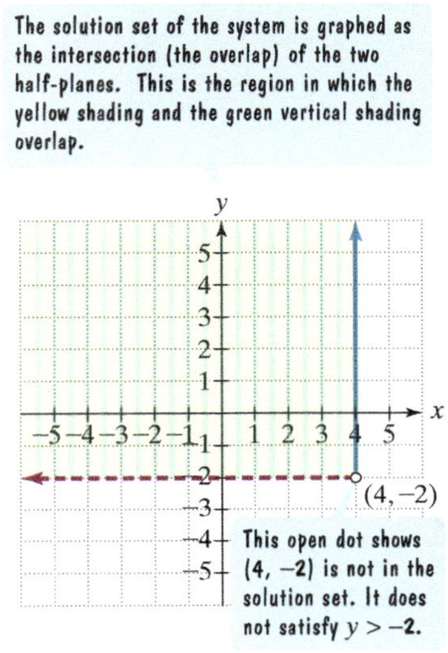

The graph of $x \le 4$ and $y > -2$

CHECK POINT 6 Graph the solution set of the system:

$$\begin{cases} x < 3 \\ y \ge -1. \end{cases}$$

Concept and Vocabulary Check

Fill in each blank so that the resulting statement is true.

1. The ordered pair (3, 2) is a/an _________ of the inequality $x + y > 1$ because when 3 is substituted for _________ and 2 is substituted for _________, the true statement _________ is obtained.
2. The set of all points that satisfy a linear inequality in two variables is called the _________ of the inequality.
3. The set of all points on one side of a line is called a/an _________.
4. True or False: The graph of $2x - 3y > 6$ includes the line $2x - 3y = 6$. _________
5. True or False: The graph of the linear equation $2x - 3y = 6$ is used to graph the linear inequality $2x - 3y > 6$. _________
6. True or False: When graphing $4x - 2y \ge 8$, to determine which side of the line to shade, choose a test point on $4x - 2y = 8$. _________
7. The solution set of the system
$$\begin{cases} x - y < 1 \\ 2x + 3y \ge 12 \end{cases}$$
is the set of ordered pairs that satisfy _________ and _________.
8. True or False: The graph of the solution set of the system
$$\begin{cases} x - 3y < 6 \\ 2x + 3y \ge -6 \end{cases}$$
includes the intersection point of $x - 3y = 6$ and $2x + 3y = -6$. _________

Exercise Set 7.4

Practice Exercises

In Exercises 1–22, graph each linear inequality.

1. $x + y \geq 2$
2. $x - y \leq 1$
3. $3x - y \geq 6$
4. $3x + y \leq 3$
5. $2x + 3y > 12$
6. $2x - 5y < 10$
7. $5x + 3y \leq -15$
8. $3x + 4y \leq -12$
9. $2y - 3x > 6$
10. $2y - x > 4$
11. $y > \frac{1}{3}x$
12. $y > \frac{1}{4}x$
13. $y \leq 3x + 2$
14. $y \leq 2x - 1$
15. $y < -\frac{1}{4}x$
16. $y < -\frac{1}{3}x$
17. $x \leq 2$
18. $x \leq -4$
19. $y > -4$
20. $y > -2$
21. $y \geq 0$
22. $x \geq 0$

In Exercises 23–38, graph the solution set of each system of inequalities.

23. $\begin{cases} 3x + 6y \leq 6 \\ 2x + y \leq 8 \end{cases}$
24. $\begin{cases} x - y \geq 4 \\ x + y \leq 6 \end{cases}$
25. $\begin{cases} 2x + y < 3 \\ x - y > 2 \end{cases}$
26. $\begin{cases} x + y < 4 \\ 4x - 2y < 6 \end{cases}$
27. $\begin{cases} 2x + y < 4 \\ x - y > 4 \end{cases}$
28. $\begin{cases} 2x - y < 3 \\ x + y < 6 \end{cases}$
29. $\begin{cases} x \geq 2 \\ y \leq 3 \end{cases}$
30. $\begin{cases} x \geq 4 \\ y \leq 2 \end{cases}$
31. $\begin{cases} x \leq 5 \\ y > -3 \end{cases}$
32. $\begin{cases} x \leq 3 \\ y > -1 \end{cases}$
33. $\begin{cases} x - y \leq 1 \\ x \geq 2 \end{cases}$
34. $\begin{cases} 4x - 5y \geq -20 \\ x \geq -3 \end{cases}$
35. $\begin{cases} y > 2x - 3 \\ y < -x + 6 \end{cases}$
36. $\begin{cases} y < -2x + 4 \\ y < x - 4 \end{cases}$
37. $\begin{cases} x + 2y \leq 4 \\ y \geq x - 3 \end{cases}$
38. $\begin{cases} x + y \leq 4 \\ y \geq 2x - 4 \end{cases}$

Practice Plus

In Exercises 39–40, write each sentence as an inequality in two variables. Then graph the inequality.

39. The y-variable is at least 4 more than the product of -2 and the x-variable.
40. The y-variable is at least 2 more than the product of -3 and the x-variable.

In Exercises 41–42, write the given sentences as a system of inequalities in two variables. Then graph the system.

41. The sum of the x-variable and the y-variable is at most 4. The y-variable added to the product of 3 and the x-variable does not exceed 6.
42. The sum of the x-variable and the y-variable is at most 3. The y-variable added to the product of 4 and the x-variable does not exceed 6.

The graphs of solution sets of systems of inequalities involve finding the intersection of the solution sets of two or more inequalities. By contrast, in Exercises 43–44, you will be graphing the union of the solution sets of two inequalities.

43. Graph the union of $y > \frac{3}{2}x - 2$ and $y < 4$.
44. Graph the union of $x - y \geq -1$ and $5x - 2y \leq 10$.

Application Exercises

The figure shows the healthy weight region for various heights for people ages 35 and older.

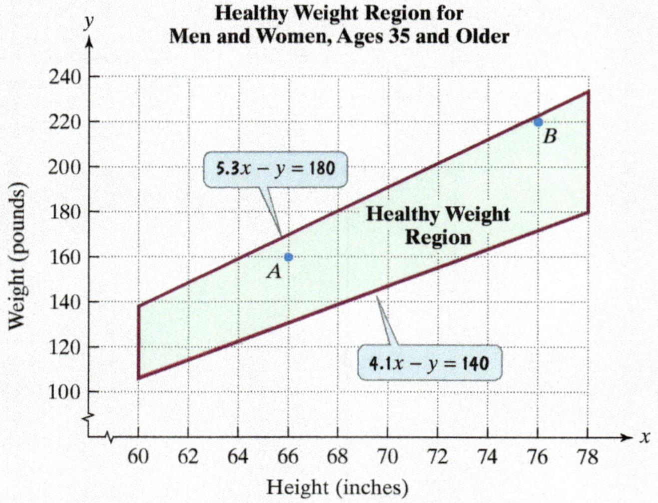

Source: U.S. Department of Health and Human Services

If x represents height, in inches, and y represents weight, in pounds, the healthy weight region can be modeled by the following system of linear inequalities:

$$\begin{cases} 5.3x - y \geq 180 \\ 4.1x - y \leq 140. \end{cases}$$

Use this information to solve Exercises 45–48.

45. Show that point A is a solution of the system of inequalities that describes healthy weight for this age group.
46. Show that point B is a solution of the system of inequalities that describes healthy weight for this age group.
47. Is a person in this age group who is 6 feet tall weighing 205 pounds within the healthy weight region?
48. Is a person in this age group who is 5 feet 8 inches tall weighing 135 pounds within the healthy weight region?
49. Many elevators have a capacity of 2000 pounds.
    a. If a child averages 50 pounds and an adult 150 pounds, write an inequality that describes when x children and y adults will cause the elevator to be overloaded.
    b. Graph the inequality. Because x and y must be positive, limit the graph to quadrant I only.
    c. Select an ordered pair satisfying the inequality. What are its coordinates and what do they represent in this situation?

50. A patient is not allowed to have more than 330 milligrams of cholesterol per day from a diet of eggs and meat. Each egg provides 165 milligrams of cholesterol. Each ounce of meat provides 110 milligrams.

a. Write an inequality that describes the patient's dietary restrictions for x eggs and y ounces of meat.

b. Graph the inequality. Because x and y must be positive, limit the graph to quadrant I only.

c. Select an ordered pair satisfying the inequality. What are its coordinates and what do they represent in this situation?

The graph of an inequality in two variables is a region in the rectangular coordinate system. Regions in coordinate systems have numerous applications. For example, the regions in the following two graphs indicate whether a person is obese, overweight, borderline overweight, normal weight, or underweight.

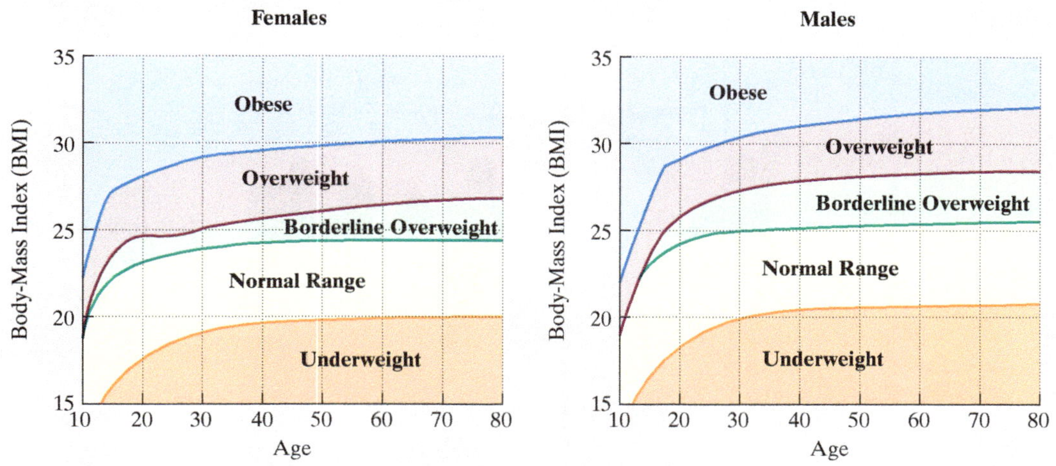

Source: Centers for Disease Control and Prevention

In these graphs, each horizontal axis shows a person's age. Each vertical axis shows that person's body-mass index (BMI), computed using the following formula:

$$\text{BMI} = \frac{703W}{H^2}.$$

The variable W represents weight, in pounds. The variable H represents height, in inches. Use this information and the graphs shown above to solve Exercises 51–52.

51. A man is 20 years old, 72 inches (6 feet) tall, and weighs 200 pounds.

a. Compute the man's BMI. Round to the nearest tenth.

b. Use the man's age and his BMI to locate this information as a point in the coordinate system for males. Is this person obese, overweight, borderline overweight, normal weight, or underweight?

52. A woman is 25 years old, 66 inches (5 feet, 6 inches) tall, and weighs 105 pounds.

a. Compute the woman's BMI. Round to the nearest tenth.

b. Use the woman's age and her BMI to locate this information as a point in the coordinate system for females. Is this person obese, overweight, borderline overweight, normal weight, or underweight?

Writing in Mathematics

53. What is a half-plane?

54. What does a dashed line mean in the graph of an inequality?

55. Explain how to graph $2x - 3y < 6$.

56. Compare the graphs of $3x - 2y > 6$ and $3x - 2y \le 6$. Discuss similarities and differences between the graphs.

57. Describe how to solve a system of linear inequalities.

Critical Thinking Exercises

Make Sense? *In Exercises 58–61, determine whether each statement makes sense or does not make sense, and explain your reasoning.*

58. When graphing a linear inequality, I should always use $(0, 0)$ as a test point because it's easy to perform the calculations when 0 is substituted for each variable.

59. When graphing $3x - 4y < 12$, it's not necessary for me to graph the linear equation $3x - 4y = 12$ because the inequality contains a $<$ symbol, in which equality is not included.

60. Systems of linear inequalities are appropriate for modeling healthy weight because guidelines give healthy weight ranges, rather than specific weights, for various heights.

61. I graphed the solution set of $y \ge x + 2$ and $x \ge 1$ without using test points.

In Exercises 62–63, write a system of inequalities for each graph.

62.

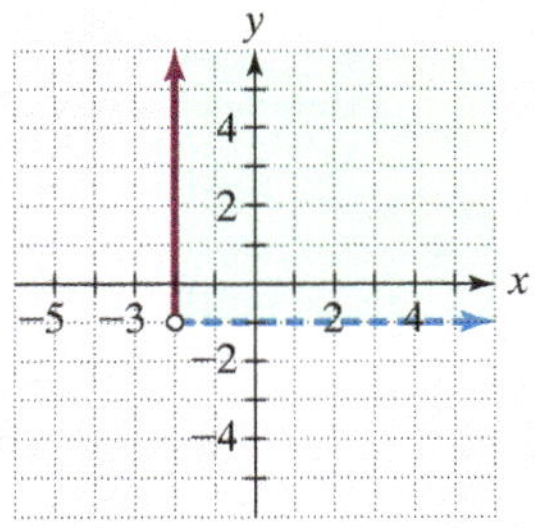

63.

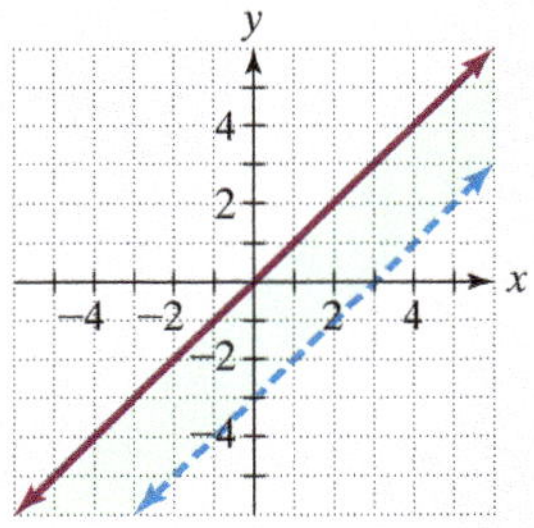

Without graphing, in Exercises 64–67, determine if each system has no solution or infinitely many solutions.

64. $\begin{cases} 3x + y < 9 \\ 3x + y > 9 \end{cases}$

65. $\begin{cases} 6x - y \le 24 \\ 6x - y > 24 \end{cases}$

66. $\begin{cases} 3x + y \le 9 \\ 3x + y \ge 9 \end{cases}$

67. $\begin{cases} 6x - y \le 24 \\ 6x - y \ge 24 \end{cases}$

7.5 Linear Programming

WHAT AM I SUPPOSED TO LEARN?

After you have read this section, you should be able to:

1. Write an objective function describing a quantity that must be maximized or minimized.
2. Use inequalities to describe limitations in a situation.
3. Use linear programming to solve problems.

West Berlin children at Tempelhof airport watch fleets of U.S. airplanes bringing in supplies to circumvent the Soviet blockade. The airlift began June 28, 1948, and continued for 15 months.

THE BERLIN AIRLIFT (1948–1949) was an operation by the United States and Great Britain in response to military action by the former Soviet Union: Soviet troops closed all roads and rail lines between West Germany and Berlin, cutting off supply routes to the city. The Allies used a mathematical technique developed during World War II to maximize the amount of supplies transported. During the 15-month airlift, 278,228 flights provided basic necessities to blockaded Berlin, saving one of the world's great cities.

In this section, we will look at an important application of systems of linear inequalities. Such systems arise in **linear programming**, a method for solving problems in which a particular quantity that must be maximized or minimized is limited by other factors. Linear programming is one of the most widely used tools in management science. It helps businesses allocate resources to manufacture products in a way that will maximize profit. Linear programming accounts for more than 50% and perhaps as much as 90% of all computing time used for management decisions in business. The Allies used linear programming to save Berlin.

Objective Functions in Linear Programming

1 Write an objective function describing a quantity that must be maximized or minimized.

Many problems involve quantities that must be maximized or minimized. Businesses are interested in maximizing profit. An operation in which bottled water and medical kits are shipped to earthquake survivors needs to maximize the number of survivors helped by this shipment. An **objective function** is an algebraic expression in two or more variables describing a quantity that must be maximized or minimized.

EXAMPLE 1 *Writing an Objective Function*

Bottled water and medical supplies are to be shipped to survivors of an earthquake by plane. Each container of bottled water will serve ten people and each medical kit will aid six people. If x represents the number of bottles of water to be shipped and y represents the number of medical kits, write the objective function that describes the number of people who can be helped.

SOLUTION

Because each bottle of water serves ten people and each medical kit aids six people, we have

The number of people helped	is	10 times the number of bottles of water	plus	6 times the number of medical kits.
	$=$	$10x$	$+$	$6y$.

Using z to represent the number of people helped, the objective function is

$$z = 10x + 6y.$$

Unlike the functions that we have seen so far, the objective function is an equation in three variables. For a value of x and a value of y, there is one and only one value of z. Thus, z is a function of x and y.

CHECK POINT 1 A company manufactures bookshelves and desks for computers. Let x represent the number of bookshelves manufactured daily and y the number of desks manufactured daily. The company's profits are \$25 per bookshelf and \$55 per desk. Write the objective function that describes the company's total daily profit, z, from x bookshelves and y desks. (Check Points 2 through 4 are also related to this situation, so keep track of your answers.)

2 Use inequalities to describe limitations in a situation.

Constraints in Linear Programming

Ideally, the number of earthquake survivors helped in Example 1 should increase without restriction so that every survivor receives water and medical kits. However, the planes that ship these supplies are subject to weight and volume restrictions. In linear programming problems, such restrictions are called **constraints**. Each constraint is expressed as a linear inequality. The list of constraints forms a system of linear inequalities.

EXAMPLE 2 *Writing a Constraint*

Each plane can carry no more than 80,000 pounds. The bottled water weighs 20 pounds per container and each medical kit weighs 10 pounds. Let x represent the number of bottles of water to be shipped and y the number of medical kits. Write an inequality that describes this constraint.

SOLUTION

Because each plane can carry no more than 80,000 pounds, we have

The total weight of the water bottles	plus	the total weight of the medical kits	must be less than or equal to	80,000 pounds.
$20x$	$+$	$10y$	$\le$	80,000.
Each bottle weighs 20 pounds.		Each kit weighs 10 pounds.		

The plane's weight constraint is described by the inequality

$$20x + 10y \le 80{,}000.$$

CHECK POINT 2 To maintain high quality, the company in Check Point 1 should not manufacture more than a total of 80 bookshelves and desks per day. Write an inequality that describes this constraint.

In addition to a weight constraint on its cargo, each plane has a limited amount of space in which to carry supplies. Example 3 demonstrates how to express this constraint.

EXAMPLE 3 *Writing a Constraint*

Each plane can carry a total volume of supplies that does not exceed 6000 cubic feet. Each water bottle is 1 cubic foot and each medical kit also has a volume of 1 cubic foot. With x still representing the number of water bottles and y the number of medical kits, write an inequality that describes this second constraint.

SOLUTION

Because each plane can carry a volume of supplies that does not exceed 6000 cubic feet, we have

The total volume of the water	plus	the total volume of the medical kits	must be less than or equal to	6000 cubic feet.
$1x$	$+$	$1y$	$\leq$	$6000.$
Each bottle is 1 cubic foot.		Each kit is 1 cubic foot.		

The plane's volume constraint is described by the inequality $x + y \leq 6000$.

In summary, here's what we have described so far in this aid-to-earthquake-survivors situation:

$$z = 10x + 6y$$

This is the objective function describing the number of people helped with x bottles of water and y medical kits.

$$\begin{cases} 20x + 10y \leq 80{,}000 \\ x + y \leq 6000. \end{cases}$$

These are the constraints based on each plane's weight and volume limitations.

CHECK POINT 3 To meet customer demand, the company in Check Point 1 must manufacture between 30 and 80 bookshelves per day, inclusive. Furthermore, the company must manufacture at least 10 and no more than 30 desks per day. Write an inequality that describes each of these sentences. Then summarize what you have described about this company by writing the objective function for its profits and the three constraints.

3 Use linear programming to solve problems.

Solving Problems with Linear Programming

The problem in the earthquake situation described previously is to maximize the number of survivors who can be helped, subject to each plane's weight and volume constraints. The process of solving this problem is called *linear programming*, based on a theorem that was proven during World War II.

SOLVING A LINEAR PROGRAMMING PROBLEM

Let $z = ax + by$ be an objective function that depends on x and y. Furthermore, z is subject to a number of constraints on x and y. If a maximum or minimum value of z exists, it can be determined as follows:

1. Graph the system of inequalities representing the constraints.
2. Find the value of the objective function at each corner, or **vertex**, of the graphed region. The maximum and minimum of the objective function occur at one or more of the corner points.

EXAMPLE 4 *Solving a Linear Programming Problem*

Determine how many bottles of water and how many medical kits should be sent on each plane to maximize the number of earthquake survivors who can be helped.

SOLUTION

We must maximize $z = 10x + 6y$ subject to the following constraints:

$$\begin{cases} 20x + 10y \le 80{,}000 \\ x + y \le 6000. \end{cases}$$

Step 1 Graph the system of inequalities representing the constraints. Because x (the number of bottles of water per plane) and y (the number of medical kits per plane) must be nonnegative, we need to graph the system of inequalities in quadrant I and its boundary only.

To graph the inequality $20x + 10y \le 80{,}000$, we graph the equation $20x + 10y = 80{,}000$ as a solid blue line (**Figure 7.39**). Setting $y = 0$, the x-intercept is 4000 and setting $x = 0$, the y-intercept is 8000. Using $(0, 0)$ as a test point, the inequality is satisfied, so we shade below the blue line, as shown in yellow in **Figure 7.39**.

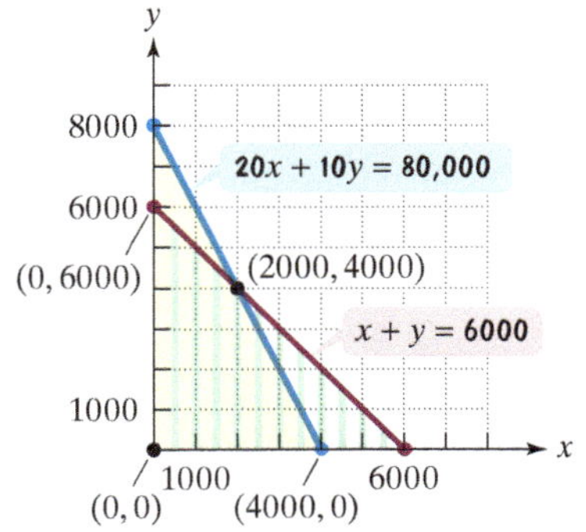

FIGURE 7.39 The region in quadrant I representing the constraints $20x + 10y \le 80{,}000$ and $x + y \le 6000$

Now we graph $x + y \le 6000$ by first graphing $x + y = 6000$ as a solid red line. Setting $y = 0$, the x-intercept is 6000. Setting $x = 0$, the y-intercept is 6000. Using $(0, 0)$ as a test point, the inequality is satisfied, so we shade below the red line, as shown using green vertical shading in **Figure 7.39**.

We use the addition method to find where the lines $20x + 10y = 80{,}000$ and $x + y = 6000$ intersect.

$$\begin{cases} 20x + 10y = 80{,}000 \\ x + y = 6000 \end{cases} \xrightarrow[\text{Multiply by } -10.]{\text{No change}} \begin{cases} 20x + 10y = 80{,}000 \\ -10x - 10y = -60{,}000 \end{cases}$$

$$\text{Add:} \quad 10x = 20{,}000$$

$$x = 2000$$

Back-substituting 2000 for x in $x + y = 6000$, we find $y = 4000$, so the intersection point is $(2000, 4000)$.

The system of inequalities representing the constraints is shown by the region in which the yellow shading and the green vertical shading overlap in **Figure 7.39**. The graph of the system of inequalities is shown again in **Figure 7.40**. The red and blue line segments are included in the graph.

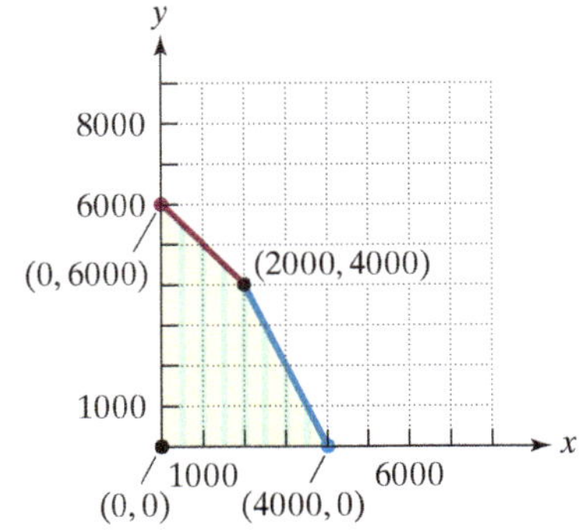

FIGURE 7.40

Step 2 Find the value of the objective function at each corner of the graphed region. The maximum and minimum of the objective function occur at one or more of the corner points. We must evaluate the objective function, $z = 10x + 6y$, at the four corners, or vertices, of the region in **Figure 7.40**.

Corner (x, y)	Objective Function $z = 10x + 6y$
$(0, 0)$	$z = 10(0) + 6(0) = 0$
$(4000, 0)$	$z = 10(4000) + 6(0) = 40{,}000$
$(2000, 4000)$	$z = 10(2000) + 6(4000) = 44{,}000$ ← maximum
$(0, 6000)$	$z = 10(0) + 6(6000) = 36{,}000$

Thus, the maximum value of z is 44,000 and this occurs when $x = 2000$ and $y = 4000$. In practical terms, this means that the maximum number of earthquake survivors who can be helped with each plane shipment is 44,000. This can be accomplished by sending 2000 water bottles and 4000 medical kits per plane.

CHECK POINT 4 For the company in Check Points 1–3, how many bookshelves and how many desks should be manufactured per day to obtain a maximum profit? What is the maximum daily profit?

Concept and Vocabulary Check

Fill in each blank so that the resulting statement is true.

1. A method for finding the maximum or minimum value of a quantity that is subject to various limitations is called ____________.
2. An algebraic expression in two or more variables describing a quantity that must be maximized or minimized is called a/an ________ function.
3. A system of linear inequalities is used to represent restrictions, or ________, on a function that must be maximized or minimized. Using the graph of such a system of inequalities, the maximum and minimum values of the function occur at one or more of the ________ points.

Exercise Set 7.5

Practice Exercises

In Exercises 1–4, find the value of the objective function at each corner of the graphed region. What is the maximum value of the objective function? What is the minimum value of the objective function?

1. Objective Function

$z = 5x + 6y$

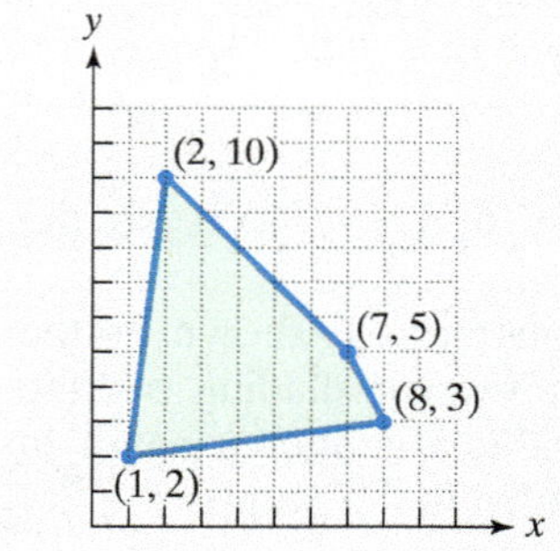

2. Objective Function

$z = 3x + 2y$

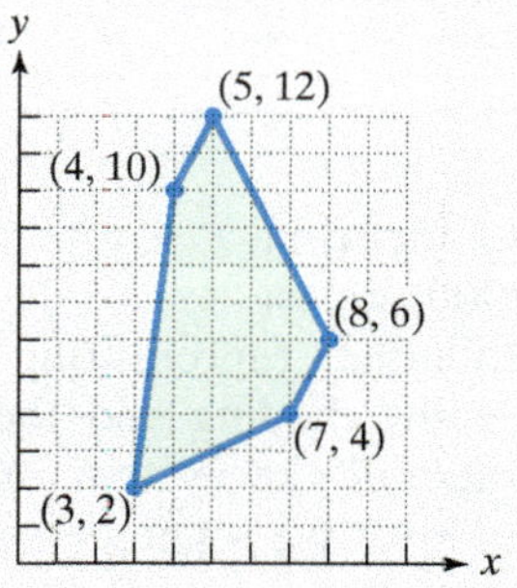

3. Objective Function

$z = 40x + 50y$

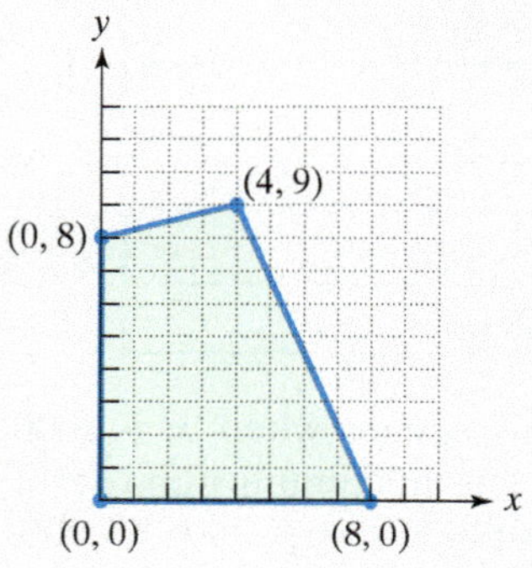

4. Objective Function

$z = 30x + 45y$

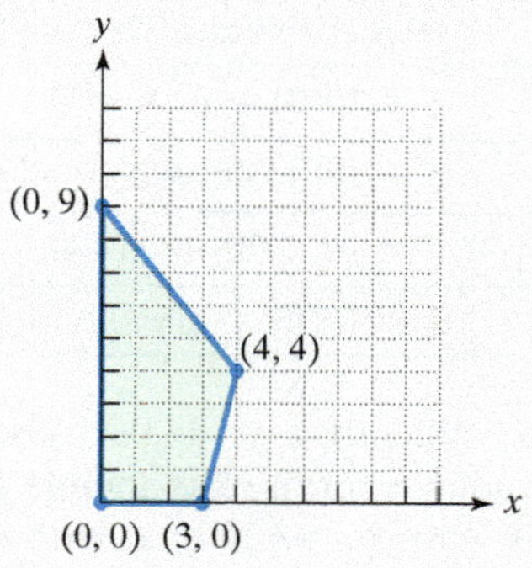

In Exercises 5–8, an objective function and a system of linear inequalities representing constraints are given.

a. *Graph the system of inequalities representing the constraints.*

b. *Find the value of the objective function at each corner of the graphed region.*

c. *Use the values in part (b) to determine the maximum value of the objective function and the values of x and y for which the maximum occurs.*

5. Objective Function
$z = x + y$
Constraints

$$\begin{cases} x \le 6 \\ y \ge 1 \\ 2x - y \ge -1 \end{cases}$$

6. Objective Function
$z = 3x - 2y$
Constraints

$$\begin{cases} x \ge 1 \\ x \le 5 \\ y \ge 2 \\ x - y \ge -3 \end{cases}$$

7. Objective Function
$z = 6x + 10y$
Constraints

$$\begin{cases} x + y \le 12 \\ x + 2y \le 20 \\ \left.\begin{matrix} x \ge 0 \\ y \ge 0 \end{matrix}\right\} \text{Quadrant I and its boundary} \end{cases}$$

8. Objective Function
$z = x + 3y$
Constraints

$$\begin{cases} x + y \ge 2 \\ x \le 6 \\ y \le 5 \\ \left.\begin{matrix} x \ge 0 \\ y \ge 0 \end{matrix}\right\} \text{Quadrant I and its boundary} \end{cases}$$

Practice Plus

Use the directions for Exercises 5–8 to solve Exercises 9–12.

9. Objective Function
$z = 5x - 2y$
Constraints

$$\begin{cases} 0 \le x \le 5 \\ 0 \le y \le 3 \\ x + y \ge 2 \end{cases}$$

10. Objective Function
$z = 3x - 2y$
Constraints

$$\begin{cases} 1 \le x \le 5 \\ y \ge 2 \\ x - y \ge -3 \end{cases}$$

11. Objective Function
$z = 10x + 12y$
Constraints
$$\begin{cases} x \geq 0, y \geq 0 \\ x + y \leq 7 \\ 2x + y \leq 10 \\ 2x + 3y \leq 18 \end{cases}$$

12. Objective Function
$z = 5x + 6y$
Constraints
$$\begin{cases} x \geq 0, y \geq 0 \\ 2x + y \geq 10 \\ x + 2y \geq 10 \\ x + y \leq 10 \end{cases}$$

Application Exercises

13. a. A student earns \$10 per hour for tutoring and \$7 per hour as a teacher's aide. Let x = the number of hours each week spent tutoring and y = the number of hours each week spent as a teacher's aide. Write the objective function that describes total weekly earnings.

b. The student is bound by the following constraints:

- To have enough time for studies, the student can work no more than 20 hours per week.
- The tutoring center requires that each tutor spend at least three hours per week tutoring.
- The tutoring center requires that each tutor spend no more than eight hours per week tutoring.

Write a system of three inequalities that describes these constraints.

c. Graph the system of inequalities in part (b). Use only the first quadrant and its boundary, because x and y are nonnegative.

d. Evaluate the objective function for total weekly earnings at each of the four vertices of the graphed region. [The vertices should occur at $(3, 0)$, $(8, 0)$, $(3, 17)$, and $(8, 12)$.]

e. Complete the missing portions of this statement: The student can earn the maximum amount per week by tutoring for __ hours per week and working as a teacher's aide for __ hours per week. The maximum amount that the student can earn each week is \$ __.

14. A television manufacturer makes LCD and plasma televisions. The profit per unit is \$125 for the LCD televisions and \$200 for the plasma televisions.

a. Let x = the number of LCD televisions manufactured in a month and y = the number of plasma televisions manufactured in a month. Write the objective function that describes the total monthly profit.

b. The manufacturer is bound by the following constraints:

- Equipment in the factory allows for making at most 450 LCD televisions in one month.
- Equipment in the factory allows for making at most 200 plasma televisions in one month.
- The cost to the manufacturer per unit is \$600 for the LCD televisions and \$900 for the plasma televisions. Total monthly costs cannot exceed \$360,000.

Write a system of three inequalities that describes these constraints.

c. Graph the system of inequalities in part (b). Use only the first quadrant and its boundary, because x and y must both be nonnegative.

d. Evaluate the objective function for total monthly profit at each of the five vertices of the graphed region. [The vertices should occur at $(0, 0)$, $(0, 200)$, $(300, 200)$, $(450, 100)$, and $(450, 0)$.]

e. Complete the missing portions of this statement: The television manufacturer will make the greatest profit by manufacturing ___ LCD televisions each month and ___ plasma televisions each month. The maximum monthly profit is \$ ____.

15. Food and clothing are shipped to survivors of a natural disaster. Each carton of food will feed 12 people, while each carton of clothing will help 5 people. Each 20-cubic-foot box of food weighs 50 pounds and each 10-cubic-foot box of clothing weighs 20 pounds. The commercial carriers transporting food and clothing are bound by the following constraints:

- The total weight per carrier cannot exceed 19,000 pounds.
- The total volume must be no more than 8000 cubic feet.

How many cartons of food and clothing should be sent with each plane shipment to maximize the number of people who can be helped?

16. You are about to take a test that contains computation problems worth 6 points each and word problems worth 10 points each. You can do a computation problem in 2 minutes and a word problem in 4 minutes. You have 40 minutes to take the test and may answer no more than 12 problems. Assuming you answer all the problems attempted correctly, how many of each type of problem must you answer to maximize your score? What is the maximum score?

17. A theater is presenting a program on drinking and driving for students and their parents. The proceeds will be donated to a local alcohol information center. Admission is \$2 for parents and \$1 for students. However, the situation has two constraints: The theater can hold no more than 150 people and every two parents must bring at least one student. How many parents and students should attend to raise the maximum amount of money?

18. On June 24, 1948, the former Soviet Union blocked all land and water routes through East Germany to Berlin. A gigantic airlift was organized using American and British planes to bring food, clothing, and other supplies to the more than 2 million people in West Berlin. The cargo capacity was 30,000 cubic feet for an American plane and 20,000 cubic feet for a British plane. To break the Soviet blockade, the Western Allies had to maximize cargo capacity, but were subject to the following restrictions:

- No more than 44 planes could be used.
- The larger American planes required 16 personnel per flight, double that of the requirement for the British planes. The total number of personnel available could not exceed 512.
- The cost of an American flight was \$9000 and the cost of a British flight was \$5000. Total weekly costs could not exceed \$300,000.

Find the number of American and British planes that were used to maximize cargo capacity.

Writing in Mathematics

19. What kinds of problems are solved using the linear programming method?

20. What is an objective function in a linear programming problem?

21. What is a constraint in a linear programming problem? How is a constraint represented?

22. In your own words, describe how to solve a linear programming problem.

23. Describe a situation in your life in which you would like to maximize something, but you are limited by at least two constraints. Can linear programming be used in this situation? Explain your answer.

Critical Thinking Exercises

Make Sense? *In Exercises 24–27, determine whether each statement makes sense or does not make sense, and explain your reasoning.*

24. In order to solve a linear programming problem, I use the graph representing the constraints and the graph of the objective function.

25. I use the coordinates of each vertex from my graph representing the constraints to find the values that maximize or minimize an objective function.

26. I need to be able to graph systems of linear inequalities in order to solve linear programming problems.

27. An important application of linear programming for businesses involves maximizing profit.

28. Suppose that you inherit \$10,000. The will states how you must invest the money. Some (or all) of the money must be invested in stocks and bonds. The requirements are that at least \$3000 be invested in bonds, with expected returns of \$0.08 per dollar, and at least \$2000 be invested in stocks, with expected returns of \$0.12 per dollar. Because the stocks are medium risk, the final stipulation requires that the investment in bonds should never be less than the investment in stocks. How should the money be invested so as to maximize your expected returns?

Group Exercises

29. Group members should choose a particular field of interest. Research how linear programming is used to solve problems in that field. If possible, investigate the solution of a specific practical problem. Present a report on your findings, including the contributions of George Dantzig, Narendra Karmarkar, and L. G. Khachion to linear programming.

30. Members of the group should interview a business executive who is in charge of deciding the product mix for a business. How are production policy decisions made? Are other methods used in conjunction with linear programming? What are these methods? What sort of academic background, particularly in mathematics, does this executive have? Present a group report addressing these questions, emphasizing the role of linear programming for the business.

7.6 Modeling Data: Exponential, Logarithmic, and Quadratic Functions

WHAT AM I SUPPOSED TO LEARN?

After you have read this section, you should be able to:

1 Graph exponential functions.
2 Use exponential models.
3 Graph logarithmic functions.
4 Use logarithmic models.
5 Graph quadratic functions.
6 Use quadratic models.
7 Determine an appropriate function for modeling data.

IS THERE A RELATIONSHIP BETWEEN LITERACY AND child mortality? As the percentage of adult females who are literate increases, does the mortality of children under five decrease? **Figure 7.41**, based on data from the United Nations, indicates that this is, indeed, the case. Each point in the figure represents one country.

Data presented in a visual form as a set of points are called a **scatter plot**. Also shown in **Figure 7.41** is a line that passes through or near the points. The line that best fits the data points in a scatter plot is called a **regression line**. We can use the line's slope and y-intercept to obtain a linear model for under-five mortality, y, per thousand, as a function of the percentage of literate adult females, x. The model is given at the top of the next page.

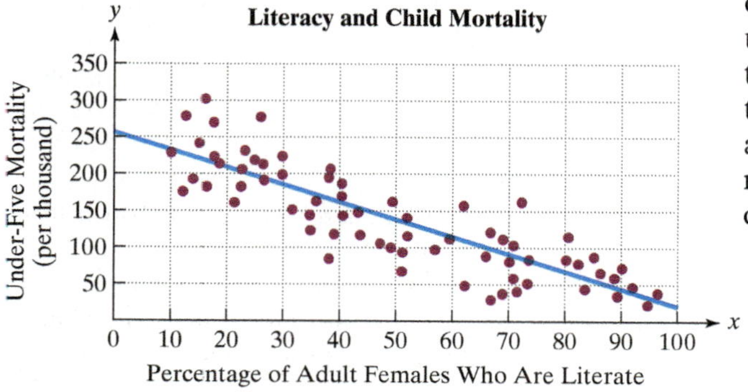

FIGURE 7.41
Source: United Nations

$$y = -2.3x + 255$$

For each percent increase in adult female literacy, under-five mortality decreases by 2.3 per thousand.

Using this model, we can make predictions about child mortality based on the percentage of literate adult females in a country.

In **Figure 7.41**, the data fall on or near a line. However, scatter plots are often curved in a way that indicates that the data do not fall near a line. In this section, we will use functions that are not linear functions to model such data and make predictions.

Modeling with Exponential Functions

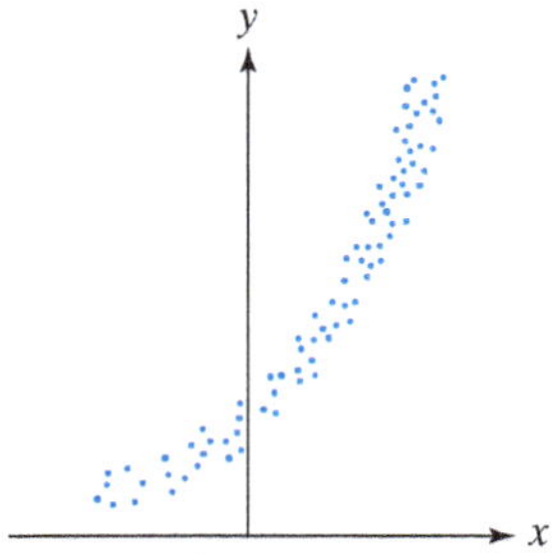

FIGURE 7.42

The scatter plot in **Figure 7.42** has a shape that indicates the data are increasing more and more rapidly. *Exponential functions* can be used to model this explosive growth, typically associated with populations, epidemics, and interest-bearing bank accounts.

DEFINITION OF THE EXPONENTIAL FUNCTION

The **exponential function with base *b*** is defined by

$$y = b^x \quad \text{or} \quad f(x) = b^x,$$

where b is a positive constant other than 1 ($b > 0$ and $b \neq 1$) and x is any real number.

Graph exponential functions.

EXAMPLE 1 *Graphing an Exponential Function*

Graph: $f(x) = 2^x$.

SOLUTION

We start by selecting numbers for x and finding the corresponding values for $f(x)$.

x	$f(x) = 2^x$	(x, y)
−3	$f(-3) = 2^{-3} = \frac{1}{8}$	$(-3, \frac{1}{8})$
−2	$f(-2) = 2^{-2} = \frac{1}{4}$	$(-2, \frac{1}{4})$
−1	$f(-1) = 2^{-1} = \frac{1}{2}$	$(-1, \frac{1}{2})$
0	$f(0) = 2^0 = 1$	$(0, 1)$
1	$f(1) = 2^1 = 2$	$(1, 2)$
2	$f(2) = 2^2 = 4$	$(2, 4)$
3	$f(3) = 2^3 = 8$	$(3, 8)$

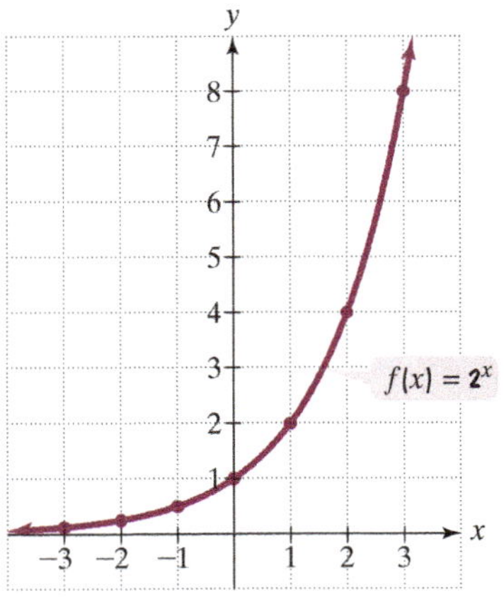

FIGURE 7.43 The graph of $f(x) = 2^x$

We plot these points, connecting them with a smooth curve. **Figure 7.43** shows the graph of $f(x) = 2^x$.

All exponential functions of the form $y = b^x$, or $f(x) = b^x$, where b is a number greater than 1, have the shape of the graph shown in **Figure 7.43**. The graph approaches, but never touches, the negative portion of the x-axis.

CHECK POINT 1 Graph: $f(x) = 3^x$.

Blitzer Bonus

Exponential Growth: The Year Humans Become Immortal

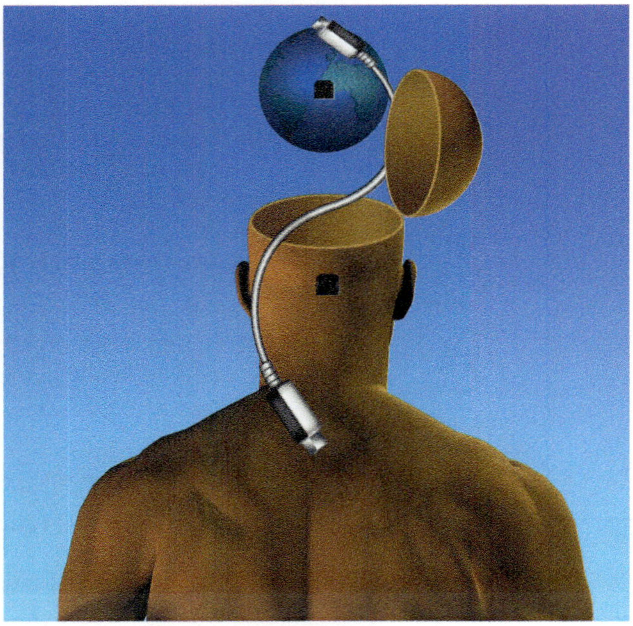

In 2011, *Jeopardy!* aired a three-night match between a personable computer named Watson and the show's two most successful players. The winner: Watson. In the time it took each human contestant to respond to one trivia question, Watson was able to scan the content of one million books. It was also trained to understand the puns and twists of phrases unique to *Jeopardy!* clues.

Watson's remarkable accomplishments can be thought of as a single data point on an exponential curve that models growth in computing power. According to inventor, author, and computer scientist Ray Kurzweil (1948–), computer technology is progressing exponentially, doubling in power each year. What does this mean in terms of the accelerating pace of the graph of $y = 2^x$ that starts slowly and then rockets skyward toward infinity? According to Kurzweil, by 2023, a supercomputer will surpass the brainpower of a human. As progress accelerates exponentially and every hour brings a century's worth of scientific breakthroughs, by 2045, computers will surpass the brainpower equivalent to that of all human brains combined. Here's where it gets exponentially weird: In that year (says Kurzweil), we will be able to scan our consciousness into computers and enter a virtual existence, or swap our bodies for immortal robots. Indefinite life extension will become a reality and people will die only if they choose to.

2 Use exponential models.

Figure 7.44(a) shows world population, in billions, for seven selected years from 1950 through 2010. A scatter plot of the data is shown in **Figure 7.44(b)**.

World Population, 1950–2010

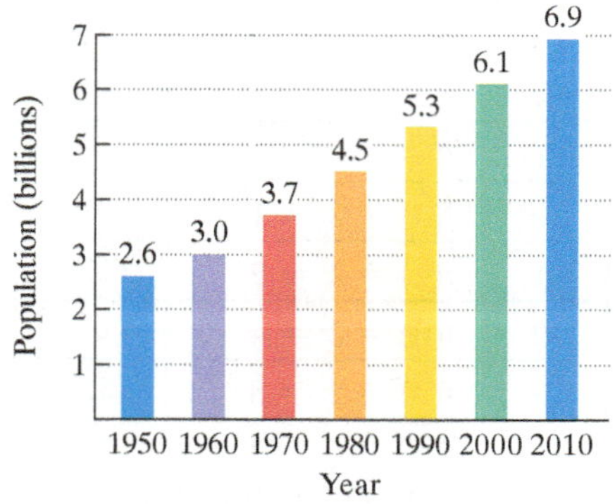

FIGURE 7.44(a)
Source: U.S. Census Bureau, International Database

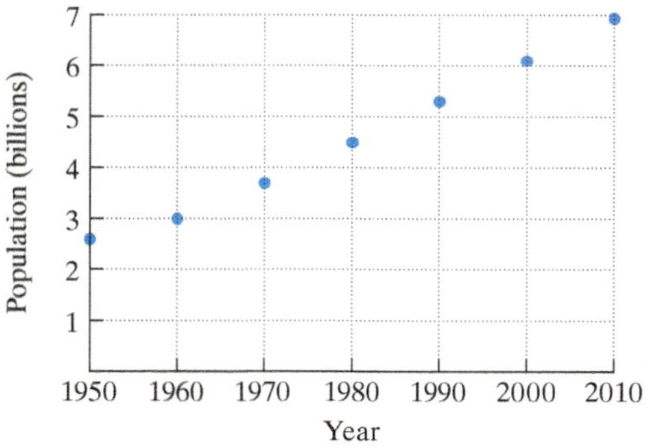

FIGURE 7.44(b)

Because the data in the scatter plot appear to increase more and more rapidly, the shape suggests that an exponential function might be a good choice for modeling the data. Furthermore, we can probably draw a line that passes through or near the seven points. Thus, a linear function would also be a good choice for a model.

EXAMPLE 2 *Comparing Linear and Exponential Models*

The data for world population are shown in **Table 7.3**. Using a graphing utility's linear regression feature and exponential regression feature, we enter the data and obtain the models shown in **Figure 7.45**.

Although the domain of $y = ab^x$ is the set of all real numbers, some graphing calculators only accept positive values for x. That's why we assigned x to represent the number of years after 1949.

TABLE 7.3

x, Number of Years after 1949	y, World Population (billions)
1 (1950)	2.6
11 (1960)	3.0
21 (1970)	3.7
31 (1980)	4.5
41 (1990)	5.3
51 (2000)	6.1
61 (2010)	6.9

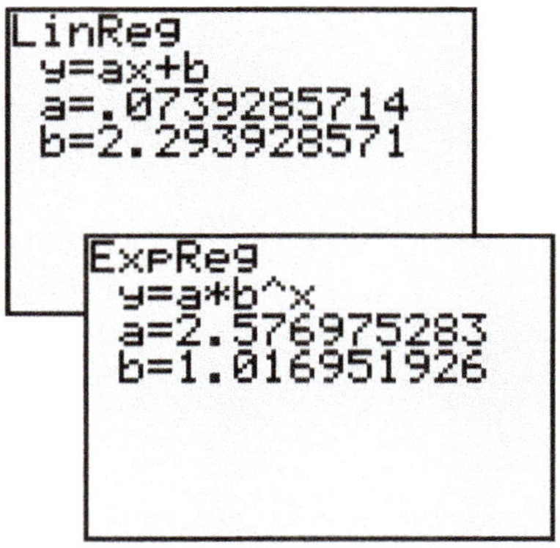

FIGURE 7.45 A linear model and an exponential model for the data in Table 7.3

a. Use **Figure 7.45** to express each model in function notation, with numbers rounded to three decimal places.

b. How well do the functions model world population in 2000?

c. By one projection, world population is expected to reach 8 billion in the year 2026. Which function serves as a better model for this prediction?

SOLUTION

a. Using **Figure 7.45** and rounding to three decimal places, the functions

$$f(x) = 0.074x + 2.294 \quad \text{and} \quad g(x) = 2.577(1.017)^x$$

model world population, in billions, x years after 1949. We named the linear function f and the exponential function g, although any letters can be used.

b. **Table 7.3** shows that world population in 2000 was 6.1 billion. The year 2000 is 51 years after 1949. Thus, we substitute 51 for x in each function's equation and then evaluate the resulting expressions with a calculator to see how well the functions describe world population in 2000.

$f(x) = 0.074x + 2.294$ This is the linear model.

$f(51) = 0.074(51) + 2.294$ Substitute 51 for x.

≈ 6.1 Use a calculator.

$g(x) = 2.577(1.017)^x$ This is the exponential model.

$g(51) = 2.577(1.017)^{51}$ Substitute 51 for x.

≈ 6.1 Use a calculator: 2.577 [×] 1.017 [y^x] (or [∧]) 51 [=].

Because 6.1 billion was the actual world population in 2000, both functions model world population in 2000 extremely well.

Blitzer Bonus

Global Population Increase

Exponential functions of the form $y = ab^x, b > 1$, model growth in which quantities increase at a rate proportional to their size. Populations that are growing exponentially grow extremely rapidly as they get larger because there are more adults to have offspring. Here's a way to put this idea into perspective:

By the time you finish reading Example 2 and working Check Point 2, more than 1000 people will have been added to our planet. By this time tomorrow, world population will have increased by more than 220,000.

c. Let's see which model comes closer to projecting a world population of 8 billion in the year 2026. Because 2026 is 77 years after 1949 $(2026 - 1949 = 77)$, we substitute 77 for x in each function's equation.

$f(x) = 0.074x + 2.294$ — This is the linear model.

$f(77) = 0.074(77) + 2.294$ — Substitute 77 for x.

≈ 8.0 — Use a calculator.

$g(x) = 2.577(1.017)^x$ — This is the exponential model.

$g(77) = 2.577(1.017)^{77}$ — Substitute 77 for x.

≈ 9.4 — Use a calculator: 2.577 [×] 1.017 [y^x] (or [∧]) 77 [=].

The linear function $f(x) = 0.074x + 2.294$ serves as a better model for a projected world population of 8 billion in 2026.

CHECK POINT 2 Use the models $f(x) = 0.074x + 2.294$ and $g(x) = 2.577(1.017)^x$ to solve this problem.

a. World population in 1970 was 3.7 billion. Which function serves as a better model for this year?

b. By one projection, world population is expected to reach 9.3 billion by 2050. Which function serves as a better model for this projection?

The Role of *e* in Applied Exponential Functions

An irrational number, symbolized by the letter e, appears as the base in many applied exponential functions. This irrational number is approximately equal to 2.72. More accurately,

$$e \approx 2.71828\ldots.$$

The number e is called the **natural base**. The function $f(x) = e^x$ is called the **natural exponential function.**

Use a scientific or graphing calculator with an [e^x] key to evaluate e to various powers. For example, to find e^2, press the following keys on most calculators:

Scientific calculator: 2 [e^x]

Graphing calculator: [e^x] 2 [ENTER].

The calculator display for e^2 should be approximately 7.389.

$$e^2 \approx 7.389$$

The number e lies between 2 and 3. Because $2^2 = 4$ and $3^2 = 9$, it makes sense that e^2, approximately 7.389, lies between 4 and 9.

Because $2 < e < 3$, the graph of $y = e^x$ lies between the graphs of $y = 2^x$ and $y = 3^x$, shown in **Figure 7.46**.

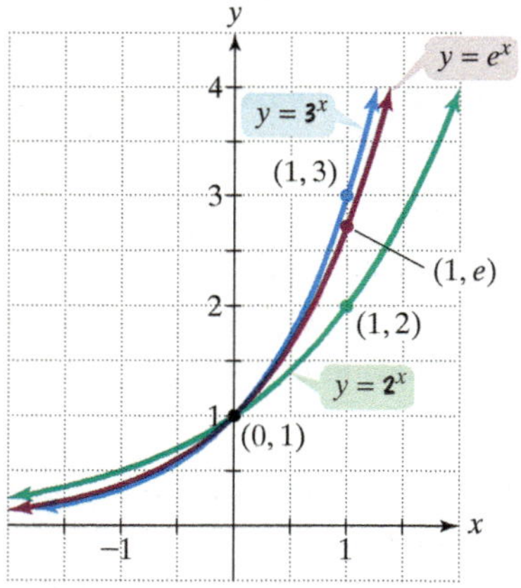

FIGURE 7.46 Graphs of three exponential functions

EXAMPLE 3 *Alcohol and Risk of a Car Accident*

Medical research indicates that the risk of having a car accident increases exponentially as the concentration of alcohol in the blood increases. The risk is modeled by

$$R = 6e^{12.77x},$$

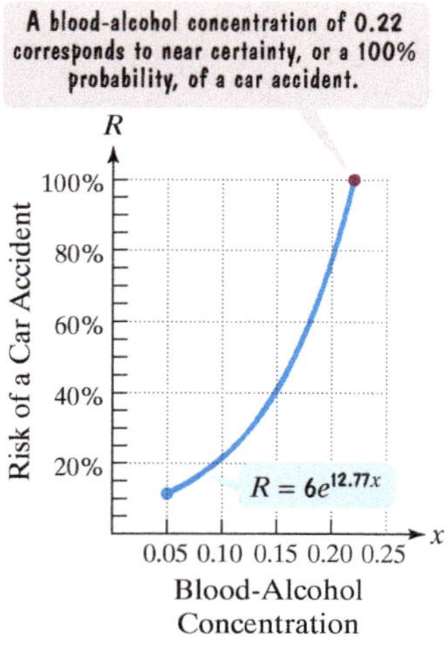

FIGURE 7.47

where x is the blood-alcohol concentration and R, given as a percent, is the risk of having a car accident. In every state, it is illegal to drive with a blood-alcohol concentration of 0.08 or greater. What is the risk of a car accident with a blood-alcohol concentration of 0.08? How is this shown on the graph of R in **Figure 7.47**?

SOLUTION

For a blood-alcohol concentration of 0.08, we substitute 0.08 for x in the exponential model's equation. Then we use a calculator to evaluate the resulting expression.

$$R = 6e^{12.77x} \quad \text{This is the given exponential model.}$$

$$R = 6e^{12.77(0.08)} \quad \text{Substitute 0.08 for } x.$$

Perform this computation on your calculator.

Scientific calculator: 6 [×] [(] 12.77 [×] .08 [)] [e^x] [=]

Graphing calculator: 6 [×] [e^x] [(] 12.77 [×] .08 [)] [ENTER]

The display should be approximately 16.665813. Rounding to one decimal place, the risk of a car accident is approximately 16.7% with a blood-alcohol concentration of 0.08. This can be visualized as the point (0.08, 16.7) on the graph of R in **Figure 7.47**. Take a moment to locate this point on the curve.

CHECK POINT 3 Use the model in Example 3 to solve this problem. In many states, it is illegal for drivers under 21 years old to drive with a blood-alcohol concentration of 0.01 or greater. What is the risk of a car accident with a blood-alcohol concentration of 0.01? Round to one decimal place.

Modeling with Logarithmic Functions

The scatter plot in **Figure 7.48** starts with rapid growth and then the growth begins to level off. This type of behavior can be modeled by *logarithmic functions*.

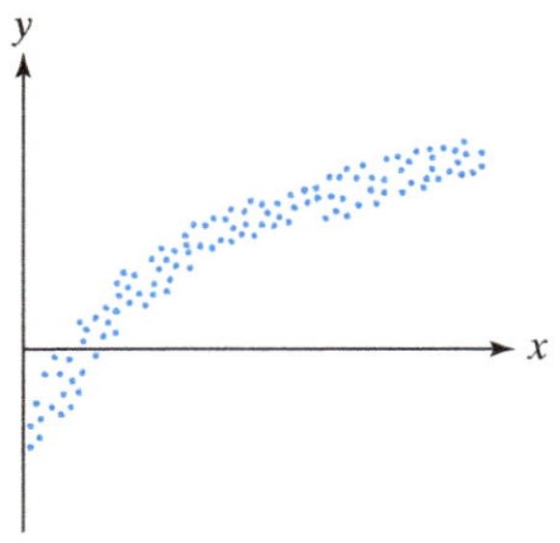

FIGURE 7.48

DEFINITION OF THE LOGARITHMIC FUNCTION

For $x > 0$ and $b > 0, b \neq 1$,

$$y = \log_b x \text{ is equivalent to } b^y = x.$$

The function $f(x) = \log_b x$ is the **logarithmic function with base *b*.**

The equations

$$y = \log_b x \quad \text{and} \quad b^y = x$$

are different ways of expressing the same thing. The first equation is in **logarithmic form**, and the second equivalent equation is in **exponential form**.

Notice that a **logarithm**, y, is an **exponent**. You should learn the location of the base and exponent in each form.

LOCATION OF BASE AND EXPONENT IN EXPONENTIAL AND LOGARITHMIC FORMS

Exponent — Exponent

Logarithmic Form: $y = \log_b x$ Exponential Form: $b^y = x$

Base — Base

3 Graph logarithmic functions.

EXAMPLE 4 *Graphing a Logarithmic Function*

Graph: $y = \log_2 x$.

SOLUTION

Because $y = \log_2 x$ means $2^y = x$, we will use the exponential form of the equation to obtain the function's graph. Using $2^y = x$, we start by selecting numbers for y and then we find the corresponding values for x.

Start with values for y.

$x = 2^y$	y	(x, y)
$2^{-2} = \frac{1}{4}$	-2	$(\frac{1}{4}, -2)$
$2^{-1} = \frac{1}{2}$	-1	$(\frac{1}{2}, -1)$
$2^0 = 1$	0	$(1, 0)$
$2^1 = 2$	1	$(2, 1)$
$2^2 = 4$	2	$(4, 2)$
$2^3 = 8$	3	$(8, 3)$

Compute x using $x = 2^y$.

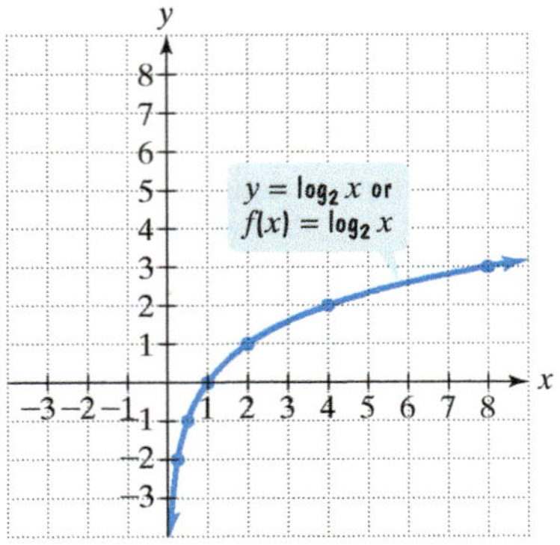

FIGURE 7.49 The graph of the logarithmic function with base 2

We plot the six ordered pairs in the table, connecting the points with a smooth curve. **Figure 7.49** shows the graph of $y = \log_2 x$.

> **GREAT QUESTION!**
>
> **I know that $y = \log_2 x$ means $2^y = x$. But what's the relationship between $y = \log_2 x$ and $y = 2^x$?**
>
> The coordinates for the logarithmic function with base 2 are the reverse of those for the exponential function with base 2. In general, $y = \log_b x$ reverses the x- and y-coordinates of $y = b^x$.

All logarithmic functions of the form $y = \log_b x$, or $f(x) = \log_b x$, where $b > 1$, have the shape of the graph shown in **Figure 7.49**. The graph approaches, but never touches, the negative portion of the y-axis. Observe that the graph is increasing from left to right. However, the rate of increase is slowing down as the graph moves to the right. This is why logarithmic functions are often used to model growing phenomena with growth that is leveling off.

CHECK POINT 4 Rewrite $y = \log_3 x$ in exponential form. Then use the exponential form of the equation to obtain the function's graph. Select integers from -2 to 2, inclusive, for y.

Scientific and graphing calculators contain keys that can be used to evaluate the logarithmic function with base 10 and the logarithmic function with base e.

Key	Function the Key Is Used to Evaluate
LOG	$y = \log_{10} x$ — This is called the **common logarithmic function**, usually expressed as $y = \log x$.
LN	$y = \log_e x$ — This is called the **natural logarithmic function**, usually expressed as $y = \ln x$.

4 Use logarithmic models.

EXAMPLE 5 *Dangerous Heat: Temperature in an Enclosed Vehicle*

When the outside air temperature is anywhere from 72° to 96° Fahrenheit, the temperature in an enclosed vehicle climbs by 43° in the first hour. The bar graph in **Figure 7.50(a)** shows the temperature increase throughout the hour. A scatter plot of the data is shown in **Figure 7.50(b)**.

Temperature Increase in an Enclosed Vehicle

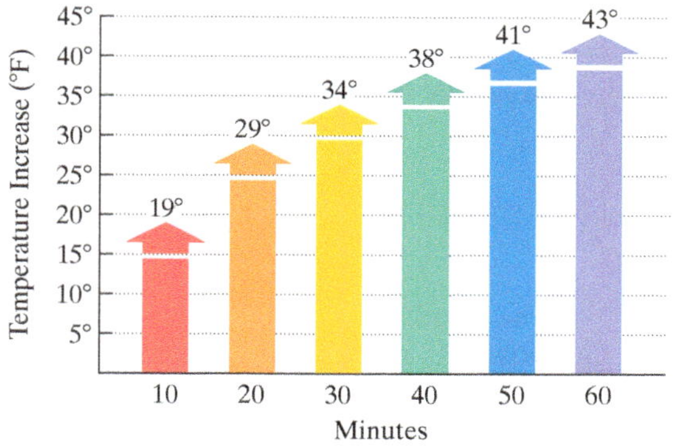

FIGURE 7.50(a)
Source: Professor Jan Null, San Francisco State University

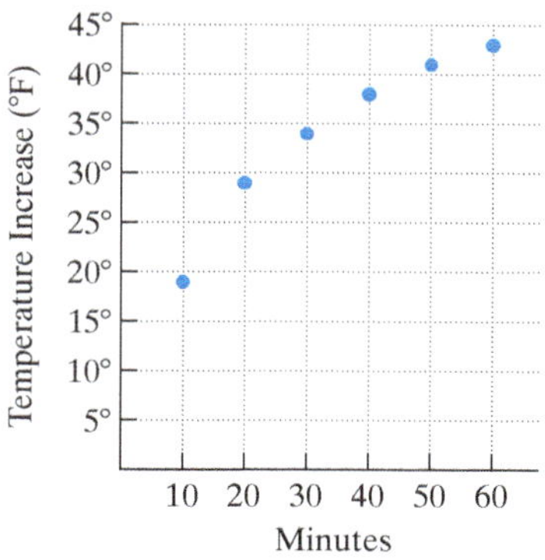

FIGURE 7.50(b)

FIGURE 7.51 Data (10, 19), (20, 29), (30, 34), (40, 38), (50, 41), (60, 43)

Because the data in the scatter plot increase rapidly at first and then begin to level off a bit, the shape suggests that a logarithmic function is a good choice for a model. After entering the data, a graphing calculator displays the logarithmic model, $y = a + b \ln x$, shown in **Figure 7.51**.

a. Express the model in function notation, with numbers rounded to one decimal place.

b. Use the function to find the temperature increase, to the nearest degree, after 50 minutes. How well does the function model the actual increase shown in **Figure 7.50(a)**?

SOLUTION

a. Using **Figure 7.51** and rounding to one decimal place, the function

$$f(x) = -11.6 + 13.4 \ln x$$

models the temperature increase, $f(x)$, in degrees Fahrenheit, after x minutes.

b. We find the temperature increase after 50 minutes by substituting 50 for x and evaluating the function at 50.

$f(x) = -11.6 + 13.4 \ln x$ This is the logarithmic model from part (a).

$f(50) = -11.6 + 13.4 \ln 50$ Substitute 50 for x.

Perform this computation on your calculator.

Scientific calculator: 11.6 [+/–] [+] 13.4 [×] 50 [LN] [=]

Graphing calculator: [(–)] 11.6 [+] 13.4 [LN] 50 [ENTER]

The display should be approximately 40.821108. Rounding to the nearest degree, the logarithmic model indicates that the temperature will have increased by approximately 41° after 50 minutes. Because the increase shown in **Figure 7.50(a)** is 41°, the function models the actual increase extremely well.

CHECK POINT 5 Use the model obtained in Example 5(a) to find the temperature increase, to the nearest degree, after 30 minutes. How well does the function model the actual increase shown in **Figure 7.50(a)**?

GREAT QUESTION!

How can I use a graphing calculator to see how well my models describe the data?

Once you have obtained one or more models for the data, you can use a graphing calculator's [TABLE] feature to numerically see how well each model describes the data. Enter the models as y_1, y_2, and so on. Create a table, scroll through the table, and compare the table values given by the models to the actual data.

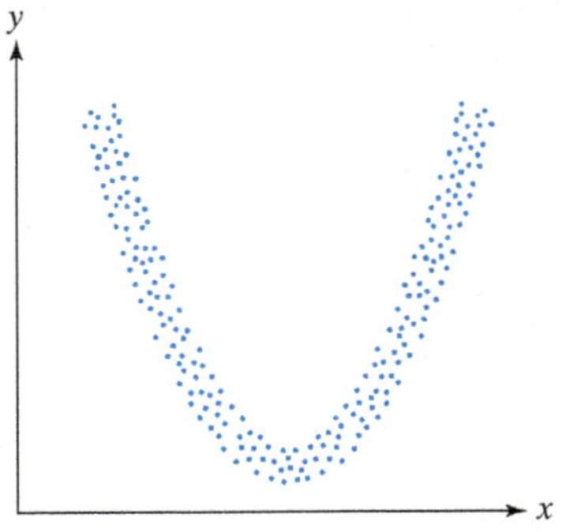

FIGURE 7.52

Modeling with Quadratic Functions

The scatter plot in **Figure 7.52** has a shape that indicates the data are first decreasing and then increasing. This type of behavior can be modeled by a *quadratic function*.

DEFINITION OF THE QUADRATIC FUNCTION

A **quadratic function** is any function of the form

$$y = ax^2 + bx + c \quad \text{or} \quad f(x) = ax^2 + bx + c,$$

where a, b, and c are real numbers, with $a \neq 0$.

The graph of any quadratic function is called a **parabola**. Parabolas are shaped like bowls or inverted bowls, as shown in **Figure 7.53**. If the coefficient of x^2 (the value of a in $ax^2 + bx + c$) is positive, the parabola opens upward. If the coefficient of x^2 is negative, the graph opens downward. The **vertex** (or turning point) of the parabola is the lowest point on the graph when it opens upward and the highest point on the graph when it opens downward.

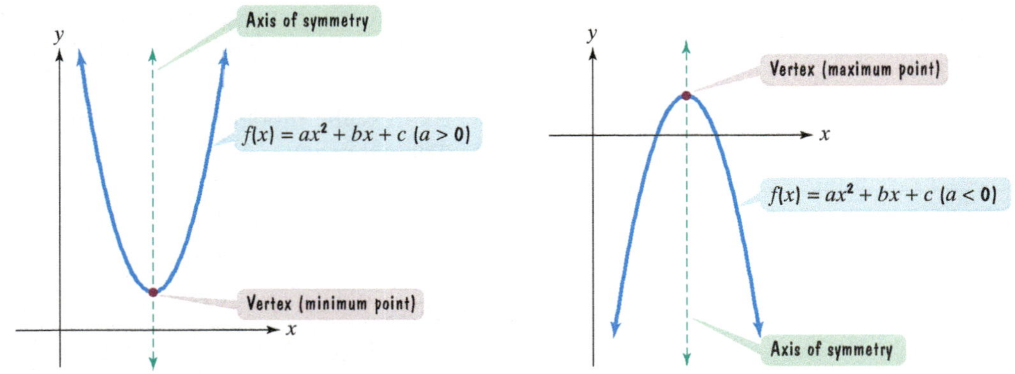

$a > 0$: Parabola opens upward. $a < 0$: Parabola opens downward.

FIGURE 7.53 Characteristics of graphs of quadratic functions

Look at the unusual image of the word *mirror* shown below. The artist, Scott Kim, has created the image so that the two halves of the whole are mirror images of each other. A parabola shares this kind of symmetry, in which a line through the vertex divides the figure in half. Parabolas are symmetric with respect to this line, called the **axis of symmetry**. If a parabola is folded along its axis of symmetry, the two halves match exactly.

© 2011 Scott Kim, scottkim.com

When graphing quadratic functions or using them as models, it is frequently helpful to determine where the vertex, or turning point, occurs.

THE VERTEX OF A PARABOLA

The vertex of a parabola whose equation is $y = ax^2 + bx + c$ occurs where

$$x = \frac{-b}{2a}.$$

5 Graph quadratic functions.

Several points are helpful when graphing a quadratic function. These points are the x-intercepts (although not every parabola has x-intercepts), the y-intercept, and the vertex.

GRAPHING QUADRATIC FUNCTIONS

The graph of $y = ax^2 + bx + c$ or $f(x) = ax^2 + bx + c$, called a parabola, can be graphed using the following steps:

1. Determine whether the parabola opens upward or downward. If $a > 0$, it opens upward. If $a < 0$, it opens downward.
2. Determine the vertex of the parabola. The x-coordinate is $\frac{-b}{2a}$.

   The y-coordinate is found by substituting the x-coordinate into the parabola's equation and evaluating.
3. Find any x-intercepts by replacing y or $f(x)$ with 0. Solve the resulting quadratic equation for x.
4. Find the y-intercept by replacing x with 0. Because $f(0) = c$ (the constant term in the function's equation), the y-intercept is c and the parabola passes through $(0, c)$.
5. Plot the intercepts and the vertex.
6. Connect these points with a smooth curve.

EXAMPLE 6 *Graphing a Parabola*

Graph the quadratic function: $y = x^2 - 2x - 3$.

SOLUTION

We can graph this function by following the steps in the box.

Step 1 Determine how the parabola opens. Note that a, the coefficient of x^2, is 1. Thus, $a > 0$; this positive value tells us that the parabola opens upward.

Step 2 Find the vertex. We know that the x-coordinate of the vertex is $\frac{-b}{2a}$. Let's identify the numbers a, b, and c in the given equation, which is in the form $y = ax^2 + bx + c$.

$$y = x^2 - 2x - 3$$

$a = 1$ $b = -2$ $c = -3$

Now we substitute the values of a and b into the expression for the x-coordinate:

$$x\text{-coordinate of vertex} = \frac{-b}{2a} = \frac{-(-2)}{2(1)} = \frac{2}{2} = 1.$$

The x-coordinate of the vertex is 1. We substitute 1 for x in the equation $y = x^2 - 2x - 3$ to find the y-coordinate:

$$y\text{-coordinate of vertex} = 1^2 - 2 \cdot 1 - 3 = 1 - 2 - 3 = -4.$$

The vertex is $(1, -4)$, shown in **Figure 7.54**.

Step 3 Find the x-intercepts. Replace y with 0 in $y = x^2 - 2x - 3$. We obtain $0 = x^2 - 2x - 3$ or $x^2 - 2x - 3 = 0$. We can solve this equation by factoring.

$$x^2 - 2x - 3 = 0$$
$$(x - 3)(x + 1) = 0$$
$$x - 3 = 0 \quad \text{or} \quad x + 1 = 0$$
$$x = 3 \qquad\qquad x = -1$$

The x-intercepts are 3 and -1. The parabola passes through $(3, 0)$ and $(-1, 0)$, shown in **Figure 7.54**.

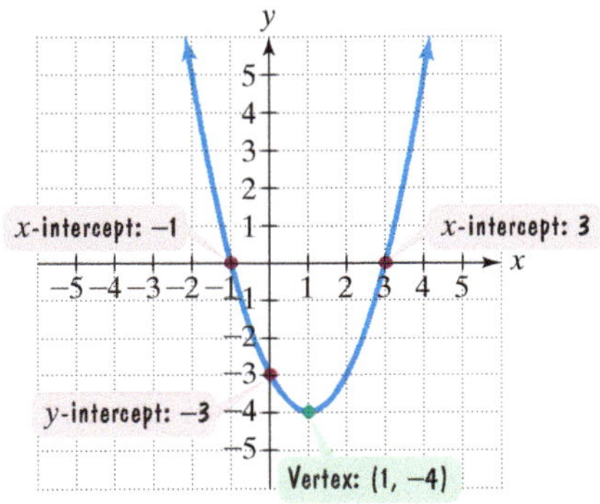

FIGURE 7.54 The graph of $y = x^2 - 2x - 3$

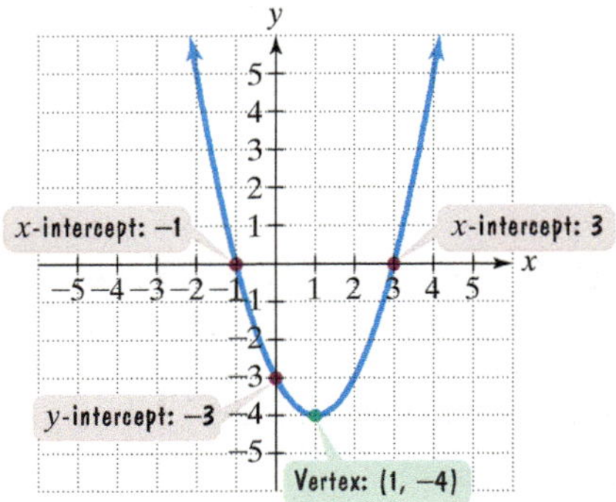

FIGURE 7.54 (repeated) The graph of $y = x^2 - 2x - 3$

Step 4 Find the y-intercept. Replace x with 0 in $y = x^2 - 2x - 3$:

$$y = 0^2 - 2 \cdot 0 - 3 = 0 - 0 - 3 = -3.$$

The y-intercept is -3. The parabola passes through $(0, -3)$, shown in **Figure 7.54**.

Steps 5 and 6 Plot the intercepts and the vertex. Connect these points with a smooth curve. The intercepts and the vertex are shown as the four labeled points in **Figure 7.54**. Also shown is the graph of the quadratic function, obtained by connecting the points with a smooth curve.

CHECK POINT 6 Graph the quadratic function: $y = x^2 + 6x + 5$.

6 Use quadratic models.

EXAMPLE 7 *Modeling the Parabolic Path of a Punted Football*

Figure 7.55 shows that when a football was kicked, the nearest defensive player was 6 feet from the point of impact with the kicker's foot. **Table 7.4** shows five measurements indicating the football's height at various horizontal distances from its point of impact. A scatter plot of the data is shown in **Figure 7.56**.

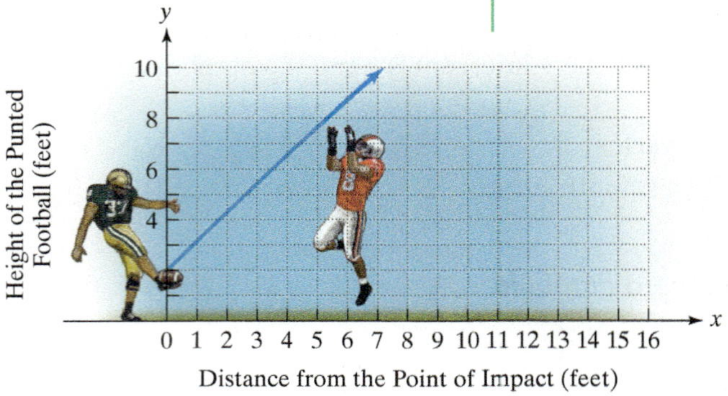

FIGURE 7.55

TABLE 7.4

x, Football's Horizontal Distance (feet)	y, Football's Height (feet)
0	2
30	28.4
60	36.8
90	27.2
110	10.8

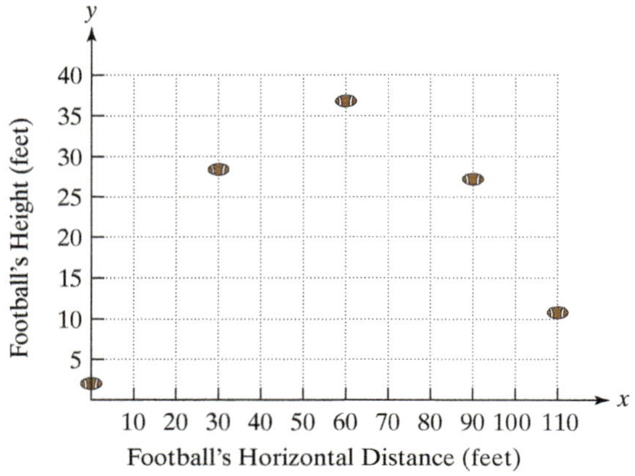

FIGURE 7.56 A scatter plot for the data in Table 7.4

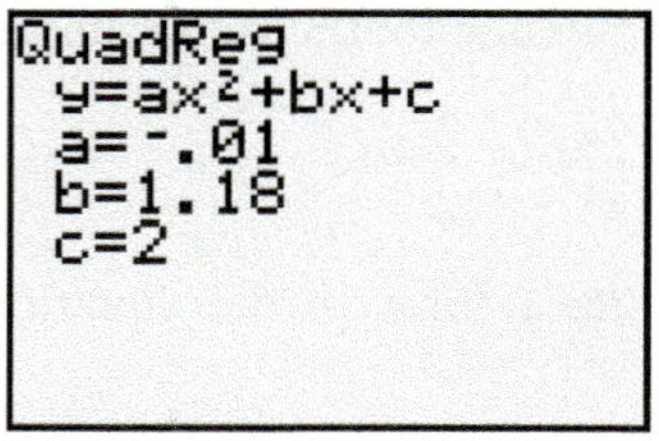

FIGURE 7.57 Data (0, 2), (30, 28.4), (60, 36.8), (90, 27.2), (110, 10.8)

Because the data in the scatter plot first increase and then decrease, the shape suggests that a quadratic function is a good choice for a model. Using the data in **Table 7.4**, a graphing calculator displays the quadratic function, $y = ax^2 + bx + c$, shown in **Figure 7.57**.

a. Express the model in function notation.

b. How far would the nearest defensive player, who was 6 feet from the kicker's point of impact, have needed to reach to block the punt?

SOLUTION

a. Using **Figure 7.57**, the function

$$f(x) = -0.01x^2 + 1.18x + 2$$

models the football's height, $f(x)$, in feet, in terms of its horizontal distance, x, in feet.

b. **Figure 7.55** shows that the defensive player was 6 feet from the point of impact. To block the punt, he needed to touch the football along its parabolic path. This means that we must find the height of the ball 6 feet from the kicker. Replace x with 6 in the function, $f(x) = -0.01x^2 + 1.18x + 2$.

$$f(6) = -0.01(6)^2 + 1.18(6) + 2 = -0.36 + 7.08 + 2 = 8.72$$

The defensive player would have needed to reach 8.72 feet above the ground to block the punt.

Assuming that the football was not blocked by the defensive player, the graph of the function that models the football's parabolic path is shown in **Figure 7.58**. The graph is shown only for $x \geq 0$, indicating horizontal distances that begin at the football's impact with the kicker's foot and end with the ball hitting the ground. Notice how the graph provides a visual story of the punted football's parabolic path.

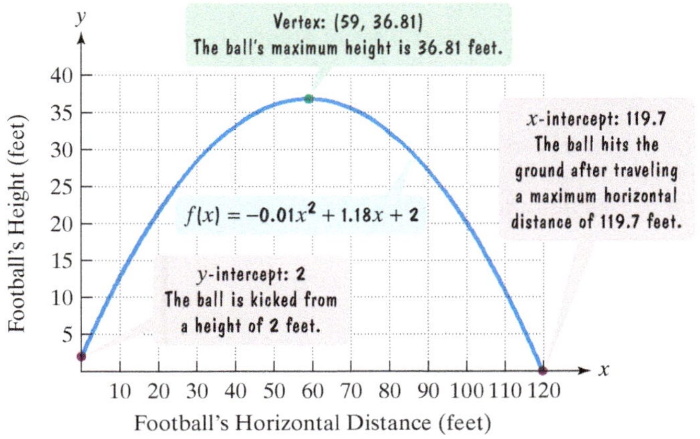

FIGURE 7.58 The parabolic path of a punted football

CHECK POINT 7 Use the model obtained in Example 7(a) to answer this question: If the defensive player had been 8 feet from the kicker's point of impact, how far would he have needed to reach to block the punt? Does this seem realistic? Identify the solution as a point on the graph in **Figure 7.58**.

---

7 Determine an appropriate function for modeling data.

Table 7.5 contains a description of the scatter plots we have encountered in this section, as well as the type of function that can serve as an appropriate model for each description.

TABLE 7.5 Modeling Data

Description of Data Points in a Scatter Plot	Model
Lie on or near a line	Linear Function: $y = mx + b$ or $f(x) = mx + b$
Increasing more and more rapidly	Exponential Function: $y = b^x$ or $f(x) = b^x, b > 1$
Increasing, although rate of increase is slowing down	Logarithmic Function: $y = \log_b x$ or $f(x) = \log_b x, b > 1$ ($y = \log_b x$ means $b^y = x$.)
Decreasing and then increasing	Quadratic Function: $y = ax^2 + bx + c$ or $f(x) = ax^2 + bx + c, a > 0$ The vertex, $\left(\frac{-b}{2a}, f\left(\frac{-b}{2a}\right)\right)$, is a minimum point on the parabola.
Increasing and then decreasing	Quadratic Function: $y = ax^2 + bx + c$ or $f(x) = ax^2 + bx + c, a < 0$ The vertex, $\left(\frac{-b}{2a}, f\left(\frac{-b}{2a}\right)\right)$, is a maximum point on the parabola.

Once the type of model has been determined, the data can be entered into a graphing calculator. The calculator's regression feature will display the specific function of the type requested that best fits the data. That, in short, is how your author obtained the algebraic models you have encountered throughout this book. In this era of technology, the process of determining models that approximate real-world situations is based on a knowledge of functions and their graphs, and has nothing to do with long and tedious computations.

Concept and Vocabulary Check

Fill in each blank so that the resulting statement is true.

1. Data presented in a visual form as a set of points are called a/an ________. The line that best fits the data points is called a/an ________ line.

For each set of points in Exercises 2–5, determine whether an exponential function, a logarithmic function, a linear function, or a quadratic function is the best choice for modeling the data.

2.

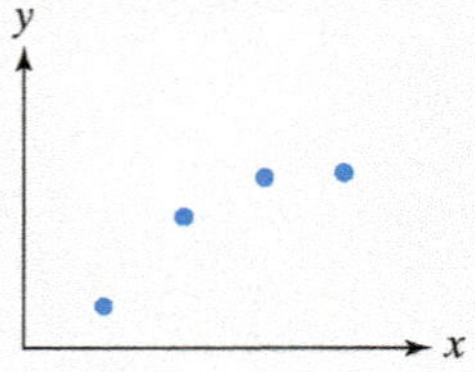

____

3.

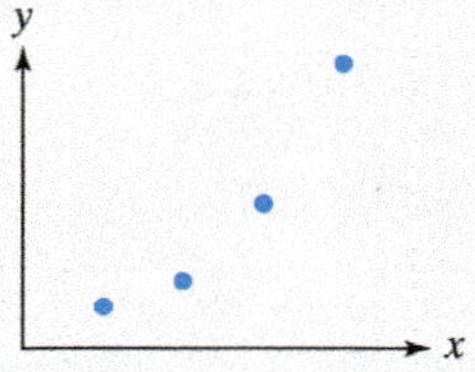

____

4.

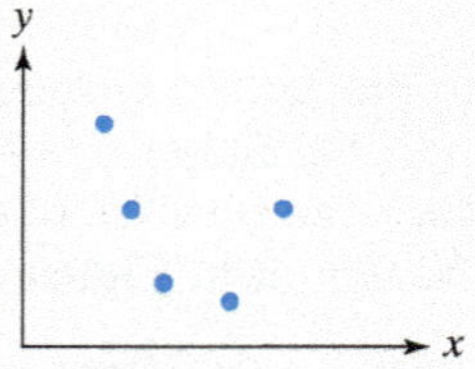

____

5.

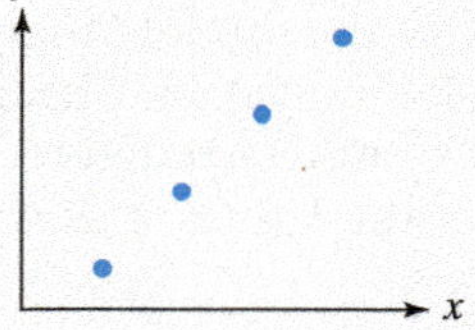

____

6. The exponential function with base b is defined by $y =$ ____, $b > 0$ and $b \neq 1$.

7. The irrational number e is approximately equal to ________. The function $y = e^x$ or $f(x) = e^x$ is called the ________ exponential function.

8. $y = \log_b x$ is equivalent to the exponential form ________, $x > 0, b > 0, b \neq 1$.

9. $y = \log_{10} x$ is usually expressed as $y =$ ______ and is called the ________ logarithmic function.

10. $y = \log_e x$ is usually expressed as $y =$______ and is called the ________ logarithmic function.

11. The function $y = ax^2 + bx + c$ or $f(x) = ax^2 + bx + c$, $a \neq 0$, is called a/an ________ function. The graph of this function is called a/an ________. The vertex, or turning point, occurs where $x =$ ____.

Exercise Set 7.6

Practice Exercises

In Exercises 1–6, use a table of coordinates to graph each exponential function. Begin by selecting −2, −1, 0, 1, and 2 for x.

1. $f(x) = 4^x$ **2.** $f(x) = 5^x$ **3.** $y = 2^{x+1}$

4. $y = 2^{x-1}$ **5.** $f(x) = 3^{x-1}$ **6.** $f(x) = 3^{x+1}$

In Exercises 7–8,

a. *Rewrite each equation in exponential form.*

b. *Use a table of coordinates and the exponential form from part (a) to graph each logarithmic function. Begin by selecting −2, −1, 0, 1, and 2 for y.*

7. $y = \log_4 x$ **8.** $y = \log_5 x$

In Exercises 9–14,

a. *Determine if the parabola whose equation is given opens upward or downward.*

b. *Find the vertex.*

c. *Find the x-intercepts.*

d. *Find the y-intercept.*

e. *Use (a)–(d) to graph the quadratic function.*

9. $y = x^2 + 8x + 7$ **10.** $y = x^2 + 10x + 9$

11. $f(x) = x^2 - 2x - 8$ **12.** $f(x) = x^2 + 4x - 5$

13. $y = -x^2 + 4x - 3$

14. $y = -x^2 + 2x + 3$

In Exercises 15–22,

a. *Create a scatter plot for the data in each table.*

b. *Use the shape of the scatter plot to determine if the data are best modeled by a linear function, an exponential function, a logarithmic function, or a quadratic function.*

15.

x	y
0	0
9	1
16	1.2
19	1.3
25	1.4

16.

x	y
0	0.3
8	1
15	1.2
18	1.3
24	1.4

17.

x	y
0	−3
1	2
2	7
3	12
4	17

18.

x	y
0	5
1	3
2	1
3	−1
4	−3

19.

x	y
0	4
1	1
2	0
3	1
4	4

20.

x	y
0	−4
1	−1
2	0
3	−1
4	−4

21.

x	y
0	−3
1	−2
2	0
3	4
4	12

22.

x	y
0	4
1	5
2	7
3	11
4	19

Practice Plus

In Exercises 23–24, use a table of coordinates to graph each exponential function. Begin by selecting −2, −1, 0, 1, and 2 for x. Based on your graph, describe the shape of a scatter plot that can be modeled by $f(x) = b^x, 0 < b < 1$.

23. $f(x) = (\frac{1}{2})^x$ (Equivalently, $y = 2^{-x}$)

24. $f(x) = (\frac{1}{3})^x$ (Equivalently, $y = 3^{-x}$)

In Exercises 25–26, use the directions for Exercises 7–8 to graph each logarithmic function. Based on your graph, describe the shape of a scatter plot that can be modeled by $f(x) = \log_b x, 0 < b < 1$.

25. $y = \log_{\frac{1}{2}} x$

26. $y = \log_{\frac{1}{3}} x$

In Exercises 27–28, use the directions for Exercises 9–14 to graph each quadratic function. Use the quadratic formula to find x-intercepts, rounded to the nearest tenth.

27. $f(x) = -2x^2 + 4x + 5$

28. $f(x) = -3x^2 + 6x - 2$

In Exercises 29–30, find the vertex for the parabola whose equation is given by writing the equation in the form $y = ax^2 + bx + c$.

29. $y = (x - 3)^2 + 2$

30. $y = (x - 4)^2 + 3$

Application Exercises

In 1900, Americans age 65 and over made up only 4.1% of the population. By 2010, that figure was 13%, or approximately 44.6 million people. Demographic projections indicate that by the year 2050, approximately 82 million Americans, or 20% of the U.S. population, will be at least 65. The bar graph shows the number of people in the United States age 65 and over, with projected figures for the year 2020 and beyond.

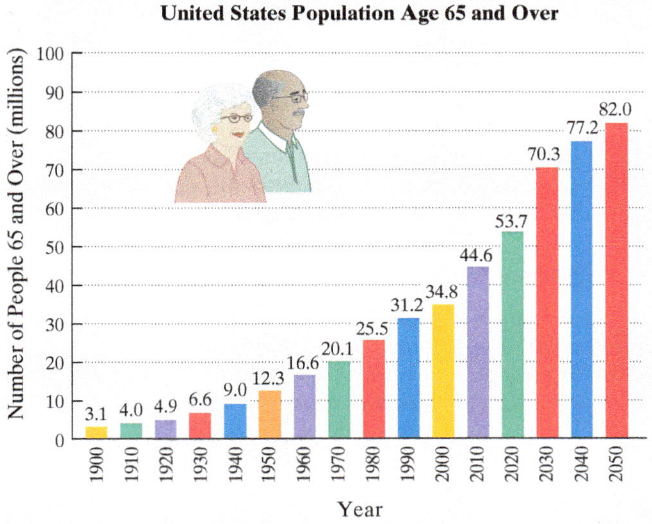

Source: U.S. Census Bureau

The graphing calculator screen displays an exponential function that models the U.S. population age 65 and over, y, in millions, x years after 1899. Use this information to solve Exercises 31–32.

31. a. Explain why an exponential function was used to model the population data.

b. Use the graphing calculator screen to express the model in function notation, with numbers rounded to three decimal places.

c. According to the model in part (b), how many Americans age 65 and over were there in 2010? Use a calculator with a y^x key or a ∧ key, and round to one decimal place. Does this rounded number overestimate or underestimate the 2010 population displayed by the bar graph? By how much?

d. According to the model in part (b), how many Americans age 65 and over will there be in 2020? Round to one decimal place. Does this rounded number overestimate or underestimate the 2020 population projection displayed by the bar graph? By how much?

32. Refer to the graph showing the U.S. population age 65 and over and the graphing calculator screen on the previous page.

a. Explain why an exponential function was used to model the population data.

b. Use the graphing calculator screen to express the model in function notation, with numbers rounded to three decimal places.

c. According to the model in part (b), how many Americans age 65 and over were there in 2000? Use a calculator with a $\boxed{y^x}$ key or a $\boxed{\wedge}$ key, and round to one decimal place. Does this rounded number overestimate or underestimate the 2000 population displayed by the bar graph? By how much?

d. According to the model in part (b), how many Americans age 65 and over will there be in 2030? Round to one decimal place. Does this rounded number overestimate or underestimate the 2030 population projection displayed by the bar graph? By how much?

Use a calculator with an $\boxed{e^x}$ *key to solve Exercises 33–34.*

Average annual premiums for employer-sponsored family health insurance policies more than doubled over 11 years. The bar graph shows the average cost of a family health insurance plan in the United States for six selected years from 2000 through 2011.

Average Cost of a Family Health Insurance Plan

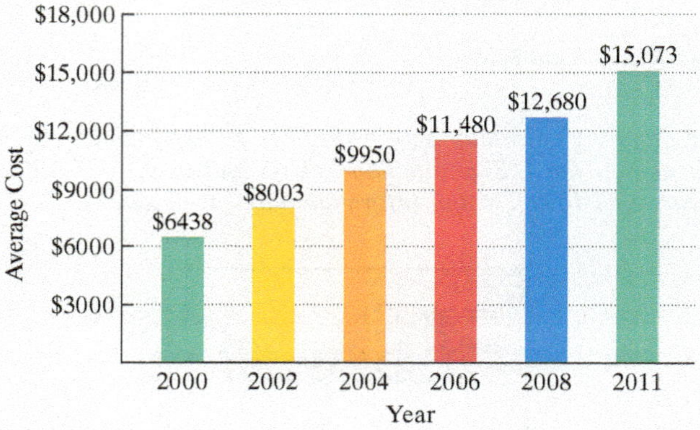

Source: Kaiser Family Foundation

The data can be modeled by

$$f(x) = 782x + 6564 \quad \text{and} \quad g(x) = 6875e^{0.077x},$$

in which f(x) and g(x) represent the average cost of a family health insurance plan x years after 2000. Use these functions to solve Exercises 33–34. Where necessary, round answers to the nearest whole dollar.

33. a. According to the linear model, what was the average cost of a family health insurance plan in 2011?

b. According to the exponential model, what was the average cost of a family health insurance plan in 2011?

c. Which function is a better model for the data in 2011?

34. a. According to the linear model, what was the average cost of a family health insurance plan in 2008?

b. According to the exponential model, what was the average cost of a family health insurance plan in 2008?

c. Which function is a better model for the data in 2008?

The data in the following table indicate that between the ages of 1 and 11, the human brain does not grow linearly, or steadily. A scatter plot for the data is shown to the right of the table.

GROWTH OF THE HUMAN BRAIN

Age	Percentage of Adult Size Brain
1	30%
2	50%
4	78%
6	88%
8	92%
10	95%
11	99%

Source: Gerrig and Zimbardo, *Psychology and Life*, 18th Edition, Allyn and Bacon, 2008.

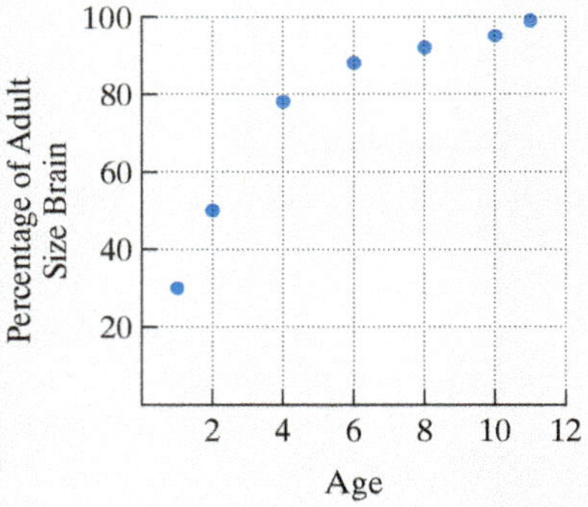

The graphing calculator screen displays the percentage of an adult size brain, y, for a child at age x, where $1 \le x \le 11$. *Use this information to solve Exercises 35–36.*

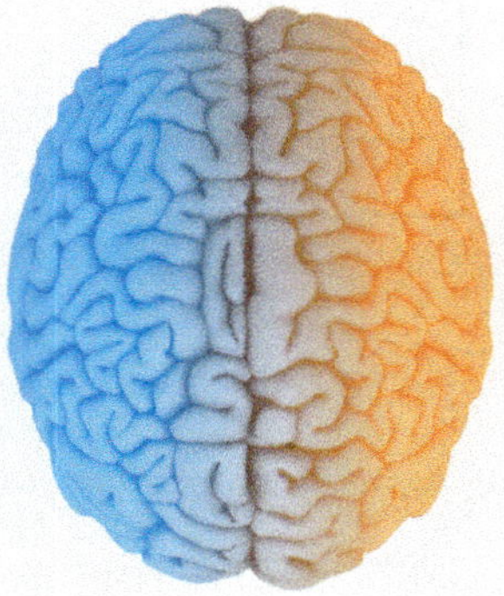

35. a. Explain why a logarithmic function was used to model the data.

b. Use the graphing calculator screen to express the model in function notation, with numbers rounded to the nearest whole number.

c. According to the model in part (b), what percentage of an adult size brain does a child have at age 10? Use a calculator with an $\boxed{\text{LN}}$ key and round to the nearest whole percent. Does this overestimate or underestimate the percent displayed by the table? By how much?

36. a. Explain why a logarithmic function was used to model the data.

b. Use the graphing calculator screen to express the model in function notation, with numbers rounded to the nearest whole number.

c. According to the model in part (b), what percentage of an adult size brain does a child have at age 8? Use a calculator with an $\boxed{\text{LN}}$ key and round to the nearest whole percent. How does this compare with the percent displayed by the table?

The percentage of adult height attained by a girl who is x years old can be modeled by

$$f(x) = 62 + 35\log(x - 4),$$

where x represents the girl's age (from 5 to 15) and f(x) represents the percentage of her adult height. Use the function to solve Exercises 37–38.

37. **a.** According to the model, what percentage of her adult height has a girl attained at age 13? Use a calculator with a LOG key and round to the nearest tenth of a percent.

b. Why was a logarithmic function used to model the percentage of adult height attained by a girl from ages 5 to 15, inclusive?

38. **a.** According to the model, what percentage of her adult height has a girl attained at age ten? Use a calculator with a LOG key and round to the nearest tenth of a percent.

b. Why was a logarithmic function used to model the percentage of adult height attained by a girl from ages 5 to 15, inclusive?

39. A ball is thrown upward and outward from a height of 6 feet. The table shows four measurements indicating the ball's height at various horizontal distances from where it was thrown. The graphing calculator screen displays a quadratic function that models the ball's height, y, in feet, in terms of its horizontal distance, x, in feet.

x, Ball's Horizontal Distance (feet)	y, Ball's Height (feet)
0	6
1	7.6
3	6
4	2.8

```
QuadReg
 y=ax²+bx+c
 a=-.8
 b=2.4
 c=6
```

a. Explain why a quadratic function was used to model the data. Why is the value of a negative?

b. Use the graphing calculator screen to express the model in function notation.

c. Use the model from part (b) to determine the x-coordinate of the quadratic function's vertex. Then complete this statement: The maximum height of the ball occurs ________ feet from where it was thrown and the maximum height is ________ feet.

40. A ball is thrown upward and outward from a height of 6 feet. The table shows four measurements indicating the ball's height at various horizontal distances from where it was thrown. The graphing calculator screen displays a quadratic function that models the ball's height, y, in feet, in terms of its horizontal distance, x, in feet.

x, Ball's Horizontal Distance (feet)	y, Ball's Height (feet)
0	6
0.5	7.4
1.5	9
4	6

```
QuadReg
 y=ax²+bx+c
 a=-.8
 b=3.2
 c=6
```

a. Explain why a quadratic function was used to model the data. Why is the value of a negative?

b. Use the graphing calculator screen to express the model in function notation.

c. Use the model from part (b) to determine the x-coordinate of the quadratic function's vertex. Then complete this statement: The maximum height of the ball occurs ________ feet from where it was thrown, and the maximum height is ________ feet.

Writing in Mathematics

41. What is a scatter plot?

42. What is an exponential function?

43. Describe the shape of a scatter plot that suggests modeling the data with an exponential function.

44. Describe the shape of a scatter plot that suggests modeling the data with a logarithmic function.

45. Would you prefer that your salary be modeled exponentially or logarithmically? Explain your answer.

46. Describe the shape of a scatter plot that suggests modeling the data with a quadratic function.

Critical Thinking Exercises

Make Sense? *In Exercises 47–50, determine whether each statement makes sense or does not make sense, and explain your reasoning.*

47. I'm looking at data that show the number of new college programs in green studies, and a linear function appears to be a better choice than an exponential function for modeling the number of new college programs from 2005 through 2009.

Number of New Programs in Green Studies at U.S. Colleges

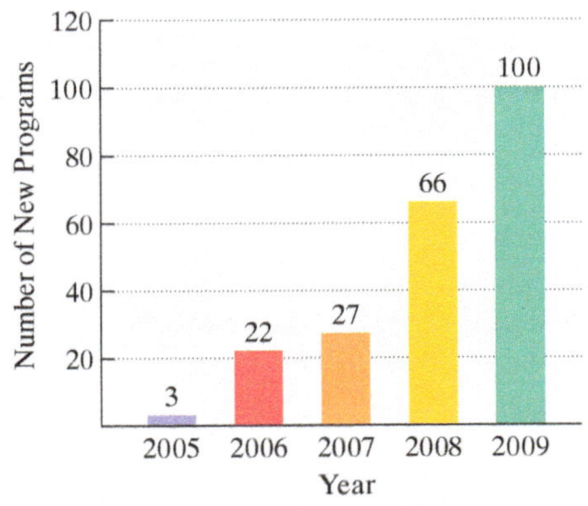

Source: Association for the Advancement of Sustainability in Higher Education

48. I used two different functions to model the data in a scatter plot.

49. Drinking and driving is extremely dangerous because the risk of a car accident increases logarithmically as the concentration of alcohol in the blood increases.

50. This work by artist Scott Kim (1955–) has the same kind of symmetry as the graph of a quadratic function.

In Exercises 51–53, the value of a in $y = ax^2 + bx + c$ and the vertex of the parabola are given. How many x-intercepts does the parabola have? Explain how you arrived at this number.

51. $a = -2$; vertex at $(4, 8)$

52. $a = 1$; vertex at $(2, 0)$

53. $a = 3$; vertex at $(3, 1)$

Technology Exercises

54. Use a graphing calculator to graph the exponential functions that you graphed by hand in Exercises 1–6. Describe similarities and differences between the graphs obtained by hand and those that appear in the calculator's viewing window.

55. Use a graphing calculator to graph the quadratic functions that you graphed by hand in Exercises 9–14.

Group Exercise

56. Each group member should consult an almanac, newspaper, magazine, or the Internet to find data that can be modeled by linear, exponential, logarithmic, or quadratic functions. Group members should select the two sets of data that are most interesting and relevant. Then consult a person who is familiar with graphing calculators to show you how to obtain a function that best fits each set of data. Once you have these functions, each group member should make one prediction based on one of the models, and then discuss a consequence of this prediction. What factors might change the accuracy of the prediction?

Chapter Summary, Review, and Test

SUMMARY – DEFINITIONS AND CONCEPTS	EXAMPLES
7.1 Graphing and Functions	
a. The rectangular coordinate system is formed using two number lines that intersect at right angles at their zero points. See Figure 7.1 on page 408. The horizontal line is the x-axis and the vertical line is the y-axis. Their point of intersection, $(0, 0)$, is the origin. Each point in the system corresponds to an ordered pair of real numbers, (x, y).	Ex. 1, p. 409
b. The graph of an equation in two variables is the set of all points whose coordinates satisfy the equation.	Ex. 2, p. 410; Ex. 3, p. 410
c. If an equation in x and y yields one value of y for each value of x, then y is a function of x, indicated by writing $f(x)$ for y.	Ex. 4, p. 412; Ex. 5, p. 413
d. The graph of a function is the graph of its ordered pairs.	Ex. 6, p. 414
e. The Vertical Line Test: If any vertical line intersects a graph in more than one point, the graph does not define y as a function of x.	Ex. 7, p. 415
7.2 Linear Functions and Their Graphs	
a. A function whose graph is a straight line is a linear function.	
b. The graph of $Ax + By = C$, a linear equation in two variables, is a straight line. The line can be graphed using intercepts and a checkpoint. To locate the x-intercept, set $y = 0$ and solve for x. To locate the y-intercept, set $x = 0$ and solve for y.	Ex. 1, p. 421
c. The slope of the line through (x_1, y_1) and (x_2, y_2) is $m = \frac{\text{Rise}}{\text{Run}} = \frac{y_2 - y_1}{x_2 - x_1} = \frac{y_1 - y_2}{x_1 - x_2}$.	Ex. 2, p. 422; Ex. 7, p. 428
d. The equation $y = mx + b$ is the slope-intercept form of the equation of a line, in which m is the slope and b is the y-intercept.	Ex. 3, p. 425; Ex. 4, p. 426; Ex. 8, p. 429
e. Horizontal and Vertical Lines **1.** The graph of $y = b$ is a horizontal line that intersects the y-axis at $(0, b)$. **2.** The graph of $x = a$ is a vertical line that intersects the x-axis at $(a, 0)$.	Ex. 5, p. 427; Ex. 6, p. 427

7.3 Systems of Linear Equations in Two Variables

a. Two equations in the form $Ax + By = C$ are called a system of linear equations. A solution of the system is an ordered pair that satisfies both equations in the system. — Ex. 1, p. 433

b. Linear systems with one solution can be solved by graphing. The coordinates of the point of intersection of the lines are the system's solution. — Ex. 2, p. 434

c. Systems of linear equations in two variables can be solved by eliminating a variable, using the substitution method (see the box on page 435) or the addition method (see the box on page 438). — Ex. 3, p. 435; Ex. 4, p. 436; Ex. 5, p. 438; Ex. 6, p. 439

d. When solving by substitution or addition, if the variable is eliminated and a false statement results, the linear system has no solution. If the variable is eliminated and a true statement results, the system has infinitely many solutions. — Ex. 7, p. 440; Ex. 8, p. 441

e. Functions of Business. A company produces and sells x units of a product. — Ex. 9, p. 442

Revenue Function: $R(x) = (\text{price per unit sold})x$

Cost Function: $C(x) = \text{fixed cost} + (\text{cost per unit produced})x$

Profit Function: $P(x) = R(x) - C(x)$

The point of intersection of the graphs of R and C is the break-even point. The x-coordinate of the point reveals the number of units that a company must produce and sell so that the money coming in, the revenue, is equal to the money going out, the cost. The y-coordinate gives the amount of money coming in and going out.

7.4 Linear Inequalities in Two Variables

a. A linear inequality in two variables can be written in the form $Ax + By > C$, $Ax + By \geq C$, $Ax + By < C$, or $Ax + By \leq C$.

b. The procedure for graphing a linear inequality in two variables is given in the box on page 448. — Ex. 1, p. 448; Ex. 2, p. 449

c. Some inequalities can be graphed without using test points, including $y > mx + b$ (the half-plane above $y = mx + b$), $y < mx + b$, $x > a$ (the half-plane to the right of $x = a$), $x < a$, $y > b$ (the half-plane above $y = b$), and $y < b$. — Ex. 3, p. 450

d. Graphing Systems of Linear Inequalities — Ex. 5, p. 452; Ex. 6, p. 453
1. Graph each inequality in the system in the same rectangular coordinate system.
2. Find the intersection of the individual graphs.

7.5 Linear Programming

a. An objective function is an algebraic expression in three variables describing a quantity that must be maximized or minimized. — Ex. 1, p. 456

b. Constraints are restrictions, expressed as linear inequalities. — Ex. 2, p. 457; Ex. 3, p. 458

c. Steps for solving a linear programming problem are given in the box on page 458. — Ex. 4, p. 459

7.6 Modeling Data: Exponential, Logarithmic, and Quadratic Functions

a. The exponential function with base b is defined by $y = b^x$ or $f(x) = b^x$, $b > 0$ and $b \neq 1$. — Ex. 1, p. 463

b. All exponential functions of the form $y = b^x$, where $b > 1$, have the shape of the graph in Figure 7.43 on page 463, making this function a good model for data points in a scatter plot that are increasing more and more rapidly. — Ex. 2, p. 465

c. The irrational number e, $e \approx 2.72$, appears in many applied exponential functions. The function $f(x) = e^x$ is called the natural exponential function. — Ex. 3, p. 466

d. The logarithmic function with base b is defined by $y = \log_b x$ or $f(x) = \log_b x$, $x > 0$, $b > 0$, and $b \neq 1$. $y = \log_b x$ means $b^y = x$, so a logarithm is an exponent. — Ex. 4, p. 468

e. $y = \log x$ means $y = \log_{10} x$, the common logarithmic function. $y = \ln x$ means $y = \log_e x$, the natural logarithmic function. Calculators contain LOG and LN keys for evaluating these functions.

f.	All logarithmic functions of the form $y = \log_b x$, where $b > 1$, have the shape of the graph in Figure 7.49 on page 468, making this function a good model for data points in a scatter plot that are increasing, but whose rate of increase is slowing down.	Ex. 5, p. 468
g.	A quadratic function is any function of the form $y = ax^2 + bx + c$ or $f(x) = ax^2 + bx + c, a \neq 0$. The graph of a quadratic function is called a parabola.	Figure 7.53, p. 470
h.	The x-coordinate of a parabola's vertex is $x = \frac{-b}{2a}$. The y-coordinate is found by substituting the x-coordinate into the parabola's equation. The vertex, or turning point, is the low point when the graph opens upward and the high point when the graph opens downward.	
i.	The six steps for graphing a parabola are given in the box on page 471.	Ex. 6, p. 471
j.	All quadratic functions have the shape of one of the graphs in Figure 7.53 on page 470, making this function a good model for data points in a scatter plot that are decreasing and then increasing, or vice versa. See Table 7.5 on page 473 .	Ex. 7, p. 472

Review Exercises

7.1

In Exercises 1–4, plot the given point in a rectangular coordinate system.

1. $(2, 5)$ **2.** $(-4, 3)$

3. $(-5, -3)$ **4.** $(2, -5)$

Graph each equation in Exercises 5–7. Let $x = -3, -2, -1, 0, 1, 2$, and 3.

5. $y = 2x - 2$ **6.** $y = |x| + 2$ **7.** $y = x$

8. If $f(x) = 4x + 11$, find $f(-2)$.

9. If $f(x) = -7x + 5$, find $f(-3)$.

10. If $f(x) = 3x^2 - 5x + 2$, find $f(4)$.

11. If $f(x) = -3x^2 + 6x + 8$, find $f(-4)$.

In Exercises 12–13, evaluate $f(x)$ for the given values of x. Then use the ordered pairs $(x, f(x))$ from your table to graph the function.

12. $f(x) = \frac{1}{2}|x|$

x	$f(x) = \frac{1}{2}\lvert x\rvert$
−6	
−4	
−2	
0	
2	
4	
6	

13. $f(x) = x^2 - 2$

x	$f(x) = x^2 - 2$
−2	
−1	
0	
1	
2	

In Exercises 14–15, use the vertical line test to identify graphs in which y is a function of x.

14.

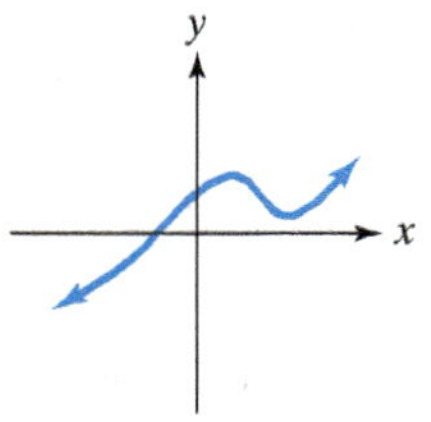

15.

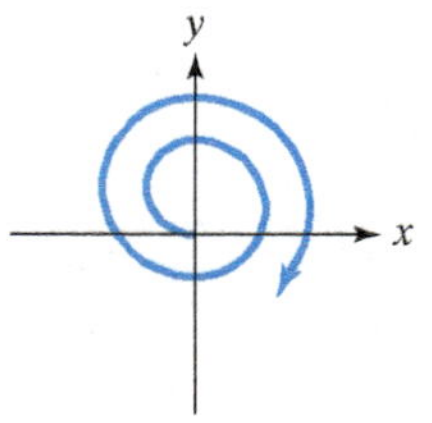

16. Whether on the slopes or at the shore, people are exposed to harmful amounts of the sun's skin-damaging ultraviolet (UV) rays. The quadratic function

$$D(x) = 0.8x^2 - 17x + 109$$

models the average time in which skin damage begins for burn-prone people, $D(x)$, in minutes, where x is the UV index, or measure of the sun's UV intensity. The graph of D is shown for a UV index from 1 (low) to 11 (high).

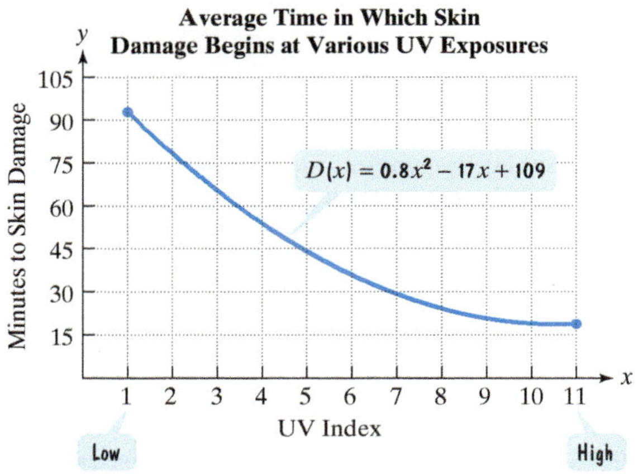

Source: National Oceanic and Atmospheric Administration

a. Find and interpret $D(1)$. How is this shown on the graph of D?

b. Find and interpret $D(10)$. How is this shown on the graph of D?

7.2

In Exercises 17–19, use the x- and y-intercepts to graph each linear equation.

17. $2x + y = 4$

18. $2x - 3y = 6$

19. $5x - 3y = 15$

In Exercises 20–23, calculate the slope of the line passing through the given points. If the slope is undefined, so state. Then indicate whether the line rises, falls, is horizontal, or is vertical.

20. $(3, 2)$ and $(5, 1)$

21. $(-1, 2)$ and $(-3, -4)$

22. $(-3, 4)$ and $(6, 4)$

23. $(5, 3)$ and $(5, -3)$

In Exercises 24–27, graph each linear function using the slope and y-intercept.

24. $y = 2x - 4$ **25.** $y = -\frac{2}{3}x + 5$

26. $f(x) = \frac{3}{4}x - 2$ **27.** $y = \frac{1}{2}x + 0$

In Exercises 28–30, **a.** *Write the equation in slope-intercept form;* **b.** *Identify the slope and the y-intercept;* **c.** *Use the slope and y-intercept to graph the line.*

28. $2x + y = 0$ **29.** $3y = 5x$ **30.** $3x + 2y = 4$

In Exercises 31–33, graph each horizontal or vertical line.

31. $x = 3$ **32.** $y = -4$ **33.** $x + 2 = 0$

34. Shown, again, is the scatter plot that indicates a relationship between the percentage of adult females in a country who are literate and the mortality of children under five. Also shown is a line that passes through or near the points.

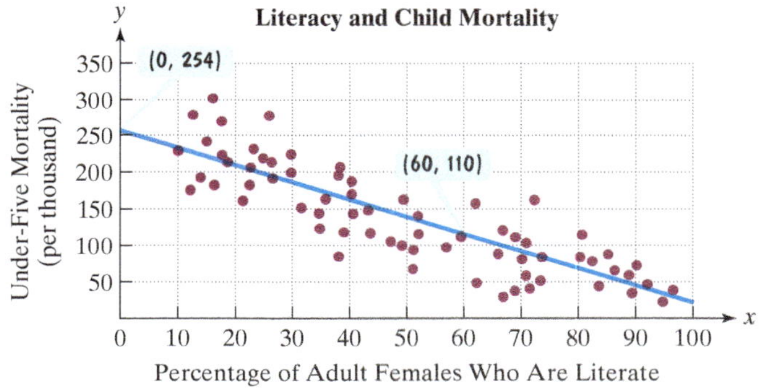

Source: United Nations

a. According to the graph, what is the y-intercept of the line? Describe what this represents in this situation.

b. Use the coordinates of the two points shown to compute the slope of the line. Describe what this means in terms of the rate of change.

c. Use the y-intercept from part (a) and the slope from part (b) to write a linear function that models child mortality, $f(x)$, per thousand, for children under five in a country where x% of adult women are literate.

d. Use the function from part (c) to predict the mortality rate of children under five in a country where 50% of adult females are literate.

7.3

In Exercises 35–37, solve each system by graphing. Check the coordinates of the intersection point in both equations.

35. $\begin{cases} x + y = 5 \\ 3x - y = 3 \end{cases}$ **36.** $\begin{cases} 2x - y = -1 \\ x + y = -5 \end{cases}$ **37.** $\begin{cases} y = -x + 5 \\ 2x - y = 4 \end{cases}$

In Exercises 38–40, solve each system by the substitution method.

38. $\begin{cases} 2x + 3y = 2 \\ x = 3y + 10 \end{cases}$ **39.** $\begin{cases} y = 4x + 1 \\ 3x + 2y = 13 \end{cases}$ **40.** $\begin{cases} x + 4y = 14 \\ 2x - y = 1 \end{cases}$

In Exercises 41–43, solve each system by the addition method.

41. $\begin{cases} x + 2y = -3 \\ x - y = -12 \end{cases}$ **42.** $\begin{cases} 2x - y = 2 \\ x + 2y = 11 \end{cases}$ **43.** $\begin{cases} 5x + 3y = 1 \\ 3x + 4y = -6 \end{cases}$

In Exercises 44–46, solve by the method of your choice. Identify systems with no solution and systems with infinitely many solutions, using set notation to express their solution sets.

44. $\begin{cases} y = -x + 4 \\ 3x + 3y = -6 \end{cases}$ **45.** $\begin{cases} 3x + y = 8 \\ 2x - 5y = 11 \end{cases}$

46. $\begin{cases} 3x - 2y = 6 \\ 6x - 4y = 12 \end{cases}$

47. A company is planning to manufacture computer desks. The fixed cost will be \$60,000 and it will cost \$200 to produce each desk. Each desk will be sold for \$450.

a. Write the cost function, C, of producing x desks.

b. Write the revenue function, R, from the sale of x desks.

c. Determine the break-even point. Describe what this means.

48. The graph shows the number of guns in private hands in the United States and the country's population, both expressed in millions, from 1995 through 2020, with projections from 2012 onward.

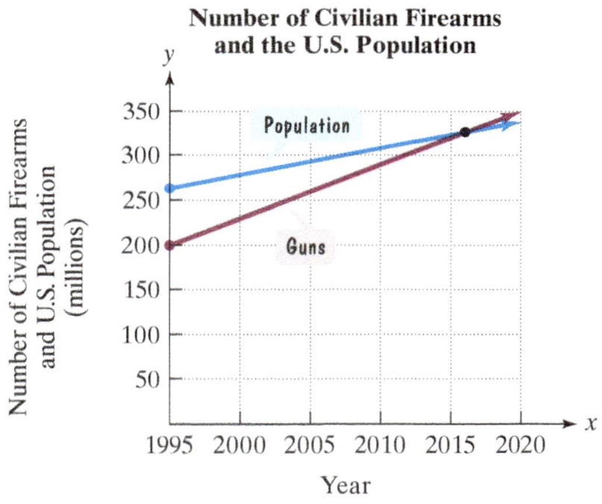

Source: *Mother Jones*, November/December 2012

a. Use the graphs to estimate the point of intersection. In what year will there be a gun for every man, woman, and child in the United States? What will be the population and the number of firearms in that year?

b. In 1995, there were an estimated 200 million firearms in private hands. This has increased at an average rate of 6 million firearms per year. Write a function that models the number of civilian firearms in the United States, y, in millions, x years after 1995.

c. The function $y - 3x = 263$ models the U.S. population, y, in millions, x years after 1995. Use this model and the model you obtained in part (b) to determine the year in which there will be a gun for every U.S. citizen. According to the models, what will be the population and the number of firearms in that year?

d. How well do the models in parts (b) and (c) describe the point of intersection of the graphs that you estimated in part (a)?

7.4

In Exercises 49–55, graph each linear inequality.

49. $x - 3y \le 6$ **50.** $2x + 3y \ge 12$ **51.** $2x - 7y > 14$

52. $y > \frac{3}{5}x$ **53.** $y \le -\frac{1}{2}x + 2$ **54.** $x \le 2$

55. $y > -3$

In Exercises 56–61, graph the solution set of each system of linear inequalities.

56. $\begin{cases} 3x - y \le 6 \\ x + y \ge 2 \end{cases}$ **57.** $\begin{cases} x + y < 4 \\ x - y < 4 \end{cases}$ **58.** $\begin{cases} x \le 3 \\ y > -2 \end{cases}$

59. $\begin{cases} 4x + 6y \le 24 \\ y > 2 \end{cases}$ **60.** $\begin{cases} x + y \le 6 \\ y \ge 2x - 3 \end{cases}$ **61.** $\begin{cases} y < -x + 4 \\ y > x - 4 \end{cases}$

7.5

62. Find the value of the objective function $z = 2x + 3y$ at each corner of the graphed region shown. What is the maximum value of the objective function? What is the minimum value of the objective function?

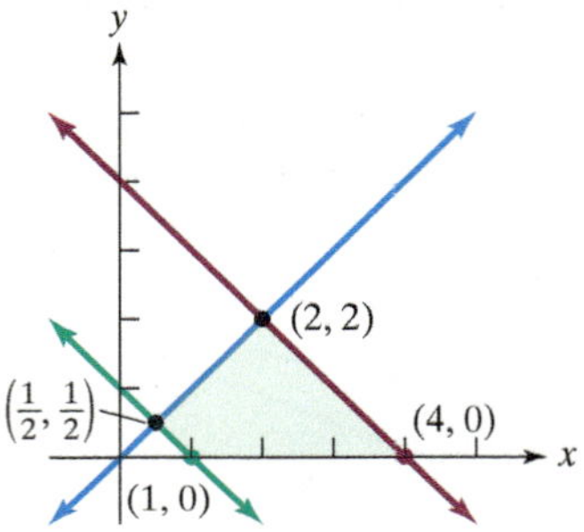

63. Consider the objective function $z = 2x + 3y$ and the following constraints:

$$x \le 6, y \le 5, x + y \ge 2, \underbrace{x \ge 0, y \ge 0}_{\text{Quadrant I and its boundary}}.$$

a. Graph the system of inequalities representing the constraints.

b. Find the value of the objective function at each corner of the graphed region.

c. Use the values in part (b) to determine the maximum and minimum values of the objective function and the values of x and y for which they occur.

64. A paper manufacturing company converts wood pulp to writing paper and newsprint. The profit on a unit of writing paper is \$500 and the profit on a unit of newsprint is \$350.

a. Let x represent the number of units of writing paper produced daily. Let y represent the number of units of newsprint produced daily. Write the objective function that models total daily profit.

b. The manufacturer is bound by the following constraints:
- Equipment in the factory allows for making at most 200 units of paper (writing paper and newsprint) in a day.
- Regular customers require at least 10 units of writing paper and at least 80 units of newsprint daily.

Write a system of inequalities that models these constraints.

c. Graph the inequalities in part (b). Use only the first quadrant and its boundary, because x and y must both be nonnegative. (*Suggestion:* Let each unit along the x- and y-axes represent 20.)

d. Evaluate the objective profit function at each of the three vertices of the graphed region.

e. Complete the missing portions of this statement: The company will make the greatest profit by producing _____ units of writing paper and _____ units of newsprint each day. The maximum daily profit is \$______.

7.6

In Exercises 65–66, use a table of coordinates to graph each exponential function. Begin by selecting −2, −1, 0, 1, and 2 for x.

65. $f(x) = 2^x$ **66.** $y = 2^{x+1}$

67. Graph $y = \log_2 x$ by rewriting the equation in exponential form. Use a table of coordinates and select −2, −1, 0, 1, and 2 for y.

In Exercises 68–69,

a. *Determine if the parabola whose equation is given opens upward or downward.*

b. *Find the vertex.*

c. *Find any x-intercepts.*

d. *Find the y-intercept.*

e. *Use (a)–(d) to graph the quadratic function.*

68. $y = x^2 - 6x - 7$ **69.** $f(x) = -x^2 - 2x + 3$

In Exercises 70–72,

a. *Create a scatter plot for the data in each table.*

b. *Use the shape of the scatter plot to determine if the data are best modeled by a linear function, an exponential function, a logarithmic function, or a quadratic function.*

70. AGE OF U.S. DRIVERS AND FATAL CRASHES

Age	Fatal Crashes per 100 Million Miles Driven
20	6.2
25	4.1
35	2.8
45	2.4
55	3.0
65	3.8
75	8.0

Source: Insurance Institute for Highway Safety

71. NUMBER OF PAGES IN THE IRS 1040 TAX INSTRUCTION BOOKLET

Year	Number of Pages
1945	4
1955	16
1965	17
1975	39
1985	52
1995	84
2005	142
2010	179

Source: National Taxpayers Union

72. INTENSITY AND LOUDNESS LEVEL OF VARIOUS SOUNDS

Intensity (watts per meter^2)	Loudness Level (decibels)
0.1 (loud thunder)	110
1 (rock concert, 2 yd from speakers)	120
10 (jackhammer)	130
100 (jet takeoff, 40 yd away)	140

73. Just browsing? Take your time. Researchers know, to the dollar, the average amount the typical consumer spends at the shopping mall. And the longer you stay, the more you spend.

Mall Browsing Time and Average Amount Spent

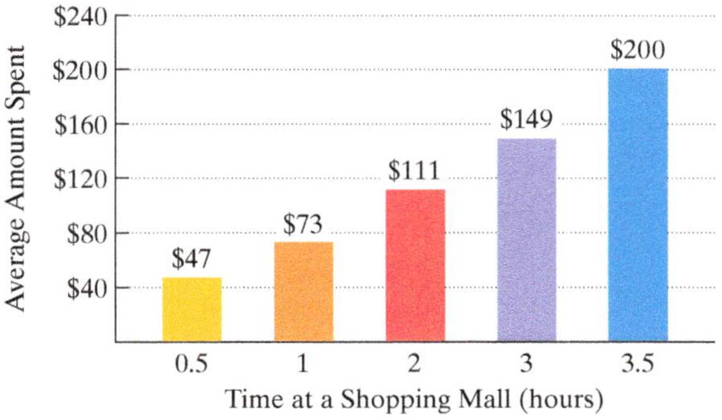

Source: International Council of Shopping Centers Research

The data in the bar graph can be modeled by the functions

$$f(x) = 47x + 22 \quad \text{and} \quad g(x) = 42.2(1.56)^x,$$

where $f(x)$ and $g(x)$ model the average amount spent, in dollars, at a shopping mall after x hours.

a. What is the slope of the linear model? What does this mean in terms of the average amount spent at a shopping mall?

b. Which function, the linear or the exponential, is a better model for the average amount spent after 3.5 hours of browsing?

74. The bar graph shows that people with lower incomes are more likely to report that their health is fair or poor.

Americans Reporting Fair or Poor Health, by Annual Income

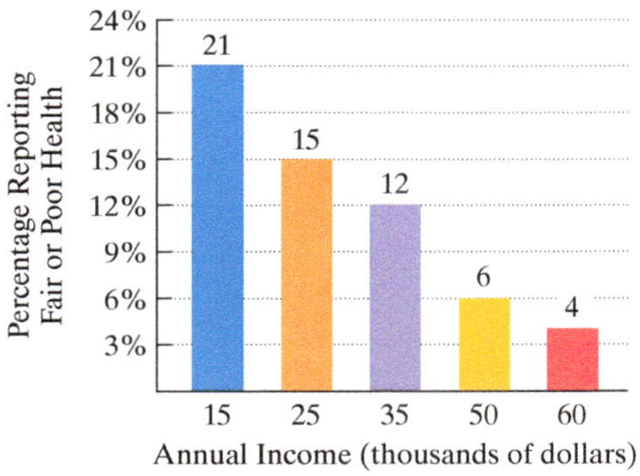

Source: William Kornblum and Joseph Julian, *Social Problems*, 12th Edition, Prentice Hall, 2007.

The data can be modeled by $f(x) = -0.4x + 25.4$ and $g(x) = 54.8 - 12.3 \ln x$, where $f(x)$ and $g(x)$ model the percentage of Americans reporting fair or poor health in terms of annual income, x, in thousands of dollars. Which function, the linear or the logarithmic, is a better model for an annual income of $60 thousand?

Chapter 7 Test

1. Graph $y = |x| - 2$. Let $x = -3, -2, -1, 0, 1, 2,$ and 3.

2. If $f(x) = 3x^2 - 7x - 5$, find $f(-2)$.

In Exercises 3–4, use the vertical line test to identify graphs in which y is a function of x.

3.

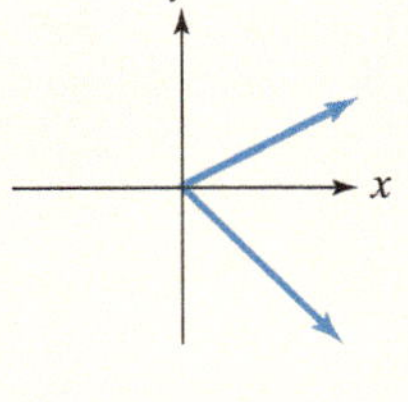

4.

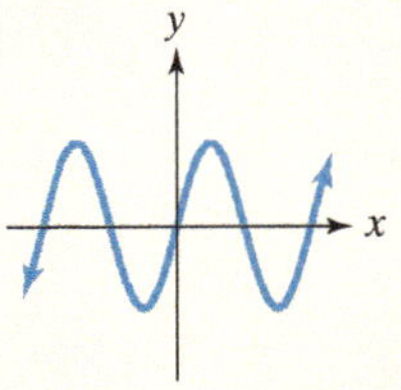

5. The graph shows the height, in meters, of an eagle in terms of its time, in seconds, in flight.

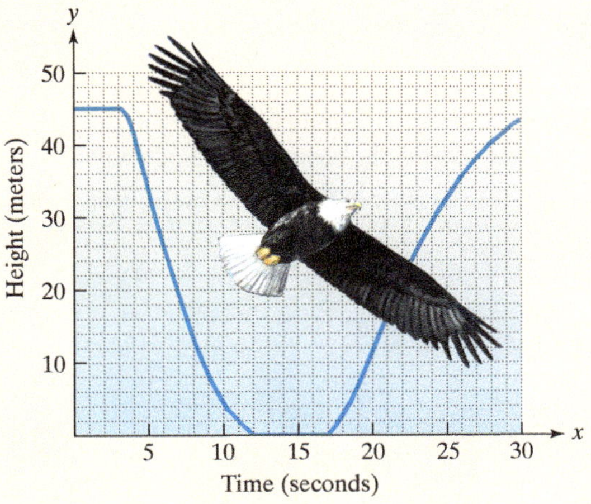

a. Is the eagle's height a function of time? Use the graph to explain why or why not.

b. Find $f(15)$. Describe what this means in practical terms.

c. What is a reasonable estimate of the eagle's maximum height?

d. During which period of time was the eagle descending?

6. Use the x- and y-intercepts to graph $4x - 2y = -8$.
7. Find the slope of the line passing through $(-3, 4)$ and $(-5, -2)$.

In Exercises 8–9, graph each linear function using the slope and y-intercept.

8. $y = \frac{2}{3}x - 1$
9. $f(x) = -2x + 3$
10. In a 2010 survey of more than 200,000 freshmen at 279 colleges, only 52% rated their emotional health high or above average, a drop from 64% in 1985.

Percentage of U.S. College Freshmen Rating Their Emotional Health High or Above Average

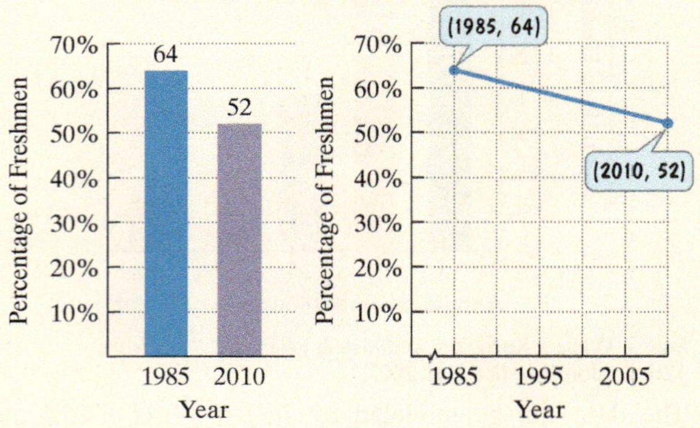

Source: UCLA Higher Education Research Institute

a. Find the slope of the line passing through the two points shown by the voice balloons.
b. Use your answer from part (a) to complete this statement:
For each year from 1985 through 2010, the percentage of U.S. college freshmen rating their emotional health high or above average decreased by ________. The rate of change was ________ per ________.

11. Studies show that texting while driving is as risky as driving with a 0.08 blood alcohol level, the standard for drunk driving. The bar graph shows the number of fatalities in the United States involving distracted driving from 2005 through 2008. Although the distracted category involves such activities as talking on cellphones, conversing with passengers, and eating, experts at the National Highway Traffic Safety Administration claim that texting while driving is the clearest menace because it requires looking away from the road. Shown to the right of the bar graph is a scatter plot with a line passing through two of the data points.

Number of Highway Fatalities in the United States Involving Distracted Driving

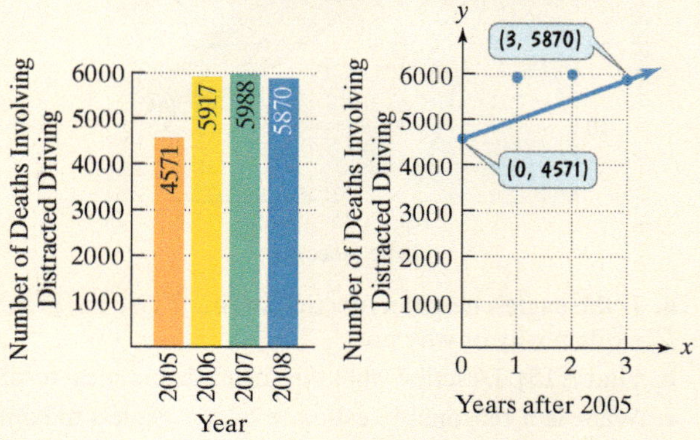

Source: National Highway Traffic Safety Administration

a. According to the scatter plot shown on the right, what is the y-intercept? Describe what this represents in this situation.
b. Use the coordinates of the two points shown in the scatter plot to compute the slope. What does this represent in terms of the rate of change in the number of highway fatalities involving distracted driving?
c. Use the y-intercept shown in the scatter plot and the slope from part (b) to write a linear function that models the number of highway fatalities involving distracted driving, $f(x)$, in the United States x years after 2005.
d. In 2010, surveys showed overwhelming public support to ban texting while driving, although at that time only 19 states and Washington, D.C., outlawed the practice. Without additional laws that penalize texting drivers, use the linear function you obtained from part (c) to project the number of fatalities in the United States in 2015 involving distracted driving.

12. Solve by graphing:
$$\begin{cases} x + y = 6 \\ 4x - y = 4. \end{cases}$$
13. Solve by substitution:
$$\begin{cases} x = y + 4 \\ 3x + 7y = -18. \end{cases}$$
14. Solve by addition:
$$\begin{cases} 5x + 4y = 10 \\ 3x + 5y = -7. \end{cases}$$
15. A company is planning to produce and sell a new line of computers. The fixed cost will be \$360,000 and it will cost \$850 to produce each computer. Each computer will be sold for \$1150.
a. Write the cost function, C, of producing x computers.
b. Write the revenue function, R, from the sale of x computers.
c. Determine the break-even point. Describe what this means.

Graph each linear inequality in Exercises 16–18.

16. $3x - 2y < 6$
17. $y \le \frac{1}{2}x - 1$
18. $y > -1$
19. Graph the system of linear inequalities:
$$\begin{cases} 2x - y \le 4 \\ 2x - y > -1. \end{cases}$$
20. Find the value of the objective function $z = 3x + 2y$ at each corner of the graphed region shown. What is the maximum value of the objective function? What is the minimum value of the objective function?

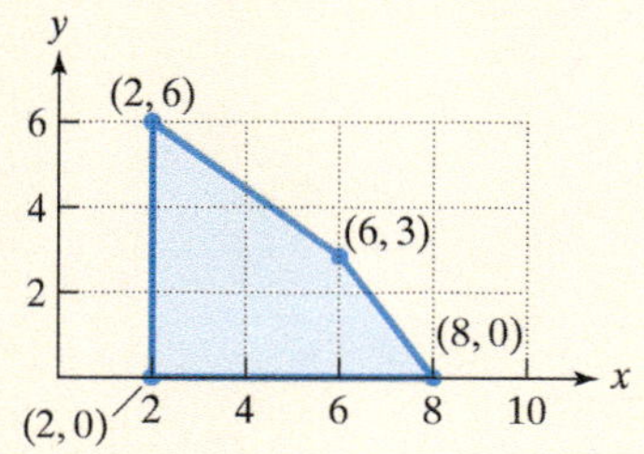

21. Find the maximum value of the objective function $z = 3x + 5y$ subject to the following constraints: $x \ge 0, y \ge 0, x + y \le 6, x \ge 2$.

22. A manufacturer makes two types of jet skis, regular and deluxe. The profit on a regular jet ski is \$200 and the profit on the deluxe model is \$250. To meet customer demand, the company must manufacture at least 50 regular jet skis per week and at least 75 deluxe models. To maintain high quality, the total number of both models of jet skis manufactured by the company should not exceed 150 per week. How many jet skis of each type should be manufactured per week to obtain maximum profit? What is the maximum weekly profit?

23. Graph $f(x) = 3^x$. Use $-2, -1, 0, 1,$ and 2 for x and find the corresponding values for y.

24. Graph $y = \log_3 x$ by rewriting the equation in exponential form. Use a table of coordinates and select $-2, -1, 0, 1,$ and 2 for y.

25. Use the vertex and intercepts to graph the quadratic function $f(x) = x^2 - 2x - 8$.

In Exercises 26–29, determine whether the values in each table belong to an exponential function, a logarithmic function, a linear function, or a quadratic function.

26.

x	y
0	3
1	1
2	-1
3	-3
4	-5

27.

x	y
$\frac{1}{3}$	-1
1	0
3	1
9	2
27	3

28.

x	y
0	1
1	5
2	25
3	125
4	625

29.

x	y
0	12
1	3
2	0
3	3
4	12

30. The bar graph and the scatter plot show what it cost the United States Mint to make a penny for five selected years from 1982 through 2012. The data can be modeled by the functions

$$f(x) = 0.03x + 0.63 \text{ and } g(x) = 0.72(1.03)^x,$$

where $f(x)$ and $g(x)$ model what it cost to make a penny x years after 1982.

The Cost of Making a Penny

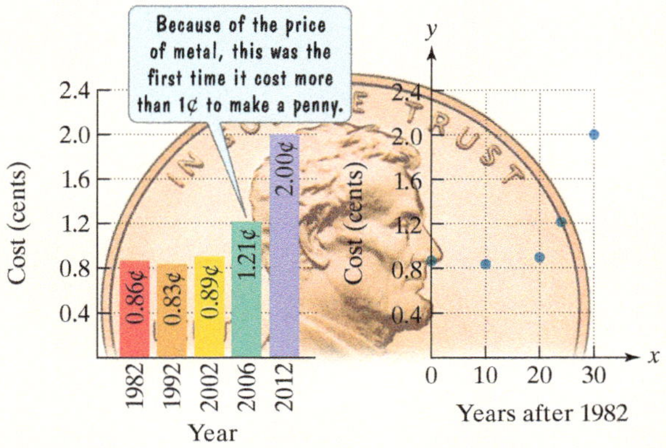

Source: U.S. Mint

a. What is the slope of the linear model? What does this mean in terms of the change in the cost of making a penny?

b. Based on the shape of the scatter plot, which function, the linear or the exponential, is a better model for the data? Explain your answer.

c. Use each function to find the cost of making a penny in 2012. Where necessary, round to two decimal places. Is either function a particularly good model for the 2012 data? Which function serves as a better model? Is this consistent with your answer in part (b)?

8

Personal Finance

"I realize, of course, that it's no shame to be poor, but it's no great honor either. So what would have been so terrible if I had a small fortune?"

—Tevye, a poor dairyman, in the musical *Fiddler on the Roof*

WE ALL WANT A WONDERFUL LIFE WITH FULFILLING WORK, GOOD HEALTH, AND loving relationships. And let's be honest: Financial security, or even a small fortune, wouldn't hurt! Achieving this goal depends on understanding basic ideas about savings, loans, and investments. A solid understanding of the topics in this chapter can pay, literally, by making your financial goals a reality.

Here's where you'll find these applications:

A number of examples illustrate how to attain fortunes ranging from over a half-million dollars to $4 million through regular savings. See Example 3 in Section 8.5 and Exercises 33–36 in Exercise Set 8.5.

8.1

WHAT AM I SUPPOSED TO LEARN?

After you have read this section, you should be able to:

1. Express a fraction as a percent.
2. Express a decimal as a percent.
3. Express a percent as a decimal.
4. Solve applied problems involving sales tax and discounts.
5. Determine percent increase or decrease.
6. Investigate some of the ways percent can be abused.

Percent, Sales Tax, and Discounts

"And if elected, it is my solemn pledge to cut your taxes by 10% for each of my first three years in office, for a total cut of 30%."

PERSONAL FINANCE INCLUDES every area of your life that involves money. It's about what you do with your money and how financial management will affect your future. Because an understanding of *percent* plays an important role in personal finance, we open the chapter with a discussion on the meaning, uses, and abuses of percent.

Basics of Percent

1 Express a fraction as a percent.

Percents are the result of expressing numbers as part of 100. The word *percent* means *per hundred*. For example, the circle graph in **Figure 8.1** shows that 55 out of every 100 college students prefer print textbooks. Thus, $\frac{55}{100} = 55\%$, indicating that 55% of college students prefer print textbooks. The percent sign, %, is used to indicate the number of parts out of 100 parts.

Kinds of Textbooks College Students Prefer: Preferences per 100 Students

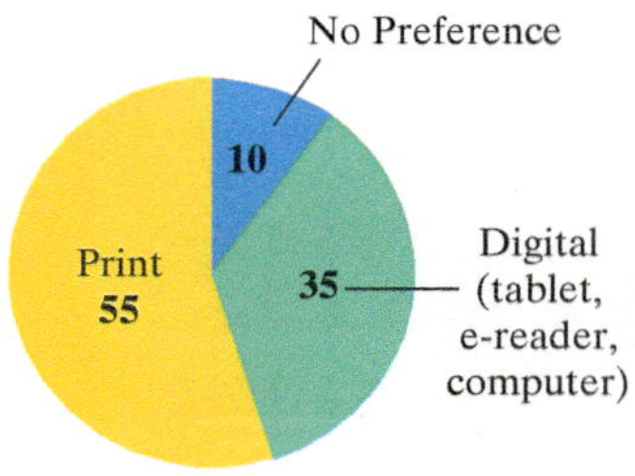

FIGURE 8.1
Source: Harris Interactive for Pearson Foundation

A fraction can be expressed as a percent using the following procedure:

EXPRESSING A FRACTION AS A PERCENT

1. Divide the numerator by the denominator.
2. Multiply the quotient by 100. This is done by moving the decimal point in the quotient two places to the right.
3. Add a percent sign.

EXAMPLE 1 *Expressing a Fraction as a Percent*

Express $\frac{5}{8}$ as a percent.

SOLUTION

Step 1 Divide the numerator by the denominator.

$$5 \div 8 = 0.625$$

Step 2 Multiply the quotient by 100.

$$0.625 \times 100 = 62.5$$

Step 3 Attach a percent sign.

$$62.5\%$$

Thus, $\frac{5}{8} = 62.5\%$.

CHECK POINT 1 Express $\frac{1}{8}$ as a percent.

---

2 Express a decimal as a percent.

Our work in Example 1 shows that $0.625 = 62.5\%$. This illustrates the procedure for expressing a decimal number as a percent.

EXPRESSING A DECIMAL NUMBER AS A PERCENT

1. Move the decimal point two places to the right.
2. Attach a percent sign.

GREAT QUESTION!

What's the difference between the word *percentage* and the word *percent*?

Dictionaries indicate that the word *percentage* has the same meaning as the word *percent*. Use the word that sounds better in the circumstance.

EXAMPLE 2 *Expressing a Decimal as a Percent*

Express 0.47 as a percent.

SOLUTION

Move decimal point two places right.

0.47 % Attach a percent sign.

Thus, 0.47 = 47%.

CHECK POINT 2 Express 0.023 as a percent.

---

3 Express a percent as a decimal.

We reverse the procedure of Example 2 to express a percent as a decimal number.

EXPRESSING A PERCENT AS A DECIMAL NUMBER

1. Move the decimal point two places to the left.
2. Remove the percent sign.

EXAMPLE 3 *Expressing Percents as Decimals*

Express each percent as a decimal:

a. 19% **b.** 180%.

SOLUTION

Use the two steps in the box.

a.

19% = 19.% = 0.19~~%~~ (The percent sign is removed.)

The decimal point starts at the far right. The decimal point is moved two places to the left.

Thus, 19% = 0.19.

b. 180% = 1.80~~%~~ = 1.80 or 1.8

CHECK POINT 3 Express each percent as a decimal:

a. 67% **b.** 250%.

If a fraction is part of a percent, as in $\frac{1}{4}\%$, begin by expressing the fraction as a decimal, retaining the percent sign. Then, express the percent as a decimal number. For example,

$$\frac{1}{4}\% = 0.25\% = 00.25\% = 0.0025.$$

GREAT QUESTION!

Can I expect to have lots of zeros when expressing a small percent as a decimal?

Yes. Be careful with the zeros. For example,

$$\frac{1}{100}\% = 0.01\% = 00.01\% = 0.0001.$$

4 Solve applied problems involving sales tax and discounts.

Percent, Sales Tax, and Discounts

Many applications involving percent are based on the following formula:

$$\underbrace{A}_{A} \underbrace{\text{is}}_{=} \underbrace{P \text{ percent}}_{P} \underbrace{\text{of}}_{\cdot} \underbrace{B.}_{B.}$$

Note that the word *of* implies multiplication.

We can use this formula to determine the **sales tax** collected by states, counties, and cities on sales of items to customers. The sales tax is a percent of the cost of an item.

Sales tax amount $=$ tax rate $\times$ item's cost

EXAMPLE 4 *Percent and Sales Tax*

Suppose that the local sales tax rate is 7.5% and you purchase a bicycle for \$894.

a. How much tax is paid?

b. What is the bicycle's total cost?

SOLUTION

a. Sales tax amount $=$ tax rate $\times$ item's cost

$= 7.5\% \times \$894 = 0.075 \times \$894 = \$67.05$

7.5% of the item's cost, or 7.5% of \$894

The tax paid is \$67.05.

b. The bicycle's total cost is the purchase price, \$894, plus the sales tax, \$67.05.

Total cost $= \$894.00 + \$67.05 = \$961.05$

The bicycle's total cost is \$961.05.

CHECK POINT 4 Suppose that the local sales tax rate is 6% and you purchase a computer for \$1260.

a. How much tax is paid?

b. What is the computer's total cost?

None of us is thrilled about sales tax, but we do like buying things that are *on sale*. Businesses reduce prices, or **discount**, to attract customers and to reduce inventory. The discount rate is a percent of the original price.

Discount amount $=$ discount rate $\times$ original price

EXAMPLE 5 *Percent and Sales Price*

A computer with an original price of \$1460 is on sale at 15% off.

a. What is the discount amount?

b. What is the computer's sale price?

TECHNOLOGY

A calculator is useful, and sometimes essential, in this chapter. The keystroke sequence that gives the sale price in Example 5 is

1460 [−] .15 [×] 1460.

Press [=] or [ENTER] to display the answer, 1241.

SOLUTION

a. Discount amount = discount rate × original price

$$= 15\% \times \$1460 = 0.15 \times \$1460 = \$219$$

15% of the original price, or 15% of $1460

The discount amount is $219.

b. The computer's sale price is the original price, $1460, minus the discount amount, $219.

$$\text{Sale price} = \$1460 - \$219 = \$1241$$

The computer's sale price is $1241.

CHECK POINT 5 A CD player with an original price of $380 is on sale at 35% off.

a. What is the discount amount?

b. What is the CD player's sale price?

GREAT QUESTION!

Do I have to determine the discount amount before finding the sale price?

No. For example, in Example 5 the computer is on sale at 15% off. This means that the sale price must be 100% − 15%, or 85%, of the original price.

$$\text{Sale price} = 85\% \times \$1460 = 0.85 \times \$1460 = \$1241$$

5 Determine percent increase or decrease.

Percent and Change

Percents are used for comparing changes, such as increases or decreases in sales, population, prices, and production. If a quantity changes, its **percent increase** or its **percent decrease** can be found as follows:

FINDING PERCENT INCREASE OR PERCENT DECREASE

1. Find the fraction for the percent increase or the percent decrease:

$$\frac{\text{amount of increase}}{\text{original amount}} \quad \text{or} \quad \frac{\text{amount of decrease}}{\text{original amount}}.$$

2. Find the percent increase or the percent decrease by expressing the fraction in step 1 as a percent.

EXAMPLE 6 *Finding Percent Increase and Decrease*

In 2000, world population was approximately 6 billion. **Figure 8.2** shows world population projections through the year 2150. The data are from the United Nations Family Planning Program and are based on optimistic or pessimistic expectations for successful control of human population growth.

a. Find the percent increase in world population from 2000 to 2150 using the high projection data.

b. Find the percent decrease in world population from 2000 to 2150 using the low projection data.

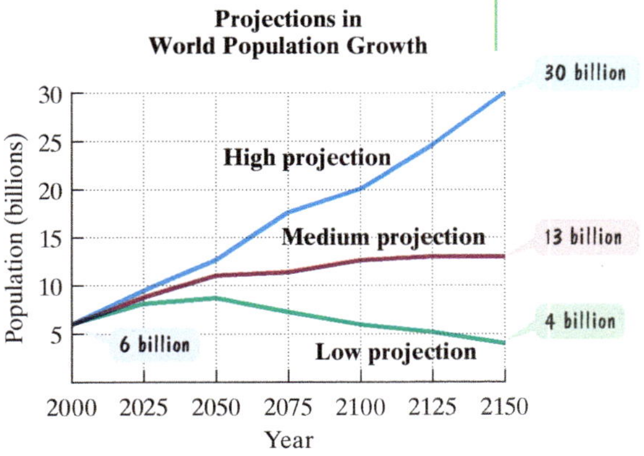

FIGURE 8.2

Source: United Nations

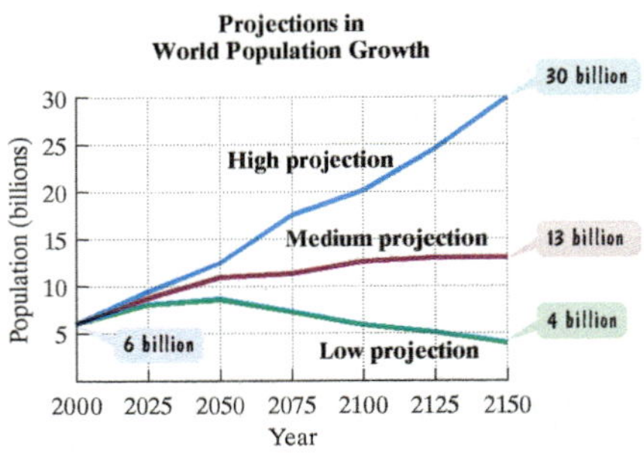

FIGURE 8.2 (repeated)

SOLUTION

a. Use the data shown on the blue, high-projection, graph.

$$\text{Percent increase} = \frac{\text{amount of increase}}{\text{original amount}}$$

$$= \frac{30-6}{6} = \frac{24}{6} = 4 = 400\%$$

The projected percent increase in world population is 400%.

b. Use the data shown on the green, low-projection, graph.

$$\text{Percent decrease} = \frac{\text{amount of decrease}}{\text{original amount}}$$

$$= \frac{6-4}{6} = \frac{2}{6} = \frac{1}{3} = 0.33\frac{1}{3} = 33\frac{1}{3}\%$$

The projected percent decrease in world population is $33\frac{1}{3}\%$.

In Example 6, we expressed the percent decrease as $33\frac{1}{3}\%$ because of the familiar conversion $\frac{1}{3} = 0.33\frac{1}{3}$. However, in many situations, rounding is needed. We suggest that you round to the nearest tenth of a percent. Carry the division in the fraction for percent increase or decrease to four places after the decimal point. Then round the decimal to three places, or to the nearest thousandth. Expressing this rounded decimal as a percent gives percent increase or decrease to the nearest tenth of a percent.

GREAT QUESTION!

I know that increasing 2 to 8 is a 300% increase. Does that mean decreasing 8 to 2 is a 300% decrease?

No. Notice the difference between the following examples:

- 2 is increased to 8.

$$\text{Percent increase} = \frac{\text{amount of increase}}{\text{original amount}} = \frac{6}{2} = 3 = 300\%$$

- 8 is decreased to 2.

$$\text{Percent decrease} = \frac{\text{amount of decrease}}{\text{original amount}} = \frac{6}{8} = \frac{3}{4} = 0.75 = 75\%$$

Although an increase from 2 to 8 is a 300% increase, a decrease from 8 to 2 is *not* a 300% decrease. **A percent decrease involving nonnegative quantities can never exceed 100%.** When a quantity is decreased by 100%, it is reduced to zero.

CHECK POINT 6

a. If 6 is increased to 10, find the percent increase.

b. If 10 is decreased to 6, find the percent decrease.

EXAMPLE 7 *Finding Percent Decrease*

A jacket regularly sells for \$135.00. The sale price is \$60.75. Find the percent decrease of the sale price from the regular price.

SOLUTION

$$\text{Percent decrease} = \frac{\text{amount of decrease}}{\text{original amount}}$$

$$= \frac{135.00 - 60.75}{135} = \frac{74.25}{135} = 0.55 = 55\%$$

The percent decrease of the sale price from the regular price is 55%. This means that the sale price of the jacket is 55% lower than the regular price.

CHECK POINT 7 A television regularly sells for $940. The sale price is $611. Find the percent decrease of the sale price from the regular price.

6 Investigate some of the ways percent can be abused.

Abuses of Percent

In our next examples, we look at a few of the many ways that percent can be used incorrectly. Confusion often arises when percent increase (or decrease) refers to a changing quantity that is itself a percent.

EXAMPLE 8 *Percents of Percents*

John Tesh, while he was still coanchoring *Entertainment Tonight*, reported that the PBS series *The Civil War* had an audience of 13% versus the usual 4% PBS audience, "an increase of more than 300%." Did Tesh report the percent increase correctly?

SOLUTION

We begin by finding the percent increase.

$$\text{Percent increase} = \frac{\text{amount of increase}}{\text{original amount}}$$

$$= \frac{13\% - 4\%}{4\%} = \frac{9\%}{4\%} = \frac{9}{4} = 2.25 = 225\%$$

The percent increase for PBS was 225%. This is not more than 300%, so Tesh did not report the percent increase correctly.

CHECK POINT 8 An episode of a television series had an audience of 12% versus its usual 10%. What was the percent increase for this episode?

EXAMPLE 9 *Promises of a Politician*

A politician states, "If you elect me to office, I promise to cut your taxes for each of my first three years in office by 10% each year, for a total reduction of 30%." Evaluate the accuracy of the politician's statement.

SOLUTION

To make things simple, let's assume that a taxpayer paid $100 in taxes in the year previous to the politician's election. A 10% reduction during year 1 is 10% of $100.

$$10\% \text{ of previous year tax} = 10\% \text{ of } \$100 = 0.10 \times \$100 = \$10$$

With a 10% reduction the first year, the taxpayer will pay only $100 − $10, or $90, in taxes during the politician's first year in office.

The following table shows how we calculate the new, reduced tax for each of the first three years in office:

Year	Tax Paid the Year Before	10% Reduction	Taxes Paid This Year
1	$100	$0.10 \times \$100 = \$10$	$\$100 - \$10 = \$90$
2	$90	$0.10 \times \$90 = \$9$	$\$90 - \$9 = \$81$
3	$81	$0.10 \times \$81 = \$8.10$	$\$81 - \$8.10 = \$72.90$

Now, we determine the percent decrease in taxes over the three years.

$$\text{Percent decrease} = \frac{\text{amount of decrease}}{\text{original amount}}$$

$$= \frac{\$100 - \$72.90}{\$100} = \frac{\$27.10}{\$100} = \frac{27.1}{100} = 0.271 = 27.1\%$$

The taxes decline by 27.1%, not by 30%. The politician is ill-informed in saying that three consecutive 10% cuts add up to a total tax cut of 30%. In our calculation, which serves as a counterexample to the promise, the total tax cut is only 27.1%.

CHECK POINT 9 Suppose you paid $1200 in taxes. During year 1, taxes decrease by 20%. During year 2, taxes increase by 20%.

a. What do you pay in taxes for year 2?

b. How do your taxes for year 2 compare with what you originally paid, namely $1200? If the taxes are not the same, find the percent increase or decrease.

Blitzer Bonus

Testing Your Financial Literacy

Scores have been falling on tests that measure financial literacy. Here are four items from a test given to high school seniors. Would you ace this one?

1. Which of the following is true about sales taxes?
- **A.** The national sales-tax percentage rate is 6%.
- **B.** The Federal Government will deduct it from your paycheck.
- **C.** You don't have to pay the tax if your income is very low.
- **D.** It makes things more expensive for you to buy.

58% of high school seniors answered incorrectly.

2. If you have caused an accident, which type of automobile insurance would cover damage to your own car?
- **A.** Comprehensive
- **B.** Liability
- **C.** Term
- **D.** Collision

63% of high school seniors answered incorrectly.

3. Which of the following types of investment would best protect the purchasing power of a family's savings in the event of a sudden increase in inflation?
- **A.** A 10-year bond issued by a corporation
- **B.** A certificate of deposit at a bank
- **C.** A 25-year corporate bond
- **D.** A house financed with a fixed-rate mortgage

64% of high school seniors answered incorrectly.

4. Sara and Joshua just had a baby. They received money as baby gifts and want to put it away for the baby's education. Which of the following tends to have the highest growth over periods of time as long as 18 years?
- **A.** A checking account
- **B.** Stocks
- **C.** A U.S. government savings bond
- **D.** A savings account

83% of high school seniors answered incorrectly.

Source: The Jump$tart Coalition's 2008 Personal Financial Survey

Answers: 1. D; 2. D; 3. D; 4. B

Concept and Vocabulary Check

Fill in each blank so that the resulting statement is true.

1. Percents are the result of expressing numbers as part of ________.
2. To express $\frac{7}{8}$ as a percent, divide ________ by ________, multiply the quotient by ________, and attach ________.
3. To express 0.1 as a percent, move the decimal point ________ places to the ________ and attach ________.
4. To express 7.5% as a decimal, move the decimal point ________ places to the ________ and remove ________.
5. To find the sales tax amount, multiply the ________ and the ________.
6. To find the discount amount, multiply the ________ and the ________.
7. The numerator of the fraction for percent increase is ________ and the denominator of the fraction for percent increase is ________.
8. The numerator of the fraction for percent decrease is ________ and the denominator of the fraction for percent decrease is ________.

Exercises 9–10 are based on items from a financial literacy survey from the Center for Economic and Entrepreneurial Literacy. Determine whether each statement is true or false. If the statement is false, make the necessary change(s) to produce a true statement.

9. Santa had to lay off 25% of his eight reindeer because of the bad economy, so only seven reindeer remained. (65% answered this question incorrectly. Santa might consider leaving *Thinking Mathematically* in stockings across the country.) ______
10. You spent 1% of your $50,000-per-year salary on gifts, so you spent $5000 on gifts for the year. ______

Exercise Set 8.1

Practice Exercises

In Exercises 1–10, express each fraction as a percent.

1. $\frac{2}{5}$
2. $\frac{3}{5}$
3. $\frac{1}{4}$
4. $\frac{3}{4}$
5. $\frac{3}{8}$
6. $\frac{7}{8}$
7. $\frac{1}{40}$
8. $\frac{3}{40}$
9. $\frac{9}{80}$
10. $\frac{13}{80}$

In Exercises 11–20, express each decimal as a percent.

11. 0.59
12. 0.96
13. 0.3844
14. 0.003
15. 2.87
16. 9.83
17. 14.87
18. 19.63
19. 100
20. 95

In Exercises 21–34, express each percent as a decimal.

21. 72%
22. 38%
23. 43.6%
24. 6.25%
25. 130%
26. 260%
27. 2%
28. 6%
29. $\frac{1}{2}\%$
30. $\frac{3}{4}\%$
31. $\frac{5}{8}\%$
32. $\frac{1}{8}\%$
33. $62\frac{1}{2}\%$
34. $87\frac{1}{2}\%$

Use the percent formula, $A = PB$: A is P percent of B, to solve Exercises 35–38.

35. What is 3% of 200?
36. What is 8% of 300?
37. What is 18% of 40?
38. What is 16% of 90?

Practice Plus

Three basic types of percent problems can be solved using the percent formula $A = PB$.

Question	Given	Percent Formula
What is P percent of B?	P and B	Solve for A.
A is P percent of what?	A and P	Solve for B.
A is what percent of B?	A and B	Solve for P.

Exercises 35–38 involved using the formula to answer the first question. In Exercises 39–46, use the percent formula to answer the second or third question.

39. 3 is 60% of what?
40. 8 is 40% of what?
41. 24% of what number is 40.8?
42. 32% of what number is 51.2?
43. 3 is what percent of 15?
44. 18 is what percent of 90?
45. What percent of 2.5 is 0.3?
46. What percent of 7.5 is 0.6?

Application Exercises

47. Suppose that the local sales tax rate is 6% and you purchase a car for $32,800.
    a. How much tax is paid?
    b. What is the car's total cost?
48. Suppose that the local sales tax rate is 7% and you purchase a graphing calculator for $96.
    a. How much tax is paid?
    b. What is the calculator's total cost?
49. An exercise machine with an original price of $860 is on sale at 12% off.
    a. What is the discount amount?
    b. What is the exercise machine's sale price?
50. A dictionary that normally sells for $16.50 is on sale at 40% off.
    a. What is the discount amount?
    b. What is the dictionary's sale price?

The circle graph shows a breakdown of spending for the average U.S. household using 365 days worked as a basis of comparison. Use this information to solve Exercises 51–52. Round answers to the nearest tenth of a percent.

Spending for the Average American Household, by 365 Days Worked

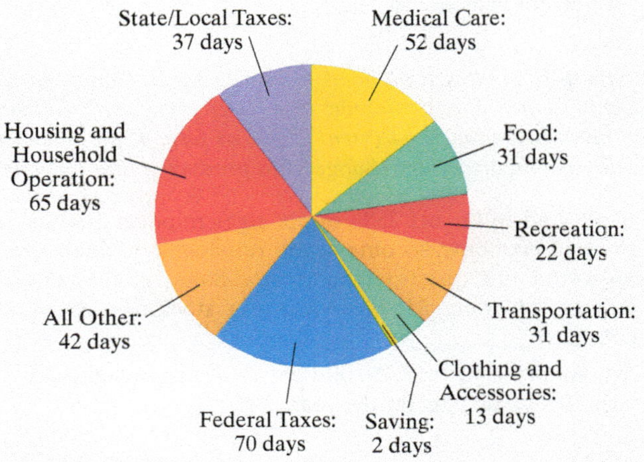

Source: The Tax Foundation

51. What percentage of work time does the average U.S. household spend paying for federal taxes?

52. What percentage of work time does the average U.S. household spend paying for state and local taxes?

Although you want to choose a career that fits your interests and abilities, it is good to have an idea of what jobs pay when looking at career options. The bar graph shows the average yearly earnings of full-time employed college graduates with only a bachelor's degree based on their college major. Use this information to solve Exercises 53–54. Round all answers to the nearest tenth of a percent.

Average Earnings, by College Major

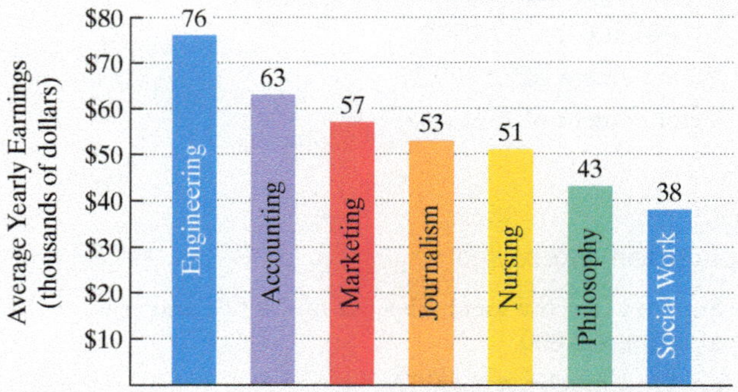

Source: Arthur J. Keown, *Personal Finance*, Fourth Edition, Pearson, 2007.

53. Find the percent increase in the average yearly earnings from students majoring in social work to students majoring in engineering.

54. Find the percent increase for the average yearly earnings from students majoring in philosophy to students majoring in accounting.

55. A sofa regularly sells for \$840. The sale price is \$714. Find the percent decrease of the sale price from the regular price.

56. A FAX machine regularly sells for \$380. The sale price is \$266. Find the percent decrease of the sale price from the regular price.

57. Suppose that you have \$10,000 in a rather risky investment recommended by your financial advisor. During the first year, your investment decreases by 30% of its original value. During the second year, your investment increases by 40% of its first-year value. Your advisor tells you that there must have been a 10% overall increase of your original \$10,000 investment. Is your financial advisor using percentages properly? If not, what is your actual percent gain or loss of your original \$10,000 investment?

58. The price of a color printer is reduced by 30% of its original price. When it still does not sell, its price is reduced by 20% of the reduced price. The salesperson informs you that there has been a total reduction of 50%. Is the salesperson using percentages properly? If not, what is the actual percent reduction from the original price?

Writing in Mathematics

59. What is a percent?

60. Describe how to express a decimal number as a percent and give an example.

61. Describe how to express a percent as a decimal number and give an example.

62. Explain how to use the sales tax rate to determine an item's total cost.

63. Describe how to find percent increase and give an example.

64. Describe how to find percent decrease and give an example.

Critical Thinking Exercises

Make Sense? *In Exercises 65–68, determine whether each statement makes sense or does not make sense, and explain your reasoning.*

65. I have \$100 and my restaurant bill comes to \$80, which is not enough to leave a 20% tip.

66. I found the percent decrease in a jacket's price to be 120%.

67. My weight increased by 1% in January and 1% in February, so my increase in weight over the two months is 2%.

68. My rent increased from 20% to 30% of my income, so the percent increase is 10%.

69. What is the total cost of a \$720 iPad that is on sale at 15% off if the local sales tax rate is 6%?

70. A condominium is taxed based on its \$78,500 value. The tax rate is \$3.40 for every \$100 of value. If the tax is paid before March 1, 3% of the normal tax is given as a discount. How much tax is paid if the condominium owner takes advantage of the discount?

71. In January, each of 60 people purchased a \$500 washing machine. In February, 10% fewer customers purchased the same washing machine that had increased in price by 20%. What was the change in sales from January to February?

72. When you buy something, it actually costs more than you may think—at least in terms of how much money you must earn to buy it. For example, if you pay 28% of your income in taxes, how much money would you have to earn to buy a used car for \$7200?

8.2

WHAT AM I SUPPOSED TO LEARN?

After you have read this section, you should be able to:

1. Determine gross income, adjustable gross income, and taxable income.
2. Calculate federal income tax.
3. Calculate FICA taxes.
4. Solve problems involving working students and taxes.

Income Tax

"THE TROUBLE WITH TRILLIONS" EPISODE of the *Simpsons* finds Homer frantically putting together his tax return two hours before the April 15th mailing deadline. In a frenzy, he shouts to his wife, "Marge, how many kids do we have, no time to count, I'll just estimate nine. If anyone asks, you need 24-hour nursing care, Lisa is a clergyman, Maggie is seven people, and Bart was wounded in Vietnam."

"Cool!" replies Bart.

It isn't only cartoon characters who are driven into states of frantic agitation over taxes. The average American pays over $10,000 per year in income tax. Yes, it's important to pay Uncle Sam what you owe, but not a penny more. People who do not understand the federal tax system often pay more than they have to. In this section, you will learn how income taxes are determined and calculated, reinforcing the role of tax planning in personal finance.

Paying Income Tax

Income tax is a percentage of your income collected by the government to fund its services and programs. The federal government collects income tax, and most, but not all, state governments do, too. (Alaska, Florida, Nevada, South Dakota, Texas, Washington, and Wyoming have no state income tax.) Tax revenue pays for our national defense, fire and police protection, road construction, schools, libraries, and parks. Without taxes, the government would not be able to conduct medical research, provide medical care for the elderly, or send astronauts into space.

Income tax is automatically withheld from your paycheck by your employer. The precise amount withheld for federal income tax depends on how you fill out your W-4 form, which you complete when you start a new job.

Although the United States Congress determines federal tax laws, the Internal Revenue Service (IRS) is the government body that enforces the laws and collects taxes. The IRS is a branch of the Treasury Department.

1 Determine gross income, adjustable gross income, and taxable income.

Determining Taxable Income

Federal income taxes are a percentage of your *taxable income*, which is based on your earnings in the calendar year—January to December. When the year is over, you have until April 15th to file your tax return.

Calculating your federal income tax begins with **gross income**, or total income for the year. This includes income from wages, tips, interest or dividends from investments, unemployment compensation, profits from a business, rental income, and even game-show winnings. It does not matter whether these winnings are in cash or in the form of items such as cars or vacations.

The next step in calculating your federal income tax is to determine your **adjusted gross income**. Adjusted gross income is figured by taking gross income and subtracting certain allowable amounts, called **adjustments**. These untaxed portions of gross income include contributions to certain retirement accounts and tax-deferred savings plans, interest paid on student loans, and alimony payments. In a traditional tax-deferred retirement plan, you get to deduct the full amount of your contribution from your gross income. You pay taxes on the money later, when you withdraw it at retirement.

$$\text{Adjusted gross income} = \text{Gross income} - \text{Adjustments}$$

IRS rules detail exactly what can be subtracted from gross income to determine your adjusted gross income.

Blitzer Bonus

Willie Nelson's Adjustments

In 1990, the IRS sent singer Willie Nelson a bill for $32 million. Egad! But as Willie said, "Thirty-two million ain't much if you say it fast." How did this happen? On bad advice, Willie got involved in a number of investments that he declared as adjustments. This reduced his adjusted gross income to a negligible amount that made paying taxes unnecessary. The IRS ruled these "adjustments" as blatant tax-avoidance schemes. Eventually, Willie and the IRS settled on a $9 million payment.

Source: Arthur J. Keown, *Personal Finance*, Fourth Edition, Pearson, 2007.

You are entitled to certain *exemptions* and *deductions*, subtracted from your adjusted gross income, before calculating your taxes. An **exemption** is a fixed amount on your return for each person supported by your income. You are entitled to this fixed amount ($3800 in 2012) for yourself and the same amount for each dependent.

A **standard deduction** is a lump-sum amount that you can subtract from your adjusted gross income. The IRS sets this amount. Most young people take the standard deduction because their financial situations are relatively simple and they are not eligible for numerous deductions associated with owning a home or making charitable contributions. **Itemized deductions** are deductions you list separately if you have incurred a large number of deductible expenses. Itemized deductions include interest on home mortgages, state income taxes, property taxes, charitable contributions, and medical expenses exceeding 7.5% of adjusted gross income. Taxpayers should choose the greater of a standard deduction or an itemized deduction.

Taxable income is figured by subtracting exemptions and deductions from adjusted gross income.

$$\text{Taxable income} = \text{Adjusted gross income} - (\text{Exemptions} + \text{Deductions})$$

EXAMPLE 1 *Gross Income, Adjusted Gross Income, and Taxable Income*

A single man earned wages of $46,500, received $1850 in interest from a savings account, received $15,000 in winnings on a television game show, and contributed $2300 to a tax-deferred savings plan. He is entitled to a personal exemption of $3800 and a standard deduction of $5950. The interest on his home mortgage was $6500, he paid $2100 in property taxes and $1855 in state taxes, and he contributed $3000 to charity.

a. Determine the man's gross income.

b. Determine the man's adjusted gross income.

c. Determine the man's taxable income.

SOLUTION

a. Gross income refers to this person's total income, which includes wages, interest from a savings account, and game-show winnings.

$$\text{Gross income} = \$46{,}500 + \$1850 + \$15{,}000 = \$63{,}350$$

Wages — Earned interest — Game-show winnings

The gross income is $63,350.

b. Adjusted gross income is gross income minus adjustments. The adjustment in this case is the contribution of $2300 to a tax-deferred savings plan. The full amount of this contribution is deducted from this year's gross income, although taxes will be paid on the money later when it is withdrawn, probably at retirement.

$$\text{Adjusted gross income} = \text{Gross income} - \text{Adjustments} = \$63{,}350 - \$2300 = \$61{,}050$$

Contribution to a tax-deferred savings plan

The adjusted gross income is $61,050.

c. We need to subtract exemptions and deductions from the adjusted gross income to determine the man's taxable income. This taxpayer is entitled to a personal exemption of $3800 and a standard deduction of $5950. However, a deduction greater than $5950 is obtained by itemizing deductions.

$$\text{Itemized deductions} = \$6500 + \$2100 + \$1855 + \$3000 = \$13{,}455$$

Interest on home mortgage — Property taxes — State taxes — Charity

We choose the itemized deductions of $13,455 because they are greater than the standard deduction of $5950.

$$\begin{aligned}\text{Taxable income} &= \text{Adjusted gross income} - (\text{Exemptions} + \text{Deductions})\\ &= \$61{,}050 - (\$3800 + \$13{,}455)\\ &= \$61{,}050 - \$17{,}255\\ &= \$43{,}795\end{aligned}$$

The taxable income is $43,795.

SUMMARY OF KINDS OF INCOME ASSOCIATED WITH FEDERAL TAXES

Gross income is total income for the year.

Adjusted gross income = Gross income − Adjustments

Taxable income = Adjusted gross income − (Exemptions + Deductions)

CHECK POINT 1 A single woman earned wages of $87,200, received $2680 in interest from a savings account, and contributed $3200 to a tax-deferred savings plan. She is entitled to a personal exemption of $3800 and a standard deduction of $5950. The interest on her home mortgage was $11,700, she paid $4300 in property taxes and $5220 in state taxes, and she contributed $15,000 to charity.

a. Determine the woman's gross income.

b. Determine the woman's adjusted gross income.

c. Determine the woman's taxable income.

2 Calculate federal income tax.

Calculating Federal Income Tax

A tax table is used to determine how much you owe based on your taxable income. However, you do not have to pay this much tax if you are entitled to any *tax credits*. **Tax credits** are sums of money that reduce the income tax owed by the full dollar-for-dollar amount of the credits.

Blitzer Bonus

Taking a Bite Out of Taxes

A tax credit is not the same thing as a tax deduction. A tax deduction reduces taxable income, saving only a percentage of the deduction in taxes. A tax credit reduces the income tax owed on the full dollar amount of the credit. There are credits available for everything from donating a kidney to buying an energy efficient dishwasher. The American Opportunity Credit, included in the economic stimulus package of 2009, provides a tax credit of up to $2500 per student. The credit can be used to lower the costs of the first four years of college. You can claim the credit for up to 100% of the first $2000 in qualified college costs and 25% of the next $2000. Significantly, 40% of this credit is refundable. This means that even if you do not have any taxable income, you could receive a check from the government for up to $1000.

Credits are awarded for a variety of activities that the government wants to encourage. Because a tax credit represents a dollar-for-dollar reduction in your tax bill, it pays to know your tax credits. You can learn more about tax credits at www.irs.gov.

Most people pay part or all of their tax bill during the year. If you are employed, your employer deducts federal taxes through *withholdings* based on a percentage of your gross pay. If you are self-employed, you pay your tax bill through *quarterly estimated taxes.*

When you file your tax return, all you are doing is settling up with the IRS over the amount of taxes you paid during the year versus the federal income tax that you owe. Many people will have paid more during the year than they owe, in which case they receive a *tax refund.* Others will not have paid enough and need to send the rest to the IRS by the deadline.

CALCULATING FEDERAL INCOME TAX

1. Determine your adjusted gross income:

$$\text{Adjusted gross income} = \text{Gross income} - \text{Adjustments}.$$

Gross income: All income for the year, including wages, tips, earnings from investments, and unemployment compensation

Adjustments: Includes payments to tax-deferred savings plans

2. Determine your taxable income:

$$\text{Taxable income} = \text{Adjusted gross income} - (\text{Exemptions} + \text{Deductions}).$$

Exemptions: A fixed amount for yourself ($3800 in 2012) and the same amount for each dependent

Deductions: Choose the greater of a standard deduction or an itemized deduction, which includes interest on home mortgages, state income taxes, property taxes, charitable contributions, and medical expenses exceeding 7.5% of adjusted gross income.

3. Determine your income tax:

$$\text{Income tax} = \text{Tax computation} - \text{Tax credits.}$$

Use your taxable income and tax rates for your filing status (single, married, etc.) to determine this amount.

Lawmakers have enacted numerous tax credits to help defray college costs.

Table 8.1 shows 2012 tax rates, standard deductions, and exemptions for the four **filing status** categories described in the voice balloons. The tax rates in the left column, called **marginal tax rates**, are assigned to various income ranges, called margins. For example, suppose you are single and your taxable income is \$25,000. The singles column of the table shows that you must pay 10% tax on the first \$8700, which is

$$10\% \text{ of } \$8700 = 0.10 \times \$8700 = \$870.$$

You must also pay 15% tax on the remaining \$16,300 (\$25,000 − \$8700 = \$16,300), which is

$$15\% \text{ of } \$16,300 = 0.15 \times \$16,300 = \$2445.$$

Your total tax is \$870 + \$2445 = \$3315. In this scenario, your *marginal rate* is 15% and you are in the 15% *tax bracket*.

Table 8.1 2012 Marginal Tax Rates, Standard Deductions, and Exemptions

	Unmarried, divorced, or legally separated	Married and each partner files a separate tax return	Married and both partners file a single tax return	Unmarried and paying more than half the cost of supporting a child or parent
Tax Rate	**Single**	**Married Filing Separately**	**Married Filing Jointly**	**Head of Household**
10%	up to \$8700	up to \$8700	up to \$17,400	up to \$12,400
15%	\$8701 to \$35,350	\$8701 to \$35,350	\$17,401 to \$70,700	\$12,401 to \$47,350
25%	\$35,351 to \$85,650	\$35,351 to \$71,350	\$70,701 to \$142,700	\$47,351 to \$122,300
28%	\$85,651 to \$178,650	\$71,351 to \$108,725	\$142,701 to \$217,450	\$122,301 to \$198,050
33%	\$178,651 to \$388,350	\$108,726 to \$194,175	\$217,451 to \$388,350	\$198,051 to \$388,350
35%	more than \$388,350	more than \$194,175	more than \$388,350	more than \$388,350
Standard Deduction	\$5950	\$5950	\$11,900	\$8700
Exemptions (per person)	\$3800	\$3800	\$3800	\$3800

SINGLE WOMAN WITH NO DEPENDENTS

Gross income: \$62,000
Adjustments: \$4000 paid to a tax-deferred IRA (Individual Retirement Account)
Deductions:
- \$7500: mortgage interest
- \$2200: property taxes
- \$2400: charitable contributions
- \$1500: medical expenses not covered by insurance

Tax credit: \$500

EXAMPLE 2 *Computing Federal Income Tax*

Calculate the federal income tax owed by a single woman with no dependents whose gross income, adjustments, deductions, and credits are given in the margin. Use the 2012 marginal tax rates in **Table 8.1**.

SOLUTION

Step 1 Determine the adjusted gross income.

$$\begin{aligned}\text{Adjusted gross income} &= \text{Gross income} - \text{Adjustments}\\ &= \$62,000 - \$4000\\ &= \$58,000\end{aligned}$$

SINGLE WOMAN WITH NO DEPENDENTS

Gross income: $62,000

Adjustments: $4000 paid to a tax-deferred IRA (Individual Retirement Account)

Deductions:

- $7500: mortgage interest
- $2200: property taxes
- $2400: charitable contributions
- $1500: medical expenses not covered by insurance

Tax credit: $500

Step 2 Determine the taxable income.

$$\text{Taxable income} = \text{Adjusted gross income} - (\text{Exemptions} + \text{Deductions})$$
$$= \$58{,}000 - (\$3800 + \text{Deductions})$$

The singles column in **Table 8.1** shows a personal exemption of $3800.

The singles column in **Table 8.1** shows a $5950 standard deduction. A greater deduction can be obtained by itemizing.

Itemized Deductions

$7500 : mortgage interest
$2200 : property taxes
$2400 : charitable contributions
~~$1500 : medical expenses~~
$12,100 : total of deductible expenditures

Can only deduct amount in excess of 7.5% of adjusted gross income: $0.075 \times \$58{,}000 = \$4350$

We substitute $12,100 for deductions in the formula for taxable income.

$$\text{Taxable income} = \text{Adjusted gross income} - (\text{Exemptions} + \text{Deductions})$$
$$= \$58{,}000 - (\$3800 + \$12{,}100)$$
$$= \$58{,}000 - \$15{,}900$$
$$= \$42{,}100$$

Tax Rate	Single
10%	up to $8700
15%	$8701 to $35,350
25%	$35,351 to $85,650

A portion of **Table 8.1** (repeated)

Step 3 Determine the income tax.

$$\text{Income tax} = \text{Tax computation} - \text{Tax credits}$$
$$= \text{Tax computation} - \$500$$

We perform the tax computation using the singles rates in **Table 8.1**, partly repeated in the margin. Our taxpayer is in the 25% tax bracket because her taxable income, $42,100, is in the $35,351 to $85,650 income range. This means that she owes 10% on the first $8700 of her taxable income, 15% on her taxable income between $8701 and $35,350, inclusive, and 25% on her taxable income above $35,350.

10% marginal rate on first $8700 of taxable income

15% marginal rate on taxable income between $8701 and $35,350

25% marginal rate on taxable income above $35,350

$$\text{Tax computation} = 0.10 \times \$8700 + 0.15 \times (\$35{,}350 - \$8700) + 0.25 \times (\$42{,}100 - \$35{,}350)$$
$$= 0.10 \times \$8700 + 0.15 \times \$26{,}650 + 0.25 \times \$6750$$
$$= \$870.00 + \$3997.50 + \$1687.50$$
$$= \$6555.00$$

We substitute $6555.00 for the tax computation in the formula for income tax.

$$\text{Income tax} = \text{Tax computation} - \text{Tax credits}$$
$$= \$6555.00 - \$500$$
$$= \$6055.00$$

The federal income tax owed is $6055.00.

CHECK POINT 2 Use the 2012 marginal tax rates in **Table 8.1** on page 501 to calculate the federal tax owed by a single man with no dependents whose gross income, adjustments, deductions, and credits are given as follows:

Gross income: $40,000
Adjustments: $1000
Deductions: $3000: charitable contributions
$1500: theft loss
$300: cost of tax preparation
Tax credit: none.

3 Calculate FICA taxes.

Social Security and Medicare (FICA)

In addition to income tax, we are required to pay the federal government **FICA** (Federal Insurance Contributions Act) taxes that are used for Social Security and Medicare benefits. **Social Security** provides payments to eligible retirees, people with health problems, eligible dependents of deceased persons, and disabled citizens. **Medicare** provides health care coverage mostly to Americans 65 and older.

The 2012 FICA tax rates are given in **Table 8.2**.

TABLE 8.2 2012 FICA Tax Rates

Employee's Rates	Matching Rates Paid by the Employer	Self-Employed Rates
• 5.65% on first $110,000 of income • 1.45% of income in excess of $110,000	• 7.65% on first $110,000 paid in wages • 1.45% of wages paid in excess of $110,000	• 13.3% on first $110,000 of net profits • 2.9% of net profits in excess of $110,000

Taxpayers are not permitted to subtract adjustments, exemptions, or deductions when determining FICA taxes.

EXAMPLE 3 *Computing FICA Tax*

If you are not self-employed and earn $150,000, what are your FICA taxes?

SOLUTION

The tax rates are 5.65% on the first $110,000 of income and 1.45% on income in excess of $110,000.

5.65% rate on the first $110,000 of income

1.45% rate on income in excess of $110,000

$$\begin{aligned}\text{FICA Tax} &= 0.0565 \times \$110{,}000 + 0.0145 \times (\$150{,}000 - \$110{,}000)\\ &= 0.0565 \times \$110{,}000 + 0.0145 \times \$40{,}000\\ &= \$6215 + \$580\\ &= \$6795\end{aligned}$$

The FICA taxes are $6795.

CHECK POINT 3 If you are not self-employed and earn $200,000, what are your FICA taxes?

4 Solve problems involving working students and taxes.

Working Students and Taxes

For those of you who work part-time, getting paid is great. However, because employers withhold federal and state taxes, as well as FICA, your paychecks probably contain less spending money than you had anticipated.

A pay stub attached to your paycheck provides a lot of information about the money you earned, including both your *gross pay* and your *net pay*. **Gross pay**, also known as **base pay**, is your salary prior to any withheld taxes for the pay period the check covers. Your gross pay is what you would receive if nothing were deducted. **Net pay** is the actual amount of your check after taxes have been withheld.

Blitzer Bonus

Percents and Tax Rates

In 1944 and 1945, the highest marginal tax rate in the United States was a staggering 94%. The tax rate on the highest-income Americans remained at approximately 90% throughout the 1950s, decreased to 70% in the 1960s and 1970s, and to 50% in the early 1980s before reaching a post–World War II low of 28% in 1998. (*Source*: IRS) In 2012, Sweden had the world's highest tax rate among industrialized countries (56.6%), followed by Nordic neighbor Denmark (55.3%) and crisis-laden Spain (52%). (*Source*: KPMG)

EXAMPLE 4 *Taxes for a Working Student*

You would like to have extra spending money, so you decide to work part-time at the local gym. The job pays \$10 per hour and you work 20 hours per week. Your employer withholds 10% of your gross pay for federal taxes, 5.65% for FICA taxes, and 3% for state taxes.

a. What is your weekly gross pay?

b. How much is withheld per week for federal taxes?

c. How much is withheld per week for FICA taxes?

d. How much is withheld per week for state taxes?

e. What is your weekly net pay?

f. What percentage of your gross pay is withheld for taxes? Round to the nearest tenth of a percent.

SOLUTION

a. Your weekly gross pay is the number of hours worked, 20, times your hourly wage, \$10 per hour.

$$\text{Gross pay} = 20\ \cancel{\text{hours}} \times \frac{\$10}{\cancel{\text{hour}}} = 20 \times \$10 = \$200$$

Your weekly gross pay, or what you would receive if nothing were deducted, is \$200.

b. Your employer withholds 10% of your gross pay for federal taxes.

$$\text{Federal taxes} = 10\% \text{ of } \$200 = 0.10 \times \$200 = \$20$$

\$20 is withheld per week for federal taxes.

c. Your employer withholds 5.65% of your gross pay for FICA taxes.

$$\text{FICA taxes} = 5.65\% \text{ of } \$200 = 0.0565 \times \$200 = \$11.30$$

\$11.30 is withheld per week for FICA taxes.

d. Your employer withholds 3% of your gross pay for state taxes.

$$\text{State taxes} = 3\% \text{ of } \$200 = 0.03 \times \$200 = \$6$$

\$6 is withheld per week for state taxes.

e. Your weekly net pay is your gross pay minus the amounts withheld for federal, FICA, and state taxes.

Gross pay | Federal taxes | FICA taxes | State taxes

$$\begin{aligned}\text{Net pay} &= \$200 - (\$20 + \$11.30 + \$6)\\ &= \$200.00 - \$37.30\\ &= \$162.70\end{aligned}$$

Your weekly net pay is \$162.70. This is the actual amount of your paycheck.

f. Our work in part (e) shows that \$37.30 is withheld for taxes. The fractional part of your gross pay that is withheld for taxes is the amount that is withheld, \$37.30, divided by your gross pay, \$200.00. We then express this fraction as a percent.

Percent of gross pay withheld for taxes

Taxes

$$= \frac{\$37.30}{\$200.00} = 0.1865 = 18.65\% \approx 18.7\%$$

Gross pay

Your employer takes \$37.30 from your weekly gross salary and sends the money to the government. This represents approximately 18.7% of your gross pay.

Figure 8.3 contains a sample pay stub for the working student in Example 4.

YOUR WORKPLACE 10 MAIN STREET ANY TOWN, STATE	YOUR NAME YOUR ADDRESS YOUR CITY, STATE, ZIP CODE SSN: 000-00-0000	Pay End Date: 0/00/12	
		Federal	State
		Single 1	Single 1

HOURS AND EARNINGS				
Current			Year to Date	
Base Rate (\$10 per hour)	Hours	Earnings	Hours	Earnings
	20	\$200	400	\$4000

TAXES		
Description:	Current	Year to Date
Federal	\$20.00	\$400.00
FICA	\$11.30	\$226.00
State	\$ 6.00	\$120.00
TOTAL	\$37.30	\$746.00
	Current	Year to Date
TOTAL GROSS	\$200.00	\$4000.00
TOTAL DEDUCTIONS	\$ 37.30	\$ 746.00
NET PAY	\$162.70	\$3254.00

FIGURE 8.3 A working student's sample pay stub

Pay stubs are attached to your paycheck and usually contain four sections:

- Personal information about the employee: This may include your name, your address, your social security number, and your marital status.
- Information about earnings: This includes your hourly wage, the number of hours worked in the current pay period, and the number of hours worked year-to-date, which is usually the total number of hours worked since January 1 of the current year (the first pay of the year may include the latter part of December), and the amount earned during the current pay period and the amount earned year-to-date.
- Information about tax deductions, summarizing withholdings for the current pay period and the withholdings year-to-date.
- Gross pay, total deductions, and net pay for the current period, and the year-to-date total for each.

CHECK POINT 4 You decide to work part-time at a local nursery. The job pays \$12 per hour and you work 15 hours per week. Your employer withholds 10% of your gross pay for federal taxes, 5.65% for FICA taxes, and 4% for state taxes.

a. What is your weekly gross pay?

b. How much is withheld per week for federal taxes?

c. How much is withheld per week for FICA taxes?

d. How much is withheld per week for state taxes?

e. What is your weekly net pay?

f. What percentage of your gross pay is withheld for taxes? Round to the nearest tenth of a percent.

Concept and Vocabulary Check

Fill in each blank so that the resulting statement is true.

1. Your ________ income is your total income for the year.
2. Subtracting certain allowable amounts from the income in Exercise 1 results in your ____________ income. These allowable amounts, or untaxed portions of income, are called __________.
3. A fixed amount deducted on your tax return for each person supported by your income, including yourself, is called a/an _________.
4. Your taxable income is your ___________ income minus the sum of your _________ and _________.
5. Sums of money that reduce federal income tax by the full dollar-for-dollar amount are called _________.
6. Taxes used for Social Security and Medicare benefits are called _________ taxes.
7. Your base pay, or _________ pay, is your salary prior to any withheld taxes.
8. The actual amount of your paycheck after taxes have been withheld is called your _________ pay.

In Exercises 9–12, determine whether each statement is true or false. If the statement is false, make the necessary change(s) to produce a true statement.

9. Federal income tax is a percentage of your gross income. ______
10. If tax credits are equal, federal tax tables show that the greater your taxable income, the more you pay. ______
11. People in some states are not required to pay state income taxes. ______
12. FICA tax is a percentage of your gross income. ______

Exercise Set 8.2

Practice and Application Exercises

In Exercises 1–2, find the gross income, the adjusted gross income, and the taxable income.

1. A taxpayer earned wages of $52,600, received $720 in interest from a savings account, and contributed $3200 to a tax-deferred retirement plan. He was entitled to a personal exemption of $3800 and had deductions totaling $7250.
2. A taxpayer earned wages of $23,500, received $495 in interest from a savings account, and contributed $1200 to a tax-deferred retirement plan. She was entitled to a personal exemption of $3800 and had deductions totaling $5450.

In Exercises 3–4, find the gross income, the adjusted gross income, and the taxable income. Base the taxable income on the greater of a standard deduction or an itemized deduction.

3. Suppose your neighbor earned wages of $86,250, received $1240 in interest from a savings account, and contributed $2200 to a tax-deferred retirement plan. She is entitled to a personal exemption of $3800 and a standard deduction of $5950. The interest on her home mortgage was $8900, she contributed $2400 to charity, and she paid $1725 in state taxes.
4. Suppose your neighbor earned wages of $319,150, received $1790 in interest from a savings account, and contributed $4100 to a tax-deferred retirement plan. He is entitled to a personal exemption of $3800 and the same exemption for each of his two children. He is also entitled to a standard deduction of $5950. The interest on his home mortgage was $51,235, he contributed $74,000 to charity, and he paid $12,760 in state taxes.

*In Exercises 5–14, use the 2012 marginal tax rates in **Table 8.1** on page 501 to compute the tax owed by each person or couple.*

5. a single man with a taxable income of $40,000
6. a single woman with a taxable income of $42,000
7. a married woman filing separately with a taxable income of $120,000
8. a married man filing separately with a taxable income of $110,000
9. a single man with a taxable income of $15,000 and a $2500 tax credit
10. a single woman with a taxable income of $12,000 and a $3500 tax credit
11. a married couple filing jointly with a taxable income of $250,000 and a $7500 tax credit
12. a married couple filing jointly with a taxable income of $400,000 and a $4500 tax credit
13. a head of household with a taxable income of $58,000 and a $6500 tax credit
14. a head of household with a taxable income of $46,000 and a $3000 tax credit

*In Exercises 15–18, use the 2012 marginal tax rates in **Table 8.1** on page 501 to calculate the income tax owed by each person.*

15. Single male, no dependents

    Gross income: $75,000

    Adjustments: $4000

    Deductions:

    $28,000 mortgage interest

    $4200 property taxes

    $3000 charitable contributions

    Tax credit: none

16. Single female, no dependents

    Gross income: $70,000

    Adjustments: $2000

    Deductions:

    $10,000 mortgage interest

    $2500 property taxes

    $1200 charitable contributions

    Tax credit: none

17. Unmarried head of household with two dependent children

Gross income: $50,000
Adjustments: none
Deductions:
$4500 state taxes
$2000 theft loss
Tax credit: $2000

18. Unmarried head of household with one dependent child

Gross income: $40,000
Adjustments: $1500
Deductions:
$3600 state taxes
$800 charitable contributions
Tax credit: $2500

In Exercises 19–24, use the 2012 FICA tax rates in **Table 8.2** *on page 503.*

19. If you are not self-employed and earn $120,000 what are your FICA taxes?

20. If you are not self-employed and earn $140,000 what are your FICA taxes?

21. If you are self-employed and earn $150,000, what are your FICA taxes?

22. If you are self-employed and earn $160,000, what are your FICA taxes?

23. To help pay for college, you worked part-time at a local restaurant, earning $20,000 in wages and tips.
  a. Calculate your FICA taxes.
  b. Use **Table 8.1** on page 501 to calculate your income tax. Assume you are single with no dependents, have no adjustments or tax credit, and you take the standard deduction.
  c. Including both FICA and income tax, what percentage of your gross income are your federal taxes? Round to the nearest tenth of a percent.

24. To help pay for college, you worked part-time at a local restaurant, earning $18,000 in wages and tips.
  a. Calculate your FICA taxes.
  b. Use **Table 8.1** on page 501 to calculate your income tax. Assume you are single with no dependents, have no adjustments or tax credit, and take the standard deduction.
  c. Including both FICA and income tax, what percentage of your gross income are your federal taxes? Round to the nearest tenth of a percent.

25. You decide to work part-time at a local supermarket. The job pays $8.50 per hour and you work 20 hours per week. Your employer withholds 10% of your gross pay for federal taxes, 5.65% for FICA taxes, and 3% for state taxes.
  a. What is your weekly gross pay?
  b. How much is withheld per week for federal taxes?
  c. How much is withheld per week for FICA taxes?
  d. How much is withheld per week for state taxes?
  e. What is your weekly net pay?
  f. What percentage of your gross pay is withheld for taxes? Round to the nearest tenth of a percent.

26. You decide to work part-time at a local veterinary hospital. The job pays $9.50 per hour and you work 20 hours per week. Your employer withholds 10% of your gross pay for federal taxes, 5.65% for FICA taxes, and 5% for state taxes.
  a. What is your weekly gross pay?
  b. How much is withheld per week for federal taxes?
  c. How much is withheld per week for FICA taxes?
  d. How much is withheld per week for state taxes?
  e. What is your weekly net pay?
  f. What percentage of your gross pay is withheld for taxes? Round to the nearest tenth of a percent

Writing in Mathematics

27. What is income tax?

28. What is gross income?

29. What is adjusted gross income?

30. What are exemptions?

31. What are deductions?

32. Under what circumstances should taxpayers itemize deductions?

33. How is taxable income determined?

34. What are tax credits?

35. What is the difference between a tax credit and a tax deduction?

36. What are FICA taxes?

37. How do you determine your net pay?

Critical Thinking Exercises

Make Sense? *In Exercises 38–42, determine whether each statement makes sense or does not make sense, and explain your reasoning.*

38. The only important thing to know about my taxes is whether I receive a refund or owe money on my return.

39. Because I am a student with a part-time job, federal tax law does not allow me to itemize deductions.

40. I'm paying less federal tax on my first dollars of earnings and more federal tax on my last dollars of earnings.

41. My employer withholds the same amount of federal tax on my first dollars of earnings and my last dollars of earnings.

42. Now that I'm a college student, I can choose a $4000 deduction or a $2500 credit to offset tuition and fees. I'll pay less federal taxes by selecting the $4000 deduction.

43. Suppose you are in the 10% tax bracket. As a college student, you can choose a $4000 deduction or a $2500 credit to offset tuition and fees. Which option will reduce your tax bill by the greater amount? What is the difference in your savings between the two options?

44. A common complaint about income tax is "I can't afford to work more because it will put me in a higher tax bracket." Is it possible that being in a higher bracket means you actually lose money? Explain your answer.

45. Because of the mortgage interest tax deduction, is it possible to save money buying a house rather than renting, even though rent payments are lower than mortgage payments? Explain your answer.

Group Exercises

The following topics are appropriate for either individual or group research projects. Use the Internet to investigate each topic.

46. Proposals to Simplify Federal Tax Laws and Filing Procedures

47. The Most Commonly Recommended Tax Saving Strategies

48. The Most Commonly Audited Tax Return Sections

49. Federal Tax Procedures Questioned over Issues of Fairness (Examples include the marriage penalty, the alternative minimum tax (AMT), and capital gains rates.)

8.3

WHAT AM I SUPPOSED TO LEARN?

After you have read this section, you should be able to:

1 Calculate simple interest.
2 Use the future value formula.

Simple Interest

IN 1626, PETER MINUIT CONVINCED the Wappinger Indians to sell him Manhattan Island for $24. If the Native Americans had put the $24 into a bank account at a 5% interest rate compounded monthly, by the year 2010 there would have been well over $5 billion in the account!

Although you may not yet understand terms such as *interest rate* and *compounded monthly*, one thing seems clear: Money in certain savings accounts grows in remarkable ways. You, too, can take advantage of such accounts with astonishing results. In the next two sections, we will show you how.

1 Calculate simple interest.

Simple Interest

Interest is the amount of money that we get paid for lending or investing money, or that we pay for borrowing money. When we deposit money in a savings institution, the institution pays us interest for its use. When we borrow money, interest is the price we pay for the privilege of using the money until we repay it.

The amount of money that we deposit or borrow is called the **principal**. For example, if you deposit $2000 in a savings account, then $2000 is the principal. The amount of interest depends on the principal, the interest **rate**, which is given as a percent and varies from bank to bank, and the length of time for which the money is deposited. In this section, the rate is assumed to be annual (per year).

Simple interest involves interest calculated only on the principal. The following formula is used to find simple interest:

CALCULATING SIMPLE INTEREST

$$\text{Interest} = \text{principal} \times \text{rate} \times \text{time}$$
$$I = Prt$$

The rate, r, is expressed as a decimal when calculating simple interest.

Throughout this section and the chapter, keep in mind that all given rates are assumed to be *per year*, unless otherwise stated.

EXAMPLE 1 *Calculating Simple Interest for a Year*

You deposit $2000 in a savings account at Hometown Bank, which has a rate of 6%. Find the interest at the end of the first year.

SOLUTION

To find the interest at the end of the first year, we use the simple interest formula.

$$I = Prt = (2000)(0.06)(1) = 120$$

Principal, or amount deposited, is $2000. Rate is 6% = 0.06. Time is 1 year.

At the end of the first year, the interest is $120. You can withdraw the $120 interest and you still have $2000 in the savings account.

CHECK POINT 1 You deposit \$3000 in a savings account at Yourtown Bank, which has a rate of 5%. Find the interest at the end of the first year.

EXAMPLE 2 *Calculating Simple Interest for More Than a Year*

A student took out a simple interest loan for \$1800 for two years at a rate of 8% to purchase a used car. What is the interest on the loan?

SOLUTION

To find the interest on the loan, we use the simple interest formula.

$$I = Prt = (1800)(0.08)(2) = 288$$

Principal, or amount borrowed, is \$1800. Rate is 8% = 0.08. Time is 2 years.

The interest on the loan is \$288.

CHECK POINT 2 A student took out a simple interest loan for \$2400 for two years at a rate of 7%. What is the interest on the loan?

Simple interest is used for many short-term loans, including automobile and consumer loans. Imagine that a short-term loan is taken for 125 days. The time of the loan is $\frac{125}{365}$ because there are 365 days in a year. However, before the modern use of calculators and computers, the **Banker's rule** allowed financial institutions to use 360 in the denominator of such a fraction because this simplified the interest calculation. Using the Banker's rule, the time, t, for a 125-day short-term loan is

$$\frac{125\text{ days}}{360\text{ days}} = \frac{125}{360}.$$

Compare the values for time, t, for a 125-day short-term loan using denominators of 360 and 365.

$$\frac{125}{360} \approx 0.347 \qquad \frac{125}{365} \approx 0.342$$

The denominator of 360 benefits the bank by resulting in a greater period of time for the loan, and consequently more interest.

With the widespread use of calculators and computers, government agencies and the Federal Reserve Bank calculate simple interest using 365 days in a year, as do many credit unions and banks. However, there are still some financial institutions that use the Banker's rule with 360 days in a year because it produces a greater amount of interest.

2 Use the future value formula.

Future Value: Principal Plus Interest

When a loan is repaid, the interest is added to the original principal to find the total amount due. In Example 2, at the end of two years, the student will have to repay

$$\text{principal} + \text{interest} = \$1800 + \$288 = \$2088.$$

In general, if a principal P is borrowed at a simple interest rate r, then after t years the amount due, A, can be determined as follows:

$$A = P + I = P + Prt = P(1 + rt).$$

The amount due, A, is called the **future value** of the loan. The principal borrowed now, P, is also known as the loan's **present value**.

CALCULATING FUTURE VALUE FOR SIMPLE INTEREST

The future value, A, of P dollars at simple interest rate r (as a decimal) for t years is given by

$$A = P(1 + rt).$$

EXAMPLE 3 *Calculating Future Value*

A loan of $1060 has been made at 6.5% for three months. Find the loan's future value.

SOLUTION

The amount borrowed, or principal, P, is $1060. The rate, r, is 6.5%, or 0.065. The time, t, is given as three months. We need to express the time in years because the rate is understood to be 6.5% per year. Because three months is $\frac{3}{12}$ of a year, $t = \frac{3}{12} = \frac{1}{4} = 0.25$.

The loan's future value, or the total amount due after three months, is

$$A = P(1 + rt) = 1060[1 + (0.065)(0.25)] \approx \$1077.23.$$

Rounded to the nearest cent, the loan's future value is $1077.23.

TECHNOLOGY

1060 [1 + (0.065)(0.25)]

On a Scientific Calculator:

1060 [×] [(] 1 [+] .065 [×] .25 [)] [=]

CHECK POINT 3 A loan of $2040 has been made at 7.5% for four months. Find the loan's future value.

EXAMPLE 4 *Earning Money by Putting Your Wallet Away Today*

Suppose you spend $4 each day, five days per week, on gourmet coffee.

a. How much do you spend on this item in a year?

b. If you invested your yearly spending on gourmet coffee in a savings account with a rate of 5%, how much would you have after one year?

SOLUTION

a. Because you are spending $4 each day, five days per week, you are spending

$$\frac{\$4}{\text{day}} \times \frac{5 \text{ days}}{\text{week}} = \frac{\$4 \times 5}{\text{week}} = \frac{\$20}{\text{week}},$$

or $20 each week on gourmet coffee. Assuming that this continues throughout the 52 weeks in the year, you are spending

$$\frac{\$20}{\text{week}} \times \frac{52 \text{ weeks}}{\text{year}} = \frac{\$20 \times 52}{\text{year}} = \frac{\$1040}{\text{year}},$$

or $1040 each year on gourmet coffee.

b. Now suppose you invest $1040 in a savings account with a rate of 5%. To find your savings after one year, we use the future value formula for simple interest.

$$A = P(1 + rt) = 1040[1 + (0.05)(1)] = 1040(1.05) = 1092$$

Giving up the day-to-day expense of gourmet coffee can result in potential savings of $1092.

CHECK POINT 4 In addition to jeopardizing your health, cigarette smoking is a costly addiction. Consider, for example, a person with a pack-a-day cigarette habit who spends $5 per day, seven days each week, on cigarettes.

a. How much is spent on this item in a year?

b. If this person invested the yearly spending on cigarettes in a savings account with a rate of 4%, how much would be saved after one year?

The formula for future value, $A = P(1 + rt)$, has four variables. If we are given values for any three of these variables, we can solve for the fourth.

EXAMPLE 5 *Determining a Simple Interest Rate*

You borrow \$2500 from a friend and promise to pay back \$2655 in six months. What simple interest rate will you pay?

SOLUTION

We use the formula for future value, $A = P(1 + rt)$. You borrow \$2500: $P = 2500$. You will pay back \$2655, so this is the future value: $A = 2655$. You will do this in six months, which must be expressed in years: $t = \frac{6}{12} = \frac{1}{2} = 0.5$. To determine the simple interest rate you will pay, we solve the future value formula for r.

$$A = P(1 + rt) \quad \text{This is the formula for future value.}$$

$$2655 = 2500[1 + r(0.5)] \quad \text{Substitute the given values.}$$

$$2655 = 2500 + 1250r \quad \text{Use the distributive property.}$$

$$155 = 1250r \quad \text{Subtract 2500 from both sides.}$$

$$\frac{155}{1250} = \frac{1250r}{1250} \quad \text{Divide both sides by 1250.}$$

$$r = 0.124 = 12.4\% \quad \text{Express } \frac{155}{1250} \text{ as a percent.}$$

You will pay a simple interest rate of 12.4%.

CHECK POINT 5 You borrow \$5000 from a friend and promise to pay back \$6800 in two years. What simple interest rate will you pay?

EXAMPLE 6 *Determining a Present Value*

You plan to save \$2000 for a trip to Europe in two years. You decide to purchase a certificate of deposit (CD) from your bank that pays a simple interest rate of 4%. How much must you put in this CD now in order to have the \$2000 in two years?

SOLUTION

We use the formula for future value, $A = P(1 + rt)$. We are interested in finding the principal, P, or the present value.

$$A = P(1 + rt) \quad \text{This is the formula for future value.}$$

$$2000 = P[1 + (0.04)(2)] \quad A\text{(future value)} = \$2000,\ r\text{(interest rate)} = 0.04, \text{ and } t = 2 \text{ (you want \$2000 in two years).}$$

$$2000 = 1.08P \quad \text{Simplify: } 1 + (0.04)(2) = 1.08.$$

$$\frac{2000}{1.08} = \frac{1.08P}{1.08} \quad \text{Divide both sides by 1.08.}$$

$$P \approx 1851.852 \quad \text{Simplify.}$$

To make sure you will have enough money for the vacation, let's round this principal *up* to \$1851.86. Thus, you should put \$1851.86 in the CD now to have \$2000 in two years.

GREAT QUESTION!

How should I round when finding present value?

When computing present value, round the principal, or present value, *up*. To round up to the nearest cent, add 1 to the hundredths digit, regardless of the digit to the right. In this way, you'll be sure to have enough money to meet future goals.

CHECK POINT 6 How much should you put in an investment paying a simple interest rate of 8% if you need $4000 in six months?

Concept and Vocabulary Check

Fill in each blank so that the resulting statement is true.

1. The formula for calculating simple interest, I, is ________, where P is the ________, r is the ________, and t is the ________.
2. The future value, A, of P dollars at simple interest rate r for t years is given by the formula ____________.
3. The Banker's rule allows using ________ days in a year.

In Exercises 4–6, determine whether each statement is true or false. If the statement is false, make the necessary change(s) to produce a true statement.

4. Interest is the amount of money we get paid for borrowing money or that we pay for investing money. ______
5. In simple interest, only the original money invested or borrowed generates interest over time. _______
6. If $4000 is borrowed at 7.6% for three months, the loan's future value is $76. ________

Exercise Set 8.3

Practice Exercises

In Exercises 1–8, the principal P is borrowed at simple interest rate r for a period of time t. Find the simple interest owed for the use of the money. Assume 360 days in a year.

1. $P = \$4000, r = 6\%, t = 1$ year
2. $P = \$7000, r = 5\%, t = 1$ year
3. $P = \$180, r = 3\%, t = 2$ years
4. $P = \$260, r = 4\%, t = 3$ years
5. $P = \$5000, r = 8.5\%, t = 9$ months
6. $P = \$18{,}000, r = 7.5\%, t = 18$ months
7. $P = \$15{,}500, r = 11\%, t = 90$ days
8. $P = \$12{,}600, r = 9\%, t = 60$ days

In Exercises 9–14, the principal P is borrowed at simple interest rate r for a period of time t. Find the loan's future value, A, or the total amount due at time t.

9. $P = \$3000, r = 7\%, t = 2$ years
10. $P = \$2000, r = 6\%, t = 3$ years
11. $P = \$26{,}000, r = 9.5\%, t = 5$ years
12. $P = \$24{,}000, r = 8.5\%, t = 6$ years
13. $P = \$9000, r = 6.5\%, t = 8$ months
14. $P = \$6000, r = 4.5\%, t = 9$ months

In Exercises 15–20, the principal P is borrowed and the loan's future value, A, at time t is given. Determine the loan's simple interest rate, r, to the nearest tenth of a percent.

15. $P = \$2000, A = \$2150, t = 1$ year
16. $P = \$3000, A = \$3180, t = 1$ year
17. $P = \$5000, A = \$5900, t = 2$ years
18. $P = \$10{,}000, A = \$14{,}060, t = 2$ years
19. $P = \$2300, A = \$2840, t = 9$ months
20. $P = \$1700, A = \$1820, t = 6$ months

In Exercises 21–26, determine the present value, P, you must invest to have the future value, A, at simple interest rate r after time t. Round answers up to the nearest cent.

21. $A = \$6000, r = 8\%, t = 2$ years
22. $A = \$8500, r = 7\%, t = 3$ years
23. $A = \$14{,}000, r = 9.5\%, t = 6$ years
24. $A = \$16{,}000, r = 11.5\%, t = 5$ years
25. $A = \$5000, r = 14.5\%, t = 9$ months
26. $A = \$2000, r = 12.6\%, t = 8$ months

Practice Plus

27. Solve for r: $A = P(1 + rt)$.
28. Solve for t: $A = P(1 + rt)$.
29. Solve for P: $A = P(1 + rt)$.
30. Solve for P: $A = P(1 + \frac{r}{n})^{nt}$. (We will be using this formula in the next section.)

Application Exercises

31. In order to start a small business, a student takes out a simple interest loan for $4000 for nine months at a rate of 8.25%.
    a. How much interest must the student pay?
    b. Find the future value of the loan.
32. In order to pay for baseball uniforms, a school takes out a simple interest loan for $20,000 for seven months at a rate of 12%.
    a. How much interest must the school pay?
    b. Find the future value of the loan.
33. You borrow $1400 from a friend and promise to pay back $2000 in two years. What simple interest rate, to the nearest tenth of a percent, will you pay?

34. Treasury bills (T-bills) can be purchased from the U.S. Treasury Department. You buy a T-bill for $981.60 that pays $1000 in 13 weeks. What simple interest rate, to the nearest tenth of a percent, does this T-bill earn?

35. To borrow money, you pawn your guitar. Based on the value of the guitar, the pawnbroker loans you $960. One month later, you get the guitar back by paying the pawnbroker $1472. What annual interest rate did you pay?

36. To borrow money, you pawn your mountain bike. Based on the value of the bike, the pawnbroker loans you $552. One month later, you get the bike back by paying the pawnbroker $851. What annual interest rate did you pay?

37. A bank offers a CD that pays a simple interest rate of 6.5%. How much must you put in this CD now in order to have $3000 for a home-entertainment center in two years?

38. A bank offers a CD that pays a simple interest rate of 5.5%. How much must you put in this CD now in order to have $8000 for a kitchen remodeling project in two years?

Writing in Mathematics

39. Explain how to calculate simple interest.

40. What is the future value of a loan and how is it determined?

Critical Thinking Exercises

Make Sense? *In Exercises 41–43, determine whether each statement makes sense or does not make sense, and explain your reasoning.*

41. After depositing $1500 in an account at a rate of 4%, my balance at the end of the first year was $(1500)(0.04).

42. I saved money on my short-term loan for 90 days by finding a financial institution that used the Banker's rule rather than one that calculated interest using 365 days in a year.

43. I planned to save $5000 in four years, computed the present value to be $3846.153, so I rounded the principal to $3846.15.

44. Use the future value formula to show that the time required for an amount of money P to double in value to $2P$ is given by

$$t = \frac{1}{r}.$$

45. You deposit $5000 in an account that earns 5.5% simple interest.

a. Express the future value in the account as a linear function of time, t.

b. Determine the slope of the function in part (a) and describe what this means. Use the phrase "rate of change" in your description.

8.4

WHAT AM I SUPPOSED TO LEARN?

After you have read this section, you should be able to:

1 Use compound interest formulas.

2 Calculate present value.

3 Understand and compute effective annual yield.

Compound Interest

SO, HOW DID THE PRESENT value of Manhattan in 1626—that is, the $24 paid to the Native Americans—attain a future value of over $5 billion in 2010, 384 years later, at a mere 5% interest rate? After all, the future value on $24 for 384 years at 5% simple interest is

$$A = P(1 + rt)$$
$$= 24[1 + (0.05)(384)] = 484.8,$$

or a paltry $484.80, compared to over $5 billion. To understand this dramatic difference in future value, we turn to the concept of *compound interest*.

1 Use compound interest formulas.

Compound Interest

Compound interest is interest computed on the original principal as well as on any accumulated interest. Many savings accounts pay compound interest. For example, suppose you deposit $1000 in a savings account at a rate of 5%. **Table 8.3** on the next page shows how the investment grows if the interest earned is automatically added on to the principal.

TABLE 8.3 Calculating the Amount in an Account Subject to Compound Interest

Use $A = P(1 + rt)$ with $r = 0.05$ and $t = 1$, or $A = P(1 + 0.05)$.

Year	Starting Balance	Amount in the Account at Year's End
1	\$1000	$A = \$1000(1 + 0.05) = \$1050$
2	\$1050 or $\$1000(1 + 0.05)$	$A = \$1050(1 + 0.05) = \$1102.50$ or $A = \$1000(1 + 0.05)(1 + 0.05) = \$1000(1 + 0.05)^2$
3	\$1102.50 or $\$1000(1 + 0.05)^2$	$A = \$1102.50(1 + 0.05) \approx \$1157.63$ or $A = \$1000(1 + 0.05)^2(1 + 0.05) = \$1000(1 + 0.05)^3$

Using inductive reasoning, the amount, A, in the account after t years is the original principal, \$1000, times $(1 + 0.05)^t$: $A = 1000(1 + 0.05)^t$.

If the original principal is P and the interest rate is r, we can use this same approach to determine the amount, A, in an account subject to compound interest.

CALCULATING THE AMOUNT IN AN ACCOUNT FOR COMPOUND INTEREST PAID ONCE A YEAR

If you deposit P dollars at rate r, in decimal form, subject to compound interest, then the amount, A, of money in the account after t years is given by

$$A = P(1 + r)^t.$$

The amount A is called the account's **future value** and the principal P is called its **present value**.

EXAMPLE 1 *Using the Compound Interest Formula*

You deposit \$2000 in a savings account at Hometown Bank, which has a rate of 6%.

a. Find the amount, A, of money in the account after three years subject to interest compounded once a year.

b. Find the interest.

TECHNOLOGY

Here are the calculator keystrokes to compute $2000(1.06)^3$:

Many Scientific Calculators

2000 [×] 1.06 [y^x] 3 [=]

Many Graphing Calculators

2000 [×] 1.06 [∧] 3 [ENTER]

SOLUTION

a. The amount deposited, or principal, P, is \$2000. The rate, r, is 6%, or 0.06. The time of the deposit, t, is three years. The amount in the account after three years is

$$A = P(1 + r)^t = 2000(1 + 0.06)^3 = 2000(1.06)^3 \approx 2382.03.$$

Rounded to the nearest cent, the amount in the savings account after three years is \$2382.03.

b. Because the amount in the account is \$2382.03 and the original principal is \$2000, the interest is \$2382.03 − \$2000, or \$382.03.

CHECK POINT 1 You deposit \$1000 in a savings account at a bank that has a rate of 4%.

a. Find the amount, A, of money in the account after five years subject to interest compounded once a year. Round to the nearest cent.

b. Find the interest.

Compound Interest Paid More Than Once a Year

The period of time between two interest payments is called the **compounding period**. When compound interest is paid once per year, the compounding period is one year. We say that the interest is **compounded annually**.

Most savings institutions have plans in which interest is paid more than once per year. If compound interest is paid twice per year, the compounding period is six months. We say that the interest is **compounded semiannually**. When compound interest is paid four times per year, the compounding period is three months and the interest is said to be **compounded quarterly**. Some plans allow for monthly compounding or daily compounding.

In general, when compound interest is paid n times a year, we say that there are ***n* compounding periods per year**. **Table 8.4** shows the three most frequently used plans in which interest is paid more than once a year.

TABLE 8.4 Interest Plans

Name	Number of Compounding Periods per Year	Length of Each Compounding Period
Semiannual Compounding	$n = 2$	6 months
Quarterly Compounding	$n = 4$	3 months
Monthly Compounding	$n = 12$	1 month

The following formula is used to calculate the amount in an account subject to compound interest with n compounding periods per year:

CALCULATING THE AMOUNT IN AN ACCOUNT FOR COMPOUND INTEREST PAID *n* TIMES A YEAR

If you deposit P dollars at rate r, in decimal form, subject to compound interest paid n times per year, then the amount, A, of money in the account after t years is given by

$$A = P\left(1 + \frac{r}{n}\right)^{nt}.$$

A is the account's **future value** and the principal P is its **present value**.

TECHNOLOGY

Here are the calculator keystrokes to compute

$$7500\left(1 + \frac{0.06}{12}\right)^{12 \cdot 5}.$$

Many Scientific Calculators

7500 [×] [(] 1 [+] .06 [÷] 12 [)] [y^x] [(] 12 [×] 5 [)] [=]

Many Graphing Calculators

7500 [(] 1 [+] .06 [÷] 12 [)] [∧] [(] 12 [×] 5 [)] [ENTER]

You may find it easier to compute the value of the exponent nt (12 × 5 or 60) before using these keystrokes. By entering 60 instead of 12 × 5, you're less likely to make a careless mistake by forgetting to enclose the exponent n times t in parentheses.

EXAMPLE 2 *Using the Compound Interest Formula*

You deposit \$7500 in a savings account that has a rate of 6%. The interest is compounded monthly.

a. How much money will you have after five years?

b. Find the interest after five years.

SOLUTION

a. The amount deposited, or principal, P, is \$7500. The rate, r, is 6%, or 0.06. Because interest is compounded monthly, there are 12 compounding periods per year, so $n = 12$. The time of the deposit, t, is five years. The amount in the account after five years is

$$A = P\left(1 + \frac{r}{n}\right)^{nt} = 7500\left(1 + \frac{0.06}{12}\right)^{12 \cdot 5} = 7500(1.005)^{60} \approx 10{,}116.38.$$

Rounded to the nearest cent, you will have \$10,116.38 after five years.

b. Because the amount in the account is \$10,116.38 and the original principal is \$7500, the interest after five years is \$10,116.38 − \$7500, or \$2616.38.

GREAT QUESTION!

Can I use the formula for compound interest paid n times a year to calculate the amount in an account that pays compound interest only once a year?

Yes. If $n = 1$ (interest paid once a year), the formula

$$A = P\left(1 + \frac{r}{n}\right)^{nt}$$

becomes

$$A = P\left(1 + \frac{r}{1}\right)^{1t}, \text{ or}$$

$$A = P(1 + r)^t.$$

This shows that the amount in an account subject to annual compounding is just one application of the general formula for compound interest paid n times a year. With this general formula, you no longer need a separate formula for annual compounding.

TABLE 8.5 As n Takes on Increasingly Large Values, the Expression $\left(1 + \frac{1}{n}\right)^n$ Approaches the Irrational Number e.

n	$\left(1 + \frac{1}{n}\right)^n$
1	2
2	2.25
5	2.48832
10	2.59374246
100	2.704813829
1000	2.716923932
10,000	2.718145927
100,000	2.718268237
1,000,000	2.718280469
1,000,000,000	2.718281827

CHECK POINT 2 You deposit \$4200 in a savings account that has a rate of 4%. The interest is compounded quarterly.

a. How much money will you have after 10 years? Round to the nearest cent.

b. Find the interest after 10 years.

Continuous Compounding

Some banks use **continuous compounding**, where the compounding periods increase infinitely (compounding interest every trillionth of a second, every quadrillionth of a second, etc.). As n, the number of compounding periods in a year, increases without bound, the expression $\left(1 + \frac{1}{n}\right)^n$ approaches the irrational number e: $e \approx 2.71828$. This is illustrated in **Table 8.5**. As a result, the formula for the balance in an account with n compounding periods per year, $A = P(1 + \frac{r}{n})^{nt}$, becomes $A = Pe^{rt}$ with continuous compounding. Although continuous compounding sounds terrific, it yields only a fraction of a percent more interest over a year than daily compounding.

FORMULAS FOR COMPOUND INTEREST

After t years, the balance, A, in an account with principal P and annual interest rate r (in decimal form) is given by the following formulas:

1. For n compounding periods per year: $A = P\left(1 + \frac{r}{n}\right)^{nt}$

2. For continuous compounding: $A = Pe^{rt}$.

TECHNOLOGY

You can compute e to a power using the $\boxed{e^x}$ key on your calculator. Use the key to enter e^1 and verify that e is approximately equal to 2.71828.

Scientific Calculators	Graphing Calculators
1 $\boxed{e^x}$	$\boxed{e^x}$ 1 $\boxed{\text{ENTER}}$

EXAMPLE 3 *Choosing between Investments*

You decide to invest \$8000 for six years and you have a choice between two accounts. The first pays 7% per year, compounded monthly. The second pays 6.85% per year, compounded continuously. Which is the better investment?

SOLUTION

The better investment is the one with the greater balance in the account after six years. Let's begin with the account with monthly compounding. We use the compound interest formula with $P = 8000$, $r = 7\% = 0.07$, $n = 12$ (monthly compounding means 12 compounding periods per year), and $t = 6$.

$$A = P\left(1 + \frac{r}{n}\right)^{nt} = 8000\left(1 + \frac{0.07}{12}\right)^{12 \cdot 6} \approx 12{,}160.84$$

The balance in this account after six years would be \$12,160.84.

TECHNOLOGY

Here are the calculator keystrokes to compute

$8000e^{0.0685(6)}$:

Many Scientific Calculators

8000 [×] [(] .0685 [×] 6 [)] [e^x] [=]

Many Graphing Calculators

8000 [e^x] [(] .0685 [×] 6 [)] [ENTER]

For the second investment option, we use the formula for continuous compounding with $P = 8000$, $r = 6.85\% = 0.0685$, and $t = 6$.

$$A = Pe^{rt} = 8000e^{0.0685(6)} \approx 12{,}066.60$$

The balance in this account after six years would be \$12,066.60, slightly less than the previous amount. Thus, the better investment is the 7% monthly compounding option.

CHECK POINT 3 A sum of \$10,000 is invested at an annual rate of 8%. Find the balance in the account after five years subject to **a.** quarterly compounding and **b.** continuous compounding.

2 Calculate present value.

Planning for the Future with Compound Interest

Just as we did in Section 8.3, we can determine P, the principal or present value, that should be deposited now in order to have a certain amount, A, in the future. If an account earns compound interest, the amount of money that should be invested today to obtain a future value of A dollars can be determined by solving the compound interest formula for P:

CALCULATING PRESENT VALUE

If A dollars are to be accumulated in t years in an account that pays rate r compounded n times per year, then the present value, P, that needs to be invested now is given by

$$P = \frac{A}{\left(1 + \frac{r}{n}\right)^{nt}}.$$

Remember to round the principal *up* to the nearest cent when computing present value so there will be enough money to meet future goals.

EXAMPLE 4 *Calculating Present Value*

How much money should be deposited today in an account that earns 6% compounded monthly so that it will accumulate to \$20,000 in five years?

SOLUTION

The amount we need today, or the present value, is determined by the present value formula. Because the interest is compounded monthly, $n = 12$. Furthermore, A (the future value) $= \$20{,}000$, r (the rate) $= 6\% = 0.06$, and t (time in years) $= 5$.

$$P = \frac{A}{\left(1 + \frac{r}{n}\right)^{nt}} = \frac{20{,}000}{\left(1 + \frac{0.06}{12}\right)^{12 \cdot 5}} \approx 14{,}827.4439$$

To make sure there will be enough money, we round the principal *up* to \$14,827.45. Approximately \$14,827.45 should be invested today in order to accumulate to \$20,000 in five years.

TECHNOLOGY

Here are the keystrokes for Example 4:

Many Scientific Calculators

20000 [÷] [(] 1 [+] .06 [÷] 12 [)]

[y^x] [(] 12 [×] 5 [)] [=]

Many Graphing Calculators

20000 [÷] [(] 1 [+] .06 [÷] 12 [)]

[∧] [(] 12 [×] 5 [)] [ENTER]

CHECK POINT 4 How much money should be deposited today in an account that earns 7% compounded weekly so that it will accumulate to \$10,000 in eight years?

Blitzer Bonus

The Time Value of Money

When you complete your education and begin earning money, it will be tempting to spend every penny earned. By doing this, you will fail to take advantage of the *time value of money*. The **time value of money** means that a dollar received today is worth more than a dollar received next year or the year after. This is because a sum of money invested today starts earning compound interest sooner than a sum of money invested some time in the future. **A significant way to increase your wealth is to spend less than you earn and invest the difference.** With time on your side, even a small amount of money can be turned into a substantial sum through the power of compounding. Make the time value of money work for you by postponing certain purchases now and investing the savings instead. Pay close attention to your spending habits as you study the time value of money.

3 Understand and compute effective annual yield.

Effective Annual Yield

As we've seen before, a common problem in financial planning is selecting the best investment from two or more investments. For example, is an investment that pays 8.25% interest compounded quarterly better than one that pays 8.3% interest compounded semiannually? Another way to answer the question is to compare the *effective rates* of the investments, also called their *effective annual yields*.

Blitzer Bonus

Doubling Your Money: The Rule of 72

Here's a shortcut for estimating the number of years it will take for your investment to double: Divide 72 by the effective annual yield without the percent sign. For example, if the effective annual yield is 6%, your money will double in approximately

$$\frac{72}{6}$$

years, or in 12 years.

EFFECTIVE ANNUAL YIELD

The **effective annual yield**, or the **effective rate**, is the simple interest rate that produces the same amount of money in an account at the end of one year as when the account is subject to compound interest at a stated rate.

EXAMPLE 5 *Understanding Effective Annual Yield*

You deposit $4000 in an account that pays 8% interest compounded monthly.

a. Find the future value after one year.

b. Use the future value formula for simple interest to determine the effective annual yield.

SOLUTION

a. We use the compound interest formula to find the account's future value after one year.

$$A = P\left(1 + \frac{r}{n}\right)^{nt} = 4000\left(1 + \frac{0.08}{12}\right)^{12 \cdot 1} \approx \$4332.00$$

Principal is $4000.

Stated rate is 8% = 0.08.

Monthly compounding: $n = 12$
Time is one year: $t = 1$.

Rounded to the nearest cent, the future value after one year is $4332.00.

b. The effective annual yield, or effective rate, is a simple interest rate. We use the future value formula for simple interest to determine the simple interest rate that produces a future value of $4332 for a $4000 deposit after one year.

$$A = P(1 + rt)$$ This is the future value formula for simple interest.

$$4332 = 4000(1 + r \cdot 1)$$ Substitute the given values.

$$4332 = 4000 + 4000r$$ Use the distributive property.

$$332 = 4000r$$ Subtract 4000 from both sides.

$$\frac{332}{4000} = \frac{4000r}{4000}$$ Divide both sides by 4000.

$$r = \frac{332}{4000} = 0.083 = 8.3\%$$ Express r as a percent.

The effective annual yield, or effective rate, is 8.3%. This means that money invested at 8.3% simple interest earns the same amount in one year as money invested at 8% interest compounded monthly.

Blitzer Bonus

High-Yield Savings

Online banks can afford to pay the best rates on savings accounts by avoiding the expense of managing a brick-and-mortar bank. To search for the best deals, go to BankRate.com for bank accounts and iMoneyNet.com for money-market funds.

In Example 5, the stated 8% rate is called the **nominal rate**. The 8.3% rate is the effective rate and is a simple interest rate.

CHECK POINT 5 You deposit \$6000 in an account that pays 10% interest compounded monthly.

a. Find the future value after one year.

b. Determine the effective annual yield.

Generalizing the procedure of Example 5 and Check Point 5 gives a formula for effective annual yield:

CALCULATING EFFECTIVE ANNUAL YIELD

Suppose that an investment has a nominal interest rate, r, in decimal form, and pays compound interest n times per year. The investment's effective annual yield, Y, in decimal form, is given by

$$Y = \left(1 + \frac{r}{n}\right)^n - 1.$$

The decimal form of Y given by the formula should then be converted to a percent.

TECHNOLOGY

Here are the keystrokes for Example 6:

Many Scientific Calculators

[(] 1 [+] .05 [÷] 360 [)] [y^x] 360

[−] 1 [=]

Many Graphing Calculators

[(] 1 [+] .05 [÷] 360 [)] [∧] 360

[−] 1 [ENTER].

Given the nominal rate and the number of compounding periods per year, some graphing calculators display the effective annual yield. The screen shows the calculation of the effective rate in Example 6 on the TI-84 Plus.

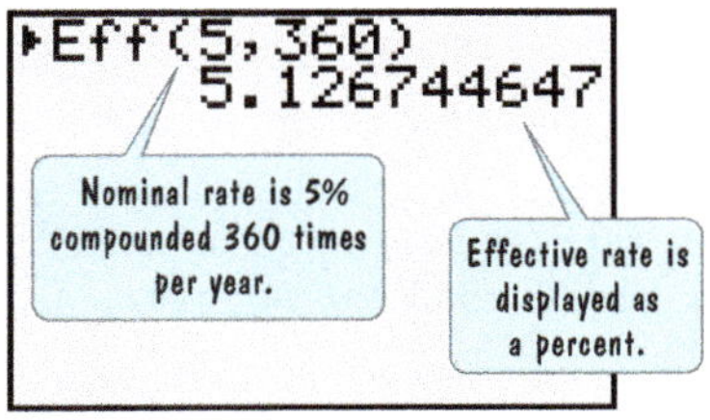

EXAMPLE 6 *Calculating Effective Annual Yield*

A passbook savings account has a nominal rate of 5%. The interest is compounded daily. Find the account's effective annual yield. (Assume 360 days in a year.)

SOLUTION

The rate, r, is 5%, or 0.05. Because interest is compounded daily and we assume 360 days in a year, $n = 360$. The account's effective annual yield is

$$Y = \left(1 + \frac{r}{n}\right)^n - 1 = \left(1 + \frac{0.05}{360}\right)^{360} - 1 \approx 0.0513 = 5.13\%.$$

The effective annual yield is 5.13%. Thus, money invested at 5.13% simple interest earns the same amount of interest in one year as money invested at 5% interest, the nominal rate, compounded daily.

CHECK POINT 6 What is the effective annual yield of an account paying 8% compounded quarterly?

The effective annual yield is often included in the information about investments or loans. Because it's the true interest rate you're earning or paying, it's the number you should pay attention to. **If you are selecting the best investment from two or more investments, the best choice is the account with the greatest effective annual yield.** However, there are differences in the types of accounts that you need to take into consideration. Some pay interest from the day of deposit to the day of withdrawal. Other accounts start paying interest the first day of the month that follows the day of deposit. Some savings institutions stop paying interest if the balance in the account falls below a certain amount.

When *borrowing money*, the effective rate or effective annual yield is usually called the **annual percentage rate**. If all other factors are equal and you are borrowing money, select the option with the least annual percentage rate.

Concept and Vocabulary Check

Fill in each blank so that the resulting statement is true.

1. Compound interest is interest computed on the original ________ as well as on any accumulated ________.

2. The formula $A = P\left(1 + \frac{r}{n}\right)^{nt}$ gives the amount of money, A, in an account after ________ years at rate ________ subject to compound interest paid ________ times per year.

3. If interest is compounded once a year, the formula in Exercise 2 becomes ________.

4. If compound interest is paid twice per year, the compounding period is ________ months and the interest is compounded ________.

5. If compound interest is paid four times per year, the compounding period is ________ months and the interest is compounded ________.

6. When the number of compounding periods in a year increases without bound, this is known as ________ compounding.

7. In the formula

$$P = \frac{A}{\left(1 + \frac{r}{n}\right)^{nt}},$$

the variable ________ represents the amount that needs to be invested now in order to have ________ dollars accumulated in ________ years in an account that pays rate ________ compounded ________ times per year.

8. If you are selecting the best investment from two or more investments, the best choice is the account with the greatest ________, which is the ________ interest rate that produces the same amount of money at the end of one year as when the account is subject to compound interest at a stated rate.

In Exercises 9–12, determine whether each statement is true or false. If the statement is false, make the necessary change(s) to produce a true statement.

9. Formulas for compound interest show that a dollar invested today is worth more than a dollar invested in the future.______

10. Formulas for compound interest show that if you make the decision to postpone certain purchases and save the money instead, small amounts of money can be turned into substantial sums over a period of years.______

11. At a given annual interest rate, your money grows faster as the compounding period becomes shorter.______

12. According to the Rule of 72 (see the Blitzer Bonus on page 518), an investment with an effective annual yield of 12% can double in six years.______

Exercise Set 8.4

Here is a list of formulas needed to solve the exercises. Be sure you understand what each formula describes and the meaning of the variables in the formulas.

$$A = P\left(1 + \frac{r}{n}\right)^{nt} \qquad P = \frac{A}{\left(1 + \frac{r}{n}\right)^{nt}}$$

$$A = Pe^{rt} \qquad Y = \left(1 + \frac{r}{n}\right)^{n} - 1$$

Practice Exercises

In Exercises 1–12, the principal represents an amount of money deposited in a savings account subject to compound interest at the given rate.

a. *Find how much money there will be in the account after the given number of years. (Assume 360 days in a year.)*

b. *Find the interest earned.*

Round answers to the nearest cent.

Principal	Rate	Compounded	Time
1. \$10,000	4%	annually	2 years
2. \$8000	6%	annually	3 years
3. \$3000	5%	semiannually	4 years
4. \$4000	4%	semiannually	5 years
5. \$9500	6%	quarterly	5 years
6. \$2500	8%	quarterly	6 years
7. \$4500	4.5%	monthly	3 years
8. \$2500	6.5%	monthly	4 years
9. \$1500	8.5%	daily	2.5 years
10. \$1200	8.5%	daily	3.5 years
11. \$20,000	4.5%	daily	20 years
12. \$25,000	5.5%	daily	20 years

Solve Exercises 13–16 using appropriate compound interest formulas. Round answers to the nearest cent.

13. Find the accumulated value of an investment of \$10,000 for five years at an interest rate of 5.5% if the money is **a.** compounded semiannually; **b.** compounded quarterly; **c.** compounded monthly; **d.** compounded continuously.

14. Find the accumulated value of an investment of \$5000 for 10 years at an interest rate of 6.5% if the money is **a.** compounded semiannually; **b.** compounded quarterly; **c.** compounded monthly; **d.** compounded continuously.

15. Suppose that you have \$12,000 to invest. Which investment yields the greater return over three years: 7% compounded monthly or 6.85% compounded continuously?

16. Suppose that you have \$6000 to invest. Which investment yields the greater return over four years: 8.25% compounded quarterly or 8.3% compounded semiannually?

In Exercises 17–20, round answers up to the nearest cent.

17. How much money should be deposited today in an account that earns 6% compounded semiannually so that it will accumulate to \$10,000 in three years?

18. How much money should be deposited today in an account that earns 7% compounded semiannually so that it will accumulate to \$12,000 in four years?

19. How much money should be deposited today in an account that earns 9.5% compounded monthly so that it will accumulate to \$10,000 in three years?

20. How much money should be deposited today in an account that earns 10.5% compounded monthly so that it will accumulate to \$22,000 in four years?

21. You deposit \$10,000 in an account that pays 4.5% interest compounded quarterly.

a. Find the future value after one year.

b. Use the future value formula for simple interest to determine the effective annual yield.

22. You deposit \$12,000 in an account that pays 6.5% interest compounded quarterly.

a. Find the future value after one year.

b. Use the future value formula for simple interest to determine the effective annual yield.

In Exercises 23–28, a passbook savings account has a rate of 6%. Find the effective annual yield, rounded to the nearest tenth of a percent, if the interest is compounded

23. semiannually.

24. quarterly.

25. monthly.

26. daily. (Assume 360 days in a year.)

27. 1000 times per year.

28. 100,000 times per year.

In Exercises 29–32, determine the effective annual yield for each investment. Then select the better investment. Assume 360 days in a year. If rounding is required, round to the nearest tenth of a percent.

29. 8% compounded monthly; 8.25% compounded annually

30. 5% compounded monthly; 5.25% compounded quarterly

31. 5.5% compounded semiannually; 5.4% compounded daily

32. 7% compounded annually; 6.85% compounded daily

Practice Plus

In Exercises 33–36, how much more would you earn in the first investment than in the second investment? Round answers to the nearest dollar.

33.
- \$25,000 invested for 40 years at 12% compounded annually
- \$25,000 invested for 40 years at 6% compounded annually

34.
- \$30,000 invested for 40 years at 10% compounded annually
- \$30,000 invested for 40 years at 5% compounded annually

35.
- \$50,000 invested for 30 years at 10% compounded annually
- \$50,000 invested for 30 years at 5% compounded monthly

36.
- \$20,000 invested for 30 years at 12% compounded annually
- \$20,000 invested for 30 years at 6% compounded monthly

Application Exercises

Assume that the accounts described in the exercises have no other deposits or withdrawals except for what is stated. Round all answers to the nearest dollar, rounding up to the nearest dollar in present-value problems. Assume 360 days in a year.

37. At the time of a child's birth, \$12,000 was deposited in an account paying 6% interest compounded semiannually. What will be the value of the account at the child's twenty-first birthday?

38. At the time of a child's birth, \$10,000 was deposited in an account paying 5% interest compounded semiannually. What will be the value of the account at the child's twenty-first birthday?

39. You deposit \$2600 in an account that pays 4% interest compounded once a year. Your friend deposits \$2200 in an account that pays 5% interest compounded monthly.

a. Who will have more money in their account after one year? How much more?

b. Who will have more money in their account after five years? How much more?

c. Who will have more money in their account after 20 years? How much more?

40. You deposit \$3000 in an account that pays 3.5% interest compounded once a year. Your friend deposits \$2500 in an account that pays 4.8% interest compounded monthly.

a. Who will have more money in their account after one year? How much more?

b. Who will have more money in their account after five years? How much more?

c. Who will have more money in their account after 20 years? How much more?

41. You deposit \$3000 in an account that pays 7% interest compounded semiannually. After 10 years, the interest rate is increased to 7.25% compounded quarterly. What will be the value of the account after 16 years?

42. You deposit \$6000 in an account that pays 5.25% interest compounded semiannually. After 10 years, the interest rate is increased to 5.4% compounded quarterly. What will be the value of the account after 18 years?

43. In 1626, Peter Minuit convinced the Wappinger Indians to sell him Manhattan Island for \$24. If the Native Americans had put the \$24 into a bank account paying compound interest at a 5% rate, how much would the investment have been worth in the year 2010 ($t = 384$ years) if interest were compounded **a.** monthly? **b.** 360 times per year?

44. In 1777, Jacob DeHaven loaned George Washington's army \$450,000 in gold and supplies. Due to a disagreement over the method of repayment (gold versus Continental money), DeHaven was never repaid, dying penniless. In 1989, his descendants sued the U.S. government over the 212-year-old debt. If the DeHavens used an interest rate of 6% and daily compounding (the rate offered by the Continental Congress in 1777), how much money did the DeHaven family demand in their suit? (*Hint:* Use the compound interest formula with $n = 360$ and $t = 212$ years.)

45. Will you earn more interest in one year by depositing \$2000 in a simple interest account that pays 6% or in an account that pays 5.9% interest compounded daily? How much more interest will you earn?

46. Will you earn more interest in one year by depositing \$1000 in a simple interest account that pays 7% or in an account that pays 6.9% interest compounded daily? How much more interest will you earn?

47. Two accounts each begin with a deposit of \$5000. Both accounts have rates of 5.5%, but one account compounds interest once a year while the other account compounds interest continuously. Make a table that shows the amount in each account and the interest earned after 1 year, 5 years, 10 years, and 20 years.

48. Two accounts each begin with a deposit of \$10,000. Both accounts have rates of 6.5%, but one account compounds interest once a year while the other account compounds interest continuously. Make a table that shows the amount in each account and the interest earned after 1 year, 5 years, 10 years, and 20 years.

49. Parents wish to have \$80,000 available for a child's education. If the child is now 5 years old, how much money must be set aside at 6% compounded semiannually to meet their financial goal when the child is 18?

50. A 30-year-old worker plans to retire at age 65. He believes that \$500,000 is needed to retire comfortably. How much should be deposited now at 7% compounded monthly to meet the \$500,000 retirement goal?

51. You would like to have \$75,000 available in 15 years. There are two options. Account A has a rate of 4.5% compounded once a year. Account B has a rate of 4% compounded daily. How much would you have to deposit in each account to reach your goal?

52. You would like to have \$150,000 available in 20 years. There are two options. Account A has a rate of 5.5% compounded once a year. Account B has a rate of 5% compounded daily. How much would you have to deposit in each account to reach your goal?

53. You invest \$1600 in an account paying 5.4% interest compounded daily. What is the account's effective annual yield? Round to the nearest hundredth of a percent.

54. You invest \$3700 in an account paying 3.75% interest compounded daily. What is the account's effective annual yield? Round to the nearest hundredth of a percent.

55. An account has a nominal rate of 4.2%. Find the effective annual yield, rounded to the nearest tenth of a percent, with quarterly compounding, monthly compounding, and daily compounding. How does changing the compounding period affect the effective annual yield?

56. An account has a nominal rate of 4.6%. Find the effective annual yield, rounded to the nearest tenth of a percent, with quarterly compounding, monthly compounding, and daily compounding. How does changing the compounding period affect the effective annual yield?

57. A bank offers a money market account paying 4.5% interest compounded semiannually. A competing bank offers a money market account paying 4.4% interest compounded daily. Which account is the better investment?

58. A bank offers a money market account paying 4.9% interest compounded semiannually. A competing bank offers a money market account paying 4.8% interest compounded daily. Which account is the better investment?

Writing in Mathematics

59. Describe the difference between simple and compound interest.

60. Give two examples that illustrate the difference between a compound interest problem involving future value and a compound interest problem involving present value.

61. What is effective annual yield?

62. Explain how to select the best investment from two or more investments.

Critical Thinking Exercises

Make Sense? *In Exercises 63–66, determine whether each statement makes sense or does not make sense, and explain your reasoning.*

63. My bank provides simple interest at 3.25% per year, but I can't determine if this is a better deal than a competing bank offering 3.25% compound interest without knowing the compounding period.

64. When choosing between two accounts, the one with the greater annual interest rate is always the better deal.

65. A bank can't increase compounding periods indefinitely without owing its customers an infinite amount of money.

66. My bank advertises a compound interest rate of 2.4%, although, without making deposits or withdrawals, the balance in my account increased by 2.43% in one year.

67. A depositor opens a new savings account with \$6000 at 5% compounded semiannually. At the beginning of year 3, an additional \$4000 is deposited. At the end of six years, what is the balance in the account?

68. A depositor opens a money market account with \$5000 at 8% compounded monthly. After two years, \$1500 is withdrawn from the account to buy a new computer. A year later, \$2000 is put in the account. What will be the ending balance if the money is kept in the account for another three years?

69. Use the future value formulas for simple and compound interest in one year to derive the formula for effective annual yield.

Group Exercise

70. This activity is a group research project intended for four or five people. Present your research in a seminar on the history of interest and banking. The seminar should last about 30 minutes. Address the following questions:

When was interest first charged on loans? How was lending money for a fee opposed historically? What is usury? What connection did banking and interest rates play in the historic European rivalries between Christians and Jews? When and where were some of the highest interest rates charged? What were the rates? Where does the word *interest* come from? What is the origin of the word *shylock*? What is the difference between usury and interest in modern times? What is the history of a national bank in the United States?

8.5 Annuities, Methods of Saving, and Investments

WHAT AM I SUPPOSED TO LEARN?

After you have read this section, you should be able to:

1. Determine the value of an annuity.
2. Determine regular annuity payments needed to achieve a financial goal.
3. Understand stocks and bonds as investments.
4. Read stock tables.
5. Understand accounts designed for retirement savings.

ACCORDING TO THE *FORBES Billionaires List*, in 2012 the two richest Americans were Bill Gates (net worth: \$61 billion) and Warren Buffett (net worth: \$44 billion). In May 1965, Buffett's new company, Berkshire Hathaway, was selling one share of stock for \$18. By the end of 2008, the price of a share had increased to \$96,600. If you had purchased one share in May 1965, your **return**, or percent increase, would be

Warren Buffett and Bill Gates

$$\frac{\text{amount of increase}}{\text{original amount}} = \frac{\$96{,}600 - \$18}{\$18} \approx 5365.67 = 536{,}567\%.$$

What does a return of nearly 540,000% mean? If you had invested \$250 in Warren Buffett's company in May 1965, your shares would have been worth over \$1.3 million by December 2008.

Of course, investments that potentially offer outrageous returns come with great risk of losing part or all of the principal. The bottom line: Is there a safe way to save regularly and have an investment worth one million dollars or more? In this section, we consider such savings plans, some of which come with special tax treatment, as well as riskier investments in stocks and bonds.

Determine the value of an annuity.

Annuities

The compound interest formula

$$A = P(1 + r)^t$$

gives the future value, A, after t years, when a fixed amount of money, P, the principal, is deposited in an account that pays an annual interest rate r (in decimal form) compounded once a year. However, money is often invested in small amounts at periodic intervals. For example, to save for retirement, you might decide to place \$1000 into an Individual Retirement Account (IRA) at the end of each year until you retire. An **annuity** is a sequence of equal payments made at equal time periods. An IRA is an example of an annuity.

The **value of an annuity** is the sum of all deposits plus all interest paid. Our first example illustrates how to find this value.

EXAMPLE 1 Determining the Value of an Annuity

You deposit \$1000 into a savings plan at the end of each year for three years. The interest rate is 8% per year compounded annually.

a. Find the value of the annuity after three years.

b. Find the interest.

SOLUTION

a. The value of the annuity after three years is the sum of all deposits made plus all interest paid over three years.

This is the \$1000 deposit at year's end.

Value at end of year 1 = \$1000

This is the first-year deposit with interest earned for a year. This is the \$1000 deposit at year's end.

Value at end of year 2 = \$1000(1 + 0.08) + \$1000

= \$1080 + \$1000 = \$2080

Use $A = P(1 + r)^t$ with $r = 0.08$ and $t = 1$, or $A = P(1 + 0.08)$.

This is the second-year balance, \$2080, with interest earned for a year. This is the \$1000 deposit at year's end.

Value at end of year 3 = \$2080(1 + 0.08) + \$1000

= \$2246.40 + \$1000 = \$3246.40

The value of the annuity at the end of three years is \$3246.40.

b. You made three payments of \$1000 each, depositing a total of 3 × \$1000, or \$3000. Because the value of the annuity is \$3246.40, the interest is \$3246.40 − \$3000, or \$246.40.

CHECK POINT 1 You deposit \$2000 into a savings plan at the end of each year for three years. The interest rate is 10% per year compounded annually.

a. Find the value of the annuity after three years.

b. Find the interest.

Suppose that you deposit P dollars into an account at the end of each year. The account pays an annual interest rate, r, compounded annually. At the end of the first year, the account contains P dollars. At the end of the second year, P dollars is deposited again. At the time of this deposit, the first deposit has received interest earned during the second year. Thus, the value of the annuity after two years is

$$P + P(1 + r).$$

Deposit of P dollars at end of second year

First-year deposit of P dollars with interest earned for a year

The value of the annuity after three years is

$$P \quad + \quad P(1 + r) \quad + \quad P(1 + r)^2.$$

Deposit of P dollars at end of third year

Second-year deposit of P dollars with interest earned for a year

First-year deposit of P dollars with interest earned over two years

The value of the annuity after t years is

$$P + P(1 + r) + P(1 + r)^2 + P(1 + r)^3 + \cdots + P(1 + r)^{t-1}.$$

Deposit of P dollars at end of year t

First-year deposit of P dollars with interest earned over $t - 1$ years

Each term in this sum is obtained by multiplying the preceding term by $(1 + r)$. Thus, the terms form a geometric sequence. Using a formula for the sum of the terms of a geometric sequence, we can obtain the following formula that gives the value of this annuity:

VALUE OF AN ANNUITY: INTEREST COMPOUNDED ONCE A YEAR

If P is the deposit made at the end of each year for an annuity that pays an annual interest rate r (in decimal form) compounded once a year, the value, A, of the annuity after t years is

$$A = \frac{P[(1 + r)^t - 1]}{r}.$$

EXAMPLE 2 Determining the Value of an Annuity

Although you are a long way from retirement, the time to begin retirement savings is when you begin earning a paycheck and can take advantage of the time value of your money.

Suppose that when you are 35, you decide to save for retirement by depositing \$1000 into an IRA at the end of each year for 30 years. If you can count on an interest rate of 10% per year compounded annually,

a. How much will you have from the IRA after 30 years?

b. Find the interest.

Round answers to the nearest dollar.

TECHNOLOGY

Here are the calculator keystrokes to compute

$$\frac{1000[(1 + 0.10)^{30} - 1]}{0.10}:$$

Many Scientific Calculators

1000 [×] [(] [(] 1 [+] .10 [)]

[y^x] 30 [−] 1 [)] [)] [÷] .10 [=]

Observe that the part of the numerator in brackets is entered as $((1 + 0.10)^{30} - 1)$ to maintain the order of operations.

Many Graphing Calculators

1000 [(] [(] 1 [+] .10 [)] [∧]

30 [−] 1 [)] [)] [÷] .10 [ENTER]

GREAT QUESTION!

Can I use the formula for the value of an annuity with interest compounded *n* times per year to calculate the value of an annuity if interest is compounded only once a year?

Yes. If $n = 1$ (interest compounded once a year), the formula in the box at the right below becomes

$$A = \frac{P[(1 + \frac{r}{1})^{1t} - 1]}{(\frac{r}{1})}, \text{ or}$$

$$A = \frac{P[(1 + r)^t - 1]}{r}.$$

This was the boxed formula on page 525, showing that the value of an annuity with interest compounded once a year is just one application of the general formula in the box on the right. With this general formula, you no longer need a separate formula for annual compounding.

SOLUTION

a. The amount that you will have from the IRA is its value after 30 years.

$$A = \frac{P[(1 + r)^t - 1]}{r}$$ Use the formula for the value of an annuity.

$$A = \frac{1000[(1 + 0.10)^{30} - 1]}{0.10}$$ The annuity involves year-end deposits of \$1000: $P = 1000$. The interest rate is 10%: $r = 0.10$. The number of years is 30: $t = 30$. The Technology box shows how this computation can be done in a single step using parentheses keys.

$$= \frac{1000[(1.10)^{30} - 1]}{0.10}$$ Add inside parentheses: $1 + 0.10 = 1.10$.

$$\approx \frac{1000(17.4494 - 1)}{0.10}$$ Use a calculator to find $(1.10)^{30}$: 1.1 [y^x] 30 [=].

$$= \frac{1000(16.4494)}{0.10}$$ Simplify inside parentheses: $17.4494 - 1 = 16.4494$.

$$= 164{,}494$$ Use a calculator: 1000 [×] 16.4494 [÷] .10 [=].

After 30 years, you will have approximately \$164,494 from the IRA.

b. You made 30 payments of \$1000 each, depositing a total of 30 × \$1000, or \$30,000. Because the value of the annuity is approximately \$164,494, the interest is approximately

$$\$164{,}494 - \$30{,}000, \text{ or } \$134{,}494.$$

The interest is nearly $4\frac{1}{2}$ times the amount of your payments, illustrating the power of compounding.

CHECK POINT 2 Suppose that when you are 25, you deposit \$3000 into an IRA at the end of each year for 40 years. If you can count on an interest rate of 8% per year compounded annually,

a. How much will you have from the IRA after 40 years?

b. Find the interest.

Round answers to the nearest dollar.

We can adjust the formula for the value of an annuity if equal payments are made at the end of each of *n* yearly compounding periods.

VALUE OF AN ANNUITY: INTEREST COMPOUNDED *n* TIMES PER YEAR

If *P* is the deposit made at the end of each compounding period for an annuity that pays an annual interest rate *r* (in decimal form) compounded *n* times per year, the value, *A*, of the annuity after *t* years is

$$A = \frac{P\left[\left(1 + \frac{r}{n}\right)^{nt} - 1\right]}{\left(\frac{r}{n}\right)}.$$

EXAMPLE 3 *Determining the Value of an Annuity*

At age 25, to save for retirement, you decide to deposit \$200 at the end of each month into an IRA that pays 7.5% compounded monthly.

a. How much will you have from the IRA when you retire at age 65?

b. Find the interest.

Round answers to the nearest dollar.

SOLUTION

a. Because you are 25, the amount that you will have from the IRA when you retire at 65 is its value after 40 years.

$$A = \frac{P\left[\left(1+\frac{r}{n}\right)^{nt}-1\right]}{\left(\frac{r}{n}\right)}$$ Use the formula for the value of an annuity.

$$A = \frac{200\left[\left(1+\frac{0.075}{12}\right)^{12\cdot 40}-1\right]}{\left(\frac{0.075}{12}\right)}$$ The annuity involves month-end deposits of \$200: $P = 200$. The interest rate is 7.5%: $r = 0.075$. The interest is compounded monthly: $n = 12$. The number of years is 40: $t = 40$.

$$= \frac{200[(1+0.00625)^{480}-1]}{0.00625}$$ Using parentheses keys, these calculations can be performed in a single step on a calculator. Answers may slightly vary if you do the calculations in stages and round along the way.

$$= \frac{200[(1.00625)^{480}-1]}{0.00625}$$ Add inside parentheses: $1 + 0.00625 = 1.00625$.

$$\approx \frac{200(19.8989-1)}{0.00625}$$ Use a calculator to find $(1.00625)^{480}$: 1.00625 [y^x] 480 [=].

$$\approx 604{,}765$$

After 40 years, you will have approximately \$604,765 when retiring at age 65.

b. Interest = Value of the IRA − Total deposits

$\approx \$604{,}765 - \$200 \cdot 12 \cdot 40$

\$200 per month × 12 months per year × 40 years

$= \$604{,}765 - \$96{,}000 = \$508{,}765$

The interest is approximately \$508,765, more than five times the amount of your contributions to the IRA.

Annuities can be categorized by when payments are made. The formula used to solve Example 3 describes **ordinary annuities**, where payments are made at the end of each period. The formula assumes the same number of yearly payments and yearly compounding periods. An annuity plan in which payments are made at the beginning of each period is called an **annuity due**. The formula for the value of this type of annuity is slightly different than the one used in Example 3.

CHECK POINT 3 At age 30, to save for retirement, you decide to deposit \$100 at the end of each month into an IRA that pays 9.5% compounded monthly.

a. How much will you have from the IRA when you retire at age 65?

b. Find the interest.

Round answers to the nearest dollar.

2 Determine regular annuity payments needed to achieve a financial goal.

Planning for the Future with an Annuity

By solving the annuity formula for P, we can determine the amount of money that should be deposited at the end of each compounding period so that an annuity has a future value of A dollars. The following formula gives the regular payments, P, needed to reach a financial goal, A:

REGULAR PAYMENTS NEEDED TO ACHIEVE A FINANCIAL GOAL

The deposit, P, that must be made at the end of each compounding period into an annuity that pays an annual interest rate r (in decimal form) compounded n times per year in order to achieve a value of A dollars after t years is

$$P = \frac{A\left(\frac{r}{n}\right)}{\left[\left(1 + \frac{r}{n}\right)^{nt} - 1\right]}.$$

When computing regular payments needed to achieve a financial goal, round the deposit made at the end of each compounding period *up*. In this way, you won't fall slightly short of being able to meet future goals. In this section, we will round annuity payments up to the nearest dollar.

EXAMPLE 4 *Using Long-Term Planning to Achieve a Financial Goal*

Suppose that once you complete your college education and begin working, you would like to save \$20,000 over five years to use as a down payment for a home. You anticipate making regular, end-of-month deposits in an annuity that pays 6% compounded monthly.

a. How much should you deposit each month? Round up to the nearest dollar.

b. How much of the \$20,000 down payment comes from deposits and how much comes from interest?

SOLUTION

a. $$P = \frac{A\left(\frac{r}{n}\right)}{\left[\left(1 + \frac{r}{n}\right)^{nt} - 1\right]}$$

Use the formula for regular payments, P, needed to achieve a financial goal, A.

$$P = \frac{20{,}000\left(\frac{0.06}{12}\right)}{\left[\left(1 + \frac{0.06}{12}\right)^{12 \cdot 5} - 1\right]}$$

Your goal is to accumulate \$20,000 ($A = 20{,}000$) over five years ($t = 5$). The interest rate is 6% ($r = 0.06$) compounded monthly ($n = 12$).

$$\approx 287$$

Use a calculator and round up to the nearest dollar to be certain you do not fall short of your goal.

You should deposit \$287 each month to be certain of having \$20,000 for a down payment on a home.

GREAT QUESTION!

Which will earn me more money: a lump-sum deposit or an annuity?

In Example 4 of Section 8.4, we saw that a lump-sum deposit of approximately \$14,828 at 6% compounded monthly would accumulate to \$20,000 in five years. In Example 4 on the right, we see that total deposits of \$17,220 are required to reach the same goal. With the same interest rate, compounding period, and time period, a lump-sum deposit will generate more interest than an annuity. If you don't have a large sum of money to open an account, an annuity is a realistic, although more expensive, option to a lump-sum deposit for achieving the same financial goal.

b. Total deposits $= \$287 \cdot 12 \cdot 5 = \$17{,}220$

\$287 per month × 12 months per year × 5 years

Interest $= \$20{,}000 - \$17{,}220 = \$2780$

We see that \$17,220 of the \$20,000 comes from your deposits and the remainder, \$2780, comes from interest.

CHECK POINT 4 Parents of a baby girl are in a financial position to begin saving for her college education. They plan to have \$100,000 in a college fund in 18 years by making regular, end-of-month deposits in an annuity that pays 9% compounded monthly.

a. How much should they deposit each month? Round up to the nearest dollar.

b. How much of the \$100,000 college fund comes from deposits and how much comes from interest?

3 Understand stocks and bonds as investments.

Investments: Risk and Return

When you deposit money into a bank account, you are making a **cash investment**. Because bank accounts up to \$250,000 are insured by the federal government, there is no risk of losing the principal you've invested. The account's interest rate guarantees a certain percent increase in your investment, called its **return**.

All investments involve a trade-off between risk and return. The different types of bank accounts carry little or no risk, so investors must be willing to accept low returns. There are other kinds of investments that are riskier, meaning that it is possible to lose all or part of your principal. These investments, including *stocks* and *bonds*, give a reasonable expectation of higher returns to attract investors.

Stocks

Investors purchase **stock**, shares of ownership in a company. The shares indicate the percent of ownership. For example, if a company has issued a total of one million shares and an investor owns 20,000 of these shares, that investor owns

$$\frac{20{,}000 \text{ shares}}{1{,}000{,}000 \text{ shares}} = 0.02$$

or 2% of the company. Any investor who owns some percentage of the company is called a **shareholder**.

Buying or selling stock is referred to as **trading**. Shares of stock need both a seller and a buyer to be traded. Stocks are traded on a **stock exchange**. The price of a share of stock is determined by the law of supply and demand. If a company is prospering, investors will be willing to pay a good price for its stock, and so the stock price goes up. If the company does not do well, investors may decide to sell, and the stock price goes down. Stock prices indicate the performance of the companies they represent, as well as the state of the national and global economies.

There are two ways to make money by investing in stock:

- You sell the shares for more money than what you paid for them, in which case you have a **capital gain** on the sale of stock. (There can also be a capital loss by selling for less than what you paid, or if the company goes bankrupt.)
- While you own the stock, the company distributes all or part of its profits to shareholders as **dividends**. Each share is paid the same dividend, so the amount you receive depends on the number of shares owned. (Some companies reinvest all profits and do not distribute dividends.)

When more and more average Americans began investing and making money in stocks in the 1990s, the federal government cut the capital-gains tax rate. Long-term capital gains (profits on items held for more than a year before being sold) and dividends are taxed at lower rates than wages and interest earnings.

Bonds

People who buy stock become part owners in a company. In order to raise money and not dilute the ownership of current stockholders, companies sell **bonds**. People who buy a bond are **lending money** to the company from which they buy the bond. Bonds are a commitment from a company to pay the price an investor pays for the bond at the time it was purchased, called the **face value**, along with interest payments at a given rate.

There are many reasons for issuing bonds. A company might need to raise money for research on a drug that has the potential for curing AIDS, so it issues bonds. The U.S. Treasury Department issues 30-year bonds at a fixed 7% annual rate to borrow money to cover federal deficits. Local governments often issue bonds to borrow money to build schools, parks, and libraries.

Bonds are traded like stock, and their price is a function of supply and demand. If a company goes bankrupt, bondholders are the first to claim the company's assets. They make their claims before the stockholders, even though (unlike stockholders) they do not own a share of the company. Generally speaking, investing in bonds is less risky than investing in stocks, although the return is lower.

Mutual Funds

It is not an easy job to determine which stocks and bonds to buy or sell, or when to do so. Even IRAs can be funded by mixing stocks and bonds. Many small investors have decided that they do not have the time to stay informed about the progress of corporations, even with the help of online industry research. Instead, they invest in a **mutual fund**. A mutual fund is a group of stocks and/or bonds managed by a professional investor. When you purchase shares in a mutual fund, you give your money to the **fund manager**. Your money is combined with the money of other investors in the mutual fund. The fund manager invests this pool of money, buying and selling shares of stocks and bonds to obtain the maximum possible returns.

Investors in mutual funds own a small portion of many different companies, which may protect them against the poor performance of a single company. When comparing mutual funds, consider both the fees charged for investing and how well the fund manager is doing with the fund's money. Newspapers publish ratings from 1 (worst) to 5 (best) of mutual fund performance based on whether the manager is doing a good job with its investors' money. Two numbers are given. The first number compares the performance of the mutual fund to a large group of similar funds. The second number compares the performance to funds that are nearly identical. The best rating a fund manager can receive is 5/5; the worst is 1/1.

A listing of all the investments that a person holds is called a **financial portfolio**. Most financial advisors recommend a portfolio with a mixture of low-risk and high-risk investments, called a **diversified portfolio**.

4 Read stock tables.

Reading Stock Tables

Daily newspapers and online services give current stock prices and other information about stocks. We will use FedEx (Federal Express) stock to learn how to read these daily stock tables. Look at the following newspaper listing of FedEx stock.

52-Week High	52-Week Low	Stock	SYM	Div	Yld %	PE	Vol 100s	Hi	Lo	Close	Net Chg
99.46	34.02	FedEx	FDX	.44	1.0	19	37701	45	43.47	44.08	−1.60

The headings indicate the meanings of the numbers across the row.

52-Week High
99.46

The heading **52-Week High** refers to the *highest price* at which FedEx stock traded during the past 52 weeks. The highest price was $99.46 per share. This means that during the past 52 weeks at least one investor was willing to pay $99.46 for a share of FedEx stock. Notice that 99.46 represents a quantity in dollars, although the stock table does not show the dollar sign.

52-Week Low
34.02

The heading **52-Week Low** refers to the *lowest price* at which FedEx stock traded during the past 52 weeks. This price is $34.02.

Stock	SYM
FedEx	FDX

The heading **Stock** is the *company name*, FedEx. The heading **SYM** is the *symbol* the company uses for trading. FedEx uses the symbol FDX.

Div
.44

The heading **Div** refers to *dividends* paid per share to stockholders during the past year. FedEx paid a dividend of $0.44 per share. Once again, the dollar symbol does not appear in the table. Thus, if you owned 100 shares, you received a dividend of $0.44 × 100, or $44.00.

Yld %
1.0

The heading **Yld %** stands for *percent yield*. In this case, the percent yield is 1.0%. (The stock table does not show the percent sign.) This means that the dividends alone gave investors an annual return of 1.0%. This is much lower than the average inflation rate. However, this percent does not take into account the fact that FedEx stock prices might rise. If an investor sells shares for more than the purchase price, the gain will probably make FedEx stock a much better investment than a bank account.

In order to understand the meaning of the heading PE, we need to understand some of the other numbers in the table. We will return to this column.

Vol 100s
37701

The heading **Vol 100s** stands for *sales volume in hundreds*. This is the number of shares traded yesterday, in hundreds. The number in the table is 37,701. This means that yesterday, a total of 37,701 × 100, or 3,770,100 shares of FedEx were traded.

Hi
45

The heading **Hi** stands for the *highest price* at which FedEx stock traded *yesterday*. This number is 45. Yesterday, FedEx's highest trading price was $45 a share.

Lo
43.47

The heading **Lo** stands for the *lowest price* at which FedEx stock traded *yesterday*. This number is 43.47. Yesterday, FedEx's lowest trading price was $43.47 a share.

Close
44.08

The heading **Close** stands for the *price* at which shares last traded *when the stock exchange closed yesterday*. This number is 44.08. Thus, the price at which shares of FedEx traded when the stock exchange closed yesterday was $44.08 per share. This is called yesterday's **closing price**.

Net Chg
−1.60

The heading **Net Chg** stands for *net change*. This is the change in price from the market close two days ago to yesterday's market close. This number is −1.60. Thus, the price of a share of FedEx stock went down by $1.60. For some stock listings, the notation. . . appears under Net Chg. This means that there was *no change in price* for a share of stock from the market close two days ago to yesterday's market close.

PE
19

Now, we are ready to return to the heading **PE**, standing for the *price-to-earnings ratio*.

$$\text{PE ratio} = \frac{\text{Yesterday's closing price per share}}{\text{Annual earnings per share}}$$

This can also be expressed as

$$\text{Annual earnings per share} = \frac{\text{Yesterday's closing price per share}}{\text{PE ratio}}.$$

The PE ratio for FedEx is given to be 19. Yesterday's closing price per share was 44.08. We can substitute these numbers into the formula to find annual earnings per share:

Close	PE
44.08	19

$$\text{Annual earnings per share} = \frac{44.08}{19} = 2.32.$$

The annual earnings per share for FedEx were $2.32. The PE ratio, 19, tells us that yesterday's closing price per share, $44.08, is 19 times greater than the earnings per share, $2.32.

EXAMPLE 5 *Reading Stock Tables*

52-Week High	52-Week Low	Stock	SYM	Div	Yld %	PE	Vol 100s	Hi	Lo	Close	Net Chg
42.38	22.50	Disney	DIS	.21	.6	43	115900	32.50	31.25	32.50	. . .

Use the stock table for Disney to answer the following questions.

a. What were the high and low prices for the past 52 weeks?

b. If you owned 3000 shares of Disney stock last year, what dividend did you receive?

c. What is the annual return for dividends alone? How does this compare to a bank account offering a 3.5% interest rate?

d. How many shares of Disney were traded yesterday?

e. What were the high and low prices for Disney shares yesterday?

f. What was the price at which Disney shares last traded when the stock exchange closed yesterday?

g. What does the value or symbol in the net change column mean?

h. Compute Disney's annual earnings per share using

$$\text{Annual earnings per share} = \frac{\text{Yesterday's closing price per share}}{\text{PE ratio}}.$$

SOLUTION

a. We find the high price for the past 52 weeks by looking under the heading **High**. The price is listed in dollars, given as 42.38. Thus, the high price for a share of stock for the past 52 weeks was \$42.38. We find the low price for the past 52 weeks by looking under the heading **Low**. This price is also listed in dollars, given as 22.50. Thus, the low price for a share of Disney stock for the past 52 weeks was \$22.50.

b. We find the dividend paid for a share of Disney stock last year by looking under the heading **Div**. The price is listed in dollars, given as .21. Thus, Disney paid a dividend of \$0.21 per share to stockholders last year. If you owned 3000 shares, you received a dividend of \$0.21 $\times$ 3000, or \$630.

c. We find the annual return for dividends alone by looking under the heading **Yld %**, standing for percent yield. The number in the table, .6, is a percent. This means that the dividends alone gave Disney investors an annual return of 0.6%. This is much lower than a bank account paying a 3.5% interest rate. However, if Disney shares increase in value, the gain might make Disney stock a better investment than the bank account.

d. We find the number of shares of Disney traded yesterday by looking under the heading **Vol 100s**, standing for sales volume in hundreds. The number in the table is 115,900. This means that yesterday, a total of 115,900 $\times$ 100, or 11,590,000 shares, were traded.

e. We find the high and low prices for Disney shares yesterday by looking under the headings **Hi** and **Lo**. Both prices are listed in dollars, given as 32.50 and 31.25. Thus, the high and low prices for Disney shares yesterday were \$32.50 and \$31.25, respectively.

f. We find the price at which Disney shares last traded when the stock exchange closed yesterday by looking under the heading **Close**. The price is listed in dollars, given as 32.50. Thus, when the stock exchange closed yesterday, the price of a share of Disney stock was \$32.50.

g. The . . . under **Net Chg** means that there was no change in price in Disney stock from the market close two days ago to yesterday's market close. In part (f), we found that the price of a share of Disney stock at yesterday's close was \$32.50, so the price at the market close two days ago was also \$32.50.

h. We are now ready to use

$$\text{Annual earnings per share} = \frac{\text{Yesterday's closing price per share}}{\text{PE ratio}}$$

to compute Disney's annual earnings per share. We found that yesterday's closing price per share was \$32.50. We find the PE ratio under the heading **PE**. The given number is 43. Thus,

$$\text{Annual earnings per share} = \frac{\$32.50}{43} \approx \$0.76.$$

The annual earnings per share for Disney were \$0.76. The PE ratio, 43, tells us that yesterday's closing price per share, \$32.50, is 43 times greater than the earnings per share, approximately \$0.76.

CHECK POINT 5 Use the stock table for Coca-Cola to solve parts (a) through (h) in Example 5 for Coca-Cola.

52-Week High	52-Week Low	Stock	SYM	Div	Yld %	PE	Vol 100s	Hi	Lo	Close	Net Chg
63.38	42.37	Coca-Cola	CocaCl	.72	1.5	37	72032	49.94	48.33	49.50	+0.03

Blitzer Bonus

The Bottom Line on Investments

Here are some investment suggestions from financial advisors:

- Do not invest money in the stock market that you will need within 10 years. Government bonds, CDs, and money-market accounts are more appropriate options for short-term goals.
- If you are a years old, approximately $(100 - a)\%$ of your investments should be in stocks. For example, at age 25, approximately $(100 - 25)\%$, or 75%, of your portfolio should be invested in stocks.
- Diversify your investments. Invest in a variety of different companies, as well as cushioning stock investments with cash investments and bonds. Diversification enables investors to take advantage of the stock market's superior returns while reducing risk to manageable levels.

Sources: Ralph Frasca, *Personal Finance*, Eighth Edition, Pearson, 2009; Eric Tyson, *Personal Finance for Dummies*, Sixth Edition, Wiley, 2010; Liz Pulliam Weston, *Easy Money*, Pearson, 2008.

5 Understand accounts designed for retirement savings.

Retirement Savings: Stashing Cash and Making Taxes Less Taxing

As you prepare for your future career, retirement probably seems very far away. However, we have seen that you can accumulate wealth much more easily if you have time to make your money work. As soon as you have a job and a paycheck, you should start putting some money away for retirement. Opening a retirement savings account early in your career is a smart way to gain more control over how you will spend a large part of your life.

You can use regular savings and investment accounts to save for retirement. There are also a variety of accounts designed specifically for retirement savings.

- A **traditional individual retirement account (IRA)** is a savings plan that allows you to set aside money for retirement, up to \$5500 per year for people under 50 and \$6500 per year for people 50 or older. You do not pay taxes on the money you deposit into the IRA. You can start withdrawing from your IRA when you are $59\frac{1}{2}$ years old. The withdrawals are taxed.
- A **Roth IRA** is a type of IRA with slightly different tax benefits. You pay taxes on the money you deposit into the IRA, but then you can withdraw your earnings tax-free when you are $59\frac{1}{2}$ years old. Although your contributions are not tax deductible, your earnings are never taxed, even after withdrawal.
- **Employer-sponsored retirement plans**, including 401(k) and 403(b) plans, are set up by the employer, who often makes some contribution to the plan on your behalf. These plans are not offered by all employers and are used to attract high-quality employees.

All accounts designed specifically for retirement savings have penalties for withdrawals before age $59\frac{1}{2}$.

EXAMPLE 6 *Dollars and Sense of Retirement Plans*

a. Suppose that between the ages of 25 and 35, you contribute \$4000 per year to a 401(k) and your employer matches this contribution dollar for dollar on your behalf. The interest rate is 8.5% compounded annually. What is the value of the 401(k) at the end of the 10 years?

b. After 10 years of working for this firm, you move on to a new job. However, you keep your accumulated retirement funds in the 401(k). How much money will you have in the plan when you reach age 65?

c. What is the difference between the amount of money you will have accumulated in the 401(k) at age 65 and the amount you contributed to the plan?

SOLUTION

a. We begin by finding the value of your 401(k) after 10 years.

$$A = \frac{P[(1+r)^t - 1]}{r}$$

Use the formula for the value of an annuity with interest compounded once a year.

$$A = \frac{8000[(1+0.085)^{10} - 1]}{0.085}$$

You contribute \$4000 and your employer matches this each year: $P = 4000 + 4000 = 8000$. The interest rate is 8.5%: $r = 0.085$. The time from age 25 to 35 is 10 years: $t = 10$.

$$\approx 118{,}681$$

Use a calculator.

The value of the 401(k) at the end of the 10 years is approximately \$118,681.

b. Now we find the value of your investment at age 65.

$$A = P(1+r)^t$$

Use the formula from Section 8.4 for future value with interest compounded once a year.

$$A = 118{,}681(1+0.085)^{30}$$

The value of the 401(k) is \$118,681: $P = 118{,}681$. The interest rate is 8.5%: $r = 0.085$. The time from age 35 to age 65 is 30 years: $t = 30$.

$$\approx 1{,}371{,}745$$

Use a calculator.

You will have approximately \$1,371,745 in the 401(k) when you reach age 65.

c. You contributed \$4000 per year to the 401(k) for 10 years, for a total of \$4000 × 10, or \$40,000. The difference between the amount you will have accumulated in the plan at age 65, \$1,371,745, and the amount you contributed, \$40,000, is

$$\$1{,}371{,}745 - \$40{,}000, \quad \text{or} \quad \$1{,}331{,}745.$$

Even when taxes are taken into consideration, we suspect you'll be quite pleased with your earnings from 10 years of savings.

CHECK POINT 6

a. Suppose that between the ages of 25 and 40, you contribute \$2000 per year to a 401(k) and your employer contributes \$1000 per year on your behalf. The interest rate is 8% compounded annually. What is the value of the 401(k), rounded to the nearest dollar, after 15 years?

b. After 15 years of working for this firm, you move on to a new job. However, you keep your accumulated retirement funds in the 401(k). How much money, to the nearest dollar, will you have in the plan when you reach age 65?

c. What is the difference between the amount of money you will have accumulated in the 401(k) and the amount you contributed to the plan?

Concept and Vocabulary Check

Fill in each blank so that the resulting statement is true.

1. A sequence of equal payments made at equal time periods is called a/an ________.

2. In the formula

$$A = \frac{P\left[\left(1 + \frac{r}{n}\right)^{nt} - 1\right]}{\left(\frac{r}{n}\right)},$$

________ is the deposit made at the end of each compounding period, ________ is the annual interest rate compounded ________ times per year, and A is the ________ after ________ years.

3. Shares of ownership in a company are called ________. If you sell shares for more money than what you paid for them, you have a/an ________ gain on the sale. Some companies distribute all or part of their profits to shareholders as ________.

4. People who buy ________ are lending money to the company from which they buy them.

5. A listing of all the investments that a person holds is called a financial ________. To minimize risk, it should be ________, containing a mixture of low-risk and high-risk investments.

6. A group of investments managed by a professional investor is called a/an ________.

7. With a/an ________ IRA, you pay taxes on the money you deposit, but you can withdraw your earnings tax-free beginning when you are ____ years old.

In Exercises 8–11, determine whether each statement is true or false. If the statement is false, make the necessary change(s) to produce a true statement.

8. With the same interest rate, compounding period, and time period, a lump-sum deposit will generate more interest than an annuity. ______

9. People who buy bonds are purchasing shares of ownership in a company. ______

10. Stocks are generally considered higher-risk investments than bonds. ______

11. A traditional IRA requires paying taxes when withdrawing money from the account at age $59\frac{1}{2}$ or older. ______

Exercise Set 8.5

Practice Exercises

Here are the formulas needed to solve the exercises. Be sure you understand what each formula describes and the meaning of the variables in the formulas.

$$A = \frac{P[(1+r)^t - 1]}{r} \qquad A = \frac{P\left[\left(1+\frac{r}{n}\right)^{nt} - 1\right]}{\left(\frac{r}{n}\right)} \qquad P = \frac{A\left(\frac{r}{n}\right)}{\left[\left(1+\frac{r}{n}\right)^{nt} - 1\right]}$$

In Exercises 1–10,

a. *Find the value of each annuity. Round to the nearest dollar.*
b. *Find the interest.*

	Periodic Deposit	Rate	Time
1.	$2000 at the end of each year	5% compounded annually	20 years
2.	$3000 at the end of each year	4% compounded annually	20 years
3.	$4000 at the end of each year	6.5% compounded annually	40 years
4.	$4000 at the end of each year	5.5% compounded annually	40 years
5.	$50 at the end of each month	6% compounded monthly	30 years

	Periodic Deposit	Rate	Time
6.	$60 at the end of each month	5% compounded monthly	30 years
7.	$100 at the end of every six months	4.5% compounded semiannually	25 years
8.	$150 at the end of every six months	6.5% compounded semiannually	25 years
9.	$1000 at the end of every three months	6.25% compounded quarterly	6 years
10.	$1200 at the end of every three months	3.25% compounded quarterly	6 years

In Exercises 11–18,

a. *Determine the periodic deposit. Round up to the nearest dollar.*
b. *How much of the financial goal comes from deposits and how much comes from interest?*

	Periodic Deposit	Rate	Time	Financial Goal
11.	$? at the end of each year	6% compounded annually	18 years	$140,000
12.	$? at the end of each year	5% compounded annually	18 years	$150,000
13.	$? at the end of each month	4.5% compounded monthly	10 years	$200,000
14.	$? at the end of each month	7.5% compounded monthly	10 years	$250,000
15.	$? at the end of each month	7.25% compounded monthly	40 years	$1,000,000
16.	$? at the end of each month	8.25% compounded monthly	40 years	$1,500,000
17.	$? at the end of every three months	3.5% compounded quarterly	5 years	$20,000
18.	$? at the end of every three months	4.5% compounded quarterly	5 years	$25,000

Exercises 19 and 20 refer to the stock tables for Goodyear (the tire company) and Dow Chemical given below. In each exercise, use the stock table to answer the following questions. Where necessary, round dollar amounts to the nearest cent.

a. *What were the high and low prices for a share for the past 52 weeks?*
b. *If you owned 700 shares of this stock last year, what dividend did you receive?*
c. *What is the annual return for the dividends alone? How does this compare to a bank offering a 3% interest rate?*
d. *How many shares of this company's stock were traded yesterday?*
e. *What were the high and low prices for a share yesterday?*
f. *What was the price at which a share last traded when the stock exchange closed yesterday?*
g. *What was the change in price for a share of stock from the market close two days ago to yesterday's market close?*
h. *Compute the company's annual earnings per share using Annual earnings per share*

$$= \frac{\text{Yesterday's closing price per share}}{\text{PE ratio}}.$$

19.

52-Week High	52-Week Low	Stock	SYM	Div	Yld %	PE	Vol 100s	Hi	Lo	Close	Net Chg
73.25	45.44	Goodyear	GT	1.20	2.2	17	5915	56.38	54.38	55.50	+1.25

20.

52-Week High	52-Week Low	Stock	SYM	Div	Yld %	PE	Vol 100s	Hi	Lo	Close	Net Chg
56.75	37.95	Dow Chemical	DOW	1.34	3.0	12	23997	44.75	44.35	44.69	+0.16

Practice Plus

Here are additional formulas that you will use to solve some of the remaining exercises. Be sure you understand what each formula describes and the meaning of the variables in the formulas.

$$A = P(1 + r)^t \qquad A = P\left(1 + \frac{r}{n}\right)^{nt}$$

In Exercises 21–22, round all answers to the nearest dollar.

21. Here are two ways of investing \$30,000 for 20 years.

-

Lump-Sum Deposit	Rate	Time
\$30,000	5% compounded annually	20 years

-

Periodic Deposit	Rate	Time
\$1500 at the end of each year	5% compounded annually	20 years

a. After 20 years, how much more will you have from the lump-sum investment than from the annuity?

b. After 20 years, how much more interest will have been earned from the lump-sum investment than from the annuity?

22. Here are two ways of investing \$40,000 for 25 years.

-

Lump-Sum Deposit	Rate	Time
\$40,000	6.5% compounded annually	25 years

-

Periodic Deposit	Rate	Time
\$1600 at the end of each year	6.5% compounded annually	25 years

a. After 25 years, how much more will you have from the lump-sum investment than from the annuity?

b. After 25 years, how much more interest will have been earned from the lump-sum investment than from the annuity?

23. Solve for P:

$$A = \frac{P[(1 + r)^t - 1]}{r}.$$

What does the resulting formula describe?

24. Solve for P:

$$A = \frac{P\left[\left(1 + \frac{r}{n}\right)^{nt} - 1\right]}{\left(\frac{r}{n}\right)}.$$

What does the resulting formula describe?

Application Exercises

In Exercises 25–30, round to the nearest dollar.

25. Suppose that you earned a bachelor's degree and now you're teaching high school. The school district offers teachers the opportunity to take a year off to earn a master's degree. To achieve this goal, you deposit \$2000 at the end of each year in an annuity that pays 7.5% compounded annually.

a. How much will you have saved at the end of 5 years?

b. Find the interest.

26. Suppose that you earned a bachelor's degree and now you're teaching high school. The school district offers teachers the opportunity to take a year off to earn a master's degree. To achieve this goal, you deposit \$2500 at the end of each year in an annuity that pays 6.25% compounded annually.

a. How much will you have saved at the end of 5 years?

b. Find the interest.

27. Suppose that at age 25, you decide to save for retirement by depositing \$50 at the end of each month in an IRA that pays 5.5% compounded monthly.

a. How much will you have from the IRA when you retire at age 65?

b. Find the interest.

28. Suppose that at age 25, you decide to save for retirement by depositing \$75 at the end of each month in an IRA that pays 6.5% compounded monthly.

a. How much will you have from the IRA when you retire at age 65?

b. Find the interest.

29. To offer scholarships to children of employees, a company invests \$10,000 at the end of every three months in an annuity that pays 10.5% compounded quarterly.

a. How much will the company have in scholarship funds at the end of 10 years?

b. Find the interest.

30. To offer scholarships to children of employees, a company invests \$15,000 at the end of every three months in an annuity that pays 9% compounded quarterly.

a. How much will the company have in scholarship funds at the end of 10 years?

b. Find the interest.

In Exercises 31–34, round up to the nearest dollar.

31. You would like to have \$3500 in four years for a special vacation following college graduation by making deposits at the end of every six months in an annuity that pays 5% compounded semiannually.

a. How much should you deposit at the end of every six months?

b. How much of the \$3500 comes from deposits and how much comes from interest?

32. You would like to have \$4000 in four years for a special vacation following college graduation by making deposits at the end of every six months in an annuity that pays 7% compounded semiannually.

a. How much should you deposit at the end of every six months?

b. How much of the \$4000 comes from deposits and how much comes from interest?

33. How much should you deposit at the end of each month into an IRA that pays 6.5% compounded monthly to have \$2 million when you retire in 45 years? How much of the \$2 million comes from interest?

34. How much should you deposit at the end of each month into an IRA that pays 8.5% compounded monthly to have \$4 million when you retire in 45 years? How much of the \$4 million comes from interest?

35. a. Suppose that between the ages of 22 and 40, you contribute \$3000 per year to a 401(k) and your employer contributes \$1500 per year on your behalf. The interest rate is 8.3% compounded annually. What is the value of the 401(k), rounded to the nearest dollar, after 18 years?

b. Suppose that after 18 years of working for this firm, you move on to a new job. However, you keep your accumulated retirement funds in the 401(k). How much money, to the nearest dollar, will you have in the plan when you reach age 65?

c. What is the difference between the amount of money you will have accumulated in the 401(k) and the amount you contributed to the plan?

36. a. Suppose that between the ages of 25 and 37, you contribute \$3500 per year to a 401(k) and your employer matches this contribution dollar for dollar on your behalf. The interest rate is 8.25% compounded annually. What is the value of the 401(k), rounded to the nearest dollar, after 12 years?

b. Suppose that after 12 years of working for this firm, you move on to a new job. However, you keep your accumulated retirement funds in the 401(k). How much money, to the nearest dollar, will you have in the plan when you reach age 65?

c. What is the difference between the amount of money you will have accumulated in the 401(k) and the amount you contributed to the plan?

Writing in Mathematics

37. What is an annuity?

38. What is meant by the value of an annuity?

39. What is stock?

40. Describe how to find the percent ownership that a shareholder has in a company.

41. Describe the two ways that investors make money with stock.

42. What is a bond? Describe the difference between a stock and a bond.

43. If an investor sees that the return from dividends for a stock is lower than the return for a no-risk bank account, should the stock be sold and the money placed in the bank account? Explain your answer.

44. What is a mutual fund?

45. What is the difference between a traditional IRA and a Roth IRA?

46. Write a problem involving the formula for regular payments needed to achieve a financial goal. The problem should be similar to Example 4 on page 528. However, the problem should be unique to your situation. Include something for which you would like to save, how much you need to save, and how long you have to achieve your goal. Then solve the problem.

Critical Thinking Exercises

Make Sense? *In Exercises 47–53, determine whether each statement makes sense or does not make sense, and explain your reasoning.*

47. By putting \$10 at the end of each month into an annuity that pays 3.5% compounded monthly, I'll be able to retire comfortably in just 30 years.

48. When I invest my money, I am making a trade-off between risk and return.

49. I have little tolerance for risk, so I must be willing to accept lower returns on my investments.

50. Diversification is like saying "don't put all your eggs in one basket."

51. Now that I've purchased bonds, I'm a shareholder in the company.

52. I've been promised a 20% return on an investment without any risk.

53. I appreciate my Roth IRA because I do not pay taxes on my deposits.

In Exercises 54–55,

a. *Determine the deposit at the end of each month. Round up to the nearest dollar.*

b. *Assume that the annuity in part (a) is a tax-deferred IRA belonging to a man whose gross income is \$50,000. Use* ***Table 8.1*** *on page 501 to calculate his taxes first with and then without the IRA. Assume the man is single with no dependents, has no tax credits, and takes the standard deduction.*

c. *What percent of his gross income are the man's federal taxes with and without the IRA? Round to the nearest tenth of a percent.*

54.

Periodic Deposit	Rate	Time	Financial Goal
\$? at the end of each month	8% compounded monthly	40 years	\$1,000,000

55.

Periodic Deposit	Rate	Time	Financial Goal
\$? at the end of each month	7% compounded monthly	40 years	\$650,000

56. How much should you deposit at the end of each month in an IRA that pays 8% compounded monthly to earn \$60,000 per year from interest alone, while leaving the principal untouched, when you retire in 30 years?

Group Exercises

57. Each group should have a newspaper with current stock quotations. Choose nine stocks that group members think would make good investments. Imagine that you invest \$10,000 in each of these nine investments. Check the value of your stock each day over the next five weeks and then sell the nine stocks after five weeks. What is the group's profit or loss over the five-week period? Compare this figure with the profit or loss of other groups in your class for this activity.

58. This activity is a group research project intended for four or five people. Use the research to present a seminar on investments. The seminar is intended to last about 30 minutes and should result in an interesting and informative presentation to the entire class. The seminar should include investment considerations, how to read the bond section of the newspaper, how to read the mutual fund section, and higher-risk investments.

59. Group members have inherited \$1 million. However, the group cannot spend any of the money for 10 years. As a group, determine how to invest this money in order to maximize the money you will make over 10 years. The money can be invested in as many ways as the group decides. Explain each investment decision. What are the risks involved in each investment plan?

8.6

WHAT AM I SUPPOSED TO LEARN?

After you have read this section, you should be able to:

1 Compute the monthly payment and interest costs for a car loan.
2 Understand the types of leasing contracts.
3 Understand the pros and cons of leasing versus buying a car.
4 Understand the different kinds of car insurance.
5 Compare monthly payments on new and used cars.
6 Solve problems related to owning and operating a car.

Cars

TO THE GUYS AT RYDELL HIGH IN THE musical *Grease!*, Kenickie's new car looks like a hunk of junk, but to him it's Greased Lightnin', a hot-rodding work of art on wheels. As with many teens, Kenickie's first car is a rite of passage—a symbol of emerging adulthood.

Our love affair with cars began in the early 1900s when Henry Ford cranked out the first Model T. Since then, we've admired cars to the point of identifying with the vehicles we drive. Cars can serve as status symbols, providing unique insights into a driver's personality.

In this section, we view cars from another vantage point—money. The money pit of owning a car ranges from financing the purchase to escalating costs of everything from fuel to tires to insurance. We open the section with the main reason people spend more money on a car than they can afford: financing.

The Mathematics of Financing a Car

A loan that you pay off with weekly or monthly payments, or payments in some other time period, is called an **installment loan**. The advantage of an installment loan is that the consumer gets to use a product immediately. The disadvantage is that the interest can add a substantial amount to the cost of a purchase.

Let's begin with car loans in which you make regular monthly payments, called **fixed installment loans**. Suppose that you borrow P dollars at interest rate r over t years.

The lender expects A dollars at the end of t years.

$$A = P\left(1 + \frac{r}{n}\right)^{nt}$$

You save the A dollars in an annuity by paying PMT dollars n times per year.

$$A = \frac{PMT\left[\left(1 + \frac{r}{n}\right)^{nt} - 1\right]}{\left(\frac{r}{n}\right)}$$

To find your regular payment amount, PMT, we set the amount the lender expects to receive equal to the amount you will save in the annuity:

$$P\left(1 + \frac{r}{n}\right)^{nt} = \frac{PMT\left[\left(1 + \frac{r}{n}\right)^{nt} - 1\right]}{\left(\frac{r}{n}\right)}.$$

Solving this equation for PMT, we obtain a formula for the loan payment for any installment loan, including payments on car loans.

Compute the monthly payment and interest costs for a car loan.

LOAN PAYMENT FORMULA FOR FIXED INSTALLMENT LOANS

The regular payment amount, PMT, required to repay a loan of P dollars paid n times per year over t years at an annual rate r is given by

$$PMT = \frac{P\left(\frac{r}{n}\right)}{\left[1 - \left(1 + \frac{r}{n}\right)^{-nt}\right]}.$$

EXAMPLE 1 *Comparing Car Loans*

Suppose that you decide to borrow \$20,000 for a new car. You can select one of the following loans, each requiring regular monthly payments:

Installment Loan A: three-year loan at 7%
Installment Loan B: five-year loan at 9%.

a. Find the monthly payments and the total interest for Loan A.

b. Find the monthly payments and the total interest for Loan B.

c. Compare the monthly payments and total interest for the two loans.

SOLUTION

For each loan, we use the loan payment formula to compute the monthly payments.

a. We first determine monthly payments and total interest for Loan A.

P, the loan amount, is \$20,000.
Rate, r, is 7%.
12 payments per year
The loan is for 3 years.

$$PMT = \frac{P\left(\frac{r}{n}\right)}{\left[1 - \left(1 + \frac{r}{n}\right)^{-nt}\right]} = \frac{20{,}000\left(\frac{0.07}{12}\right)}{\left[1 - \left(1 + \frac{0.07}{12}\right)^{-12(3)}\right]} \approx 618$$

The monthly payments are approximately \$618.

Now we calculate the interest over three years, or 36 months.

Total interest over 3 years = Total of all monthly payments minus amount of the loan.

$$= \$618 \times 36 - \$20{,}000$$
$$= \$2248$$

The total interest paid over three years is approximately \$2248.

b. Next, we determine monthly payments and total interest for Loan B.

P, the loan amount, is \$20,000.
Rate, r, is 9%.
12 payments per year
The loan is for 5 years.

$$PMT = \frac{P\left(\frac{r}{n}\right)}{\left[1 - \left(1 + \frac{r}{n}\right)^{-nt}\right]} = \frac{20{,}000\left(\frac{0.09}{12}\right)}{\left[1 - \left(1 + \frac{0.09}{12}\right)^{-12(5)}\right]} \approx 415$$

The monthly payments are approximately \$415.

TECHNOLOGY

Here are the calculator keystrokes to compute

$$\frac{20{,}000\left(\frac{0.07}{12}\right)}{\left[1 - \left(1 + \frac{0.07}{12}\right)^{-12(3)}\right]}.$$

Begin by simplifying the exponent, $-12(3)$, to -36 to avoid possible errors with parentheses:

$$\frac{20{,}000\left(\frac{0.07}{12}\right)}{\left[1 - \left(1 + \frac{0.07}{12}\right)^{-36}\right]}.$$

Scientific and graphing calculator keystrokes require placing parentheses around the expressions in both the numerator and the denominator.

Many Scientific Calculators

[(] 20000 [×] .07 [÷] 12 [)] [÷]

[(] 1 [−] [(] 1 [+] .07 [÷] 12 [)]

[y^x] 36 [+/−] [)] [=]

Many Graphing Calculators

[(] 20000 [×] .07 [÷] 12 [)] [÷]

[(] 1 [−] [(] 1 [+] .07 [÷] 12 [)]

[∧] [(−)] 36 [)] [ENTER]

Answers may vary if you do calculations in stages and round along the way.

Now we calculate the interest over five years, or 60 months.

$$\underbrace{\text{Total interest over 5 years}} = \underbrace{\text{Total of all monthly payments}} \text{ minus } \underbrace{\text{amount of the loan.}}$$

$$= \$415 \times 60 - \$20{,}000$$

$$= \$4900$$

The total interest paid over five years is approximately $4900.

c. **Table 8.6** compares the monthly payments and total interest for the two loans.

Table 8.6 Comparing Car Loans

$20,000 loan	Monthly Payment	Total Interest
3-year loan at 7%	$618	$2248
5-year loan at 9%	$415	$4900

Monthly payments are less with the longer-term loan.

Interest is more with the longer-term loan.

CHECK POINT 1 Suppose that you decide to borrow $15,000 for a new car. You can select one of the following loans, each requiring regular monthly payments:

Installment Loan A: four-year loan at 8%
Installment Loan B: six-year loan at 10%.

a. Find the monthly payments and the total interest for Loan A.

b. Find the monthly payments and the total interest for Loan B.

c. Compare the monthly payments and total interest for the two loans.

Blitzer Bonus

Financing Your Car

- Check out financing options. It's a good idea to get preapproved for a car loan through a bank or credit union before going to the dealer. You can then compare the loan offered by the dealer to your preapproved loan. Furthermore, with more money in hand, you'll have more negotiating power.
- Dealer financing often costs 1% or 2% more than a bank or credit union. Shop around for interest rates. Credit unions traditionally offer the best rates on car loans, more than 1.5% less on average than a bank loan.
- Put down as much money as you can. Interest rates generally decrease as the money you put down toward the car increases. Furthermore, you'll be borrowing less money, thereby paying less interest.
- A general rule is that you should spend no more than 20% of your net monthly income on a car payment.

2 Understand the types of leasing contracts.

The Leasing Alternative

Leasing is the practice of paying a specified amount of money over a specified time for the use of a product. Leasing is essentially a long-term rental agreement.

Leasing a car instead of buying one has become increasingly popular over the past several years. There are two types of leasing contracts:

- **A closed-end lease:** Each month, you make a fixed payment based on estimated usage. When the lease ends, you return the car and pay for mileage in excess of your estimate.

- **An open-end lease:** Each month you make a fixed payment based on the car's *residual value*. **Residual value** is the estimated resale value of the car at the end of the lease and is determined by the dealer. When the lease ends, you return the car and make a payment based on its appraised value at that time compared to its residual value. If the appraised value is less than the residual value stated in the lease, you pay all or a portion of the difference. If the appraised value is greater than or equal to the residual value, you owe nothing and you may receive a refund.

3 Understand the pros and cons of leasing versus buying a car.

Leasing a car offers both advantages and disadvantages over buying one.

Advantages of Leasing

- Leases require only a small down payment, or no down payment at all.
- Lease payments for a new car are lower than loan payments for the same car. Most people can lease a more expensive car than they would be able to buy.
- When the lease ends, you return the car to the dealer and do not have to be concerned about selling the car.

Disadvantages of Leasing

- When the lease ends, you do not own the car.
- Most lease agreements have mileage limits: 12,000 to 15,000 miles per year is common. If you exceed the number of miles allowed, there can be considerable charges.
- When mileage penalties and other costs at the end of the leasing period are taken into consideration, the total cost of leasing is almost always more expensive than financing a car.
- While leasing the car, you are responsible for keeping it in perfect condition. You are liable for any damage to the car.
- Leasing does not cover maintenance.
- There are penalties for ending the lease early.

Car leases tend to be extremely complicated. It can appear that there are as many lease deals as there are kinds of cars. A helpful pamphlet entitled "Keys to Vehicle Leasing" is published by the Federal Reserve Board. Copies are available on the Internet. Additional information can be found at websites such as home.autos.msn.com or intellichoice.com.

4 Understand the different kinds of car insurance.

The Importance of Auto Insurance

Who needs auto insurance? The simple answer is that if you own or lease a car, you do.

When you purchase **insurance**, you buy protection against loss associated with unexpected events. Different types of coverage are associated with auto insurance, but the one required by nearly every state is *liability*. There are two components of **liability coverage**:

- **Bodily injury liability** covers the costs of lawsuits if someone is injured or killed in an accident in which you are at fault.
- **Property damage liability** covers damage to other cars and property from negligent operation of your vehicle.

If you have a car loan or lease a car, you will also need *collision* and *comprehensive* coverage:

- **Collision coverage** pays for damage or loss of your car if you're in an accident.
- **Comprehensive coverage** protects your car from perils such as fire, theft, falling objects, acts of nature, and collision with an animal.

There is a big difference in auto insurance rates, so be sure to shop around. Insurance can be very expensive for younger drivers with limited driving experience. A poor driving record dramatically increases your insurance rates. Other factors that impact your insurance premium include where you live, the number of miles you drive each year, and the value of your car.

5 Compare monthly payments on new and used cars.

New or Used?

Who insists you need a new car? A new car loses an average of 12% of its value the moment it is driven off the dealer's lot. It's already a used car and you haven't even arrived home.

Used cars are a good option for many people. Your best buy is typically a two- to three-year-old car because the annual depreciation in price is greatest over the first few years. Furthermore, many sources of financing for used cars will loan money only on newer models that are less than five years old. Reputable car dealerships offer a good selection of used cars, with extended warranties and other perks.

The two most commonly used sources of pricing information for used cars are the *National Automobile Dealers Association Official Used Car Guide* (www.nada.com) and the *Kelley Blue Book Used Car Guide* (www.kbb.com). They contain the average retail price for many different makes of used cars.

EXAMPLE 2 *Saving Money with a Used Car*

Suppose that you are thinking about buying a car and have narrowed down your choices to two options:

- The new-car option: The new car costs $25,000 and can be financed with a four-year loan at 7.9%.
- The used-car option: A three-year-old model of the same car costs $14,000 and can be financed with a four-year loan at 8.45%.

What is the difference in monthly payments between financing the new car and financing the used car?

SOLUTION

We first determine the monthly payments for the new car that costs $25,000, financed with a four-year loan at 7.9%.

P, the loan amount, is $25,000. Rate, *r*, is 7.9%. 12 payments per year. The loan is for 4 years.

$$PMT = \frac{P\left(\frac{r}{n}\right)}{\left[1-\left(1+\frac{r}{n}\right)^{-nt}\right]} = \frac{25{,}000\left(\frac{0.079}{12}\right)}{\left[1-\left(1+\frac{0.079}{12}\right)^{-12(4)}\right]} \approx 609$$

The monthly payments for the new car are approximately $609. Now we determine the monthly payments for the used car that costs $14,000, financed with a four-year loan at 8.45%.

P, the loan amount, is $14,000. Rate, *r*, is 8.45%. 12 payments per year. The loan is for 4 years.

$$PMT = \frac{P\left(\frac{r}{n}\right)}{\left[1-\left(1+\frac{r}{n}\right)^{-nt}\right]} = \frac{14{,}000\left(\frac{0.0845}{12}\right)}{\left[1-\left(1+\frac{0.0845}{12}\right)^{-12(4)}\right]} \approx 345$$

The monthly payments for the used car are approximately $345. The difference in monthly payments between the new-car loan, $609, and the used-car loan, $345, is

$609 − $345, or $264.

You save $264 each month over a period of four years with the used-car option.

CHECK POINT 2 Suppose that you are thinking about buying a car and have narrowed down your choices to two options:

The new-car option: The new car costs \$19,000 and can be financed with a three-year loan at 6.18%.

The used-car option: A two-year-old model of the same car costs \$11,500 and can be financed with a three-year loan at 7.5%.

What is the difference in monthly payments between financing the new car and financing the used car?

6 Solve problems related to owning and operating a car.

The Money Pit of Car Ownership

Buying a car is a huge expense. To make matters worse, the car continues costing money after you purchase it. These costs include operating expenses such as fuel, maintenance, tires, tolls, parking, and cleaning. The costs also include ownership expenses such as insurance, license fees, registration fees, taxes, and interest on loans.

The significant expense of owning and operating a car is shown in **Table 8.7**. According to the American Automobile Association (AAA), the average yearly cost of owning and operating a car is just under \$9000.

TABLE 8.7 Annual Costs of Owning and Operating a Car in 2012*

Type of Car	Small Sedan	Medium Sedan	Minivan	Large Sedan	SUV 4WD
Cost per mile	44.9¢	58.5¢	63.4¢	75.5¢	75.7¢
Cost per year	\$6735	\$8780	\$9504	\$11,324	\$11,360

* Based on driving 15,000 miles per year.
Source: AAA

A large portion of a car's operating expenses involves the cost of gasoline. As the luster of big gas-guzzlers becomes less appealing, many people are turning to fuel-efficient hybrid cars that use a combination of gasoline and rechargeable batteries as power sources.

Our next example compares fuel expenses for a gas-guzzler and a hybrid. You can estimate the annual fuel expense for a vehicle if you know approximately how many miles the vehicle will be driven each year, how many miles the vehicle can be driven per gallon of gasoline, and how much a gallon of gasoline will cost.

THE COST OF GASOLINE

$$\text{Annual fuel expense} = \frac{\text{annual miles driven}}{\text{miles per gallon}} \times \text{price per gallon}$$

EXAMPLE 3 *Comparing Fuel Expenses*

Suppose that you drive 24,000 miles per year and gas averages \$4 per gallon.

a. What will you save in annual fuel expenses by owning a hybrid car averaging 50 miles per gallon rather than an SUV (sport utility vehicle) averaging 12 miles per gallon?

b. If you deposit your monthly fuel savings at the end of each month into an annuity that pays 7.3% compounded monthly, how much will you have saved at the end of six years?

SOLUTION

a. We use the formula for annual fuel expense.

$$\text{Annual fuel expense} = \frac{\text{annual miles driven}}{\text{miles per gallon}} \times \text{price per gallon}$$

$$\text{Annual fuel expense for the hybrid} = \frac{24{,}000}{50} \times \$4 = 480 \times \$4 = \$1920$$

The hybrid averages 50 miles per gallon.

$$\text{Annual fuel expense for the SUV} = \frac{24{,}000}{12} \times \$4 = 2000 \times \$4 = \$8000$$

The SUV averages 12 miles per gallon.

Your annual fuel expense is \$1920 for the hybrid and \$8000 for the SUV. By owning the hybrid rather than the SUV, you save

$$\$8000 - \$1920, \text{ or } \$6080$$

in annual fuel expenses.

b. Because you save \$6080 per year, you save

$$\frac{\$6080}{12} \approx \$507,$$

or approximately \$507 per month. Now you deposit \$507 at the end of each month into an annuity that pays 7.3% compounded monthly. We use the formula for the value of an annuity to determine your savings at the end of six years.

$$A = \frac{P\left[\left(1 + \frac{r}{n}\right)^{nt} - 1\right]}{\left(\frac{r}{n}\right)}$$

Use the formula for the value of an annuity.

$$A = \frac{507\left[\left(1 + \frac{0.073}{12}\right)^{12\cdot 6} - 1\right]}{\left(\frac{0.073}{12}\right)}$$

The annuity involves month-end deposits of \$507: $P = 507$. The interest rate is 7.3%: $r = 0.073$. The interest is compounded monthly: $n = 12$. The number of years is 6: $t = 6$.

$$\approx 45{,}634$$

Use a calculator.

You will have saved approximately \$45,634 at the end of six years. This illustrates how driving a car that consumes less gas can yield significant savings for your future.

CHECK POINT 3 Suppose that you drive 36,000 miles per year and gas averages \$3.50 per gallon.

a. What will you save in annual fuel expenses by owning a hybrid car averaging 40 miles per gallon rather than an SUV averaging 15 miles per gallon?

b. If you deposit your monthly fuel savings at the end of each month into an annuity that pays 7.25% compounded monthly, how much will you have saved at the end of seven years? Round all computations to the nearest dollar.

Concept and Vocabulary Check

Fill in each blank so that the resulting statement is true.

1. In the formula

$$PMT = \frac{P\left(\frac{r}{n}\right)}{\left[1 - \left(1 + \frac{r}{n}\right)^{-nt}\right]},$$

________ is the regular payment amount required to repay a loan of ________ dollars paid ________ times per year over ________ years at an annual interest rate ________.

2. The two types of contracts involved with leasing a car are called a/an ________ lease and a/an ________ lease.

3. The estimated resale value of a car at the end of its lease is called the car's ________.

4. There are two components of liability insurance. The component that covers costs if someone is injured or killed in an accident in which you are at fault is called ________ liability. The component that covers damage to other cars if you are at fault is called ________ liability.

5. The type of car insurance that pays for damage or loss of your car if you're in an accident is called ________ coverage.

6. The type of insurance that pays for damage to your car due to fire, theft, or falling objects is called ________ coverage.

In Exercises 7–12, determine whether each statement is true or false. If the statement is false, make the necessary changes(s) to produce a true statement.

7. The interest on a car loan can be determined by taking the difference between the total of all monthly payments and the amount of the loan. ______

8. When an open-end lease terminates and the car's appraised value is less than the residual value stated in the lease, you owe nothing. ______

9. One advantage to leasing a car is that you are not responsible for any damage to the car. ______

10. One disadvantage to leasing a car is that most lease agreements have mileage limits. ______

11. Collision coverage pays for damage to another car if you cause an accident. ______

12. Due to operating and ownership expenses, a car continues costing money after you buy it. ______

Exercise Set 8.6

Practice and Application Exercises

In Exercises 1–10, use

$$PMT = \frac{P\left(\frac{r}{n}\right)}{\left[1 - \left(1 + \frac{r}{n}\right)^{-nt}\right]}.$$

Round answers to the nearest dollar.

1. Suppose that you borrow \$10,000 for four years at 8% toward the purchase of a car. Find the monthly payments and the total interest for the loan.

2. Suppose that you borrow \$30,000 for four years at 8% for the purchase of a car. Find the monthly payments and the total interest for the loan.

3. Suppose that you decide to borrow \$15,000 for a new car. You can select one of the following loans, each requiring regular monthly payments:

   Installment Loan A: three-year loan at 5.1%
   Installment Loan B: five-year loan at 6.4%.

   a. Find the monthly payments and the total interest for Loan A.
   b. Find the monthly payments and the total interest for Loan B.
   c. Compare the monthly payments and the total interest for the two loans.

4. Suppose that you decide to borrow \$40,000 for a new car. You can select one of the following loans, each requiring regular monthly payments:

   Installment Loan A: three-year loan at 6.1%
   Installment Loan B: five-year loan at 7.2%.

   a. Find the monthly payments and the total interest for Loan A.
   b. Find the monthly payments and the total interest for Loan B.
   c. Compare the monthly payments and the total interest for the two loans.

5. Suppose that you are thinking about buying a car and have narrowed down your choices to two options:

   The new-car option: The new car costs \$28,000 and can be financed with a four-year loan at 6.12%.

   The used-car option: A three-year old model of the same car costs \$16,000 and can be financed with a four-year loan at 6.86%.

   What is the difference in monthly payments between financing the new car and financing the used car?

6. Suppose that you are thinking about buying a car and have narrowed down your choices to two options:

   The new-car option: The new car costs \$68,000 and can be financed with a four-year loan at 7.14%.

   The used-car option: A three-year old model of the same car costs \$28,000 and can be financed with a four-year loan at 7.92%.

   What is the difference in monthly payments between financing the new car and financing the used car?

7. Suppose that you decide to buy a car for $29,635, including taxes and license fees. You saved $9000 for a down payment and can get a five-year car loan at 6.62%. Find the monthly payment and the total interest for the loan.

8. Suppose that you decide to buy a car for $37,925, including taxes and license fees. You saved $12,000 for a down payment and can get a five-year loan at 6.58%. Find the monthly payment and the total interest for the loan.

9. Suppose that you are buying a car for $60,000, including taxes and license fees. You saved $10,000 for a down payment. The dealer is offering you two incentives:

Incentive A is $5000 off the price of the car, followed by a five-year loan at 7.34%.

Incentive B does not have a cash rebate, but provides free financing (no interest) over five years.

What is the difference in monthly payments between the two offers? Which incentive is the better deal?

10. Suppose that you are buying a car for $56,000, including taxes and license fees. You saved $8000 for a down payment. The dealer is offering you two incentives:

Incentive A is $10,000 off the price of the car, followed by a four-year loan at 12.5%.

Incentive B does not have a cash rebate, but provides free financing (no interest) over four years.

What is the difference in monthly payments between the two offers? Which incentive is the better deal?

In Exercises 11–14, use the formula

$$A = \frac{P\left[\left(1 + \frac{r}{n}\right)^{nt} - 1\right]}{\left(\frac{r}{n}\right)}.$$

Round all computations to the nearest dollar.

11. Suppose that you drive 40,000 miles per year and gas averages $4 per gallon.

a. What will you save in annual fuel expenses by owning a hybrid car averaging 40 miles per gallon rather than an SUV averaging 16 miles per gallon?

b. If you deposit your monthly fuel savings at the end of each month into an annuity that pays 5.2% compounded monthly, how much will you have saved at the end of six years?

12. Suppose that you drive 15,000 miles per year and gas averages $3.50 per gallon.

a. What will you save in annual fuel expenses by owning a hybrid car averaging 60 miles per gallon rather than an SUV averaging 15 miles per gallon?

b. If you deposit your monthly fuel savings at the end of each month into an annuity that pays 5.7% compounded monthly, how much will you have saved at the end of six years?

The table shows the expense of operating and owning four selected cars, by average costs per mile. Use the appropriate information in the table to solve Exercises 13–16.

AVERAGE ANNUAL COSTS OF OWNING AND OPERATING A CAR

	Average Costs per Mile		
Make and Model	**Operating**	**Ownership**	**Total**
Cadillac STS	$0.26	$0.72	$0.98
Mercury Grand Marquis GS	$0.23	$0.42	$0.65
Honda Accord LX	$0.21	$0.34	$0.55
Toyota Corolla CE	$0.15	$0.25	$0.40

Source: Runzheimer International

13. a. If you drive 20,000 miles per year, what is the total annual expense for a Cadillac STS?

b. If the total annual expense for a Cadillac STS is deposited at the end of each year into an IRA paying 8.5% compounded yearly, how much will be saved at the end of six years?

14. a. If you drive 14,000 miles per year, what is the total annual expense for a Toyota Corolla CE?

b. If the total annual expense for a Toyota Corolla CE is deposited at the end of each year into an IRA paying 8.2% compounded yearly, how much will be saved at the end of six years?

15. If you drive 30,000 miles per year, by how much does the total annual expense for a Cadillac STS exceed that of a Toyota Corolla CE over six years?

16. If you drive 25,000 miles per year, by how much does the total annual expense for a Mercury Grand Marquis GS exceed that of a Honda Accord LX over six years?

Writing in Mathematics

17. If a three-year car loan has the same interest rate as a six-year car loan, how do the monthly payments and the total interest compare for the two loans?

18. What is the difference between a closed-end car lease and an open-end car lease?

19. Describe two advantages of leasing a car over buying one.

20. Describe two disadvantages of leasing a car over buying one.

21. What are the two components of liability coverage and what is covered by each component?

22. What does collision coverage pay for?

23. What does comprehensive coverage pay for?

24. How can you estimate a car's annual fuel expense?

Critical Thinking Exercises

Make Sense? *In Exercises 25–30, determine whether each statement makes sense or does not make sense, and explain your reasoning.*

25. If I purchase a car using money that I've saved, I can eliminate paying interest on a car loan, but then I have to give up the interest income I could have earned on my savings.

26. The problem with my car lease is that when it ends, I have to be concerned about selling the car.

27. Although lease payments for a new car are lower than loan payments for the same car, once I take mileage penalties and other costs into consideration, the total cost of leasing is more expensive than financing the car.

28. I've paid off my car loan, so I am not required to have liability coverage.

29. Buying a used car or a fuel-efficient car can yield significant savings for my future.

30. Because it is extremely expensive to own and operate a car, I plan to look closely at whether or not a car is essential and consider other modes of transportation.

31. Use the discussion at the bottom of page 539 to prove the loan payment formula shown in the box at the top of page 540. Work with the equation in which the amount the lender expects to receive is equal to the amount saved in the annuity. Multiply both sides of this equation by $\frac{r}{n}$ and then solve for *PMT* by dividing both sides by the appropriate expression. Finally, divide the numerator and the denominator of the resulting formula for *PMT* by $(1 + \frac{r}{n})^{nt}$ to obtain the form of the loan payment formula shown in the box.

32. The unpaid balance of an installment loan is equal to the present value of the remaining payments. The unpaid balance, *P*, is given by

$$P = PMT\frac{\left[1-\left(1+\frac{r}{n}\right)^{-nt}\right]}{\left(\frac{r}{n}\right)},$$

where *PMT* is the regular payment amount, *r* is the annual interest rate, *n* is the number of payments per year, and *t* is the number of years remaining in the loan.

a. Use the loan payment formula to derive the unpaid balance formula.

b. The price of a car is \$24,000. You have saved 20% of the price as a down payment. After the down payment, the balance is financed with a 5-year loan at 9%. Determine the unpaid balance after three years. Round all calculations to the nearest dollar.

Group Exercises

33. Group members should go to the Internet and select a car that they might like to buy. Price the car and its options. Then find two loans with the best rates, but with different terms. For each loan, calculate the monthly payments and total interest.

34. Student Loans

Group members should present a report on federal loans to finance college costs, including Stafford loans, Perkins loans, and PLUS loans. Also include a discussion of grants that do not have to be repaid, such as Pell Grants and National Merit Scholarships. Refer to *Funding Education Beyond High School*, published by the Department of Education and available at studentaid.ed.gov. Use the loan repayment formula that we applied to car loans to determine regular payments and interest on some of the loan options presented in your report.

8.7

WHAT AM I SUPPOSED TO LEARN?

After you have read this section, you should be able to:

1 Compute the monthly payment and interest costs for a mortgage.

2 Prepare a partial loan amortization schedule.

3 Solve problems involving what you can afford to spend for a mortgage.

4 Understand the pros and cons of renting versus buying.

The Cost of Home Ownership

THE BIGGEST SINGLE PURCHASE THAT MOST PEOPLE make in their lives is the purchase of a home. If you choose home ownership at some point in the future, it is likely that you will finance the purchase with an installment loan. Knowing the unique issues surrounding the purchase of a home, and whether or not this aspect of the American dream is right for you, can play a significant role in your financial future.

Mortgages

A **mortgage** is a long-term installment loan (perhaps up to 30, 40, or even 50 years) for the purpose of buying a home, and for which the property is pledged as security for payment. If payments are not made on the loan, the lender may take possession of the property. The **down payment** is the portion of the sale price of the home that the buyer initially pays to the seller. The minimum required down payment is computed as a percentage of the sale price. For example, suppose you decide to buy a \$220,000 home. The lender requires you to pay the seller 10% of the sale price. You must pay 10% of \$220,000, which is $0.10 \times 220{,}000$ or \$22,000, to the seller. Thus, \$22,000 is the down payment. The **amount of the mortgage** is the difference between the sale price and the down payment. For your \$220,000 home, the amount of the mortgage is \$220,000 − \$22,000, or \$198,000.

Monthly payments for a mortgage depend on the amount of the mortgage (the principal), the interest rate, and the duration of the mortgage. Mortgages can have a fixed interest rate or a variable interest rate. **Fixed-rate mortgages** have the same monthly principal and interest payment during the entire time of the loan. A loan like this that has a schedule for paying a fixed amount each period is called a **fixed installment loan**. **Variable-rate mortgages**, also known as **adjustable-rate mortgages** (ARMs), have payment amounts that change from time to time depending on changes in the interest rate. ARMs are less predictable than fixed-rate mortgages. They start out at lower rates than fixed-rate mortgages. Caps limit how high rates can go over the term of the loan.

Compute the monthly payment and interest costs for a mortgage.

Computations Involved with Buying a Home

Although monthly payments for a mortgage depend on the amount of the mortgage, the duration of the loan, and the interest rate, the interest is not the only cost of a mortgage. Most lending institutions require the buyer to pay one or more **points** at the time of closing—that is, the time at which the mortgage begins. A point is a one-time charge that equals 1% of the loan amount. For example, two points means that the buyer must pay 2% of the loan amount at closing. Often, a buyer can pay fewer points in exchange for a higher interest rate or more points for a lower rate. A document, called the **Truth-in-Lending Disclosure Statement**, shows the buyer the APR, or the annual percentage rate, for the mortgage. The APR takes into account the interest rate and points.

A monthly mortgage payment is used to repay the principal plus interest. In addition, lending institutions can require monthly deposits into an **escrow account**, an account used by the lender to pay real estate taxes and insurance. These deposits increase the amount of the monthly payment.

In the previous section, we used the loan payment formula for fixed installment loans to determine payments on car loans. Because a fixed-rate mortgage is a fixed installment loan, we use the same formula to compute the monthly payment for a mortgage.

LOAN PAYMENT FORMULA FOR FIXED INSTALLMENT LOANS

The regular payment amount, PMT, required to repay a loan of P dollars paid n times per year over t years at an annual rate r is given by

$$PMT = \frac{P\left(\frac{r}{n}\right)}{\left[1 - \left(1 + \frac{r}{n}\right)^{-nt}\right]}.$$

EXAMPLE 1 Computing the Monthly Payment and Interest Costs for a Mortgage

The price of a home is \$195,000. The bank requires a 10% down payment and two points at the time of closing. The cost of the home is financed with a 30-year fixed-rate mortgage at 7.5%.

a. Find the required down payment.

b. Find the amount of the mortgage.

c. How much must be paid for the two points at closing?

d. Find the monthly payment (excluding escrowed taxes and insurance).

e. Find the total interest paid over 30 years.

SOLUTION

a. The required down payment is 10% of \$195,000 or

$$0.10 \times \$195{,}000 = \$19{,}500.$$

b. The amount of the mortgage is the difference between the price of the home and the down payment.

Amount of the mortgage = sale price − down payment

$$= \$195{,}000 - \$19{,}500$$

$$= \$175{,}500$$

c. To find the cost of two points on a mortgage of \$175,500, find 2% of \$175,500.

$$0.02 \times \$175{,}500 = \$3510$$

The down payment (\$19,500) is paid to the seller and the cost of two points (\$3510) is paid to the lending institution.

d. We are interested in finding the monthly payment for a \$175,500 mortgage at 7.5% for 30 years. We use the loan payment formula for installment loans.

P, the mortgage amount, is \$175,500. Fixed rate, r, is 7.5%. 12 payments per year. The mortgage time, t, is 30 years.

$$PMT = \frac{P\left(\frac{r}{n}\right)}{\left[1 - \left(1 + \frac{r}{n}\right)^{-nt}\right]} = \frac{175{,}500\left(\frac{0.075}{12}\right)}{\left[1 - \left(1 + \frac{0.075}{12}\right)^{-12(30)}\right]}$$

$$\approx 1227$$

The monthly mortgage payment for principal and interest is approximately \$1227. (Keep in mind that this payment does not include escrowed taxes and insurance.)

e. The total cost of interest over 30 years is equal to the difference between the total of all monthly payments and the amount of the mortgage. The total of all monthly payments is equal to the amount of the monthly payment multiplied by the number of payments. We found the amount of each monthly payment in (d): \$1227. The number of payments is equal to the number of months in a year, 12, multiplied by the number of years in the mortgage, 30: $12 \times 30 = 360$. Thus, the total of all monthly payments $= \$1227 \times 360$.

Now we can calculate the interest over 30 years.

Total interest paid = total of all monthly payments minus amount of the mortgage.

$$= \$1227 \times 360 - \$175{,}500$$

$$= \$441{,}720 - \$175{,}500 = \$266{,}220$$

The total interest paid over 30 years is approximately \$266,220.

TECHNOLOGY

Here are the calculator keystrokes to compute

$$\frac{175{,}500\left(\frac{0.075}{12}\right)}{\left[1 - \left(1 + \frac{0.075}{12}\right)^{-12(30)}\right]}.$$

Begin by simplifying the exponent, −12(30), to −360 to avoid possible errors with parentheses:

$$\frac{175{,}500\left(\frac{0.075}{12}\right)}{\left[1 - \left(1 + \frac{0.075}{12}\right)^{-360}\right]}.$$

Many Scientific Calculators

[( ] 175500 [×] .075 [÷] 12 [ )] [÷]

[( ] 1 [−] [( ] 1 [+] .075 [÷] 12 [ )]

[y^x] 360 [+/−] [ )] [=]

Many Graphing Calculators

[( ] 175500 [×] .075 [÷] 12 [ )] [÷]

[( ] 1 [−] [( ] 1 [+] .075 [÷] 12 [ )]

[∧] [(−)] 360 [ )] [ENTER]

CHECK POINT 1 In Example 1, the \$175,500 mortgage was financed with a 30-year fixed rate at 7.5%. The total interest paid over 30 years was approximately \$266,220.

a. Use the loan payment formula for installment loans to find the monthly payment if the time of the mortgage is reduced to 15 years. Round to the nearest dollar.

b. Find the total interest paid over 15 years.

c. How much interest is saved by reducing the mortgage from 30 to 15 years?

2 Prepare a partial loan amortization schedule.

Loan Amortization Schedules

When a mortgage loan is paid off through a series of regular payments, it is said to be **amortized**, which literally means "killed off." In working Check Point 1(c), were you surprised that nearly \$150,000 was saved when the mortgage was amortized over 15 years rather than over 30 years? What adds to the interest cost is the long period over which the loan is financed. **Although each payment is the same, with each successive payment the interest portion decreases and the portion applied toward paying off the principal increases.** The interest is computed using the simple interest formula $I = Prt$. The principal, P, is equal to the balance of the loan, which decreases each month. The rate, r, is the annual interest rate of the mortgage loan. Because a payment is made each month, the time, t, is

$$\frac{1 \text{ month}}{12 \text{ months}} = \frac{1 \cancel{\text{month}}}{12 \cancel{\text{months}}}$$

or $\frac{1}{12}$ of a year.

A document showing how the payment each month is split between interest and principal is called a **loan amortization schedule**. Typically, for each payment, this document includes the payment number, the interest for the payment, the amount of the payment applied to the principal, and the balance of the loan after the payment is applied.

Blitzer Bonus

The Mortgage Crisis

In 2006, the median U.S. home price jumped to \$206,000, up a stunning 15% in just one year and 55% over five years. This rise in home values made real estate an attractive investment to many people, including those with poor credit records and low incomes. Credit standards for mortgages were lowered and loans were made to high-risk borrowers. By 2008, America's raucous house party was over. A brief period of easy lending, especially lax mortgage practices from 2002 through 2006, exploded into the worst financial crisis since the Great Depression. The plunge in home prices wiped out trillions of dollars in home equity, setting off fears that foreclosures and tight credit could send home prices falling to the point that millions of families and thousands of banks might be thrust into insolvency.

EXAMPLE 2 *Preparing a Loan Amortization Schedule*

Prepare a loan amortization schedule for the first two months of the mortgage loan shown in the table below. Round entries to the nearest cent.

LOAN AMORTIZATION SCHEDULE

Annual % Rate: 9.5%			
Amount of Mortgage: \$130,000		**Monthly Payment: \$1357.50**	
Number of Monthly Payments: 180		**Term: Years 15, Months 0**	
Payment Number	**Interest Payment**	**Principal Payment**	**Balance of Loan**
1			
2			

SOLUTION

We begin with payment number 1.

$$\text{Interest for the month} = Prt = \$130{,}000 \times 0.095 \times \frac{1}{12} \approx \$1029.17$$

$$\text{Principal payment} = \text{Monthly payment} - \text{Interest payment}$$
$$= \$1357.50 - \$1029.17 = \$328.33$$

$$\text{Balance of loan} = \text{Principal balance} - \text{Principal payment}$$
$$= \$130{,}000 - \$328.33 = \$129{,}671.67$$

Now, starting with a loan balance of \$129,671.67, we repeat these computations for the second payment.

$$\text{Interest for the month} = Prt = \$129{,}671.67 \times 0.095 \times \frac{1}{12} \approx \$1026.57$$

$$\text{Principal payment} = \text{Monthly payment} - \text{Interest payment}$$
$$= \$1357.50 - \$1026.57 = \$330.93$$

$$\text{Balance of loan} = \text{Principal balance} - \text{Principal payment}$$
$$= \$129{,}671.67 - \$330.93 = \$129{,}340.74$$

The results of these computations are included in **Table 8.8** on the next page, a partial loan amortization schedule. By using the simple interest formula month-to-month on the loan's balance, a complete loan amortization schedule for all 180 payments can be calculated.

TABLE 8.8 Loan Amortization Schedule

Annual % Rate: 9.5% Amount of Mortgage: $130,000 Number of Monthly Payments: 180		Monthly Payment: $1357.50 Term: Years 15, Months 0	
Payment Number	**Interest Payment**	**Principal Payment**	**Balance of Loan**
1	$1029.17	$ 328.33	$129,671.67
2	$1026.57	$ 330.93	$129,340.74
3	$1023.96	$ 333.54	$129,007.22
4	$1021.32	$ 336.18	$128,671.04
30	$ 944.82	$ 412.68	$118,931.35
31	$ 941.55	$ 415.95	$118,515.52
125	$ 484.62	$ 872.88	$ 60,340.84
126	$ 477.71	$ 879.79	$ 59,461.05
179	$ 21.26	$1336.24	$ 1347.74
180	$ 9.76	$1347.74	

Many lenders supply a loan amortization schedule like the one in Example 2 at the time of closing. Such a schedule shows how the buyer pays slightly less in interest and more in principal for each payment over the entire life of the loan.

CHECK POINT 2 Prepare a loan amortization schedule for the first two months of the mortgage loan shown in the following table. Round entries to the nearest cent.

Annual % Rate: 7.0% Amount of Mortgage: $200,000 Number of Monthly Payments: 240		Monthly Payment: $1550.00 Term: Years 20, Months 0	
Payment Number	**Interest Payment**	**Principal Payment**	**Balance of Loan**
1			
2			

Blitzer Bonus

Bittersweet Interest

Looking at amortization tables, you could get discouraged by how much of your early mortgage payments goes toward interest and how little goes toward paying off the principal. Although you get socked with tons of interest in the early years of a loan, the one bright side to the staggering cost of a mortgage is the **mortgage interest tax deduction**. To make the cost of owning a home more affordable, the tax code permits deducting all the mortgage interest (but not the principal) that you pay per year on the loan. **Table 8.9** illustrates how this tax loophole reduces the cost of the mortgage.

TABLE 8.9 Tax Deductions for a $100,000 Mortgage at 7% for a Taxpayer in the 28% Tax Bracket

Year	Interest	Tax Savings	Net Cost of Mortgage
1	$6968	$1951	$5017
2	$6895	$1931	$4964
3	$6816	$1908	$4908
4	$6732	$1885	$4847
5	$6641	$1859	$4782

3 Solve problems involving what you can afford to spend for a mortgage.

Determining What You Can Afford

Here's the bottom line from most financial advisers:

- Spend no more than 28% of your gross monthly income for your mortgage payment.
- Spend no more than 36% of your gross monthly income for your total monthly debt, including mortgage payments, car payments, credit card bills, student loans, and medical debt.

Using these guidelines, **Table 8.10** shows the maximum monthly amount you could afford for mortgage payments and total credit obligations for a variety of income levels.

TABLE 8.10 Maximum Amount You Can Afford

Gross Annual Income	Monthly Mortgage Payment	Total Monthly Credit Obligations
\$20,000	\$467	\$600
\$30,000	\$700	\$900
\$40,000	\$933	\$1200
\$50,000	\$1167	\$1500
\$60,000	\$1400	\$1800
\$70,000	\$1633	\$2100
\$80,000	\$1867	\$2400
\$90,000	\$2100	\$2700
\$100,000	\$2333	\$3000

Source: Fannie Mae

EXAMPLE 3 *What Can You Afford?*

Suppose that your gross annual income is \$25,000.

a. What is the maximum amount you should spend each month on a mortgage payment?

b. What is the maximum amount you should spend each month for total credit obligations?

c. If your monthly mortgage payment is 80% of the maximum amount you can afford, what is the maximum amount you should spend each month for all other debt?

Round all computations to the nearest dollar.

SOLUTION

With a gross annual income of \$25,000, your gross monthly income is

$$\frac{\$25{,}000}{12},$$

or approximately \$2083.

a. You should spend no more than 28% of your gross monthly income, \$2083, on a mortgage payment.

$$28\% \text{ of } \$2083 = 0.28 \times \$2083 \approx \$583.$$

Your monthly mortgage payment should not exceed \$583.

b. You should spend no more than 36% of your gross monthly income, \$2083, for total monthly debt.

$$36\% \text{ of } \$2083 = 0.36 \times \$2083 \approx \$750.$$

Your total monthly credit obligations should not exceed \$750.

c. The problem's conditions state that your monthly mortgage payment is 80% of the maximum you can afford, which is \$583. This means that your monthly mortgage payment is 80% of \$583.

$$80\% \text{ of } \$583 = 0.8 \times \$583 \approx \$466.$$

In part (b), we saw that your total monthly debt should not exceed \$750. Because you are paying \$466 for your mortgage payment, this leaves \$750 − \$466, or \$284, for all other debt. Your monthly credit obligations, excluding mortgage payments, should not exceed \$284.

CHECK POINT 3 Suppose that your gross annual income is \$240,000.

a. What is the maximum amount you should spend each month on a mortgage payment?

b. What is the maximum amount you should spend each month for total credit obligations?

c. If your monthly mortgage payment is 90% of the maximum amount you can afford, what is the maximum amount you should spend each month for all other debt?

Round all computations to the nearest dollar.

4 Understand the pros and cons of renting versus buying.

Renting versus Buying

Nearly everyone is faced at some stage in life with the dilemma "should I rent or should I buy a home?" The rent-or-buy decision can be highly complex and is often based on lifestyle rather than finances. Aside from a changing economic climate, there are many factors to consider. Here are some advantages of both renting and buying to help smooth the way:

Benefits of Renting

- No down payment or points are required. You generally have a security deposit that is returned at the end of your lease.
- Very mobile: You can easily relocate, moving as often as you like and as your lease permits.
- Does not tie up hundreds of thousands of dollars that might be invested more safely and lucratively elsewhere. Most financial advisers agree that you should buy a home because you want to live in it, not because you want to fund your retirement.
- Does not clutter what you can afford for your total monthly debt with mortgage payments.
- May involve lower monthly expenses. You pay rent, whereas a homeowner pays the mortgage, taxes, insurance, and upkeep.
- Can provide amenities like swimming pools, tennis courts, and health clubs.
- Avoids the risk of falling housing prices.
- Does not require home repair, maintenance, and groundskeeping.
- There are no property taxes.
- Generally less costly than buying a home when staying in it for fewer than three years.

Benefits of Home Ownership

- Peace of mind and stability.
- Provides significant tax advantages, including deduction of mortgage interest and property taxes.
- There is no chance of rent increasing over time.
- Allows for freedom to remodel, landscape, and redecorate.
- You can build up **equity**, the difference between the home's value and what you owe on the mortgage, as the mortgage is paid off. The possibility of home appreciation is a potential source of cash in the form of home equity loans.
- When looking at seven-year time frames, the total cost of renting (monthly rent, renter's insurance, loss of potential interest on a security deposit) is more than twice the total cost of buying for home owners who itemize their tax deductions.

Source: Arthur J. Keown, *Personal Finance*, Fourth Edition, Pearson, 2007.

Blitzer Bonus

Reducing Rental Costs

Let's assume that one of your long-term financial goals involves home ownership. It's still likely that you'll be renting for a while once you complete your education and begin your first job. Other than living in a tent, here are some realistic suggestions for reducing costs on your first rental:

- Select a lower-cost rental. Who says you should begin your career in a large apartment with fancy amenities, private parking spots, and lakefront views? The less you spend renting, the more you can save for a down payment toward buying your own place. You will ultimately qualify for the most favorable mortgage terms by making a down payment of at least 20% of the purchase price of the property.
- Negotiate rental increases. Landlords do not want to lose good tenants who are respectful of their property and pay rent on time. Filling vacancies can be time consuming and costly.
- Rent a larger place with roommates. By sharing a rental, you will decrease rental costs and get more home for your rental dollars.

Concept and Vocabulary Check

Fill in each blank so that the resulting statement is true.

1. A long-term installment loan for the purpose of buying a home is called a/an ________. The portion of the sale price of the home that the buyer initially pays to the seller is called the ___________.
2. A document showing how each monthly installment payment is split between interest and principal is called a/an ____________________.

In Exercises 3–6, determine whether each statement is true or false. If the statement is false, make the necessary change(s) to produce a true statement.

3. Over the life of an installment loan, the interest portion increases and the portion applied to paying off the principal decreases with each successive payment. ______
4. Financial advisors suggest spending no more than 5% of your gross monthly income for your mortgage payment. ______
5. Renters are not required to pay property taxes. ______
6. Renters can build up equity as the rent is paid each month. ______

Exercise Set 8.7

Practice and Application Exercises

In Exercises 1–10, use

$$PMT = \frac{P\left(\frac{r}{n}\right)}{\left[1 - \left(1 + \frac{r}{n}\right)^{-nt}\right]}$$

to determine the regular payment amount, rounded to the nearest dollar.

1. The price of a home is \$220,000. The bank requires a 20% down payment and three points at the time of closing. The cost of the home is financed with a 30-year fixed-rate mortgage at 7%.
   a. Find the required down payment.
   b. Find the amount of the mortgage.
   c. How much must be paid for the three points at closing?

d. Find the monthly payment (excluding escrowed taxes and insurance).

e. Find the total cost of interest over 30 years.

2. The price of a condominium is \$180,000. The bank requires a 5% down payment and one point at the time of closing. The cost of the condominium is financed with a 30-year fixed-rate mortgage at 8%.

a. Find the required down payment.

b. Find the amount of the mortgage.

c. How much must be paid for the one point at closing?

d. Find the monthly payment (excluding escrowed taxes and insurance).

e. Find the total cost of interest over 30 years.

3. The price of a small cabin is \$100,000. The bank requires a 5% down payment. The buyer is offered two mortgage options: 20-year fixed at 8% or 30-year fixed at 8%. Calculate the amount of interest paid for each option. How much does the buyer save in interest with the 20-year option?

4. The price of a home is \$160,000. The bank requires a 15% down payment. The buyer is offered two mortgage options: 15-year fixed at 8% or 30-year fixed at 8%. Calculate the amount of interest paid for each option. How much does the buyer save in interest with the 15-year option?

5. In terms of paying less in interest, which is more economical for a \$150,000 mortgage: a 30-year fixed-rate at 8% or a 20-year fixed-rate at 7.5%? How much is saved in interest?

6. In terms of paying less in interest, which is more economical for a \$90,000 mortgage: a 30-year fixed-rate at 8% or a 15-year fixed-rate at 7.5%? How much is saved in interest?

In Exercises 7–8, which mortgage loan has the greater total cost (closing costs + the amount paid for points + total cost of interest)? By how much?

7. A \$120,000 mortgage with two loan options:

Mortgage A: 30-year fixed at 7% with closing costs of \$2000 and one point

Mortgage B: 30-year fixed at 6.5% with closing costs of \$1500 and four points

8. A \$250,000 mortgage with two loan options:

Mortgage A: 30-year fixed at 7.25% with closing costs of \$2000 and one point

Mortgage B: 30-year fixed at 6.25% with closing costs of \$350 and four points

9. The cost of a home is financed with a \$120,000 30-year fixed-rate mortgage at 4.5%.

a. Find the monthly payments and the total interest for the loan.

b. Prepare a loan amortization schedule for the first three months of the mortgage. Round entries to the nearest cent.

Payment Number	Interest	Principal	Loan Balance
1			
2			
3			

10. The cost of a home is financed with a \$160,000 30-year fixed-rate mortgage at 4.2%.

a. Find the monthly payments and the total interest for the loan.

b. Prepare a loan amortization schedule for the first three months of the mortgage. Round entries to the nearest cent.

Payment Number	Interest	Principal	Loan Balance
1			
2			
3			

Use this advice from most financial advisers to solve Exercises 11–12.

- *Spend no more than 28% of your gross monthly income for your mortgage payment.*
- *Spend no more than 36% of your gross monthly income for your total monthly debt.*

Round all computations to the nearest dollar.

11. Suppose that your gross annual income is \$36,000.

a. What is the maximum amount you should spend each month on a mortgage payment?

b. What is the maximum amount you should spend each month for total credit obligations?

c. If your monthly mortgage payment is 70% of the maximum you can afford, what is the maximum amount you should spend each month for all other debt?

12. Suppose that your gross annual income is \$62,000.

a. What is the maximum amount you should spend each month on a mortgage payment?

b. What is the maximum amount you should spend each month for total credit obligations?

c. If your monthly mortgage payment is 90% of the maximum you can afford, what is the maximum amount you should spend each month for all other debt?

Writing in Mathematics

13. What is a mortgage?

14. What is a down payment?

15. How is the amount of a mortgage determined?

16. Describe why a buyer would select a 30-year fixed-rate mortgage instead of a 15-year fixed-rate mortage if interest rates are $\frac{1}{4}$% to $\frac{1}{2}$% lower on a 15-year mortgage.

17. Describe one advantage and one disadvantage of an adjustable-rate mortage over a fixed-rate mortgage.

18. What is a loan amortization schedule?

19. Describe what happens to the portions of payments going to principal and interest over the life of an installment loan.

20. Describe how to determine what you can afford for your monthly mortgage payment.

21. Describe two advantages of renting over home ownership.

22. Describe two advantages of home ownership over renting.

Critical Thinking Exercises

Make Sense? *In Exercises 23–26, determine whether each statement makes sense or does not make sense, and explain your reasoning.*

23. I use the same formula to determine mortgage payments and payments for car loans.

24. There must be an error in the loan amortization schedule for my mortgage because the annual interest rate is only 3.5%, yet the schedule shows that I'm paying more on interest than on the principal for many of my payments.

25. My landlord required me to pay 2 points when I signed my rental lease.

26. I include rental payments among my itemized tax deductions.

27. If your gross annual income is $75,000, use appropriate computations to determine whether you could afford a $200,000 30-year fixed-rate mortgage at 5.5%.

28. The partial loan amortization schedule shows payments 50–54. Although payment 50 is correct, there are errors in one or more of the payments from 51 through 54. Find the errors and correct them.

LOAN AMORTIZATION SCHEDULE

Annual % Rate: 6.0%
Amount of Mortgage: $120,000 — **Monthly Payment: $1012.63**
Number of Monthly Payments: 180 — **Term: Years 15, Months 0**

Payment Number	Interest Payment	Principal Payment	Balance of Loan
50	$485.77	$526.86	$96,626.51
51	$483.13	$529.50	$96,097.01
52	$477.82	$534.81	$95,030.06
53	$480.49	$532.14	$95,564.87
54	$495.15	$537.48	$94,492.58

8.8

WHAT AM I SUPPOSED TO LEARN?

After you have read this section, you should be able to:

1. Find the interest, the balance due, and the minimum monthly payment for credit card loans.
2. Understand the pros and cons of using credit cards.
3. Understand the difference between credit cards and debit cards.
4. Know what is contained in a credit report.
5. Understand credit scores as measures of creditworthiness.

Credit Cards

WOULD YOU LIKE TO BUY PRODUCTS WITH A CREDIT card? Although the card will let you use a product while paying for it, the costs associated with such cards, including their high interest rates, fees, and penalties, stack the odds in favor of your getting hurt by them. In 2012, the average credit-card debt per U.S. household was $15,374. One advantage of making a purchase with a credit card is that the consumer gets to use a product immediately. In this section, we will see that a significant disadvantage is that it can add a substantial amount to the cost of a purchase. When it comes to using a credit card, consumer beware!

Open-End Installment Loans

Using a credit card is an example of an open-end installment loan, commonly called **revolving credit**. Open-end loans differ from fixed installment loans such as car loans and mortgages in that there is no schedule for paying a fixed amount each period. Credit card loans require users to make only a minimum monthly payment that depends on the unpaid balance and the interest rate. Credit cards have high interest rates compared to other kinds of loans. The interest on credit cards is computed using the simple interest formula $I = Prt$. However, r represents the *monthly* interest rate and t is time in months rather than in years. A typical interest rate is 1.57% monthly. This is equivalent to a yearly rate of $12 \times 1.57\%$, or 18.84%. With such a high annual percentage rate, credit card balances should be paid off as quickly as possible.

Most credit card customers are billed every month. A typical billing period is May 1 through May 31, but it can also run from, say, May 5 through June 4. Customers receive a statement, called an **itemized billing**, that includes the unpaid balance on the first day of the billing period, the total balance owed on the last day of the billing period, a list of purchases and cash advances made during the billing period, any finance charges or other fees incurred, the date of the last day of the billing period, the payment due date, and the minimum payment required.

Customers who make a purchase during the billing period and pay the entire amount of the purchase by the payment due date are not charged interest. By contrast, customers who make cash advances using their credit cards must pay interest from the day the money is advanced until the day it is repaid.

Find the interest, the balance due, and the minimum monthly payment for credit card loans.

Interest on Credit Cards: The Average Daily Balance Method

Methods for calculating interest, or finance charges, on credit cards may vary and the interest can differ on credit cards that show the same annual percentage rate, or APR. The method used for calculating interest on most credit cards is called the *average daily balance method.*

THE AVERAGE DAILY BALANCE METHOD

Interest is calculated using $I = Prt$, where r is the monthly rate and t is one month. The principal, P, is the *average daily balance*. The **average daily balance** is the sum of the unpaid balances for each day in the billing period divided by the number of days in the billing period.

$$\text{Average daily balance} = \frac{\text{Sum of the unpaid balances for each day in the billing period}}{\text{Number of days in the billing period}}$$

In Example 1, we illustrate how to determine the average daily balance. At the conclusion of the example, we summarize the steps used in the computation.

EXAMPLE 1 *Balance Due on a Credit Card*

The issuer of a particular VISA card calculates interest using the average daily balance method. The monthly interest rate is 1.3% of the average daily balance. The following transactions occurred during the May 1–May 31 billing period.

Transaction Description	Transaction Amount
Previous balance, \$1350.00	
May 1 Billing date	
May 8 Payment	\$250.00 credit
May 10 Charge: Airline Tickets	\$375.00
May 20 Charge: Books	\$ 57.50
May 28 Charge: Restaurant	\$ 65.30
May 31 End of billing period	
Payment Due Date: June 9	

a. Find the average daily balance for the billing period. Round to the nearest cent.

b. Find the interest to be paid on June 1, the next billing date. Round to the nearest cent.

c. Find the balance due on June 1.

d. This credit card requires a \$10 minimum monthly payment if the balance due at the end of the billing period is less than \$360. Otherwise, the minimum monthly payment is $\frac{1}{36}$ of the balance due at the end of the billing period, rounded up to the nearest whole dollar. What is the minimum monthly payment due by June 9?

SOLUTION

a. We begin by finding the average daily balance for the billing period. First make a table that shows the beginning date of the billing period, each transaction date, and the unpaid balance for each date.

Date	Unpaid Balance	
May 1	\$1350.00	previous balance
May 8	\$1350.00 − \$250.00 = \$1100.00	\$250.00 payment
May 10	\$1100.00 + \$375.00 = \$1475.00	\$375.00 charge
May 20	\$1475.00 + \$57.50 = \$1532.50	\$57.50 charge
May 28	\$1532.50 + \$65.30 = \$1597.80	\$65.30 charge

We now extend our table by adding two columns. One column shows the number of days at each unpaid balance. The final column shows each unpaid balance multiplied by the number of days that the balance is outstanding.

Date	Unpaid Balance	Number of Days at Each Unpaid Balance	(Unpaid Balance) · (Number of Days)
May 1	\$1350.00	7	(\$1350.00)(7) = \$9450.00
May 8	\$1100.00	2	(\$1100.00)(2) = \$2200.00
May 10	\$1475.00	10	(\$1475.00)(10) = \$14,750.00
May 20	\$1532.50	8	(\$1532.50)(8) = \$12,260.00
May 28	\$1597.80	4	(\$1597.80)(4) = \$6391.20
		Total: 31	Total: \$45,051.20

There are 4 days at this unpaid balance, May 28, 29, 30, and 31, before the beginning of the next billing period, June 1.

This is the number of days in the billing period.

This is the sum of the unpaid balances for each day in the billing period.

Notice that we found the sum of the products in the final column of the table. This dollar amount, \$45,051.20, gives the sum of the unpaid balances for each day in the billing period.

Now we divide the sum of the unpaid balances for each day in the billing period, \$45,051.20, by the number of days in the billing period, 31. This gives the average daily balance.

Average daily balance

$$= \frac{\text{Sum of the unpaid balances for each day in the billing period}}{\text{Number of days in the billing period}}$$

$$= \frac{\$45{,}051.20}{31} \approx \$1453.26$$

The average daily balance is approximately \$1453.26.

b. Now we find the interest to be paid on June 1, the next billing date. The monthly interest rate is 1.3% of the average daily balance. The interest due is computed using $I = Prt$.

$$I = Prt = (\$1453.26)\,(0.013)\,(1) \approx \$18.89$$

The average daily balance serves as the principal.

Time, t, is measured in months, and $t = 1$ month.

The interest, or finance charge, for the June 1 billing will be \$18.89.

GREAT QUESTION!

How do credit card companies round when determining a minimum monthly payment?

They round *up* to the nearest dollar. For example, the quotient in part (d) is approximately 44.908, which rounds to 45. Because the minimum monthly payment is rounded up, $45 would still be the payment if the approximate quotient had been 44.098.

c. The balance due on June 1, the next billing date, is the unpaid balance on May 31 plus the interest.

$$\text{Balance due} = \$1597.80 + \$18.89 = \$1616.69$$

Unpaid balance on May 31, obtained from the second table on the previous page

Interest, or finance charge, obtained from part (b)

The balance due on June 1 is $1616.69.

d. Because the balance due, $1616.69, exceeds $360, the customer must pay a minimum of $\frac{1}{36}$ of the balance due.

$$\text{Minimum monthly payment} = \frac{\text{balance due}}{36} = \frac{\$1616.69}{36} \approx \$45$$

Rounded up to the nearest whole dollar, the minimum monthly payment due by June 9 is $45.

The following box summarizes the steps used in Example 1 to determine the average daily balance. Calculating the average daily balance can be quite tedious when there are numerous transactions during a billing period.

DETERMINING THE AVERAGE DAILY BALANCE

Step 1 Make a table that shows the beginning date of the billing period, each transaction date, and the unpaid balance for each date.
Step 2 Add a column to the table that shows the number of days at each unpaid balance.
Step 3 Add a final column to the table that shows each unpaid balance multiplied by the number of days that the balance is outstanding.
Step 4 Find the sum of the products in the final column of the table. This dollar amount is the sum of the unpaid balances for each day in the billing period.
Step 5 Compute the average daily balance.

$$\text{Average daily balance} = \frac{\text{Sum of the unpaid balances for each day in the billing period}}{\text{Number of days in the billing period}}$$

CHECK POINT 1 A credit card company calculates interest using the average daily balance method. The monthly interest rate is 1.6% of the average daily balance. The following transactions occurred during the May 1–May 31 billing period.

Transaction Description	Transaction Amount
Previous balance, $8240.00	
May 1 Billing date	
May 7 Payment	$ 350.00 credit
May 15 Charge: Computer	$ 1405.00
May 17 Charge: Restaurant	$ 45.20
May 30 Charge: Clothing	$ 180.72
May 31 End of billing period	
Payment Due Date: June 9	

Answer parts (a) through (d) in Example 1 on page 558 using this information.

2 Understand the pros and cons of using credit cards.

Credit Cards: Marvelous Tools or Snakes in Your Wallet?

Credit cards are convenient. Pay the entire balance by the due date for each monthly billing and you avoid interest charges. Carry over a balance and interest charges quickly add up. With this in mind, let's consider the positives and the negatives involved with credit card usage.

Advantages of Using Credit Cards

- Get to use a product before actually paying for it.
- No interest charges by paying the balance due at the end of each billing period.
- Responsible use is an effective way to build a good credit score. (See page 562 for a discussion of credit scores.)
- No need to carry around large amounts of cash.
- More convenient to use than checks.
- Offer consumer protections: If there is a disputed or fraudulent charge on your credit card statement, let the card issuer know and the amount is generally removed.
- Provide a source of temporary emergency funds.
- Extend shopping opportunities to purchases over the phone or the Internet.
- Simple tasks like renting a car or booking a hotel room can be difficult or impossible without a credit card.
- Monthly statements can help keep track of spending. Some card issuers provide an annual statement that aids in tax preparation.
- May provide amenities such as free miles toward air travel.
- Useful as identification when multiple pieces of identification are needed.

Credit Card Woes

- High interest rates on unpaid balances. In 2009, interest rates were as high as 30%.
- No cap on interest rates. In 2009, the U.S. Senate defeated an amendment that would have imposed a 15% cap on credit card interest rates. (The Credit Card Act, passed by Congress in 2009, does restrict when issuers can raise rates on existing unpaid balances.) Your initial credit card interest rate is unlikely to go down, but it can sure go up.
- No cap on fees. *Consumer Reports* (October 2008) cited a credit card with an enticing 9.9% annual interest rate. But the fine print revealed a $29 account-setup fee, a $95 program fee, a $48 annual fee, and a $7 monthly servicing fee. Nearly 40% of the $40 billion in profits that U.S. card issuers earned in 2008 came from fees. Furthermore, issuers can hike fees at any time, for any reason. Read the fine print of a credit card agreement before you sign up.
- Easy to overspend. Purchases with credit cards can create the illusion that you are not actually spending money.
- Can serve as a tool for financial trouble. Using a credit card to buy more than you can afford and failing to pay the bill in full each month can result in serious debt. Fees and interest charges are added to the balance, which continues to grow, even if there are no new purchases.
- The minimum-payment trap: Credit-card debt is made worse by paying only the required minimum, a mistake made by 11% of credit-card debtors. Pay the minimum and most of it goes to interest charges.

Credit card statements now include a Minimum Payment Warning: "If you make only the minimum payment each period, you will pay more in interest and it will take you longer to pay off your balance."

Debit Cards

Credit cards have been around since the 1950s. Their plastic lookalikes, debit cards, were introduced in the mid-1970s. Although debit cards look like credit cards, the big difference is that debit cards are linked to your bank account. When you use a debit card,

3 Understand the difference between credit cards and debit cards.

the money you spend is deducted electronically from your bank account. It's similar to writing an electronic check, but there's no paper involved and the check gets "cashed" instantly.

Debit cards offer the convenience of making purchases with a piece of plastic without the temptation or ability to run up credit card debt. You can't spend money you don't have because the card won't work if the money isn't in your checking or savings account.

Debit cards have drawbacks. They may not offer the protection a credit card does for disputed purchases. It's easy to rack up overdraft charges if your bank enrolls you in an "overdraft protection" program. That means your card won't be turned down if you do not have sufficient funds in your account to cover your purchase. You are spared the embarrassment of having your card rejected, but it will cost you fees of approximately $27 per overdraft.

Debit card purchases should be treated like those for which you use a check. Record all transactions and their amounts in your checkbook, including any cash received from an ATM. Always keep track of how much money is available in your account. Your balance can be checked at an ATM or online.

4 Know what is contained in a credit report.

Credit Reports and Credit Scores

As a college student, it is unlikely that you have a credit history. Once you apply for your first credit card, your personal *credit report* will begin. A **credit report** contains the following information:

- **Identifying Information:** This includes your name, social security number, current address, and previous addresses.
- **Record of Credit Accounts:** This includes details about all open or closed credit accounts, such as when each account was opened, the latest balance, and the payment history.
- **Public Record Information:** Any of your public records, such as bankruptcy information, appears in this section of the credit report.
- **Collection Agency Account Information:** Unpaid accounts are turned over to collection agencies. Information about such actions appear in this section of the credit report.
- **Inquiries:** Companies that have asked for your credit information because you applied for credit are listed here.

Organizations known as **credit bureaus** collect credit information on individual consumers and provide credit reports to potential lenders, employers, and others upon request. The three main credit bureaus are Equifax, Experian, and TransUnion. You can get your credit report from the three bureaus free at www.annualcreditreport.com.

5 Understand credit scores as measures of creditworthiness.

Credit bureaus use data from your credit report to create a *credit score*, which is used to measure your creditworthiness. **Credit scores**, or **FICO scores**, range from 300 to 850, with a higher score indicating better credit. **Table 8.11** contains ranges of credit scores and their measures of creditworthiness.

TABLE 8.11 Credit Scores and Their Significance

Scores	Creditworthiness
720–850	Very good to excellent; Best interest rates on loans
650–719	Good; Likely to get credit, but not the best interest rates on loans
630–649	Fair; May get credit, but only at higher rates
580–629	Poor; Likely to be denied credit by all but a high-interest lender
300–579	Bad; Likely to be denied credit

Your credit score will have an enormous effect on your financial life. Individuals with higher credit scores get better interest rates on loans because they are considered to have a lower risk of defaulting. A good credit score can save you thousands of dollars in interest charges over your lifetime.

Blitzer Bonus

College Students and Credit Cards

If you have no credit history but are at least 18 years old with a job, you may be able to get a card with a limited amount of credit, usually $500 to $1000. Your chances of getting a credit card increase if you apply through a bank with which you have an account.

Prior to 2010, it was not necessary for college students to have a job to get a credit card. Credit card companies viewed college students as good and responsible customers who would continue to have a lifelong need for credit. The issuers anticipated retaining these students after graduation when their accounts would become more valuable. Many resorted to aggressive marketing tactics, offering everything from T-shirts to iPods to students who signed up.

Times have changed. In May 2009, President Obama signed legislation that prohibits issuing credit cards to college students younger than 21 unless they can prove they are able to make payments or get a parent or guardian to co-sign. Because college students do not have much money, most won't be able to get a credit card without permission from their parents. The bill also requires lenders to get permission from the co-signer before increasing the card's credit limit.

Before credit card reform swiped easy plastic from college students, those who fell behind on their credit card bills often left college with blemished credit reports. This made it more difficult for them to rent an apartment, get a car loan, or even find a job. "A lot of kids got themselves in trouble," said Adam Levin, founder of Credit.com, a consumer Web site. "As much as college students are obsessed with GPAs, their credit score is the most important number they're going to have to deal with after graduation."

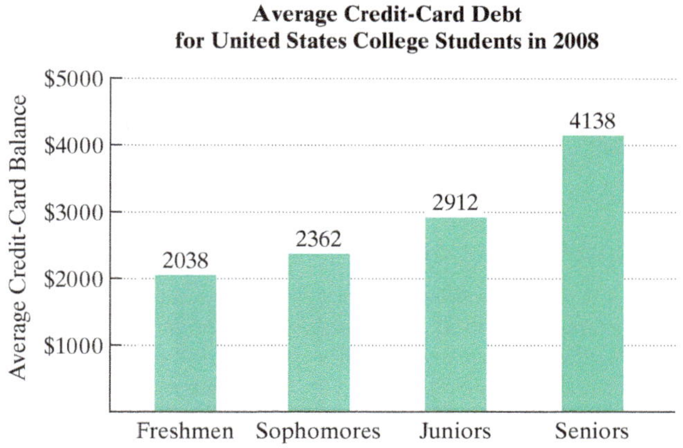

Source: Nellie Mae

GREAT QUESTION!

What's the bottom line on responsible credit card use?

A critical component of financial success involves demonstrating that you can handle the responsibility of using a credit card. Responsible credit card use means:

- You pay the entire balance by the due date for each monthly billing.
- You only use the card to make purchases that you can afford.
- You save all receipts for credit card purchases and check each itemized billing carefully for any errors.
- You use your credit card sometimes, but also use cash, checks, or your debit card.

When you decide to apply for your first credit card, you should look for a card with no annual fee and a low interest rate. You can compare rates and fees at www.creditcards.com, www.bankrate.com, and www.indexcreditcards.com.

Concept and Vocabulary Check

Fill in each blank so that the resulting statement is true.

1. Using a credit card is an example of a/an ________ installment loan for which there is no schedule for paying a fixed amount each period.
2. The average daily balance for a credit card's billing period is ________ divided by ________________________.
3. When you use a/an ________ card, the money you spend is deducted electronically from your bank account.
4. Details about when all your credit accounts were opened, their latest balance, and their payment history are included in a/an ________.
5. Credit scores range from ________ to ________, with a higher score indicating ______________.

In Exercises 6–9, determine whether each statement is true or false. If the statement is false, make the necessary change(s) to produce a true statement.

6. Interest on a credit card is calculated using $I = Prt$, where r is the monthly rate, t is one month, and P is the balance due. _______

7. When using a credit card, the money spent is deducted electronically from the user's bank account. _______

8. Credit reports contain bankruptcy information. _______

9. Higher credit scores indicate better credit. _______

Exercise Set 8.8

Practice and Application Exercises

Exercises 1–2 involve credit cards that calculate interest using the average daily balance method. The monthly interest rate is 1.5% of the average daily balance. Each exercise shows transactions that occurred during the March 1–March 31 billing period. In each exercise,

a. *Find the average daily balance for the billing period. Round to the nearest cent.*

b. *Find the interest to be paid on April 1, the next billing date. Round to the nearest cent.*

c. *Find the balance due on April 1.*

d. *This credit card requires a \$10 minimum monthly payment if the balance due at the end of the billing period is less than \$360. Otherwise, the minimum monthly payment is $\frac{1}{36}$ of the balance due at the end of the billing period, rounded up to the nearest whole dollar. What is the minimum monthly payment due by April 9?*

1.

Transaction Description	Transaction Amount
Previous balance, \$6240.00	
March 1 Billing date	
March 5 Payment	\$300 credit
March 7 Charge: Restaurant	\$ 40
March 12 Charge: Groceries	\$ 90
March 21 Charge: Car Repairs	\$230
March 31 End of billing period	
Payment Due Date: April 9	

2.

Transaction Description	Transaction Amount
Previous balance, \$7150.00	
March 1 Billing date	
March 4 Payment	\$ 400 credit
March 6 Charge: Furniture	\$ 1200
March 15 Charge: Gas	\$ 40
March 30 Charge: Groceries	\$ 50
March 31 End of billing period	
Payment Due Date: April 9	

Exercises 3–4 involve credit cards that calculate interest using the average daily balance method. The monthly interest rate is 1.2% of the average daily balance. Each exercise shows transactions that occurred during the June 1–June 30 billing period. In each exercise,

a. *Find the average daily balance for the billing period. Round to the nearest cent.*

b. *Find the interest to be paid on July 1, the next billing date. Round to the nearest cent.*

c. *Find the balance due on July 1.*

d. *This credit card requires a \$30 minimum monthly payment if the balance due at the end of the billing period is less than \$400. Otherwise, the minimum monthly payment is $\frac{1}{25}$ of the balance due at the end of the billing period, rounded up to the nearest whole dollar. What is the minimum monthly payment due by July 9?*

3.

Transaction Description	Transaction Amount
Previous balance, \$2653.48	
June 1 Billing date	
June 6 Payment	\$1000.00 credit
June 8 Charge: Gas	\$ 36.25
June 9 Charge: Groceries	\$ 138.43
June 17 Charge: Gas Charge: Groceries	\$ 42.36 \$ 127.19
June 27 Charge: Clothing	\$ 214.83
June 30 End of billing period	
Payment Due Date: July 9	

4.

Transaction Description	Transaction Amount
Previous balance, \$4037.93	
June 1 Billing date	
June 5 Payment	\$ 350.00 credit
June 10 Charge: Gas	\$ 31.17
June 15 Charge: Prescriptions	\$ 42.50
June 22 Charge: Gas Charge: Groceries	\$ 43.86 \$ 112.91
June 29 Charge: Clothing	\$ 96.73
June 30 End of billing period	
Payment Due Date: July 9	

In Exercises 5–10, use

$$PMT = \frac{P\left(\frac{r}{n}\right)}{\left[1 - \left(1 + \frac{r}{n}\right)^{-nt}\right]}$$

to determine the regular payment amount, rounded to the nearest dollar.

5. Suppose your credit card has a balance of \$4200 and an annual interest rate of 18%. You decide to pay off the balance over two years. If there are no further purchases charged to the card,

a. How much must you pay each month?

b. How much total interest will you pay?

6. Suppose your credit card has a balance of \$3600 and an annual interest rate of 16.5%. You decide to pay off the balance over two years. If there are no further purchases charged to the card,

a. How much must you pay each month?

b. How much total interest will you pay?

7. To pay off the \$4200 credit card balance in Exercise 5, suppose that you can get a bank loan at 10.5% with a term of three years.

a. How much will you pay each month? How does this compare with your credit card payment in Exercise 5?

b. How much total interest will you pay? How does this compare with your total credit card interest in Exercise 5?

8. To pay off the \$3600 credit card balance in Exercise 6, suppose that you can get a bank loan at 9.5% with a term of three years.

a. How much will you pay each month? How does this compare with your credit card payment in Exercise 6?

b. How much total interest will you pay? How does this compare with your total credit card interest in Exercise 6?

9. Rework Exercise 5 assuming you decide to pay off the balance over one year rather than two. How much more must you pay each month and how much less will you pay in total interest?

10. Rework Exercise 6 assuming you decide to pay off the balance over one year rather than two. How much more must you pay each month and how much less will you pay in total interest?

Writing in Mathematics

11. Describe the difference between a fixed installment loan and an open-end installment loan.

12. For a credit card billing period, describe how the average daily balance is determined. Why is this computation somewhat tedious when done by hand?

13. Describe two advantages of using credit cards.

14. Describe two disadvantages of using credit cards.

15. What is a debit card?

16. Describe what is contained in a credit report.

17. What are credit scores?

18. Describe two aspects of responsible credit card use.

Critical Thinking Exercises

Make Sense? *In Exercises 19–25, determine whether each statement makes sense or does not make sense, and explain your reasoning.*

19. I like to keep all my money, so I pay only the minimum required payment on my credit card.

20. One advantage of using credit cards is that there are caps on interest rates and fees.

21. The balance due on my credit card from last month to this month increased even though I made no new purchases.

22. My debit card offers the convenience of making purchases with a piece of plastic without the ability to run up credit-card debt.

23. My debit score is 630, so I anticipate an offer on a car loan at a very low interest rate.

24. In order to achieve financial success, I should consistently pay my entire credit card balance by the due date.

25. As a college student, I'm frequently enticed to sign up for credit cards with offers of free T-shirts or iPods.

26. A bank bills its credit card holders on the first of each month for each itemized billing. The card provides a 20-day period in which to pay the bill before charging interest. If the card holder wants to buy an expensive gift for a September 30 wedding but can't pay for it until November 5, explain how this can be done without adding an interest charge.

Group Exercises

27. Cellphone Plans

If credit cards can cause financial woes, cellphone plans are not far behind. Group members should present a report on cellphone plans, addressing each of the following questions: What are the monthly fees for these plans and what features are included? What happens if you use the phone more than the plan allows? Are there higher rates for texting and Internet access? What additional charges are imposed by the carrier on top of the monthly fee? What are the termination fees if you default on the plan? What can happen to your credit report and your credit score in the event of early termination? Does the carrier use free T-shirts, phones, and other items to entice new subscribers into binding contracts? What suggestions can the group offer to avoid financial difficulties with these plans?

28. Risky Credit Arrangements

Group members should present a report on the characteristics and financial risks associated with payday lending, tax refund loans, and pawn shops.

Chapter Summary, Review, and Test

SUMMARY – DEFINITIONS AND CONCEPTS

SUMMARY – DEFINITIONS AND CONCEPTS	EXAMPLES
8.1 Percent, Sales Tax, and Discounts	
a. Percent means per hundred. Thus, $97\% = \frac{97}{100}$.	
b. To express a fraction as a percent, divide the numerator by the denominator, move the decimal point in the quotient two places to the right, and add a percent sign.	Ex. 1, p. 488
c. To express a decimal number as a percent, move the decimal point two places to the right and add a percent sign.	Ex. 2, p. 489
d. To express a percent as a decimal number, move the decimal point two places to the left and remove the percent sign.	Ex. 3, p. 489
e. The percent formula, $A = PB$, means A is P percent of B.	
f. Sales tax amount = tax rate × item's cost	Ex. 4, p. 490
g. Discount amount = discount rate × original price	Ex. 5, p. 490
h. The fraction for percent increase (or decrease) is $\frac{\text{amount of increase (or decrease)}}{\text{original amount}}$. Find the percent increase (or decrease) by expressing this fraction as a percent.	Ex. 6, p. 491; Ex. 7, p. 492; Ex. 8, p. 493; Ex. 9, p. 493
8.2 Income Tax	
a. Calculating Income Tax **1.** Determine adjusted gross income: Adjusted gross income = Gross income − Adjustments. **2.** Determine taxable income: Taxable income = Adjusted gross income − (Exemptions + Deductions). **3.** Determine the income tax: Income tax = Tax computation − Tax credits. See details in the box on pages 500 – 501.	Ex. 1, p. 498; Ex. 2, p. 501
b. FICA taxes are used for Social Security and Medicare benefits. FICA tax rates are given in Table 8.2 on page 503.	Ex. 3, p. 503
c. Gross pay is your salary prior to any withheld taxes for a pay period. Net pay is the actual amount of your check after taxes have been withheld.	Ex. 4, p. 504
8.3 Simple Interest	
a. Interest is the amount of money that we get paid for lending or investing money, or that we pay for borrowing money. The amount deposited or borrowed is the principal. The charge for interest, given as a percent, is the rate, assumed to be per year.	
b. Simple interest involves interest calculated only on the principal and is computed using $I = Prt$.	Ex. 1, p. 508; Ex. 2, p. 509
c. The future value, A, of P dollars at simple interest rate r for t years is $A = P(1 + rt)$.	Ex. 3, p. 510; Ex. 4, p. 510; Ex. 5, p. 511; Ex. 6, p. 511
8.4 Compound Interest	
a. Compound interest involves interest computed on the original principal as well as on any accumulated interest. The amount in an account for one compounding period per year is $A = P(1 + r)^t$. For n compounding periods per year, the amount is $A = P\left(1 + \frac{r}{n}\right)^{nt}$. For continuous compounding, the amount is $A = Pe^{rt}$, where $e \approx 2.72$.	Ex. 1, p. 514; Ex. 2, p. 515; Ex. 3, p. 516

b. Calculating Present Value
If A dollars are to be accumulated in t years in an account that pays rate r compounded n times per year, then the present value, P, that needs to be invested now is given by

$$P = \frac{A}{\left(1 + \frac{r}{n}\right)^{nt}}.$$

Ex. 4, p. 517

c. Effective Annual Yield
Effective annual yield is defined in the box on page 518. The effective annual yield, Y, for an account that pays rate r compounded n times per year is given by

$$Y = \left(1 + \frac{r}{n}\right)^{n} - 1.$$

Ex. 5, p. 518; Ex. 6, p. 519

8.5 Annuities, Methods of Saving, and Investments

a. An annuity is a sequence of equal payments made at equal time periods. The value of an annuity is the sum of all deposits plus all interest paid.

Ex. 1, p. 524

b. The value of an annuity after t years is

$$A = \frac{P\left[\left(1 + \frac{r}{n}\right)^{nt} - 1\right]}{\left(\frac{r}{n}\right)},$$

where interest is compounded n times per year. See the box on page 526.

Ex. 2, p. 525; Ex. 3, p. 527

c. The formula

$$P = \frac{A\left(\frac{r}{n}\right)}{\left[\left(1 + \frac{r}{n}\right)^{nt} - 1\right]}$$

gives the deposit, P, into an annuity at the end of each compounding period needed to achieve a value of A dollars after t years. See the box on page 528.

Ex. 4, p. 528

d. The return on an investment is the percent increase in the investment.

e. Investors purchase stock, shares of ownership in a company. The shares indicate the percent of ownership. Trading refers to buying and selling stock. Investors make money by selling a stock for more money than they paid for it. They can also make money while they own stock if a company distributes all or part of its profits as dividends. Each share of stock is paid the same dividend.

f. Investors purchase a bond, lending money to the company from which they purchase the bond. The company commits itself to pay the price an investor pays for the bond at the time it was purchased, called its face value, along with interest payments at a given rate.

A listing of all the investments a person holds is called a financial portfolio. A portfolio with a mixture of low-risk and high-risk investments is called a diversified portfolio.

A mutual fund is a group of stocks and/or bonds managed by a professional investor, called the fund manager.

g. Reading stock tables is explained on pages 530–531.

Ex. 5, p. 532

h. Accounts designed for retirement savings include traditional IRAs (requires paying taxes when withdrawing money at age $59\frac{1}{2}$ or older), Roth IRAs (requires paying taxes on deposits, but not withdrawals/earnings at age $59\frac{1}{2}$ or older), and employer-sponsored plans, such as 401(k) and 403(b) plans.

Ex. 6, p. 534

8.6 Cars

a. A fixed installment loan is paid off with a series of equal periodic payments. A car loan is an example of a fixed installment loan.
Loan Payment Formula for Fixed Installment Loans

$$PMT = \frac{P\left(\frac{r}{n}\right)}{\left[1 - \left(1 + \frac{r}{n}\right)^{-nt}\right]}$$

PMT is the regular payment amount required to repay a loan of P dollars paid n times per year over t years at an annual interest rate r.

Ex. 1, p. 540; Ex. 2, p. 543

b. Leasing is a long-term rental agreement. Leasing a car offers some advantages over buying one (small down payment; lower monthly payments; no concerns about selling the car when the lease ends). Leasing also offers some disadvantages over buying (mileage penalties and other costs at the end of the leasing period often make the total cost of leasing more expensive than buying; penalties for ending the lease early; not owning the car when the lease ends; liability for damage to the car). Ex. 3, p. 544

Auto insurance includes liability coverage: Bodily injury liability covers the costs of lawsuits if someone is injured or killed in an accident in which you are at fault. Property damage liability covers damage to other cars and property from negligent operation of your vehicle. If you have a car loan or lease a car, you also need collision coverage (pays for damage or loss of your car if you're in an accident) and comprehensive coverage (protects your car from fire, theft, and acts of nature).

The Cost of Gasoline

$$\text{Annual fuel expense} = \frac{\text{annual miles driven}}{\text{miles per gallon}} \times \text{price per gallon}$$

8.7 The Cost of Home Ownership

a. A mortgage is a long-term loan for the purpose of buying a home, and for which the property is pledged as security for payment. The term of the mortgage is the number of years until final payoff. The down payment is the portion of the sale price of the home that the buyer initially pays. The amount of the mortgage is the difference between the sale price and the down payment.

b. Fixed-rate mortgages have the same monthly payment during the entire time of the loan. Variable-rate mortgages, or adjustable-rate mortgages, have payment amounts that change from time to time depending on changes in the interest rate.

c. A point is a one-time charge that equals 1% of the amount of a mortgage loan.

d. The loan payment formula for fixed installment loans can be used to determine the monthly payment for a mortgage. Ex. 1, p. 549

e. Amortizing a mortgage loan is the process of making regular payments on the principal and interest until the loan is paid off. A loan amortization schedule is a document showing the following information for each mortgage payment: the payment number, the interest paid from the payment, the amount of the payment applied to the principal, and the balance of the loan after the payment. Such a schedule shows how the buyer pays slightly less in interest and more in principal for each payment over the entire life of the loan. Ex. 2, p. 551

f. Here are the guidelines for what you can spend for a mortgage: Ex. 3, p. 553

- Spend no more than 28% of your gross monthly income for your mortgage payment.
- Spend no more than 36% of your gross monthly income for your total monthly debt.

g. Home ownership provides significant tax advantages, including deduction of mortgage interest and property taxes. Renting does not provide tax benefits, although renters do not pay property taxes.

8.8 Credit Cards

a. A fixed installment loan is paid off with a series of equal periodic payments. An open-end installment loan is paid off with variable monthly payments. Credit card loans are open-end installment loans.

Most credit cards calculate interest using the average daily balance method. Interest is calculated using $I = Prt$, where P is the average daily balance, r is the monthly rate, and t is one month. Ex. 1, p. 558

Average daily balance

$$= \frac{\text{Sum of the unpaid balances for each day in the billing period}}{\text{Number of days in the billing period}}$$

The steps needed to determine the average daily balance are given in the box on page 560.

b. One advantage of using a credit card is that there are no interest charges by paying the balance due at the end of each billing period. A disadvantage is the high interest rate on unpaid balances. Failing to pay the bill in full each month can result in serious debt.

c. The difference between debit cards and credit cards is that debit cards are linked to your bank account. When you use a debit card, the money you spend is deducted electronically from your account balance.

d. Credit reports include details about all open or closed credit accounts, such as when each account was opened, the latest balance, and the payment history. They also contain bankruptcy information and information about unpaid accounts that were turned over to collection agencies. Credit scores range from 300 to 850, with a higher score indicating better credit.

PERSONAL FINANCE FORMULAS

Simple Interest

$$I = Prt$$
$$A = P(1 + rt)$$

Compound Interest

$$A = P\left(1 + \frac{r}{n}\right)^{nt}$$
$$P = \frac{A}{\left(1 + \frac{r}{n}\right)^{nt}}$$
$$Y = \left(1 + \frac{r}{n}\right)^{n} - 1$$

Annuities

$$A = \frac{P\left[\left(1 + \frac{r}{n}\right)^{nt} - 1\right]}{\left(\frac{r}{n}\right)}$$
$$P = \frac{A\left(\frac{r}{n}\right)}{\left[\left(1 + \frac{r}{n}\right)^{nt} - 1\right]}$$

Amortization

$$PMT = \frac{P\left(\frac{r}{n}\right)}{\left[1 - \left(1 + \frac{r}{n}\right)^{-nt}\right]}$$

Be sure you understand what each formula in the box describes and the meaning of the variables in the formulas. Select the appropriate formula or formulas as you work the exercises in the Review Exercises and the Chapter 8 Test.

Review Exercises

8.1

In Exercises 1–3, express each fraction as a percent.

1. $\frac{4}{5}$ **2.** $\frac{1}{8}$ **3.** $\frac{3}{4}$

In Exercises 4–6, express each decimal as a percent.

4. 0.72 **5.** 0.0035 **6.** 4.756

In Exercises 7–12, express each percent as a decimal.

7. 65% **8.** 99.7% **9.** 150%

10. 3% **11.** 0.65% **12.** $\frac{1}{4}$%

13. What is 8% of 120?

14. Suppose that the local sales-tax rate is 6% and you purchase a backpack for $24.
- **a.** How much tax is paid?
- **b.** What is the backpack's total cost?

15. A television with an original price of $850 is on sale at 35% off.
- **a.** What is the discount amount?
- **b.** What is the television's sale price?

16. A college that had 40 students for each lecture course increased the number to 45 students. What is the percent increase in the number of students in a lecture course?

17. A dictionary regularly sells for $56.00. The sale price is $36.40. Find the percent decrease of the sale price from the regular price.

18. Consider the following statement:

> My investment portfolio fell 10% last year, but then it rose 10% this year, so at least I recouped my losses.

Is this statement true? In particular, suppose you invested $10,000 in the stock market last year. How much money would be left in your portfolio with a 10% fall and then a 10% rise? If there is a loss, what is the percent decrease, to the nearest tenth of a percent, in your portfolio?

8.2

In Exercises 19–20, find the gross income, the adjusted gross income, and the taxable income. In Exercise 20, base the taxable income on the greater of a standard deduction or an itemized deduction.

19. Your neighbor earned wages of $30,200, received $130 in interest from a savings account, and contributed $1100 to a tax-deferred retirement plan. He was entitled to a personal exemption of $3800 and had deductions totaling $5450.

20. Your neighbor earned wages of $86,400, won $350,000 on a television game show, and contributed $50,000 to a tax-deferred savings plan. She is entitled to a personal exemption of $3800 and a standard deduction of $5950. The interest on her home mortgage was $9200 and she contributed $95,000 to charity.

In Exercises 21–22, use the 2012 marginal tax rates in ***Table 8.1*** *on page 501 to compute the tax owed by each person or couple.*

21. A single woman with a taxable income of $600,000

22. A married couple filing jointly with a taxable income of $82,000 and a $7500 tax credit

23. Use the 2012 marginal tax rates in **Table 8.1** to calculate the income tax owed by the following person:
- Single, no dependents
- Gross income: $40,000
- $2500 paid to a tax-deferred IRA
- $6500 mortgage interest
- $1800 property taxes
- No tax credits

Use the 2012 FICA tax rates in ***Table 8.2*** *on page 503 to solve Exercises 24–25.*

24. If you are not self-employed and earn $86,000, what are your FICA taxes?

25. If you are self-employed and earn $260,000, what are your FICA taxes?

26. You decide to work part-time at a local clothing store. The job pays $8.50 per hour and you work 16 hours per week. Your employer withholds 10% of your gross pay for federal taxes, 5.65% for FICA taxes, and 4% for state taxes.
a. What is your weekly gross pay?
b. How much is withheld per week for federal taxes?
c. How much is withheld per week for FICA taxes?
d. How much is withheld per week for state taxes?
e. What is your weekly net pay?
f. What percentage of your gross pay is withheld for taxes? Round to the nearest tenth of a percent.

8.3

In Exercises 27–30, find the simple interest. (Assume 360 days in a year.)

	Principal	Rate	Time
27.	$6000	3%	1 year
28.	$8400	5%	6 years
29.	$20,000	8%	9 months
30.	$36,000	15%	60 days

31. In order to pay for tuition and books, a college student borrows $3500 for four months at 10.5% interest.
a. How much interest must the student pay?
b. Find the future value of the loan.

In Exercises 32–34, use the formula for future value with simple interest to find the missing quantity. Round dollar amounts to the nearest cent and rates to the nearest tenth of a percent.

32. $A = ?, P = \$12{,}000, r = 8.2\%, t = 9$ months

33. $A = \$5750, P = \$5000, r = ?, t = 2$ years

34. $A = \$16{,}000, P = ?, r = 6.5\%, t = 3$ years

35. You plan to buy a $12,000 sailboat in four years. How much should you invest now, at 7.3% simple interest, to have enough for the boat in four years? (Round up to the nearest cent.)

36. You borrow $1500 from a friend and promise to pay back $1800 in six months. What simple interest rate will you pay?

8.4

In Exercises 37–39, the principal represents an amount of money deposited in a savings account that provides the lender compound interest at the given rate.

a. *Find how much money, to the nearest cent, there will be in the account after the given number of years.*
b. *Find the interest earned.*

	Principal	Rate	Compounding Periods per Year	Time
37.	$7000	3%	1	5 years
38.	$30,000	2.5%	4	10 years
39.	$2500	4%	12	20 years

40. Suppose that you have $14,000 to invest. Which investment yields the greater return over 10 years: 7% compounded monthly or 6.85% compounded continuously? How much more (to the nearest dollar) is yielded by the better investment?

In Exercises 41–42, round answers up to the nearest cent.

41. How much money should parents deposit today in an account that earns 7% compounded monthly so that it will accumulate to $100,000 in 18 years for their child's college education?

42. How much money should be deposited today in an account that earns 5% compounded quarterly so that it will accumulate to $75,000 in 35 years for retirement?

43. You deposit $2000 in an account that pays 6% interest compounded quarterly.
a. Find the future value, to the nearest cent, after one year.
b. Use the future value formula for simple interest to determine the effective annual yield. Round to the nearest tenth of a percent.

44. What is the effective annual yield, to the nearest hundredth of a percent, of an account paying 5.5% compounded quarterly? What does your answer mean?

45. Which investment is the better choice: 6.25% compounded monthly or 6.3% compounded annually?

8.5

In Exercises 46–48, round the value of each annuity to the nearest dollar.

46. A person who does not understand probability theory (see Chapter 11) wastes $10 per week on lottery tickets, averaging $520 per year. Instead of buying tickets, if this person deposits the $520 at the end of each year in an annuity paying 6% compounded annually,
a. How much would he or she have after 20 years?
b. Find the interest.

47. To save for retirement, you decide to deposit $100 at the end of each month in an IRA that pays 5.5% compounded monthly.
a. How much will you have from the IRA after 30 years?
b. Find the interest.

48. Suppose that you would like to have \$25,000 to use as a down payment for a home in five years by making regular deposits at the end of every three months in an annuity that pays 7.25% compounded quarterly.
   a. Determine the amount of each deposit. Round up to the nearest dollar.
   b. How much of the \$25,000 comes from deposits and how much comes from interest?

For Exercises 49–56, refer to the stock table for Harley Davidson (the motorcycle company). Where necessary, round dollar amounts to the nearest cent.

52-Week High	52-Week Low	Stock	SYM	Div	Yld %	PE
64.06	26.13	Harley Dav	HOG	.16	.3	41

Vol 100s	Hi	Lo	Close	Net Chg
5458	61.25	59.25	61	+1.75

49. What were the high and low prices for a share for the past 52 weeks?
50. If you owned 900 shares of this stock last year, what dividend did you receive?
51. What is the annual return for the dividends alone?
52. How many shares of this company's stock were traded yesterday?
53. What were the high and low prices for a share yesterday?
54. What was the price at which a share last traded when the stock exchange closed yesterday?
55. What was the change in price for a share of stock from the market close two days ago to yesterday's market close?
56. Compute the company's annual earnings per share using

$$\frac{\text{Yesterday's closing price per share}}{\text{PE ratio}}.$$

57. Explain the difference between investing in a stock and investing in a bond.
58. What is the difference between tax benefits for a traditional IRA and a Roth IRA?

8.6

59. Suppose that you decide to take a \$15,000 loan for a new car. You can select one of the following loans, each requiring regular monthly payments:
   Loan A: three-year loan at 7.2%
   Loan B: five-year loan at 8.1%.
   a. Find the monthly payments and the total interest for Loan A.
   b. Find the monthly payments and the total interest for Loan B.
   c. Compare the monthly payments and interest for the longer-term loan to the monthly payments and interest for the shorter-term loan.

60. Describe two advantages of leasing a car.
61. Describe two disadvantages of leasing a car.
62. Two components of auto insurance are property damage liability and collision. What is the difference between these types of coverage?
63. Suppose that you drive 36,000 miles per year and gas averages \$3.60 per gallon.
   a. What will you save in annual fuel expenses by owning a hybrid car averaging 40 miles per gallon rather than an SUV averaging 12 miles per gallon?
   b. If you deposit your monthly fuel savings at the end of each month into an annuity that pays 5.2% compounded monthly, how much will you have saved at the end of six years? Round all computations to the nearest dollar.

8.7

In Exercises 64–66, round to the nearest dollar.

64. The price of a home is \$240,000. The bank requires a 20% down payment and two points at the time of closing. The cost of the home is financed with a 30-year fixed-rate mortgage at 7%.
   a. Find the required down payment.
   b. Find the amount of the mortgage.
   c. How much must be paid for the two points at closing?
   d. Find the monthly payment (excluding escrowed taxes and insurance).
   e. Find the total cost of interest over 30 years.
65. In terms of paying less in interest, which is more economical for a \$70,000 mortgage: a 30-year fixed-rate at 8.5% or a 20-year fixed-rate at 8%? How much is saved in interest? Discuss one advantage and one disadvantage for each mortgage option.
66. Suppose that you need a loan of \$100,000 to buy a home. Here are your options:
   Option A: 30-year fixed-rate at 8.5% with no closing costs and no points
   Option B: 30-year fixed-rate at 7.5% with closing costs of \$1300 and three points.
   a. Determine your monthly payments for each option and discuss how you would decide between the two options.
   b. Which mortgage loan has the greater total cost (closing costs + the amount paid for points + total cost of interest)? By how much?
67. The cost of a home is financed with a \$300,000 30-year fixed rate mortgage at 6.5%.
   a. Find the monthly payments, rounded to the nearest dollar, for the loan.
   b. Prepare a loan amortization schedule for the first three months of the mortgage. Round entries to the nearest cent.

Payment Number	Interest	Principal	Loan Balance
1			
2			
3			

68. Use these guidelines to solve this exercise: Spend no more than 28% of your gross monthly income for your mortgage payment and no more than 36% for your total monthly debt. Round all computations to the nearest dollar. Suppose that your gross annual income is $54,000.
  a. What is the maximum amount you should spend each month on a mortgage payment?
  b. What is the maximum amount you should spend each month for total credit obligations?
  c. If your monthly mortgage payment is 80% of the maximum amount you can afford, what is the maximum amount you should spend each month for all other debt?

69. Describe three benefits of renting over home ownership.

70. Describe three benefits of home ownership over renting.

8.8

71. A credit card issuer calculates interest using the average daily balance method. The monthly interest rate is 1.1% of the average daily balance. The following transactions occurred during the November 1–November 30 billing period.

Transaction Description		Transaction Amount
Previous balance, $4620.80		
November 1	Billing date	
November 7	Payment	$650.00 credit
November 11	Charge: Airline Tickets	$350.25
November 25	Charge: Groceries	$125.70
November 28	Charge: Gas	$ 38.25
November 30	End of billing period	
Payment Due Date: December 9		

  a. Find the average daily balance for the billing period. Round to the nearest cent.
  b. Find the interest to be paid on December 1, the next billing date. Round to the nearest cent.
  c. Find the balance due on December 1.
  d. This credit card requires a $10 minimum monthly payment if the balance due at the end of the billing period is less than $360. Otherwise, the minimum monthly payment is $\frac{1}{36}$ of the balance due at the end of the billing period, rounded up to the nearest whole dollar. What is the minimum monthly payment due by December 9?

72. In 2012, the average credit-card debt was $15,374. Suppose your card has this balance and an annual interest rate of 18%. You decide to pay off the balance over two years. If there are no further purchases charged to the card,
  a. How much must you pay each month?
  b. How much total interest will you pay?
  Round answers to the nearest dollar.

73. Describe two advantages of using credit cards.

74. Describe two disadvantages of using credit cards.

75. How does a debit card differ from a credit card?

76. Is a credit report the same thing as a credit score? If not, what is the difference?

77. Describe two ways to demonstrate that you can handle the responsibility of using a credit card.

Chapter 8 Test

The box on page 569 summarizes the finance formulas you have worked with throughout the chapter. Where applicable, use the appropriate formula to solve an exercise in this test. Unless otherwise stated, round dollar amounts to the nearest cent and rates to the nearest tenth of a percent.

1. A CD player with an original price of $120 is on sale at 15% off.
  a. What is the amount of the discount?
  b. What is the sale price of the CD player?

2. You purchased shares of stock for $2000 and sold them for $3500. Find the percent increase, or your return, on this investment.

3. You earned wages of $46,500, received $790 in interest from a savings account, and contributed $1100 to a tax-deferred savings plan. You are entitled to a personal exemption of $3800 and a standard deduction of $5950. The interest on your home mortgage was $7300, you contributed $350 to charity, and you paid $1395 in state taxes.
  a. Find your gross income.
  b. Find your adjusted gross income.
  c. Find your taxable income. Base your taxable income on the greater of the standard deduction or an itemized deduction.

4. Use the 2012 marginal tax rates in **Table 8.1** on page 501 to calculate the federal income tax owed by the following person:
- Single, no dependents
- Gross income: $36,500
- $2000 paid to a tax-deferred IRA
- $4700 mortgage interest
- $1300 property taxes
- No tax credits

5. Use FICA tax rates for people who are not self-employed, 5.65% on the first $110,000 of income and 1.45% on income in excess of $110,000, to answer this question: If a person is not self-employed and earns $150,000, what are that person's FICA taxes?

6. You decide to work part-time at a local stationery store. The job pays $10 per hour and you work 15 hours per week. Your employer withholds 10% of your gross pay for federal taxes, 5.65% for FICA taxes, and 3% for state taxes.
   a. What is your weekly gross pay?
   b. How much is withheld per week for federal taxes?
   c. How much is withheld per week for FICA taxes?
   d. How much is withheld per week for state taxes?
   e. What is your weekly net pay?
   f. What percentage of your gross pay is withheld for taxes? Round to the nearest tenth of a percent.
7. You borrow $2400 for three months at 12% simple interest. Find the amount of interest paid and the future value of the loan.
8. You borrow $2000 from a friend and promise to pay back $3000 in two years. What simple interest rate will you pay?
9. In six months, you want to have $7000 worth of remodeling done to your home. How much should you invest now, at 9% simple interest, to have enough money for the project? (Round up to the nearest cent.)
10. Find the effective annual yield, to the nearest hundredth of a percent, of an account paying 4.5% compounded quarterly. What does your answer mean?
11. You receive an inheritance of $20,000 and invest it in an account that pays 6.5% compounded monthly.
    a. How much, to the nearest dollar, will you have after 40 years?
    b. Find the interest.
12. You would like to have $3000 in four years for a special vacation by making a lump-sum investment in an account that pays 9.5% compounded semiannually. How much should you deposit now? Round up to the nearest dollar.
13. Suppose that you save money for a down payment to buy a home in five years by depositing $6000 in an account that pays 6.5% compounded monthly.
    a. How much, to the nearest dollar, will you have as a down payment after five years?
    b. Find the interest.
14. Instead of making the lump-sum deposit of $6000 described in Exercise 13, suppose that you decide to deposit $100 at the end of each month in an annuity that pays 6.5% compounded monthly.
    a. How much, to the nearest dollar, will you have as a down payment after five years?
    b. Find the interest.
    c. Why is less interest earned from this annuity than from the lump-sum deposit in Exercise 13? With less interest earned, why would one select the annuity rather than the lump-sum deposit?
15. Suppose that you want to retire in 40 years. How much should you deposit at the end of each month in an IRA that pays 6.25% compounded monthly to have $1,500,000 in 40 years? Round up to the nearest dollar. How much of the $1.5 million comes from interest?

Use the stock table for AT&T to solve Exercises 16–18.

52-Week High	52-Week Low	Stock	SYM	Div	Yld %	PE
26.50	24.25	AT&T	PNS	2.03	7.9	18

Vol 100s	Hi	Lo	Close	Net Chg
961	25.75	25.50	25.75	+0.13

16. What were the high and low prices for a share yesterday?
17. If you owned 1000 shares of this stock last year, what dividend did you receive?
18. Suppose that you bought 600 shares of AT&T, paying the price per share at which a share traded when the stock exchange closed yesterday. If the broker charges 2.5% of the price paid for all 600 shares, find the broker's commission.
19. Suppose that you drive 30,000 miles per year and gas averages $3.80 per gallon. What will you save in annual fuel expense by owning a hybrid car averaging 50 miles per gallon rather than a pickup truck averaging 15 miles per gallon?

Use this information to solve Exercises 20–25. The price of a home is $120,000. The bank requires a 10% down payment and two points at the time of closing. The cost of the home is financed with a 30-year fixed-rate mortgage at 8.5%.

20. Find the required down payment.
21. Find the amount of the mortgage.
22. How much must be paid for the two points at closing?
23. Find the monthly payment (excluding escrowed taxes and insurance). Round to the nearest dollar.
24. Find the total cost of interest over 30 years.
25. Prepare a loan amortization schedule for the first two months of the mortgage. Round entries to the nearest cent.

Payment Number	Interest	Principal	Loan Balance
1			
2			

26. Use these guidelines to solve this exercise. Spend no more than 28% of your gross monthly income for your mortgage payment and no more than 36% for your total monthly debt. Round all computations to the nearest dollar. Suppose that your gross annual income is $66,000.
    a. What is the maximum amount you should spend each month on a mortgage payment?
    b. What is the maximum amount you should spend each month for total credit obligations?
    c. If your monthly mortgage payment is 90% of the maximum amount you can afford, what is the maximum amount you should spend each month for all other debt?

27. A credit card issuer calculates interest using the average daily balance method. The monthly interest rate is 2% of the average daily balance. The following transactions occurred during the September 1–September 30 billing period.

Transaction Description		Transaction Amount
Previous balance, $3800.00		
September 1	Billing date	
September 5	Payment	$800.00 credit
September 9	Charge: Gas	$ 40.00
September 19	Charge: Clothing	$160.00
September 27	Charge: Airline Ticket	$200.00
September 30	End of billing period	
Payment Due Date: October 9		

a. Find the average daily balance for the billing period. Round to the nearest cent.
b. Find the interest to be paid on October 1, the next billing date. Round to the nearest cent.
c. Find the balance due on October 1.
d. Terms for the credit card require a $10 minimum monthly payment if the balance due is less than $360. Otherwise, the minimum monthly payment is $\frac{1}{36}$ of the balance due, rounded up to the nearest whole dollar. What is the minimum monthly payment due by October 9?

In Exercises 28–34, determine whether each statement is true or false. If the statement is false, make the necessary change(s) to produce a true statement.

28. By buying bonds, you purchase shares of ownership in a company.

29. A traditional IRA requires paying taxes when withdrawing money from the account at age $59\frac{1}{2}$ or older.

30. One advantage to leasing a car is that there are no penalties for ending the lease early.

31. If you cause an accident, collision coverage pays for damage to the other car.

32. Home ownership provides significant tax advantages, including deduction of mortgage interest.

33. Money spent using a credit card is deducted electronically from your bank account.

34. Credit scores range from 100 to 1000, with a higher score indicating better credit.

9

Measurement

YOU ARE FEELING CROWDED IN. PERHAPS IT WOULD BE A GOOD TIME TO MOVE TO A STATE WITH MORE ELBOW room. But which state? You can look up the population of each state, but that does not take into account the amount of land the population occupies. How is this land measured and how can you use this measure to select a place where wildlife outnumber humans?

In this chapter, we explore ways of measuring things in our English system, as well as in the metric system. Knowing how units of measure are used to describe your world can help you make decisions on issues ranging from where to live to alcohol consumption to determining proper dosages of medication.

Here's where you'll find these applications:

- Finding a state with lots of room to spread out is explored in Example 2 of Section 9.2.
- Measuring alcohol consumption is addressed in the Blitzer Bonus on page 600.
- Dosages of medication form the basis of Example 8 in Section 9.2 and Example 4 in Section 9.3.

9.1 Measuring Length; The Metric System

WHAT AM I SUPPOSED TO LEARN?

After you have read this section, you should be able to:

1. Use dimensional analysis to change units of measurement.
2. Understand and use metric prefixes.
3. Convert units within the metric system.
4. Use dimensional analysis to change to and from the metric system.

HAVE YOU SEEN EITHER OF THE *JURASSIC PARK* FILMS? The popularity of these movies reflects our fascination with dinosaurs and their incredible size. From end to end, the largest dinosaur from the Jurassic period, which lasted from 208 to 146 million years ago, was about 88 feet. To **measure** an object such as a dinosaur is to assign a number to its size. The number representing its measure from end to end is called its **length**. Measurements are used to describe properties of length, area, volume, weight, and temperature. Over the centuries, people have developed systems of measurement that are now accepted in most of the world.

Linear units of measure were originally based on parts of the body. The Egyptians used the palm (equal to four fingers), the span (a handspan), and the cubit (length of forearm).

Length

Every measurement consists of two parts: a number and a unit of measure. For example, if the length of a dinosaur is 88 feet, the number is 88 and the unit of measure is the foot. Many different units are commonly used in measuring length. The foot is from a system of measurement called the **English system**, which is generally used in the United States. In this system of measurement, length is expressed in such units as inches, feet, yards, and miles.

The result obtained from measuring length is called a **linear measurement** and is stated in **linear units**.

LINEAR UNITS OF MEASURE: THE ENGLISH SYSTEM

12 inches (in.) = 1 foot (ft)

3 feet = 1 yard (yd)

36 inches = 1 yard

5280 feet = 1 mile (mi)

1 Use dimensional analysis to change units of measurement.

Because many of us are familiar with the measures in the box, we find it simple to change from one measure to another, say from feet to inches. We know that there are 12 inches in a foot. To convert from 5 feet to a measure in inches, we multiply by 12. Thus, 5 feet = 5×12 inches = 60 inches.

Another procedure used to convert from one unit of measurement to another is called **dimensional analysis**. Dimensional analysis uses *unit fractions*. A **unit fraction** has two properties: The numerator and denominator contain different units and the value of the unit fraction is 1. Here are some examples of unit fractions:

$$\frac{12\text{ in.}}{1\text{ ft}};\quad \frac{1\text{ ft}}{12\text{ in.}};\quad \frac{3\text{ ft}}{1\text{ yd}};\quad \frac{1\text{ yd}}{3\text{ ft}};\quad \frac{5280\text{ ft}}{1\text{ mi}};\quad \frac{1\text{ mi}}{5280\text{ ft}}.$$

In each unit fraction, the numerator and denominator are equal measures, making the value of the fraction 1.

Let's see how to convert 5 feet to inches using dimensional analysis.

$$5\text{ ft} = ?\text{ in.}$$

We need to eliminate feet and introduce inches. The unit we need to introduce, inches, must appear in the numerator of the fraction. The unit we need to eliminate, feet, must appear in the denominator. Therefore, we choose the unit fraction with inches in the numerator and feet in the denominator. The units divide out as follows:

$$5 \text{ ft} = \frac{5 \text{ ft}}{1} \cdot \frac{12 \text{ in.}}{1 \text{ ft}} = 5 \cdot 12 \text{ in.} = 60 \text{ in.}$$

unit fraction

DIMENSIONAL ANALYSIS

To convert a measurement to a different unit, multiply by a unit fraction (or by unit fractions). The given unit of measurement should appear in the denominator of the unit fraction so that this unit cancels upon multiplication. The unit of measurement that needs to be introduced should appear in the numerator of the fraction so that this unit will be retained upon multiplication.

EXAMPLE 1 *Using Dimensional Analysis to Change Units of Measurement*

Convert:

a. 40 inches to feet **b.** 13,200 feet to miles **c.** 9 inches to yards.

SOLUTION

a. Because we want to convert 40 inches to feet, feet should appear in the numerator and inches in the denominator. We use the unit fraction

$$\frac{1 \text{ ft}}{12 \text{ in.}}$$

and proceed as follows:

$$40 \text{ in.} = \frac{40 \text{ in.}}{1} \cdot \frac{1 \text{ ft}}{12 \text{ in.}} = \frac{40}{12} \text{ ft} = 3\tfrac{1}{3} \text{ ft or } 3.\overline{3} \text{ ft.}$$

This period ends the sentence and is not part of the abbreviated unit.

b. To convert 13,200 feet to miles, miles should appear in the numerator and feet in the denominator. We use the unit fraction

$$\frac{1 \text{ mi}}{5280 \text{ ft}}$$

and proceed as follows:

$$13{,}200 \text{ ft} = \frac{13{,}200 \text{ ft}}{1} \cdot \frac{1 \text{ mi}}{5280 \text{ ft}} = \frac{13{,}200}{5280} \text{ mi} = 2\frac{1}{2} \text{ mi or } 2.5 \text{ mi.}$$

c. To convert 9 inches to yards, yards should appear in the numerator and inches in the denominator. We use the unit fraction

$$\frac{1 \text{ yd}}{36 \text{ in.}}$$

and proceed as follows:

$$9 \text{ in.} = \frac{9 \text{ in.}}{1} \cdot \frac{1 \text{ yd}}{36 \text{ in.}} = \frac{9}{36} \text{ yd} = \frac{1}{4} \text{ yd or } 0.25 \text{ yd.}$$

GREAT QUESTION!

When should I use a period for abbreviated units of measurement?

The abbreviations for units of measurement are written without a period, such as ft for feet, except for the abbreviation for inches (in.).

In each part of Example 1, we converted from a smaller unit to a larger unit. Did you notice that this results in a smaller number in the converted unit of measure? **Converting to a larger unit always produces a smaller number. Converting to a smaller unit always produces a larger number.**

CHECK POINT 1 Convert:

a. 78 inches to feet

b. 17,160 feet to miles

c. 3 inches to yards.

2 Understand and use metric prefixes.

Length and the Metric System

Although the English system of measurement is most commonly used in the United States, most industrialized countries use the metric system of measurement. One of the advantages of the metric system is that units are based on powers of ten, making it much easier than the English system to change from one unit of measure to another.

The basic unit for linear measure in the metric system is the meter (m). A meter is slightly longer than a yard, approximately 39 inches. Prefixes are used to denote a multiple or part of a meter. **Table 9.1** summarizes the more commonly used metric prefixes and their meanings.

kilodollar

hectodollar

dekadollar

dollar

decidollar

centidollar

Like our system of money, the metric system is based on powers of ten.

TABLE 9.1 Commonly Used Metric Prefixes

Prefix	Symbol	Meaning
kilo	k	$1000 \times$ base unit
hecto	h	$100 \times$ base unit
deka	da	$10 \times$ base unit
deci	d	$\frac{1}{10}$ of base unit
centi	c	$\frac{1}{100}$ of base unit
milli	m	$\frac{1}{1000}$ of base unit

GREAT QUESTION!

Does the "i" in the prefixes ending in "i" (deci, centi, milli) have any significance?

Prefixes ending in "i" are all fractional parts of one unit.

The prefixes kilo, centi, and milli are used more frequently than hecto, deka, and deci. **Table 9.2** applies all six prefixes to the meter. The first part of the symbol indicates the prefix and the second part (m) indicates meter.

TABLE 9.2 Commonly Used Units of Linear Measure in the Metric System

Symbol	Unit	Meaning
km	kilometer	1000 meters
hm	hectometer	100 meters
dam	dekameter	10 meters
m	meter	1 meter
dm	decimeter	0.1 meter
cm	centimeter	0.01 meter
mm	millimeter	0.001 meter

Kilometer is pronounced kil'-oh-met-er with the accent on the FIRST syllable. If pronounced correctly, kilometers should sound something like "kill all meters."

In the metric system, the kilometer is used to measure distances comparable to those measured in miles in the English system. One kilometer is approximately 0.6 mile, and one mile is approximately 1.6 kilometers.

Metric units of centimeters and millimeters are used to measure what the English system measures in inches. **Figure 9.1** on the next page shows that a centimeter is less than half an inch; there are 2.54 centimeters in an inch. The smaller markings on the bottom scale are millimeters. A millimeter is approximately the thickness of a dime. The length of a bee or a fly may be measured in millimeters.

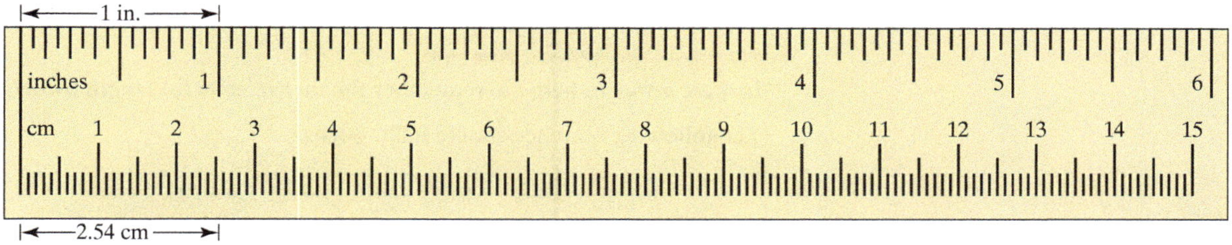

FIGURE 9.1

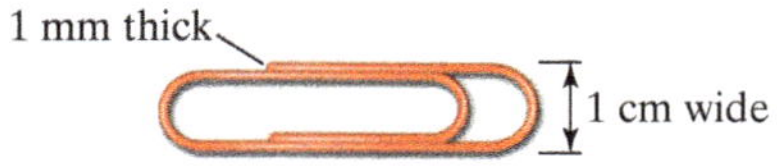

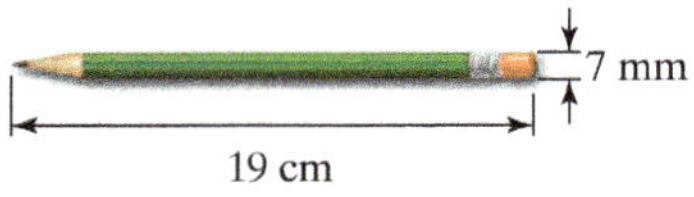

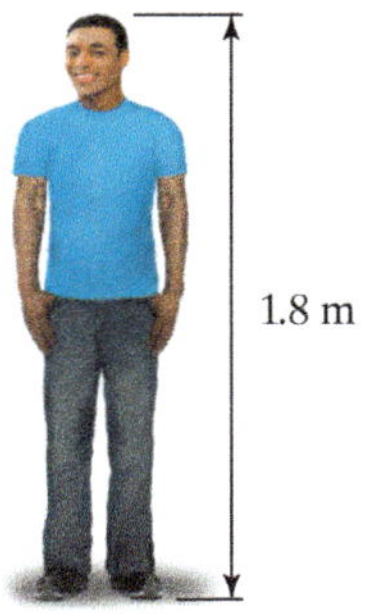

Those of us born in the United States have a good sense of what a length in the English system tells us about an object. An 88-foot dinosaur is huge, about 15 times the height of a 6-foot man. But what sense can we make of knowing that a whale is 25 meters long? The following lengths and the given approximations can help give you a feel for metric units of linear measure.

(1 meter ≈ 39 inches 1 kilometer ≈ 0.6 mile)

Item	Approximate Length
Width of lead in pencil	2 mm or 0.08 in.
Width of an adult's thumb	2 cm or 0.8 in.
Height of adult male	1.8 m or 6 ft
Typical room height	2.5 m or 8.3 ft
Length of medium-size car	5 m or 16.7 ft
Height of Empire State Building	381 m or 1270 ft
Average depth of ocean	4 km or 2.5 mi
Length of Manhattan Island	18 km or 11.25 mi
Distance from New York City to San Francisco	4800 km or 3000 mi
Radius of Earth	6378 km or 3986 mi
Distance from Earth to the moon	384,401 km or 240,251 mi

3 Convert units within the metric system.

Although dimensional analysis can be used to convert from one unit to another within the metric system, there is an easier, faster way to accomplish this conversion. The procedure is based on the observation that successively smaller units involve division by 10 and successively larger units involve multiplication by 10.

CHANGING UNITS WITHIN THE METRIC SYSTEM

Use the following chart to find equivalent measures of length:

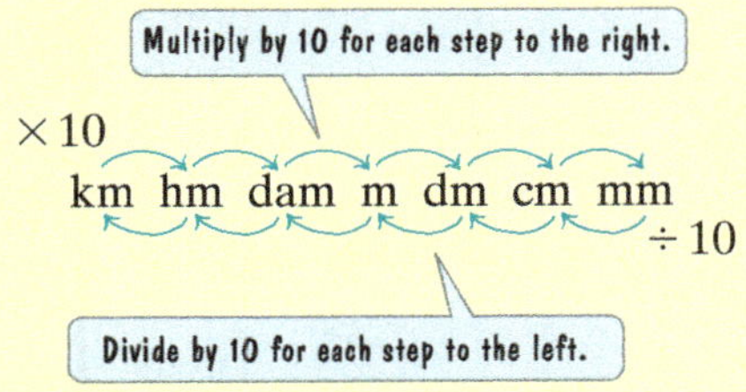

1. To change from a larger unit to a smaller unit (moving to the right in the diagram), multiply by 10 for each step to the right. Thus, move the decimal point in the given quantity one place to the right for each smaller unit until the desired unit is reached.
2. To change from a smaller unit to a larger unit (moving to the left in the diagram), divide by 10 for each step to the left. Thus, move the decimal point in the given quantity one place to the left for each larger unit until the desired unit is reached.

GREAT QUESTION!

Is there a way to help me remember the metric units for length from largest to smallest?

The following sentence should help:

King	Henry	died	Monday	drinking	chocolate	milk.
km	hm	dam	m	dm	cm	mm

Blitzer Bonus

The First Meter

The French first defined the meter in 1791, calculating its length as a romantic one ten-millionth of a line running from the Equator through Paris to the North Pole. Today's meter, officially accepted in 1983, is equal to the length of the path traveled by light in a vacuum during the time interval of 1/299,794,458 of a second.

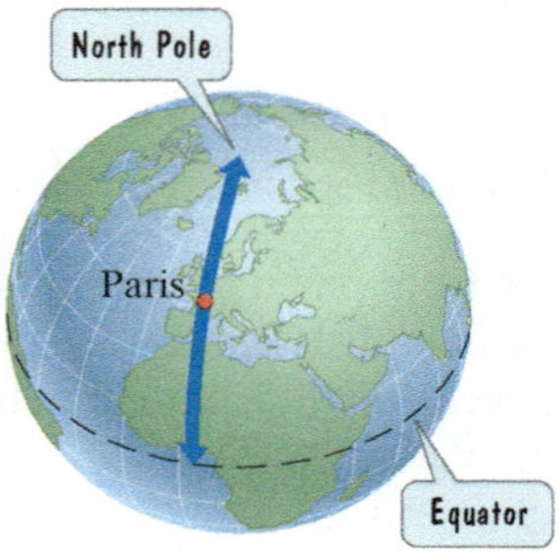

1 meter = 1/10,000,000 of the distance from the North Pole to the Equator on the meridian through Paris

EXAMPLE 2 *Changing Units within the Metric System*

a. Convert 504.7 meters to kilometers.

b. Convert 27 meters to centimeters.

c. Convert 704 mm to hm.

d. Convert 9.71 dam to dm.

SOLUTION

a. To convert from meters to kilometers, we start at meters and move three steps to the left to obtain kilometers:

km hm dam m dm cm mm.

Hence, we move the decimal point three places to the left:

504.7 m = 0.5047 km.

Thus, 504.7 meters converts to 0.5047 kilometer. Changing from a smaller unit of measurement (meter) to a larger unit of measurement (kilometer) results in an answer with a smaller number of units.

b. To convert from meters to centimeters, we start at meters and move two steps to the right to obtain centimeters:

km hm dam m dm cm mm.

Hence, we move the decimal point two places to the right:

27 m = 2700 cm.

Thus, 27 meters converts to 2700 centimeters. Changing from a larger unit of measurement (meter) to a smaller unit of measurement (centimeter) results in an answer with a larger number of units.

c. To convert from mm (millimeters) to hm (hectometers), we start at mm and move five steps to the left to obtain hm:

km hm dam m dm cm mm.

Hence, we move the decimal point five places to the left:

704 mm = 0.00704 hm.

d. To convert from dam (dekameters) to dm (decimeters), we start at dam and move two places to the right to obtain dm:

km hm dam m dm cm mm.

Hence, we move the decimal point two places to the right:

9.71 dam = 971 dm.

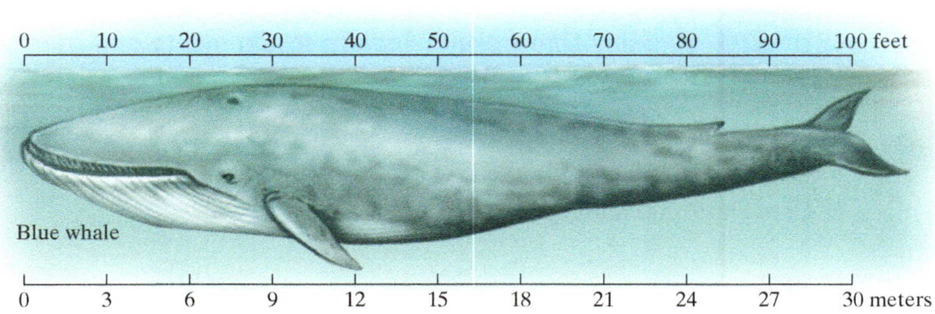

In Example 2(b), we showed that 27 meters converts to 2700 centimeters. This is the average length of the California blue whale, the longest of the great whales. Blue whales can have lengths that exceed 30 meters, making them over 100 feet long.

CHECK POINT 2

a. Convert 8000 meters to kilometers.
b. Convert 53 meters to millimeters.
c. Convert 604 cm to hm.
d. Convert 6.72 dam to cm.

Blitzer Bonus

Viruses and Metric Prefixes

Viruses are measured in attometers. An attometer is one quintillionth of a meter, or 10^{-18} meter, symbolized am. If a virus measures 1 am, you can place 10^{15} of them across a penciled 1 millimeter dot. If you were to enlarge each of these viruses to the size of the dot, they would stretch far into space, almost reaching Saturn.

Here is a list of all 20 metric prefixes. When applied to the meter, they range from the yottameter (10^{24} meters) to the yoctometer (10^{-24} meter).

LARGER THAN THE BASIC UNIT

Prefix	Symbol	Power of Ten	English Name
yotta-	Y	+24	septillion
zetta-	Z	+21	sextillion
exa-	E	+18	quintillion
peta-	P	+15	quadrillion
tera-	T	+12	trillion
giga-	G	+9	billion
mega-	M	+6	million
kilo-	k	+3	thousand
hecto-	h	+2	hundred
deca-	da	+1	ten

SMALLER THAN THE BASIC UNIT

Prefix	Symbol	Power of Ten	English Name
deci-	d	−1	tenth
centi-	c	−2	hundredth
milli-	m	−3	thousandth
micro-	μ	−6	millionth
nano-	n	−9	billionth
pico-	p	−12	trillionth
femto-	f	−15	quadrillionth
atto-	a	−18	quintillionth
zepto-	z	−21	sextillionth
yocto-	y	−24	septillionth

4 Use dimensional analysis to change to and from the metric system.

Although dimensional analysis is not necessary when changing units within the metric system, it is a useful tool when converting to and from the metric system. Some conversions are given in **Table 9.3**.

TABLE 9.3 English and Metric Equivalents

1 inch (in.) = 2.54 centimeters (cm)
1 foot (ft) = 30.48 centimeters (cm)
1 yard (yd) ≈ 0.9 meter (m)
1 mile (mi) ≈ 1.6 kilometers (km)

These conversions are exact.

These conversions are approximate.

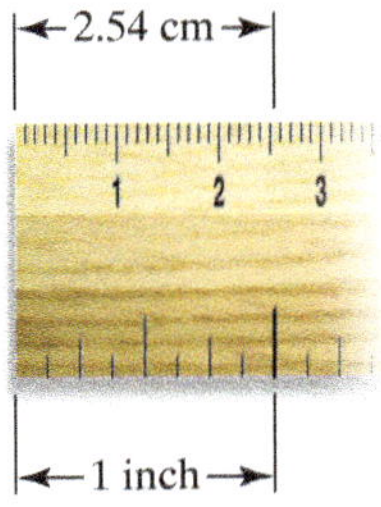

1 in. = 2.54 cm

TABLE 9.3 English and Metric Equivalents
1 inch (in.) = 2.54 centimeters (cm)
1 foot (ft) = 30.48 centimeters (cm)
1 yard (yd) ≈ 0.9 meter (m)
1 mile (mi) ≈ 1.6 kilometers (km)

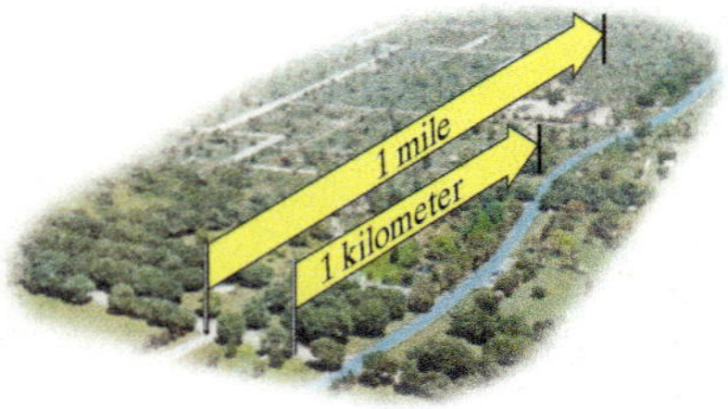

1 mi ≈ 1.6 km

GREAT QUESTION!

Is the conversion in part (b) approximate or exact?

It's approximate. More accurately,

$$1 \text{ mi} \approx 1.61 \text{ km}.$$

Thus,

$$125 \text{ mi} \approx 125(1.61) \text{ km} \approx 201.25 \text{ km}.$$

EXAMPLE 3 *Using Dimensional Analysis to Change to and from the Metric System*

a. Convert 8 inches to centimeters.
b. Convert 125 miles to kilometers.
c. Convert 26,800 millimeters to inches.

SOLUTION

a. To convert 8 inches to centimeters, we use a unit fraction with centimeters in the numerator and inches in the denominator:

$$\frac{2.54 \text{ cm}}{1 \text{ in.}}.$$ Table 9.3 shows that 1 in. = 2.54 cm.

We proceed as follows.

$$8 \text{ in.} = \frac{8 \text{ in.}}{1} \cdot \frac{2.54 \text{ cm}}{1 \text{ in.}} = 8(2.54) \text{ cm} = 20.32 \text{ cm}$$

b. To convert 125 miles to kilometers, we use a unit fraction with kilometers in the numerator and miles in the denominator:

$$\frac{1.6 \text{ km}}{1 \text{ mi}}.$$ Table 9.3 shows that 1 mi ≈ 1.6 km.

Thus,

$$125 \text{ mi} \approx \frac{125 \text{ mi}}{1} \cdot \frac{1.6 \text{ km}}{1 \text{ mi}} = 125(1.6) \text{ km} = 200 \text{ km}.$$

c. To convert 26,800 millimeters to inches, we observe that **Table 9.3** has only a conversion factor between inches and centimeters. We begin by changing millimeters to centimeters:

$$26{,}800 \text{ mm} = 2680.0 \text{ cm}$$

Now we need to convert 2680 centimeters to inches. We use a unit fraction with inches in the numerator and centimeters in the denominator:

$$\frac{1 \text{ in.}}{2.54 \text{ cm}}.$$

Thus,

$$26{,}800 \text{ mm} = 2680 \text{ cm} = \frac{2680 \text{ cm}}{1} \cdot \frac{1 \text{ in.}}{2.54 \text{ cm}} = \frac{2680}{2.54} \text{ in.} \approx 1055 \text{ in.}$$

This measure is equivalent to about 88 feet, the length of the largest dinosaur from the Jurassic period. The diplodocus, a plant eater, was 26.8 meters, approximately 88 feet, long.

CHECK POINT 3

a. Convert 8 feet to centimeters.
b. Convert 20 meters to yards.
c. Convert 30 meters to inches.

So far, we have used dimensional analysis to change units of length. Dimensional analysis may also be used to convert other kinds of measures, such as speed.

Blitzer Bonus

The Mars Climate Orbiter

If you think that using dimensional analysis to change to and from the metric system is no big deal, consider this: The Mars Climate Orbiter, launched on December 11, 1998, was lost when it crashed into Mars about a year later due to failure to properly convert English units into metric units. The price tag: $125 million!

EXAMPLE 4 *Using Dimensional Analysis*

a. The speed limit on many highways in the United States is 55 miles per hour (mi/hr). How many kilometers per hour (km/hr) is this?

b. If a high-speed train in Japan is capable of traveling at 200 kilometers per hour, how many miles per hour is this?

SOLUTION

a. To change miles per hour to kilometers per hour, we need to concentrate on changing miles to kilometers, so we need a unit fraction with kilometers in the numerator and miles in the denominator:

$$\frac{1.6 \text{ km}}{1 \text{ mi}}. \qquad \text{Table 9.3 shows that 1 mi} \approx \text{1.6 km.}$$

Thus,

$$\frac{55 \text{ mi}}{\text{hr}} \approx \frac{55 \cancel{\text{mi}}}{\text{hr}} \cdot \frac{1.6 \text{ km}}{1 \cancel{\text{mi}}} = 55(1.6)\frac{\text{km}}{\text{hr}} = 88 \text{ km/hr}.$$

This shows that 55 miles per hour is approximately 88 kilometers per hour.

b. To change 200 kilometers per hour to miles per hour, we must convert kilometers to miles. We need a unit fraction with miles in the numerator and kilometers in the denominator:

$$\frac{1 \text{ mi}}{1.6 \text{ km}}. \qquad \text{Table 9.3 shows that 1 mi} \approx \text{1.6 km.}$$

Thus,

$$\frac{200 \text{ km}}{\text{hr}} \approx \frac{200 \cancel{\text{km}}}{\text{hr}} \cdot \frac{1 \text{ mi}}{1.6 \cancel{\text{km}}} = \frac{200}{1.6}\frac{\text{mi}}{\text{hr}} = 125 \text{ mi/hr}.$$

A train capable of traveling at 200 kilometers per hour can therefore travel at about 125 miles per hour.

CHECK POINT 4 A road in Europe has a speed limit of 60 kilometers per hour. Approximately how many miles per hour is this?

Concept and Vocabulary Check

Fill in each blank so that the resulting statement is true.

1. The result obtained from measuring length is called a/an ________ measurement and is stated in ________ units.
2. In the English system, ________ in. = 1 ft, ________ ft = 1 yd, ________ in. = 1 yd, and ________ ft = 1 mi.
3. Fractions such as $\frac{12 \text{ in.}}{1 \text{ ft}}$ and $\frac{1 \text{ yd}}{3 \text{ ft}}$ are called ________ fractions. The value of such fractions is ________.
4. In the metric system, 1 km = ________ m, 1 hm = ________ m, 1 dam = ________ m, 1 dm = ________ m, 1 cm = ________ m, and 1 mm = ________ m.

In Exercises 5–8, determine whether each statement is true or false. If the statement is false, make the necessary change(s) to produce a true statement.

5. Dimensional analysis uses powers of 10 to convert from one unit of measurement to another. ______
6. One of the advantages of the English system is that units are based on powers of 10. ______
7. There are 2.54 inches in a centimeter. ______
8. The height of an adult male is approximately 6 meters. ______

Exercise Set 9.1

Practice Exercises

In Exercises 1–16, use dimensional analysis to convert the quantity to the indicated unit. If necessary, round the answer to two decimal places.

1. 30 in. to ft
2. 100 in. to ft
3. 30 ft to in.
4. 100 ft to in.
5. 6 in. to yd
6. 21 in. to yd
7. 6 yd to in.
8. 21 yd to in.
9. 6 yd to ft
10. 12 yd to ft
11. 6 ft to yd
12. 12 ft to yd
13. 23,760 ft to mi
14. 19,800 ft to mi
15. 0.75 mi to ft
16. 0.25 mi to ft

In Exercises 17–26, use the diagram in the box on page 579 to convert the given measurement to the unit indicated.

17. 5 m to cm
18. 8 dam to m
19. 16.3 hm to m
20. 0.37 hm to m
21. 317.8 cm to hm
22. 8.64 hm to cm
23. 0.023 mm to m
24. 0.00037 km to cm
25. 2196 mm to dm
26. 71 dm to km

In Exercises 27–44, use the following English and metric equivalents, along with dimensional analysis, to convert the given measurement to the unit indicated.

English and Metric Equivalents

1 in. = 2.54 cm
1 ft = 30.48 cm
1 yd ≈ 0.9 m
1 mi ≈ 1.6 km

27. 14 in. to cm
28. 26 in. to cm
29. 14 cm to in.
30. 26 cm to in.
31. 265 mi to km
32. 776 mi to km
33. 265 km to mi
34. 776 km to mi
35. 12 m to yd
36. 20 m to yd
37. 14 dm to in.
38. 1.2 dam to in.
39. 160 in. to dam
40. 180 in. to hm
41. 5 ft to m
42. 8 ft to m
43. 5 m to ft
44. 8 m to ft

Use 1 mi ≈ 1.6 km *to solve Exercises 45–48.*

45. Express 96 kilometers per hour in miles per hour.
46. Express 104 kilometers per hour in miles per hour.
47. Express 45 miles per hour in kilometers per hour.
48. Express 50 miles per hour in kilometers per hour.

Practice Plus

In Exercises 49–52, use the unit fractions

$$\frac{36 \text{ in.}}{1 \text{ yd}} \text{ and } \frac{2.54 \text{ cm}}{1 \text{ in.}}.$$

49. Convert 5 yd to cm.
50. Convert 8 yd to cm.
51. Convert 762 cm to yd.
52. Convert 1016 cm to yd.

In Exercises 53–54, use the unit fractions

$$\frac{5280 \text{ ft}}{1 \text{ mi}}, \frac{12 \text{ in.}}{1 \text{ ft}}, \text{ and } \frac{2.54 \text{ cm}}{1 \text{ in.}}.$$

53. Convert 30 mi to km.
54. Convert 50 mi to km.
55. Use unit fractions to express 120 miles per hour in feet per second.
56. Use unit fractions to express 100 miles per hour in feet per second.

Application Exercises

In Exercises 57–66, selecting from millimeter, meter, and kilometer, determine the best unit of measure to express the given length.

57. A person's height
58. The length of a football field
59. The length of a bee
60. The distance from New York City to Washington, D.C.
61. The distance around a one-acre lot
62. The length of a car
63. The width of a book
64. The altitude of an airplane
65. The diameter of a screw
66. The width of a human foot

In Exercises 67–74, select the best estimate for the measure of the given item.

67. The length of a pen
a. 30 cm **b.** 19 cm **c.** 19 mm

68. The length of this page
a. 2.5 mm **b.** 25 mm **c.** 250 mm

69. The height of a skyscraper
a. 325 m **b.** 32.5 km **c.** 325 km **d.** 3250 km

70. The length of a pair of pants
a. 700 cm **b.** 70 cm **c.** 7 cm

71. The height of a room
a. 4 mm **b.** 4 cm **c.** 4 m **d.** 4 dm

72. The length of a rowboat
a. 4 cm **b.** 4 dm **c.** 4 m **d.** 4 dam

73. The width of an electric cord

a. 4 mm **b.** 4 cm **c.** 4 dm **d.** 4 m

74. The dimensions of a piece of typing paper

a. 22 mm by 28 mm **b.** 22 cm by 28 cm

c. 22 dm by 28 dm **d.** 22 m by 28 m

75. A baseball diamond measures 27 meters along each side. If a batter scored two home runs in a game, how many kilometers did the batter run?

76. If you jog six times around a track that is 700 meters long, how many kilometers have you covered?

77. The distance from the Earth to the sun is about 93 million miles. What is this distance in kilometers?

78. The distance from New York City to Los Angeles is 4690 kilometers. What is the distance in miles?

Exercises 79–80 give the approximate length of some of the world's longest rivers. In each exercise, determine which is the longer river and by how many kilometers.

79. Nile: 4130 miles; Amazon: 6400 kilometers

80. Yangtze: 3940 miles; Mississippi: 6275 kilometers

Exercises 81–82 give the approximate height of some of the world's tallest mountains. In each exercise, determine which is the taller mountain and by how many meters. Round to the nearest meter.

81. K2: 8611 meters; Everest: 29,035 feet

82. Lhotse: 8516 meters; Kangchenjunga: 28,170 feet

Exercises 83–84 give the average rainfall of some of the world's wettest places. In each exercise, determine which location has the greater average rainfall and by how many inches. Round to the nearest inch.

83. Debundscha (Cameroon): 10,280 millimeters; Waialeale (Hawaii): 451 inches

84. Mawsynram (India): 11,870 millimeters; Cherrapunji (India): 498 inches

(Source for Exercises 79–84: Russell Ash, *The Top 10 of Everything 2009*)

Writing in Mathematics

85. Describe the two parts of a measurement.

86. Describe how to use dimensional analysis to convert 20 inches to feet.

87. Describe advantages of the metric system over the English system.

88. Explain how to change units within the metric system.

89. You jog 500 meters in a given period of time. The next day, you jog 500 yards over the same time period. On which day was your speed faster? Explain your answer.

90. What kind of difficulties might arise if the United States immediately eliminated all units of measure in the English system and replaced the system by the metric system?

91. The United States is the only Westernized country that does not use the metric system as its primary system of measurement. What reasons might be given for continuing to use the English system?

Critical Thinking Exercises

Make Sense? *In Exercises 92–95, determine whether each statement makes sense or does not make sense, and explain your reasoning.*

92. I can run 4000 meters in approximately one hour.

93. I ran 2000 meters and you ran 2000 yards in the same time, so I ran at a faster rate.

94. The most frequent use of dimensional analysis involves changing units within the metric system.

95. When multiplying by a unit fraction, I put the unit of measure that needs to be introduced in the denominator.

In Exercises 96–100, convert to an appropriate metric unit so that the numerical expression in the given measure does not contain any zeros.

96. 6000 cm **97.** 900 m

98. 7000 dm **99.** 11,000 mm

100. 0.0002 km

9.2 Measuring Area and Volume

WHAT AM I SUPPOSED TO LEARN?

After you have read this section, you should be able to:

1. Use square units to measure area.
2. Use dimensional analysis to change units for area.
3. Use cubic units to measure volume.
4. Use English and metric units to measure capacity.

ARE YOU FEELING A BIT CROWDED in? Although there are more people on the East Coast of the United States than there are bears, there are places in the Northwest where bears outnumber humans. The most densely populated state is New Jersey, averaging 1195.5 people per square mile. The least densely populated state is Alaska, averaging 1.2 persons per square mile. The U.S. average is 87.4 persons per square mile.

A square mile is one way of measuring the **area** of a state. A state's area is the region within its boundaries. Its **population density** is its population divided by its area. In this section, we discuss methods for measuring both area and volume.

Use square units to measure area.

Measuring Area

In order to measure a region that is enclosed by boundaries, we begin by selecting a *square unit*. A **square unit** is a square, each of whose sides is one unit in length, illustrated in **Figure 9.2**. The region in **Figure 9.2** is said to have an area of **one square unit**. The side of the square can be 1 inch, 1 centimeter, 1 meter, 1 foot, or one of any linear unit of measure. The corresponding units of area are the square inch (in.^{2}), the square centimeter (cm^2), the square meter (m^2), the square foot (ft^2), and so on. **Figure 9.3** illustrates 1 square inch and 1 square centimeter, drawn to actual size.

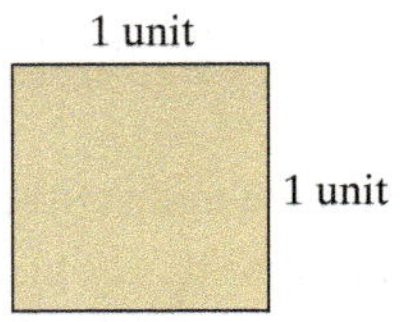

FIGURE 9.2 One square unit

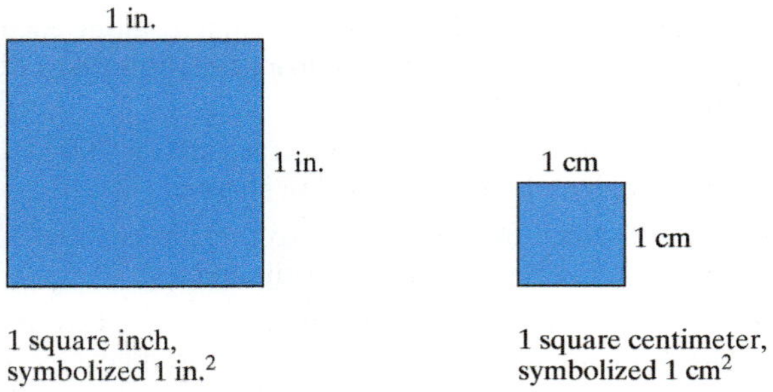

FIGURE 9.3 Common units of measurement for area, drawn to actual size

EXAMPLE 1 *Measuring Area*

What is the area of the region shown in **Figure 9.4**?

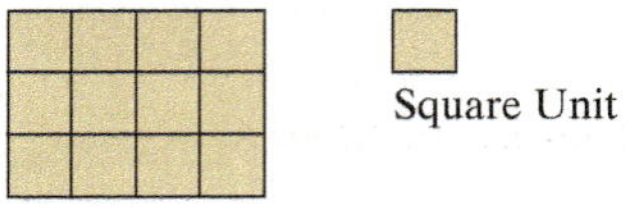

FIGURE 9.4

SOLUTION

We can determine the area of the region by counting the number of square units contained within the region. There are 12 such units. Therefore, the area of the region is 12 square units.

CHECK POINT 1 What is the area of the region represented by the first two rows in **Figure 9.4**?

Although there are 12 inches in one foot and 3 feet in one yard, these numerical relationships are not the same for square units.

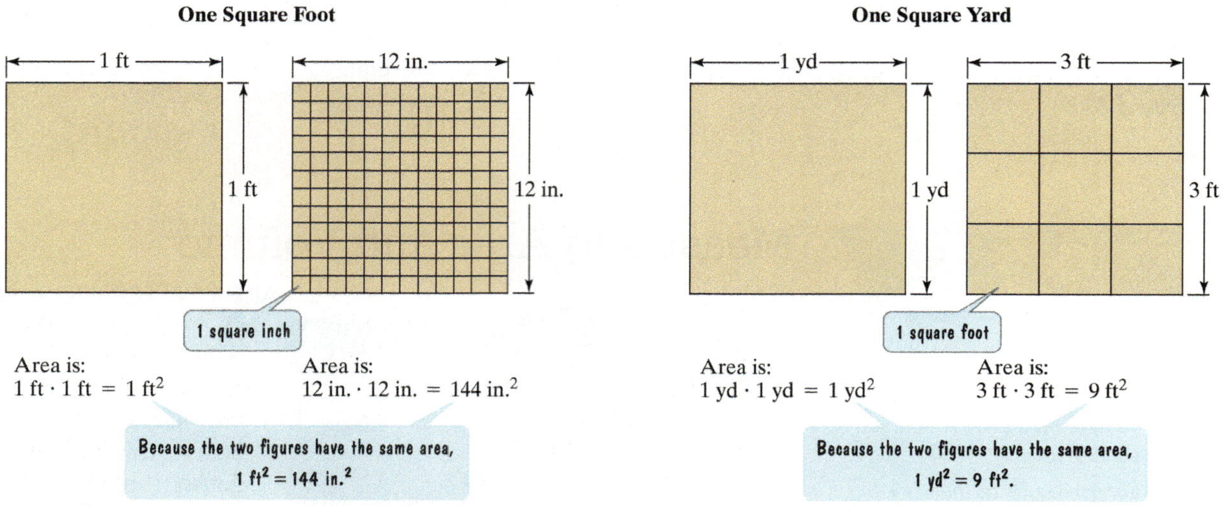

SQUARE UNITS OF MEASURE: THE ENGLISH SYSTEM

1 square foot (ft^2) = 144 square inches (in.^{2})
1 square yard (yd^2) = 9 square feet (ft^2)
1 acre (a) = 43,560 ft^2 or 4840 yd^2
1 square mile (mi^2) = 640 acres

GREAT QUESTION!

Which square units of measure are frequently used in everyday situations?

A small plot of land is usually measured in square feet, rather than a small fraction of an acre. Curiously, square yards are rarely used. However, square yards are commonly used for carpet and flooring measures.

EXAMPLE 2 *Using Square Units to Compute Population Density*

After Alaska, Wyoming is the least densely populated state. The population of Wyoming is 568,158 and its area is 97,814 square miles. What is Wyoming's population density?

SOLUTION

We compute the population density by dividing Wyoming's population by its area.

$$\text{population density} = \frac{\text{population}}{\text{area}} = \frac{568{,}158 \text{ people}}{97{,}814 \text{ square miles}}$$

Using a calculator and rounding to the nearest tenth, we obtain a population density of 5.8 people per square mile. This means that there is an average of only 5.8 people for each square mile of area.

CHECK POINT 2 The population of California is 37,691,912 and its area is 158,633 square miles. What is California's population density? Round to the nearest tenth.

2 Use dimensional analysis to change units for area.

EXAMPLE 3 *Using Dimensional Analysis on Units of Area*

Your author wrote *Thinking Mathematically* in Point Reyes National Seashore, 40 miles north of San Francisco. The national park consists of 75,000 acres with miles of pristine surf-washed beaches, forested ridges, and bays bordered by white cliffs. How large is the national park in square miles?

SOLUTION

We use the fact that 1 square mile = 640 acres to set up our unit fraction:

$$\frac{1 \text{ mi}^2}{640 \text{ acres}}.$$

Thus,

$$75{,}000 \text{ acres} = \frac{75{,}000 \text{ acres}}{1} \cdot \frac{1 \text{ mi}^2}{640 \text{ acres}} = \frac{75{,}000}{640} \text{ mi}^2 \approx 117 \text{ mi}^2.$$

The area of Point Reyes National Seashore is approximately 117 square miles.

CHECK POINT 3 The National Park Service administers approximately 84,000,000 acres of national parks. How large is this in square miles?

In Section 9.1, we saw that in most other countries, the system of measurement that is used is the metric system. In the metric system, the square centimeter is used instead of the square inch. The square meter replaces the square foot and the square yard.

The English system uses the acre and the square mile to measure large land areas, where one square mile = 640 acres. The metric system uses the hectare (symbolized ha and pronounced "hectair"). A hectare is about the area of two football fields placed side by side, approximately 2.5 acres. One square mile of land consists of approximately 260 hectares. Just as the hectare replaces the acre, the square kilometer is used instead of the square mile. One square kilometer is approximately 0.38 square mile.

Some basic approximate conversions for units of area are given in **Table 9.4**.

TABLE 9.4 English and Metric Equivalents for Area

1 square inch (in.^{2})	$\approx$ 6.5 square centimeters (cm^2)
1 square foot (ft^2)	$\approx$ 0.09 square meter (m^2)
1 square yard (yd^2)	$\approx$ 0.8 square meter (m^2)
1 square mile (mi^2)	$\approx$ 2.6 square kilometers (km^2)
1 acre	$\approx$ 0.4 hectare (ha)

EXAMPLE 4 *Using Dimensional Analysis on Units of Area*

A property in Italy is advertised at $545,000 for 6.8 hectares.

a. Find the area of the property in acres.

b. What is the price per acre?

SOLUTION

a. Using **Table 9.4**, we see that 1 acre $\approx$ 0.4 hectare. To convert 6.8 hectares to acres, we use a unit fraction with acres in the numerator and hectares in the denominator.

$$6.8 \text{ ha} \approx \frac{6.8 \text{ ha}}{1} \cdot \frac{1 \text{ acre}}{0.4 \text{ ha}} = \frac{6.8}{0.4} \text{ acres} = 17 \text{ acres}$$

The area of the property is approximately 17 acres.

b. The price per acre is the total price, $545,000, divided by the number of acres, 17.

$$\text{price per acre} = \frac{\$545{,}000}{17 \text{ acres}} \approx \$32{,}059/\text{acre}$$

The price is approximately $32,059 per acre.

CHECK POINT 4 A property in northern California is on the market at $415,000 for 1.8 acres.

a. Find the area of the property in hectares.

b. What is price per hectare?

3 Use cubic units to measure volume.

Measuring Volume

A shoe box and a basketball are examples of three-dimensional figures. **Volume** refers to the amount of space occupied by such figures. In order to measure this space, we begin by selecting a *cubic unit*. Two such cubic units are shown in **Figure 9.5**.

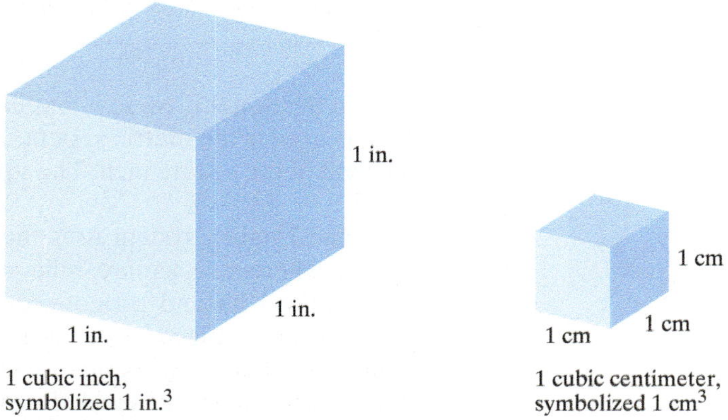

FIGURE 9.5 Common units of measurement for volume

The edges of a cube all have the same length. Other cubic units used to measure volume include 1 cubic foot (1 ft^3) and 1 cubic meter (1 m^3). One way to measure the volume of a solid is to calculate the number of cubic units contained in its interior.

EXAMPLE 5 *Measuring Volume*

What is the volume of the solid shown in **Figure 9.6**?

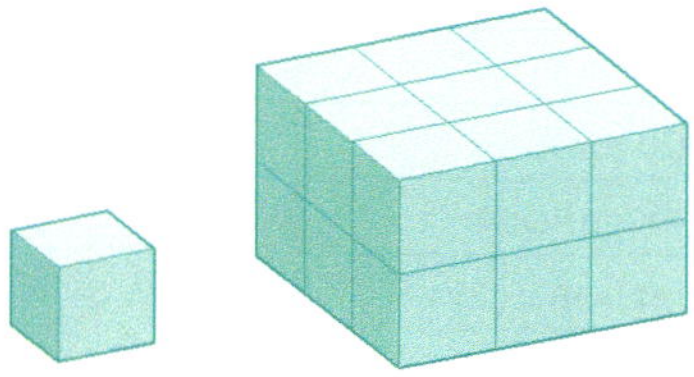

FIGURE 9.6

SOLUTION

We can determine the volume of the solid by counting the number of cubic units contained within the region. Because we have drawn a solid three-dimensional figure on a flat two-dimensional page, some of the small cubic units in the back, right are hidden. The figures below show how the cubic units are used to fill the inside of the solid.

Do these figures help you to see that there are 18 cubic units inside the solid? The volume of the solid is 18 cubic units.

CHECK POINT 5 What is the volume of the region represented by the bottom row of blocks in **Figure 9.6**?

We have seen that there are 3 feet in a yard, but 9 square feet in a square yard. Neither of these relationships holds for cubic units. **Figure 9.7** illustrates that there are 27 cubic feet in a cubic yard. Furthermore, there are 1728 cubic inches in a cubic foot.

GREAT QUESTION!

I'm having difficulty seeing the detail in Figure 9.7. Can you help me out?

Cubing numbers is helpful:

$$3 \text{ ft} = 1 \text{ yd}$$
$$(3 \text{ ft})^3 = (1 \text{ yd})^3$$
$$3 \cdot 3 \cdot 3 \text{ ft}^3 = 1 \cdot 1 \cdot 1 \text{ yd}^3.$$

Conclusion: $27 \text{ ft}^3 = 1 \text{ yd}^3$

$$12 \text{ in.} = 1 \text{ ft}$$
$$(12 \text{ in.})^3 = (1 \text{ ft})^3$$
$$12 \cdot 12 \cdot 12 \text{ in.}^3 = 1 \cdot 1 \cdot 1 \text{ ft}^3$$

Conclusion: $1728 \text{ in.}^3 = 1 \text{ ft}^3$

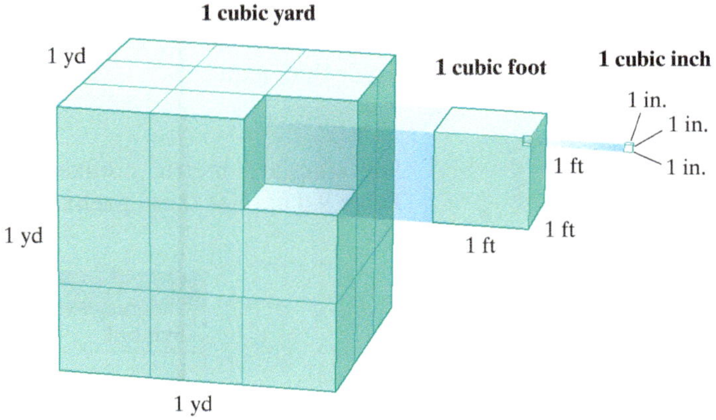

FIGURE 9.7 $27 \text{ ft}^3 = 1 \text{ yd}^3$ and $1728 \text{ in.}^3 = 1 \text{ ft}^3$

The measure of volume also includes the amount of fluid that a three-dimensional object can hold. This is often called the object's **capacity**. For example, we often refer to the capacity, in gallons, of a gas tank. A cubic yard has a capacity of about 200 gallons and a cubic foot has a capacity of about 7.48 gallons.

4 Use English and metric units to measure capacity.

Table 9.5 contains information about standard units of capacity in the English system.

TABLE 9.5 English Units for Capacity

2 pints (pt)	= 1 quart (qt)
4 quarts	= 1 gallon (gal)
1 gallon	= 128 fluid ounces (fl oz)
1 cup (c)	= 8 fluid ounces
Volume in Cubic Units	**Capacity**
1 cubic yard	about 200 gallons
1 cubic foot	about 7.48 gallons
231 cubic inches	about 1 gallon

EXAMPLE 6 *Volume and Capacity in the English System*

A swimming pool has a volume of 22,500 cubic feet. How many gallons of water does the pool hold?

SOLUTION

We use the fact that 1 cubic foot has a capacity of about 7.48 gallons to set up our unit fraction:

$$\frac{7.48 \text{ gal}}{1 \text{ ft}^3}.$$

We use this unit fraction to find the capacity of the 22,500 cubic feet.

$$22{,}500 \text{ ft}^3 \approx \frac{22{,}500 \cancel{\text{ft}^3}}{1} \cdot \frac{7.48 \text{ gal}}{1 \cancel{\text{ft}^3}} = 22{,}500(7.48) \text{ gal} = 168{,}300 \text{ gal}$$

The pool holds approximately 168,300 gallons of water.

CHECK POINT 6 A pool has a volume of 10,000 cubic feet. How many gallons of water does the pool hold?

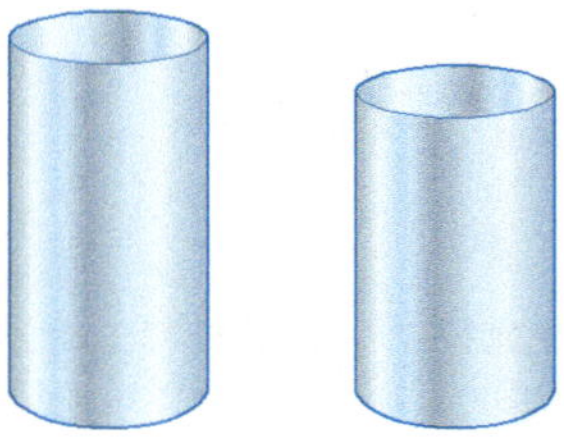

1 liter 1 quart

As we have come to expect, things are simpler when the metric system is used to measure capacity. The basic unit is the **liter**, symbolized by L. A liter is slightly larger than a quart.

$$1 \text{ liter} \approx 1.0567 \text{ quarts}$$

The standard metric prefixes are used to denote a multiple or part of a liter. **Table 9.6** applies these prefixes to the liter.

TABLE 9.6 Units of Capacity in the Metric System

Symbol	Unit	Meaning
kL	kiloliter	1000 liters
hL	hectoliter	100 liters
daL	dekaliter	10 liters
L	liter	1 liter ≈ 1.06 quarts
dL	deciliter	0.1 liter
cL	centiliter	0.01 liter
mL	milliliter	0.001 liter

The following list should help give you a feel for capacity in the metric system.

Item	Capacity
Average cup of coffee	250 mL
12-ounce can of soda	355 mL
One quart of fruit juice	0.95 L
One gallon of milk	3.78 L
Average gas capacity of a car (about 18.5 gallons)	70 L

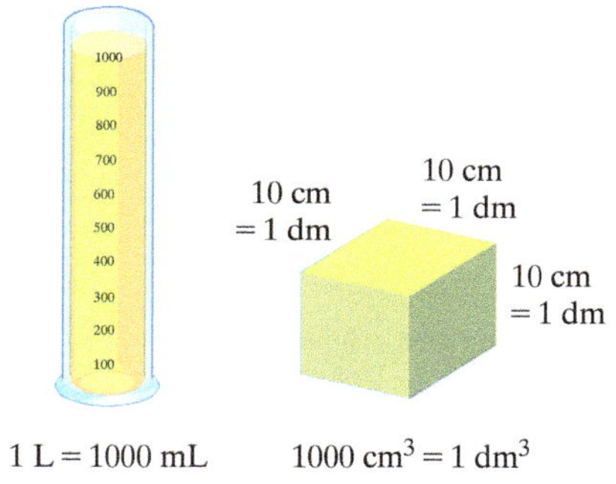

FIGURE 9.8

Figure 9.8 shows a 1-liter container filled with water. The water in the liter container will fit exactly into the cube shown to its right. The volume of this cube is 1000 cubic centimeters, or equivalently, 1 cubic decimeter. Thus,

$$1000 \text{ cm}^3 = 1 \text{ dm}^3 = 1 \text{ L}.$$

Table 9.7 expands on this relationship between volume and capacity in the metric system.

TABLE 9.7 Volume and Capacity in the Metric System

Volume in Cubic Units		Capacity
1 cm^3	=	1 mL
$1 \text{ dm}^3 = 1000 \text{ cm}^3$	=	1 L
1 m^3	=	1 kL

A milliliter is the capacity of a cube measuring 1 centimeter on each side.

A liter is the capacity of a cube measuring 10 centimeters on each side.

EXAMPLE 7 *Volume and Capacity in the Metric System*

An aquarium has a volume of 36,000 cubic centimeters. How many liters of water does the aquarium hold?

SOLUTION

We use the fact that 1000 cubic centimeters corresponds to a capacity of 1 liter to set up our unit fraction:

$$\frac{1 \text{ L}}{1000 \text{ cm}^3}.$$

We use this unit fraction to find the capacity of the 36,000 cubic centimeters.

$$36{,}000 \text{ cm}^3 = \frac{36{,}000 \text{ cm}^3}{1} \cdot \frac{1 \text{ L}}{1000 \text{ cm}^3} = \frac{36{,}000}{1000} \text{ L} = 36 \text{ L}$$

The aquarium holds 36 liters of water.

CHECK POINT 7 A fish pond has a volume of 220,000 cubic centimeters. How many liters of water does the pond hold?

Table 9.8 can be used to convert units of capacity to and from the metric system.

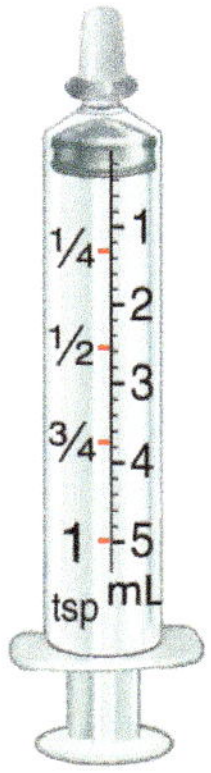

TABLE 9.8 English and Metric Equivalents for Capacity

1 teaspoon (tsp)	≈	5 milliliters (mL)
1 tablespoon (tbsp)	≈	15 milliliters (mL)
1 fluid ounce (fl oz)	≈	30 milliliters (mL)
1 cup (c)	≈	0.24 liter (L)
1 pint (pt)	≈	0.47 liter (L)
1 quart (qt)	≈	0.95 liter (L)
1 gallon (gal)	≈	3.8 liters (L)

Our next example involves measuring dosages of medicine in liquid form. We have seen that

$$1 \text{ cm}^3 = 1 \text{ mL}.$$

Dosages of liquid medication are measured using cubic centimeters, or milliliters as they are also called. In the United States, cc, rather than cm^3, denotes cubic centimeters.

EXAMPLE 8 *Measuring Dosages of Medicine in Liquid Form*

A physician orders 10 cc of the drug Lexapro (used to treat depression and anxiety) to be administered to a patient in liquid form.

a. How many milliliters of the drug should be administered?

b. How many fluid ounces of the drug should be administered?

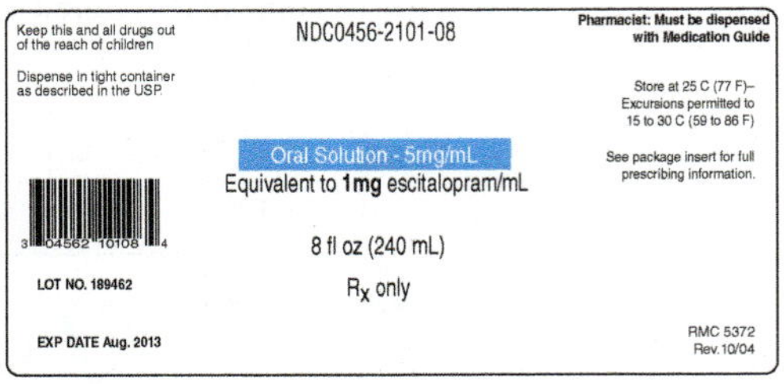

SOLUTION

a. We use a relationship between volume and capacity in the metric system: 1 cubic centimeter (cc) = 1 milliliter (mL). Because 10 cubic centimeters (cc) of the drug is to be administered, this is equivalent to 10 milliliters (mL) of Lexapro.

b. We now need to convert 10 milliliters to fluid ounces. **Table 9.8** on the previous page shows that 1 fluid ounce (fl oz) ≈ 30 milliliters (mL). We use

$$\frac{1 \text{ fl oz}}{30 \text{ mL}}$$

as our unit fraction.

$$10 \text{ mL} \approx \frac{10 \cancel{\text{mL}}}{1} \cdot \frac{1 \text{ fl oz}}{30 \cancel{\text{mL}}} = \frac{10}{30} \text{ fl oz} \approx 0.33 \text{ fl oz}$$

Approximately 0.33 fluid ounce (fl oz) of Lexapro should be administered.

CHECK POINT 8 A physician orders 20 cc of the antibiotic Omnicef to be administered every 12 hours.

a. How many milliliters of the drug should be administered?

b. How many fluid ounces of the drug should be administered? Round to two decimal places.

Concept and Vocabulary Check

Fill in each blank so that the resulting statement is true.

1. Area is measured in ________ units and volume is measured in ________ units.
2. In the English system, 1 ft^2 = ____________ $in.^2$ and 1 yd^2 = ____________ ft^2.
3. Because 1 mi^2 = 640 acres, the unit fraction needed to convert from acres to square miles is ________ and the unit fraction needed to convert from square miles to acres is ________.
4. In the English system, ____________ pt = 1 qt and ____________ qt = 1 gal.
5. The amount of fluid that a three-dimensional object can hold is called the object's ________, measured in the metric system using a basic unit called the ________.
6. A state's population density is its population divided by its ________.
7. 1 cm^3, or 1 cc, = ________ mL

In Exercises 8–10, determine whether each statement is true or false. If the statement is false, make the necessary change(s) to produce a true statement.

8. Because there are 3 feet in one yard, there are also 3 square feet in one square yard. ________
9. The English system uses $in.^2$ to measure large land areas. ________
10. One quart is approximately 1.06 liters. ________

Exercise Set 9.2

Practice Exercises

In Exercises 1–4, use the given figure to find its area in square units.

1.

2.

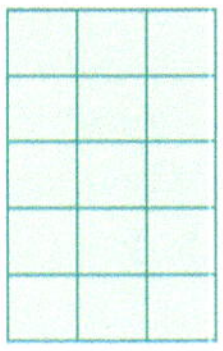

3.

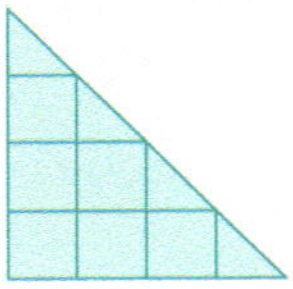

4.

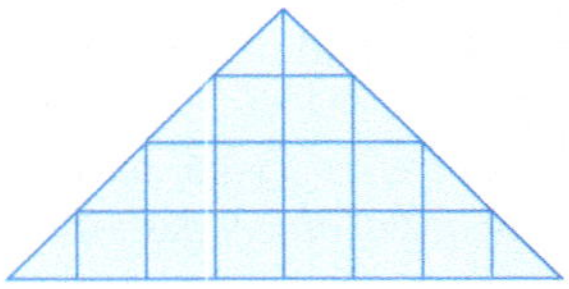

In Exercises 5–12, use ***Table 9.4*** *on page 588, along with dimensional analysis, to approximately convert the given square unit to the square unit indicated. Where necessary, round answers to two decimal places.*

5. 14 cm^2 to $in.^2$
6. 20 m^2 to ft^2
7. 30 m^2 to yd^2
8. 14 mi^2 to km^2
9. 10.2 ha to acres
10. 20.6 ha to acres
11. 14 $in.^2$ to cm^2
12. 20 $in.^2$ to cm^2

In Exercises 13–14, use the given figure to find its volume in cubic units.

13.

14.

In Exercises 15–22, use ***Table 9.5*** *on page 590, along with dimensional analysis, to approximately convert the given unit to the unit indicated. Where necessary, round answers to two decimal places.*

15. 10,000 ft^3 to gal
16. 25,000 ft^3 to gal
17. 8 yd^3 to gal
18. 35 yd^3 to gal
19. 2079 $in.^3$ to gal
20. 6237 $in.^3$ to gal
21. 2700 gal to yd^3
22. 1496 gal to ft^3

In Exercises 23–32, use ***Table 9.7*** *on page 591, along with dimensional analysis, to convert the given unit to the unit indicated.*

23. 45,000 cm^3 to L
24. 75,000 cm^3 to L
25. 17 cm^3 to mL
26. 19 cm^3 to mL
27. 1.5 L to cm^3
28. 4.5 L to cm^3
29. 150 mL to cm^3
30. 250 mL to cm^3
31. 12 kL to dm^3
32. 16 kL to dm^3

In Exercises 33–48, use ***Table 9.8*** *on page 591, along with dimensional analysis, to convert the given unit to the unit indicated. Where necessary, round to two decimal places.*

33. 12 mL to tsp
34. 14 mL to tsp
35. 3 tbsp to mL

36. 4 tbsp to mL

37. 70 mL to fl oz

38. 80 mL to fl oz

39. 1.4 L to c

40. 2.6 L to c

41. 6 pt to L

42. 9 pt to L

43. 4 L to qt

44. 7 L to qt

45. 3 gal to L

46. 5 gal to L

47. 2000 mL to qt

48. 5000 mL to qt

Practice Plus

The bar graph shows the resident population and the land area of the United States for selected years from 1800 through 2010. Use the information shown by the graph to solve Exercises 49–52.

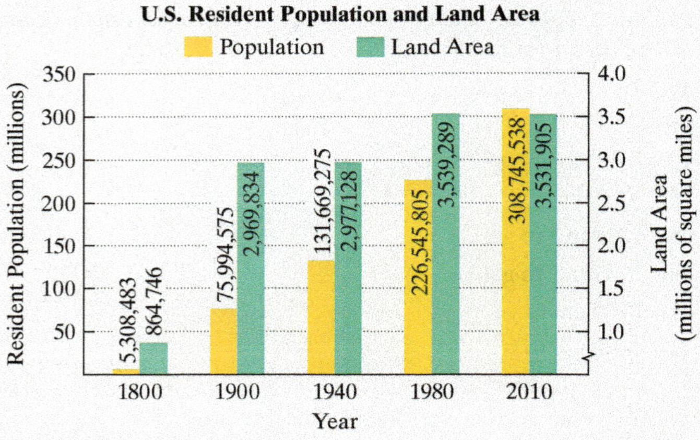

Source: U.S. Bureau of the Census

49. a. Find the population density of the United States, to the nearest tenth, in 1900 and in 2010.

b. Find the percent increase in population density, to the nearest tenth of a percent, from 1900 to 2010.

50. a. Find the population density of the United States, to the nearest tenth, in 1800 and in 2010.

b. Find the percent increase in population density, to the nearest tenth of a percent, from 1800 to 2010.

51. Find the population density of the United States, to the nearest tenth, expressed in people per square kilometer, in 1940.

52. Find the population density of the United States, to the nearest tenth, expressed in people per square kilometer, in 1980.

Application Exercises

In Exercises 53–54, find the population density, to the nearest tenth, for each state. Which state has the greater population density? How many more people per square mile inhabit the state with the greater density than inhabit the state with the lesser density?

53. Illinois population: 12,869,257 area: 57,914 mi^2

Ohio population: 11,544,951 area: 44,826 mi^2

54. New York population: 19,465,197 area: 54,555 mi^2

Rhode Island population: 1,051,302 area: 1545 mi^2

In Exercises 55–56, use the fact that 1 square mile = 640 acres to find the area of each national park to the nearest square mile.

55. Everglades National Park (Florida): 1,509,154 acres

56. Yosemite National Park (California): 761,268 acres

57. A property that measures 8 hectares is for sale.

a. How large is the property in acres?

b. If the property is selling for $250,000, what is the price per acre?

58. A property that measures 100 hectares is for sale.

a. How large is the property in acres?

b. If the property is selling for $350,000, what is the price per acre?

In Exercises 59–62, selecting from square centimeters, square meters, or square kilometers, determine the best unit of measure to express the area of the object described.

59. The top of a desk

60. A dollar bill

61. A national park

62. The wall of a room

In Exercises 63–66, select the best estimate for the measure of the area of the object described.

63. The area of the floor of a room

a. 25 cm^2 **b.** 25 m^2 **c.** 25 km^2

64. The area of a television screen

a. 2050 mm^2 **b.** 2050 cm^2 **c.** 2050 dm^2

65. The area of the face of a small coin

a. 6 mm^2 **b.** 6 cm^2 **c.** 6 dm^2

66. The area of a parcel of land in a large metropolitan area on which a house can be built

a. 900 cm^2 **b.** 900 m^2 **c.** 900 ha

67. A swimming pool has a volume of 45,000 cubic feet. How many gallons of water does the pool hold?

68. A swimming pool has a volume of 66,000 cubic feet. How many gallons of water does the pool hold?

69. A container of grapefruit juice has a volume of 4000 cubic centimeters. How many liters of juice does the container hold?

70. An aquarium has a volume of 17,500 cubic centimeters. How many liters of water does the aquarium hold?

Exercises 71–72 give the approximate area of some of the world's largest island countries. In each exercise, determine which country has the greater area and by how many square kilometers.

71. Philippines: 300,000 km^2; Japan: 145,900 mi^2

72. Iceland: 103,000 km^2; Cuba: 42,800 mi^2

Exercises 73–74 give the approximate area of some of the world's largest islands. In each exercise, determine which island has the greater area and by how many square miles. Round to the nearest square mile.

73. Baffin Island, Canada: 194,574 mi^2;
Sumatra, Indonesia: 443,070 km^2

74. Honshu, Japan: 87,805 mi^2;
Victoria Island, Canada: 217,300 km^2

(Source for Exercises 71–74: Russell Ash, *The Top 10 of Everything 2009*)

Exercises 75–76 involve dosages of the anti-inflammatory drug indomethacin, administered in liquid form. Refer to the appropriate entries in ***Table 9.8*** *on page 591 to convert given units to the unit indicated. Where necessary, round to two decimal places.*

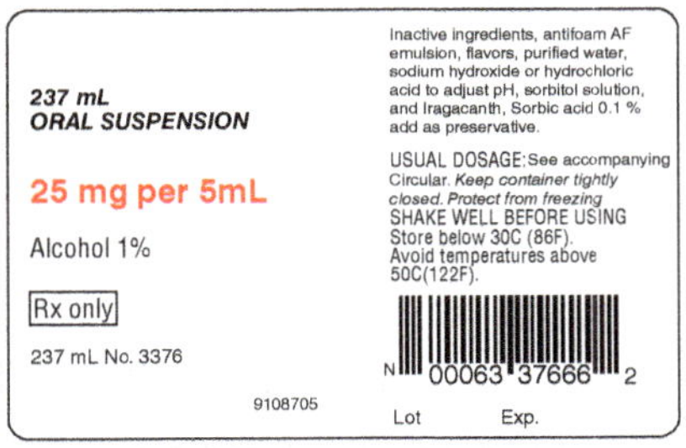

75. A physician orders 3 teaspoons daily of indomethacin in three equally divided doses.

a. How many milliliters of the drug should be administered daily?

b. How many cc of the drug should be administered daily?

c. How many fluid ounces of the drug should be administered daily?

d. How many fluid ounces of the drug should the patient receive in each divided dose?

76. A physician orders 4 teaspoons daily of indomethacin in four equally divided doses.

a. How many milliliters of the drug should be administered daily?

b. How many cc of the drug should be administered daily?

c. How many fluid ounces of the drug should be administered daily?

d. How many fluid ounces of the drug should the patient receive in each divided dose?

Writing in Mathematics

77. Describe how area is measured. Explain why linear units cannot be used.

78. New Mexico has a population density of 17 people per square mile. Describe what this means. Use an almanac or the Internet to compare this population density to that of the state in which you are now living.

79. Describe the difference between the following problems: How much fencing is needed to enclose a garden? How much fertilizer is needed for the garden?

80. Describe how volume is measured. Explain why linear or square units cannot be used.

81. For a swimming pool, what is the difference between the following units of measure: cubic feet and gallons? For each unit, write a sentence about the pool that makes the use of the unit appropriate.

82. If there are 10 decimeters in a meter, explain why there are not 10 cubic decimeters in a cubic meter.

Critical Thinking Exercises

Make Sense? *In Exercises 83–86, determine whether each statement makes sense or does not make sense, and explain your reasoning.*

83. The gas capacity of my car is approximately 40 meters.

84. The population density of Montana is approximately 6.5 people per mile.

85. Yesterday I drank 2000 cm^3 of water.

86. I read that one quart is approximately 0.94635295 liter.

87. Singapore has the highest population density of any country: 46,690 people per 1000 hectares. How many people are there per square mile?

88. Nebraska has a population density of 23.8 people per square mile and a population of 1,842,641. What is the area of Nebraska? Round to the nearest square mile.

89. A high population density is a condition common to extremely poor and extremely rich locales. Explain why this is so.

90. Although Alaska is the least densely populated state, over 90% of its land is protected federal property that is off-limits to settlement. A resident of Anchorage, Alaska, might feel hemmed in. In terms of "elbow room," what other important factor must be considered when calculating a state's population density?

91. Does an adult's body contain approximately 6.5 liters, kiloliters, or milliliters of blood? Explain your answer.

92. Is the volume of a coin approximately 1 cubic centimeter, 1 cubic millimeter, or 1 cubic decimeter? Explain your answer.

Group Exercise

93. If you could select any place in the world, where would you like to live? Look up the population and area of your ideal place, and compute its population density. Group members should share places to live and population densities. What trend, if any, does the group observe?

9.3

WHAT AM I SUPPOSED TO LEARN?

After you have read this section, you should be able to:

1 Apply metric prefixes to units of weight.
2 Convert units of weight within the metric system.
3 Use relationships between volume and weight within the metric system.
4 Use dimensional analysis to change units of weight to and from the metric system.
5 Understand temperature scales.

Measuring Weight and Temperature

YOU ARE WATCHING CNN INTERNATIONAL on cable television. The temperature in Honolulu, Hawaii, is reported as 30°C. Are Honolulu's tourists running around in winter jackets? In this section, we will make sense of Celsius temperature readings, as we discuss methods for measuring temperature and weight.

Measuring Weight

You step on the scale at the doctor's office to check your weight, discovering that you are 150 pounds. Compare this to your weight on the moon: 25 pounds. Why the difference? **Weight** is the measure of the gravitational pull on an object. The gravitational pull on the moon is only about one-sixth the gravitational pull on Earth. Although your weight varies depending on the force of gravity, your mass is exactly the same in all locations. **Mass** is a measure of the quantity of matter in an object, determined by its molecular structure. On Earth, as your weight increases, so does your mass. In this section, measurements are assumed to involve everyday situations on the surface of Earth. Thus, we will treat weight and mass as equivalent, and refer strictly to weight.

UNITS OF WEIGHT: THE ENGLISH SYSTEM

16 ounces (oz) = 1 pound (lb)

2000 pounds (lb) = 1 ton (T)

1 Apply metric prefixes to units of weight.

The basic metric unit of weight is the **gram** (g), used for very small objects such as a coin, a candy bar, or a teaspoon of salt. A nickel has a weight of about 5 grams.

As with meters, prefixes are used to denote a multiple or part of a gram. **Table 9.9** applies the common metric prefixes to the gram. The first part of the symbol indicates the prefix and the second part (g) indicates gram.

Weight of pineapple is 1kg, or 1000g.

TABLE 9.9 Commonly Used Units of Weight in the Metric System

Symbol	Unit	Meaning
kg	kilogram	1000 grams
hg	hectogram	100 grams
dag	dekagram	10 grams
g	gram	1 gram
dg	decigram	0.1 gram
cg	centigram	0.01 gram
mg	milligram	0.001 gram

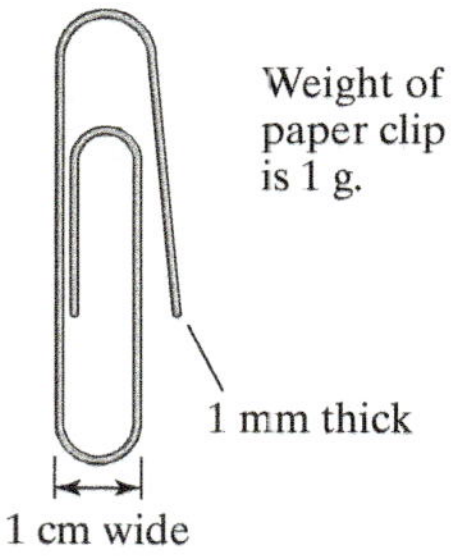

Weight of paper clip is 1 g.

In the metric system, the kilogram is the comparable unit to the pound in the English system. **A weight of 1 kilogram is approximately 2.2 pounds.** Thus, an average man has a weight of about 75 kilograms. Objects that we measure in pounds are measured in kilograms in most countries.

A milligram, equivalent to 0.001 gram, is an extremely small unit of weight and is used extensively in the pharmaceutical industry. If you look at the label on a bottle of tablets, you will see that the amounts of different substances in each tablet are expressed in milligrams.

The weight of a very heavy object is expressed in terms of the metric tonne (t), which is equivalent to 1000 kilograms, or about 2200 pounds. This is 10 percent more than the English ton (T) of 2000 pounds.

We change units of weight within the metric system exactly the same way that we changed units of length.

2 Convert units of weight within the metric system.

CHANGING UNITS OF WEIGHT WITHIN THE METRIC SYSTEM

Use the following diagram to find equivalent measures of weight:

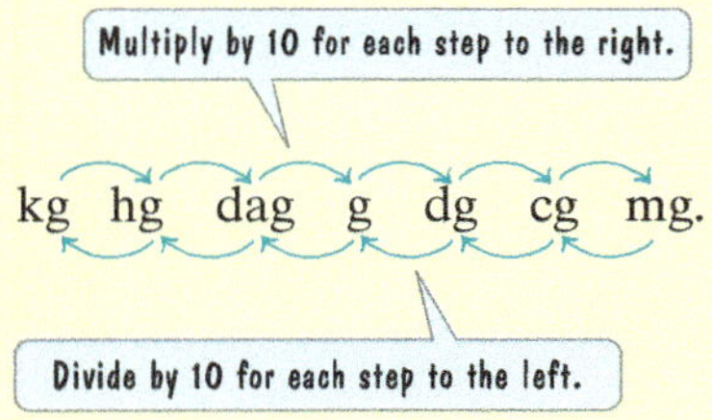

EXAMPLE 1 *Changing Units within the Metric System*

a. Convert 8.7 dg to mg.

b. Convert 950 mg to g.

SOLUTION

a. To convert from dg (decigrams) to mg (milligrams), we start at dg and move two steps to the right:

kg hg dag g dg cg mg.

Hence, we move the decimal point two places to the right:

$$8.7 \text{ dg} = 870 \text{ mg}.$$

b. To convert from mg (milligrams) to g (grams), we start at mg and move three steps to the left:

kg hg dag g dg cg mg.

Hence, we move the decimal point three places to the left:

$$950 \text{ mg} = 0.950 \text{ g}.$$

CHECK POINT 1

a. Convert 4.2 dg to mg.

b. Convert 620 cg to g.

---

3 Use relationships between volume and weight within the metric system.

We have seen a convenient relationship in the metric system between volume and capacity:

$$1000 \text{ cm}^3 = 1 \text{ dm}^3 = 1 \text{ L}.$$

This relationship can be extended to include weight based on the following:

One kilogram of water has a volume of 1 liter.

Thus,

$$1000 \text{ cm}^3 = 1 \text{ dm}^3 = 1 \text{ L} = 1 \text{ kg}.$$

Table 9.10 shows the relationships between volume and weight of water in the metric system.

TABLE 9.10 Volume and Weight of Water in the Metric System

Volume		Capacity		Weight
1 cm^3	=	1 mL	=	1 g
1 dm^3 = 1000 cm^3	=	1 L	=	1 kg
1 m^3	=	1 kL	=	1000 kg = 1 t

EXAMPLE 2 *Volume and Weight in the Metric System*

An aquarium holds 0.25 m^3 of water. How much does the water weigh?

SOLUTION

We use the fact that 1 m^3 of water = 1000 kg of water to set up our unit fraction:

$$\frac{1000 \text{ kg}}{1 \text{ m}^3}.$$

Thus,

$$0.25 \text{ m}^3 = \frac{0.25 \text{ m}^3}{1} \cdot \frac{1000 \text{ kg}}{1 \text{ m}^3} = 250 \text{ kg}.$$

The water weighs 250 kilograms.

CHECK POINT 2 An aquarium holds 0.145 m^3 of water. How much does the water weigh?

4 Use dimensional analysis to change units of weight to and from the metric system.

A problem like Example 2 involves more awkward computation in the English system. For example, if you know the aquarium's volume in cubic feet, you must also know that 1 cubic foot of water weighs about 62.5 pounds to determine the water's weight.

Dimensional analysis is a useful tool when converting units of weight between the English and metric systems. Some basic approximate conversions are given in **Table 9.11**.

1-ounce coin ≈ 28 grams

1-pound lobster ≈ 0.45 kilogram

TABLE 9.11 Weight: English and Metric Equivalents

1 ounce (oz)	≈ 28 grams (g)
1 pound (lb)	≈ 0.45 kilogram (kg)
1 ton (T)	≈ 0.9 tonne (t)

1-ton bison ≈ 0.9 tonne

EXAMPLE 3 *Using Dimensional Analysis*

a. Convert 160 pounds to kilograms.

b. Convert 300 grams to ounces.

Blitzer Bonus

The SI System

The metric system is officially called the Système International d'Unités, abbreviated the SI system. Yemen, Brunei, and the United States are the only countries in the world not using the SI system.

In 1975, the United States prepared to adopt the SI system with the Metric Conversion Act. However, due to concerns in grassroots America, Congress made the conversion voluntary, effectively chasing it off. But it's not as bleak as you think; from metric tools to milligram measures of medication, we have been edging, inching, even centimetering toward the metric system.

SOLUTION

a. To convert 160 pounds to kilograms, we use a unit fraction with kilograms in the numerator and pounds in the denominator:

$$\frac{0.45 \text{ kg}}{1 \text{ lb}}.$$

Table 9.11 shows that 1 lb ≈ 0.45 kg.

Thus,

$$160 \text{ lb} \approx \frac{160 \text{ lb}}{1} \cdot \frac{0.45 \text{ kg}}{1 \text{ lb}} = 160(0.45) \text{ kg} = 72 \text{ kg}.$$

b. To convert 300 grams to ounces, we use a unit fraction with ounces in the numerator and grams in the denominator:

$$\frac{1 \text{ oz}}{28 \text{ g}}.$$

Table 9.11 shows that 1 oz ≈ 28 g.

Thus,

$$300 \text{ g} \approx \frac{300 \text{ g}}{1} \cdot \frac{1 \text{ oz}}{28 \text{ g}} = \frac{300}{28} \text{ oz} \approx 10.7 \text{ oz}.$$

CHECK POINT 3

a. Convert 120 pounds to kilograms.

b. Convert 500 grams to ounces.

EXAMPLE 4 *Dosages of Medication and Weight in the Metric System*

Drug dosage is frequently based on a patient's weight, measured in kilograms. For example, 20 milligrams of the drug Didronel (used to treat irregular bone formation) should be administered daily for each kilogram of a patient's weight. This can be expressed as 20 mg/kg. How many 400-milligram tablets should be given each day to a patient who weighs 180 pounds?

SOLUTION

- Convert 180 pounds to kilograms, using 1 lb ≈ 0.45 kg to set up the unit fraction.

$$180 \text{ lb} \approx \frac{180 \text{ lb}}{1} \cdot \frac{0.45 \text{ kg}}{1 \text{ lb}} = 180(0.45) \text{ kg} = 81 \text{ kg}$$

- Determine the dosage. We are given 20 mg/kg, meaning that 20 milligrams of the drug is to be given for each kilogram of a patient's weight. Multiply the patient's weight, 81 kilograms, by 20 to determine the dosage.

$$\text{Dosage} = \frac{81 \text{ kg}}{1} \cdot \frac{20 \text{ mg}}{1 \text{ kg}} = 81(20) \text{ mg} = 1620 \text{ mg}$$

- Determine the number of tablets that should be given each day. The patient should receive 1620 milligrams of Didronel daily. We are given that each tablet contains 400 milligrams.

$$\text{Number of tablets} = \frac{1620 \text{ mg}}{400 \text{ mg}} = 4.05$$

The patient should receive 4 tablets of Didronel daily.

CHECK POINT 4

The prescribed dosage of a drug is 6 mg/kg daily. How many 200-milligram tablets should be given each day to a patient who weighs 150 pounds?

Blitzer Bonus

Using the Metric System to Measure Blood-Alcohol Concentration

In Chapter 6, we presented a formula for determining blood-alcohol concentration. Blood-alcohol concentration (BAC) is measured in grams of alcohol per 100 milliliters of blood. To put this measurement into perspective, a 175-pound man has approximately 5 liters (5000 milliliters) of blood and a 12-ounce can of 6%-alcohol-by-volume beer contains about 15 grams of alcohol. Based on the time it takes for alcohol to be absorbed into the bloodstream, as well as its elimination at a rate of 10–15 grams per hour, **Figure 9.9** shows the effect on BAC of a number of drinks on individuals within weight classes. It is illegal to drive with a BAC at 0.08 g/100 mL, or 0.08 gram of alcohol per 100 milliliters of blood, or greater.

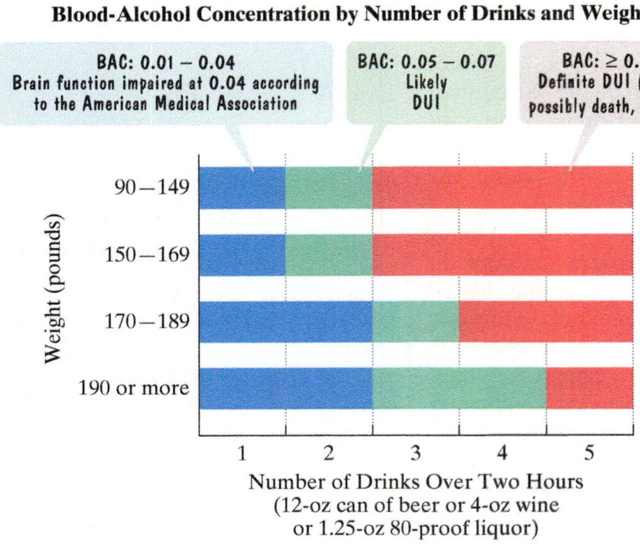

FIGURE 9.9
Source: Patrick McSharry, *Everyday Numbers*, Random House, 2002.

5 Understand temperature scales.

Measuring Temperature

You'll be leaving the cold of winter for a vacation to Hawaii. CNN International reports a temperature in Hawaii of 30°C. Should you pack a winter coat?

The idea of changing from Celsius readings—or is it Centigrade?—to familiar Fahrenheit readings can be disorienting. Reporting a temperature of 30°C doesn't have the same impact as the Fahrenheit equivalent of 86 degrees (don't pack the coat). Why these annoying temperature scales?

The **Fahrenheit temperature scale**, the one we are accustomed to, was established in 1714 by the German physicist Gabriel Daniel Fahrenheit. He took a mixture of salt and ice, then thought to be the coldest possible temperature, and called it 0 degrees. He called the temperature of the human body 96 degrees, dividing the space between 0 and 96 into 96 parts. Fahrenheit was wrong about body temperature. It was later found to be 98.6 degrees. On his scale, water froze (without salt) at 32 degrees and boiled at 212 degrees. The symbol ° was used to replace the word *degree*.

Twenty years later, the Swedish scientist Anders Celsius introduced another temperature scale. He set the freezing point of water at 0° and its boiling point at 100°, dividing the space into 100 parts. Degrees were called centigrade until 1948, when the name was officially changed to honor its inventor. However, *centigrade* is still commonly used in the United States.

Figure 9.10 shows a thermometer that measures temperatures in both degrees Celsius (°C is the scale on the left) and degrees Fahrenheit (°F is the scale on the right). The thermometer should help orient you if you need to know what a temperature in °C means. For example, if it is 40°C, find the horizontal line representing this temperature on the left. Now read across to the °F scale on the right. The reading is above 100°, indicating heat wave conditions.

The following formulas can be used to convert from one temperature scale to the other:

FIGURE 9.10 The Celsius scale is on the left and the Fahrenheit scale is on the right.

FROM CELSIUS TO FAHRENHEIT

$$F = \frac{9}{5}C + 32$$

FROM FAHRENHEIT TO CELSIUS

$$C = \frac{5}{9}(F - 32)$$

GREAT QUESTION!

Because $\frac{9}{5} = 1.8$, can I use 1.8 instead of $\frac{9}{5}$ in the formula $F = \frac{9}{5}C + 32$?

Yes. The formula used to convert from Celsius to Fahrenheit can be expressed without the use of fractions:

$$F = 1.8C + 32.$$

Some students find this form of the formula easier to memorize.

EXAMPLE 5 *Converting from Celsius to Fahrenheit*

The bills from your European vacation have you feeling a bit feverish, so you decide to take your temperature. The thermometer reads 37°C. Should you panic?

SOLUTION

Use the formula

$$F = \frac{9}{5}C + 32$$

to convert 37°C from °C to °F. Substitute 37 for C in the formula and find the value of F.

$$F = \frac{9}{5}(37) + 32 = 66.6 + 32 = 98.6$$

No need to panic! Your temperature is 98.6°F, which is perfectly normal.

CHECK POINT 5 Convert 50°C from °C to °F.

EXAMPLE 6 *Converting from Fahrenheit to Celsius*

The temperature on a warm spring day is 77°F. Find the equivalent temperature on the Celsius scale.

SOLUTION

Use the formula

$$C = \frac{5}{9}(F - 32)$$

to convert 77°F from °F to °C. Substitute 77 for F in the formula and find the value of C.

$$C = \frac{5}{9}(77 - 32) = \frac{5}{9}(45) = 25$$

Thus, 77°F is equivalent to 25°C.

CHECK POINT 6 Convert 59°F from °F to °C.

Lake Baikal, Siberia, is one of the coldest places on Earth, reaching −76°F (−60°C) in winter. The lowest temperature possible is absolute zero. Scientists have cooled atoms to a few millionths of a degree above absolute zero.

Because temperature is a measure of heat, scientists do not find negative temperatures meaningful in their work. In 1948, the British physicist Lord Kelvin introduced a third temperature scale. He put 0 degrees at absolute zero, the coldest possible temperature, at which there is no heat and molecules stop moving. **Figure 9.11** on the next page illustrates the three temperature scales.

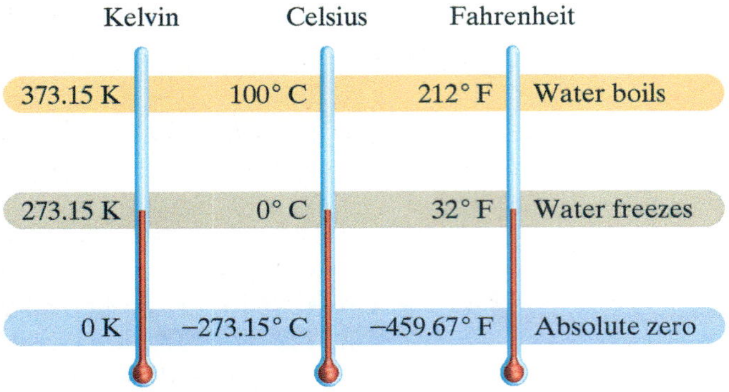

FIGURE 9.11 The three temperature scales

Figure 9.11 shows that water freezes at 273.15 K (read "K" or "Kelvins," not "degrees Kelvin") and boils at 373.15 K. The Kelvin scale is the same as the Celsius scale, except in its starting (zero) point. This makes it easy to go back and forth from Celsius to Kelvin.

FROM CELSIUS TO KELVIN

$$K = C + 273.15$$

FROM KELVIN TO CELSIUS

$$C = K - 273.15$$

Kelvin's scale was embraced by the scientific community. Today, it is the final authority, as scientists define Celsius and Fahrenheit in terms of Kelvins.

Blitzer Bonus

Running a 5 K Race?

A 5 K race means a race at 5 Kelvins. This is a race so cold that no one would be able to move because all the participants would be frozen solid! The proper symbol for a race five kilometers long is a 5 km race.

Concept and Vocabulary Check

Fill in each blank so that the resulting statement is true.

1. In the English system, ________ oz = 1 lb and ________ lb = 1 T.
2. The basic metric unit of weight is the ________, used for very small objects such as a coin, a candy bar, or a teaspoon of salt.
3. A weight of 1 kilogram is approximately ________ pounds.
4. One kilogram of water has a volume of ________ liter(s).
5. On the Fahrenheit temperature scale, water freezes at ________ degrees and boils at ________ degrees.
6. On the Celsius temperature scale, water freezes at ________ degrees and boils at ________ degrees.

In Exercises 7–10, determine whether each statement is true or false. If the statement is false, make the necessary change(s) to produce a true statement.

7. 1 gram = 1000 kilograms ______
8. 1 gram ≈ 28 ounces ______
9. The formula $F = \frac{9}{5}C + 32$ is used to convert from Fahrenheit to Celsius. ______
10. The formula $C = \frac{1}{9}(F - 32)$ is used to convert from Fahrenheit to Celsius. ______

Exercise Set 9.3

Practice Exercises

In Exercises 1–10, convert the given unit of weight to the unit indicated.

1. 7.4 dg to mg
2. 6.9 dg to mg
3. 870 mg to g
4. 640 mg to g
5. 8 g to cg
6. 7 g to cg
7. 18.6 kg to g
8. 0.37 kg to g
9. 0.018 mg to g
10. 0.029 mg to g

*In Exercises 11–18, use **Table 9.10** on page 598 to convert the given measurement of water to the unit indicated.*

11. 0.05 m^3 to kg
12. 0.02 m^3 to kg
13. 4.2 kg to cm^3
14. 5.8 kg to cm^3

15. 1100 m^3 to t

16. 1500 t to m^3

17. 0.04 kL to g

18. 0.03 kL to g

In Exercises 19–30, use the following equivalents, along with dimensional analysis, to convert the given measurement to the unit indicated. When necessary, round answers to two decimal places.

$$16 \text{ oz} = 1 \text{ lb}$$
$$2000 \text{ lb} = 1 \text{ T}$$
$$1 \text{ oz} \approx 28 \text{ g}$$
$$1 \text{ lb} \approx 0.45 \text{ kg}$$
$$1 \text{ T} \approx 0.9 \text{ t}$$

19. 36 oz to lb

20. 26 oz to lb

21. 36 oz to g

22. 26 oz to g

23. 540 lb to kg

24. 220 lb to kg

25. 80 lb to g

26. 150 lb to g

27. 540 kg to lb

28. 220 kg to lb

29. 200 t to T

30. 100 t to T

In Exercises 31–38, convert the given Celsius temperature to its equivalent temperature on the Fahrenheit scale. Where appropriate, round to the nearest tenth of a degree.

31. 10°C

32. 20°C

33. 35°C

34. 45°C

35. 57°C

36. 98°C

37. −5°C

38. −10°C

In Exercises 39–50, convert the given Fahrenheit temperature to its equivalent temperature on the Celsius scale. Where appropriate, round to the nearest tenth of a degree.

39. 68°F

40. 86°F

41. 41°F

42. 50°F

43. 72°F

44. 90°F

45. 23°F

46. 14°F

47. 350°F

48. 475°F

49. −22°F

50. −31°F

Practice Plus

51. The nine points shown below represent Celsius temperatures and their equivalent Fahrenheit temperatures. Also shown is a line that passes through the points.

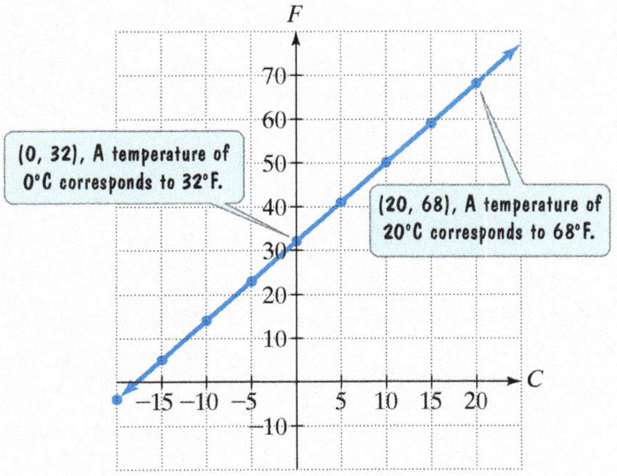

a. Use the coordinates of the two points identified by the voice balloons to compute the line's slope. Express the answer as a fraction reduced to lowest terms. What does this mean about the change in Fahrenheit temperature for each degree change in Celsius temperature?

b. Use the slope-intercept form of the equation of a line, $y = mx + b$, the slope from part (a), and the y-intercept shown by the graph to derive the formula used to convert from Celsius to Fahrenheit.

52. Solve the formula used to convert from Celsius to Fahrenheit for C and derive the other temperature conversion formula.

Application Exercises

In Exercises 53–59, selecting from milligram, gram, kilogram, and tonne, determine the best unit of measure to express the given item's weight.

53. A bee

54. This book

55. A tablespoon of salt

56. A Boeing 747

57. A stacked washer-dryer

58. A pen

59. An adult male

In Exercises 60–66, select the best estimate for the weight of the given item.

60. A newborn infant's weight

a. 3000 kg **b.** 300 kg
c. 30 kg **d.** 3 kg

61. The weight of a nickel

a. 5 kg **b.** 5 g **c.** 5 mg

62. A person's weight

a. 60 kg **b.** 60 g **c.** 60 dag

63. The weight of a box of cereal

a. 0.5 kg **b.** 0.5 g **c.** 0.5 t

64. The weight of a glass of water

a. 400 dg **b.** 400 g
c. 400 dag **d.** 400 hg

65. The weight of a regular-size car

a. 1500 dag **b.** 1500 hg **c.** 1500 kg **d.** 15,000 kg

66. The weight of a bicycle

a. 140 kg **b.** 140 hg **c.** 140 dag **d.** 140 g

67. Six items purchased at a grocery store weigh 14 kilograms. One of the items is detergent weighing 720 grams. What is the total weight, in kilograms, of the other five items?

68. If a nickel weighs 5 grams, how many nickels are there in 4 kilograms of nickels?

69. If the cost to mail a letter is 44 cents for mail weighing up to one ounce and 24 cents for each additional ounce or fraction of an ounce, find the cost of mailing a letter that weighs 85 grams.

70. Using the information given below the pictured finback whale, estimate the weight, in tons and kilograms, of the killer whale.

71. Which is more economical: purchasing the economy size of a detergent at 3 kilograms for \$3.15 or purchasing the regular size at 720 grams for 60¢?

Exercises 72–73 ask you to determine drug dosage by a patient's weight. Use the fact that 1 lb ≈ 0.45 kg.

72. The prescribed dosage of a drug is 10 mg/kg daily, meaning that 10 milligrams of the drug should be administered daily for each kilogram of a patient's weight. How many 400-milligram tablets should be given each day to a patient who weighs 175 pounds?

73. The prescribed dosage of a drug is 15 mg/kg daily, meaning that 15 milligrams of the drug should be administered daily for each kilogram of a patient's weight. How many 200-milligram tablets should be given each day to a patient who weighs 120 pounds?

The label on a bottle of Emetrol ("for food or drink indiscretions") reads

Each 5 mL teaspoonful contains glucose, 1.87 g; levulose, 1.87 g; and phosphoric acid, 21.5 mg.

Use this information to solve Exercises 74–75.

74. **a.** Find the amount of glucose in the recommended dosage of two teaspoons.

b. If the bottle contains 4 ounces, find the quantity of glucose in the bottle. (1 oz ≈ 30 mL)

75. **a.** Find the amount of phosphoric acid in the recommended dosage of two teaspoons.

b. If the bottle contains 4 ounces, find the quantity of phosphoric acid in the bottle. (1 oz ≈ 30 mL)

In Exercises 76–79, select the best estimate of the Celsius temperature of

76. A very hot day.

a. 85°C **b.** 65°C **c.** 35°C **d.** 20°C

77. A warm winter day in Washington, D.C.

a. 10°C **b.** 30°C **c.** 50°C **d.** 70°C

78. A setting for a home thermostat.

a. 20°C **b.** 40°C **c.** 60°C **d.** 80°C

79. The oven temperature for cooking a roast.

a. 80°C **b.** 100°C **c.** 175°C **d.** 350°C

Exercises 80–81 give the average temperature of some of the world's hottest places. In each exercise, determine which location has the hotter average temperature and by how many degrees Fahrenheit. Round to the nearest tenth of a degree.

80. Assab, Eritrea: 86.8°F; Dalol, Ethiopia: 34.6°C

81. Berbera, Somalia: 86.2°F; Néma, Mauritania: 30.3°C

Exercises 82–83 give the average temperature of some of the world's coldest places. In each exercise, determine which location has the colder average temperature and by how many degrees Celsius. Round to the nearest tenth of a degree.

82. Plateau, Antarctica: −56.7°C;
Amundsen-Scott, Antarctica: −56.2°F

83. Eismitte, Greenland: −29.2°C;
Resolute, Canada: −11.6°F

(Source for Exercises 80–83: Russell Ash, *The Top 10 of Everything 2009*)

Writing in Mathematics

84. Describe the difference between weight and mass.

85. Explain how to use dimensional analysis to convert 200 pounds to kilograms.

86. Why do you think that countries using the metric system prefer the Celsius scale over the Fahrenheit scale?

87. Describe in words how to convert from Celsius to Fahrenheit.

88. Describe in words how to convert from Fahrenheit to Celsius.

89. If you decide to travel outside the United States, which one of the two temperature conversion formulas should you take? Explain your answer.

Critical Thinking Exercises

Make Sense? *In Exercises 90–93, determine whether each statement makes sense or does not make sense, and explain your reasoning.*

90. When I played high school basketball, I weighed 250 kilograms.

91. It's not realistic to convert 500 mg to g because measurements should not involve fractional units.

92. Both the English system and the metric system enable me to easily see relationships among volume, capacity, and weight.

93. A comfortable classroom temperature is 20°C.

In Exercises 94–101, determine whether each statement is true or false. If the statement is false, make the necessary change(s) to produce a true statement.

94. A 4-pound object weighs more than a 2000-gram object.

95. A 100-milligram object weighs more than a 2-ounce object.

96. A 50-gram object weighs more than a 2-ounce object.

97. A 10-pound object weighs more than a 4-kilogram object.

98. Flour selling at 3¢ per gram is a better buy than flour selling at 55¢ per pound.

99. The measures

32,600 g, 32.1 kg, 4 lb, 36 oz

are arranged in order, from greatest to least weight.

100. If you are taking aspirin to relieve cold symptoms, a reasonable dose is 2 kilograms four times a day.

101. A large dog weighs about 350 kilograms.

Group Exercise

102. Present a group report on the current status of the metric system in the United States. At present, does it appear that the United States will convert to the metric system? Who supports the conversion and who opposes it? Summarize each side's position. Give examples of how our current system of weights and measures is an economic liability. What are the current obstacles to metric conversion?

Chapter Summary, Review, and Test

SUMMARY – DEFINITIONS AND CONCEPTS

9.1 Measuring Length; The Metric System

SUMMARY – DEFINITIONS AND CONCEPTS	EXAMPLES
a. The result obtained from measuring length is called a linear measurement, stated in linear units.	
b. Linear Units: The English System 12 in. = 1 ft, 3 ft = 1 yd, 36 in. = 1 yd, 5280 ft = 1 mi	
c. Dimensional Analysis Multiply the given measurement by a unit fraction with the unit of measurement that needs to be introduced in the numerator and the unit of measurement that needs to be eliminated in the denominator.	Ex. 1, p. 577
d. Linear Units: The Metric System The basic unit is the meter (m), approximately 39 inches. 1 km = 1000 m, 1 hm = 100 m, 1 dam = 10 m, 1 dm = 0.1 m, 1 cm = 0.01 m, 1 mm = 0.001 m	
e. Changing Linear Units within the Metric System $\times 10$ km hm dam m dm cm mm $\div 10$	Ex. 2, p. 580
f. English and metric equivalents for length are given in **Table 9.3** on page 581.	Ex. 3, p. 582; Ex. 4, p. 583

9.2 Measuring Area and Volume

SUMMARY – DEFINITIONS AND CONCEPTS	EXAMPLES
a. The area measure of a plane region is the number of square units contained in the given region.	Ex. 1, p. 586
b. Square Units: The English System $1\text{ ft}^2 = 144\text{ in.}^2$, $1\text{ yd}^2 = 9\text{ ft}^2$, $1\text{ a} = 43{,}560\text{ ft}^2 = 4840\text{ yd}^2$, $1\text{ mi}^2 = 640\text{ a}$	Ex. 2, p. 587; Ex. 3, p. 587
c. English and metric equivalents for area are given in **Table 9.4** on page 588.	Ex. 4, p. 588
d. The volume measure of a three-dimensional figure is the number of cubic units contained in its interior.	Ex. 5, p. 589
e. Capacity refers to the amount of fluid that a three-dimensional object can hold. English units for capacity include pints, quarts, and gallons: 2 pt = 1 qt; 4 qt = 1 gal; one cubic yard has a capacity of about 200 gallons. See **Table 9.5** on page 590.	Ex. 6, p. 590

f. The basic unit for capacity in the metric system is the liter (L). One liter is about 1.06 quarts. Prefixes for the liter are the same as throughout the metric system.

g. **Table 9.7** on page 591 shows relationships between volume and capacity in the metric system. Ex. 7, p. 591

$$1\text{ dm}^3 = 1000\text{ cm}^3 = 1\text{ L}$$

h. English and metric equivalents for capacity are given in **Table 9.8** on page 591. Ex. 8, p. 592

9.3 Measuring Weight and Temperature

a. Weight: The English System

$$16\text{ oz} = 1\text{ lb}, 2000\text{ lb} = 1\text{ T}$$

b. Units of Weight: The Metric System
The basic unit is the gram (g).

$$1000\text{ grams (1 kg)} \approx 2.2\text{ lb},$$
$$1\text{ kg} = 1000\text{ g}, 1\text{ hg} = 100\text{ g}, 1\text{ dag} = 10\text{ g},$$
$$1\text{ dg} = 0.1\text{ g}, 1\text{ cg} = 0.01\text{ g}, 1\text{ mg} = 0.001\text{ g}$$

c. Changing Units of Weight within the Metric System Ex. 1, p. 597

$\times 10$

kg hg dag g dg cg mg

$\div 10$

d. One kilogram of water has a volume of 1 liter. **Table 9.10** on page 598 shows relationships between volume and weight of water in the metric system. Ex. 2, p. 598

e. English and metric equivalents for weight are given in **Table 9.11** on page 598. Ex. 3, p. 598; Ex. 4, p. 599

f. Temperature Scales: Celsius to Fahrenheit: $F = \frac{9}{5}C + 32$ Ex. 5, p. 601

g. Temperature Scales: Fahrenheit to Celsius: $C = \frac{5}{9}(F - 32)$ Ex. 6, p. 601

Review Exercises

9.1

In Exercises 1–4, use dimensional analysis to convert the quantity to the indicated unit.

1. 69 in. to ft
2. 9 in. to yd
3. 21 ft to yd
4. 13,200 ft to mi

In Exercises 5–10, convert the given linear measurement to the metric unit indicated.

5. 22.8 m to cm
6. 7 dam to m
7. 19.2 hm to m
8. 144 cm to hm
9. 0.5 mm to m
10. 18 cm to mm

In Exercises 11–16, use the given English and metric equivalents, along with dimensional analysis, to convert the given measurement to the unit indicated. Where necessary, round answers to two decimal places.

1 in. = 2.54 cm
1 ft = 30.48 cm
1 yd ≈ 0.9 m
1 mi ≈ 1.6 km

11. 23 in. to cm
12. 19 cm to in.
13. 330 mi to km
14. 600 km to mi
15. 14 m to yd
16. 12 m to ft

17. Express 45 kilometers per hour in miles per hour.
18. Express 60 miles per hour in kilometers per hour.
19. Arrange from smallest to largest: 0.024 km, 2400 m, 24,000 cm.
20. If you jog six times around a track that is 800 meters long, how many kilometers have you covered?

9.2

21. Use the given figure to find its area in square units.

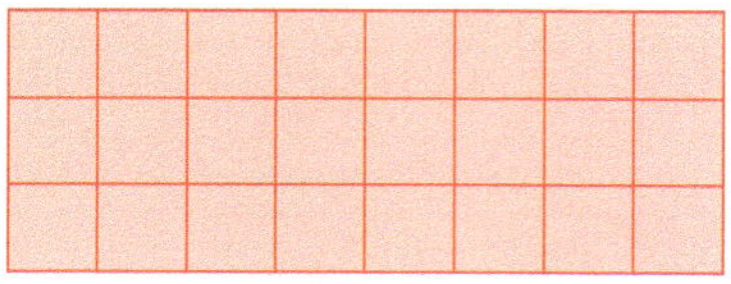

22. Singapore, with an area of 268 square miles and a population of 4,425,700, is one of the world's most densely populated countries. Find Singapore's population density, to the nearest tenth. Describe what this means.

23. Acadia National Park on the coast of Maine consists of 47,453 acres. How large is the national park in square miles? Round to the nearest square mile. (1 mi^2 = 640 a)

24. Given 1 acre ≈ 0.4 hectare, use dimensional analysis to find the size of a property in acres measured at 7.2 hectares.

25. Using 1 ft^2 ≈ 0.09 m^2, convert 30 m^2 to ft^2.

26. Using 1 mi^2 ≈ 2.6 km^2, convert 12 mi^2 to km^2.

27. Which one of the following is a reasonable measure for the area of a flower garden in a person's yard?

a. 100 m^2 **b.** 0.4 ha **c.** 0.01 km^2

28. Use the given figure to find its volume in cubic units.

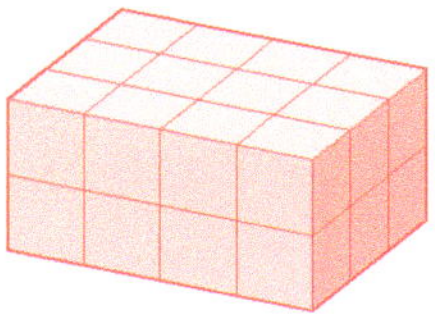

29. A swimming pool has a volume of 33,600 cubic feet. Given that 1 cubic foot has a capacity of about 7.48 gallons, how many gallons of water does the pool hold?

30. An aquarium has a volume of 76,000 cubic centimeters. How many liters of water does the aquarium hold?

In Exercises 31–35, use the given English and metric equivalents, along with dimensional analysis, to convert the given measurement to the unit indicated. Where necessary, round answers to two decimal places.

31. 22 mL to tsp

32. 5.4 L to c

33. 8 L to qt

34. 6 gal to L

1 tsp	≈ 5 mL
1 tbsp	≈ 15 mL
1 fl oz	≈ 30 mL
1 c	≈ 0.24 L
1 pt	≈ 0.47 L
1 qt	≈ 0.95 L
1 gal	≈ 3.8 L

35. A physician orders 15 cc of medication.

a. How many milliliters of the drug should be administered?

b. How many fluid ounces of the drug should be administered?

36. The capacity of a one-quart container of juice is approximately

a. 0.1 kL **b.** 0.5 L **c.** 1 L **d.** 1 mL

37. There are 3 feet in a yard. Explain why there are not 3 square feet in a square yard. If helpful, illustrate your explanation with a diagram.

38. Explain why the area of Texas could not be measured in cubic miles.

9.3

In Exercises 39–42, convert the given unit of weight to the unit indicated.

39. 12.4 dg to mg

40. 12 g to cg

41. 0.012 mg to g

42. 450 mg to kg

*In Exercises 43–44, use **Table 9.10** on page 598 to convert the given measurement of water to the unit or units indicated.*

43. 50 kg to cm^3

44. 4 kL to dm^3 to g

45. Using 1 lb ≈ 0.45 kg, convert 210 pounds to kilograms.

46. Using 1 oz ≈ 28 g, convert 392 grams to ounces.

47. The prescribed dosage of a drug is 12 mg/kg daily. How many 400-milligram tablets should be given each day to a patient who weighs 220 pounds? (1 lb ≈ 0.45 kg)

48. If you are interested in your weight in the metric system, would it be best to report it in milligrams, grams, or kilograms? Explain why the unit you selected would be most appropriate. Explain why each of the other two units is not the best choice for reporting your weight.

49. Given 16 oz = 1 lb, use dimensional analysis to convert 36 ounces to pounds.

In Exercises 50–51, select the best estimate for the weight of the given item.

50. A dollar bill:

a. 1 g **b.** 10 g **c.** 1 kg **d.** 4 kg

51. A hamburger:

a. 3 kg **b.** 1 kg **c.** 200 g **d.** 5 g

In Exercises 52–56, convert the given Celsius temperature to its equivalent temperature on the Fahrenheit scale.

52. 15°C **53.** 100°C

54. 5°C **55.** 0°C

56. −25°C

In Exercises 57–62, convert the given Fahrenheit temperature to its equivalent temperature on the Celsius scale.

57. 59°F **58.** 41°F

59. 212°F **60.** 98.6°F

61. 0°F **62.** 14°F

63. Is a decrease of 15° Celsius more or less than a decrease of 15° Fahrenheit? Explain your answer.

Chapter 9 Test

1. Change 807 mm to hm.

2. Given 1 inch = 2.54 centimeters, use dimensional analysis to change 635 centimeters to inches.

3. If you jog eight times around a track that is 600 meters long, how many kilometers have you covered?

In Exercises 4–6, write the most reasonable metric unit for length in each blank. Select from mm, cm, m, and km.

4. A human thumb is 20 ______ wide.

5. The height of the table is 45 ______.

6. The towns are 60 ______ apart.

7. If 1 mile ≈ 1.6 kilometers, express 80 miles per hour in kilometers per hour.

8. How many times greater is a square yard than a square foot?

9. Australia has a population of 20,090,400 and an area of 2,967,908 square miles. Find Australia's population density to the nearest tenth. Describe what this means.

10. Given 1 acre ≈ 0.4 hectare, use dimensional analysis to find the area of a property measured at 18 hectares.

11. The area of a dollar bill is approximately

a. 10 cm^2 **b.** 100 cm^2 **c.** 1000 cm^2 **d.** 1 m^2

12. There are 10 decimeters in a meter. Explain why there are not 10 cubic decimeters in a cubic meter. How many times greater is a cubic meter than a cubic decimeter?

13. The label on a bottle of Pepto Bismol indicates that the dosage should not exceed 16 tablespoons in 24 hours. How many fluid ounces is the maximum daily dose? (1 tbsp ≈ 15 mL; 1 fl oz ≈ 30 mL)

14. A swimming pool has a volume of 10,000 cubic feet. Given that 1 cubic foot has a capacity of about 7.48 gallons, how many gallons of water does the pool hold?

15. The capacity of a pail used to wash floors is approximately

a. 3 L **b.** 12 L **c.** 80 L **d.** 2 kL

16. Change 137 g to kg.

17. The prescribed dosage of a drug is 10 mg/kg daily. How many 200-milligram tablets should be given each day to a patient who weighs 130 pounds? (1 lb ≈ 0.45 kg)

In Exercises 18–19, write the most reasonable metric unit for weight in each blank. Select from mg, g, and kg.

18. My suitcase weighs 20 ______.

19. I took a 350 ______ aspirin.

20. Convert 30°C to Fahrenheit.

21. Convert 176°F to Celsius.

22. Comfortable room temperature is approximately

a. 70°C **b.** 50°C **c.** 30°C **d.** 20°C

10

Geometry

GEOMETRY IS THE STUDY OF THE SPACE YOU LIVE IN AND THE SHAPES THAT SURROUND YOU. YOU'RE EVEN MADE OF IT! THE HUMAN lung consists of nearly 300 spherical air sacs, geometrically designed to provide the greatest surface area within the limited volume of our bodies. Viewed in this way, geometry becomes an intimate experience.

For thousands of years, people have studied geometry in some form to obtain a better understanding of the world in which they live. A study of the shape of your world will provide you with many practical applications and perhaps help to increase your appreciation of its beauty.

Here's where you'll find these applications:

- A relationship between geometry and the visual arts is developed in Section 10.3 (Tessellations: pages 633–635).
- Using geometry to describe nature's complexity is discussed in Section 10.7 (Fractals: pages 674–675).

10.1

WHAT AM I SUPPOSED TO LEARN?

After you have read this section, you should be able to:

1 Understand points, lines, and planes as the basis of geometry.
2 Solve problems involving angle measures.
3 Solve problems involving angles formed by parallel lines and transversals.

Points, Lines, Planes, and Angles

THE SAN FRANCISCO MUSEUM OF MODERN ART WAS constructed in 1995 to illustrate how art and architecture can enrich one another. The exterior involves geometric shapes, symmetry, and unusual facades. Although there are no windows, natural light streams in through a truncated cylindrical skylight that crowns the building. The architect worked with a scale model of the museum at the site and observed how light hit it during different times of the day. These observations were used to cut the cylindrical skylight at an angle that maximizes sunlight entering the interior.

Angles play a critical role in creating modern architecture. They are also fundamental in the study of geometry. The word "geometry" means "earth measure." Because it involves the mathematics of shapes, geometry connects mathematics to art and architecture. It also has many practical applications. You can use geometry at home when you buy carpet, build a fence, tile a floor, or determine whether a piece of furniture will fit through your doorway. In this chapter, we look at the shapes that surround us and their applications.

Points, Lines, and Planes

1 Understand points, lines, and planes as the basis of geometry.

Points, lines, and planes make up the basis of all geometry. Stars in the night sky look like points of light. Long stretches of overhead power lines that appear to extend endlessly look like lines. The top of a flat table resembles part of a plane. However, stars, power lines, and tabletops only approximate points, lines, and planes. Points, lines, and planes do not exist in the physical world. Representations of these forms are shown in **Figure 10.1**. A **point**, represented as a small dot, has no length, width, or thickness. No object in the real world has zero size. A **line**, connecting two points along the shortest possible path, has no thickness and extends infinitely in both directions. However, no familiar everyday object is infinite in length. A **plane** is a flat surface with no thickness and no boundaries. This page resembles a plane, although it does not extend indefinitely and it does have thickness.

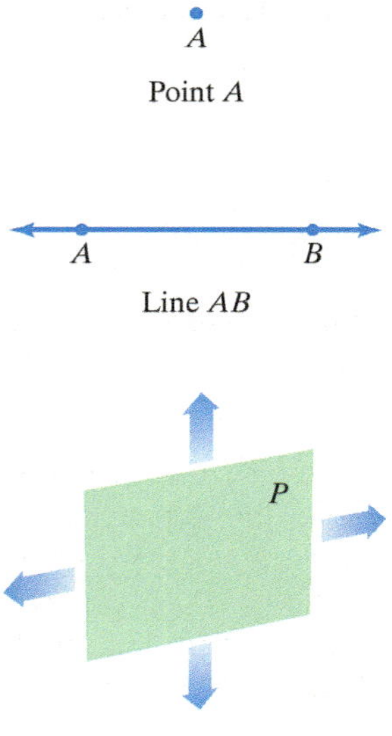

FIGURE 10.1 Representing a point, a line, and a plane

A line may be named using any two of its points. In **Figure 10.2(a)**, line AB can be symbolized $\overleftrightarrow{AB}$ or $\overleftrightarrow{BA}$. Any point on the line divides the line into three parts—the point and two **half-lines**. **Figure 10.2(b)** illustrates half-line AB, symbolized $\overset{\circ\!\!\rightarrow}{AB}$. The open circle above the A in the symbol and in the diagram indicates that point A is not included in the half-line. A **ray** is a half-line with its endpoint included. **Figure 10.2(c)** illustrates ray AB, symbolized $\overrightarrow{AB}$. The closed dot above the A in the diagram shows that point A is included in the ray. A portion of a line joining two points and including the endpoints is called a **line segment**. **Figure 10.2(d)** illustrates line segment AB, symbolized $\overline{AB}$ or $\overline{BA}$.

(a) Line AB — $\overleftrightarrow{AB}$ or $\overleftrightarrow{BA}$ **(b)** Half-line AB — $\overset{\circ\!\!\rightarrow}{AB}$ **(c)** Ray AB — $\overrightarrow{AB}$ **(d)** Line Segment AB — $\overline{AB}$ or $\overline{BA}$

FIGURE 10.2 Lines, half-lines, rays, and line segments

Angles

An **angle**, symbolized $\measuredangle$, is formed by the union of two rays that have a common endpoint. One ray is called the **initial side** and the other the **terminal side**.

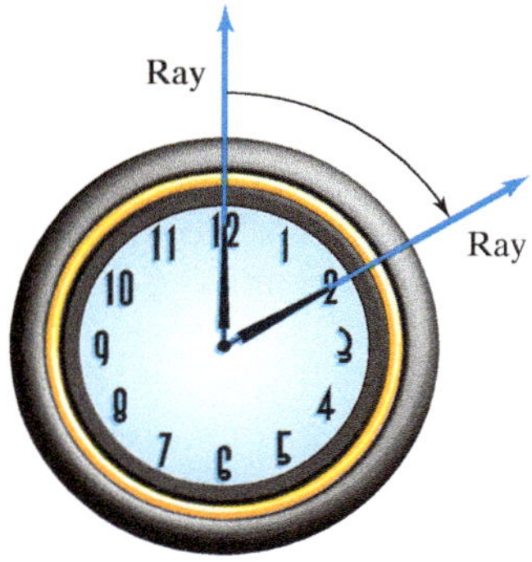

FIGURE 10.3 Clock with hands forming an angle

A rotating ray is often a useful way to think about angles. The ray in **Figure 10.3** rotates from 12 to 2. The ray pointing to 12 is the initial side and the ray pointing to 2 is the terminal side. The common endpoint of an angle's initial side and terminal side is the **vertex** of the angle.

Figure 10.4 shows an angle. The common endpoint of the two rays, B, is the vertex. The two rays that form the angle, $\overrightarrow{BA}$ and $\overrightarrow{BC}$, are the sides. The four ways of naming the angle are shown to the right of **Figure 10.4**.

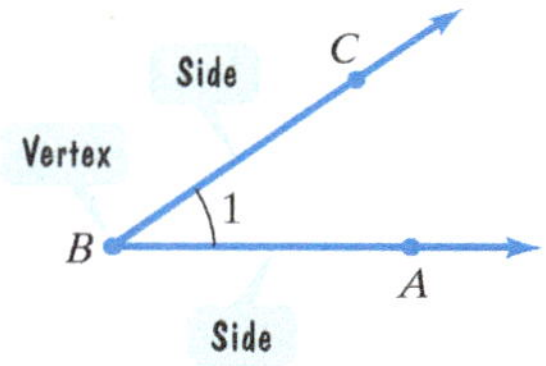

FIGURE 10.4 An angle: two rays with a common endpoint

Naming the Angle

$\measuredangle 1$ $\measuredangle B$ $\measuredangle ABC$ $\measuredangle CBA$

Vertex alone — Vertex letter in the middle

2 Solve problems involving angle measures.

Measuring Angles Using Degrees

Angles are measured by determining the amount of rotation from the initial side to the terminal side. One way to measure angles is in **degrees**, symbolized by a small, raised circle °. Think of the hour hand of a clock. From 12 noon to 12 midnight, the hour hand moves around in a complete circle. By definition, the ray has rotated through 360 degrees, or 360°, shown in **Figure 10.5**. Using 360° as the amount of rotation of a ray back onto itself, **a degree, 1°, is $\frac{1}{360}$ of a complete rotation.**

FIGURE 10.5 A complete 360° rotation

EXAMPLE 1 *Using Degree Measure*

The hand of a clock moves from 12 to 2 o'clock, shown in **Figure 10.3**. Through how many degrees does it move?

SOLUTION

We know that one complete rotation is 360°. Moving from 12 to 2 o'clock is $\frac{2}{12}$, or $\frac{1}{6}$, of a complete revolution. Thus, the hour hand moves

$$\frac{1}{6} \times 360° = \frac{360°}{6} = 60°$$

in going from 12 to 2 o'clock.

CHECK POINT 1 The hand of a clock moves from 12 to 1 o'clock. Through how many degrees does it move?

Figure 10.6 shows angles classified by their degree measurement. An **acute angle** measures less than 90° [see **Figure 10.6(a)**]. A **right angle**, one-quarter of a complete rotation, measures 90° [**Figure 10.6(b)**]. Examine the right angle—do you see a small square at the vertex? This symbol is used to indicate a right angle. An **obtuse angle** measures more than 90°, but less than 180° [**Figure 10.6(c)**]. Finally, a **straight angle**, one-half a complete rotation, measures 180° [**Figure 10.6(d)**]. The two rays in a straight angle form a straight line.

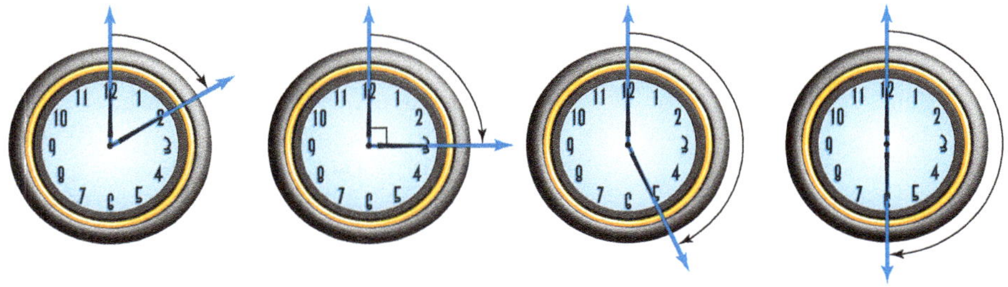

(a) Acute angle Less than 90° — **(b) Right angle** 90° — **(c) Obtuse angle** More than 90° but less than 180° — **(d) Straight angle** 180°

FIGURE 10.6 Classifying angles by their degree measurements

Figure 10.7 illustrates a **protractor**, used for finding the degree measure of an angle. As shown in the figure, we measure an angle by placing the center point of the protractor on the vertex of the angle and the straight side of the protractor along one side of the angle. The measure of $\angle ABC$ is then read as 50°. Observe that the measure is not 130° because the angle is obviously less than 90°. We indicate the angle's measure by writing $m\angle ABC = 50°$, read "the measure of angle ABC is 50°."

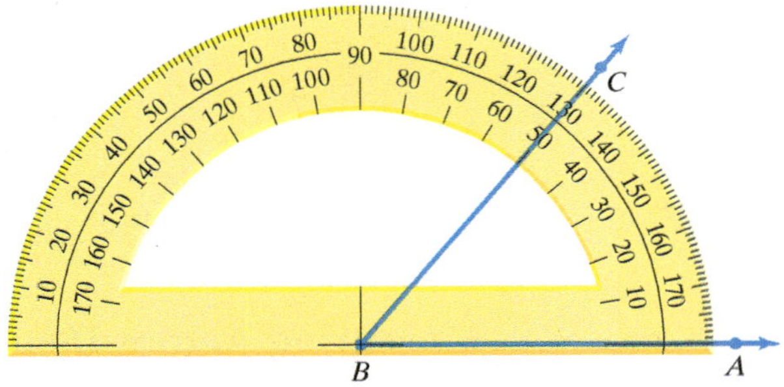

FIGURE 10.7 Using a protractor to measure an angle: $m\angle ABC = 50°$

Two angles whose measures have a sum of 90° are called **complementary angles**. For example, angles measuring 70° and 20° are complementary angles because $70° + 20° = 90°$. For angles such as those measuring 70° and 20°, each angle is the **complement** of the other: The 70° angle is the complement of the 20° angle and the 20° angle is the complement of the 70° angle. **The measure of the complement can be found by subtracting the angle's measure from 90°.** For example, we can find the complement of a 25° angle by subtracting 25° from 90°: $90° - 25° = 65°$. Thus, an angle measuring 65° is the complement of one measuring 25°.

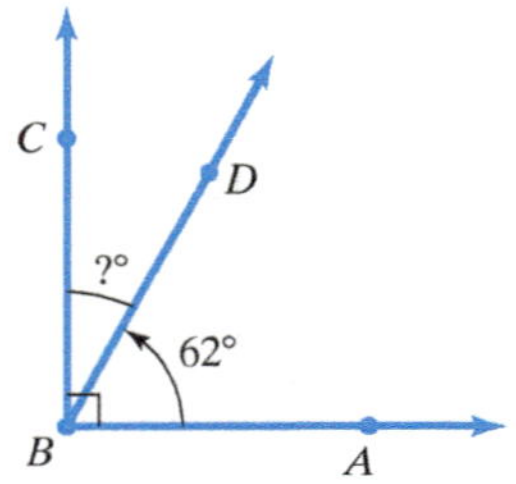

FIGURE 10.8

EXAMPLE 2 *Angle Measures and Complements*

Use **Figure 10.8** to find $m\angle DBC$.

SOLUTION

The measure of $\angle DBC$ is not yet known. It is shown as ?° in **Figure 10.8**. The acute angles $\angle ABD$, which measures 62°, and $\angle DBC$ form a right angle, indicated by the square at the vertex. This means that the measures of the acute angles add up to 90°. Thus, $\angle DBC$ is the complement of the angle measuring 62°. The measure of $\angle DBC$ is found by subtracting 62° from 90°:

$$m\angle DBC = 90° - 62° = 28°.$$

The measure of $\angle DBC$ can also be found using an algebraic approach.

$m\angle ABD + m\angle DBC = 90°$	The sum of the measures of complementary angles is 90°.
$62° + m\angle DBC = 90°$	We are given $m\angle ABD = 62°$.
$m\angle DBC = 90° - 62° = 28°$	Subtract 62° from both sides of the equation.

CHECK POINT 2 In **Figure 10.8**, let $m\angle DBC = 19°$. Find $m\angle DBA$.

Two angles whose measures have a sum of 180° are called **supplementary angles**. For example, angles measuring 110° and 70° are supplementary angles because $110° + 70° = 180°$. For angles such as those measuring 110° and 70°, each angle is the **supplement** of the other: The 110° angle is the supplement of the 70° angle, and the 70° angle is the supplement of the 110° angle. **The measure of the supplement can be found by subtracting the angle's measure from 180°.** For example, we can find the supplement of a 25° angle by subtracting 25° from 180°: $180° - 25° = 155°$. Thus, an angle measuring 155° is the supplement of one measuring 25°.

EXAMPLE 3 *Angle Measures and Supplements*

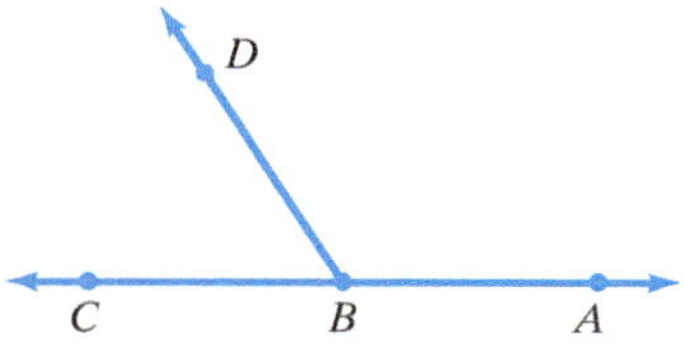

FIGURE 10.9

Figure 10.9 shows that $\measuredangle ABD$ and $\measuredangle DBC$ are supplementary angles. If $m\measuredangle ABD$ is 66° greater than $m\measuredangle DBC$, find the measure of each angle.

SOLUTION

Let $m\measuredangle DBC = x$. Because $m\measuredangle ABD$ is 66° greater than $m\measuredangle DBC$, then $m\measuredangle ABD = x + 66°$. We are given that these angles are supplementary.

$m\measuredangle DBC + m\measuredangle ABD = 180°$	The sum of the measures of supplementary angles is 180°.
$x + (x + 66°) = 180°$	Substitute the variable expressions for the measures.
$2x + 66° = 180°$	Combine like terms: $x + x = 2x$.
$2x = 114°$	Subtract 66° from both sides.
$x = 57°$	Divide both sides by 2.

Thus, $m\measuredangle DBC = 57°$ and $m\measuredangle ABD = 57° + 66° = 123°$.

CHECK POINT 3 In **Figure 10.9**, if $m\measuredangle ABD$ is 88° greater than $m\measuredangle DBC$, find the measure of each angle.

FIGURE 10.10

Figure 10.10 illustrates a highway sign that warns of a railroad crossing. When two lines intersect, the opposite angles formed are called **vertical angles**.

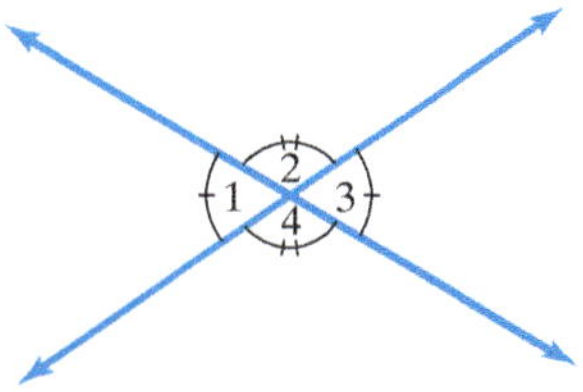

FIGURE 10.11

In **Figure 10.11**, there are two pairs of vertical angles. Angles 1 and 3 are vertical angles. Angles 2 and 4 are also vertical angles.

We can use **Figure 10.11** to show that vertical angles have the same measure. Let's concentrate on angles 1 and 3, each denoted by one tick mark. Can you see that each of these angles is supplementary to angle 2?

$m\measuredangle 1 + m\measuredangle 2 = 180°$	The sum of the measures of supplementary angles is 180°.
$m\measuredangle 2 + m\measuredangle 3 = 180°$	
$m\measuredangle 1 + m\measuredangle 2 = m\measuredangle 2 + m\measuredangle 3$	Substitute $m\measuredangle 2 + m\measuredangle 3$ for 180° in the first equation.
$m\measuredangle 1 = m\measuredangle 3$	Subtract $m\measuredangle 2$ from both sides.

Using a similar approach, we can show that $m\measuredangle 2 = m\measuredangle 4$, each denoted by two tick marks in **Figure 10.11**.

Vertical angles have the same measure.

EXAMPLE 4 *Using Vertical Angles*

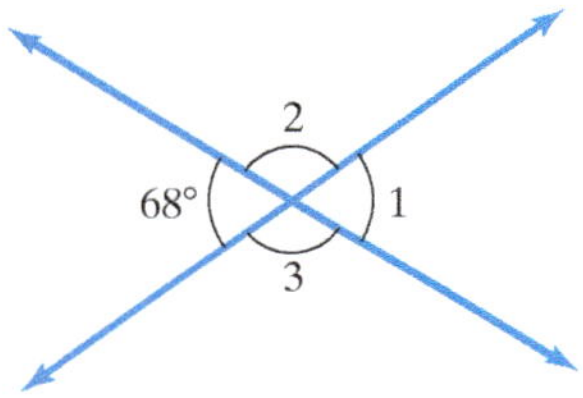

FIGURE 10.12

Figure 10.12 shows that the angle on the left measures 68°. Find the measures of the other three angles.

SOLUTION

Angle 1 and the angle measuring 68° are vertical angles. Because vertical angles have the same measure,

$$m\measuredangle 1 = 68°.$$

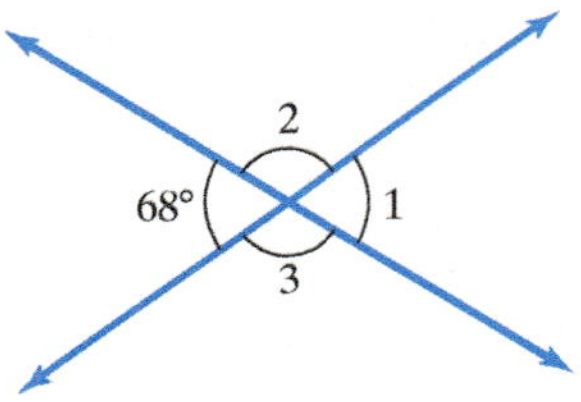

FIGURE 10.12 (repeated)

Angle 2 and the angle measuring 68° form a straight angle and are supplementary. Because their measures add up to 180°,

$$m\measuredangle 2 = 180° - 68° = 112°.$$

Angle 2 and angle 3 are also vertical angles, so they have the same measure. Because the measure of angle 2 is 112°,

$$m\measuredangle 3 = 112°.$$

CHECK POINT 4 In **Figure 10.12**, assume that the angle on the left measures 57°. Find the measures of the other three angles.

Parallel Lines

Parallel lines are lines that lie in the same plane and have no points in common. If two different lines in the same plane are not parallel, they have a single point in common and are called **intersecting lines**. If the lines intersect at an angle of 90°, they are called **perpendicular lines**.

If we intersect a pair of parallel lines with a third line, called a **transversal**, eight angles are formed, as shown in **Figure 10.13**. Certain pairs of these angles have special names, as well as special properties. These names and properties are summarized in **Table 10.1**.

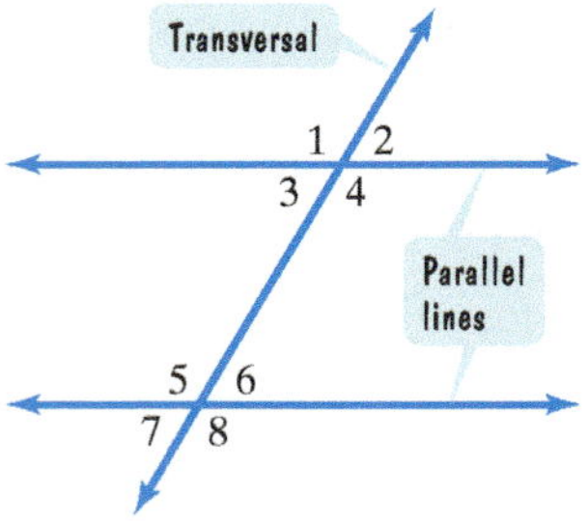

FIGURE 10.13

TABLE 10.1 Names of Angle Pairs Formed by a Transversal Intersecting Parallel Lines

Name	Description	Sketch	Angle Pairs Described	Property
Alternate interior angles	Interior angles that do not have a common vertex on alternate sides of the transversal	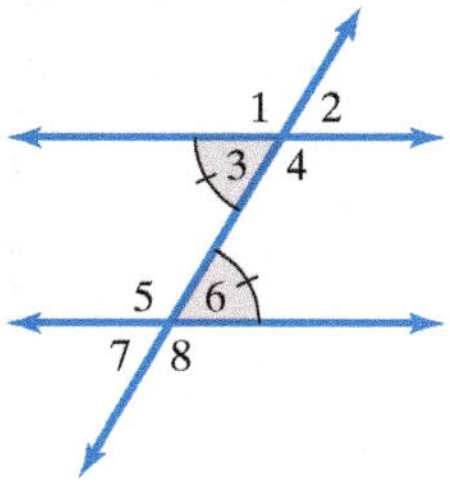	$\measuredangle 3$ and $\measuredangle 6$ $\measuredangle 4$ and $\measuredangle 5$	Alternate interior angles have the same measure. $m\measuredangle 3 = m\measuredangle 6$ $m\measuredangle 4 = m\measuredangle 5$
Alternate exterior angles	Exterior angles that do not have a common vertex on alternate sides of the transversal	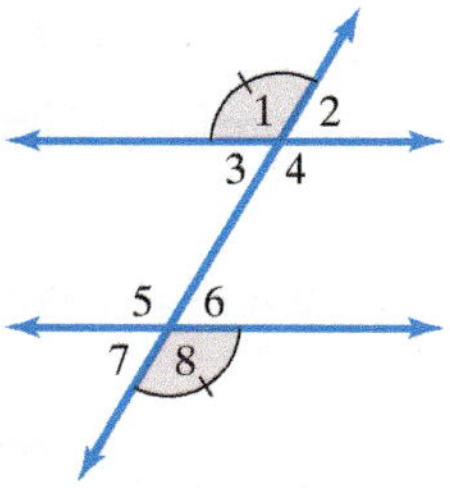	$\measuredangle 1$ and $\measuredangle 8$ $\measuredangle 2$ and $\measuredangle 7$	Alternate exterior angles have the same measure. $m\measuredangle 1 = m\measuredangle 8$ $m\measuredangle 2 = m\measuredangle 7$
Corresponding angles	One interior and one exterior angle on the same side of the transversal	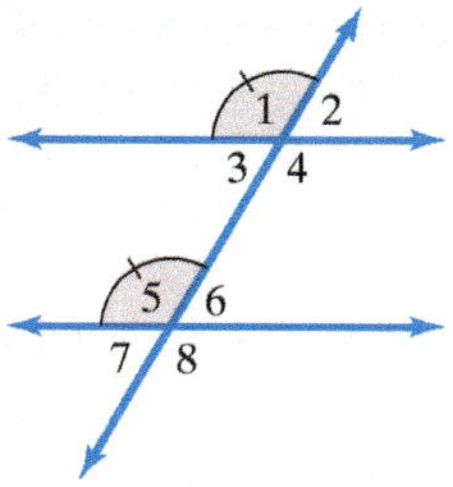	$\measuredangle 1$ and $\measuredangle 5$ $\measuredangle 2$ and $\measuredangle 6$ $\measuredangle 3$ and $\measuredangle 7$ $\measuredangle 4$ and $\measuredangle 8$	Corresponding angles have the same measure. $m\measuredangle 1 = m\measuredangle 5$ $m\measuredangle 2 = m\measuredangle 6$ $m\measuredangle 3 = m\measuredangle 7$ $m\measuredangle 4 = m\measuredangle 8$

3 Solve problems involving angles formed by parallel lines and transversals.

When two parallel lines are intersected by a transversal, the following relationships are true:

PARALLEL LINES AND ANGLE PAIRS

If parallel lines are intersected by a transversal,

- alternate interior angles have the same measure,
- alternate exterior angles have the same measure, and
- corresponding angles have the same measure.

Conversely, if two lines are intersected by a third line and a pair of alternate interior angles or a pair of alternate exterior angles or a pair of corresponding angles have the same measure, then the two lines are parallel.

EXAMPLE 5 *Finding Angle Measures When Parallel Lines Are Intersected by a Transversal*

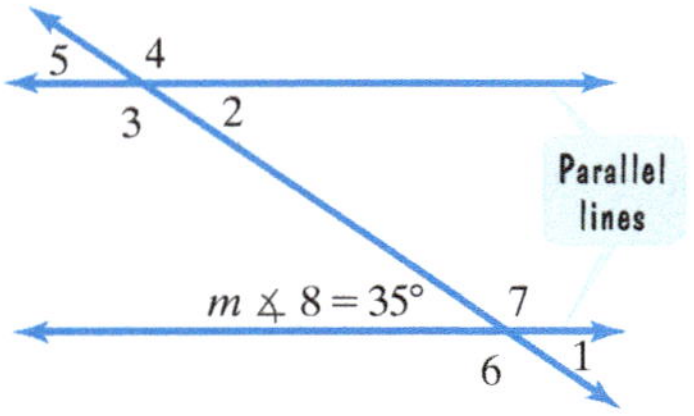

FIGURE 10.14

In **Figure 10.14**, two parallel lines are intersected by a transversal. One of the angles (∡ 8) has a measure of 35°. Find the measure of each of the other seven angles.

SOLUTION

Look carefully at **Figure 10.14** and fill in the angle measures as you read each line in this solution.

$m\measuredangle 1 = 35°$	∡ 8 and ∡ 1 are vertical angles and vertical angles have the same measure.
$m\measuredangle 6 = 180° - 35° = 145°$	∡ 8 and ∡ 6 are supplementary.
$m\measuredangle 7 = 145°$	∡ 6 and ∡ 7 are vertical angles, so they have the same measure.
$m\measuredangle 2 = 35°$	∡ 8 and ∡ 2 are alternate interior angles, so they have the same measure.
$m\measuredangle 3 = 145°$	∡ 7 and ∡ 3 are alternate interior angles. Thus, they have the same measure.
$m\measuredangle 5 = 35°$	∡ 8 and ∡ 5 are corresponding angles. Thus, they have the same measure.
$m\measuredangle 4 = 180° - 35° = 145°$	∡ 4 and ∡ 5 are supplementary.

GREAT QUESTION!

Is there more than one way to solve Example 5?

Yes. For example, once you know that $m\measuredangle 2 = 35°$, $m\measuredangle 5$ is also 35° because ∡ 2 and ∡ 5 are vertical angles.

CHECK POINT 5 In **Figure 10.14**, assume that $m\measuredangle 8 = 29°$. Find the measure of each of the other seven angles.

Concept and Vocabulary Check

Fill in each blank so that the resulting statement is true.

1. $\overleftrightarrow{AB}$ symbolizes __________ AB, $\overrightarrow{AB}$ symbolizes __________ AB, $\overleftrightarrow{AB}$ symbolizes __________ AB, and $\overline{AB}$ symbolizes __________ AB.
2. A/an __________ angle measures less than 90°, a/an __________ angle measures 90°, a/an __________ angle measures more than 90° and less than 180°, and a/an __________ angle measures 180°.
3. Two angles whose measures have a sum of 90° are called __________ angles. Two angles whose measures have a sum of 180° are called __________ angles.
4. When two lines intersect, the opposite angles are called __________ angles.
5. Lines that lie in the same plane and have no points in common are called __________ lines. If these lines are intersected by a third line, called a __________, eight angles are formed.
6. Lines that intersect at an angle of 90° are called __________ lines.

In Exercises 7–12, determine whether each statement is true or false. If the statement is false, make the necessary change(s) to produce a true statement.

7. A ray extends infinitely in both directions. ______
8. An angle is formed by the union of two lines that have a common endpoint. ______
9. A degree, 1°, is $\frac{1}{90}$ of a complete rotation. ______
10. A ruler is used for finding the degree measure of an angle. ______
11. The measure of an angle's complement is found by subtracting the angle's measure from 90°. ______
12. Vertical angles have the same measure. ______

Exercise Set 10.1

Practice Exercises

1. The hour hand of a clock moves from 12 to 5 o'clock. Through how many degrees does it move?
2. The hour hand of a clock moves from 12 to 4 o'clock. Through how many degrees does it move?
3. The hour hand of a clock moves from 1 to 4 o'clock. Through how many degrees does it move?
4. The hour hand of a clock moves from 1 to 7 o'clock. Through how many degrees does it move?

In Exercises 5–10, use the protractor to find the measure of each angle. Then classify the angle as acute, right, straight, or obtuse.

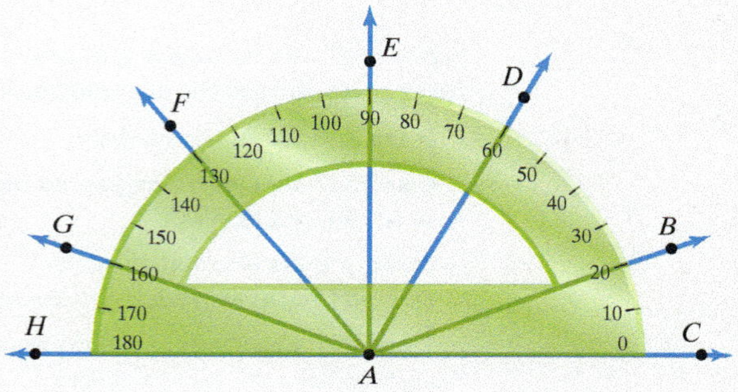

5. $\measuredangle CAB$
6. $\measuredangle CAF$
7. $\measuredangle HAB$
8. $\measuredangle HAF$
9. $\measuredangle CAH$
10. $\measuredangle HAE$

In Exercises 11–14, find the measure of the angle in which a question mark with a degree symbol appears.

11.

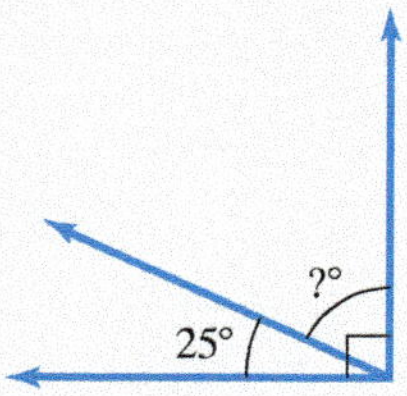

12.

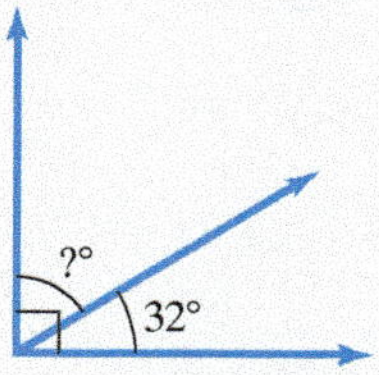

13.

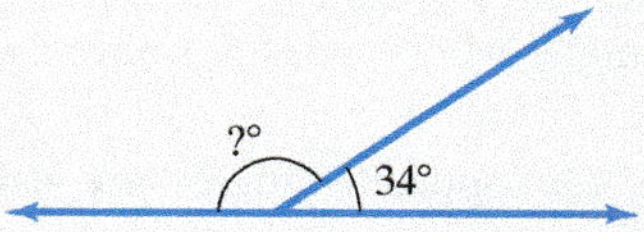

14.

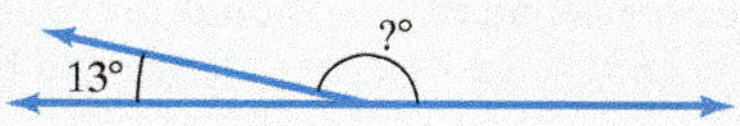

In Exercises 15–20, find the measure of the complement and the supplement of each angle.

15. 48°	**16.** 52°	**17.** 89°
18. 1°	**19.** 37.4°	**20.** $15\frac{1}{3}°$

In Exercises 21–24, use an algebraic equation to find the measures of the two angles described. Begin by letting x represent the degree measure of the angle's complement or its supplement.

21. The measure of the angle is 12° greater than its complement.
22. The measure of the angle is 56° greater than its complement.
23. The measure of the angle is three times greater than its supplement.
24. The measure of the angle is 81° more than twice that of its supplement.

In Exercises 25–28, find the measures of angles 1, 2, and 3.

25.

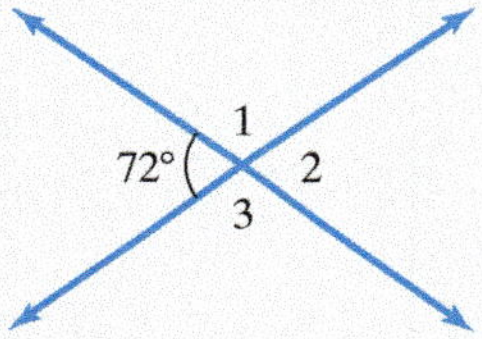

26.

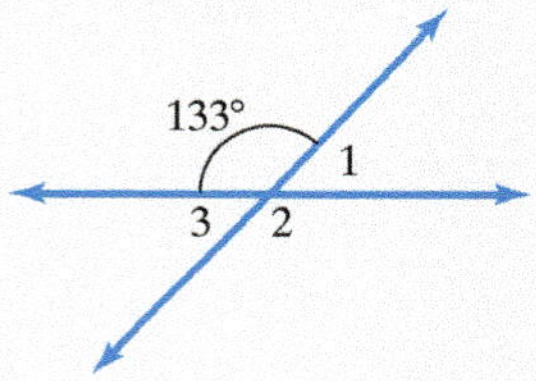

27.

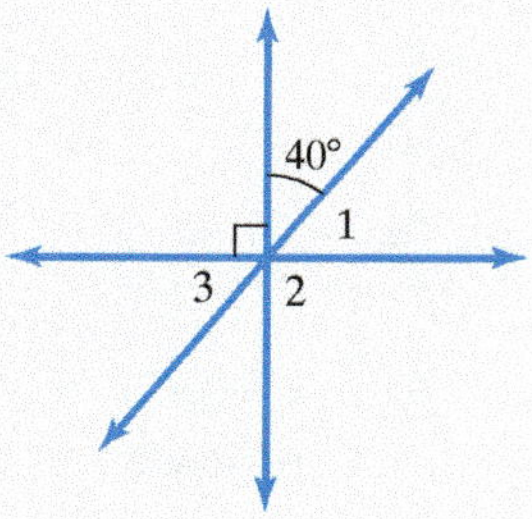

28.

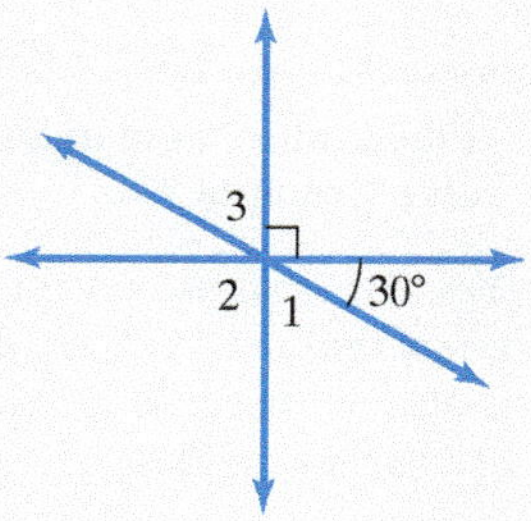

The figures for Exercises 29–30 show two parallel lines intersected by a transversal. One of the angle measures is given. Find the measure of each of the other seven angles.

29.

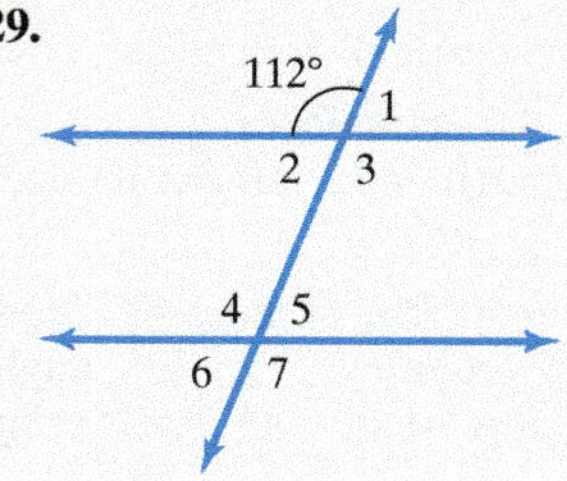

30.

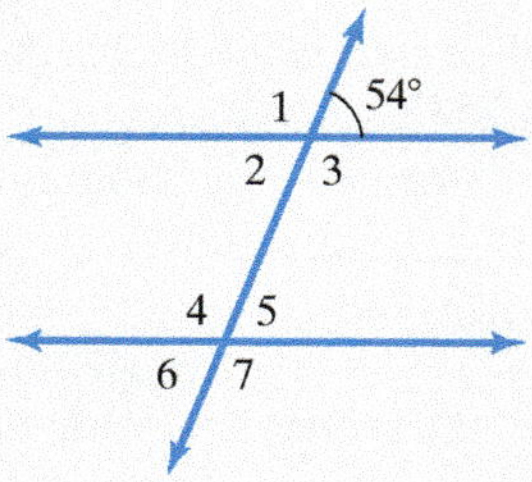

The figures for Exercises 31–34 show two parallel lines intersected by more than one transversal. Two of the angle measures are given. Find the measures of angles 1, 2, and 3.

31.

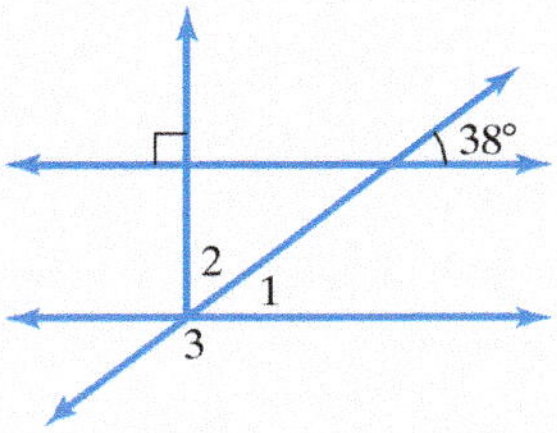

32.

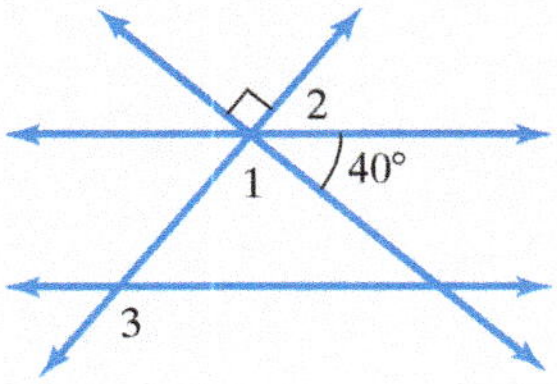

33.

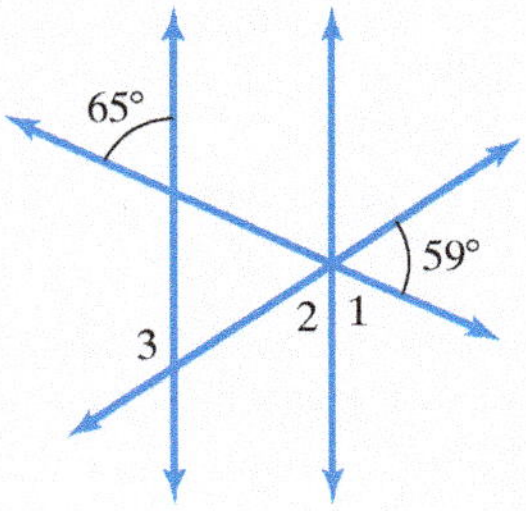

34.

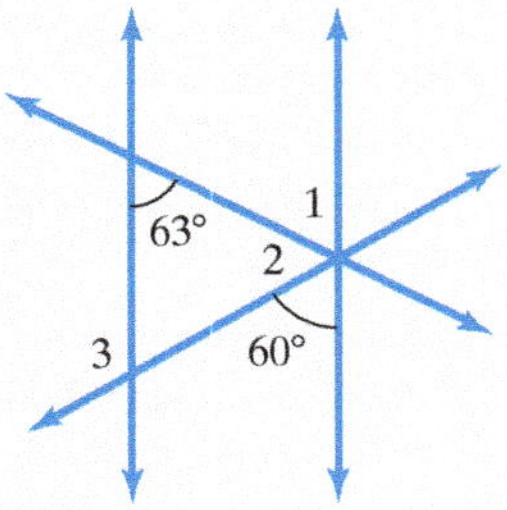

Use the following figure to determine whether each statement in Exercises 35–38 is true or false.

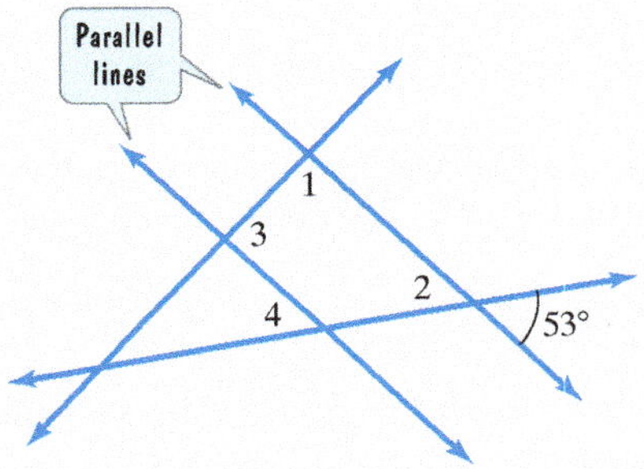

35. $m\measuredangle 2 = 37°$

36. $m\measuredangle 1 = m\measuredangle 2$

37. $m\measuredangle 4 = 53°$

38. $m\measuredangle 3 = m\measuredangle 4$

Use the following figure to determine whether each statement in Exercises 39–42 is true or false.

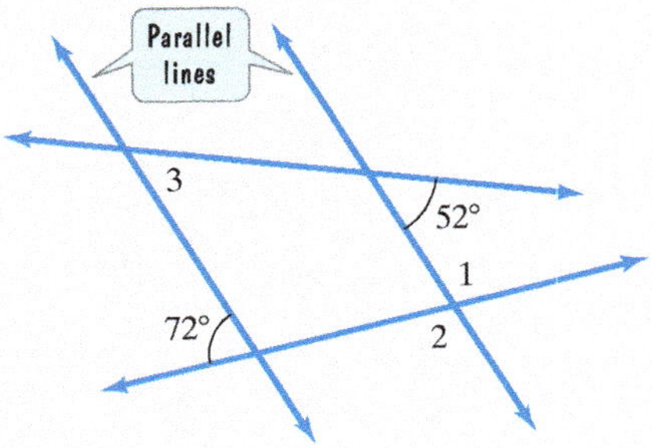

39. $m\measuredangle 1 = 38°$

40. $m\measuredangle 1 = 108°$

41. $m\measuredangle 2 = 52°$

42. $m\measuredangle 3 = 72°$

Practice Plus

In Exercises 43–46, use an algebraic equation to find the measure of each angle that is represented in terms of x.

43.

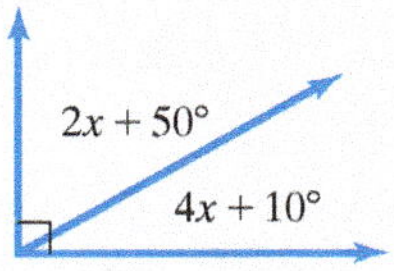

44.

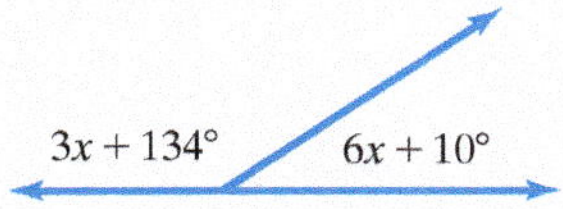

45.

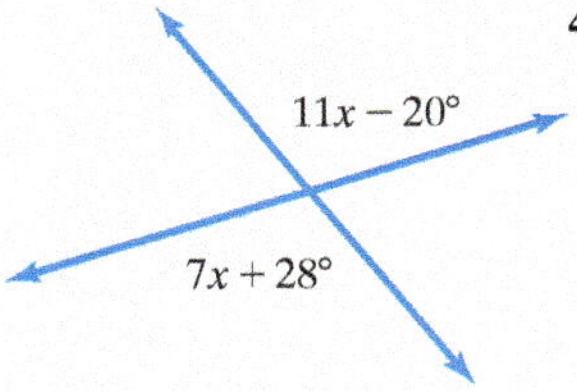

46.

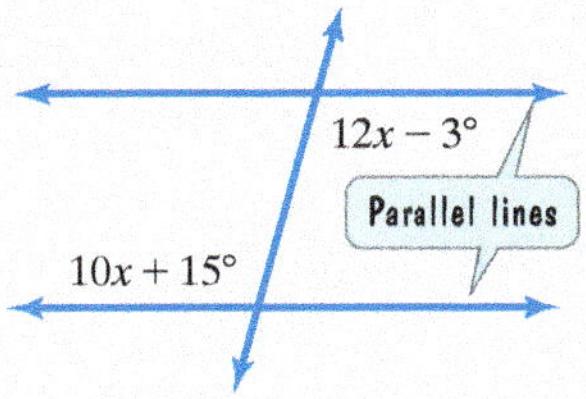

Because geometric figures consist of sets of points, we can apply set operations to obtain the union, ∪, or the intersection, ∩, of such figures. The union of two geometric figures is the set of points that belongs to either of the figures or to both figures. The intersection of two geometric figures is the set of points common to both figures. In Exercises 47–54, use the line shown to find each set of points.

47. $\overline{AC} \cap \overline{BD}$

48. $\overline{AB} \cap \overline{BC}$

49. $\overline{AC} \cup \overline{BD}$

50. $\overline{AB} \cup \overline{BC}$

51. $\overrightarrow{BA} \cup \overrightarrow{BC}$

52. $\overrightarrow{CB} \cup \overrightarrow{CD}$

53. $\overline{AD} \cap \overrightarrow{DB}$

54. $\overline{AC} \cap \overrightarrow{CB}$

Application Exercises

55. The picture shows the top of an umbrella in which all the angles formed by the spokes have the same measure. Find the measure of each angle.

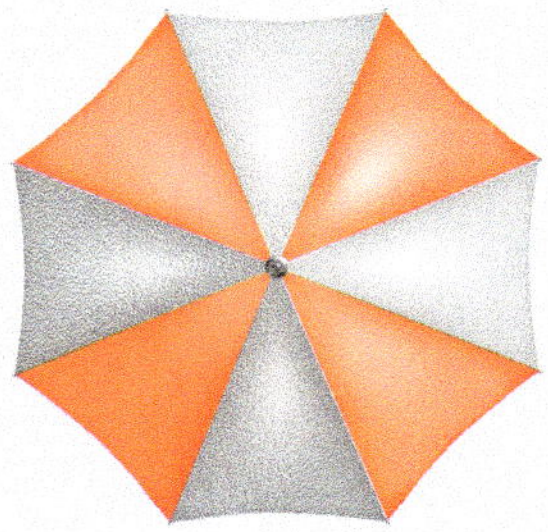

56. In the musical *Company*, composer Stephen Sondheim describes the marriage between two of the play's characters as "parallel lines who meet." What is the composer saying about this relationship?

57. The picture shows a window with parallel framing in which snow has collected in the corners. What property of parallel lines is illustrated by where the snow has collected?

In Exercises 58–59, consider the following uppercase letters from the English alphabet:

A E F H N T X Z.

58. Which letters contain parallel line segments?

59. Which letters contain perpendicular line segments?

Angles play an important role in custom bikes that are properly fitted to the biking needs of cyclists. One of the angles to help find the perfect fit is called the hip angle. The figure indicates that the hip angle is created when you're sitting on the bike, gripping the handlebars, and your leg is fully extended. Your hip is the vertex, with one ray extending to your shoulder and the other ray extending to the front-bottom of your foot.

The table indicates hip angles for various biking needs. Use this information to pedal through Exercises 60–63.

Hip Angle	Used For
$85° \le$ hip angle $\le 89°$	short-distance aggressive racing
$91° \le$ hip angle $\le 115°$	long-distance riding
$116° \le$ hip angle $\le 130°$	mountain biking

60. Which type or types of biking require an acute hip angle?

61. Which type or types of biking require an obtuse hip angle?

62. A racer who had an 89° hip angle decides to switch to long-distance riding. What is the maximum difference in hip angle for the two types of biking?

63. A racer who had an 89° hip angle decides to switch to mountain biking. What is the minimum difference in hip angle for the two types of biking?

(Source for Exercises 60–63: *Scholastic Math*, January 11, 2010)

Writing in Mathematics

64. Describe the differences among lines, half-lines, rays, and line segments.

65. What is an angle and what determines its size?

66. Describe each type of angle: acute, right, obtuse, and straight.

67. What are complementary angles? Describe how to find the measure of an angle's complement.

68. What are supplementary angles? Describe how to find the measure of an angle's supplement.

69. Describe the difference between perpendicular and parallel lines.

70. If two parallel lines are intersected by a transversal, describe the location of the alternate interior angles, the alternate exterior angles, and the corresponding angles.

71. Describe everyday objects that approximate points, lines, and planes.

72. If a transversal is perpendicular to one of two parallel lines, must it be perpendicular to the other parallel line as well? Explain your answer.

Critical Thinking Exercises

Make Sense? *In Exercises 73–76, determine whether each statement makes sense or does not make sense, and explain your reasoning.*

73. I drew two lines that are not parallel and that intersect twice.

74. I used the length of an angle's sides to determine that it was obtuse.

75. I'm working with two angles that are not complementary, so I can conclude that they are supplementary.

76. The rungs of a ladder are perpendicular to each side, so the rungs are parallel to each other.

77. Use the figure to select a pair of complementary angles.

a. $\measuredangle 1$ and $\measuredangle 4$

b. $\measuredangle 3$ and $\measuredangle 6$

c. $\measuredangle 2$ and $\measuredangle 5$

d. $\measuredangle 1$ and $\measuredangle 5$

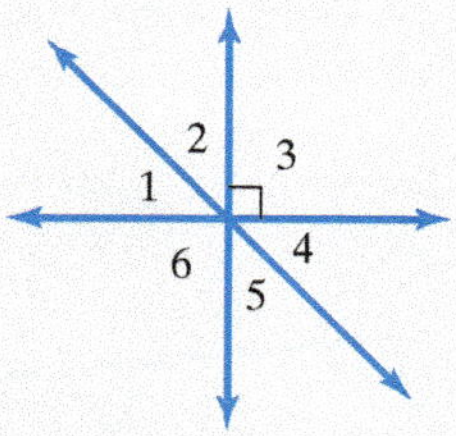

78. If $m\measuredangle AGB = m\measuredangle BGC$, and $m\measuredangle CGD = m\measuredangle DGE$, find $m\measuredangle BGD$.

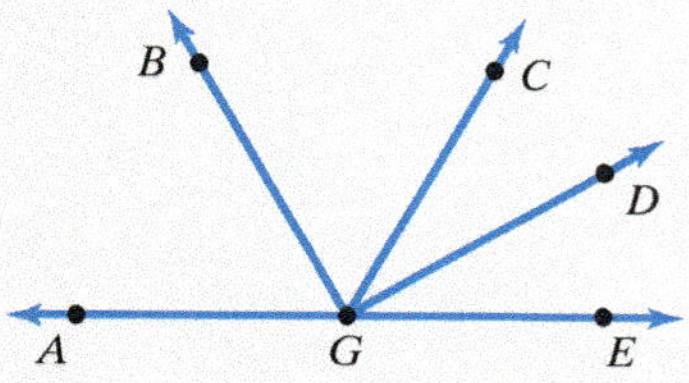

10.2

WHAT AM I SUPPOSED TO LEARN?

After you have read this section, you should be able to:

1 Solve problems involving angle relationships in triangles.
2 Solve problems involving similar triangles.
3 Solve problems using the Pythagorean Theorem.

Triangles

The Walter Pyramid, California State University, Long Beach

IN CHAPTER 1, WE DEFINED *deductive reasoning* as the process of proving a specific conclusion from one or more general statements. A conclusion that is proved to be true through deductive reasoning is called a **theorem**. The Greek mathematician Euclid, who lived more than 2000 years ago, used deductive reasoning. In his 13-volume book, *Elements,* Euclid proved over 465 theorems about geometric figures. Euclid's work established deductive reasoning as a fundamental tool of mathematics. Here's looking at Euclid!

A **triangle** is a geometric figure that has three sides, all of which lie on a flat surface or plane. If you start at any point along the triangle and trace along the entire figure exactly once, you will end at the same point at which you started. Because the beginning point and ending point are the same, the triangle is called a **closed** geometric figure. Euclid used parallel lines to prove one of the most important properties of triangles: The sum of the measures of the three angles of any triangle is 180°. Here is how he did it. He began with the following general statement:

EUCLID'S ASSUMPTION ABOUT PARALLEL LINES

Given a line and a point not on the line, one and only one line can be drawn through the given point parallel to the given line.

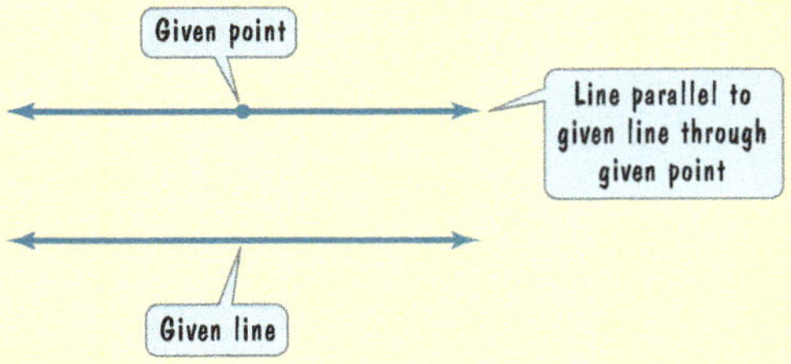

In **Figure 10.15**, triangle ABC represents any triangle. Using the general assumption given above, we draw a line through point B parallel to line AC.

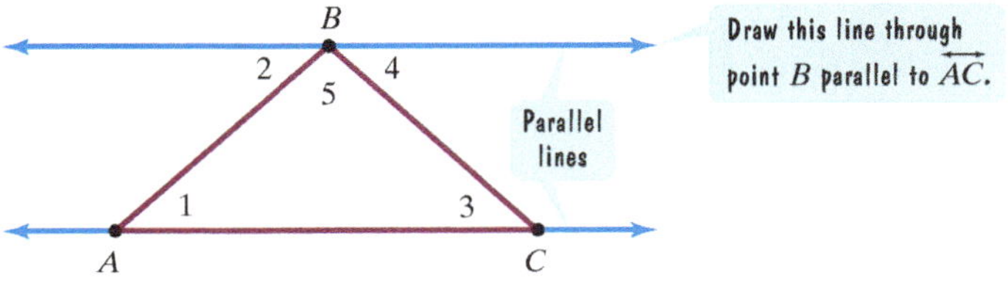

FIGURE 10.15

Because the lines are parallel, alternate interior angles have the same measure.

$$m\measuredangle 1 = m\measuredangle 2 \quad \text{and} \quad m\measuredangle 3 = m\measuredangle 4$$

Also observe that angles 2, 5, and 4 form a straight angle.

$$m\measuredangle 2 + m\measuredangle 5 + m\measuredangle 4 = 180°$$

Because $m\measuredangle 1 = m\measuredangle 2$, replace $m\measuredangle 2$ with $m\measuredangle 1$.

Because $m\measuredangle 3 = m\measuredangle 4$, replace $m\measuredangle 4$ with $m\measuredangle 3$.

$$m\measuredangle 1 + m\measuredangle 5 + m\measuredangle 3 = 180°$$

Because $\measuredangle 1$, $\measuredangle 5$, and $\measuredangle 3$ are the three angles of the triangle, this last equation shows that the measures of the triangle's three angles have a 180° sum.

THE ANGLES OF A TRIANGLE

The sum of the measures of the three angles of any triangle is 180°.

Solve problems involving angle relationships in triangles.

EXAMPLE 1 Using Angle Relationships in Triangles

Find the measure of angle A for triangle ABC in **Figure 10.16**.

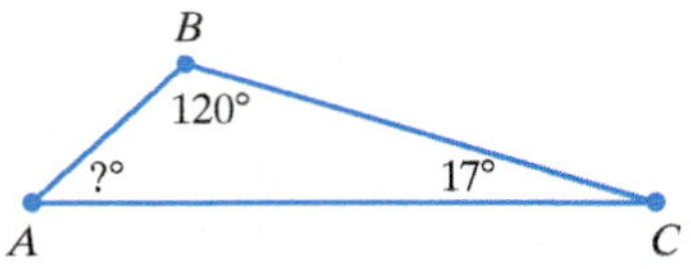

FIGURE 10.16

SOLUTION

Because $m\measuredangle A + m\measuredangle B + m\measuredangle C = 180°$, we obtain

$$m\measuredangle A + 120° + 17° = 180°$$ The sum of the measures of a triangle's three angles is 180°.

$$m\measuredangle A + 137° = 180°$$ Simplify: 120° + 17° = 137°.

$$m\measuredangle A = 180° - 137°$$ Find the measure of A by subtracting 137° from both sides of the equation.

$$m\measuredangle A = 43°$$ Simplify.

CHECK POINT 1 In **Figure 10.16**, suppose that $m\measuredangle B = 116°$ and $m\measuredangle C = 15°$. Find $m\measuredangle A$.

EXAMPLE 2 Using Angle Relationships in Triangles

Find the measures of angles 1 through 5 in **Figure 10.17**.

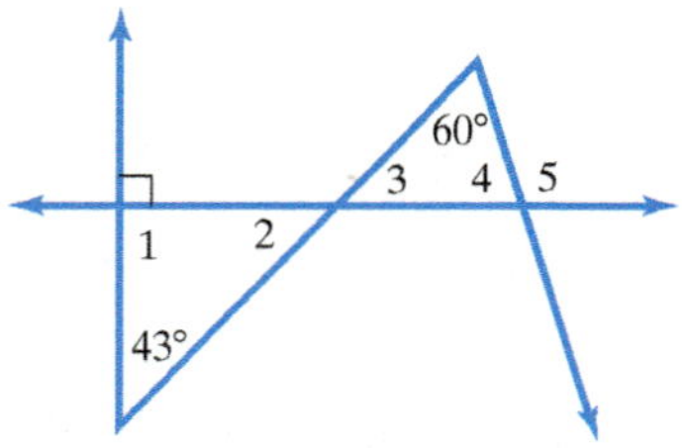

FIGURE 10.17

SOLUTION

Because $\measuredangle 1$ is supplementary to the right angle, $m\measuredangle 1 = 90°$.

$m\measuredangle 2$ can be found using the fact that the sum of the measures of the angles of a triangle is 180°.

$$m\measuredangle 1 + m\measuredangle 2 + 43° = 180°$$ The sum of the measures of a triangle's three angles is 180°.

$$90° + m\measuredangle 2 + 43° = 180°$$ We previously found that $m\measuredangle 1 = 90°$.

$$m\measuredangle 2 + 133° = 180°$$ Simplify: 90° + 43° = 133°.

$$m\measuredangle 2 = 180° - 133°$$ Subtract 133° from both sides.

$$m\measuredangle 2 = 47°$$ Simplify.

$m\measuredangle 3$ can be found using the fact that vertical angles have equal measures: $m\measuredangle 3 = m\measuredangle 2$. Thus, $m\measuredangle 3 = 47°$.

$m\measuredangle 4$ can be found using the fact that the sum of the measures of the angles of a triangle is 180°. Refer to the triangle at the top of **Figure 10.17**.

$$m\measuredangle 3 + m\measuredangle 4 + 60° = 180°$$ The sum of the measures of a triangle's three angles is 180°.

$$47° + m\measuredangle 4 + 60° = 180°$$ We previously found that $m\measuredangle 3 = 47°$.

$$m\measuredangle 4 + 107° = 180°$$ Simplify: 47° + 60° = 107°.

$$m\measuredangle 4 = 180° - 107°$$ Subtract 107° from both sides.

$$m\measuredangle 4 = 73°$$ Simplify.

Finally, we can find $m\angle 5$ by observing that angles 4 and 5 form a straight angle.

$$m\angle 4 + m\angle 5 = 180^\circ \quad \text{A straight angle measures } 180^\circ.$$
$$73^\circ + m\angle 5 = 180^\circ \quad \text{We previously found that } m\angle 4 = 73^\circ.$$
$$m\angle 5 = 180^\circ - 73^\circ \quad \text{Subtract } 73^\circ \text{ from both sides.}$$
$$m\angle 5 = 107^\circ \quad \text{Simplify.}$$

CHECK POINT 2 In **Figure 10.17**, suppose that the angle shown to measure 43° measures, instead, 36°. Further suppose that the angle shown to measure 60° measures, instead, 58°. Under these new conditions, find the measures of angles 1 through 5 in the figure.

Triangles can be described using characteristics of their angles or their sides.

TRIANGLES AND THEIR CHARACTERISTICS

Classification by Angles

Acute Triangle
All angles are acute.

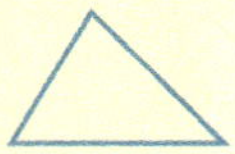

Right Triangle
One angle measures 90°.

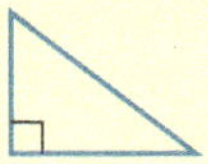

Obtuse Triangle
One angle is obtuse.

Classification by Sides

Isosceles Triangle
Two sides have equal length. (Angles opposite these sides have the same measure.)

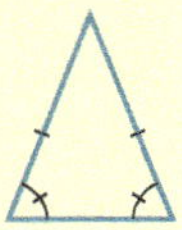

Equilateral Triangle
All sides have equal length. (Each angle measures 60°.)

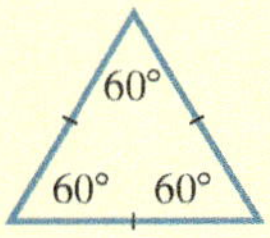

Scalene Triangle
No two sides are equal in length.

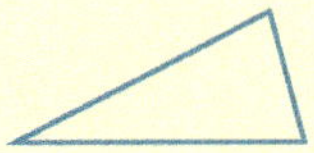

GREAT QUESTION!

Does an isosceles triangle have exactly two equal sides or at least two equal sides?

We checked a variety of math textbooks and found two slightly different definitions for an isosceles triangle. One definition states that an isosceles triangle has *exactly* two sides of equal length. A second definition asserts that an isosceles triangle has *at least* two sides of equal length. (This makes an equilateral triangle, with three sides of the same length, an isosceles triangle.) In this book, we assume that an isosceles triangle has exactly two sides, but not three sides, of the same length.

2 Solve problems involving similar triangles.

Pedestrian crossing

Similar Triangles

Shown in the margin is an international road sign. This sign is shaped just like the actual sign, although its size is smaller. Figures that have the same shape, but not the same size, are used in **scale drawings**. A scale drawing always pictures the exact shape of the object that the drawing represents. Architects, engineers, landscape gardeners, and interior decorators use scale drawings in planning their work.

Figures that have the same shape, but not necessarily the same size, are called **similar figures**. In **Figure 10.18**, triangles ABC and DEF are similar. Angles A and D measure the same number of degrees and are called **corresponding angles**. Angles C and F are corresponding angles, as are angles B and E. Angles with the same number of tick marks in **Figure 10.18** are the corresponding angles.

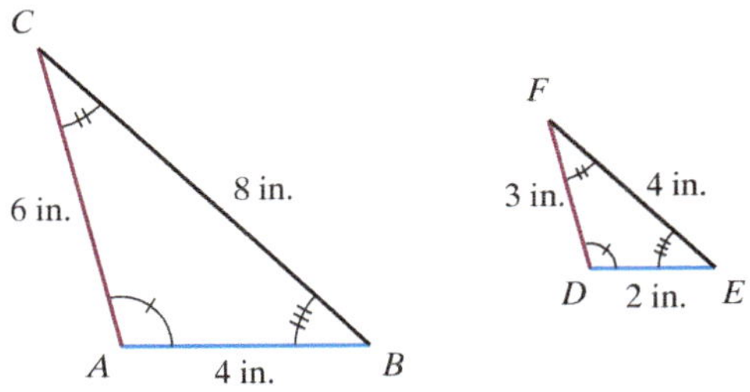

FIGURE 10.18

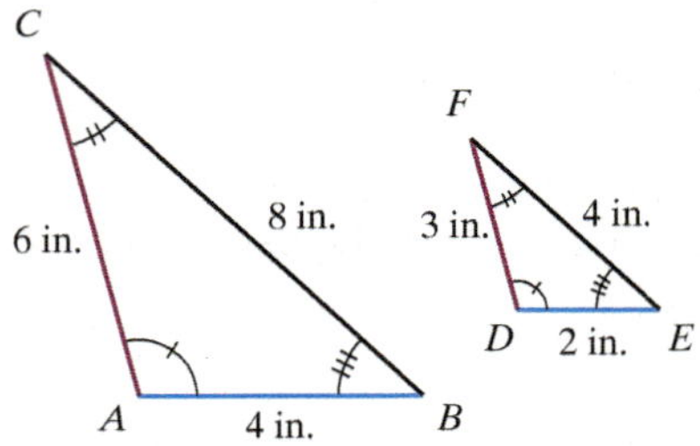

FIGURE 10.18 (repeated)

The sides opposite the corresponding angles are called **corresponding sides.** Thus, $\overline{CB}$ and $\overline{FE}$ are corresponding sides. $\overline{AB}$ and $\overline{DE}$ are also corresponding sides, as are $\overline{AC}$ and $\overline{DF}$. Corresponding angles measure the same number of degrees, but corresponding sides may or may not be the same length. For the triangles in **Figure 10.18**, each side in the smaller triangle is half the length of the corresponding side in the larger triangle.

The triangles in **Figure 10.18** illustrate what it means to be **similar triangles. Corresponding angles have the same measure and the ratios of the lengths of the corresponding sides are equal.**

$$\frac{\text{length of } \overline{AC}}{\text{length of } \overline{DF}} = \frac{6 \text{ in.}}{3 \text{ in.}} = \frac{2}{1}; \quad \frac{\text{length of } \overline{CB}}{\text{length of } \overline{FE}} = \frac{8 \text{ in.}}{4 \text{ in.}} = \frac{2}{1}; \quad \frac{\text{length of } \overline{AB}}{\text{length of } \overline{DE}} = \frac{4 \text{ in.}}{2 \text{ in.}} = \frac{2}{1}$$

In similar triangles, the lengths of the corresponding sides are proportional. Thus,

AC represents the length of $\overline{AC}$, DF the length of $\overline{DF}$, and so on.

$$\frac{AC}{DF} = \frac{CB}{FE} = \frac{AB}{DE}.$$

How can we quickly determine if two triangles are similar? **If the measures of two angles of one triangle are equal to those of two angles of a second triangle, then the two triangles are similar.** If the triangles are similar, then their corresponding sides are proportional.

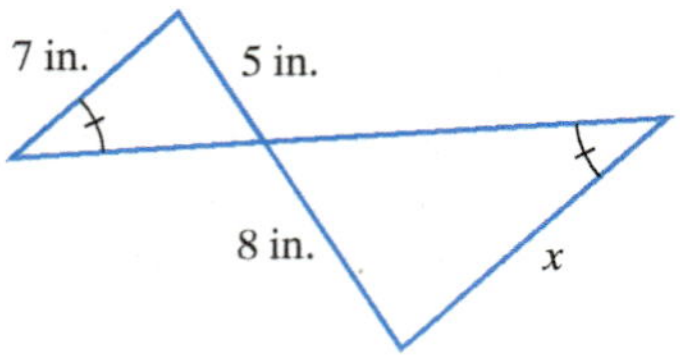

FIGURE 10.19

EXAMPLE 3 *Using Similar Triangles*

Explain why the triangles in **Figure 10.19** are similar. Then find the missing length, x.

SOLUTION

Figure 10.20 shows that two angles of the small triangle are equal in measure to two angles of the large triangle. One angle pair is given to have the same measure. Another angle pair consists of vertical angles with the same measure. Thus, the triangles are similar and their corresponding sides are proportional.

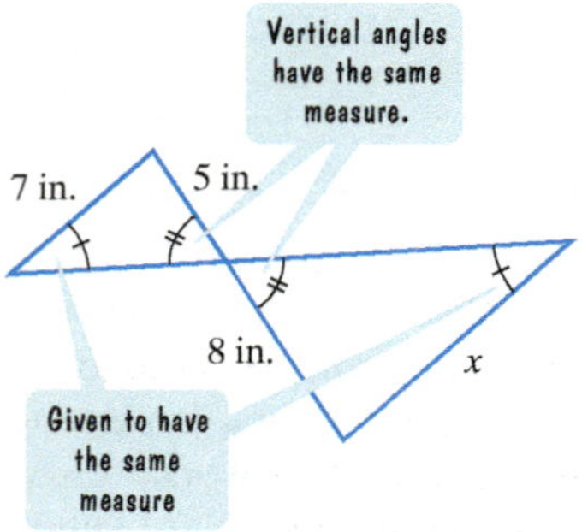

FIGURE 10.20

Opposite the angle with one tick in the small △ / Opposite the angle with one tick in the large △

$$\frac{5}{8} = \frac{7}{x}$$

Opposite the angle with two ticks in the small △ / Opposite the angle with two ticks in the large △

We solve $\frac{5}{8} = \frac{7}{x}$ for x by applying the cross-products principle for proportions that we discussed in Section 6.2: If $\frac{a}{b} = \frac{c}{d}$, then $ad = bc$.

$5x = 8 \cdot 7$ Apply the cross-products principle.

$5x = 56$ Multiply: $8 \cdot 7 = 56$.

$\frac{5x}{5} = \frac{56}{5}$ Divide both sides by 5.

$x = 11.2$ Simplify.

The missing length, x, is 11.2 inches.

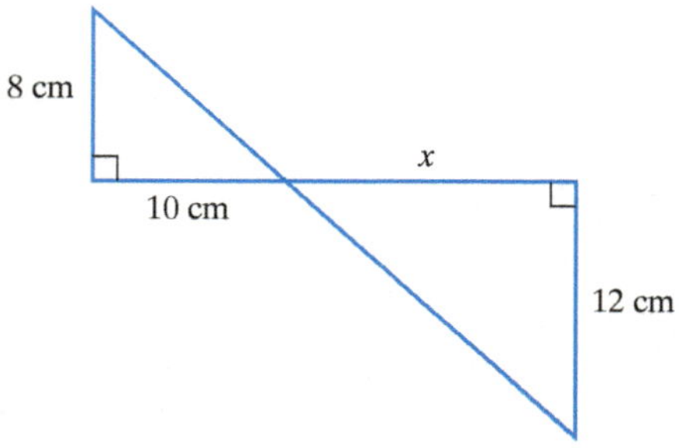

FIGURE 10.21

CHECK POINT 3 Explain why the triangles in **Figure 10.21** are similar. Then find the missing length, x.

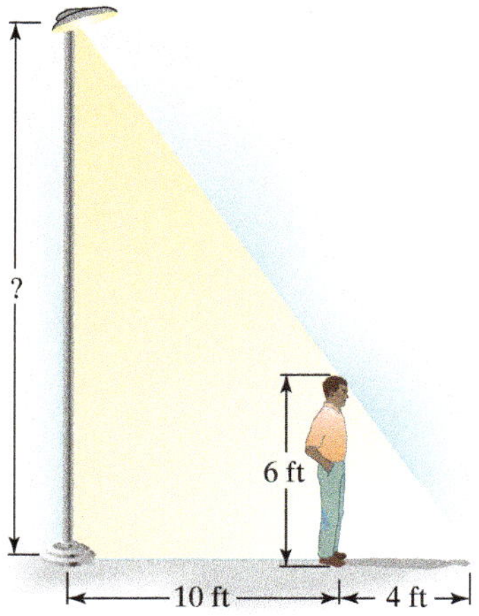

FIGURE 10.22

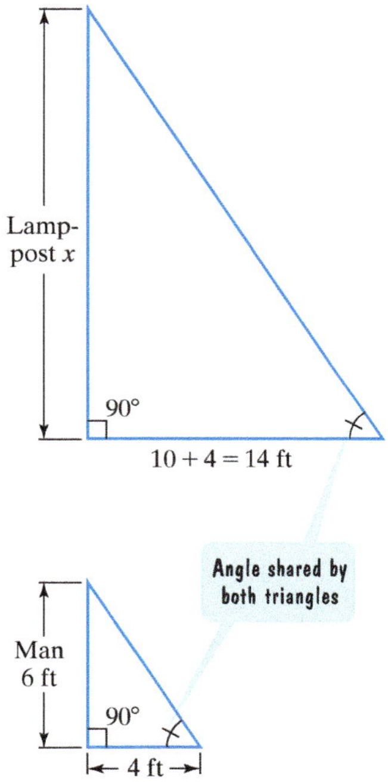

FIGURE 10.23

EXAMPLE 4 *Problem Solving Using Similar Triangles*

A man who is 6 feet tall is standing 10 feet from the base of a lamppost (see **Figure 10.22**). The man's shadow has a length of 4 feet. How tall is the lamppost?

SOLUTION

The drawing in **Figure 10.23** makes the similarity of the triangles easier to see. The large triangle with the lamppost on the left and the small triangle with the man on the left both contain 90° angles. They also share an angle. Thus, two angles of the large triangle are equal in measure to two angles of the small triangle. This means that the triangles are similar and their corresponding sides are proportional. We begin by letting x represent the height of the lamppost, in feet. Because corresponding sides of the two similar triangles are proportional,

$$\frac{x}{6} = \frac{14}{4}.$$

(Numerator x: Opposite the angle with one tick in the large △; denominator 6: Opposite the angle with one tick in the small △; numerator 14: Opposite the unmarked angle in the large △; denominator 4: Opposite the unmarked angle in the small △)

We solve for x by applying the cross-products principle.

$$4x = 6 \cdot 14 \quad \text{Apply the cross-products principle.}$$
$$4x = 84 \quad \text{Multiply: } 6 \cdot 14 = 84.$$
$$\frac{4x}{4} = \frac{84}{4} \quad \text{Divide both sides by 4.}$$
$$x = 21 \quad \text{Simplify.}$$

The lamppost is 21 feet tall.

CHECK POINT 4 Find the height of the lookout tower shown in **Figure 10.24** using the figure that lines up the top of the tower with the top of a stick that is 2 yards long and 3.5 yards from the line to the top of the tower.

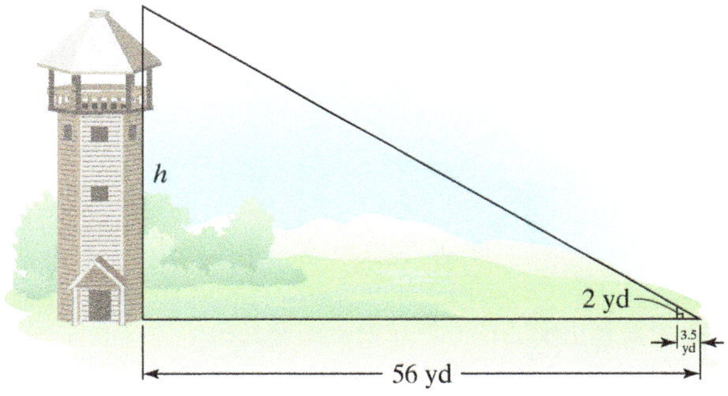

FIGURE 10.24

3 Solve problems using the Pythagorean Theorem.

The Pythagorean Theorem

The ancient Greek philosopher and mathematician Pythagoras (approximately 582–500 B.C.) founded a school whose motto was "All is number." Pythagoras is best remembered for his work with the **right triangle**, a triangle with one angle measuring 90°. The side opposite the 90° angle is called the **hypotenuse**. The other sides are called **legs**. Pythagoras found that if he constructed squares on each of the legs, as well as a larger square on the hypotenuse, the sum of the areas of the smaller squares is equal to the area of the larger square. This is illustrated in **Figure 10.25**.

This relationship is usually stated in terms of the lengths of the three sides of a right triangle and is called the **Pythagorean Theorem**.

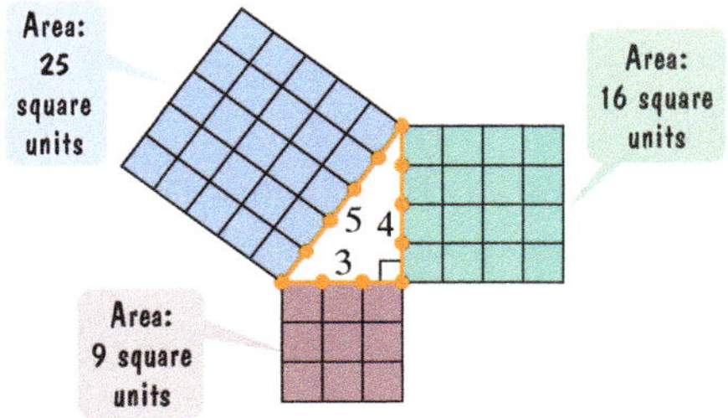

FIGURE 10.25 The area of the large square equals the sum of the areas of the smaller squares.

THE PYTHAGOREAN THEOREM

The sum of the squares of the lengths of the legs of a right triangle equals the square of the length of the hypotenuse.

If the legs have lengths a and b and the hypotenuse has length c, then

$$a^2 + b^2 = c^2.$$

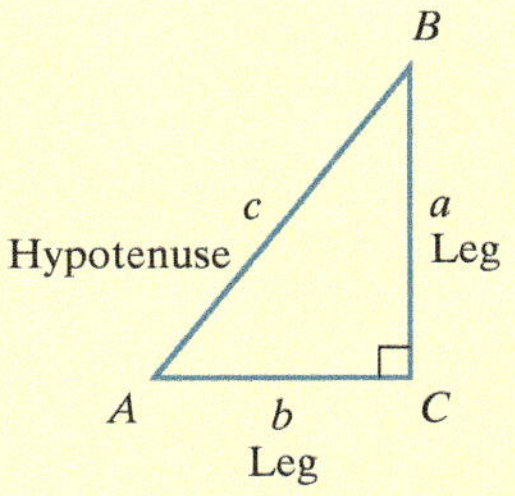

Blitzer Bonus

The Universality of Mathematics

Is mathematics discovered or invented? The "Pythagorean" Theorem, credited to Pythagoras, was known in China, from land surveying, and in Egypt, from pyramid building, centuries before Pythagoras was born. The same mathematics is often discovered/invented by independent researchers separated by time, place, and culture.

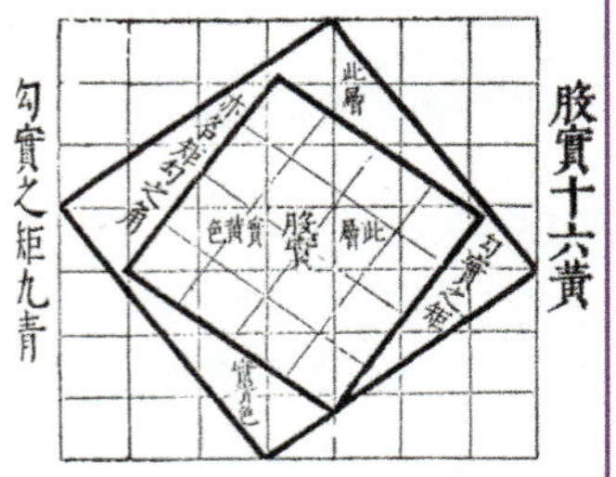

This diagram of the Pythagorean Theorem is from a Chinese manuscript dated as early as 1200 B.C.

EXAMPLE 5 *Using the Pythagorean Theorem*

Find the length of the hypotenuse c in the right triangle shown in **Figure 10.26**.

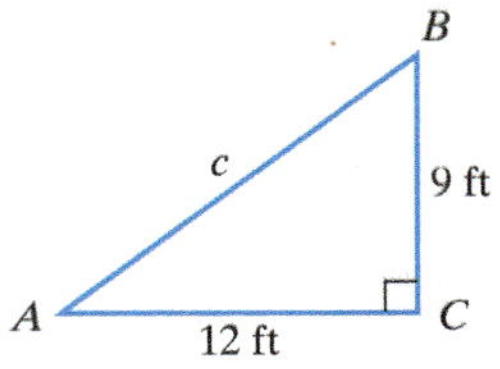

FIGURE 10.26

SOLUTION

Let $a = 9$ and $b = 12$. Substituting these values into $c^2 = a^2 + b^2$ enables us to solve for c.

$c^2 = a^2 + b^2$ — Use the symbolic statement of the Pythagorean Theorem.

$c^2 = 9^2 + 12^2$ — Let $a = 9$ and $b = 12$.

$c^2 = 81 + 144$ — $9^2 = 9 \cdot 9 = 81$ and $12^2 = 12 \cdot 12 = 144$.

$c^2 = 225$ — Add.

$c = \sqrt{225} = 15$ — Solve for c by taking the positive square root of 225.

The length of the hypotenuse is 15 feet.

CHECK POINT 5 Find the length of the hypotenuse in a right triangle whose legs have lengths 7 feet and 24 feet.

EXAMPLE 6 *Using the Pythagorean Theorem*

a. A wheelchair ramp with a length of 122 inches has a horizontal distance of 120 inches. What is the ramp's vertical distance?

b. Construction laws are very specific when it comes to access ramps for the disabled. Every vertical rise of 1 inch requires a horizontal run of 12 inches. Does this ramp satisfy the requirement?

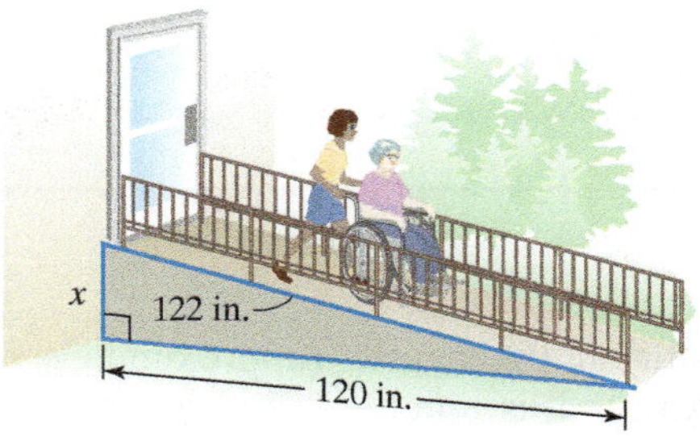

FIGURE 10.27

SOLUTION

a. The problem's conditions state that the wheelchair ramp has a length of 122 inches and a horizontal distance of 120 inches. **Figure 10.27** shows the right triangle that is formed by the ramp, the wall, and the ground. We can find x, the ramp's vertical distance, using the Pythagorean Theorem.

(leg)^2 plus (leg)^2 equals (hypotenuse)^2

$$x^2 + 120^2 = 122^2$$

$x^2 + 120^2 = 122^2$ — This is the equation resulting from the Pythagorean Theorem.

$x^2 + 14{,}400 = 14{,}884$ — Square 120 and 122.

$x^2 = 484$ — Isolate x^2 by subtracting 14,400 from both sides.

$x = \sqrt{484} = 22$ — Solve for x by taking the positive square root of 484.

The ramp's vertical distance is 22 inches.

b. Every vertical rise of 1 inch requires a horizontal run of 12 inches. Because the ramp has a vertical distance of 22 inches, it requires a horizontal distance of 22(12) inches, or 264 inches. The horizontal distance is only 120 inches, so this ramp does not satisfy construction laws for access ramps for the disabled.

CHECK POINT 6 A radio tower is supported by two wires that are each 130 yards long and attached to the ground 50 yards from the base of the tower. How far from the ground are the wires attached to the tower?

Blitzer Bonus

Curiosities Associated with the Pythagorean Theorem

- *Tarzan raised Desi Arnaz' rat.*
  O, had I nine more hero-men in Idaho!
  These sentences read the same from left to right and from right to left. We call such sentences **palindromes**. Here are some examples of lengths of sides of right triangles for which the lengths of two legs are palindromic numbers:

a	b	c
363	484	605
313	48,984	48,985
29,192	60,006	66,730
34,104,840,143	42,002,820,024	54,105,240,145

In each case, $a^2 + b^2 = c^2$, and a and b read the same in either direction.

- The lengths of the sides of a right triangle can display unusual symmetric properties.

Palindromic number (a) — Reverses of one another (b and c)

a	b	c
33	56	65
3333	5656	6565

Reverses of one another (a and b)

a	b	c
88,209	90,288	126,225

In each case, $a^2 + b^2 = c^2$.

- Some right triangles have unusual areas.

a	b	c	Area (square units)
443,531	6660	443,581	1,476,958,230 Each of the ten digits, 0, 1, 2, 3, …, 9 appears once.
693	1924	2045	666,666 See the Blitzer Bonus (Integers, Karma, Exponents) on page 266.

- There is a surprising relationship between the irrational number π and the Pythagorean Theorem.

$$\pi \approx 3.14159265$$

Let $a = 159$. Let $c = 265$.

Introducing the palindromic number $b = 212$, the resulting three numbers satisfy the Pythagorean Theorem.

a	b	c	$a^2 + b^2 = c^2$
159	212	265	$159^2 + 212^2 = 265^2$

Furthermore,

$$\frac{666}{212} \approx 3.1415 \approx \pi$$

See the Blitzer Bonus on page 266.

This is the number we introduced above to form the sides of a right triangle.

Source: Alfred Posamentier, *The Pythagorean Theorem—The Story of Its Power and Beauty,* Prometheus Books, 2010.

Concept and Vocabulary Check

Fill in each blank so that the resulting statement is true.

1. The sum of the measures of the three angles of any triangle is __________.
2. A triangle in which each angle measures less than 90° is called a/an __________ triangle.
3. A triangle in which one angle measures more than 90° is called a/an __________ triangle.
4. A triangle with exactly two sides of the same length is called a/an __________ triangle.
5. A triangle whose sides are all the same length is called a/an __________ triangle.
6. A triangle that has no sides of the same length is called a/an __________ triangle.
7. Triangles that have the same shape, but not necessarily the same size, are called __________ triangles. For such triangles, corresponding angles have __________ and the lengths of the corresponding sides are __________.
8. The Pythagorean Theorem states that in any __________ triangle, the sum of the squares of the lengths of the __________ equals __________.

In Exercises 9–13, determine whether each statement is true or false. If the statement is false, make the necessary change(s) to produce a true statement.

9. Euclid's assumption about parallel lines states that given a line and a point not on the line, one and only one line can be drawn through the given point parallel to the given line. ______
10. A triangle cannot have both a right angle and an obtuse angle. ______
11. Each angle of an isosceles triangle measures 60°. ______
12. If the measures of two angles of one triangle are equal to those of two angles of a second triangle, then the two triangles have same size and shape. ______
13. In any triangle, the sum of the squares of the lengths of the two shorter sides equals the square of the length of the longest side. ______

Exercise Set 10.2

Practice Exercises

In Exercises 1–4, find the measure of angle A for the triangle shown.

1.

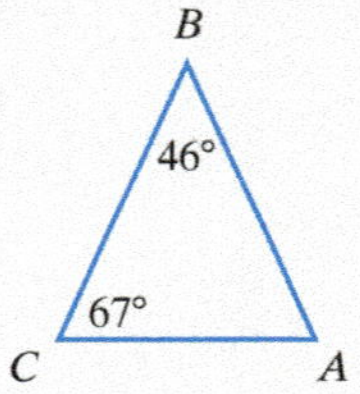

2.

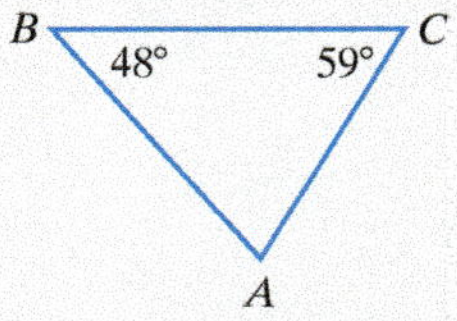

3.

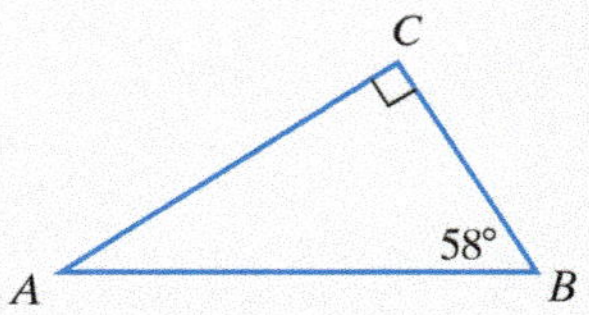

4.

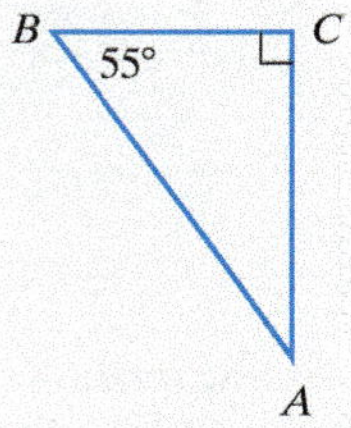

In Exercises 5–6, find the measures of angles 1 through 5 in the figure shown.

5.

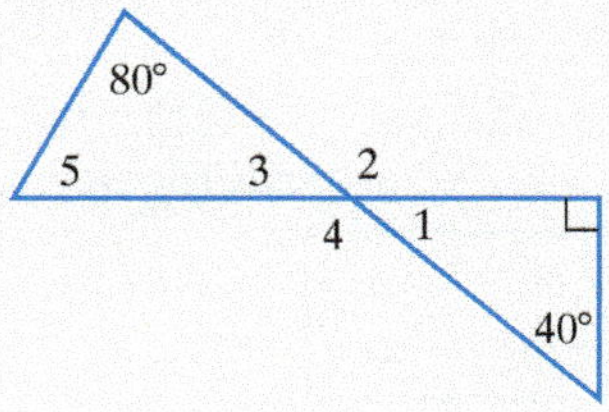

6.

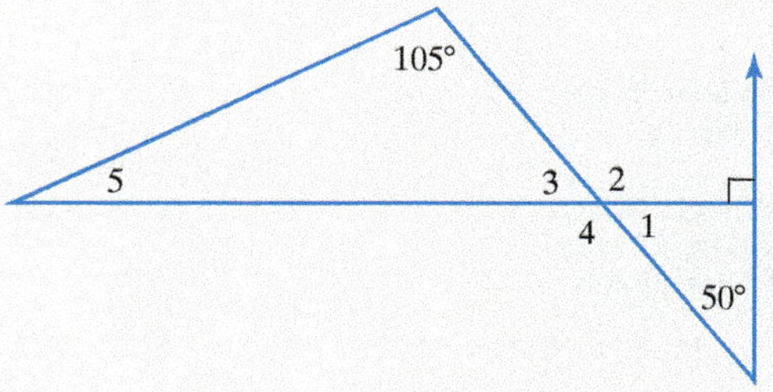

We have seen that isosceles triangles have two sides of equal length. The angles opposite these sides have the same measure. In Exercises 7–8, use this information to help find the measure of each numbered angle.

7.

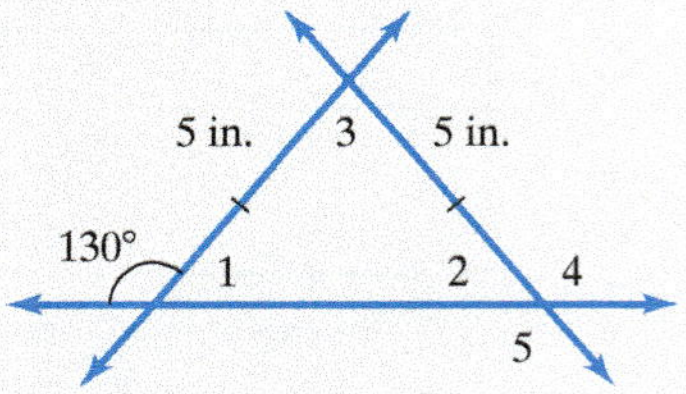

8.

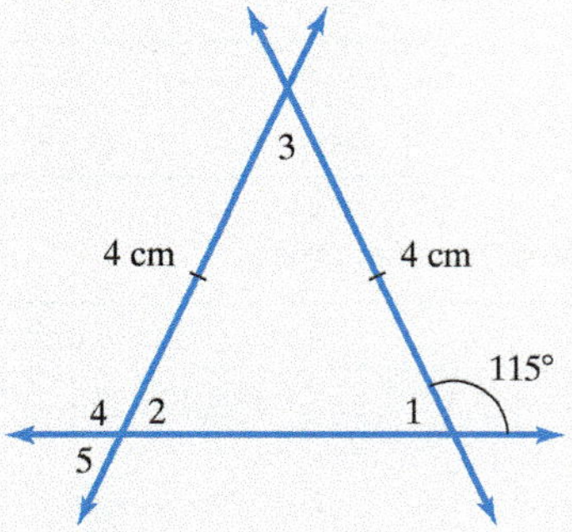

In Exercises 9–10, lines l and m are parallel. Find the measure of each numbered angle.

9.

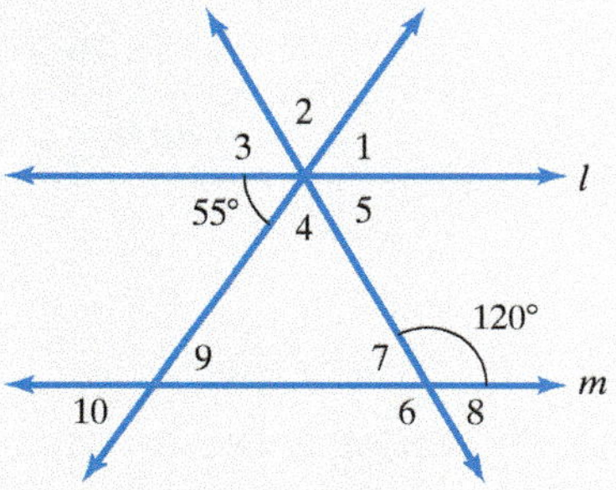

10.

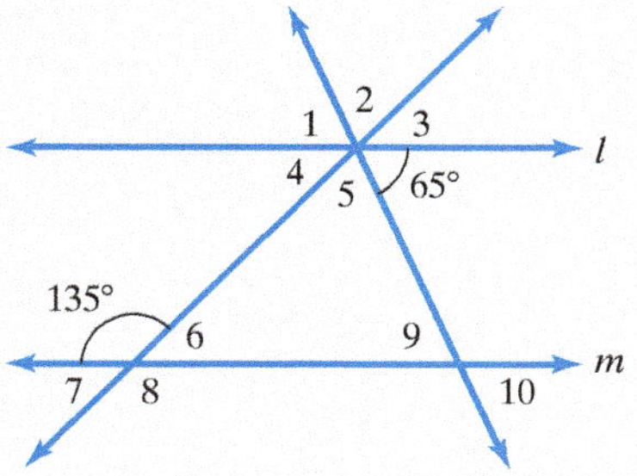

In Exercises 11–16, explain why the triangles are similar. Then find the missing length, x.

11.

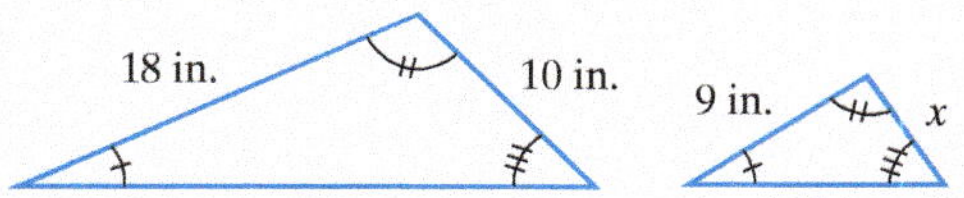

12.

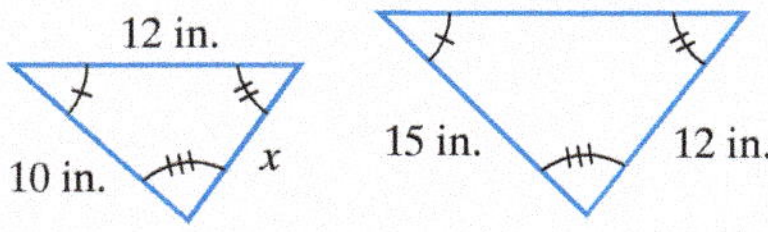

13.

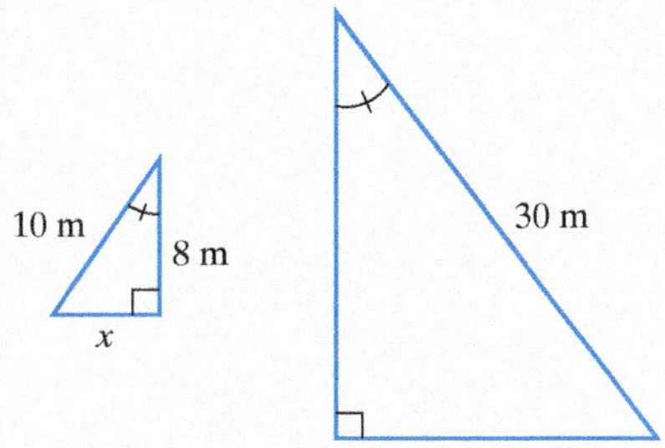

14.

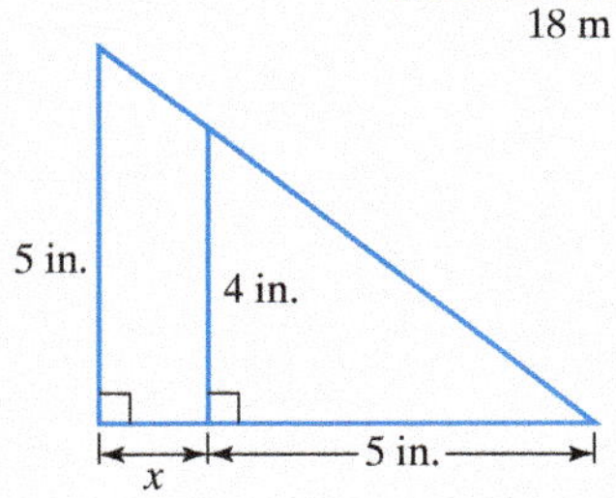

15.

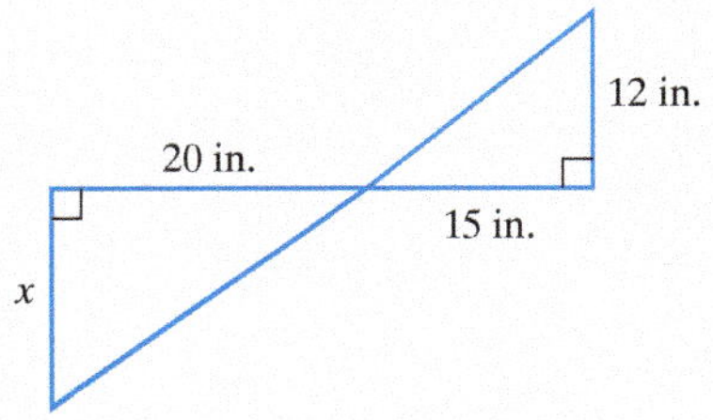

16.

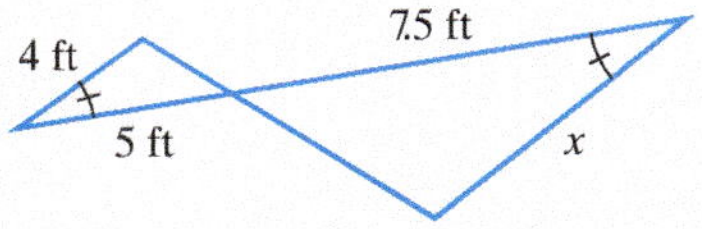

In Exercises 17–19, ΔABC and ΔADE are similar. Find the length of the indicated side.

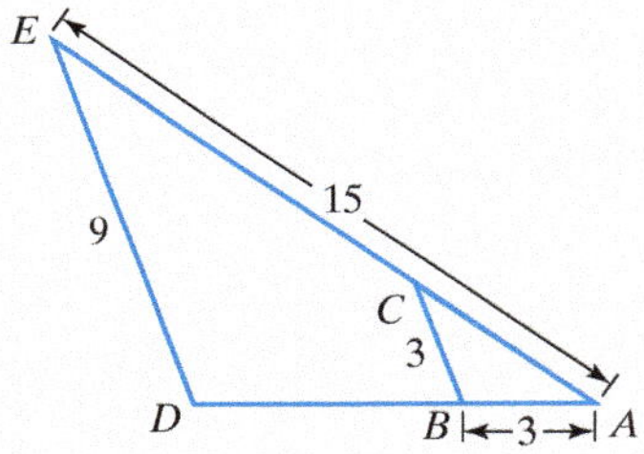

17. $\overline{CA}$ **18.** $\overline{DB}$ **19.** $\overline{DA}$

20. In the diagram for Exercises 17–19, suppose that you are not told that ΔABC and ΔADE are similar. Instead, you are given that $\overleftrightarrow{ED}$ and $\overleftrightarrow{CB}$ are parallel. Under these conditions, explain why the triangles must be similar.

In Exercises 21–26, use the Pythagorean Theorem to find the missing length in each right triangle. Use your calculator to find square roots, rounding, if necessary, to the nearest tenth.

21.

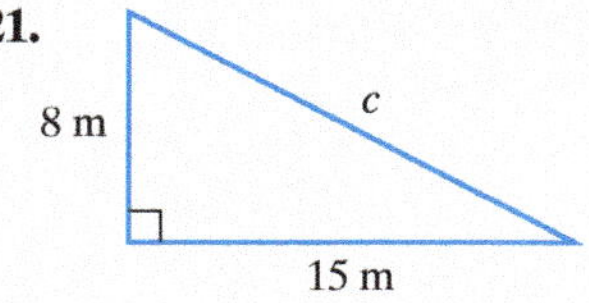

22.

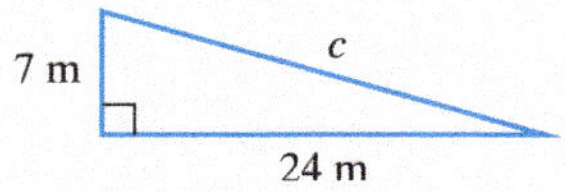

23.

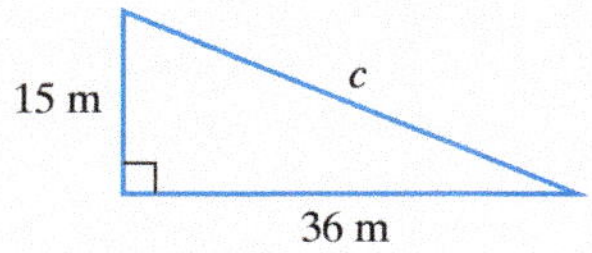

24.

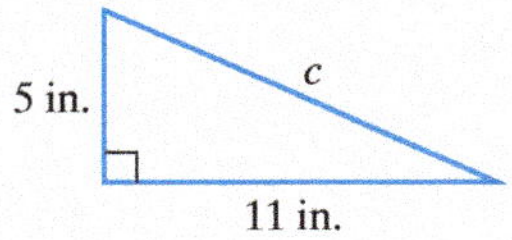

25.

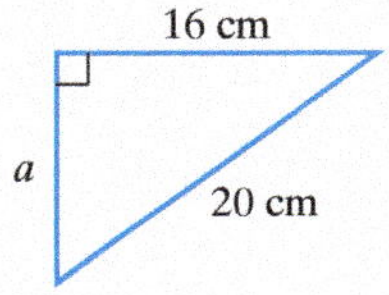

26.

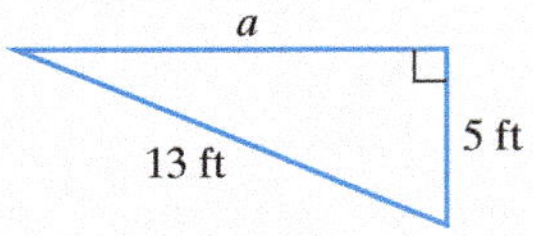

Practice Plus

*Two triangles are **congruent** if they have the same shape and the same size. In congruent triangles, the measures of corresponding angles are equal and the corresponding sides have the same length. The following triangles are congruent:*

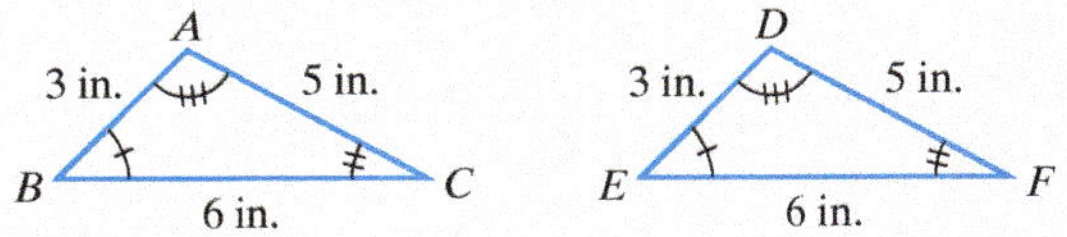

Any one of the following may be used to determine if two triangles are congruent.

Determining Congruent Triangles

1. Side-Side-Side (SSS)

If the lengths of three sides of one triangle equal the lengths of the corresponding sides of a second triangle, then the two triangles are congruent.

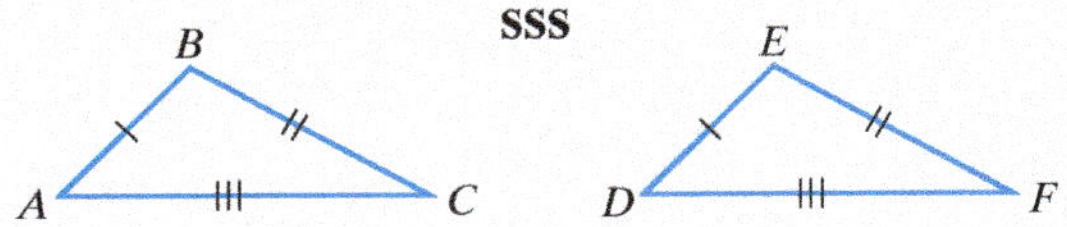

2. Side-Angle-Side (SAS)
If the lengths of two sides of one triangle equal the lengths of the corresponding sides of a second triangle and the measures of the angles between each pair of sides are equal, then the two triangles are congruent.

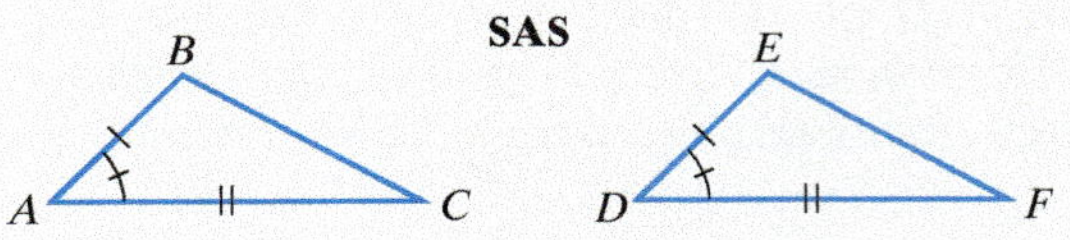

3. Angle-Side-Angle (ASA)
If the measures of two angles of one triangle equal the measures of two angles of a second triangle and the lengths of the sides between each pair of angles are equal, then the two triangles are congruent.

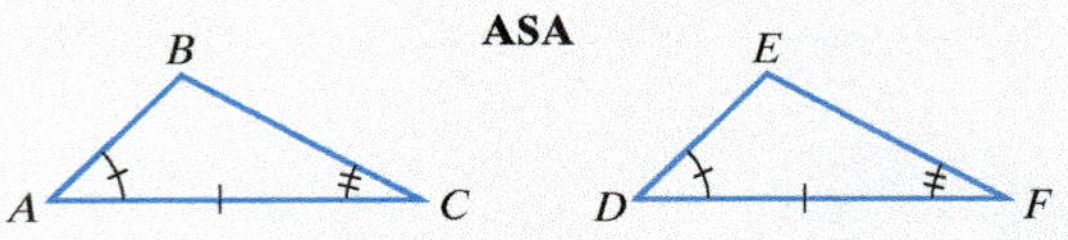

In Exercises 27–36, determine whether ΔI and ΔII are congruent. If the triangles are congruent, state the reason why, selecting from SSS, SAS, or ASA. (More than one reason may be possible.)

27.

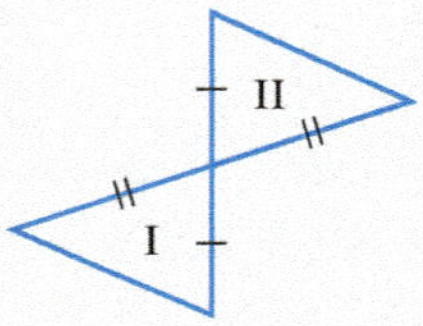

28.

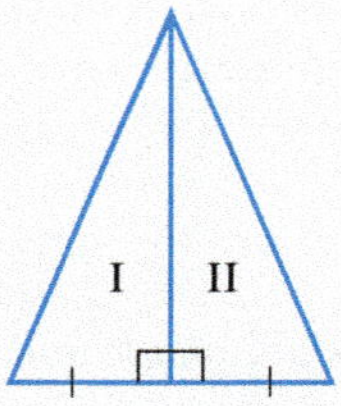

29.

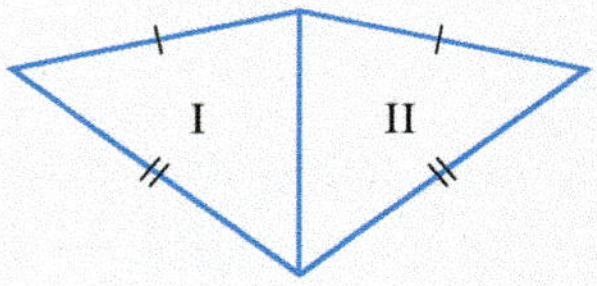

30.

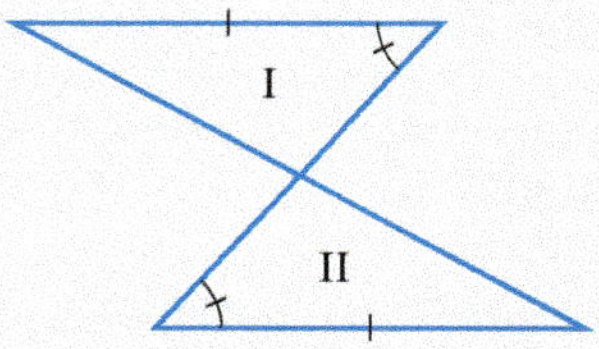

31.

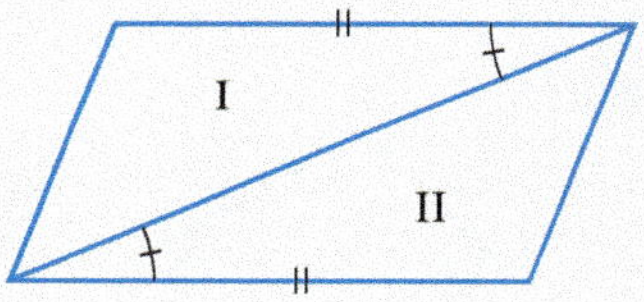

32.

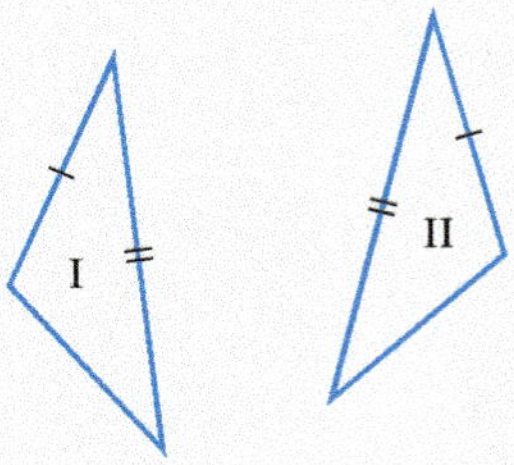

33.

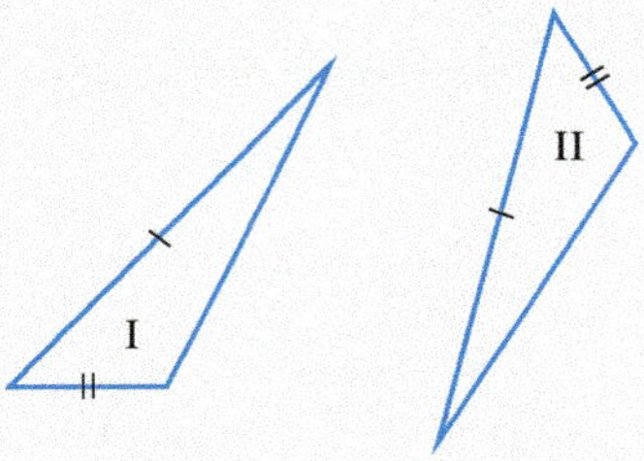

34.

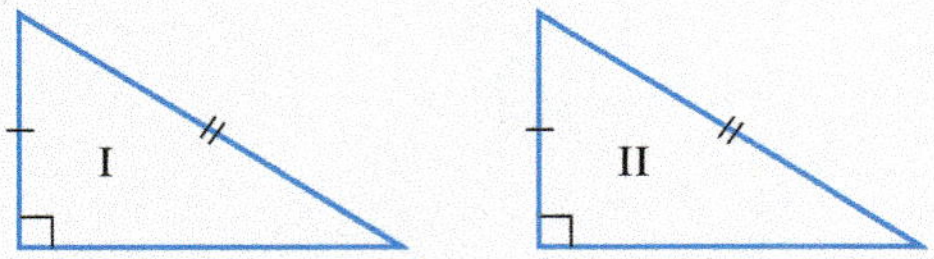

35.

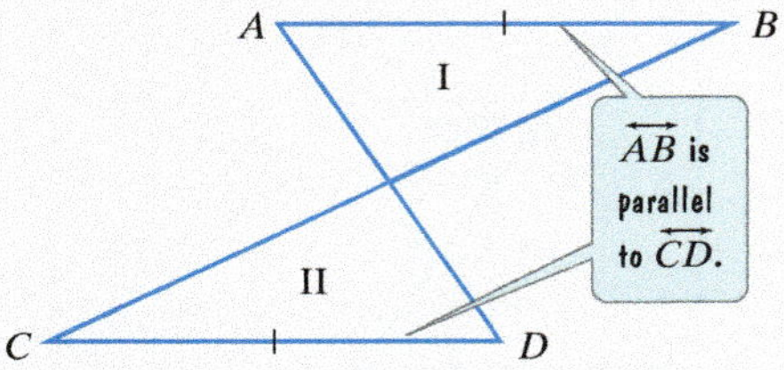

36.

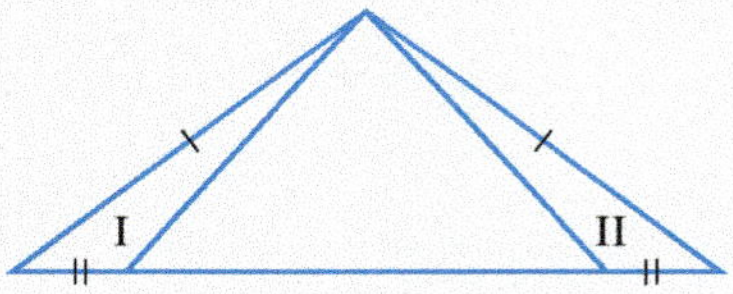

Application Exercises

Use similar triangles to solve Exercises 37–38.

37. A person who is 5 feet tall is standing 80 feet from the base of a tree and the tree casts an 86-foot shadow. The person's shadow is 6 feet in length. What is the tree's height?

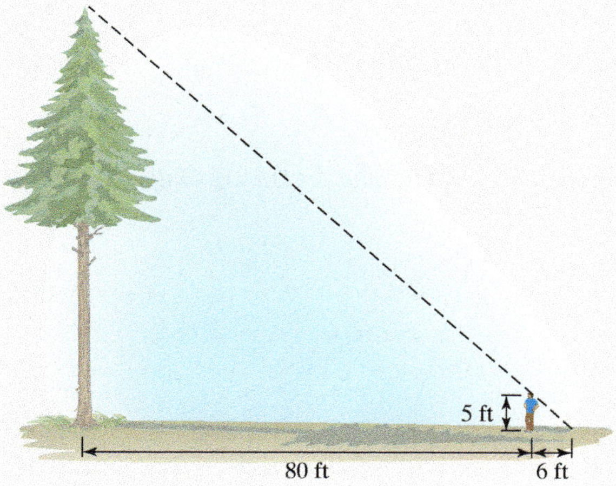

38. A tree casts a shadow 12 feet long. At the same time, a vertical rod 8 feet high casts a shadow that is 6 feet long. How tall is the tree?

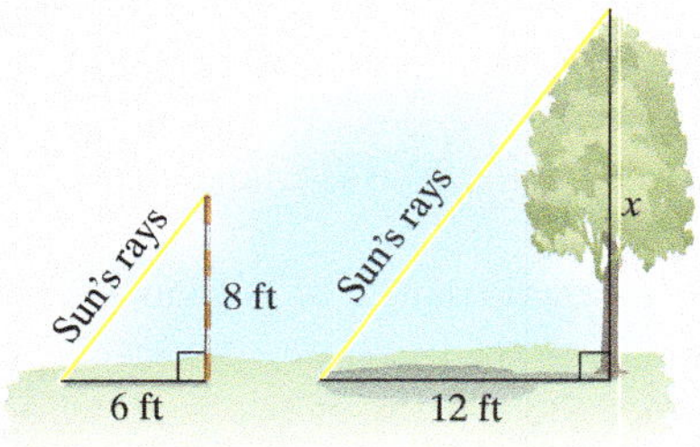

Use the Pythagorean Theorem to solve Exercises 39–46. Use your calculator to find square roots, rounding, if necessary, to the nearest tenth.

39. A baseball diamond is actually a square with 90-foot sides. What is the distance from home plate to second base?

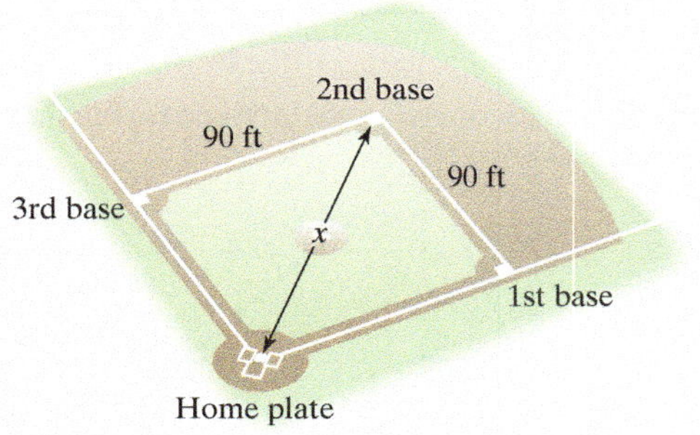

40. The base of a 20-foot ladder is 15 feet from the house. How far up the house does the ladder reach?

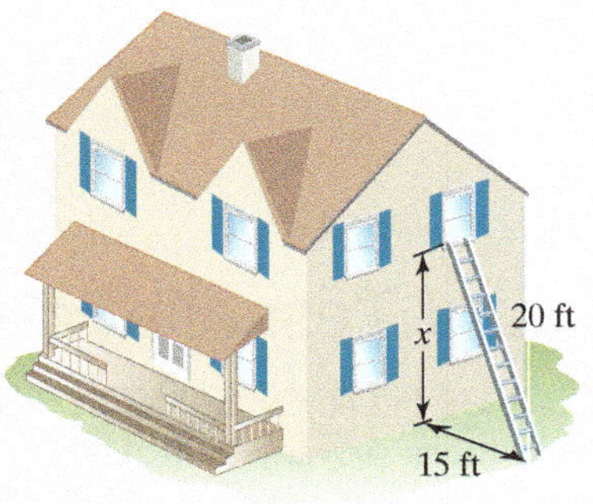

41. A flagpole has a height of 16 yards. It will be supported by three cables, each of which is attached to the flagpole at a point 4 yards below the top of the pole and attached to the ground at a point that is 9 yards from the base of the pole. Find the total number of yards of cable that will be required.

42. A flagpole has a height of 10 yards. It will be supported by three cables, each of which is attached to the flagpole at a point 4 yards below the top of the pole and attached to the ground at a point that is 8 yards from the base of the pole. Find the total number of yards of cable that will be required.

43. A rectangular garden bed measures 5 feet by 12 feet. A water faucet is located at one corner of the garden bed. A hose will be connected to the water faucet. The hose must be long enough to reach the opposite corner of the garden bed when stretched straight. Find the required length of hose.

44. A rocket ascends vertically after being launched from a location that is midway between two ground-based tracking stations. When the rocket reaches an altitude of 4 kilometers, it is 5 kilometers from each of the tracking stations. Assuming that this is a locale where the terrain is flat, how far apart are the two tracking stations?

45. If construction costs are $150,000 per *kilometer,* find the cost of building the new road in the figure shown.

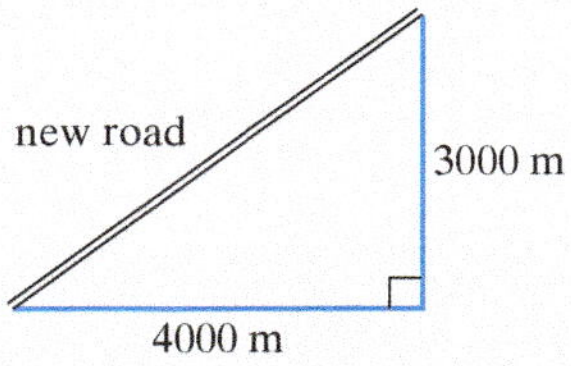

46. **Picky, Picky, Picky** This problem appeared on a high school exit exam:

Alex is building a ramp for a bike competition. He has two rectangular boards. One board is six meters and the other is five meters long. If the ramp has to form a right triangle, what should its height be?

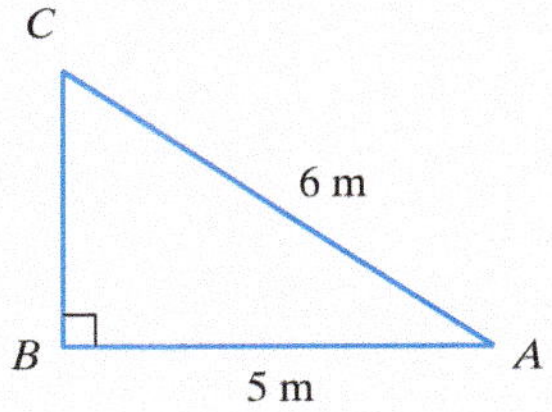

Students were asked to select the correct answer from the following options:

3 meters; 4 meters; 3.3 meters; 7.8 meters.

a. Among the available choices, which option best expresses the ramp's height? How many feet, to the nearest tenth of a foot, is this? Does a bike competition that requires riders to jump off these heights seem realistic? (ouch!)

b. Express the ramp's height to the nearest hundredth of a meter. By how many centimeters does this differ from the "correct" answer on the test? How many inches, to the nearest half inch, is this? Is it likely that a carpenter with a tape measure would make this error?

c. According to the problem, Alex has boards that measure 5 meters and 6 meters. A 6-meter board? How many feet, to the nearest tenth of a foot, is this? When was the last time you found a board of this length at Home Depot? (*Source: The New York Times,* April 24, 2005)

Writing in Mathematics

47. If the measures of two angles of a triangle are known, explain how to find the measure of the third angle.

48. Can a triangle contain two right angles? Explain your answer.

49. What general assumption did Euclid make about a point and a line in order to prove that the sum of the measures of the angles of a triangle is 180°?

50. What are similar triangles?

51. If the ratio of the corresponding sides of two similar triangles is 1 to 1 $\left(\frac{1}{1}\right)$, what must be true about the triangles?

52. What are corresponding angles in similar triangles?

53. Describe how to identify the corresponding sides in similar triangles.

54. In your own words, state the Pythagorean Theorem.

55. In the 1939 movie *The Wizard of Oz,* upon being presented with a Th.D. (Doctor of Thinkology), the Scarecrow proudly exclaims, "The sum of the square roots of any two sides of an isosceles triangle is equal to the square root of the remaining side." Did the Scarecrow get the Pythagorean Theorem right? In particular, describe four errors in the Scarecrow's statement.

Critical Thinking Exercises

Make Sense? *In Exercises 56–59, determine whether each statement makes sense or does not make sense, and explain your reasoning.*

56. I'm fencing off a triangular plot of land that has two right angles.

57. Triangle I is equilateral, as is triangle II, so the triangles are similar.

58. Triangle I is a right triangle, as is triangle II, so the triangles are similar.

59. If I am given the lengths of two sides of a triangle, I can use the Pythagorean Theorem to find the length of the third side.

60. Find the measure of angle R.

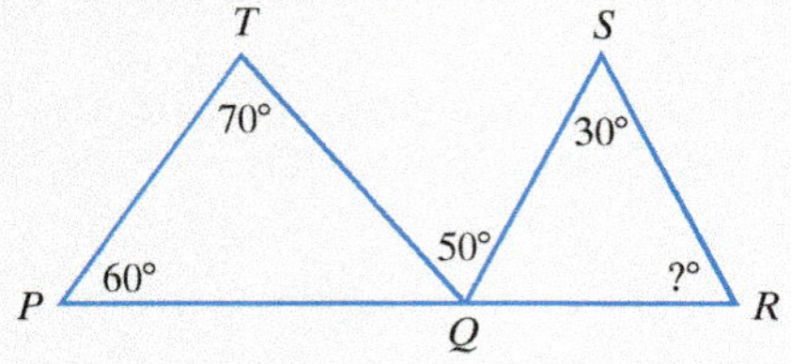

61. What is the length of $\overline{AB}$ in the accompanying figure?

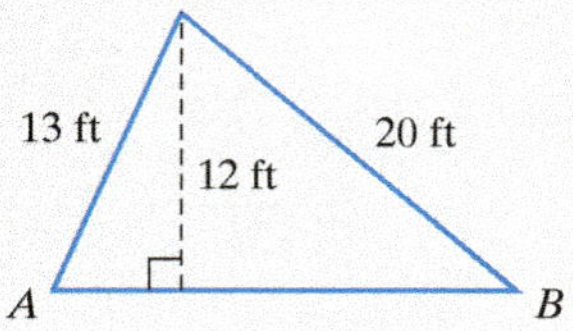

62. Use a quadratic equation to solve this problem. The width of a rectangular carpet is 7 meters shorter than the length, and the diagonal is 1 meter longer than the length. What are the carpet's dimensions?

10.3

WHAT AM I SUPPOSED TO LEARN?

After you have read this section, you should be able to:

1 Name certain polygons according to the number of sides.

2 Recognize the characteristics of certain quadrilaterals.

3 Solve problems involving a polygon's perimeter.

4 Find the sum of the measures of a polygon's angles.

5 Understand tessellations and their angle requirements.

Polygons, Perimeter, and Tessellations

YOU HAVE JUST PURCHASED A BEAUTIFUL PLOT OF LAND IN THE COUNTRY, shown in **Figure 10.28**. In order to have more privacy, you decide to put fencing along each of its four sides. The cost of this project depends on the distance around the four outside edges of the plot, called its **perimeter**, as well as the cost for each foot of fencing.

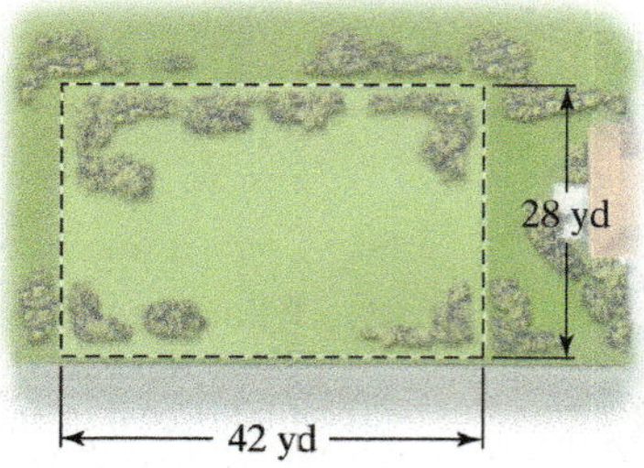

FIGURE 10.28

Your plot of land is a geometric figure: It has four straight sides that are line segments. The plot is on level ground, so that the four sides all lie on a flat surface, or plane. The plot is an example of a *polygon*. Any closed shape in the plane formed by three or more line segments that intersect only at their endpoints is a **polygon**.

1 Name certain polygons according to the number of sides.

A polygon is named according to the number of sides it has. We know that a three-sided polygon is called a **triangle**. A four-sided polygon is called a **quadrilateral**.

A polygon whose sides are all the same length and whose angles all have the same measure is called a **regular polygon**. **Table 10.2** provides the names of six polygons. Also shown are illustrations of regular polygons.

Table 10.2 Illustrations of Regular Polygons

Name	Picture	Name	Picture
Triangle 3 sides	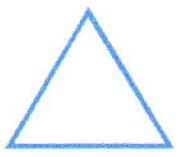	Hexagon 6 sides	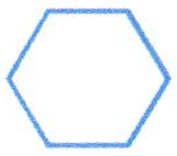
Quadrilateral 4 sides		Heptagon 7 sides	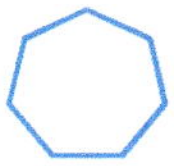
Pentagon 5 sides	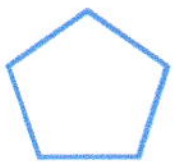	Octagon 8 sides	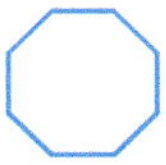

2 Recognize the characteristics of certain quadrilaterals.

Blitzer Bonus

Bees and Regular Hexagons

Bees use honeycombs to store honey and house larvae. They construct honey storage cells from wax. Each cell has the shape of a regular hexagon. The cells fit together perfectly, preventing dirt or predators from entering. Squares or equilateral triangles would fit equally well, but regular hexagons provide the largest storage space for the amount of wax used.

Quadrilaterals

The plot of land in **Figure 10.28** is a four-sided polygon, or a quadrilateral. However, when you first looked at the figure, perhaps you thought of the plot as a rectangular field. A **rectangle** is a special kind of quadrilateral in which both pairs of opposite sides are parallel, have the same measure, and whose angles are right angles. **Table 10.3** presents some special quadrilaterals and their characteristics.

TABLE 10.3 Types of Quadrilaterals

Name	Characteristics	Representation
Parallelogram	Quadrilateral in which both pairs of opposite sides are parallel and have the same measure. Opposite angles have the same measure.	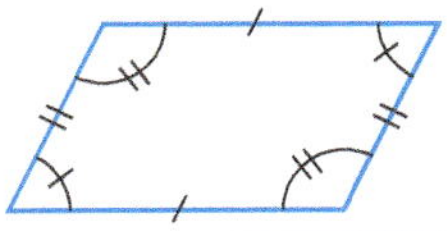
Rhombus	Parallelogram with all sides having equal length.	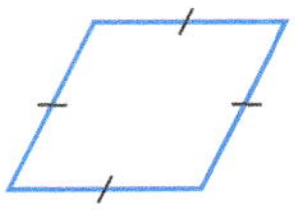
Rectangle	Parallelogram with four right angles. Because a rectangle is a parallelogram, opposite sides are parallel and have the same measure.	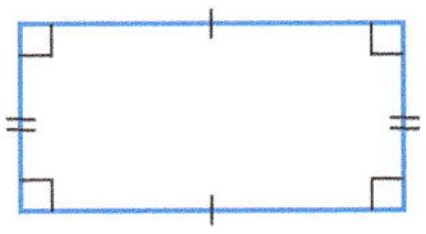
Square	A rectangle with all sides having equal length. Each angle measures 90°, and the square is a regular quadrilateral.	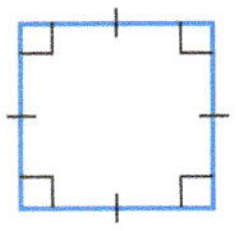
Trapezoid	A quadrilateral with exactly one pair of parallel sides.	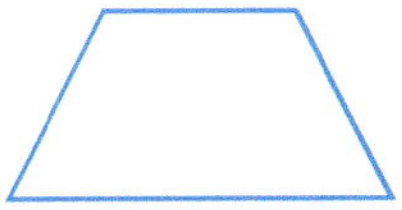

3 Solve problems involving a polygon's perimeter.

Perimeter

The **perimeter**, P, of a polygon is the sum of the lengths of its sides. Perimeter is measured in linear units, such as inches, feet, yards, meters, or kilometers.

Example 1 involves the perimeter of a rectangle. Because perimeter is the sum of the lengths of the sides, the perimeter of the rectangle shown in **Figure 10.29** is $l + w + l + w$. This can be expressed as

$$P = 2l + 2w.$$

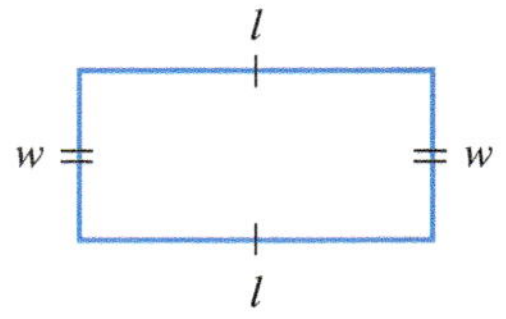

FIGURE 10.29 A rectangle with length l and width w

EXAMPLE 1 *An Application of Perimeter*

The rectangular field we discussed at the beginning of this section (see **Figure 10.28**) has a length of 42 yards and a width of 28 yards. If fencing costs \$5.25 per foot, find the cost to enclose the field with fencing.

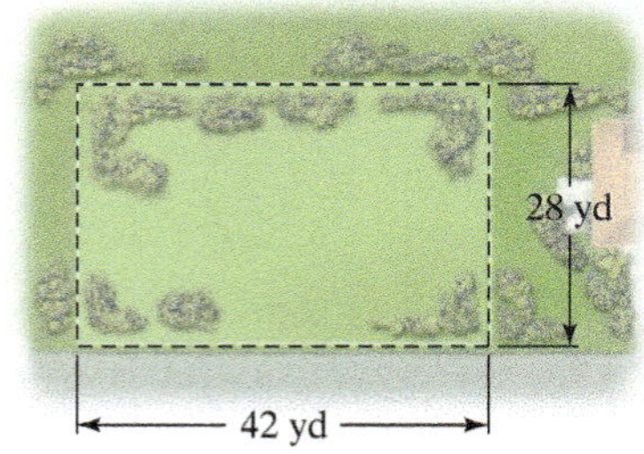

FIGURE 10.28 (repeated)

SOLUTION

We begin by finding the perimeter of the rectangle in yards. Using 3 ft = 1 yd and dimensional analysis, we express the perimeter in feet. Finally, we multiply the perimeter, in feet, by \$5.25 because the fencing costs \$5.25 per foot.

The length, l, is 42 yards and the width, w, is 28 yards. The perimeter of the rectangle is determined using the formula $P = 2l + 2w$.

$$P = 2l + 2w = 2 \cdot 42 \text{ yd} + 2 \cdot 28 \text{ yd} = 84 \text{ yd} + 56 \text{ yd} = 140 \text{ yd}$$

Because 3 ft = 1 yd, we use the unit fraction $\frac{3 \text{ ft}}{1 \text{ yd}}$ to convert from yards to feet.

$$140 \text{ yd} = \frac{140 \text{ yd}}{1} \cdot \frac{3 \text{ ft}}{1 \text{ yd}} = 140 \cdot 3 \text{ ft} = 420 \text{ ft}$$

The perimeter of the rectangle is 420 feet. Now we are ready to find the cost of the fencing. We multiply 420 feet by \$5.25, the cost per foot.

$$\text{Cost} = \frac{420 \text{ feet}}{1} \cdot \frac{\$5.25}{\text{foot}} = 420(\$5.25) = \$2205$$

The cost to enclose the field with fencing is \$2205.

CHECK POINT 1 A rectangular field has a length of 50 yards and a width of 30 yards. If fencing costs \$6.50 per foot, find the cost to enclose the field with fencing.

4 Find the sum of the measures of a polygon's angles.

The Sum of the Measures of a Polygon's Angles

We know that the sum of the measures of the three angles of any triangle is 180°. We can use inductive reasoning to find the sum of the measures of the angles of any polygon. Start by drawing line segments from a single point where two sides meet so that nonoverlapping triangles are formed. This is done in **Figure 10.30**.

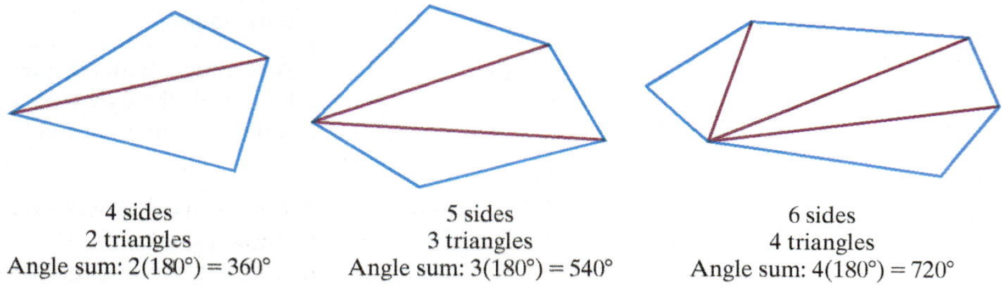

FIGURE 10.30

In each case, the number of triangles is two less than the number of sides of the polygon. Thus, for an n-sided polygon, there are $n - 2$ triangles. Because each triangle has an angle-measure sum of 180°, the sum of the measures for the angles in the $n - 2$ triangles is $(n - 2)180°$. Thus, the sum of the measures of the angles of an n-sided polygon is $(n - 2)180°$.

THE ANGLES OF A POLYGON

The sum of the measures of the angles of a polygon with n sides is

$$(n - 2)180°.$$

EXAMPLE 2 *Using the Formula for the Angles of a Polygon*

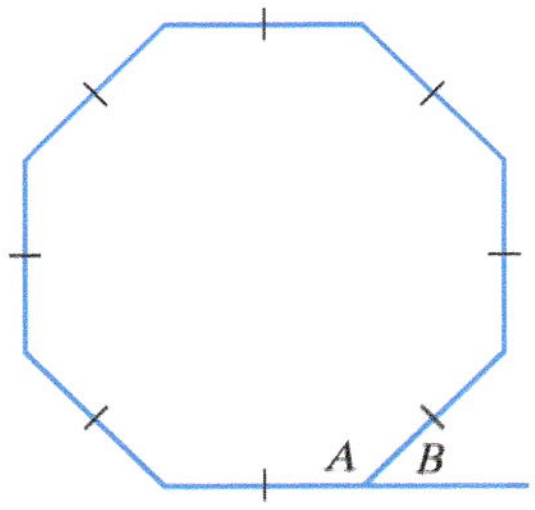

FIGURE 10.31 A regular octagon

a. Find the sum of the measures of the angles of an octagon.

b. **Figure 10.31** shows a regular octagon. Find the measure of angle A.

c. Find the measure of exterior angle B.

SOLUTION

a. An octagon has eight sides. Using the formula $(n - 2)180°$ with $n = 8$, we can find the sum of the measures of its eight angles.
The sum of the measures of an octagon's angles is

$$\begin{aligned} &(n - 2)180° \\ &= (8 - 2)180° \\ &= 6 \cdot 180° \\ &= 1080°. \end{aligned}$$

b. Examine the regular octagon in **Figure 10.31**. Note that all eight sides have the same length. Likewise, all eight angles have the same degree measure. Angle A is one of these eight angles. We find its measure by taking the sum of the measures of all eight angles, 1080°, and dividing by 8.

$$m\measuredangle A = \frac{1080°}{8} = 135°$$

c. Because $\measuredangle B$ is the supplement of $\measuredangle A$,

$$m\measuredangle B = 180° - 135° = 45°.$$

CHECK POINT 2

a. Find the sum of the measures of the angles of a 12-sided polygon.

b. Find the measure of an angle of a regular 12-sided polygon.

5 Understand tessellations and their angle requirements.

Tessellations

A relationship between geometry and the visual arts is found in an art form called *tessellations*. A **tessellation**, or **tiling**, is a pattern consisting of the repeated use of the same geometric figures to completely cover a plane, leaving no gaps and no overlaps. **Figure 10.32** on the next page shows eight tessellations, each consisting of the repeated use of two or more regular polygons.

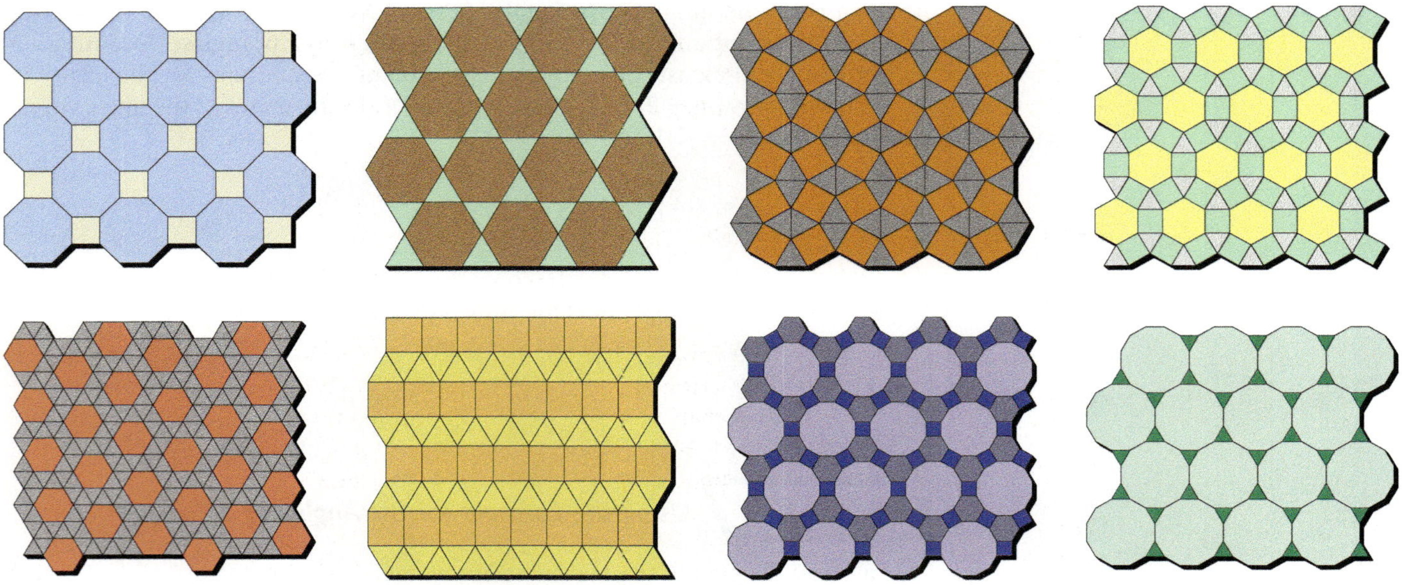

FIGURE 10.32 Eight tessellations formed by two or more regular polygons

In each tessellation in **Figure 10.32**, the same types of regular polygons surround every vertex (intersection point). Furthermore, **the sum of the measures of the angles that come together at each vertex is 360°, a requirement for the formation of a tessellation**. This is illustrated in the enlarged version of the tessellation in **Figure 10.33**. If you select any vertex and count the sides of the polygons that touch it, you'll see that vertices are surrounded by two regular octagons and a square. Can you see why the sum of the angle measures at every vertex is 360°?

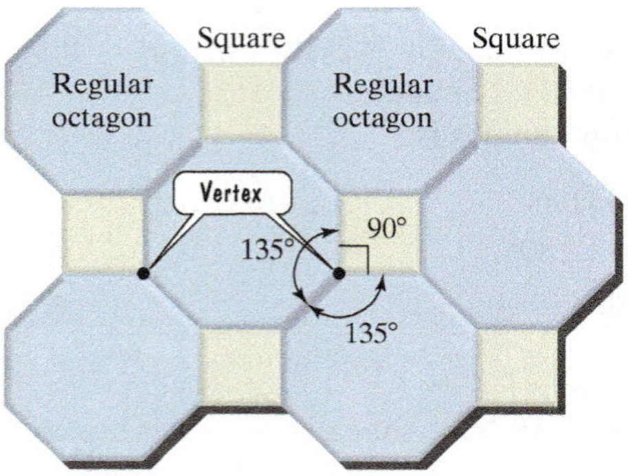

FIGURE 10.33

$$135° + 135° + 90° = 360°$$

In Example 2, we found that each angle of a regular octagon measures 135°.

Each angle of a square measures 90°.

The most restrictive condition in creating tessellations is that just *one type of regular polygon* may be used. With this restriction, there are only three possible tessellations, made from equilateral triangles or squares or regular hexagons, as shown in **Figure 10.34**.

FIGURE 10.34 The three tessellations formed using one regular polygon

Each tessellation is possible because the angle sum at every vertex is 360°:

Six equilateral triangles at each vertex

$$60° + 60° + 60° + 60° + 60° + 60° = 360°$$

Four squares at each vertex

$$90° + 90° + 90° + 90° = 360°$$

Three regular hexagons at each vertex

$$120° + 120° + 120° = 360°.$$

Each angle measures $\frac{(n-2)180°}{n} = \frac{(6-2)180°}{6} = 120°$.

EXAMPLE 3 *Angle Requirements of Tessellations*

Explain why a tessellation cannot be created using only regular pentagons.

SOLUTION

Let's begin by applying $(n - 2)180°$ to find the measure of each angle of a regular pentagon. Each angle measures

$$\frac{(5-2)180°}{5} = \frac{3(180°)}{5} = 108°.$$

A requirement for the formation of a tessellation is that the measures of the angles that come together at each vertex is 360°. With each angle of a regular pentagon measuring 108°, **Figure 10.35** shows that three regular pentagons fill in $3 \cdot 108° = 324°$ and leave a $360° - 324°$, or a 36°, gap. Because the 360° angle requirement cannot be met, no tessellation by regular pentagons is possible.

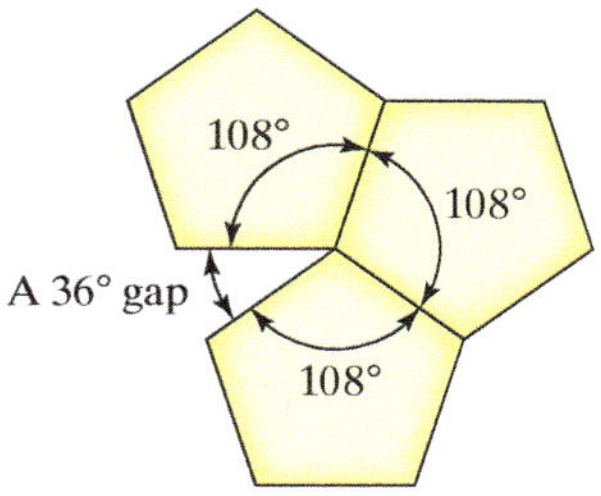

FIGURE 10.35

"The regular division of the plane into congruent figures evoking an association in the observer with a familiar natural object is one of these hobbies or problems . . . I have embarked on this geometric problem again and again over the years, trying to throw light on different aspects each time. I cannot imagine what my life would be like if this problem had never occurred to me; one might say that I am head over heels in love with it, and I still don't know why."

—M. C. Escher

CHECK POINT 3 Explain why a tessellation cannot be created using only regular octagons.

Tessellations that are not restricted to the repeated use of regular polygons are endless in number. They are prominent in Islamic art, Italian mosaics, quilts, and ceramics. The Dutch artist M. C. Escher (1898–1972) created a dazzling array of prints, drawings, and paintings using tessellations composed of stylized interlocking animals. Escher's art reflects the mathematics that underlies all things, while creating surreal manipulations of space and perspective that make gentle fun of consensus reality.

Symmetry Drawing E85 (1952), M.C. Escher.

Concept and Vocabulary Check

Fill in each blank so that the resulting statement is true.

1. The distance around the sides of a polygon is called its ____________.
2. A four-sided polygon is a/an ____________, a five-sided polygon is a/an ____________, a six-sided polygon is a/an ____________, a seven-sided polygon is a/an ____________, and an eight-sided polygon is a/an ____________.
3. A polygon whose sides are all the same length and whose angles all have the same measure is called a/an ____________ polygon.
4. Opposite sides of a parallelogram are ____________ and ____________.
5. A parallelogram with all sides of equal length without any right angles is called a/an ____________.
6. A parallelogram with four right angles without all sides of equal length is called a/an ____________.
7. A parallelogram with four right angles and all sides of equal length is called a/an ____________.
8. A four-sided figure with exactly one pair of parallel sides is called a/an ____________.
9. The perimeter, P, of a rectangle with length l and width w is given by the formula ____________.
10. The sum of the measures of the angles of a polygon with n sides is ____________.
11. A pattern consisting of the repeated use of the same geometric figures to completely cover a plane, leaving no gaps and no overlaps, is called a/an ____________.

In Exercises 12–18, determine whether each statement is true or false. If the statement is false, make the necessary change(s) to produce a true statement.

12. Every parallelogram is a rhombus. ______
13. Every rhombus is a parallelogram. ______
14. All squares are rectangles. ______
15. Some rectangles are not squares. ______
16. No triangles are polygons. ______
17. Every rhombus is a regular polygon. ______
18. A requirement for the formation of a tessellation is that the sum of the measures of the angles that come together at each vertex is 360°. ______

Exercise Set 10.3

Practice Exercises

In Exercises 1–4, use the number of sides to name the polygon.

1.

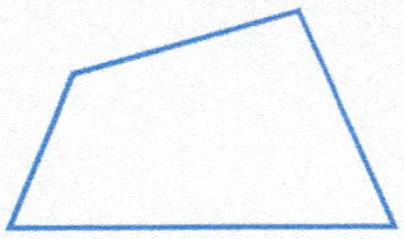

2.

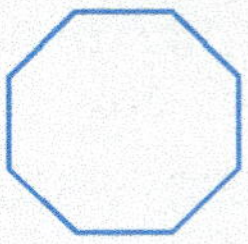

3.

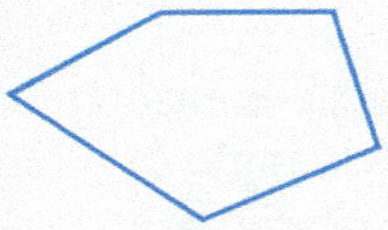

4.

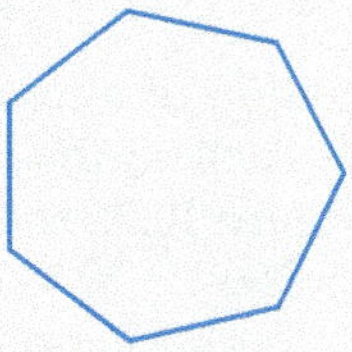

Use these quadrilaterals to solve Exercises 5–10.

a.

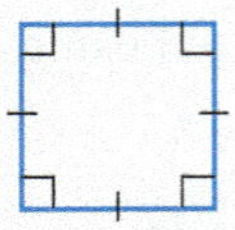

b.

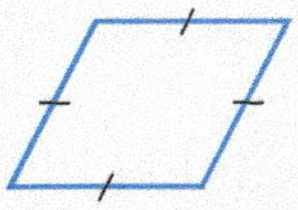

c.

d.

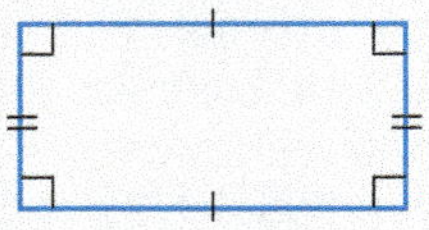

e.

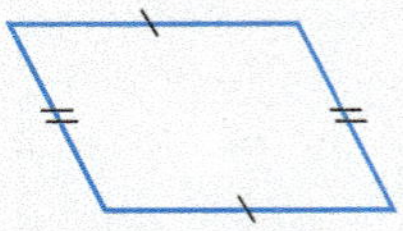

5. Which of these quadrilaterals have opposite sides that are parallel? Name these quadrilaterals.

6. Which of these quadrilaterals have sides of equal length that meet at a vertex? Name these quadrilaterals.

7. Which of these quadrilaterals have right angles? Name these quadrilaterals.

8. Which of these quadrilaterals do not have four sides of equal length? Name these quadrilaterals.

9. Which of these quadrilaterals is not a parallelogram? Name this quadrilateral.

10. Which of these quadrilaterals is/are a regular polygon? Name this/these quadrilateral(s).

In Exercises 11–20, find the perimeter of the figure named and shown. Express the perimeter using the same unit of measure that appears on the given side or sides.

11. Rectangle

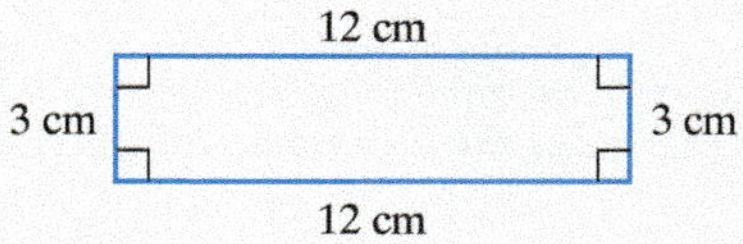

12. Parallelogram

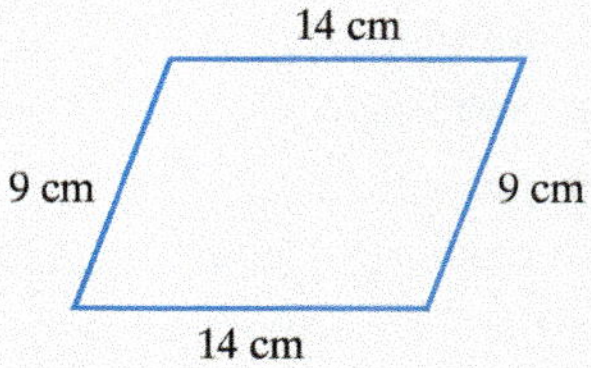

13. Rectangle

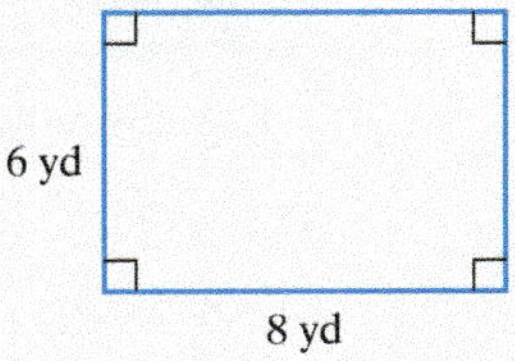

14. Rectangle

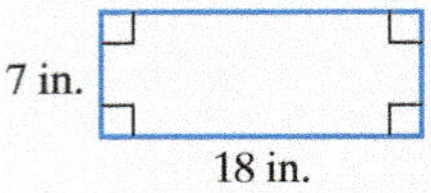

15. Square

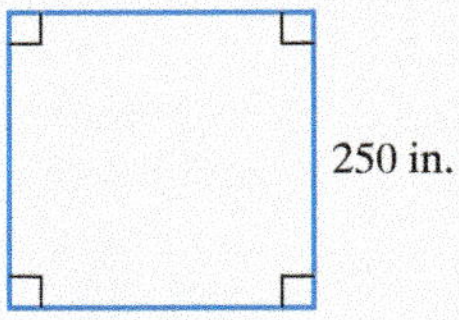

16. Square

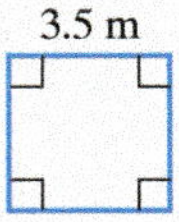

17. Triangle

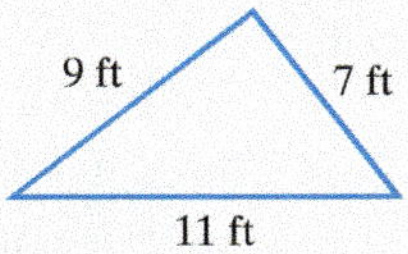

18. Triangle

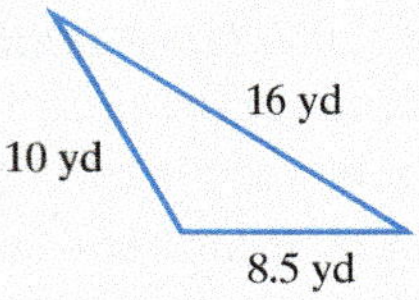

19. Equilateral triangle

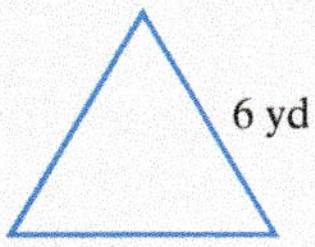

20. Regular hexagon

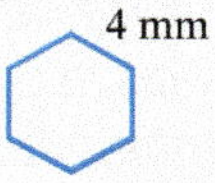

In Exercises 21–24, find the perimeter of the figure shown. Express the perimeter using the same unit of measure that appears on the given side or sides.

21.

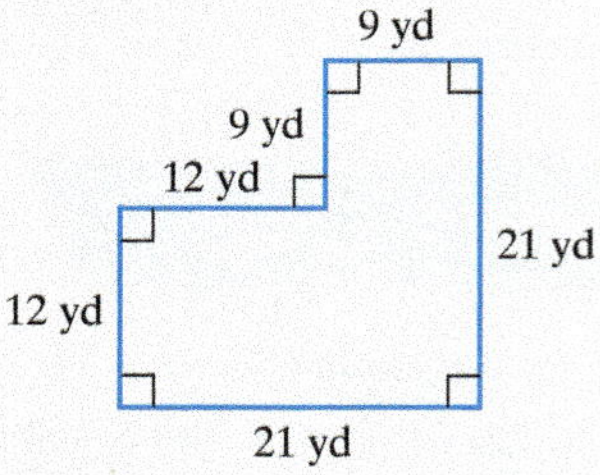

22.

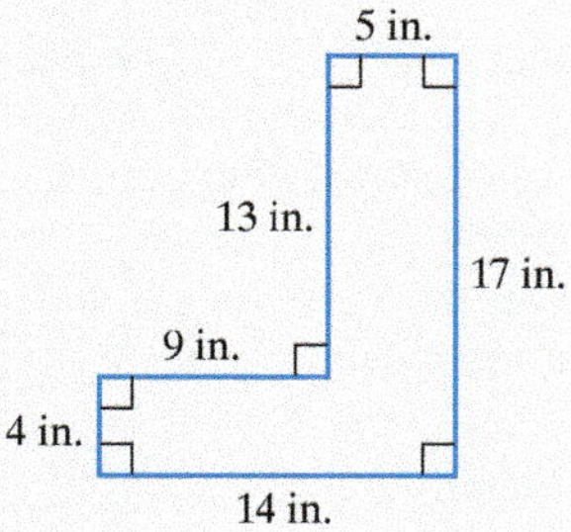

23.

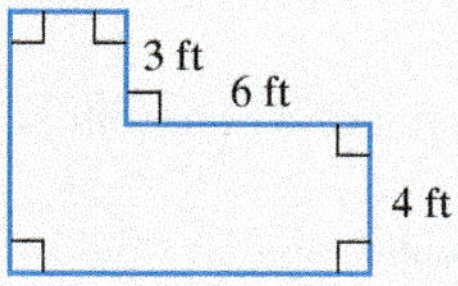

24.

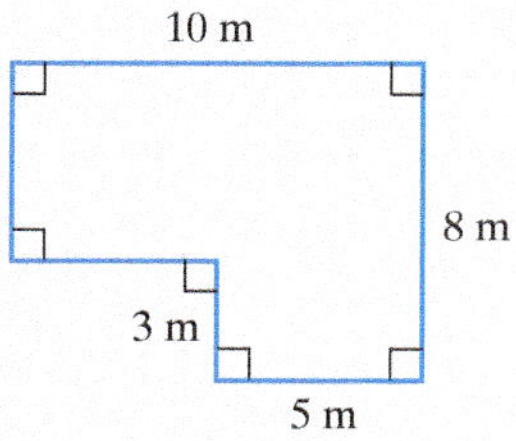

25. Find the sum of the measures of the angles of a five-sided polygon.

26. Find the sum of the measures of the angles of a six-sided polygon.

27. Find the sum of the measures of the angles of a quadrilateral.

28. Find the sum of the measures of the angles of a heptagon.

In Exercises 29–30, each figure shows a regular polygon. Find the measures of angle A and angle B.

29.

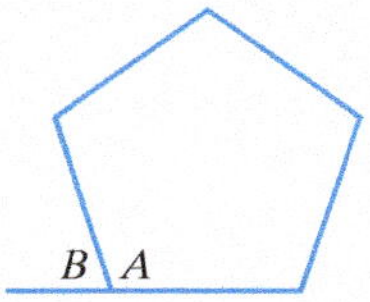

30.

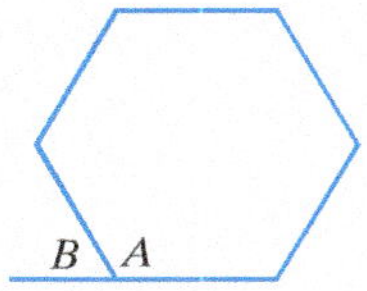

In Exercises 31–32, **a.** *Find the sum of the measures of the angles for the figure given;* **b.** *Find the measures of angle A and angle B.*

31.

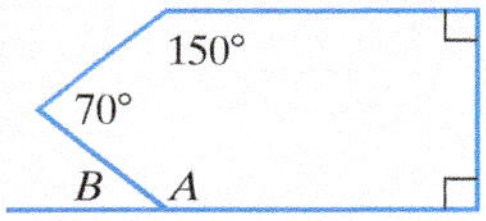

32.

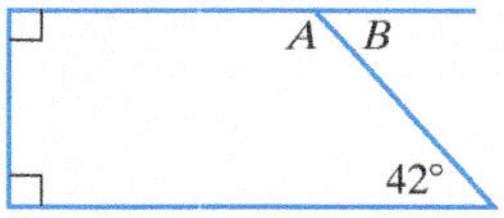

In Exercises 33–36, tessellations formed by two or more regular polygons are shown.

a. *Name the types of regular polygons that surround each vertex.*

b. *Determine the number of angles that come together at each vertex, as well as the measures of these angles.*

c. *Use the angle measures from part (b) to explain why the tessellation is possible.*

33.

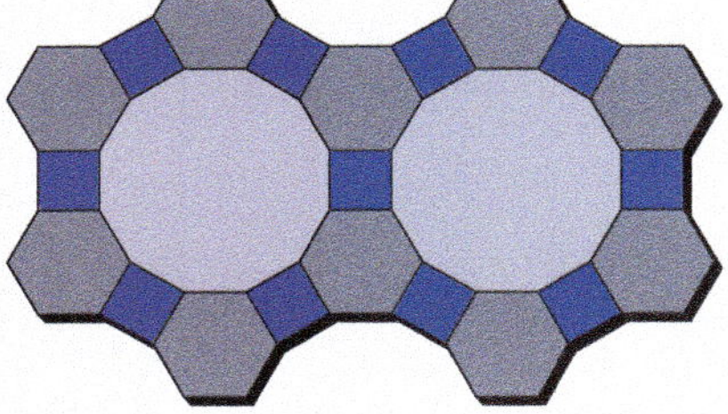

34.

35.

36.

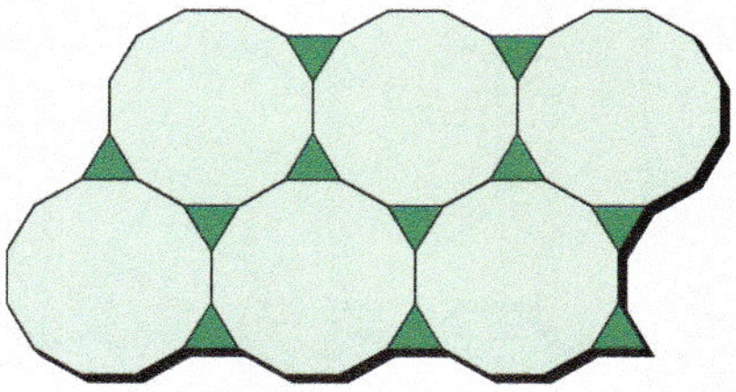

37. Can a tessellation be created using only regular nine-sided polygons? Explain your answer.

38. Can a tessellation be created using only regular ten-sided polygons? Explain your answer.

Practice Plus

In Exercises 39–42, use an algebraic equation to determine each rectangle's dimensions.

39. A rectangular field is four times as long as it is wide. If the perimeter of the field is 500 yards, what are the field's dimensions?

40. A rectangular field is five times as long as it is wide. If the perimeter of the field is 288 yards, what are the field's dimensions?

41. An American football field is a rectangle with a perimeter of 1040 feet. The length is 200 feet more than the width. Find the width and length of the rectangular field.

42. A basketball court is a rectangle with a perimeter of 86 meters. The length is 13 meters more than the width. Find the width and length of the basketball court.

In Exercises 43–44, use algebraic equations to find the measure of each angle that is represented in terms of x.

43.

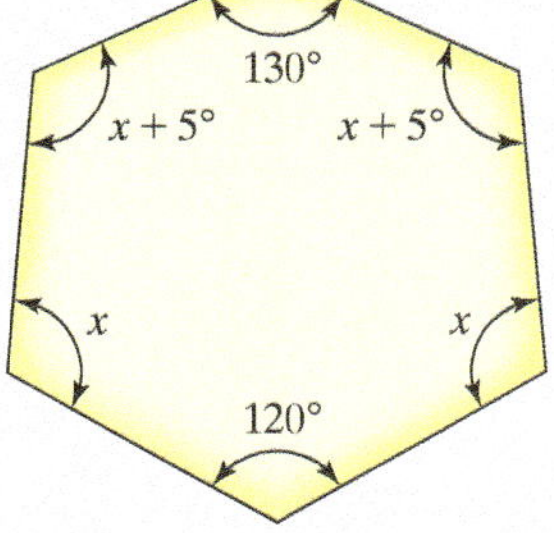

44.

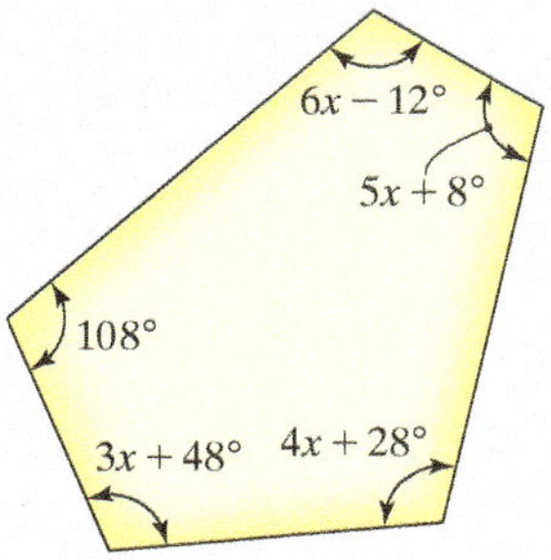

In the figure shown, the artist has cunningly distorted the "regular" polygons to create a fraudulent tessellation with discrepancies that are too subtle for the eye to notice. In Exercises 45–46, you will use mathematics, not your eyes, to observe the irregularities.

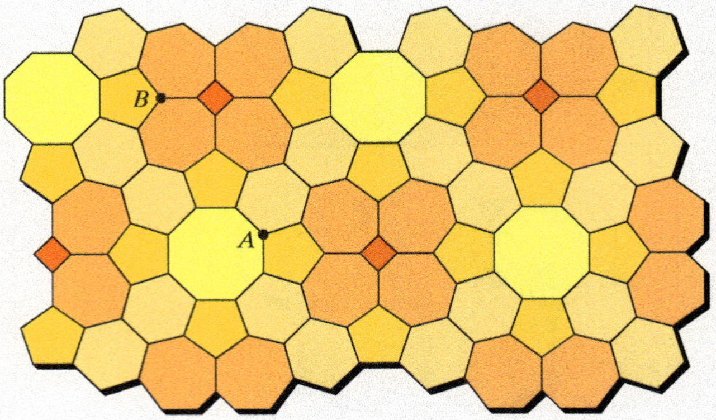

45. Find the sum of the angle measures at vertex A. Then explain why the tessellation is a fake.

46. Find the sum of the angle measures at vertex B. Then explain why the tessellation is a fake.

Application Exercises

47. A school playground is in the shape of a rectangle 400 feet long and 200 feet wide. If fencing costs \$14 per yard, what will it cost to place fencing around the playground?

48. A rectangular field is 70 feet long and 30 feet wide. If fencing costs \$8 per yard, how much will it cost to enclose the field?

49. One side of a square flower bed is 8 feet long. How many plants are needed if they are to be spaced 8 inches apart around the outside of the bed?

50. What will it cost to place baseboard around the region shown if the baseboard costs \$0.25 per foot? No baseboard is needed for the 2-foot doorway.

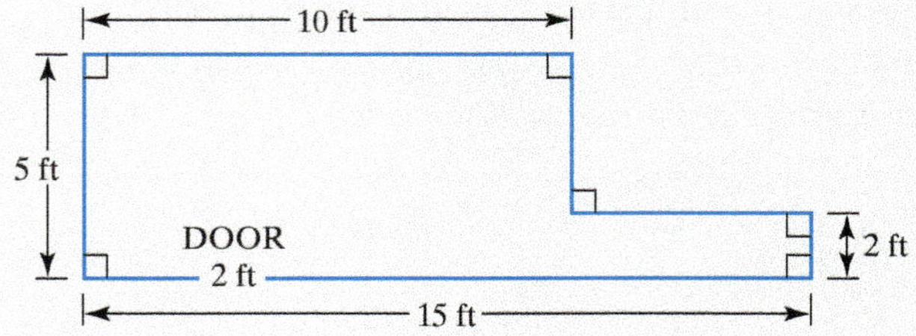

Writing in Mathematics

51. What is a polygon?

52. Explain why rectangles and rhombuses are also parallelograms.

53. Explain why every square is a rectangle, a rhombus, a parallelogram, a quadrilateral, and a polygon.

54. Explain why a square is a regular polygon, but a rhombus is not.

55. Using words only, describe how to find the perimeter of a rectangle.

56. Describe a practical situation in which you needed to apply the concept of a geometric figure's perimeter.

57. Describe how to find the measure of an angle of a regular pentagon.

Critical Thinking Exercises

Make Sense? *In Exercises 58–61, determine whether each statement makes sense or does not make sense, and explain your reasoning.*

58. I drew a polygon having two sides that intersect at their midpoints.

59. I find it helpful to think of a polygon's perimeter as the length of its boundary.

60. If a polygon is not regular, I can determine the sum of the measures of its angles, but not the measure of any one of its angles.

61. I used floor tiles in the shape of regular pentagons to completely cover my kitchen floor.

In Exercises 62–63, write an algebraic expression that represents the perimeter of the figure shown.

62.

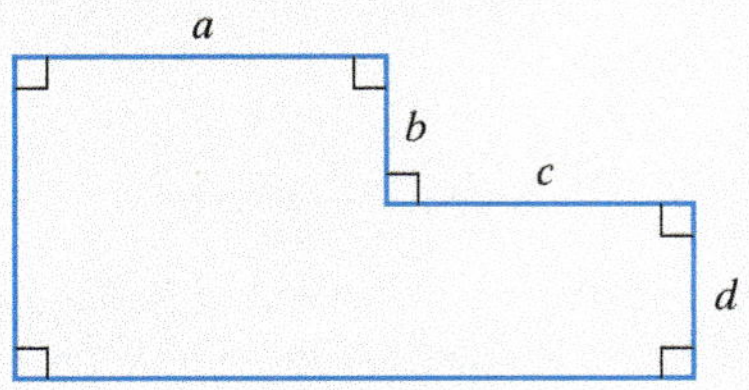

63.

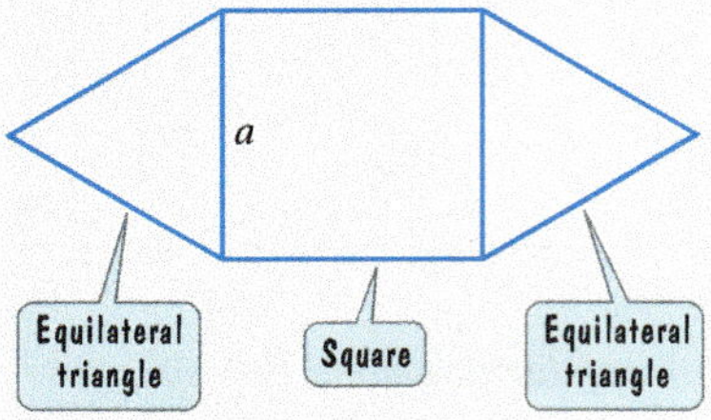

64. Find $m\angle 1$ in the figure shown.

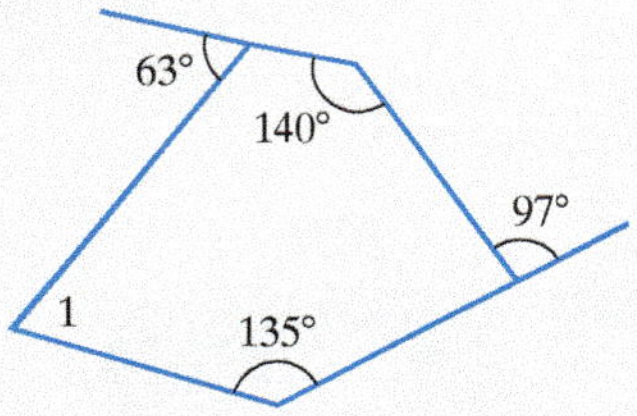

Group Exercise

65. Group members should consult sites on the Internet devoted to tessellations, or tilings, and present a report that expands upon the information in this section. Include a discussion of cultures that have used tessellations on fabrics, wall coverings, baskets, rugs, and pottery, with examples. Include the Alhambra, a fourteenth-century palace in Granada, Spain, in the presentation, as well as works by the artist M. C. Escher. Discuss the various symmetries (translations, rotations, reflections) associated with tessellations. Demonstrate how to create unique tessellations, including Escher-type patterns. Other than creating beautiful works of art, are there any practical applications of tessellations?

10.4

WHAT AM I SUPPOSED TO LEARN?

After you have read this section, you should be able to:

1. Use area formulas to compute the areas of plane regions and solve applied problems.
2. Use formulas for a circle's circumference and area.

Area and Circumference

The size of a house is described in square feet. But how do you know from the real estate ad whether the 1200-square-foot home with the backyard pool is large enough to warrant a visit? Faced with hundreds of ads, you need some way to sort out the best bets. What does 1200 square feet mean and how is this area determined? In this section, we discuss how to compute the areas of plane regions.

1 Use area formulas to compute the areas of plane regions and solve applied problems.

Formulas for Area

In Section 9.2, we saw that the area of a two-dimensional figure is the number of square units, such as square inches or square miles, it takes to fill the interior of the figure. For example, **Figure 10.36** shows that there are 12 square units contained within the rectangular region. The area of the region is 12 square units. Notice that the area can be determined in the following manner:

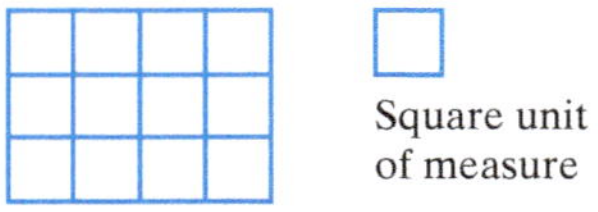

FIGURE 10.36 The area of the region on the left is 12 square units.

Distance across   Distance down

$$4 \text{ units} \times 3 \text{ units} = 4 \times 3 \times \text{units} \times \text{units}$$
$$= 12 \text{ square units.}$$

The area of a rectangular region, usually referred to as the area of a rectangle, is the product of the distance across (length) and the distance down (width).

AREA OF A RECTANGLE AND A SQUARE

The area, A, of a rectangle with length l and width w is given by the formula

$$A = lw.$$

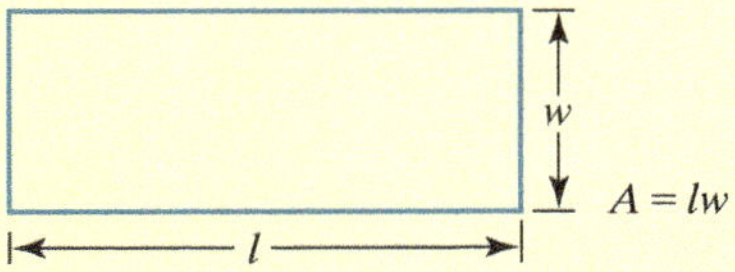

The area, A, of a square with one side measuring s linear units is given by the formula

$$A = s^2.$$

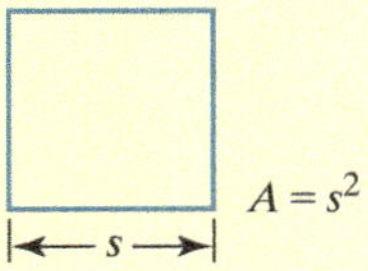

EXAMPLE 1 *Solving an Area Problem*

You decide to cover the path shown in **Figure 10.37** with bricks.

a. Find the area of the path.

b. If the path requires four bricks for every square foot, how many bricks are needed for the project?

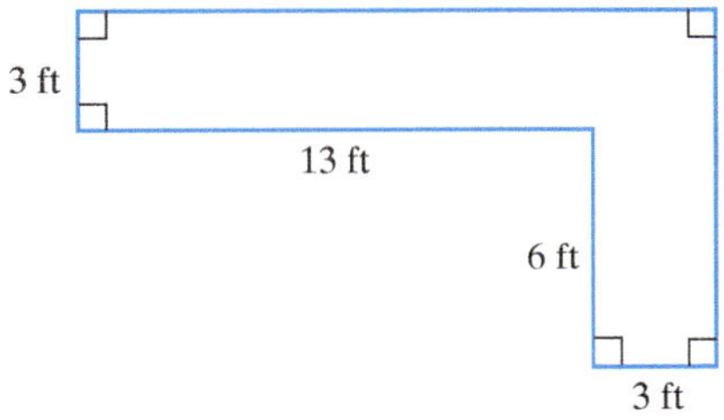

FIGURE 10.37

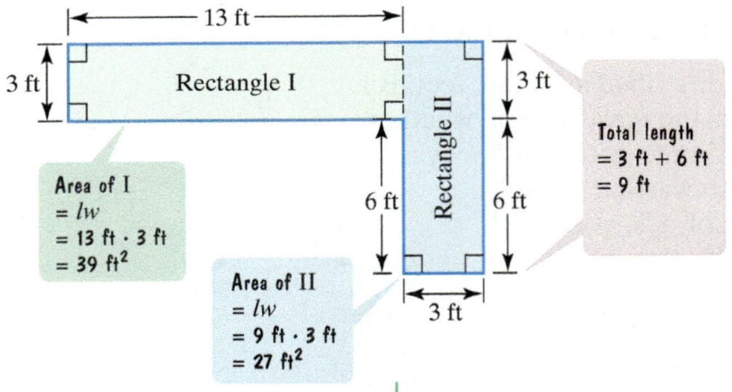

SOLUTION

a. Because we have a formula for the area of a rectangle, we begin by drawing a dashed line that divides the path into two rectangles. One way of doing this is shown at the left. We then use the length and width of each rectangle to find its area. The computations for area are shown in the green and blue voice balloons.

The area of the path is found by adding the areas of the two rectangles.

$$\text{Area of path} = 39 \text{ ft}^2 + 27 \text{ ft}^2 = 66 \text{ ft}^2$$

b. The path requires 4 bricks per square foot. The number of bricks needed for the project is the number of square feet in the path, its area, times 4.

$$\text{Number of bricks needed} = 66 \cancel{\text{ft}^2} \cdot \frac{4 \text{ bricks}}{\cancel{\text{ft}^2}} = 66 \cdot 4 \text{ bricks} = 264 \text{ bricks}$$

Thus, 264 bricks are needed for the project.

CHECK POINT 1 Find the area of the path described in Example 1, rendered on the right as a green region, by first measuring off a large rectangle as shown. The area of the path is the area of the large rectangle (the blue and green regions combined) minus the area of the blue rectangle. Do you get the same answer as we did in Example 1(a)?

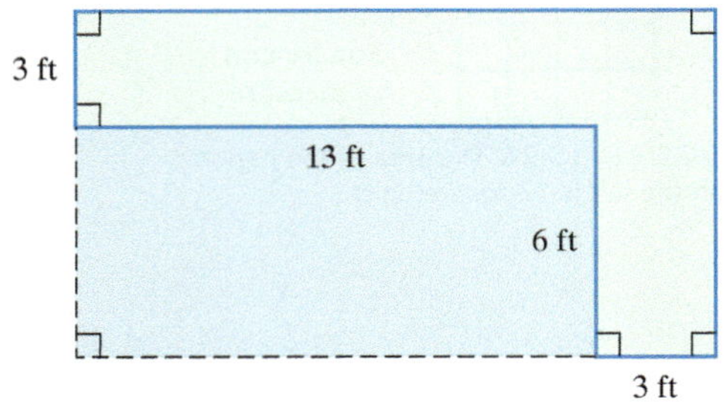

Blitzer Bonus

Appraising a House

A house is measured by an appraiser hired by a bank to help establish its value. The appraiser works from the outside, measuring off a rectangle. Then the appraiser adds the living spaces that lie outside the rectangle and subtracts the empty areas inside the rectangle. The final figure, in square feet, includes all the finished floor space in the house. Not included are the garage, outside porches, decks, or an unfinished basement.

A 1000-square-foot house is considered small, one with 2000 square feet average, and one with more than 2500 square feet pleasantly large. If a 1200–square-foot house has three bedrooms, the individual rooms might seem snug and cozy. With only one bedroom, the space may feel palatial!

Average Size of New U.S. Single-Family Homes

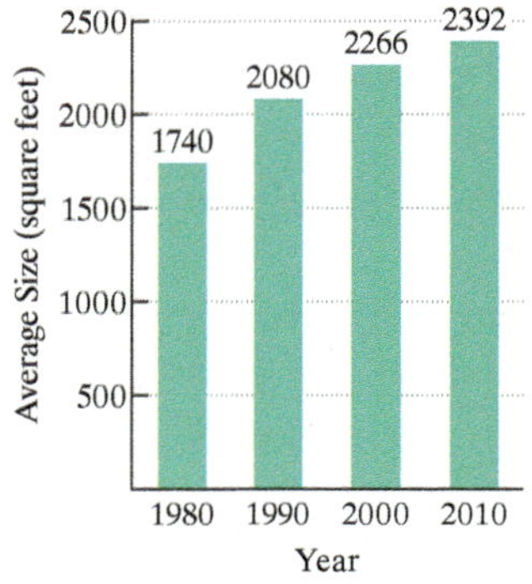

Source: U.S. Census Bureau

In Section 9.2, we saw that although there are 3 linear feet in a linear yard, there are 9 square feet in a square yard. If a problem requires measurement of area in square yards and the linear measures are given in feet, to avoid errors, first convert feet to yards. Then apply the area formula. This idea is illustrated in Example 2.

EXAMPLE 2 *Solving an Area Problem*

What will it cost to carpet a rectangular floor measuring 12 feet by 15 feet if the carpet costs \$18.50 per square yard?

SOLUTION

We begin by converting the linear measures from feet to yards.

$$12 \text{ ft} = \frac{12 \cancel{\text{ft}}}{1} \cdot \frac{1 \text{ yd}}{3 \cancel{\text{ft}}} = \frac{12}{3} \text{ yd} = 4 \text{ yd}$$

$$15 \text{ ft} = \frac{15 \cancel{\text{ft}}}{1} \cdot \frac{1 \text{ yd}}{3 \cancel{\text{ft}}} = \frac{15}{3} \text{ yd} = 5 \text{ yd}$$

Next, we find the area of the rectangular floor in square yards.

$$A = lw = 5 \text{ yd} \cdot 4 \text{ yd} = 20 \text{ yd}^2$$

Finally, we find the cost of the carpet by multiplying the cost per square yard, \$18.50, by the number of square yards in the floor, 20.

$$\text{Cost of carpet} = \frac{\$18.50}{\text{yd}^2} \cdot \frac{20\ \text{yd}^2}{1} = \$18.50(20) = \$370$$

It will cost \$370 to carpet the floor.

CHECK POINT 2 What will it cost to carpet a rectangular floor measuring 18 feet by 21 feet if the carpet costs \$16 per square yard?

We can use the formula for the area of a rectangle to develop formulas for areas of other polygons. We begin with a parallelogram, a quadrilateral with opposite sides equal and parallel. The **height** of a parallelogram is the perpendicular distance between two of the parallel sides. Height is denoted by h in **Figure 10.38**. The **base**, denoted by b, is the length of either of these parallel sides.

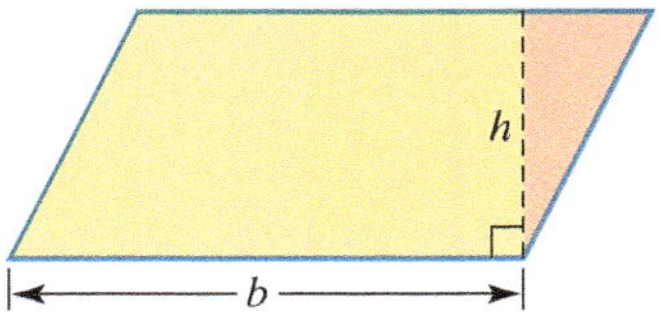

FIGURE 10.38

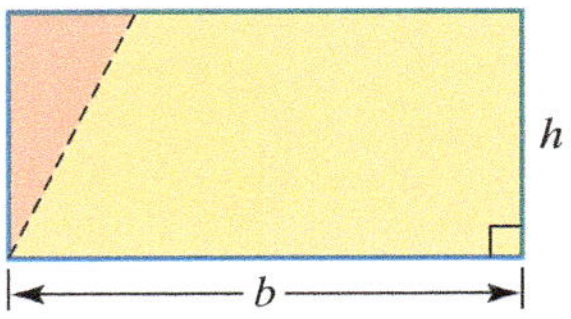

FIGURE 10.39

In **Figure 10.39**, the orange triangular region has been cut off from the right of the parallelogram and attached to the left. The resulting figure is a rectangle with length b and width h. Because bh is the area of the rectangle, it also represents the area of the parallelogram.

GREAT QUESTION!

Is the height of a parallelogram the length of one of its sides?

No. The height of a parallelogram is the perpendicular distance between two of the parallel sides. It is *not* the length of a side.

AREA OF A PARALLELOGRAM

The area, A, of a parallelogram with height h and base b is given by the formula

$$A = bh.$$

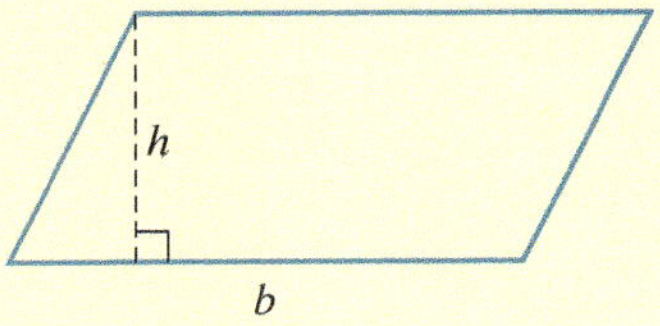

EXAMPLE 3 *Using the Formula for a Parallelogram's Area*

Find the area of the parallelogram in **Figure 10.40**.

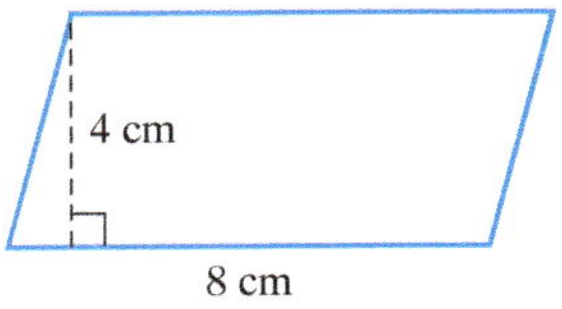

FIGURE 10.40

SOLUTION

As shown in the figure, the base is 8 centimeters and the height is 4 centimeters. Thus, $b = 8$ and $h = 4$.

$$A = bh$$

$$A = 8\text{ cm} \cdot 4\text{ cm} = 32\text{ cm}^2$$

The area is 32 square centimeters.

CHECK POINT 3 Find the area of a parallelogram with a base of 10 inches and a height of 6 inches.

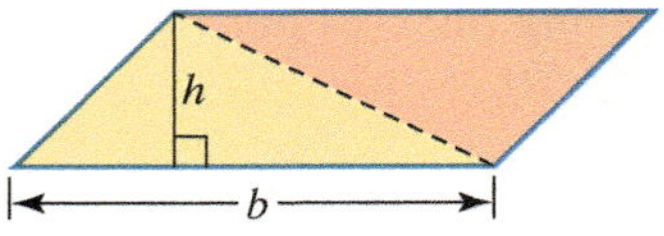

(a)

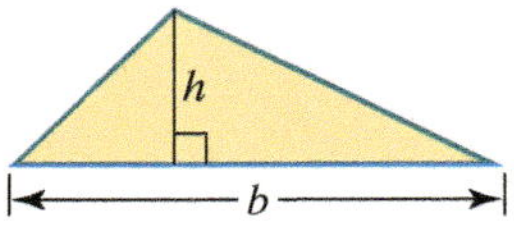

(b)

FIGURE 10.41

Figure 10.41 demonstrates how we can use the formula for the area of a parallelogram to obtain a formula for the area of a triangle. The area of the parallelogram in **Figure 10.41(a)** is given by $A = bh$. The diagonal shown in the parallelogram divides it into two triangles with the same size and shape. This means that the area of each triangle is one-half that of the parallelogram. Thus, the area of the triangle in **Figure 10.41(b)** is given by $A = \frac{1}{2}bh$.

AREA OF A TRIANGLE

The area, A, of a triangle with height h and base b is given by the formula

$$A = \frac{1}{2}bh.$$

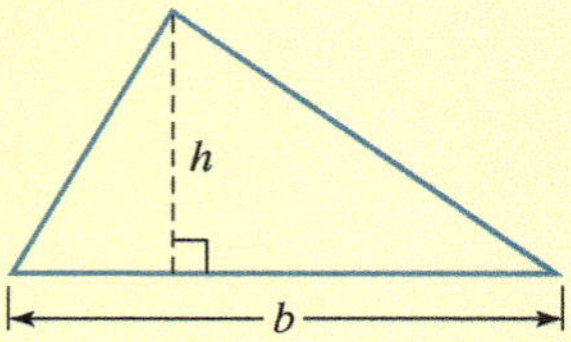

EXAMPLE 4 *Using the Formula for a Triangle's Area*

Find the area of each triangle in **Figure 10.42**.

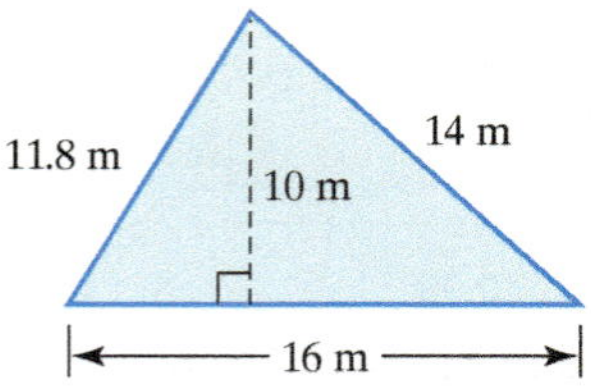

(a)

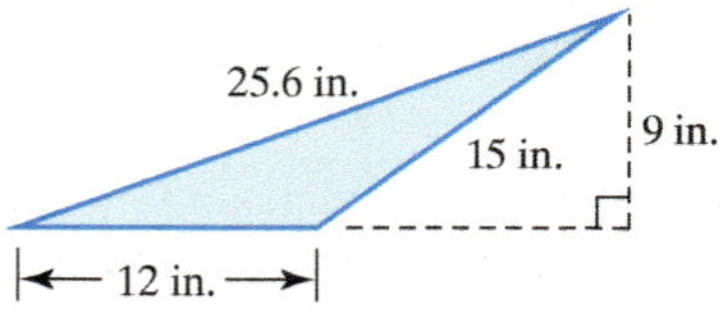

(b)

FIGURE 10.42

SOLUTION

a. In **Figure 10.42(a)**, the base is 16 meters and the height is 10 meters, so $b = 16$ and $h = 10$. We do not need the 11.8 meters or the 14 meters to find the area. The area of the triangle is

$$A = \frac{1}{2}bh = \frac{1}{2}\cdot 16\text{ m}\cdot 10\text{ m} = 80\text{ m}^2.$$

The area is 80 square meters.

b. In **Figure 10.42(b)**, the base is 12 inches. The base line needs to be extended to draw the height. However, we still use 12 inches for b in the area formula. The height, h, is given to be 9 inches. The area of the triangle is

$$A = \frac{1}{2}bh = \frac{1}{2}\cdot 12\text{ in.}\cdot 9\text{ in.} = 54\text{ in.}^2.$$

The area of the triangle is 54 square inches.

CHECK POINT 4 A sailboat has a triangular sail with a base of 12 feet and a height of 5 feet. Find the area of the sail.

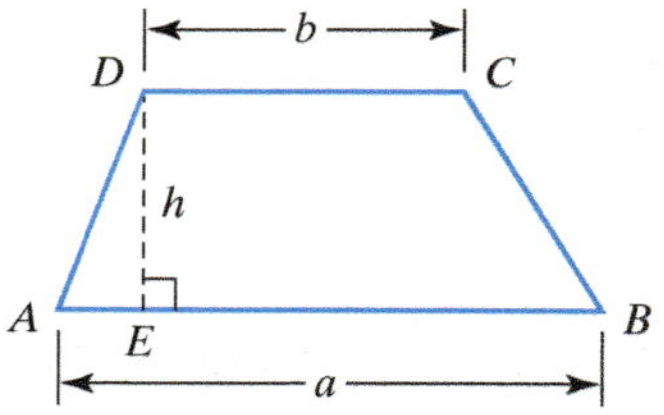

FIGURE 10.43

The formula for the area of a triangle can be used to obtain a formula for the area of a trapezoid. Consider the trapezoid shown in **Figure 10.43**. The lengths of the two parallel sides, called the **bases**, are represented by a (the lower base) and b (the upper base). The trapezoid's height, denoted by h, is the perpendicular distance between the two parallel sides.

In **Figure 10.44**, we have drawn line segment BD, dividing the trapezoid into two triangles, shown in yellow and orange. The area of the trapezoid is the sum of the areas of these triangles.

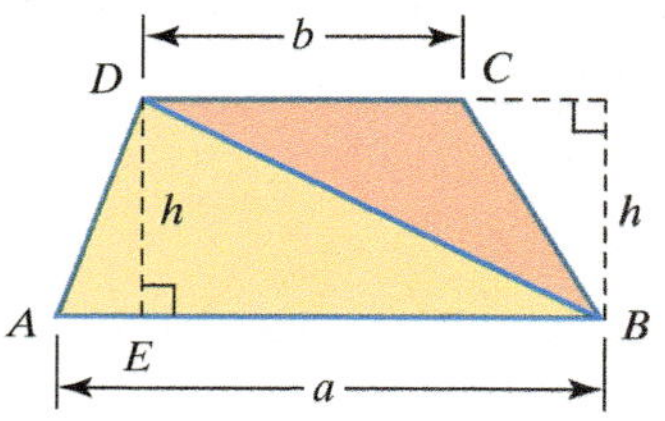

FIGURE 10.44

Area of trapezoid	=	Area of yellow Δ	plus	Area of orange Δ
A	$=$	$\frac{1}{2}ah$	$+$	$\frac{1}{2}bh$
	$=$	$\frac{1}{2}h(a+b)$		Factor out $\frac{1}{2}h$.

AREA OF A TRAPEZOID

The area, A, of a trapezoid with parallel bases a and b and height h is given by the formula

$$A = \frac{1}{2}h(a + b).$$

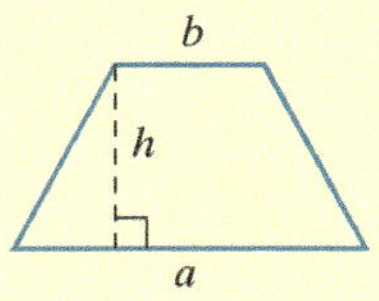

EXAMPLE 5 *Finding the Area of a Trapezoid*

Find the area of the trapezoid in **Figure 10.45**.

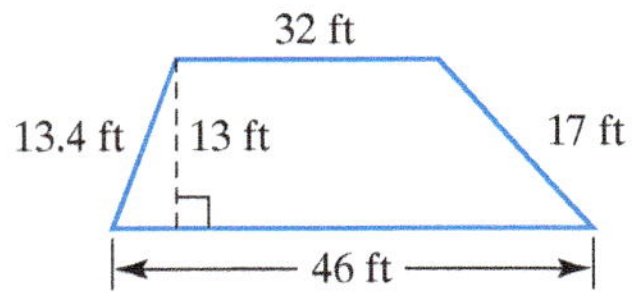

FIGURE 10.45

SOLUTION

The height, h, is 13 feet. The lower base, a, is 46 feet, and the upper base, b, is 32 feet. We do not use the 17-foot and 13.4-foot sides in finding the trapezoid's area.

$$A = \frac{1}{2}h(a + b) = \frac{1}{2} \cdot 13 \text{ ft} \cdot (46 \text{ ft} + 32 \text{ ft})$$

$$= \frac{1}{2} \cdot 13 \text{ ft} \cdot 78 \text{ ft} = 507 \text{ ft}^2$$

The area of the trapezoid is 507 square feet.

CHECK POINT 5 Find the area of a trapezoid with bases of length 20 feet and 10 feet and height 7 feet.

---

2 Use formulas for a circle's circumference and area.

The point at which a pebble hits a flat surface of water becomes the center of a number of circular ripples.

It's a good idea to know your way around a circle. Clocks, angles, maps, and compasses are based on circles. Circles occur everywhere in nature: in ripples on water, patterns on a butterfly's wings, and cross sections of trees. Some consider the circle to be the most pleasing of all shapes.

A **circle** is a set of points in the plane equally distant from a given point, its **center**. **Figure 10.46** shows two circles. The **radius** (plural: radii), r, is a line segment from the center to any point on the circle. For a given circle, all radii have the same length. The **diameter**, d, is a line segment through the center whose endpoints both lie on the circle. For a given circle, all diameters have the same length. In any circle, the **length of the diameter is twice the length of the radius.**

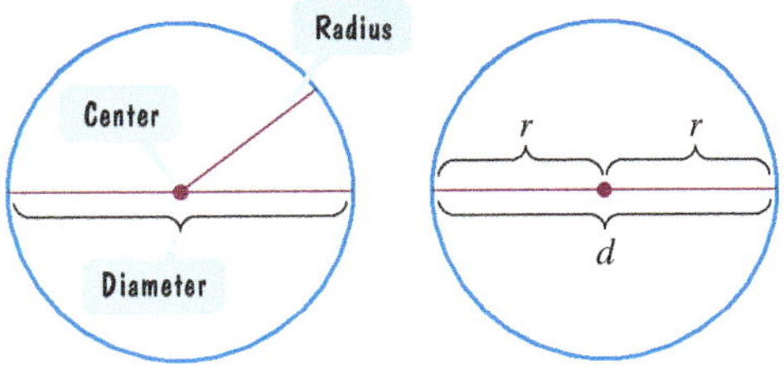

FIGURE 10.46

The words *radius* and *diameter* refer to both the line segments in **Figure 10.46** as well as to their linear measures. The distance around a circle (its perimeter) is called its **circumference**, C. For all circles, if you divide the circumference by the diameter, or by twice the radius, you will get the same number. This ratio is the irrational number π and is approximately equal to 3.14:

$$\frac{C}{d} = \pi \quad \text{or} \quad \frac{C}{2r} = \pi.$$

Thus,

$$C = \pi d \quad \text{or} \quad C = 2\pi r.$$

FINDING THE DISTANCE AROUND A CIRCLE

The circumference, C, of a circle with diameter d and radius r is

$$C = \pi d \quad \text{or} \quad C = 2\pi r.$$

When computing a circle's circumference by hand, round π to 3.14. When using a calculator, use the $\boxed{\pi}$ key, which gives the value of π rounded to approximately 11 decimal places. In either case, calculations involving π give approximate answers. These answers can vary slightly depending on how π is rounded. The symbol $\approx$ (is approximately equal to) will be written in these calculations.

EXAMPLE 6 *Finding a Circle's Circumference*

Find the circumference of the circle in **Figure 10.47**.

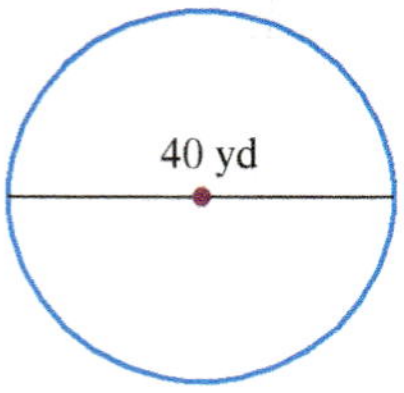

FIGURE 10.47

SOLUTION

The diameter is 40 yards, so we use the formula for circumference with d in it.

$$C = \pi d = \pi(40 \text{ yd}) = 40\pi \text{ yd} \approx 125.7 \text{ yd}$$

The distance around the circle is approximately 125.7 yards.

CHECK POINT 6 Find the circumference of a circle whose diameter measures 10 inches. Express the answer in terms of π and then round to the nearest tenth of an inch.

EXAMPLE 7 *Using the Circumference Formula*

How much trim, to the nearest tenth of a foot, is needed to go around the window shown in **Figure 10.48**?

FIGURE 10.48

SOLUTION

The trim covers the 6-foot bottom of the window, the two 8-foot sides, and the half-circle (called a semicircle) on top. The length needed is

$$6 \text{ ft} + 8 \text{ ft} + 8 \text{ ft} + \text{circumference of the semicircle.}$$

The circumference of the semicircle is half the circumference of a circle whose diameter is 6 feet.

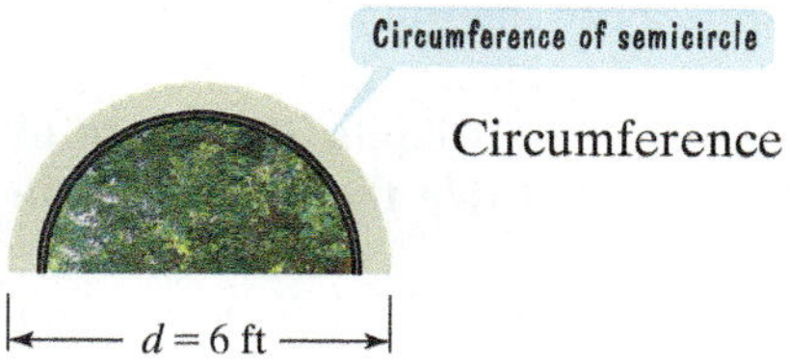

$$\begin{aligned}\text{Circumference of semicircle} &= \frac{1}{2}\pi d \\ &= \frac{1}{2}\pi(6 \text{ ft}) = 3\pi \text{ ft} \approx 9.4 \text{ ft}\end{aligned}$$

Rounding the circumference to the nearest tenth (9.4 feet), the length of trim that is needed is approximately

$$6 \text{ ft} + 8 \text{ ft} + 8 \text{ ft} + 9.4 \text{ ft},$$

or 31.4 feet.

CHECK POINT 7 In **Figure 10.48**, suppose that the dimensions are 10 feet and 12 feet for the window's bottom and side, respectively. How much trim, to the nearest tenth of a foot, is needed to go around the window?

The irrational number π is also used to find the area of a circle in square units. This is because the ratio of a circle's area to the square of its radius is π:

$$\frac{A}{r^2} = \pi.$$

Multiplying both sides of this equation by r^2 gives a formula for determining a circle's area.

FINDING THE AREA OF A CIRCLE

The area, A, of a circle with radius r is

$$A = \pi r^2.$$

EXAMPLE 8 *Problem Solving Using the Formula for a Circle's Area*

Which one of the following is the better buy: a large pizza with a 16-inch diameter for \$15.00 or a medium pizza with an 8-inch diameter for \$7.50?

SOLUTION

The better buy is the pizza with the lower price per square inch. The radius of the large pizza is $\frac{1}{2} \cdot 16$ inches, or 8 inches, and the radius of the medium pizza is $\frac{1}{2} \cdot 8$ inches, or 4 inches. The area of the surface of each circular pizza is determined using the formula for the area of a circle.

$$\text{Large pizza: } A = \pi r^2 = \pi(8 \text{ in.})^2 = 64\pi \text{ in.}^2 \approx 201 \text{ in.}^2$$

$$\text{Medium pizza: } A = \pi r^2 = \pi(4 \text{ in.})^2 = 16\pi \text{ in.}^2 \approx 50 \text{ in.}^2$$

For each pizza, the price per square inch is found by dividing the price by the area:

$$\text{Price per square inch for large pizza} = \frac{\$15.00}{64\pi \text{ in.}^2} \approx \frac{\$15.00}{201 \text{ in.}^2} \approx \frac{\$0.07}{\text{in.}^2}$$

$$\text{Price per square inch for medium pizza} = \frac{\$7.50}{16\pi \text{ in.}^2} \approx \frac{\$7.50}{50 \text{ in.}^2} = \frac{\$0.15}{\text{in.}^2}$$

The large pizza costs approximately \$0.07 per square inch and the medium pizza costs approximately \$0.15 per square inch. Thus, the large pizza is the better buy.

TECHNOLOGY

You can use your calculator to obtain the price per square inch for each pizza in Example 8. The price per square inch for the large pizza, $\frac{15}{64\pi}$, is approximated by one of the following sequences of keystrokes:

Many Scientific Calculators

15 [÷] [(] 64 [×] [π] [)] [=]

Many Graphing Calculators

15 [÷] [(] 64 [π] [)] [ENTER]

In Example 8, did you at first think that the price per square inch would be the same for the large and the medium pizzas? After all, the radius of the large pizza is twice that of the medium pizza, and the cost of the large is twice that of the medium. However, the large pizza's area, 64π square inches, is *four times the area* of the medium pizza's, 16π square inches. Doubling the radius of a circle increases its area by a factor of 2^2, or 4. In general, if the radius of a circle is increased by k times its original linear measure, the area is multiplied by k^2. The same principle is true for any two-dimensional figure: If the shape of the figure is kept the same while linear dimensions are increased k times, the area of the larger, similar, figure is k^2 times greater than the area of the original figure.

CHECK POINT 8 Which one of the following is the better buy: a large pizza with an 18-inch diameter for \$20 or a medium pizza with a 14-inch diameter for \$14?

Concept and Vocabulary Check

Fill in each blank so that the resulting statement is true.

1. The area, A, of a rectangle with length l and width w is given by the formula ______________.
2. The area, A, of a square with one side measuring s linear units is given by the formula ______________.
3. The area, A, of a parallelogram with height h and base b is given by the formula ______________.
4. The area, A, of a triangle with height h and base b is given by the formula ______________.
5. The area, A, of a trapezoid with parallel bases a and b and height h is given by the formula ______________.
6. The circumference, C, of a circle with diameter d is given by the formula ______________.
7. The circumference, C, of a circle with radius r is given by the formula ______________.
8. The area, A, of a circle with radius r is given by the formula ______________.

In Exercises 9–13, determine whether each statement is true or false. If the statement is false, make the necessary change(s) to produce a true statement.

9. The area, A, of a rectangle with length l and width w is given by the formula $A = 2l + 2w$. ____________
10. The height of a parallelogram is the perpendicular distance between two of the parallel sides. ____________
11. The area of either triangle formed by drawing a diagonal in a parallelogram is one-half that of the parallelogram. ____________
12. In any circle, the length of the radius is twice the length of the diameter. ____________
13. The ratio of a circle's circumference to its diameter is the irrational number π. ____________

Exercise Set 10.4

Practice Exercises

In Exercises 1–14, use the formulas developed in this section to find the area of each figure.

1.

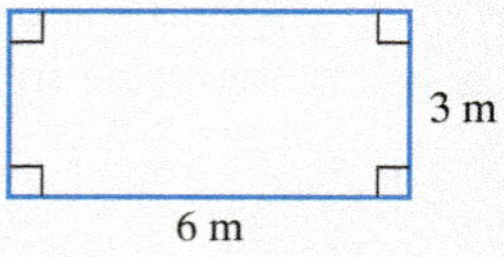

2.

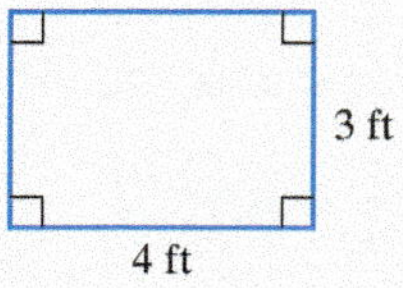

3.

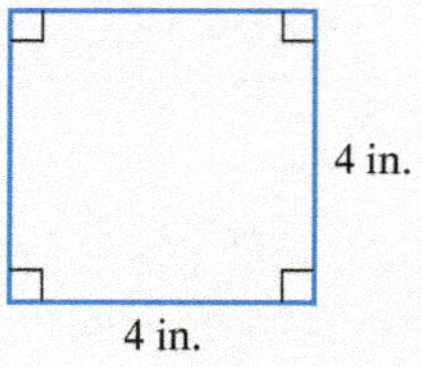

4.

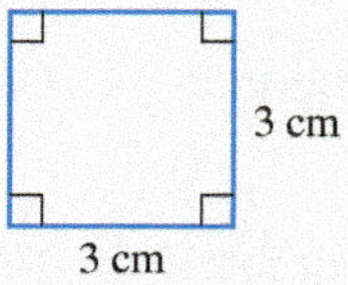

5.

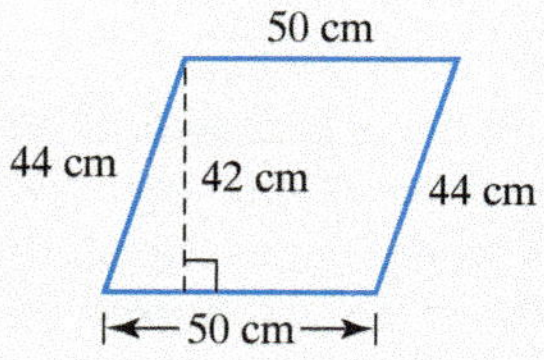

6.

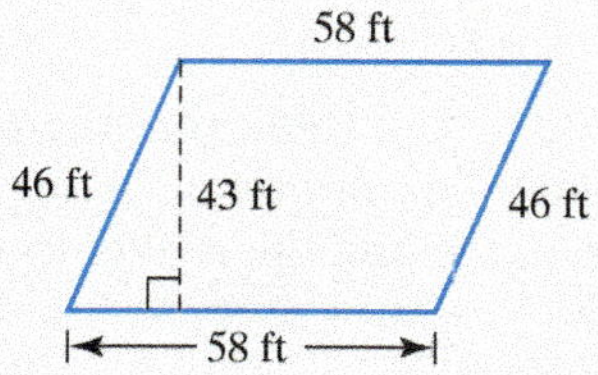

7.

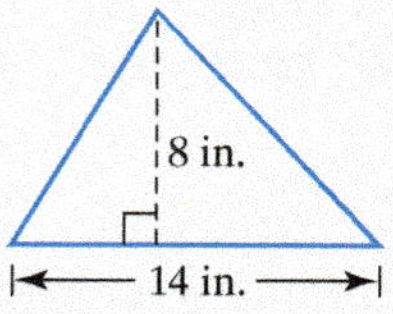

8.

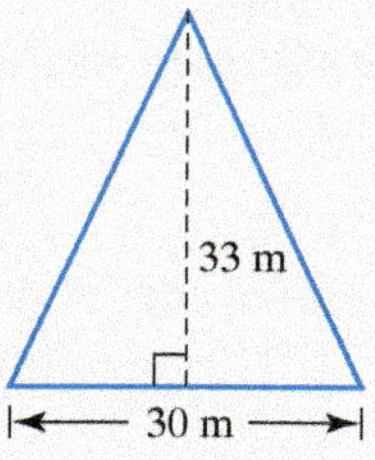

9.

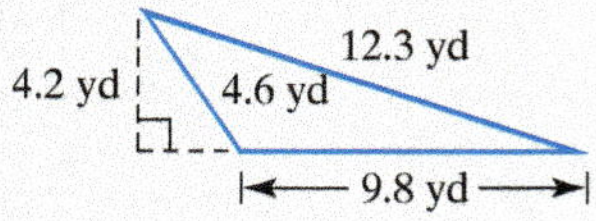

10.

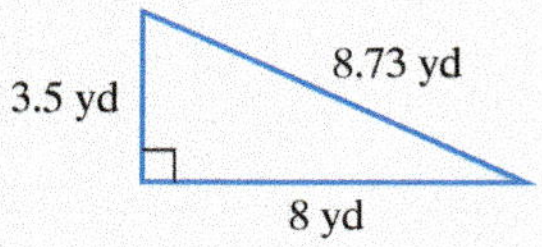

11.

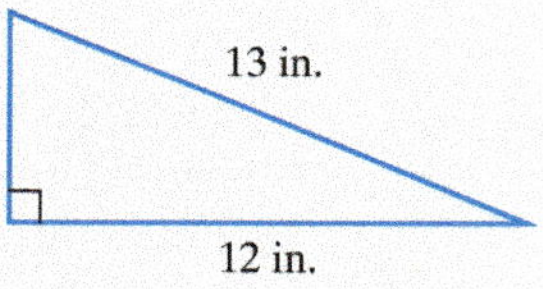

12.

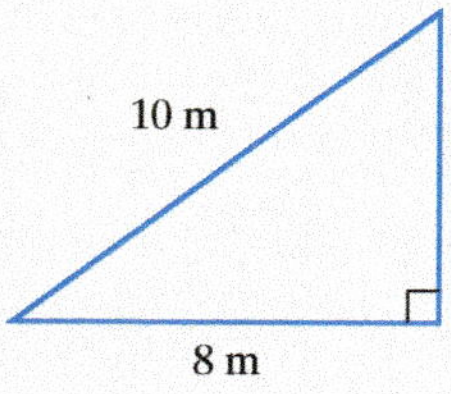

13.

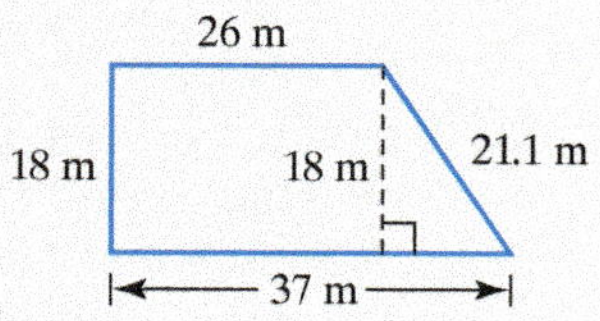

14.

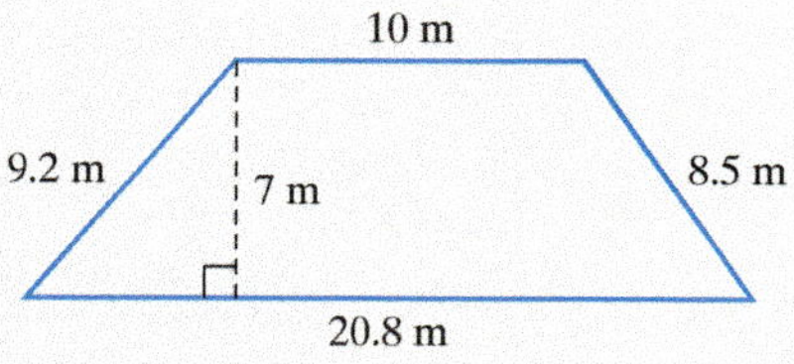

In Exercises 15–18, find the circumference and area of each circle. Express answers in terms of π and then round to the nearest tenth.

15.

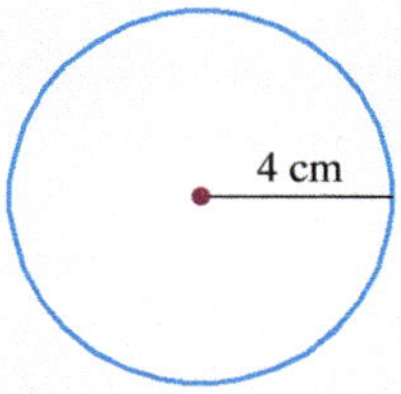

16.

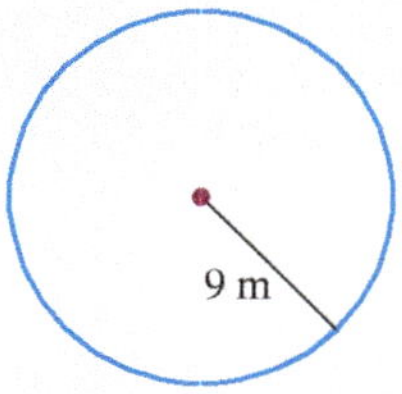

17.

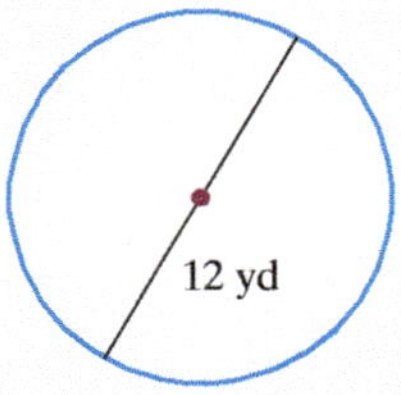

18.

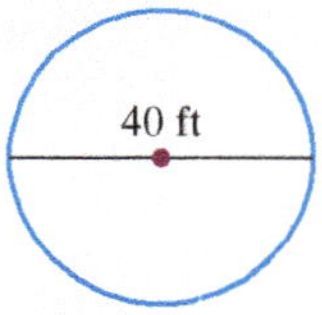

Find the area of each figure in Exercises 19–24. Where necessary, express answers in terms of π and then round to the nearest tenth.

19.

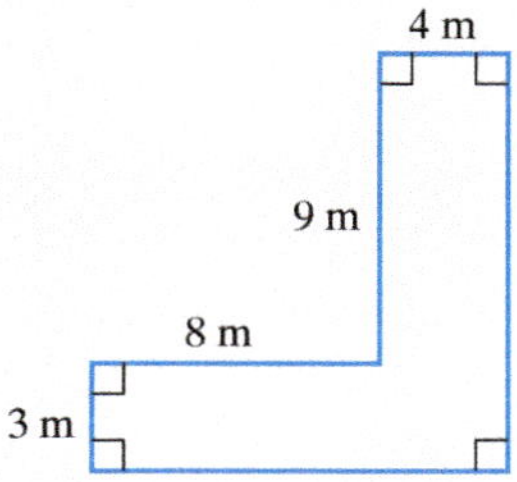

20.

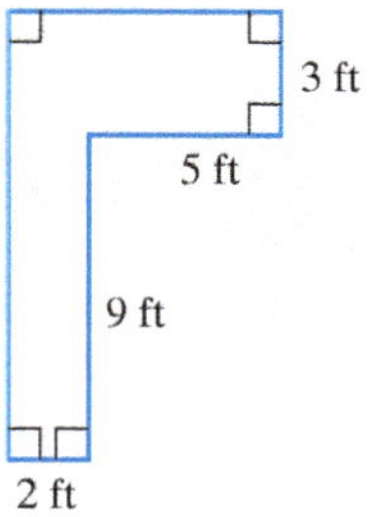

21.

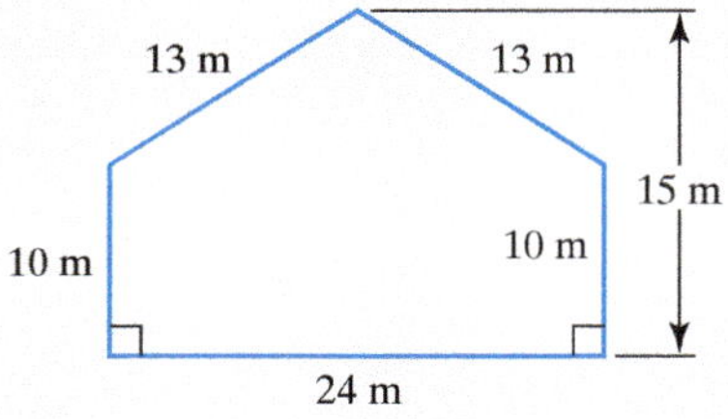

22.

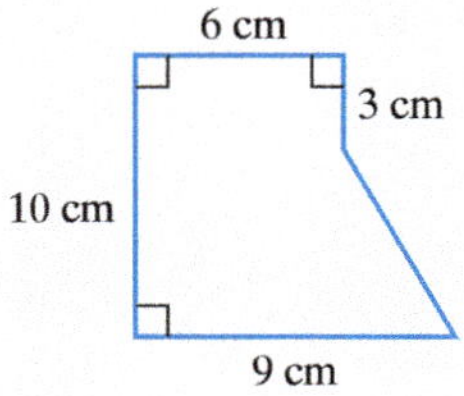

23.

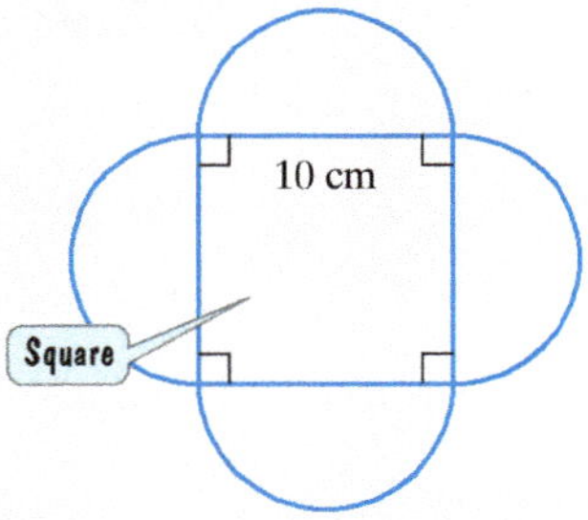

24.

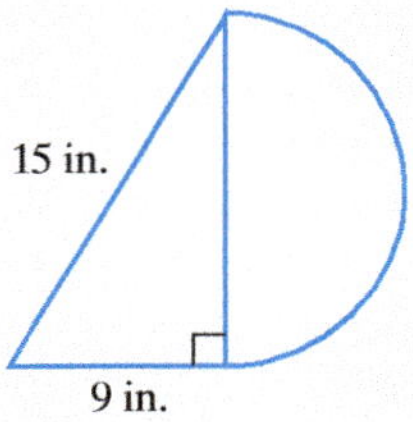

In Exercises 25–28, find a formula for the total area, A, of each figure in terms of the variable(s) shown. Where necessary, use π in the formula.

25.

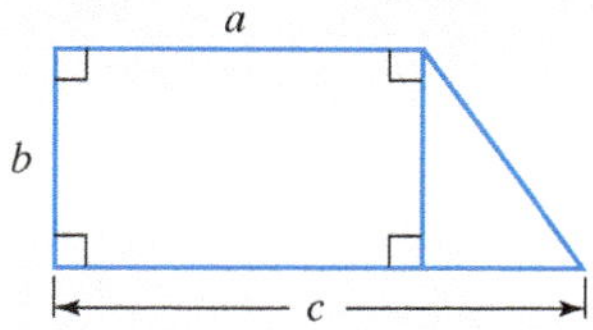

26.

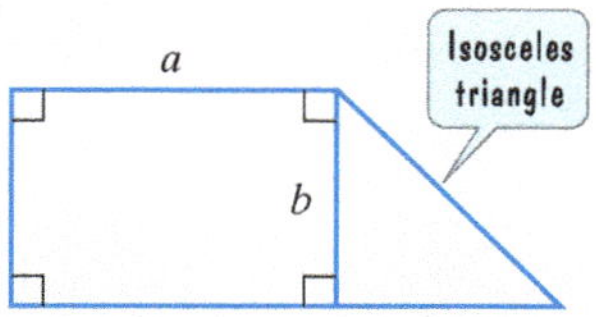

27.

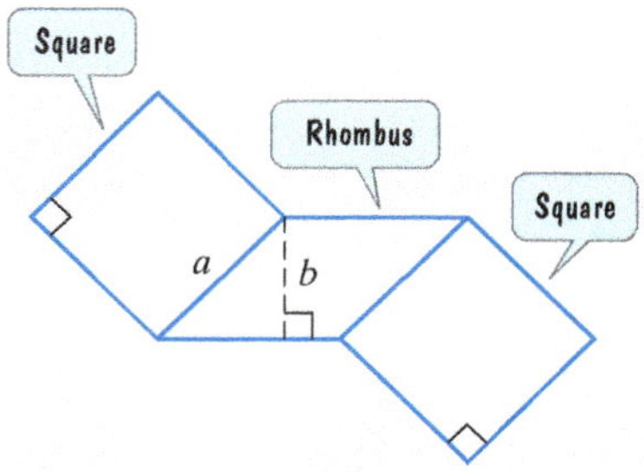

28.

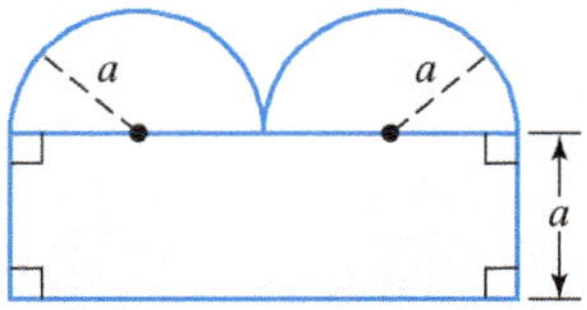

Practice Plus

In Exercises 29–30, find the area of each shaded region.

29.

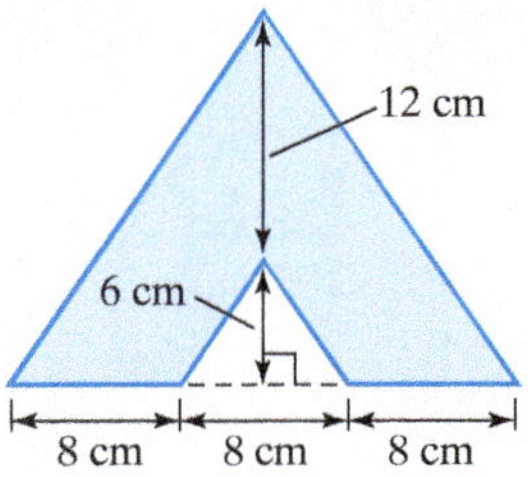

30.

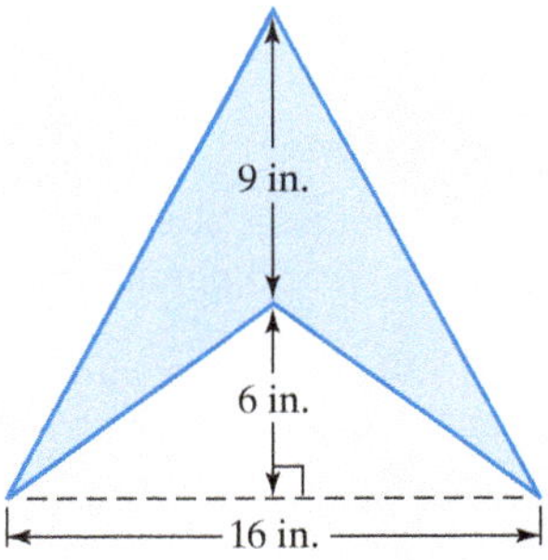

In Exercises 31–34, find the area of each shaded region in terms of π.

31.

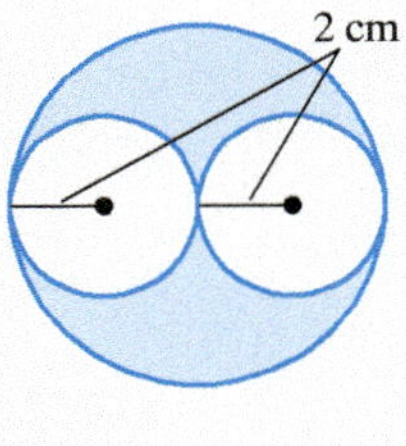

32.

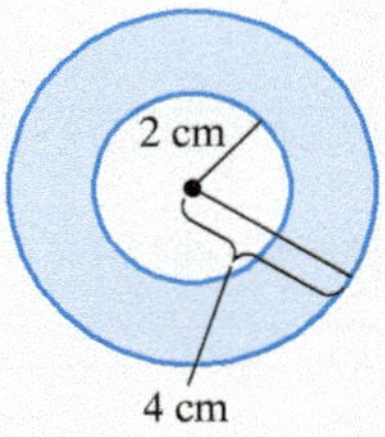

33.

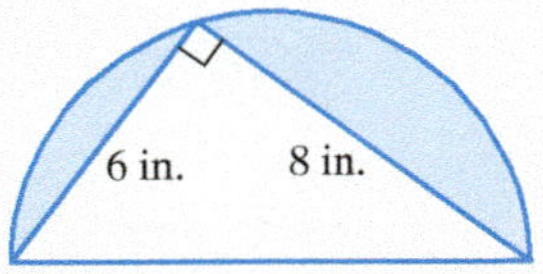

34.

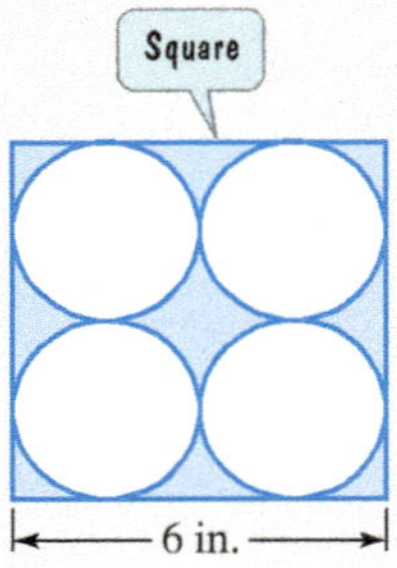

In Exercises 35–36, find the perimeter and the area of each figure. Where necessary, express answers in terms of π and round to the nearest tenth.

35.

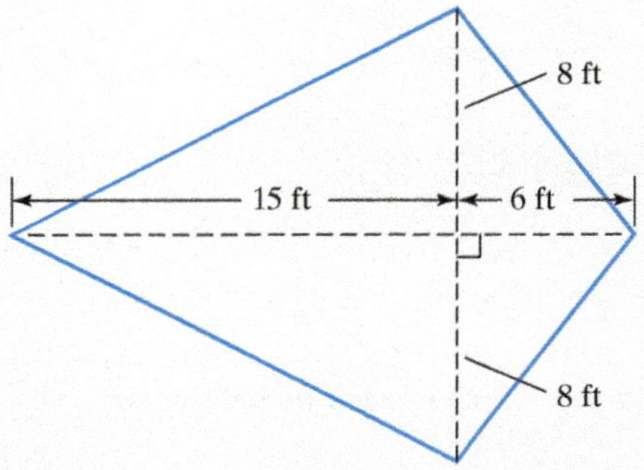

36.

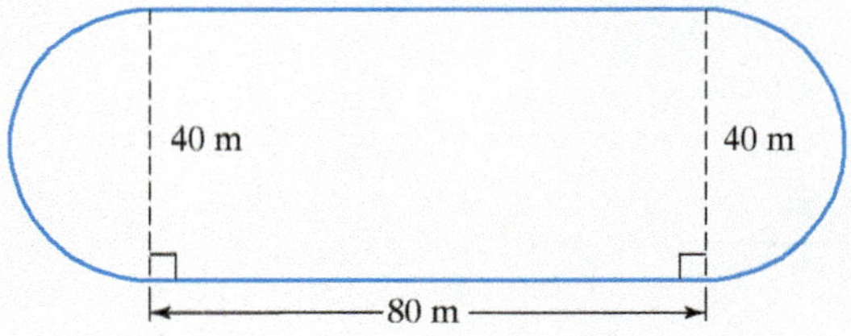

Application Exercises

37. What will it cost to carpet a rectangular floor measuring 9 feet by 21 feet if the carpet costs $26.50 per square yard?

38. A plastering contractor charges $18 per square yard. What is the cost of plastering 60 feet of wall in a house with a 9-foot ceiling?

39. A rectangular kitchen floor measures 12 feet by 15 feet. A stove on the floor has a rectangular base measuring 3 feet by 4 feet, and a refrigerator covers a rectangular area of the floor measuring 4 feet by 5 feet. How many square feet of tile will be needed to cover the kitchen floor not counting the area used by the stove and the refrigerator?

40. A rectangular room measures 12 feet by 15 feet. The entire room is to be covered with rectangular tiles that measure 3 inches by 2 inches. If the tiles are sold at ten for $1.20, what will it cost to tile the room?

41. The lot in the figure shown, except for the house, shed, and driveway, is lawn. One bag of lawn fertilizer costs $25 and covers 4000 square feet.

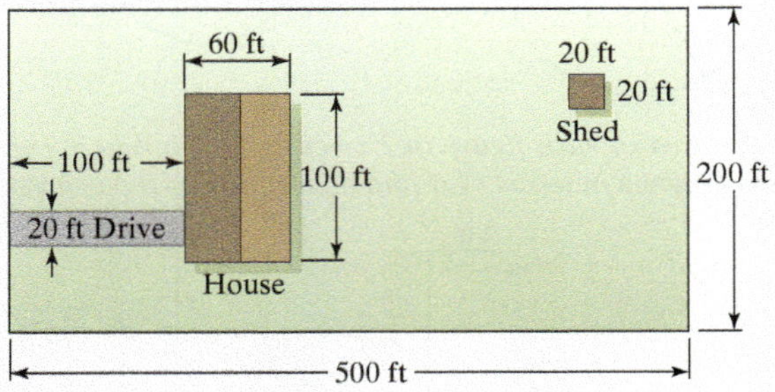

a. Determine the minimum number of bags of fertilizer needed for the lawn.

b. Find the total cost of the fertilizer.

42. Taxpayers with an office in their home may deduct a percentage of their home-related expenses. This percentage is based on the ratio of the office's area to the area of the home. A taxpayer with an office in a 2200-square-foot home maintains a 20 foot by 16 foot office. If the yearly utility bills for the home come to $4800, how much of this is deductible?

43. You are planning to paint the house whose dimensions are shown in the figure.

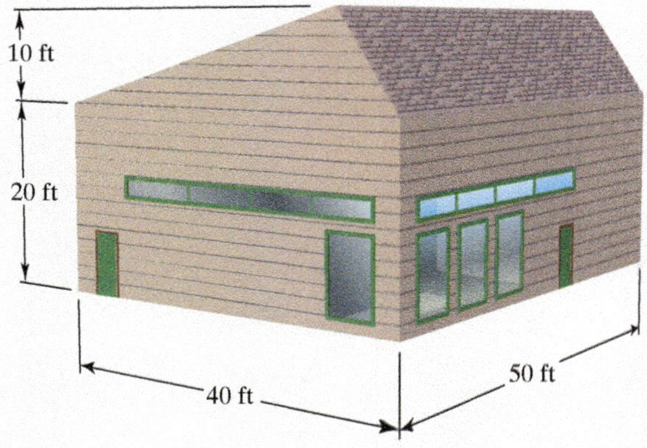

a. How many square feet will you need to paint? (There are four windows, each 8 feet by 5 feet; two windows, each 30 feet by 2 feet; and two doors, each 80 inches by 36 inches, that do not require paint.)

b. The paint that you have chosen is available in gallon cans only. Each can covers 500 square feet. If you want to use two coats of paint, how many cans will you need for the project?

c. If the paint you have chosen sells for $26.95 per gallon, what will it cost to paint the house?

The diagram shows the floor plan for a one-story home. Use the given measurements to solve Exercises 44–46. (A calculator will be helpful in performing the necessary computations.)

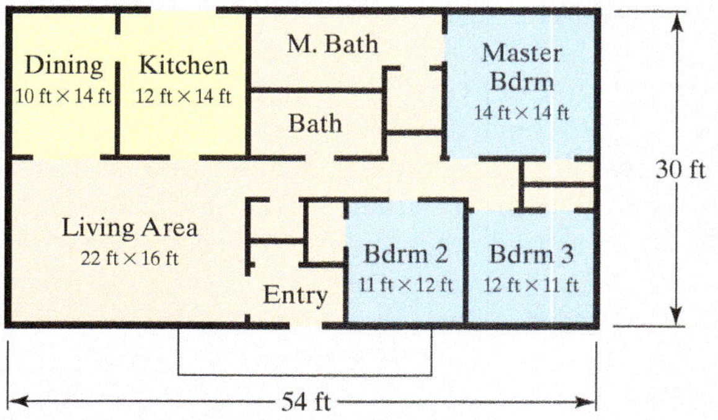

44. If construction costs \$95 per square foot, find the cost of building the home.

45. If carpet costs \$17.95 per square yard and is available in whole square yards only, find the cost of carpeting the three bedroom floors.

46. If ceramic tile costs \$26.95 per square yard and is available in whole square yards only, find the cost of installing ceramic tile on the kitchen and dining room floors.

In Exercises 47–48, express the required calculation in terms of π and then round to the nearest tenth.

47. How much fencing is required to enclose a circular garden whose radius is 20 meters?

48. A circular rug is 6 feet in diameter. How many feet of fringe is required to edge this rug?

49. How many plants spaced every 6 inches are needed to surround a circular garden with a 30-foot radius?

50. A stained glass window is to be placed in a house. The window consists of a rectangle, 6 feet high by 3 feet wide, with a semicircle at the top. Approximately how many feet of stripping, to the nearest tenth of a foot, will be needed to frame the window?

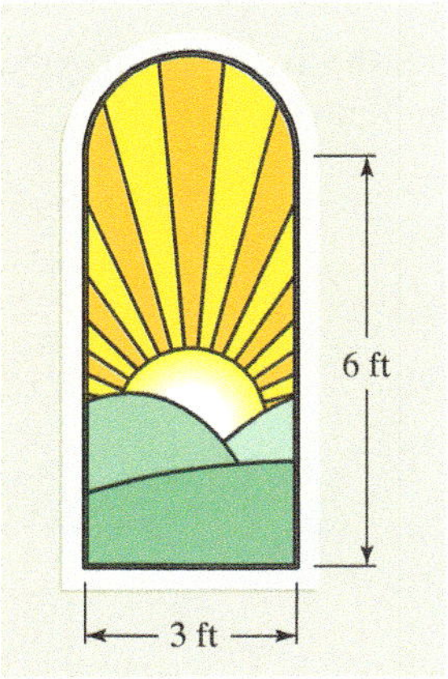

51. Which one of the following is a better buy: a large pizza with a 14-inch diameter for \$12 or a medium pizza with a 7-inch diameter for \$5?

52. Which one of the following is a better buy: a large pizza with a 16-inch diameter for \$12 or two small pizzas, each with a 10-inch diameter, for \$12?

Writing in Mathematics

53. Using the formula for the area of a rectangle, explain how the formula for the area of a parallelogram ($A = bh$) is obtained.

54. Using the formula for the area of a parallelogram ($A = bh$), explain how the formula for the area of a triangle $\left(A = \frac{1}{2}bh\right)$ is obtained.

55. Using the formula for the area of a triangle, explain how the formula for the area of a trapezoid is obtained.

56. Explain why a circle is not a polygon.

57. Describe the difference between the following problems: How much fencing is needed to enclose a circular garden? How much fertilizer is needed for a circular garden?

Critical Thinking Exercises

Make Sense? *In Exercises 58–61, determine whether each statement makes sense or does not make sense, and explain your reasoning.*

58. The house is a 1500-square-foot mansion with six bedrooms.

59. Because a parallelogram can be divided into two triangles with the same size and shape, the area of a triangle is one-half that of a parallelogram.

60. I used $A = \pi r^2$ to determine the amount of fencing needed to enclose my circular garden.

61. I paid \$10 for a pizza, so I would expect to pay approximately \$20 for the same kind of pizza with twice the radius.

62. You need to enclose a rectangular region with 200 feet of fencing. Experiment with different lengths and widths to determine the maximum area you can enclose. Which quadrilateral encloses the most area?

63. Suppose you know the cost for building a rectangular deck measuring 8 feet by 10 feet. If you decide to increase the dimensions to 12 feet by 15 feet, by how much will the cost increase?

64. A rectangular swimming pool measures 14 feet by 30 feet. The pool is surrounded on all four sides by a path that is 3 feet wide. If the cost to resurface the path is \$2 per square foot, what is the total cost of resurfacing the path?

65. A proposed oil pipeline will cross 16.8 miles of national forest. The width of the land needed for the pipeline is 200 feet. If the U.S. Forest Service charges the oil company \$32 per acre, calculate the total cost. (1 mile = 5280 feet and 1 acre = 43,560 square feet.)

10.5

WHAT AM I SUPPOSED TO LEARN?

After you have read this section, you should be able to:

1. Use volume formulas to compute the volumes of three-dimensional figures and solve applied problems.
2. Compute the surface area of a three-dimensional figure.

Volume and Surface Area

YOU ARE CONSIDERING GOING TO JUDGE JUDY'S WEB site and filling out a case submission form to appear on her TV show. The case involves your contractor, who promised to install a water tank that holds 500 gallons of water. Upon delivery, you noticed that capacity was not printed anywhere, so you decided to do some measuring. The tank is shaped like a giant tuna can, with a circular top and bottom. You measured the radius of each circle to be 3 feet and you measured the tank's height to be 2 feet 4 inches. You know that 500 gallons is the capacity of a solid figure with a volume of about 67 cubic feet. Now you need some sort of method to compute the volume of the water tank. In this section, we discuss how to compute the volumes of various solid, three-dimensional figures. Using a formula you will learn in the section, you can determine whether the evidence indicates you can win a case against the contractor if you appear on *Judge Judy*. Or do you risk joining a cast of bozos who entertain television viewers by being loudly castigated by the judge? (Before a possible ear-piercing "Baloney, sir, you're a geometric idiot!", we suggest working Exercise 45 in Exercise Set 10.5.)

1 Use volume formulas to compute the volumes of three-dimensional figures and solve applied problems.

Formulas for Volume

In Section 9.2, we saw that **volume** refers to the amount of space occupied by a solid object, determined by the number of cubic units it takes to fill the interior of that object. For example, **Figure 10.49** shows that there are 18 cubic units contained within the box. The volume of the box, called a **rectangular solid**, is 18 cubic units. The box has a length of 3 units, a width of 3 units, and a height of 2 units. The volume, 18 cubic units, may be determined by finding the product of the length, the width, and the height:

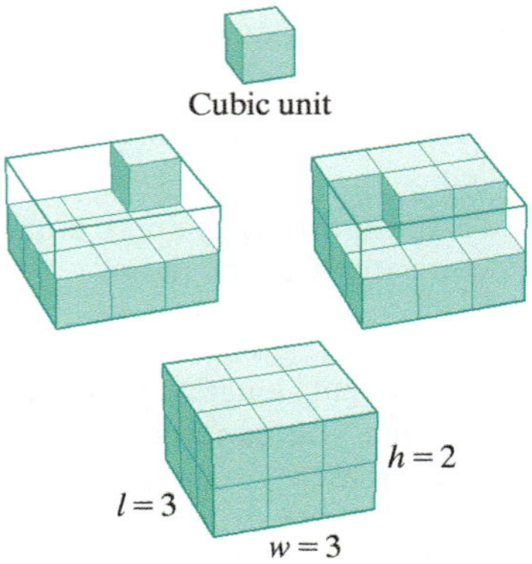

FIGURE 10.49 Volume = 18 cubic units

$$\text{Volume} = 3 \text{ units} \cdot 3 \text{ units} \cdot 2 \text{ units} = 18 \text{ units}^3.$$

In general, the volume, V, of a rectangular solid is the product of its length, l, its width, w, and its height, h:

$$V = lwh.$$

If the length, width, and height are the same, the rectangular solid is called a **cube**. Formulas for these boxlike shapes are given below.

VOLUMES OF BOXLIKE SHAPES

Rectangular solid

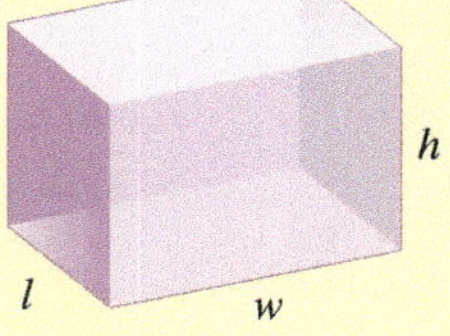

$V = lwh$

Cube

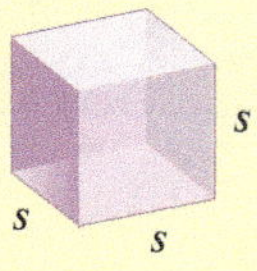

$V = sss = s^3$

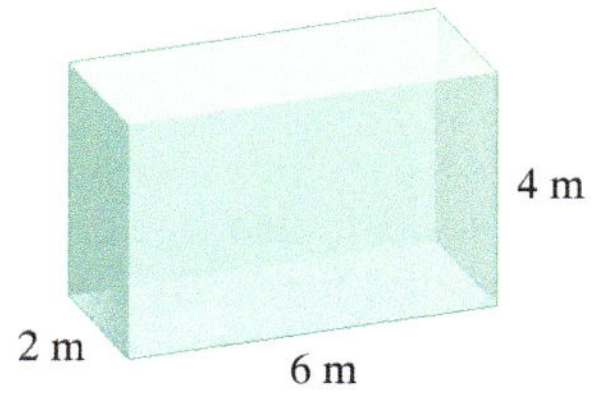

FIGURE 10.50

EXAMPLE 1 Finding the Volume of a Rectangular Solid

Find the volume of the rectangular solid in **Figure 10.50**.

SOLUTION

As shown in the figure, the length is 6 meters, the width is 2 meters, and the height is 4 meters. Thus, $l = 6$, $w = 2$, and $h = 4$.

$$V = lwh = 6\text{ m}\cdot 2\text{ m}\cdot 4\text{ m} = 48\text{ m}^3$$

The volume of the rectangular solid is 48 cubic meters.

CHECK POINT 1 Find the volume of a rectangular solid with length 5 feet, width 3 feet, and height 7 feet.

In Section 9.2, we saw that although there are 3 feet in a yard, there are 27 cubic feet in a cubic yard. If a problem requires measurement of volume in cubic yards and the linear measures are given in feet, to avoid errors, first convert feet to yards. Then apply the volume formula. This idea is illustrated in Example 2.

EXAMPLE 2 Solving a Volume Problem

You are about to begin work on a swimming pool in your yard. The first step is to have a hole dug that is 90 feet long, 60 feet wide, and 6 feet deep. You will use a truck that can carry 10 cubic yards of dirt and charges \$35 per load. How much will it cost you to have all the dirt hauled away?

SOLUTION

We begin by converting feet to yards:

$$90\text{ ft} = \frac{90\text{ ft}}{1}\cdot\frac{1\text{ yd}}{3\text{ ft}} = \frac{90}{3}\text{ yd} = 30\text{ yd}.$$

Similarly, $60\text{ ft} = 20\text{ yd}$ and $6\text{ ft} = 2\text{ yd}$. Next, we find the volume of dirt that needs to be dug out and hauled off.

$$V = lwh = 30\text{ yd}\cdot 20\text{ yd}\cdot 2\text{ yd} = 1200\text{ yd}^3$$

Now, we find the number of loads that the truck needs to haul off all the dirt. Because the truck carries 10 cubic yards, divide the number of cubic yards of dirt by 10.

$$\text{Number of truckloads} = \frac{1200\text{ yd}^3}{\frac{10\text{ yd}^3}{\text{trip}}} = \frac{1200\text{ yd}^3}{1}\cdot\frac{\text{trip}}{10\text{ yd}^3} = \frac{1200}{10}\text{ trips} = 120\text{ trips}$$

Because the truck charges \$35 per trip, the cost to have all the dirt hauled away is the number of trips, 120, times the cost per trip, \$35.

$$\text{Cost to haul all dirt away} = \frac{120\text{ trips}}{1}\cdot\frac{\$35}{\text{trip}} = 120(\$35) = \$4200$$

The dirt-hauling phase of the pool project will cost you \$4200.

CHECK POINT 2 Find the volume, in cubic yards, of a cube whose edges each measure 6 feet.

FIGURE 10.51 The volume of a pyramid is $\frac{1}{3}$ the volume of a rectangular solid having the same base and the same height.

A rectangular solid is an example of a **polyhedron**, a solid figure bounded by polygons. A rectangular solid is bounded by six rectangles, called faces. By contrast, a **pyramid** is a polyhedron whose base is a polygon and whose sides are triangles. **Figure 10.51** shows a pyramid with a rectangular base drawn inside a rectangular solid. The contents of three pyramids with rectangular bases exactly fill a rectangular solid of the same base and height. Thus, the formula for the volume of the pyramid is $\frac{1}{3}$ that of the rectangular solid.

VOLUME OF A PYRAMID

The volume, V, of a pyramid is given by the formula

$$V = \frac{1}{3}Bh,$$

where B is the area of the base and h is the height (the perpendicular distance from the top to the base).

Pyramid

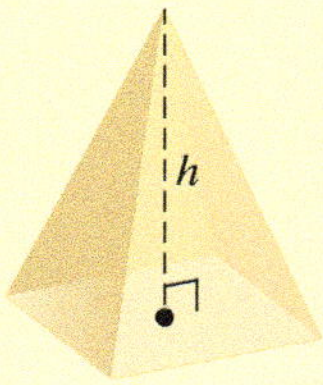

EXAMPLE 3 *Using the Formula for a Pyramid's Volume*

Capped with a pointed spire on top of its 48 stories, the Transamerica Tower in San Francisco is a pyramid with a square base. The pyramid is 256 meters (853 feet) tall. Each side of the square base has a length of 52 meters. Although San Franciscans disliked it when it opened in 1972, they have since accepted it as part of the skyline. Find the volume of the building.

The Transamerica Tower's 3678 windows take cleaners one month to wash. Its foundation is sunk 15.5 m (52 ft) into the ground and is designed to move with earth tremors.

SOLUTION

First find the area of the square base, represented as B in the volume formula. Because each side of the square base is 52 meters, the area of the square base is

$$B = 52\text{ m} \cdot 52\text{ m} = 2704\text{ m}^2.$$

The area of the square base is 2704 square meters. Because the pyramid is 256 meters tall, its height, h, is 256 meters. Now we apply the formula for the volume of a pyramid:

$$V = \frac{1}{3}Bh = \frac{1}{3} \cdot \frac{2704\text{ m}^2}{1} \cdot \frac{256\text{ m}}{1} = \frac{2704 \cdot 256}{3}\text{ m}^3 \approx 230{,}741\text{ m}^3.$$

The volume of the building is approximately 230,741 cubic meters.

The San Francisco pyramid is relatively small compared to the Great Pyramid outside Cairo, Egypt. Built in about 2550 B.C. by a labor force of 100,000, the Great Pyramid is approximately 11 times the volume of San Francisco's pyramid.

CHECK POINT 3 A pyramid is 4 feet tall. Each side of the square base has a length of 6 feet. Find the pyramid's volume.

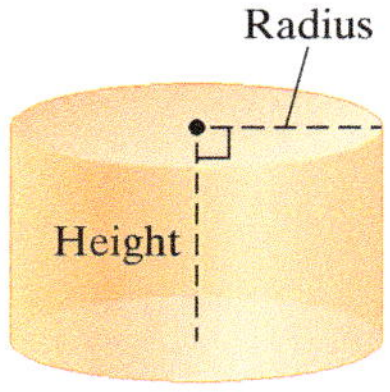

FIGURE 10.52

Not every three-dimensional figure is a polyhedron. Take, for example, the right circular cylinder shown in **Figure 10.52**. Its shape should remind you of a soup can or a stack of coins. The right circular cylinder is so named because the top and bottom are circles, and the side forms a right angle with the top and bottom. The formula for the volume of a right circular cylinder is given as follows:

VOLUME OF A RIGHT CIRCULAR CYLINDER

The volume, V, of a right circular cylinder is given by the formula

$$V = \pi r^2 h,$$

where r is the radius of the circle at either end and h is the height.

Right circular cylinder

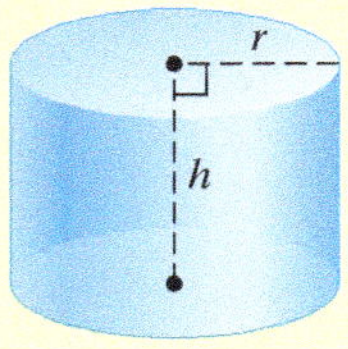

EXAMPLE 4 *Finding the Volume of a Cylinder*

Find the volume of the cylinder in **Figure 10.53**.

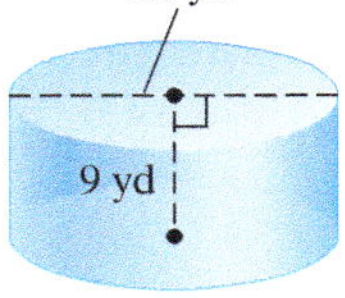

FIGURE 10.53

SOLUTION

In order to find the cylinder's volume, we need both its radius and its height. Because the diameter is 20 yards, the radius is half this length, or 10 yards. The height of the cylinder is given to be 9 yards. Thus, $r = 10$ and $h = 9$. Now we apply the formula for the volume of a cylinder.

$$V = \pi r^2 h = \pi(10\text{ yd})^2 \cdot 9\text{ yd} = 900\pi\text{ yd}^3 \approx 2827\text{ yd}^3$$

The volume of the cylinder is approximately 2827 cubic yards.

CHECK POINT 4 Find the volume, to the nearest cubic inch, of a cylinder with a diameter of 8 inches and a height of 6 inches.

FIGURE 10.54

Figure 10.54 shows a **right circular cone** inside a cylinder, sharing the same circular base as the cylinder. The height of the cone, the perpendicular distance from the top to the circular base, is the same as that of the cylinder. Three such cones can occupy the same amount of space as the cylinder. Therefore, the formula for the volume of the cone is $\frac{1}{3}$ the volume of the cylinder.

VOLUME OF A CONE

The volume, V, of a right circular cone that has height h and radius r is given by the formula

$$V = \frac{1}{3}\pi r^2 h.$$

Cone

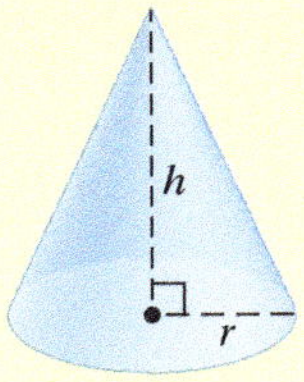

EXAMPLE 5 *Finding the Volume of a Cone*

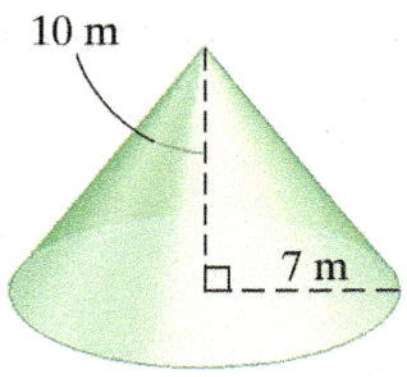

FIGURE 10.55

Find the volume of the cone in **Figure 10.55**.

SOLUTION

The radius of the cone is 7 meters and the height is 10 meters. Thus, $r = 7$ and $h = 10$. Now we apply the formula for the volume of a cone.

$$V = \frac{1}{3}\pi r^2 h = \frac{1}{3}\pi(7 \text{ m})^2 \cdot 10 \text{ m} = \frac{490\pi}{3} \text{ m}^3 \approx 513 \text{ m}^3$$

The volume of the cone is approximately 513 cubic meters.

CHECK POINT 5 Find the volume, to the nearest cubic inch, of a cone with a radius of 4 inches and a height of 6 inches.

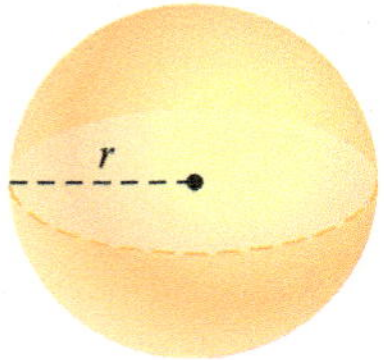

FIGURE 10.56

Figure 10.56 shows a *sphere*. Its shape may remind you of a basketball. The Earth is not a perfect sphere, but it's close. A **sphere** is the set of points in space equally distant from a given point, its **center**. Any line segment from the center to a point on the sphere is a **radius** of the sphere. The word *radius* is also used to refer to the length of this line segment. A sphere's volume can be found by using π and its radius.

VOLUME OF A SPHERE

The volume, V, of a sphere of radius r is given by the formula

$$V = \frac{4}{3}\pi r^3.$$

Sphere

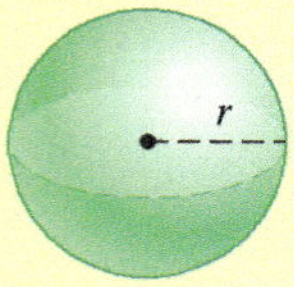

EXAMPLE 6 *Applying Volume Formulas*

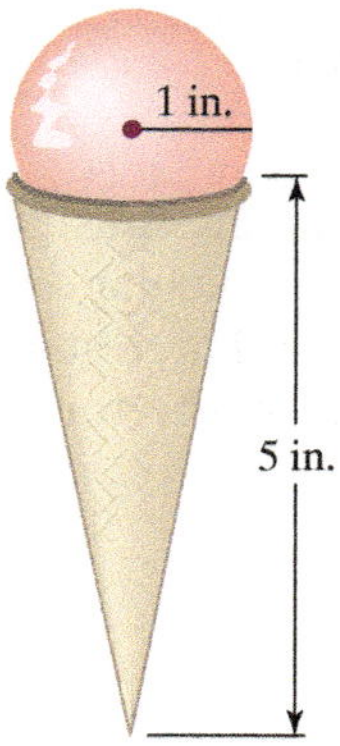

FIGURE 10.57

An ice cream cone is 5 inches deep and has a radius of 1 inch. A spherical scoop of ice cream also has a radius of 1 inch. (See **Figure 10.57**.) If the ice cream melts into the cone, will it overflow?

SOLUTION

The ice cream will overflow if the volume of the ice cream, a sphere, is greater than the volume of the cone. Find the volume of each.

$$V_{\text{cone}} = \frac{1}{3}\pi r^2 h = \frac{1}{3}\pi(1 \text{ in.})^2 \cdot 5 \text{ in.} = \frac{5\pi}{3} \text{ in.}^3 \approx 5 \text{ in.}^3$$

$$V_{\text{sphere}} = \frac{4}{3}\pi r^3 = \frac{4}{3}\pi(1 \text{ in.})^3 = \frac{4\pi}{3} \text{ in.}^3 \approx 4 \text{ in.}^3$$

The volume of the spherical scoop of ice cream is less than the volume of the cone, so there will be no overflow.

CHECK POINT 6 A basketball has a radius of 4.5 inches. If the ball is filled with 350 cubic inches of air, is this enough air to fill it completely?

2 Compute the surface area of a three-dimensional figure.

Surface Area

In addition to volume, we can also measure the area of the outer surface of a three-dimensional object, called its **surface area**. Like area, surface area is measured in square units. For example, the surface area of the rectangular solid in **Figure 10.58** is the sum of the areas of the six outside rectangles of the solid.

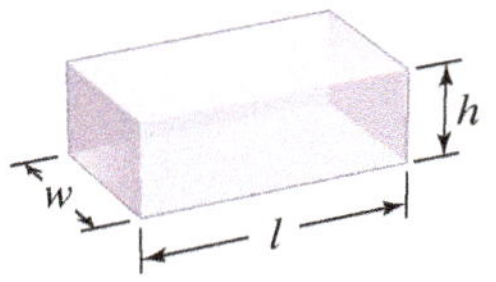

FIGURE 10.58

$$\text{Surface Area} = lw + lw + lh + lh + wh + wh$$

Areas of top and bottom rectangles — Areas of front and back rectangles — Areas of rectangles on left and right sides

$$= 2lw + 2lh + 2wh$$

Formulas for the surface area, abbreviated SA, of three-dimensional figures are given in **Table 10.4**.

Table 10.4 Common Formulas for Surface Area

Cube	Rectangular Solid	Circular Cylinder
$SA = 6s^2$	$SA = 2lw + 2lh + 2wh$	$SA = 2\pi r^2 + 2\pi rh$
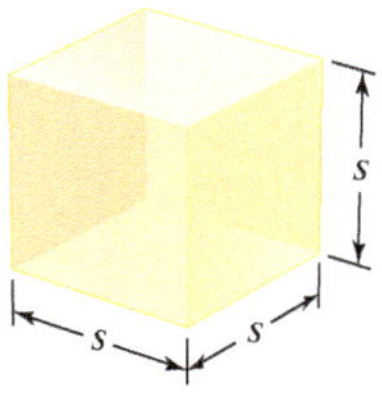	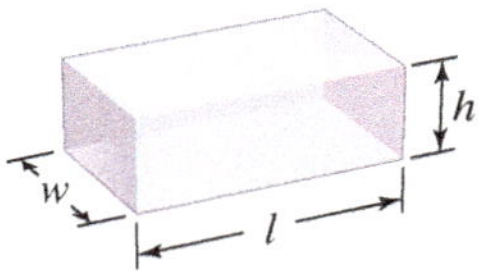	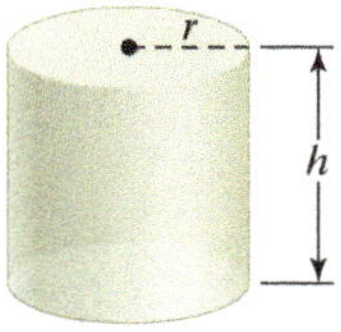

EXAMPLE 7 *Finding the Surface Area of a Solid*

Find the surface area of the rectangular solid in **Figure 10.59**.

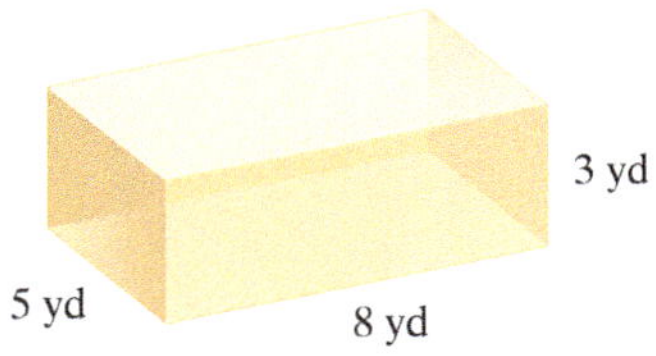

FIGURE 10.59

SOLUTION

As shown in the figure, the length is 8 yards, the width is 5 yards, and the height is 3 yards. Thus, $l = 8$, $w = 5$, and $h = 3$.

$$\begin{aligned} SA &= 2lw + 2lh + 2wh \\ &= 2 \cdot 8 \text{ yd} \cdot 5 \text{ yd} + 2 \cdot 8 \text{ yd} \cdot 3 \text{ yd} + 2 \cdot 5 \text{ yd} \cdot 3 \text{ yd} \\ &= 80 \text{ yd}^2 + 48 \text{ yd}^2 + 30 \text{ yd}^2 = 158 \text{ yd}^2 \end{aligned}$$

The surface area is 158 square yards.

CHECK POINT 7 If the length, width, and height shown in **Figure 10.59** are each doubled, find the surface area of the resulting rectangular solid.

Concept and Vocabulary Check

Fill in each blank so that the resulting statement is true.

1. The volume, V, of a rectangular solid with length l, width w, and height h is given by the formula ________.
2. The volume, V, of a cube with an edge that measures s linear units is given by the formula ________.
3. A solid figure bounded by polygons is called a/an ________.
4. The volume, V, of a pyramid with base area B and height h is given by the formula ________.
5. The volume, V, of a right circular cylinder with height h and radius r is given by the formula ________.
6. The volume, V, of a right circular cone with height h and radius r is given by the formula ________.
7. The volume, V, of a sphere of radius r is given by the formula ________.

In Exercises 8–14, determine whether each statement is true or false. If the statement is false, make the necessary change(s) to produce a true statement.

8. A cube is a rectangular solid with the same length, width, and height. ______

9. A cube is an example of a polyhedron. ______

10. The volume of a pyramid is $\frac{1}{2}$ the volume of a rectangular solid having the same base and the same height. ______

11. Some three-dimensional figures are not polyhedrons. ______

12. A sphere is the set of points in space equally distant from its center. ______

13. Surface area refers to the area of the outer surface of a three-dimensional object. ______

14. The surface area, SA, of a rectangular solid with length l, width w, and height h is given by the formula $SA = lw + lh + wh$. ______

Exercise Set 10.5

Practice Exercises

In Exercises 1–20, find the volume of each figure. If necessary, express answers in terms of π and then round to the nearest whole number.

1.

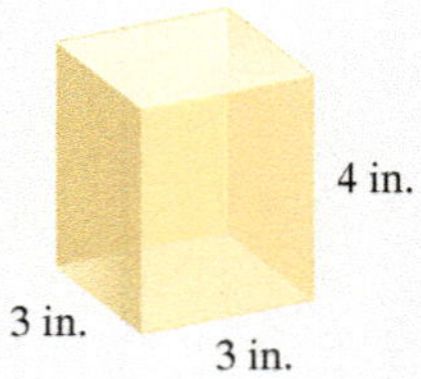

2.

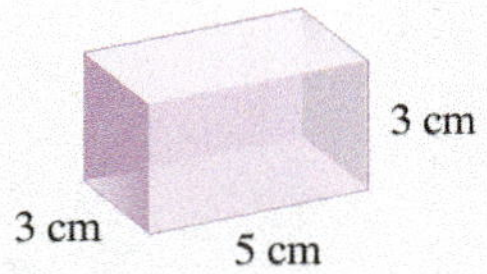

3.

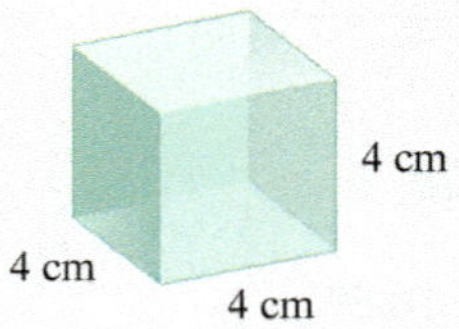

4.

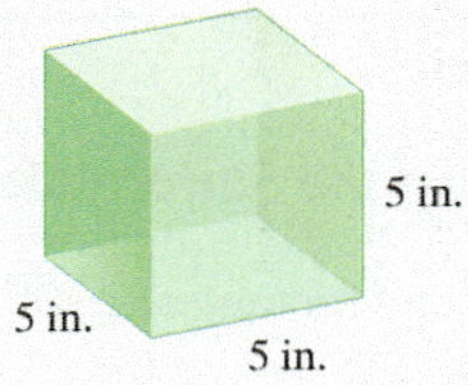

5.

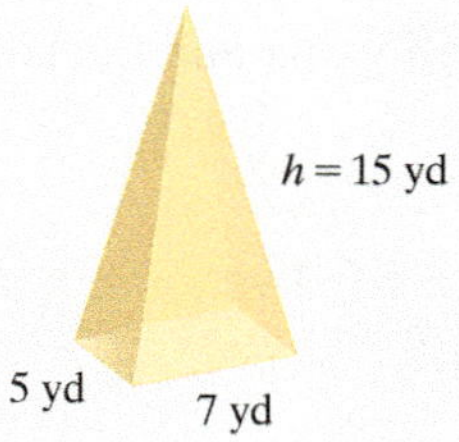

6.

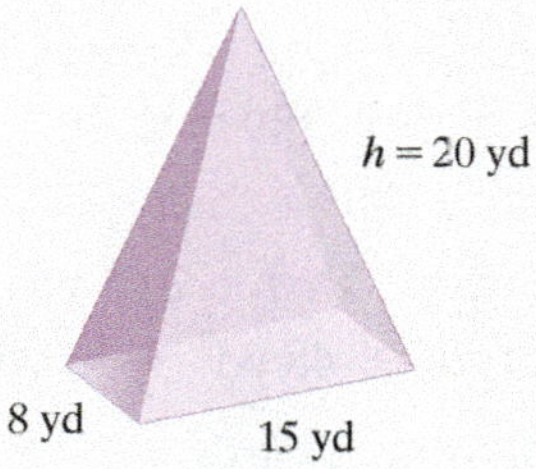

7.

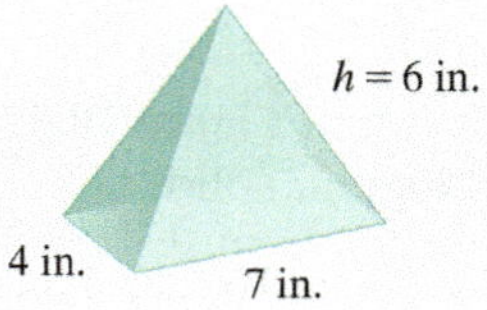

8.

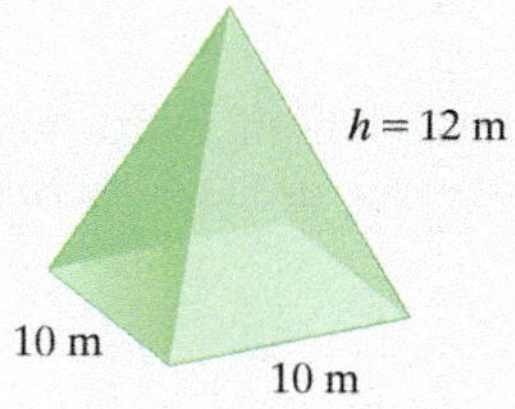

9.

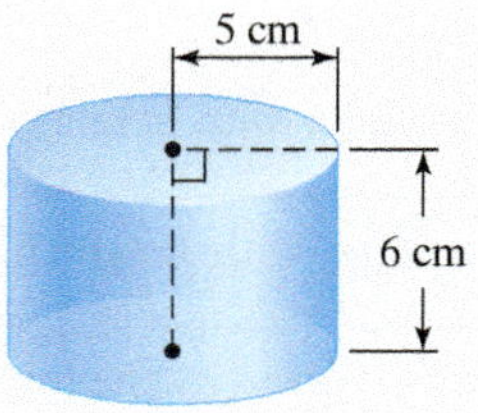

10.

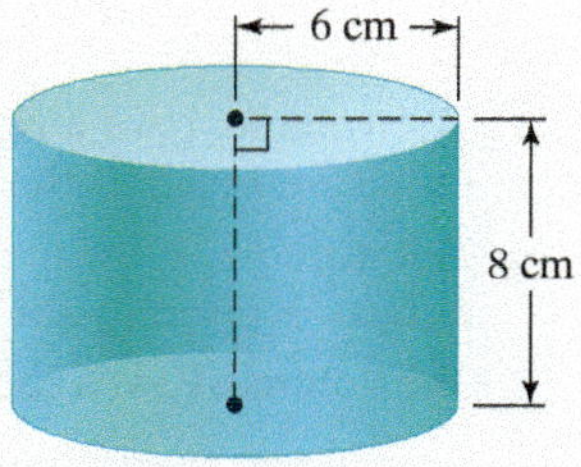

11.

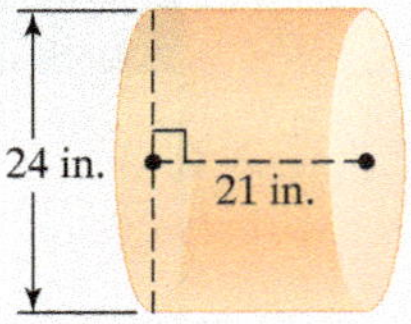

12.

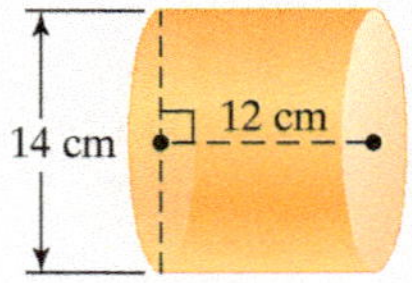

13.

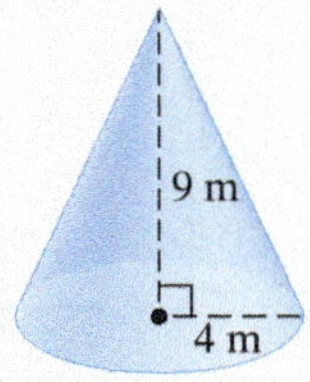

14.

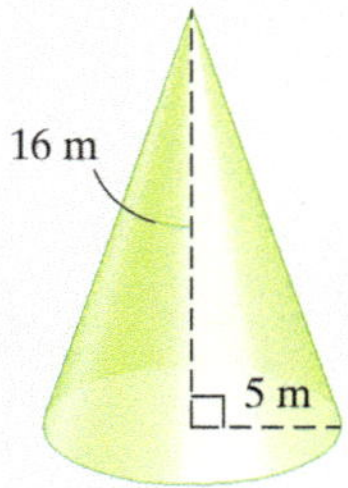

15.

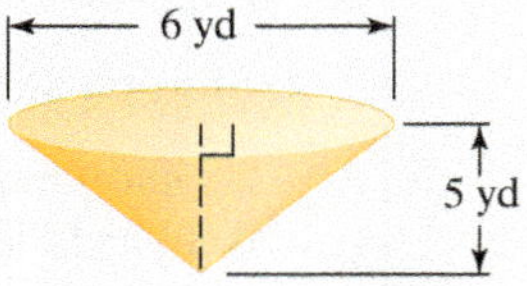

16.

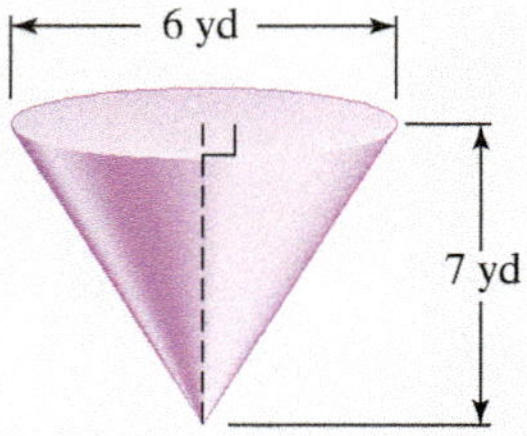

17.

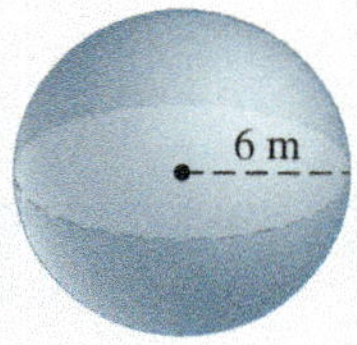

18.

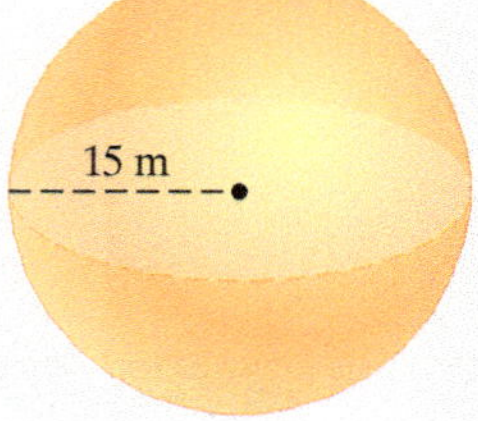

19.

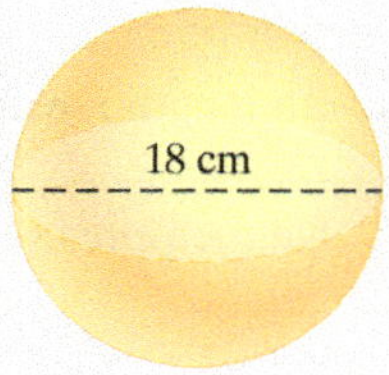

20.

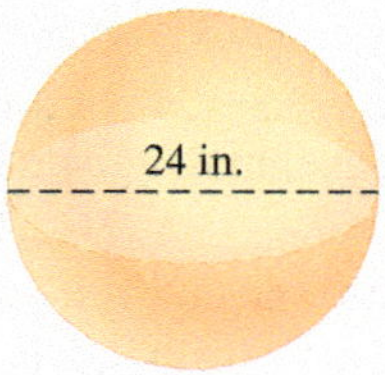

In Exercises 21–24, find the surface area of each figure.

21.

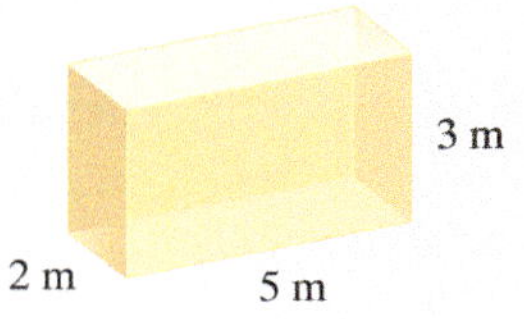

22.

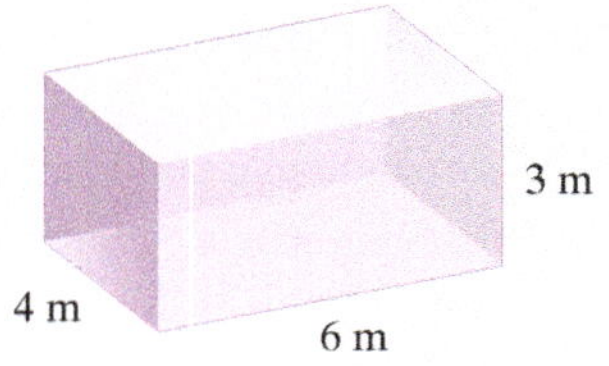

23.

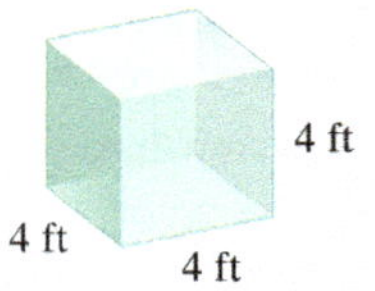

24.

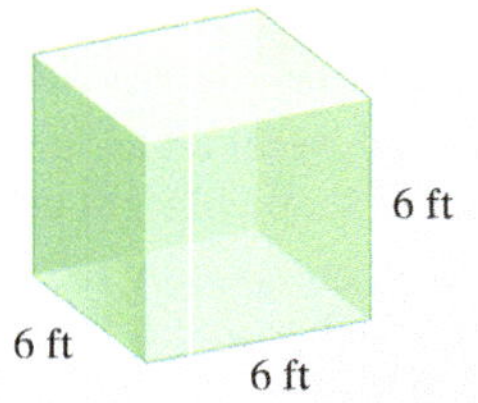

In Exercises 25–30, use two formulas for volume to find the volume of each figure. Express answers in terms of π and then round to the nearest whole number.

25.

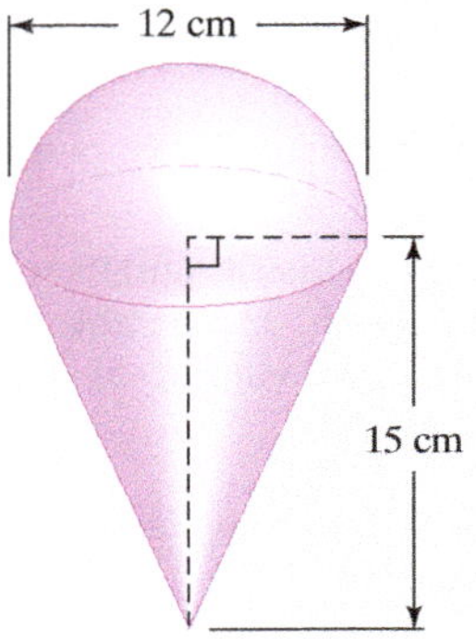

26.

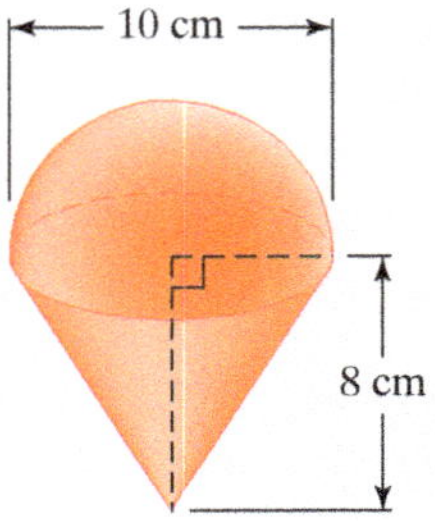

27.

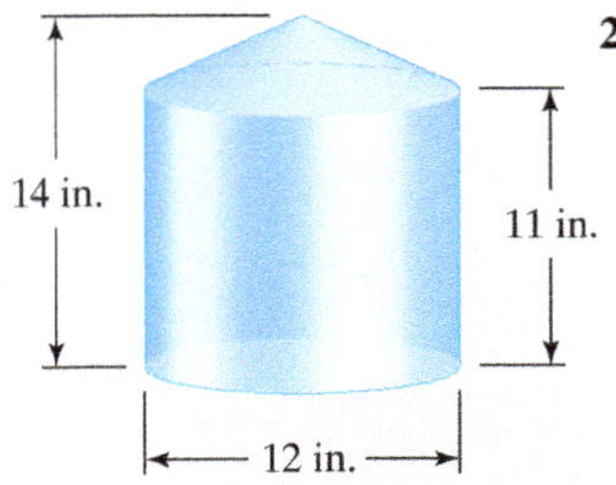

28.

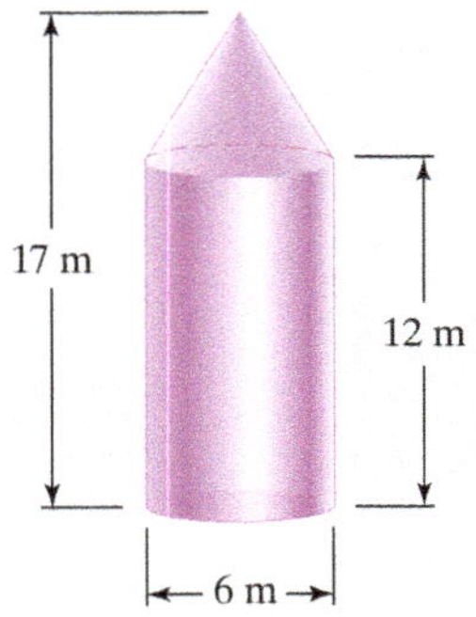

29.

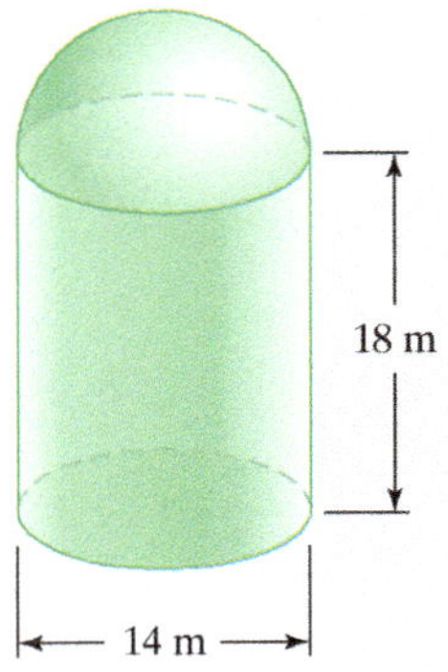

30.

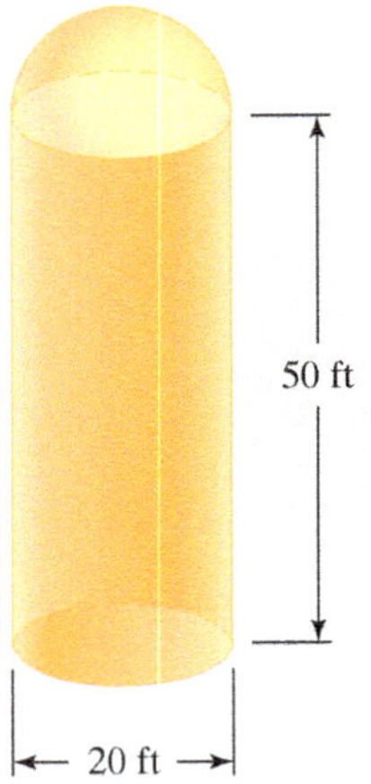

Practice Plus

31. Find the surface area and the volume of the figure shown.

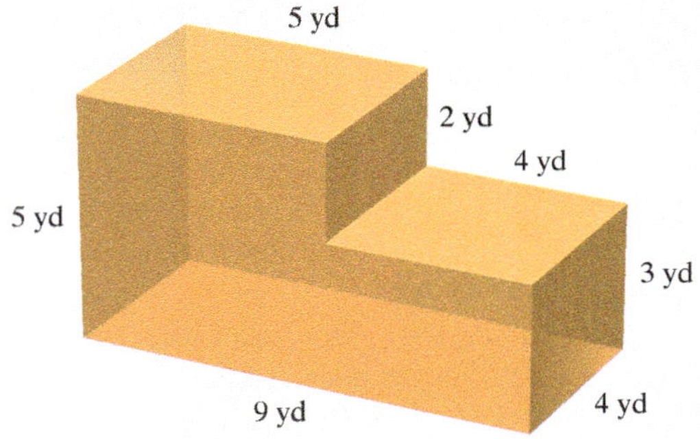

32. Find the surface area and the volume of the cement block in the figure shown.

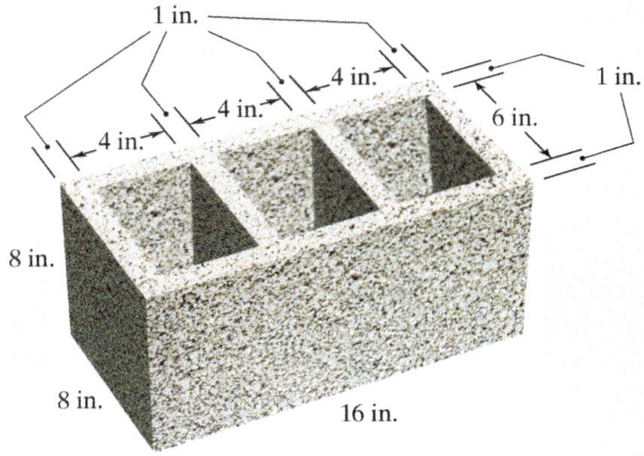

33. Find the surface area of the figure shown.

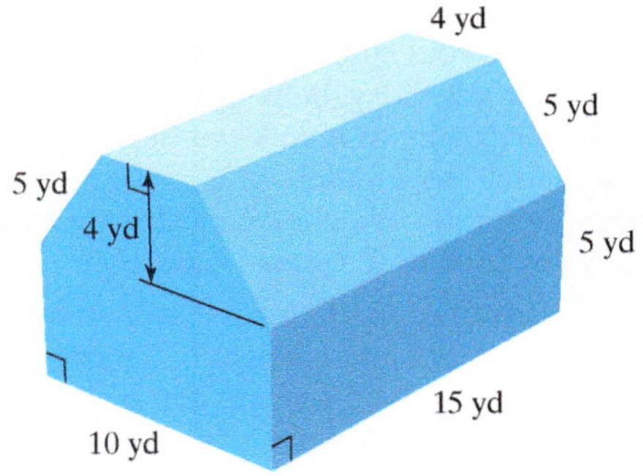

34. A machine produces open boxes using square sheets of metal measuring 12 inches on each side. The machine cuts equal-sized squares whose sides measure 2 inches from each corner. Then it shapes the metal into an open box by turning up the sides. Find the volume of the box.

35. Find the ratio, reduced to lowest terms, of the volume of a sphere with a radius of 3 inches to the volume of a sphere with a radius of 6 inches.

36. Find the ratio, reduced to lowest terms, of the volume of a sphere with a radius of 3 inches to the volume of a sphere with a radius of 9 inches.

37. A cylinder with radius 3 inches and height 4 inches has its radius tripled. How many times greater is the volume of the larger cylinder than the smaller cylinder?

38. A cylinder with radius 2 inches and height 3 inches has its radius quadrupled. How many times greater is the volume of the larger cylinder than the smaller cylinder?

Application Exercises

39. A building contractor is to dig a foundation 12 feet long, 9 feet wide, and 6 feet deep for a toll booth. The contractor pays $85 per load for trucks to remove the dirt. Each truck holds 6 cubic yards. What is the cost to the contractor to have all the dirt hauled away?

40. What is the cost of concrete for a walkway that is 15 feet long, 8 feet wide, and 9 inches deep if the concrete costs $30 per cubic yard?

41. A furnace is designed to heat 10,000 cubic feet. Will this furnace be adequate for a 1400-square-foot house with a 9-foot ceiling?

42. A water reservoir is shaped like a rectangular solid with a base that is 50 yards by 30 yards, and a vertical height of 20 yards. At the start of a three-month period of no rain, the reservoir was completely full. At the end of this period, the height of the water was down to 6 yards. How much water was used in the three-month period?

43. The Great Pyramid outside Cairo, Egypt, has a square base measuring 756 feet on a side and a height of 480 feet.

a. What is the volume of the Great Pyramid, in cubic yards?

b. The stones used to build the Great Pyramid were limestone blocks with an average volume of 1.5 cubic yards. How many of these blocks were needed to construct the Great Pyramid?

44. Although the Eiffel Tower in Paris is not a solid pyramid, its shape approximates that of a pyramid with a square base measuring 120 feet on a side and a height of 980 feet. If it were a solid pyramid, what would be the Eiffel Tower's volume, in cubic yards?

45. You are about to sue your contractor who promised to install a water tank that holds 500 gallons of water. You know that 500 gallons is the capacity of a tank that holds 67 cubic feet. The cylindrical tank has a radius of 3 feet and a height of 2 feet 4 inches. Does the evidence indicate you can win the case against the contractor if it goes to court?

46. Two cylindrical cans of soup sell for the same price. One can has a diameter of 6 inches and a height of 5 inches. The other has a diameter of 5 inches and a height of 6 inches. Which can contains more soup and, therefore, is the better buy?

47. A circular backyard pool has a diameter of 24 feet and is 4 feet deep. One cubic foot of water has a capacity of approximately 7.48 gallons. If water costs $2 per thousand gallons, how much, to the nearest dollar, will it cost to fill the pool?

48. The tunnel under the English Channel that connects England and France is the world's longest tunnel. There are actually three separate tunnels built side by side. Each is a half-cylinder that is 50,000 meters long and 4 meters high. How many cubic meters of dirt had to be removed to build the tunnel?

Writing in Mathematics

49. Explain the following analogy:

In terms of formulas used to compute volume, a pyramid is to a rectangular solid just as a cone is to a cylinder.

50. Explain why a cylinder is not a polyhedron.

Critical Thinking Exercises

Make Sense? *In Exercises 51–54, determine whether each statement makes sense or does not make sense, and explain your reasoning.*

51. The physical education department ordered new basketballs in the shape of right circular cylinders.

52. When completely full, a cylindrical soup can with a diameter of 3 inches and a height of 4 inches holds more soup than a cylindrical can with a diameter of 4 inches and a height of 3 inches.

53. I found the volume of a rectangular solid in cubic inches and then divided by 12 to convert the volume to cubic feet.

54. Because a cylinder is a solid figure, I use cubic units to express its surface area.

55. What happens to the volume of a sphere if its radius is doubled?

56. A scale model of a car is constructed so that its length, width, and height are each $\frac{1}{10}$ the length, width, and height of the actual car. By how many times does the volume of the car exceed its scale model?

In Exercises 57–58, find the volume of the darkly shaded region. If necessary, round to the nearest whole number.

57.

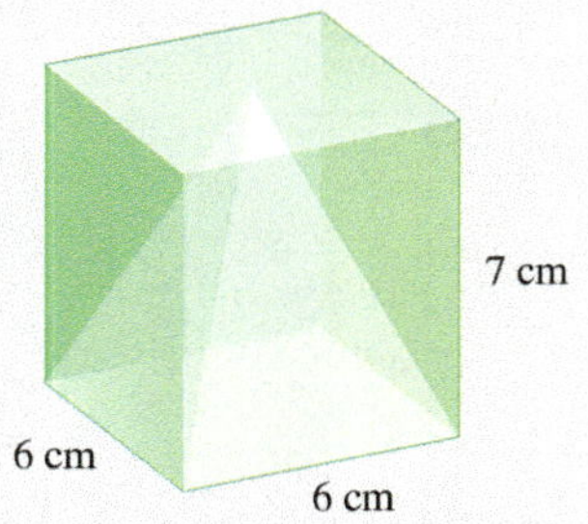

58.

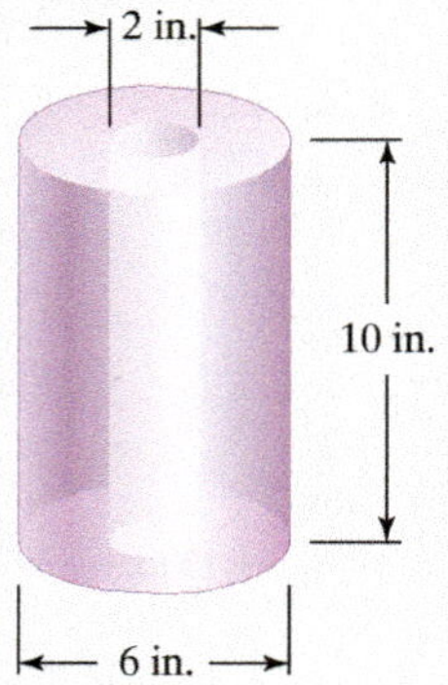

59. Find the surface area of the figure shown.

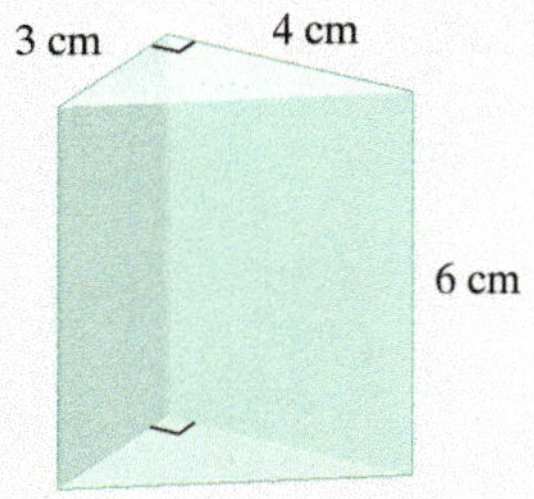

10.6

WHAT AM I SUPPOSED TO LEARN?

After you have read this section, you should be able to:

1. Use the lengths of the sides of a right triangle to find trigonometric ratios.
2. Use trigonometric ratios to find missing parts of right triangles.
3. Use trigonometric ratios to solve applied problems.

Right Triangle Trigonometry

MOUNTAIN CLIMBERS HAVE FOREVER BEEN fascinated by reaching the top of Mount Everest, sometimes with tragic results. The mountain, on Asia's Tibet-Nepal border, is Earth's highest, peaking at an incredible 29,035 feet. The heights of mountains can be found using *trigonometry*. The word **trigonometry** means *measurement of triangles.* Trigonometry is used in navigation, building, and engineering. For centuries, Muslims used trigonometry and the stars to navigate across the Arabian desert to Mecca, the birthplace of the prophet Muhammad, the founder of Islam. The ancient Greeks used trigonometry to record the locations of thousands of stars and worked out the motion of the Moon relative to Earth. Today, trigonometry is used to study the structure of DNA, the master molecule that determines how we grow from a single cell to a complex, fully developed adult.

Ratios in Right Triangles

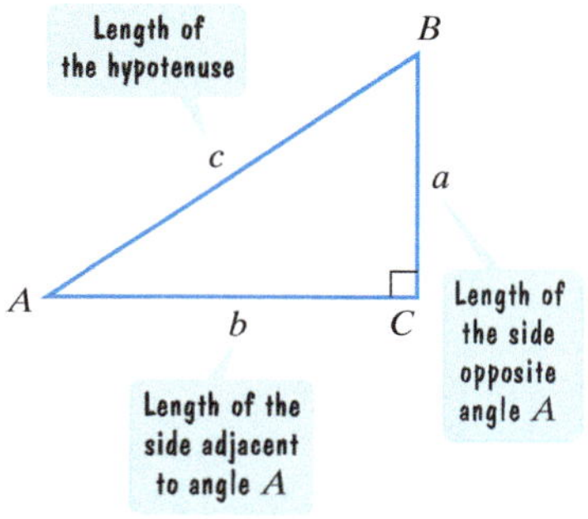

FIGURE 10.60 Naming a right triangle's sides from the point of view of an acute angle

The right triangle forms the basis of trigonometry. If either acute angle of a right triangle stays the same size, the shape of the triangle does not change even if it is made larger or smaller. Because of properties of similar triangles, this means that the ratios of certain lengths stay the same regardless of the right triangle's size. These ratios have special names and are defined in terms of the **side opposite** an acute angle, the **side adjacent** to the acute angle, and the **hypotenuse**. In **Figure 10.60**, the length of the hypotenuse, the side opposite the 90° angle, is represented by c. The length of the side opposite angle A is represented by a. The length of the side adjacent to angle A is represented by b.

The three fundamental trigonometric ratios, **sine** (abbreviated sin), **cosine** (abbreviated cos), and **tangent** (abbreviated tan), are defined as ratios of the lengths of the sides of a right triangle. In the box that follows, when a side of a triangle is mentioned, we are referring to the *length* of that side.

1 Use the lengths of the sides of a right triangle to find trigonometric ratios.

TRIGONOMETRIC RATIOS

Let A represent an acute angle of a right triangle, with right angle C, shown in **Figure 10.60**. For angle A, the trigonometric ratios are defined as follows:

sine of A
$$\sin A = \frac{\text{side opposite angle } A}{\text{hypotenuse}} = \frac{a}{c}$$

cosine of A
$$\cos A = \frac{\text{side adjacent to angle } A}{\text{hypotenuse}} = \frac{b}{c}$$

tangent of A
$$\tan A = \frac{\text{side opposite angle } A}{\text{side adjacent to angle } A} = \frac{a}{b}.$$

GREAT QUESTION!

Is there a way to help me remember the definitions of the trigonometric ratios?

The word

SOHCAHTOA (pronounced: so-cah-tow-ah)

may be helpful in remembering the definitions of the three trigonometric ratios.

S	O H	C	A H	T	O A
Sine	$\frac{\text{opp}}{\text{hyp}}$	Cosine	$\frac{\text{adj}}{\text{hyp}}$	Tangent	$\frac{\text{opp}}{\text{adj}}$

"Some Old Hog Came Around Here and Took Our Apples."

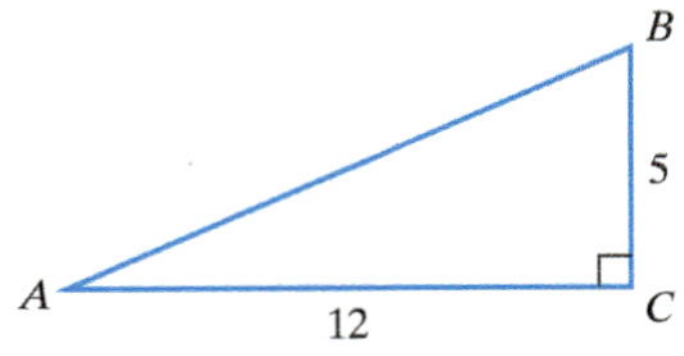

FIGURE 10.61

EXAMPLE 1 *Becoming Familiar with the Trigonometric Ratios*

Find the sine, cosine, and tangent of A in **Figure 10.61**.

SOLUTION

We begin by finding the measure of the hypotenuse c using the Pythagorean Theorem.

$$c^2 = a^2 + b^2 = 5^2 + 12^2 = 25 + 144 = 169$$

$$c = \sqrt{169} = 13$$

Now, we apply the definitions of the trigonometric ratios.

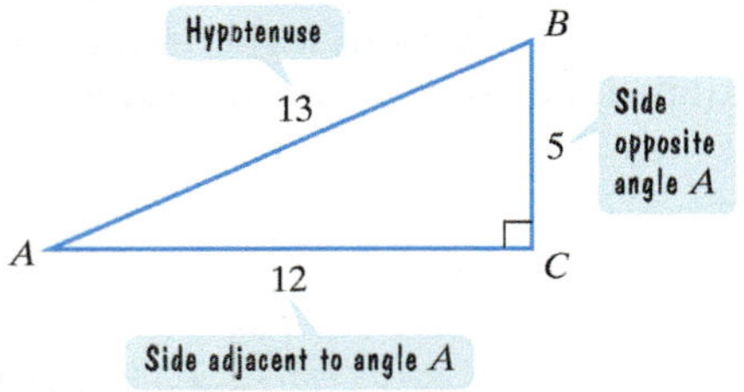

$$\sin A = \frac{\text{side opposite angle } A}{\text{hypotenuse}} = \frac{5}{13}$$

$$\cos A = \frac{\text{side adjacent to angle } A}{\text{hypotenuse}} = \frac{12}{13}$$

$$\tan A = \frac{\text{side opposite angle } A}{\text{side adjacent to angle } A} = \frac{5}{12}$$

GREAT QUESTION!

Can you clarify which is the opposite side and which is the adjacent side?

We know that the longest side of the right triangle is the hypotenuse. The two legs of the triangle are described by their relationship to each acute angle. The adjacent leg "touches" the acute angle by forming one of the angle's sides. The opposite leg is not a side of the acute angle because it lies directly opposite the angle.

CHECK POINT 1 Find the sine, cosine, and tangent of A in the figure shown.

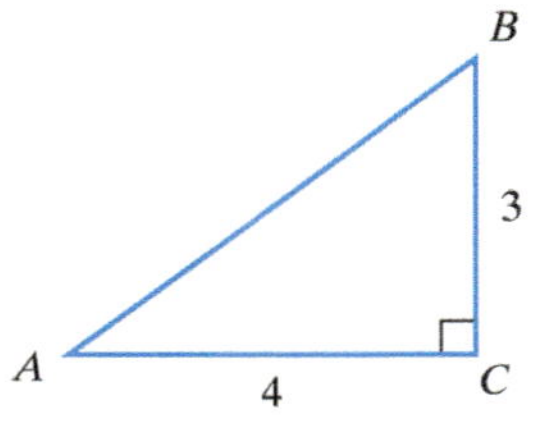

A BRIEF REVIEW *Equations with Fractions*

Finding missing parts of right triangles involves solving equations with fractions.

- The variable can appear in a fraction's numerator.

Example

Solve for a: $0.3 = \frac{a}{150}$

$150(0.3) = 150\left(\frac{a}{150}\right)$ Clear fractions by multiplying both sides by 150.

$45 = a$ Simplify: $150(0.3) = 45$ and $150\left(\frac{a}{150}\right) = \frac{\overset{1}{\cancel{150}}}{1}\cdot\frac{a}{\underset{1}{\cancel{150}}} = a.$

- The variable can appear in a fraction's denominator.

Example

Solve for c: $0.8 = \frac{150}{c}$

$0.8c = \left(\frac{150}{c}\right)c$ Clear fractions by multiplying both sides by c.

$0.8c = 150$ Simplify: $\left(\frac{150}{c}\right)c = \frac{150}{\underset{1}{\cancel{c}}}\cdot\frac{\overset{1}{\cancel{c}}}{1} = 150.$

We're still not done. We have not isolated c.

$\frac{0.8c}{0.8} = \frac{150}{0.8}$ Divide both sides by 0.8.

$c = 187.5$ Simplify.

2 Use trigonometric ratios to find missing parts of right triangles.

A scientific or graphing calculator in the degree mode will give you decimal approximations for the trigonometric ratios of any angle. For example, to find an approximation for tan 37°, the tangent of an angle measuring 37°, a keystroke sequence similar to one of the following can be used:

Many Scientific Calculators: 37 [TAN]

Many Graphing Calculators: [TAN] 37 [ENTER].

The tangent of 37°, rounded to four decimal places, is 0.7536.

If we are given the length of one side and the measure of an acute angle of a right triangle, we can use trigonometry to solve for the length of either of the other two sides. Example 2 illustrates how this is done.

EXAMPLE 2 *Finding a Missing Leg of a Right Triangle*

Find a in the right triangle in **Figure 10.62**.

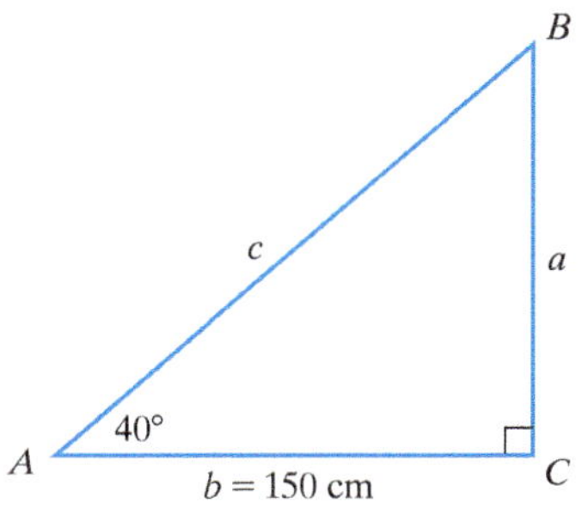

FIGURE 10.62

SOLUTION

We need to identify a trigonometric ratio that will make it possible to find a. Because we have a known angle, 40°, an unknown opposite side, a, and a known adjacent side, 150 cm, we use the tangent ratio.

$\tan 40° = \frac{a}{150}$ (Side opposite the 40° angle / Side adjacent to the 40° angle)

Now we solve for a by multiplying both sides by 150.

$a = 150 \tan 40° \approx 126$

The tangent ratio reveals that a is approximately 126 centimeters.

TECHNOLOGY

Here is the keystroke sequence for 150 tan 40°:

Many Scientific Calculators

150 [×] 40 [TAN] [=]

Many Graphing Calculators

150 [TAN] 40 [ENTER]

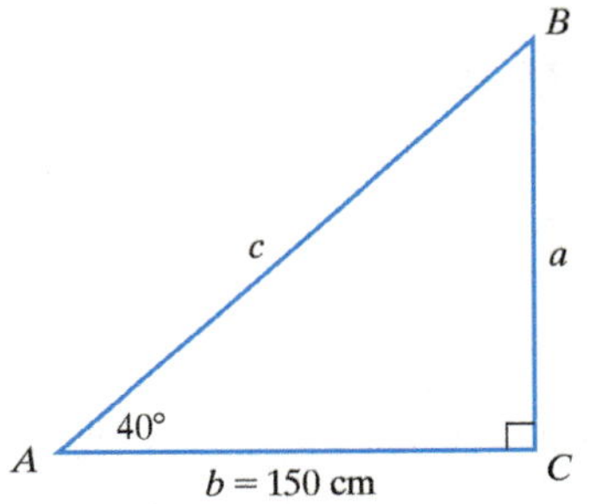

FIGURE 10.62 (repeated)

CHECK POINT 2 In **Figure 10.62**, let $m\measuredangle A = 62°$ and $b = 140$ cm. Find a to the nearest centimeter.

EXAMPLE 3 *Finding a Missing Hypotenuse of a Right Triangle*

Find c in the right triangle in **Figure 10.62**.

SOLUTION

In Example 2, we found a: $a \approx 126$. Because we are given that $b = 150$, it is possible to find c using the Pythagorean Theorem: $c^2 = a^2 + b^2$. However, if we made an error in computing a, we will perpetuate our mistake using this approach.

Instead, we will use the quantities given and identify a trigonometric ratio that will make it possible to find c. Refer to **Figure 10.62**. Because we have a known angle, 40°, a known adjacent side, 150 cm, and an unknown hypotenuse, c, we use the cosine ratio.

$$\cos 40° = \frac{150}{c}$$

150: Side adjacent to the 40° angle
c: Hypotenuse

$$c \cos 40° = 150 \qquad \text{Multiply both sides by } c.$$

$$c = \frac{150}{\cos 40°} \qquad \text{Divide both sides by } \cos 40°.$$

$$c \approx 196 \qquad \text{Use a calculator.}$$

The cosine ratio reveals that the hypotenuse is approximately 196 centimeters.

TECHNOLOGY

Here is the keystroke sequence for $\frac{150}{\cos 40°}$:

Many Scientific Calculators

150 [÷] 40 [COS] [=]

Many Graphing Calculators

150 [÷] [COS] 40 [ENTER]

CHECK POINT 3 In **Figure 10.62**, let $m\measuredangle A = 62°$ and $b = 140$ cm. Find c to the nearest centimeter.

3 Use trigonometric ratios to solve applied problems.

Applications of the Trigonometric Ratios

Trigonometry was first developed to determine heights and distances that are inconvenient or impossible to measure. These applications often involve the angle made with an imaginary horizontal line. As shown in **Figure 10.63**, an angle formed by a horizontal line and the line of sight to an object that is above the horizontal line is called the **angle of elevation**. The angle formed by a horizontal line and the line of sight to an object that is below the horizontal line is called the **angle of depression**. Transits and sextants are instruments used to measure such angles.

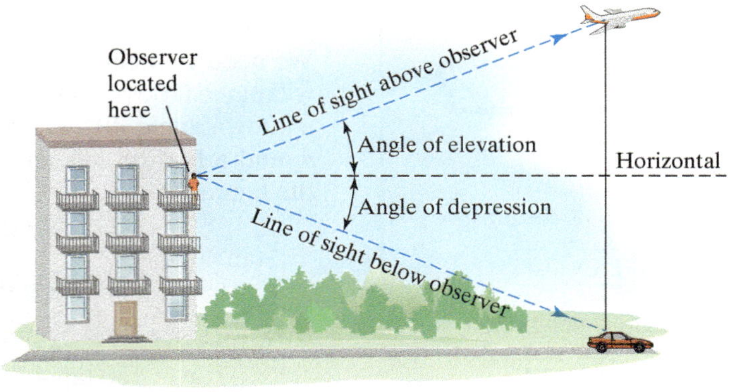

FIGURE 10.63

EXAMPLE 4 *Problem Solving Using an Angle of Elevation*

From a point on level ground 125 feet from the base of a tower, the angle of elevation to the top of the tower is 57.2°. Approximate the height of the tower to the nearest foot.

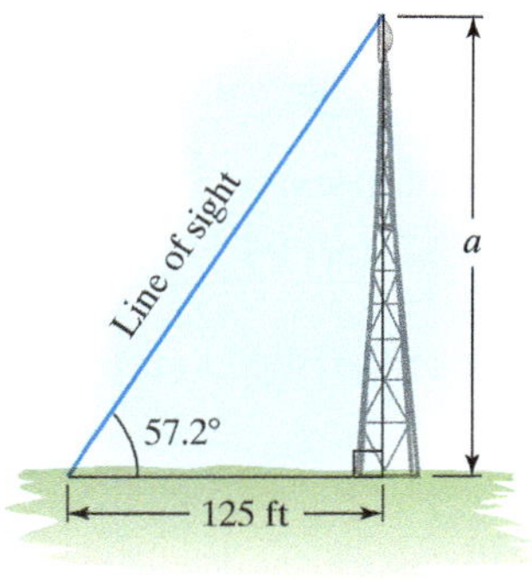

FIGURE 10.64 Determining height without using direct measurement

SOLUTION

A sketch is shown in **Figure 10.64**, where a represents the height of the tower. In the right triangle, we have a known angle, an unknown opposite side, and a known adjacent side. Therefore, we use the tangent ratio.

$$\tan 57.2° = \frac{a}{125}$$

Side opposite the 57.2° angle

Side adjacent to the 57.2° angle

We solve for a by multiplying both sides of this equation by 125:

$$a = 125 \tan 57.2° \approx 194.$$

The tower is approximately 194 feet high.

CHECK POINT 4 From a point on level ground 80 feet from the base of the Eiffel Tower, the angle of elevation is 85.4°. Approximate the height of the Eiffel Tower to the nearest foot.

If the measures of two sides of a right triangle are known, the measures of the two acute angles can be found using the **inverse trigonometric keys** on a calculator. For example, suppose that $\sin A = 0.866$. We can find the measure of angle A by using the *inverse sine* key, usually labeled [SIN^{-1}]. The key [SIN^{-1}] is not a button you will actually press; it is the secondary function for the button labeled [SIN].

Many Scientific Calculators:

.866 [2nd] [SIN]

Many Graphing Calculators:

[2nd] [SIN] .866 [ENTER]

Pressing [2nd] [SIN] accesses the inverse sine key, [SIN^{-1}].

The display should show approximately 59.99°, which can be rounded to 60°. Thus, if $\sin A = 0.866$, then $m\measuredangle A \approx 60°$.

EXAMPLE 5 *Determining the Angle of Elevation*

A building that is 21 meters tall casts a shadow 25 meters long. Find the angle of elevation of the sun.

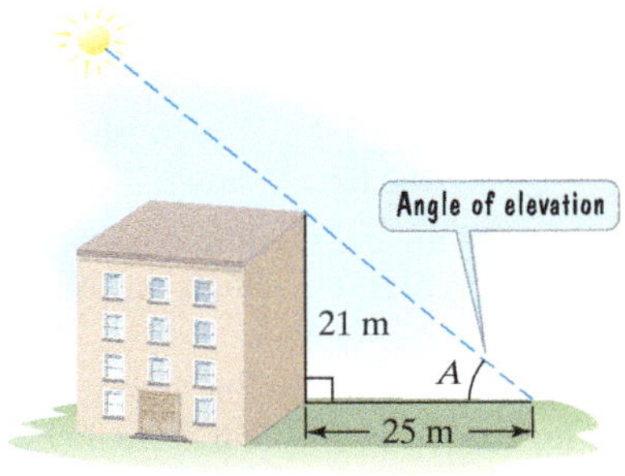

FIGURE 10.65

SOLUTION

The situation is illustrated in **Figure 10.65**. We are asked to find $m\measuredangle A$. We begin with the tangent ratio.

$$\tan A = \frac{\text{side opposite } A}{\text{side adjacent to } A} = \frac{21}{25}$$

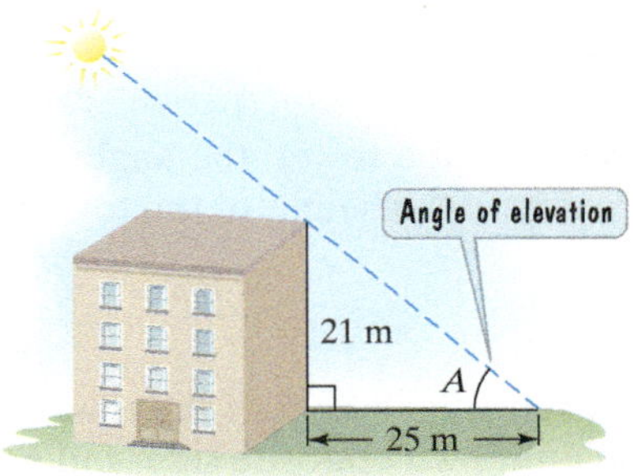

FIGURE 10.65 (repeated)

With $\tan A = \frac{21}{25}$, we use the **inverse tangent** key, $\boxed{\text{TAN}^{-1}}$, to find A.

Pressing [2nd][TAN] accesses the inverse tangent key, [TAN^{-1}].

Many Scientific Calculators:

[(] 21 [÷] 25 [)] [2nd] [TAN]

Many Graphing Calculators:

[2nd] [TAN] [(] 21 [÷] 25 [)] [ENTER]

The display should show approximately 40. Thus, the angle of elevation of the sun is approximately 40°.

CHECK POINT 5 A flagpole that is 14 meters tall casts a shadow 10 meters long. Find the angle of elevation of the sun to the nearest degree.

Blitzer Bonus

The Mountain Man

In the 1930s, a *National Geographic* team headed by Brad Washburn used trigonometry to create a map of the 5000-square-mile region of the Yukon, near the Canadian border. The team started with aerial photography. By drawing a network of angles on the photographs, the approximate locations of the major mountains and their rough heights were determined. The expedition then spent three months on foot to find the exact heights. Team members established two base points a known distance apart, one directly under the mountain's peak. By measuring the angle of elevation from one of the base points to the peak, the tangent ratio was used to determine the peak's height. The Yukon expedition was a major advance in the way maps are made.

Concept and Vocabulary Check

Fill in each blank so that the resulting statement is true. Exercises 1–3 are based on the following right triangle:

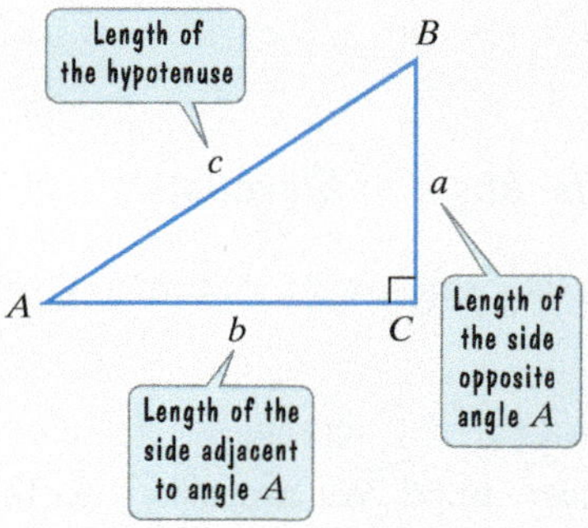

1. sin A, which represents the ____________ of angle A, is defined by the length of the side ____________ angle A divided by the length of the ____________. This is represented by ____________ in the figure shown.

2. cos A, which represents the ____________ of angle A, is defined by the length of the side ____________ angle A divided by the length of the ____________. This is represented by ____________ in the figure shown.

3. tan A, which represents the ____________ of angle A, is defined by the length of the side ____________ angle A divided by the length of the side ____________ angle A. This is represented by ____________ in the figure shown.

4. An angle formed by a horizontal line and the line of sight to an object that is above the horizontal line is called the angle of ____________.

5. An angle formed by a horizontal line and the line of sight to an object that is below the horizontal line is called the angle of ____________.

In Exercises 6–9, determine whether each statement is true or false. If the statement is false, make the necessary change(s) to produce a true statement.

6. sin 30° increases as the size of a right triangle with an acute angle of 30° grows larger. ______

7. The side of a right triangle opposite an acute angle forms one of the acute angle's sides. ______

8. On a scientific calculator, the keystroke 47 [TAN] gives a decimal approximation for tan 47°. ______

9. The equation $\cos 40° = \frac{150}{c}$ is solved for c by multiplying both sides by 150. ______

Exercise Set 10.6

Practice Exercises

In Exercises 1–8, use the given right triangles to find ratios, in reduced form, for sin A, cos A, and tan A.

1.

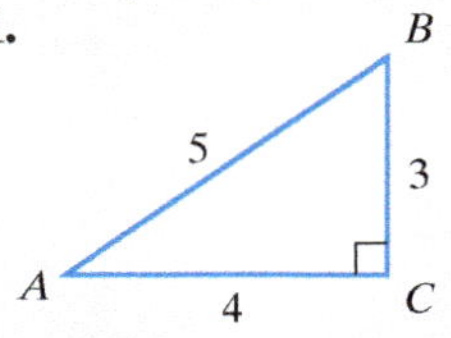

2.

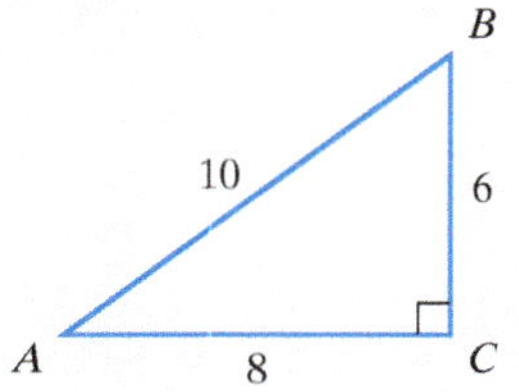

3.

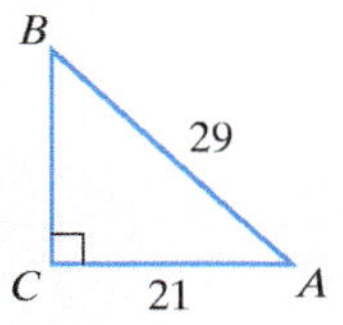

4.

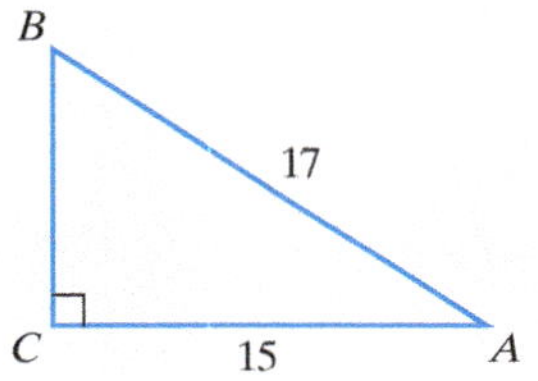

5.

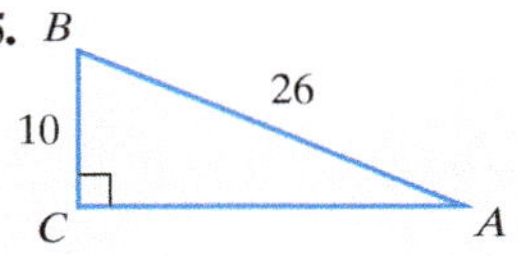

6.

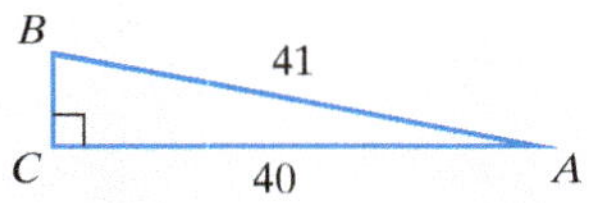

7.

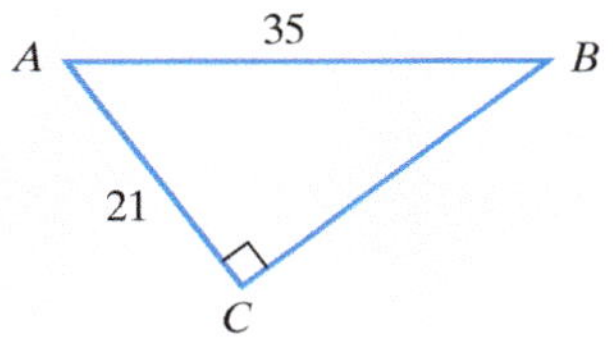

8.

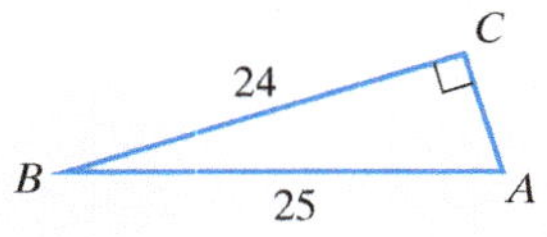

In Exercises 9–18, find the measure of the side of the right triangle whose length is designated by a lowercase letter. Round answers to the nearest whole number.

9.

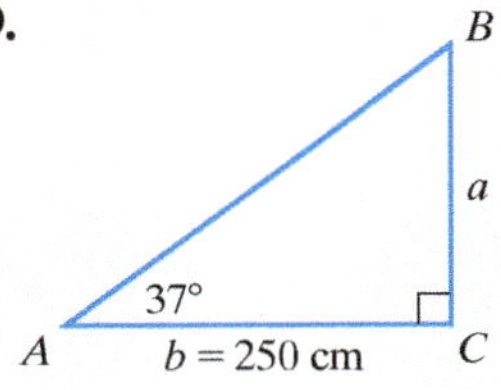

10.

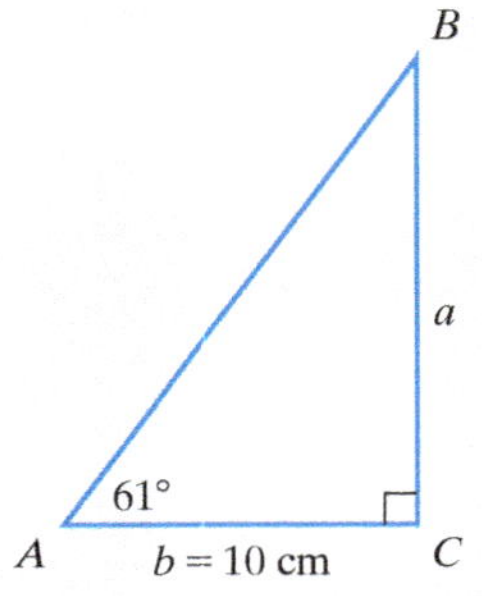

11.

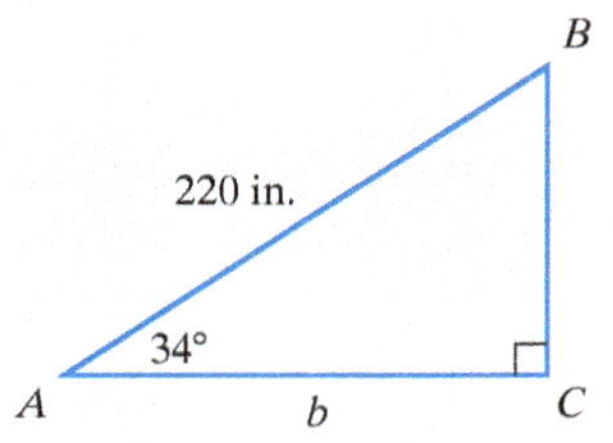

12.

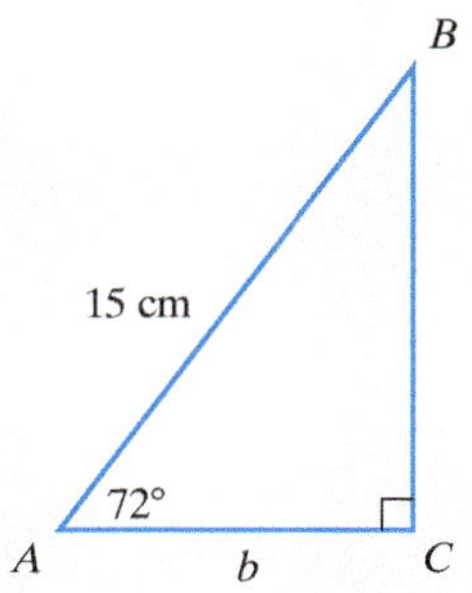

13.

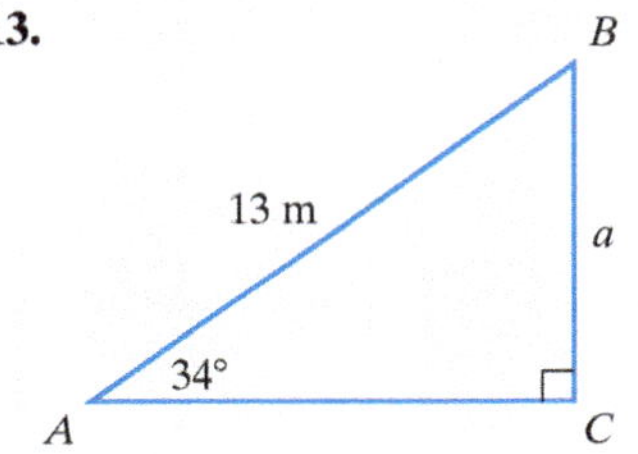

14.

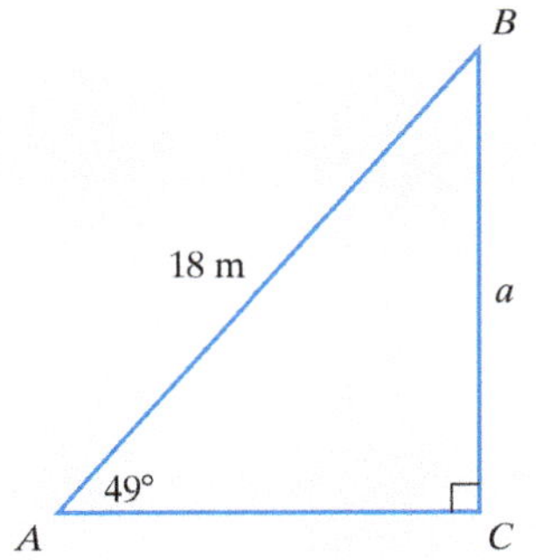

15.

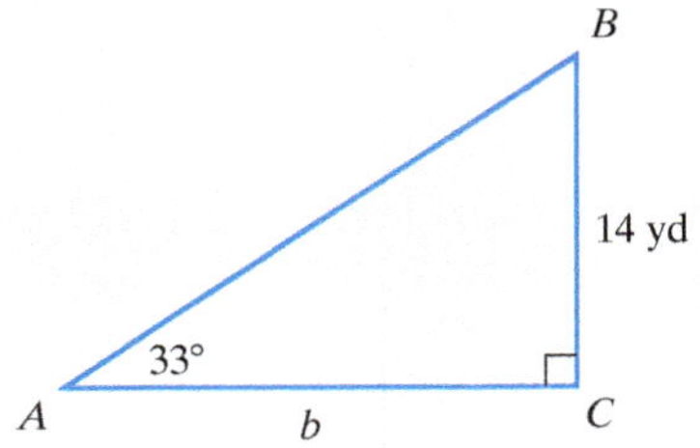

16.

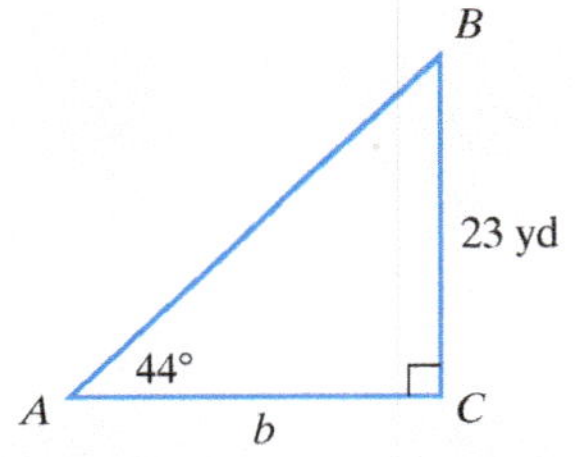

17.

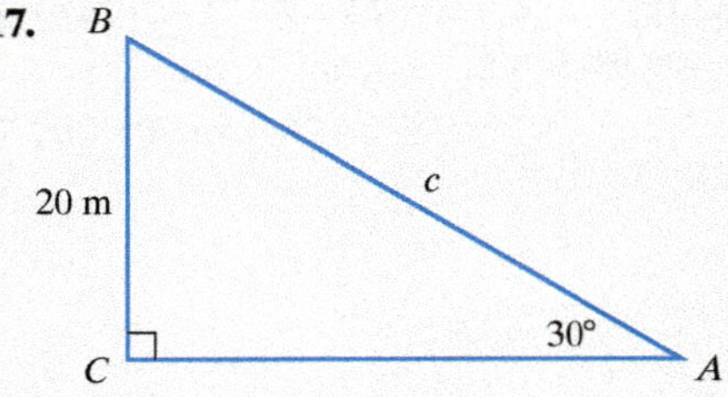

18.

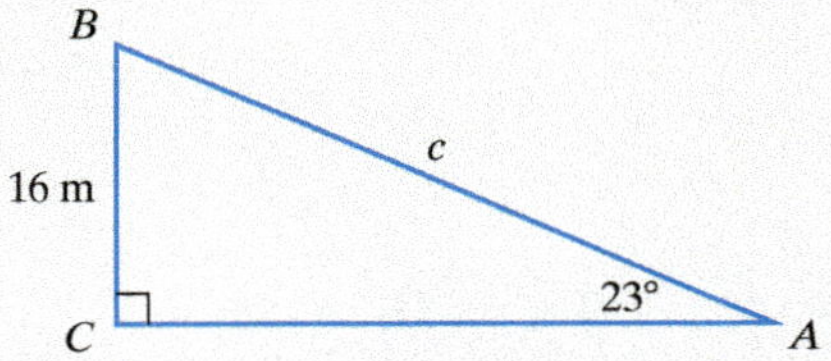

In Exercises 19–22, find the measures of the parts of the right triangle that are not given. Round all answers to the nearest whole number.

19.

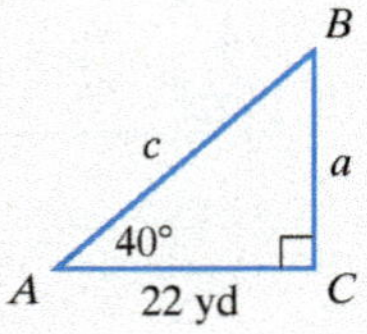

20.

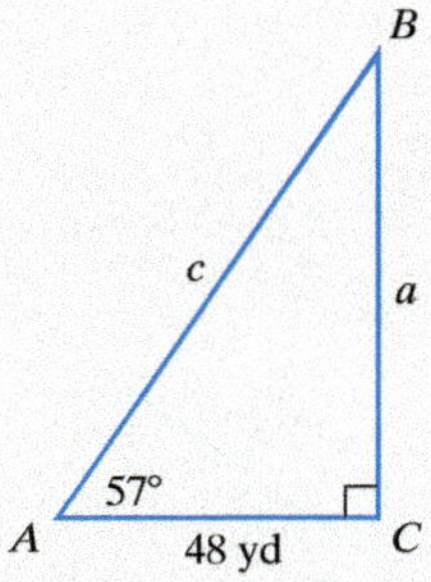

21.

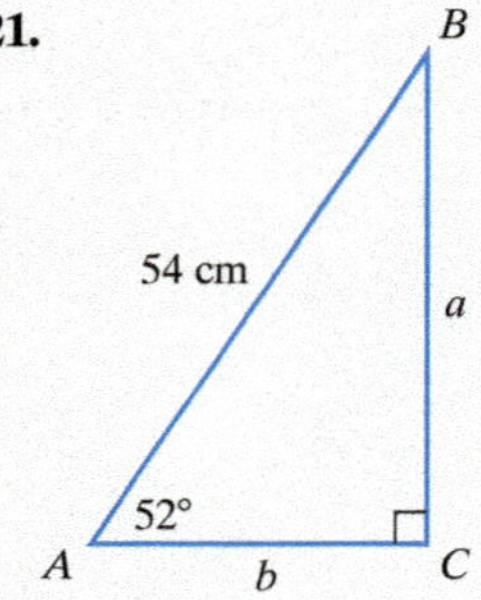

22.

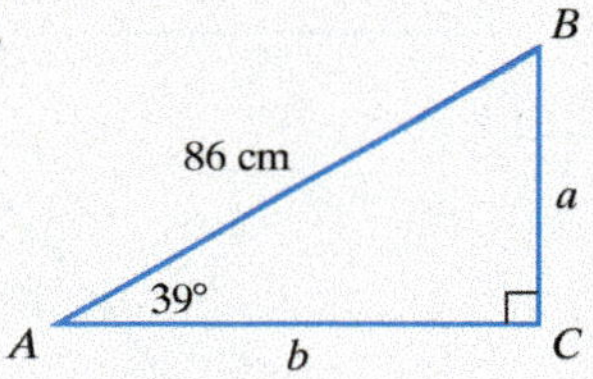

In Exercises 23–26, use the inverse trigonometric keys on a calculator to find the measure of angle A, rounded to the nearest whole degree.

23.

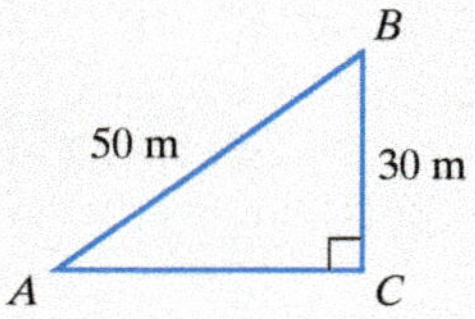

24.

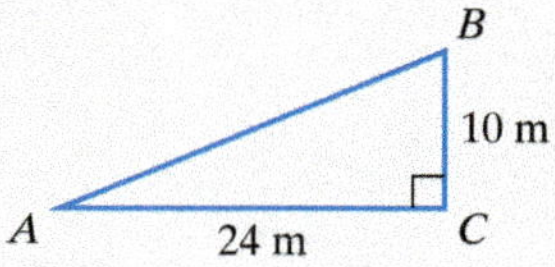

25.

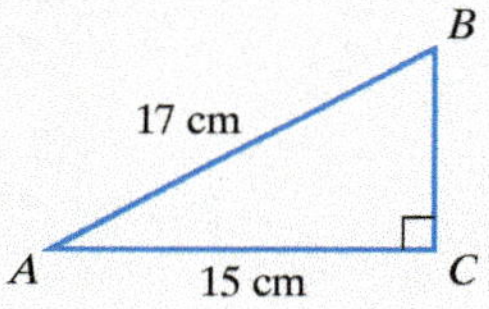

26.

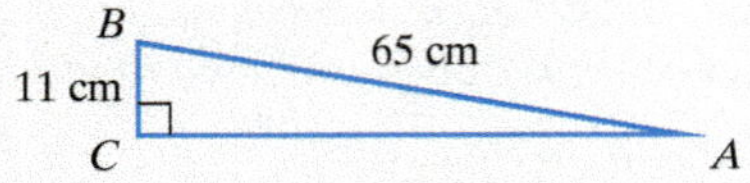

Practice Plus

In Exercises 27–34, find the length x to the nearest whole number.

27.

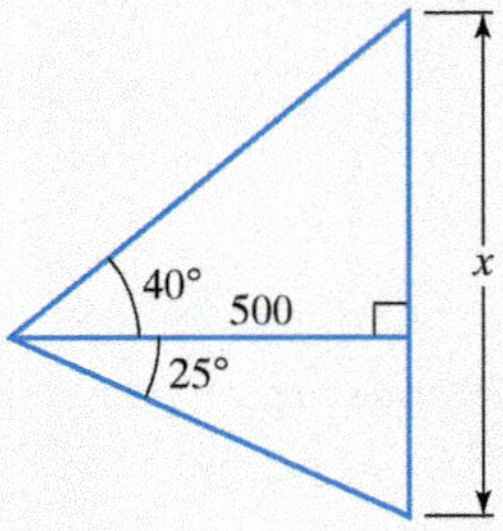

28.

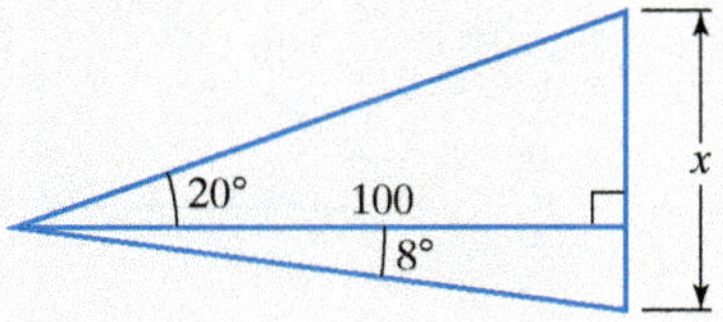

29.

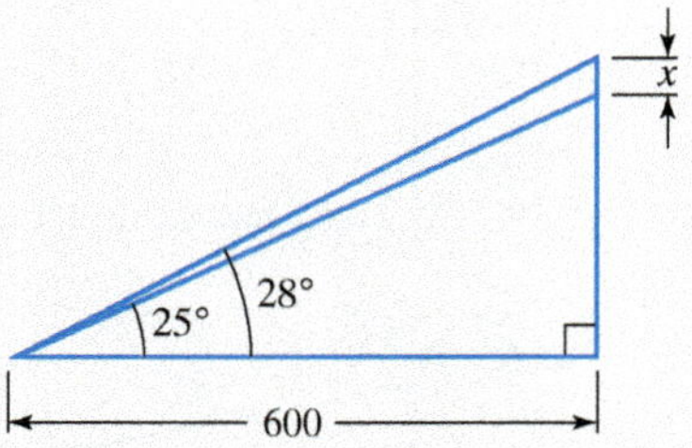

30.

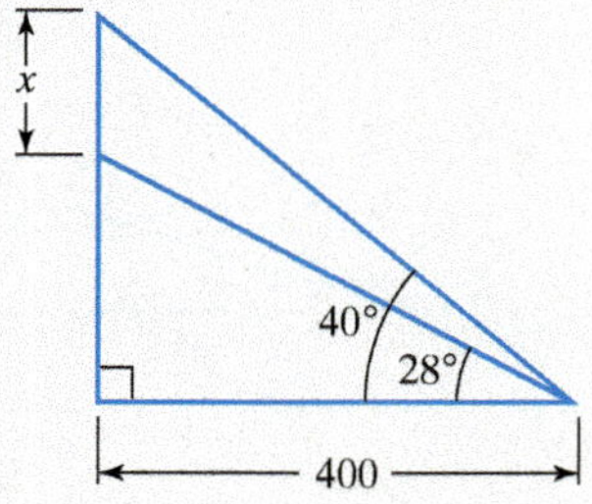

31.

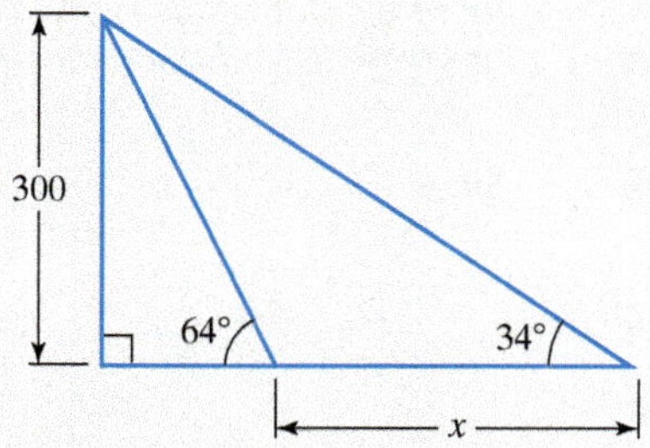

32.

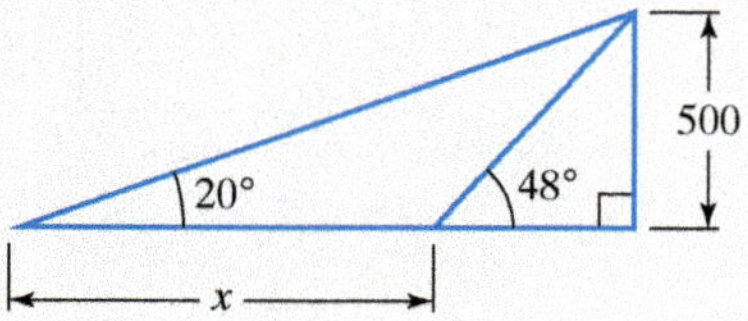

33.

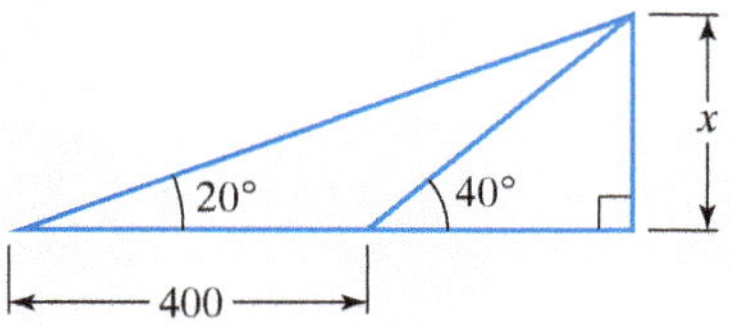

34.

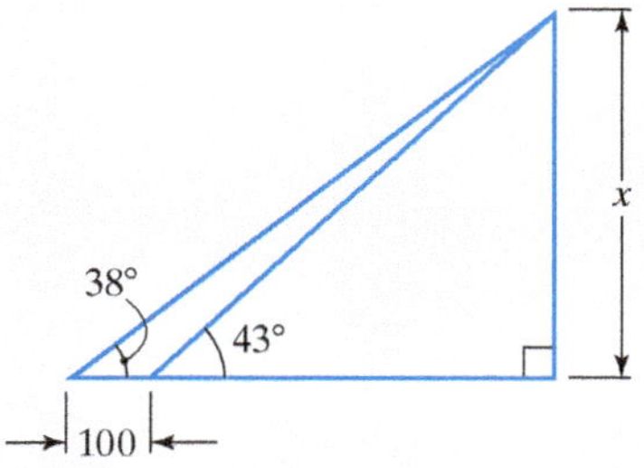

Application Exercises

35. To find the distance across a lake, a surveyor took the measurements shown in the figure. Use these measurements to determine how far it is across the lake. Round to the nearest yard.

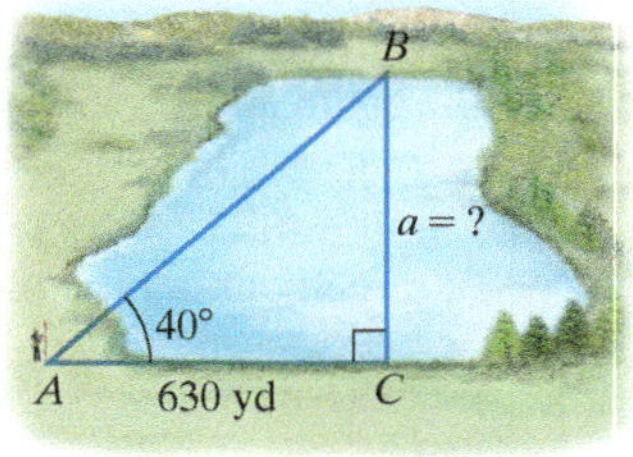

36. At a certain time of day, the angle of elevation of the sun is 40°. To the nearest foot, find the height of a tree whose shadow is 35 feet long.

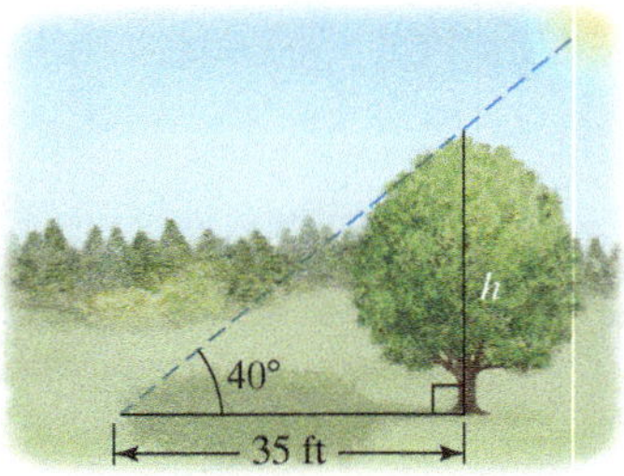

37. A plane rises from take-off and flies at an angle of 10° with the horizontal runway. When it has gained 500 feet in altitude, find the distance, to the nearest foot, the plane has flown.

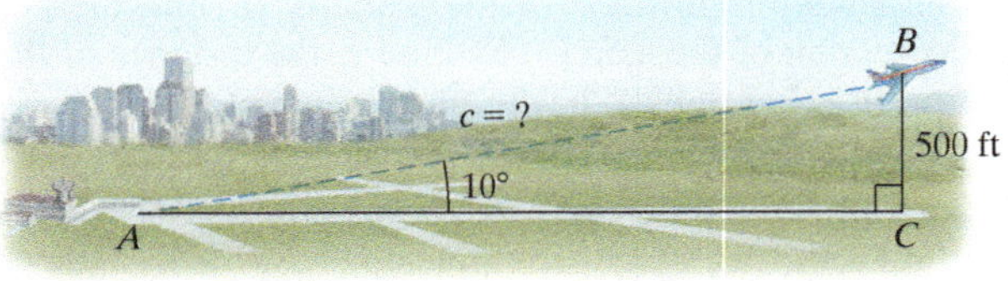

38. A road is inclined at an angle of 5°. After driving 5000 feet along this road, find the driver's increase in altitude. Round to the nearest foot.

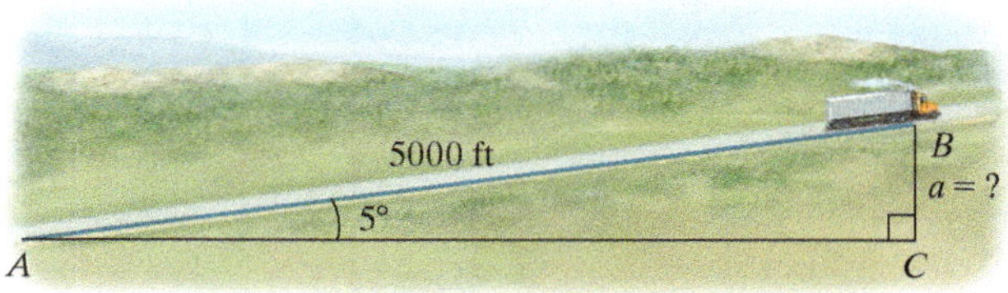

39. The tallest television transmitting tower in the world is in North Dakota. From a point on level ground 5280 feet (one mile) from the base of the tower, the angle of elevation to the top of the tower is 21.3°. Approximate the height of the tower to the nearest foot.

40. From a point on level ground 30 yards from the base of a building, the angle of elevation to the top of the building is 38.7°. Approximate the height of the building to the nearest foot.

41. The Statue of Liberty is approximately 305 feet tall. If the angle of elevation of a ship to the top of the statue is 23.7°, how far, to the nearest foot, is the ship from the statue's base?

42. A 200-foot cliff drops vertically into the ocean. If the angle of elevation of a ship to the top of the cliff is 22.3°, how far off shore, to the nearest foot, is the ship?

43. A tower that is 125 feet tall casts a shadow 172 feet long. Find the angle of elevation of the sun to the nearest degree.

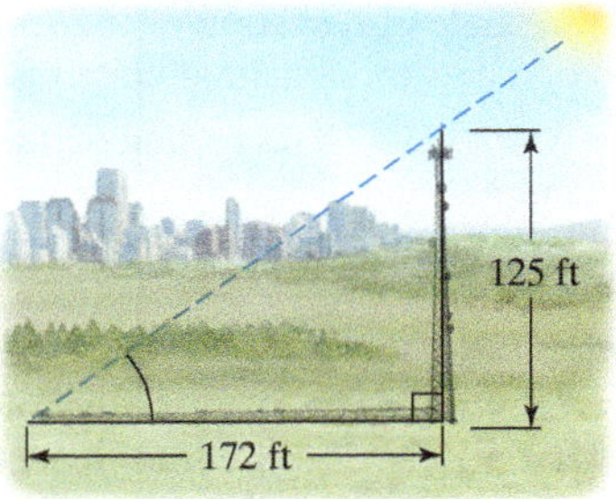

44. The Washington Monument is 555 feet high. If you stand one quarter of a mile, or 1320 feet, from the base of the monument and look to the top, find the angle of elevation to the nearest degree.

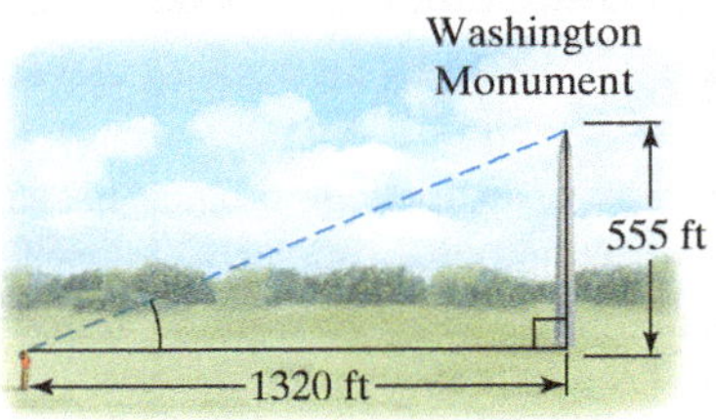

45. A helicopter hovers 1000 feet above a small island. The figure below shows that the angle of depression from the helicopter to point P is 36°. How far off the coast, to the nearest foot, is the island?

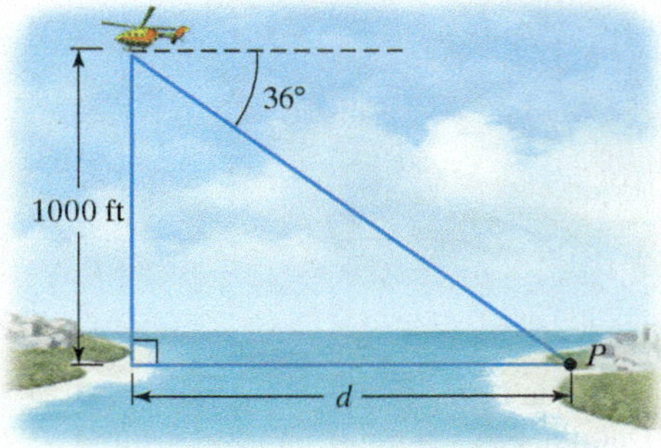

46. A police helicopter is flying at 800 feet. A stolen car is sighted at an angle of depression of 72°. Find the distance of the stolen car, to the nearest foot, from a point directly below the helicopter.

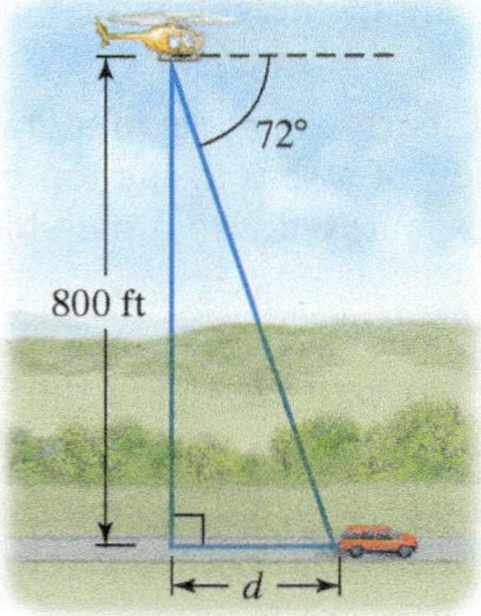

47. A wheelchair ramp is to be built beside the steps to the campus library. Find the angle of elevation of the 23-foot ramp, to the nearest tenth of a degree, if its final height is 6 feet.

48. A kite flies at a height of 30 feet when 65 feet of string is out. If the string is in a straight line, find the angle that it makes with the ground. Round to the nearest tenth of a degree.

Writing in Mathematics

49. If you are given the lengths of the sides of a right triangle, describe how to find the sine of either acute angle.

50. Describe one similarity and one difference between the sine ratio and the cosine ratio in terms of the sides of a right triangle.

51. If one of the acute angles of a right triangle is 37°, explain why the sine ratio does not increase as the size of the triangle increases.

52. If the measure of one of the acute angles and the hypotenuse of a right triangle are known, describe how to find the measure of the remaining parts of the triangle.

53. Describe what is meant by an angle of elevation and an angle of depression.

54. Give an example of an applied problem that can be solved using one or more trigonometric ratios. Be as specific as possible.

55. Use a calculator to find each of the following: sin 32° and cos 58°; sin 17° and cos 73°; sin 50° and cos 40°; sin 88° and cos 2°. Describe what you observe. Based on your observations, what do you think the *co* in *cosine* stands for?

56. Stonehenge, the famous "stone circle" in England, was built between 2750 B.C. and 1300 B.C. using solid stone blocks weighing over 99,000 pounds each. It required 550 people to pull a single stone up a ramp inclined at a 9° angle. Describe how right triangle trigonometry can be used to determine the distance the 550 workers had to drag a stone in order to raise it to a height of 30 feet.

Critical Thinking Exercises

Make Sense? *In Exercises 57–60, determine whether each statement makes sense or does not make sense, and explain your reasoning.*

57. For a given angle A, I found a slight increase in sin A as the size of the triangle increased.

58. I'm working with a right triangle in which the hypotenuse is the side adjacent to the acute angle.

59. Standing under this arch, I can determine its height by measuring the angle of elevation to the top of the arch and my distance to a point directly under the arch.

Delicate Arch in Arches National Park, Utah

60. A wheelchair ramp must be constructed so that the slope is not more than 1 inch of rise for every 1 foot of run, so I used the tangent ratio to determine the maximum angle that the ramp can make with the ground.

61. Explain why the sine or cosine of an acute angle cannot be greater than or equal to 1.

62. Describe what happens to the tangent of an angle as the measure of the angle gets close to 90°.

63. From the top of a 250-foot lighthouse, a plane is sighted overhead and a ship is observed directly below the plane. The angle of elevation of the plane is 22° and the angle of depression of the ship is 35°. Find **a.** the distance of the ship from the lighthouse; **b.** the plane's height above the water. Round to the nearest foot.

64. Sighting the top of a building, a surveyor measured the angle of elevation to be 22°. The transit is 5 feet above the ground and 300 feet from the building. Find the building's height. Round to the nearest foot.

10.7

Beyond Euclidean Geometry

WHAT AM I SUPPOSED TO LEARN?

After you have read this section, you should be able to:

1 Gain an understanding of some of the general ideas of other kinds of geometries.

The Family (1963), Marisol Escobar.

"I love Euclidean geometry, but it is quite clear that it does not give a reasonable presentation of the world. Mountains are not cones, clouds are not spheres, trees are not cylinders, neither does lightning travel in a straight line. Almost everything around us is non-Euclidean."

—Benoit Mandelbrot

THINK OF YOUR FAVORITE NATURAL SETTING. DURING THE LAST QUARTER OF the twentieth century, mathematicians developed a new kind of geometry that uses the computer to produce the diverse forms of nature that we see around us. These forms are not polygons, polyhedrons, circles, or cylinders. In this section, we move beyond Euclidean geometry to explore ideas which have extended geometry beyond the boundaries first laid down by the ancient Greek scholars.

1 Gain an understanding of some of the general ideas of other kinds of geometries.

The Geometry of Graphs

In the early 1700s, the city of Königsberg, Germany, was connected by seven bridges, shown in **Figure 10.66**. Many people in the city were interested in finding if it were possible to walk through the city so as to cross each bridge exactly once. After a few trials, you may be convinced that the answer is no. However, it is not easy to prove your answer by trial and error because there are a large number of ways of taking such a walk.

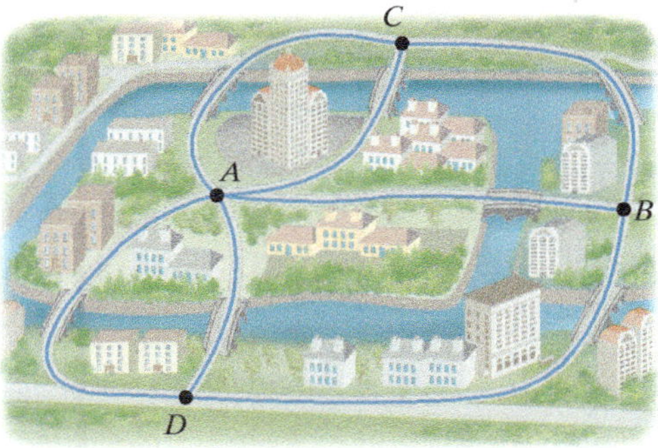

FIGURE 10.66

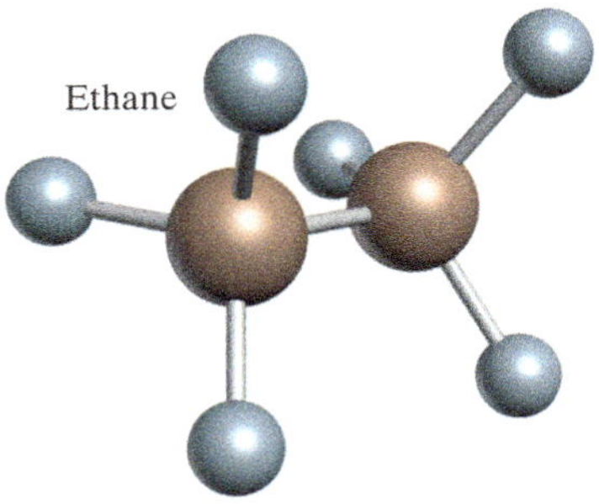

Graph theory is used to show how atoms are linked to form molecules.

The problem was taken to the Swiss mathematician Leonhard Euler (1707–1783). In the year 1736, Euler (pronounced "oil er") proved that it is not possible to stroll through the city and cross each bridge exactly once. His solution of the problem opened up a new kind of geometry called **graph theory**. Graph theory is now used to design city streets, analyze traffic patterns, and find the most efficient routes for public transportation.

Euler solved the problem by introducing the following definitions:

A **vertex** is a point. An **edge** is a line segment or curve that starts and ends at a vertex.

Vertices and edges form a **graph**. A vertex with an odd number of attached edges is an **odd vertex**. A vertex with an even number of attached edges is an **even vertex**.

Using these definitions, **Figure 10.67** shows that the graph in the Königsberg bridge problem (**Figure 10.66**) has four odd vertices.

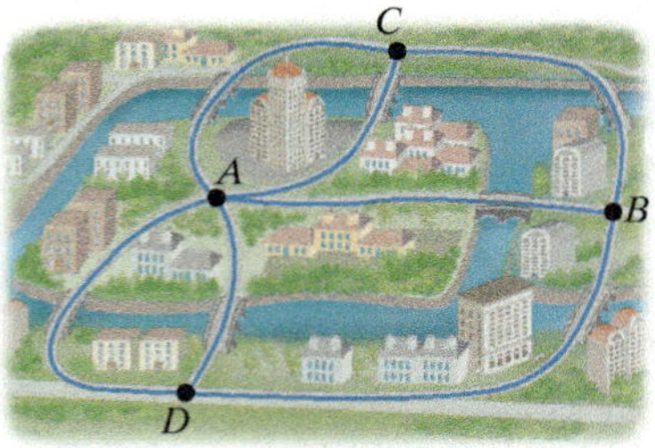

FIGURE 10.66 (repeated)

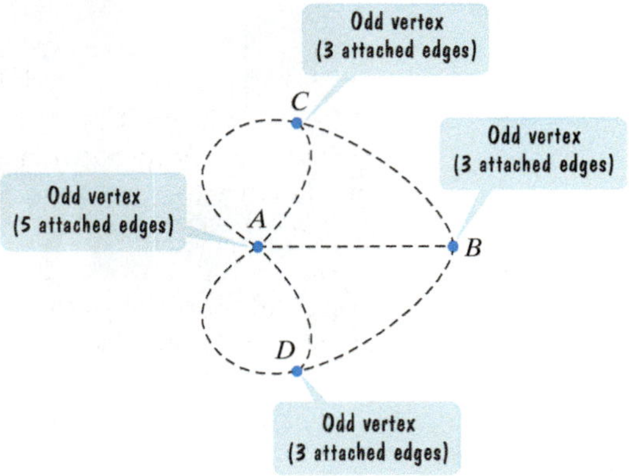

FIGURE 10.67

A graph is **traversable** if it can be traced without lifting the pencil from the paper and without tracing an edge more than once. Euler proved the following rules of traversability in solving the problem:

RULES OF TRAVERSABILITY

1. A graph with all even vertices is traversable. One can start at any vertex and end where one began.
2. A graph with two odd vertices is traversable. One must start at either of the odd vertices and finish at the other.
3. A graph with more than two odd vertices is not traversable.

Because the graph in the Königsberg bridge problem has four odd vertices, it cannot be traversed.

EXAMPLE 1 *To Traverse or Not to Traverse?*

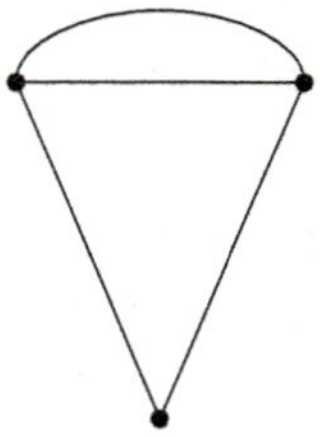
FIGURE 10.68

Consider the graph in **Figure 10.68**.

a. Is this graph traversable?

b. If it is, describe a path that will traverse it.

SOLUTION

a. Begin by determining whether each vertex is even or odd.

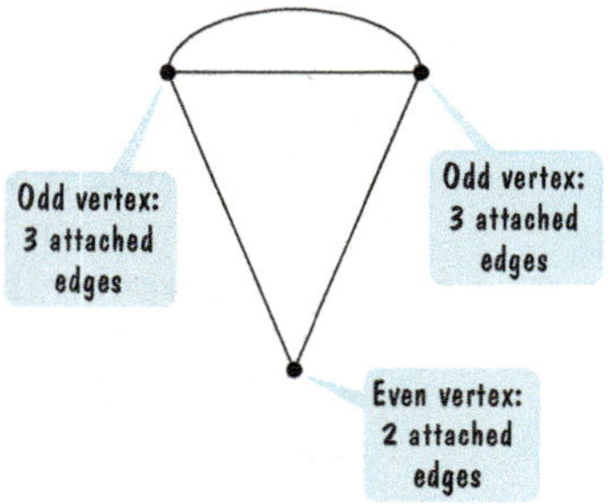

Because this graph has exactly two odd vertices, by Euler's second rule, it is traversable.

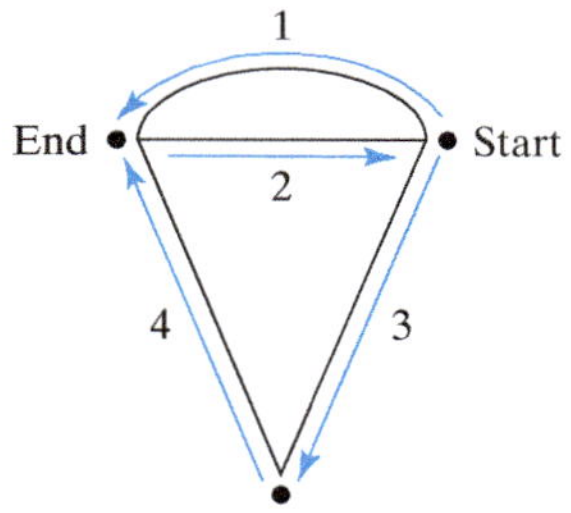

FIGURE 10.69

b. In order to find a path that traverses this graph, we again use Euler's second rule. It states that we must start at either of the odd vertices and finish at the other. One possible path is shown in **Figure 10.69**.

CHECK POINT 1 Create a graph with two even and two odd vertices. Then describe a path that will traverse it.

Graph theory is used to solve many kinds of practical problems. The first step is to create a graph that represents, or models, the problem. For example, consider a UPS driver trying to find the best way to deliver packages around town. By modeling the delivery locations as vertices and roads between locations as edges, graph theory reveals the driver's most efficient path. Graphs are powerful tools because they illustrate the important aspects of a problem and leave out unnecessary detail. A more detailed presentation on the geometry of graphs, called *graph theory*, can be found in Chapter 14.

Topology

A branch of modern geometry called **topology** looks at shapes in a completely new way. In Euclidean geometry, shapes are rigid and unchanging. In topology, shapes can be twisted, stretched, bent, and shrunk. Total flexibility is the rule.

A topologist does not know the difference between a doughnut and a coffee cup! This is because, topologically speaking, there is no difference. They both have one hole, and in this strange geometry of transformations, a doughnut can be flattened, pulled, and pushed to form a coffee cup.

In topology, objects are classified according to the number of holes in them, called their **genus**. The genus gives the largest number of complete cuts that can be made in the object without cutting the object into two pieces. Objects with the same genus are **topologically equivalent**.

The three shapes shown below (sphere, cube, irregular blob) have the same genus: 0. No complete cuts can be made without cutting these objects into two pieces.

GREAT QUESTION!

Can you remind me of the relationship between genus and holes?

The genus of an object is the same as the number of holes in the object.

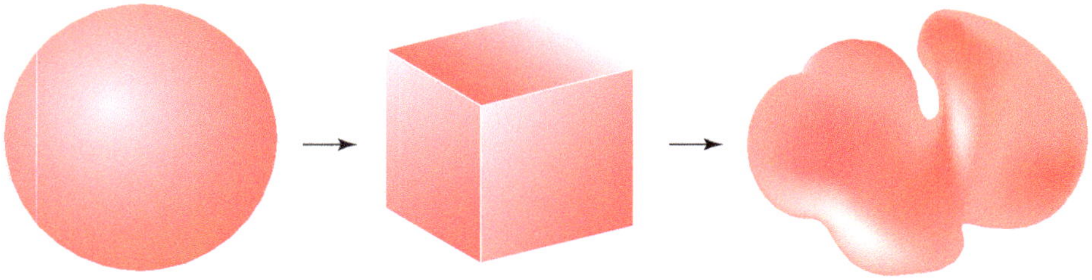

How can a doughnut be transformed into a coffee cup? It can be stretched and deformed to make a bowl part of the surface. Both the doughnut and the coffee cup have genus 1. One complete cut can be made without cutting these objects into two pieces.

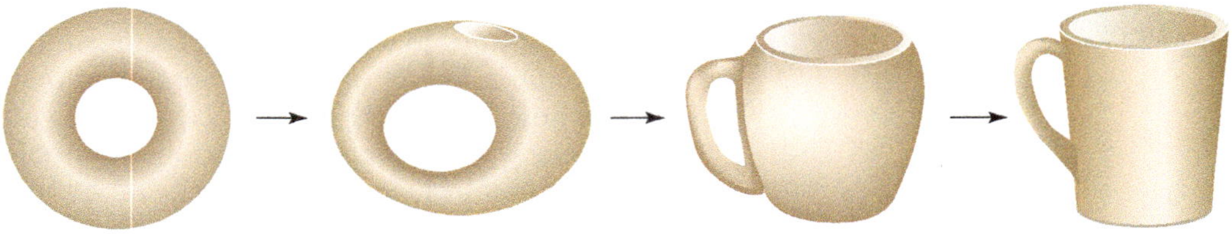

To a topologist, a lump with two holes in it is no different than a sugar bowl with two handles. Both have genus 2. Two complete cuts can be made without cutting these objects into two pieces.

Shown below is a transformation that results in a most unusual topological figure.

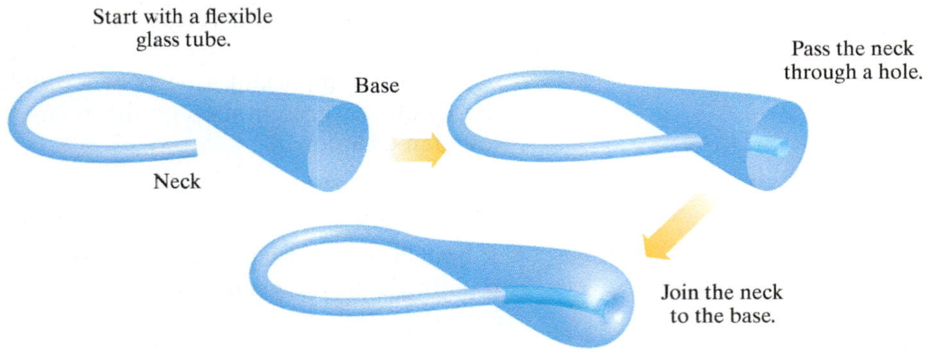

The figure that results is called a **Klein bottle**. You couldn't use one to carry water on a long hike. Because the inside surface loops back on itself to merge with the outside, it has neither an outside nor an inside and cannot hold water. A Klein bottle passes through itself without the existence of a hole, which is impossible in three-dimensional space. A true Klein bottle is visible only when generated on a computer.

Topology is more than an excursion in geometric fantasy. Topologists study knots and ways in which they are topologically equivalent. Knot theory is used to identify viruses and understand the ways in which they invade our cells. Such an understanding is a first step in finding vaccines for viruses ranging from the common cold to HIV.

Non-Euclidean Geometries

Imagine this: Earth is squeezed and compressed to the size of a golf ball. It becomes a **black hole**, a region in space where everything seems to disappear. Its pull of gravity is now so strong that the space near the golf ball can be thought of as a funnel with the black hole sitting at the bottom. Any object that comes near the funnel will be pulled into it by the immense gravity of the golf-ball-sized Earth. Then the object disappears! Although this sounds like science fiction, some physicists believe that there are billions of black holes in space.

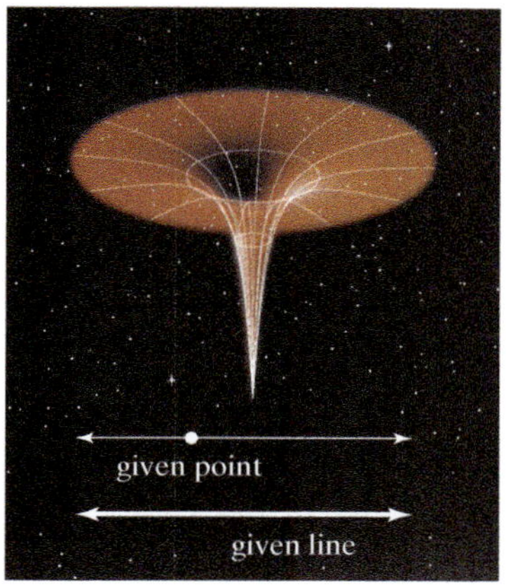

FIGURE 10.70

Is there a geometry that describes this unusual phenomenon? Yes, although it's not the geometry of Euclid. Euclidean geometry is based on the assumption that given a point not on a line, there is one line that can be drawn through the point parallel to the given line, shown as the thinner white line in **Figure 10.70**. We used this assumption to prove that the sum of the measures of the three angles of any triangle is 180°.

Hyperbolic geometry was developed independently by the Russian mathematician Nikolay Lobachevsky (1792–1856) and the Hungarian mathematician Janos Bolyai (1802–1860). It is based on the assumption that given a point not on a line, there are an infinite number of lines that can be drawn through the point parallel to the given line. The shapes in this non-Euclidean geometry are not represented on a plane. Instead, they are drawn on a funnel-like surface, called a *pseudosphere,* shown in **Figure 10.71**. On such a surface, the shortest distance between two points is a curved line. A triangle's sides are composed of arcs, and the sum of the measures of its angles is less than 180°. The shape of the pseudosphere looks like the distorted space near a black hole.

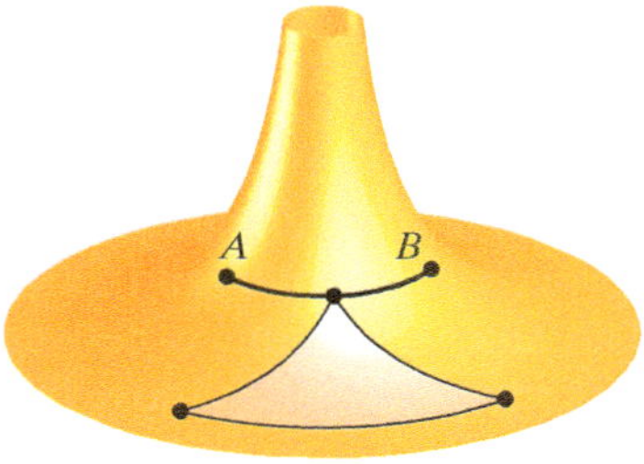

FIGURE 10.71

Once the ice had been broken by Lobachevsky and Bolyai, mathematicians were stimulated to set up other non-Euclidean geometries. The best known of these was **elliptic geometry**, proposed by the German mathematician Bernhard Riemann (1826–1866). Riemann began his geometry with the assumption that there are no parallel lines. As shown in **Figure 10.72**, elliptic geometry is on a sphere, and the sum of the measures of the angles of a triangle is greater than 180°.

Riemann's elliptic geometry was used by Albert Einstein in his theory of the universe. One aspect of the theory states that if you set off on a journey through space and keep going in the same direction, you will eventually come back to your starting point. The same thing happens on the surface of a sphere. This means that space itself is "curved," although the idea is difficult to visualize. It is interesting that Einstein used ideas of non-Euclidean geometry more than 50 years after the system was logically developed. Mathematics often moves ahead of our understanding of the physical world.

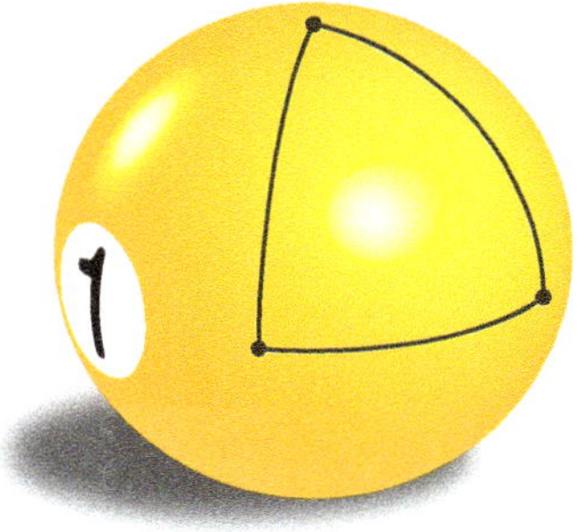

FIGURE 10.72

Blitzer Bonus

Colorful Puzzles

How many different colors are needed to color a plane map so that any two regions that share a common border are colored differently? Even though no map could be found that required more than four colors, a proof that four colors would work for every map remained elusive to mathematicians for over 100 years. The conjecture that at most four colors are needed, stated in 1852, was proved by American mathematicians Kenneth Appel and Wolfgang Haken of the University of Illinois in 1976. Their proof translated maps into graphs that represented each region as a vertex and each shared common border as an edge between corresponding vertices. The proof relied heavily on computer computations that checked approximately 1500 special cases, requiring 1200 hours of computer time. Although there are recent rumors about a flaw in the proof, no one has found a proof that does not enlist the use of computer calculations.

If you take a strip of paper, give it a single half-twist, and paste the ends together, the result is a one-sided surface called a Möbius strip. A map on a Möbius strip requires six colors to make sure that no adjacent areas are the same color.

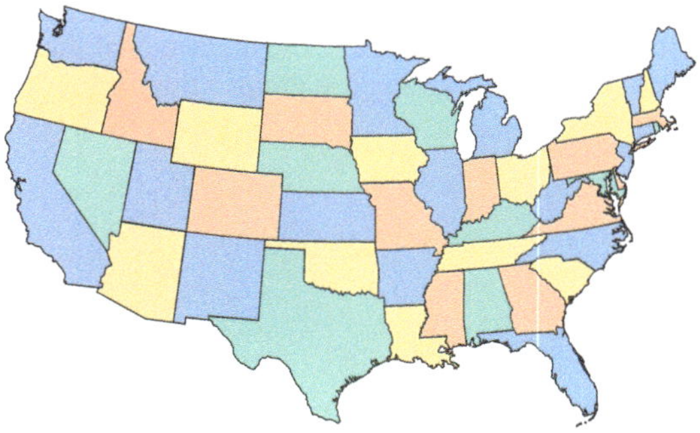

At most four colors are needed on a plane map.

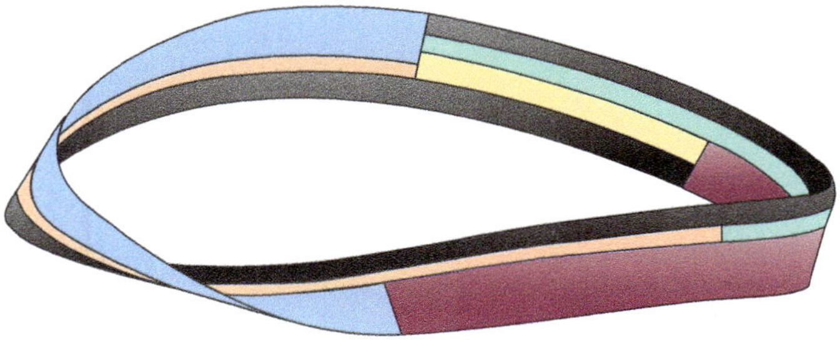

At most six colors are needed on a map drawn on a Möbius strip.

Circle Limits III (1959), M.C. Escher.

We have seen that Dutch artist M. C. Escher is known for combining art and geometry. Many of his images are based on non-Euclidean geometry. In the woodcut shown above, things enlarge as they approach the center and shrink proportionally as they approach the boundary. The shortest distance between two points is a curved line, and the triangles that are formed look like those from hyperbolic geometry.

Table 10.5 compares Euclidean geometry with hyperbolic and elliptic geometry.

TABLE 10.5 Comparing Three Systems of Geometry

<table>
<tr><th>Euclidean Geometry
Euclid (300 B.C.)</th><th>Hyperbolic Geometry
Lobachevsky, Bolyai (1830)</th><th>Elliptic Geometry
Riemann (1850)</th></tr>
<tr><td>Given a point not on a line, there is one and only one line through the point parallel to the given line.</td><td>Given a point not on a line, there are an infinite number of lines through the point that do not intersect the given line.</td><td>There are no parallel lines.</td></tr>
<tr><td>Geometry is on a plane:
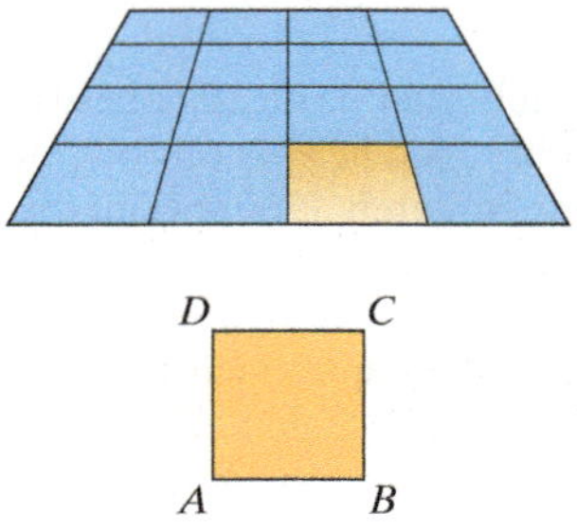</td><td>Geometry is on a pseudosphere:
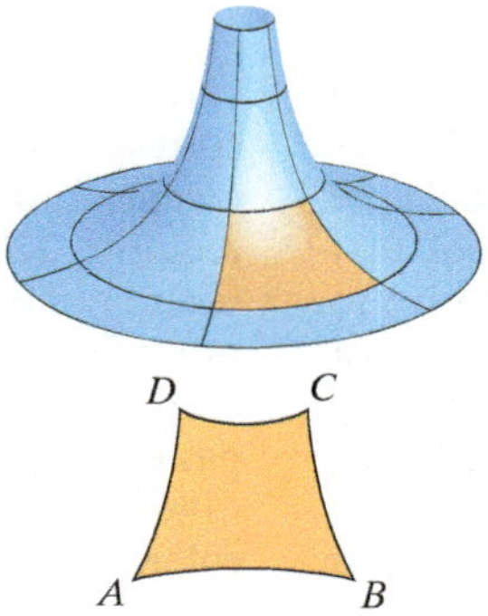</td><td>Geometry is on a sphere:
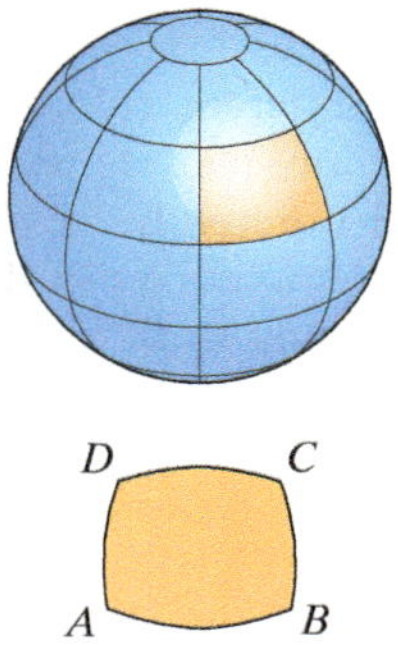</td></tr>
<tr><td>The sum of the measures of the angles of a triangle is 180°.
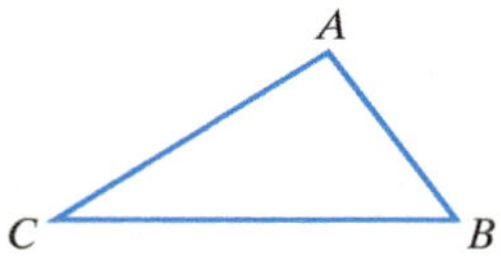</td><td>The sum of the measures of the angles of a triangle is less than 180°.
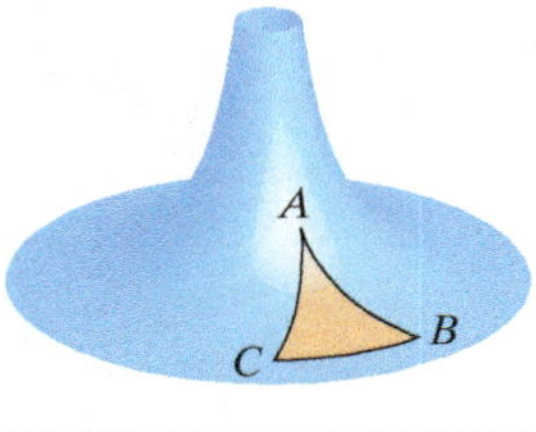</td><td>The sum of the measures of the angles of a triangle is greater than 180°.
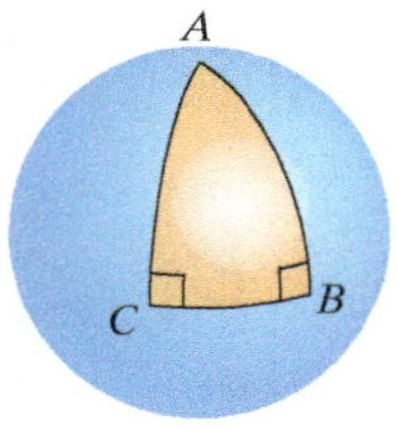</td></tr>
</table>

Fractal Geometry

How can geometry describe nature's complexity in objects such as ferns, coastlines, or the human circulatory system? Any magnified portion of a fern repeats much of the pattern of the whole fern, as well as new and unexpected patterns. This property is called **self-similarity**.

Through the use of computers, a new geometry of natural shapes, called **fractal geometry**, has arrived. The word *fractal* is from the Latin word *fractus*, meaning "broken up," or "fragmented." The fractal shape shown in **Figure 10.73** is, indeed, broken up. This shape was obtained by repeatedly subtracting triangles from within triangles. From the midpoints of the sides of triangle ABC, remove triangle $A'B'C'$. Repeat the process for the three remaining triangles, then with each of the triangles left after that, and so on ad infinitum.

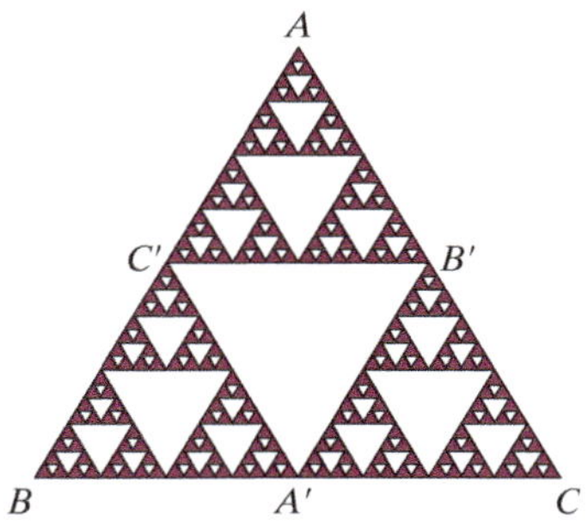

FIGURE 10.73

The process of repeating a rule again and again to create a self-similar fractal like the broken-up triangle is called **iteration**. The self-similar fractal shape in **Figure 10.74** was generated on a computer by iteration that involved subtracting pyramids within pyramids. Computers can generate fractal shapes by carrying out thousands or millions of iterations.

FIGURE 10.74 Daryl H. Hepting and Allan N. Snider, "Desktop Tetrahedron" 1990. Computer generated at the University of Regina in Canada.

In 1975, Benoit Mandelbrot (1924–2010), the mathematician who developed fractal geometry, published *The Fractal Geometry of Nature*. This book of beautiful computer graphics shows images that look like actual mountains, coastlines, underwater coral gardens, and flowers, all imitating nature's forms. Many of the realistic-looking landscapes that you see in movies are actually computer-generated fractal images.

These images are computer-generated fractals.

Concept and Vocabulary Check

Fill in each blank so that the resulting statement is true.

1. In the geometry of graphs, a point is called a/an ____________. A line segment or curve that starts and ends at a point is called a/an ____________. A set of points connected by line segments or curves is called a/an ____________.
2. In the geometry of graphs, if you can trace a graph without lifting your pencil from the paper and without tracing an edge more than once, then the graph is called ____________.
3. In topology, objects are classified according to the number of holes in them, called their ____________.

4. In Euclidean geometry, a basic assumption states that given a line and a point not on the line, one and only one line may be drawn through the given point __________ to the given line.

5. By changing the basic assumption in Exercise 4, we obtain __________ geometries. In one of these geometries, called elliptic geometry, there are no __________ lines.

6. In fractal geometry, magnified portions of an object repeat much of the pattern of the whole object, as well as new and unexpected patterns. This property is called __________. The process of repeating a rule again and again to create such a fractal image is called __________.

In Exercises 7–10, determine whether each statement is true or false. If the statement is false, make the necessary change(s) to produce a true statement.

7. A graph that contains exactly three vertices, each of which is even, is traversable. ______

8. A graph that contains exactly three vertices, each of which is odd, is traversable. ______

9. A doughnut and a compact disc both have genus 1 and are topologically equivalent. ______

10. In non-Euclidean geometries, the sum of the measures of the angles of a triangle is not necessarily 180°. ______

Exercise Set 10.7

Practice and Application Exercises

*For each graph in Exercises 1–6, **a.** Is the graph traversable? **b.** If it is, describe a path that will traverse it.*

1.

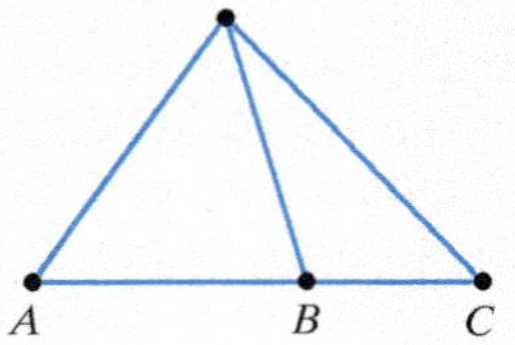

2.

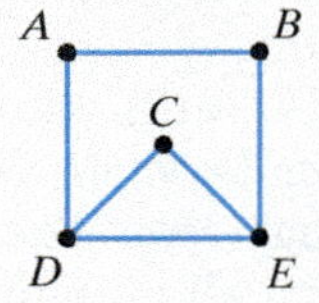

3.

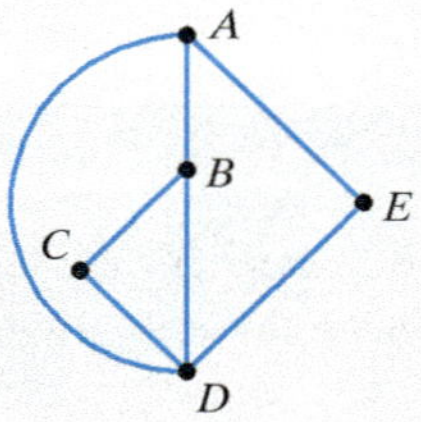

4.

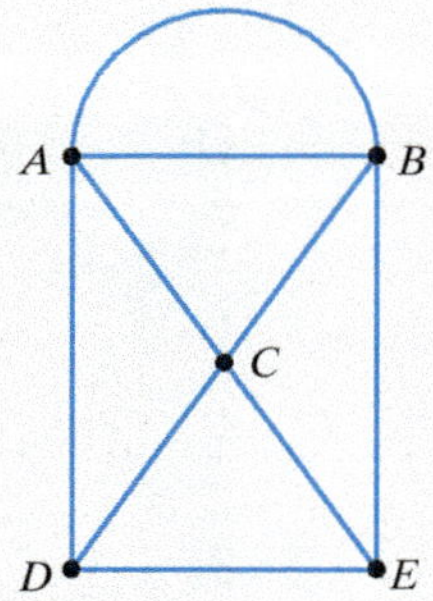

5.

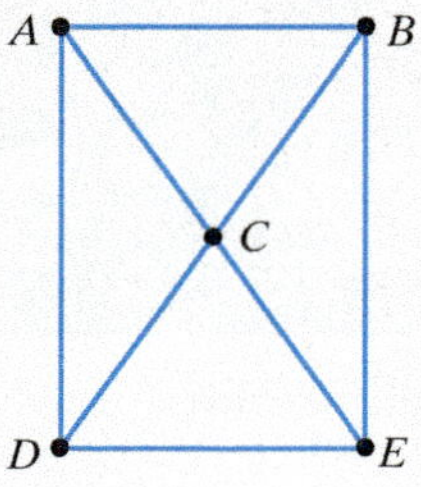

6.

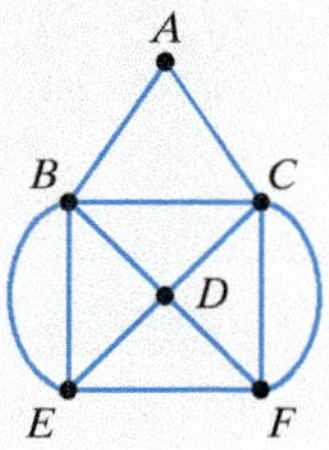

The model shows the way that carbon atoms (red) and hydrogen atoms (blue) link together to form a propane molecule. Use the model to answer Exercises 7–9.

Propane

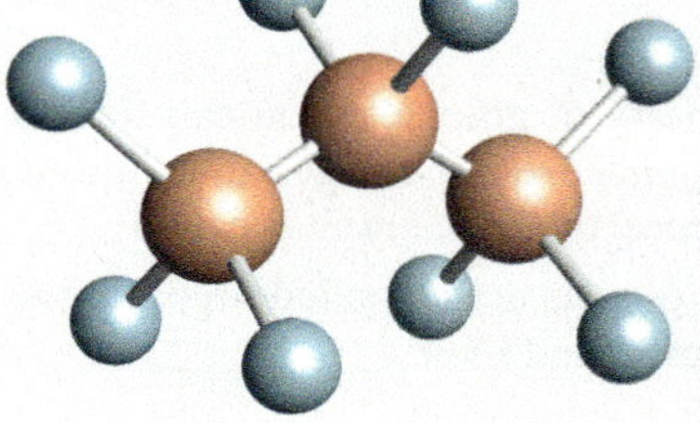

7. Using each of the three red carbon atoms as a vertex and each of the eight blue hydrogen atoms as a vertex, draw a graph for the propane molecule.

8. Determine if each vertex of the graph is even or odd.

9. Is the graph traversable?

The figure below on the left shows the floor plan of a four-room house. By representing rooms as vertices, the outside, E, as a vertex, and doors as edges, the figure below on the right is a graph that models the floor plan. Use these figures to solve Exercises 10–13.

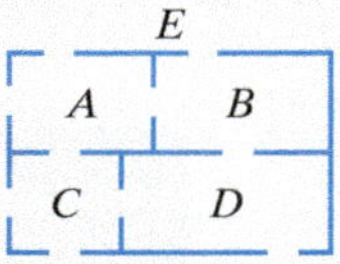

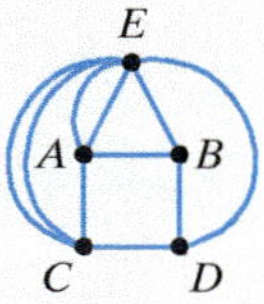

10. Use the floor plan to determine how many doors connect the outside, *E*, to room *A*. How is this shown in the graph?

11. Use the floor plan to determine how many doors connect the outside, *E*, to room *C*. How is this shown in the graph?

12. Is the graph traversable? Explain your answer.

13. If the graph is traversable, determine a path to walk through every room, using each door only once.

In Exercises 14–17, give the genus of each object.

14.

Pretzel

15.

Pitcher

16.

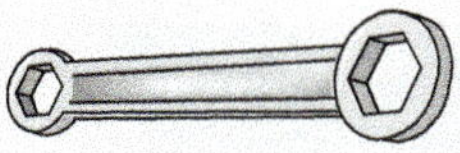

Wrench

17.

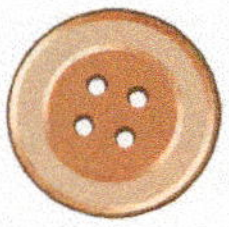

Button

18. In Exercises 14–17, which objects are topologically equivalent?

19. Draw (or find and describe) an object of genus 4 or more.

20. What does the figure shown illustrate about angles in a triangle when the triangle is drawn on a sphere?

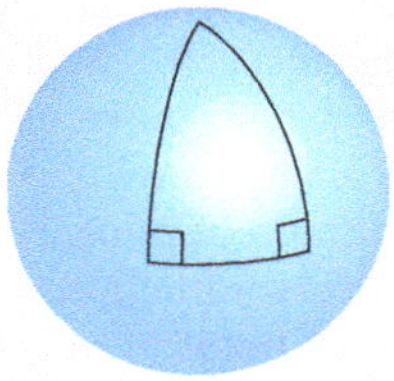

21. The figure shows a quadrilateral that was drawn on the surface of a sphere. Angles C and D are obtuse angles. What does the figure illustrate about the sum of the measures of the four angles in a quadrilateral in elliptic geometry?

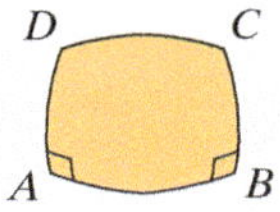

In Exercises 22–24, describe whether each figure shown exhibits self-similarity.

22.

23.

24.

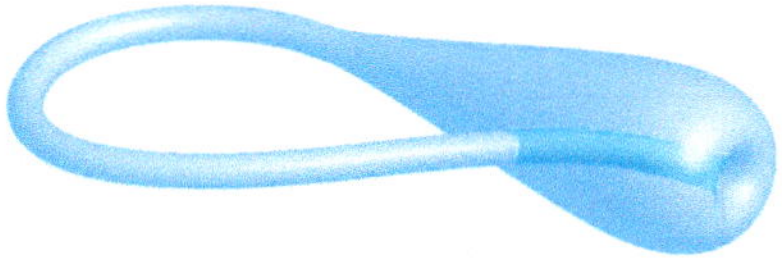

25. Consult an Internet site devoted to fractals and download two fractal images that imitate nature's forms. How is self-similarity shown in each image?

Writing in Mathematics

26. What is a graph?

27. What does it mean if a graph is traversable?

28. How do you determine whether or not a graph is traversable?

29. Describe one way in which topology is different than Euclidean geometry.

30. What is the genus of an object?

31. Describe how a rectangular solid and a sphere are topologically equivalent.

32. State the assumption that Euclid made about parallel lines that was altered in both hyperbolic and elliptic geometry.

33. How does hyperbolic geometry differ from Euclidean geometry?

34. How does elliptic geometry differ from Euclidean geometry?

35. What is self-similarity? Describe an object in nature that has this characteristic.

36. Some suggest that nature produces its many forms through a combination of both iteration and a touch of randomness. Describe what this means.

37. Find an Internet site devoted to fractals. Use the site to write a paper on a specific use of fractals.

38. Did you know that laid end to end, the veins, arteries, and capillaries of your body would reach over 40,000 miles? However, your vascular system occupies a very small fraction of your body's volume. Describe a self-similar object in nature, such as your vascular system, with enormous length but relatively small volume.

39. Describe a difference between the shapes of man-made objects and objects that occur in nature.

40. Explain what this short poem by Jonathan Swift has to do with fractal images:

So Nat'ralists observe, A Flea
Hath Smaller Fleas that on him prey
and these have smaller Fleas to bite'em
And so proceed, ad infinitum

41. In Tom Stoppard's play *Arcadia,* the characters talk about mathematics, including ideas from fractal geometry. Read the play and write a paper on one of the ideas that the characters discuss that is related to something presented in this section.

Critical Thinking Exercises

Make Sense? *In Exercises 42–45, determine whether each statement makes sense or does not make sense, and explain your reasoning.*

42. The most effective way to determine if a graph is traversable is by trial and error.

43. Objects that appear to be quite different can be topologically equivalent.

44. Non-Euclidean geometries can be used to prove that the sum of the measures of a triangle's angles is 180°.

45. Euclidean geometry serves as an excellent model for describing the diverse forms that arise in nature.

Group Exercises

46. This activity is suggested for two or three people. Research some of the practical applications of topology. Present the results of your research in a seminar to the entire class. You may also want to include a discussion of some of topology's more unusal figures, such as the Klein bottle and the Möbius strip.

47. Research non-Euclidean geometry and plan a seminar based on your group's research. Each group member should research one of the following five areas:

a. Present an overview of the history of the people who developed non-Euclidean geometry. Who first used the term and why did he never publish his work?

b. Present an overview of the connection between Saccheri quadrilaterals and non-Euclidean geometry. Describe the work of Girolamo Saccheri.

c. Describe how Albert Einstein applied the ideas of Gauss and Riemann. Discuss the notion of curved space and a fourth dimension.

d. Present examples of the work of M. C. Escher that provide ways of visualizing hyperbolic and elliptic geometry.

e. Describe how non-Euclidean geometry changed the direction of subsequent research in mathematics.

After all research has been completed, the group should plan the order in which each group member will speak. Each person should plan on taking about five minutes for his or her portion of the presentation.

48. Albert Einstein's theory of general relativity is concerned with the structure, or the geometry, of the universe. In order to describe the universe, Einstein discovered that he needed four variables: three variables to locate an object in space and a fourth variable describing time. This system is known as space-time.

Because we are three-dimensional beings, how can we imagine four dimensions? One interesting approach to visualizing four dimensions is to consider an analogy of a two-dimensional being struggling to understand three dimensions. This approach first appeared in a book called *Flatland* by Edwin Abbott, written around 1884.

Flatland describes an entire civilization of beings who are two dimensional, living on a flat plane, unaware of the existence of anything outside their universe. A house in Flatland would look like a blueprint or a line drawing to us. If we were to draw a closed circle around Flatlanders, they would be imprisoned in a cell with no way to see out or escape because there is no way to move up and over the circle. For a two-dimensional being moving only on a plane, the idea of up would be incomprehensible. We could explain that up means moving in a new direction, perpendicular to the two dimensions they know, but it would be similar to telling us we can move in the fourth dimension by traveling perpendicular to our three dimensions.

Group members should obtain copies of or excerpts from Edwin Abbott's *Flatland*. We especially recommend *The Annotated Flatland*, Perseus Publishing, 2002, with fascinating commentary by mathematician and author Ian Stewart. Once all group members have read the story, the following questions are offered for group discussion.

a. How does the sphere, the visitor from the third dimension, reflect the same narrow perspective as the Flatlanders?

b. What are some of the sociological problems raised in the story?

c. What happens when we have a certain way of seeing the world that is challenged by coming into contact with something quite different? Be as specific as possible, citing either personal examples or historical examples.

d. How are A. Square's difficulties in visualizing three dimensions similar to those of a three-dimensional dweller trying to visualize four dimensions?

e. How does the author reflect the overt sexism of his time?

f. What "upward not northward" ideas do you hold that, if shared, would result in criticism, rejection, or a fate similar to that of the narrator of *Flatland*?

Chapter Summary, Review, and Test

SUMMARY – DEFINITIONS AND CONCEPTS

10.1 Points, Lines, Planes, and Angles

SUMMARY – DEFINITIONS AND CONCEPTS	EXAMPLES
a. Line AB ($\overleftrightarrow{AB}$ or $\overleftrightarrow{BA}$), half-line AB ($\overset{\circ}{\overrightarrow{AB}}$), ray AB ($\overrightarrow{AB}$), and line segment AB ($\overline{AB}$ or $\overline{BA}$) are represented in Figure 10.2 on page 610.	
b. Angles are measured in degrees. A degree, $1°$, is $\frac{1}{360}$ of a complete rotation. Acute angles measure less than $90°$, right angles $90°$, obtuse angles more than $90°$ but less than $180°$, and straight angles $180°$.	Ex. 1, p. 611
c. Complementary angles are two angles whose measures have a sum of $90°$. Supplementary angles are two angles whose measures have a sum of $180°$.	Ex. 2, p. 612; Ex. 3, p. 613
d. Vertical angles have the same measure.	Ex. 4, p. 613
e. If parallel lines are intersected by a transversal, alternate interior angles, alternate exterior angles, and corresponding angles have the same measure.	Ex. 5, p. 615

10.2 Triangles

SUMMARY – DEFINITIONS AND CONCEPTS	EXAMPLES
a. The sum of the measures of the three angles of any triangle is $180°$.	Ex. 1, p. 620; Ex. 2, p. 620
b. Triangles can be classified by angles (acute, right, obtuse) or by sides (isosceles, equilateral, scalene). See the box on page 621.	

c. Similar triangles have the same shape, but not necessarily the same size. Corresponding angles have the same measure and corresponding sides are proportional. If the measures of two angles of one triangle are equal to those of two angles of a second triangle, then the two triangles are similar.	Ex. 3, p. 622; Ex. 4, p. 623
d. The Pythagorean Theorem: The sum of the squares of the lengths of the legs of a right triangle equals the square of the length of the hypotenuse.	Ex. 5, p. 624; Ex. 6, p. 624

10.3 Polygons, Perimeter, and Tessellations

a. A polygon is a closed geometric figure in a plane formed by three or more line segments. Names of some polygons, given in Table 10.2 on page 631, include triangles (three sides), quadrilaterals (four sides), pentagons (five sides), hexagons (six sides), heptagons (seven sides), and octagons (eight sides). A regular polygon is one whose sides are all the same length and whose angles all have the same measure. The perimeter of a polygon is the sum of the lengths of its sides.	Ex. 1, p. 632
b. Types of quadrilaterals, including the parallelogram, rhombus, rectangle, square, and trapezoid, and their characteristics, are given in Table 10.3 on page 631.	
c. The sum of the measures of the angles of an n-sided polygon is $(n - 2)180°$. If the n-sided polygon is a regular polygon, then each angle measures $\frac{(n - 2)180°}{n}$.	Ex. 2, p. 633
d. A tessellation is a pattern consisting of the repeated use of the same geometric figures to completely cover a plane, leaving no gaps and having no overlaps. The angle requirement for the formation of a tessellation is that the sum of the measures of the angles at each vertex must be 360°.	Ex. 3, p. 635

10.4 Area and Circumference

a. Formulas for Area Rectangle: $A = lw$; Square: $A = s^2$; Parallelogram: $A = bh$; Triangle: $A = \frac{1}{2}bh$; Trapezoid: $A = \frac{1}{2}h(a + b)$	Ex. 1, p. 639; Ex. 2, p. 640; Ex. 3, p. 641; Ex. 4, p. 642; Ex. 5, p. 643
b. Circles Circumference: $C = 2\pi r$ or $C = \pi d$ Area: $A = \pi r^2$	Ex. 6, p. 644; Ex. 7, p. 644; Ex. 8, p. 645

10.5 Volume and Surface Area

a. Formulas for Volume Rectangular Solid: $V = lwh$; Cube: $V = s^3$; Pyramid: $V = \frac{1}{3}Bh$; Cylinder: $V = \pi r^2 h$; Cone: $V = \frac{1}{3}\pi r^2 h$ Sphere: $V = \frac{4}{3}\pi r^3$	Ex. 1, p. 651; Ex. 2, p. 651; Ex. 3, p. 652; Ex. 4, p. 653; Ex. 5, p. 654; Ex. 6, p. 654
b. Formulas for surface area are given in Table 10.4 on page 655.	Ex. 7, p. 655

10.6 Right Triangle Trigonometry

a. Trigonometric ratios, sin A, cos A, and tan A, are defined in the box on page 659.	Ex. 1, p. 660
b. Given one side and the measure of an acute angle of a right triangle, the trigonometric ratios can be used to find the measures of the other parts of the right triangle.	Ex. 2, p. 661; Ex. 3, p. 662; Ex. 4, p. 663
c. Given the measures of two sides of a right triangle, the measures of the acute angles can be found using the inverse trigonometric ratio keys $\boxed{SIN^{-1}}$, $\boxed{COS^{-1}}$, and $\boxed{TAN^{-1}}$, on a calculator.	Ex. 5, p. 663

10.7 Beyond Euclidean Geometry

a. A vertex is a point. An edge is a line segment or curve that starts and ends at a vertex. Vertices and edges form a graph. A vertex with an odd number of edges is odd; one with an even number of edges is even. A traversable graph is one that can be traced without removing the pencil from the paper and without tracing an edge more than once. A graph with all even vertices or one with exactly two odd vertices is traversable. With more than two odd vertices, the graph cannot be traversed.	Ex. 1, p. 670

b. In topology, shapes are twisted, stretched, bent, and shrunk. The genus of an object refers to the number of holes in the object. Objects with the same genus are topologically equivalent.

c. Non-Euclidean geometries, including hyperbolic geometry and elliptic geometry, are outlined in Table 10.5 on page 674.

d. Fractal geometry includes self-similar forms; magnified portions of such forms repeat much of the pattern of the whole form.

Review Exercises

10.1

In the figure shown, lines l and m are parallel. In Exercises 1–7, match each term with the numbered angle or angles in the figure.

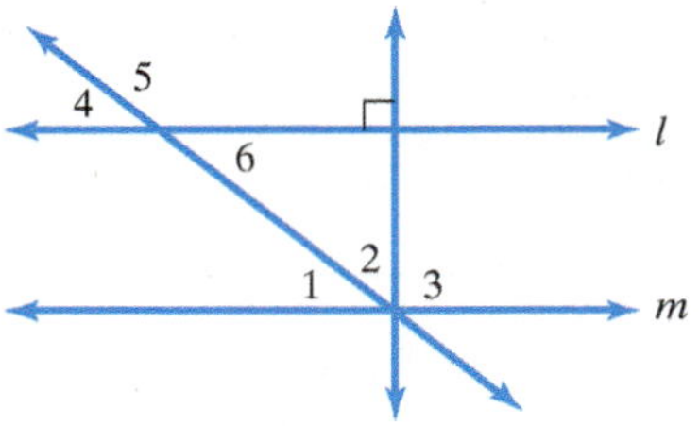

1. right angle

2. obtuse angle

3. vertical angles

4. alternate interior angles

5. corresponding angles

6. the complement of ∡1

7. the supplement of ∡6

In Exercises 8–9, find the measure of the angle in which a question mark with a degree symbol appears.

8.

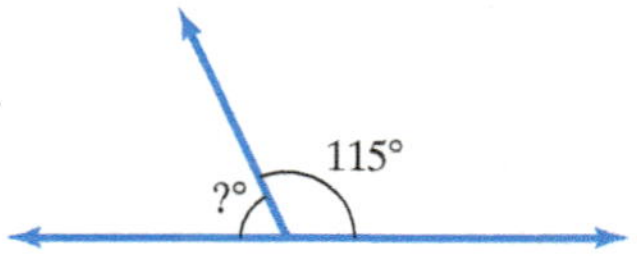

9.

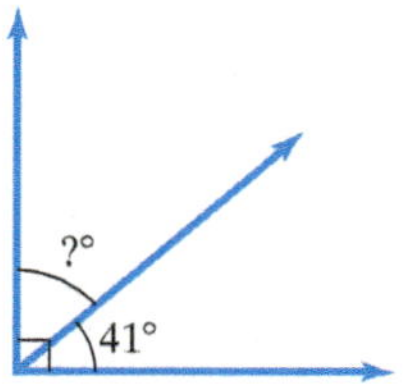

10. If an angle measures 73°, find the measure of its complement.

11. If an angle measures 46°, find the measure of its supplement.

12. In the figure shown, find the measures of angles 1, 2, and 3.

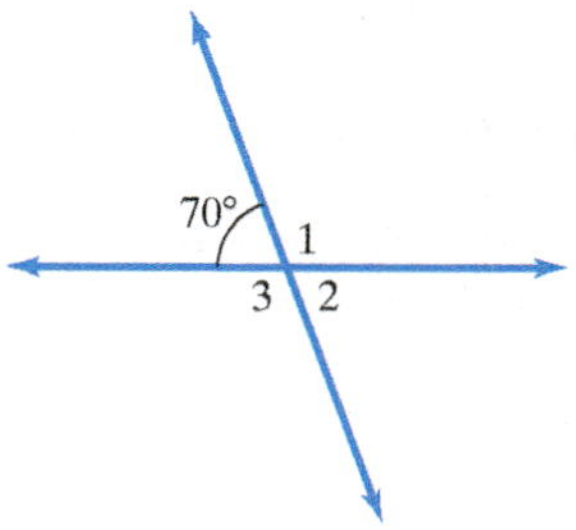

13. In the figure shown, two parallel lines are intersected by a transversal. One of the angle measures is given. Find the measure of each of the other seven angles.

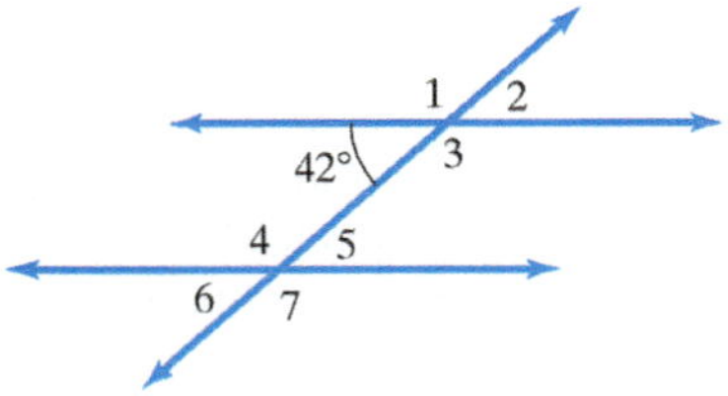

10.2

In Exercises 14–15, find the measure of angle A for the triangle shown.

14.

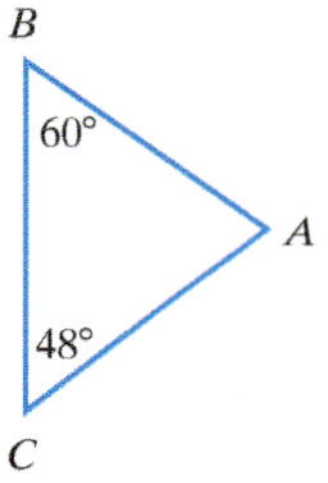

15.

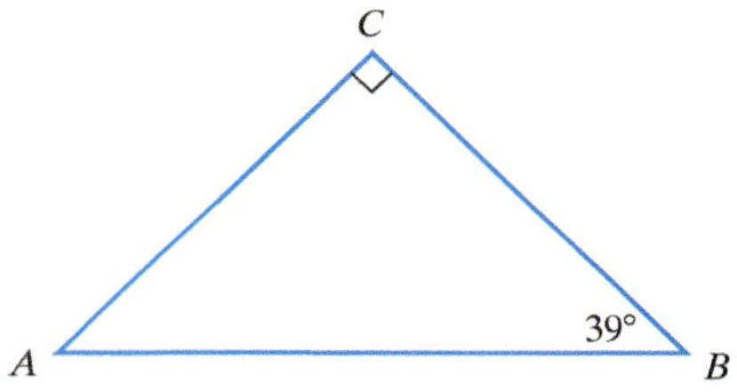

16. Find the measures of angles 1 through 5 in the figure shown.

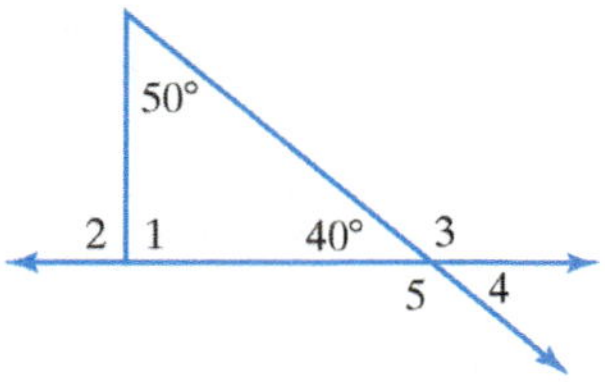

17. In the figure shown, lines l and m are parallel. Find the measure of each numbered angle.

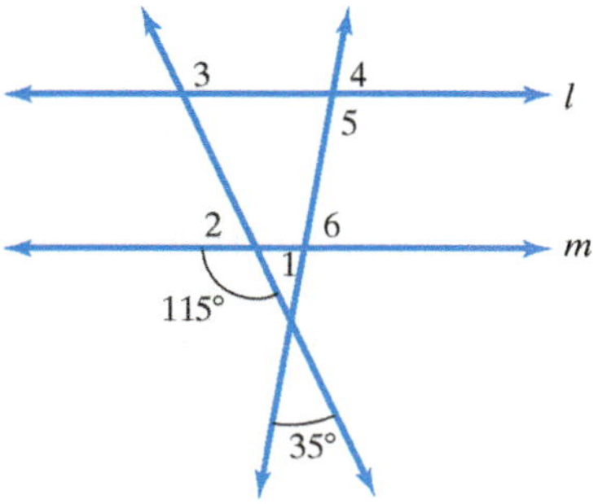

In Exercises 18–19, use similar triangles and the fact that corresponding sides are proportional to find the length of each side marked with an x.

18.

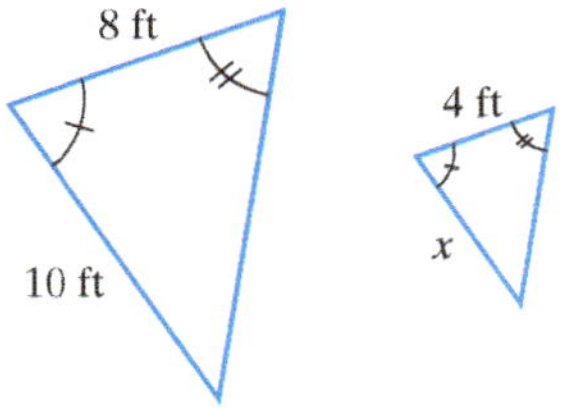

19.

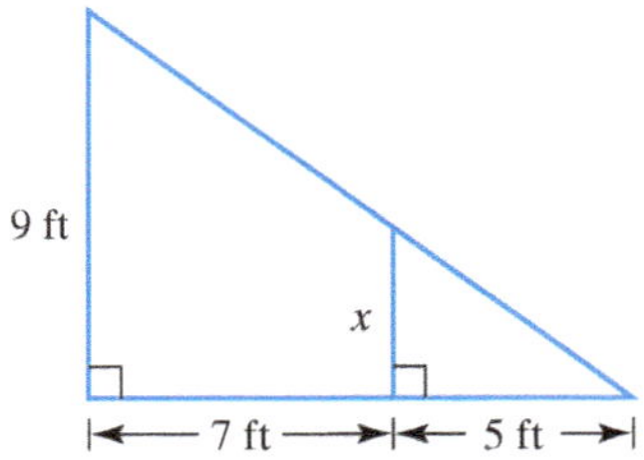

In Exercises 20–22, use the Pythagorean Theorem to find the missing length in each right triangle. Round, if necessary, to the nearest tenth.

20.

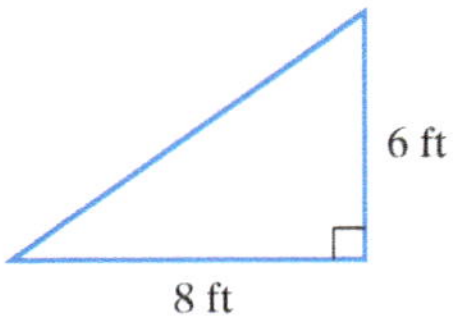

21.

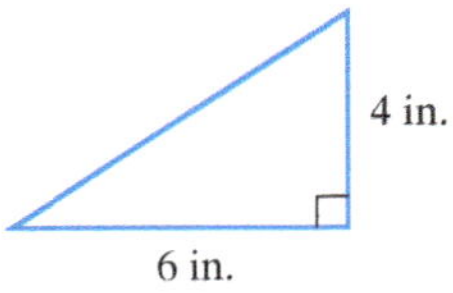

22.

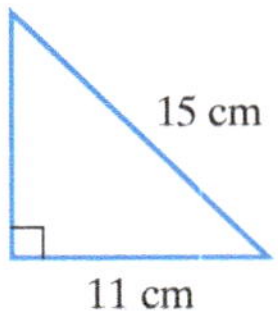

23. Find the height of the lamppost in the figure.

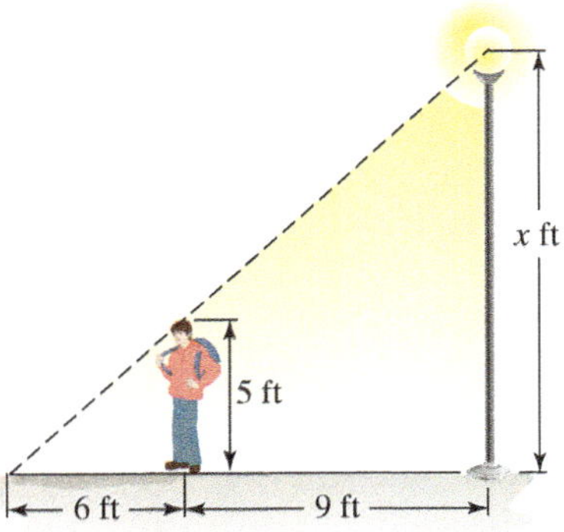

24. How far away from the building in the figure shown is the bottom of the ladder?

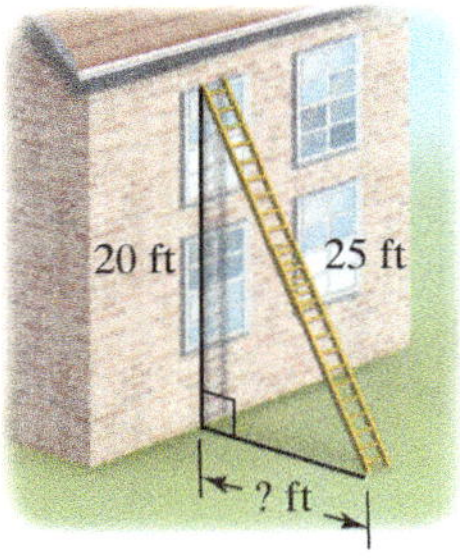

25. A vertical pole is to be supported by three wires. Each wire is 13 yards long and is anchored 5 yards from the base of the pole. How far up the pole will the wires be attached?

10.3

26. Write the names of all quadrilaterals that always have four right angles.

27. Write the names of all quadrilaterals with four sides always having the same measure.

28. Write the names of all quadrilaterals that do not always have four angles with the same measure.

In Exercises 29–31, find the perimeter of the figure shown. Express the perimeter using the same unit of measure that appears in the figure.

29.

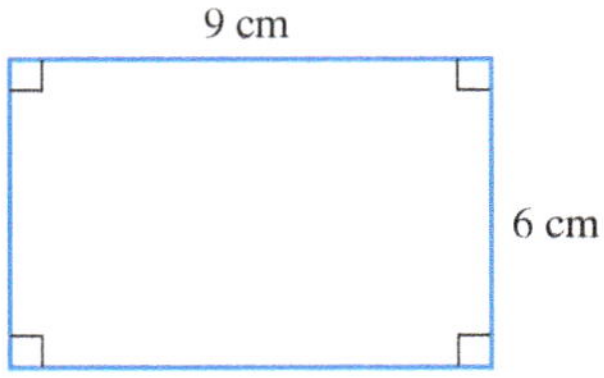

30.

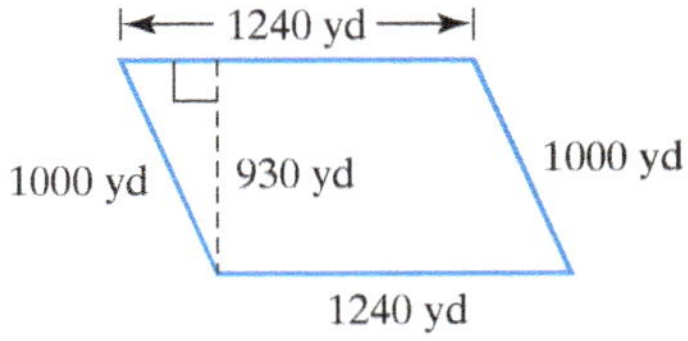

31.

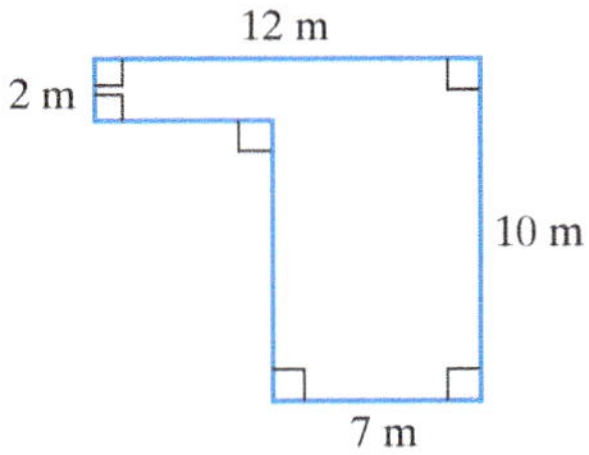

32. Find the sum of the measures of the angles of a 12-sided polygon.

33. Find the sum of the measures of the angles of an octagon.

34. The figure shown is a regular polygon. Find the measures of angle 1 and angle 2.

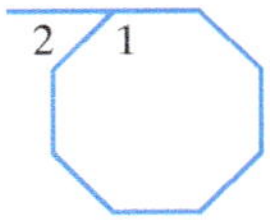

35. A carpenter is installing a baseboard around a room that has a length of 35 feet and a width of 15 feet. The room has four doorways and each doorway is 3 feet wide. If no baseboard is to be put across the doorways and the cost of the baseboard is $1.50 per foot, what is the cost of installing the baseboard around the room?

36. Use the following tessellation to solve this exercise.

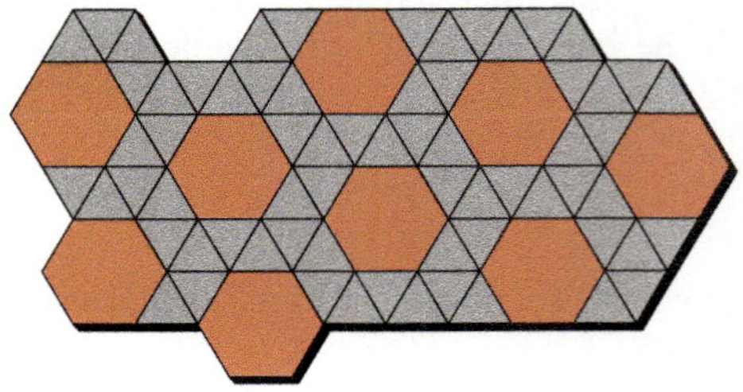

a. Name the types of regular polygons that surround each vertex.
b. Determine the number of angles that come together at each vertex, as well as the measures of these angles.
c. Use the angle measures from part (b) to explain why the tessellation is possible.

37. Can a tessellation be created using only regular hexagons? Explain your answer.

10.4

In Exercises 38–41, find the area of each figure.

38.

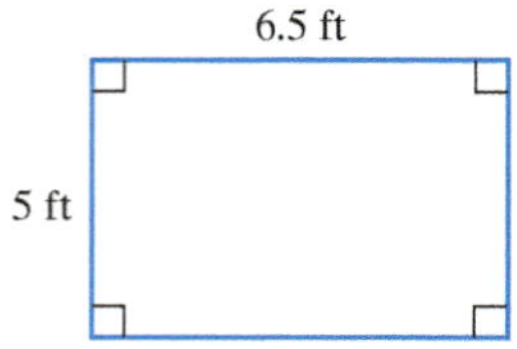

39.

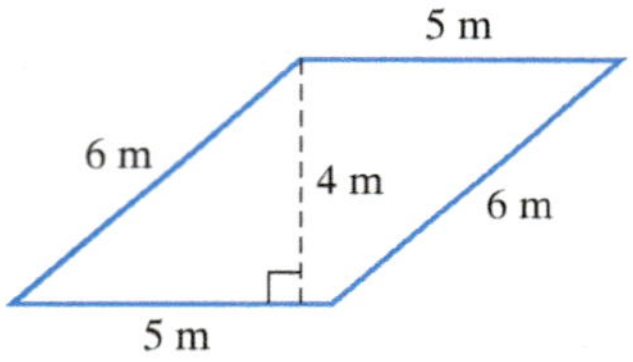

40.

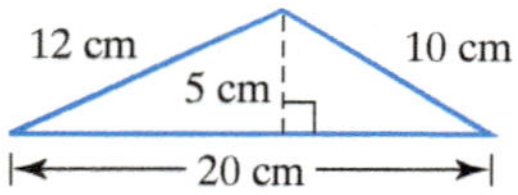

41.

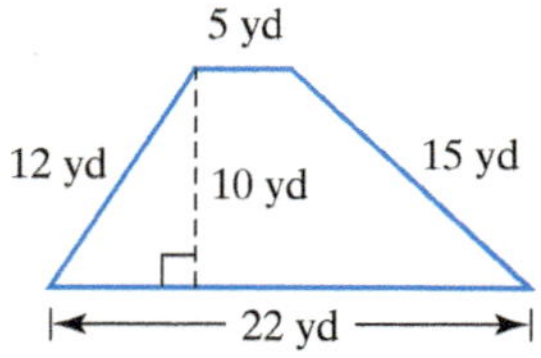

42. Find the circumference and the area of a circle with a diameter of 20 meters. Express answers in terms of π and then round to the nearest tenth.

In Exercises 43–44, find the area of each figure.

43.

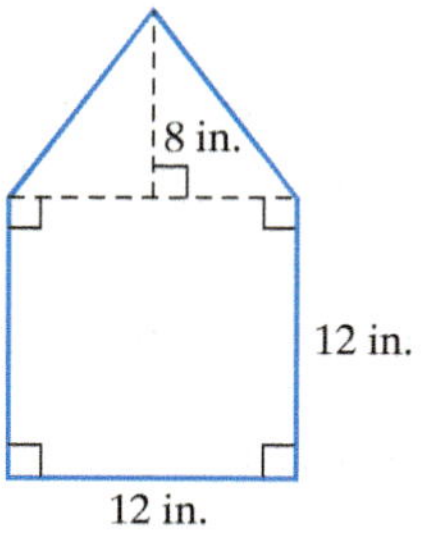

44.

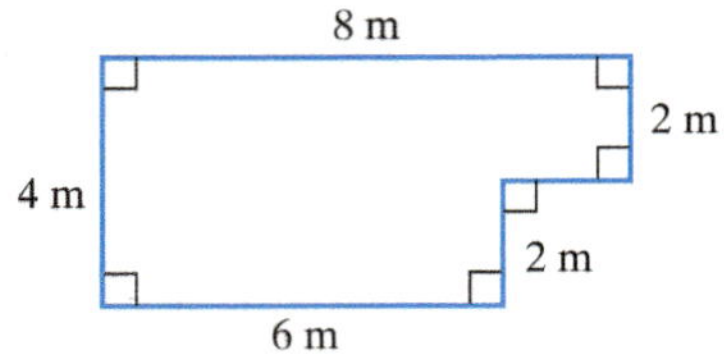

In Exercises 45–46, find the area of each shaded region. Where necessary, express answers in terms of π and then round to the nearest tenth.

45.

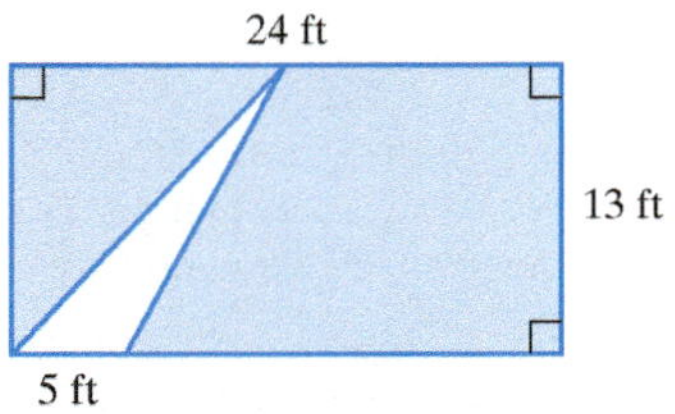

46.

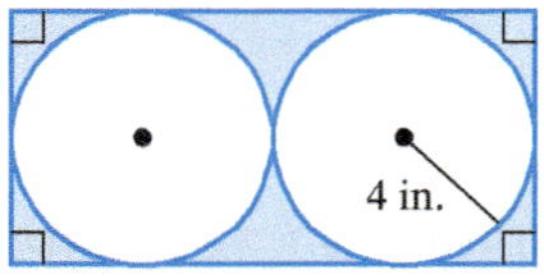

47. What will it cost to carpet a rectangular floor measuring 15 feet by 21 feet if the carpet costs $22.50 per square yard?

48. What will it cost to cover a rectangular floor measuring 40 feet by 50 feet with square tiles that measure 2 feet on each side if a package of 10 tiles costs $13?

49. How much fencing, to the nearest whole yard, is needed to enclose a circular garden that measures 10 yards across?

10.5

In Exercises 50–54, find the volume of each figure. Where necessary, express answers in terms of π and then round to the nearest whole number.

50.

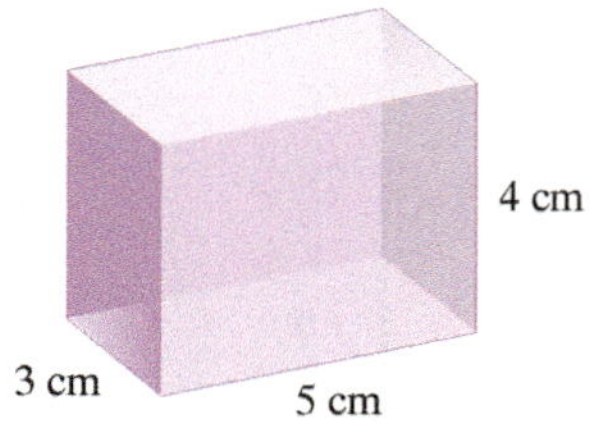

51.

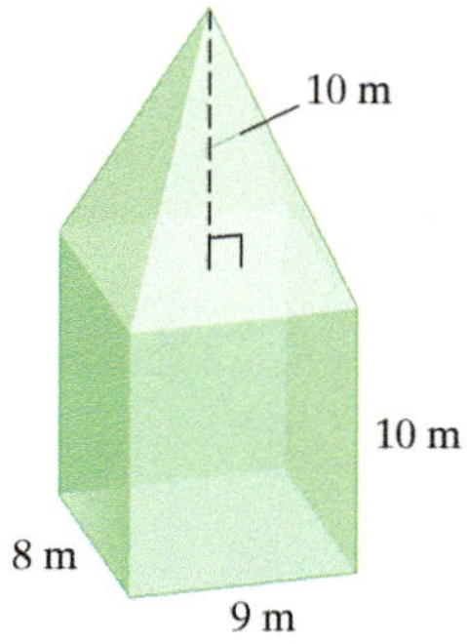

52.

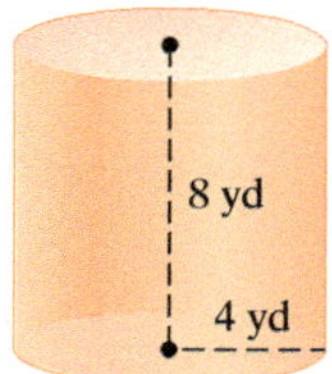

53.

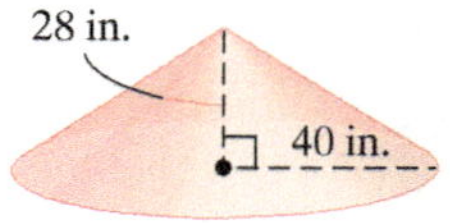

54.

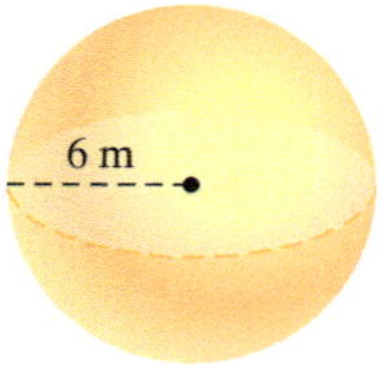

55. Find the surface area of the figure shown.

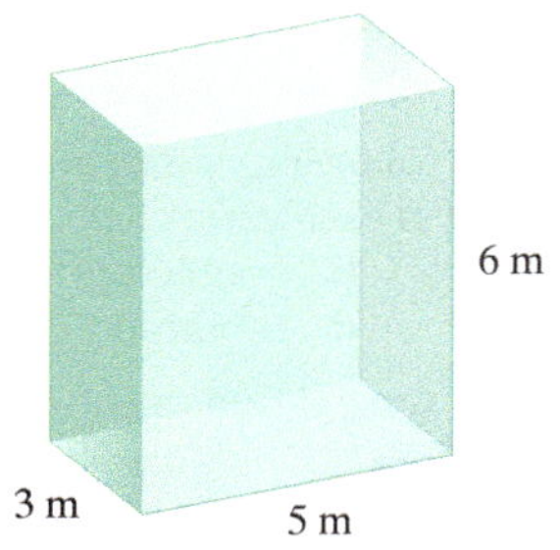

56. A train is being loaded with shipping boxes. Each box is 8 meters long, 4 meters wide, and 3 meters high. If there are 50 shipping boxes, how much space is needed?

57. An Egyptian pyramid has a square base measuring 145 meters on each side. If the height of the pyramid is 93 meters, find its volume.

58. What is the cost of concrete for a walkway that is 27 feet long, 4 feet wide, and 6 inches deep if the concrete is $40 per cubic yard?

10.6

59. Use the right triangle shown to find ratios, in reduced form, for sin A, cos A, and tan A.

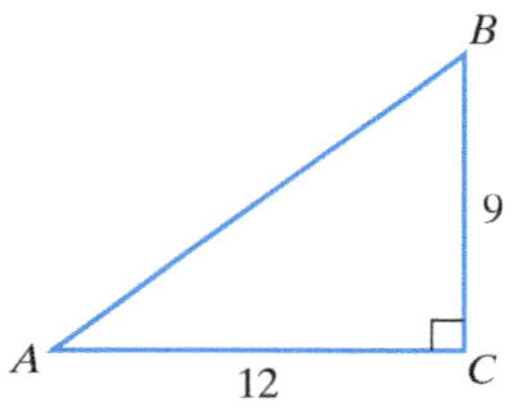

In Exercises 60–62, find the measure of the side of the right triangle whose length is designated by a lowercase letter. Round answers to the nearest whole number.

60.

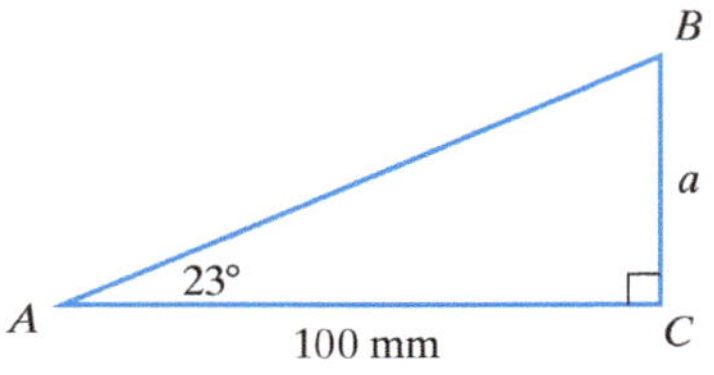

61.

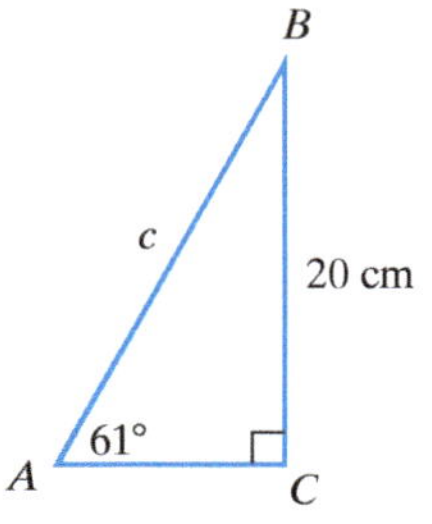

62.

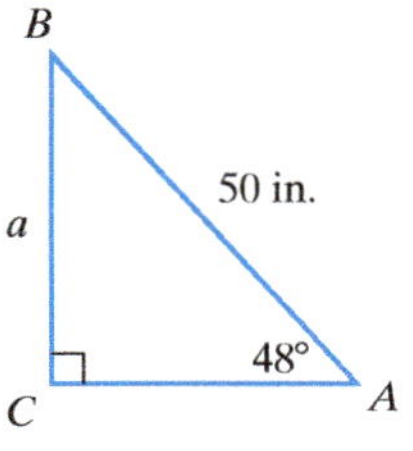

63. Find the measure of angle A in the right triangle shown. Round to the nearest whole degree.

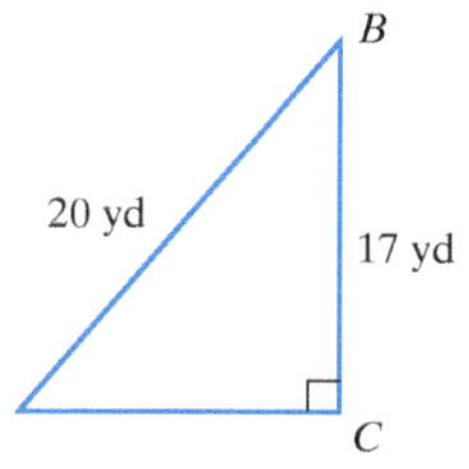

64. A hiker climbs for a half mile (2640 feet) up a slope whose inclination is 17°. How many feet of altitude, to the nearest foot, does the hiker gain?

65. To find the distance across a lake, a surveyor took the measurements in the figure shown. What is the distance across the lake? Round to the nearest meter.

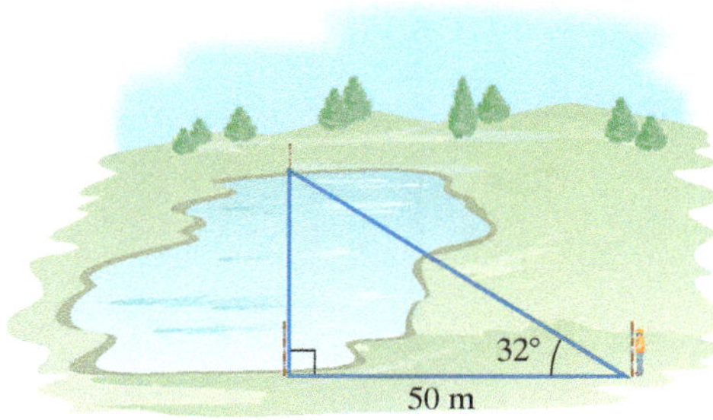

66. When a six-foot pole casts a four-foot shadow, what is the angle of elevation of the sun? Round to the nearest whole degree.

10.7

For each graph in Exercises 67–68, determine whether it is traversable. If it is, describe a path that will traverse it.

67.

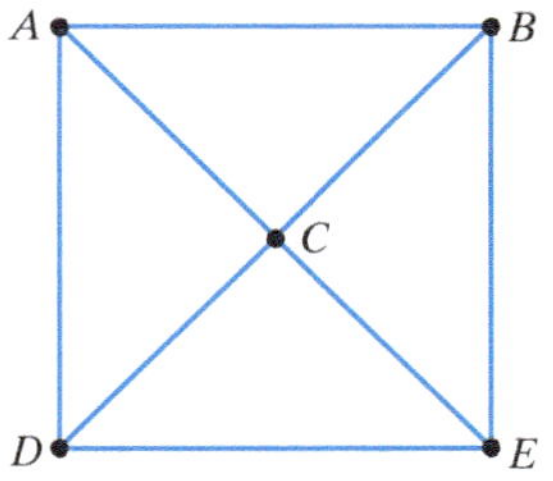

68.

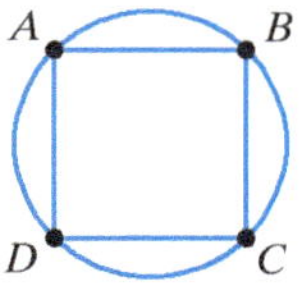

In Exercises 69–72, give the genus of each object. Which objects are topologically equivalent?

69.

70.

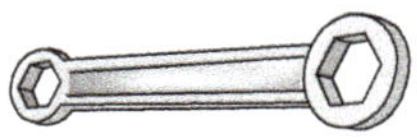

71.

72.

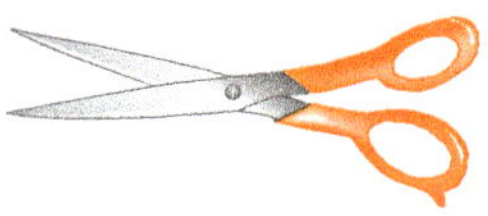

73. State Euclid's assumption about parallel lines that no longer applies in hyperbolic and elliptic geometry.

74. What is self-similarity? Describe an object in nature that has this characteristic.

Chapter 10 Test

1. If an angle measures 54°, find the measure of its complement and supplement.

In Exercises 2–4, use the figure shown to find the measure of angle 1.

2.

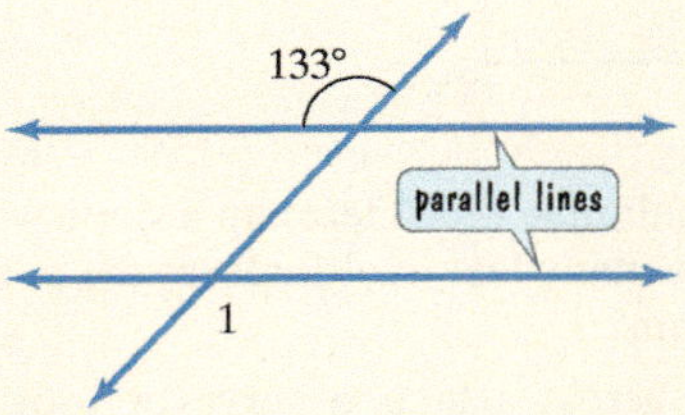

3.

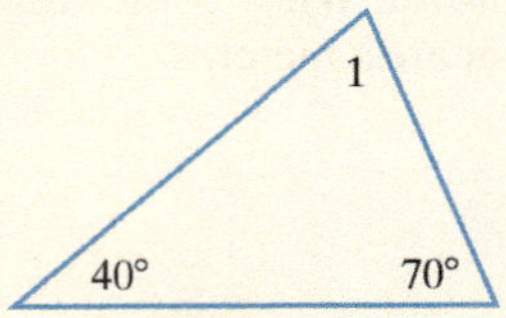

4.

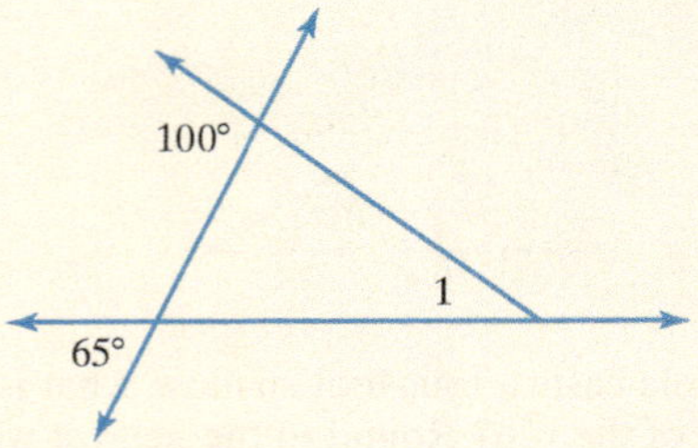

5. The triangles in the figure are similar. Find the length of the side marked with an x.

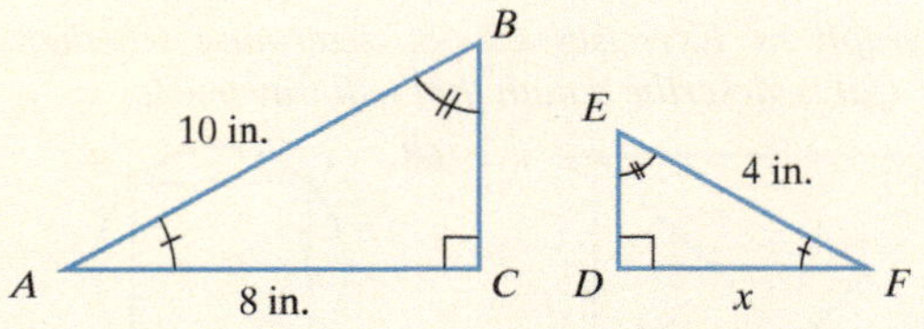

6. A vertical pole is to be supported by three wires. Each wire is 26 feet long and is anchored 24 feet from the base of the pole. How far up the pole should the wires be attached?

7. Find the sum of the measures of the angles of a ten-sided polygon.

8. Find the perimeter of the figure shown.

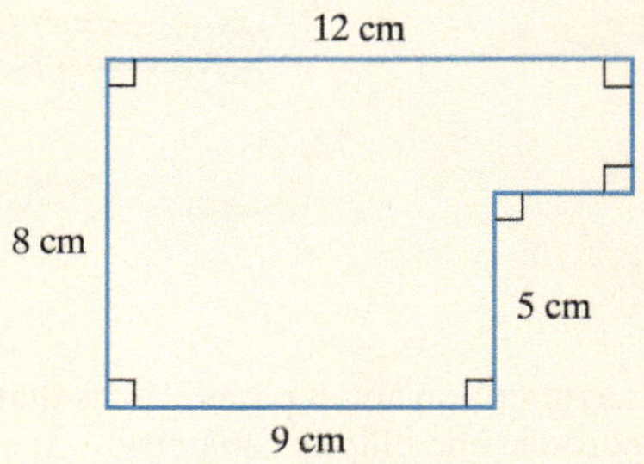

9. Which one of the following names a quadrilateral in which the sides that meet at each vertex have the same measure?
   a. rectangle **b.** parallelogram
   c. trapezoid **d.** rhombus

10. Use the following tessellation to solve this exercise.

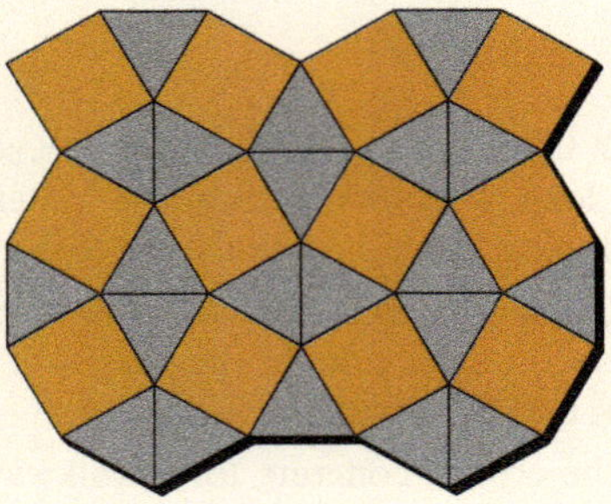

   a. Name the types of regular polygons that surround each vertex.
   b. Determine the number of angles that come together at each vertex, as well as the measures of these angles.
   c. Use the angle measures from part (b) to explain why the tessellation is possible.

In Exercises 11–12, find the area of each figure.

11.

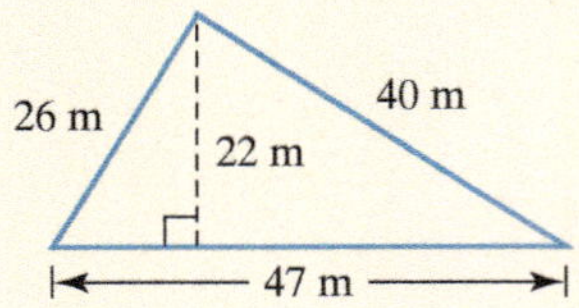

12.

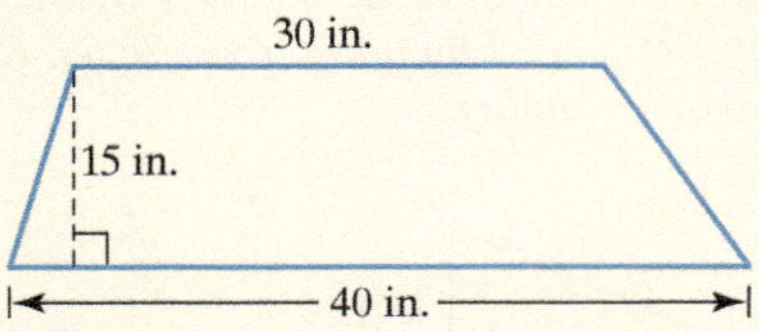

13. The right triangle shown has one leg of length 5 centimeters and a hypotenuse of length 13 centimeters.

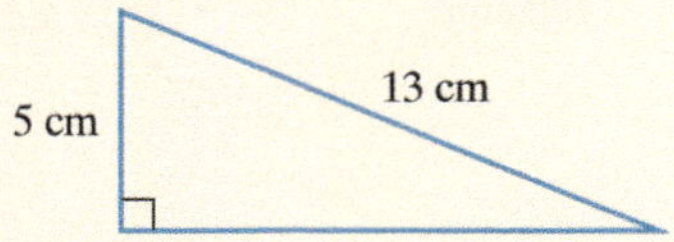

   a. Find the length of the other leg.
   b. What is the perimeter of the triangle?
   c. What is the area of the triangle?

14. Find the circumference and area of a circle with a diameter of 40 meters. Express answers in terms of π and then round to the nearest tenth.

15. A rectangular floor measuring 8 feet by 6 feet is to be completely covered with square tiles measuring 8 inches on each side. How many tiles are needed to completely cover the floor?

In Exercises 16–18, find the volume of each figure. If necessary, express the answer in terms of π and then round to the nearest whole number.

16.

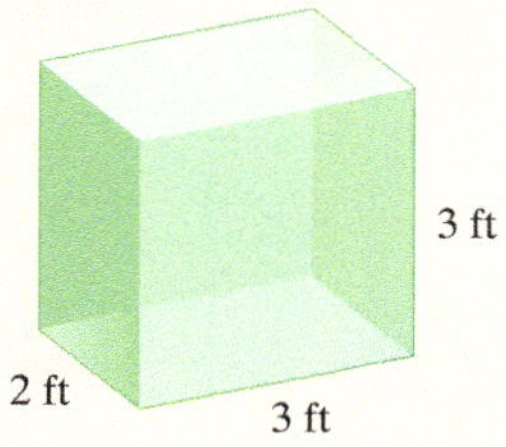

17.

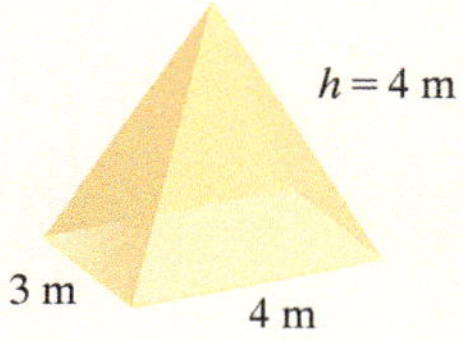

18.

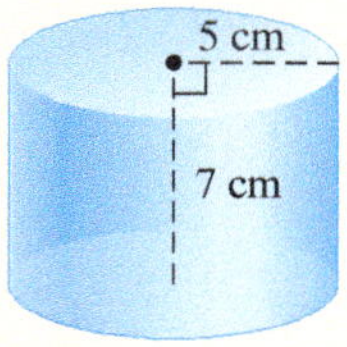

19. Find the measure, to the nearest whole number, of the side of the right triangle whose length is designated by c.

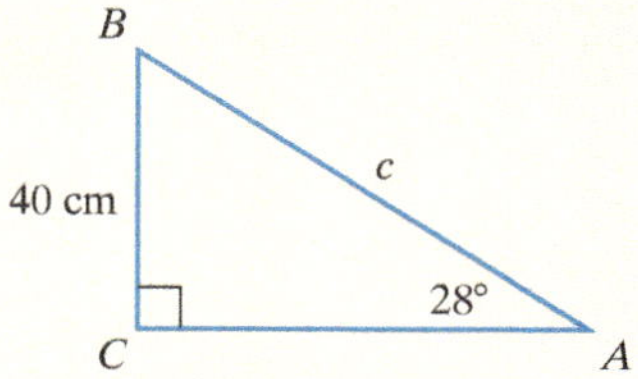

20. At a certain time of day, the angle of elevation of the sun is 34°. If a building casts a shadow measuring 104 feet, find the height of the building to the nearest foot.

21. Determine if the graph shown is traversable. If it is, describe a path that will traverse it.

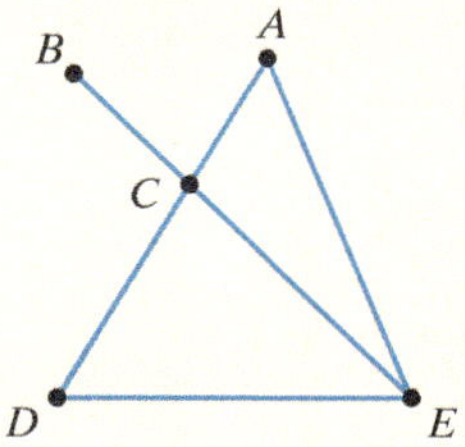

22. Describe a difference between the shapes of fractal geometry and those of Euclidean geometry.

12

Statistics

SOME RANDOM STATISTICAL FACTOIDS:

- 28% of liberals have insomnia, compared with 16% of conservatives. (*Mother Jones*)
- 17% of American workers would reveal company secrets for money, and 8% have done it already. (Monster.com)
- 31% of American adults find giving up their smartphone for a day more difficult than giving up their significant other. (Microsoft)
- Between the ages of 18 and 24, 73% of Americans have used text messages to send suggestive pictures, compared with 55% ages 25–29, 52% ages 30–34, 42% ages 35–44, 26% ages 45–54, 10% ages 55–64, and 7% ages 65 and older. (*Time*)
- 49% of Americans cite a "lot" of stress at age 22, compared with 45% at 42, 35% at 58, 29% at 62, and 20% by 70. (*Proceedings of the National Academy of Sciences*)
- 34% of American adults believe in ghosts. (AP/Ipsos)
- Unwillingness to eat sushi correlates nearly perfectly with disapproval of marriage equality. (Pew Research Center)

Statisticians collect numerical data from subgroups of populations to find out everything imaginable about the population as a whole, including whom they favor in an election, what they watch on TV, how much money they make, or what worries them. Comedians and statisticians joke that 62.38% of all statistics are made up on the spot. Because statisticians both record and influence our behavior, it is important to distinguish between good and bad methods for collecting, presenting, and interpreting data.

Here's where you'll find these applications:

Throughout this chapter, you will gain an understanding of where data come from and how these numbers are used to make decisions. We'll return to the bizarre sushi/marriage equality correlation in Exercises 5 and 35 of Exercise Set 12.6.

12.1 Sampling, Frequency Distributions, and Graphs

WHAT AM I SUPPOSED TO LEARN?

After you have read this section, you should be able to:

1. Describe the population whose properties are to be analyzed.
2. Select an appropriate sampling technique.
3. Organize and present data.
4. Identify deceptions in visual displays of data.

*M*A*S*H* took place in the early 1950s, during the Korean War. By the final episode, the show had lasted four times as long as the Korean War.

AT THE END OF THE twentieth century, there were 94 million households in the United States with television sets. The television program viewed by the greatest percentage of such households in that century was the final episode of *M*A*S*H*. Over 50 million American households watched this program.

Numerical information, such as the information about the top three TV shows of the twentieth century, shown in **Table 12.1**, is called **data**. The word **statistics** is often used when referring to data. However, statistics has a second meaning: Statistics is also a method for collecting, organizing, analyzing, and interpreting data, as well as drawing conclusions based on the data. This methodology divides statistics into two main areas. **Descriptive statistics** is concerned with collecting, organizing, summarizing, and presenting data. **Inferential statistics** has to do with making generalizations about and drawing conclusions from the data collected.

TABLE 12.1 TV Programs with the Greatest U.S. Audience Viewing Percentage of the Twentieth Century

Program	Total Households	Viewing Percentage
1. *M*A*S*H* Feb. 28, 1983	50,150,000	60.2%
2. *Dallas* Nov. 21, 1980	41,470,000	53.3%
3. *Roots Part 8* Jan. 30, 1977	36,380,000	51.1%

Source: Nielsen Media Research

1 Describe the population whose properties are to be analyzed.

Populations and Samples

Consider the set of all American TV households. Such a set is called the *population*. In general, a **population** is the set containing all the people or objects whose properties are to be described and analyzed by the data collector.

The population of American TV households is huge. At the time of the *M*A*S*H* conclusion, there were nearly 84 million such households. Did over 50 million American TV households really watch the final episode of *M*A*S*H*? A friendly phone call to each household ("So, how are you? What's new? Watch any good television last night? If so, what?") is, of course, absurd. A **sample**, which is a subset or subgroup of the population, is needed. In this case, it would be appropriate to have a sample of a few thousand TV households to draw conclusions about the population of all TV households.

Blitzer Bonus

A Sampling Fiasco

In 1936, the *Literary Digest* mailed out over ten million ballots to voters throughout the country. The results poured in, and the magazine predicted a landslide victory for Republican Alf Landon over Democrat Franklin Roosevelt. However, the prediction of the *Literary Digest* was wrong. Why? The mailing lists the editors used included people from their own subscriber list, directories of automobile owners, and telephone books. As a result, its sample was anything but random. It excluded most of the poor, who were unlikely to subscribe to the *Literary Digest,* or to own a car or telephone in the heart of the Depression. Prosperous people in 1936 were more likely to be Republican than the poor. Thus, although the sample was massive, it included a higher percentage of affluent individuals than the population as a whole did. A victim of both the Depression and the 1936 sampling fiasco, the *Literary Digest* folded in 1937.

EXAMPLE 1 *Populations and Samples*

A group of hotel owners in a large city decide to conduct a survey among citizens of the city to discover their opinions about casino gambling.

a. Describe the population.

b. One of the hotel owners suggests obtaining a sample by surveying all the people at six of the largest nightclubs in the city on a Saturday night. Each person will be asked to express his or her opinion on casino gambling. Does this seem like a good idea?

SOLUTION

a. The population is the set containing all the citizens of the city.

b. Questioning people at six of the city's largest nightclubs is a terrible idea. The nightclub subset is probably more likely to have a positive attitude toward casino gambling than the population of all the city's citizens.

CHECK POINT 1 A city government wants to conduct a survey among the city's homeless to discover their opinions about required residence in city shelters from midnight until 6 A.M.

a. Describe the population.

b. A city commissioner suggests obtaining a sample by surveying all the homeless people at the city's largest shelter on a Sunday night. Does this seem like a good idea? Explain your answer.

Random Sampling

There is a way to use a small sample to make generalizations about a large population: Guarantee that every member of the population has an equal chance to be selected for the sample. Surveying people at six of the city's largest nightclubs does not provide this guarantee. Unless it can be established that all citizens of the city frequent these clubs, which seems unlikely, this sampling scheme does not permit each citizen an equal chance of selection.

RANDOM SAMPLES

A **random sample** is a sample obtained in such a way that every element in the population has an equal chance of being selected for the sample.

Suppose that you are elated with the quality of one of your courses. Although it's an auditorium section with 120 students, you feel that the professor is lecturing right to you. During a wonderful lecture, you look around the auditorium to see if any of the other students are sharing your enthusiasm. Based on body language, it's hard to tell. You really want to know the opinion of the population of 120 students taking this course. You think about asking students to grade the course on an A to F scale, anticipating a unanimous A. You cannot survey everyone. Eureka! Suddenly you have an idea on how to take a sample. Place cards numbered from 1 through 120, one number per card, in a box. Because the course has assigned seating by number, each numbered card corresponds to a student in the class. Reach in and randomly select six cards. Each card, and therefore each student, has an equal chance of being selected. Then use the opinions about the course from the six randomly selected students to generalize about the course opinion for the entire 120-student population.

2 Select an appropriate sampling technique.

Your idea is precisely how random samples are obtained. In random sampling, each element in the population must be identified and assigned a number. The numbers are generally assigned in order. The way to sample from the larger numbered population is to generate random numbers using a computer or calculator. Each numbered element from the population that corresponds to one of the generated random numbers is selected for the sample.

Call-in polls on radio and television are not reliable because those polled do not represent the larger population. A person who calls in is likely to have feelings about an issue that are consistent with the politics of the show's host. For a poll to be accurate, the sample must be chosen randomly from the larger population. The A. C. Nielsen Company uses a random sample of approximately 5000 TV households to measure the percentage of households tuned in to a television program.

EXAMPLE 2 *Selecting an Appropriate Sampling Technique*

We return to the hotel owners in the large city who are interested in how the city's citizens feel about casino gambling. Which of the following would be the most appropriate way to select a random sample?

a. Randomly survey people who live in the oceanfront condominiums in the city.

b. Survey the first 200 people whose names appear in the city's telephone directory.

c. Randomly select neighborhoods of the city and then randomly survey people within the selected neighborhoods.

SOLUTION

Keep in mind that the population is the set containing all the city's citizens. A random sample must give each citizen an equal chance of being selected.

a. Randomly selecting people who live in the city's oceanfront condominiums is not a good idea. Many hotels lie along the oceanfront, and the oceanfront property owners might object to the traffic and noise as a result of casino gambling. Furthermore, this sample does not give each citizen of the city an equal chance of being selected.

b. If the hotel owners survey the first 200 names in the city's telephone directory, all citizens do not have an equal chance of selection. For example, individuals whose last name begins with a letter toward the end of the alphabet have no chance of being selected.

c. Randomly selecting neighborhoods of the city and then randomly surveying people within the selected neighborhoods is an appropriate technique. Using this method, each citizen has an equal chance of being selected.

In summary, given the three options, the sampling technique in part (c) is the most appropriate.

Surveys and polls involve data from a sample of some population. Regardless of the sampling technique used, the sample should exhibit characteristics typical of those possessed by the target population. This type of sample is called a **representative sample**.

CHECK POINT 2 Explain why the sampling technique described in Check Point 1(b) on page 767 is not a random sample. Then describe an appropriate way to select a random sample of the city's homeless.

Blitzer Bonus

The United States Census

A census is a survey that attempts to include the entire population. The U.S. Constitution requires a census of the American population every ten years. When the Founding Fathers invented American democracy, they realized that if you are going to have government by the people, you need to know who and where they are. Nowadays about $400 billion per year in federal aid is distributed based on the Census numbers, for everything from jobs to bridges to schools. For every 100 people not counted, states and communities could lose as much as $130,000 annually, or $1300 per person each year, so this really matters.

Although the Census generates volumes of statistics, its main purpose is to give the government block-by-block population figures. The U.S. Census is not foolproof. The 1990 Census missed 1.6% of the American population, including an estimated 4.4% of the African-American population, largely in inner cities. Only 67% of households responded to the 2000 Census, even after door-to-door canvassing. About 6.4 million people were missed and 3.1 million were counted twice. Although the 2010 Census was one of the shortest forms in history, counting each person was not an easy task, particularly with concerns about immigration status and privacy of data.

Of course, there would be more than $400 billion to spread around if it didn't cost so much to count us in the first place: about $15 billion for the 2010 Census. That included $338 million for ads in 28 languages, a Census-sponsored NASCAR entry, and $2.5 million for a Super Bowl ad. The ads were meant to boost the response rate, since any household that did not mail back its form got visited by a Census worker, another pricey item. In all, the cost of the 2010 Census worked out to appoximately $49 per person.

3 Organize and present data.

Frequency Distributions

After data have been collected from a sample of the population, the next task facing the statistician is to present the data in a condensed and manageable form. In this way, the data can be more easily interpreted.

Suppose, for example, that researchers are interested in determining the age at which adolescent males show the greatest rate of physical growth. A random sample of 35 ten-year-old boys is measured for height and then remeasured each year until they reach 18. The age of maximum yearly growth for each subject is as follows:

12, 14, 13, 14, 16, 14, 14, 17, 13, 10, 13, 18, 12, 15, 14, 15, 15, 14, 14, 13, 15, 16, 15, 12, 13, 16, 11, 15, 12, 13, 12, 11, 13, 14, 14.

A piece of data is called a **data item**. This list of data has 35 data items. Some of the data items are identical. Two of the data items are 11 and 11. Thus, we can say that the **data value** 11 occurs twice. Similarly, because five of the data items are 12, 12, 12, 12, and 12, the data value 12 occurs five times.

Collected data can be presented using a **frequency distribution**. Such a distribution consists of two columns. The data values are listed in one column. Numerical data are generally listed from smallest to largest. The adjacent column is labeled **frequency** and indicates the number of times each value occurs.

EXAMPLE 3 Constructing a Frequency Distribution

Construct a frequency distribution for the data of the age of maximum yearly growth for 35 boys:

12, 14, 13, 14, 16, 14, 14, 17, 13, 10, 13, 18, 12, 15, 14, 15, 15, 14, 14, 13, 15, 16, 15, 12, 13, 16, 11, 15, 12, 13, 12, 11, 13, 14, 14.

SOLUTION

It is difficult to determine trends in the data above in their current format. Perhaps we can make sense of the data by organizing them into a frequency distribution. Let us create two columns. One lists all possible data values, from smallest (10) to largest (18). The other column indicates the number of times the value occurs in the sample. The frequency distribution is shown in **Table 12.2**.

The frequency distribution indicates that one subject had maximum growth at age 10, two at age 11, five at age 12, seven at age 13, and so on. The maximum growth for most of the subjects occurred between the ages of 12 and 15. Nine boys experienced maximum growth at age 14, more than at any other age within the sample. The sum of the frequencies, 35, is equal to the original number of data items.

The trend shown by the frequency distribution in **Table 12.2** indicates that the number of boys who attain their maximum yearly growth at a given age increases until age 14 and decreases after that. This trend is not evident in the data in their original format.

TABLE 12.2 A Frequency Distribution for a Boy's Age of Maximum Yearly Growth

Age of Maximum Growth	Number of Boys (Frequency)
10	1
11	2
12	5
13	7
14	9
15	6
16	3
17	1
18	1
Total:	$n = 35$

35 is the sum of the frequencies.

CHECK POINT 3 Construct a frequency distribution for the data showing final course grades for students in a precalculus course, listed alphabetically by student name in a grade book:

F, A, B, B, C, C, B, C, A, A, C, C, D, C, B, D, C, C, B, C.

A frequency distribution that lists all possible data items can be quite cumbersome when there are many such items. For example, consider the following data items. These are statistics test scores for a class of 40 students.

82	47	75	64	57	82	63	93
76	68	84	54	88	77	79	80
94	92	94	80	94	66	81	67
75	73	66	87	76	45	43	56
57	74	50	78	71	84	59	76

It's difficult to determine how well the group did when the grades are displayed like this. Because there are so many data items, one way to organize these data so that the results are more meaningful is to arrange the grades into groups, or **classes**, based on something that interests us. Many grading systems assign an A to grades in the 90–100 class, B to grades in the 80–89 class, C to grades in the 70–79 class, and so on. These classes provide one way to organize the data.

Looking at the 40 statistics test scores, we see that they range from a low of 43 to a high of 94. We can use classes that run from 40 through 49, 50 through 59, 60 through 69, and so on up to 90 through 99, to organize the scores. In Example 4, we go through the data and tally each item into the appropriate class. This method for organizing data is called a **grouped frequency distribution**.

EXAMPLE 4 *Constructing a Grouped Frequency Distribution*

Use the classes 40–49, 50–59, 60–69, 70–79, 80–89, and 90–99 to construct a grouped frequency distribution for the 40 test scores on the previous page.

SOLUTION

We use the 40 given scores and tally the number of scores in each class.

Tallying Statistics Test Scores

Test Scores (Class)	Tally	Number of Students (Frequency)
40–49	|||	3
50–59	~~||||~~ |	6
60–69	~~||||~~ |	6
70–79	~~||||~~ ~~||||~~ |	11
80–89	~~||||~~ ||||	9
90–99	~~||||~~	5

The second score in the list, 47, is shown as the first tally in this row.

The first score in the list, 82, is shown as the first tally in this row.

TABLE 12.3 A Grouped Frequency Distribution for Statistics Test Scores

Class	Frequency
40–49	3
50–59	6
60–69	6
70–79	11
80–89	9
90–99	5
Total:	$n = 40$

40, the sum of the frequencies, is the number of data items.

Omitting the tally column results in the grouped frequency distribution in **Table 12.3**. The distribution shows that the greatest frequency of students scored in the 70–79 class. The number of students decreases in classes that contain successively lower and higher scores. The sum of the frequencies, 40, is equal to the original number of data items.

The leftmost number in each class of a grouped frequency distribution is called the **lower class limit**. For example, in **Table 12.3**, the lower limit of the first class is 40 and the lower limit of the third class is 60. The rightmost number in each class is called the **upper class limit**. In **Table 12.3**, 49 and 69 are the upper limits for the first and third classes, respectively. Notice that if we take the difference between any two consecutive lower class limits, we get the same number:

$$50 - 40 = 10,\ 60 - 50 = 10,\ 70 - 60 = 10,\ 80 - 70 = 10,\ 90 - 80 = 10.$$

The number 10 is called the **class width**.

When setting up class limits, each class, with the possible exception of the first or last, should have the same width. Because each data item must fall into exactly one class, it is sometimes helpful to vary the width of the first or last class to allow for items that fall far above or below most of the data.

CHECK POINT 4 Use the classes in **Table 12.3** to construct a grouped frequency distribution for the following 37 exam scores:

73	58	68	75	94	79	96	79
87	83	89	52	99	97	89	58
95	77	75	81	75	73	73	62
69	76	77	71	50	57	41	98
77	71	69	90	75.			

TABLE 12.2 A Frequency Distribution for a Boy's Age of Maximum Yearly Growth

Age of Maximum Growth	Number of Boys (Frequency)
10	1
11	2
12	5
13	7
14	9
15	6
16	3
17	1
18	1
Total:	$n = 35$

35 is the sum of the frequencies.

Histograms and Frequency Polygons

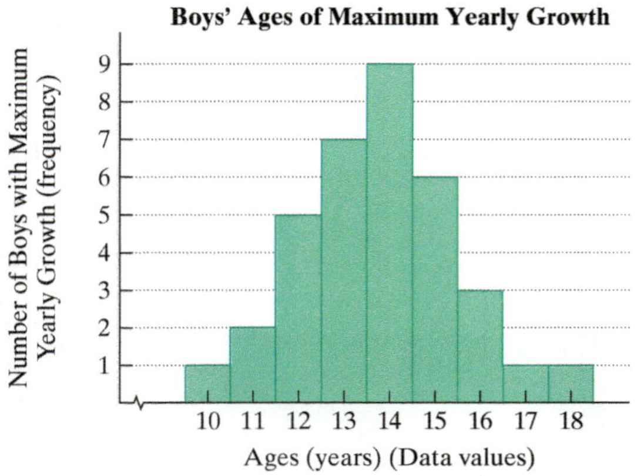

FIGURE 12.1 A histogram for a boy's age of maximum yearly growth

Take a second look at the frequency distribution for the age of a boy's maximum yearly growth in **Table 12.2**. A bar graph with bars that touch can be used to visually display the data. Such a graph is called a **histogram**. **Figure 12.1** illustrates a histogram that was constructed using the frequency distribution in **Table 12.2**. A series of rectangles whose heights represent the frequencies are placed next to each other. For example, the height of the bar for the data value 10, shown in **Figure 12.1**, is 1. This corresponds to the frequency for 10 given in **Table 12.2**. The higher the bar, the more frequent the age. The break along the horizontal axis, symbolized by ⩘, eliminates listing the ages 1 through 9.

A line graph called a **frequency polygon** can also be used to visually convey the information shown in **Figure 12.1**. The axes are labeled just like those in a histogram. Thus, the horizontal axis shows data values and the vertical axis shows frequencies. Once a histogram has been constructed, it's fairly easy to draw a frequency polygon. **Figure 12.2** shows a histogram with a dot at the top of each rectangle at its midpoint. Connect each of these midpoints with a straight line. To complete the frequency polygon at both ends, the lines should be drawn down to touch the horizontal axis. The completed frequency polygon is shown in **Figure 12.3**.

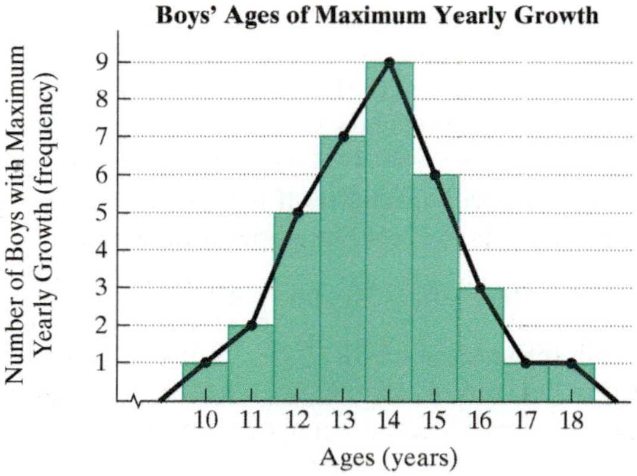

FIGURE 12.2 A histogram with a superimposed frequency polygon

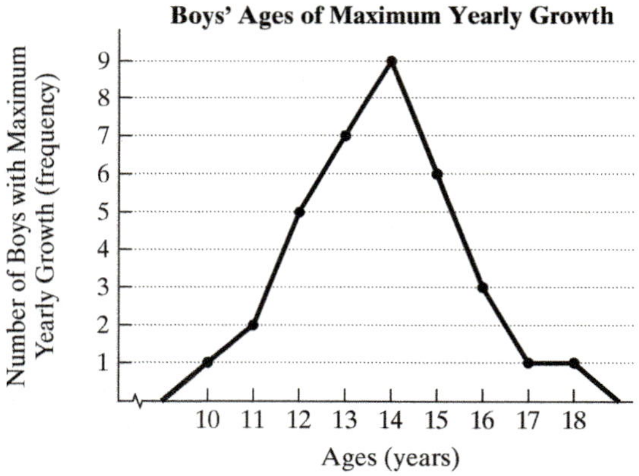

FIGURE 12.3 A frequency polygon

Stem-and-Leaf Plots

A unique way of displaying data uses a tool called a **stem-and-leaf plot**. Example 5 illustrates how we sort the data, revealing the same visual impression created by a histogram.

EXAMPLE 5 *Constructing a Stem-and-Leaf Plot*

Use the data showing statistics test scores for 40 students to construct a stem-and-leaf plot:

82	47	75	64	57	82	63	93
76	68	84	54	88	77	79	80
94	92	94	80	94	66	81	67
75	73	66	87	76	45	43	56
57	74	50	78	71	84	59	76.

SOLUTION

The plot is constructed by separating each data item into two parts. The first part is the *stem*. The **stem** consists of the tens digit. For example, the stem for the score of 82 is 8. The second part is the *leaf*. The **leaf** consists of the units digit for a given value. For the score of 82, the leaf is 2. The possible stems for the 40 scores are 4, 5, 6, 7, 8, and 9, entered in the left column of the plot.

Begin by entering each data item in the first row:

82 47 75 64 57 82 63 93.

Entering 82:

Stems	Leaves
4	
5	
6	
7	
8	2
9	

Adding 47:

Stems	Leaves
4	7
5	
6	
7	
8	2
9	

Adding 75:

Stems	Leaves
4	7
5	
6	
7	5
8	2
9	

Adding 64:

Stems	Leaves
4	7
5	
6	4
7	5
8	2
9	

Adding 57:

Stems	Leaves
4	7
5	7
6	4
7	5
8	2
9	

Adding 82:

Stems	Leaves
4	7
5	7
6	4
7	5
8	2 2
9	

Adding 63:

Stems	Leaves
4	7
5	7
6	4 3
7	5
8	2 2
9	

Adding 93:

Stems	Leaves
4	7
5	7
6	4 3
7	5
8	2 2
9	3

We continue in this manner and enter all the data items. **Figure 12.4** shows the completed stem-and-leaf plot. If you turn the page so that the left margin is on the bottom and facing you, the visual impression created by the enclosed leaves is the same as that created by a histogram. An advantage over the histogram is that the stem-and-leaf plot preserves exact data items. The enclosed leaves extend farthest to the right when the stem is 7. This shows that the greatest frequency of students scored in the 70s.

A Stem-and-Leaf Plot for 40 Test Scores

Tens digit Units digit

Stems	Leaves
4	7 5 3
5	7 4 6 7 0 9
6	4 3 8 6 7 6
7	5 6 7 9 5 3 6 4 8 1 6
8	2 2 4 8 0 0 1 7 4
9	3 4 2 4 4

FIGURE 12.4 A stem-and-leaf plot displaying 40 test scores

CHECK POINT 5 Construct a stem-and-leaf plot for the data in Check Point 4 on page 771.

4 Identify deceptions in visual displays of data.

Deceptions in Visual Displays of Data

Benjamin Disraeli, Queen Victoria's prime minister, stated that there are "lies, damned lies, and statistics." The problem is not that statistics lie, but rather that liars use statistics. Graphs can be used to distort the underlying data, making it difficult for the viewer to learn the truth. One potential source of misunderstanding is the scale on the vertical axis used to draw the graph. This scale is important because it lets a researcher "inflate" or "deflate" a trend. For example, both graphs in **Figure 12.5** present identical data for the percentage of people in the United States living below the poverty level from 2001 through 2005. The graph on the left stretches the scale on the vertical axis to create an overall impression of a poverty rate increasing rapidly over time. The graph on the right compresses the scale on the vertical axis to create an impression of a poverty rate that is slowly increasing, and beginning to level off, over time.

Percentage of People in the United States Living below the Poverty Level, 2001–2005

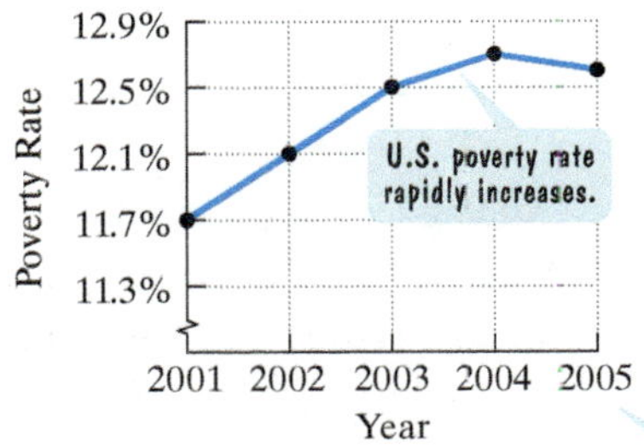

Year	Poverty Rate
2001	11.7%
2002	12.1%
2003	12.5%
2004	12.7%
2005	12.6%

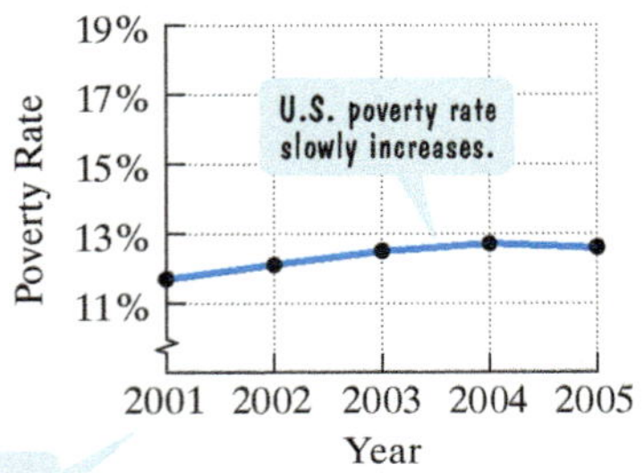

The graph in both figures present this data.

FIGURE 12.5
Source: U.S. Census Bureau

There is another problem with the data in **Figure 12.5**. Look at **Table 12.4** that shows the poverty rate from 2001 through 2011. Depending on the time frame chosen, the data can be interpreted in various ways. Carefully choosing a time frame can help represent data trends in the most positive or negative light.

TABLE 12.4 U.S. Poverty Rate from 2001 to 2011

Year	Poverty Rate
2001	11.7%
2002	12.1%
2003	12.5%
2004	12.7%
2005	12.6%
2006	12.3%
2007	12.5%
2008	13.2%
2009	14.3%
2010	15.1%
2011	15.0%

THINGS TO WATCH FOR IN VISUAL DISPLAYS OF DATA

1. Is there a title that explains what is being displayed?
2. Are numbers lined up with tick marks on the vertical axis that clearly indicate the scale? Has the scale been varied to create a more or less dramatic impression than shown by the actual data?
3. Do too many design and cosmetic effects draw attention from or distort the data?
4. Has the wrong impression been created about how the data are changing because equally spaced time intervals are not used on the horizontal axis? Furthermore, has a time interval been chosen that allows the data to be interpreted in various ways?
5. Are bar sizes scaled proportionately in terms of the data they represent?
6. Is there a source that indicates where the data in the display came from? Do the data come from an entire population or a sample? Was a random sample used and, if so, are there possible differences between what is displayed in the graph and what is occurring in the entire population? (We'll discuss these *margins of error* in Section 12.4.) Who is presenting the visual display, and does that person have a special case to make for or against the trend shown by the graph?

Table 12.5 contains two examples of misleading visual displays.

TABLE 12.5 Examples of Misleading Visual Displays

Graphic Display	Presentation Problems
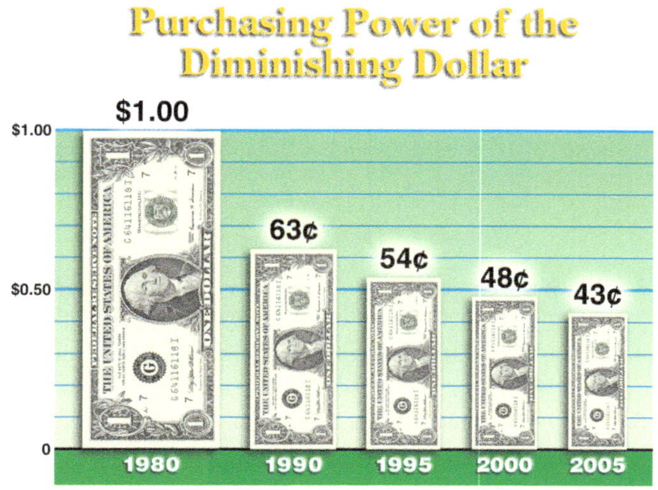 *Source:* Bureau of Labor Statistics	Although the length of each dollar bill is proportional to its spending power, the visual display varies both the length *and width* of the bills to show the diminishing power of the dollar over time. Because our eyes focus on the *areas* of the dollar-shaped bars, this creates the impression that the purchasing power of the dollar diminished even more than it really did. If the area of the dollar were drawn to reflect its purchasing power, the 2005 dollar would be approximately twice as large as the one shown in the graphic display.
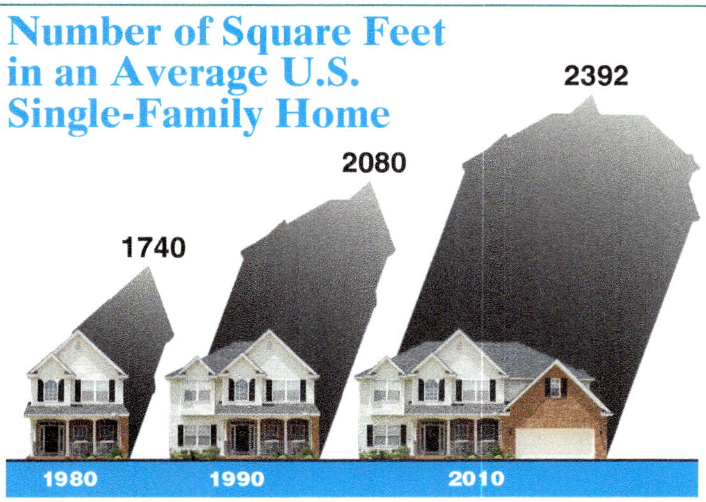 *Source:* U.S. Census Bureau	Cosmetic effects of homes with equal heights, but different frontal additions and shadow lengths, make it impossible to tell if they proportionately depict the given areas. Time intervals on the horizontal axis are not uniform in size, making it appear that dwelling swelling has been linear from 1980 through 2010. The data indicate that this is not the case. There was a greater increase in area from 1980 through 1990, averaging 34 square feet per year, than from 1990 through 2010, averaging 15.6 square feet per year.

Concept and Vocabulary Check

Fill in each blank so that the resulting statement is true.

1. A sample obtained in such a way that every member of the population has an equal chance of being selected is called a/an __________ sample.
2. If data values are listed in one column and the adjacent column indicates the number of times each value occurs, the data presentation is called a/an ________________.
3. If the data presentation in Exercise 2 is varied by organizing the data into classes, the data presentation is called a/an ______________________. If one class in such a distribution is 80–89, the lower class limit is __________ and the upper class limit is __________.
4. Data can be displayed using a bar graph with bars that touch each other. This visual presentation of the data is called a/an __________. The heights of the bars represent the __________ of the data values.
5. If the midpoints of the tops of the bars for the data presentation in Exercise 4 are connected with straight lines, the resulting line graph is a data presentation called a/an ____________. To complete such a graph at both ends, the lines are drawn down to touch the ____________.
6. A data presentation that separates each data item into two parts is called a/an ______________.

In Exercises 7–10, determine whether each statement is true or false. If the statement is false, make the necessary change(s) to produce a true statement.

7. A sample is the set of all the people or objects whose properties are to be described and analyzed by the data collector. ______

8. A call-in poll on radio or television is not reliable because the sample is not chosen randomly from a larger population. ______

9. One disadvantage of a stem-and-leaf plot is that it does not display the data items. ______

10. A deception in the visual display of data can result by stretching or compressing the scale on a graph's vertical axis. ______

Exercise Set 12.1

Practice and Application Exercises

1. The government of a large city needs to determine whether the city's residents will support the construction of a new jail. The government decides to conduct a survey of a sample of the city's residents. Which one of the following procedures would be most appropriate for obtaining a sample of the city's residents?
   a. Survey a random sample of the employees and inmates at the old jail.
   b. Survey every fifth person who walks into City Hall on a given day.
   c. Survey a random sample of persons within each geographic region of the city.
   d. Survey the first 200 people listed in the city's telephone directory.

2. The city council of a large city needs to know whether its residents will support the building of three new schools. The council decides to conduct a survey of a sample of the city's residents. Which procedure would be most appropriate for obtaining a sample of the city's residents?
   a. Survey a random sample of teachers who live in the city.
   b. Survey 100 individuals who are randomly selected from a list of all people living in the state in which the city in question is located.
   c. Survey a random sample of persons within each neighborhood of the city.
   d. Survey every tenth person who enters City Hall on a randomly selected day.

A questionnaire was given to students in an introductory statistics class during the first week of the course. One question asked, "How stressed have you been in the last $2\frac{1}{2}$ weeks, on a scale of 0 to 10, with 0 being not at all stressed and 10 being as stressed as possible?" The students' responses are shown in the frequency distribution. Use this frequency distribution to solve Exercises 3–6.

Stress Rating	Frequency
0	2
1	1
2	3
3	12
4	16
5	18

Stress Rating	Frequency
6	13
7	31
8	26
9	15
10	14

Source: Journal of Personality and Social Psychology, 69, 1102–1112

3. Which stress rating describes the greatest number of students? How many students responded with this rating?

4. Which stress rating describes the least number of students? How many responded with this rating?

5. How many students were involved in this study?

6. How many students had a stress rating of 8 or more?

7. A random sample of 30 college students is selected. Each student is asked how much time he or she spent on homework during the previous week. The following times (in hours) are obtained:

   16, 24, 18, 21, 18, 16, 18, 17, 15, 21, 19, 17, 17, 16, 19, 18, 15, 15, 20, 17, 15, 17, 24, 19, 16, 20, 16, 19, 18, 17.

   Construct a frequency distribution for the data.

8. A random sample of 30 male college students is selected. Each student is asked his height (to the nearest inch). The heights are as follows:

   72, 70, 68, 72, 71, 71, 71, 69, 73, 71, 73, 75, 66, 67, 75, 74, 73, 71, 72, 67, 72, 68, 67, 71, 73, 71, 72, 70, 73, 70.

   Construct a frequency distribution for the data.

A college professor had students keep a diary of their social interactions for a week. Excluding family and work situations, the number of social interactions of ten minutes or longer over the week is shown in the following grouped frequency distribution. Use this information to solve Exercises 9–16.

Number of Social Interactions	Frequency
0–4	12
5–9	16
10–14	16
15–19	16
20–24	10
25–29	11
30–34	4
35–39	3
40–44	3
45–49	3

Source: Society for Personality and Social Psychology

9. Identify the lower class limit for each class.

10. Identify the upper class limit for each class.

11. What is the class width?

12. How many students were involved in this study?

13. How many students had at least 30 social interactions for the week?

14. How many students had at most 14 social interactions for the week?

15. Among the classes with the greatest frequency, which class has the least number of social interactions?

16. Among the classes with the smallest frequency, which class has the least number of social interactions?

17. As of 2011, the following are the ages, in chronological order, at which U.S. presidents were inaugurated:

57, 61, 57, 57, 58, 57, 61, 54, 68, 51, 49, 64, 50, 48, 65, 52, 56, 46, 54, 49, 50, 47, 55, 55, 54, 42, 51, 56, 55, 51, 54, 51, 60, 62, 43, 55, 56, 61, 52, 69, 64, 46, 54, 47.

Source: Time Almanac

Construct a grouped frequency distribution for the data. Use 41–45 for the first class and use the same width for each subsequent class.

18. The IQ scores of 70 students enrolled in a liberal arts course at a college are as follows:

102, 100, 103, 86, 120, 117, 111, 101, 93, 97, 99, 95, 95, 104, 104, 105, 106, 109, 109, 89, 94, 95, 99, 99, 103, 104, 105, 109, 110, 114, 124, 123, 118, 117, 116, 110, 114, 114, 96, 99, 103, 103, 104, 107, 107, 110, 111, 112, 113, 117, 115, 116, 100, 104, 102, 94, 93, 93, 96, 96, 111, 116, 107, 109, 105, 106, 97, 106, 107, 108.

Construct a grouped frequency distribution for the data. Use 85–89 for the first class and use the same width for each subsequent class.

19. Construct a histogram and a frequency polygon for the data involving stress ratings in Exercises 3–6.

20. Construct a histogram and a frequency polygon for the data in Exercise 7.

21. Construct a histogram and a frequency polygon for the data in Exercise 8.

The histogram shows the distribution of starting salaries (rounded to the nearest thousand dollars) for college graduates based on a random sample of recent graduates.

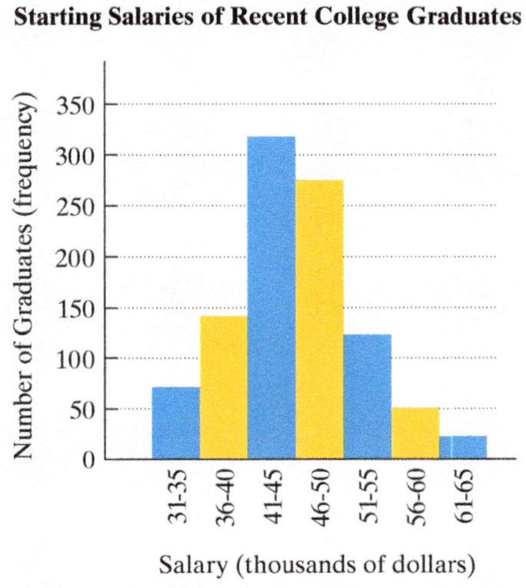

In Exercises 22–25, determine whether each statement is true or false according to the graph at the bottom of the previous column.

22. The graph is based on a sample of approximately 500 recent college graduates.

23. More college graduates had starting salaries in the \$51,000–\$55,000 range than in the \$36,000–\$40,000 range.

24. If the sample is truly representative, then for a group of 400 college graduates, we can expect about 28 of them to have starting salaries in the \$31,000–\$35,000 range.

25. The percentage of starting salaries falling above those shown by any rectangular bar is equal to the percentage of starting salaries falling below that bar.

The frequency polygon shows a distribution of IQ scores.

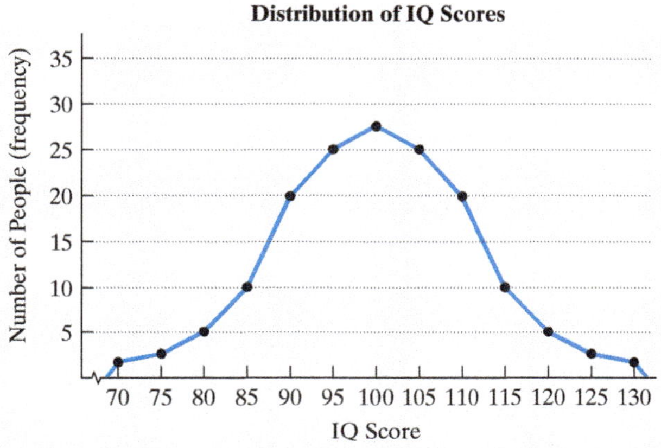

In Exercises 26–29, determine whether each statement is true or false according to the graph.

26. The graph is based on a sample of approximately 50 people.

27. More people had an IQ score of 100 than any other IQ score, and as the deviation from 100 increases or decreases, the scores fall off in a symmetrical manner.

28. More people had an IQ score of 110 than a score of 90.

29. The percentage of scores above any IQ score is equal to the percentage of scores below that score.

30. Construct a stem-and-leaf plot for the data in Exercise 17 showing the ages at which U.S. presidents were inaugurated.

31. A random sample of 40 college professors is selected from all professors at a university. The following list gives their ages:

63, 48, 42, 42, 38, 59, 41, 44, 45, 28, 54, 62, 51, 44, 63, 66, 59, 46, 51, 28, 37, 66, 42, 40, 30, 31, 48, 32, 29, 42, 63, 37, 36, 47, 25, 34, 49, 30, 35, 50.

Construct a stem-and-leaf plot for the data. What does the shape of the display reveal about the ages of the professors?

32. In "Ages of Oscar-Winning Best Actors and Actresses" (*Mathematics Teacher* magazine) by Richard Brown and Gretchen Davis, the stem-and-leaf plots shown on the right compare the ages of 30 actors and 30 actresses at the time they won the award.

a. What is the age of the youngest actor to win an Oscar?

b. What is the age difference between the oldest and the youngest actress to win an Oscar?

c. What is the oldest age shared by two actors to win an Oscar?

d. What differences do you observe between the two stem-and-leaf plots? What explanations can you offer for these differences?

Actors	Stems	Actresses
	2	146667
98753221	3	00113344455778
88776543322100	4	11129
6651	5	
210	6	011
6	7	4
	8	0

In Exercises 33–37, describe what is misleading in each visual display of data.

33.

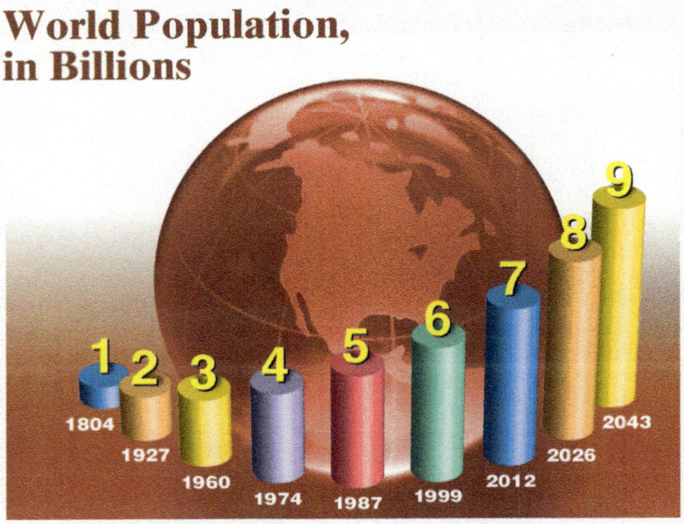

Source: U.S. Census Bureau

34.

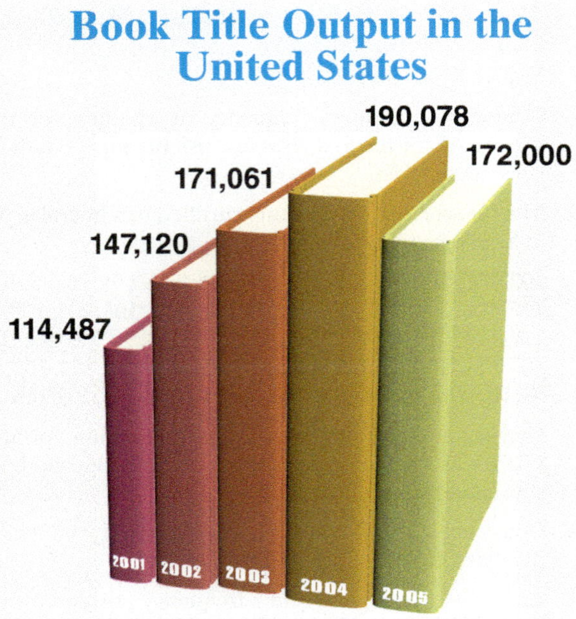

Source: R. R. Bowker

35.

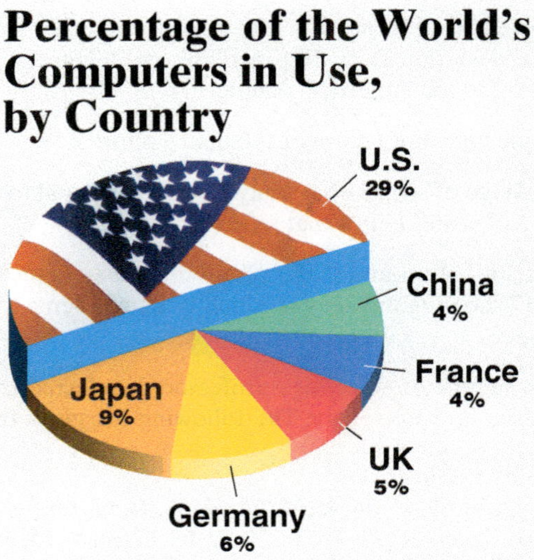

Source: Computer Industry Almanac

36.

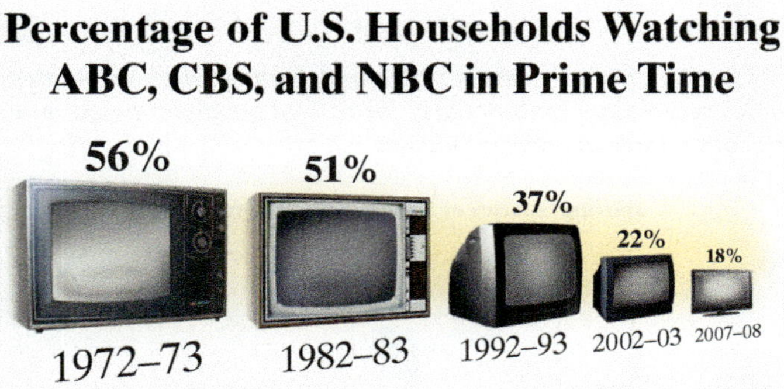

Source: Nielsen Media Research

37.

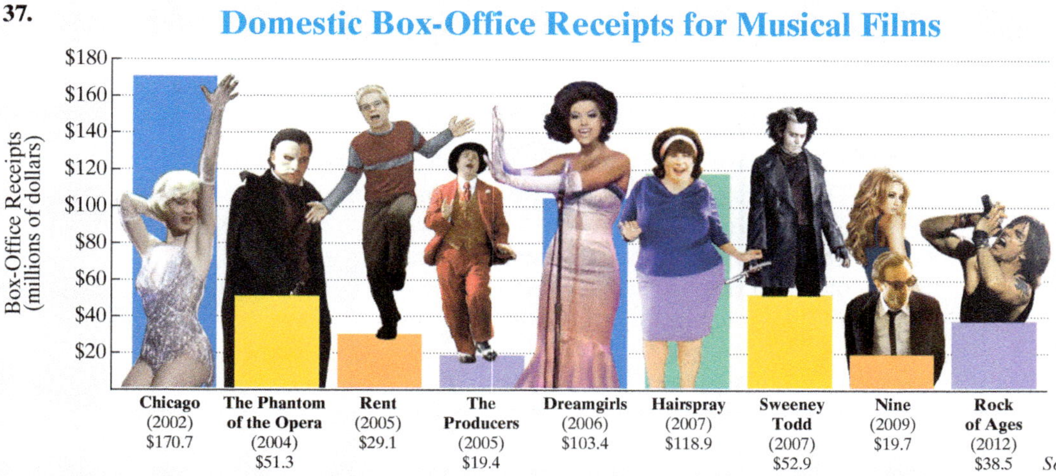

Source: Entertainment Weekly

Writing in Mathematics

38. What is a population? What is a sample?

39. Describe what is meant by a random sample.

40. Suppose you are interested in whether or not the students at your college would favor a grading system in which students may receive final grades of A+, A, A−, B+, B, B−,C+, C, C−, and so on. Describe how you might obtain a random sample of 100 students from the entire student population.

41. For Exercise 40, would questioning every fifth student as he or she is leaving the campus library until 100 students are interviewed be a good way to obtain a random sample? Explain your answer.

42. What is a frequency distribution?

43. What is a histogram?

44. What is a frequency polygon?

45. Describe how to construct a frequency polygon from a histogram.

46. Describe how to construct a stem-and-leaf plot from a set of data.

47. Describe two ways that graphs can be misleading.

Critical Thinking Exercises

Make Sense? *In Exercises 48–51, determine whether each statement makes sense or does not make sense, and explain your reasoning.*

48. The death rate from this new strain of flu is catastrophic because 25% of the people hospitalized with the disease have died.

49. The following graph indicates that for the period from 2000 through 2010, the percentage of female college freshmen describing their health as "above average" has rapidly decreased.

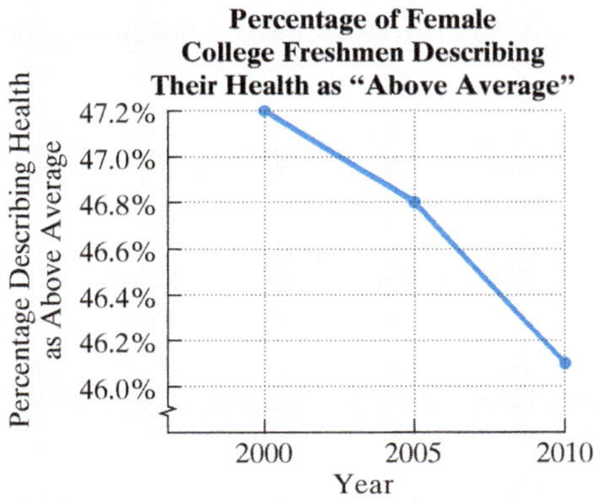

Source: John Macionis, *Sociology, Fourteenth Edition*, Pearson, 2012.

50. A public radio station needs to survey its contributors to determine their programming interests, so they should select a random sample of 100 of their largest contributors.

51. Improperly worded questions can steer respondents toward answers that are not their own.

52. Construct a grouped frequency distribution for the following data, showing the length, in miles, of the 25 longest rivers in the United States. Use five classes that have the same width.

2540	2340	1980	1900	1900
1460	1450	1420	1310	1290
1280	1240	1040	990	926
906	886	862	800	774
743	724	692	659	649

Source: U.S. Department of the Interior

Group Exercises

53. The classic book on distortion using statistics is *How to Lie with Statistics* by Darrell Huff. This activity is designed for five people. Each person should select two chapters from Huff's book and then present to the class the common methods of statistical manipulation and distortion that Huff discusses.

54. Each group member should find one example of a graph that presents data with integrity and one example of a graph that is misleading. Use newspapers, magazines, the Internet, books, and so forth. Once graphs have been collected, each member should share his or her graphs with the entire group. Be sure to explain why one graph depicts data in a forthright manner and how the other graph misleads the viewer.

12.2

WHAT AM I SUPPOSED TO LEARN?

After you have read this section, you should be able to:

1 Determine the mean for a data set.
2 Determine the median for a data set.
3 Determine the mode for a data set.
4 Determine the midrange for a data set.

Measures of Central Tendency

DURING A LIFETIME, AMERICANS AVERAGE TWO WEEKS KISSING.

But wait, there's more:

- 130: The average number of "Friends" for a Facebook user
- 12: The average number of cars an American owns during a lifetime
- 300: The average number of times a 6-year-old child laughs each day
- 550: The average number of hairs in the human eyebrow
- 28: The average number of years in the lifespan of a citizen during the Roman Empire
- 6,000,000: The average number of dust mites living in a U.S. bed.

Source: Listomania, Harper Design

These numbers represent what is "average" or "typical" in a variety of situations. In statistics, such values are known as **measures of central tendency** because they are generally located toward the center of a distribution. Four such measures are discussed in this section: the mean, the median, the mode, and the midrange. Each measure of central tendency is calculated in a different way. Thus, it is better to use a specific term (mean, median, mode, or midrange) than to use the generic descriptive term "average."

1 Determine the mean for a data set.

The Mean

By far the most commonly used measure of central tendency is the *mean*. The **mean** is obtained by adding all the data items and then dividing the sum by the number of items. The Greek letter sigma, Σ, called a **symbol of summation**, is used to indicate the sum of data items. The notation Σx, read "the sum of x," means to add all the data items in a given data set. We can use this symbol to give a formula for calculating the mean.

THE MEAN

The **mean** is the sum of the data items divided by the number of items.

$$\text{Mean} = \frac{\Sigma x}{n},$$

where Σx represents the sum of all the data items and n represents the number of items.

The mean of a sample is symbolized by $\bar{x}$ (read "x bar"), while the mean of an entire population is symbolized by μ (the lowercase Greek letter *mu*). Unless otherwise indicated, the data sets throughout this chapter represent samples, so we will use $\bar{x}$ for the mean: $\bar{x} = \frac{\Sigma x}{n}$.

EXAMPLE 1 *Calculating the Mean*

Table 12.6 at the top of the next page shows the ten highest-earning TV actors and the ten highest-earning TV actresses for the 2010–2011 television season. Find the mean earnings, in millions of dollars, for the ten highest-earning actors.

Blitzer Bonus

Using Means to Compare How the U.S. Measures Up

- Mean Life Expectancy

U.S.	ITALY
78	82

- Mean Cost of an Angiogram

U.S.	CANADA
$798	$35

- Mean Number of Hours of TV Watched Daily

U.S.	ENGLAND
2.7	2.8

- Mean Number of Working Hours per Week

U.S.	GERMANY
35.1	27.8

- Mean Size of a Steak Served at Restaurants

U.S.	ENGLAND
13 ounces	8 ounces

Source: Time, USA Today

TABLE 12.6 Highest-Earning TV Actors and Actresses, 2010–2011

Actor	Earnings (millions of dollars)	Actress	Earnings (millions of dollars)
Charlie Sheen	$40	Eva Longoria	$13
Ray Romano	$20	Tina Fey	$13
Steve Carell	$15	Marcia Cross	$10
Mark Harmon	$13	Mariska Hargitay	$10
Jon Cryer	$11	Marg Helgenberger	$10
Laurence Fishburne	$11	Teri Hatcher	$9
Patrick Dempsey	$10	Felicity Huffman	$9
Simon Baker	$9	Courteney Cox	$7
Hugh Laurie	$9	Ellen Pompeo	$7
Chris Meloni	$9	Julianna Margulies	$7

Source: Forbes

SOLUTION

We find the mean, $\bar{x}$, by adding the earnings for the actors and dividing this sum by 10, the number of data items.

$$\bar{x} = \frac{\sum x}{n} = \frac{40 + 20 + 15 + 13 + 11 + 11 + 10 + 9 + 9 + 9}{10} = \frac{147}{10} = 14.7$$

The mean earnings of the ten highest-earning actors is $14.7 million.

One and only one mean can be calculated for any group of numerical data. The mean may or may not be one of the actual data items. In Example 1, the mean was 14.7, although no data item is 14.7.

CHECK POINT 1 Use **Table 12.6** to find the mean earnings, $\bar{x}$, in millions of dollars, for the ten highest-earning actresses.

In Example 1, some of the data items were identical. We can use multiplication when computing the sum for these identical items.

$$\bar{x} = \frac{40 + 20 + 15 + 13 + 11 + 11 + 10 + 9 + 9 + 9}{10}$$

$$= \frac{40 \cdot 1 + 20 \cdot 1 + 15 \cdot 1 + 13 \cdot 1 + 11 \cdot 2 + 10 \cdot 1 + 9 \cdot 3}{10}$$

The data value 11 has a frequency of 2.
The data value 9 has a frequency of 3.

When many data values occur more than once and a frequency distribution is used to organize the data, we can use the following formula to calculate the mean:

CALCULATING THE MEAN FOR A FREQUENCY DISTRIBUTION

$$\text{Mean} = \bar{x} = \frac{\Sigma xf}{n},$$

where

x represents a data value.
f represents the frequency of that data value.
Σxf represents the sum of all the products obtained by multiplying each data value by its frequency.
n represents the *total frequency* of the distribution.

TABLE 12.7 Students' Stress-Level Ratings

Stress Rating x	Frequency f
0	2
1	1
2	3
3	12
4	16
5	18
6	13
7	31
8	26
9	15
10	14

Source: Journal of Personality and Social Psychology, 69, 1102–1112

EXAMPLE 2 *Calculating the Mean for a Frequency Distribution*

In the previous Exercise Set, we mentioned a questionnaire given to students in an introductory statistics class during the first week of the course. One question asked, "How stressed have you been in the last $2\frac{1}{2}$ weeks, on a scale of 0 to 10, with 0 being not at all stressed and 10 being as stressed as possible?" **Table 12.7** shows the students' responses. Use this frequency distribution to find the mean of the stress-level ratings.

SOLUTION

We use the formula for the mean, $\bar{x}$:

$$\bar{x} = \frac{\Sigma xf}{n}.$$

First, we must find xf, obtained by multiplying each data value, x, by its frequency, f. Then, we need to find the sum of these products, Σxf. We can use the frequency distribution to organize these computations. Add a third column in which each data value is multiplied by its frequency. This column, shown on the right, is headed xf. Then, find the sum of the values, Σxf, in this column.

x	f	xf
0	2	$0 \cdot 2 = 0$
1	1	$1 \cdot 1 = 1$
2	3	$2 \cdot 3 = 6$
3	12	$3 \cdot 12 = 36$
4	16	$4 \cdot 16 = 64$
5	18	$5 \cdot 18 = 90$
6	13	$6 \cdot 13 = 78$
7	31	$7 \cdot 31 = 217$
8	26	$8 \cdot 26 = 208$
9	15	$9 \cdot 15 = 135$
10	14	$10 \cdot 14 = 140$

Totals: $n = 151$ $\quad \Sigma xf = 975$

Σxf is the sum of the numbers in the third column.

This value, the sum of the numbers in the second column, is the total frequency of the distribution.

Now, substitute these values into the formula for the mean, $\bar{x}$. Remember that n is the *total frequency* of the distribution, or 151.

$$\bar{x} = \frac{\Sigma xf}{n} = \frac{975}{151} \approx 6.46$$

The mean of the 0 to 10 stress-level ratings is approximately 6.46. Notice that the mean is greater than 5, the middle of the 0 to 10 scale.

CHECK POINT 2 Find the mean, $\bar{x}$, for the data items in the frequency distribution. (In order to save space, we've written the frequency distribution horizontally.)

Score, x	30	33	40	50
Frequency, f	3	4	4	1

2 Determine the median for a data set.

The Median

The *median* age in the United States is 37.2. The oldest state by median age is Maine (42.7) and the youngest state is Utah (29.2). To find these values, researchers begin with appropriate random samples. The data items—that is, the ages—are arranged from youngest to oldest. The median age is the data item in the middle of each set of ranked, or ordered, data.

THE MEDIAN

To find the **median** of a group of data items,

1. Arrange the data items in order, from smallest to largest.
2. If the number of data items is odd, the median is the data item in the middle of the list.
3. If the number of data items is even, the median is the mean of the two middle data items.

EXAMPLE 3 *Finding the Median*

Find the median for each of the following groups of data:

a. 84, 90, 98, 95, 88

b. 68, 74, 7, 13, 15, 25, 28, 59, 34, 47.

GREAT QUESTION!

What exactly does the median do with the data?

The median splits the data items down the middle, like the median strip in a road.

SOLUTION

a. Arrange the data items in order, from smallest to largest. The number of data items in the list, five, is odd. Thus, the median is the middle number.

84, 88, 90, 95, 98

Middle data item

The median is 90. Notice that two data items lie above 90 and two data items lie below 90.

b. Arrange the data items in order, from smallest to largest. The number of data items in the list, ten, is even. Thus, the median is the mean of the two middle data items.

7, 13, 15, 25, 28, 34, 47, 59, 68, 74

Middle data items are 28 and 34.

$$\text{Median} = \frac{28 + 34}{2} = \frac{62}{2} = 31$$

The median is 31. Five data items lie above 31 and five data items lie below 31.

7 13 15 25 28 | 34 47 59 68 74

Five data items lie below 31. Five data items lie above 31.

Median is 31.

CHECK POINT 3 Find the median for each of the following groups of data:

a. 28, 42, 40, 25, 35

b. 72, 61, 85, 93, 79, 87.

If a relatively long list of data items is arranged in order, it may be difficult to identify the item or items in the middle. In cases like this, the median can be found by determining its position in the list of items.

GREAT QUESTION!

Does the formula

$$\frac{n+1}{2}$$

give the value of the median?

No. The formula gives the *position* of the median, and not the actual value of the median. When finding the median, be sure to first arrange the data items in order from smallest to largest.

POSITION OF THE MEDIAN

If n data items are arranged in order, from smallest to largest, the median is the value in the

$$\frac{n+1}{2}$$

position.

EXAMPLE 4 *Finding the Median Using the Position Formula*

Table 12.8 gives the nine longest words in the English language. Find the median number of letters for the nine longest words.

TABLE 12.8 The Nine Longest Words in the English Language

Word	Number of Letters
Pneumonoultramicroscopicsilicovolcanoconiosis A lung disease caused by breathing in volcanic dust	45
Supercalifragilisticexpialidocious Meaning "wonderful", from song of this title in the movie *Mary Poppins*	34
Floccinaucinihilipilification Meaning "the action or habit of estimating as worthless"	29
Trinitrophenylmethylnitramine A chemical compound used as a detonator in shells	29
Antidisestablishmentarianism Meaning "opposition to the disestablishment of the Church of England"	28
Electroencephalographically Relating to brain waves	27
Microspectrophotometrically Relating to the measurement of light waves	27
Immunoelectrophoretically Relating to measurement of immunoglobulin	25
Spectroheliokinematograph A 1930s' device for monitoring and filming solar activity	25

Source: Chris Cole, rec.puzzles archive

SOLUTION

We begin by listing the data items from smallest to largest.

25, 25, 27, 27, 28, 29, 29, 34, 45

There are nine data items, so $n = 9$. The median is the value in the

$$\frac{n+1}{2}\text{ position} = \frac{9+1}{2}\text{ position} = \frac{10}{2}\text{ position} = \text{fifth position.}$$

We find the median by selecting the data item in the fifth position.

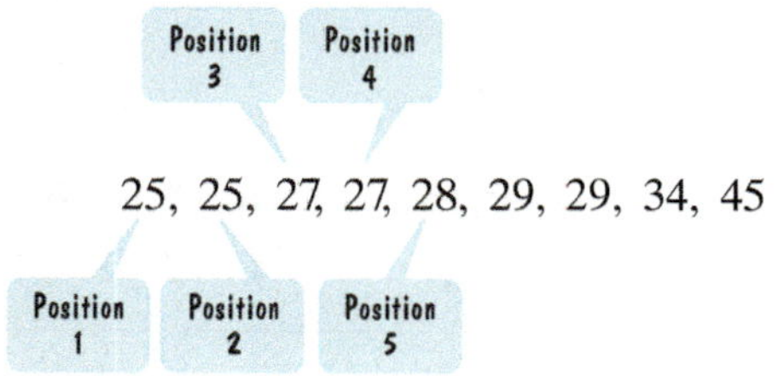

The median is 28. Notice that four data items lie above 28 and four data items lie below it. The median number of letters for the nine longest words in the English language is 28.

CHECK POINT 4 Find the median for the following group of data items:

1, 2, 2, 2, 3, 3, 3, 3, 3, 5, 6, 7, 7, 10, 11, 13, 19, 24, 26.

TABLE 12.9 Hours and Minutes per Day Spent Sleeping and Eating in Selected Countries

Country	Sleeping	Eating
France	8:50	2:15
U.S.	8:38	1:14
Spain	8:34	1:46
New Zealand	8:33	2:10
Australia	8:32	1:29
Turkey	8:32	1:29
Canada	8:29	1:09
Poland	8:28	1:34
Finland	8:27	1:21
Belgium	8:25	1:49
United Kingdom	8:23	1:25
Mexico	8:21	1:06
Italy	8:18	1:54
Germany	8:12	1:45
Sweden	8:06	1:34
Norway	8:03	1:22
Japan	7:50	1:57
S. Korea	7:49	1:36

Source: Organization for Economic Cooperation and Development

EXAMPLE 5 *Finding the Median Using the Position Formula*

Table 12.9 gives the mean number of hours and minutes per day spent sleeping and eating in 18 selected countries. Find the median number of hours and minutes per day spent sleeping for these countries.

SOLUTION

Reading from the bottom to the top of **Table 12.9**, the data items for sleeping appear from smallest to largest. There are 18 data items, so $n = 18$. The median is the value in the

$$\frac{n+1}{2}\text{ position} = \frac{18+1}{2}\text{ position} = \frac{19}{2}\text{ position} = 9.5\text{ position}.$$

This means that the median is the mean of the data items in positions 9 and 10.

7:49, 7:50, 8:03, 8:06, 8:12, 8:18, 8:21, 8:23, [8:25, 8:27,] 8:28, 8:29, 8:32, 8:32, 8:33, 8:34, 8:38, 8:50

$$\text{Median} = \frac{8{:}25 + 8{:}27}{2} = \frac{16{:}52}{2} = 8{:}26$$

The median number of hours per day spent sleeping for the 18 countries is 8 hours, 26 minutes.

CHECK POINT 5 Arrange the data items for eating in **Table 12.9** from smallest to largest. Then find the median number of hours and minutes per day spent eating for the 18 countries.

When individual data items are listed from smallest to largest, you can find the median by identifying the item or items in the middle or by using the $\frac{n+1}{2}$ formula for its position. However, the formula for the position of the median is more useful when data items are organized in a frequency distribution.

EXAMPLE 6 Finding the Median for a Frequency Distribution

The frequency distribution for the stress-level ratings of 151 students is repeated below using a horizontal format. Find the median stress-level rating.

Stress rating

Number of college students

x	0	1	2	3	4	5	6	7	8	9	10
f	2	1	3	12	16	18	13	31	26	15	14

Total: $n = 151$

SOLUTION

There are 151 data items, so $n = 151$. The median is the value in the

$$\frac{n+1}{2} \text{ position} = \frac{151+1}{2} \text{ position} = \frac{152}{2} \text{ position} = 76\text{th position.}$$

We find the median by selecting the data item in the 76th position. The frequency distribution indicates that the data items begin with

$$0, 0, 1, 2, 2, 2, \ldots .$$

We can write the data items all out and then select the median, the 76th data item. A more efficient way to proceed is to count down the frequency column in the distribution until we identify the 76th data item:

x	f
0	2
1	1
2	3
3	12
4	16
5	18
6	13
7	31
8	26
9	15
10	14

We count down the frequency column.
1, 2
3
4, 5, 6
7, 8, 9, 10, 11, 12, 13, 14, 15, 16, 17, 18
19, 20, 21, 22, 23, 24, 25, 26, 27, 28, 29, 30, 31, 32, 33, 34
35, 36, 37, 38, 39, 40, 41, 42, 43, 44, 45, 46, 47, 48, 49, 50, 51, 52
53, 54, 55, 56, 57, 58, 59, 60, 61, 62, 63, 64, 65
66, 67, 68, 69, 70, 71, 72, 73, 74, 75, 76

Stop counting. We've reached the 76th data item.

The 76th data item is 7. The median stress-level rating is 7.

CHECK POINT 6 Find the median for the following frequency distribution.

Age at presidential inauguration

Number of U.S. presidents assuming office in the 20th century with the given age

x	42	43	46	51	52	54	55	56	60	61	64	69
f	1	1	1	3	1	2	2	2	1	2	1	1

Statisticians generally use the median, rather than the mean, when reporting income. Why? Our next example will help to answer this question.

EXAMPLE 7 Comparing the Median and the Mean

Five employees in the assembly section of a television manufacturing company earn salaries of \$19,700, \$20,400, \$21,500, \$22,600, and \$23,000 annually. The section manager has an annual salary of \$95,000.

a. Find the median annual salary for the six people.

b. Find the mean annual salary for the six people.

SOLUTION

a. To compute the median, first arrange the salaries in order:

$$\$19{,}700,\quad \$20{,}400,\quad \$21{,}500,\quad \$22{,}600,\quad \$23{,}000,\quad \$95{,}000.$$

Because the list contains an even number of data items, six, the median is the mean of the two middle items.

$$\text{Median} = \frac{\$21{,}500 + \$22{,}600}{2} = \frac{\$44{,}100}{2} = \$22{,}050$$

The median annual salary is $22,050.

b. We find the mean annual salary by adding the six annual salaries and dividing by 6.

$$\text{Mean} = \frac{\$19{,}700 + \$20{,}400 + \$21{,}500 + \$22{,}600 + \$23{,}000 + \$95{,}000}{6}$$

$$= \frac{\$202{,}200}{6} = \$33{,}700$$

The mean annual salary is $33,700.

In Example 7, the median annual salary is $22,050 and the mean annual salary is $33,700. Why such a big difference between these two measures of central tendency? The relatively high annual salary of the section manager, $95,000, pulls the mean salary to a value considerably higher than the median salary. When one or more data items are much greater than the other items, these extreme values can greatly influence the mean. In cases like this, the median is often more representative of the data.

This is why the median, rather than the mean, is used to summarize the incomes, by gender and race, shown in **Figure 12.6**. Because no one can earn less than $0, the distribution of income must come to an end at $0 for each of these eight groups. By contrast, there is no upper limit on income on the high side. In the United States, the wealthiest 20% of the population earn about 50% of the total income. The relatively few people with very high annual incomes tend to pull the mean income to a value considerably greater than the median income. Reporting mean incomes in **Figure 12.6** would inflate the numbers shown, making them nonrepresentative of the millions of workers in each of the eight groups.

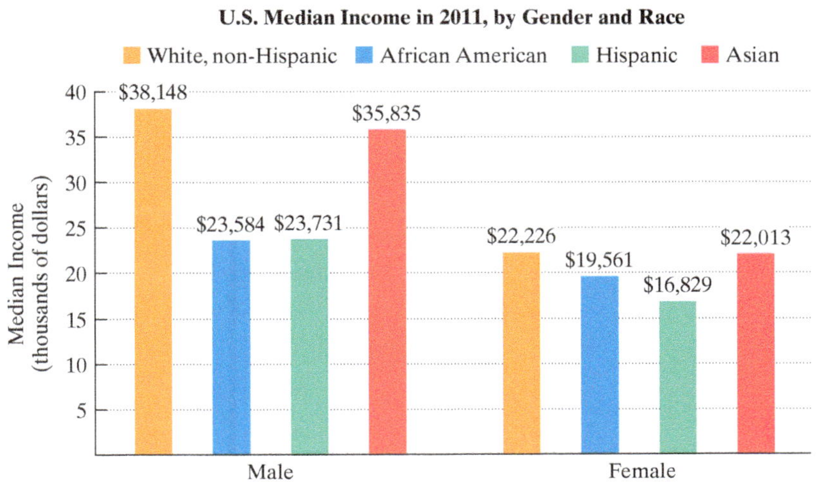

FIGURE 12.6
Source: U.S. Census Bureau

CHECK POINT 7 **Table 12.10** shows the net worth, in millions of 2010 dollars, for ten U.S. presidents from Kennedy through Obama.

TABLE 12.10 Net Worth for Ten U.S. Presidents

President	Net Worth (millions of dollars)
Kennedy	$1000 (i.e. $1 billion)
Johnson	$98
Nixon	$15
Ford	$7
Carter	$7
Reagan	$13
Bush	$23
Clinton	$38
Bush	$20
Obama	$5

Source: Time

a. Find the mean net worth, in millions of dollars, for the ten presidents.
b. Find the median net worth, in millions of dollars, for the ten presidents.
c. Describe why one of the measures of central tendency is greater than the other.

3 Determine the mode for a data set.

The Mode

Let's take one final look at the frequency distribution for the stress-level ratings of 151 college students.

Stress rating

Number of college students

x	0	1	2	3	4	5	6	7	8	9	10
f	2	1	3	12	16	18	13	31	26	15	14

7 is the stress rating with the greatest frequency.

The data value that occurs most often in this distribution is 7, the stress rating for 31 of the 151 students. We call 7 the *mode* of this distribution.

THE MODE

The **mode** is the data value that occurs most often in a data set. If more than one data value has the highest frequency, then each of these data values is a mode. If there is no data value that occurs most often, then the data set has no mode.

EXAMPLE 8 Finding the Mode

Find the mode for each of the following groups of data:
a. 7, 2, 4, 7, 8, 10 **b.** 2, 1, 4, 5, 3 **c.** 3, 3, 4, 5, 6, 6.

SOLUTION

a. 7, 2, 4, 7, 8, 10

7 occurs most often.

The mode is 7.

b. 2, 1, 4, 5, 3

Each data item occurs the same number of times.

There is no mode.

c. 3, 3, 4, 5, 6, 6

Both 3 and 6 occur most often.

The modes are 3 and 6. The data set is said to be **bimodal**.

CHECK POINT 8 Find the mode for each of the following groups of data:

a. 3, 8, 5, 8, 9, 10

b. 3, 8, 5, 8, 9, 3

c. 3, 8, 5, 6, 9, 10.

4 Determine the midrange for a data set.

The Midrange

Table 12.11 shows the ten hottest cities in the United States. Because temperature is constantly changing, you might wonder how the mean temperatures shown in the table are obtained.

First, we need to find a representative daily temperature. This is obtained by adding the lowest and highest temperatures for the day and then dividing this sum by 2. Next, we take the representative daily temperatures for all 365 days, add them, and divide the sum by 365. These are the mean temperatures that appear in **Table 12.11**.

Representative daily temperature,

$$\frac{\text{lowest daily temperature} + \text{highest daily temperature}}{2},$$

is an example of a measure of central tendency called the *midrange*.

TABLE 12.11 Ten Hottest U.S. Cities

City	Mean Temperature
Key West, FL	77.8°
Miami, FL	75.9°
West Palm Beach, FL	74.7°
Fort Myers, FL	74.4°
Yuma, AZ	74.2°
Brownsville, TX	73.8°
Phoenix, AZ	72.6°
Vero Beach, FL	72.4°
Orlando, FL	72.3°
Tampa, FL	72.3°

Source: National Oceanic and Atmospheric Administration

THE MIDRANGE

The **midrange** is found by adding the lowest and highest data values and dividing the sum by 2.

$$\text{Midrange} = \frac{\text{lowest data value} + \text{highest data value}}{2}$$

EXAMPLE 9 *Finding the Midrange*

Newsweek magazine examined factors that affect women's lives, including justice, health, education, economics, and politics. Using these five factors, the magazine graded each of 165 countries on a scale from 0 to 100. The 12 best places to be a woman and the 12 worst places to be a woman are shown in **Table 12.12**.

TABLE 12.12 Women in the World

Best Places to Be a Woman		Worst Places to Be a Woman	
Country	**Score**	**Country**	**Score**
Iceland	100.0	Chad	0.0
Canada	99.6	Afghanistan	2.0
Sweden	99.2	Yemen	12.1
Denmark	95.3	Democratic Republic of the Congo	13.6
Finland	92.8	Mali	17.6
Switzerland	91.9	Solomon Islands	20.8
Norway	91.3	Niger	21.2
United States	89.8	Pakistan	21.4
Australia	88.2	Ethiopia	23.7
Netherlands	87.7	Sudan	26.1
New Zealand	87.2	Guinea	28.5
France	87.2	Sierra Leone	29.0

Source: Newsweek

Find the midrange score among the 12 best countries to be a woman.

SOLUTION

Refer to **Table 12.12** on the previous page.

$$\text{Midrange} = \frac{\text{best place with the lowest score} + \text{best place with the highest score}}{2}$$

$$= \frac{87.2 + 100.0}{2} = \frac{187.2}{2} = 93.6$$

The midrange score among the 12 best countries to be a woman is 93.6.

We can find the mean score among the 12 best countries to be a woman by adding up the 12 scores and then dividing the sum by 12. By doing so, we can determine that the mean score is approximately 92.5. It is much faster to calculate the midrange, which is often used as an estimate for the mean.

CHECK POINT 9 Use **Table 12.12** on the previous page to find the midrange score among the 12 worst countries to be a woman.

EXAMPLE 10 *Finding the Four Measures of Central Tendency*

Suppose your six exam grades in a course are

52, 69, 75, 86, 86, and 92.

Compute your final course grade (90–100 = A, 80–89 = B, 70–79 = C, 60–69 = D, below 60 = F) using the

a. mean. **b.** median. **c.** mode. **d.** midrange.

SOLUTION

a. The mean is the sum of the data items divided by the number of items, 6.

$$\text{Mean} = \frac{52 + 69 + 75 + 86 + 86 + 92}{6} = \frac{460}{6} \approx 76.67$$

Using the mean, your final course grade is C.

b. The six data items, 52, 69, 75, 86, 86, and 92, are arranged in order. Because the number of data items is even, the median is the mean of the two middle items.

$$\text{Median} = \frac{75 + 86}{2} = \frac{161}{2} = 80.5$$

Using the median, your final course grade is B.

c. The mode is the data value that occurs most frequently. Because 86 occurs most often, the mode is 86. Using the mode, your final course grade is B.

d. The midrange is the mean of the lowest and highest data values.

$$\text{Midrange} = \frac{52 + 92}{2} = \frac{144}{2} = 72$$

Using the midrange, your final course grade is C.

CHECK POINT 10 *Consumer Reports* magazine gave the following data for the number of calories in a meat hot dog for each of 17 brands:

173, 191, 182, 190, 172, 147, 146, 138, 175, 136, 179, 153, 107, 195, 135, 140, 138.

Find the mean, median, mode, and midrange for the number of calories in a meat hot dog for the 17 brands. If necessary, round answers to the nearest tenth of a calorie.

Concept and Vocabulary Check

Fill in each blank so that the resulting statement is true.

1. $\frac{\Sigma x}{n}$, the sum of all the data items divided by the number of data items, is the measure of central tendency called the ________.
2. The measure of central tendency that is the data item in the middle of ranked, or ordered, data is called the ________.
3. If n data items are arranged in order, from smallest to largest, the data item in the middle is the value in ________ position.
4. A data value that occurs most often in a data set is the measure of central tendency called the ________.
5. The measure of central tendency that is found by adding the lowest and highest data values and dividing the sum by 2 is called the ________.

In Exercises 6–9, determine whether each statement is true or false. If the statement is false, make the necessary change(s) to produce a true statement.

6. Numbers representing what is average or typical about a data set are called measures of central tendency. ______
7. When finding the mean, it is necessary to arrange the data items in order. ______
8. If one or more data items are much greater than the other items, the mean, rather than the median, is more representative of the data. ______
9. A data set can contain more than one median, or no median at all. ______

Exercise Set 12.2

Practice Exercises

In Exercises 1–8, find the mean for each group of data items.

1. 7, 4, 3, 2, 8, 5, 1, 3
2. 11, 6, 4, 0, 2, 1, 12, 0, 0
3. 91, 95, 99, 97, 93, 95
4. 100, 100, 90, 30, 70, 100
5. 100, 40, 70, 40, 60
6. 1, 3, 5, 10, 8, 5, 6, 8
7. 1.6, 3.8, 5.0, 2.7, 4.2, 4.2, 3.2, 4.7, 3.6, 2.5, 2.5
8. 1.4, 2.1, 1.6, 3.0, 1.4, 2.2, 1.4, 9.0, 9.0, 1.8

In Exercises 9–12, find the mean for the data items in the given frequency distribution.

9.

Score x	Frequency f
1	1
2	3
3	4
4	4
5	6
6	5
7	3
8	2

10.

Score x	Frequency f
1	2
2	4
3	5
4	7
5	6
6	4
7	3

11.

Score x	Frequency f
1	1
2	1
3	2
4	5
5	7
6	9
7	8
8	6
9	4
10	3

12.

Score x	Frequency f
1	3
2	4
3	6
4	8
5	9
6	7
7	5
8	2
9	1
10	1

In Exercises 13–20, find the median for each group of data items.

13. 7, 4, 3, 2, 8, 5, 1, 3
14. 11, 6, 4, 0, 2, 1, 12, 0, 0
15. 91, 95, 99, 97, 93, 95
16. 100, 100, 90, 30, 70, 100
17. 100, 40, 70, 40, 60
18. 1, 3, 5, 10, 8, 5, 6, 8
19. 1.6, 3.8, 5.0, 2.7, 4.2, 4.2, 3.2, 4.7, 3.6, 2.5, 2.5
20. 1.4, 2.1, 1.6, 3.0, 1.4, 2.2, 1.4, 9.0, 9.0, 1.8

Find the median for the data items in the frequency distribution in

21. Exercise 9.
22. Exercise 10.
23. Exercise 11.
24. Exercise 12.

In Exercises 25–32, find the mode for each group of data items. If there is no mode, so state.

25. 7, 4, 3, 2, 8, 5, 1, 3
26. 11, 6, 4, 0, 2, 1, 12, 0, 0
27. 91, 95, 99, 97, 93, 95
28. 100, 100, 90, 30, 70, 100
29. 100, 40, 70, 40, 60
30. 1, 3, 5, 10, 8, 5, 6, 8
31. 1.6, 3.8, 5.0, 2.7, 4.2, 4.2, 3.2, 4.7, 3.6, 2.5, 2.5
32. 1.4, 2.1, 1.6, 3.0, 1.4, 2.2, 1.4, 9.0, 9.0, 1.8

Find the mode for the data items in the frequency distribution in

33. Exercise 9.
34. Exercise 10.
35. Exercise 11.
36. Exercise 12.

In Exercises 37–44, find the midrange for each group of data items.

37. 7, 4, 3, 2, 8, 5, 1, 3
38. 11, 6, 4, 0, 2, 1, 12, 0, 0
39. 91, 95, 99, 97, 93, 95
40. 100, 100, 90, 30, 70, 100
41. 100, 40, 70, 40, 60
42. 1, 3, 5, 10, 8, 5, 6, 8
43. 1.6, 3.8, 5.0, 2.7, 4.2, 4.2, 3.2, 4.7, 3.6, 2.5, 2.5
44. 1.4, 2.1, 1.6, 3.0, 1.4, 2.2, 1.4, 9.0, 9.0, 1.8

Find the midrange for the data items in the frequency distribution in

45. Exercise 9.
46. Exercise 10.
47. Exercise 11.
48. Exercise 12.

Practice Plus

In Exercises 49–54, use each display of data items to find the mean, median, mode, and midrange.

49.

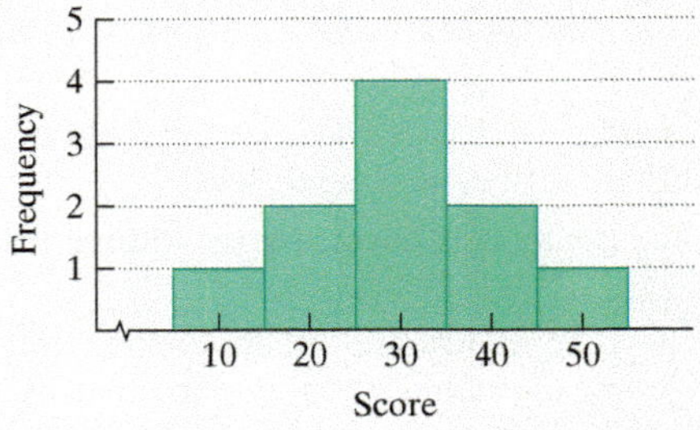

50.

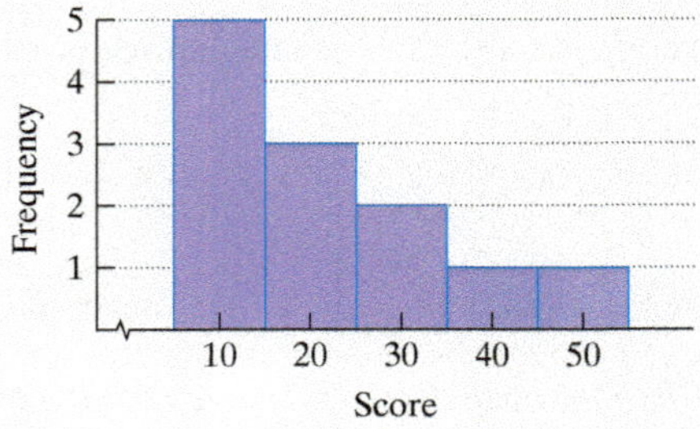

51.

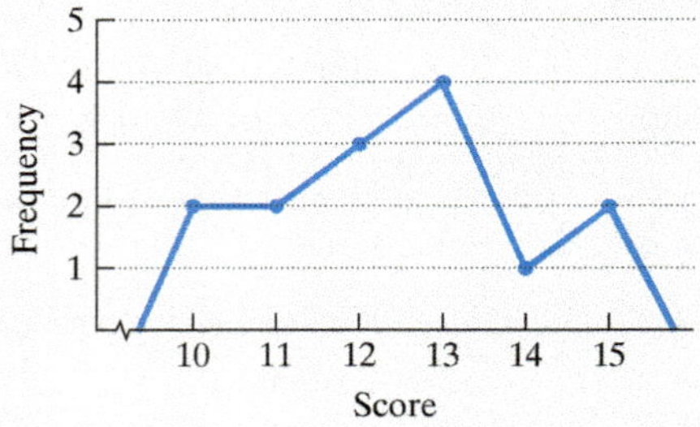

52.

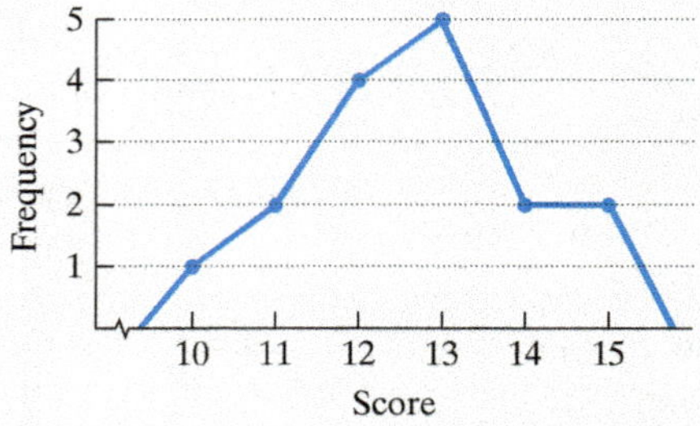

53.

Stems	Leaves			
2	1	4	5	
3	0	1	1	3
4	2	5		

54.

Stems	Leaves			
2	8			
3	2	4	4	9
4	0	1	5	7

Application Exercises

Exercises 55–57 present data on a variety of topics. For each data set described in boldface, find the

a. *mean.* **b.** *median.*

c. *mode (or state that there is no mode).*

d. *midrange.*

55. Top Cities with New College Graduates

Metro Area	Number of Recent College Graduates Who Moved to the Area from 2000–2011 (thousands)
New York	200
Chicago	97
Washington, D.C.	92
Los Angeles	92
San Francisco	64
Houston	51
Boston	51
Dallas-Fort Worth	50
Philadelphia	49
Denver	37
Seattle	34
Minneapolis-St. Paul	32
San Jose	27

Source: USA Today

56. Net Worth for the First 13 U.S. Presidents

President	Net Worth (millions of 2010 dollars)
Washington	$525
Adams	$19
Jefferson	$212
Madison	$101
Monroe	$27
Adams	$21
Jackson	$119
Van Buren	$26
Harrison	$5
Tyler	$51
Polk	$10
Taylor	$6
Fillmore	$4

Source: Time

57. Number of Social Interactions of College Students In Exercise Set 12.1, we presented a grouped frequency distribution showing the number of social interactions of ten minutes or longer over a one-week period for a group of college students. (These interactions excluded family and work situations.) Use the frequency distribution shown to solve this exercise. (This distribution was obtained by replacing the classes in the grouped frequency distribution previously shown with the midpoints of the classes.)

Social interactions in a week → x

Number of college students → f

x	2	7	12	17	22	27	32	37	42	47
f	12	16	16	16	10	11	4	3	3	3

The weights (to the nearest five pounds) of 40 randomly selected male college students are organized in a histogram with a superimposed frequency polygon. Use the graph to answer Exercises 58–61.

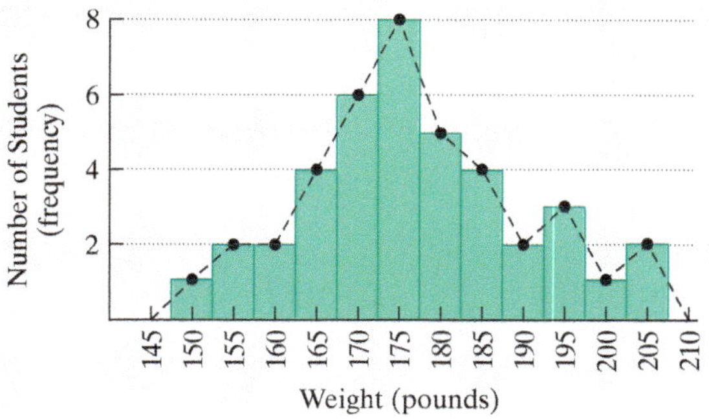

58. Find the mean weight.

59. Find the median weight.

60. Find the modal weight.

61. Find the midrange weight.

62. An advertisement for a speed-reading course claimed that the "average" reading speed for people completing the course was 1000 words per minute. Shown below are the actual data for the reading speeds per minute for a sample of 24 people who completed the course.

1000	900	800	1000	900	850
650	1000	1050	800	1000	850
700	750	800	850	900	950
600	1100	950	700	750	650

a. Find the mean, median, mode, and midrange. (If you prefer, first organize the data in a frequency distribution.)

b. Which measure of central tendency was given in the advertisement?

c. Which measure of central tendency is the best indicator of the "average" reading speed in this situation? Explain your answer.

63. In one common system for finding a grade-point average, or GPA,

$$A = 4, B = 3, C = 2, D = 1, F = 0.$$

The GPA is calculated by multiplying the number of credit hours for a course and the number assigned to each grade, and then adding these products. Then divide this sum by the total number of credit hours. Because each course grade is weighted according to the number of credits of the course, GPA is called a *weighted mean*. Calculate the GPA for this transcript:

Sociology: 3 cr. A; Biology: 3.5 cr. C; Music: 1 cr. B; Math: 4 cr. B; English: 3 cr. C.

Writing in Mathematics

64. What is the mean and how is it obtained?

65. What is the median and how is it obtained?

66. What is the mode and how is it obtained?

67. What is the midrange and how is it obtained?

68. The "average" income in the United States can be given by the mean or the median.

a. Which measure would be used in anti-U.S. propaganda? Explain your answer.

b. Which measure would be used in pro-U.S. propaganda? Explain your answer.

69. In a class of 40 students, 21 have examination scores of 77%. Which measure or measures of central tendency can you immediately determine? Explain your answer.

70. You read an article that states, "Of the 411 players in the National Basketball Association, only 138 make more than the average salary of \$3.12 million." Is \$3.12 million the mean or the median salary? Explain your answer.

71. A student's parents promise to pay for next semester's tuition if an A average is earned in chemistry. With examination grades of 97%, 97%, 75%, 70%, and 55%, the student reports that an A average has been earned. Which measure of central tendency is the student reporting as the average? How is this student misrepresenting the course performance with statistics?

72. According to the National Oceanic and Atmospheric Administration, the coldest city in the United States is International Falls, Minnesota, with a mean Fahrenheit temperature of 36.8°. Explain how this mean is obtained.

Critical Thinking Exercises

Make Sense? *In Exercises 73–76, determine whether each statement makes sense or does not make sense, and explain your reasoning.*

73. I'm working with a data set for which neither the mean nor the median is one of the data items.

74. I made a distribution of the heights of the 12 players on our basketball team. Because one player is much taller than the others, the team's median height is greater than its mean height.

75. Although the data set 1, 1, 2, 3, 3, 3, 4, 4 has a number of repeated items, there is only one mode.

76. If professors use the same test scores for a particular student and calculate measures of central tendency correctly, they will always agree on the student's final course grade.

77. Give an example of a set of six examination grades (from 0 to 100) with each of the following characteristics:

a. The mean and the median have the same value, but the mode has a different value.

b. The mean and the mode have the same value, but the median has a different value.

c. The mean is greater than the median.

d. The mode is greater than the mean.

e. The mean, median, and mode have the same value.

f. The mean and mode have values of 72.

78. On an examination given to 30 students, no student scored below the mean. Describe how this occurred.

Group Exercises

79. Select a characteristic, such as shoe size or height, for which each member of the group can provide a number. Choose a characteristic of genuine interest to the group. For this characteristic, organize the data collected into a frequency distribution and a graph. Compute the mean, median, mode, and midrange. Discuss any differences among these values. What happens if the group is divided (men and women, or people under a certain age and people over a certain age) and these measures of central tendency are computed for each of the subgroups? Attempt to use measures of central tendency to discover something interesting about the entire group or the subgroups.

80. A book on spotting bad statistics and learning to think critically about these influential numbers is *Damn Lies and Statistics* by Joel Best (University of California Press, 2001). This activity is designed for six people. Each person should select one chapter from Best's book. The group report should include examples of the use, misuse, and abuse of statistical information. Explain exactly how and why bad statistics emerge, spread, and come to shape policy debates. What specific ways does Best recommend to detect bad statistics?

12.3

WHAT AM I SUPPOSED TO LEARN?

After you have read this section, you should be able to:

1 Determine the range for a data set.
2 Determine the standard deviation for a data set.

Measures of Dispersion

WHEN YOU THINK OF HOUSTON, TEXAS, AND Honolulu, Hawaii, do balmy temperatures come to mind? Both cities have a mean temperature of 75°. However, the mean temperature does not tell the whole story. The temperature in Houston differs seasonally from a low of about 40° in January to a high of close to 100° in July and August. By contrast, Honolulu's temperature varies less throughout the year, usually ranging between 60° and 90°.

Measures of dispersion are used to describe the spread of data items in a data set. Two of the most common measures of dispersion, the *range* and the *standard deviation,* are discussed in this section.

1 Determine the range for a data set.

The Range

A quick but rough measure of dispersion is the **range**, the difference between the highest and lowest data values in a data set. For example, if Houston's hottest annual temperature is 103° and its coldest annual temperature is 33°, the range in temperature is

$$103° - 33°, \quad \text{or} \quad 70°.$$

If Honolulu's hottest day is 89° and its coldest day 61°, the range in temperature is

$$89° - 61°, \quad \text{or} \quad 28°.$$

THE RANGE

The **range**, the difference between the highest and lowest data values in a data set, indicates the total spread of the data.

$$\text{Range} = \text{highest data value} - \text{lowest data value}$$

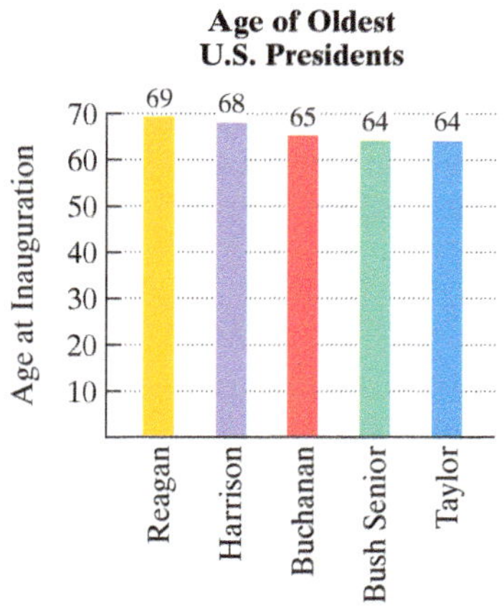

FIGURE 12.7
Source: Internet Public Library

EXAMPLE 1 *Computing the Range*

Figure 12.7 shows the age of the five oldest U.S presidents at the start of their first term. Find the age range for the five oldest presidents.

SOLUTION

$$\text{Range} = \text{highest data value} - \text{lowest data value}$$
$$= 69 - 64 = 5$$

The range is 5 years.

CHECK POINT 1 Find the range for the following group of data items:

4, 2, 11, 7.

The Standard Deviation

A second measure of dispersion, and one that is dependent on *all* of the data items, is called the **standard deviation**. The standard deviation is found by determining how much each data item differs from the mean.

In order to compute the standard deviation, it is necessary to find by how much each data item deviates from the mean. First compute the mean, $\bar{x}$. Then subtract the mean from each data item, $x - \bar{x}$. Example 2 shows how this is done. In Example 3, we will use this skill to actually find the standard deviation.

EXAMPLE 2 *Preparing to Find the Standard Deviation; Finding Deviations from the Mean*

Find the deviations from the mean for the five data items 69, 68, 65, 64, and 64, shown in **Figure 12.7**.

SOLUTION

First, calculate the mean, $\bar{x}$.

$$\bar{x} = \frac{\Sigma x}{n} = \frac{69 + 68 + 65 + 64 + 64}{5} = \frac{330}{5} = 66$$

The mean age for the five oldest U.S. presidents is 66 years. Now, let's find by how much each of the five data items in **Figure 12.7** differs from 66, the mean. For Reagan, who was 69 at the start of his first term, the computation is shown as follows:

$$\text{Deviation from mean} = \text{data item} - \text{mean}$$
$$= x - \bar{x}$$
$$= 69 - 66 = 3.$$

This indicates that Reagan's inaugural age exceeds the mean by three years.

The computation for Buchanan, who was 65 at the start of his first term, is given by

$$\text{Deviation from mean} = \text{data item} - \text{mean}$$
$$= x - \bar{x}$$
$$= 65 - 66 = -1.$$

This indicates that Buchanan's inaugural age is one year below the mean.

The deviations from the mean for each of the five given data items are shown in **Table 12.13**.

TABLE 12.13 Deviations from the Mean

Data item x	Deviation: data item − mean $x - \bar{x}$
69	$69 - 66 = 3$
68	$68 - 66 = 2$
65	$65 - 66 = -1$
64	$64 - 66 = -2$
64	$64 - 66 = -2$

CHECK POINT 2 Compute the mean for the following group of data items:

2, 4, 7, 11.

Then find the deviations from the mean for the four data items. Organize your work in table form just like **Table 12.13**. Keep track of these computations. You will be using them in Check Point 3.

TABLE 12.13 (repeated)
Deviations from the Mean

Data item x	Deviation: data item − mean $x - \bar{x}$
69	$69 - 66 = 3$
68	$68 - 66 = 2$
65	$65 - 66 = -1$
64	$64 - 66 = -2$
64	$64 - 66 = -2$

The sum of the deviations from the mean for a set of data is always zero: $\Sigma(x - \bar{x}) = 0$. For the deviations from the mean shown in **Table 12.13**,

$$3 + 2 + (-1) + (-2) + (-2) = 5 + (-5) = 0.$$

This shows that we cannot find a measure of dispersion by finding the mean of the deviations, because this value is always zero. However, a kind of average of the deviations from the mean, called the **standard deviation**, can be computed. We do so by squaring each deviation and later introducing a square root in the computation. Here are the details on how to find the standard deviation for a set of data:

2 Determine the standard deviation for a data set.

COMPUTING THE STANDARD DEVIATION FOR A DATA SET

1. Find the mean of the data items.
2. Find the deviation of each data item from the mean:
$$\text{data item} - \text{mean}.$$
3. Square each deviation:
$$(\text{data item} - \text{mean})^2.$$
4. Sum the squared deviations:
$$\Sigma(\text{data item} - \text{mean})^2.$$
5. Divide the sum in step 4 by $n - 1$, where n represents the number of data items:
$$\frac{\Sigma(\text{data item} - \text{mean})^2}{n - 1}.$$
6. Take the square root of the quotient in step 5. This value is the standard deviation for the data set.
$$\text{Standard deviation} = \sqrt{\frac{\Sigma(\text{data item} - \text{mean})^2}{n - 1}}$$

The standard deviation of a sample is symbolized by s, while the standard deviation of an entire population is symbolized by σ (the lowercase Greek letter *sigma*). Unless otherwise indicated, data sets represent samples, so we will use s for the standard deviation:

$$s = \sqrt{\frac{\Sigma(x - \bar{x})^2}{n - 1}}.$$

The computation of the standard deviation can be organized using a table with three columns:

Data item x	Deviation: $x - \bar{x}$ Data item − mean	(Deviation)^2: $(x - \bar{x})^2$ (Data item − mean)^2

In Example 2, we worked out the first two columns of such a table. Let's continue working with the data for the ages of the five oldest U.S. presidents and compute the standard deviation.

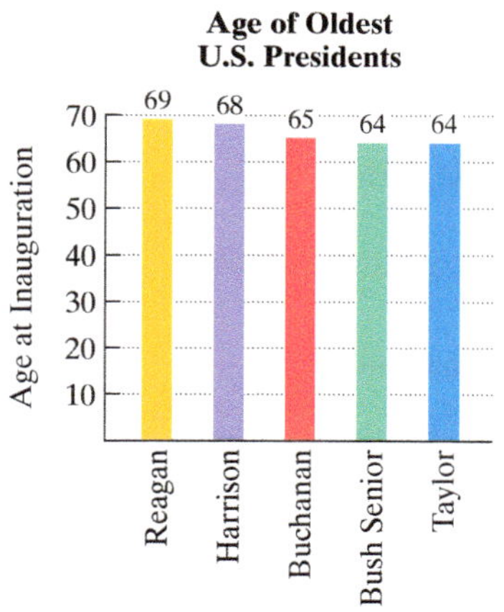

FIGURE 12.7 (repeated)

EXAMPLE 3 *Computing the Standard Deviation*

Figure 12.7, showing the age of the five oldest U.S. presidents at the start of their first term, is repeated in the margin. Find the standard deviation for the ages of the five presidents.

SOLUTION

Step 1 Find the mean. From our work in Example 2, the mean is 66: $\bar{x} = 66$.

Step 2 Find the deviation of each data item from the mean: data item − mean or $x - \bar{x}$. This, too, was done in Example 2 for each of the five data items.

Step 3 Square each deviation: (data item − mean)^2 or $(x - \bar{x})^2$. We square each of the numbers in the (data item − mean) column, shown in **Table 12.14**. Notice that squaring the difference always results in a nonnegative number.

TABLE 12.14 Computing the Standard Deviation

Data item x	Deviation: data item − mean $x - \bar{x}$	(Deviation)^2: (data item − mean)^2 $(x - \bar{x})^2$
69	$69 - 66 = 3$	$3^2 = 3 \cdot 3 = 9$
68	$68 - 66 = 2$	$2^2 = 2 \cdot 2 = 4$
65	$65 - 66 = -1$	$(-1)^2 = (-1) \cdot (-1) = 1$
64	$64 - 66 = -2$	$(-2)^2 = (-2) \cdot (-2) = 4$
64	$64 - 66 = -2$	$(-2)^2 = (-2) \cdot (-2) = 4$
Totals:	$\Sigma(x - \bar{x}) = 0$	$\Sigma(x - \bar{x})^2 = 22$

The sum of the deviations for a set of data is always zero.

Adding the five numbers in the third column gives the sum of the squared deviations: Σ(data item − mean)^2.

Step 4 Sum the squared deviations: Σ(data item − mean)^2. This step is shown in **Table 12.14**. The squares in the third column were added, resulting in a sum of 22: $\Sigma(x - \bar{x})^2 = 22$.

Step 5 Divide the sum in step 4 by $n - 1$, where n represents the number of data items. The number of data items is 5 so we divide by 4.

$$\frac{\Sigma(x - \bar{x})^2}{n - 1} = \frac{\Sigma(\text{data item} - \text{mean})^2}{n - 1} = \frac{22}{5 - 1} = \frac{22}{4} = 5.5$$

Step 6 The standard deviation, s, is the square root of the quotient in step 5.

$$s = \sqrt{\frac{\Sigma(x - \bar{x})^2}{n - 1}} = \sqrt{\frac{\Sigma(\text{data item} - \text{mean})^2}{n - 1}} = \sqrt{5.5} \approx 2.35$$

The standard deviation for the five oldest U.S. presidents is approximately 2.35 years.

CHECK POINT 3 Find the standard deviation for the group of data items in Check Point 2 on page 796. Round to two decimal places.

TECHNOLOGY

Almost all scientific and graphing calculators compute the standard deviation of a set of data. Using the data items in Example 3,

69, 68, 65, 64, 64,

the keystrokes for obtaining the standard deviation on many scientific calculators are as follows:

69 [Σ+] 68 [Σ+] 65 [Σ+]

64 [Σ+] 64 [Σ+] [2nd] [$\sigma n - 1$].

Graphing calculators require that you specify if data items are from an entire population or a sample of the population.

Example 4 illustrates that as the spread of data items increases, the standard deviation gets larger.

EXAMPLE 4 *Computing the Standard Deviation*

Find the standard deviation of the data items in each of the samples shown below.

Sample A	**Sample B**
17, 18, 19, 20, 21, 22, 23	5, 10, 15, 20, 25, 30, 35

SOLUTION

Begin by finding the mean for each sample.

Sample A:

$$\text{Mean} = \frac{17 + 18 + 19 + 20 + 21 + 22 + 23}{7} = \frac{140}{7} = 20$$

Sample B:

$$\text{Mean} = \frac{5 + 10 + 15 + 20 + 25 + 30 + 35}{7} = \frac{140}{7} = 20$$

Although both samples have the same mean, the data items in sample B are more spread out. Thus, we would expect sample B to have the greater standard deviation. The computation of the standard deviation requires that we find $\Sigma(\text{data item} - \text{mean})^2$, shown in **Table 12.15**.

TABLE 12.15 Computing Standard Deviations for Two Samples

Sample A

Data item x	Deviation: data item − mean $x - \bar{x}$	(Deviation)^2: (data item − mean)^2 $(x - \bar{x})^2$
17	$17 - 20 = -3$	$(-3)^2 = 9$
18	$18 - 20 = -2$	$(-2)^2 = 4$
19	$19 - 20 = -1$	$(-1)^2 = 1$
20	$20 - 20 = 0$	$0^2 = 0$
21	$21 - 20 = 1$	$1^2 = 1$
22	$22 - 20 = 2$	$2^2 = 4$
23	$23 - 20 = 3$	$3^2 = 9$
Totals:		$\Sigma(x - \bar{x})^2 = 28$

Sample B

Data item x	Deviation: data item − mean $x - \bar{x}$	(Deviation)^2: (data item − mean)^2 $(x - \bar{x})$
5	$5 - 20 = -15$	$(-15)^2 = 225$
10	$10 - 20 = -10$	$(-10)^2 = 100$
15	$15 - 20 = -5$	$(-5)^2 = 25$
20	$20 - 20 = 0$	$0^2 = 0$
25	$25 - 20 = 5$	$5^2 = 25$
30	$30 - 20 = 10$	$10^2 = 100$
35	$35 - 20 = 15$	$15^2 = 225$
		$\Sigma(x - \bar{x})^2 = 700$

Each sample contains seven data items, so we compute the standard deviation by dividing the sums in **Table 12.15**, 28 and 700, by $7 - 1$, or 6. Then we take the square root of each quotient.

$$\text{Standard deviation} = \sqrt{\frac{\Sigma(x - \bar{x})^2}{n - 1}} = \sqrt{\frac{\Sigma(\text{data item} - \text{mean})^2}{n - 1}}$$

Sample A:

$$\text{Standard deviation} = \sqrt{\frac{28}{6}} \approx 2.16$$

Sample B:

$$\text{Standard deviation} = \sqrt{\frac{700}{6}} \approx 10.80$$

Sample A has a standard deviation of approximately 2.16 and sample B has a standard deviation of approximately 10.80. The data in sample B are more spread out than those in sample A.

CHECK POINT 4 Find the standard deviation of the data items in each of the samples shown below. Round to two decimal places.

Sample A: 73, 75, 77, 79, 81, 83

Sample B: 40, 44, 92, 94, 98, 100

Figure 12.8 illustrates four sets of data items organized in histograms. From left to right, the data items are

Figure 12.8(a): 4, 4, 4, 4, 4, 4, 4

Figure 12.8(b): 3, 3, 4, 4, 4, 5, 5

Figure 12.8(c): 3, 3, 3, 4, 5, 5, 5

Figure 12.8(d): 1, 1, 1, 4, 7, 7, 7.

Each data set has a mean of 4. However, as the spread of the data items increases, the standard deviation gets larger. Observe that when all the data items are the same, the standard deviation is 0.

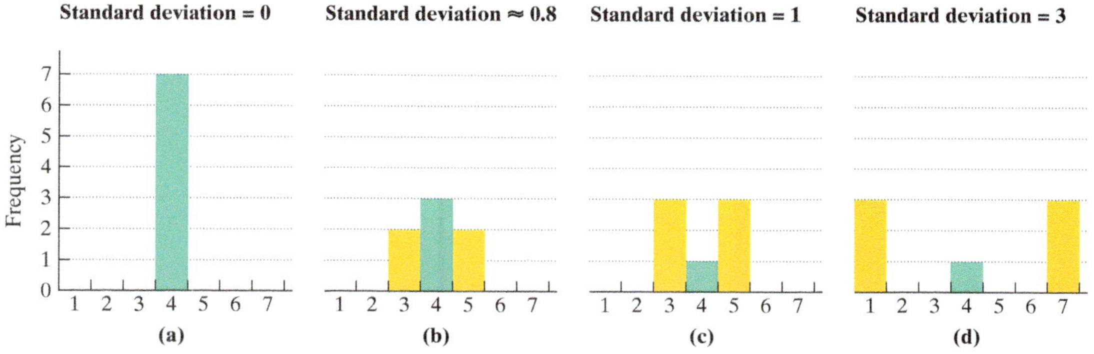

FIGURE 12.8 The standard deviation gets larger with increased dispersion among data items. In each case, the mean is 4.

EXAMPLE 5 *Interpreting Standard Deviation*

Two fifth-grade classes have nearly identical mean scores on an aptitude test, but one class has a standard deviation three times that of the other. All other factors being equal, which class is easier to teach, and why?

SOLUTION

The class with the smaller standard deviation is easier to teach because there is less variation among student aptitudes. Course work can be aimed at the average student without too much concern that the work will be too easy for some or too difficult for others. By contrast, the class with greater dispersion poses a greater challenge. By teaching to the average student, the students whose scores are significantly above the mean will be bored; students whose scores are significantly below the mean will be confused.

CHECK POINT 5 Shown below are the means and standard deviations of the yearly returns on two investments from 1926 through 2004.

Investment	Mean Yearly Return	Standard Deviation
Small-Company Stocks	17.5%	33.3%
Large-Company Stocks	12.4%	20.4%

Source: Summary Statistics of Annual Total Returns 1926 to 2004 Yearbook, Ibbotson Associates, Chicago

a. Use the means to determine which investment provided the greater yearly return.

b. Use the standard deviations to determine which investment had the greater risk. Explain your answer.

Concept and Vocabulary Check

Fill in each blank so that the resulting statement is true.

1. The difference between the highest and lowest data values in a data set is called the ____________.
2. The formula

$$\sqrt{\frac{\Sigma(\text{data item} - \text{mean})^2}{n-1}}$$

gives the value of the ______________ for a data set.

In Exercises 3–5, determine whether each statement is true or false. If the statement is false, make the necessary change(s) to produce a true statement.

3. Measures of dispersion are used to describe the spread of data items in a data set. ______
4. The sum of the deviations from the mean for a data set is always zero. ______
5. Measures of dispersion get smaller as the spread of data items increases. ______

Exercise Set 12.3

Practice Exercises

In Exercises 1–6, find the range for each group of data items.

1. 1, 2, 3, 4, 5
2. 16, 17, 18, 19, 20
3. 7, 9, 9, 15
4. 11, 13, 14, 15, 17
5. 3, 3, 4, 4, 5, 5
6. 3, 3, 3, 4, 5, 5, 5

In Exercises 7–10, a group of data items and their mean are given.

a. *Find the deviation from the mean for each of the data items.*

b. *Find the sum of the deviations in part (a).*

7. 3, 5, 7, 12, 18, 27; Mean = 12
8. 84, 88, 90, 95, 98; Mean = 91
9. 29, 38, 48, 49, 53, 77; Mean = 49
10. 60, 60, 62, 65, 65, 65, 66, 67, 70, 70; Mean = 65

In Exercises 11–16, find a. the mean; b. the deviation from the mean for each data item; and c. the sum of the deviations in part (b).

11. 85, 95, 90, 85, 100
12. 94, 62, 88, 85, 91
13. 146, 153, 155, 160, 161
14. 150, 132, 144, 122
15. 2.25, 3.50, 2.75, 3.10, 1.90
16. 0.35, 0.37, 0.41, 0.39, 0.43

In Exercises 17–26, find the standard deviation for each group of data items. Round answers to two decimal places.

17. 1, 2, 3, 4, 5
18. 16, 17, 18, 19, 20
19. 7, 9, 9, 15
20. 11, 13, 14, 15, 17
21. 3, 3, 4, 4, 5, 5
22. 3, 3, 3, 4, 5, 5, 5
23. 1, 1, 1, 4, 7, 7, 7
24. 6, 6, 6, 6, 7, 7, 7, 4, 8, 3
25. 9, 5, 9, 5, 9, 5, 9, 5
26. 6, 10, 6, 10, 6, 10, 6, 10

In Exercises 27–28, compute the mean, range, and standard deviation for the data items in each of the three samples. Then describe one way in which the samples are alike and one way in which they are different.

27. Sample A: 6, 8, 10, 12, 14, 16, 18
    Sample B: 6, 7, 8, 12, 16, 17, 18
    Sample C: 6, 6, 6, 12, 18, 18, 18
28. Sample A: 8, 10, 12, 14, 16, 18, 20
    Sample B: 8, 9, 10, 14, 18, 19, 20
    Sample C: 8, 8, 8, 14, 20, 20, 20

Practice Plus

In Exercises 29–36, use each display of data items to find the standard deviation. Where necessary, round answers to two decimal places.

29.

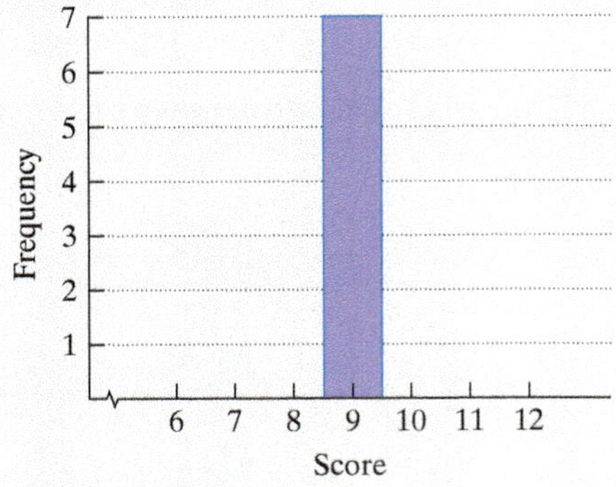

30.

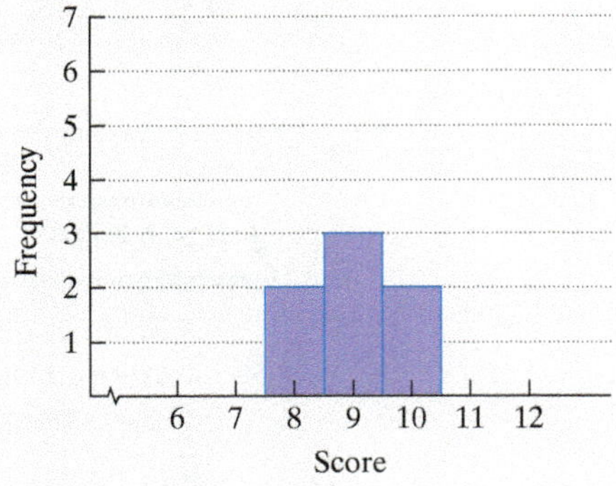

31.

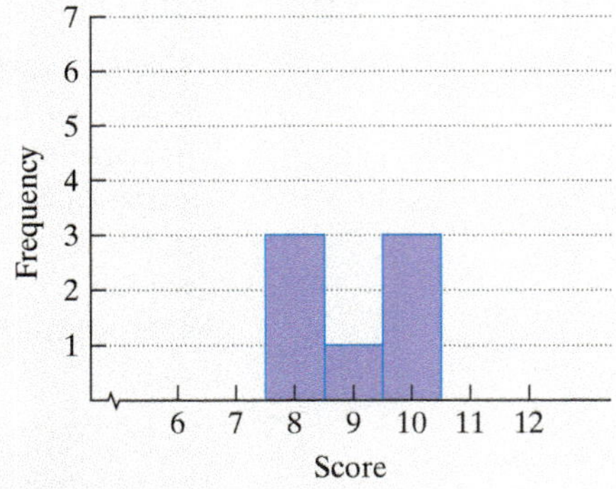

32.

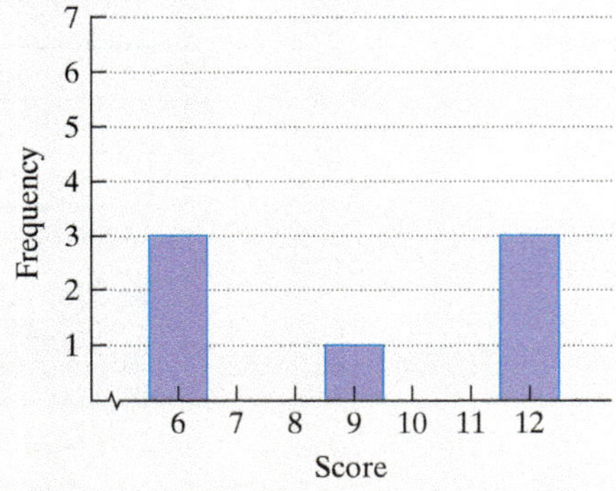

33.

Stems	Leaves
0	5
1	0 5
2	0 5

34.

Stems	Leaves
0	4 8
1	2 6
2	0

35.

Stems	Leaves
1	8 9 9 8 7 8
2	0 1 0 2

36.

Stems	Leaves
1	3 5 3 8 3 4
2	3 0 0 4

Application Exercises

37. The data sets give the number of platinum albums for the five male artists and the five female artists in the United States with the most platinum albums. (Platinum albums sell one million units or more.)

MALE ARTISTS WITH THE MOST PLATINUM ALBUMS

Artist	Platinum Albums
Garth Brooks	145
Elvis Presley	104
Billy Joel	80
Michael Jackson	71
Elton John	65

FEMALE ARTISTS WITH THE MOST PLATINUM ALBUMS

Artist	Platinum Albums
Mariah Carey	64
Madonna	63
Barbra Streisand	61
Whitney Houston	54
Celine Dion	48

Source: RIAA

a. Without calculating, which data set has the greater mean number of platinum albums? Explain your answer.

b. Verify your conjecture from part (a) by calculating the mean number of platinum albums for each data set.

c. Without calculating, which data set has the greater standard deviation? Explain your answer.

d. Verify your conjecture from part (c) by calculating the standard deviation for each data set. Round answers to two decimal places.

38. The data sets give the ages of the first six U.S. presidents and the last six U.S. presidents (through Barack Obama).

AGE OF FIRST SIX U.S. PRESIDENTS AT INAUGURATION

President	Age
Washington	57
J. Adams	61
Jefferson	57
Madison	57
Monroe	58
J. Q. Adams	57

AGE OF LAST SIX U.S. PRESIDENTS AT INAUGURATION

President	Age
Carter	52
Reagan	69
G. H. W. Bush	64
Clinton	46
G. W. Bush	54
Obama	47

Source: *Time Almanac*

a. Without calculating, which set has the greater standard deviation? Explain your answer.

b. Verify your conjecture from part (b) by calculating the standard deviation for each data set. Round answers to two decimal places.

Writing in Mathematics

39. Describe how to find the range of a data set.

40. Describe why the range might not be the best measure of dispersion.

41. Describe how the standard deviation is computed.

42. Describe what the standard deviation reveals about a data set.

43. If a set of test scores has a standard deviation of zero, what does this mean about the scores?

44. Two classes took a statistics test. Both classes had a mean score of 73. The scores of class A had a standard deviation of 5 and those of class B had a standard deviation of 10. Discuss the difference between the two classes' performance on the test.

45. A sample of cereals indicates a mean potassium content per serving of 93 milligrams and a standard deviation of 2 milligrams. Write a description of what this means for a person who knows nothing about statistics.

46. Over a one-month period, stock A had a mean daily closing price of 124.7 and a standard deviation of 12.5. By contrast, stock B had a mean daily closing price of 78.2 and a standard deviation of 6.1. Which stock was more volatile? Explain your answer.

Critical Thinking Exercises

Make Sense? *In Exercises 47–50, determine whether each statement makes sense or does not make sense, and explain your reasoning.*

47. The mean can be misleading if you don't know the spread of data items.

48. The standard deviation for the weights of college students is greater than the standard deviation for the weights of 3-year-old children.

49. I'm working with data sets with different means and the same standard deviation.

50. I'm working with data sets with the same mean and different standard deviations.

51. Describe a situation in which a relatively large standard deviation is desirable.

52. If a set of test scores has a large range but a small standard deviation, describe what this means about students' performance on the test.

53. Use the data 1, 2, 3, 5, 6, 7. Without actually computing the standard deviation, which of the following best approximates the standard deviation?

a. 2 **b.** 6 **c.** 10 **d.** 20

54. Use the data 0, 1, 3, 4, 4, 6. Add 2 to each of the numbers. How does this affect the mean? How does this affect the standard deviation?

Group Exercises

55. As a follow-up to Group Exercise 79 on page 794, the group should reassemble and compute the standard deviation for each data set whose mean you previously determined. Does the standard deviation tell you anything new or interesting about the entire group or subgroups that you did not discover during the previous group activity?

56. Group members should consult a current almanac or the Internet and select intriguing data. The group's function is to use statistics to tell a story. Once "intriguing" data are identified, as a group

a. Summarize the data. Use words, frequency distributions, and graphic displays.

b. Compute measures of central tendency and dispersion, using these statistics to discuss the data.

12.4

WHAT AM I SUPPOSED TO LEARN?

After you have read this section, you should be able to:

1 Recognize characteristics of normal distributions.
2 Understand the 68–95–99.7 Rule.
3 Find scores at a specified standard deviation from the mean.
4 Use the 68–95–99.7 Rule.
5 Convert a data item to a z-score.
6 Understand percentiles and quartiles.
7 Use and interpret margins of error.
8 Recognize distributions that are not normal.

The Normal Distribution

OUR HEIGHTS ARE ON THE RISE! IN ONE million B.C., the mean height for men was 4 feet 6 inches. The mean height for women was 4 feet 2 inches. Because of improved diets and medical care, the mean height for men is now 5 feet 10 inches and for women it is 5 feet 5 inches. Mean adult heights are expected to plateau by 2050.

Mean Adult Heights

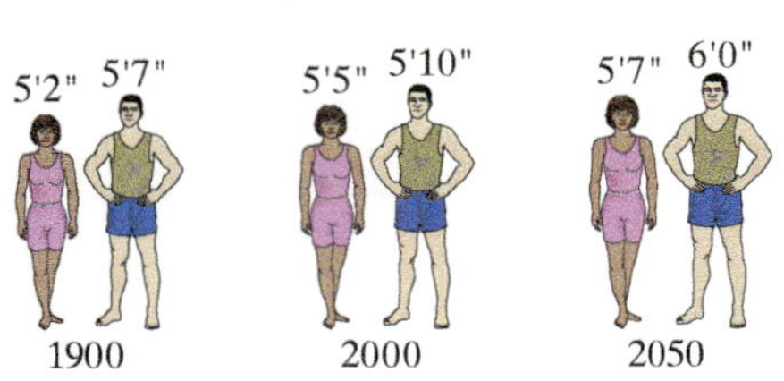

Source: National Center for Health Statistics

1 Recognize characteristics of normal distributions.

Suppose that a researcher selects a random sample of 100 adult men, measures their heights, and constructs a histogram for the data. The graph is shown in **Figure 12.9(a)** below. **Figure 12.9(b)** and **(c)** illustrate what happens as the sample size increases. In **Figure 12.9(c)**, if you were to fold the graph down the middle, the left side would fit the right side. As we move out from the middle, the heights of the bars are the same to the left and right. Such a histogram is called **symmetric**. As the sample size increases, so does the graph's symmetry. If it were possible to measure the heights of all adult males, the entire population, the histogram would approach what is called the **normal distribution**, shown in **Figure 12.9(d)**. This distribution is also called the **bell curve** or the **Gaussian distribution**, named for the German mathematician Carl Friedrich Gauss (1777–1855).

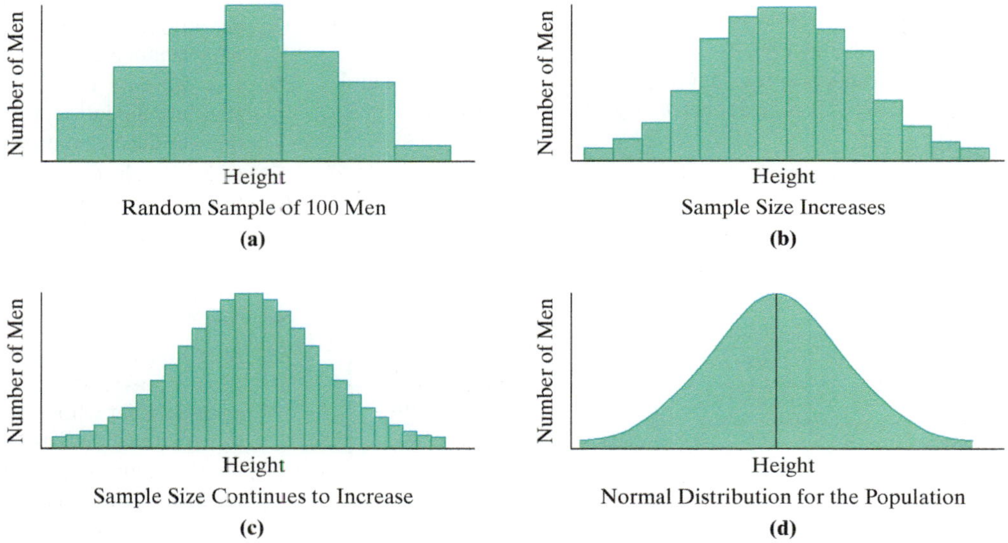

FIGURE 12.9 Heights of adult males

Figure 12.9(d) illustrates that the normal distribution is bell shaped and symmetric about a vertical line through its center. Furthermore, **the mean, median, and mode of a normal distribution are all equal** and located at the center of the distribution.

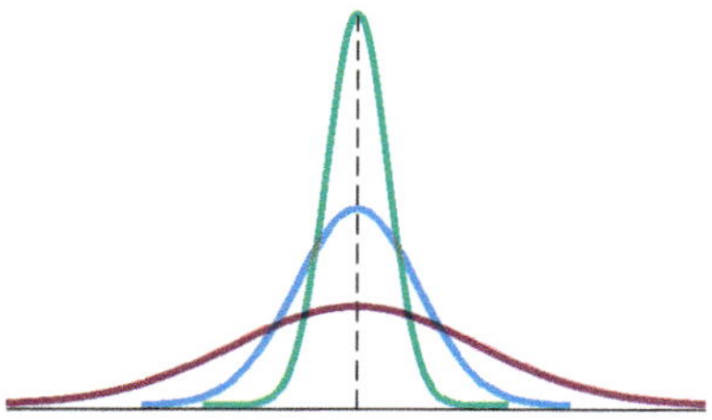

FIGURE 12.10

The shape of the normal distribution depends on the mean and the standard deviation. **Figure 12.10** illustrates three normal distributions with the same mean, but different standard deviations. As the standard deviation increases, the distribution becomes more dispersed, or spread out, but retains its symmetric bell shape.

The normal distribution provides a wonderful model for all kinds of phenomena because many sets of data items closely resemble this population distribution. Examples include heights and weights of adult males, intelligence quotients, SAT scores, prices paid for a new car model, and life spans of light bulbs. In these distributions, the data items tend to cluster around the mean. The more an item differs from the mean, the less likely it is to occur.

The normal distribution is used to make predictions about an entire population using data from a sample. In this section, we focus on the characteristics and applications of the normal distribution.

2 Understand the 68–95–99.7 Rule.

The Standard Deviation and z-Scores in Normal Distributions

The standard deviation plays a crucial role in the normal distribution, summarized by the **68–95–99.7 Rule**. This rule is illustrated in **Figure 12.11**.

THE 68–95–99.7 RULE FOR THE NORMAL DISTRIBUTION

1. Approximately 68% of the data items fall within 1 standard deviation of the mean (in both directions).
2. Approximately 95% of the data items fall within 2 standard deviations of the mean.
3. Approximately 99.7% of the data items fall within 3 standard deviations of the mean.

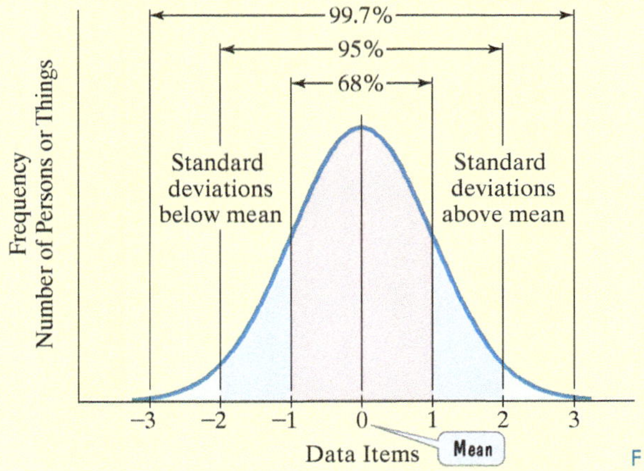

FIGURE 12.11

Blitzer Bonus

Well-Worn Steps and the Normal Distribution

These ancient steps each take on the shape of a normal distribution when the picture is viewed upside down. The center of each step is more worn than the outer edges. The greatest number of people have walked in the center, making this the mean, median, and mode for where people have walked.

Figure 12.11 illustrates that a very small percentage of the data in a normal distribution lies more than 3 standard deviations above or below the mean. As we move from the mean, the curve falls rapidly, and then more and more gradually, toward the horizontal axis. The tails of the curve approach, but never touch, the horizontal axis, although they are quite close to the axis at 3 standard deviations from the mean. The range of the normal distribution is infinite. No matter how far out from the mean we move, there is always the probability (although very small) of a data item occurring even farther out.

3 Find scores at a specified standard deviation from the mean.

EXAMPLE 1 *Finding Scores at a Specified Standard Deviation from the Mean*

Male adult heights in North America are approximately normally distributed with a mean of 70 inches and a standard deviation of 4 inches. Find the height that is

a. 2 standard deviations above the mean.

b. 3 standard deviations below the mean.

SOLUTION

a. First, let us find the height that is 2 standard deviations above the mean.

$$\text{Height} = \text{mean} + 2 \cdot \text{standard deviation}$$
$$= 70 + 2 \cdot 4 = 70 + 8 = 78$$

A height of 78 inches is 2 standard deviations above the mean.

b. Next, let us find the height that is 3 standard deviations below the mean.

$$\text{Height} = \text{mean} - 3 \cdot \text{standard deviation}$$
$$= 70 - 3 \cdot 4 = 70 - 12 = 58$$

A height of 58 inches is 3 standard deviations below the mean.

The distribution of male adult heights in North America is illustrated as a normal distribution in **Figure 12.12**.

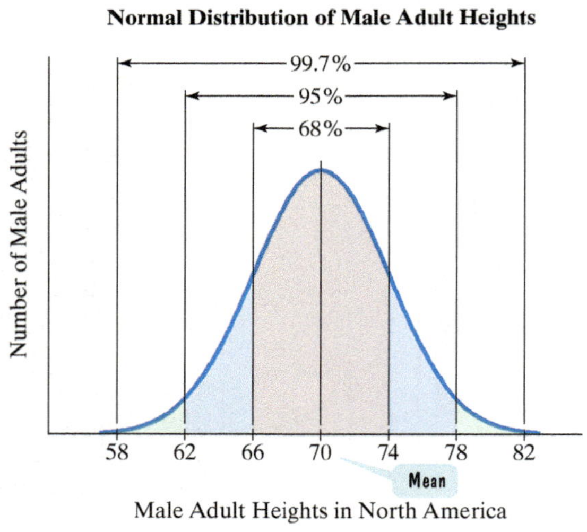

FIGURE 12.12

CHECK POINT 1 Female adult heights in North America are approximately normally distributed with a mean of 65 inches and a standard deviation of 3.5 inches. Find the height that is

a. 3 standard deviations above the mean.

b. 2 standard deviations below the mean.

4 Use the 68–95–99.7 Rule.

EXAMPLE 2 *Using the 68–95–99.7 Rule*

Use the distribution of male adult heights in **Figure 12.12** on the previous page to find the percentage of men in North America with heights

a. between 66 inches and 74 inches. **b.** between 70 inches and 74 inches.

c. above 78 inches.

SOLUTION

a. The 68–95–99.7 Rule states that approximately 68% of the data items fall within 1 standard deviation, 4, of the mean, 70.

$$\text{mean} - 1 \cdot \text{standard deviation} = 70 - 1 \cdot 4 = 70 - 4 = 66$$
$$\text{mean} + 1 \cdot \text{standard deviation} = 70 + 1 \cdot 4 = 70 + 4 = 74$$

Figure 12.12 shows that 68% of male adults have heights between 66 inches and 74 inches.

b. The percentage of men with heights between 70 inches and 74 inches is not given directly in **Figure 12.12**. Because of the distribution's symmetry, the percentage with heights between 66 inches and 70 inches is the same as the percentage with heights between 70 and 74 inches. **Figure 12.13** indicates that 68% have heights between 66 inches and 74 inches. Thus, half of 68%, or 34%, of men have heights between 70 inches and 74 inches.

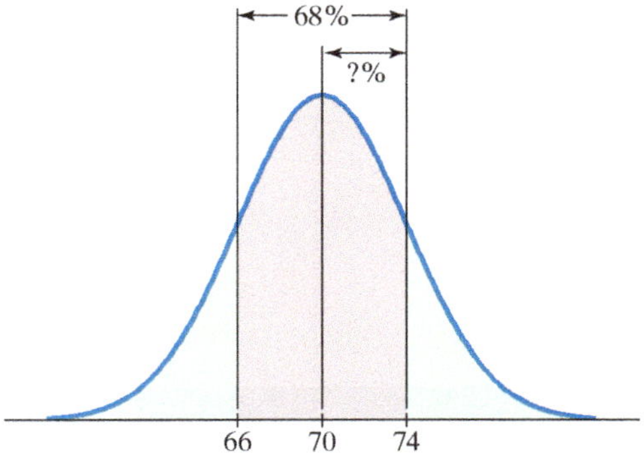

FIGURE 12.13 What percentage have heights between 70 inches and 74 inches?

c. The percentage of men with heights above 78 inches is not given directly in **Figure 12.12**. A height of 78 inches is 2 standard deviations, $2 \cdot 4$, or 8 inches, above the mean, 70 inches. The 68–95–99.7 Rule states that approximately 95% of the data items fall within 2 standard deviations of the mean. Thus, approximately 100% − 95%, or 5%, of the data items are farther than 2 standard deviations from the mean. The 5% of the data items are represented by the two shaded green regions in **Figure 12.14**. Because of the distribution's symmetry, half of 5%, or 2.5%, of the data items are more than 2 standard deviations above the mean. This means that 2.5% of men have heights above 78 inches.

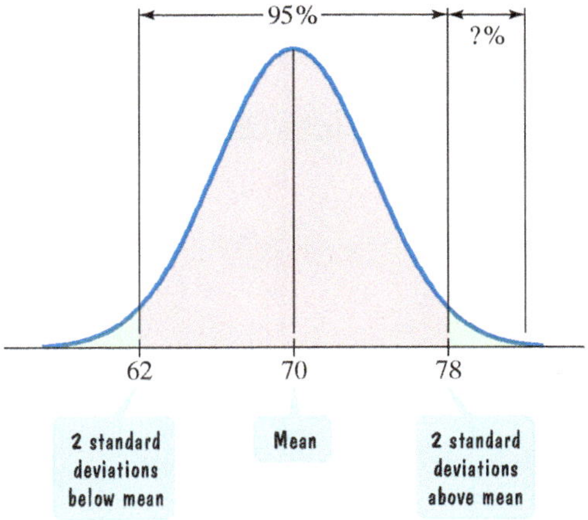

FIGURE 12.14 What percentage have heights above 78 inches?

CHECK POINT 2 Use the distribution of male adult heights in North America in **Figure 12.12** to find the percentage of men with heights

a. between 62 inches and 78 inches.

b. between 70 inches and 78 inches. **c.** above 74 inches.

Because the normal distribution of male adult heights in North America has a mean of 70 inches and a standard deviation of 4 inches, a height of 78 inches lies 2 standard deviations above the mean. In a normal distribution, a **z-score** describes how many standard deviations a particular data item lies above or below the mean. Thus, the z-score for the data item 78 is 2.

The following formula can be used to express a data item in a normal distribution as a z-score:

5 Convert a data item to a z-score.

COMPUTING z-SCORES

A z-score describes how many standard deviations a data item in a normal distribution lies above or below the mean. The z-score can be obtained using

$$z\text{-score} = \frac{\text{data item} - \text{mean}}{\text{standard deviation}}.$$

Data items above the mean have positive z-scores. Data items below the mean have negative z-scores. The z-score for the mean is 0.

EXAMPLE 3 *Computing z-Scores*

The mean weight of newborn infants is 7 pounds, and the standard deviation is 0.8 pound. The weights of newborn infants are normally distributed. Find the z-score for a weight of

a. 9 pounds. **b.** 7 pounds. **c.** 6 pounds.

SOLUTION

We compute the z-score for each weight by using the z-score formula. The mean is 7 and the standard deviation is 0.8.

a. The z-score for a weight of 9 pounds, written z_9, is

$$z_9 = \frac{\text{data item} - \text{mean}}{\text{standard deviation}} = \frac{9-7}{0.8} = \frac{2}{0.8} = 2.5.$$

The z-score of a data item greater than the mean is always positive. A 9-pound infant is a chubby little tyke, with a weight that is 2.5 standard deviations above the mean.

b. The z-score for a weight of 7 pounds is

$$z_7 = \frac{\text{data item} - \text{mean}}{\text{standard deviation}} = \frac{7-7}{0.8} = \frac{0}{0.8} = 0.$$

The z-score for the mean is always 0. A 7-pound infant is right at the mean, deviating 0 pounds above or below it.

c. The z-score for a weight of 6 pounds is

$$z_6 = \frac{\text{data item} - \text{mean}}{\text{standard deviation}} = \frac{6-7}{0.8} = \frac{-1}{0.8} = -1.25.$$

The z-score of a data item less than the mean is always negative. A 6-pound infant's weight is 1.25 standard deviations below the mean.

Figure 12.15 shows the normal distribution of weights of newborn infants. The horizontal axis is labeled in terms of weights and z-scores.

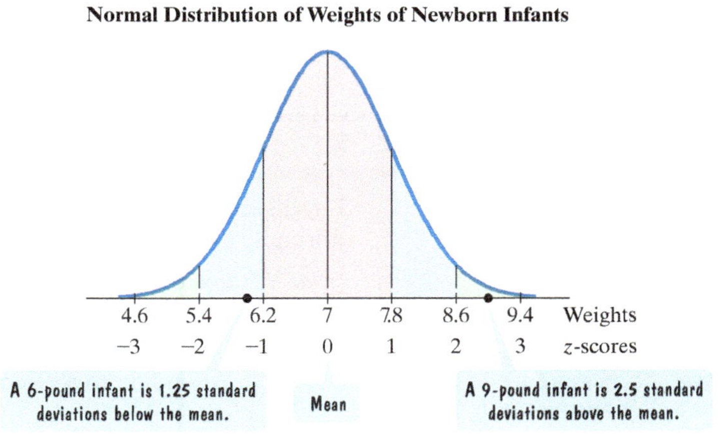

FIGURE 12.15 Infants' weights are normally distributed.

CHECK POINT 3 The length of horse pregnancies from conception to birth is normally distributed with a mean of 336 days and a standard deviation of 3 days. Find the z-score for a horse pregnancy of

a. 342 days. **b.** 336 days. **c.** 333 days.

In Example 4, we consider two normally distributed sets of test scores, in which a higher score generally indicates a better result. To compare scores on two different tests in relation to the mean on each test, we can use z-scores. The better score is the item with the greater z-score.

EXAMPLE 4 *Using and Interpreting z-Scores*

A student scores 70 on an arithmetic test and 66 on a vocabulary test. The scores for both tests are normally distributed. The arithmetic test has a mean of 60 and a standard deviation of 20. The vocabulary test has a mean of 60 and a standard deviation of 2. On which test did the student have the better score?

SOLUTION

To answer the question, we need to find the student's z-score on each test, using

$$z = \frac{\text{data item} - \text{mean}}{\text{standard deviation}}.$$

The arithmetic test has a mean of 60 and a standard deviation of 20.

$$z\text{-score for } 70 = z_{70} = \frac{70 - 60}{20} = \frac{10}{20} = 0.5$$

The vocabulary test has a mean of 60 and a standard deviation of 2.

$$z\text{-score for } 66 = z_{66} = \frac{66 - 60}{2} = \frac{6}{2} = 3$$

The arithmetic score, 70, is half a standard deviation above the mean, whereas the vocabulary score, 66, is 3 standard deviations above the mean. The student did much better than the mean on the vocabulary test.

CHECK POINT 4 The SAT (Scholastic Aptitude Test) has a mean of 500 and a standard deviation of 100. The ACT (American College Test) has a mean of 18 and a standard deviation of 6. Both tests measure the same kind of ability, with scores that are normally distributed. Suppose that you score 550 on the SAT and 24 on the ACT. On which test did you have the better score?

Blitzer Bonus

The IQ Controversy

Is intelligence something we are born with or is it a quality that can be manipulated through education? Can it be measured accurately and is IQ the way to measure it? There are no clear answers to these questions.

In a study by Carolyn Bird (*Pygmalion in the Classroom*), a group of third-grade teachers was told that they had classes of students with IQs well above the mean. These classes made incredible progress throughout the year. In reality, these were not gifted kids, but, rather, a random sample of all third-graders. It was the teachers' expectations, and not the IQs of the students, that resulted in increased performance.

EXAMPLE 5 *Understanding z-Scores*

Intelligence quotients (IQs) on the Stanford-Binet intelligence test are normally distributed with a mean of 100 and a standard deviation of 16.

a. What is the IQ corresponding to a z-score of -1.5?

b. Mensa is a group of people with high IQs whose members have z-scores of 2.05 or greater on the Stanford-Binet intelligence test. What is the IQ corresponding to a z-score of 2.05?

SOLUTION

a. We begin with the IQ corresponding to a z-score of -1.5. The negative sign in -1.5 tells us that the IQ is $1\frac{1}{2}$ standard deviations below the mean.

$$\begin{aligned} \text{IQ} &= \text{mean} - 1.5 \cdot \text{standard deviation} \\ &= 100 - 1.5(16) = 100 - 24 = 76 \end{aligned}$$

The IQ corresponding to a z-score of -1.5 is 76.

b. Next, we find the IQ corresponding to a z-score of 2.05. The positive sign implied in 2.05 tells us that the IQ is 2.05 standard deviations above the mean.

$$\begin{aligned} \text{IQ} &= \text{mean} + 2.05 \cdot \text{standard deviation} \\ &= 100 + 2.05(16) = 100 + 32.8 = 132.8 \end{aligned}$$

The IQ corresponding to a z-score of 2.05 is 132.8. (An IQ score of at least 133 is required to join Mensa.)

CHECK POINT 5 Use the information in Example 5 to find the IQ corresponding to a z-score of

a. -2.25.

b. 1.75.

6 Understand percentiles and quartiles.

Percentiles and Quartiles

A z-score measures a data item's position in a normal distribution. Another measure of a data item's position is its **percentile**. Percentiles are often associated with scores on standardized tests. If a score is in the 45th percentile, this means that 45% of the scores are less than this score. If a score is in the 95th percentile, this indicates that 95% of the scores are less than this score.

PERCENTILES

If n% of the items in a distribution are less than a particular data item, we say that the data item is in the ***n*th percentile** of the distribution.

EXAMPLE 6 *Interpreting Percentile*

The cutoff IQ score for Mensa membership, 132.8, is in the 98th percentile. What does this mean?

SOLUTION

Because 132.8 is in the 98th percentile, this means that 98% of IQ scores fall below 132.8. Caution: A score in the 98th percentile does *not* mean that 98% of the answers are correct. Nor does it mean that the score was 98%.

CHECK POINT 6 A student scored in the 75th percentile on the SAT. What does this mean?

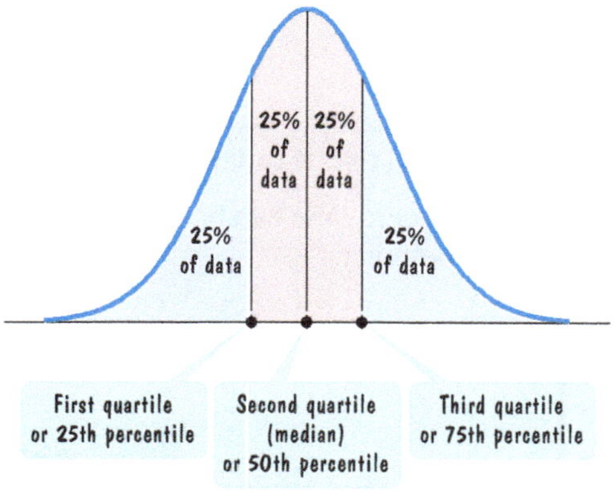

FIGURE 12.16 Quartiles

Three commonly encountered percentiles are the *quartiles*. **Quartiles** divide data sets into four equal parts. The 25th percentile is the **first quartile**: 25% of the data fall below the first quartile. The 50th percentile is the **second quartile**: 50% of the data fall below the second quartile, so the second quartile is equivalent to the median. The 75th percentile is the **third quartile**: 75% of the data fall below the third quartile. **Figure 12.16** illustrates the concept of quartiles for the normal distribution.

7 Use and interpret margins of error.

Polls and Margins of Error

What activities do you dread? Reading math textbooks with cows posing for photos on the cover? (Be kind!) No, that's not America's most-dreaded activity. In a random sample of 1000 U.S. adults, 46% of those questioned responded, "Public speaking." The problem is that this is a single random sample. Do 46% of adults in the entire U.S. population dread public speaking?

Statisticians use properties of the normal distribution to estimate the probability that a result obtained from a single sample reflects what is truly happening in the population. If you look at the results of a poll like the one shown in **Figure 12.17**, you will observe that a *margin of error* is reported. Surveys and opinion polls often give a margin of error. Let's use our understanding of the normal distribution to see how to calculate and interpret margins of error.

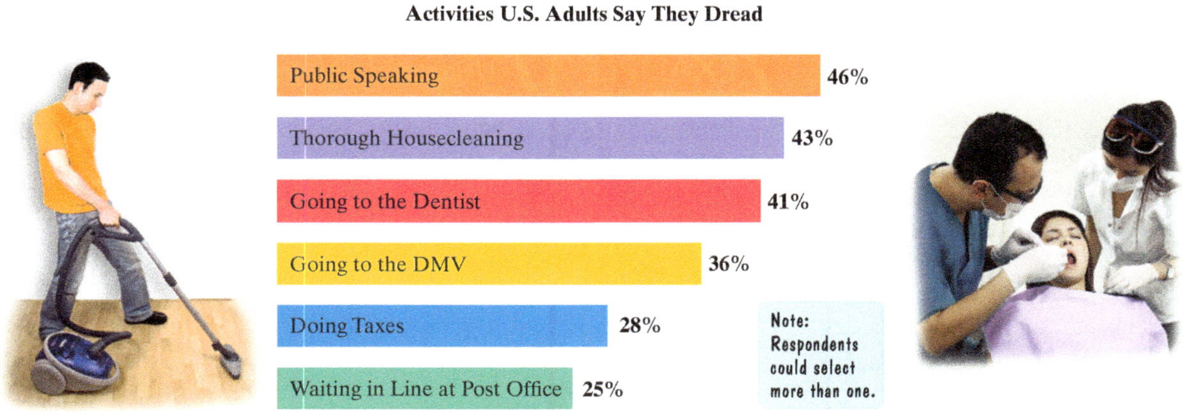

FIGURE 12.17 *Source:* TNS survey of 1000 adults, March 2010, Margin of error: ±3.2%

Note the margin of error.

Suppose that $p\%$ of the population of U.S. adults dread public speaking. Instead of taking only one random sample of 1000 adults, we repeat the process of selecting a random sample of 1000 adults hundreds of times. Then, we calculate the percentage of adults for each sample who dread public speaking. With random sampling, we expect to find the percentage in many of the samples close to $p\%$, with relatively few samples having percentages far from $p\%$. **Figure 12.18** shows that the percentages of U.S adults from the hundreds of samples can be modeled by a normal distribution. The mean of this distribution is the actual population percent, $p\%$, and is the most frequent result from the samples.

Mathematicians have shown that the standard deviation of a normal distribution of samples like the one in **Figure 12.18** is approximately $\frac{1}{2\sqrt{n}} \times 100\%$, where n is the sample size. Using the 68–95–99.7 Rule, approximately 95% of the samples have a percentage within 2 standard deviations of the true population percentage, $p\%$:

$$2 \text{ standard deviations} = 2 \cdot \frac{1}{2\sqrt{n}} \times 100\% = \frac{1}{\sqrt{n}} \times 100\%.$$

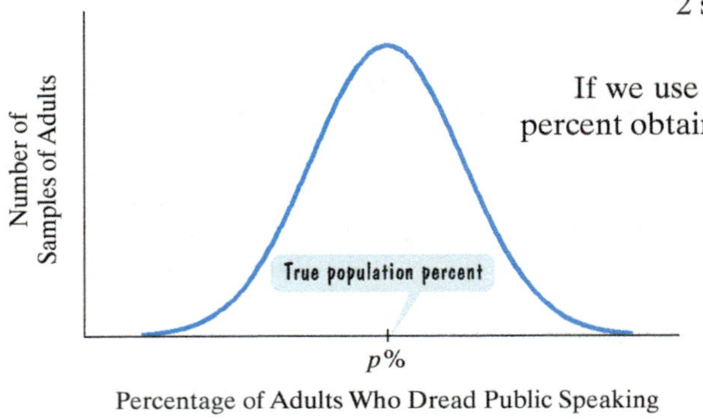

FIGURE 12.18 Percentage of U.S. adults who dread public speaking

If we use a single random sample of size n, there is a 95% probability that the percent obtained will lie within two standard deviations, or $\frac{1}{\sqrt{n}} \times 100\%$, of the true population percent. We can be 95% confident that the true population percent lies between

$$\text{the sample percent} - \frac{1}{\sqrt{n}} \times 100\%$$

and

$$\text{the sample percent} + \frac{1}{\sqrt{n}} \times 100\%.$$

We call $\pm \frac{1}{\sqrt{n}} \times 100\%$ the **margin of error.**

MARGIN OF ERROR IN A SURVEY

If a statistic is obtained from a random sample of size n, there is a 95% probability that it lies within $\frac{1}{\sqrt{n}} \times 100\%$ of the true population percent, where $\pm\frac{1}{\sqrt{n}} \times 100\%$ is called the **margin of error.**

TABLE 12.16 Activities U.S. Adults Dread

Activity	Percentage Who Dread the Activity
Public speaking	46%
Thorough house-cleaning	43%
Going to the dentist	41%
Going to the DMV	36%
Doing taxes	28%
Waiting in line at the post office	25%

Source: TNS survey of 1000 adults, March 2010,

EXAMPLE 7 *Using and Interpreting Margin of Error*

Table 12.16 shows that in a random sample of 1000 U.S adults, 46% of those questioned said that they dread public speaking.

a. Verify the margin of error that was given for this survey.

b. Write a statement about the percentage of adults in the U.S. population who dread public speaking.

SOLUTION

a. The sample size is $n = 1000$. The margin of error is

$$\pm\frac{1}{\sqrt{n}} \times 100\% = \pm\frac{1}{\sqrt{1000}} \times 100\% \approx \pm 0.032 \times 100\% = \pm 3.2\%.$$

b. There is a 95% probability that the true population percentage lies between

$$\text{the sample percent} - \frac{1}{\sqrt{n}} \times 100\% = 46\% - 3.2\% = 42.8\%$$

and

$$\text{the sample percent} + \frac{1}{\sqrt{n}} \times 100\% = 46\% + 3.2\% = 49.2\%.$$

We can be 95% confident that between 42.8% and 49.2% of all U.S adults dread public speaking.

Blitzer Bonus

A Caveat Giving a True Picture of a Poll's Accuracy

Unlike the precise calculation of a poll's margin of error, certain polling imperfections cannot be determined exactly. One problem is that people do not always respond to polls honestly and accurately. Some people are embarrassed to say "undecided," so they make up an answer. Other people may try to respond to questions in the way they think will make the pollster happy, just to be "nice." Perhaps the following caveat, applied to the poll in Example 7, would give the public a truer picture of its accuracy:

> The poll results are 42.8% to 49.2% at the 95% confidence level, but it's only under ideal conditions that we can be 95% confident that the true numbers are within 3.2% of the poll's results. The true error span is probably greater than 3.2% due to limitations that are inherent in this and every poll, but, unfortunately, this additional error amount cannot be calculated precisely. Warning: Five percent of the time—that's one time out of 20—the error will be greater than 3.2%. We remind readers of the poll that things occurring "only" 5% of the time do, indeed, happen.

We suspect that the public would tire of hearing this.

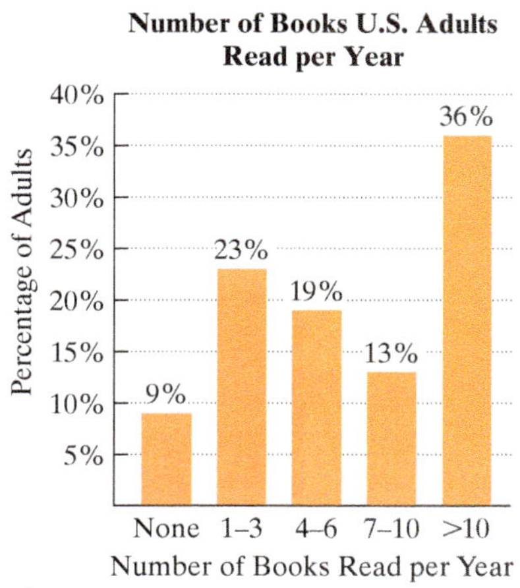

FIGURE 12.19
Source: Harris Poll of 2513 U.S. adults ages 18 and older conducted March 11 and 18, 2008

CHECK POINT 7 A Harris Poll of 2513 U.S. adults ages 18 and older asked the question

How many books do you typically read in a year?

The results of the poll are shown in **Figure 12.19**.

a. Find the margin of error for this survey. Round to the nearest tenth of a percent.

b. Write a statement about the percentage of U.S. adults who read more than ten books per year.

c. Why might some people not respond honestly and accurately to the question in this poll?

8 Recognize distributions that are not normal.

Other Kinds of Distributions

Although the normal distribution is the most important of all distributions in terms of analyzing data, not all data can be approximated by this symmetric distribution with its mean, median, and mode all having the same value.

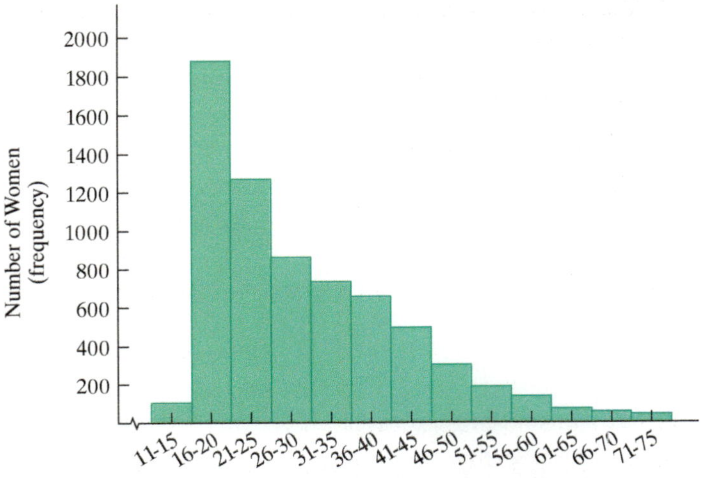

FIGURE 12.20 Histogram of the ages of females interviewed by Kinsey and his associates

The histogram in **Figure 12.20** represents the frequencies of the ages of women interviewed by Kinsey and his colleagues in their study of female sexual behavior. This distribution is not symmetric. The greatest frequency of women interviewed was in the 16–20 age range. The bars get shorter and shorter after this. The shorter bars fall on the right, indicating that relatively few older women were included in Kinsey's interviews.

In our discussion of measures of central tendency, we mentioned that the median, rather than the mean, is used to summarize income. **Figure 12.21** illustrates the population distribution of weekly earnings in the United States. There is no upper limit on weekly earnings. The relatively few people with very high weekly incomes tend to pull the mean income to a value greater than the median. The most frequent income, the mode, occurs toward the low end of the data items. The mean, median, and mode do not have the same value, and a normal distribution is not an appropriate model for describing weekly earnings in the United States.

The distribution in **Figure 12.21** is called a *skewed distribution.* A distribution of data is **skewed** if a large number of data items are piled up at one end or the other, with a "tail" at the opposite end. In the distribution of weekly earnings in **Figure 12.21**, the tail is to the right. Such a distribution is said to be **skewed to the right**.

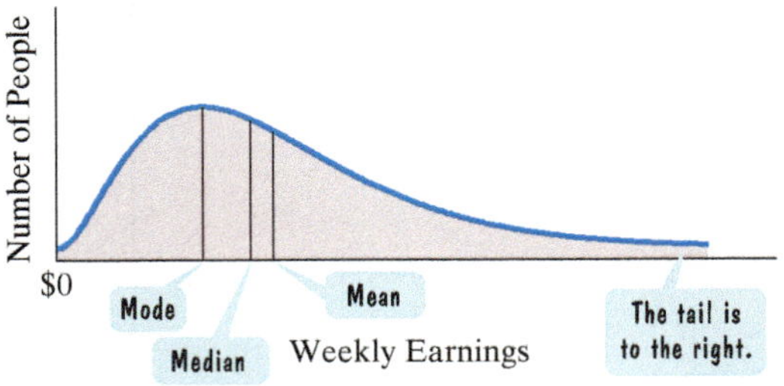

FIGURE 12.21 Skewed to the right

By contrast to the distribution of weekly earnings, the distribution in **Figure 12.22** has more data items at the high end of the scale than at the low end. The tail of this distribution is to the left. The distribution is said to be **skewed to the left**. In many colleges, an example of a distribution skewed to the left is based on the student ratings of faculty teaching performance. Most professors are given rather high ratings, while only a few are rated as terrible. These low ratings pull the value of the mean lower than the median.

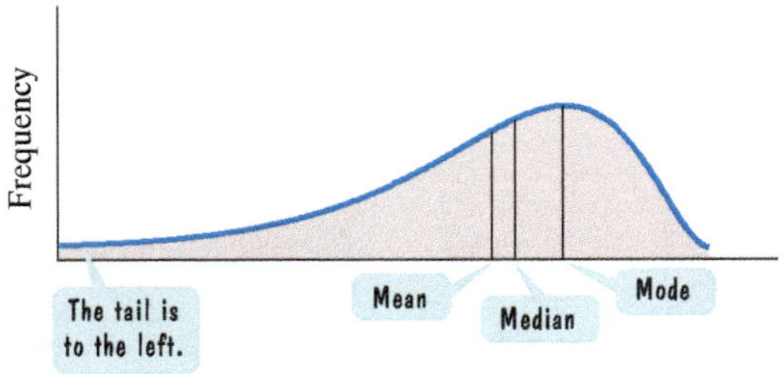

FIGURE 12.22 Skewed to the left

GREAT QUESTION!

What's the bottom line on the relationship between the mean and the median for skewed distributions?

If the data are skewed to the right, the mean is greater than the median. If the data are skewed to the left, the mean is less than the median.

Concept and Vocabulary Check

Fill in each blank so that the resulting statement is true.

1. In a normal distribution, approximately ____________% of the data items fall within 1 standard deviation of the mean, approximately ____________% of the data items fall within 2 standard deviations of the mean, and approximately ____________% of the data items fall within 3 standard deviations of the mean.

2. A z-score describes how many standard deviations a data item in a normal distribution lies above or below the ____________.

3. If n% of the items in a distribution are less than a particular data item, we say that the data item is in the nth ____________ of the distribution.

4. If a statistic is obtained from a random sample of size n, there is a 95% probability that it lies within $\frac{1}{\sqrt{n}} \times 100\%$ of the true population percent, where $\pm \frac{1}{\sqrt{n}} \times 100\%$ is called the ____________.

In Exercises 5–8, determine whether each statement is true or false. If the statement is false, make the necessary change(s) to produce a true statement.

5. The mean, median, and mode of a normal distribution are all equal. ______

6. In a normal distribution, the z-score for the mean is 0. ______

7. The z-score for a data item in a normal distribution is obtained using

$$z\text{-score} = \frac{\text{data item} - \text{standard deviation}}{\text{mean}}.$$ ______

8. A score in the 50th percentile on a standardized test is the median. ______

Exercise Set 12.4

Practice and Application Exercises

The scores on a test are normally distributed with a mean of 100 and a standard deviation of 20. In Exercises 1–10, find the score that is

1. 1 standard deviation above the mean.

2. 2 standard deviations above the mean.

3. 3 standard deviations above the mean.

4. $1\frac{1}{2}$ standard deviations above the mean.

5. $2\frac{1}{2}$ standard deviations above the mean.

6. 1 standard deviation below the mean.

7. 2 standard deviations below the mean.

8. 3 standard deviations below the mean.

9. one-half a standard deviation below the mean.

10. $2\frac{1}{2}$ standard deviations below the mean.

Not everyone pays the same price for the same model of a car. The figure illustrates a normal distribution for the prices paid for a particular model of a new car. The mean is \$17,000 and the standard deviation is \$500.

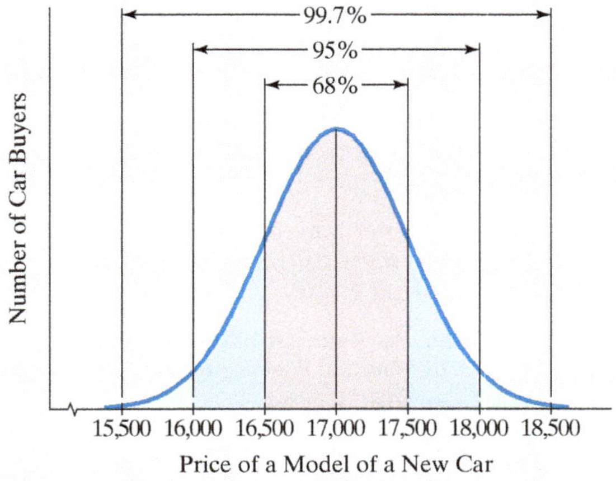

In Exercises 11–22, use the 68–95–99.7 Rule, illustrated in the figure, to find the percentage of buyers who paid

11. between \$16,500 and \$17,500.

12. between \$16,000 and \$18,000.

13. between \$17,000 and \$17,500.

14. between \$17,000 and \$18,000.

15. between \$16,000 and \$17,000.

16. between \$16,500 and \$17,000.

17. between \$15,500 and \$17,000.

18. between \$17,000 and \$18,500.

19. more than \$17,500.

20. more than \$18,000.

21. less than \$16,000.

22. less than \$16,500.

Intelligence quotients (IQs) on the Stanford-Binet intelligence test are normally distributed with a mean of 100 and a standard deviation of 16. In Exercises 23–32, use the 68–95–99.7 Rule to find the percentage of people with IQs

23. between 68 and 132.

24. between 84 and 116.

25. between 68 and 100.

26. between 84 and 100.

27. above 116.

28. above 132.

29. below 68.

30. below 84.

31. above 148.

32. below 52.

A set of data items is normally distributed with a mean of 60 and a standard deviation of 8. In Exercises 33–48, convert each data item to a z-score.

33. 68 **34.** 76 **35.** 84

36. 92 **37.** 64 **38.** 72

39. 74 **40.** 78 **41.** 60

42. 100 **43.** 52 **44.** 44

45. 48 **46.** 40 **47.** 34

48. 30

Scores on a dental anxiety scale range from 0 (no anxiety) to 20 (extreme anxiety). The scores are normally distributed with a mean of 11 and a standard deviation of 4. In Exercises 49–56, find the z-score for the given score on this dental anxiety scale.

49. 17 **50.** 18

51. 20 **52.** 12

53. 6 **54.** 8

55. 5 **56.** 1

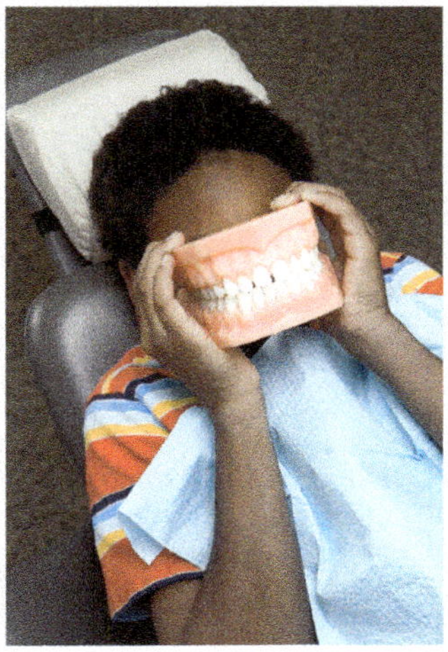

Intelligence quotients on the Stanford-Binet intelligence test are normally distributed with a mean of 100 and a standard deviation of 16. Intelligence quotients on the Wechsler intelligence test are normally distributed with a mean of 100 and a standard deviation of 15. Use this information to solve Exercises 57–58.

57. Use z-scores to determine which person has the higher IQ: an individual who scores 128 on the Stanford-Binet or an individual who scores 127 on the Wechsler.

58. Use z-scores to determine which person has the higher IQ: an individual who scores 150 on the Stanford-Binet or an individual who scores 148 on the Wechsler.

A set of data items is normally distributed with a mean of 400 and a standard deviation of 50. In Exercises 59–66, find the data item in this distribution that corresponds to the given z-score.

59. $z = 2$

60. $z = 3$

61. $z = 1.5$

62. $z = 2.5$

63. $z = -3$

64. $z = -2$

65. $z = -2.5$

66. $z = -1.5$

67. Reducing Gun Violence The data in the bar graph are from a random sample of 814 American adults. The graph shows four proposals to reduce gun violence in the United States and the percentage of surveyed adults who favored each of these proposals.

Proposals to Reduce Gun Violence in the United States

Source: Time/CNN poll using a sample of 814 American adults, January 14–15, 2013

a. Find the margin of error, to the nearest tenth of a percent, for this survey.

b. Write a statement about the percentage of adults in the U.S population who favor required gun registration to reduce gun violence.

68. How to Blow Your Job Interview The data in the bar graph at the top of the next column are from a random sample of 1910 job interviewers. The graph shows the top interviewer turnoffs and the percentage of surveyed interviewers who were offended by each of these behaviors.

Top Interviewer Turnoffs

Source: Scott Erker, PhD., and Kelli Buczynski, "Are You Failing the Interview? 2009 Survey of Global Interviewing Practices and Perceptions." Development Dimensions International.

a. Find the margin of error, to the nearest tenth of a percent, for this survey.

b. Write a statement about the percentage of interviewers in the population who are turned off by a job applicant being arrogant.

69. Using a random sample of 4000 TV households, Nielsen Media Research found that 60.2% watched the final episode of *M*A*S*H*.

a. Find the margin of error in this percent.

b. Write a statement about the percentage of TV households in the population that tuned into the final episode of *M*A*S*H*.

70. Using a random sample of 4000 TV households, Nielsen Media Research found that 51.1% watched *Roots, Part 8*.

a. Find the margin of error in this percent.

b. Write a statement about the percentage of TV households in the population that tuned into *Roots, Part 8*.

71. In 1997, Nielsen Media Research increased its random sample to 5000 TV households. By how much, to the nearest tenth of a percent, did this improve the margin of error over that in Exercises 69 and 70?

72. If Nielsen Media Research were to increase its random sample from 5000 to 10,000 TV households, by how much, to the nearest tenth of a percent, would this improve the margin of error?

The histogram shows murder rates per 100,000 residents and the number of U.S. states that had these rates for a recent year. Use this histogram to solve Exercises 73–74.

U.S. Murder Rates per 100,000 Residents, by State and Washington, D.C.

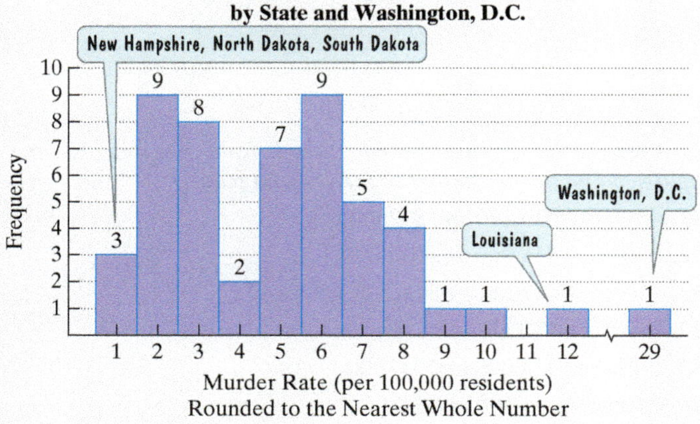

Source: FBI, *Crime in the United States*

73. a. Is the shape of this distribution best classified as normal, skewed to the right, or skewed to the left?

b. Calculate the mean murder rate per 100,000 residents for the 50 states and Washington, D.C.

c. Find the median murder rate per 100,000 residents for the 50 states and Washington, D.C.

d. Are the mean and median murder rates consistent with the shape of the distribution that you described in part (a)? Explain your answer.

e. The standard deviation for the data is approximately 4.2. If the distribution were roughly normal, what would be the z-score, rounded to one decimal place, for Washington, D.C.? Does this seem unusually high? Explain your answer.

74. a. Find the median murder rate per 100,000 residents for the 50 states and Washington, D.C.

b. Find the first quartile by determining the median of the lower half of the data. (This is the median of the items that lie below the median that you found in part (a).)

c. Find the third quartile by determining the median of the upper half of the data. (This is the median of the items that lie above the median that you found in part (a).)

d. Use the following numerical scale:

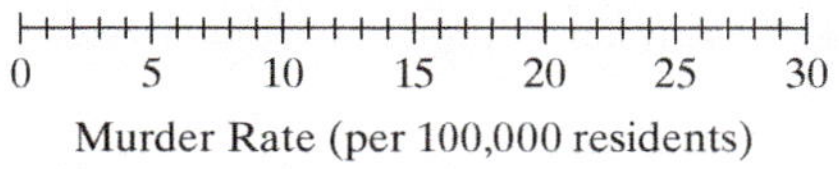

Above this scale, show five points, each at the same height. (The height is arbitrary.) Each point should represent one of the following numbers:

lowest data value, first quartile, median, third quartile, highest data value.

e. A **box-and-whisker plot** consists of a rectangular box extending from the first quartile to the third quartile, with a dashed line representing the median, and line segments (or whiskers) extending outward from the box to the lowest and highest data values:

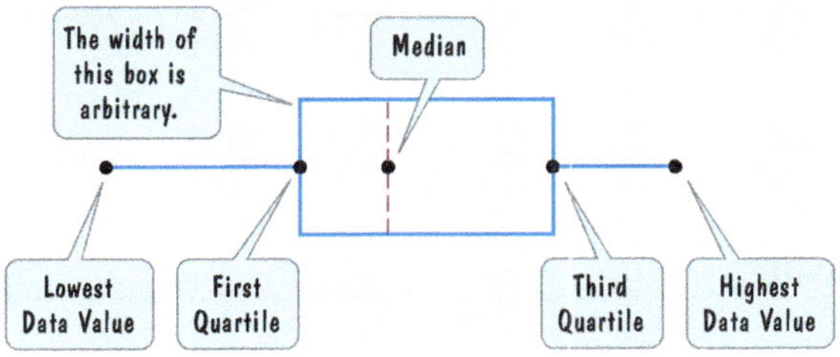

Use your graph from part (d) to create a box-and-whisker plot for U.S. murder rates per 100,000 residents.

f. If one of the whiskers in a box-and-whisker plot is clearly longer, the distribution is usually skewed in the direction of the longer whisker. Based on this observation, does your box-and-whisker plot in part (e) indicate that the distribution is skewed to the right or skewed to the left?

g. Is the shape of the distribution of scores shown by the given histogram consistent with your observation in part (f)?

Writing in Mathematics

75. What is a symmetric histogram?

76. Describe the normal distribution and discuss some of its properties.

77. Describe the 68–95–99.7 Rule.

78. Describe how to determine the z-score for a data item in a normal distribution.

79. What does a z-score measure?

80. Give an example of both a commonly occurring and an infrequently occurring z-score. Explain how you arrived at these examples.

81. Describe when a z-score is negative.

82. If you score in the 83rd percentile, what does this mean?

83. If your weight is in the third quartile, what does this mean?

84. Two students have scores with the same percentile, but for different administrations of the SAT. Does this mean that the students have the same score on the SAT? Explain your answer.

85. Give an example of a phenomenon that is normally distributed. Explain why. (Try to be creative and not use one of the distributions discussed in this section.) Estimate what the mean and the standard deviation might be and describe how you determined these estimates.

86. Give an example of a phenomenon that is not normally distributed and explain why.

Critical Thinking Exercises

Make Sense? *In Exercises 87–90, determine whether each statement makes sense or does not make sense, and explain your reasoning.*

87. The heights of the men on our college basketball team are normally distributed with a mean of 6 feet 3 inches and a standard deviation of 1 foot 2 inches.

88. I scored in the 50th percentile on a standardized test, so my score is the median.

89. A poll administered to a random sample of 1150 voters shows 51% in favor of candidate A, so I'm 95% confident that candidate A will win the election.

90. My math teacher gave a very difficult exam for which the distribution of scores was skewed to the right.

Group Exercise

91. For this activity, group members will conduct interviews with a random sample of students on campus. Each student is to be asked, "What is the worst thing about being a student?" One response should be recorded for each student.

a. Each member should interview enough students so that there are at least 50 randomly selected students in the sample.

b. After all responses have been recorded, the group should organize the four most common answers. For each answer, compute the percentage of students in the sample who felt that this is the worst thing about being a student.

c. Find the margin of error for your survey.

d. For each of the four most common answers, write a statement about the percentage of all students on your campus who feel that this is the worst thing about being a student.

12.5

Problem Solving with the Normal Distribution

WHAT AM I SUPPOSED TO LEARN?

After you have read this section, you should be able to:

1 Solve applied problems involving normal distributions.

WE HAVE SEEN THAT MALE HEIGHTS IN NORTH AMERICA ARE approximately normally distributed with a mean of 70 inches and a standard deviation of 4 inches. Suppose we are interested in the percentage of men with heights below 80 inches:

$$z_{80} = \frac{\text{data item} - \text{mean}}{\text{standard deviation}} = \frac{80 - 70}{4} = \frac{10}{4} = 2.5.$$

Because this z-score is not an integer, the 68–95–99.7 Rule is not helpful in finding the percentage of data items that fall below 2.5 standard deviations of the mean. In this section, we will use a table that contains numerous z-scores and their percentiles to solve a variety of problems involving the normal distribution.

1 Solve applied problems involving normal distributions.

Problem Solving Using z-Scores and Percentiles

Table 12.17 gives a percentile interpretation for z-scores.

TABLE 12.17 z-Scores and Percentiles

z-Score	Percentile	z-Score	Percentile	z-Score	Percentile	z-Score	Percentile
−4.0	0.003	−1.0	15.87	0.0	50.00	1.1	86.43
−3.5	0.02	−0.95	17.11	0.05	51.99	1.2	88.49
−3.0	0.13	−0.90	18.41	0.10	53.98	1.3	90.32
−2.9	0.19	−0.85	19.77	0.15	55.96	1.4	91.92
−2.8	0.26	−0.80	21.19	0.20	57.93	1.5	93.32
−2.7	0.35	−0.75	22.66	0.25	59.87	1.6	94.52
−2.6	0.47	−0.70	24.20	0.30	61.79	1.7	95.54
−2.5	0.62	−0.65	25.78	0.35	63.68	1.8	96.41
−2.4	0.82	−0.60	27.43	0.40	65.54	1.9	97.13
−2.3	1.07	−0.55	29.12	0.45	67.36	2.0	97.72
−2.2	1.39	−0.50	30.85	0.50	69.15	2.1	98.21
−2.1	1.79	−0.45	32.64	0.55	70.88	2.2	98.61
−2.0	2.28	−0.40	34.46	0.60	72.57	2.3	98.93
−1.9	2.87	−0.35	36.32	0.65	74.22	2.4	99.18
−1.8	3.59	−0.30	38.21	0.70	75.80	2.5	99.38
−1.7	4.46	−0.25	40.13	0.75	77.34	2.6	99.53
−1.6	5.48	−0.20	42.07	0.80	78.81	2.7	99.65
−1.5	6.68	−0.15	44.04	0.85	80.23	2.8	99.74
−1.4	8.08	−0.10	46.02	0.90	81.59	2.9	99.81
−1.3	9.68	−0.05	48.01	0.95	82.89	3.0	99.87
−1.2	11.51	0.0	50.00	1.0	84.13	3.5	99.98
−1.1	13.57					4.0	99.997

TWO ENTRIES FROM TABLE 12.17

z-Score	Percentile
2.5	99.38
0.0	50.00

The portion of the table in the margin indicates that the corresponding percentile for a z-score of 2.5 is 99.38. This tells us that 99.38% of North American men have heights that are less than 80 inches, or $z = 2.5$.

In a normal distribution, the mean, median, and mode all have a corresponding z-score of 0. **Table 12.17** shows that the percentile for a z-score of 0 is 50.00. Thus,

50% of the data items in a normal distribution are less than the mean, median, and mode. Consequently, 50% of the data items are greater than or equal to the mean, median, and mode.

Table 12.17 can be used to find the percentage of data items that are less than any data item in a normal distribution. Begin by converting the data item to a z-score. Then, use the table to find the percentile for this z-score. This percentile is the percentage of data items that are less than the data item in question.

EXAMPLE 1 *Finding the Percentage of Data Items Less Than a Given Data Item*

According to the Department of Health and Education, cholesterol levels are normally distributed. For men between 18 and 24 years, the mean is 178.1 (measured in milligrams per 100 milliliters) and the standard deviation is 40.7. What percentage of men in this age range have a cholesterol level less than 239.15?

SOLUTION

If you are familiar with your own cholesterol level, you probably recognize that a level of 239.15 is fairly high for a young man. Because of this, we would expect most young men to have a level less than 239.15. Let's see if this is so. **Table 12.17** requires that we use z-scores. We compute the z-score for a 239.15 cholesterol level by using the z-score formula.

$$z_{239.15} = \frac{\text{data item} - \text{mean}}{\text{standard deviation}} = \frac{239.15 - 178.1}{40.7} = \frac{61.05}{40.7} = 1.5$$

A PORTION OF TABLE 12.17

z-Score	Percentile
1.4	91.92
1.5	93.32
1.6	94.52

A man between 18 and 24 with a 239.15 cholesterol level is 1.5 standard deviations above the mean, illustrated in **Figure 12.23(a)**. The question mark indicates that we must find the percentage of men with a cholesterol level less than $z = 1.5$, the z-score for a 239.15 cholesterol level. **Table 12.17** gives this percentage as a percentile. Find 1.5 in the z-score column in the right portion of the table. The percentile given to the right of 1.5 is 93.32. Thus, 93.32% of men between 18 and 24 have a cholesterol level less than 239.15, shown in **Figure 12.23(b)**.

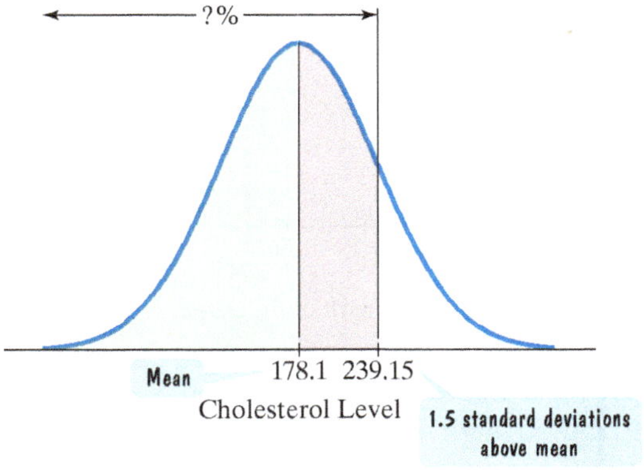

FIGURE 12.23(a)

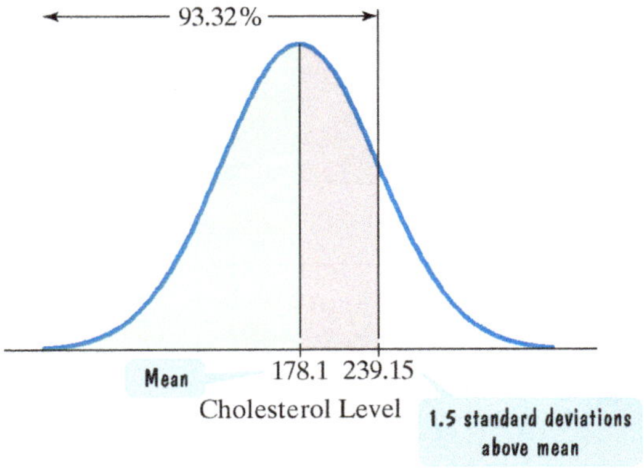

FIGURE 12.23(b)

CHECK POINT 1 The distribution of monthly charges for cellphone plans in the United States is approximately normal with a mean of \$62 and a standard deviation of \$18. What percentage of plans have charges that are less than \$83.60?

The normal distribution accounts for all data items, meaning 100% of the scores. This means that **Table 12.17** can also be used to find the percentage of data items that are greater than any data item in a normal distribution. Use the percentile in the table to determine the percentage of data items less than the data item in question. Then subtract this percentage from 100% to find the percentage of data items greater than the item in question. In using this technique, we will treat the phrases "greater than" and "greater than or equal to" as equivalent.

EXAMPLE 2 *Finding the Percentage of Data Items Greater Than a Given Data Item*

Lengths of pregnancies of women are normally distributed with a mean of 266 days and a standard deviation of 16 days. What percentage of children are born from pregnancies lasting more than 274 days?

SOLUTION

A PORTION OF TABLE 12.17

z-Score	Percentile
0.45	67.36
0.50	69.15
0.55	70.88

Table 12.17 requires that we use z-scores. We compute the z-score for a 274-day pregnancy by using the z-score formula.

$$z_{274} = \frac{\text{data item} - \text{mean}}{\text{standard deviation}} = \frac{274 - 266}{16} = \frac{8}{16} = 0.5$$

A 274-day pregnancy is 0.5 standard deviation above the mean. **Table 12.17** gives the percentile corresponding to 0.50 as 69.15. This means that 69.15% of pregnancies last less than 274 days, illustrated in **Figure 12.24**. We must find the percentage of pregnancies lasting more than 274 days by subtracting 69.15% from 100%.

$$100\% - 69.15\% = 30.85\%$$

Thus, 30.85% of children are born from pregnancies lasting more than 274 days.

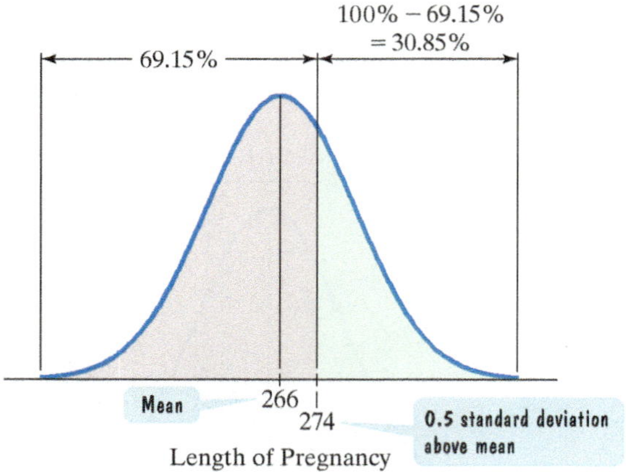

FIGURE 12.24

CHECK POINT 2 Female adult heights in North America are approximately normally distributed with a mean of 65 inches and a standard deviation of 3.5 inches. What percentage of North American women have heights that exceed 69.9 inches?

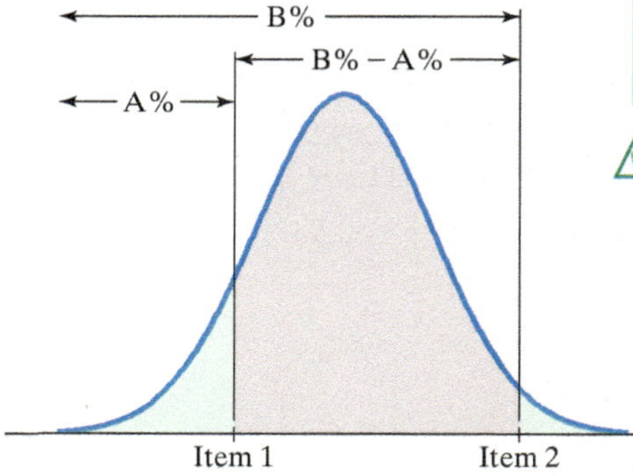

FIGURE 12.25 The percentile for data item 1 is A. The percentile for data item 2 is B. The percentage of data items between item 1 and item 2 is B% − A%.

We have seen how **Table 12.17** is used to find the percentage of data items that are less than or greater than any given item. The table can also be used to find the percentage of data items *between* two given items. Because the percentile for each item is the percentage of data items less than the given item, the percentage of data between the two given items is found by subtracting the lesser percent from the greater percent. This is illustrated in **Figure 12.25**.

FINDING THE PERCENTAGE OF DATA ITEMS BETWEEN TWO GIVEN ITEMS IN A NORMAL DISTRIBUTION

1. Convert each given data item to a z-score:

$$z = \frac{\text{data item} - \text{mean}}{\text{standard deviation}}.$$

2. Use **Table 12.17** to find the percentile corresponding to each z-score in step 1.
3. Subtract the lesser percentile from the greater percentile and attach a % sign.

EXAMPLE 3 *Finding the Percentage of Data Items between Two Given Data Items*

A PORTION OF TABLE 12.17

z-Score	Percentile
−0.55	29.12
−0.50	30.85
−0.45	32.64

A PORTION OF TABLE 12.17

z-Score	Percentile
2.4	99.18
2.5	99.38
2.6	99.53

The amount of time that self-employed Americans work each week is normally distributed with a mean of 44.6 hours and a standard deviation of 14.4 hours. What percentage of self-employed individuals in the United States work between 37.4 and 80.6 hours per week?

SOLUTION

Step 1 Convert each given data item to a z-score.

$$z_{37.4} = \frac{\text{data item} - \text{mean}}{\text{standard deviation}} = \frac{37.4 - 44.6}{14.4} = \frac{-7.2}{14.4} = -0.5$$

$$z_{80.6} = \frac{\text{data item} - \text{mean}}{\text{standard deviation}} = \frac{80.6 - 44.6}{14.4} = \frac{36}{14.4} = 2.5$$

Step 2 Use Table 12.17 to find the percentile corresponding to these z-scores. The percentile given to the right of −0.50 is 30.85. This means that 30.85% of self-employed Americans work less than 37.4 hours per week.

Table 12.17 also gives the percentile corresponding to $z = 2.5$. Find 2.5 in the z-score column in the far-right portion of the table. The percentile given to the right of 2.5 is 99.38. This means that 99.38% of self-employed Americans work less than 80.6 hours per week.

Step 3 Subtract the lesser percentile from the greater percentile and attach a % sign. Subtracting percentiles, we obtain

$$99.38 - 30.85 = 68.53.$$

Thus, 68.53% of self-employed Americans work between 37.4 and 80.6 hours per week. The solution is illustrated in **Figure 12.26**.

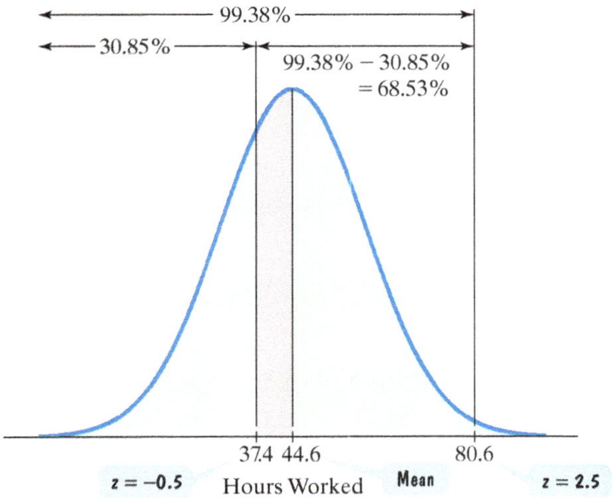

FIGURE 12.26

CHECK POINT 3 The distribution for the life of refrigerators is approximately normal with a mean of 14 years and a standard deviation of 2.5 years. What percentage of refrigerators have lives between 11 years and 18 years?

Our work in Examples 1 through 3 is summarized as follows:

COMPUTING PERCENTAGE OF DATA ITEMS FOR NORMAL DISTRIBUTIONS

Description of Percentage	Graph	Computation of Percentage
Percentage of data items less than a given data item with $z = b$	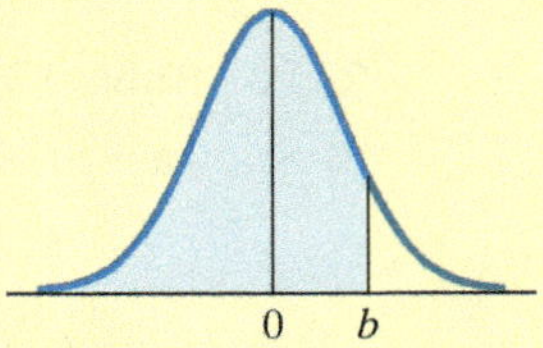	Use the table percentile for $z = b$ and add a % sign.
Percentage of data items greater than a given data item with $z = a$	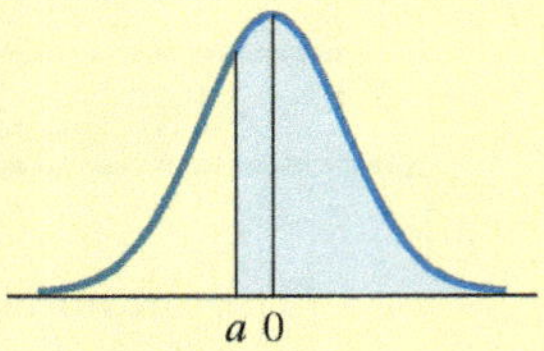	Subtract the table percentile for $z = a$ from 100 and add a % sign.
Percentage of data items between two given data items with $z = a$ and $z = b$	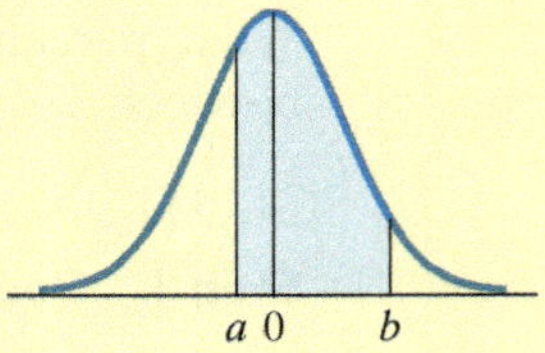	Subtract the table percentile for $z = a$ from the table percentile for $z = b$ and add a % sign.

Concept and Vocabulary Check

Use the information shown below to fill in each blank so that the resulting statement is true.

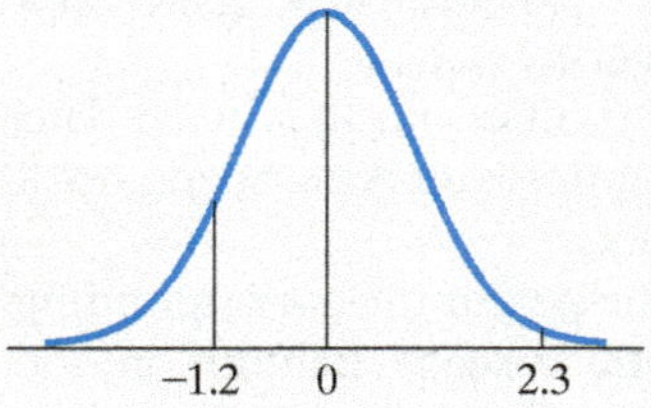

z-Score	Percentile
−1.2	11.51
2.3	98.93

1. The percentage of scores less than $z = 2.3$ is __________.
2. The percentage of scores greater than $z = 2.3$ is __________.
3. The percentage of scores greater than $z = -1.2$ is __________.
4. The percentage of scores between $z = -1.2$ and $z = 2.3$ is __________.
5. True or False: The 68–95–99.7 Rule cannot be used if z-scores are not integers. ______

Exercise Set 12.5

Practice and Application Exercises

Use **Table 12.17** *on page 816 to solve Exercises 1–16.*

In Exercises 1–8, find the percentage of data items in a normal distribution that lie **a.** *below and* **b.** *above the given z-score.*

1. $z = 0.6$ **2.** $z = 0.8$

3. $z = 1.2$ **4.** $z = 1.4$

5. $z = -0.7$ **6.** $z = -0.4$

7. $z = -1.2$ **8.** $z = -1.8$

In Exercises 9–16, find the percentage of data items in a normal distribution that lie between

9. $z = 0.2$ and $z = 1.4$. **10.** $z = 0.3$ and $z = 2.1$.

11. $z = 1$ and $z = 3$. **12.** $z = 2$ and $z = 3$.

13. $z = -1.5$ and $z = 1.5$. **14.** $z = -1.2$ and $z = 1.2$.

15. $z = -2$ and $z = -0.5$. **16.** $z = -2.2$ and $z = -0.3$.

Systolic blood pressure readings are normally distributed with a mean of 121 and a standard deviation of 15. (A reading above 140 is considered to be high blood pressure.) In Exercises 17–26, begin by converting any given blood pressure reading or readings into z-scores. Then use **Table 12.17** *on page 816 to find the percentage of people with blood pressure readings*

17. below 142. **18.** below 148.

19. above 130. **20.** above 133.

21. above 103. **22.** above 100.

23. between 142 and 154. **24.** between 145 and 157.

25. between 112 and 130. **26.** between 109 and 133.

The weights for 12-month-old baby boys are normally distributed with a mean of 22.5 pounds and a standard deviation of 2.2 pounds. In Exercises 27–30, use **Table 12.17** *on page 816 to find the percentage of 12-month-old baby boys who weigh*

27. more than 25.8 pounds.

28. more than 23.6 pounds.

29. between 19.2 and 21.4 pounds.

30. between 18.1 and 19.2 pounds.

Practice Plus

The table shows selected ages of licensed drivers in the United States and the corresponding percentiles.

AGES OF U.S. DRIVERS

Age	Percentile
75	98
65	88
55	77
45	60
35	37
25	14
20	5

Source: Department of Transportation

In Exercises 31–36, use the information given by the table to find the percentage of U.S. drivers who are

31. younger than 55.

32. younger than 45.

33. at least 25.

34. at least 35.

35. at least 65 and younger than 75.

36. at least 20 and younger than 65.

Writing in Mathematics

37. Explain when it is necessary to use a table showing z-scores and percentiles rather than the 68–95–99.7 Rule to determine the percentage of data items less than a given data item.

38. Explain how to use a table showing z-scores and percentiles to determine the percentage of data items between two z-scores.

Critical Thinking Exercises

Make Sense? *In Exercises 39–42, determine whether each statement makes sense or does not make sense, and explain your reasoning.*

39. I'm using a table showing z-scores and percentiles that has positive percentiles corresponding to positive z-scores and negative percentiles corresponding to negative z-scores.

40. My table showing z-scores and percentiles displays the percentage of data items less than a given value of z.

41. My table showing z-scores and percentiles does not display the percentage of data items greater than a given value of z.

42. I can use a table showing z-scores and percentiles to verify the three approximate numbers given by the 68–95–99.7 Rule.

43. Find two z-scores so that 40% of the data in the distribution lies between them. (More than one answer is possible.)

44. A woman insists that she will never marry a man as short or shorter than she, knowing that only one man in 400 falls into this category. Assuming a mean height of 69 inches for men with a standard deviation of 2.5 inches (and a normal distribution), approximately how tall is the woman?

45. The placement test for a college has scores that are normally distributed with a mean of 500 and a standard deviation of 100. If the college accepts only the top 10% of examinees, what is the cutoff score on the test for admission?

12.6 Scatter Plots, Correlation, and Regression Lines

WHAT AM I SUPPOSED TO LEARN?

After you have read this section, you should be able to:

1. Make a scatter plot for a table of data items.
2. Interpret information given in a scatter plot.
3. Compute the correlation coefficient.
4. Write the equation of the regression line.
5. Use a sample's correlation coefficient to determine whether there is a correlation in the population.

THESE PHOTOS OF PRESIDENTIAL PUFFING INDICATE that the White House was not always a no-smoking zone. According to *Cigar Aficionado*, nearly half of U.S. presidents have had a nicotine habit, from cigarettes to pipes to cigars. Franklin Roosevelt's stylish way with a cigarette holder was part of his mystique. Although Dwight Eisenhower quit his wartime four-pack-a-day habit before taking office, smoking in the residence was still common, with ashtrays on the tables at state dinners and free cigarettes for guests. In 1993, Hillary Clinton banned smoking in the White House, although Bill Clinton's cigars later made a sordid cameo in the Lewinsky scandal. Barack Obama quit smoking before entering the White House, but had "fallen off the wagon occasionally" as he admitted in a *Meet the Press* interview.

Changing attitudes toward smoking, both inside and outside the White House, date back to 1964 and an equation in two variables. To understand the mathematics behind this turning point in public health, we need to explore situations involving data collected on two variables.

Up to this point in the chapter, we have studied situations in which data sets involve a single variable, such as height, weight, cholesterol level, and length of pregnancies. By contrast, the 1964 study involved data collected on two variables from 11 countries—annual cigarette consumption for each adult male and deaths per million males from lung cancer. In this section, we consider situations in which there are two data items for each randomly selected person or thing. Our interest is in determining whether or not there is a relationship between the two variables and, if so, the strength of that relationship.

Make a scatter plot for a table of data items.

Scatter Plots and Correlation

Is there a relationship between education and prejudice? With increased education, does a person's level of prejudice tend to decrease? Notice that we are interested in two quantities—years of education and level of prejudice. For each person in our sample, we will record the number of years of school completed and the score on a test measuring prejudice. Higher scores on this 1-to-10 test indicate greater prejudice. Using x to represent years of education and y to represent scores on a test measuring prejudice, **Table 12.18** shows these two quantities for a random sample of ten people.

TABLE 12.18 Recording Two Quantities in a Sample of Ten People

Respondent	A	B	C	D	E	F	G	H	I	J
Years of education (x)	12	5	14	13	8	10	16	11	12	4
Score on prejudice test (y)	1	7	2	3	5	4	1	2	3	10

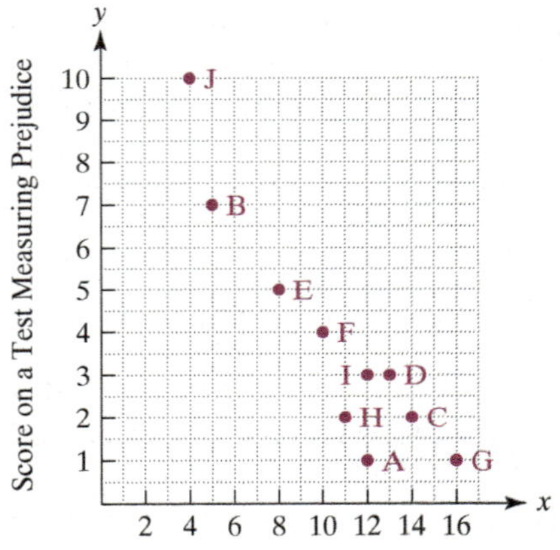

FIGURE 12.27 A scatter plot for education-prejudice data

When two data items are collected for every person or object in a sample, the data items can be visually displayed using a *scatter plot*. A **scatter plot** is a collection of data points, one data point per person or object. We can make a scatter plot of the data in **Table 12.18** by drawing a horizontal axis to represent years of education and a vertical axis to represent scores on a test measuring prejudice. We then represent each of the ten respondents with a single point on the graph. For example, the dot for respondent A is located to represent 12 years of education on the horizontal axis and 1 on the prejudice test on the vertical axis. Plotting each of the ten pieces of data in a rectangular coordinate system results in the scatter plot shown in **Figure 12.27**.

A scatter plot like the one in **Figure 12.27** can be used to determine whether two quantities are related. If there is a clear relationship, the quantities are said to be **correlated**. The scatter plot shows a downward trend among the data points, although there are a few exceptions. People with increased education tend to have a lower score on the test measuring prejudice. **Correlation** is used to determine if there is a relationship between two variables and, if so, the strength and direction of that relationship.

Correlation and Causal Connections

Correlations can often be seen when data items are displayed on a scatter plot. Although the scatter plot in **Figure 12.27** indicates a correlation between education and prejudice, we cannot conclude that increased education causes a person's level of prejudice to decrease. There are at least three possible explanations:

1. The correlation between increased education and decreased prejudice is simply a coincidence.
2. Education usually involves classrooms with a variety of different kinds of people. Increased exposure to diversity in the classroom setting, which accompanies increased levels of education, might be an underlying cause for decreased prejudice.
3. Education, the process of acquiring knowledge, requires people to look at new ideas and see things in different ways. Thus, education causes one to be more tolerant and less prejudiced.

This list represents three possibilities. Perhaps you can provide a better explanation about decreasing prejudice with increased education.

Establishing that one thing causes another is extremely difficult, even if there is a strong correlation between these things. For example, as the air temperature increases, there is an increase in the number of people stung by jellyfish at the beach. This does not mean that an increase in air temperature causes more people to be stung. It might mean that because it is hotter, more people go into the water. With an increased number of swimmers, more people are likely to be stung. In short, correlation is not necessarily causation.

2 Interpret information given in a scatter plot.

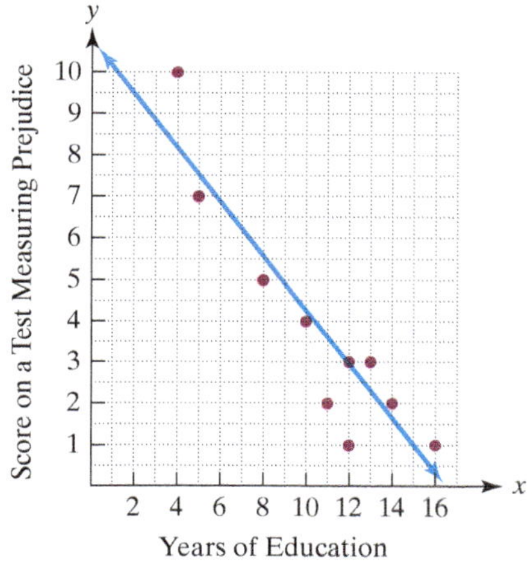

FIGURE 12.28 A scatter plot with a regression line

Regression Lines and Correlation Coefficients

Figure 12.28 shows the scatter plot for the education-prejudice data. Also shown is a straight line that seems to approximately "fit" the data points. Most of the data points lie either near or on this line. A line that best fits the data points in a scatter plot is called a **regression line**. The regression line is the particular line in which the spread of the data points around it is as small as possible.

A measure that is used to describe the strength and direction of a relationship between variables whose data points lie on or near a line is called the **correlation coefficient,** designated by r. **Figure 12.29** shows scatter plots and correlation coefficients. Variables are **positively correlated** if they tend to increase or decrease together, as in **Figure 12.29(a)**, **(b)**, and **(c)**. By contrast, variables are **negatively correlated** if one variable tends to decrease while the other increases, as in **Figure 12.29(e)**, **(f)**, and **(g)**. **Figure 12.29** illustrates that a correlation coefficient, r, is a number between -1 and 1, inclusive. **Figure 12.29(a)** shows a value of 1. This indicates a **perfect positive correlation** in which all points in the scatter plot lie precisely on the regression line that rises from left to right. **Figure 12.29(g)** shows a value of -1. This indicates a **perfect negative correlation** in which all points in the scatter plot lie precisely on the regression line that falls from left to right.

Take another look at **Figure 12.29**. If r is between 0 and 1, as in **(b)** and **(c)**, the two variables are positively correlated, but not perfectly. Although all the data points will not lie on the regression line, as in **(a)**, an increase in one variable tends to be accompanied by an increase in the other. Negative correlations are also illustrated in **Figure 12.29**. If r is between 0 and -1, as in **(e)** and **(f)**, the two variables are negatively correlated, but not perfectly. Although all the data points will not lie on the regression line, as in **(g)**, an increase in one variable tends to be accompanied by a decrease in the other.

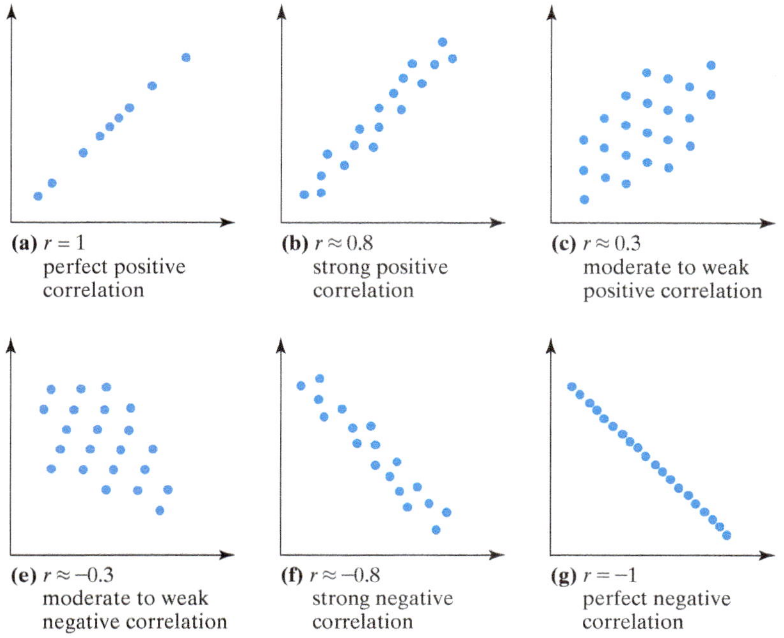

FIGURE 12.29 Scatter plots and correlation coefficients

Blitzer Bonus

Beneficial Uses of Correlation Coefficients

- A Florida study showed a high positive correlation between the number of powerboats and the number of manatee deaths. Many of these deaths were seen to be caused by boats' propellers gashing into the manatees' bodies. Based on this study, Florida set up coastal sanctuaries where powerboats are prohibited so that these large gentle mammals that float just below the water's surface could thrive.
- Researchers studied how psychiatric patients readjusted to their community after their release from a mental hospital. A moderate positive correlation ($r = 0.38$) was found between patients' attractiveness and their postdischarge social adjustment. The better-looking patients were better off. The researchers suggested that physical attractiveness plays a role in patients' readjustment to community living because good-looking people tend to be treated better by others than homely people are.

EXAMPLE 1 *Interpreting a Correlation Coefficient*

In a 1971 study involving 232 subjects, researchers found a relationship between the subjects' level of stress and how often they became ill. The correlation coefficient in this study was 0.32. Does this indicate a strong relationship between stress and illness?

SOLUTION

The correlation coefficient $r = 0.32$ means that as stress increases, frequency of illness also tends to increase. However, 0.32 is only a moderate correlation, illustrated in **Figure 12.29(c)** on the previous page. There is not, based on this study, a strong relationship between stress and illness. In this study, the relationship is somewhat weak.

CHECK POINT 1 In a 1996 study involving obesity in mothers and daughters, researchers found a relationship between a high body-mass index for the girls and their mothers. (Body-mass index is a measure of weight relative to height. People with a high body-mass index are overweight or obese.) The correlation coefficient in this study was 0.51. Does this indicate a weak relationship between the body-mass index of daughters and the body-mass index of their mothers?

How to Obtain the Correlation Coefficient and the Equation of the Regression Line

The easiest way to find the correlation coefficient and the equation of the regression line is to use a graphing or statistical calculator. Graphing calculators have statistical menus that enable you to enter the x and y data items for the variables. Based on this information, you can instruct the calculator to display a scatter plot, the equation of the regression line, and the correlation coefficient.

We can also compute the correlation coefficient and the equation of the regression line by hand using formulas. First, we compute the correlation coefficient.

COMPUTING THE CORRELATION COEFFICIENT BY HAND

The following formula is used to calculate the correlation coefficient, r:

$$r = \frac{n(\Sigma xy) - (\Sigma x)(\Sigma y)}{\sqrt{n(\Sigma x^2) - (\Sigma x)^2}\sqrt{n(\Sigma y^2) - (\Sigma y)^2}}.$$

In the formula,

n = the number of data points, (x, y)
Σx = the sum of the x-values
Σy = the sum of the y-values
Σxy = the sum of the product of x and y in each pair
Σx^2 = the sum of the squares of the x-values
Σy^2 = the sum of the squares of the y-values
$(\Sigma x)^2$ = the square of the sum of the x-values
$(\Sigma y)^2$ = the square of the sum of the y-values

When computing the correlation coefficient by hand, organize your work in five columns:

x	y	xy	x^2	y^2

Find the sum of the numbers in each column. Then, substitute these values into the formula for r. Example 2 illustrates computing the correlation coefficient for the education-prejudice test data.

3 Compute the correlation coefficient.

EXAMPLE 2 *Computing the Correlation Coefficient*

Shown below are the data involving the number of years of school, x, completed by ten randomly selected people and their scores on a test measuring prejudice, y. Recall that higher scores on the measure of prejudice (1 to 10) indicate greater levels of prejudice. Determine the correlation coefficient between years of education and scores on a prejudice test.

Respondent	**A**	**B**	**C**	**D**	**E**	**F**	**G**	**H**	**I**	**J**
Years of education (x)	12	5	14	13	8	10	16	11	12	4
Score on prejudice test (y)	1	7	2	3	5	4	1	2	3	10

SOLUTION

As suggested, organize the work in five columns.

x	y	xy	x^2	y^2
12	1	12	144	1
5	7	35	25	49
14	2	28	196	4
13	3	39	169	9
8	5	40	64	25
10	4	40	100	16
16	1	16	256	1
11	2	22	121	4
12	3	36	144	9
4	10	40	16	100
$\Sigma x = 105$	$\Sigma y = 38$	$\Sigma xy = 308$	$\Sigma x^2 = 1235$	$\Sigma y^2 = 218$
Add all values in the x-column.	Add all values in the y-column.	Add all values in the xy-column.	Add all values in the x^2-column.	Add all values in the y^2-column.

We use these five sums to calculate the correlation coefficient.

Another value in the formula for r that we have not yet determined is n, the number of data points (x, y). Because there are ten items in the x-column and ten items in the y-column, the number of data points (x, y) is ten. Thus, $n = 10$.

In order to calculate r, we also need to find the square of the sum of the x-values and the y-values:

$$(\Sigma x)^2 = (105)^2 = 11{,}025 \quad \text{and} \quad (\Sigma y)^2 = (38)^2 = 1444.$$

TECHNOLOGY

Graphing Calculators, Scatter Plots, and Regression Lines

You can use a graphing calculator to display a scatter plot and the regression line. After entering the x and y data items for years of education and scores on a prejudice test, the calculator shows the scatter plot of the data and the regression line.

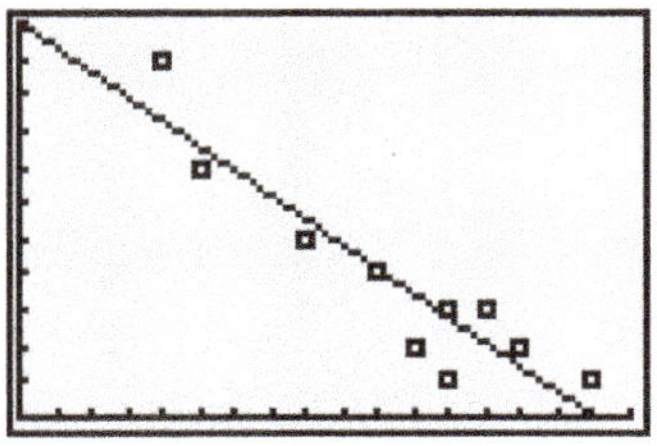

Also displayed below is the regression line's equation and the correlation coefficient, r. The slope shown below is approximately -0.69. The negative slope reinforces the fact that there is a negative correlation between the variables in Example 2.

We are ready to determine the value for r. We use the sums obtained on the previous page, with $n = 10$

$$r = \frac{n(\Sigma xy) - (\Sigma x)(\Sigma y)}{\sqrt{n(\Sigma x^2) - (\Sigma x)^2}\sqrt{n(\Sigma y^2) - (\Sigma y)^2}}$$

$$= \frac{10(308) - 105(38)}{\sqrt{10(1235) - 11{,}025}\sqrt{10(218) - 1444}}$$

$$= \frac{-910}{\sqrt{1325}\sqrt{736}}$$

$$\approx -0.92$$

The value for r, approximately -0.92, is fairly close to -1 and indicates a strong negative correlation. This means that the more education a person has, the less prejudiced that person is (based on scores on the test measuring levels of prejudice).

CHECK POINT 2 The points in the scatter plot in **Figure 12.30** show the number of firearms per 100 persons and the number of deaths per 100,000 persons for the ten industrialized countries with the highest death rates. Use the data displayed by the voice balloons to determine the correlation coefficient between these variables. Round to two decimal places. What does the correlation coefficient indicate about the strength and direction of the relationship between firearms per 100 persons and deaths per 100,000 persons?

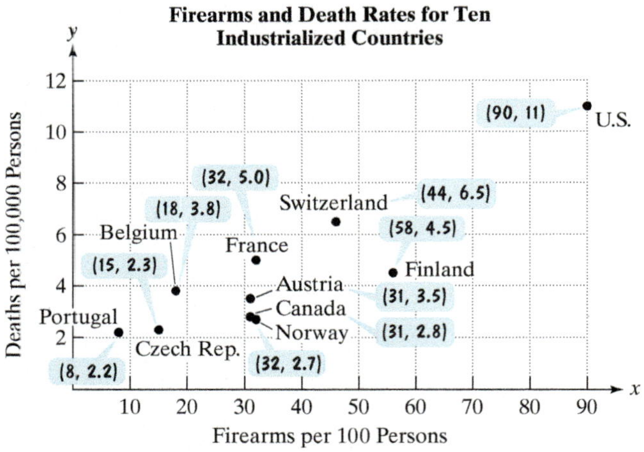

FIGURE 12.30

Source: International Action Network on Small Arms

Once we have determined that two variables are related, we can use the equation of the regression line to determine the exact relationship. Here is the formula for writing the equation of the line that best fits the data:

4 Write the equation of the regression line.

WRITING THE EQUATION OF THE REGRESSION LINE BY HAND

The equation of the regression line is

$$y = mx + b,$$

where

$$m = \frac{n(\Sigma xy) - (\Sigma x)(\Sigma y)}{n(\Sigma x^2) - (\Sigma x)^2} \quad \text{and} \quad b = \frac{\Sigma y - m(\Sigma x)}{n}.$$

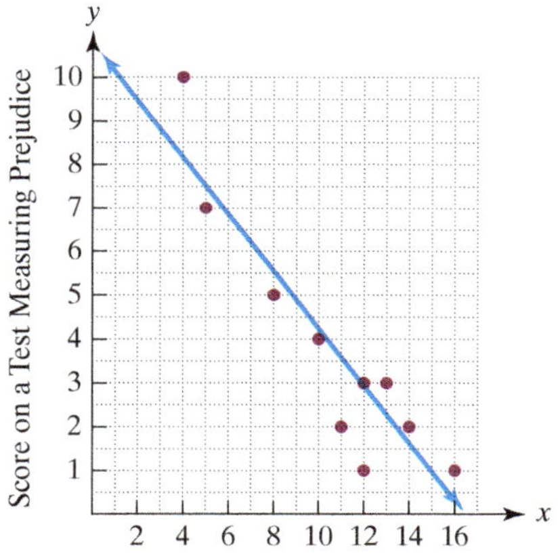

FIGURE 12.28 (repeated)

GREAT QUESTION!

Why is $b \approx 11.05$ in Example 3, but $b \approx 11.01$ in the Technology box on page 825?

In Example 3, we rounded the value of m when we calculated b. The value of b on the calculator screen on page 825 is more accurate.

EXAMPLE 3 *Writing the Equation of the Regression Line*

a. Shown, again, in **Figure 12.28** is the scatter plot and the regression line for the data in Example 2. Use the data to find the equation of the regression line that relates years of education and scores on a prejudice test.

b. Approximately what score on the test can be anticipated by a person with nine years of education?

SOLUTION

a. We use the sums obtained in Example 2. We begin by computing m.

$$m = \frac{n(\Sigma xy) - (\Sigma x)(\Sigma y)}{n(\Sigma x^2) - (\Sigma x)^2} = \frac{10(308) - 105(38)}{10(1235) - (105)^2} = \frac{-910}{1325} \approx -0.69$$

With a negative correlation coefficient, it makes sense that the slope of the regression line is negative. This line falls from left to right, indicating a negative correlation.

Now, we find the y-intercept, b.

$$b = \frac{\Sigma y - m(\Sigma x)}{n} = \frac{38 - (-0.69)(105)}{10} = \frac{110.45}{10} \approx 11.05$$

Using $m \approx -0.69$ and $b \approx 11.05$, the equation of the regression line, $y = mx + b$, is

$$y = -0.69x + 11.05,$$

where x represents the number of years of education and y represents the score on the prejudice test.

b. To anticipate the score on the prejudice test for a person with nine years of education, substitute 9 for x in the regression line's equation.

$$y = -0.69x + 11.05$$
$$y = -0.69(9) + 11.05 = 4.84$$

A person with nine years of education is anticipated to have a score close to 5 on the prejudice test.

CHECK POINT 3 Use the data in **Figure 12.30** of Check Point 2 on page 826 to find the equation of the regression line. Round m and b to one decimal place. Then use the equation to project the number of deaths per 100,000 persons in a country with 80 firearms per 100 persons.

5 Use a sample's correlation coefficient to determine whether there is a correlation in the population.

The Level of Significance of r

In Example 2, we found a strong negative correlation between education and prejudice, computing the correlation coefficient, r, to be -0.92. However, the sample size ($n = 10$) was relatively small. With such a small sample, can we truly conclude that a correlation exists in the population? Or could it be that education and prejudice are not related? Perhaps the results we obtained were simply due to sampling error and chance.

Mathematicians have identified values to determine whether r, the correlation coefficient for a sample, can be attributed to a relationship between variables in the population. These values are shown in the second and third columns of

TABLE 12.19 Values for Determining Correlations in a Population

n	$\alpha = 0.05$	$\alpha = 0.01$
4	0.950	0.990
5	0.878	0.959
6	0.811	0.917
7	0.754	0.875
8	0.707	0.834
9	0.666	0.798
10	0.632	0.765
11	0.602	0.735
12	0.576	0.708
13	0.553	0.684
14	0.532	0.661
15	0.514	0.641
16	0.497	0.623
17	0.482	0.606
18	0.468	0.590
19	0.456	0.575
20	0.444	0.561
22	0.423	0.537
27	0.381	0.487
32	0.349	0.449
37	0.325	0.418
42	0.304	0.393
47	0.288	0.372
52	0.273	0.354
62	0.250	0.325
72	0.232	0.302
82	0.217	0.283
92	0.205	0.267
102	0.195	0.254

The larger the sample size, n, the smaller is the value of r needed for a correlation in the population.

Table 12.19. They depend on the sample size, n, listed in the left column. If $|r|$, the absolute value of the correlation coefficient computed for the sample, is greater than the value given in the table, a correlation exists between the variables in the population. The column headed $\alpha = 0.05$ denotes a **significance level of 5%**, meaning that there is a 0.05 probability that, when the statistician says the variables are correlated, they are actually not related in the population. The column on the right, headed $\alpha = 0.01$, denotes a **significance level of 1%**, meaning that there is a 0.01 probability that, when the statistician says the variables are correlated, they are actually not related in the population. Values in the $\alpha = 0.01$ column are greater than those in the $\alpha = 0.05$ column. Because of the possibility of sampling error, there is always a probability that when we say the variables are related, there is actually not a correlation in the population from which the sample was randomly selected.

EXAMPLE 4 *Determining a Correlation in the Population*

In Example 2, we computed $r = -0.92$ for $n = 10$. Can we conclude that there is a negative correlation between education and prejudice in the population?

SOLUTION

Begin by taking the absolute value of the calculated correlation coefficient.

$$|r| = |-0.92| = 0.92$$

Now, look to the right of $n = 10$ in **Table 12.19**. Because 0.92 is greater than both of these values (0.632 and 0.765), we may conclude that a correlation does exist between education and prejudice in the population. (There is a probability of at most 0.01 that the variables are not really correlated in the population and our results could be attributed to chance.)

Blitzer Bonus

Cigarettes and Lung Cancer

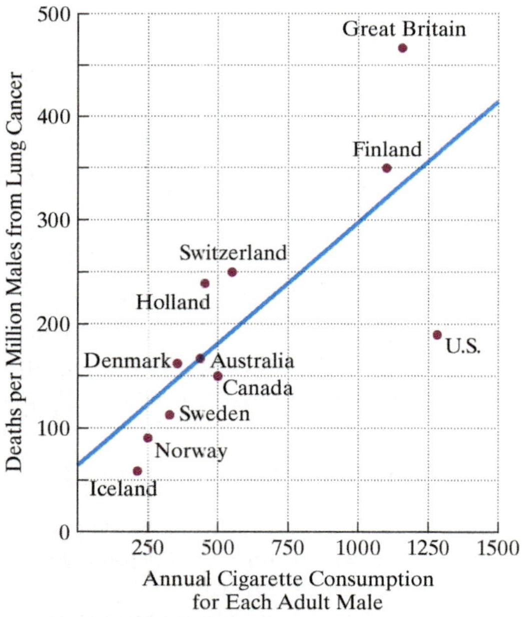

Source: Smoking and Health, Washington, D.C., 1964

This scatter plot shows a relationship between cigarette consumption among males and deaths due to lung cancer per million males. The data are from 11 countries and date back to a 1964 report by the U.S. Surgeon General. The scatter plot can be modeled by a line whose slope indicates an increasing death rate from lung cancer with increased cigarette consumption. At that time, the tobacco industry argued that in spite of this regression line, tobacco use is not the cause of cancer. Recent data do, indeed, show a causal effect between tobacco use and numerous diseases.

CHECK POINT 4 If you worked Check Point 2 correctly, you should have found that $r \approx 0.89$ for $n = 10$. Can you conclude that there is a positive correlation for all industrialized countries between firearms per 100 persons and deaths per 100,000 persons?

Concept and Vocabulary Check

Fill in each blank so that the resulting statement is true.

1. A set of points representing data is called a/an ____________.
2. The line that best fits a set of points is called a/an ____________.
3. A measure that is used to describe the strength and direction of a relationship between variables whose data points lie on or near a line is called the ________________, ranging from $r =$ ____________ to $r =$ ____________.

In Exercises 4–7, determine whether each statement is true or false. If the statement is false, make the necessary change(s) to produce a true statement.

4. If $r = 0$, there is no correlation between two variables. ______
5. If $r = 1$, changes in one variable cause changes in the other variable. ______
6. If $r = -0.1$, there is a strong negative correlation between two variables. ______
7. A significance level of 5% means that there is a 0.05 probability that when a statistician says that variables are correlated, they are actually not related in the population. ______

Exercise Set 12.6

Practice and Application Exercises

In Exercises 1–8, make a scatter plot for the given data. Use the scatter plot to describe whether or not the variables appear to be related.

1.

x	1	6	4	3	7	2
y	2	5	3	3	4	1

2.

x	2	1	6	3	4
y	4	5	10	8	9

3.

x	8	6	1	5	4	10	3
y	2	4	10	5	6	2	9

4.

x	4	5	2	1
y	1	3	5	4

5. HAMACHIPHOBIA

Generation	Percentage Who Won't Try Sushi x	Percentage Who Don't Approve of Marriage Equality y
Millennials	42	36
Gen X	52	49
Boomers	60	59
Silent/Greatest Generation	72	66

Source: Pew Research Center

6. TREASURED CHEST: FILMS OF MATTHEW MCCONAUGHEY

Film	Minutes Shirtless x	Opening Weekend Gross (millions of dollars) y
We Are Marshall	0	6.1
EDtv	0.8	8.3
Reign of Fire	1.6	15.6
Sahara	1.8	18.1
Fool's Gold	14.6	21.6

Source: Entertainment Weekly

7. TEENAGE DRUG USE

Country	Percentage Who Have Used Marijuana x	Percentage Who Have Used Other Illegal Drugs y
Czech Republic	22	4
Denmark	17	3
England	40	21
Finland	5	1
Ireland	37	16
Italy	19	8
Northern Ireland	23	14
Norway	6	3
Portugal	7	3
Scotland	53	31
United States	34	24

Source: De Veaux et.al., *Intro Stats,* Pearson, 2009.

8. LITERACY AND HUNGER

	Percentage Who Are	
	Literate	**Undernourished**
Country	x	y
Cuba	100	2
Egypt	71	4
Ethiopia	36	46
Grenada	96	7
Italy	98	2
Jamaica	80	9
Jordan	91	6
Pakistan	50	24
Russia	99	3
Togo	53	24
Uganda	67	19

Source: The Penguin State of the World Atlas, 2008

The scatter plot in the figure shows the relationship between the percentage of married women of child-bearing age using contraceptives and births per woman in selected countries. Use the scatter plot to determine whether each of the statements in Exercises 9–18 is true or false.

Contraceptive Prevalence and Average Number of Births per Woman, Selected Countries

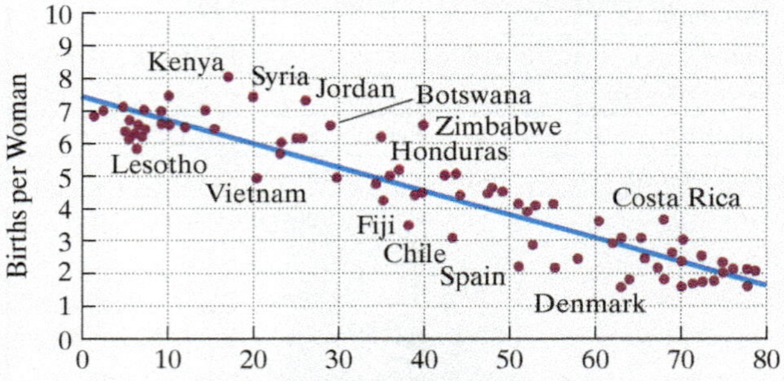

Source: Population Reference Bureau

9. There is a strong positive correlation between contraceptive use and births per woman.

10. There is no correlation between contraceptive use and births per woman.

11. There is a strong negative correlation between contraceptive use and births per woman.

12. There is a causal relationship between contraceptive use and births per woman.

13. With approximately 43% of women of child-bearing age using contraceptives, there are three births per woman in Chile.

14. With 20% of women of child-bearing age using contraceptives, there are six births per woman in Vietnam.

15. No two countries have a different number of births per woman with the same percentage of married women using contraceptives.

16. The country with the greatest number of births per woman also has the smallest percentage of women using contraceptives.

17. Most of the data points do not lie on the regression line.

18. The number of selected countries shown in the scatter plot is approximately 20.

Just as money doesn't buy happiness for individuals, the two don't necessarily go together for countries either. However, the scatter plot does show a relationship between a country's annual per capita income and the percentage of people in that country who call themselves "happy." Use the scatter plot to determine whether each of the statements in Exercises 19–26 is true or false.

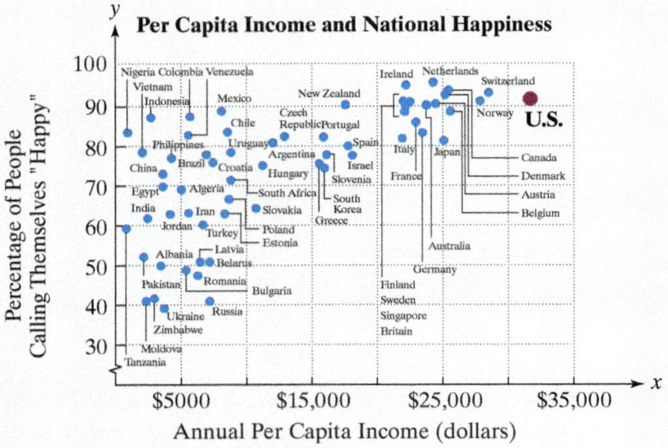

Source: Richard Layard, *Happiness: Lessons from a New Science*, Penguin, 2005

19. There is no correlation between per capita income and the percentage of people who call themselves "happy."

20. There is an almost-perfect positive correlation between per capita income and the percentage of people who call themselves "happy."

21. There is a positive correlation between per capita income and the percentage of people who call themselves "happy."

22. As per capita income decreases, the percentage of people who call themselves "happy" also tends to decrease.

23. The country with the lowest per capita income has the least percentage of people who call themselves "happy."

24. The country with the highest per capita income has the greatest percentage of people who call themselves "happy."

25. A reasonable estimate of the correlation coefficient for the data is 0.8.

26. A reasonable estimate of the correlation coefficient for the data is -0.3.

Use the scatter plots shown, labeled (a)–(f), to solve Exercises 27–30.

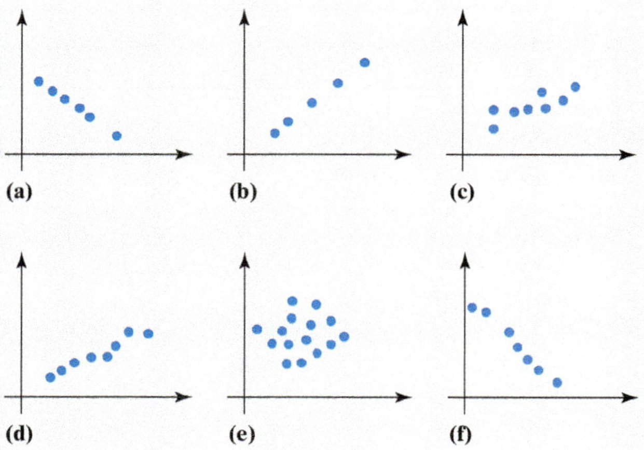

27. Which scatter plot indicates a perfect negative correlation?

28. Which scatter plot indicates a perfect positive correlation?

29. In which scatter plot is $r = 0.9$?

30. In which scatter plot is $r = 0.01$?

Compute r, the correlation coefficient, rounded to two decimal places, for the data in

31. Exercise 1. **32.** Exercise 2.

33. Exercise 3. **34.** Exercise 4.

35. Use the data in Exercise 5 to solve this exercise.

a. Determine the correlation coefficient, rounded to two decimal places, between the percentage of people who won't try sushi and the percentage who don't approve of marriage equality.

b. What explanations can you offer for the correlation coefficient in part (a)?

c. Find the equation of the regression line for the percentage who won't try sushi and the percentage who don't approve of marriage equality. Round m and b to two decimal places.

d. What percentage of people, to the nearest percent, can we anticipate do not approve of marriage equality in a generation where 30% won't try sushi?

36. Use the data in Exercise 6 to solve this exercise.

a. Determine the correlation coefficient, rounded to two decimal places, between the minutes Matthew McConaughey appeared shirtless in a film and the film's opening weekend gross.

b. Find the equation of the regression line for the minutes McConaughey appeared shirtless in a film and the film's opening weekend gross. Round m and b to two decimal places.

c. What opening weekend gross, to the nearest tenth of a million dollars, can we anticipate in a McConaughey film in which he appears shirtless for 20 minutes?

37. Use the data in Exercise 7 to solve this exercise.

a. Determine the correlation coefficient, rounded to two decimal places, between the percentage of teenagers who have used marijuana and the percentage who have used other drugs.

b. Find the equation of the regression line for the percentage of teenagers who have used marijuana and the percentage who have used other drugs. Round m and b to two decimal places.

c. What percentage of teenagers, to the nearest percent, can we anticipate using illegal drugs other than marijuana in a country where 10% of teenagers have used marijuana?

38. Use the data in Exercise 8 to solve this exercise.

a. Determine the correlation coefficient, rounded to two decimal places, between the percentage of people in a country who are literate and the percentage who are undernourished.

b. Find the equation of the regression line for the percentage who are literate and the percentage who are undernourished. Round m and b to two decimal places.

c. What percentage of people, to the nearest percent, can we anticipate are undernourished in a country where 60% of the people are literate?

In Exercises 39–45, the correlation coefficient, r, is given for a sample of n data points. Use the $\alpha = 0.05$ column in **Table 12.19** *on page 828 to determine whether or not we may conclude that a correlation does exist in the population. (Using the $\alpha = 0.05$ column, there is a probability of 0.05 that the variables are not really correlated in the population and our results could be attributed to chance. Ignore this possibility when concluding whether or not there is a correlation in the population.)*

39. $n = 20, r = 0.5$ **40.** $n = 27, r = 0.4$

41. $n = 12, r = 0.5$ **42.** $n = 22, r = 0.04$

43. $n = 72, r = -0.351$ **44.** $n = 37, r = -0.37$

45. $n = 20, r = -0.37$

46. In the 1964 study on cigarette consumption and deaths due to lung cancer (see the Blitzer Bonus on page 828), $n = 11$ and $r = 0.73$. What can you conclude using the $\alpha = 0.05$ column in **Table 12.19** on page 828?

Writing in Mathematics

47. What is a scatter plot?

48. How does a scatter plot indicate that two variables are correlated?

49. Give an example of two variables with a strong positive correlation and explain why this is so.

50. Give an example of two variables with a strong negative correlation and explain why this is so.

51. What is meant by a regression line?

52. When all points in a scatter plot fall on the regression line, what is the value of the correlation coefficient? Describe what this means.

For the pairs of quantities in Exercises 53–56, describe whether a scatter plot will show a positive correlation, a negative correlation, or no correlation. If there is a correlation, is it strong, moderate, or weak? Explain your answers.

53. Height and weight

54. Number of days absent and grade in a course

55. Height and grade in a course

56. Hours of television watched and grade in a course

57. Explain how to use the correlation coefficient for a sample to determine if there is a correlation in the population.

Critical Thinking Exercises

Make Sense? *In Exercises 58–61, determine whether each statement makes sense or does not make sense, and explain your reasoning.*

58. I found a strong positive correlation for the data in Exercise 7 relating the percentage of teenagers in various countries who have used marijuana and the percentage who have used other drugs. I concluded that using marijuana leads to the use of other drugs.

59. I found a strong negative correlation for the data in Exercise 8 relating the percentage of people in various countries who are literate and the percentage who are undernourished. I concluded that an increase in literacy causes a decrease in undernourishment.

60. I'm working with a data set for which the correlation coefficient and the slope of the regression line have opposite signs.

61. I read that there is a correlation of 0.72 between IQ scores of identical twins reared apart, so I would expect a significantly lower correlation, approximately 0.52, between IQ scores of identical twins reared together.

62. Give an example of two variables with a strong correlation, where each variable is not the cause of the other.

Technology Exercise

63. Use the linear regression feature of a graphing calculator to verify your work in any two exercises from Exercises 35–38, parts (a) and (b).

Group Exercises

64. The group should select two variables related to people on your campus that it believes have a strong positive or negative correlation. Once these variables have been determined,

- **a.** Collect at least 30 ordered pairs of data (x, y) from a sample of people on your campus.
- **b.** Draw a scatter plot for the data collected.
- **c.** Does the scatter plot indicate a positive correlation, a negative correlation, or no relationship between the variables?
- **d.** Calculate r. Does the value of r reinforce the impression conveyed by the scatter plot?
- **e.** Find the equation of the regression line.
- **f.** Use the regression line's equation to make a prediction about a y-value given an x-value.
- **g.** Are the results of this project consistent with the group's original belief about the correlation between the variables, or are there some surprises in the data collected?

65. What is the opinion of students on your campus about . . .? Group members should begin by deciding on some aspect of college life around which student opinion can be polled. The poll should consist of the question, "What is your opinion of . . .?" Be sure to provide options such as excellent, good, average, poor, horrible, or a 1-to-10 scale, or possibly grades of A, B, C, D, F. Use a random sample of students on your campus and conduct the opinion survey. After collecting the data, present and interpret it using as many of the skills and techniques learned in this chapter as possible.

Chapter Summary, Review, and Test

SUMMARY – DEFINITIONS AND CONCEPTS	EXAMPLES
12.1 Sampling, Frequency Distributions, and Graphs	
a. A population is the set containing all objects whose properties are to be described and analyzed. A sample is a subset of the population.	Ex. 1, p. 767
b. Random samples are obtained in such a way that each member of the population has an equal chance of being selected.	Ex. 2, p. 768
c. Data can be organized and presented in frequency distributions, grouped frequency distributions, histograms, frequency polygons, and stem-and-leaf plots.	Ex. 3, p. 770; Ex. 4, p. 771; Figures 12.2 and 12.3, p. 772; Ex. 5, p. 772
d. The box on page 774 lists some things to watch for in visual displays of data.	Table 12.5, p. 775
12.2 Measures of Central Tendency	
a. The mean, $\bar{x}$, is the sum of the data items divided by the number of items: $\bar{x} = \frac{\Sigma x}{n}$.	Ex. 1, p. 780
b. The mean, $\bar{x}$, of a frequency distribution is computed using $\bar{x} = \frac{\Sigma xf}{n}$, where x is a data value, f is its frequency, and n is the total frequency of the distribution.	Ex. 2, p. 782
c. The median of ranked data is the item in the middle or the mean of the two middlemost items. The median is the value in the $\frac{n+1}{2}$ position in the list of ranked data.	Ex. 3, p. 783; Ex. 4, p. 784; Ex. 5, p. 785; Ex. 6, p. 786

d. When one or more data items are much greater than or much less than the other items, these extreme values greatly influence the mean, often making the median more representative of the data.	Ex. 7, p. 786
e. The mode of a data set is the value that occurs most often. If there is no such value, there is no mode. If more than one data value has the highest frequency, then each of these data values is a mode.	Ex. 8, p. 788
f. The midrange is computed using $\dfrac{\text{lowest data value} + \text{highest data value}}{2}$.	Ex. 9, p. 789; Ex. 10, p. 790

12.3 Measures of Dispersion

a. Range = highest data value − lowest data value	Ex. 1, p. 795
b. Standard deviation $= \sqrt{\dfrac{\Sigma(\text{data item} - \text{mean})^2}{n-1}}$ This is symbolized by $s = \sqrt{\dfrac{\Sigma(x-\bar{x})^2}{n-1}}$.	Ex. 2, p. 795; Ex. 3, p. 797; Ex. 4, p. 798
c. As the spread of data items increases, the standard deviation gets larger.	Ex. 5, p. 799

12.4 The Normal Distribution

a. The normal distribution is a theoretical distribution for the entire population. The distribution is bell shaped and symmetric about a vertical line through its center, where the mean, median, and mode are located.	
b. The 68–95–99.7 Rule Approximately 68% of the data items fall within 1 standard deviation of the mean. Approximately 95% of the data items fall within 2 standard deviations of the mean. Approximately 99.7% of the data items fall within 3 standard deviations of the mean.	Ex. 1, p. 804; Ex. 2, p. 805
c. A z-score describes how many standard deviations a data item in a normal distribution lies above or below the mean. $z\text{-score} = \dfrac{\text{data item} - \text{mean}}{\text{standard deviation}}$	Ex. 3, p. 806; Ex. 4, p. 807; Ex. 5, p. 808
d. If n% of the items in a distribution are less than a particular data item, that data item is in the nth percentile of the distribution. The 25th percentile is the first quartile, the 50th percentile, or the median, is the second quartile, and the 75th percentile is the third quartile.	Ex. 6, p. 809; Figure 12.16, p. 809
e. If a statistic is obtained from a random sample of size n, there is a 95% probability that it lies within $\dfrac{1}{\sqrt{n}} \times 100\%$ of the true population statistic. $\pm\dfrac{1}{\sqrt{n}} \times 100\%$ is called the margin of error.	Ex. 7, p. 810
f. A distribution of data is skewed if a large number of data items are piled up at one end or the other, with a "tail" at the opposite end.	Figure 12.21, p. 812; Figure 12.22, p. 812

12.5 Problem Solving with the Normal Distribution

a. A table showing z-scores and their percentiles can be used to find the percentage of data items less than or greater than a given data item in a normal distribution, as well as the percentage of data items between two given items. See the boxed summary on computing percentage of data items on page 820.	Ex. 1, p. 817; Ex. 2, p. 818; Ex. 3, p. 819

12.6 Scatter Plots, Correlation, and Regression Lines

a. A plot of data points is called a scatter plot. If the points lie approximately along a line, the line that best fits the data is called a regression line.	
b. A correlation coefficient, r, measures the strength and direction of a possible relationship between variables. If $r = 1$, there is a perfect positive correlation, and if $r = -1$, there is a perfect negative correlation. If $r = 0$, there is no relationship between the variables. Table 12.19 on page 828 indicates whether r denotes a correlation in the population.	Ex. 1, p. 824; Ex. 4, p. 828
c. The formula for computing the correlation coefficient, r, is given in the box on page 824. The equation of the regression line is given in the box on page 826.	Ex. 2, p. 825; Ex. 3, p. 827

Review Exercises

12.1

1. The government of a large city wants to know if its citizens will support a three-year tax increase to provide additional support to the city's community college system. The government decides to conduct a survey of the city's residents before placing a tax increase initiative on the ballot. Which one of the following is most appropriate for obtaining a sample of the city's residents?
   a. Survey a random sample of persons within each geographic region of the city.
   b. Survey a random sample of community college professors living in the city.
   c. Survey every tenth person who walks into the city's government center on two randomly selected days of the week.
   d. Survey a random sample of persons within each geographic region of the state in which the city is located.

A random sample of ten college students is selected and each student is asked how much time he or she spent on homework during the previous weekend. The following times, in hours, are obtained:

8, 10, 9, 7, 9, 8, 7, 6, 8, 7.

Use these data items to solve Exercises 2–4.

2. Construct a frequency distribution for the data.
3. Construct a histogram for the data.
4. Construct a frequency polygon for the data.

The 50 grades on a physiology test are shown. Use the data to solve Exercises 5–6.

44	24	54	81	18
34	39	63	67	60
72	36	91	47	75
57	74	87	49	86
59	14	26	41	90
13	29	13	31	68
63	35	29	70	22
95	17	50	42	27
73	11	42	31	69
56	40	31	45	51

5. Construct a grouped frequency distribution for the data. Use 0–39 for the first class, 40–49 for the second class, and make each subsequent class width the same as the second class.
6. Construct a stem-and-leaf plot for the data.

7. Describe what is misleading about the size of the barrels in the following visual display.

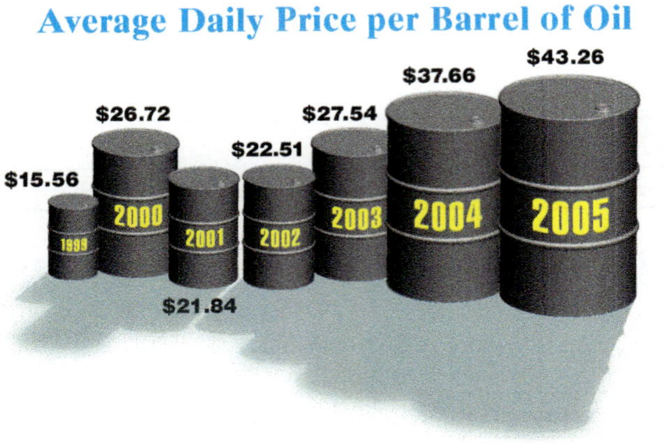

Source: U.S. Department of Energy

12.2

In Exercises 8–9, find the mean for each group of data items.

8. 84, 90, 95, 89, 98
9. 33, 27, 9, 10, 6, 7, 11, 23, 27
10. Find the mean for the data items in the given frequency distribution.

Score x	Frequency f
1	2
2	4
3	3
4	1

In Exercises 11–12, find the median for each group of data items.

11. 33, 27, 9, 10, 6, 7, 11, 23, 27
12. 28, 16, 22, 28, 34
13. Find the median for the data items in the frequency distribution in Exercise 10.

In Exercises 14–15, find the mode for each group of data items. If there is no mode, so state.

14. 33, 27, 9, 10, 6, 7, 11, 23, 27
15. 582, 585, 583, 585, 587, 587, 589
16. Find the mode for the data items in the frequency distribution in Exercise 10.

In Exercises 17–18, find the midrange for each group of data items.

17. 84, 90, 95, 88, 98
18. 33, 27, 9, 10, 6, 7, 11, 23, 27
19. Find the midrange for the data items in the frequency distribution in Exercise 10.

20. A student took seven tests in a course, scoring between 90% and 95% on three of the tests, between 80% and 89% on three of the tests, and below 40% on one of the tests. In this distribution, is the mean or the median more representative of the student's overall performance in the course? Explain your answer.

21. The data items below are the ages of U.S. presidents at the time of their first inauguration.

57 61 57 57 58 57 61 54 68 51 49 64 50 48
65 52 56 46 54 49 51 47 55 55 54 42 51 56
55 51 54 51 60 62 43 55 56 61 52 69 64 46 54 47

a. Organize the data in a frequency distribution.
b. Use the frequency distribution to find the mean age, median age, modal age, and midrange age of the presidents when they were inaugurated.

12.3

In Exercises 22–23, find the range for each group of data items.

22. 28, 34, 16, 22, 28

23. 312, 783, 219, 312, 426, 219

24. The mean for the data items 29, 9, 8, 22, 46, 51, 48, 42, 53, 42 is 35. Find **a.** the deviation from the mean for each data item and **b.** the sum of the deviations in part (a).

25. Use the data items 36, 26, 24, 90, and 74 to find **a.** the mean, **b.** the deviation from the mean for each data item, and **c.** the sum of the deviations in part (b).

In Exercises 26–27, find the standard deviation for each group of data items.

26. 3, 3, 5, 8, 10, 13

27. 20, 27, 23, 26, 28, 32, 33, 35

28. A test measuring anxiety levels is administered to a sample of ten college students with the following results. (High scores indicate high anxiety.)

10, 30, 37, 40, 43, 44, 45, 69, 86, 86

Find the mean, range, and standard deviation for the data.

29. Compute the mean and the standard deviation for each of the following data sets. Then, write a brief description of similarities and differences between the two sets based on each of your computations.

Set A: 80, 80, 80, 80   Set B: 70, 70, 90, 90

30. Describe how you would determine
a. which of the two groups, men or women, at your college has a higher mean grade point average.
b. which of the groups is more consistently close to its mean grade point average.

12.4

The scores on a test are normally distributed with a mean of 70 and a standard deviation of 8. In Exercises 31–33, find the score that is

31. 2 standard deviations above the mean.

32. $3\frac{1}{2}$ standard deviations above the mean.

33. $1\frac{1}{4}$ standard deviations below the mean.

The ages of people living in a retirement community are normally distributed with a mean age of 68 years and a standard deviation of 4 years. In Exercises 34–40, use the 68–95–99.7 Rule to find the percentage of people in the community whose ages

34. are between 64 and 72.

35. are between 60 and 76.

36. are between 68 and 72.

37. are between 56 and 80.

38. exceed 72.

39. are less than 72.

40. exceed 76.

A set of data items is normally distributed with a mean of 50 and a standard deviation of 5. In Exercises 41–45, convert each data item to a z-score.

41. 50

42. 60

43. 58

44. 35

45. 44

46. A student scores 60 on a vocabulary test and 80 on a grammar test. The data items for both tests are normally distributed. The vocabulary test has a mean of 50 and a standard deviation of 5. The grammar test has a mean of 72 and a standard deviation of 6. On which test did the student have the better score? Explain why this is so.

The number of miles that a particular brand of car tires lasts is normally distributed with a mean of 32,000 miles and a standard deviation of 4000 miles. In Exercises 47–49, find the data item in this distribution that corresponds to the given z-score.

47. $z = 1.5$

48. $z = 2.25$

49. $z = -2.5$

50. Using a random sample of 2281 American adults ages 18 and older, an Adecco survey asked respondents if they would be willing to sacrifice a percentage of their salary in order to work for an environmentally friendly company. The poll indicated that 31% of the respondents said "yes," 39% said "no," and 30% declined to answer.
a. Find the margin of error, to the nearest tenth of a percent, for this survey.
b. Write a statement about the percentage of American adults who would be willing to sacrifice a percentage of their salary in order to work for an environmentally friendly company.

51. The histogram indicates the frequencies of the number of syllables per word for 100 randomly selected words in Japanese.

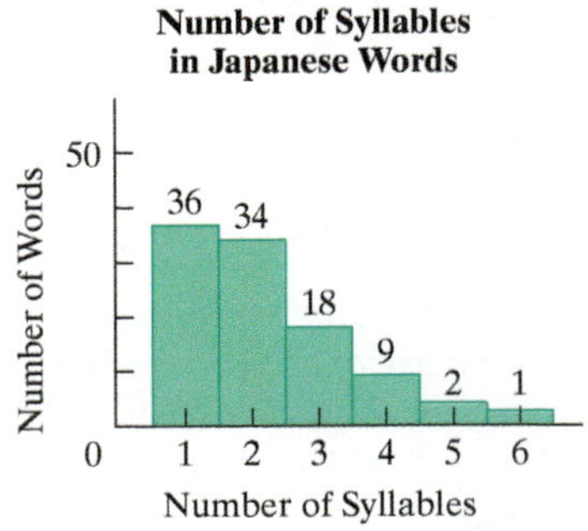

a. Is the shape of this distribution best classified as normal, skewed to the right, or skewed to the left?

b. Find the mean, median, and mode for the number of syllables in the sample of Japanese words.

c. Are the measures of central tendency from part (b) consistent with the shape of the distribution that you described in part (a)? Explain your answer.

12.5

The mean cholesterol level for all men in the United States is 200 and the standard deviation is 15. In Exercises 52–55, use **Table 12.17** *on page 816 to find the percentage of U.S. men whose cholesterol level*

52. is less than 221.

53. is greater than 173.

54. is between 173 and 221.

55. is between 164 and 182.

Use the percentiles for the weights of adult men over 40 to solve Exercises 56–58.

Weight	Percentile
235	86
227	third quartile
180	second quartile
173	first quartile

Find the percentage of men over 40 who weigh

56. less than 227 pounds.

57. more than 235 pounds.

58. between 227 and 235 pounds.

12.6

In Exercises 59–60, make a scatter plot for the given data. Use the scatter plot to describe whether or not the variables appear to be related.

59.

x	1	3	4	6	8	9
y	1	2	3	3	5	5

60.

Country	Canada	U.S.	Mexico	Brazil	Costa Rica	Denmark	China	Egypt	Pakistan	Bangladesh	Australia	Japan	Russia
Life expectancy in years, x	81	78	76	72	77	78	73	72	64	63	82	82	66
Infant deaths per 1000 births, y	5.1	6.3	19.0	23.3	9.0	4.4	21.2	28.4	66.9	57.5	4.8	2.8	10.8

Source: U.S. Bureau of the Census International Database

The scatter plot shows the relationship between the percentage of adult females in a country who are literate and the mortality of children under five. Also shown is the regression line. Use this information to determine whether each of the statements in Exercises 61–67 is true or false.

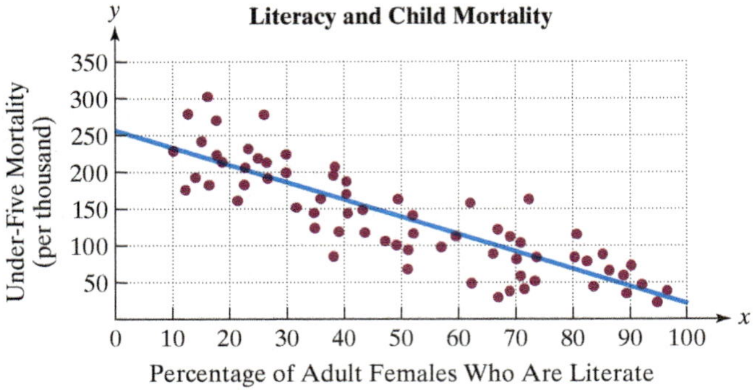

Source: United Nations

61. There is a perfect negative correlation between the percentage of adult females who are literate and under-five mortality.

62. As the percentage of adult females who are literate increases, under-five mortality tends to decrease.

63. The country with the least percentage of adult females who are literate has the greatest under-five mortality.

64. No two countries have the same percentage of adult females who are literate but different under-five mortalities.

65. There are more than 20 countries in this sample.

66. There is no correlation between the percentage of adult females who are literate and under-five mortality.

67. The country with the greatest percentage of adult females who are literate has an under-five mortality rate that is less than 50 children per thousand.

68. Which one of the following scatter plots indicates a correlation coefficient of approximately -0.9?

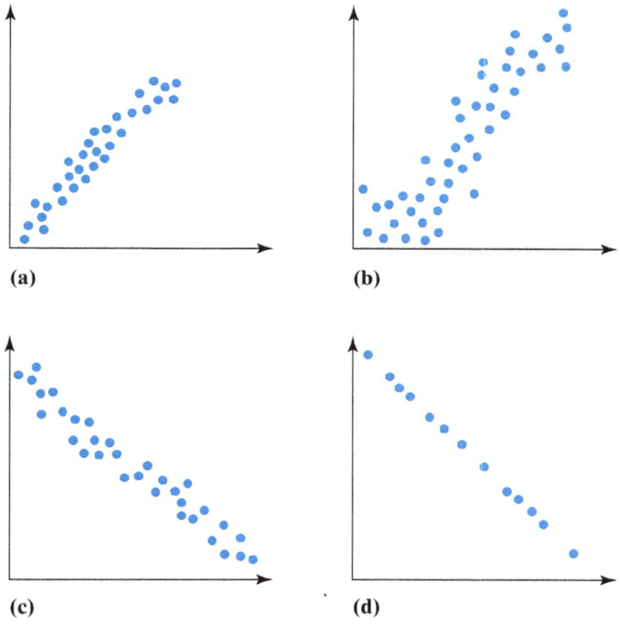

69. Use the data in Exercise 59 to solve this exercise.

a. Compute r, the correlation coefficient, rounded to the nearest thousandth.

b. Find the equation of the regression line.

70. The graph, based on Nielsen Media Research data taken from random samples of Americans at various ages, indicates that as we get older, we watch more television.

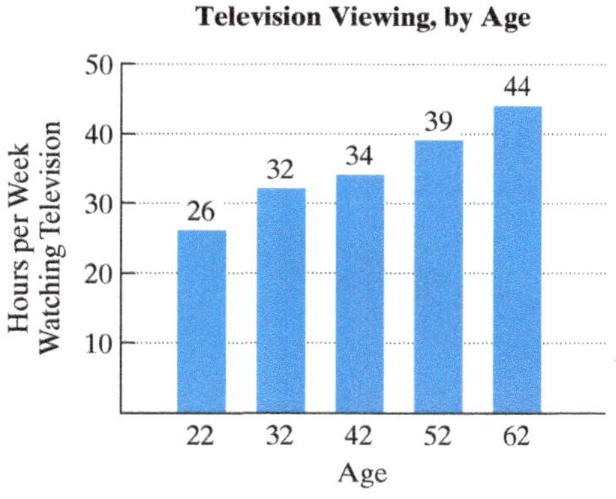

Source: Nielsen Media Research

a. Let x represent one's age and let y represent hours per week watching television. Calculate the correlation coefficient.

b. Using **Table 12.19** on page 828 and the $\alpha = 0.05$ column, determine whether there is a correlation between age and time spent watching television in the American population.

Chapter 12 Test

1. Politicians in the Florida Keys need to know if the residents of Key Largo think the amount of money charged for water is reasonable. The politicians decide to conduct a survey of a sample of Key Largo's residents. Which procedure would be most appropriate for a sample of Key Largo's residents?

a. Survey all water customers who pay their water bills at Key Largo City Hall on the third day of the month.

b. Survey a random sample of executives who work for the water company in Key Largo.

c. Survey 5000 individuals who are randomly selected from a list of all people living in Georgia and Florida.

d. Survey a random sample of persons within each neighborhood of Key Largo.

Use these scores on a ten-point quiz to solve Exercises 2–4.

8, 5, 3, 6, 5, 10, 6, 9, 4, 5, 7, 9, 7, 4, 8, 8

2. Construct a frequency distribution for the data.

3. Construct a histogram for the data.

4. Construct a frequency polygon for the data.

Use the 30 test scores listed below to solve Exercises 5–6.

79	51	67	50	78
62	89	83	73	80
88	48	60	71	79
89	63	55	93	71
41	81	46	50	61
59	50	90	75	61

5. Construct a grouped frequency distribution for the data. Use 40–49 for the first class and use the same width for each subsequent class.

6. Construct a stem-and-leaf display for the data.

7. The graph shows the percentage of students in the United States through grade 12 who were home-schooled in 1999 and 2007. What impression does the roofline in the visual display imply about what occurred in 2000 through 2006? How might this be misleading?

Source: National Center for Education Statistics

Use the six data items listed below to solve Exercises 8–11.

3, 6, 2, 1, 7, 3

8. Find the mean.
9. Find the median.
10. Find the midrange.
11. Find the standard deviation.

Use the frequency distribution shown to solve Exercises 12–14.

Score x	Frequency f
1	3
2	5
3	2
4	2

12. Find the mean.
13. Find the median.
14. Find the mode.

15. The annual salaries of four salespeople and the owner of a bookstore are

\$17,500, \$19,000, \$22,000, \$27,500, \$98,500.

Is the mean or the median more representative of the five annual salaries? Briefly explain your answer.

According to the American Freshman, *the number of hours that college freshmen spend studying each week is normally distributed with a mean of 7 hours and a standard deviation of 5.3 hours. In Exercises 16–17, use the 68–95–99.7 Rule to find the percentage of college freshmen who study*

16. between 7 and 12.3 hours each week.

17. more than 17.6 hours each week.

18. IQ scores are normally distributed in the population. Who has a higher IQ: a student with a 120 IQ on a scale where 100 is the mean and 10 is the standard deviation, or a professor with a 128 IQ on a scale where 100 is the mean and 15 is the standard deviation? Briefly explain your answer.

19. Use the z-scores and the corresponding percentiles shown at the top of the next column to solve this exercise. Test scores are normally distributed with a mean of 74 and a standard deviation of 10. What percentage of the scores are above 88?

z-Score	Percentile
1.1	86.43
1.2	88.49
1.3	90.32
1.4	91.92
1.5	93.32

20. Use the percentiles in the table shown below to find the percentage of scores between 630 and 690.

Score	Percentile
780	99
750	87
720	72
690	49
660	26
630	8
600	1

21. Using a random sample of 100 students from a campus of approximately 12,000 students, 60% of the students in the sample said they were very satisfied with their professors.

a. Find the margin of error in this percent.

b. Write a statement about the percentage of the entire population of students from this campus who are very satisfied with their professors.

22. Make a scatter plot for the given data. Use the scatter plot to describe whether or not the variables appear to be related.

x	1	4	3	5	2
y	5	2	2	1	4

The scatter plot shows the number of minutes each of 16 people exercise per week and the number of headaches per month each person experiences. Use the scatter plot to determine whether each of the statements in Exercises 23–25 is true or false.

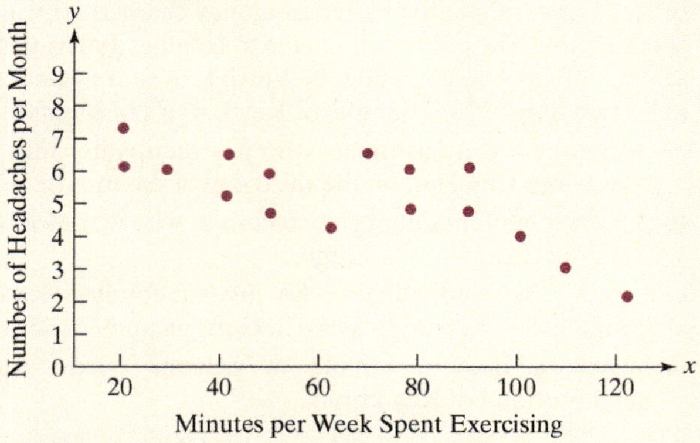

23. An increase in the number of minutes devoted to exercise causes a decrease in headaches.

24. There is a perfect negative correlation between time spent exercising and number of headaches.

25. The person who exercised most per week had the least number of headaches per month.

26. Is the relationship between the price of gas and the number of people visiting our national parks a positive correlation, a negative correlation, or is there no correlation? Explain your answer.

Answers to Selected Exercises

CHAPTER 1

Section 1.1

Check Point Exercises

1. Answers will vary; an example is $40 \times 40 = 1600$. **2. a.** Each number in the list is obtained by adding 6 to the previous number.; 33 **b.** Each number in the list is obtained by multiplying the previous number by 5.; 1250 **c.** To get the second number, multiply the previous number by 2. Then multiply by 3 and then by 4. Then multiply by 2, then by 3, and then by 4, repeatedly.; 3456 **d.** To get the second number, add 8 to the previous number. Then add 8 and then subtract 14. Then add 8, then add 8, and then subtract 14, repeatedly.; 7 **3. a.** Starting with the third number, each number is the sum of the previous two numbers.; 76 **b.** Starting with the second number, each number is one less than twice the previous number.; 257
4. The figures alternate between rectangles and triangles, and the number of appendages follows the pattern: one, two, three, one, two, three, etc.;

5. a. The result of the process is two times the original number selected.
b. Using n to represent the original number, we have

Select a number:	n
Multiply the number by 4:	$4n$
Add 6 to the product:	$4n + 6$
Divide this sum by 2:	$\frac{4n + 6}{2} = 2n + 3$
Subtract 3 from the quotient:	$2n + 3 - 3 = 2n.$

Blitzer Bonus: Are You Smart Enough to Work at Google?

1. SSSS **2.** 3 1 2 2 1 1

Concept and Vocabulary Check

1. counterexample **2.** deductive **3.** inductive **4.** true

Exercise Set 1.1

1. Answers will vary; an example is: Barack Obama was younger than 65 at the time of his inauguration. **3.** Answers will vary; an example is: 3 multiplied by itself is 9, which is not even.
5. Answers will vary; an example is: Adding 1 to the numerator and denominator of $\frac{1}{2}$ results in $\frac{2}{3}$, which is not equal to $\frac{1}{2}$.
7. Answers will vary; an example is: When -1 is added to itself, the result is -2, which is less than -1. **9.** Each number in the list is obtained by adding 4 to the previous number.; 28 **11.** Each number in the list is obtained by subtracting 5 from the previous number.; 12 **13.** Each number in the list is obtained by multiplying the previous number by 3.; 729 **15.** Each number in the list is obtained by multiplying the previous number by 2.; 32 **17.** The numbers in the list alternate between 1 and numbers obtained by multiplying the number prior to the previous number by 2.; 32
19. Each number in the list is obtained by subtracting 2 from the previous number.; -6
21. Each number in the list is obtained by adding 4 to the denominator of the previous fraction.; $\frac{1}{22}$
23. Each number in the list is obtained by multiplying the previous number by $\frac{1}{3}$.; $\frac{1}{81}$
25. The second number is obtained by adding 4 to the first number. The third number is obtained by adding 5 to the second number. The number being added to the previous number increases by 1 each time.; 42 **27.** The second number is obtained by adding 3 to the first number. The third number is obtained by adding 5 to the second number. The number being added to the previous number increases by 2 each time.; 51 **29.** Starting with the third number, each number is the sum of the previous two numbers.; 71 **31.** To get the second number, add 5 to the previous number. Then add 5 and then subtract 7. Then add 5, then add 5, and then subtract 7, repeatedly.; 18 **33.** The second number is obtained by multiplying the first number by 2. The third number is obtained by subtracting 1 from the second number. Then multiply by 2 and then subtract 1, repeatedly.; 33
35. Each number in the list is obtained by multiplying the previous number by $-\frac{1}{4}$.; $\frac{1}{4}$
37. For each pair in the list, the second number is obtained by subtracting 4 from the first number.; -1
39. The pattern is: square, triangle, circle, square, triangle, circle, etc.;

41. Each figure contains the letter of the alphabet following the letter in the previous figure with one more occurrence than in the previous figure.;

d	d	d
d	d	

43. a. The result of the process is two times the original number selected.
b. Using n to represent the original number, we have

Select a number:	n
Multiply the number by 4:	$4n$
Add 8 to the product:	$4n + 8$
Divide this sum by 2:	$\frac{4n + 8}{2} = 2n + 4$
Subtract 4 from the quotient:	$2n + 4 - 4 = 2n.$

45. a. The result of the process is 3.
b. Using n to represent the original number as we have

Select a number:	n
Add 5:	$n + 5$
Double the result:	$2(n + 5) = 2n + 10$
Subtract 4:	$2n + 10 - 4 = 2n + 6$
Divide by 2:	$\frac{2n + 6}{2} = n + 3$
Subtract n:	$n + 3 - n = 3.$

47. $1 + 2 + 3 + 4 + 5 + 6 = \frac{6 \times 7}{2}$; $21 = 21$ **49.** $1 + 3 + 5 + 7 + 9 + 11 = 6 \times 6$; $36 = 36$ **51.** $98{,}765 \times 9 + 3 = 888{,}888$; correct

53. $165 \times 3367 = 555{,}555$; correct **55.** b **57.** deductive reasoning; Answers will vary. **59.** inductive reasoning; Answers will vary. **61. a.** 28, 36, 45, 55, 66 **b.** 36, 49, 64, 81, 100 **c.** 35, 51, 70, 92, 117 **d.** square **67.** makes sense **69.** makes sense

71. a.

16	3	11
5	10	15
9	17	4

The sums are all 30. **b.**

17	5	14
9	12	15
10	19	7

The sums are all 36. **c.** For any values of a, b, and c, the sums of all rows, all columns, and both diagonals are the same. **d.** The sums of the expressions in each row, each column, and each diagonal is $3a$. **e.** Add the variable expressions in a, b, and c, in each row, each column, and each diagonal. The sum is always $3a$.

73. a. The result is a three-or four-digit number in which the thousands and hundreds places represent the month of the birthday and the tens and ones places represent the day of the birthday. **b.** $5[4(5M + 6) + 9] + D - 165 = 100M + D$ **75. a.** 10,101; 20,202; 30,303; 40,404 **b.** In the multiplications, the first factor is always 3367, and the second factors are consecutive multiples of 3, beginning with $3 \times 1 = 3$.; The second and fourth digits of the products are always 0; the first, third, and last digits are the same within each product; this digit is 1 in the first product and increases by 1 in each subsequent product. **c.** $3367 \times 15 = 50{,}505$; $3367 \times 18 = 60{,}606$ **d.** inductive reasoning; Answers will vary.

Section 1.2

Check Point Exercises

1. a. 7,000,000,000 **b.** 7,058,750,000 **2. a.** 3.1 **b.** 3.1416 **3. a.** \$3, \$2, \$6, \$5, \$3, \$3, and \$4; ≈ \$26 **b.** no **4. a.** ≈ \$2000 per wk **b.** ≈ \$100,000 per yr **5. a.** 0.48×2148.72 **b.** $0.5 \times 2100 = 1050$; Your family spent approximately \$1050 on heating and cooling last year. **6. a.** ≈ 0.17 year for each subsequent birth year **b.** ≈ 88.1 years **7. a.** 22% **b.** 1994 through 1998 **c.** 1982 and 1994 **8. a.** \$1076 **b.** $T = 15{,}518 + 1076x$ **c.** \$30,582

Concept and Vocabulary Check

1. estimation **2.** circle graph **3.** mathematical model **4.** true **5.** true **6.** false

Exercise Set 1.2

1. a. 19,465,200 **b.** 19,465,000 **c.** 19,470,000 **d.** 19,500,000 **e.** 19,000,000 **f.** 20,000,000 **3.** 2.718 **5.** 2.71828 **7.** 2.718281828 **9.** $350 + 600 = 950$; 955; reasonably well **11.** $9 + 1 + 19 = 29$; 29.23; quite well **13.** $32 - 11 = 21$; 20.911; quite well **15.** $40 \times 6 = 240$; 218.185; not so well **17.** $0.8 \times 400 = 320$; 327.06; reasonably well **19.** $48 \div 3 = 16$; ≈ 16.49; quite well **21.** 30% of 200,000 is 60,000.; 59,920.96; quite well **23.** ≈ \$43 **25.** ≈ \$40,000 per yr **27.** ≈ \$24,000 **29.** ≈ \$1000 **31.** ≈ \$30 per hr **33.** ≈ 700,800 hr **35.** ≈ 40; ≈ 42.03; quite reasonable **37.** b **39.** c **41.** ≈ 3 hr **43.** ≈ $0.10 \times 16{,}000{,}000 = 1{,}600{,}000$ high school teenagers **45. a.** ≈ 85 people per 100 **b.** ≈ 5400 people **47. a.** ≈ 0.5% per year **b.** 29.7% **49. a.** ≈ 38% **b.** 2008 and 2009 **c.** 2003 **51. a.** 1.33 ppm per year **b.** $C = 310 + 1.33x$ **c.** 443 ppm **67.** does not make sense **69.** does not make sense **71.** a **73.** b **75.** ≈ 667 days; ≈ 1.8 yr

Section 1.3

Check Point Exercises

1. the amount of money given to the cashier **2.** The 128-ounce bottle at approximately 4¢ an ounce is the better value. **3.** 14 months **4.** 5 combinations **5.** 6 outfits **6.** A route that will cost less than \$1460 is A, D, E, C, B, A.

Trick Questions

1. 12 **2.** 12 **3.** sister and brother **4.** match

Concept and Vocabulary Check

1. understand **2.** devise a plan **3.** false **4.** false

Exercise Set 1.3

1. the price of the computer **3.** the number of words on the page **5.** unnecessary information: weekly salary of \$350; extra pay: \$180 **7.** unnecessary information: \$20 given to the parking attendant; charge: \$4.50 **9. a.** 24-ounce box for \$4.59 **b.** 22¢ per ounce for the 15.3-ounce box and \$3.06 per pound for the 24-ounce box **c.** no; Answers will vary. **11.** \$3000 **13.** \$50 **15.** \$90 **17.** \$4525 **19.** \$565 **21.** 4 mi **23.** \$104 **25.** \$14,300 **27.** 5 ways **29.** 6 ways **31.** 9 ways **33.** 10 different total scores **35.** B owes \$18 and C owes \$2.; A is owed \$14, D is owed \$4, and E is owed \$2.; B should give A \$14 and D \$4, while C should give E \$2. **37.** 4 ways **39.** Andy, Darnell, Caleb, Beth, Ella **41.** Home, Bank, Post Office, Dry Cleaners, Home

43. Sample answer:
CO, WY, UT, AZ, NM, CO, UT

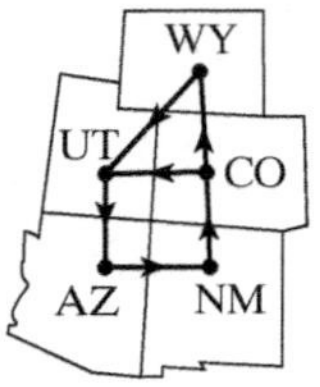

45. Bob's major is psychology.

47. a.

5	22	18
28	15	2
12	8	25

b.

4	9	8
11	7	3
6	5	10

49.

9	6	7
8	1	4
3	2	5

51.

```
     156
28)4368
   28
   156
   140
    168
    168
      0
```

57. makes sense **59.** does not make sense **61.** the dentist with poor dental work **63.** Friday
65. Sample answer:

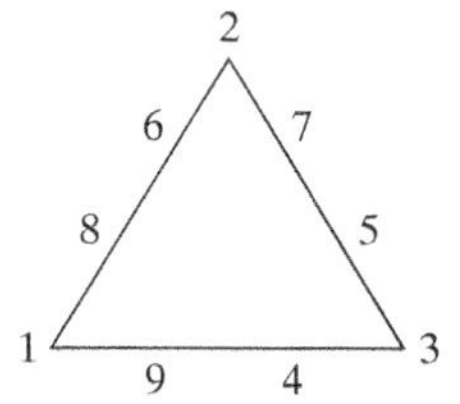

67. There is no missing dollar; in the end, the customers paid a total of $27 of which $25 went to the restaurant and $2 was stolen by the waiter.
69. Answers will vary; an example is

State	A	B	C	D
Congressional Seats	4	6	8	12

Chapter 1 Review Exercises

1. deductive reasoning: Answers will vary. **2.** inductive reasoning; Answers will vary. **3.** Each number in the list is obtained by adding 5 to the previous number.; 24 **4.** Each number in the list is obtained by multiplying the previous number by 2.; 112 **5.** The successive differences are consecutive counting numbers beginning with 2.; 21 **6.** Each number in the list is obtained by writing a fraction with a denominator that is one more than the denominator of the previous fraction before it is reduced to lowest terms.; $\frac{3}{8}$ **7.** Each number in the list is obtained by multiplying the previous number by $-\frac{1}{2}$.; $\frac{5}{2}$ **8.** Each number in the list is obtained by subtracting 60 from the previous number; −200. **9.** Each number beginning with the third number is the sum of the two previous numbers.; 42 **10.** To get the second number, multiply the first number by 3. Then multiply the second number by 2 to get the third number. Then multiply by 3 and then by 2, repeatedly.; 432 **11.** The figures alternate between squares and circles, and in each figure the tick mark has been rotated 90° clockwise from its position in the previous figure.;

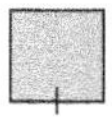

12. $2 + 4 + 8 + 16 + 32 = 64 - 2$; correct **13.** $444 \div 12 = 37$; correct
14. a. The result is the original number.
b. Using n to represent the original number, we have

Select a number:	n
Double the number:	$2n$
Add 4 to the product:	$2n + 4$
Divide the sum by 2:	$\frac{2n+4}{2} = n + 2$

Subtract 2 from the quotient: $n + 2 - 2 = n$.

15. a. 923,187,500 **b.** 923,187,000 **c.** 923,200,000 **d.** 923,000,000 **e.** 900,000,000 **16. a.** 1.5 **b.** 1.51 **c.** 1.507 **d.** 1.5065917 **17.** ≈ 16; 15.71; quite reasonable **18.** ≈ 450; 432.67; somewhat reasonable **19.** ≈ 5; ≈ 4.76; quite reasonable **20.** ≈ 2400; 2397.0548; quite reasonable **21.** ≈ $18.00 **22.** ≈ $560 **23.** ≈ $60 **24.** ≈ 5,100,000 students **25.** b **26.** c **27. a.** Asian; ≈ 122 **b.** ≈ 990 million **28. a.** 0.4% per yr **b.** 34% **29. a.** 115 beats per min; 10 min **b.** 64 beats per min; 8 min **c.** between 9 and 10 minutes **d.** 9 min **30. a.** 2.65 million **b.** $p = 203.3 + 2.65x$ **c.** 335.8 million **31.** the weight of the child **32.** unnecessary information: $20 given to driver; cost of trip: $8.00 **33.** 8 lb **34.** $885 **35.** Plan A; $90 **36.** 5.75 hr or 5 hr 45 min **37.** $15,500 **38.** 6 combinations

Chapter 1 Test

1. deductive **2.** inductive **3.** Each number in the list is obtained by adding 5 to the previous number.; 20
4. Each number in the list is obtained by multiplying the previous number by $\frac{1}{2}$.; $\frac{1}{96}$ **5.** $3367 \times 15 = 50{,}505$
6. The outer figure is always a square; the inner figure follows the pattern: triangle, circle, square, triangle, circle, square, etc.; the number of appendages on the outer square alternates between 1 and 2.;

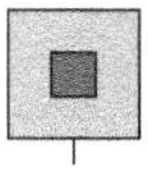

7. a. The original number is doubled.
b. Using n to represent the original number, we have

Select a number:	n
Multiply the number by 4:	$4n$
Add 8 to the product:	$4n + 8$
Divide the sum by 2:	$\frac{4n+8}{2} = 2n + 4$

Subtract 4 from the quotient: $2n + 4 - 4 = 2n$.

8. 3,300,000 **9.** 706.38 **10.** ≈ $90 **11.** ≈ $25,000 per person **12.** ≈ 60 **13.** ≈ 50 billion pounds **14.** a **15. a.** 2001; ≈ 1275 discharges **b.** 2010; 275 discharges **c.** between 2001 and 2002 **d.** 1997 **16. a.** 0.7% per year **b.** $p = 17.6 + 0.7x$ **c.** 54%
17. Estes Rental; $12 **18.** $14,080 **19.** 26 weeks **20.** Belgium; 160,000 people

CHAPTER 2

Section 2.1

Check Point Exercises

1. L is the set of the first six lowercase letters of the alphabet. **2.** $M = \{\text{April, August}\}$ **3.** $O = \{1, 3, 5, 7, 9\}$ **4. a.** not the empty set **b.** empty set **c.** not the empty set **d.** not the empty set **5. a.** true **b.** true **c.** false **6. a.** $A = \{1, 2, 3\}$ **b.** $B = \{15, 16, 17, 18, \ldots\}$ **c.** $O = \{1, 3, 5, 7, \ldots\}$ **7. a.** $\{1, 2, 3, 4, \ldots, 199\}$ **b.** $\{51, 52, 53, 54, \ldots, 200\}$
8. a. $n(A) = 5$ **b.** $n(B) = 1$ **c.** $n(C) = 8$ **d.** $n(D) = 0$ **9.** No; the sets do not contain the same number of distinct elements.
10. a. true **b.** false

Concept and Vocabulary Check

1. roster; set-builder **2.** empty; $\varnothing$ **3.** is an element **4.** natural numbers **5.** cardinal; $n(A)$ **6.** equivalent **7.** equal

Exercise Set 2.1

1. well defined; set **3.** not well defined; not a set **5.** well defined; set **7.** the set of planets in our solar system **9.** the set of months that begin with J **11.** the set of natural numbers greater than 5 **13.** the set of natural numbers between 6 and 20, inclusive **15.** {winter, spring, summer, fall} **17.** {September, October, November, December} **19.** {1, 2, 3} **21.** {1, 3, 5, 7, 9, 11} **23.** {1, 2, 3, 4, 5} **25.** {6, 7, 8, 9, . . . } **27.** {7, 8, 9, 10} **29.** {10, 11, 12, 13, . . . , 79} **31.** {2} **33.** not the empty set **35.** empty set **37.** not the empty set **39.** empty set **41.** empty set **43.** not the empty set **45.** not the empty set **47.** true **49.** true **51.** false **53.** true **55.** false **57.** false **59.** true **61.** false **63.** false **65.** true **67.** 5 **69.** 15 **71.** 0 **73.** 1 **75.** 4 **77.** 5 **79.** 0 **81. a.** not equivalent; Answers will vary. **b.** not equal; Answers will vary. **83. a.** equivalent; Answers will vary. **b.** not equal; Answers will vary. **85. a.** equivalent; Answers will vary. **b.** equal; Answers will vary. **87. a.** equivalent; Answers will vary. **b.** not equal; Answers will vary. **89. a.** equivalent; Answers will vary. **b.** equal; Answers will vary. **91.** infinite **93.** finite **95.** finite **97.** $\{x \mid x \in \mathbf{N} \text{ and } x \geq 61\}$ **99.** $\{x \mid x \in \mathbf{N} \text{ and } 61 \leq x \leq 89\}$ **101.** Answers will vary; an example is: {0, 1, 2, 3} and {1, 2, 3, 4}. **103.** impossible; Answers will vary. **105.** {New Zealand, Australia, United States} **107.** {Australia, United States, United Kingdom, Switzerland, Ireland} **109.** {United Kingdom, Switzerland, Ireland} **111.** $\varnothing$ or { } **113.** {12, 19} **115.** {20, 21} **117.** no one-to-one correspondence; not equivalent **125.** does not make sense **127.** makes sense **129.** false **131.** true **133.** false **135.** false

Section 2.2

Check Point Exercises

1. a. $\not\subseteq$ **b.** $\subseteq$ **c.** $\subseteq$ **2. a.** $\subseteq$, $\subset$ **b.** $\subseteq$, $\subset$ **3.** yes **4. a.** 16; 15 **b.** 64; 63

Concept and Vocabulary Check

1. $A \subseteq B$; every element in set A is also an element in set B **2.** $A \subset B$; sets A and B are not equal **3.** the empty set; subset **4.** 2^n **5.** $2^n - 1$

Exercise Set 2.2

1. $\subseteq$ **3.** $\not\subseteq$ **5.** $\not\subseteq$ **7.** $\not\subseteq$ **9.** $\subseteq$ **11.** $\not\subseteq$ **13.** $\subseteq$ **15.** $\not\subseteq$ **17.** $\subseteq$ **19.** both **21.** $\subseteq$ **23.** neither **25.** both **27.** $\subseteq$ **29.** $\subseteq$ **31.** both **33.** both **35.** both **37.** neither **39.** $\subseteq$ **41.** true **43.** false; Explanations will vary. **45.** true **47.** false; Explanations will vary. **49.** true **51.** false; Explanations will vary. **53.** true **55.** $\varnothing$, {border collie}, {poodle}, {border collie, poodle} **57.** $\varnothing$, {t}, {a}, {b}, {t, a}, {t, b}, {a, b}, {t, a, b} **59.** $\varnothing$ and {0} **61.** 16; 15 **63.** 64; 63 **65.** 128; 127 **67.** 8; 7 **69.** false; The set {1, 2, 3, . . . , 1000} has $2^{1000} - 1$ proper subsets. **71.** true **73.** false; $\varnothing \subseteq \{\varnothing, \{\varnothing\}\}$ **75.** true **77.** true **79.** true **81.** false; The set of subsets of {a, e, i, o, u} contains 32 elements. **83.** false; $D \subseteq T$ **85.** true **87.** false; If $x \in W$, then $x \in D$. **89.** true **91.** true **93.** 32 **95.** 64 **97.** 256 **105.** does not make sense **107.** does not make sense **109.** false **111.** false **113.** \$0.00, \$0.05, \$0.10, \$0.15, \$0.25, \$0.30, \$0.35, and \$0.40

Section 2.3

Check Point Exercises

1. a. {1, 5, 6, 7, 9} **b.** {1, 5, 6} **c.** {7, 9} **2. a.** {a, b, c, d} **b.** {e} **c.** {e, f, g} **d.** {f, g} **3.** {b, c, e} **4. a.** {7, 10} **b.** $\varnothing$ **c.** $\varnothing$ **5. a.** {1, 3, 5, 6, 7, 10, 11} **b.** {1, 2, 3, 4, 5, 6, 7} **c.** {1, 2, 3} **6. a.** {a, d} **b.** {a, d} **7. a.** {5} **b.** {2, 3, 7, 11, 13, 17, 19} **c.** {2, 3, 5, 7, 11, 13} **d.** {17, 19} **e.** {5, 7, 11, 13, 17, 19} **f.** {2, 3} **8.** 28

Concept and Vocabulary Check

1. Venn diagrams **2.** complement; A' **3.** intersection; $A \cap B$ **4.** union; $A \cup B$ **5.** $n(A) + n(B) - n(A \cap B)$ **6.** true **7.** false **8.** true **9.** false

Exercise Set 2.3

1. the set of all composers **3.** the set of all brands of soft drinks **5.** {c, d, e} **7.** {b, c, d, e, f} **9.** {6, 7, 8, 9, . . . , 20} **11.** {2, 4, 6, 8, . . . , 20} **13.** {21, 22, 23, 24, . . . } **15.** {1, 3, 5, 7, . . . } **17.** {1, 3} **19.** {1, 2, 3, 5, 7} **21.** {2, 4, 6} **23.** {4, 6} **25.** {1, 3, 5, 7} or A **27.** {1, 2, 4, 6, 7} **29.** {1, 2, 4, 6, 7} **31.** {4, 6} **33.** {1, 3, 5, 7} or A **35.** $\varnothing$ **37.** {1, 2, 3, 4, 5, 6, 7} or U **39.** {1, 3, 5, 7} or A **41.** {g, h} **43.** {a, b, g, h} **45.** {b, c, d, e, f} or C **47.** {c, d, e, f} **49.** {a, g, h} or A **51.** {a, b, c, d, e, f, g, h} or U **53.** {a, b, c, d, e, f, g, h} or U **55.** {c, d, e, f} **57.** {a, g, h} or A **59.** $\varnothing$ **61.** {a, b, c, d, e, f, g, h} or U **63.** {a, g, h} or A **65.** {a, c, d, e, f, g, h} **67.** {1, 3, 4, 7} **69.** {1, 2, 3, 4, 5, 6, 7, 8, 9} **71.** {3, 7} **73.** {1, 4, 8, 9} **75.** {8, 9} **77.** {1, 4} **79.** {Δ, two, four, six} **81.** {Δ, #, \$, two, four, six} **83.** 6 **85.** 5 **87.** {#, \$, two, four, six, 10, 01} **89.** {two, four, six} **91.** 4 **93.** 31 **95.** 27 **97.** {1, 2, 3, 4, 5, 6, 7, 8} or U **99.** {1, 3, 5, 7} or A **101.** {1, 7} **103.** {1, 3, 5, 6, 7, 8} **105.** {23, 29, 31, 37, 41, 43, 53, 59, 61, 67, 71} **107.** 60 **109.** {Ashley, Mike, Josh} **111.** {Ashley, Mike, Josh, Emily, Hanna, Ethan} **113.** {Ashley} **115.** {Jacob} **117.** III **119.** I **121.** II **123.** I **125.** IV **127.** II **129.** III **131.** I **133.** {1980, 1990} **135.** {1980, 1990, 2000} **137.** $\varnothing$ **139.** 283 people **153.** makes sense **155.** makes sense **157.** true **159.** false **161.** false **163.** true

165.

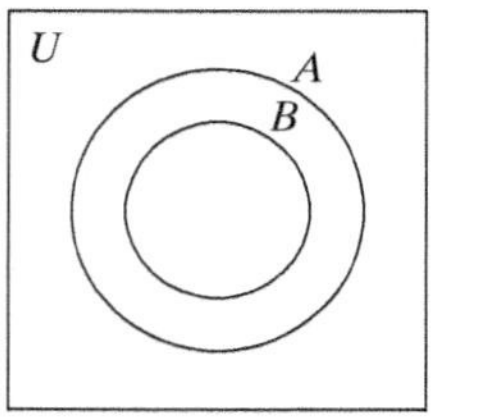

167.

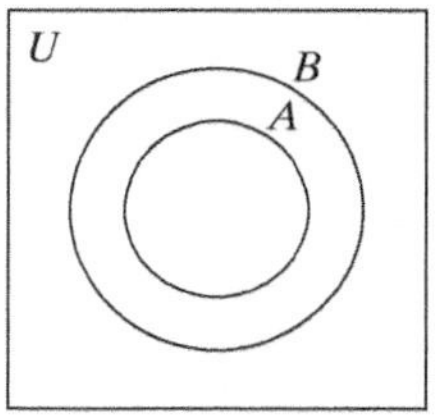

Section 2.4

Check Point Exercises

1. a. {a, b, c, d, f} **b.** {a, b, c, d, f} **c.** {a, b, d}
2. a. {5, 6, 7, 8, 9} **b.** {1, 2, 5, 6, 7, 8, 9, 10, 12} **c.** {5, 6, 7} **d.** {3, 4, 6, 8, 11} **e.** {1, 2, 3, 5, 6, 7, 8, 9, 10, 11, 12}

3.
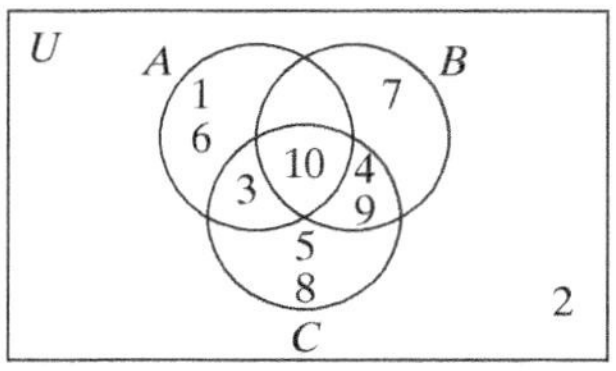

4. a. IV **b.** IV **c.** $(A \cup B)' = A' \cap B'$ **5. a.** II, IV, and V **b.** II, IV, and V **c.** $A \cap (B \cup C) = (A \cap B) \cup (A \cap C)$

Concept and Vocabulary Check

1. inside parentheses **2.** eight **3.** false **4.** true

Exercise Set 2.4

1. {1, 2, 3, 5, 7} **3.** {1, 2, 3, 5, 7} **5.** {2} **7.** {2} **9.** ∅ **11.** {4, 6} **13.** {a, b, g, h} **15.** {a, b, g, h} **17.** {b}
19. {b} **21.** ∅ **23.** {c, d, e, f} **25.** II, III, V, and VI **27.** I, II, IV, V, VI, and VII **29.** II and V **31.** I, IV, VII, and VIII
33. {1, 2, 3, 4, 5, 6, 7, 8} **35.** {1, 2, 3, 4, 5, 6, 7, 8, 9, 10, 11} **37.** {12, 13} **39.** {4, 5, 6} **41.** {6} **43.** {1, 2, 3, 4, 5, 7, 8, 9, 10, 11, 12, 13}

45.
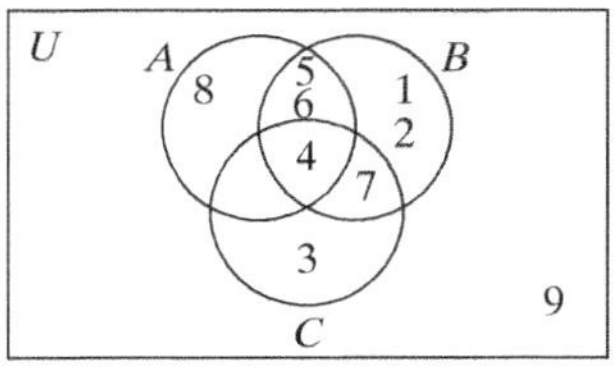

47.
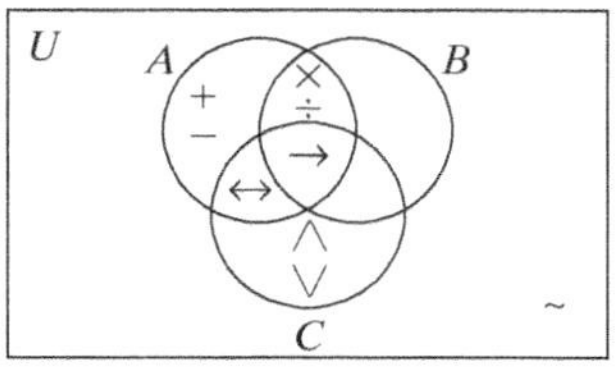

49. a. II **b.** II **c.** $A \cap B = B \cap A$ **51. a.** I, III, and IV **b.** IV **c.** no; Answers will vary. **53.** not equal **55.** not equal
57. equal **59. a.** II, IV, V, VI, and VII **b.** II, IV, V, VI, and VII **c.** $(A \cap B) \cup C = (A \cup C) \cap (B \cup C)$ **61. a.** II, IV, and V
b. I, II, IV, V, and VI **c.** no; Answers will vary. **63.** not true **65.** true; theorem **67.** true; theorem **69. a.** {c, e, f}; {c, e, f}
b. {1, 3, 5, 7, 8}; {1, 3, 5, 7, 8} **c.** $A \cup (B' \cap C') = (A \cup B') \cap (A \cup C')$ **d.** theorem **71.** $(A \cap B)' \cap (A \cup B)$ **73.** $A' \cup B$
75. $(A \cap B) \cup C$ **77.** $A' \cap (B \cup C)$ **79.** {Ann, Jose, Al, Gavin, Amy, Ron, Grace} **81.** {Jose} **83.** {Lily, Emma}
85. {Lily, Emma, Ann, Jose, Lee, Maria, Fred, Ben, Sheila, Ellen, Gary} **87.** {Lily, Emma, Al, Gavin, Amy, Lee, Maria}
89. {Al, Gavin, Amy} **91.** The set of students who scored 90% or above on exam 1 and exam 3 but not on exam 2 **93.** I **95.** V
97. VI **99.** III **101.** IV **103.** VI

105.
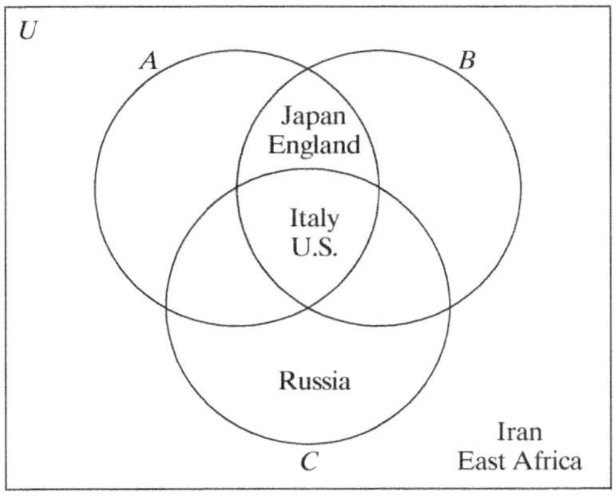

109. does not make sense **111.** makes sense **113.** AB^+ **115.** no

Section 2.5

Check Point Exercises

1. a. 75 **b.** 90 **c.** 20 **d.** 145 **e.** 55 **f.** 70 **g.** 30 **h.** 175 **2. a.** 490 men **b.** 510 men

3.
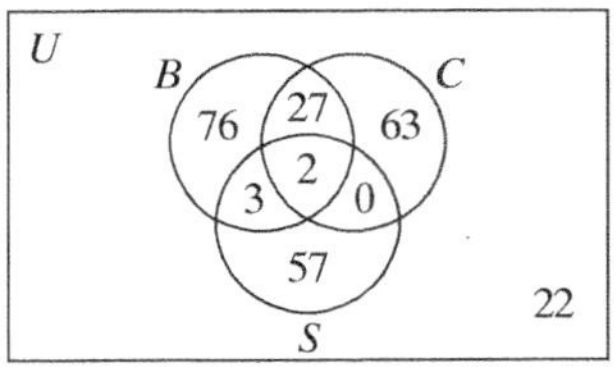

4. a. 63 **b.** 3 **c.** 136 **d.** 30 **e.** 228 **f.** 22

Concept and Vocabulary Check

1. *and/but* **2.** *or* **3.** *not* **4.** innermost; subtraction **5.** true **6.** true **7.** false **8.** true

Exercise Set 2.5

1. 26 **3.** 17 **5.** 37 **7.** 7 **9.** I: 14; III: 22; IV: 5 **11.** 17 **13.** 6 **15.** 28 **17.** 9 **19.** 3 **21.** 19 **23.** 21 **25.** 34 **27.** I: 5; II: 1; III: 4; IV: 3; VI: 1; VII: 8; VIII: 6 **29.** I: 5; II: 10; III: 3; IV: 4; V: 7; VI: 1; VII: 6; VIII: 2 **31.** impossible; There are only 10 elements in set *A* but there are 13 elements in set *A* that are also in sets *B* or *C*. A similar problem exists for set *C*. **33.** 18 **35.** 9 **37.** 9

39. a–d.

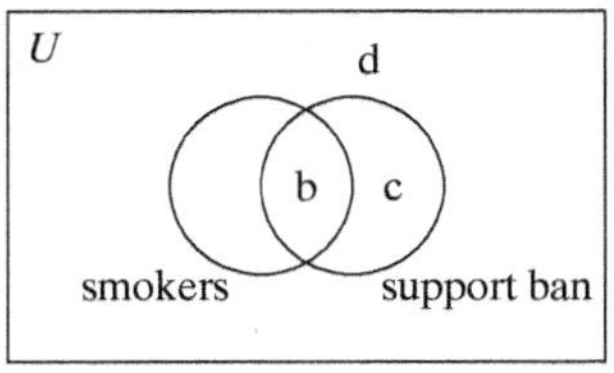

41. a–d.

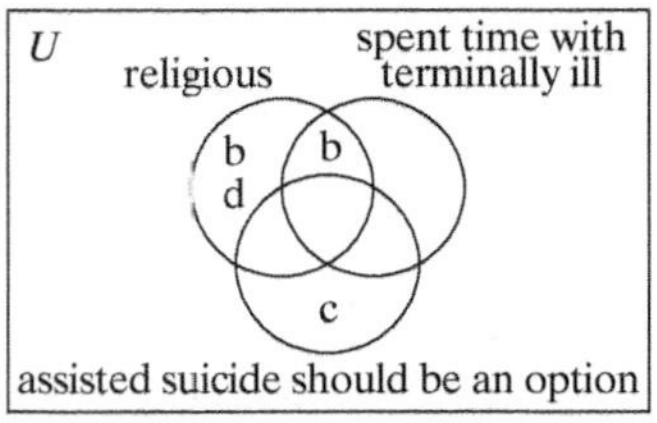

e. Answers will vary.

43.

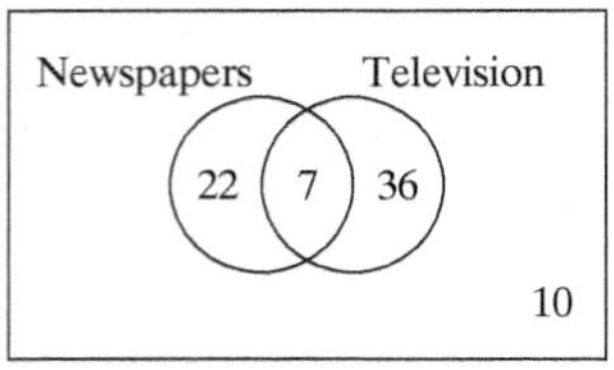

a. 22 **b.** 36 **c.** 65 **d.** 10

45.

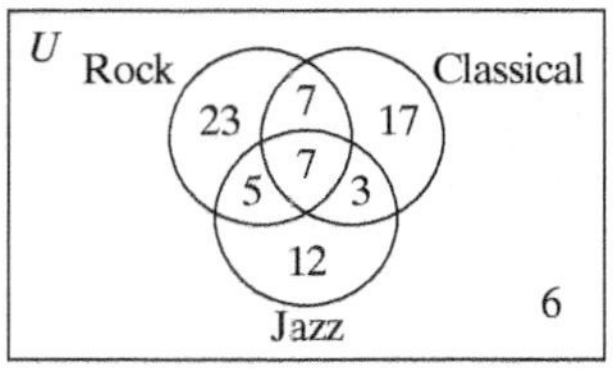

a. 23 **b.** 3 **c.** 32 **d.** 52 **e.** 22 **f.** 6

47. a. 1500 **b.** 1135 **c.** 56 **d.** 327 **e.** 526 **f.** 366 **g.** 1191 **51.** does not make sense **53.** does not make sense **55.** false **57.** false **59. a.** 0 **b.** 30 **c.** 60

Chapter 2 Review Exercises

1. the set of days of the week beginning with the letter T **2.** the set of natural numbers between 1 and 10, inclusive **3.** $\{m, i, s\}$ **4.** $\{8, 9, 10, 11, 12\}$ **5.** $\{1, 2, 3, \ldots, 30\}$ **6.** not empty **7.** empty set **8.** $\in$ **9.** $\notin$ **10.** 12 **11.** 15 **12.** $\neq$ **13.** $\neq$ **14.** equivalent **15.** both **16.** finite **17.** infinite **18.** $\subseteq$ **19.** $\not\subseteq$ **20.** $\subseteq$ **21.** $\subseteq$ **22.** both **23.** false; Answers will vary. **24.** false; Answers will vary. **25.** true **26.** false; Answers will vary. **27.** true **28.** false; It has $2^1 = 2$ subsets. **29.** true **30.** $\varnothing, \{1\}, \{5\}, \{1, 5\}; \{1, 5\}$ **31.** 32; 31 **32.** 8; 7 **33.** $\{1, 2, 4\}$ **34.** $\{1, 2, 3, 4, 6, 7, 8\}$ **35.** $\{5\}$ **36.** $\{6, 7, 8\}$ **37.** $\{6, 7, 8\}$ **38.** $\{4, 5, 6\}$ **39.** $\{2, 3, 6, 7\}$ **40.** $\{1, 4, 5, 6, 8, 9\}$ **41.** $\{4, 5\}$ **42.** $\{1, 2, 3, 6, 7, 8, 9\}$ **43.** $\{2, 3, 7\}$ **44.** $\{6\}$ **45.** $\{1, 2, 3, 4, 5, 6, 7, 8, 9\}$ **46.** 33 **47.** $\{1, 2, 3, 4, 5\}$ **48.** $\{1, 2, 3, 4, 5, 6, 7, 8\}$ or U **49.** $\{c, d, e, f, k, p, r\}$ **50.** $\{f, p\}$ **51.** $\{c, d, f, k, p, r\}$ **52.** $\{c, d, e\}$ **53.** $\{a, b, c, d, e, g, h, p, r\}$ **54.** $\{f\}$

55.

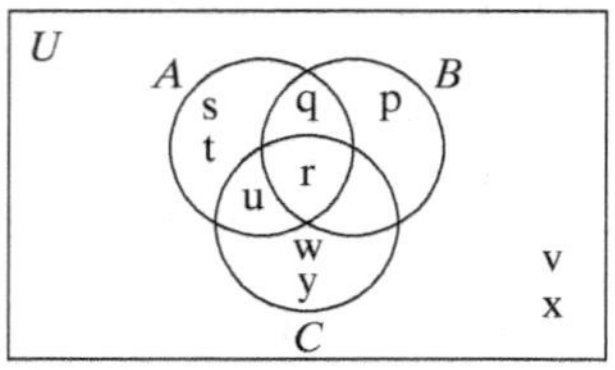

56. Use Figure 2.22.

Set	Regions in the Venn Diagram
A	I, II
B	II, III
$A \cup B$	I, II, III
$(A \cup B)'$	IV

Set	Regions in the Venn Diagram
A'	III, IV
B'	I, IV
$A' \cap B'$	IV

Since $(A \cup B)'$ and $A' \cap B'$ are represented by the same region, $(A \cup B)' = A' \cap B'$.

57. Use Figure 2.23.

Set	Regions in the Venn Diagram
A	I, II, IV, V
B	II, III, V, VI
C	IV, V, VI, VII
$B \cup C$	II, III, IV, V, VI, VII
$A \cap (B \cup C)$	II, IV, V

Set	Regions in the Venn Diagram
A	I, II, IV, V
B	II, III, V, VI
C	IV, V, VI, VII
$B \cap C$	V, VI
$A \cup (B \cap C)$	I, II, IV, V, VI

Since $A \cap (B \cup C)$ and $A \cup (B \cap C)$ are not represented by the same regions, $A \cap (B \cup C) = A \cup (B \cap C)$ is not a theorem.

58. United States: V; Italy: IV; Turkey: VIII; Norway: V; Pakistan: VIII; Iceland: V; Mexico: I

59. a–c

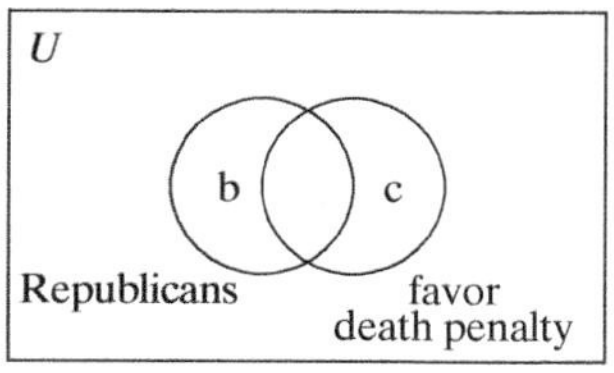

60. a. 250 **b.** 800 **c.** 200 **61. a.** 50 **b.** 26 **c.** 130 **d.** 46 **e.** 0

Chapter 2 Test

1. {18, 19, 20, 21, 22, 23, 24} **2.** false; Answers will vary. **3.** true **4.** true **5.** false; Answers will vary. **6.** true **7.** false; Answers will vary. **8.** false; Answers will vary. **9.** false; Answers will vary. **10.** ∅, {6}, {9}, {6, 9}; {6, 9} **11.** {a, b, c, d, e, f} **12.** {a, b, c, d, f, g} **13.** {b, c, d} **14.** {a, e} **15.** 5 **16.** {b, c, d, i, j, k} **17.** {a} **18.** {a, f, h}

19.

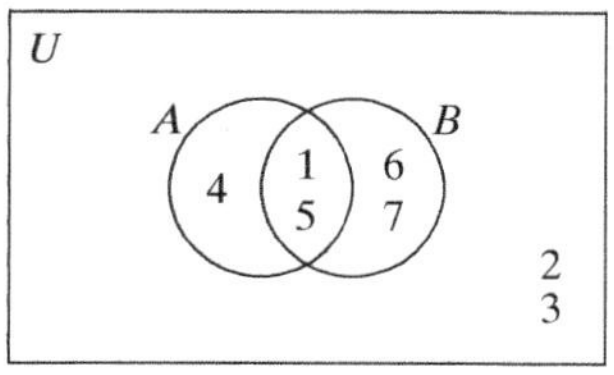

20. theorem **21. a.** V **b.** VII **c.** IV **d.** I **e.** VI

22. a.

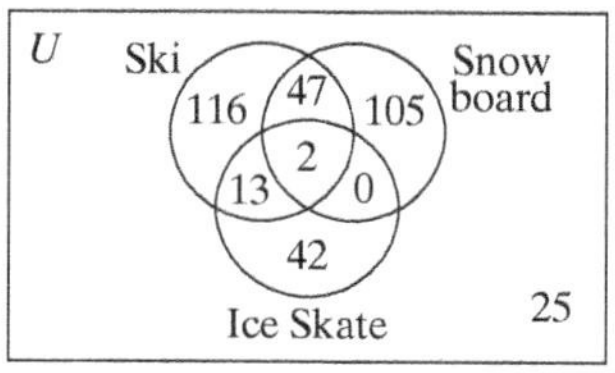

b. 263 **c.** 25 **d.** 62 **e.** 0 **f.** 147 **g.** 116

CHAPTER 3

Section 3.1

Check Point Exercises

1. a. Paris is not the capital of Spain. **b.** July is a month. **2. a.** $\sim p$ **b.** $\sim q$ **3.** Chicago O'Hare is not the world's busiest airport. **4.** Some new tax dollars will not be used to improve education.; At least one new tax dollar will not be used to improve education.

Concept and Vocabulary Check

1. true; false **2.** false; true **3.** $\sim p$; not p **4.** quantified **5.** There are no *A* that are not *B*. **6.** There exists at least one *A* that is a *B*. **7.** All *A* are not *B*. **8.** Not all *A* are *B*. **9.** Some *A* are not *B*. **10.** No *A* are *B*.

Exercise Set 3.1

1. statement **3.** statement **5.** not a statement **7.** statement **9.** not a statement **11.** statement **13.** statement **15.** It is not raining. **17.** Facts cease to exist when they are ignored. **19.** Chocolate in moderation is good for the heart. **21.** $\sim p$ **23.** $\sim r$ **25.** Listening to classical music does not make infants smarter. **27.** Sigmund Freud's father was 20 years older than his mother. **29. a.** There are no whales that are not mammals. **b.** Some whales are not mammals. **31. a.** At least one student is a business major. **b.** No students are business majors. **33. a.** At least one thief is not a criminal. **b.** All thieves are criminals. **35. a.** All Democratic presidents have not been impeached. **b.** Some Democratic presidents have been impeached. **37. a.** All seniors graduated. **b.** Some seniors did not graduate. **39. a.** Some parrots are not pets. **b.** All parrots are pets. **41. a.** No atheist is a churchgoer. **b.** Some atheists are churchgoers. **43.** Some Africans have Jewish ancestry. **45.** Some rap is not hip-hop. **47. a.** All birds are parrots. **b.** false; Some birds are not parrots. **49. a.** No college students are business majors. **b.** false; Some college students are business majors. **51. a.** All people like Sara Lee. **b.** Some people don't like Sara Lee. **53. a.** No safe thing is exciting. **b.** Some safe things are exciting. **55. a.** Some great actors are not a Tom Hanks. **b.** All great actors are a Tom Hanks. **57.** b **59.** true **61.** false; Some college students in the United States are willing to marry without romantic love. **63.** true **65.** false; The sentence "5% of college students in Australia are willing to marry without romantic love" is a statement. **75.** does not make sense **77.** does not make sense

Section 3.2

Check Point Exercises

1. a. $q \wedge p$ **b.** $\sim p \wedge q$ **2. a.** $p \vee q$ **b.** $q \vee \sim p$ **3. a.** $\sim p \rightarrow \sim q$ **b.** $q \rightarrow \sim p$ **4.** $\sim q \rightarrow p$ **5. a.** $q \leftrightarrow p$ **b.** $\sim p \leftrightarrow \sim q$ **6. a.** It is not true that he earns $105,000 yearly and that he is often happy. **b.** He is not often happy and he earns $105,000 yearly. **c.** It is not true that if he is often happy then he earns $105,000 yearly. **7. a.** If the plant is fertilized and watered, then it will not wilt. **b.** The plant is fertilized, and if it is watered then it will not wilt. **8.** p: There is too much homework.; q: A teacher is boring.; r: I take the class. **a.** $(p \vee q) \rightarrow \sim r$ **b.** $p \vee (q \rightarrow \sim r)$

Concept and Vocabulary Check

1. $p \wedge q$; conjunction **2.** $p \vee q$; disjunction **3.** $p \rightarrow q$; conditional **4.** $p \leftrightarrow q$; biconditional **5.** true **6.** false **7.** true **8.** false **9.** true **10.** true **11.** true **12.** true **13.** false

Exercise Set 3.2

1. $p \wedge q$ **3.** $q \wedge \sim p$ **5.** $\sim q \wedge p$ **7.** $p \vee q$ **9.** $p \vee \sim q$ **11.** $p \rightarrow q$ **13.** $\sim p \rightarrow \sim q$ **15.** $\sim q \rightarrow \sim p$ **17.** $p \rightarrow q$ **19.** $p \rightarrow \sim q$ **21.** $q \rightarrow \sim p$ **23.** $p \rightarrow \sim q$ **25.** $q \rightarrow \sim p$ **27.** $p \leftrightarrow q$ **29.** $\sim q \leftrightarrow \sim p$ **31.** $q \leftrightarrow p$ **33.** The heater is not working and the house is cold. **35.** The heater is working or the house is not cold. **37.** If the heater is working, then the house is not cold. **39.** The heater is working if and only if the house is not cold. **41.** It is July 4th and we are not having a barbecue. **43.** It is not July 4th or we are having a barbecue. **45.** If we are having a barbecue, then it is not July 4th. **47.** It is not July 4th if and only if we are having a barbecue. **49.** It is not true that Romeo loves Juliet and Juliet loves Romeo. **51.** Romeo does not love Juliet but Juliet loves Romeo. **53.** It is not true that Juliet loves Romeo or Romeo loves Juliet; Neither Juliet loves Romeo nor Romeo loves Juliet. **55.** Juliet does not love Romeo or Romeo loves Juliet. **57.** Romeo does not love Juliet and Juliet does not love Romeo. **59.** $(p \wedge q) \vee r$ **61.** $(p \vee \sim q) \rightarrow r$ **63.** $r \leftrightarrow (p \wedge \sim q)$ **65.** $(p \wedge \sim q) \rightarrow r$ **67.** If the temperature is above 85° and we have finished studying, then we will go to the beach. **69.** The temperature is above 85°, and if we have finished studying then we will go to the beach. **71.** $\sim r \rightarrow (\sim p \vee \sim q)$; If we do not go to the beach, then the temperature is not above 85° or we have not finished studying. **73.** If we do not go to the beach then we have not finished studying, or the temperature is above 85°. **75.** $r \leftrightarrow (p \wedge q)$; We will go to the beach if and only if the temperature is above 85° and we have finished studying. **77.** The temperature is above 85° if and only if we have finished studying, and we go to the beach. **79.** If we do not go to the beach, then it is not true that both the temperature is above 85° and we have finished studying. **81.** p: I like the teacher.; q: The course is interesting.; r: I miss class.; $(p \vee q) \rightarrow \sim r$ **83.** p: I like the teacher.; q: The course is interesting.; r: I miss class.; $p \vee (q \rightarrow \sim r)$ **85.** p: I like the teacher.; q: The course is interesting.; r: I miss class.; $r \leftrightarrow \sim (p \wedge q)$ **87.** p: I like the teacher.; q: The course is interesting.; r: I miss class.; $(p \rightarrow \sim r) \leftrightarrow q$ **89.** p: I like the teacher.; q: The course is interesting.; r: I miss class.; s: I spend extra time reading the book.; $(\sim p \wedge r) \rightarrow (\sim q \vee s)$ **91.** p: Being French is necessary for being a Parisian.; q: Being German is necessary for being a Berliner.; $\sim p \rightarrow \sim q$ **93.** p: You file an income tax report.; q: You file a complete statement of earnings.; r: You are a taxpayer.; s: You are an authorized tax preparer.; $(r \vee s) \rightarrow (p \wedge q)$ **95.** p: You are wealthy.; q: You are happy.; r: You live contentedly.; $\sim (p \rightarrow (q \wedge r))$ **97.** $[p \rightarrow (q \vee r)] \leftrightarrow (p \wedge r)$ **99.** $(p \rightarrow p) \leftrightarrow [(p \wedge p) \rightarrow \sim p]$ **101.** $\sim p \rightarrow q$ **103.** $(p \rightarrow q) \wedge \sim q$ **105.** $((p \rightarrow q) \wedge \sim p) \rightarrow \sim q$ **115.** makes sense **117.** does not make sense **119.** p: Shooting unarmed civilians is morally justifiable.; q: Bombing unarmed civilians is morally justifiable.; $((p \leftrightarrow q) \wedge \sim p) \rightarrow \sim q$

Section 3.3

Check Point Exercises

1. a. false **b.** true **c.** false **d.** true

2.

p	q	$p \vee q$	$\sim(p \vee q)$
T	T	T	F
T	F	T	F
F	T	T	F
F	F	F	T

$\sim(p \vee q)$ is true when both p and q are false; otherwise, $\sim(p \vee q)$ is false.

3.

p	q	$\sim p$	$\sim q$	$\sim p \wedge \sim q$
T	T	F	F	F
T	F	F	T	F
F	T	T	F	F
F	F	T	T	T

$\sim p \wedge \sim q$ is true when both p and q are false; otherwise, $\sim p \wedge \sim q$ is false.

4.

p	q	$\sim q$	$p \wedge \sim q$	$\sim p$	$(p \wedge \sim q) \vee \sim p$
T	T	F	F	F	F
T	F	T	T	F	T
F	T	F	F	T	T
F	F	T	F	T	T

$(p \wedge \sim q) \vee \sim p$ is false when both p and q are true; otherwise, $(p \wedge \sim q) \vee \sim p$ is true.

5.

p	$\sim p$	$p \wedge \sim p$
T	F	F
F	T	F

$p \wedge \sim p$ is false in all cases.

6. p: I study hard.; q: I ace the final.; r: I fail the course.

a.

p	q	r	$q \vee r$	$p \wedge (q \vee r)$
T	T	T	T	T
T	T	F	T	T
T	F	T	T	T
T	F	F	F	F
F	T	T	T	F
F	T	F	T	F
F	F	T	T	F
F	F	F	F	F

b. false **7.** true

Concept and Vocabulary Check

1. opposite **2.** both p and q are true **3.** p and q are false **4.** true **5.** true **6.** true **7.** false **8.** true

Exercise Set 3.3

1. true **3.** false **5.** false **7.** false **9.** true **11.** true **13.** true **15.** true

17.

p	$\sim p$	$\sim p \wedge p$
T	F	F
F	T	F

19.

p	q	$\sim p$	$\sim p \wedge q$
T	T	F	F
T	F	F	F
F	T	T	T
F	F	T	F

21.

p	q	$p \vee q$	$\sim(p \vee q)$
T	T	T	F
T	F	T	F
F	T	T	F
F	F	F	T

23.

p	q	$\sim p$	$\sim q$	$\sim p \wedge \sim q$
T	T	F	F	F
T	F	F	T	F
F	T	T	F	F
F	F	T	T	T

25.

p	q	$\sim q$	$p \vee \sim q$
T	T	F	T
T	F	T	T
F	T	F	F
F	F	T	T

27.

p	q	$\sim p$	$\sim p \vee q$	$\sim(\sim p \vee q)$
T	T	F	T	F
T	F	F	F	T
F	T	T	T	F
F	F	T	T	F

29.

p	q	$\sim p$	$p \vee q$	$(p \vee q) \wedge \sim p$
T	T	F	T	F
T	F	F	T	F
F	T	T	T	T
F	F	T	F	F

31.

p	q	$\sim p$	$\sim q$	$p \wedge \sim q$	$\sim p \vee (p \wedge \sim q)$
T	T	F	F	F	F
T	F	F	T	T	T
F	T	T	F	F	T
F	F	T	T	F	T

33.

p	q	$\sim p$	$\sim q$	$p \vee q$	$\sim p \vee \sim q$	$(p \vee q) \wedge (\sim p \vee \sim q)$
T	T	F	F	T	F	F
T	F	F	T	T	T	T
F	T	T	F	T	T	T
F	F	T	T	F	T	F

35.

p	q	$\sim q$	$p \wedge \sim q$	$p \wedge q$	$(p \wedge \sim q) \vee (p \wedge q)$
T	T	F	F	T	T
T	F	T	T	F	T
F	T	F	F	F	F
F	F	T	F	F	F

37.

p	q	r	$\sim q$	$\sim q \vee r$	$p \wedge (\sim q \vee r)$
T	T	T	F	T	T
T	T	F	F	F	F
T	F	T	T	T	T
T	F	F	T	T	T
F	T	T	F	T	F
F	T	F	F	F	F
F	F	T	T	T	F
F	F	F	T	T	F

39.

p	q	r	$\sim p$	$\sim q$	$r \wedge \sim p$	$(r \wedge \sim p) \vee \sim q$
T	T	T	F	F	F	F
T	T	F	F	F	F	F
T	F	T	F	T	F	T
T	F	F	F	T	F	T
F	T	T	T	F	T	T
F	T	F	T	F	F	F
F	F	T	T	T	T	T
F	F	F	T	T	F	T

41.

p	q	r	$p \vee q$	$\sim(p \vee q)$	$\sim r$	$\sim(p \vee q) \wedge \sim r$
T	T	T	T	F	F	F
T	T	F	T	F	T	F
T	F	T	T	F	F	F
T	F	F	T	F	T	F
F	T	T	T	F	F	F
F	T	F	T	F	T	F
F	F	T	F	T	F	F
F	F	F	F	T	T	T

43. a. p: You did the dishes.; q: You left the room a mess.; $\sim p \wedge q$ **b.** See truth table for Exercise 19. **c.** The statement is true when p is false and q is true. **45. a.** p: I bought a meal ticket.; q: I used it.; $\sim(p \wedge \sim q)$

b.

p	q	$\sim q$	$p \wedge \sim q$	$\sim(p \wedge \sim q)$
T	T	F	F	T
T	F	T	T	F
F	T	F	F	T
F	F	T	F	T

c. Answers will vary; an example is: The statement is true when p and q are true.

47. a. p: The student is intelligent.; q: The student is an overachiever.; $(p \vee q) \wedge \sim q$

b.

p	q	$\sim q$	$p \vee q$	$(p \vee q) \wedge \sim q$
T	T	F	T	F
T	F	T	T	T
F	T	F	T	F
F	F	T	F	F

c. The statement is true when p is true and q is false.

49. a. p: Married people are healthier than single people.; q: Married people are more economically stable than single people.; r: Children of married people do better on a variety of indicators.; $(p \wedge q) \wedge r$

b.

p	q	r	$p \wedge q$	$(p \wedge q) \wedge r$
T	T	T	T	T
T	T	F	T	F
T	F	T	F	F
T	F	F	F	F
F	T	T	F	F
F	T	F	F	F
F	F	T	F	F
F	F	F	F	F

c. The statement is true when p, q, and r are all true.

51. a. p: I go to office hours.; q: I ask questions.; r: My professor remembers me.; $(p \wedge q) \vee {\sim}r$

b.

p	q	r	${\sim}r$	$p \wedge q$	$(p \wedge q) \vee {\sim}r$
T	T	T	F	T	T
T	T	F	T	T	T
T	F	T	F	F	F
T	F	F	T	F	T
F	T	T	F	F	F
F	T	F	T	F	T
F	F	T	F	F	F
F	F	F	T	F	T

c. Answers will vary; an example is: The statement is true when p, q, and r are all true.

53. false **55.** true **57.** true **59.** false **61.** true

63.

p	q	${\sim}[{\sim}(p \wedge {\sim}q) \vee {\sim}({\sim}p \vee q)]$
T	T	F
T	F	F
F	T	F
F	F	F

65.

p	q	r	$[(p \wedge {\sim}r) \vee (q \wedge {\sim}r)] \wedge {\sim}({\sim}p \vee r)$
T	T	T	F
T	T	F	T
T	F	T	F
T	F	F	T
F	T	T	F
F	T	F	F
F	F	T	F
F	F	F	F

67. p: You notice this notice.; q: You notice this notice is not worth noticing.; $(p \vee {\sim}p) \wedge q$

p	q	${\sim}p$	$p \vee {\sim}p$	$(p \vee {\sim}p) \wedge q$
T	T	F	T	T
T	F	F	T	F
F	T	T	T	T
F	F	T	T	F

The statement is true when q is true.

69. $p: x \leq 3$; $q: x \geq 7$; $\sim(p \vee q) \wedge (\sim p \wedge \sim q)$

p	q	$\sim p$	$\sim q$	$p \vee q$	$\sim(p \vee q)$	$\sim p \wedge \sim q$	$\sim(p \vee q) \wedge (\sim p \wedge \sim q)$
T	T	F	F	T	F	F	F
T	F	F	T	T	F	F	F
F	T	T	F	T	F	F	F
F	F	T	T	F	T	T	T

The statement is true when both p and q are false.

71. The percent body fat in women peaks at age 55 and the percent body fat in men does not peak at age 65.; false **73.** The percent body fat in women does not peak at age 55 and men have more than 24% body fat at age 25.; false **75.** The percent body fat in women peaks at age 55 or the percent body fat in men does not peak at age 65.; true **77.** The percent body fat in women does not peak at age 55 or men have more than 24% body fat at age 25.; false **79.** The percent body fat in women peaks at age 55 and the percent body fat in men peaks at age 65, or men have more than 24% body fat at age 25.; true **81.** p: More than 10% named business.; q: 9% named engineering.; $p \vee \sim q$; true **83.** p: 7.5% named teaching.; q: 12% named business.; $p \vee \sim q$; true **85. a.** Hora Gershwin **b.** Bolera Mozart does not have a master's degree in music. Cha-Cha Bach does not either play three instruments or have five years experience playing with a symphony orchestra. **95.** does not make sense **97.** does not make sense

101.

p	q	$p \veebar q$
T	T	F
T	F	T
F	T	T
F	F	F

Section 3.4

Check Point Exercises

1.

p	q	$\sim p$	$\sim q$	$\sim p \rightarrow \sim q$
T	T	F	F	T
T	F	F	T	T
F	T	T	F	F
F	F	T	T	T

2. The table shows that $[(p \rightarrow q) \wedge \sim q] \rightarrow \sim p$ is always true; therefore, it is a tautology.

p	q	$\sim p$	$\sim q$	$p \rightarrow q$	$(p \rightarrow q) \wedge \sim q$	$[(p \rightarrow q) \wedge \sim q] \rightarrow \sim p$
T	T	F	F	T	F	T
T	F	F	T	F	F	T
F	T	T	F	T	F	T
F	F	T	T	T	T	T

3. a. p: You use Hair Grow.; q: You apply it daily.; r: You go bald.

p	q	r	$\sim r$	$p \wedge q$	$(p \wedge q) \rightarrow \sim r$
T	T	T	F	T	F
T	T	F	T	T	T
T	F	T	F	F	T
T	F	F	T	F	T
F	T	T	F	F	T
F	T	F	T	F	T
F	F	T	F	F	T
F	F	F	T	F	T

b. no

4.

p	q	$p \vee q$	$\sim p$	$\sim p \rightarrow q$	$(p \vee q) \leftrightarrow (\sim p \rightarrow q)$
T	T	T	F	T	T
T	F	T	F	T	T
F	T	T	T	T	T
F	F	F	T	F	T

Because all cases are true, the statement is a tautology.

5. true

Concept and Vocabulary Check

1. p is true and q is false **2.** tautology; implications; self-contradiction **3.** p and q have the same truth value **4.** false **5.** false **6.** true **7.** true

Exercise Set 3.4

1.

p	q	$\sim q$	$p \rightarrow \sim q$
T	T	F	F
T	F	T	T
F	T	F	T
F	F	T	T

3.

p	q	$q \rightarrow p$	$\sim(q \rightarrow p)$
T	T	T	F
T	F	T	F
F	T	F	T
F	F	T	F

5.

p	q	$p \wedge q$	$p \vee q$	$(p \wedge q) \rightarrow (p \vee q)$
T	T	T	T	T
T	F	F	T	T
F	T	F	T	T
F	F	F	F	T

7.

p	q	$p \rightarrow q$	$\sim q$	$(p \rightarrow q) \wedge \sim q$
T	T	T	F	F
T	F	F	T	F
F	T	T	F	F
F	F	T	T	T

9.

p	q	r	$p \vee q$	$(p \vee q) \rightarrow r$
T	T	T	T	T
T	T	F	T	F
T	F	T	T	T
T	F	F	T	F
F	T	T	T	T
F	T	F	T	F
F	F	T	F	T
F	F	F	F	T

11.

p	q	r	$p \wedge q$	$r \rightarrow (p \wedge q)$
T	T	T	T	T
T	T	F	T	T
T	F	T	F	F
T	F	F	F	T
F	T	T	F	F
F	T	F	F	T
F	F	T	F	F
F	F	F	F	T

13.

p	q	r	$\sim q$	$\sim r$	$\sim q \rightarrow p$	$\sim r \wedge (\sim q \rightarrow p)$
T	T	T	F	F	T	F
T	T	F	F	T	T	T
T	F	T	T	F	T	F
T	F	F	T	T	T	T
F	T	T	F	F	T	F
F	T	F	F	T	T	T
F	F	T	T	F	F	F
F	F	F	T	T	F	F

15.

p	q	r	$\sim q$	$p \wedge r$	$\sim (p \wedge r)$	$\sim q \vee r$	$\sim (p \wedge r) \rightarrow (\sim q \vee r)$
T	T	T	F	T	F	T	T
T	T	F	F	F	T	F	F
T	F	T	T	T	F	T	T
T	F	F	T	F	T	T	T
F	T	T	F	F	T	T	T
F	T	F	F	F	T	F	F
F	F	T	T	F	T	T	T
F	F	F	T	F	T	T	T

17.

p	q	$\sim q$	$p \leftrightarrow \sim q$
T	T	F	F
T	F	T	T
F	T	F	T
F	F	T	F

19.

p	q	$p \leftrightarrow q$	$\sim (p \leftrightarrow q)$
T	T	T	F
T	F	F	T
F	T	F	T
F	F	T	F

21.

p	q	$p \leftrightarrow q$	$(p \leftrightarrow q) \rightarrow p$
T	T	T	T
T	F	F	T
F	T	F	T
F	F	T	F

23.

p	q	$\sim p$	$\sim p \leftrightarrow q$	$\sim p \rightarrow q$	$(\sim p \leftrightarrow q) \rightarrow (\sim p \rightarrow q)$
T	T	F	F	T	T
T	F	F	T	T	T
F	T	T	T	T	T
F	F	T	F	F	T

25.

p	q	$p \wedge q$	$q \rightarrow p$	$(p \wedge q) \wedge (q \rightarrow p)$	$[(p \wedge q) \wedge (q \rightarrow p)] \leftrightarrow (p \wedge q)$
T	T	T	T	T	T
T	F	F	T	F	T
F	T	F	F	F	T
F	F	F	T	F	T

27.

p	q	r	$\sim r$	$p \leftrightarrow q$	$(p \leftrightarrow q) \rightarrow \sim r$
T	T	T	F	T	F
T	T	F	T	T	T
T	F	T	F	F	T
T	F	F	T	F	T
F	T	T	F	F	T
F	T	F	T	F	T
F	F	T	F	T	F
F	F	F	T	T	T

29.

p	q	r	$p \wedge r$	$q \vee r$	$\sim(q \vee r)$	$(p \wedge r) \leftrightarrow \sim(q \vee r)$
T	T	T	T	T	F	F
T	T	F	F	T	F	T
T	F	T	T	T	F	F
T	F	F	F	F	T	F
F	T	T	F	T	F	T
F	T	F	F	T	F	T
F	F	T	F	T	F	T
F	F	F	F	F	T	F

31.

p	q	r	$\sim q$	$\sim q \wedge p$	$r \vee (\sim q \wedge p)$	$\sim p$	$[r \vee (\sim q \wedge p)] \leftrightarrow \sim p$
T	T	T	F	F	T	F	F
T	T	F	F	F	F	F	T
T	F	T	T	T	T	F	F
T	F	F	T	T	T	F	F
F	T	T	F	F	T	T	T
F	T	F	F	F	F	T	F
F	F	T	T	F	T	T	T
F	F	F	T	F	F	T	F

33. neither

p	q	$p \rightarrow q$	$(p \rightarrow q) \wedge q$	$[(p \rightarrow q) \wedge q] \rightarrow p$
T	T	T	T	T
T	F	F	F	T
F	T	T	T	F
F	F	T	F	T

35. tautology

p	q	$\sim p$	$\sim q$	$p \rightarrow q$	$(p \rightarrow q) \wedge \sim q$	$[(p \rightarrow q) \wedge \sim q] \rightarrow \sim p$
T	T	F	F	T	F	T
T	F	F	T	F	F	T
F	T	T	F	T	F	T
F	F	T	T	T	T	T

37. neither

p	q	$\sim q$	$p \vee q$	$(p \vee q) \wedge p$	$[(p \vee q) \wedge p] \rightarrow \sim q$
T	T	F	T	T	F
T	F	T	T	T	T
F	T	F	T	F	T
F	F	T	F	F	T

39. tautology

p	q	$\sim p$	$p \rightarrow q$	$\sim p \vee q$	$(p \rightarrow q) \rightarrow (\sim p \vee q)$
T	T	F	T	T	T
T	F	F	F	F	T
F	T	T	T	T	T
F	F	T	T	T	T

41. self-contradiction

p	q	$\sim p$	$\sim q$	$p \wedge q$	$\sim p \vee \sim q$	$(p \wedge q) \wedge (\sim p \vee \sim q)$
T	T	F	F	T	F	F
T	F	F	T	F	T	F
F	T	T	F	F	T	F
F	F	T	T	F	T	F

43. neither

p	q	$\sim p$	$\sim q$	$p \wedge q$	$\sim(p \wedge q)$	$\sim p \wedge \sim q$	$\sim(p \wedge q) \leftrightarrow (\sim p \wedge \sim q)$
T	T	F	F	T	F	F	T
T	F	F	T	F	T	F	F
F	T	T	F	F	T	F	F
F	F	T	T	F	T	T	T

45. neither

p	q	$p \rightarrow q$	$q \rightarrow p$	$(p \rightarrow q) \leftrightarrow (q \rightarrow p)$
T	T	T	T	T
T	F	F	T	F
F	T	T	F	F
F	F	T	T	T

47. tautology

p	q	$\sim p$	$p \rightarrow q$	$\sim p \vee q$	$(p \rightarrow q) \leftrightarrow (\sim p \vee q)$
T	T	F	T	T	T
T	F	F	F	F	T
F	T	T	T	T	T
F	F	T	T	T	T

49. tautology

p	q	$p \leftrightarrow q$	$p \rightarrow q$	$q \rightarrow p$	$(q \rightarrow p) \wedge (p \rightarrow q)$	$(p \leftrightarrow q) \leftrightarrow [(q \rightarrow p) \wedge (p \rightarrow q)]$
T	T	T	T	T	T	T
T	F	F	F	T	F	T
F	T	F	T	F	F	T
F	F	T	T	T	T	T

51. neither

p	q	r	$\sim p$	$p \wedge q$	$\sim p \vee r$	$(p \wedge q) \leftrightarrow (\sim p \vee r)$
T	T	T	F	T	T	T
T	T	F	F	T	F	F
T	F	T	F	F	T	F
T	F	F	F	F	F	T
F	T	T	T	F	T	F
F	T	F	T	F	T	F
F	F	T	T	F	T	F
F	F	F	T	F	T	F

53. tautology

p	q	r	$p \rightarrow q$	$q \rightarrow r$	$(p \rightarrow q) \wedge (q \rightarrow r)$	$p \rightarrow r$	$[(p \rightarrow q) \wedge (q \rightarrow r)] \rightarrow (p \rightarrow r)$
T	T	T	T	T	T	T	T
T	T	F	T	F	F	F	T
T	F	T	F	T	F	T	T
T	F	F	F	T	F	F	T
F	T	T	T	T	T	T	T
F	T	F	T	F	F	T	T
F	F	T	T	T	T	T	T
F	F	F	T	T	T	T	T

55. neither

p	q	r	$[(q \rightarrow r) \wedge (r \rightarrow \sim p)] \leftrightarrow (q \wedge p)$
T	T	T	F
T	T	F	F
T	F	T	T
T	F	F	F
F	T	T	F
F	T	F	T
F	F	T	F
F	F	F	F

57. a. p: You do homework right after class.; q: You fall behind.; $(p \rightarrow \sim q) \wedge (\sim p \rightarrow q)$

b.

p	q	$\sim p$	$\sim q$	$p \rightarrow \sim q$	$\sim p \rightarrow q$	$(p \rightarrow \sim q) \wedge (\sim p \rightarrow q)$
T	T	F	F	F	T	F
T	F	F	T	T	T	T
F	T	T	F	T	T	T
F	F	T	T	T	F	F

c. Answers will vary; an example is: The statement is false when p and q are both true.

59. a. p: You cut and paste from the Internet.; q: You cite the source.; r: You are charged with plagiarism.; $(p \wedge \sim q) \rightarrow r$

b.

p	q	r	$\sim q$	$p \wedge \sim q$	$(p \wedge \sim q) \rightarrow r$
T	T	T	F	F	T
T	T	F	F	F	T
T	F	T	T	T	T
T	F	F	T	T	F
F	T	T	F	F	T
F	T	F	F	F	T
F	F	T	T	F	T
F	F	F	T	F	T

c. The statement is false when p is true and q and r are false.

61. a. p: You are comfortable in your room.; q: You are honest with your roommate.; r: You enjoy the college experience.; $(p \leftrightarrow q) \vee \sim r$

b.

p	q	r	$\sim r$	$p \leftrightarrow q$	$(p \leftrightarrow q) \vee \sim r$
T	T	T	F	T	T
T	T	F	T	T	T
T	F	T	F	F	F
T	F	F	T	F	T
F	T	T	F	F	F
F	T	F	T	F	T
F	F	T	F	T	T
F	F	F	T	T	T

c. Answers will vary; an example is: The statement is false when p and r are true and q is false.

63. a. p: I enjoy the course.; q: I choose the class based on the professor.; r: I choose the class based on the course description.; $p \leftrightarrow (q \wedge \sim r)$

b.

p	q	r	$\sim r$	$q \wedge \sim r$	$p \leftrightarrow (q \wedge \sim r)$
T	T	T	F	F	F
T	T	F	T	T	T
T	F	T	F	F	F
T	F	F	T	F	F
F	T	T	F	F	T
F	T	F	T	T	F
F	F	T	F	F	T
F	F	F	T	F	T

c. Answers will vary; an example is: The statement is false when p, q, and r are all true.

65. false **67.** true **69.** false **71.** true **73.** true

75. $(p \rightarrow q) \leftrightarrow [(p \wedge q) \rightarrow \sim p]$

p	q	$(p \rightarrow q) \leftrightarrow [(p \wedge q) \rightarrow \sim p]$
T	T	F
T	F	F
F	T	T
F	F	T

77. $[p \rightarrow (\sim q \vee r)] \leftrightarrow (p \wedge r)$

p	q	r	$[p \rightarrow (\sim q \vee r)] \leftrightarrow (p \wedge r)$
T	T	T	T
T	T	F	T
T	F	T	T
T	F	F	F
F	T	T	F
F	T	F	F
F	F	T	F
F	F	F	F

79. p: You love a person.; q: You marry that person.; $(q \rightarrow p) \wedge (\sim p \rightarrow \sim q)$

p	q	$\sim p$	$\sim q$	$q \rightarrow p$	$\sim p \rightarrow \sim q$	$(q \rightarrow p) \wedge (\sim p \rightarrow \sim q)$
T	T	F	F	T	T	T
T	F	F	T	T	T	T
F	T	T	F	F	F	F
F	F	T	T	T	T	T

The statement is false when p is false and q is true.

81. p: You are happy.; q: You live contentedly.; r: You are wealthy.; $\sim [r \rightarrow (p \wedge q)]$

p	q	r	$p \wedge q$	$r \rightarrow (p \wedge q)$	$\sim [r \rightarrow (p \wedge q)]$
T	T	T	T	T	F
T	T	F	T	T	F
T	F	T	F	F	T
T	F	F	F	T	F
F	T	T	F	F	T
F	T	F	F	T	F
F	F	T	F	F	T
F	F	F	F	T	F

Answers will vary; an example is: The statement is false when p, q, and r are all true.

83. p: There was an increase in the percentage who believed in God.; q: There was a decrease in the percentage who believed in Heaven.; r: There was an increase in the percentage who believed in the devil.; $(p \wedge q) \rightarrow r$; true **85.** p: There was a decrease in the percentage who believed in God.; q: There was an increase in the percentage who believed in Heaven.; r: The percentage believing in the devil decreased.; $(p \leftrightarrow q) \vee r$; true **87.** p: Fifteen percent are capitalists.; q: Thirty-four percent are members of the upper-middle class.; r: The number of working poor exceeds the number belonging to the working class.; $(p \vee \sim q) \leftrightarrow r$; false **89.** p: There are more people in the lower-middle class than in the capitalist and upper-middle classes combined.; q: One percent are capitalists.; r: Thirty-four percent belong to the upper-middle class.; $p \rightarrow (q \wedge r)$; false **97.** makes sense **99.** does not make sense **101.** Answers will vary; examples are $p \rightarrow q$, $\sim p$, $(p \rightarrow q) \vee \sim p$, and $\sim p \rightarrow [(p \rightarrow q) \vee \sim p]$.

Section 3.5

Check Point Exercises

1. a.

p	q	$\sim q$	$p \vee q$	$\sim q \rightarrow p$
T	T	F	T	T
T	F	T	T	T
F	T	F	T	T
F	F	T	F	F

The statements are equivalent since their truth values are the same.

b. If I don't lose my scholarship, then I attend classes.

2.

p	$\sim p$	$\sim(\sim p)$	$\sim[\sim(\sim p)]$
T	F	T	F
F	T	F	T

Since the truth values are the same, $\sim[\sim(\sim p)] \equiv \sim p$.

3. c **4. a.** If you're not driving too closely, then you can't read this. **b.** If it's not time to do the laundry, then you have clean underwear. **c.** If supervision during exams is required, then some students are not honest. **d.** $q \rightarrow (p \vee r)$ **5.** Converse: If you don't see a Club Med, then you are in Iran.; Inverse: If you are not in Iran, then you see a Club Med.; Contrapositive: If you see a Club Med, then you are not in Iran.

Concept and Vocabulary Check

1. equivalent; $\equiv$ **2.** $\sim q \rightarrow \sim p$ **3.** $q \rightarrow p$ **4.** $\sim p \rightarrow \sim q$ **5.** equivalent; converse; inverse **6.** false **7.** false

Exercise Set 3.5

1. a.

p	q	$\sim p$	$p \vee q$	$\sim p \rightarrow q$
T	T	F	T	T
T	F	F	T	T
F	T	T	T	T
F	F	T	F	F

b. The United States supports the development of solar-powered cars or it will suffer increasing atmospheric pollution.

3. not equivalent **5.** equivalent **7.** equivalent **9.** not equivalent **11.** not equivalent **13.** equivalent **15.** a **17.** c **19.** Converse: If I am in Illinois, then I am in Chicago.; Inverse: If I am not in Chicago, then I am not in Illinois.; Contrapositive: If I am not in Illinois, then I am not in Chicago. **21.** Converse: If I cannot hear you, then the stereo is playing.; Inverse: If the stereo is not playing, then I can hear you.; Contrapositive: If I can hear you, then the stereo is not playing. **23.** Converse: If you die, you don't laugh.; Inverse: If you laugh, you don't die.; Contrapositive: If you don't die, you laugh. **25.** Converse: If all troops were withdrawn, then the president is telling the truth.; Inverse: If the president is not telling the truth, then some troops were not withdrawn.; Contrapositive: If some troops were not withdrawn, then the president was not telling the truth. **27.** Converse: If some people suffer, then all institutions place profit above human need.; Inverse: If some institutions do not place profit above human need, then no people suffer.; Contrapositive: If no people suffer, then some institutions do not place profit above human need. **29.** Converse: $\sim r \rightarrow \sim q$; Inverse: $q \rightarrow r$; Contrapositive: $r \rightarrow q$ **31.** If a person diets, then he or she loses weight.; Converse: If a person loses weight, then he or she is dieting.; Inverse: If a person is not dieting, then he or she is not losing weight.; Contrapositive: If a person is not losing weight, then he or she is not dieting. **33.** If a vehicle has no flashing light on top, then it is not an ambulance.; Converse: If a vehicle is not an ambulance, then it has no flashing light on top.; Inverse: If a vehicle has a flashing light on top, then it is an ambulance.; Contrapositive: If a vehicle is an ambulance, then it has a flashing light on top. **35.** If a person is an attorney, then he or she has passed the bar exam.; Converse: If a person has passed the bar exam, then he or she is an attorney.; Inverse: If a person is not an attorney, then he or she has not passed the bar exam.; Contrapositive: If a person has not passed the bar exam, then he or she is not an attorney. **37.** If a person is a pacifist, then he or she is not a warmonger.; Converse: If a person is not a warmonger, then he or she is a pacifist.; Inverse: If a person is not a pacifist, then he or she is a warmonger.; Contrapositive: If a person is a warmonger, then he or she is not a pacifist. **39. a.** true **b.** Converse: If the age of sexual consent is 14, then the country is Italy.; Inverse: If the country is not Italy, then the age of sexual consent is not 14.; Contrapositive: If the age of sexual consent is not 14, then the country is not Italy.; Converse and inverse are not necessarily true. Contrapositive is true. **47.** does not make sense **49.** makes sense

Section 3.6

Check Point Exercises

1. You do not have a fever and you have the flu. **2.** Bart Simpson is not a cartoon character or Tony Soprano is not a cartoon character. **3.** You do not leave by 5 P.M. and you arrive home on time. **4. a.** Some horror movies are not scary or none are funny. **b.** Your workouts are not strenuous and you get stronger. **5.** If we cannot swim or we can sail, it is windy.

Concept and Vocabulary Check

1. $p \wedge \sim q$; antecedent; and; consequent **2.** $\sim p \vee \sim q$; $\sim p \wedge \sim q$ **3.** or **4.** and **5.** false

Exercise Set 3.6

1. I am in Los Angeles and not in California. **3.** It is purple and it is a carrot. **5.** He doesn't and I won't. **7.** There is a blizzard and some schools are not closed. **9.** $\sim q \wedge r$ **11.** Australia is not an island or China is not an island. **13.** My high school did not encourage creativity or it did not encourage diversity. **15.** Jewish scripture does not give a clear indication of a heaven and it does not give a clear indication of an afterlife. **17.** The United States has eradicated neither poverty nor racism. **19.** $p \vee \sim q$ **21.** If you do not succeed, you did not attend lecture or did not study. **23.** If his wife does not cook and his child does not cook, then he does. **25.** $(\sim q \wedge r) \rightarrow \sim p$ **27.** I'm going to neither Seattle nor San Francisco. **29.** I do not study and I pass. **31.** I am going or he is not going. **33.** A bill does not become law or it receives majority approval. **35.** $\sim p \wedge q$ **37.** $\sim p \vee (\sim q \wedge \sim r)$ **39.** none **41.** none **43.** a and b **45.** a and b **47.** If there is no pain, there is no gain.; Converse: If there is no gain, then there is no pain.; Inverse: If there is pain, then there is gain.; Contrapositive: If there is gain, then there is pain.; Negation: There is no pain and there is gain. **49.** If you follow Buddha's "Middle Way," then you are neither hedonistic nor ascetic.; Converse: If you are neither hedonistic nor ascetic, then you follow Buddha's "Middle Way."; Inverse: If you do not follow Buddha's "Middle Way," then you are either hedonistic or ascetic.; Contrapositive: If you are either hedonistic or ascetic, then you do not follow Buddha's "Middle Way."; Negation: You follow Buddha's "Middle Way" and you are either hedonistic or ascetic. **51.** $p \wedge (\sim r \vee s)$ **53.** $\sim p \vee (r \wedge s)$ **55. a.** false **b.** Smoking does not reduce life expectancy by 2370 days or heart disease does not reduce life expectancy by 1247 days. **c.** true **57. a.** true **b.** Homicide does not reduce life expectancy by 74 days and fire reduces life expectancy by 25 days. **c.** false **59. a.** true **b.** Drowning reduces life expectancy by 10 times the number of days as airplane accidents and drowning reduces life expectancy by 24 days. **c.** false **65.** makes sense **67.** makes sense **69.** Contrapositive: If no one is eating turkey, then it is not Thanksgiving.; Negation: It is Thanksgiving and no one is eating turkey.

Section 3.7

Check Point Exercises

1. valid **2.** valid **3.** invalid **4. a.** valid **b.** invalid **c.** valid **5.** valid **6.** Some people do not lead.

Concept and Vocabulary Check

1. valid **2.** q; valid; $[(p \rightarrow q) \wedge p] \rightarrow q$ **3.** $\sim p$; valid; $[(p \rightarrow q) \wedge \sim q] \rightarrow \sim p$ **4.** $p \rightarrow r$; valid; $[(p \rightarrow q) \wedge (q \rightarrow r)] \rightarrow (p \rightarrow r)$ **5.** q; valid; $[(p \vee q) \wedge \sim p] \rightarrow q$ **6.** p **7.** $\sim q$ **8.** false **9.** true **10.** false

Exercise Set 3.7

1. invalid **3.** valid **5.** valid **7.** invalid **9.** invalid **11.** valid **13.** valid

15. $\begin{array}{l} p \rightarrow \sim q \\ q \\ \hline \therefore \sim p \end{array}$ valid **17.** $\begin{array}{l} p \vee q \\ q \\ \hline \therefore \sim p \end{array}$ invalid **19.** $\begin{array}{l} p \rightarrow q \\ q \\ \hline \therefore p \end{array}$ invalid **21.** $\begin{array}{l} p \rightarrow q \\ \sim p \rightarrow q \\ \hline \therefore q \end{array}$ valid **23.** $\begin{array}{l} p \vee q \\ \sim q \\ \hline \therefore p \end{array}$ valid **25.** $\begin{array}{l} p \rightarrow q \\ \sim p \\ \hline \therefore \sim q \end{array}$ invalid **27.** $\begin{array}{l} p \rightarrow q \\ q \rightarrow r \\ \hline \therefore p \rightarrow r \end{array}$ valid **29.** $\begin{array}{l} p \rightarrow q \\ q \rightarrow r \\ \hline \therefore r \rightarrow p \end{array}$ invalid

31. $\begin{array}{l} (p \wedge q) \rightarrow r \\ p \wedge \sim r \\ \hline \therefore \sim q \end{array}$ valid **33.** $\begin{array}{l} (p \vee q) \rightarrow r \\ \sim r \\ \hline \therefore \sim p \wedge \sim q \end{array}$ valid **35.** $\begin{array}{l} (p \vee q) \rightarrow r \\ r \\ \hline \therefore p \vee q \end{array}$ invalid **37.** $\begin{array}{l} (p \wedge q) \rightarrow r \\ \sim p \vee \sim q \\ \hline \therefore \sim r \end{array}$ invalid **39.** $\begin{array}{l} p \rightarrow q \\ \sim p \rightarrow r \\ \hline \therefore q \vee r \end{array}$ valid **41.** $\begin{array}{l} p \rightarrow q \\ q \rightarrow \sim r \\ r \\ \hline \therefore \sim p \end{array}$ valid

43. My best friend is not a chemist. **45.** They were dropped from prime time. **47.** Some electricity is not off.
49. If I vacation in Paris, I gain weight.

51. $\begin{array}{l} p \rightarrow q \\ \sim p \\ \hline \therefore \sim q \end{array}$ invalid **53.** $\begin{array}{l} p \vee q \\ \sim p \\ \hline \therefore q \end{array}$ valid **55.** $\begin{array}{l} p \rightarrow q \\ q \\ \hline \therefore p \end{array}$ invalid **57.** $\begin{array}{l} p \rightarrow q \\ \sim q \\ \hline \therefore \sim p \end{array}$ valid **59.** $\begin{array}{l} p \rightarrow q \\ q \rightarrow r \\ \hline \therefore p \rightarrow r \end{array}$ valid **61.** $\begin{array}{l} p \rightarrow q \\ \sim q \\ \hline \therefore \sim p \end{array}$ valid **63.** h **65.** i **67.** c **69.** a **71.** j

73. d **83.** does not make sense **85.** does not make sense **87.** People sometimes speak without being spoken to. **89.** The doctor either destroys the base on which the placebo rests or jeopardizes a relationship built on trust.

Section 3.8

Check Point Exercises

1. valid **2.** invalid **3.** valid **4.** invalid **5.** invalid **6.** invalid

Concept and Vocabulary Check

1. All *A* are *B*. **2.** No *A* are *B*. **3.** Some *A* are *B*. **4.** Some *A* are not *B*. **5.** false **6.** false

Exercise Set 3.8

1. valid **3.** invalid **5.** valid **7.** invalid **9.** invalid **11.** valid **13.** valid **15.** invalid **17.** invalid **19.** valid **21.** valid **23.** valid **25.** invalid **27.** invalid **29.** valid **31.** invalid **33.** invalid **35.** invalid **37.** valid **39.** valid **43.** makes sense **45.** does not make sense **47.** b **49.** Some teachers are amusing people.

Chapter 3 Review Exercises

1. $(p \wedge q) \rightarrow r$; If the temperature is below 32° and we have finished studying, we will go to the movies. **2.** $\sim r \rightarrow (\sim p \vee \sim q)$; If we do not go to the movies, then the temperature is not below 32° or we have not finished studying. **3.** The temperature is below 32°, and if we finish studying, we will go to the movies. **4.** We will go to the movies if and only if the temperature is below 32° and we have finished studying. **5.** It is not true that both the temperature is below 32° and we have finished studying. **6.** We will not go to the movies if and only if the temperature is not below 32° or we have not finished studying. **7.** $(p \wedge q) \vee r$ **8.** $(p \vee \sim q) \rightarrow r$ **9.** $q \rightarrow (p \leftrightarrow r)$ **10.** $r \leftrightarrow (p \wedge \sim q)$ **11.** $p \rightarrow r$ **12.** $q \rightarrow \sim r$ **13.** Some houses are not made with wood. **14.** Some students major in business. **15.** No crimes are motivated by passion. **16.** All Democrats are registered voters. **17.** Some new taxes will not be used for education.

18. neither

p	q	$\sim p$	$\sim p \wedge q$	$p \wedge (\sim p \wedge q)$
T	T	F	F	T
T	F	F	F	T
F	T	T	T	T
F	F	T	F	F

19. neither

p	q	$\sim p$	$\sim q$	$\sim p \vee \sim q$
T	T	F	F	F
T	F	F	T	T
F	T	T	F	T
F	F	T	T	T

20. neither

p	q	$\sim p$	$\sim p \vee q$	$p \rightarrow (\sim p \vee q)$
T	T	F	T	T
T	F	F	F	F
F	T	T	T	T
F	F	T	T	T

21. neither

p	q	$\sim q$	$p \leftrightarrow \sim q$
T	T	F	F
T	F	T	T
F	T	F	T
F	F	T	F

22. tautology

p	q	$\sim p$	$\sim q$	$p \vee q$	$\sim(p \vee q)$	$\sim p \wedge \sim q$	$\sim(p \vee q) \rightarrow (\sim p \wedge \sim q)$
T	T	F	F	T	F	F	T
T	F	F	T	T	F	F	T
F	T	T	F	T	F	F	T
F	F	T	T	F	T	T	T

23. neither

p	q	r	$\sim r$	$p \vee q$	$(p \vee q) \rightarrow \sim r$
T	T	T	F	T	F
T	T	F	T	T	T
T	F	T	F	T	F
T	F	F	T	T	T
F	T	T	F	T	F
F	T	F	T	T	T
F	F	T	F	F	T
F	F	F	T	F	T

24. neither

p	q	r	$p \wedge q$	$p \wedge r$	$(p \wedge q) \leftrightarrow (p \wedge r)$
T	T	T	T	T	T
T	T	F	T	F	F
T	F	T	F	T	F
T	F	F	F	F	T
F	T	T	F	F	T
F	T	F	F	F	T
F	F	T	F	F	T
F	F	F	F	F	T

25. neither

p	q	r	$r \rightarrow p$	$q \vee (r \rightarrow p)$	$p \wedge [q \vee (r \rightarrow p)]$
T	T	T	T	T	T
T	T	F	T	T	T
T	F	T	T	T	T
T	F	F	T	T	T
F	T	T	F	T	F
F	T	F	T	T	F
F	F	T	F	F	F
F	F	F	T	T	F

26. a. p: I'm in class.; q: I'm studying.; $(p \vee q) \wedge \sim p$

b.

p	q	$\sim p$	$p \vee q$	$(p \vee q) \wedge \sim p$
T	T	F	T	F
T	F	F	T	F
F	T	T	T	T
F	F	T	F	F

c. The statement is true when p is false and q is true.

27. **a.** p: You spit from a truck.; q: It's legal.; r: You spit from a car.; $(p \rightarrow q) \wedge (r \rightarrow \sim q)$

b.

p	q	r	$\sim q$	$p \rightarrow q$	$r \rightarrow \sim q$	$(p \rightarrow q) \wedge (r \rightarrow \sim q)$
T	T	T	F	T	F	F
T	T	F	F	T	T	T
T	F	T	T	F	T	F
T	F	F	T	F	T	F
F	T	T	F	T	F	F
F	T	F	F	T	T	T
F	F	T	T	T	T	T
F	F	F	T	T	T	T

c. The statement is true when p and q are both false.

28. false **29.** true **30.** true **31.** false **32.** p: The 2000 diversity index was 47.; q: The index increased from 2000 to 2010.; $p \wedge \sim q$; false **33.** p: The diversity index decreased from 1980 through 2010.; q: The index was 55 in 1980.; r: The index was 34 in 2010.; $p \rightarrow (q \wedge r)$; true **34.** p: The diversity increased by 6 from 1980 to 1990.; q: The diversity index increased by 7 from 1990 to 2000.; r: The index was at a maximum in 2010.; $(p \leftrightarrow q) \vee \sim r$; true

35. **a.**

p	q	$\sim p$	$\sim p \vee q$	$p \rightarrow q$
T	T	F	T	T
T	F	F	F	F
F	T	T	T	T
F	F	T	T	T

b. If the triangle is isosceles, then it has two equal sides.

36. c **37.** not equivalent **38.** equivalent **39.** Converse: If I am in the South, then I am in Atlanta.; Inverse: If I am not in Atlanta, then I am not in the South.; Contrapositive: If I am not in the South, then I am not in Atlanta. **40.** Converse: If today is not a holiday, then I am in class.; Inverse: If I am not in class, then today is a holiday.; Contrapositive: If today is a holiday, then I am not in class. **41.** Converse: If I pass all courses, then I worked hard.; Inverse: If I don't work hard, then I don't pass some courses.; Contrapositive: If I do not pass some course, then I did not work hard. **42.** Converse: $\sim q \rightarrow \sim p$; Inverse: $p \rightarrow q$.; Contrapositive: $q \rightarrow p$ **43.** An argument is sound and it is not valid. **44.** I do not work hard and I succeed. **45.** $\sim r \wedge \sim p$ **46.** Chicago is not a city or Maine is not a city. **47.** Ernest Hemingway was neither a musician nor an actor. **48.** If a number is not 0, the number is positive or negative. **49.** I do not work hard and I succeed. **50.** She is using her car or she is not taking a bus. **51.** $p \wedge \sim q$ **52.** a and c **53.** a and b **54.** a and c **55.** none **56.** invalid **57.** valid

58. $\begin{array}{l} p \rightarrow q \\ \underline{q} \\ \therefore p \end{array}$ invalid **59.** $\begin{array}{l} p \vee q \\ \underline{q} \\ \therefore \sim p \end{array}$ invalid **60.** $\begin{array}{l} p \vee q \\ \underline{\sim p} \\ \therefore q \end{array}$ valid **61.** $\begin{array}{l} p \rightarrow q \\ \underline{\sim q} \\ \therefore \sim p \end{array}$ valid **62.** $\begin{array}{l} p \rightarrow \sim q \\ \underline{\sim p \rightarrow q} \\ \therefore p \leftrightarrow \sim q \end{array}$ valid **63.** $\begin{array}{l} p \rightarrow \sim q \\ \underline{r \rightarrow q} \\ \therefore \sim r \rightarrow p \end{array}$ invalid

64. invalid **65.** valid **66.** valid **67.** invalid **68.** invalid **69.** valid

Chapter 3 Test

1. If I'm registered and I'm a citizen, then I vote. **2.** I don't vote if and only if I'm not registered or I'm not a citizen. **3.** I'm neither registered nor a citizen. **4.** $(p \wedge q) \vee \sim r$ **5.** $(\sim p \vee \sim q) \rightarrow \sim r$ **6.** $r \rightarrow q$ **7.** Some numbers are not divisible by 5. **8.** No people wear glasses.

9.

p	q	$\sim p$	$\sim p \vee q$	$p \wedge (\sim p \vee q)$
T	T	F	T	T
T	F	F	F	F
F	T	T	T	F
F	F	T	T	F

10.

p	q	$\sim p$	$\sim q$	$p \wedge q$	$\sim(p \wedge q)$	$(\sim p \vee \sim q)$	$\sim(p \wedge q) \leftrightarrow (\sim p \vee \sim q)$
T	T	F	F	T	F	F	T
T	F	F	T	F	T	T	T
F	T	T	F	F	T	T	T
F	F	T	T	F	T	T	T

11.

p	q	r	$q \vee r$	$p \leftrightarrow (q \vee r)$
T	T	T	T	T
T	T	F	T	T
T	F	T	T	T
T	F	F	F	F
F	T	T	T	F
F	T	F	T	F
F	F	T	T	F
F	F	F	F	T

12. p: You break the law.; q: You change the law.; $(p \wedge q) \rightarrow \sim p$

p	q	$\sim p$	$p \wedge q$	$(p \wedge q) \rightarrow \sim p$
T	T	F	T	F
T	F	F	F	T
F	T	T	F	T
F	F	T	F	T

The statement is false when p and q are both true.

13. true **14.** true **15.** p: There was an increase in the percentage spent on food.; q: There was an increase in the percentage spent on health care.; r: By 2010, the percentage spent on health care was more than triple the percentage spent on food.; $\sim p \vee (q \wedge r)$; true **16.** b **17.** If it snows, then it is not August. **18.** Converse: If I cannot concentrate, then the radio is playing.; Inverse: If the radio is not playing, then I can concentrate. **19.** It is cold and we use the pool. **20.** The test is not today and the party is not tonight. **21.** The banana is not green or it is ready to eat. **22.** a and b **23.** a and c **24.** invalid **25.** valid **26.** invalid **27.** invalid **28.** valid **29.** invalid

CHAPTER 4

Section 4.1

Check Point Exercises

1. a. $(4 \times 10^3) + (0 \times 10^2) + (2 \times 10^1) + (6 \times 1)$ or $(4 \times 1000) + (0 \times 100) + (2 \times 10) + (6 \times 1)$
b. $(2 \times 10^4) + (4 \times 10^3) + (2 \times 10^2) + (3 \times 10^1) + (2 \times 1)$ or $(2 \times 10{,}000) + (4 \times 1000) + (2 \times 100) + (3 \times 10) + (2 \times 1)$
2. a. 6073 **b.** 80,900 **3. a.** 12,031 **b.** 468,721 **4. a.** 80,293 **b.** 290,490

Concept and Vocabulary Check

1. numeral **2.** Hindu-Arabic **3.** 7; 10,000,000 **4.** expanded; positional **5.** 10^4; 10^3; 10^2; 10^1; 1 **6.** 10; 60
7. $(10 + 1) \times 60^2 + (1 + 1) \times 60^1 + (10 + 10 + 1 + 1) \times 1$ **8.** $(10 \times 60^3) + (1 \times 60^2) + (2 \times 60^1) + (11 \times 1)$
9. 18×20^4 **10.** $2 \times 18 \times 20 = 720$; $3 \times 20 = 60$; $9 \times 1 = 9$; 789; 789

Exercise Set 4.1

1. 25 **3.** 8 **5.** 81 **7.** 100,000 **9.** $(3 \times 10^1) + (6 \times 1)$ **11.** $(2 \times 10^2) + (4 \times 10^1) + (9 \times 1)$
13. $(7 \times 10^2) + (0 \times 10^1) + (3 \times 1)$ **15.** $(4 \times 10^3) + (8 \times 10^2) + (5 \times 10^1) + (6 \times 1)$
17. $(3 \times 10^3) + (0 \times 10^2) + (7 \times 10^1) + (0 \times 1)$ **19.** $(3 \times 10^4) + (4 \times 10^3) + (5 \times 10^2) + (6 \times 10^1) + (9 \times 1)$
21. $(2 \times 10^8) + (3 \times 10^7) + (0 \times 10^6) + (0 \times 10^5) + (0 \times 10^4) + (7 \times 10^3) + (0 \times 10^2) + (0 \times 10^1) + (4 \times 1)$ **23.** 73
25. 385 **27.** 528,743 **29.** 7002 **31.** 600,002,007 **33.** 23 **35.** 1262 **37.** 1833 **39.** 11,523 **41.** 75,851 **43.** 2,416,271
45. 655,261 **47.** 14 **49.** 6846 **51.** 3048 **53.** 14,411 **55.** 75,610 **57.** 15,842,203 **59.** 29,520,224
61. $(4 \times 10^4) + (6 \times 10^3) + (2 \times 10^2) + (2 \times 10^1) + (5 \times 1)$ **63.** $(2 \times 10^3) + (2 \times 10^2) + (9 \times 10^1) + (9 \times 1)$
65. 0.4759 **67.** 0.700203 **69.** 5000.03 **71.** 30,700.05809 **73.** 9734 **75.** 8097 **77.** 365 is the number of days in a non–leap year.
87. does not make sense **89.** makes sense **91.** <<<<VVVV <<VVVVVVV

Section 4.2

Check Point Exercises

1. 487 **2.** 51 **3.** 2772 **4.** 11_{five} **5.** 1031_{seven} **6.** 110011_{two} **7.** 42023_{five}

Concept and Vocabulary Check

1. 5; 0, 1, 2, 3, and 4 **2.** 5^2; 5^1; 1 **3.** 2; 0 and 1 **4.** 2^3; 2^2; 2^1; 1 **5.** 13 **6.** 731 **7.** two; eight; sixteen

Exercise Set 4.2

1. 23 **3.** 42 **5.** 30 **7.** 11 **9.** 455 **11.** 28,909 **13.** 8342 **15.** 53 **17.** 44,261 **19.** 12_{five} **21.** 14_{seven} **23.** 10_{two}
25. 101_{two} **27.** 1000_{two} **29.** 31_{four} **31.** 101_{six} **33.** 322_{five} **35.** 1230_{four} **37.** 10011_{two} **39.** 111001_{two} **41.** 1011010_{two}
43. 12010_{three} **45.** 1442_{six} **47.** 4443_{seven} **49.** <<<<< <<<<<VV **51.** VVVVVV <<<VV <<VVVVVV

53. 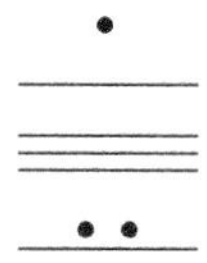 **55.** 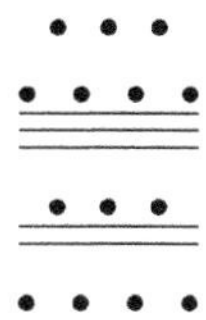 **57.** 25_{seven} **59.** 623_{eight} **61.** 1000110 **63.** 1101101 **65.** PAL **67.** 1001101110111111101101

73. does not make sense **75.** makes sense **77.** 887_{nine}; 1000_{nine} **79.** 11111011_{two}; 673_{eight}; $3A6_{twelve}$

Section 4.3

Check Point Exercises

1. 131_{five} **2.** 1110_{two} **3.** 13_{five} **4.** 1605_{seven} **5.** 201_{seven} **6.** 23_{four}

Concept and Vocabulary Check

1. 1; 1; 11 **2.** 1; 1; 11 **3.** 1; 5; 15; 1; 5 **4.** 7; 7; 10; 5; 10; 5; 3; 1; 15 **5.** 2; 1; 21; 1; 2; 2; 2; 0; 20; 201 **6.** 2; 20 **7.** true

Exercise Set 4.3

1. 102_{four} **3.** 110_{two} **5.** 1310_{five} **7.** 1302_{seven} **9.** 15760_{nine} **11.** 23536_{seven} **13.** 13_{four} **15.** 4_{five} **17.** 206_{eight} **19.** 366_{seven} **21.** 10_{two} **23.** 111_{three} **25.** 152_{six} **27.** 11_{two} **29.** 4011_{seven} **31.** 3114_{eight} **33.** 312_{four} **35.** 20_{four} **37.** 41_{five} remainder of 1_{five} **39.** 1000110_{two} **41.** 110000_{two} **43.** 110111_{two} **45.** $71BE_{sixteen}$ **47.** 01100; 11101 **49.** 10001; 11001 **51.** 01100; 00110; 00100; 11011 **53.** The circuit in Exercise 47 **57.** makes sense **59.** makes sense **61.** 8 hours, 13 minutes, 36 seconds **63.** 56_{seven}

Section 4.4

Check Point Exercises

1. 300,222 **2.**  **3.** 1361 **4.** 1447 **5.** CCCXCIX **6.** 二千六百九十三 **7.** 885

Concept and Vocabulary Check

1. 1000 + 100 + 100 + 10 + 1 + 1 = 1212 **2.** 10 + 10 + 1 + 1 + 1; dccccbbaaa **3.** true **4.** add; +; 110 **5.** subtract; –; 40 **6.** 1000; 50 × 1000 = 50,000 **7.** true **8.** H Y D X F **9.** true **10.** VPC **11.** 6000; 500; 30; 5; 36,535 **12.** true

Exercise Set 4.4

1. 322 **3.** 300,423 **5.** 132 **7.**  **9.** 

11.  **13.** 11 **15.** 16 **17.** 40 **19.** 59 **21.** 146 **23.** 1621 **25.** 2677

27. 9466 **29.** XLIII **31.** CXXIX **33.** MDCCCXCVI **35.** $\overline{\text{VI}}$DCCCXCII **37.** 88 **39.** 527 **41.** 2776

43. 四十三 **45.** 五百八十三 **47.** 四千八百七十 **49.** 12 **51.** 234 **53.** $\mu\gamma$ **55.** $\upsilon\pi\gamma$ **57.** MMCCCXXIV; 二千三百二十四

59. a.  **b.** 一千七百四十一 **61.** 3104_{five} **63.** 1232_{five} **65.** 

67. 1776; Declaration of Independence **69.** 4,640,224 **77.** does not make sense **79.** makes sense
81. Preceding: 𓍢𓍢𓎆𓎆𓎆𓎆𓎆𓎆𓎆𓎆𓎆𓏺𓏺𓏺𓏺𓏺𓏺𓏺𓏺𓏺

Following: 𓍢𓍢𓍢

Chapter 4 Review Exercises

1. 121 **2.** 343 **3.** $(4 \times 10^2) + (7 \times 10^1) + (2 \times 1)$ **4.** $(8 \times 10^3) + (0 \times 10^2) + (7 \times 10^1) + (6 \times 1)$
5. $(7 \times 10^4) + (0 \times 10^3) + (3 \times 10^2) + (2 \times 10^1) + (9 \times 1)$ **6.** 706,953 **7.** 740,000,306 **8.** 673 **9.** 8430 **10.** 2331
11. 65,536 **12.** Each position represents a particular value. The symbol in each position tells how many of that value are represented.
13. 19 **14.** 6 **15.** 325 **16.** 805 **17.** 4051 **18.** 560 **19.** 324_{five} **20.** 10101_{two} **21.** 122112_{three} **22.** 26452_{seven}
23. 111140_{six} **24.** $3A2_{\text{twelve}}$ **25.** 132_{seven} **26.** 1401_{eight} **27.** 110000_{two} **28.** $AD0_{\text{sixteen}}$ **29.** 5_{six} **30.** 345_{seven} **31.** 11_{two}
32. 2304_{five} **33.** 222_{four} **34.** 354_{seven} **35.** 1102_{five} **36.** 133_{four} **37.** 12_{five} **38.** 1246 **39.** 12,432
40. 𓆼𓆼𓍢𓍢𓍢𓍢𓎆𓎆𓎆𓎆𓎆𓎆𓎆𓎆𓎆𓏺𓏺𓏺𓏺𓏺𓏺 **41.** 𓂭𓂭𓂭𓆼𓆼𓆼𓆼𓍢𓍢𓍢𓍢𓎆𓎆𓎆𓎆𓎆𓎆𓎆𓏺𓏺𓏺 **42.** 2314
43. DDDDDCCCCBBBBBBBBBAA **44.** Answers will vary. **45.** 163 **46.** 1034 **47.** 1990 **48.** XLIX **49.** MMCMLXV
50. If symbols increase in value from left to right, subtract the value of the symbol on the left from the symbol on the right. Answers will vary.

51. 554	**52.** 8253	**53.** 二 百 七 十 四	**54.** 三 千 五 百 八 十 七	**55.** 365	**56.** 4520	**57.** G Y I X C	**58.** F Z H Y E X D	**59.** Answers will vary.	**60.** 653	

61. 678 **62.** $\upsilon\nu\gamma$ **63.** $\pi\beta$ **64.** 357 **65.** 37,894 **66.** 80,618 **67.** WRG **68.** lfVQC **69.** Answers will vary.

Chapter 4 Test

1. 729 **2.** $(5 \times 10^2) + (6 \times 10^1) + (7 \times 1)$ **3.** $(6 \times 10^4) + (3 \times 10^3) + (0 \times 10^2) + (2 \times 10^1) + (8 \times 1)$ **4.** 7493 **5.** 400,206
6. A number represents "How many?" whereas a numeral is a symbol used to write a number.
7. A symbol for zero is needed for a place holder when there are no values for a position.
8. 72,731 **9.** 1560 **10.** 113 **11.** 223 **12.** 53 **13.** 2212_{three} **14.** 111000_{two} **15.** 24334_{five} **16.** 1212_{five} **17.** 414_{seven}
18. 250_{six} **19.** 221_{five} **20.** 20,303 **21.** 𓂭𓂭𓂭𓆼𓆼𓍢𓍢𓍢𓍢𓍢𓍢𓎆𓎆𓎆𓏺𓏺𓏺𓏺 **22.** 1994 **23.** CDLIX **24.** Answers will vary.

CHAPTER 5

Section 5.1

Check Point Exercises

1. b **2.** $2^3 \cdot 3 \cdot 5$ **3.** 75 **4.** 96 **5.** 90 **6.** 5:00 P.M.

Concept and Vocabulary Check

1. prime **2.** composite **3.** greatest common divisor **4.** least common multiple **5.** false **6.** true **7.** false **8.** false

Exercise Set 5.1

1. a. yes **b.** no **c.** yes **d.** no **e.** no **f.** yes **g.** no **h.** no **i.** no **3. a.** yes **b.** yes **c.** yes **d.** no **e.** yes **f.** yes **g.** no **h.** no **i.** yes **5. a.** yes **b.** no **c.** yes **d.** no **e.** no **f.** no **g.** no **h.** no **i.** no **7. a.** yes **b.** yes **c.** yes **d.** no **e.** yes **f.** yes **g.** yes **h.** no **i.** yes **9. a.** yes **b.** yes **c.** yes **d.** yes **e.** yes **f.** yes **g.** yes **h.** yes **i.** yes **11.** true; $5 + 9 + 5 + 8 = 27$ which is divisible by 3. **13.** true; the last two digits of 10,612 form the number 12 and 12 is divisible by 4. **15.** false **17.** true; 104,538 is an even number so it is divisible by 2. $1 + 0 + 4 + 5 + 3 + 8 = 21$ which is divisible by 3 so it is divisible by 3. Any number divisible by 2 and 3 is divisible by 6. **19.** true; the last three digits of 20,104 form the number 104 and 104 is divisible by 8. **21.** false **23.** true; $5 + 1 + 7 + 8 + 7 + 2 = 30$ which is divisible by 3 so it is divisible by 3. The last two digits of 517,872 form the number 72 and 72 is divisible by 4 so it is divisible by 4. Any number divisible by 3 and 4 is divisible by 12. **25.** $3 \cdot 5^2$ **27.** $2^3 \cdot 7$ **29.** $3 \cdot 5 \cdot 7$ **31.** $2^2 \cdot 5^3$ **33.** $3 \cdot 13 \cdot 17$ **35.** $3 \cdot 5 \cdot 59$ **37.** $2^5 \cdot 3^2 \cdot 5$ **39.** $2^2 \cdot 499$ **41.** $3 \cdot 5^2 \cdot 7^2$ **43.** $2^3 \cdot 3 \cdot 5^2 \cdot 11 \cdot 13$ **45.** 14 **47.** 2 **49.** 12 **51.** 24 **53.** 38 **55.** 15 **57.** 168 **59.** 336 **61.** 540 **63.** 360 **65.** 3420 **67.** 4560 **69.** 8 **71.** 6 **73.** 2, 6 **75.** perfect **77.** not perfect **79.** not an emirp **81.** emirp **83.** not a Germain prime **85.** not a Germain prime **87.** 648; 648; Answers will vary. An example is: The product of the greatest common divisor and least common multiple of two numbers equals the product of the two numbers. **89.** The numbers are the prime numbers between 2 and 97, inclusive. **91.** 12 **93.** 10 **95.** July 1 **97.** 90 min or $1\frac{1}{2}$ hr **111.** does not make sense **113.** does not make sense **115.** GCD: $2^{14} \cdot 3^{25} \cdot 5^{30}$; LCM: $2^{17} \cdot 3^{37} \cdot 5^{31}$ **117.** 2:20 A.M. on the third day **119.** yes **121.** yes

Section 5.2

Check Point Exercises

1. 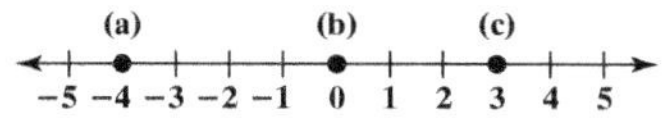 **2. a.** $>$ **b.** $<$ **c.** $<$ **d.** $<$ **3. a.** 8 **b.** 6 **c.** -8 **4. a.** 37 **b.** -4 **c.** -24 **5. a.** 8 years **b.** 14 years **6. a.** 25 **b.** -25 **c.** -64 **d.** 81 **7.** 36 **8.** 82

Concept and Vocabulary Check

1. $\{\ldots,-3,-2,-1,0,1,2,3,\ldots\}$ **2.** left **3.** the distance from 0 to a **4.** additive inverses **5.** false **6.** true **7.** true **8.** false

Exercise Set 5.2

1. 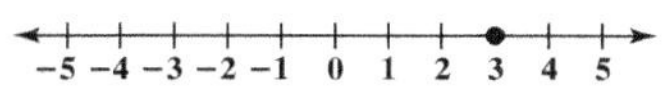 **3.** 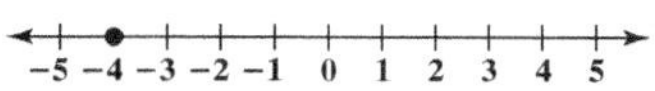 **5.** $<$ **7.** $<$ **9.** $>$ **11.** $<$ **13.** 14 **15.** 14 **17.** 300,000 **19.** -12 **21.** 4 **23.** -3 **25.** -5 **27.** -18 **29.** 0 **31.** 5 **33.** -7 **35.** 14 **37.** 11 **39.** -9 **41.** -28 **43.** -54 **45.** 21 **47.** -12 **49.** 13 **51.** 0 **53.** 25 **55.** 25 **57.** 64 **59.** -125 **61.** 625 **63.** -81 **65.** 81 **67.** -3 **69.** -7 **71.** 30 **73.** 0 **75.** undefined **77.** -20 **79.** -31 **81.** 25 **83.** 13 **85.** 13 **87.** 33 **89.** 32 **91.** -24 **93.** 0 **95.** -32 **97.** 45 **99.** 14 **101.** 14 **103.** -36 **105.** 88 **107.** -29 **109.** -10 **111.** $-10-(-2)^3; -2$ **113.** $[2(7-10)]^2; 36$ **115.** 20,602 ft **117.** shrink by 5 years **119.** shrink by 11 years **121.** no change **123.** 25 years **125.** 1 year **127. a.** \$128 billion; surplus **b.** –\$1299; deficit **c.** \$1427 billion **129.** \$1138 billion **131.** 7°F **133.** -2°F **145.** makes sense **147.** makes sense **149.** $(8-2)\cdot 3-4=14$ **151.** -36

Section 5.3

Check Point Exercises

1. $\frac{4}{5}$ **2.** $\frac{21}{8}$ **3.** $1\frac{2}{3}$ **4. a.** 0.375 **b.** $0.\overline{45}$ **5. a.** $\frac{9}{10}$ **b.** $\frac{43}{50}$ **c.** $\frac{53}{1000}$ **6.** $\frac{2}{9}$ **7.** $\frac{79}{99}$ **8. a.** $\frac{8}{33}$ **b.** $\frac{3}{2}$ or $1\frac{1}{2}$ **c.** $\frac{51}{10}$ or $5\frac{1}{10}$ **9. a.** $\frac{36}{55}$ **b.** $-\frac{4}{3}$ or $-1\frac{1}{3}$ **c.** $\frac{3}{2}$ or $1\frac{1}{2}$ **10. a.** $\frac{2}{3}$ **b.** $\frac{3}{2}$ or $1\frac{1}{2}$ **c.** $-\frac{9}{4}$ or $-2\frac{1}{4}$ **11.** $\frac{19}{20}$ **12.** $-\frac{17}{60}$ **13.** $\frac{3}{4}$ **14.** $\frac{5}{12}$ **15.** $2\frac{4}{5}$ eggs; 3 eggs

Concept and Vocabulary Check

1. rational; integers; zero **2.** improper; the numerator is greater than the denominator **3.** terminate/stop; have repeating digits **4.** reciprocal/multiplicative inverse **5.** false **6.** false **7.** true **8.** false

Exercise Set 5.3

1. $\frac{2}{3}$ **3.** $\frac{5}{6}$ **5.** $\frac{4}{7}$ **7.** $\frac{5}{9}$ **9.** $\frac{9}{10}$ **11.** $\frac{14}{19}$ **13.** $\frac{19}{8}$ **15.** $-\frac{38}{5}$ **17.** $\frac{199}{16}$ **19.** $4\frac{3}{5}$ **21.** $-8\frac{4}{9}$ **23.** $35\frac{11}{20}$ **25.** 0.75 **27.** 0.35 **29.** 0.875 **31.** $0.\overline{81}$ **33.** $3.\overline{142857}$ **35.** $0.\overline{285714}$ **37.** $\frac{3}{10}$ **39.** $\frac{2}{5}$ **41.** $\frac{39}{100}$ **43.** $\frac{41}{50}$ **45.** $\frac{29}{40}$ **47.** $\frac{5399}{10{,}000}$ **49.** $\frac{7}{9}$ **51.** 1 **53.** $\frac{4}{11}$ **55.** $\frac{257}{999}$ **57.** $\frac{21}{88}$ **59.** $-\frac{7}{120}$ **61.** $\frac{3}{2}$ **63.** 6 **65.** $\frac{10}{3}$ **67.** $-\frac{14}{15}$ **69.** 6 **71.** $\frac{5}{11}$ **73.** $\frac{2}{3}$ **75.** $\frac{2}{3}$ **77.** $\frac{7}{10}$ **79.** $\frac{9}{10}$ **81.** $\frac{53}{120}$ **83.** $\frac{1}{2}$ **85.** $\frac{7}{12}$ **87.** $-\frac{71}{150}$ **89.** $\frac{53}{12}$ or $4\frac{5}{12}$ **91.** $\frac{7}{6}$ or $1\frac{1}{6}$ **93.** $-\frac{5}{2}$ or $-2\frac{1}{2}$ **95.** $\frac{11}{14}$ **97.** $\frac{4}{15}$ **99.** $-\frac{9}{40}$ **101.** $-1\frac{1}{36}$ **103.** $-19\frac{3}{4}$ **105.** $\frac{7}{24}$ **107.** $\frac{7}{12}$ **109.** $-\frac{3}{4}$ **111.** Both are equal to $\frac{169}{36}$. **113.** $\frac{1}{2^2\cdot 3}$ **115.** $-\frac{289}{2^4\cdot 5^4\cdot 7}$ **117.** $0.\overline{54}; 0.58\overline{3}; <$ **119.** $-0.8\overline{3}; -0.\overline{8}; >$ **121. a.** $\frac{560}{1600}=\frac{7}{20}$ **b.** $0.35=35\%$ **c.** $\frac{720}{1600}=\frac{9}{20}$ **d.** $0.45=45\%$ **e.** 10% **123.** $\frac{1}{3}$ cup butter, $\frac{5}{2}=2.5$ ounces unsweetened chocolate, $\frac{3}{4}$ cup sugar, 1 teaspoon vanilla, 1 egg, $\frac{1}{2}$ cup flour **125.** $\frac{5}{6}$ cup butter, $\frac{25}{4}=6.25$ ounces unsweetened chocolate, $\frac{15}{8}=1\frac{7}{8}$ cups sugar, $\frac{5}{2}=2\frac{1}{2}$ teaspoons vanilla, $\frac{5}{2}=2\frac{1}{2}$ eggs, $\frac{5}{4}=1\frac{1}{4}$ cups flour **127.** 24 brownies **129.** $3\frac{2}{3}$ c **131. a.** D, E, G, A, B **b.** There are black keys to the left of the keys for the notes D, E, G, A, and B. **133.** $16\frac{7}{16}$ in. **135.** $\frac{1}{3}$ of the business **137.** $1\frac{3}{20}$ mi; $\frac{7}{20}$ mi **139.** $\frac{1}{10}$ **153.** makes sense **155.** makes sense **157.** $\frac{1}{1\cdot 2}+\frac{1}{2\cdot 3}+\frac{1}{3\cdot 4}+\frac{1}{4\cdot 5}+\frac{1}{5\cdot 6}=\frac{5}{6}$

Section 5.4

Check Point Exercises

1. a. $2\sqrt{3}$ **b.** $2\sqrt{15}$ **c.** cannot be simplified **2. a.** $\sqrt{30}$ **b.** 10 **c.** $2\sqrt{3}$ **3. a.** 4 **b.** $2\sqrt{2}$ **4. a.** $18\sqrt{3}$ **b.** $-5\sqrt{13}$ **c.** $8\sqrt{10}$ **5. a.** $3\sqrt{3}$ **b.** $-13\sqrt{2}$ **6. a.** $\frac{5\sqrt{10}}{2}$ **b.** $\frac{\sqrt{14}}{7}$ **c.** $\frac{5\sqrt{2}}{6}$

Concept and Vocabulary Check

1. terminating; repeating **2.** π **3.** $\sqrt{n}$; n **4.** $\sqrt{49}\cdot\sqrt{6} = 7\sqrt{6}$ **5.** coefficient **6.** $(8 + 10)\sqrt{3} = 18\sqrt{3}$ **7.** $5\sqrt{2} + 4\sqrt{2} = 9\sqrt{2}$ **8.** rationalizing the denominator **9.** $\sqrt{7}$ **10.** $\sqrt{3}$

Exercise Set 5.4

1. 3 **3.** 5 **5.** 8 **7.** 11 **9.** 13 **11. a.** 13.2 **b.** 13.15 **c.** 13.153 **13. a.** 133.3 **b.** 133.27 **c.** 133.270 **15. a.** 1.8 **b.** 1.77 **c.** 1.772 **17.** $2\sqrt{5}$ **19.** $4\sqrt{5}$ **21.** $5\sqrt{10}$ **23.** $14\sqrt{7}$ **25.** $\sqrt{42}$ **27.** 6 **29.** $3\sqrt{2}$ **31.** $2\sqrt{13}$ **33.** 3 **35.** $3\sqrt{5}$ **37.** $-4\sqrt{3}$ **39.** $13\sqrt{3}$ **41.** $-2\sqrt{13}$ **43.** $2\sqrt{5}$ **45.** $7\sqrt{2}$ **47.** $3\sqrt{5}$ **49.** $2\sqrt{2}$ **51.** $34\sqrt{2}$ **53.** $-\frac{3}{2}\sqrt{3}$ **55.** $11\sqrt{3}$ **57.** $\frac{5\sqrt{3}}{3}$ **59.** $3\sqrt{7}$ **61.** $\frac{2\sqrt{30}}{5}$ **63.** $\frac{5\sqrt{3}}{2}$ **65.** $\frac{\sqrt{10}}{5}$ **67.** $20\sqrt{2} - 5\sqrt{3}$ **69.** $-7\sqrt{7}$ **71.** $\frac{43\sqrt{2}}{35}$ **73.** $\frac{5\sqrt{6}}{6}$ **75.** $6\sqrt{3}$ miles; 10.4 miles **77.** 70 mph; He was speeding. **79. a.** 41 in. **b.** 40.6 in.; quite well **81. a.** 36 cm **b.** 44.7 cm **c.** 46.9 cm **d.** The model describes healthy children. **83.** Paige Fox is bad at math. **85.** $0.44R_f$; 44 weeks **95.** does not make sense **97.** does not make sense **99.** false; $\sqrt{9} + \sqrt{16} = 3 + 4 = 7 \neq \sqrt{25}$ **101.** false; $\frac{\sqrt{64}}{2} = \frac{8}{2} = 4$ **103.** > **105.** The square root is multiplied by $\sqrt{2}$. **107.** $\frac{3\sqrt{2}}{2}$

Section 5.5

Check Point Exercises

1. a. $\sqrt{9}$ **b.** 0, $\sqrt{9}$ **c.** $-9, 0, \sqrt{9}$ **d.** $-9, -1.3, 0, 0.\bar{3}, \sqrt{9}$ **e.** $\frac{\pi}{2}, \sqrt{10}$ **f.** $-9, -1.3, 0, 0.\bar{3}, \frac{\pi}{2}, \sqrt{9}, \sqrt{10}$ **2. a.** associative property of multiplication **b.** commutative property of addition **c.** distributive property of multiplication over addition **d.** commutative property of multiplication **e.** identity property of addition **f.** inverse property of multiplication **3. a.** yes **b.** no **4. a.** The entries in the body of the table are all elements of the set. **b.**

$$\begin{aligned}(2 \oplus 2) \oplus 3 &= 2 \oplus (2 \oplus 3)\\ 0 \oplus 3 &= 2 \oplus 1\\ 3 &= 3\end{aligned}$$

c. 0 **d.** The inverse of 0 is 0, the inverse of 1 is 3, the inverse of 2 is 2, and the inverse of 3 is 1. **e.** $1 \oplus 3 = 3 \oplus 1$; $3 \oplus 2 = 2 \oplus 3$
$0 = 0 \qquad 1 = 1$

Concept and Vocabulary Check

1. rational; irrational **2.** closure **3.** $ab = ba$ **4.** $(a + b) + c = a + (b + c)$ **5.** $a(b + c) = ab + ac$ **6.** identity **7.** identity; 1 **8.** multiplicative inverse; multiplicative identity **9.** $1 \oplus 4 = 0$; $3 \oplus 3 = 1$; $4 \oplus 2 = 1$ **10.** true

Exercise Set 5.5

1. a. $\sqrt{100}$ **b.** 0, $\sqrt{100}$ **c.** $-9, 0, \sqrt{100}$ **d.** $-9, -\frac{4}{5}, 0, 0.25, 9.2, \sqrt{100}$ **e.** $\sqrt{3}$ **f.** $-9, -\frac{4}{5}, 0, 0.25, \sqrt{3}, 9.2, \sqrt{100}$ **3. a.** $\sqrt{64}$ **b.** 0, $\sqrt{64}$ **c.** $-11, 0, \sqrt{64}$ **d.** $-11, -\frac{5}{6}, 0, 0.75, \sqrt{64}$ **e.** $\sqrt{5}, \pi$ **f.** $-11, -\frac{5}{6}, 0, 0.75, \sqrt{5}, \pi, \sqrt{64}$ **5.** 0 **7.** Answers will vary; an example is: $\frac{1}{2}$. **9.** Answers will vary; an example is: 1. **11.** Answers will vary; an example is: $\sqrt{2}$. **13.** 4 **15.** 6 **17.** 4; 3 **19.** 7 **21.** $30 + 5\sqrt{2}$ **23.** $3\sqrt{7} + \sqrt{14}$ **25.** $5\sqrt{3} + 3$ **27.** $2\sqrt{3} + 6$ **29.** commutative property of addition **31.** associative property of addition **33.** commutative property of addition **35.** distributive property of multiplication over addition **37.** associative property of multiplication **39.** identity property of multiplication **41.** inverse property of addition **43.** inverse property of multiplication **45.** Answers will vary; an example is: $1 - 2 = -1$. **47.** Answers will vary; an example is: $4 \div 8 = \frac{1}{2}$. **49.** Answers will vary; an example is: $\sqrt{2}\cdot\sqrt{2} = 2$. **51. a.** The entries in the body of the table are all elements of the set. **b.**

$$\begin{aligned}(4 \oplus 6) \oplus 7 &= 4 \oplus (6 \oplus 7)\\ 2 \oplus 7 &= 4 \oplus 5\\ 1 &= 1\end{aligned}$$

c. 0 **d.** The inverse of 0 is 0, the inverse of 1 is 7, the inverse of 2 is 6, the inverse of 3 is 5, the inverse of 4 is 4, the inverse of 5 is 3, the inverse of 6 is 2, and the inverse of 7 is 1. **e.** $5 \oplus 6 = 6 \oplus 5$; $4 \oplus 7 = 7 \oplus 4$
$3 = 3 \qquad 3 = 3$

53. true **55.** false **57.** distributive property; commutative property of addition; associative property of addition; distributive property; commutative property of addition **59. a.** $c \triangle (d \square e) = c \triangle c = e$; $(c \triangle d) \square (c \triangle e) = b \square d = e$ **b.** distributive property **61.** *b* **63.** *c* **65.** *d* **67.** vampire **69.** not a vampire **71.** narcissistic **73.** not narcissistic **75. a.** distributive property **b.** approximately 108 mg; Answers will vary. **77.** sevenfold rotational symmetry **89.** makes sense **91.** makes sense **93.** false **95.** false **97.** false **99.** false

Section 5.6

Check Point Exercises

1. a. 1 **b.** 1 **c.** 1 **d.** −1 **2. a.** $\frac{1}{81}$ **b.** $\frac{1}{216}$ **c.** $\frac{1}{12}$ **3. a.** 7,400,000,000 **b.** 0.000003017 **4. a.** 7.41×10^9 **b.** 9.2×10^{-8} **5.** 4.1×10^9 **6.** 520,000 **7.** 0.0000023 **8. a.** 1.872×10^4 **b.** 18,720 **9.** \$8300

Concept and Vocabulary Check

1. add **2.** multiply **3.** subtract **4.** one **5.** 10 to an integer power; a number greater than or equal to 1 and less than 10 **6.** false **7.** false **8.** false **9.** true **10.** false

Exercise Set 5.6

1. $2^5 = 32$ **3.** $4^3 = 64$ **5.** $2^6 = 64$ **7.** $1^{20} = 1$ **9.** $4^2 = 16$ **11.** $2^4 = 16$ **13.** 1 **15.** 1 **17.** −1 **19.** $\frac{1}{4}$ **21.** $\frac{1}{64}$ **23.** $\frac{1}{32}$ **25.** $3^2 = 9$ **27.** $3^{-2} = \frac{1}{9}$ **29.** $2^{-4} = \frac{1}{16}$ **31.** $\frac{1}{x^{16}}$ **33.** $\frac{1}{x^2}$ **35.** $\frac{1}{x^{12}}$ **37.** $\frac{2}{5}$ **39.** $-\frac{6x^2}{y^3}$ **41.** $-\frac{5y^8}{x^6}$ **43.** 270 **45.** 912,000 **47.** 80,000,000 **49.** 100,000 **51.** 0.79 **53.** 0.0215 **55.** 0.000786 **57.** 0.00000318 **59.** 3.7×10^2 **61.** 3.6×10^3 **63.** 3.2×10^4 **65.** 2.2×10^8 **67.** 2.7×10^{-2} **69.** 3.7×10^{-3} **71.** 2.93×10^{-6} **73.** 8.2×10^7 **75.** 4.1×10^5 **77.** 2.1×10^{-6} **79.** 600,000 **81.** 60,000 **83.** 0.123 **85.** 30,000 **87.** 3,000,000 **89.** 0.03 **91.** 0.0021 **93.** $(8.2 \times 10^7)(3.0 \times 10^9) = 2.46 \times 10^{17}$ **95.** $(5.0 \times 10^{-4})(6.0 \times 10^6) = 3 \times 10^3$ **97.** $\frac{9.5 \times 10^6}{5 \times 10^2} = 1.9 \times 10^4$ **99.** $\frac{8 \times 10^{-5}}{2 \times 10^2} = 4 \times 10^{-7}$ **101.** $\frac{4.8 \times 10^{11}}{1.2 \times 10^{-4}} = 4 \times 10^{15}$ **103.** $\frac{11}{18}$ **105.** $\frac{99}{25} = 3\frac{24}{25}$ **107.** 2.5×10^{-3} **109.** 8×10^{-5} **111. a.** 2.17×10^{12} **b.** 3.09×10^8 **c.** $\$7.02 \times 10^3$; \$7020 **113.** $\$1.0586 \times 10^{10}$ **115.** 1.06×10^{-18} g **117.** 4.064×10^9 **129.** makes sense **131.** does not make sense **133.** false **135.** false **137.** false **139.** false

Section 5.7

Check Point Exercises

1. 100, 120, 140, 160, 180, and 200 **2.** 8, 5, 2, −1, −4, and −7 **3.** −34 **4. a.** $a_n = 0.35n + 15.65$ **b.** 23% **5.** 12, −6, 3, $-\frac{3}{2}, \frac{3}{4}$, and $-\frac{3}{8}$ **6.** 3645 **7.** $a_n = 3(2)^{n-1}$; 384

Concept and Vocabulary Check

1. arithmetic; common difference **2.** $a_n = a_1 + (n - 1)d$; the first term; the common difference **3.** geometric; common ratio **4.** $a_n = a_1 r^{n-1}$; first term; the common ratio

Exercise Set 5.7

1. 8, 10, 12, 14, 16, and 18 **3.** 200, 220, 240, 260, 280, and 300 **5.** −7, −3, 1, 5, 9, and 13 **7.** −400, −100, 200, 500, 800, and 1100 **9.** 7, 4, 1, −2, −5, and −8 **11.** 200, 140, 80, 20, −40, and −100 **13.** $\frac{5}{2}, 3, \frac{7}{2}, 4, \frac{9}{2}$, and 5 **15.** $\frac{3}{2}, \frac{7}{4}, 2, \frac{9}{4}, \frac{5}{2}$, and $\frac{11}{4}$ **17.** 4.25, 4.55, 4.85, 5.15, 5.45, and 5.75 **19.** 4.5, 3.75, 3, 2.25, 1.5, and 0.75 **21.** 33 **23.** 252 **25.** 67 **27.** 955 **29.** −82 **31.** −142 **33.** −43 **35.** −248 **37.** $\frac{23}{2}$ **39.** 1.75 **41.** $a_n = 1 + (n - 1)4$; 77 **43.** $a_n = 7 + (n - 1)(-4)$; −69 **45.** $a_n = 9 + (n - 1)2$; 47 **47.** $a_n = -20 + (n - 1)(-4)$; −96 **49.** 4, 8, 16, 32, 64, and 128 **51.** 1000, 1000, 1000, 1000, 1000, and 1000 **53.** 3, −6, 12, −24, 48, and −96 **55.** 10, −40, 160, −640, 2560, and −10,240 **57.** 2000, −2000, 2000, −2000, 2000, and −2000 **59.** −2, 6, −18, 54, −162, and 486 **61.** −6, 30, −150, 750, −3750, and 18,750 **63.** $\frac{1}{4}, \frac{1}{2}$, 1, 2, 4, and 8 **65.** $\frac{1}{4}, \frac{1}{8}, \frac{1}{16}, \frac{1}{32}, \frac{1}{64}$, and $\frac{1}{128}$ **67.** $-\frac{1}{16}, \frac{1}{4}$, −1, 4, −16, and 64 **69.** 2, 0.2, 0.02, 0.002, 0.0002, and 0.00002 **71.** 256 **73.** $2{,}324{,}522{,}934 \approx 2.32 \times 10^9$ **75.** 50 **77.** 320 **79.** −2 **81.** 486 **83.** $\frac{3}{64}$ **85.** $-\frac{2}{27}$ **87.** $\approx -1.82 \times 10^{-9}$ **89.** 0.1 **91.** $a_n = 3(4)^{n-1}$; 12,288 **93.** $a_n = 18\left(\frac{1}{3}\right)^{n-1}$; $\frac{2}{81}$ **95.** $a_n = 1.5(-2)^{n-1}$; 96 **97.** $a_n = 0.0004(-10)^{n-1}$; 400 **99.** arithmetic; 18 and 22 **101.** geometric; 405 and 1215 **103.** arithmetic; 13 and 18 **105.** geometric; $\frac{3}{16}$ and $\frac{3}{32}$ **107.** arithmetic; $\frac{5}{2}$ and 3 **109.** geometric; 7 and −7 **111.** arithmetic; −49 and −63 **113.** geometric; $25\sqrt{5}$ and 125 **115.** arithmetic; 310 **117.** geometric; 59,048 **119.** geometric; −1023 **121.** arithmetic; 80 **123.** 5050 **125. a.** $a_n = 0.6n + 17.8$ **b.** 35.8% **127.** Company A; \$1400 **129.** \$16,384 **131.** \$3,795,957 **133. a.** approximately 1.01 for all but one division **b.** $a_n = 33.87(1.01)^{n-1}$ **c.** 41.33 million **143.** makes sense **145.** makes sense **147.** false **149.** true **151.** false **153.** true

Chapter 5 Review Exercises

1. 2: yes; 3: yes; 4: yes; 5: no; 6: yes; 8: yes; 9: no; 10: no; 12: yes **2.** 2: yes; 3: yes; 4: no; 5: yes; 6: yes; 8: no; 9: yes; 10: yes; 12: no **3.** $3 \cdot 5 \cdot 47$ **4.** $2^6 \cdot 3 \cdot 5$ **5.** $3 \cdot 5^2 \cdot 7 \cdot 13$ **6.** GCD: 6; LCM: 240 **7.** GCD: 6; LCM: 900 **8.** GCD: 2; LCM: 27,432 **9.** 12 **10.** 11:48 A.M. **11.** $<$ **12.** $>$ **13.** 860 **14.** 53 **15.** 0 **16.** −3 **17.** −11 **18.** −15 **19.** 1 **20.** 99 **21.** −15 **22.** 9 **23.** −4 **24.** −16 **25.** −16 **26.** 10 **27.** −2 **28.** 17 **29.** \$658 billion **30.** $\frac{8}{15}$ **31.** $\frac{6}{25}$ **32.** $\frac{11}{12}$ **33.** $\frac{64}{11}$ **34.** $-\frac{23}{7}$ **35.** $5\frac{2}{5}$ **36.** $-1\frac{8}{9}$ **37.** 0.8 **38.** $0.\overline{428571}$ **39.** 0.625 **40.** 0.5625 **41.** $\frac{3}{5}$ **42.** $\frac{17}{25}$ **43.** $\frac{147}{250}$ **44.** $\frac{21}{2500}$ **45.** $\frac{5}{9}$ **46.** $\frac{34}{99}$ **47.** $\frac{113}{999}$ **48.** $\frac{21}{50}$ **49.** $\frac{35}{6}$ **50.** $\frac{8}{3}$ **51.** $-\frac{1}{4}$ **52.** $\frac{2}{3}$ **53.** $\frac{43}{36}$ **54.** $\frac{37}{60}$ **55.** $\frac{11}{15}$ **56.** $\frac{5}{16}$ **57.** $-\frac{2}{5}$ **58.** $-\frac{20}{3}$ or $-6\frac{2}{3}$ **59.** $\frac{15}{112}$ **60.** $\frac{27}{40}$ **61.** $11\frac{1}{4}$ or about 11 pounds **62.** $\frac{5}{12}$ of the tank **63.** $2\sqrt{7}$ **64.** $6\sqrt{2}$ **65.** $5\sqrt{6}$ **66.** $10\sqrt{3}$ **67.** $4\sqrt{3}$ **68.** $5\sqrt{2}$ **69.** $2\sqrt{3}$ **70.** 3 **71.** $5\sqrt{5}$ **72.** $-6\sqrt{11}$ **73.** $7\sqrt{2}$ **74.** $-17\sqrt{3}$ **75.** $12\sqrt{2}$ **76.** $6\sqrt{5}$ **77.** $\frac{\sqrt{6}}{3}$ **78.** $8\sqrt{3} \approx 13.9$ feet per second **79. a.** $\sqrt{81}$ **b.** 0, $\sqrt{81}$ **c.** −17, 0, $\sqrt{81}$ **d.** −17, $-\frac{9}{13}$, 0, 0.75, $\sqrt{81}$ **e.** $\sqrt{2}$, π **f.** −17, $-\frac{9}{13}$, 0, 0.75, $\sqrt{2}$, π, $\sqrt{81}$ **80.** Answers will vary; an example is: 0. **81.** Answers will vary; an example is: $\frac{1}{2}$. **82.** Answers will vary; an example is: $\sqrt{2}$. **83.** commutative property of addition **84.** associative property of multiplication **85.** distributive property of multiplication over addition **86.** commutative property of multiplication **87.** commutative property of multiplication **88.** commutative property of addition

89. inverse property of multiplication **90.** identity property of multiplication **91.** Answers will vary; an example is: $2 \div 6 = \frac{1}{3}$. **92.** Answers will vary; an example is: $0 - 2 = -2$. **93. a.** The entries in the body of the table are all elements of the set.
b. $(4 \oplus 2) \oplus 3 = 4 \oplus (2 \oplus 3)$
$1 \oplus 3 = 4 \oplus 0$
$4 = 4$
c. 0 **d.** The inverse of 0 is 0, the inverse of 1 is 4, the inverse of 2 is 3, the inverse of 3 is 2, and the inverse of 4 is 1.
e. $3 \oplus 4 = 4 \oplus 3$; $3 \oplus 2 = 2 \oplus 3$
$2 = 2$ $\quad 0 = 0$

94. 216 **95.** 64 **96.** 16 **97.** 729 **98.** 25 **99.** 1 **100.** 1 **101.** $\frac{1}{216}$ **102.** $\frac{1}{16}$ **103.** $\frac{1}{49}$ **104.** 27 **105.** 460 **106.** 37,400 **107.** 0.00255 **108.** 0.0000745 **109.** 7.52×10^3 **110.** 3.59×10^6 **111.** 7.25×10^{-3} **112.** 4.09×10^{-7} **113.** 4.2×10^{13} **114.** 9.7×10^{-5} **115.** 390 **116.** 1,150,000 **117.** 0.023 **118.** 40 **119.** $(6.0 \times 10^4)(5.4 \times 10^5) = 3.24 \times 10^{10}$ **120.** $(9.1 \times 10^4)(4 \times 10^{-4}) = 3.64 \times 10^1$ **121.** $\frac{8.4 \times 10^6}{4 \times 10^3} = 2.1 \times 10^3$ **122.** $\frac{3 \times 10^{-6}}{6 \times 10^{-8}} = 5 \times 10^1$ **123.** 1.3×10^{12} **124.** 3.2×10^7 **125.** 40,625 years **126.** 2.88×10^{13} **127.** 7, 11, 15, 19, 23, and 27 **128.** $-4, -9, -14, -19, -24$, and -29 **129.** $\frac{3}{2}, 1, \frac{1}{2}, 0, -\frac{1}{2}$, and -1 **130.** 20 **131.** -30 **132.** -38 **133.** $a_n = -7 + (n - 1)4$; 69 **134.** $a_n = 200 + (n - 1)(-20)$; -180 **135.** 3, 6, 12, 24, 48, and 96 **136.** $\frac{1}{2}, \frac{1}{4}, \frac{1}{8}, \frac{1}{16}, \frac{1}{32}$, and $\frac{1}{64}$ **137.** 16, -8, 4, -2, 1, and $-\frac{1}{2}$ **138.** 54 **139.** $\frac{1}{2}$ **140.** -48 **141.** $a_n = 1(2)^{n-1}$; 128 **142.** $a_n = 100\left(\frac{1}{10}\right)^{n-1}$; $\frac{1}{100{,}000}$ **143.** arithmetic; 24 and 29 **144.** geometric; 162 and 486 **145.** geometric; $\frac{1}{256}$ and $\frac{1}{1024}$ **146.** arithmetic; -28 and -35 **147. a.** $a_n = 34.8 - 0.3n$ **b.** 18 hours per week **148. a.** approximately 1.02 for each division **b.** $a_n = 15.98(1.02)^{n-1}$ **c.** 28.95 million

Chapter 5 Test

1. 2, 3, 4, 6, 8, 9, and 12 **2.** $2^2 \cdot 3^2 \cdot 7$ **3.** GCD: 24; LCM: 144 **4.** 1 **5.** -4 **6.** -32 **7.** $0.58\overline{3}$ **8.** $\frac{64}{99}$ **9.** $\frac{1}{5}$ **10.** $\frac{37}{60}$ **11.** $-\frac{19}{2}$ **12.** $\frac{7}{12}$ **13.** $5\sqrt{2}$ **14.** $9\sqrt{2}$ **15.** $3\sqrt{2}$ **16.** $-7, -\frac{4}{5}, 0, 0.25, \sqrt{4}$, and $\frac{22}{7}$ **17.** commutative property of addition **18.** distributive property of multiplication over addition **19.** 243 **20.** 64 **21.** $\frac{1}{64}$ **22.** 7500 **23.** $\frac{4.9 \times 10^4}{7 \times 10^{-3}} = 7 \times 10^6$ **24.** $\$5.36 \times 10^{10}$ **25.** 3.07×10^8 **26.** $\approx \$175$ **27.** 1, -4, -9, -14, -19, and -24 **28.** 22 **29.** 16, 8, 4, 2, 1, and $\frac{1}{2}$ **30.** 320

CHAPTER 6

Section 6.1

Check Point Exercises

1. 608 **2.** -2 **3.** -94 **4.** 2662 calories; underestimates by 38 calories **5.** $3x - 21$ **6.** $38x^2 + 23x$ **7.** $2x + 36$

Concept and Vocabulary Check

1. evaluating **2.** equation **3.** terms **4.** coefficient **5.** factors **6.** like terms

Exercise Set 6.1

1. 27 **3.** 23 **5.** 29 **7.** -2 **9.** -21 **11.** -10 **13.** 140 **15.** 217 **17.** 176 **19.** 30 **21.** 44 **23.** 22 **25.** 27 **27.** -12 **29.** 69 **31.** -33 **33.** -8 **35.** 10°C **37.** 60 ft **39.** distributive property; commutative property of addition; associative property of addition; commutative property of addition **41.** $17x$ **43.** $-3x^2$ **45.** $3x + 15$ **47.** $8x - 12$ **49.** $15x + 16$ **51.** $27x - 10$ **53.** $29y - 29$ **55.** $8y - 12$ **57.** $-8x^2 + 10x$ **59.** $16y - 25$ **61.** $-10x + 6$ **63.** $12x^2 + 11$ **65.** $-2x^2 - 9$ **67. a.** 140 beats per minute **b.** 160 beats per minute **69.** 1924; underestimates by 76 calories **71. a.** 22% **b.** 18.7%; less than the estimate **81.** makes sense **83.** makes sense **85.** false **87.** false **89.** false **91.** true **93. a.** \$50.50, \$5.50, and \$1.00 per clock, respectively **b.** no; Answers will vary.

Section 6.2

Check Point Exercises

1. {6} **2.** {−2} **3.** {2} **4.** {5} **5.** {6} **6.** $\frac{37}{10}$ or 3.7; If a horizontal line is drawn from 10 on the scale for level of depression until it touches the blue line graph for the low-humor group and then a vertical line is drawn from that point on the blue line graph to the scale for the intensity of the negative life event, the vertical line will touch the scale at 3.7. **7. a.** {15} **b.** {5} **8.** \$5880 **9.** 720 deer **10.** $\varnothing$ **11.** $\{x \mid x \text{ is a real number}\}$

Concept and Vocabulary Check

1. linear **2.** equivalent **3.** $b + c$ **4.** bc **5.** apply the distributive property **6.** simplified; solved **7.** least common denominator; 12 **8.** proportion **9.** $ad = bc$ **10.** no; $\varnothing$ **11.** true; $\{x \mid x \text{ is a real number}\}$ **12.** false **13.** false **14.** false **15.** false

Exercise Set 6.2

1. {10} **3.** {−17} **5.** {12} **7.** {9} **9.** {−3} **11.** $\left\{-\frac{1}{4}\right\}$ **13.** {3} **15.** {11} **17.** {−17} **19.** {11} **21.** $\left\{\frac{7}{5}\right\}$ **23.** $\left\{\frac{25}{3}\right\}$ **25.** {8} **27.** {−3} **29.** {2} **31.** {−4} **33.** $\left\{-\frac{1}{5}\right\}$ **35.** {−4} **37.** {5} **39.** {6} **41.** {1} **43.** {−57} **45.** {18} **47.** {1} **49.** {24} **51.** {−6} **53.** {20} **55.** {5} **57.** {−7} **59.** {14} **61.** {27} **63.** {−15} **65.** {−9} **67.** {34} **69.** {−49} **71.** {10} **73.** ∅ **75.** $\{x \mid x$ is a real number$\}$ **77.** $\left\{\frac{2}{3}\right\}$ **79.** $\{x \mid x$ is a real number$\}$ **81.** ∅ **83.** {0} **85.** ∅ **87.** {0} **89.** $\left\{\frac{7}{2}\right\}$ **91.** {0} **93.** 2 **95.** 161 **97.** {−2} **99.** ∅ **101.** {10} **103.** {−2} **105.** 142 lb; 13 lb **107. a.** underestimates by 2% **b.** 2030 **109.** 6.25 qt **111.** 5.4 ft **113.** 20,489 fur seal pups **125.** makes sense **127.** does not make sense **131.** yes; Answers will vary.

Section 6.3

Check Point Exercises

1. computer science: \$56 thousand; economics; \$49 thousand; education; \$35 thousand **2.** by 67 years after 1969, or in 2036 **3.** 300 text messages **4.** \$1200 **5.** $w = \frac{P - 2l}{2}$ **6.** $m = \frac{T - D}{p}$

Concept and Vocabulary Check

1. $x + 658.6$ **2.** $31 + 2.4x$ **3.** $4 + 0.15x$ **4.** $x - 0.15x$ or $0.85x$ **5.** isolated on one side **6.** subtract b; divide by m

Exercise Set 6.3

1. 6 **3.** 25 **5.** 120 **7.** 320 **9.** 19 and 45 **11.** $x - (x + 4); -4$ **13.** $6(-5x); -30x$ **15.** $5x - 2x; 3x$ **17.** $8x - (3x + 6); 5x - 6$ **19.** TV: 9 years; sleeping: 28 years **21.** bachelor's: \$55 thousand; master's: \$61 thousand **23.** by 2015 **25. a.** births: 384,000; deaths: 156,000 **b.** 83 million **c.** approximately 4 years **27.** after 5 years **29.** after 5 months; \$165 **31.** 30 times **33.** \$600 of merchandise; \$580 **35.** 2019; 22,300 students **37.** \$420 **39.** \$150 **41.** \$467.20 **43.** $L = \frac{A}{W}$ **45.** $b = \frac{2A}{h}$ **47.** $P = \frac{I}{rt}$ **49.** $m = \frac{E}{c^2}$ **51.** $m = \frac{y - b}{x}$ **53.** $a = \frac{2A}{h} - b$ **55.** $r = \frac{S - P}{Pt}$ **57.** $x = \frac{C - By}{A}$ **59.** $n = \frac{a_n - a_1}{d} + 1$ or $n = \frac{a_n - a_1 + d}{d}$ **65.** does not make sense **67.** does not make sense **69.** \$200 **71.** 10 problems **73.** 36 plants **75.** $x = \frac{bc}{ad}$

Section 6.4

Check Point Exercises

1. a. 

b. 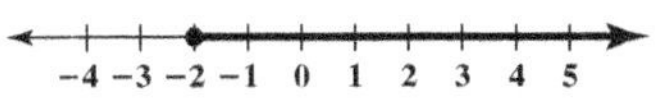

c. 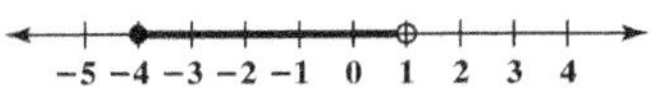

2. $\{x \mid x \leq 4\}$; 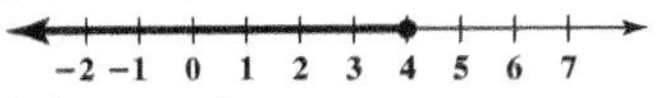

3. a. $\{x \mid x < 8\}$; 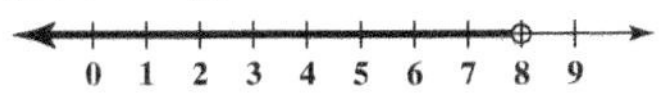

b. $\{x \mid x > -3\}$; 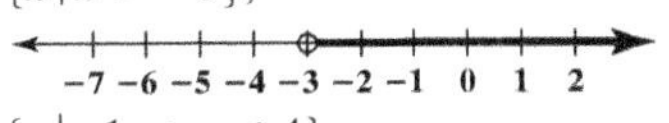

4. $\{x \mid x < -6\}$; 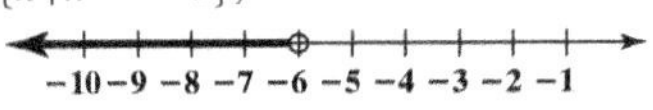

5. $\{x \mid x \geq 1\}$; 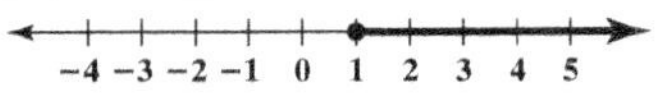

6. $\{x \mid -1 \leq x < 4\}$; 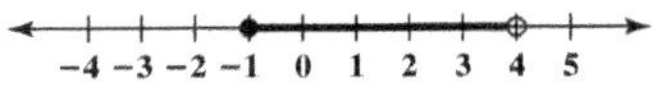

7. at least 83%

Concept and Vocabulary Check

1. is not included; is included **2.** $\{x \mid x < a\}$ **3.** $\{x \mid x < a \leq b\}$ **4.** negative **5.** false **6.** false **7.** false **8.** true

Exercise Set 6.4

1. 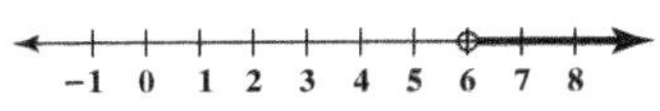

3. 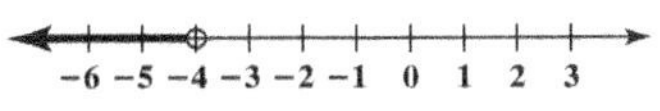

5. 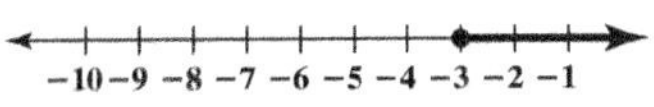

7. 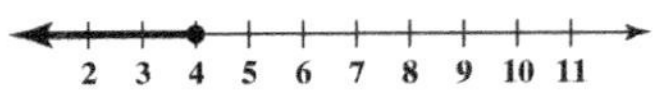

9. 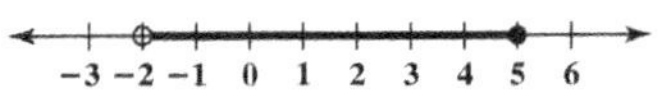

11. 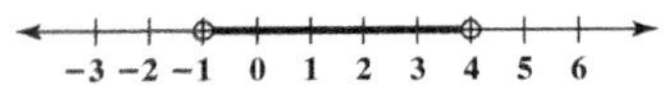

13. $\{x \mid x > 5\}$; 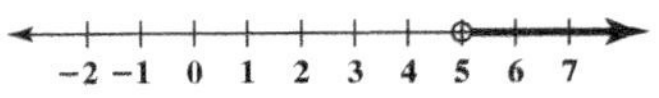

15. $\{x \mid x \leq 5\}$; 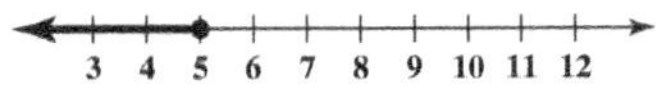

17. $\{x \mid x < 3\}$; 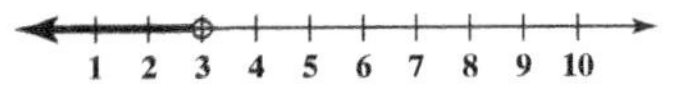

19. $\{x \mid x < 5\}$; 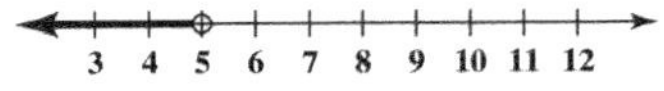

21. $\{x \mid x \geq -5\}$; 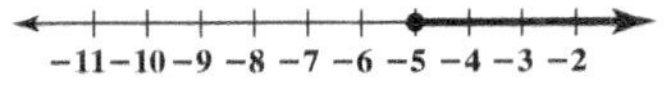

23. $\{x \mid x > 5\}$; 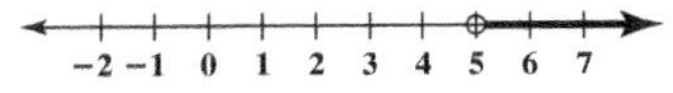

25. $\{x \mid x < 5\}$;

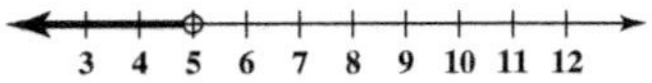

27. $\{x \mid x < 8\}$;

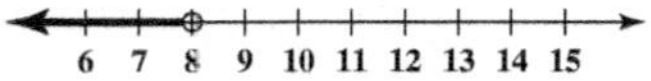

29. $\{x \mid x > -6\}$;

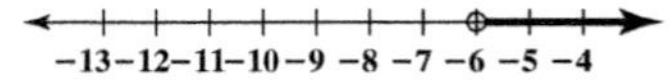

31. $\{x \mid x > -5\}$;

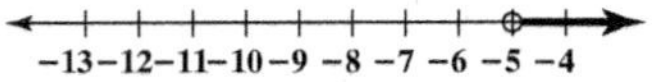

33. $\{x \mid x \le 5\}$;

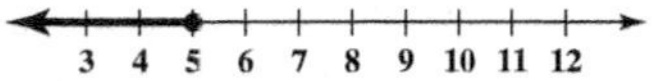

35. $\{x \mid x \le 3\}$;

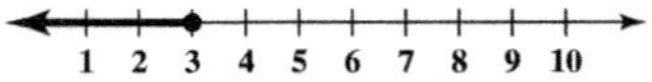

37. $\{x \mid x < 16\}$;

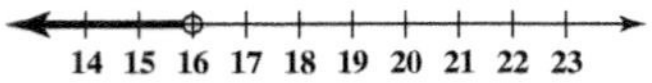

39. $\{x \mid x > -3\}$;

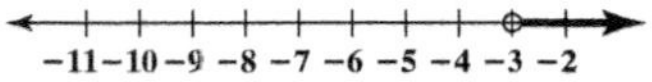

41. $\{x \mid x \ge -2\}$;

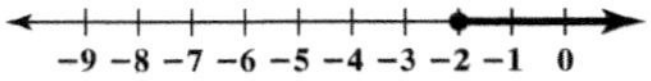

43. $\{x \mid x > -4\}$;

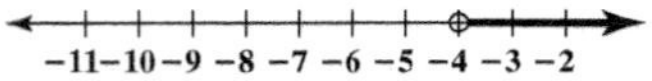

45. $\{x \mid x \ge 4\}$;

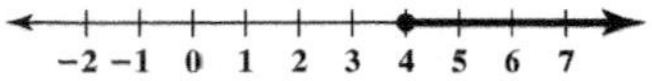

47. $\left\{x \mid x > \frac{11}{3}\right\}$;

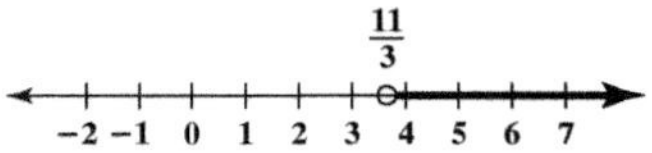

49. $\{x \mid x > 2\}$;

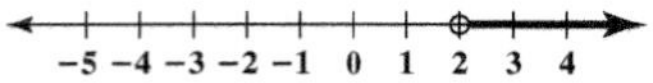

51. $\{x \mid x < 3\}$;

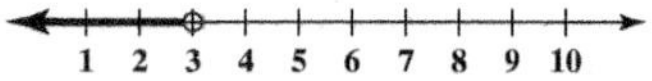

53. $\left\{x \mid x > \frac{5}{3}\right\}$;

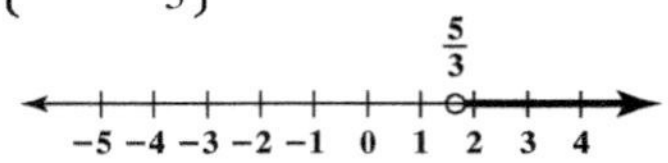

55. $\{x \mid x \ge -10\}$;

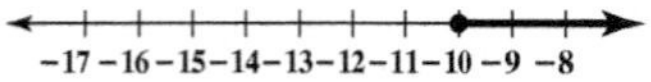

57. $\{x \mid x < -6\}$;

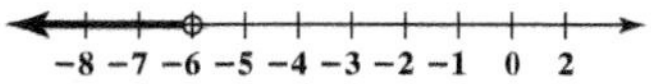

59. $\{x \mid 3 < x < 5\}$;

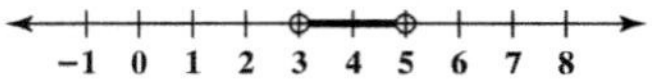

61. $\{x \mid -1 \le x < 3\}$;

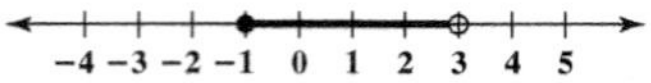

63. $\{x \mid -5 < x \le -2\}$;

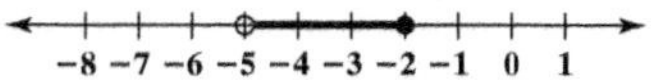

65. $\{x \mid 3 \le x < 6\}$;

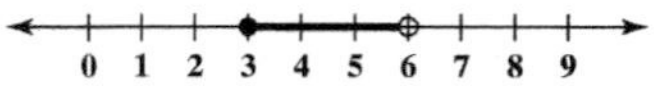

67. $x > \frac{C - By}{A}$ **69.** $x < \frac{C - By}{A}$ **71.** $\{x \mid x \le 5\}$ **73.** $\{x \mid x \le 14\}$ **75.** $\{x \mid x \le 50\}$ **77.** $\{x \mid 0 < x < 4\}$
79. intimacy $\ge$ passion or passion $\le$ intimacy **81.** commitment $>$ passion or passion $<$ commitment **83.** 9; after 3 years
85. **a.** years after 2016 **b.** years after 2020 **c.** years after 2020 **87.** **a.** at least 96 **b.** less than 66 **89.** at most 1280 mi
91. at most 29 bags **93.** between 80 and 110 minutes, inclusive **99.** makes sense **101.** makes sense **103.** more than 720 miles

Section 6.5

Check Point Exercises

1. $x^2 + 11x + 30$ **2.** $28x^2 - x - 15$ **3.** $(x + 2)(x + 3)$ **4.** $(x + 5)(x - 2)$ **5.** $(5x - 4)(x - 2)$ **6.** $(3y - 1)(2y + 7)$
7. $\{-6, 3\}$ **8.** $\{-2, 8\}$ **9.** $\left\{-4, \frac{1}{2}\right\}$ **10.** $\left\{-\frac{1}{2}, \frac{1}{4}\right\}$ **11.** $\left\{\frac{3 + \sqrt{7}}{2}, \frac{3 - \sqrt{7}}{2}\right\}$ **12.** approximately 26 years old

Concept and Vocabulary Check

1. $2x^2$; $3x$; $10x$; 15 **2.** $(x + 3)(x + 10)$ **3.** $(x - 3)(x - 6)$ **4.** $(x - 6)(x + 5)$ **5.** $(x + 2)(x - 7)$ **6.** $(4x + 1)(2x - 3)$
7. $(4x + 5)(3x - 4)$ **8.** $(x - 1)(2x - 3)$ **9.** $(2x + 3)(3x + 4)$ **10.** quadratic **11.** $A = 0$ or $B = 0$ **12.** subtracting 18
13. $x = \frac{-b \pm \sqrt{b^2 - 4ac}}{2a}$; quadratic formula **14.** false **15.** false **16.** false **17.** false **18.** false

Exercise Set 6.5

1. $x^2 + 8x + 15$ **3.** $x^2 - 2x - 15$ **5.** $2x^2 + 3x - 2$ **7.** $12x^2 - 43x + 35$ **9.** $(x + 2)(x + 3)$ **11.** $(x - 5)(x + 3)$
13. $(x - 5)(x - 3)$ **15.** $(x - 12)(x + 3)$ **17.** prime **19.** $(x + 1)(x + 16)$ **21.** $(2x + 1)(x + 3)$ **23.** $(2x - 5)(x - 6)$
25. $(3x + 2)(x - 1)$ **27.** $(3x - 28)(x + 1)$ **29.** $(3x - 4)(2x - 1)$ **31.** $(2x + 3)(2x + 5)$ **33.** $\{-3, 8\}$ **35.** $\left\{-\frac{5}{4}, 2\right\}$
37. $\{-5, -3\}$ **39.** $\{-3, 5\}$ **41.** $\{-3, 7\}$ **43.** $\{-8, -1\}$ **45.** $\{6\}$ **47.** $\left\{-\frac{1}{2}, 4\right\}$ **49.** $\left\{-2, \frac{9}{5}\right\}$ **51.** $\left\{-\frac{7}{2}, -\frac{1}{3}\right\}$ **53.** $\{-5, -3\}$
55. $\left\{\frac{-5 + \sqrt{13}}{2}, \frac{-5 - \sqrt{13}}{2}\right\}$ **57.** $\{-2 + \sqrt{10}, -2 - \sqrt{10}\}$ **59.** $\{-2 + \sqrt{11}, -2 - \sqrt{11}\}$ **61.** $\{-3, 6\}$ **63.** $\left\{-\frac{2}{3}, \frac{3}{2}\right\}$
65. $\{1 + \sqrt{11}, 1 - \sqrt{11}\}$ **67.** $\left\{\frac{1 + \sqrt{57}}{2}, \frac{1 - \sqrt{57}}{2}\right\}$ **69.** $\left\{\frac{-3 + \sqrt{3}}{6}, \frac{-3 - \sqrt{3}}{6}\right\}$ **71.** $\left\{\frac{3}{2}\right\}$ **73.** $\left\{-\frac{2}{3}, 4\right\}$ **75.** $\{-3, 1\}$
77. $\left\{1, \frac{5}{2}\right\}$ **79.** $\{\pm\sqrt{6}\}$ **81.** $1 + \sqrt{7}$ **83.** 9 teams **85.** **a.** 10.8%; overestimates by 0.4 **b.** 2020 **87.** **a.** $\frac{1}{\Phi - 1}$
b. $\Phi = \frac{1 + \sqrt{5}}{2}$ **c.** $\frac{1 + \sqrt{5}}{2}$ **93.** does not make sense **95.** does not make sense **99.** 3 and 4 **101.** $\{-3\sqrt{3}, \sqrt{3}\}$

Chapter 6 Review Exercises

1. 33 **2.** 15 **3.** −48 **4.** 55; It's the same. **5.** $17x - 15$ **6.** $5y - 13$ **7.** $14x^2 + x$ **8.** {6} **9.** {2} **10.** {12} **11.** $\left\{-\frac{13}{3}\right\}$ **12.** {2} **13.** ∅ **14.** $\{x \mid x \text{ is a real number}\}$ **15.** {5} **16.** {−65} **17.** $\left\{\frac{2}{5}\right\}$ **18.** {3} **19.** 324 teachers **20.** 287 trout **21. a.** \$15,000 **b.** \$15,180; reasonably well **c.** \$15,154; reasonably well **d.** 2025 **22.** *Scream:* 6; *Scream 2:* 8; *Scream 3:* 9 **23.** by 40 years after 1980, or in 2020 **24.** 500 text messages **25.** \$60 **26.** \$10,000 in sales **27.** $x = \frac{By + C}{A}$ **28.** $h = \frac{2A}{b}$ **29.** $B = 2A - C$ **30.** $g = \frac{s - vt}{t^2}$

31. $\{x \mid x < 4\}$;

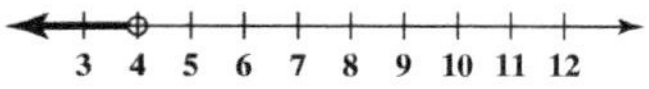

32. $\{x \mid x > -8\}$;

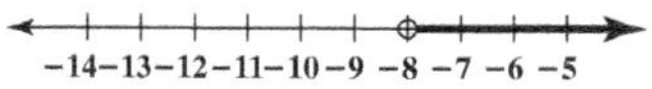

33. $\{x \mid x \geq -3\}$;

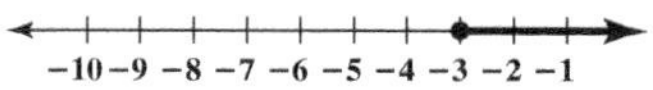

34. $\{x \mid x > 6\}$;

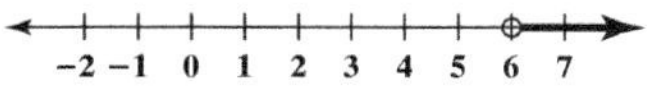

35. $\{x \mid x \geq 4\}$;

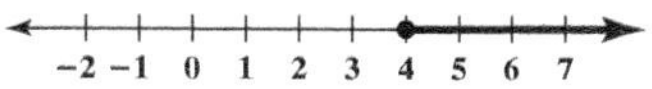

36. $\{x \mid x \leq 2\}$;

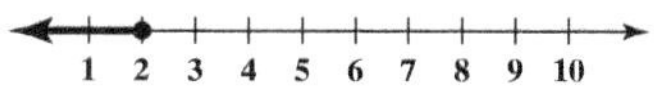

37. $\left\{x \middle| -\frac{3}{4} < x \leq 1\right\}$;

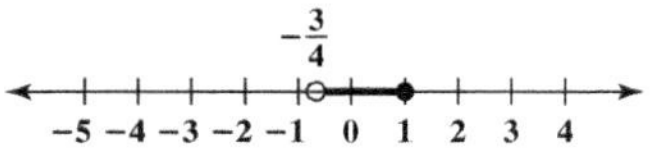

38. at least 64 **39.** $x^2 + 4x - 45$ **40.** $12x^2 - 13x - 14$ **41.** $(x - 4)(x + 3)$ **42.** $(x - 5)(x - 3)$ **43.** prime **44.** $(3x - 2)(x - 5)$ **45.** $(2x - 5)(3x + 2)$ **46.** prime **47.** {−7, 2} **48.** {−4, 8} **49.** $\left\{-8, \frac{1}{2}\right\}$ **50.** {−5, −2} **51.** {1, 3} **52.** $\left\{\frac{5 + \sqrt{41}}{2}, \frac{5 - \sqrt{41}}{2}\right\}$ **53.** $\left\{-3, \frac{1}{2}\right\}$ **54.** $\left\{\frac{3 + 2\sqrt{6}}{3}, \frac{3 - 2\sqrt{6}}{3}\right\}$ **55. a.** 182.8 ≈ 183; overestimates by 3 **b.** 2023

Chapter 6 Test

1. −44 **2.** $14x - 4$ **3.** {−5} **4.** {2} **5.** ∅ **6.** {−15} **7.** $y = \frac{Ax + A}{B}$ **8. a.** 21%; overestimates by 0.2 **b.** 20.7%; underestimates by 0.1 **c.** 2021 **9.** $\left\{\frac{15}{2}\right\}$ **10.** {3} **11.** 6000 tule elk **12.** love: 630; thanks: 243; sorry: 211 **13.** after 7 years **14.** 200 text messages **15.** \$50

16. $\{x \mid x \leq -3\}$;

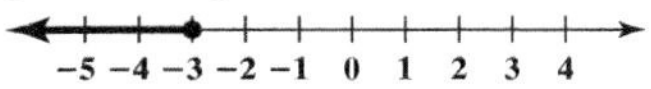

17. $\{x \mid x > 7\}$;

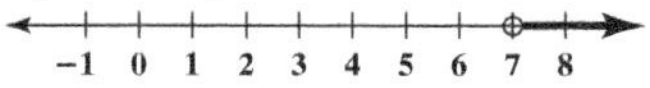

18. $\left\{x \middle| -2 \leq x < \frac{5}{2}\right\}$;

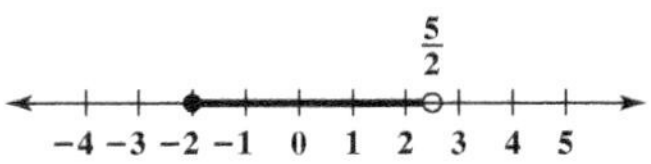

19. at least 92 **20.** $6x^2 - 7x - 20$ **21.** $(2x - 5)(x - 2)$ **22.** {−9, 4} **23.** $\left\{\frac{-2 + \sqrt{2}}{2}, \frac{-2 - \sqrt{2}}{2}\right\}$ **24.** 2018 **25.** 2018 **26.** quite well

CHAPTER 7

Section 7.1

Check Point Exercises

1.

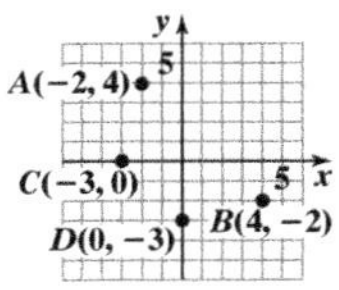

2.

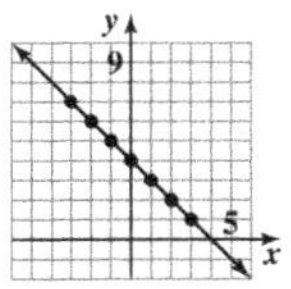

3. a. $y = 2x$

x	(x, y)
0	(0, 0)
2	(2, 4)
4	(4, 8)
6	(6, 12)
8	(8, 16)
10	(10, 20)
12	(12, 24)

$y = 10 + x$

x	(x, y)
0	(0, 10)
2	(2, 12)
4	(4, 14)
6	(6, 16)
8	(8, 18)
10	(10, 20)
12	(12, 22)

b.

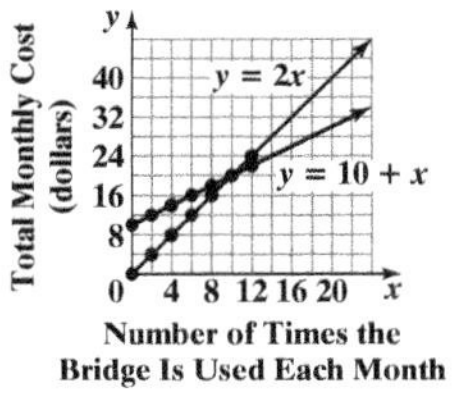

c. (10, 20); When the bridge is used 10 times during a month, the cost is \$20 with or without the discount pass.

4. a. 29 **b.** 65 **c.** 46

5. a. 190 ft **b.** 191 ft

6.

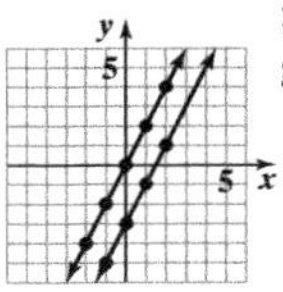

; The graph of g is the graph of f shifted down 3 units.

7. a. function **b.** function **c.** not a function **8. a.** 0 to 3 hours **b.** 3 to 13 hours **c.** 0.05 mg per 100 ml; after 3 hours **d.** None of the drug is left in the body. **e.** No vertical line intersects the graph in more than one point.

Concept and Vocabulary Check

1. x-axis **2.** y-axis **3.** origin **4.** quadrants; four **5.** x-coordinate; y-coordinate **6.** solution; satisfies **7.** y; x; function **8.** x; 6 **9.** more than once; function

Exercise Set 7.1

1.
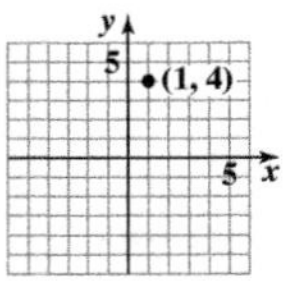

3.
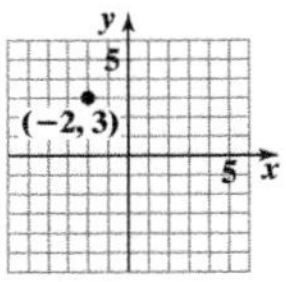

5.
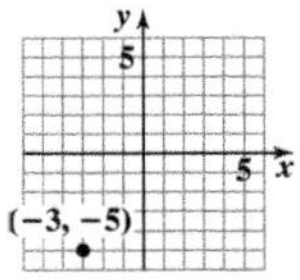

7.
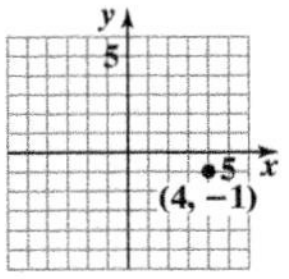

9.
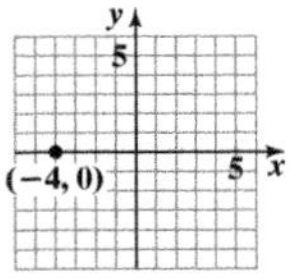

11.
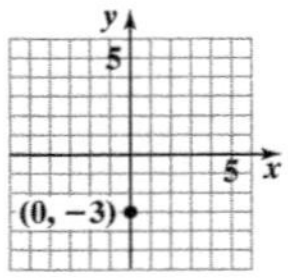

13.
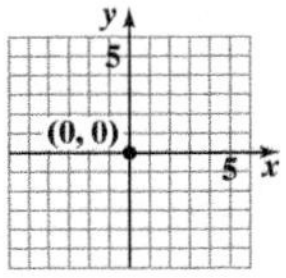

15.
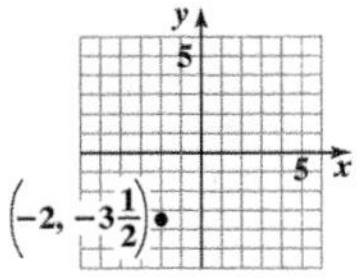

17.
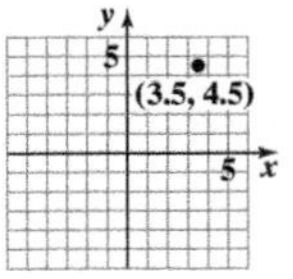

19.
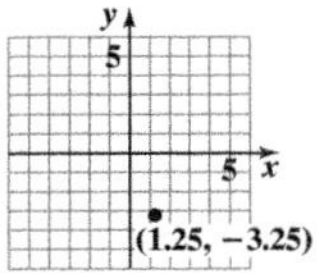

21.
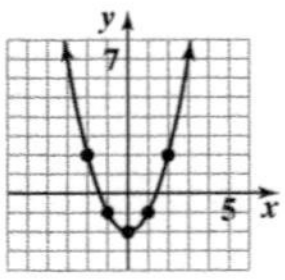

23.
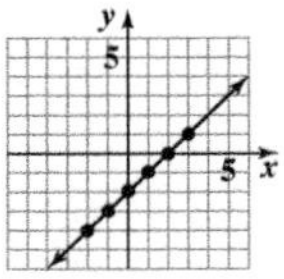

25.
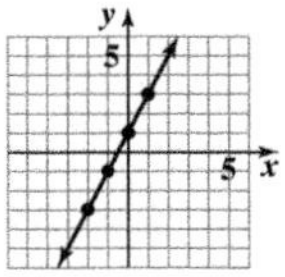

27.
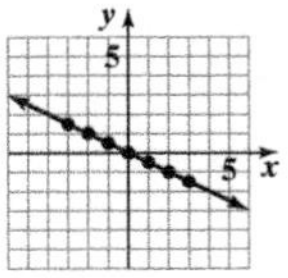

29.
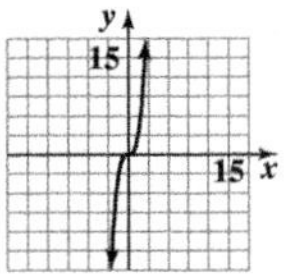

31.
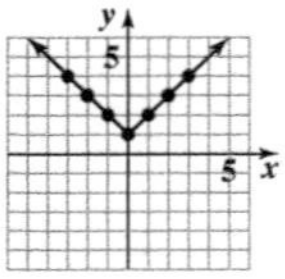

33. a. 4 **b.** −3 **35. a.** 19 **b.** −2 **37. a.** 5 **b.** 5 **39. a.** −14 **b.** −7 **41. a.** 53 **b.** 8 **43. a.** 26 **b.** 19 **45. a.** 1 **b.** −1

47.

x	$f(x) = x^2 - 1$
−2	3
−1	0
0	−1
1	0
2	3

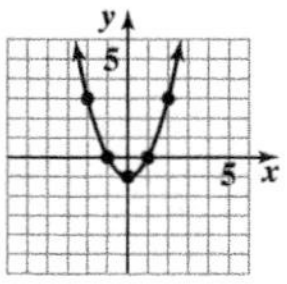

49.

x	$f(x) = x - 1$
−2	−3
−1	−2
0	−1
1	0
2	1

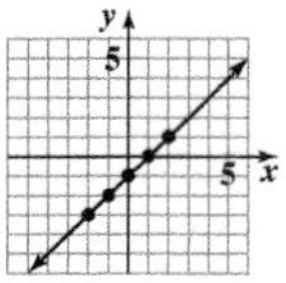

51.

x	$f(x) = (x - 2)^2$
0	4
1	1
2	0
3	1
4	4

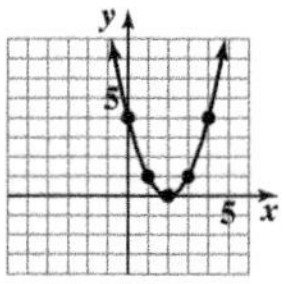

53.

x	$f(x) = x^3 + 1$
−3	−26
−2	−7
−1	0
0	1
1	2

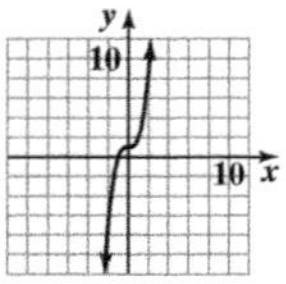

55. function **57.** function **59.** not a function **61.** function **63.** −2; 10 **65.** −38

67. $y = 2x + 4$

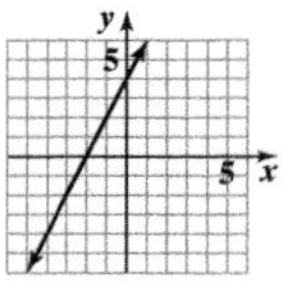

69. $y = 3 - x^2$

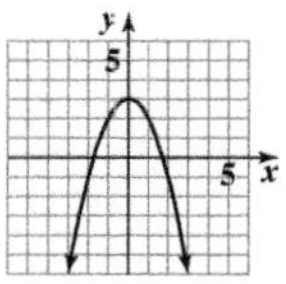

71. (2, 7); The football is 7 feet above the ground when it is 2 yards from the quarterback. **73.** (6, 9.25) **75.** 12 feet; 15 yards **77. a.** 81; According to the function, women's earnings were 81% of men's earning 30 years after 1980, or in 2010.; (30, 81) **b.** underestimates by 2% **79.** 440; For 20-year-old drivers, there are 440 accidents per 50 million miles driven.; (20, 440) **81.** $x = 45$; $y = 190$; The minimum number of accidents is 190 per 50 million miles driven and is attributed to 45-year-old drivers. **89.** makes sense **91.** makes sense **93.** −2 **95.** 1

Section 7.2

Check Point Exercises

1. 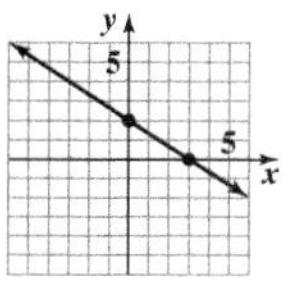 **2. a.** $m = 6$ **b.** $m = -\frac{7}{5}$ **3.** 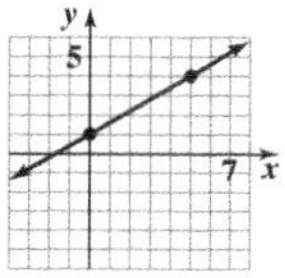 **4.** 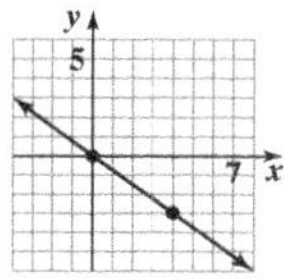 **5.** 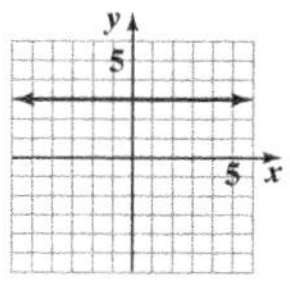 **6.** 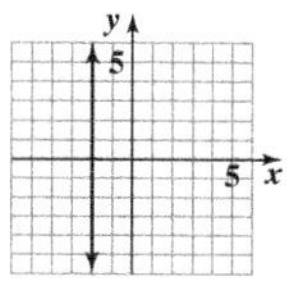

7. slope: −0.85; For the period from 1970 through 2010, the percentage of married men ages 20 to 24 decreased by 0.85 per year. The rate of change is −0.85% per year. **8. a.** $C(x) = 0.32x + 8$ **b.** 27.2%

Concept and Vocabulary Check

1. x-intercept **2.** y-intercept **3.** $\frac{y_2 - y_1}{x_2 - x_1}$ **4.** $y = mx + b$; slope; y-intercept **5.** −4; 3 **6.** (0, 3); 2; 5 **7.** horizontal **8.** vertical

Exercise Set 7.2

1. 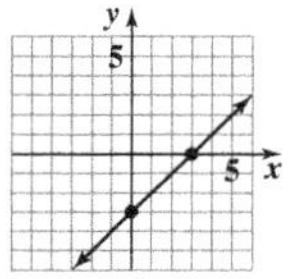 **3.** 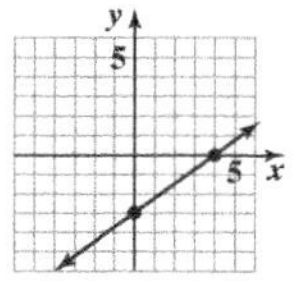 **5.** 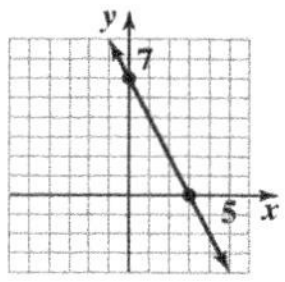 **7.** 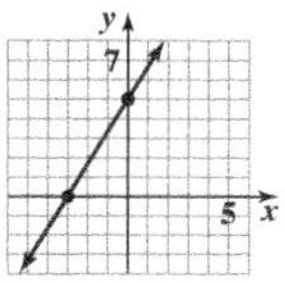

9. −1; falls **11.** $\frac{1}{4}$; rises **13.** −5; falls **15.** undefined; vertical **17.** −4; falls **19.** 0; horizontal

21. 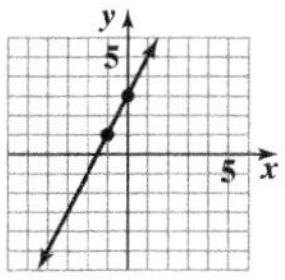 **23.** 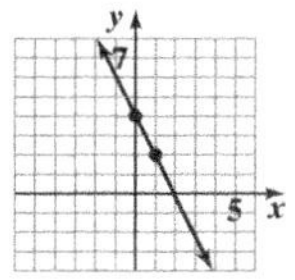 **25.** 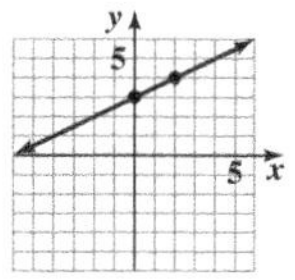 **27.** 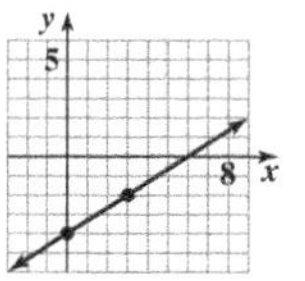 **29.** 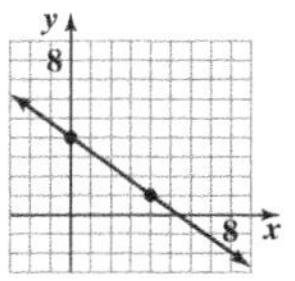 **31.** 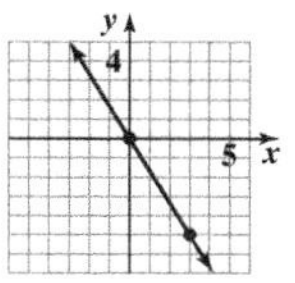

33. a. $y = -3x$ or $y = -3x + 0$
b. slope $= -3$; y-intercept $= 0$
c. 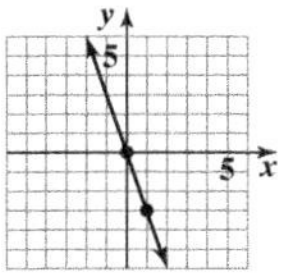

35. a. $y = \frac{4}{3}x$ or $y = \frac{4}{3}x + 0$
b. slope $= \frac{4}{3}$; y-intercept $= 0$
c. 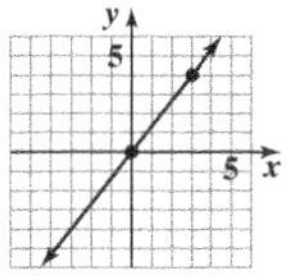

37. a. $y = -2x + 3$
b. slope $= -2$; y-intercept $= 3$
c. 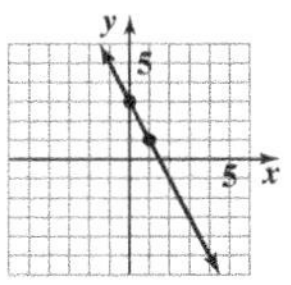

39. a. $y = -\frac{7}{2}x + 7$
b. slope $= -\frac{7}{2}$; y-intercept $= 7$
c. 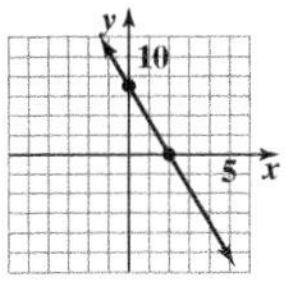

41. 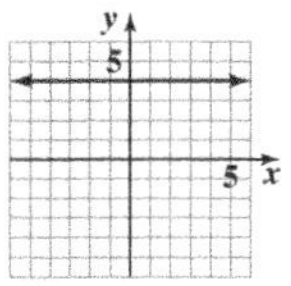 **43.** 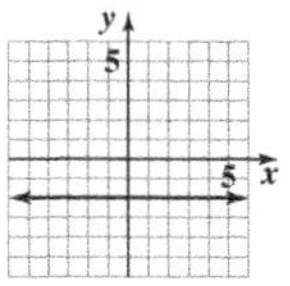 **45.** 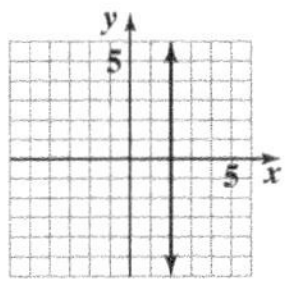 **47.** 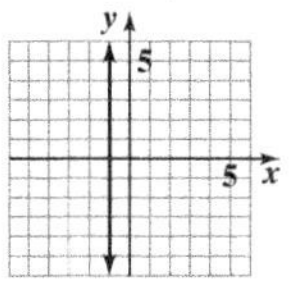

49. $m = -\frac{a}{b}$; falls **51.** undefined slope; vertical **53.** $m = -\frac{A}{B}$; $b = \frac{C}{B}$ **55.** −2 **57.** m_1, m_3, m_2, m_4 **59. a.** −0.5 **b.** 0.5; 0.5; year of aging **61.** $P(x) = 0.725x + 18$ **63. a.** $W(x) = 16x + 460$ **b.** 1100 thousand, or 1,100,000 **75.** does not make sense **77.** makes sense **79.** false **81.** false

Section 7.3

Check Point Exercises

1. not a solution **2.** 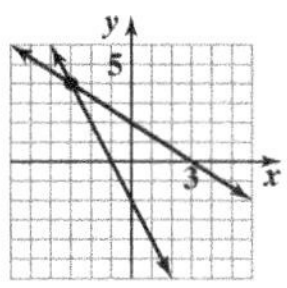 ; $\{(-3, 4)\}$ **3.** $\{(6, 11)\}$ **4.** $\{(1, -2)\}$ **5.** $\{(2, -1)\}$ **6.** $\{(2, -1)\}$ **7.** $\varnothing$

8. $\{(x, y) \mid y = 4x - 4\}$ or $\{(x, y) \mid 8x - 2y = 8\}$ **9. a.** $C(x) = 300{,}000 + 30x$ **b.** $R(x) = 80x$ **c.** (6000, 480,000); The company will break even if it produces and sells 6000 pairs of shoes.

Concept and Vocabulary Check

1. satisfies both equations in the system **2.** the intersection point **3.** $\left\{\left(\frac{1}{3}, -2\right)\right\}$ **4.** -2 **5.** -3 **6.** $\varnothing$; parallel

7. $\{(x, y) \mid x = 3y + 2$ or $\{(x, y) \mid 5x - 15y = 10$; are identical or coincide **8.** revenue; profit **9.** break-even point

Exercise Set 7.3

1. solution **3.** not a solution

5. 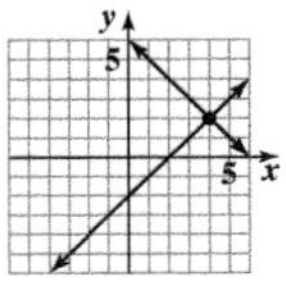

$\{(4, 2)\}$

7. 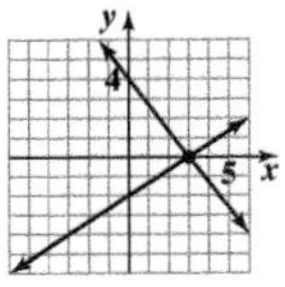

$\{(3, 0)\}$

9. 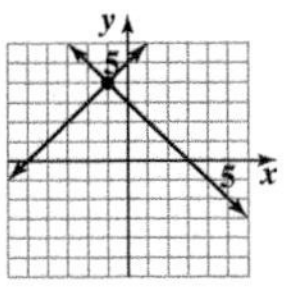

$\{(-1, 4)\}$

11. 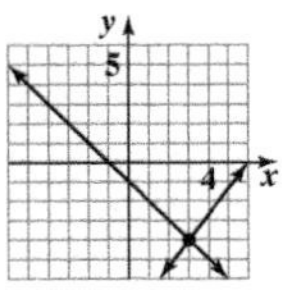

$\{(3, -4)\}$

13. $\{(1, 3)\}$ **15.** $\{(5, 1)\}$ **17.** $\{(2, 1)\}$ **19.** $\{(-1, 3)\}$ **21.** $\{(4, 5)\}$ **23.** $\left\{\left(-4, \frac{5}{4}\right)\right\}$ **25.** $\{(2, -1)\}$ **27.** $\{(3, 0)\}$

29. $\{(-4, 3)\}$ **31.** $\{(3, 1)\}$ **33.** $\{(1, -2)\}$ **35.** $\{(-5, -2)\}$ **37.** $\varnothing$ **39.** $\{(x, y) \mid y = 3x - 5\}$ or $\{(x, y) \mid 21x - 35 = 7y\}$ **41.** $\{(1, 4)\}$

43. $\{(x, y) \mid x + 3y = 2\}$ or $\{(x, y) \mid 3x + 9y = 6\}$ **45.** $\begin{cases} y = x - 4 \\ y = -\frac{1}{3}x + 4 \end{cases}$ **47.** $\left\{\left(\frac{1}{a}, 3\right)\right\}$ **49.** $m = -4, b = 3$ **51.** 500 radios

53. -6000; When the company produces and sells 200 radios, the loss is \$6000. **55. a.** $P(x) = 20x - 10{,}000$ **b.** \$190,000

57. a. $C(x) = 18{,}000 + 20x$ **b.** $R(x) = 80x$ **c.** (300, 24,000); When 300 canoes are produced and sold, both revenue and cost are \$24,000.

59. a. $C(x) = 30{,}000 + 2500x$ **b.** $R(x) = 3125x$ **c.** (48, 150,000); For 48 sold-out performances, both cost and revenue are \$150,000.

61. a. 4 million workers; \$4.50 per hour **b.** \$4.50; 4; 4 **c.** 2 million **d.** 5.7 million **e.** 3.7 million **63. a.** $y = 0.45x + 0.8$

b. $y = 0.15x + 2.6$ **c.** week 6; 3.5 symptoms; by the intersection point (6, 3.5) **75.** makes sense **77.** does not make sense

81. the twin who always lies

Section 7.4

Check Point Exercises

1. 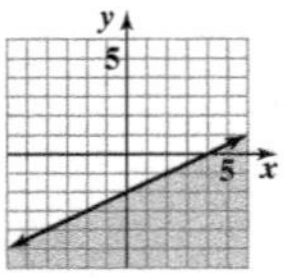

2. 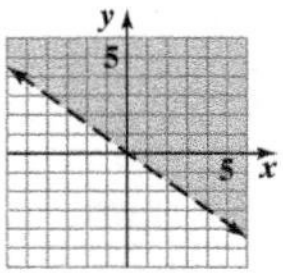

3. a. 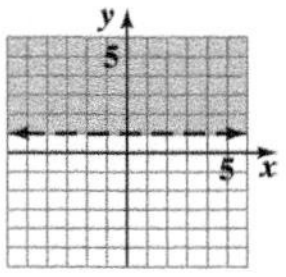

b. 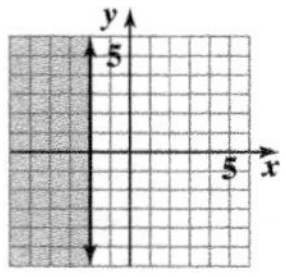

4. Point $B = (66, 130)$; $4.9(66) - 130 \geq 165$, or $193.4 \geq 165$, is true; $3.7(66) - 130 \leq 125$, or $114.2 \leq 125$, is true.

5. 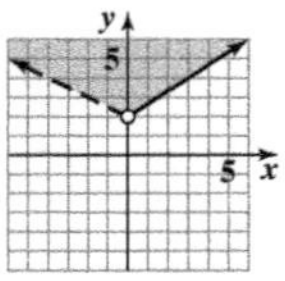

6. 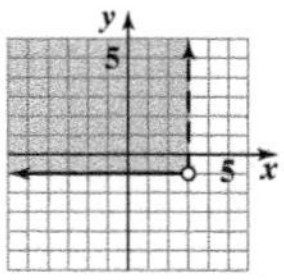

Concept and Vocabulary Check

1. solution; x; y; $5 > 1$ **2.** graph **3.** half-plane **4.** false **5.** true **6.** false **7.** $x - y < 1$; $2x + 3y \geq 12$ **8.** false

Exercise Set 7.4

1. 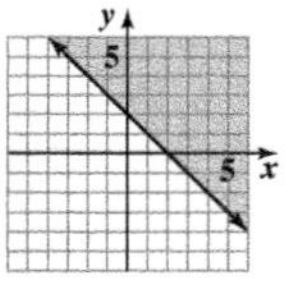

3. 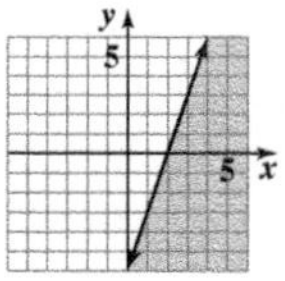

5. 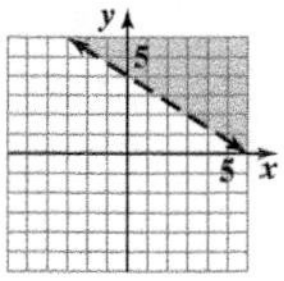

7. 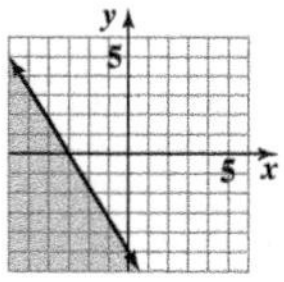

9. 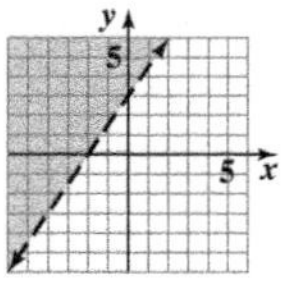

11. 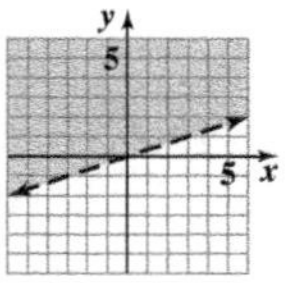

13. 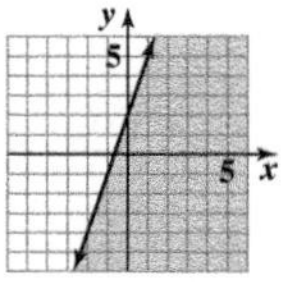

15. 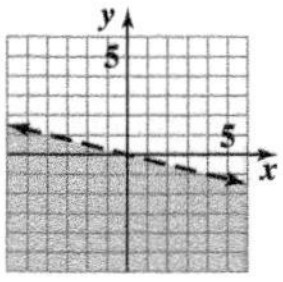

17. 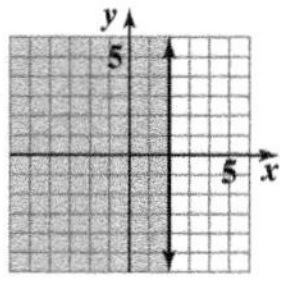

19. 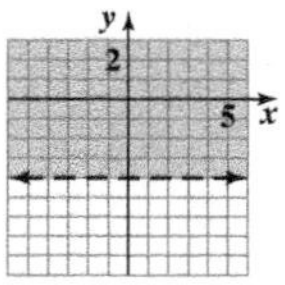

21. 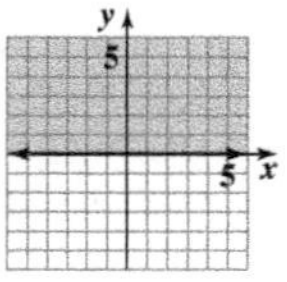

23. 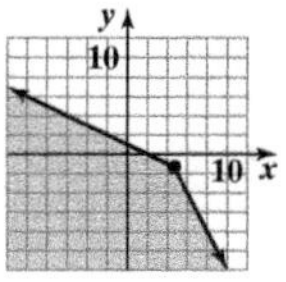

25. 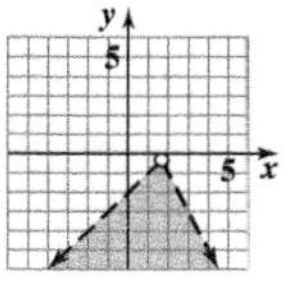

27. 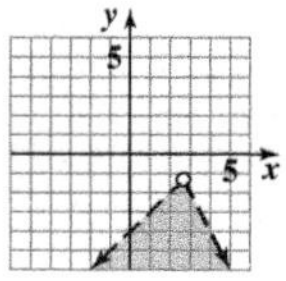

29. 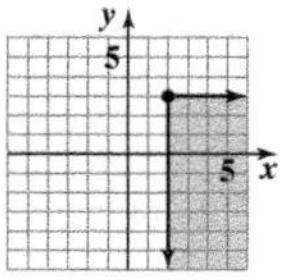

31. 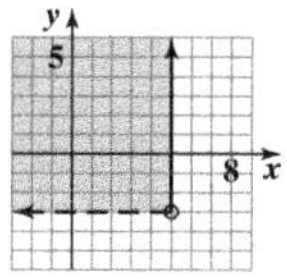

33. 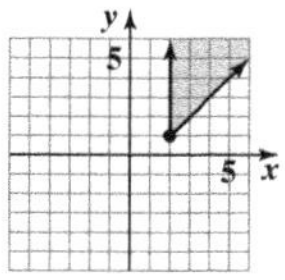

35. 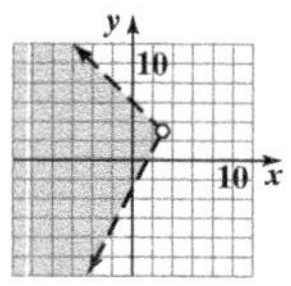

37. 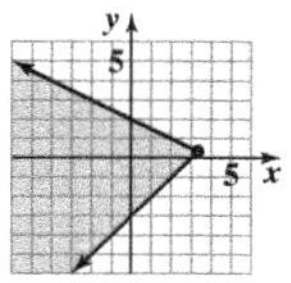

39. $y \geq -2x + 4$

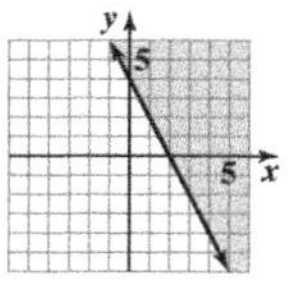

41. $\begin{cases} x + y \leq 4 \\ 3x + y \leq 6 \end{cases}$

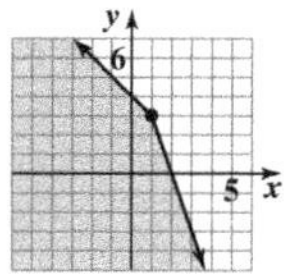

43. 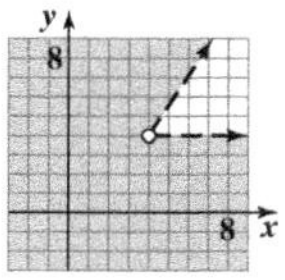

45. Point $A = (66, 160)$; $5.3(66) - 160 \geq 180$, or $189.8 \geq 180$, is true; $4.1(66) - 160 \leq 140$, or $110.6 \leq 140$, is true. **47.** no

49. a. $50x + 150y > 2000$

b. 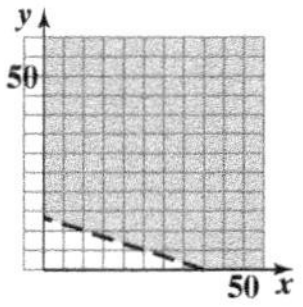

51. a. 27.1 **b.** overweight **59.** does not make sense **61.** makes sense **63.** $\begin{cases} y > x - 3 \\ y \leq x \end{cases}$ **65.** no solution **67.** infinitely many solutions

Section 7.5

Check Point Exercises

1. $z = 25x + 55y$ **2.** $x + y \leq 80$ **3.** $30 \leq x \leq 80$; $10 \leq y \leq 30$; objective function: $z = 25x + 55y$; constraints: $\begin{cases} x + y \leq 80 \\ 30 \leq x \leq 80 \\ 10 \leq y \leq 30 \end{cases}$

4. 50 bookshelves and 30 desks; \$2900

Concept and Vocabulary Check

1. linear programming **2.** objective **3.** constraints; corner

Exercise Set 7.5

1. (1, 2): 17; (2, 10): 70; (7, 5): 65; (8, 3): 58; maximum: $z = 70$; minimum: $z = 17$

3. (0, 0): 0; (0, 8): 400; (4, 9): 610; (8, 0): 320; maximum: $z = 610$; minimum: $z = 0$

5. a. 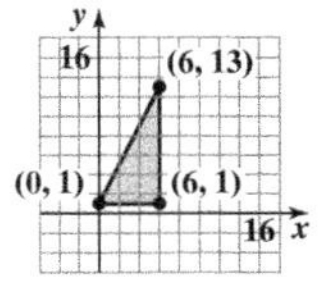

b. (0, 1): 1; (6, 13): 19; (6, 1): 7
c. maximum of 19 at $x = 6$ and $y = 13$

7. a. 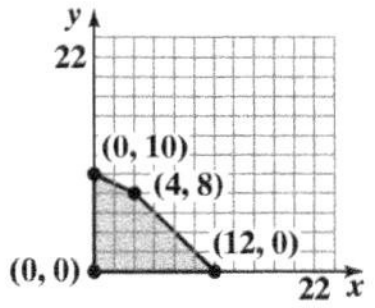

b. (0, 10): 100; (4, 8): 104; (12, 0): 72; (0, 0): 0
c. maximum of 104 at $x = 4$ and $y = 8$

9. a. 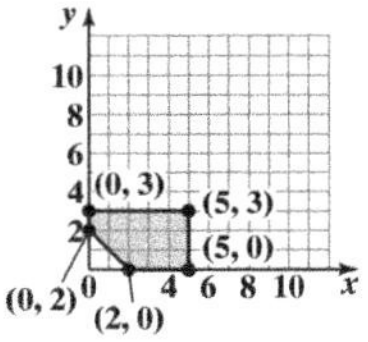

b. (0, 3): −6; (0, 2): −4; (2, 0): 10; (5, 0): 25; (5, 3): 19
c. maximum value: 25 at $x = 5$ and $y = 0$

11. a. 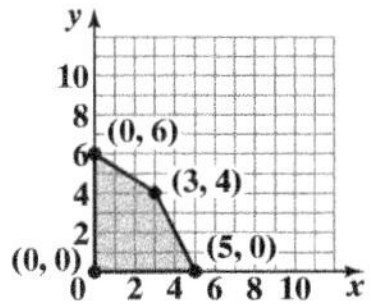

b. (0, 6): 72; (0, 0): 0; (5, 0): 50; (3, 4): 78
c. maximum value: 78 at $x = 3$ and $y = 4$

13. a. $z = 10x + 7y$

b. $\begin{cases} x + y \leq 20 \\ x \geq 3 \\ x \leq 8 \end{cases}$

c. 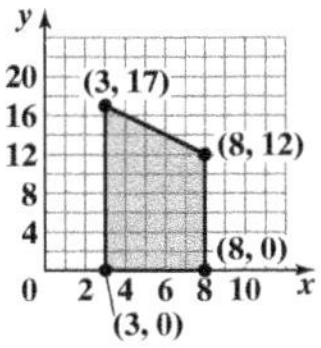

d. (3, 0): 30; (8, 0): 80; (3, 17): 149; (8, 12): 164
e. 8; 12; \$164

15. 300 cartons of food and 200 cartons of clothing **17.** 50 students and 100 parents **25.** makes sense **27.** makes sense

Section 7.6

Check Point Exercises

1.
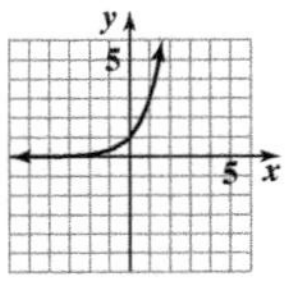

2. a. the exponential function g
b. the linear function f
3. 6.8%

4. $x = 3^y$;
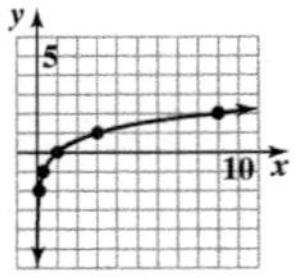

5. 34°; extremely well

6.
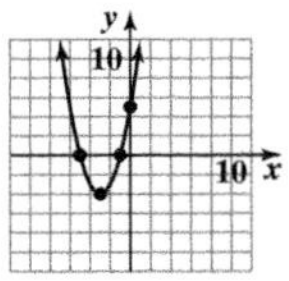

7. 10.8 ft; Answers will vary.; (8, 10.8)

Concept and Vocabulary Check

1. scatter plot; regression **2.** logarithmic **3.** exponential **4.** quadratic **5.** linear **6.** b^x **7.** 2.72; natural **8.** $b^y = x$ **9.** log x; common **10.** ln x; natural **11.** quadratic; parabola; $\frac{-b}{2a}$

Exercise Set 7.6

1.
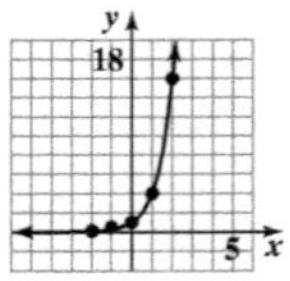

3.
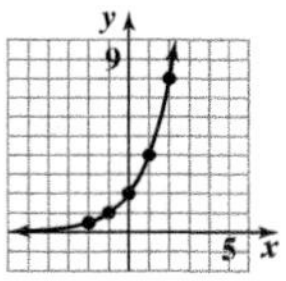

5.
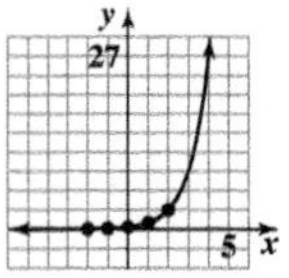

7. a. $x = 4^y$ **b.**
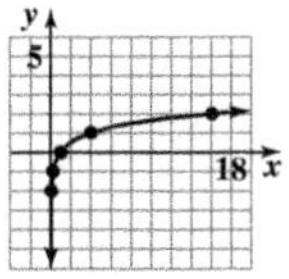

9. a. upward
b. $(-4, -9)$
c. $(-7, 0)$ and $(-1, 0)$
d. $(0, 7)$
e.
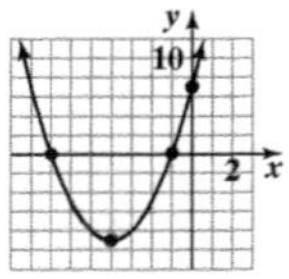

11. a. upward
b. $(1, -9)$
c. $(-2, 0)$ and $(4, 0)$
d. $(0, -8)$
e.
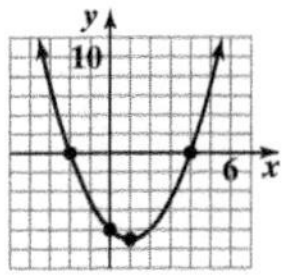

13. a. downward
b. $(2, 1)$
c. $(1, 0)$ and $(3, 0)$
d. $(0, -3)$
e.
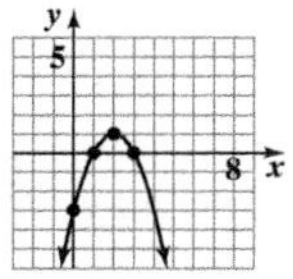

15. a.
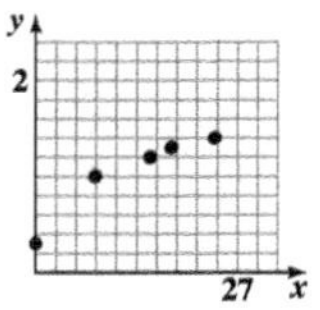

b. logarithmic

17. a.
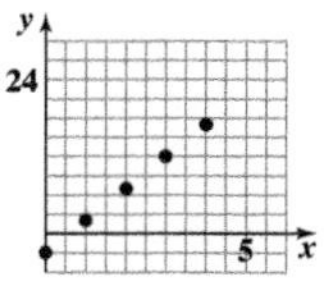

b. linear

19. a.
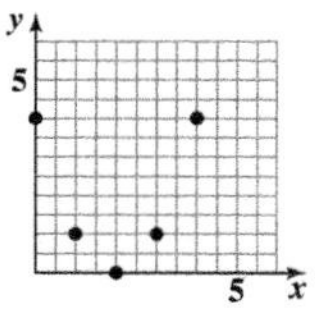

b. quadratic

21. a.
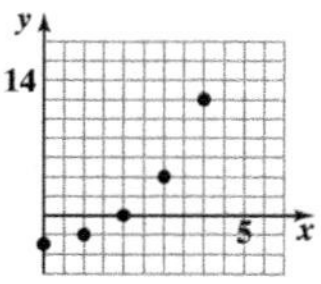

b. exponential

23.
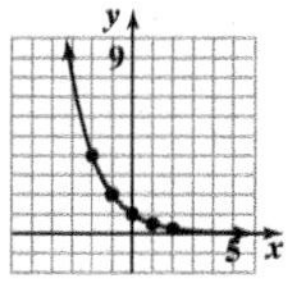

decreasing, although rate of decrease is slowing down

25.
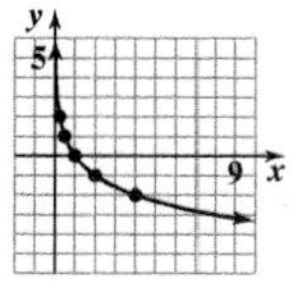

decreasing, although rate of decrease is slowing down

27. a. downward
b. $(1, 7)$
c. $(-0.9, 0)$ and $(2.9, 0)$
d. $(0, 5)$
e.
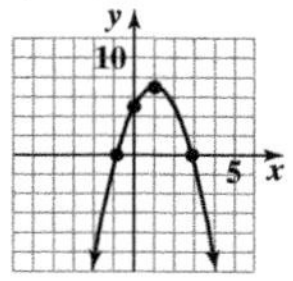

29. (3, 2) **31. a.** The data are increasing more and more rapidly. **b.** $f(x) = 3.476(1.023)^x$ **c.** 43.4 million; underestimates by 1.2 million **d.** 54.5 million; overestimates by 0.8 million **33. a.** \$15,166 **b.** \$16,037 **c.** the linear model **35. a.** The data increase rapidly and then begin to level off. **b.** $f(x) = 32 + 29 \ln x$ **c.** 99%; overestimates by 4% **37. a.** 95.4% **b.** Height increases rapidly at first and then more slowly. **39. a.** The data increase and then decrease.; The graph of the quadratic function modeling the data opens down. **b.** $f(x) = -0.8x^2 + 2.4x + 6$ **c.** 1.5; 7.8 **47.** does not make sense **49.** does not make sense **51.** 2 **53.** 0

Chapter 7 Review Exercises

1. 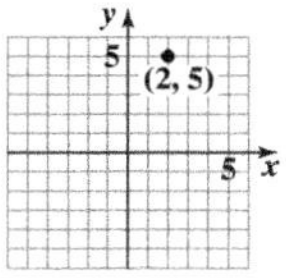

2. 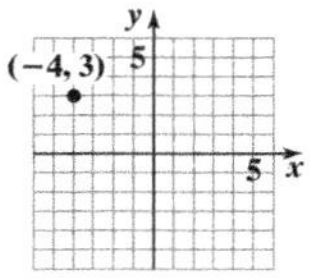

3. 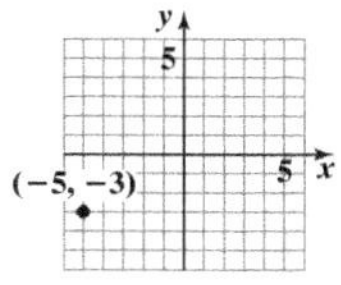

4. 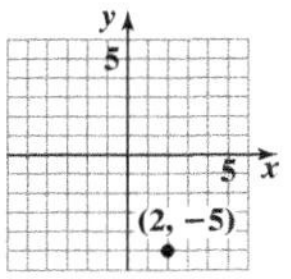

5. 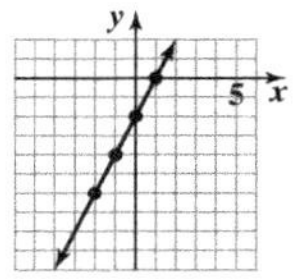

6. 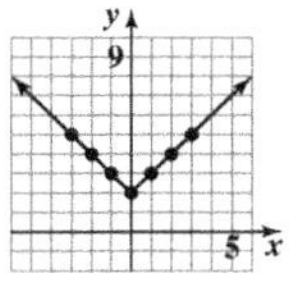

7. 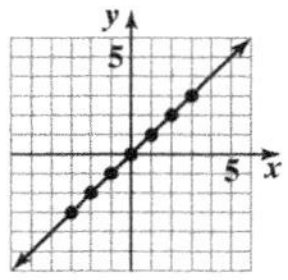

8. 3 **9.** 26 **10.** 30 **11.** −64

12.

x	$f(x) = \frac{1}{2}\lvert x\rvert$
−6	3
−4	2
−2	1
0	0
2	1
4	2
6	3

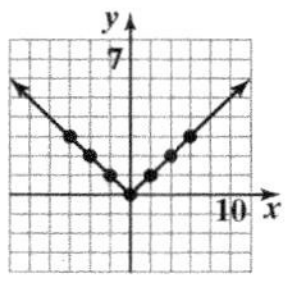

13.

x	$f(x) = x^2 - 2$
−2	2
−1	−1
0	−2
1	−1
2	2

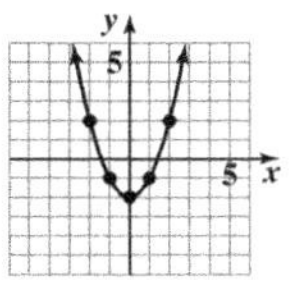

14. y is a function of x. **15.** y is not a function of x.

16. a. $D(1) = 92.8$; Skin damage begins for burn-prone people after 92.8 minutes when the sun's UV index is 1.; by the point (1, 92.8)
b. $D(10) = 19$; Skin damage begins for burn-prone people after 19 minutes when the sun's UV index is 10.; by the point (10, 19)

17. 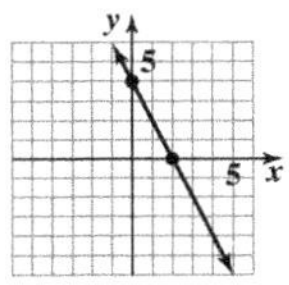

18. 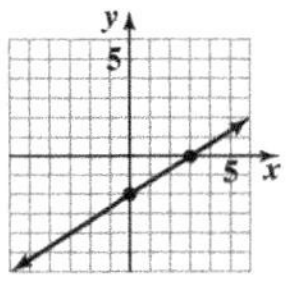

19. 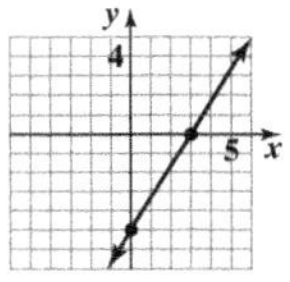

20. $-\frac{1}{2}$; falls **21.** 3; rises **22.** 0; horizontal **23.** undefined; vertical

24. 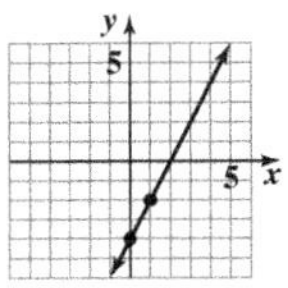

25. 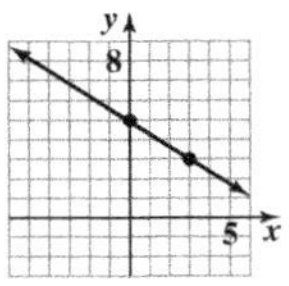

26. 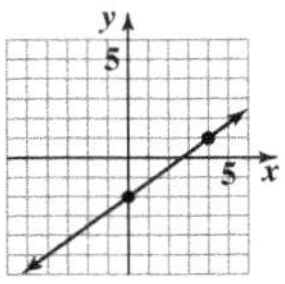

27. 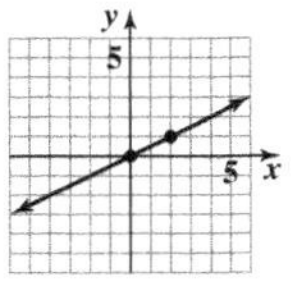

28. a. $y = -2x$
b. slope: −2; y-intercept: 0
c. 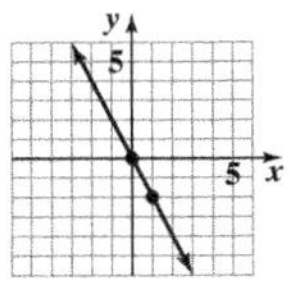

29. a. $y = \frac{5}{3}x$
b. slope: $\frac{5}{3}$; y-intercept: 0
c. 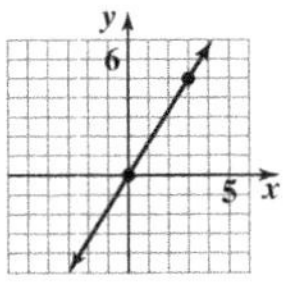

30. a. $y = -\frac{3}{2}x + 2$
b. slope: $-\frac{3}{2}$; y-intercept: 2
c. 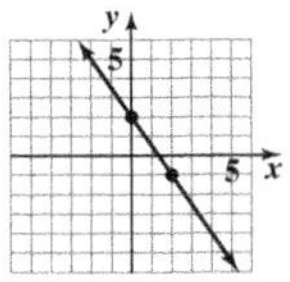

31. 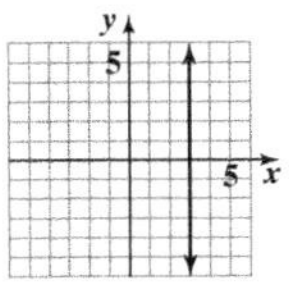

32. 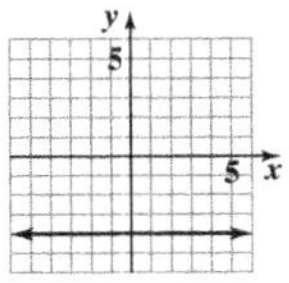

33. 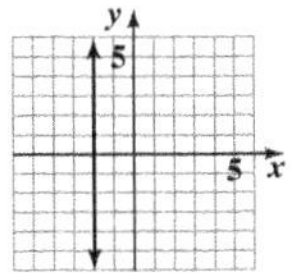

34. a. 254; If no women in a country are literate, the mortality rate of children under 5 is 254 per thousand. **b.** −2.4; For each 1% of adult females who are literate, the mortality rate of children under 5 decreases by 2.4 per thousand. **c.** $f(x) = -2.4x + 254$ **d.** 134 per thousand **35.** $\{(2, 3)\}$ **36.** $\{(-2, -3)\}$ **37.** $\{(3, 2)\}$ **38.** $\{(4, -2)\}$ **39.** $\{(1, 5)\}$ **40.** $\{(2, 3)\}$ **41.** $\{(-9, 3)\}$ **42.** $\{(3, 4)\}$ **43.** $\{(2, -3)\}$ **44.** $\varnothing$ **45.** $\{(3, -1)\}$ **46.** $\{(x, y) \mid 3x - 2y = 6\}$ or $\{(x, y) \mid 6x - 4y = 12\}$ **47. a.** $C(x) = 60{,}000 + 200x$ **b.** $R(x) = 450x$ **c.** (240, 108,000); This means the company will break even if it produces and sells 240 desks. **48. a.** Answers will vary.; approximately (2016, 325); 2016; 325 million **b.** $y = 6x + 200$ **c.** 2016; 326 million **d.** quite well, although answers may vary

49. 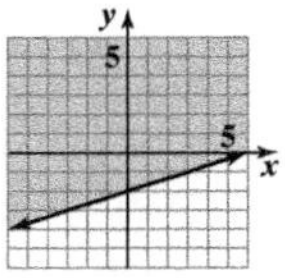

50. 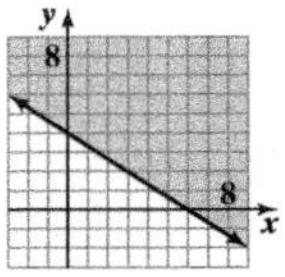

51. 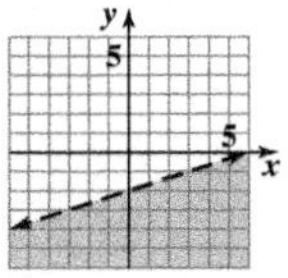

52. 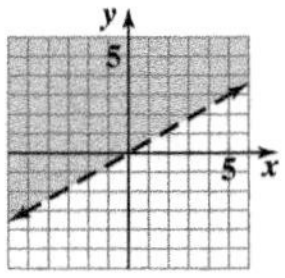

53. 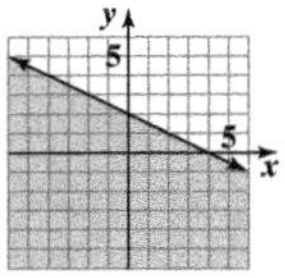

54. 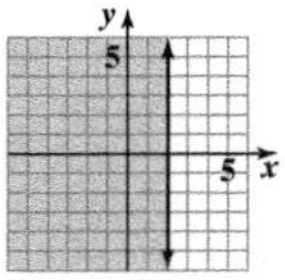

55. 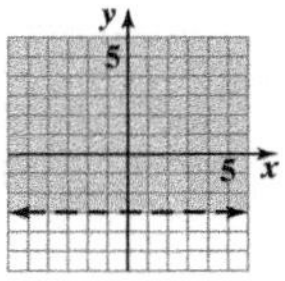

56. 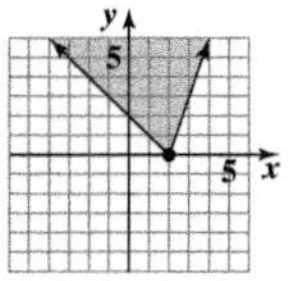

57. 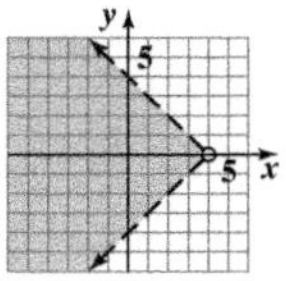

58. 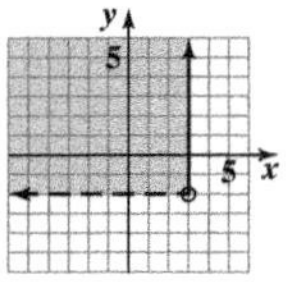

59. 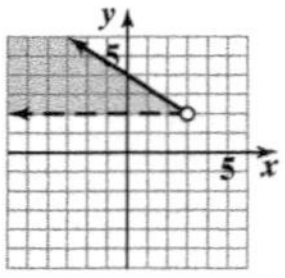

60. 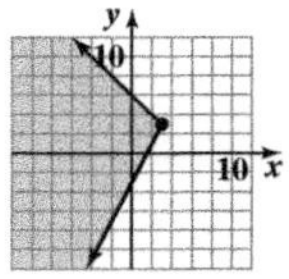

61. 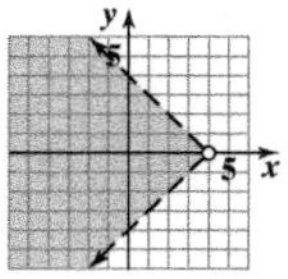

62. $(1, 0): 2; \left(\frac{1}{2}, \frac{1}{2}\right): \frac{5}{2}$; (2, 2): 10; (4, 0): 8; maximum: 10; minimum: 2

63. a. 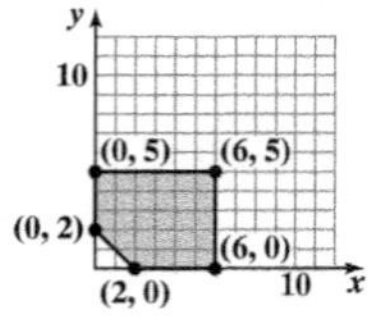

b. (0, 2): 6; (0, 5): 15; (6, 5): 27; (6, 0): 12; (2, 0): 4
c. maximum of 27 at $x = 6$ and $y = 5$; minimum of 4 at $x = 2$ and $y = 0$

64. a. $z = 500x + 350y$
b. $\begin{cases} x + y \le 200 \\ x \ge 10 \\ y \ge 80 \end{cases}$
c. 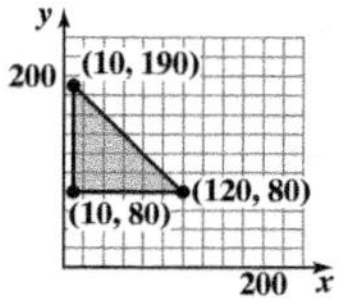

d. (10, 80): 33,000; (10, 190): 71,500; (120, 80): 88,000
e. 120; 80; 88,000

65. 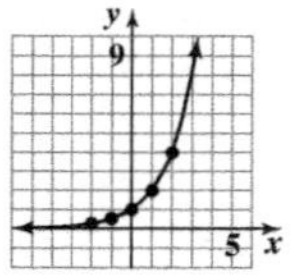

66. 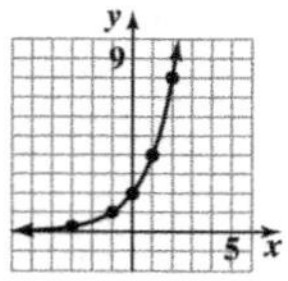

67. $x = 2^y$; 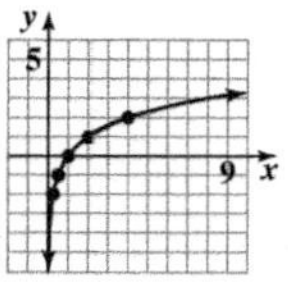

68. a. upward
b. (3, −16)
c. (−1, 0) and (7, 0)
d. (0, −7)
e. 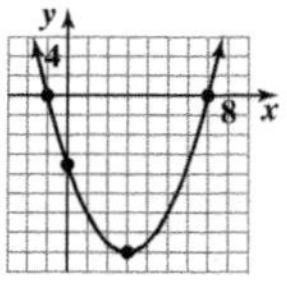

69. a. downward
b. (−1, 4)
c. (−3, 0) and (1, 0)
d. (0, 3)
e. 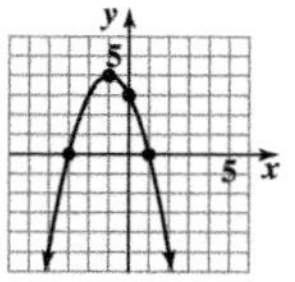

70. a. 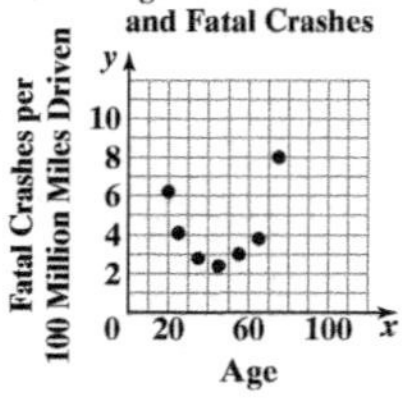

b. quadratic

71. a. 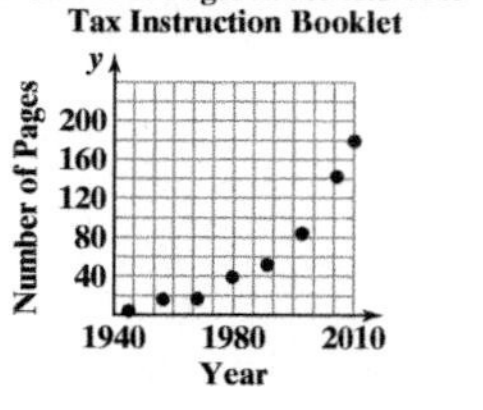

b. exponential

72. a. 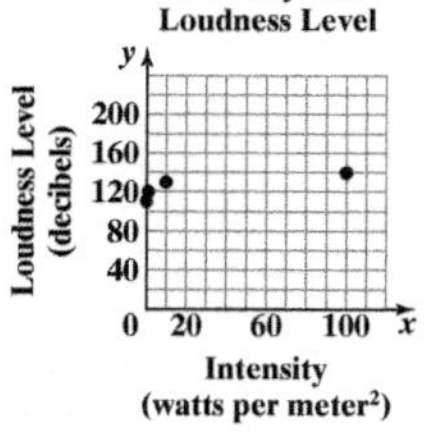

b. logarithmic

73. a. 47; For each additional hour spent at a shopping mall, the average amount spent increases by \$47. The rate of change is \$47 per hour. **b.** the exponential function **74.** the logarithmic function

Chapter 7 Test

1. 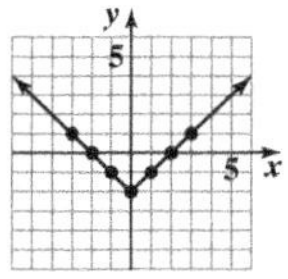 **2.** 21 **3.** y is not a function of x. **4.** y is a function of x. **5. a.** Yes; No vertical line intersects the graph in more than one point. **b.** 0; The eagle was on the ground after 15 seconds. **c.** 45 meters **d.** between 3 and 12 seconds

6. 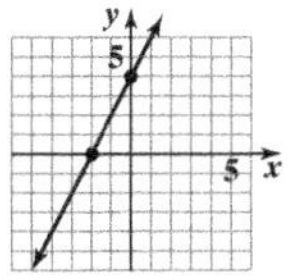 **7.** $m = 3$ **8.** 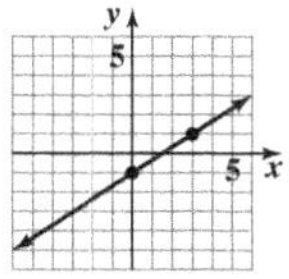 **9.** 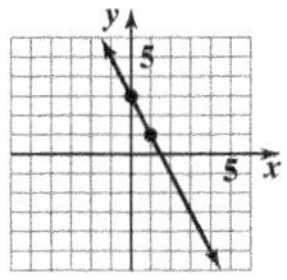 **10. a.** −0.48 **b.** 0.48; −0.48%; year **11. a.** 4571; There were 4571 deaths involving distracted driving 0 years after 2005, or in 2005. **b.** 433; The rate of change in the number of deaths involving distracted driving is 433 deaths per year. **c.** $f(x) = 433x + 4571$ **d.** 8901 deaths

12. $\{(2, 4)\}$ **13.** $\{(1, -3)\}$ **14.** $\{(6, -5)\}$ **15. a.** $C(x) = 360{,}000 + 850x$ **b.** $R(x) = 1150x$ **c.** (1200, 1,380,000); The company will break even if it produces and sells 1200 computers.

16. 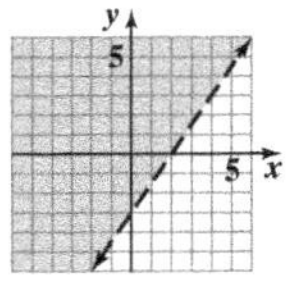 **17.** 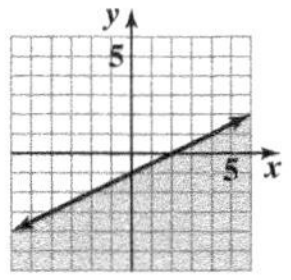 **18.** 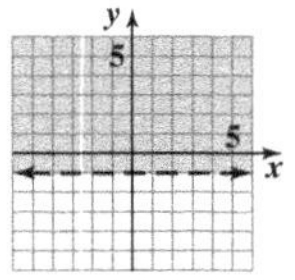 **19.** 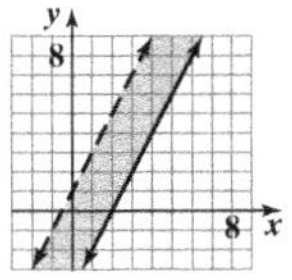

20. (2, 0): 6; (2, 6): 18; (6, 3): 24; (8, 0): 24; maximum: 24; minimum: 6 **21.** 26 **22.** 50 regular jet skis, 100 deluxe jet skis; \$35,000

23. 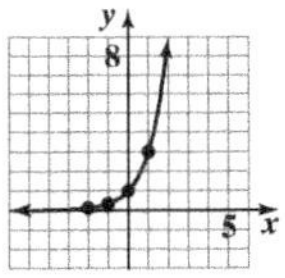 **24.** $x = 3^y$; 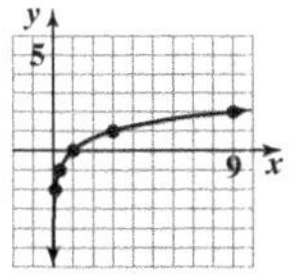 **25.** 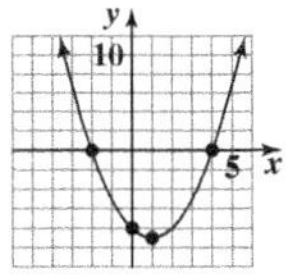

26. linear **27.** logarithmic **28.** exponential **29.** quadratic **30. a.** 0.03; The cost of making a penny is increasing an average of 0.03 cents per year. **b.** exponential; The cost is rising more and more rapidly. **c.** linear: 1.53¢; exponential: 1.75¢; not really; exponential; yes

CHAPTER 8

Section 8.1

Check Point Exercises

1. 12.5% **2.** 2.3% **3. a.** 0.67 **b.** 2.5 **4. a.** \$75.60 **b.** \$1335.60 **5. a.** \$133 **b.** \$247 **6. a.** $66\frac{2}{3}\%$ **b.** 40% **7.** 35% **8.** 20% **9. a.** \$1152 **b.** 4% decrease

Concept and Vocabulary Check

1. 100 **2.** 7; 8; 100; a percent sign **3.** two; right; a percent sign **4.** two; left; the percent sign **5.** tax rate; item's cost **6.** discount rate; original price **7.** the amount of increase; the original amount **8.** the amount of decrease; the original amount **9.** false; Changes will vary. **10.** false; Changes will vary.

Exercise Set 8.1

1. 40% **3.** 25% **5.** 37.5% **7.** 2.5% **9.** 11.25% **11.** 59% **13.** 38.44% **15.** 287% **17.** 1487% **19.** 10,000% **21.** 0.72 **23.** 0.436 **25.** 1.3 **27.** 0.02 **29.** 0.005 **31.** 0.00625 **33.** 0.625 **35.** 6 **37.** 7.2 **39.** 5 **41.** 170 **43.** 20% **45.** 12% **47. a.** \$1968 **b.** \$34,768 **49. a.** \$103.20 **b.** \$756.80 **51.** 19.2% **53.** 100% **55.** 15% **57.** no; 2% loss **65.** does not make sense **67.** does not make sense **69.** \$648.72 **71.** \$2400 increase

Section 8.2

Check Point Exercises

1. a. \$89,880 **b.** \$86,680 **c.** \$46,660 **2.** \$3952.50 **3.** \$7520 **4. a.** \$180 **b.** \$18 **c.** \$10.17 **d.** \$7.20 **e.** \$144.63 **f.** 19.7%

Concept and Vocabulary Check

1. gross **2.** adjusted gross; adjustments **3.** exemption **4.** adjusted gross; exemptions; deductions **5.** credits **6.** FICA **7.** gross **8.** net **9.** false; Changes will vary. **10.** true **11.** true **12.** true

Exercise Set 8.2

1. gross income: \$53,320; adjusted gross income: \$50,120; taxable income: \$39,070 **3.** gross income: \$87,490; adjusted gross income: \$85,290; taxable income: \$68,465 **5.** \$6030 **7.** \$28,053.25 **9.** −\$685, or a \$685 refund **11.** \$51,906.50 **13.** \$2645 **15.** \$4365 **17.** \$1865 **19.** \$6360 **21.** \$15,790 **23. a.** \$1130 **b.** \$1102.50 **c.** 11.2% **25. a.** \$170 **b.** \$17 **c.** \$9.61 **d.** \$5.10 **e.** \$138.29 **f.** 18.7% **39.** does not make sense **41.** makes sense **43.** The \$2500 credit reduces taxes by the greater amount; the difference is \$2100. **45.** yes; Answers will vary.

Section 8.3

Check Point Exercises

1. \$150 **2.** \$336 **3.** \$2091 **4. a.** \$1820 or \$1825 **b.** \$1892.80 or \$1898 **5.** 18% **6.** \$3846.16

Concept and Vocabulary Check

1. $I = Prt$; principal; rate; time **2.** $A = P(1 + rt)$ **3.** 360 **4.** false; Changes will vary. **5.** true **6.** false; Changes will vary.

Exercise Set 8.3

1. \$240 **3.** \$10.80 **5.** \$318.75 **7.** \$426.25 **9.** \$3420 **11.** \$38,350 **13.** \$9390 **15.** 7.5% **17.** 9% **19.** 31.3% **21.** \$5172.42 **23.** \$8917.20 **25.** \$4509.59 **27.** $r = \frac{A - P}{Pt}$ **29.** $P = \frac{A}{1 + rt}$ **31. a.** \$247.50 **b.** \$4247.50 **33.** 21.4% **35.** 640% **37.** \$2654.87 **41.** does not make sense **43.** does not make sense **45. a.** $A = 275t + 5000$ **b.** slope: 275; Answers will vary; an example is: The rate of change of the future value per year is \$275.

Section 8.4

Check Point Exercises

1. a. \$1216.65 **b.** \$216.65 **2. a.** \$6253.23 **b.** \$2053.23 **3. a.** \$14,859.47 **b.** \$14,918.25 **4.** \$5714.25 **5. a.** \$6628.28 **b.** 10.5% **6.** ≈ 8.24%

Concept and Vocabulary Check

1. principal; interest **2.** t; r; n **3.** $A = P(1 + r)^t$ **4.** six; semiannually **5.** three; quarterly **6.** continuous **7.** P; A; t; r; n **8.** effective annual yield; simple **9.** true **10.** true **11.** true **12.** true

Exercise Set 8.4

1. a. \$10,816 **b.** \$816 **3. a.** \$3655.21 **b.** \$655.21 **5. a.** \$12,795.12 **b.** \$3295.12 **7. a.** \$5149.12 **b.** \$649.12 **9. a.** \$1855.10 **b.** \$355.10 **11. a.** \$49,189.30 **b.** \$29,189.30 **13. a.** \$13,116.51 **b.** \$13,140.67 **c.** \$13,157.04 **d.** \$13,165.31 **15.** 7% compounded monthly **17.** \$8374.85 **19.** \$7528.59 **21. a.** \$10,457.65 **b.** 4.6% **23.** 6.1% **25.** 6.2% **27.** 6.2% **29.** 8.3%; 8.25%; 8% compounded monthly **31.** 5.6%; 5.5%; 5.5% compounded semiannually **33.** \$2,069,131 **35.** \$649,083 **37.** \$41,528 **39. a.** you; \$391 **b.** you; \$340 **c.** your friend; \$271 **41.** \$9186 or \$9187 **43. a.** ≈ \$5,027,400,000 **b.** ≈ \$5,225,000,000 **45.** 5.9% compounded daily; \$2

47.

	Once a Year		Continuous	
Years	**Amount**	**Interest**	**Amount**	**Interest**
1	\$5275	\$275	\$5283	\$283
5	\$6535	\$1535	\$6583	\$1583
10	\$8541	\$3541	\$8666	\$3666
20	\$14,589	\$9589	\$15,021	\$10,021

49. \$37,096 **51.** account A: \$38,755; account B: \$41,163 **53.** 5.55% **55.** 4.3%; 4.3%; 4.3%; As the number of compounding periods increases, the effective annual yield increases slightly. However, with the rates rounded to the nearest tenth of a percent, this increase is not evident.

57. 4.5% compounded semiannually **63.** does not make sense **65.** does not make sense **67.** \$12,942.94 or \$12,942.95 **69.** $A = P\left(1 + \frac{r}{n}\right)^{nt}$

Substitute this value of A into the future value formula for simple interest. Since the interest rates are two different values, use Y, the effective yield, for the interest rate on the right side of the equation.

$$P\left(1 + \frac{r}{n}\right)^{nt} = P\,(1 + Yt)$$

$$\left(1 + \frac{r}{n}\right)^{nt} = (1 + Yt) \qquad \text{Divide each side by } P.$$

$$\left(1 + \frac{r}{n}\right)^{n} = 1 + Y \qquad \text{Let } t = 1.$$

$$Y = \left(1 + \frac{r}{n}\right)^{n} - 1 \qquad \text{Subtract 1 from each side and interchange the sides.}$$

Section 8.5

Check Point Exercises

1. a. \$6620 **b.** \$620 **2. a.** \$777,169 or \$777,170 **b.** \$657,169 or \$657,170 **3. a.** \$333,946 or \$334,288 **b.** \$291,946 or \$292,288 **4. a.** \$187 **b.** deposits: \$40,392; interest: \$59,608 **5. a.** high: \$63.38; low: \$42.37 **b.** \$2160 **c.** 1.5%; This is much lower than a bank account paying 3.5%. **d.** 7,203,200 shares **e.** high: \$49.94; low: \$48.33 **f.** \$49.50 **g.** The closing price is up \$0.03 from the previous day's closing price. **h.** ≈ \$1.34 **6. a.** \$81,456 **b.** \$557,849 **c.** \$527,849

Concept and Vocabulary Check

1. annuity **2.** P; r; n; value of the annuity; t **3.** stock; capital; dividends **4.** bonds **5.** portfolio; diversified **6.** mutual fund **7.** Roth; $59\frac{1}{2}$ **8.** true **9.** false; Changes will vary. **10.** true **11.** true

Exercise Set 8.5

1. a. \$66,132 **b.** \$26,132 **3. a.** \$702,528 **b.** \$542,528 **5. a.** \$50,226 **b.** \$32,226 **7. a.** \$9076 **b.** \$4076 **9. a.** \$28,850 **b.** \$4850 **11. a.** \$4530 **b.** deposits: \$81,540; interest: \$58,460 **13. a.** \$1323 **b.** deposits: \$158,760; interest: \$41,240 **15. a.** \$356 **b.** deposits: \$170,880; interest: \$829,120 **17. a.** \$920 **b.** deposits: \$18,400; interest: \$1600 **19. a.** high: \$73.25; low: \$45.44 **b.** \$840 **c.** 2.2%; Answers will vary. **d.** 591,500 shares **e.** high: \$56.38; low: \$54.38 **f.** \$55.50 **g.** \$1.25 increase **h.** ≈ \$3.26 **21. a.** \$30,000 **b.** \$30,000 **23.** $P = \dfrac{Ar}{[(1+r)^t - 1]}$; the deposit at the end of each year that yields A dollars after t years with interest rate r compounded annually **25. a.** \$11,617 **b.** \$1617 **27. a.** \$87,052 **b.** \$63,052 **29. a.** \$693,031 **b.** \$293,031 **31. a.** \$401 **b.** deposits: \$3208; interest: \$292 **33.** \$620; \$1,665,200 **35. a.** \$173,527 **b.** \$1,273,733 **c.** \$1,219,733 **47.** does not make sense **49.** makes sense **51.** does not make sense **53.** does not make sense **55. a.** \$248 **b.** with: \$5862.25; without: \$6606.25 **c.** with: 11.7%; without: 13.2%

Section 8.6

Check Point Exercises

1. a. monthly payment: \$366; interest: \$2568 **b.** monthly payment: \$278; interest: \$5016 **c.** Monthly payments are lower with the longer-term loan, but there is more interest with this loan. **2.** \$222 **3. a.** \$5250 **b.** \$47,746

Concept and Vocabulary Check

1. PMT; P; n; t; r **2.** closed-end; open-end **3.** residual value **4.** bodily injury; property damage **5.** collision **6.** comprehensive **7.** true **8.** false; Changes will vary. **9.** false; Changes will vary. **10.** true **11.** false; Changes will vary. **12.** true

Exercise Set 8.6

1. monthly payment: \$244; interest: \$1712 **3. a.** monthly payment: \$450; interest: \$1200 **b.** monthly payment: \$293; interest: \$2580 **c.** Monthly payments are lower with the longer-term loan, but there is more interest with this loan. **5.** \$277 **7.** monthly payment: \$405; interest: \$3665 **9.** \$65; Incentive B is the better deal. **11. a.** \$6000 **b.** \$42,142 **13. a.** \$19,600 **b.** \$145,609 **15.** \$104,400 **25.** makes sense **27.** makes sense **29.** makes sense

31.

$$P\left(1+\frac{r}{n}\right)^{nt} = \frac{PMT\left[\left(1+\frac{r}{n}\right)^{nt}-1\right]}{\left(\frac{r}{n}\right)}$$

$$P\left(\frac{r}{n}\right)\left(1+\frac{r}{n}\right)^{nt} = PMT\left[\left(1+\frac{r}{n}\right)^{nt}-1\right]$$

$$\frac{P\left(\frac{r}{n}\right)\left(1+\frac{r}{n}\right)^{nt}}{\left[\left(1+\frac{r}{n}\right)^{nt}-1\right]} = PMT$$

$$\frac{P\left(\frac{r}{n}\right)}{\left[1-\left(1+\frac{r}{n}\right)^{-nt}\right]} = PMT$$

Section 8.7

Check Point Exercises

1. a. \$1627 **b.** \$117,360 **c.** \$148,860

2.

Payment Number	Interest Payment	Principal Payment	Balance of Loan
1	\$1166.67	\$383.33	\$199,616.67
2	\$1164.43	\$385.57	\$199,231.10

3. a. \$5600 **b.** \$7200 **c.** \$2160

Concept and Vocabulary Check

1. mortgage; down payment **2.** loan amortization schedule **3.** false; Changes will vary. **4.** false; Changes will vary. **5.** true **6.** false; Changes will vary.

Exercise Set 8.7

1. a. \$44,000 **b.** \$176,000 **c.** \$5280 **d.** \$1171 **e.** \$245,560 **3.** \$60,120 **5.** 20-year at 7.5%; \$106,440 **7.** Mortgage A; \$11,300 **9. a.** monthly payment: \$608; total interest: \$98,880

b.

Payment Number	Interest	Principal	Loan Balance
1	$450.00	$158.00	$119,842.00
2	$449.41	$158.59	$119,683.41
3	$448.81	$159.19	$119,524.22

11. a. $840 **b.** $1080 **c.** $492 **23.** makes sense **25.** does not make sense **27.** Yes

Section 8.8

Check Point Exercises

1. a. $8761.76 **b.** $140.19 **c.** $9661.11 **d.** $269

Concept and Vocabulary Check

1. open-end **2.** the sum of the unpaid balances for each day in the billing period; the number of days in the billing period **3.** debit **4.** credit report **5.** 300; 850; better credit(worthiness) **6.** false; Changes will vary. **7.** false; Changes will vary **8.** true **9.** true

Exercise Set 8.8

1. a. $6150.65 **b.** $92.26 **c.** $6392.26 **d.** $178 **3. a.** $2057.22 **b.** $24.69 **c.** $2237.23 **d.** $90 **5. a.** $210 **b.** $840 **7. a.** $137; lower monthly payment **b.** $732; less interest **9.** monthly payment: $385; total interest: $420; $175 more each month; $420 less interest **19.** does not make sense **21.** makes sense **23.** does not make sense **25.** does not make sense

Chapter 8 Review Exercises

1. 80% **2.** 12.5% **3.** 75% **4.** 72% **5.** 0.35% **6.** 475.6% **7.** 0.65 **8.** 0.997 **9.** 1.50 **10.** 0.03 **11.** 0.0065 **12.** 0.0025 **13.** 9.6 **14. a.** $1.44 **b.** $25.44 **15. a.** $297.50 **b.** $552.50 **16.** 12.5% increase **17.** 35% decrease **18.** no; $9900; 1% decrease **19.** gross income: $30,330; adjusted gross income: $29,230; taxable income: $19,480 **20.** gross income: $436,400; adjusted gross income: $386,400; taxable income: $278,400 **21.** $186,761 **22.** $5060 **23.** $3375 **24.** $4859 **25.** $18,980 **26. a.** $136 **b.** $13.60 **c.** $7.68 **d.** $5.44 **e.** $109.28 **f.** 19.6% **27.** $180 **28.** $2520 **29.** $1200 **30.** $900 **31. a.** $122.50 **b.** $3622.50 **32.** $12,738 **33.** 7.5% **34.** $13,389.12 **35.** $9287.93 **36.** 40% **37. a.** $8114.92 **b.** $1114.92 **38. a.** $38,490.80 **b.** $8490.80 **39. a.** $5556.46 **b.** $3056.46 **40.** 7% compounded monthly; $362 **41.** $28,469,44 **42.** $13,175.19 **43. a.** $2122.73 **b.** 6.1% **44.** 5.61%; Answers will vary; an example is: The same amount of money would earn 5.61% in a simple interest account for a year. **45.** 6.25% compounded monthly **46. a.** $19,129 **b.** $8729 **47. a.** $91,361 **b.** $55,361 **48. a.** $1049 **b.** $20,980; $4020 **49.** high: $64.06; low: $26.13 **50.** $144 **51.** 0.3% **52.** 545,800 shares **53.** high: $61.25; low: $59.25 **54.** $61.00 **55.** $1.75 increase **56.** $1.49 **59. a.** monthly payment: $465; total interest: $1740 **b.** monthly payment: $305; total interest: $3300 **c.** Longer term has lower monthly payment but greater total interest. **63. a.** $7560 **b.** $53,099 **64. a.** $48,000 **b.** $192,000 **c.** $3840 **d.** $1277 **e.** $267,720 **65. a.** 20-year at 8%; $53,040; Answers will vary. **66. a.** option A: $769: option B: $699: Answers will vary. **b.** Mortgage A: $20,900 **67. a.** $1896

b.

Payment Number	Interest	Principal	Loan Balance
1	$1625.00	$271.00	$299,729.00
2	$1623.53	$272.47	$299,456.53
3	$1622.06	$273.94	$299,182.59

68. a. $1260 **b.** $1620 **c.** $612 **71. a.** $4363.27 **b.** $48.00 **c.** $4533.00 **d.** $126 **72. a.** $768 **b.** $3058

Chapter 8 Test

1. a. $18 **b.** $102 **2.** 75% **3. a.** $47,290 **b.** $46,190 **c.** $33,345 **4.** $3270 **5.** $6795 **6. a.** $150 **b.** $15 **c.** $8.48 **d.** $4.50 **e.** $122.02 **f.** 18.7% **7.** $72; $2472 **8.** 25% **9.** $6698.57 **10.** 4.58%; Answers will vary; an example is: The same amount of money would earn 4.58% in a simple interest account for a year. **11. a.** $267,392 **b.** $247,392 **12.** $2070 **13. a.** $8297 **b.** $2297 **14. a.** $7067 **b.** $1067 **c.** Only part of the $6000 is invested for the entire five years.; Answers will vary. **15.** $704 per month; $1,162,080 interest **16.** high: $25.75; low: $25.50 **17.** $2030 **18.** $386.25 **19.** $5320 **20.** $12,000 **21.** $108,000 **22.** $2160 **23.** $830 **24.** $190,800

25.

Payment Number	Interest	Principal	Loan Balance
1	$765.00	$65.00	$107,935.00
2	$764.54	$65.46	$107,869.54

26. a. $1540 **b.** $1980 **c.** $594 **27. a.** $3226.67 **b.** $64.53 **c.** $3464.53 **d.** $97 **28.** false; Changes will vary. **29.** true **30.** false; Changes will vary. **31.** false; Changes will vary. **32.** true **33.** false; Changes will vary. **34.** false; Changes will vary.

CHAPTER 9

Section 9.1

Check Point Exercises

1. a. 6.5 ft **b.** 3.25 mi **c.** $\frac{1}{12}$ yd **2. a.** 8 km **b.** 53,000 mm **c.** 0.0604 hm **d.** 6720 cm **3. a.** 243.84 cm **b.** ≈22.22 yd **c.** ≈1181.1 in. **4.** ≈37.5 mi/hr

Concept and Vocabulary Check

1. linear; linear **2.** 12; 3; 36; 5280 **3.** unit; 1 **4.** 1000; 100; 10; 0.1; 0.01; 0.001 **5.** false; Changes will vary. **6.** false; Changes will vary. **7.** false; Changes will vary. **8.** false; Changes will vary.

Exercise Set 9.1

1. 2.5 ft **3.** 360 in. **5.** $\frac{1}{6}$ yd ≈0.17 yd **7.** 216 in. **9.** 18 ft **11.** 2 yd **13.** 4.5 mi **15.** 3960 ft **17.** 500 cm **19.** 1630 m **21.** 0.03178 hm **23.** 0.000023 m **25.** 21.96 dm **27.** 35.56 cm **29.** ≈5.51 in. **31.** ≈424 km **33.** ≈165.625 mi **35.** ≈13.33 yd **37.** ≈55.12 in. **39.** 0.4064 dam **41.** 1.524 m **43.** ≈16.40 ft **45.** ≈60 mi/hr **47.** ≈72 km/hr **49.** 457.2 cm **51.** $8\frac{1}{3}$yd **53.** 48.28032 km **55.** 176 ft/sec **57.** meter **59.** millimeter **61.** meter **63.** millimeter **65.** millimeter **67.** b **69.** a **71.** c **73.** a **75.** 0.216 km **77.** ≈148.8 million km **79.** Nile; ≈208 km **81.** Everest; ≈239 m **83.** Waialeale; ≈46 in. **93.** makes sense **95.** does not make sense **97.** 9 hm **99.** 11 m or 1.1 dam

Section 9.2

Check Point Exercises

1. 8 square units **2.** 237.6 people per square mile **3.** 131,250 mi^2 **4. a.** ≈0.72 ha **b.** ≈$576,389 per hectare **5.** 9 cubic units **6.** ≈74,800 gal **7.** 220 L **8. a.** 20 mL **b.** ≈0.67 fl oz

Concept and Vocabulary Check

1. square; cubic **2.** 144; 9 **3.** $\frac{1\text{ mi}^2}{640\text{ acres}}$; $\frac{640\text{ acres}}{1\text{ mi}^2}$ **4.** 2; 4 **5.** capacity; liter **6.** area **7.** 1 **8.** false; Changes will vary. **9.** false; Changes will vary. **10.** false; Changes will vary.

Exercise Set 9.2

1. 16 square units **3.** 8 square units **5.** ≈2.15 in.^{2} **7.** ≈37.5 yd^2 **9.** ≈25.5 acres **11.** ≈91 cm^2 **13.** 24 cubic units **15.** ≈74,800 gal **17.** ≈1600 gal **19.** ≈9 gal **21.** ≈13.5 yd^3 **23.** 45 L **25.** 17 mL **27.** 1500 cm^3 **29.** 150 cm^3 **31.** 12,000 dm^3 **33.** ≈2.4 tsp **35.** ≈45 mL **37.** ≈2.33 fl oz **39.** ≈5.83 c **41.** ≈2.82 L **43.** ≈4.21 qt **45.** ≈11.4 L **47.** ≈2.11 qt **49. a.** 1900: 25.6 people per square mile; 2010: 87.4 people per square mile **b.** 241.4% **51.** 17.0 people per square kilometer **53.** Illinois: 222.2 people per square mile; Ohio: 257.6 people per square mile; Ohio has the greater population density by 35.4 people per square mile. **55.** ≈2358 mi^2 **57. a.** ≈20 acres **b.** ≈$12,500 per acre **59.** square centimeters **61.** square kilometers **63.** b **65.** b **67.** ≈336,600 gal **69.** 4 L **71.** Japan; ≈79,000 km^2 **73.** Baffin Island; ≈24,162 mi^2 **75. a.** ≈15 mL **b.** ≈15 cc **c.** 0.5 fl oz **d.** ≈0.17 fl oz **83.** does not make sense **85.** makes sense **87.** ≈11,952.64 people per square mile **89.** Answers will vary. **91.** 6.5 liters; Answers will vary.

Section 9.3

Check Point Exercises

1. a. 420 mg **b.** 6.2 g **2.** 145 kg **3. a.** ≈54 kg **b.** ≈17.9 oz **4.** 2 tablets **5.** 122 °F **6.** 15°C

Concept and Vocabulary Check

1. 16; 2000 **2.** gram **3.** 2.2 **4.** 1 **5.** 32; 212 **6.** 0; 100 **7.** false; Changes will vary. **8.** false; Changes will vary. **9.** false; Changes will vary. **10.** false; Changes will vary.

Exercise Set 9.3

1. 740 mg **3.** 0.87 g **5.** 800 cg **7.** 18,600 g **9.** 0.000018 g **11.** 50 kg **13.** 4200 cm^3 **15.** 1100 t **17.** 40,000 g **19.** 2.25 lb **21.** ≈1008 g **23.** ≈243 kg **25.** ≈36,000 g **27.** ≈1200 lb **29.** ≈222.22 T **31.** 50°F **33.** 95°F **35.** 134.6°F **37.** 23°F **39.** 20°C **41.** 5°C **43.** 22.2°C **45.** −5°C **47.** 176.7°C **49.** −30°C **51. a.** $\frac{9}{5}$; Fahrenheit temperature increases by $\frac{9}{5}°$ for each 1° change in Celsius temperature. **b.** $F = \frac{9}{5}C + 32$ **53.** milligram **55.** gram **57.** kilogram **59.** kilogram **61.** b **63.** a **65.** c **67.** 13.28 kg **69.** $1.16 **71.** Purchasing the regular size is more economical. **73.** 4 tablets **75. a.** 43 mg **b.** ≈516 mg **77.** a **79.** c **81.** Néma; ≈0.3 °F **83.** Eismitte; ≈5°C **91.** does not make sense **93.** makes sense **95.** false **97.** true **99.** true **101.** false

Chapter 9 Review Exercises

1. 5.75 ft **2.** 0.25 yd **3.** 7 yd **4.** 2.5 mi **5.** 2280 cm **6.** 70 m **7.** 1920 m **8.** 0.0144 hm **9.** 0.0005 m **10.** 180 mm **11.** 58.42 cm **12.** ≈7.48 in. **13.** ≈528 km **14.** ≈375 mi **15.** ≈15.56 yd **16.** ≈39.37 ft **17.** ≈28.13 mi/hr **18.** ≈96 km/hr **19.** 0.024 km, 24,000 cm, 2400 m **20.** 4.8 km **21.** 24 square units **22.** 16,513.8 people per square mile; In Singapore, there are an average of 16,513.8 people for each square mile. **23.** 74 mi^2 **24.** ≈18 acres **25.** ≈333.33 ft^2 **26.** ≈31.2 km^2 **27.** a **28.** 24 cubic units **29.** ≈251,328 gal **30.** 76 L **31.** ≈4.4 tsp **32.** ≈22.5 c **33.** ≈8.42 qt **34.** ≈22.8 L **35. a.** 15 mL **b.** ≈0.5 fl oz **36.** c **39.** 1240 mg **40.** 1200 cg **41.** 0.000012 g **42.** 0.00045 kg **43.** 50,000 cm^3 **44.** 4000 dm^3, 4,000,000 g **45.** 94.5 kg **46.** 14 oz **47.** 3 tablets **48.** kilograms; Answers will vary. **49.** 2.25 lb **50.** a **51.** c **52.** 59 °F **53.** 212°F **54.** 41°F **55.** 32°F **56.** −13°F **57.** 15°C **58.** 5°C **59.** 100°C **60.** 37°C **61.** ≈−17.8°C **62.** −10°C **63.** more; Answers will vary.

Chapter 9 Test

1. 0.00807 hm **2.** 250 in. **3.** 4.8 km **4.** mm **5.** cm **6.** km **7.** ≈128 km/hr **8.** 9 times **9.** 6.8 people per square mile; In Australia, there is an average of 6.8 people for each square mile. **10.** ≈45 acres **11.** b **12.** Answers will vary.; 1000 times **13.** 8 fl oz **14.** ≈74,800 gal **15.** b **16.** 0.137 kg **17.** 3 tablets **18.** kg **19.** mg **20.** 86°F **21.** 80°C **22.** d

CHAPTER 10

Section 10.1

Check Point Exercises

1. 30° **2.** 71° **3.** 46°, 134° **4.** $m\angle 1 = 57°, m\angle 2 = 123°$, and $m\angle 3 = 123°$ **5.** $m\angle 1 = 29°, m\angle 2 = 29°, m\angle 3 = 151°, m\angle 4 = 151°$, $m\angle 5 = 29°, m\angle 6 = 151°$, and $m\angle 7 = 151°$

Concept and Vocabulary Check

1. line; half-line; ray; line segment **2.** acute; right; obtuse; straight **3.** complementary; supplementary **4.** vertical **5.** parallel; transversal **6.** perpendicular **7.** false; Changes will vary. **8.** false; Changes will vary. **9.** false; Changes will vary. **10.** false; Changes will vary. **11.** true **12.** true

Exercise Set 10.1

1. 150° **3.** 90° **5.** 20°; acute **7.** 160°; obtuse **9.** 180°; straight **11.** 65° **13.** 146° **15.** 42°; 132° **17.** 1°; 91° **19.** 52.6°; 142.6° **21.** $x + x + 12° = 90°$; 39°, 51° **23.** $x + 3x = 180°$; 45°, 135° **25.** $m\angle 1 = 108°; m\angle 2 = 72°; m\angle 3 = 108°$ **27.** $m\angle 1 = 50°; m\angle 2 = 90°$; $m\angle 3 = 50°$ **29.** $m\angle 1 = 68°; m\angle 2 = 68°; m\angle 3 = 112°; m\angle 4 = 112°; m\angle 5 = 68°; m\angle 6 = 68°; m\angle 7 = 112°$ **31.** $m\angle 1 = 38°; m\angle 2 = 52°$; $m\angle 3 = 142°$ **33.** $m\angle 1 = 65°; m\angle 2 = 56°; m\angle 3 = 124°$ **35.** false **37.** true **39.** false **41.** false **43.** 60°, 30° **45.** 112°, 112° **47.** $\overline{BC}$ **49.** $\overline{AD}$ **51.** $\overleftrightarrow{AD}$ **53.** $\overline{AD}$ **55.** 45° **57.** When two parallel lines are intersected by a transversal, corresponding angles have the same measure. **59.** E, F, H, and T **61.** long-distance riding and mountain biking **63.** 27° **73.** does not make sense **75.** does not make sense **77.** d

Section 10.2

Check Point Exercises

1. 49° **2.** $m\angle 1 = 90°; m\angle 2 = 54°; m\angle 3 = 54°; m\angle 4 = 68°; m\angle 5 = 112°$ **3.** Two angles of the small triangle are equal to two angles of the large triangle. One angle pair is given to have the same measure (right angles). Another angle pair consists of vertical angles with the same measure.; 15 cm **4.** 32 yd **5.** 25 ft **6.** 120 yd

Concept and Vocabulary Check

1. 180° **2.** acute **3.** obtuse **4.** isosceles **5.** equilateral **6.** scalene **7.** similar; the same measure; proportional **8.** right; legs; the square of the length of the hypotenuse **9.** true **10.** true **11.** false; Changes will vary. **12.** false; Changes will vary. **13.** false; Changes will vary.

Exercise Set 10.2

1. 67° **3.** 32° **5.** $m\angle 1 = 50°; m\angle 2 = 130°; m\angle 3 = 50°; m\angle 4 = 130°; m\angle 5 = 50°$ **7.** $m\angle 1 = 50°; m\angle 2 = 50°; m\angle 3 = 80°; m\angle 4 = 130°$; $m\angle 5 = 130°$ **9.** $m\angle 1 = 55°; m\angle 2 = 65°; m\angle 3 = 60°; m\angle 4 = 65°; m\angle 5 = 60°; m\angle 6 = 120°; m\angle 7 = 60°; m\angle 8 = 60°; m\angle 9 = 55°$; $m\angle 10 = 55°$ **11.** The three angles of the large triangle are given to have the same measures as the three angles of the small triangle.; 5 in. **13.** Two angles of the large triangle are given to have the same measures as two angles of the small triangle.; 6 m **15.** One angle pair is given to have the same measure (right angles). Another angle pair consists of vertical angles with the same measure.; 16 in. **17.** 5 **19.** 9 **21.** 17 m **23.** 39 m **25.** 12 cm **27.** congruent; SAS **29.** congruent; SSS **31.** congruent; SAS **33.** not necessarily congruent **35.** congruent; ASA **37.** 71.7 ft **39.** $90\sqrt{2}$ ft $\approx$ 127.3 ft **41.** 45 yd **43.** 13 ft **45.** $750,000 **57.** makes sense **59.** does not make sense **61.** 21 ft

Section 10.3

Check Point Exercises

1. $3120 **2.** **a.** 1800° **b.** 150° **3.** Each angle of a regular octagon measures 135°. 360° is not a multiple of 135°.

Concept and Vocabulary Check

1. perimeter **2.** quadrilateral; pentagon; hexagon; heptagon; octagon **3.** regular **4.** equal in measure; parallel **5.** rhombus **6.** rectangle **7.** square **8.** trapezoid **9.** $P = 2l + 2w$ **10.** $(n - 2)180°$ **11.** tessellation **12.** false; Changes will vary. **13.** true **14.** true **15.** true **16.** false; Changes will vary. **17.** false; Changes will vary. **18.** true

Exercise Set 10.3

1. quadrilateral **3.** pentagon **5.** a: square; b: rhombus; d: rectangle; e: parallelogram **7.** a: square; d: rectangle **9.** c: trapezoid **11.** 30 cm **13.** 28 yd **15.** 1000 in. **17.** 27 ft **19.** 18 yd **21.** 84 yd **23.** 32 ft **25.** 540° **27.** 360° **29.** 108°; 72° **31.** **a.** 540° **b.** $m\angle A = 140°; m\angle B = 40°$ **33.** **a.** squares, hexagons, dodecagons (12 sided polygons) **b.** 3 angles; 90°, 120°, 150° **c.** The sum of the measures is 360°. **35.** **a.** triangles, hexagons **b.** 4 angles; 60°, 60°, 120°, 120° **c.** The sum of the measures is 360°. **37.** no; Each angle of the polygon measures 140°. 360° is not a multiple of 140°. **39.** 50 yd by 200 yd **41.** 160 ft by 360 ft **43.** 115°, 115°, 120°, 120° **45.** If the polygons were all regular polygons, the sum would be 363°. The sum is not 360°. **47.** $5600 **49.** 48 **59.** makes sense **61.** does not make sense **63.** $6a$

Section 10.4

Check Point Exercises

1. 66 ft^2; yes **2.** $672 **3.** 60 $in.^2$ **4.** 30 ft^2 **5.** 105 ft^2 **6.** 10π in.; 31.4 in. **7.** 49.7 ft **8.** large pizza

Concept and Vocabulary Check

1. $A = lw$ **2.** $A = s^2$ **3.** $A = bh$ **4.** $A = \frac{1}{2}bh$ **5.** $A = \frac{1}{2}h(a + b)$ **6.** $C = \pi d$ **7.** $C = 2\pi r$ **8.** $A = \pi r^2$ **9.** false; Changes will vary. **10.** true **11.** true **12.** false; Changes will vary. **13.** true

Exercise Set 10.4

1. 18 m^2 **3.** 16 $in.^2$ **5.** 2100 cm^2 **7.** 56 $in.^2$ **9.** 20.58 yd^2 **11.** 30 $in.^2$ **13.** 567 m^2 **15.** C: 8π cm, $\approx$ 25.1 cm; A: $16\pi\ cm^2$, $\approx$ 50.3 cm^2 **17.** C: 12π yd, $\approx$ 37.7 yd; A: $36\pi\ yd^2$, $\approx$ 113.1 yd^2 **19.** 72 m^2 **21.** 300 m^2 **23.** $100 + 50\pi$; 257.1 cm^2 **25.** $A = ab + \frac{1}{2}(c - a)b$ or $A = \frac{1}{2}(a + c)b$ **27.** $A = 2a^2 + ab$ **29.** 192 cm^2 **31.** $8\pi\ cm^2$ **33.** $(12.5\pi - 24)in.^2$ **35.** perimeter: 54 ft; area: 168 ft^2 **37.** \$556.50 **39.** 148 ft^2 **41. a.** 23 bags **b.** \$575 **43. a.** 3680 ft^2 **b.** 15 cans of paint **c.** \$404.25 **45.** \$933.40 **47.** 40π m; 125.7 m **49.** 377 plants **51.** large pizza **59.** makes sense **61.** does not make sense **63.** by a factor of $\frac{9}{4}$ **65.** \$13,032.73

Section 10.5

Check Point Exercises

1. 105 ft^3 **2.** 8 yd^3 **3.** 48 ft^3 **4.** 302 $in.^3$ **5.** 101 $in.^3$ **6.** no **7.** 632 yd^2

Concept and Vocabulary Check

1. $V = lwh$ **2.** $V = s^3$ **3.** polyhedron **4.** $V = \frac{1}{3}Bh$ **5.** $V = \pi r^2 h$ **6.** $V = \frac{1}{3}\pi r^2 h$ **7.** $V = \frac{4}{3}\pi r^3$ **8.** true **9.** true **10.** false; Changes will vary. **11.** true **12.** true **13.** true **14.** false; Changes will vary.

Exercise set 10.5

1. 36 $in.^3$ **3.** 64 cm^3 **5.** 175 yd^3 **7.** 56 $in.^3$ **9.** $150\pi\ cm^3$, 471 cm^3 **11.** $3024\pi\ in.^3$, 9500 $in.^3$ **13.** $48\pi\ m^3$, 151 m^3 **15.** $15\pi\ yd^3$, 47 yd^3 **17.** $288\pi\ m^3$, 905 m^3 **19.** $972\pi\ cm^3$, 3054 cm^3 **21.** 62 m^2 **23.** 96 ft^2 **25.** $324\pi\ cm^3$, 1018 cm^3 **27.** $432\pi\ in.^3$, 1357 $in.^3$ **29.** $\frac{3332}{3}\pi\ m^3$, 3489 m^3 **31.** surface area: 186 yd^2; volume: 148 yd^3 **33.** 666 yd^2 **35.** $\frac{1}{8}$ **37.** 9 times **39.** \$340 **41.** no **43. a.** 3,386,880 yd^3 **b.** 2,257,920 blocks **45.** yes **47.** \$27 **51.** does not make sense **53.** does not make sense **55.** The volume is multiplied by 8. **57.** 168 cm^3 **59.** 84 cm^2

Section 10.6

Check Point Exercises

1. $\sin A = \frac{3}{5}$; $\cos A = \frac{4}{5}$; $\tan A = \frac{3}{4}$ **2.** 263 cm **3.** 298 cm **4.** 994 ft **5.** 54°

Concept and Vocabulary Check

1. sine; opposite; hypotenuse; $\frac{a}{c}$ **2.** cosine; adjacent to; hypotenuse; $\frac{b}{c}$ **3.** tangent; opposite; adjacent to; $\frac{a}{b}$ **4.** elevation **5.** depression **6.** false; Changes will vary. **7.** false; Changes will vary. **8.** true **9.** false; Changes will vary.

Exercise Set 10.6

1. $\sin A = \frac{3}{5}$; $\cos A = \frac{4}{5}$; $\tan A = \frac{3}{4}$ **3.** $\sin A = \frac{20}{29}$; $\cos A = \frac{21}{29}$; $\tan A = \frac{20}{21}$ **5.** $\sin A = \frac{5}{13}$; $\cos A = \frac{12}{13}$; $\tan A = \frac{5}{12}$ **7.** $\sin A = \frac{4}{5}$; $\cos A = \frac{3}{5}$; $\tan A = \frac{4}{3}$ **9.** 188 cm **11.** 182 in. **13.** 7 m **15.** 22 yd **17.** 40 m **19.** $m\measuredangle B = 50°$, $a = 18$yd, $c = 29$yd **21.** $m\measuredangle B = 38°$, $a = 43$ cm, $b = 33$ cm **23.** 37° **25.** 28° **27.** 653 units **29.** 39 units **31.** 298 units **33.** 257 units **35.** 529 yd **37.** 2879 ft **39.** 2059 ft **41.** 695 ft **43.** 36° **45.** 1376 ft **47.** 15.1° **57.** does not make sense **59.** makes sense **63. a.** 357 ft **b.** 394 ft

Section 10.7

Check Point Exercises

1. Answers will vary.

Concept and Vocabulary Check

1. vertex; edge; graph **2.** traversable **3.** genus **4.** parallel **5.** non-Euclidean; parallel **6.** self-similarity; iteration **7.** true **8.** false; Changes will vary. **9.** true **10.** true

Exercise Set 10.7

1. a. traversable **b.** sample path: D, A, B, D, C, B **3. a.** traversable **b.** sample path: A, D, C, B, D, E, A, B **5.** not traversable

7.

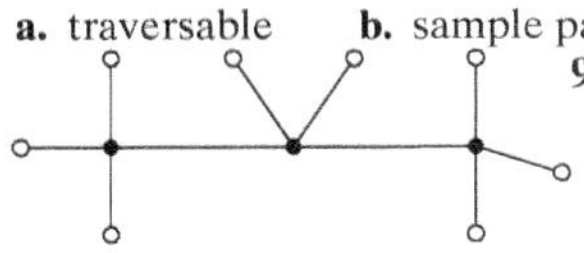

9. no

11. 2 **13.** sample path: $B, E, A, B, D, C, A, E, C, E, D$ **15.** 2 **17.** 4 **19.** Answers will vary. **21.** The sum of the angles of such a quadrilateral is greater than 360°. **23.** yes **43.** makes sense **45.** does not make sense

Chapter 10 Review Exercises

1. $\measuredangle 3$ **2.** $\measuredangle 5$ **3.** $\measuredangle 4$ and $\measuredangle 6$ **4.** $\measuredangle 1$ and $\measuredangle 6$ **5.** $\measuredangle 4$ and $\measuredangle 1$ **6.** $\measuredangle 2$ **7.** $\measuredangle 5$ **8.** 65° **9.** 49° **10.** 17° **11.** 134° **12.** $m\measuredangle 1 = 110°$; $m\measuredangle 2 = 70°$; $m\measuredangle 3 = 110°$ **13.** $m\measuredangle 1 = 138°$; $m\measuredangle 2 = 42°$; $m\measuredangle 3 = 138°$; $m\measuredangle 4 = 138°$; $m\measuredangle 5 = 42°$; $m\measuredangle 6 = 42°$; $m\measuredangle 7 = 138°$ **14.** 72° **15.** 51° **16.** $m\measuredangle 1 = 90°$; $m\measuredangle 2 = 90°$; $m\measuredangle 3 = 140°$; $m\measuredangle 4 = 40°$; $m\measuredangle 5 = 140°$ **17.** $m\measuredangle 1 = 80°$; $m\measuredangle 2 = 65°$; $m\measuredangle 3 = 115°$; $m\measuredangle 4 = 80°$; $m\measuredangle 5 = 100°$; $m\measuredangle 6 = 80°$ **18.** 5 ft **19.** 3.75 ft **20.** 10 ft **21.** 7.2 in. **22.** 10.2 cm **23.** 12.5 ft **24.** 15 ft **25.** 12 yd **26.** rectangle, square **27.** rhombus, square **28.** parallelogram, rhombus, trapezoid **29.** 30 cm **30.** 4480 yd **31.** 44 m **32.** 1800° **33.** 1080° **34.** $m\measuredangle 1 = 135°$; $m\measuredangle 2 = 45°$ **35.** $132 **36. a.** triangles, hexagons **b.** 5 angles; 60°, 60°, 60°, 60°, 120° **c.** The sum of the measures is 360°. **37.** yes; Each angle of a regular hexagon measures 120°, and 360° is a multiple of 120°. **38.** 32.5 ft^2 **39.** 20 m^2 **40.** 50 cm^2 **41.** 135 yd^2 **42.** $C = 20\pi$ m ≈ 62.8 m; $A = 100\pi$ m^2 ≈ 314.2 m^2 **43.** 192 in.^{2} **44.** 28 m^2 **45.** 279.5 ft^3 **46.** $(128 - 32\pi)$ in.^{2}; 27.5 in.^{2} **47.** $787.50 **48.** $650 **49.** 31yd **50.** 60 cm^3 **51.** 960 m^3 **52.** 128π yd^3, 402 yd^3 **53.** $\frac{44,800\pi}{3}$ in.^{3}, 46,914 in.^{3} **54.** 288π m^3, 905 m^3 **55.** 126 m^2 **56.** 4800 m^3 **57.** 651,775 m^3 **58.** $80 **59.** $\sin A = \frac{3}{5}$; $\cos A = \frac{4}{5}$; $\tan A = \frac{3}{4}$ **60.** 42 mm **61.** 23 cm **62.** 37 in. **63.** 58° **64.** 772 ft **65.** 31 m **66.** 56° **67.** not traversable **68.** traversable; sample path: *A, B, C, D, A, B, C, D, A* **69.** 0 **70.** 2 **71.** 1 **72.** 2 **73.** Answers will vary. **74.** Answers will vary.

Chapter 10 Test

1. 36°;126° **2.** 133° **3.** 70° **4.** 35° **5.** 3.2 in. **6.** 10 ft **7.** 1440° **8.** 40 cm **9.** d **10. a.** triangles, squares **b.** 5 angles; 60°, 60°, 60°, 90°, 90° **c.** The sum of the measures is 360°. **11.** 517 m^2 **12.** 525 in.^{2} **13. a.** 12 cm **b.** 30 cm **c.** 30 cm^2 **14.** *C*: 40π m, 125.7 m; *A*: 400π m^2, 1256.6 m^2 **15.** 108 tiles **16.** 18 ft^3 **17.** 16 m^3 **18.** 175π cm^3, 550 cm^3 **19.** 85 cm **20.** 70 ft **21.** traversable; sample path: *B, C, A, E, C, D, E* **22.** Answers will vary.

CHAPTER 11

Section 11.1

Check Point Exercises

1. 150 **2.** 40 **3.** 30 **4.** 160 **5.** 729 **6.** 90,000

Concept and Vocabulary Check

1. $M \cdot N$ **2.** multiplying; Fundamental Counting **3.** false; Changes will vary. **4.** true

Exercise Set 11.1

1. 80 **3.** 12 **5.** 6 **7.** 40 **9.** 144; Answers will vary. **11.** 8 **13.** 96 **15.** 243 **17.** 144 **19.** 676,000 **21.** 2187 **27.** makes sense **29.** makes sense **31.** 720 hr

Section 11.2

Check Point Exercises

1. 600 **2.** 120 **3. a.** 504 **b.** 524,160 **c.** 100 **4.** 840 **5.** 15,120 **6.** 420

Concept and Vocabulary Check

1. factorial; 5; 1; 1 **2.** $\frac{n!}{(n-r)!}$ **3.** $\frac{n!}{p!q!}$ **4.** false; Changes will vary. **5.** false; Changes will vary. **6.** true **7.** false; Changes will vary.

Exercise Set 11.2

1. 720 **3.** 120 **5.** 120 **7.** 362,880 **9.** 6 **11.** 4 **13.** 504 **15.** 570,024 **17.** 3,047,466,240 **19.** 600 **21.** 10,712 **23.** 5034 **25.** 24 **27.** 6 **29.** 42 **31.** 1716 **33.** 3024 **35.** 6720 **37.** 720 **39.** 1 **41.** 720 **43.** 8,648,640 **45.** 120 **47.** 15,120 **49.** 180 **51.** 831,600 **53.** 105 **55.** 280 **65.** makes sense **67.** does not make sense **69.** 360 **71.** 14,400 **73.** $\frac{n(n-1)\cdots 3 \cdot 2 \cdot 1}{2} = n(n-1)\cdots 3$

Section 11.3

Check Point Exercises

1. a. combinations **b.** permutations **2.** 35 **3.** 1820 **4.** 525

Concept and Vocabulary Check

1. $\frac{n!}{(n-r)!r!}$ **2.** $r!$ **3.** false; Changes will vary. **4.** false; Changes will vary.

Exercise Set 11.3

1. combinations **3.** permutations **5.** 6 **7.** 126 **9.** 330 **11.** 8 **13.** 1 **15.** 4060 **17.** 1 **19.** 7 **21.** 0 **23.** $\frac{3}{4}$ **25.** −9499 **27.** $\frac{3}{68}$ **29.** 20 **31.** 495 **33.** 24,310 **35.** 22,957,480 **37.** 735 **39.** 4,516,932,420 **41.** 360 **43.** 1716 **45.** 1140 **47.** 840 **49.** 2730 **51.** 10 **53.** 12 **55.** 10 **57.** 60 **59.** 792 **61.** 720 **63.** 20 **65.** 24 **67.** 12 **73.** does not make sense **75.** makes sense **77.** The 5/36 lottery is easier to win.; Answers will vary. **79.** 570 sec or 9.5 min; 2340 sec or 39 min

Section 11.4

Check Point Exercises

1. a. $\frac{1}{6}$ **b.** $\frac{1}{2}$ **c.** 0 **d.** 1 **2. a.** $\frac{1}{13}$ **b.** $\frac{1}{2}$ **c.** $\frac{1}{26}$ **3.** $\frac{1}{2}$ **4. a.** 0.31 **b.** 0.49

Concept and Vocabulary Check

1. sample space **2.** $P(E)$; number of outcomes in E; total number of possible outcomes **3.** 52; hearts; diamonds; clubs; spades **4.** empirical **5.** true **6.** false; Changes will vary. **7.** true **8.** false; Changes will vary.

Exercise Set 11.4

1. $\frac{1}{6}$ **3.** $\frac{1}{2}$ **5.** $\frac{1}{3}$ **7.** 1 **9.** 0 **11.** $\frac{1}{13}$ **13.** $\frac{1}{4}$ **15.** $\frac{3}{13}$ **17.** $\frac{1}{52}$ **19.** 0 **21.** $\frac{1}{4}$ **23.** $\frac{1}{2}$ **25.** $\frac{1}{2}$ **27.** $\frac{3}{8}$ **29.** $\frac{3}{8}$ **31.** $\frac{7}{8}$ **33.** 0 **35.** $\frac{1}{4}$ **37.** $\frac{1}{9}$ **39.** 0 **41.** $\frac{3}{10}$ **43.** $\frac{1}{5}$ **45.** $\frac{1}{2}$ **47.** 0 **49.** $\frac{1}{4}$ **51.** $\frac{1}{4}$ **53.** $\frac{1}{2}$ **55.** 0.43 **57.** 0.13 **59.** 0.03 **61.** 0.38 **63.** 0.78 **65.** 0.12 **67.** $\frac{1}{6}$ **69.** $\frac{11}{87}$ **79.** does not make sense **81.** makes sense **83.** $\frac{3}{8} = 0.375$

Section 11.5

Check Point Exercises

1. $\frac{2}{15}$ **2.** $\frac{27}{17{,}522{,}351} \approx 1.541 \times 10^{-6} \approx 0.000001541$ **3. a.** $\frac{1}{6}$ **b.** $\frac{1}{2}$

Concept and Vocabulary Check

1. permutations; the total number of possible permutations **2.** 1; combinations **3.** true **4.** false; Changes will vary.

Exercise Set 11.5

1. a. 120 **b.** 6 **c.** $\frac{1}{20}$ **3. a.** $\frac{1}{6}$ **b.** $\frac{1}{30}$ **c.** $\frac{1}{720}$ **d.** $\frac{1}{3}$ **5. a.** 84 **b.** 10 **c.** $\frac{5}{42}$ **7.** $\frac{1}{175{,}711{,}536}$ **9.** $\frac{2125}{29{,}285{,}256}$ **11. a.** $\frac{1}{177{,}100} \approx 0.00000565$ **b.** $\frac{27{,}132}{177{,}100} \approx 0.153$ **13.** $\frac{3}{10} = 0.3$ **15. a.** 2,598,960 **b.** 1287 **c.** $\frac{1287}{2{,}598{,}960} \approx 0.000495$ **17.** $\frac{11}{1105} \approx 0.00995$ **19.** $\frac{36}{270{,}725} \approx 0.000133$ **25.** does not make sense **27.** makes sense **29.** The prize is shared among all winners. You are guaranteed to win but not to win $500 million. **31.** $\frac{235{,}620}{2{,}598{,}960} \approx 0.0907$

Section 11.6

Check Point Exercises

1. $\frac{3}{4}$ **2. a.** $\frac{160}{191}$ **b.** $\frac{182}{191}$ **3.** $\frac{1}{3}$ **4.** $\frac{27}{50}$ **5.** $\frac{3}{4}$ **6. a.** $\frac{189}{242} \approx 0.78$ **b.** $\frac{19}{121} \approx 0.16$ **7. a.** 2:50 or 1:25 **b.** 50:2 or 25:1 **8.** 199:1 **9.** 1:15; $\frac{1}{16}$

Concept and Vocabulary Check

1. $1 - P(E)$; $1 - P(\text{not } E)$ **2.** mutually exclusive; $P(A) + P(B)$ **3.** $P(A) + P(B) - P(A \text{ and } B)$ **4.** E will occur; E will not occur **5.** $P(E) = \frac{a}{a+b}$ **6.** false; Changes will vary. **7.** false; Changes will vary. **8.** true **9.** false; Changes will vary.

Exercise Set 11.6

1. $\frac{12}{13}$ **3.** $\frac{3}{4}$ **5.** $\frac{10}{13}$ **7.** $\frac{2{,}598{,}924}{2{,}598{,}960} \approx 0.999986$ **9.** $\frac{2{,}595{,}216}{2{,}598{,}960} \approx 0.998559$ **11. a.** 0.10 **b.** 0.90 **13.** $\frac{16}{25}$ **15.** $\frac{47}{50}$ **17.** $\frac{2}{13}$ **19.** $\frac{1}{13}$ **21.** $\frac{1}{26}$ **23.** $\frac{9}{22}$ **25.** $\frac{5}{6}$ **27.** $\frac{7}{13}$ **29.** $\frac{11}{26}$ **31.** $\frac{3}{4}$ **33.** $\frac{5}{8}$ **35.** $\frac{33}{40}$ **37.** $\frac{4}{5}$ **39.** $\frac{7}{10}$ **41.** $\frac{43}{58}$ **43.** $\frac{50}{87}$ **45.** $\frac{113}{174}$ **47.** 15:43; 43:15 **49.** $\frac{85}{142}$ **51.** $\frac{127}{142}$ **53.** $\frac{53}{142}$ **55.** 16:55; 55:16 **57.** 1:141; 141:1 **59.** 61:10; 10:61 **61.** 2:1 **63.** 1:2 **65. a.** 9:91 **b.** 91:9 **67.** 1:3 **69.** 1:1 **71.** 12:1 **73.** 25:1 **75.** 47:5 **77.** 49:1 **79.** 9:10 **81.** 14:5 **83.** 14:5 **85.** 9:10 **87.** $\frac{3}{7}$ **89.** $\frac{4}{25}$; 84 **91.** $\frac{1}{1000}$ **101.** does not make sense **103.** does not make sense **105.** 0.06

Section 11.7

Check Point Exercises

1. $\frac{1}{361} \approx 0.00277$ **2.** $\frac{1}{16}$ **3. a.** $\frac{625}{130{,}321} \approx 0.005$ **b.** $\frac{38{,}416}{130{,}321} \approx 0.295$ **c.** $\frac{91{,}905}{130{,}321} \approx 0.705$ **4.** $\frac{1}{221} \approx 0.00452$ **5.** $\frac{11}{850} \approx 0.0129$ **6.** $\frac{2}{5}$ **7. a.** 1 **b.** $\frac{1}{2}$ **8. a.** $\frac{9}{10} = 0.9$ **b.** $\frac{45}{479} \approx 0.094$

Concept and Vocabulary Check

1. independent; $P(A) \cdot P(B)$ **2.** the event does not occur **3.** dependent; $P(A) \cdot P(B$ given that A occurred$)$ **4.** conditional; $P(B|A)$ **5.** false; Changes will vary. **6.** false; Changes will vary. **7.** true **8.** true

Exercise Set 11.7

1. $\frac{1}{6}$ **3.** $\frac{1}{36}$ **5.** $\frac{1}{4}$ **7.** $\frac{1}{36}$ **9.** $\frac{1}{8}$ **11.** $\frac{1}{36}$ **13.** $\frac{1}{3}$ **15.** $\frac{3}{52}$ **17.** $\frac{1}{169}$ **19.** $\frac{1}{4}$ **21.** $\frac{1}{64}$ **23.** $\frac{1}{6}$ **25. a.** $\frac{1}{256} \approx 0.00391$ **b.** $\frac{1}{4096} \approx 0.000244$ **c.** ≈ 0.524 **d.** ≈ 0.476 **27.** 0.0144 **29.** 0.008 **31.** 0.4686 **33.** $\frac{7}{29}$ **35.** $\frac{5}{87}$ **37.** $\frac{2}{21}$ **39.** $\frac{4}{35}$ **41.** $\frac{11}{21}$ **43.** $\frac{1}{57}$ **45.** $\frac{8}{855}$ **47.** $\frac{11}{57}$ **49.** $\frac{1}{5}$ **51.** $\frac{2}{3}$ **53.** $\frac{3}{4}$ **55.** $\frac{3}{4}$ **57.** $\frac{412{,}368}{412{,}878} \approx 0.999$ **59.** $\frac{412{,}368}{574{,}895} \approx 0.717$ **61.** $\frac{109}{121} \approx 0.90$ **63.** $\frac{19}{121} \approx 0.16$ **65.** $\frac{6}{11} \approx 0.55$ **67.** $\frac{5}{12} \approx 0.42$ **69.** $\frac{11}{124} \approx 0.09$ **71.** $\frac{105}{118} \approx 0.89$ **73. a.** Answers will vary. **b.** $\frac{365}{365} \cdot \frac{364}{365} \cdot \frac{363}{365} \approx 0.992$ **c.** ≈ 0.008 **d.** 0.411 **e.** 23 People, $1 - \frac{365}{365} \cdot \frac{364}{365} \cdot \frac{363}{365} \cdot \cdots \cdot \frac{343}{365} \approx 0.507$ **83.** does not make sense **85.** does not make sense **87.** $\frac{25}{7776} \approx 0.00322$ **89.** $\frac{11}{36}$

Section 11.8

Check Point Exercises

1. 2.5 **2.** 2 **3. a.** \$8000; In the long run, the average cost of a claim is \$8000. **b.** \$8000 **4.** 0; no; Answers will vary. **5.** table entries: \$998, \$48, and −\$2; expected value: −\$0.90; In the long run, a person can expect to lose an average of \$0.90 for each ticket purchased.; Answers will vary. **6.** −\$0.20; In the long run, a person can expect to lose an average of \$0.20 for each card purchased.

Concept and Vocabulary Check

1. expected; probability; add **2.** loss; probability; add **3.** false; Changes will vary. **4.** true

Exercise Set 11.8

1. 1.75 **3. a.** \$29,000; In the long run, the average cost of a claim is \$29,000. **b.** \$29,000 **c.** \$29,050 **5.** \$0; Answers will vary. **7.** $\approx$ \$0.73 **9.** $\frac{1}{16} = 0.0625$; yes **11.** the second mall **13. a.** \$140,000 **b.** no **15.** $\approx$ −\$0.17; In the long run, a person can expect to lose an average of about \$0.17 for each game played. **17.** $\approx$ −\$0.05; In the long run, a person can expect to lose an average of about \$0.05 for each game played. **19.** −\$0.50; In the long run, a person can expect to lose an average of \$0.50 for each game played. **27.** makes sense **29.** does not make sense **31.** \$160

Chapter 11 Review Exercises

1. 800 **2.** 20 **3.** 9900 **4.** 125 **5.** 243 **6.** 60 **7.** 240 **8.** 800 **9.** 114 **10.** 990 **11.** 151,200 **12.** 9900 **13.** 330 **14.** 2002 **15.** combinations **16.** permutations **17.** combinations **18.** 720 **19.** 32,760 **20.** 210 **21.** 420 **22.** 1140 **23.** 120 **24.** 1,860,480 **25.** 120 **26.** 1287 **27.** 55,440 **28.** 60 **29.** $\frac{1}{6}$ **30.** $\frac{2}{3}$ **31.** 1 **32.** 0 **33.** $\frac{1}{13}$ **34.** $\frac{3}{13}$ **35.** $\frac{3}{13}$ **36.** $\frac{1}{52}$ **37.** $\frac{1}{26}$ **38.** $\frac{1}{2}$ **39.** $\frac{1}{3}$ **40.** $\frac{1}{6}$ **41. a.** $\frac{1}{2}$ **b.** 0 **42.** $\frac{7}{12}$ **43.** $\frac{31}{60}$ **44.** $\frac{1}{30}$ **45.** $\frac{1}{24}$ **46.** $\frac{1}{5}$ **47.** $\frac{1}{30}$ **48.** $\frac{1}{720}$ **49.** $\frac{1}{3}$ **50. a.** $\frac{1}{15{,}504} \approx 0.0000645$ **b.** $\frac{100}{15{,}504} \approx 0.00645$ **51. a.** $\frac{1}{14}$ **b.** $\frac{3}{7}$ **52.** $\frac{3}{26}$ **53.** $\frac{5}{6}$ **54.** $\frac{1}{2}$ **55.** $\frac{1}{3}$ **56.** $\frac{2}{3}$ **57.** 1 **58.** $\frac{10}{13}$ **59.** $\frac{3}{4}$ **60.** $\frac{2}{13}$ **61.** $\frac{1}{13}$ **62.** $\frac{7}{13}$ **63.** $\frac{11}{26}$ **64.** $\frac{5}{6}$ **65.** $\frac{5}{6}$ **66.** $\frac{1}{2}$ **67.** $\frac{2}{3}$ **68.** 1 **69.** $\frac{5}{6}$ **70.** $\frac{4}{5}$ **71.** $\frac{3}{4}$ **72.** $\frac{18}{25}$ **73.** $\frac{6}{7}$ **74.** $\frac{3}{5}$ **75.** $\frac{12}{35}$ **76.** in favor: 1:12; against: 12:1 **77.** 99:1 **78.** $\frac{3}{4}$ **79.** $\frac{2}{9}$ **80.** $\frac{1}{36}$ **81.** $\frac{1}{9}$ **82.** $\frac{1}{36}$ **83.** $\frac{8}{27}$ **84.** $\frac{1}{32}$ **85. a.** 0.04 **b.** 0.008 **c.** 0.4096 **d.** 0.5904 **86.** $\frac{1}{9}$ **87.** $\frac{1}{12}$ **88.** $\frac{1}{196}$ **89.** $\frac{1}{3}$ **90.** $\frac{3}{10}$ **91. a.** $\frac{1}{2}$ **b.** $\frac{2}{7}$ **92.** $\frac{27}{29}$ **93.** $\frac{4}{29}$ **94.** $\frac{144}{145}$ **95.** $\frac{11}{20}$ **96.** $\frac{11}{135}$ **97.** $\frac{1}{125}$ **98.** $\frac{1}{232}$ **99.** $\frac{19}{1044}$ **100.** $\frac{27{,}336}{31{,}593} \approx 0.865$ **101.** $\frac{11{,}161}{31{,}593} \approx 0.353$ **102.** $\frac{29{,}214}{31{,}593} \approx 0.925$ **103.** $\frac{15{,}256}{31{,}593} \approx 0.483$ **104.** $\frac{4312}{31{,}593} \approx 0.136$ **105.** $\frac{3684}{27{,}336} \approx 0.135$ **106.** $\frac{2169}{2379} \approx 0.912$ **107.** 3.125 **108. a.** \$0.50; In the long run, the average cost of a claim is \$0.50. **b.** \$10.00 **109.** \$4500; Answers will vary. **110.** −\$0.25; In the long run, a person can expect to lose an average of \$0.25 for each game played.

Chapter 11 Test

1. 240 **2.** 24 **3.** 720 **4.** 990 **5.** 210 **6.** 420 **7.** $\frac{6}{25}$ **8.** $\frac{17}{25}$ **9.** $\frac{11}{25}$ **10.** $\frac{5}{13}$ **11.** $\frac{1}{210}$ **12.** $\frac{10}{1001} \approx 0.00999$ **13.** $\frac{1}{2}$ **14.** $\frac{1}{16}$ **15.** $\frac{1}{8000} = 0.000125$ **16.** $\frac{8}{13}$ **17.** $\frac{3}{5}$ **18.** $\frac{1}{19}$ **19.** $\frac{1}{256}$ **20.** 3:4 **21. a.** 4:1 **b.** $\frac{4}{5}$ **22.** $\frac{3}{5}$ **23.** $\frac{39}{50}$ **24.** $\frac{3}{5}$ **25.** $\frac{9}{19}$ **26.** $\frac{7}{150}$ **27.** \$1000; Answers will vary. **28.** −\$12.75; In the long run, a person can expect to lose an average of \$12.75 for each game played.

CHAPTER 12

Section 12.1

Check Point Exercises

1. a. the set containing all the city's homeless **b.** no: People already in the shelters are probably less likely to be against mandatory residence in the shelters. **2.** By selecting people from a shelter, homeless people who do not go to the shelters have no chance of being selected. An appropriate method would be to randomly select neighborhoods of the city and then randomly survey homeless people within the selected neighborhood.

3.

Grade	Frequency
A	3
B	5
C	9
D	2
F	1
	20

4.

Class	Frequency
40–49	1
50–59	5
60–69	4
70–79	15
80–89	5
90–99	7
	37

5.

Stem	Leaves
4	1
5	8 2 8 0 7
6	8 2 9 9
7	3 5 9 9 7 5 5 3 3 6 7 1 7 1 5
8	7 3 9 9 1
9	4 6 9 7 5 8 0

Concept and Vocabulary Check

1. random **2.** frequency distribution **3.** grouped frequency distribution; 80; 89 **4.** histogram; frequencies **5.** frequency polygon; horizontal axis **6.** stem-and-leaf plot **7.** false; Changes will vary. **8.** true **9.** false; Changes will vary. **10.** true

Exercise Set 12.1

1. c **3.** 7; 31 **5.** 151

7.

Time Spent on Homework (in hours)	Number of Students
15	4
16	5
17	6
18	5
19	4
20	2
21	2
22	0
23	0
24	2
	30

9. 0, 5, 10, . . ., 40, 45 **11.** 5 **13.** 13 **15.** the 5–9 class

17.

Age at Inauguration	Number of Presidents
41–45	2
46–50	9
51–55	15
56–60	9
61–65	7
66–70	2
	44

19. a.

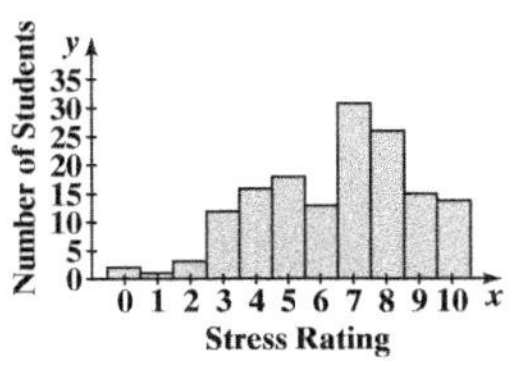

b.

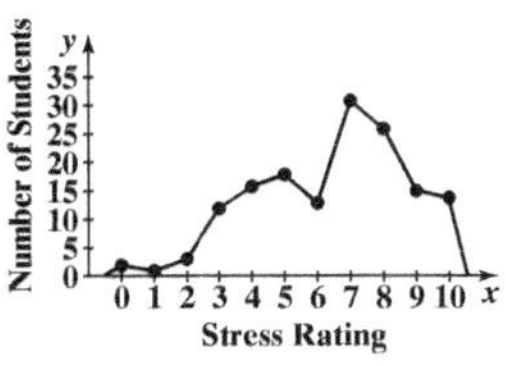

21.

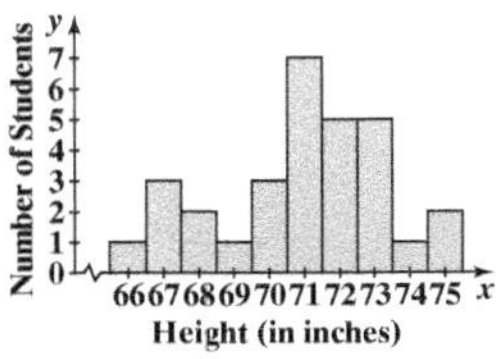

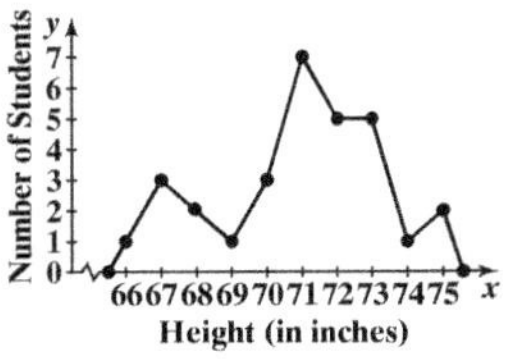

23. false **25.** false **27.** true **29.** false

31.

Stem	Leaves
2	8 8 9 5
3	8 7 0 1 2 7 6 4 0 5
4	8 2 2 1 4 5 4 6 2 0 8 2 7 9
5	9 4 1 9 1 0
6	3 2 3 6 6 3

The greatest number of college professors are in their 40s.

33. Time intervals on the horizontal axis do not represent equal amounts of time. **35.** Percentages do not add up to 100%. **37.** It is not clear whether the bars or the actors represent the box-office receipts. **49.** does not make sense **51.** makes sense

Section 12.2

Check Point Exercises

1. \$9.5 million **2.** 36 **3. a.** 35 **b.** 82 **4.** 5 **5.** 1:06, 1:09, 1:14, 1:21, 1:22, 1:25, 1:29, 1:29, 1:34, 1:34, 1:36, 1:45, 1:46, 1:49, 1:54, 1:57, 2:10, 2:15; median: 1 hour, 34 minutes **6.** 54.5 **7. a.** \$122.6 million **b.** \$17.5 million **c.** Kennedy's net worth was much greater than the other presidents'. **8. a.** 8 **b.** 3 and 8 **c.** no mode **9.** 14.5 **10.** mean: 158.6 cal; median: 153 cal; mode: 138 cal; midrange: 151 cal

Concept and Vocabulary Check

1. mean **2.** median **3.** $\frac{n+1}{2}$ **4.** mode **5.** midrange **6.** true **7.** false; Changes will vary. **8.** false; Changes will vary. **9.** false; Changes will vary.

Exercise Set 12.2

1. 4.125 **3.** 95 **5.** 62 **7.** ≈3.45 **9.** ≈4.71 **11.** ≈6.26 **13.** 3.5 **15.** 95 **17.** 60 **19.** 3.6 **21.** 5 **23.** 6 **25.** 3 **27.** 95 **29.** 40 **31.** 2.5, 4.2 (bimodal) **33.** 5 **35.** 6 **37.** 4.5 **39.** 95 **41.** 70 **43.** 3.3 **45.** 4.5 **47.** 5.5 **49.** mean: 30; median: 30; mode: 30; midrange: 30 **51.** mean: approximately 12.4; median: 12.5; mode: 13; midrange: 12.5 **53.** mean: approximately 31.3; median: 31; mode: 31; midrange: 33 **55. a.** ≈67.4 thousand **b.** 51 thousand **c.** 51 thousand and 92 thousand **d.** 113.5 thousand **57. a.** ≈ 17.27 **b.** 17 **c.** 7, 12, 17 **d.** 24.5 **59.** 175 lb **61.** 177.5 lb **63.** ≈2.76 **73.** makes sense **75.** makes sense **77.** Sample answers: **77. a.** 75, 80, 80, 90, 91, 94 **b.** 50, 80, 80, 85, 90, 95 **c.** 70, 75, 80, 85, 90, 100 **d.** 75, 80, 85, 90, 95, 95 **e.** 75, 80, 85, 85, 90, 95 **f.** 68, 70, 72, 72, 74, 76

Section 12.3

Check Point Exercises

1. 9 **2.** mean: 6;

Data item	Deviation
2	−4
4	−2
7	1
11	5

3. ≈3.92 **4.** sample A: 3.74; sample B: 28.06 **5. a.** small-company stocks **b.** small-company stocks; Answers will vary.

Concept and Vocabulary Check

1. range **2.** standard deviation **3.** true **4.** true **5.** false; Changes will vary.

Exercise Set 12.3

1. 4 **3.** 8 **5.** 2

7. a.

Data item	Deviation
3	−9
5	−7
7	−5
12	0
18	6
27	15

b. 0

9. a.

Data item	Deviation
29	−20
38	−11
48	−1
49	0
53	4
77	28

b. 0

11. a. 91

b.

Data item	Deviation
85	−6
95	4
90	−1
85	−6
100	9

c. 0

13. a. 155

b.

Data item	
146	−9
153	−2
155	0
160	5
161	6

c. 0

15. a. 2.70

b.

Data item	Deviation
2.25	−0.45
3.50	0.80
2.75	0.05
3.10	0.40
1.90	−0.80

c. 0

17. ≈1.58 **19.** ≈3.46 **21.** ≈0.89 **23.** 3 **25.** ≈2.14

27. *Sample A*: mean: 12; range: 12; standard deviation: ≈4.32
Sample B: mean: 12; range: 12; standard deviation: ≈5.07
Sample C: mean: 12; range: 12; standard deviation: 6
The samples have the same mean and range, but different standard deviations.

29. 0 **31.** 1 **33.** 7.91 **35.** 1.55 **37. a.** male artists; All of the data items for the men are greater than the greatest data item for the women. **b.** male artists: 93; female artists: 58 **c.** male artists; There is greater spread in the data for the men. **d.** male artists: 32.64; female artists: 6.82 **47.** makes sense **49.** makes sense **53.** a

Section 12.4

Check Point Exercises

1. a. 75.5 in. **b.** 58 in. **2. a.** 95% **b.** 47.5% **c.** 16% **3. a.** 2 **b.** 0 **c.** −1 **4.** ACT **5. a.** 64 **b.** 128 **6.** 75% of the scores on the SAT are less than this student's score. **7. a.** ± 2.0% **b.** We can be 95% confident that between 34% and 38% of Americans read more than ten books per year. **c.** Sample answer: Some people may be embarrassed to admit that they read few or no books per year.

Concept and Vocabulary Check

1. 68; 95; 99.7 **2.** mean **3.** percentile **4.** margin of error **5.** true **6.** true **7.** false; Changes will vary. **8.** true

Exercise Set 12.4

1. 120 **3.** 160 **5.** 150 **7.** 60 **9.** 90 **11.** 68% **13.** 34% **15.** 47.5% **17.** 49.85% **19.** 16% **21.** 2.5% **23.** 95% **25.** 47.5% **27.** 16% **29.** 2.5% **31.** 0.15% **33.** 1 **35.** 3 **37.** 0.5 **39.** 1.75 **41.** 0 **43.** −1 **45.** −1.5 **47.** −3.25 **49.** 1.5 **51.** 2.25 **53.** −1.25 **55.** −1.5 **57.** The person who scores 127 on the Wechsler has the higher IQ. **59.** 500 **61.** 475 **63.** 250 **65.** 275 **67. a.** ±3.5% **b.** We can be 95% confident that between 65.5% and 72.5% of the population favor required gun registration as a means to reduce gun violence. **69. a.** ±1.6% **b.** We can be 95% confident that between 58.6% and 61.8% of all TV households watched the final episode of *M*A*S*H*. **71.** 0.2% **73. a.** skewed to the right **b.** 5.3 murders per 100,000 residents **c.** 5 murders per 100,000 residents **d.** yes; The mean is greater than the median, which is consistent with a distribution skewed to the right. **e.** 5.6; yes; For a normal distribution, almost 100% of the z-scores are between −3 and 3. **87.** does not make sense **89.** does not make sense

Section 12.5

Check Point Exercises

1. 88.49% **2.** 8.08% **3.** 83.01%

Concept and Vocabulary Check

1. 98.93% **2.** 1.07% **3.** 88.49% **4.** 87.42% **5.** true

Exercise Set 12.5

1. a. 72.57% **b.** 27.43% **3. a.** 88.49% **b.** 11.51% **5. a.** 24.2% **b.** 75.8% **7. a.** 11.51% **b.** 88.49% **9.** 33.99% **11.** 15.74% **13.** 86.64% **15.** 28.57% **17.** 91.92% **19.** 27.43% **21.** 88.49% **23.** 6.69% **25.** 45.14% **27.** 6.68% **29.** 24.17% **31.** 77% **33.** 86% **35.** 10% **39.** does not make sense **41.** makes sense **45.** 630

Section 12.6

Check Point Exercises

1. This indicates a moderate relationship. **2.** 0.89; There is a moderately strong positive relationship between the two quantities. **3.** $y = 0.1x + 0.8$; 8.8 deaths per 100,000 people **4.** yes

Concept and Vocabulary Check

1. scatter plot **2.** regression line **3.** correlation coefficient; −1; 1 **4.** true **5.** false; Changes will vary. **6.** false; Changes will vary. **7.** true

Exercise Set 12.6

1.

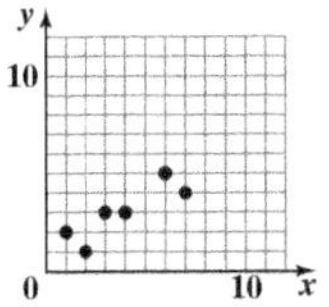

There appears to be a positive correlation.

3.

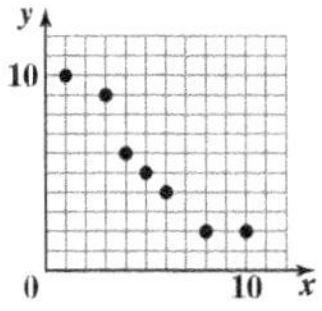

There appears to be a negative correlation.

5.

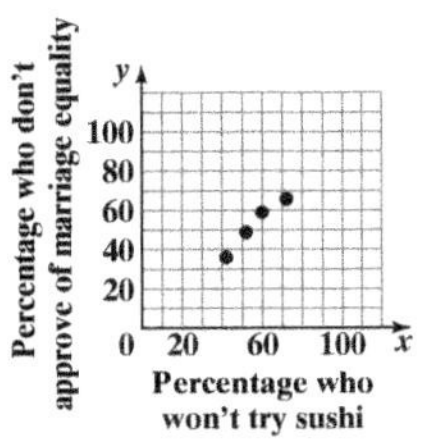

There appears to be a positive correlation.

7.

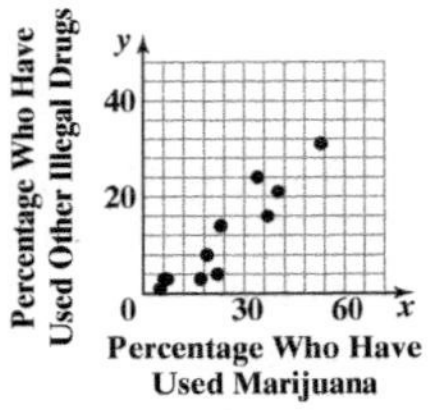

There appears to be a positive correlation.

9. false **11.** true **13.** true **15.** false **17.** true **19.** false **21.** true **23.** false **25.** false **27.** a **29.** d **31.** 0.85 **33.** −0.95 **35. a.** 0.98 **b.** Answers will vary. **c.** $y = 1.01x - 4.47$ **d.** 26% **37. a.** 0.93 **b.** $y = 0.62x - 3.07$ **c.** 3% **39.** A correlation does exist. **41.** A correlation does not exist. **43.** A correlation does exist. **45.** A correlation does not exist. **59.** does not make sense **61.** does not make sense

Chapter 12 Review Exercises

1. a

2.

Time Spent on Homework (in hours)	Number of Students
6	1
7	3
8	3
9	2
10	1
	10

3.

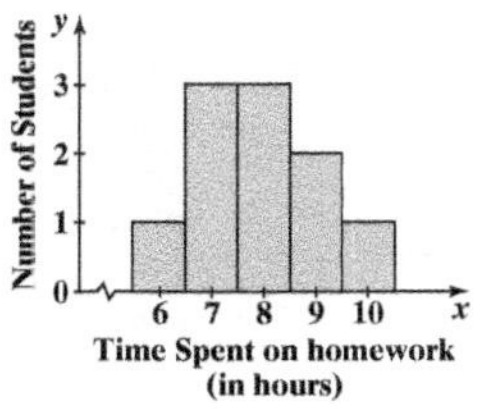

4.

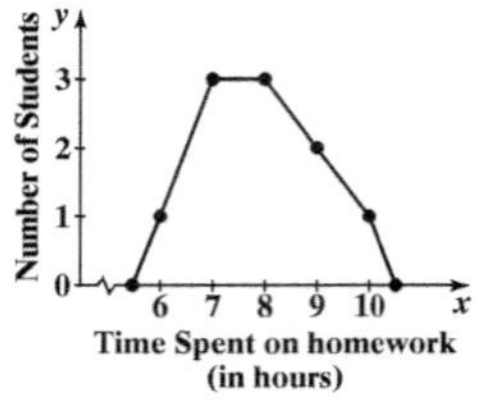

5.

Grades	Number of Students
0–39	19
40–49	8
50–59	6
60–69	6
70–79	5
80–89	3
90–99	3
	50

6.

Stem	Leaves
1	8 4 3 3 7 1
2	4 6 9 9 2 7
3	4 9 6 1 5 1 1
4	4 7 9 1 2 2 0 5
5	4 7 9 0 6 1
6	3 7 0 8 3 9
7	2 5 4 0 3
8	1 7 6
9	1 0 5

7. Sizes of barrels are not scaled proportionally in terms of the data they represent. **8.** 91.2 **9.** 17 **10.** 2.3 **11.** 11 **12.** 28 **13.** 2 **14.** 27 **15.** 585, 587 (bimodal) **16.** 2 **17.** 91 **18.** 19.5 **19.** 2.5

21. a.

Age at First Inauguration	Number of Presidents
42	1
43	1
44	0
45	0
46	2
47	2
48	1
49	2
50	1
51	5
52	2
53	0
54	5
55	4
56	3
57	4
58	1
59	0
60	1
61	3
62	1
63	0
64	2
65	1
66	0
67	0
68	1
69	1
	44

b. mean: ≈54.66 yr; median: 54.5 yr; mode: 51 yr, 54 yr (bimodal); midrange: 55.5 yr **22.** 18
23. 564

24. a.

Data item	Deviation
29	−6
9	−26
8	−27
22	−13
46	11
51	16
48	13
42	7
53	18
42	7

b. 0

25. a. 50

b.

Data item	Deviation
36	−14
26	−24
24	−26
90	40
74	24

c. 0

26. ≈ 4.05 **27.** ≈ 5.13 **28.** mean: 49; range: 76; standard deviation: ≈ 24.32 **29.** Set A: mean: 80; standard deviation: 0; Set B: mean: 80; standard deviation: ≈ 11.55; Answers will vary. **30.** Answers will vary. **31.** 86 **32.** 98 **33.** 60 **34.** 68% **35.** 95% **36.** 34% **37.** 99.7% **38.** 16% **39.** 84% **40.** 2.5% **41.** 0 **42.** 2 **43.** 1.6 **44.** −3 **45.** −1.2 **46.** vocabulary test **47.** 38,000 miles **48.** 41,000 miles **49.** 22,000 miles **50. a.** ± 2.1% **b.** We can be 95% confident that between 28.9% and 33.1% of American adults would be willing to sacrifice a percentage of their salary to work for an environmentally friendly company. **51. a.** skewed to the right **b.** mean: 2.1 syllables; median: 2 syllables; mode: 1 syllable **c.** yes; The mean is greater than the median, which is consistent with a distribution skewed to the right. **52.** 91.92% **53.** 96.41% **54.** 88.33% **55.** 10.69% **56.** 75% **57.** 14% **58.** 11%

59.

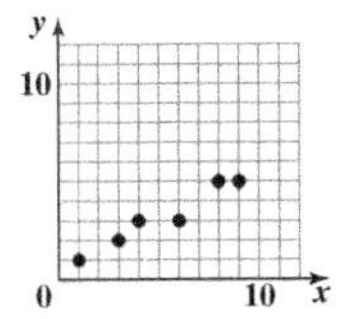

There appears to be a positive correlation.

60.

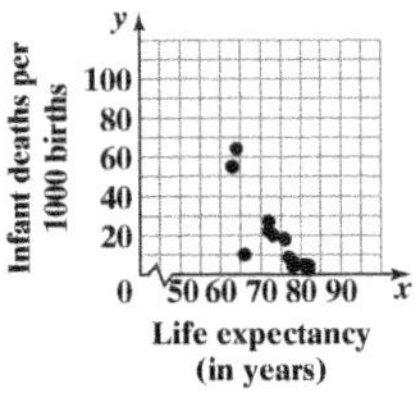

There appears to be a negative correlation.

61. false **62.** true **63.** false **64.** false **65.** true **66.** false **67.** true **68.** c **69. a.** 0.972 **b.** $y = 0.509x + 0.537$
70. a. 0.99 **b.** There is a correlation.

Chapter 12 Test

1. d

2.

Score	Frequency
3	1
4	2
5	3
6	2
7	2
8	3
9	2
10	1
	16

3.

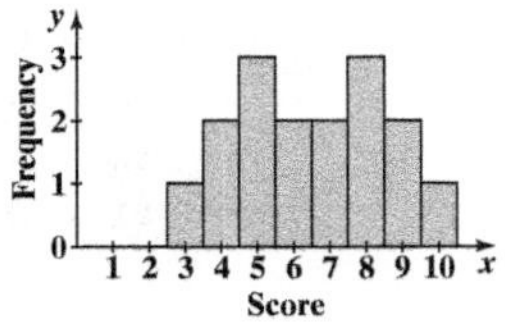

4.

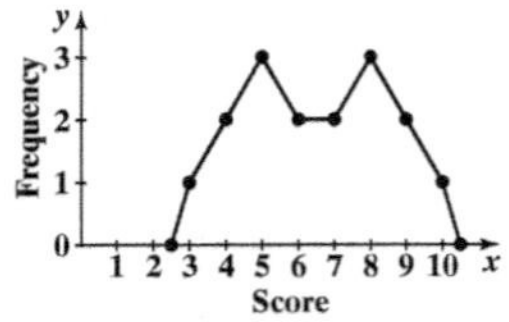

5.

Class	Frequency
40–49	3
50–59	6
60–69	6
70–79	7
80–89	6
90–99	2
	30

6.

Stem	Leaves
4	8 1 6
5	1 0 5 0 9 0
6	7 2 0 3 1 1
7	9 8 3 1 9 1 5
8	9 3 0 8 9 1
9	3 0

7. The roofline gives the impression that the percentage of home-schooled students grew at the same rate each year between the years shown which might not have happened. **8.** ≈ 3.67 **9.** 3 **10.** 4 **11.** ≈ 2.34 **12.** 2.25 **13.** 2 **14.** 2 **15.** Answers will vary. **16.** 34% **17.** 2.5% **18.** student **19.** 8.08% **20.** 41% **21. a.** $\pm 10\%$ **b.** We can be 95% confident that between 50% and 70% of all students are very satisfied with their professors.

22.

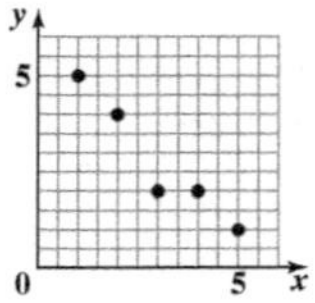

;There appears to be a strong negative correlation. **23.** false **24.** false **25.** true **26.** Answers will vary.

CHAPTER 13

Section 13.1

Check Point Exercises

1. a. 4210 **b.** 40 **c.** 2865 **2.** Donna **3.** Bob **4.** Carmen **5.** Bob

Concept and Vocabulary Check

1. preference; preference **2.** majority **3.** pairwise comparison; 1; $\frac{1}{2}$; most points **4.** $\frac{n(n-1)}{2}$ **5.** plurality **6.** plurality **7.** Borda count; most points **8.** True

Exercise Set 13.1

1. 7; 5; 4

3.

Number of Votes	5	1	4	2
First Choice	A	B	C	C
Second Choice	B	D	B	B
Third Choice	C	C	D	A
Fourth Choice	D	A	A	D

5. a. 26 **b.** 8 **c.** 22 **d.** 3 **7.** musical **9.** Darwin **11.** Comedy **13.** Freud **15.** Comedy **17.** Einstein **19.** 10 **21.** 28 **23.** Comedy **25.** Hawking **27.** A **29.** B **31.** Rent **33.** Rent **35. a.** C **b.** A **37.** C **49.** does not make sense **51.** makes sense

Section 13.2

Check Point Exercises

1. a. A **b.** B **2. a.** B **b.** A **3. a.** A **b.** C **c.** yes **4. a.** A **b.** D **c.** yes

Concept and Vocabulary Check

1. majority **2.** head-to-head **3.** monotonicity **4.** irrelevant alternatives **5.** Borda count **6.** True

Exercise Set 13.2

1. a. D **b.** A **c.** no **3. a.** A **b.** V **c.** no **5. a.** A **b.** C **c.** no **7. a.** C **b.** B **c.** no

9. a. S **b.**

Number of Votes	15	8
First Choice	S	L
Second Choice	L	S

;S **c.** yes **11. a.** A **b.** no **13. a.** C **b.** no **15. a.** B **b.** no **c.** no **d.** A; no **17. a.** B **b.** no **c.** no

19. a. A **b.** yes **c.** yes **d.**

Number of Votes	7	3	2
First Choice	A	B	A
Second Choice	B	C	C
Third Choice	C	A	B

;A **e.** yes **f.** no

29. makes sense **31.** does not make sense

Section 13.3

Check Point Exercises

1. a. 50 **b.** 22.24; 22.36; 26.4; 30.3; 98.7 **2.** 22; 22; 27; 30; 99 **3.** 22; 22; 26; 30; 100 **4.** 22; 23; 27; 30; 98 **5.** 22; 22; 27; 30; 99

Concept and Vocabulary Check

1. divisor; quota **2.** quotas **3.** lower; upper; lower quota; upper quota **4.** Hamilton's; decimal **5.** modified; Jefferson's; Adams's; Webster's **6.** quota; Jefferson's; Adams's; Webster's **7.** True

Exercise Set 13.3

1. a. 20; 20,000 **b.** 6.9; 13.3; 26.7; 33.1 **c.**

State	Lower Quota	Upper Quota
A	6	7
B	13	14
C	26	27
D	33	34

3. 7; 13; 27; 33 **5.** 30; 41; 53; 73; 103 **7.** 3; 5; 12; 15; 22 **9.** 17; 55; 24; 54 **11.** 28; 33; 53; 15; 22; 29 **13.** 12; 10; 8 **15.** 4; 10; 6 **17.** 20; 24; 29; 29; 98 **19.** 57; 81; 68; 44 **21.** 57; 81; 68; 44 **23.** 7; 2; 2; 2; 8; 14; 4; 5; 10; 10; 13; 2; 6; 2; 18 **25.** 7; 2; 2; 2; 8; 14; 4; 5; 10; 10; 12; 2; 6; 3; 18 **43.** does not make sense **45.** does not make sense

Section 13.4

Check Point Exercises

1. B's apportionment decreases from 11 to 10. **2. a.** 10; 19; 71 **b.** State A: 1.005%; State B: 0.998% **c.** State A loses a seat to State B, even though the population of State A is increasing at a faster rate. **3. a.** 21; 79 **b.** With North High added to the district, West High loses a counselor to East High.

Concept and Vocabulary Check

1. Alabama **2.** population **3.** new-states **4.** True

Exercise Set 13.4

1. a. 16; 8; 6 **b.** 17; 9; 5; Liberal Arts Math loses a teaching assistant when the total number of teaching assistants is raised from 30 to 31. This is an example of the Alabama paradox. **3.** State A loses a seat when the total number of seats increases from 40 to 41. **5. a.** 4; 6; 14 **b.** 28.3%; 26.3%; 14.7% **c.** 3; 7; 14; A loses a seat while B gains, even though A has a faster increasing population. The population paradox does occur. **7.** C loses a truck to B even though C increased in population faster than B. **9. a.** 10; 90 **b.** Branch B loses a promotion when branch C is added. This means the new-states paradox has occurred. **11. a.** 9; 91 **b.** State B loses a seat when state C is added. **13. a.** 6; 13; 31 **b.** The new-states paradox does not occur. As long as the modified divisor, *d*, remains the same, adding a new state cannot change the number of seats held by existing states. **19.** makes sense **21.** does not make sense

Chapter 13 Review Exercises

1.

Number of Votes	4	3	3	2
First Choice	A	B	C	C
Second Choice	B	D	B	B
Third Choice	C	C	D	A
Fourth Choice	D	A	A	D

2. 23 **3.** 4 **4.** 16 **5.** 14 **6.** Musical **7.** Comedy **8.** Musical **9.** Musical **10.** A **11.** B **12.** A **13.** A **14.** B **15.** A; no **16.** A; no **17.** B **18.** B; yes **19.** A **20.** B **21.** A **22.** A **23.** A; Borda count method **24.** B **25.** A; no **26.** A

27.

Number of Votes	400	450
First Choice	A	C
Second Choice	C	A

C; no **28.** B **29.** B still wins. The irrelevant alternatives criterion is satisfied. **30.** 50.05 **31.** 5.49; 7.83; 12.21; 14.47 **32.** A: 5, 6; B: 7, 8; C: 12, 13; D: 14, 15 **33.** 6; 8; 12; 14 **34.** 5; 8; 12; 15 **35.** 6; 8; 12; 14 **36.** 6; 8; 12; 14 **37.** 14; 42; 62; 82 **38.** 13; 42; 63; 82 **39.** 14; 42; 63; 81 **40.** 14; 42; 62; 82 **41. a.** 8; 67; 75 **b.** The Alabama paradox occurs. **42. a.** 72; 20; 8 **b.** 0.8%; 1.3% **c.** 72; 21; 7; The population paradox occurs. **43. a.** 7; 26 **b.** The new-states paradox does not occur. **44.** False. Answers will vary.

Chapter 13 Test

1. 3600 **2.** 600 **3.** 1500 **4.** 1500 **5.** B **6.** B **7.** A **8.** A **9.** A **10.** A; yes **11.** A **12.** B; no **13.** C **14.** B; no **15.** no **16.** 50 **17.** 2.38; 3.3; 4.32 **18.** A: 2, 3; B: 3, 4; C; 4, 5 **19.** 3; 3; 4 **20.** 2; 3; 5 **21.** 3; 3; 4 **22.** 2; 3; 5 **23.** yes **24.** The new-states paradox does not occur. **25.** Answers will vary.

CHAPTER 14

Section 14.1

Check Point Exercises

1. The two graphs have the same number of vertices connected to each other in the same way.

2.
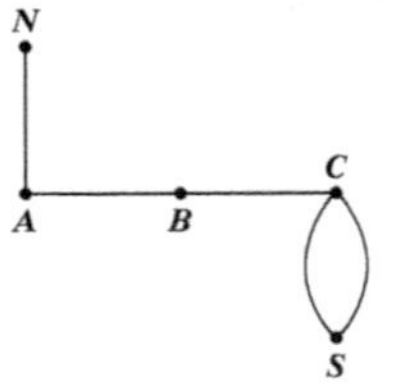

3.
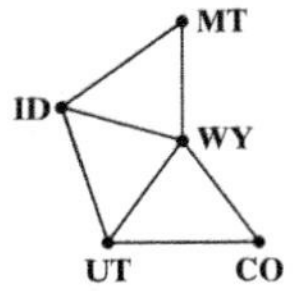

4.
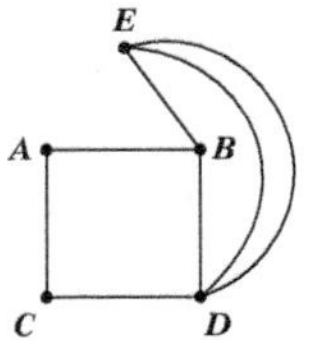

5.
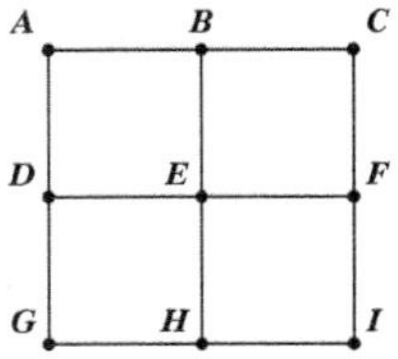

6. *A* and *B*, *A* and *C*, *A* and *D*, *A* and *E*, *B* and *C*, and *E* and *E*

Concept and Vocabulary Check

1. graph; vertices; edges; loop **2.** equivalent **3.** degree **4.** adjacent; path; circuit **5.** bridge **6.** true **7.** true **8.** true **9.** false

Exercise Set 14.1

1. 6; St. Louis, Chicago, Philadelphia, Montreal; 1, 1, 2, 2 **3.** No; no; Answers will vary.

5. Possible answers:
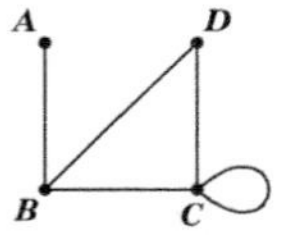

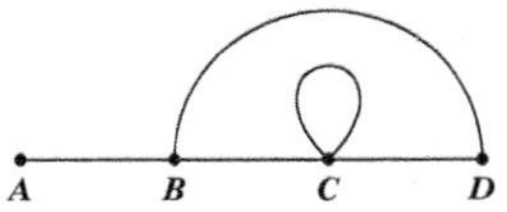

7. The two graphs have the same number of vertices connected in the same way, so they are the equivalent. Possible answer:
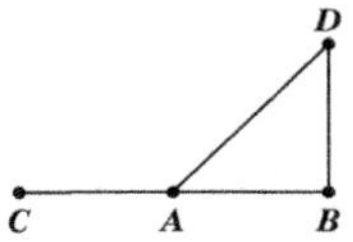

9.
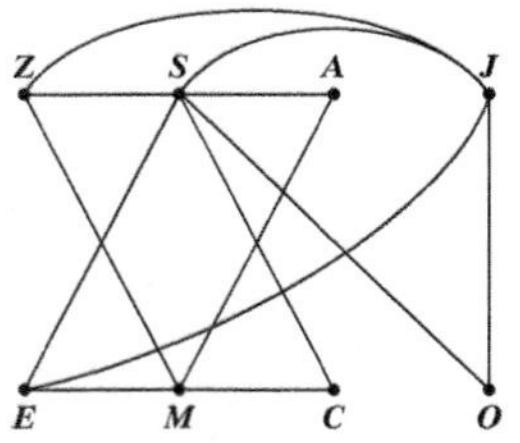

11.
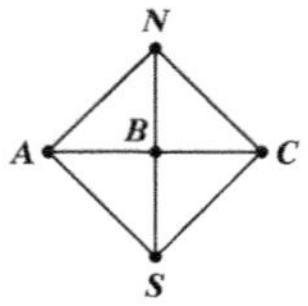

13.
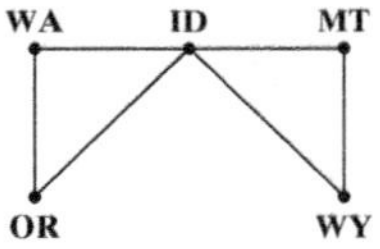

15.
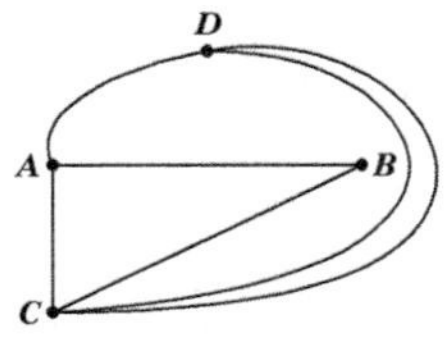

17.
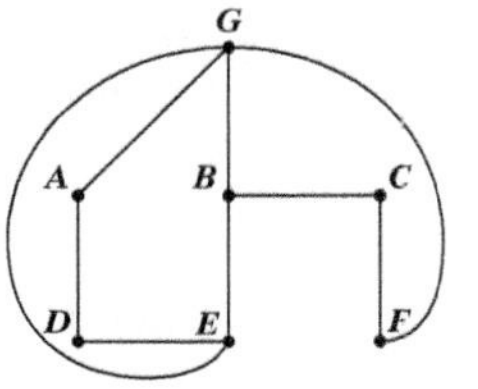

19.
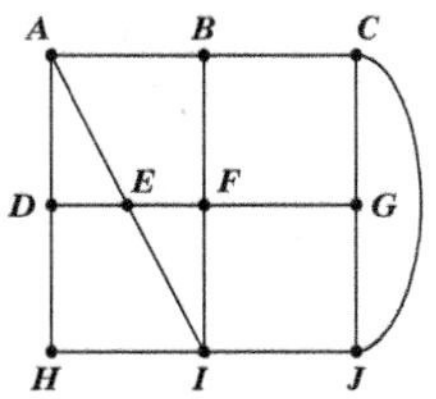

21.
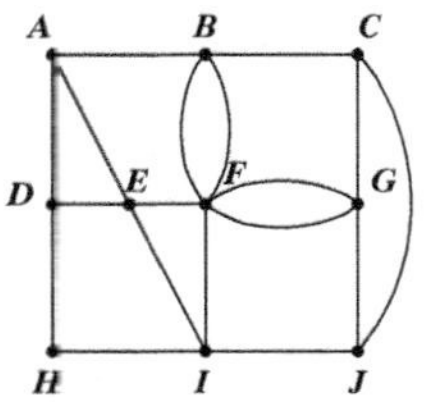

23. *A*: 2; *B*: 2; *C*: 3; *D*: 3; *E*: 3; *F*: 1 **25.** *B* and *C* **27.** *A*, *C*, *D*; *A*, *B*, *C*, *D* **29.** *AC* and *DF* **31.** While edge *CD* is included, the graph is connected. If we remove *CD*, the graph will be disconnected. **33.** *DF* **35.** Even: *A*, *B*, *G*, *H*, *I*; odd: *C*, *D*, *E*, *F* **37.** *D*, *G*, and *I* **39.** Possible answers: *B*, *C*, *D*, *F*; *B*, *A*, *C*, *D*, *F* **41.** *G*, *F*, *I*, *H*, *G* **43.** *A*, *B*, *C*, *C*, *D*, *F*, *G*, *H*, *I* **45.** *G*, *F*, *D*, *E*, *D* requires that edge *DE* be traversed twice. This is not allowed within a path. **47.** No edge connects *F* and *E*.

49. Possible answer:

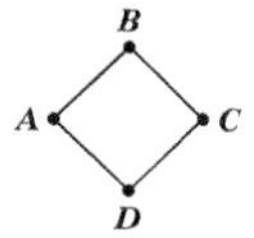

51. Possible answer:

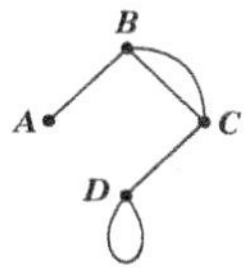

67. makes sense **69.** does not make sense **71.**

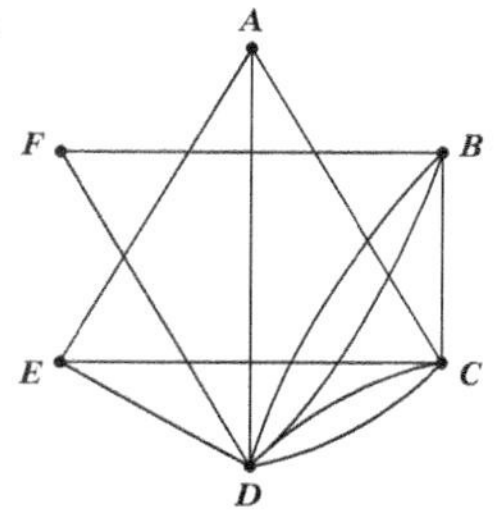

Section 14.2

Check Point Exercises

1.

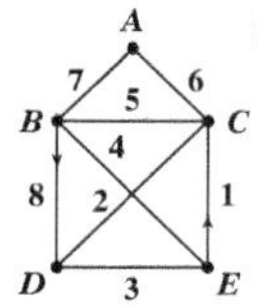

; *E, C, D, E, B, C, A, B, D*

2.

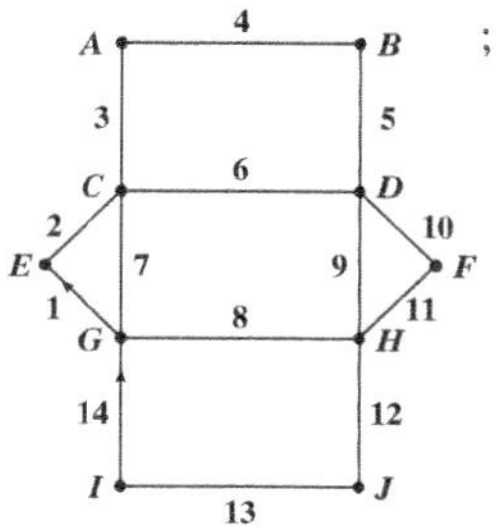

; *G, E, C, A, B, D, C, G, H, D, F, H, J, I, G*

3. a. yes **b.**

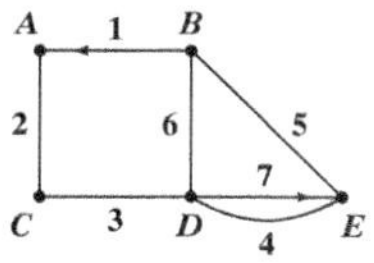

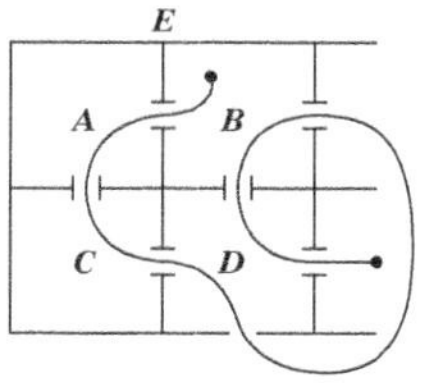

4.

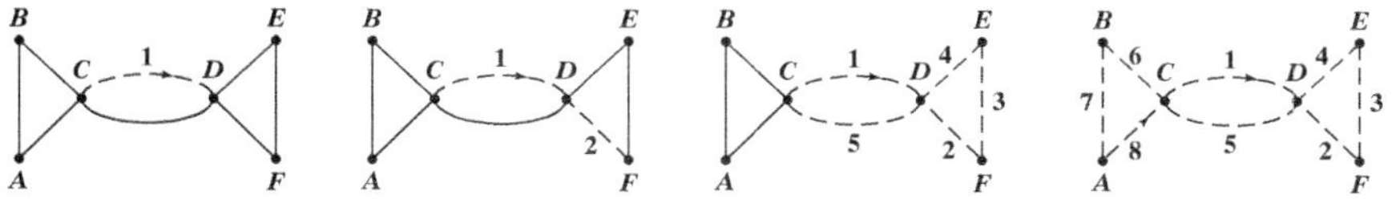

Concept and Vocabulary Check

1. Euler **2.** Euler **3.** two **4.** no; any **5.** odd **6.** *E; E; D* **7.** Fleury's; bridge **8.** true **9.** false **10.** false

Exercise Set 14.2

1.

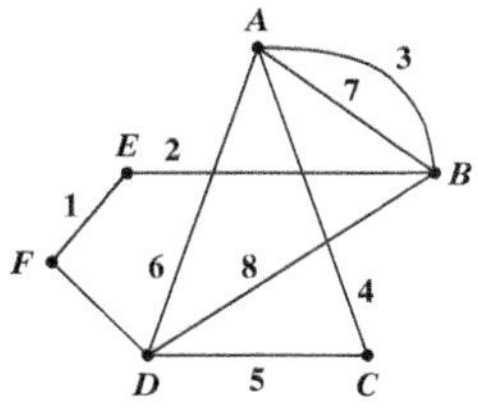

neither

3.

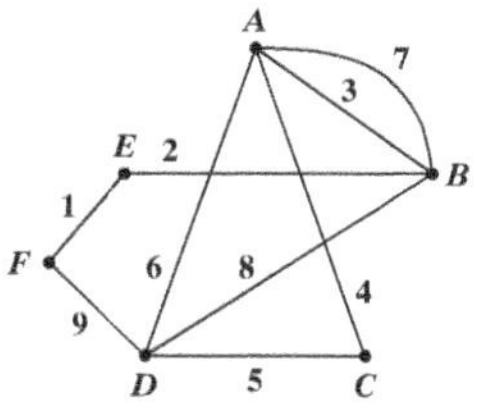

Euler circuit

5.

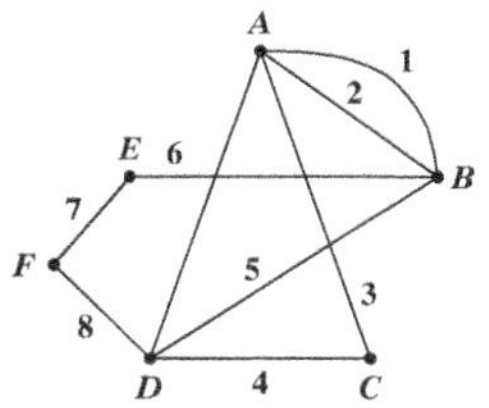

neither

7. a. There are two odd vertices.
b. Possible answer:

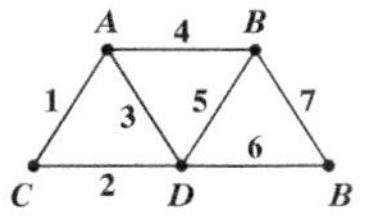

9. a. There are no odd vertices.
b. Possible answer:

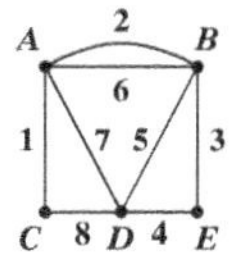

11. There are more than two odd vertices. **13.** Euler circuit **15.** Euler path, but no Euler circuit **17.** neither an Euler path nor an Euler circuit

19. a. Euler circuit

b. Possible answer:

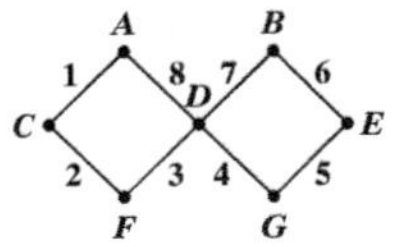

21. a. Euler path

b. Possible answer:

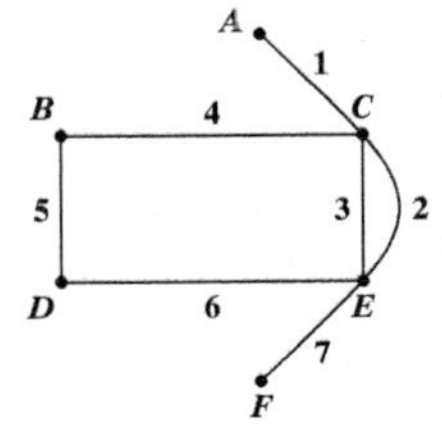

23. a. neither

25. a. Euler path

b. Possible answer:

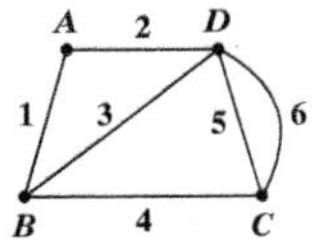

27. a. Euler circuit

b. Possible answer:

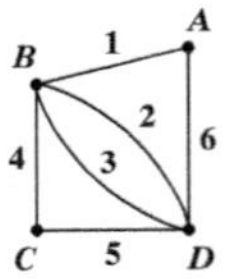

29. a. neither

31. a. Euler path

b. Possible answer:

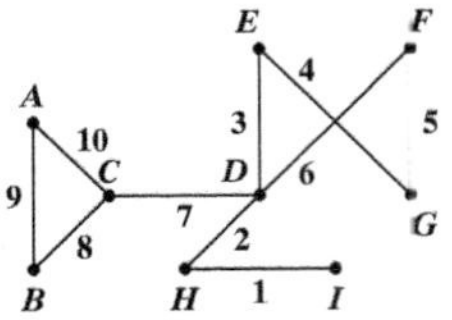

33. Possible answer:

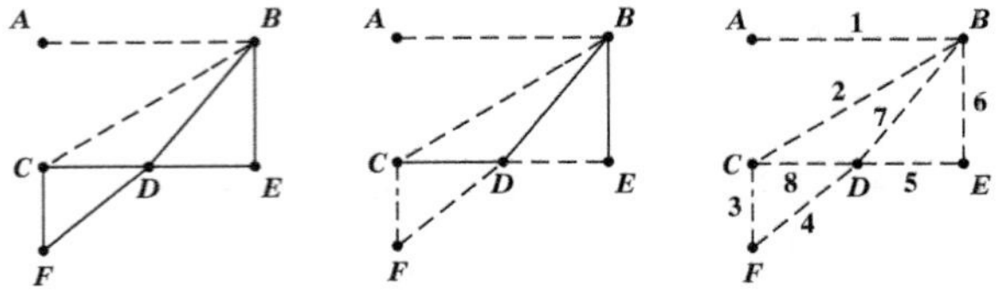

35. Possible answer:

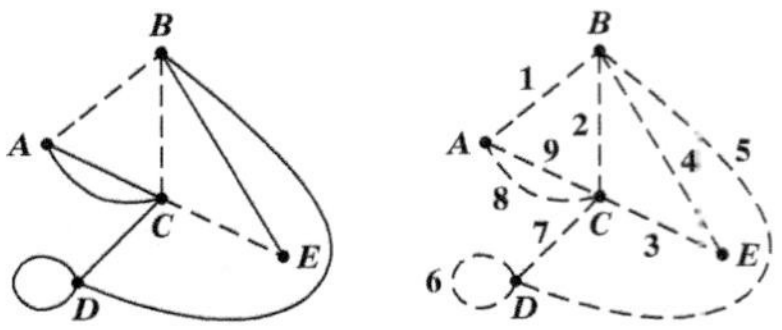

37. Possible answer:

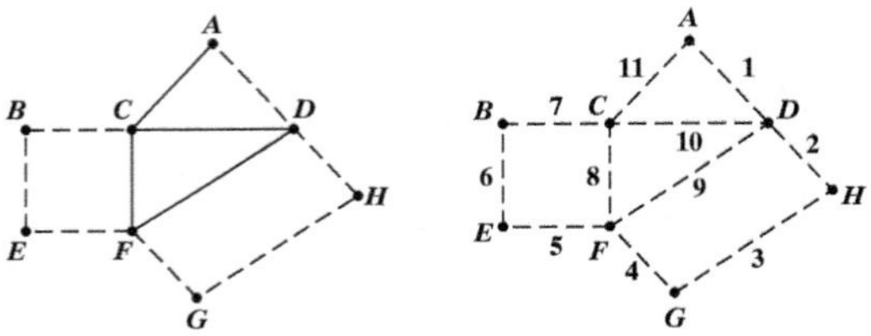

39. Possible answer:

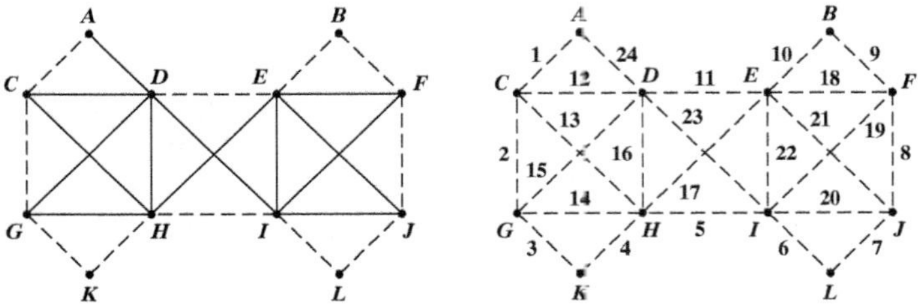

41. a. Remove *FG*. **b.** Sample answer: *EC, CB, BD, DF, FA, AD, DG, GH, HC, CG, GB, BH, HE*

43. a. Remove *BA* and *FJ*. **b.** Sample answer: *CA, AD, DI, IH, HG, GF, FC, CD, DE, EH, HJ, JG, GB, BC*

45. The graph has an Euler circuit.

47.

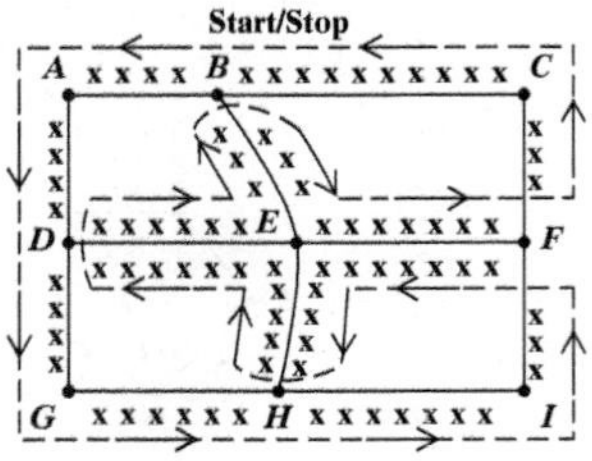

49. a.

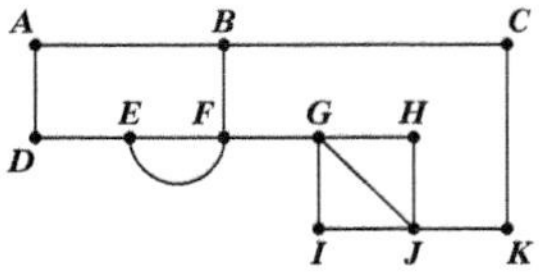

b. Yes; begin at *B* or *E*.

51. a.

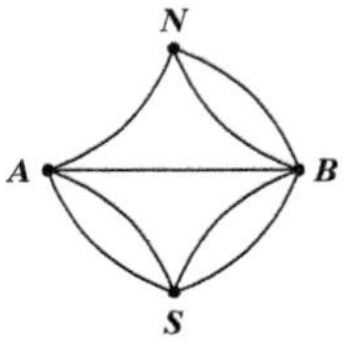

b. Yes, they can.

c. Possible answer:

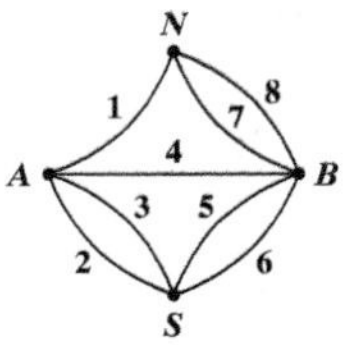

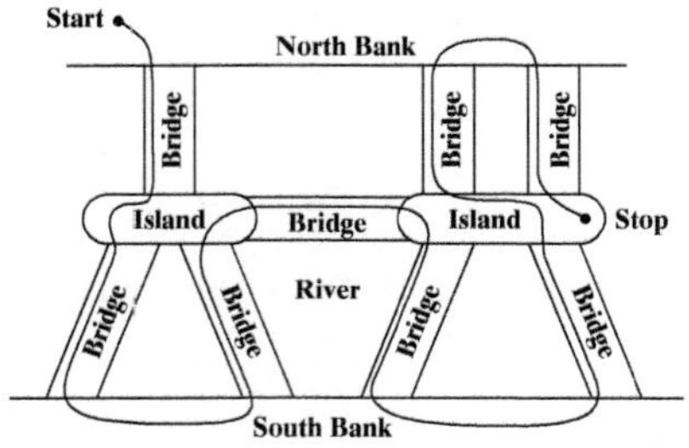

53.

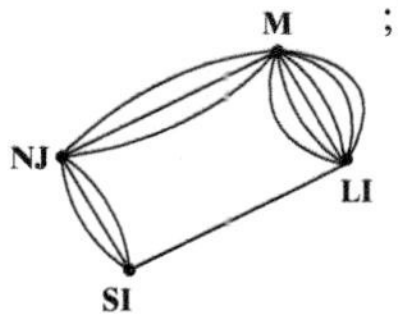

; It is possible.

55. a.

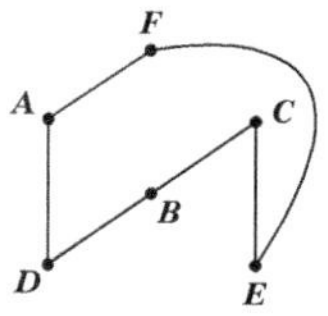

b. It is possible.

c. Possible answer:

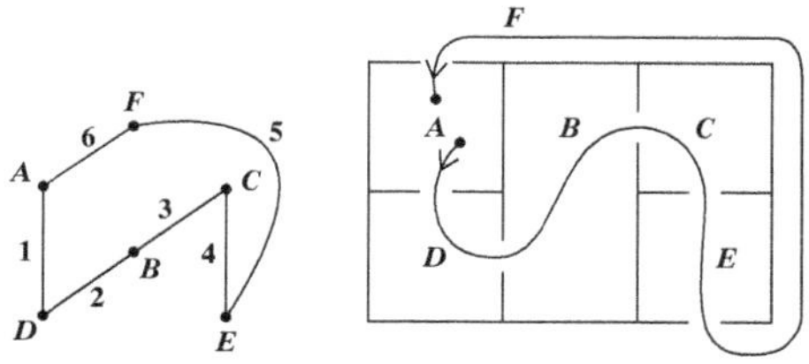

57. Possible answer:

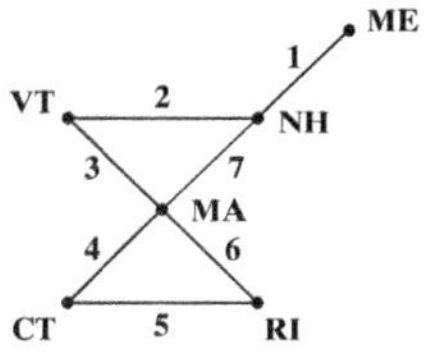

59. a.

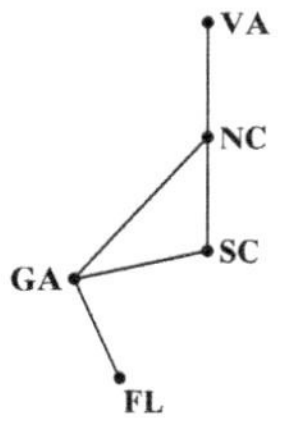

b. not possible **d.** not possible **69.** makes sense **71.** does not make sense

Section 14.3

Check Point Exercises

1. a. Possible answer: *E*, *C*, *D*, *G*, *B*, *A*, *F* **b.** Possible answer: *E*, *C*, *D*, *G*, *B*, *F*, *A*, *E* **2. a.** 2 **b.** 120 **c.** 362,880 **3.** $562

4.

Hamilton Circuit	Sum of the Weights of the Edges	=	Total Cost
A, *B*, *C*, *D*, *A*	20 + 15 + 50 + 30	=	$115
A, *B*, *D*, *C*, *A*	20 + 10 + 50 + 70	=	$150
A, *C*, *B*, *D*, *A*	70 + 15 + 10 + 30	=	$125
A, *C*, *D*, *B*, *A*	70 + 50 + 10 + 20	=	$150
A, *D*, *B*, *C*, *A*	30 + 10 + 15 + 70	=	$125
A, *D*, *C*, *B*, *A*	30 + 50 + 15 + 20	=	$115

A, *B*, *C*, *D*, *A*; *A*, *D*, *C*, *B*, *A*

5. *A*, *B*, *C*, *D*, *E*, *A*; 198

Concept and Vocabulary Check

1. Hamilton; Hamilton **2.** complete; $(n-1)!$ **3.** weighted; weights; traveling; optimal **4.** Brute Force **5.** Nearest Neighbor; weight **6.** false **7.** false **8.** false

Exercise Set 14.3

1. Possible answer: *A*, *G*, *C*, *F*, *E*, *D*, *B* **3.** Possible answer: *A*, *B*, *G*, *C*, *F*, *E*, *D*, *A* **5.** Possible answer: *A*, *F*, *G*, *E*, *C*, *B*, *D* **7.** Possible answer: *A*, *B*, *C*, *E*, *G*, *F*, *D*, *A* **9. a.** no **11. a.** yes **b.** 120 **13. a.** no **15.** 2 **17.** 39,916,800 **19.** 11 **21.** 36 **23.** 36 **25.** 88 **27.** 70 **29.** 70 **31.** *A*, *C*, *B*, *D*, *A* and *A*, *D*, *B*, *C*, *A* **33.** *B*, *D*, *C*, *A*, *B*; 82 **35. a.** Add *AB*.; 6 **b.** Sample answers: *AD*, *DB*, *BC*, *CA*; *CA*, *AB*, *BD*, *DC* **c.** Remove *CD*. **d.** Sample answer: *AC*, *CB*, *BD*, *DA* **37. a.** Add *AB*, *AC*, *BC*, and *DE*.; 24 **b.** Sample answers: *AB*, *BC*, *CE*, *ED*, *DA*; *AB*, *BC*, *CD*, *DE*, *EA* **c.** Remove *BD* and *BE*. **d.** Sample answer: *AE*, *EC*, *CD*, *DA* **39.** *A*, *D*, *B*, *C*, *E*, *A*

41.

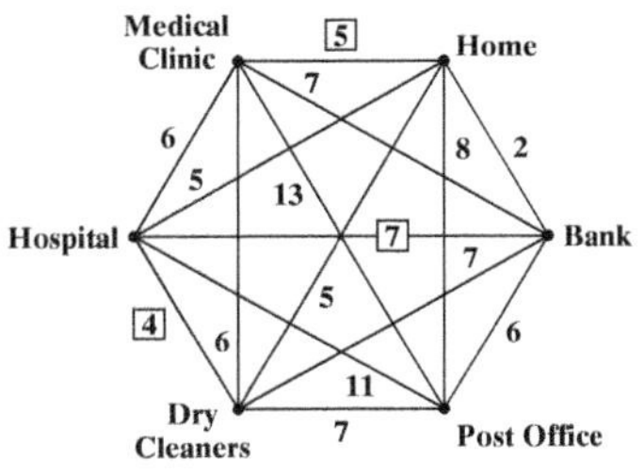

43. 30 **45.**

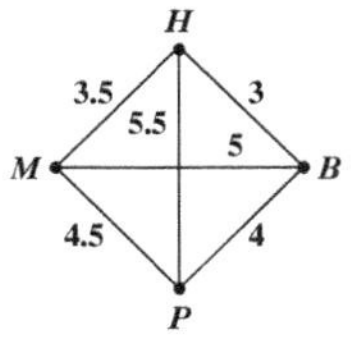

47. *H*, *B*, *P*, *M*, *H*; 15 mi; same

61. does not make sense **63.** makes sense

Section 14.4

Check Point Exercises

1. Figure 14.51(c) **2.** Possible answer:

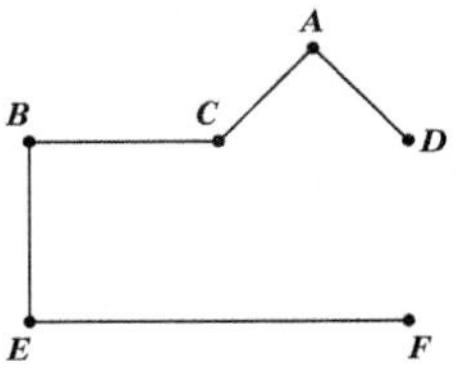

3.

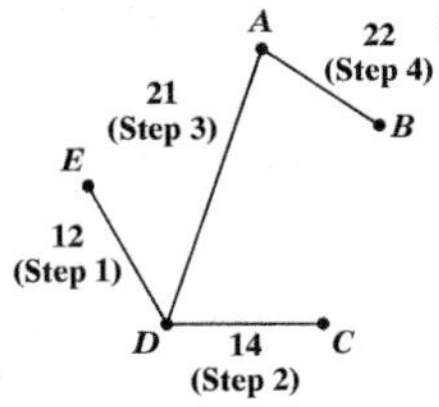

; 69

Concept and Vocabulary Check

1. tree; bridge; $n-1$ **2.** spanning **3.** minimum spanning **4.** Kruskal's; weight; circuits **5.** false **6.** true

Exercise Set 14.4

1. yes **3.** no **5.** yes **7.** no **9.** yes **11.** i **13.** ii **15.** iii

17.

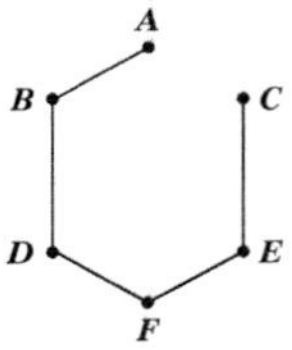

19.

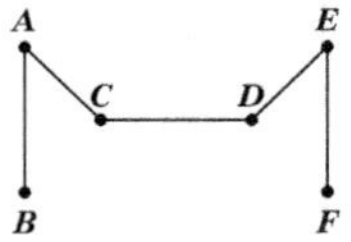

21.

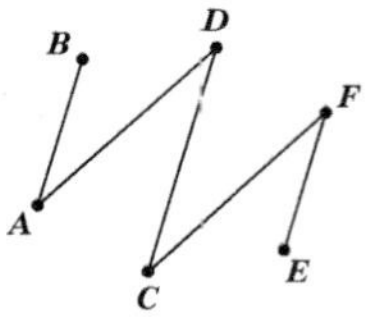

23.

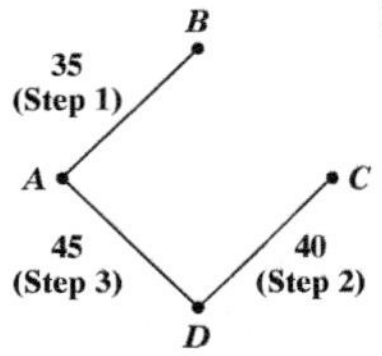

; 120

25.

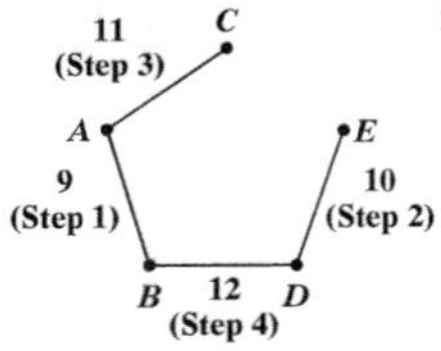

; 42

27.

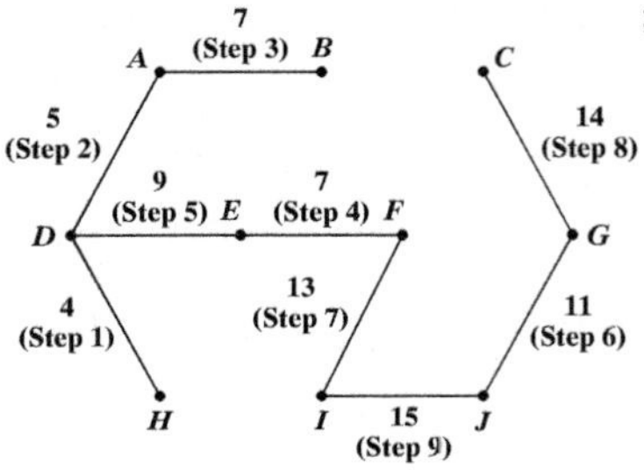

; 85

29.

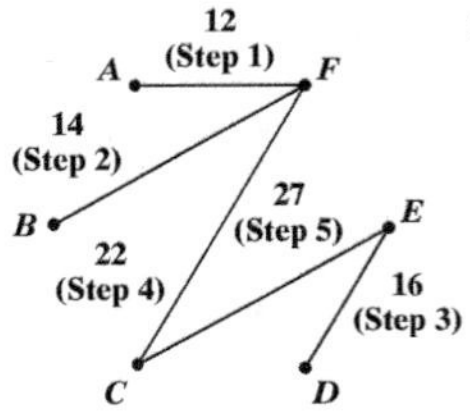

; 91

31. Sample answers: AB, AC, CD; AB, BC, CD; AB, AD, CD; AB, AD, BC **33.** AE, BC, CD, CE; 64

35. Sample answer: $AE, BC, BE, CF, DE, EH, FG, FJ, HI$; 141

37.

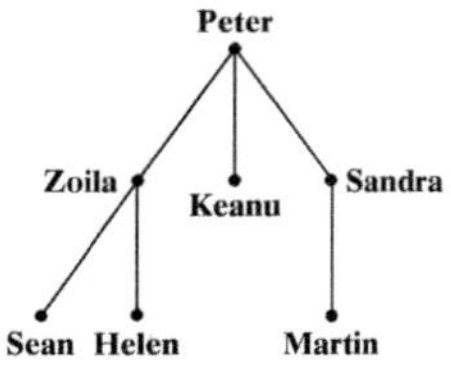

39. a.

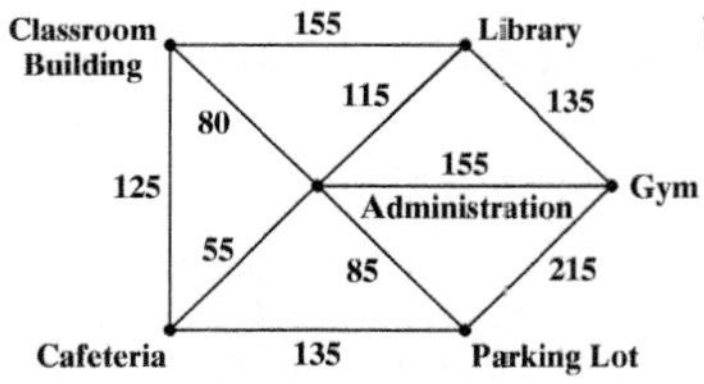

b.

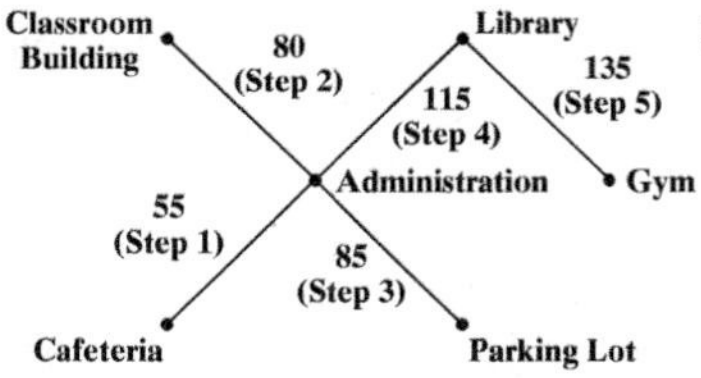

; 470 ft

41.

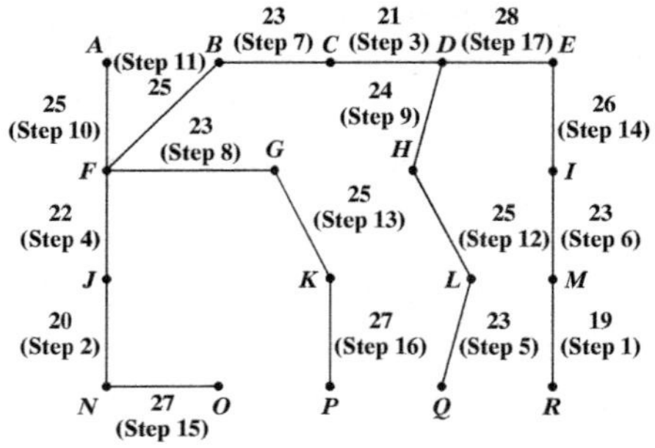

; 406 ft

53. does not make sense **55.** makes sense

Chapter 14 Review Exercises

1. Both graphs have the same number of vertices, and these vertices are connected in the same ways.
Possible answer:

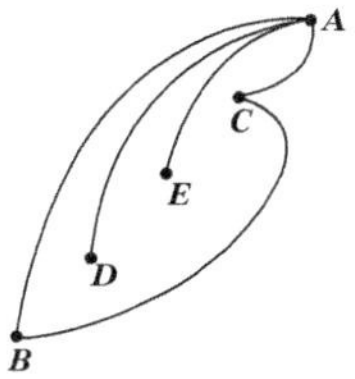

2. *A*: 5; *B*: 4; *C*: 5; *D*: 4; *E*: 2 *E*, *D*, *B*, *A* and *E*, *C*, *A*

3. Even: *B*, *D*, *E*; odd: *A*, *C*

4. *B*, *C*, and *E*

5. Possible answer:

6. Possible answer: *E*, *D*, *C*, *E*

7. yes

8. no

9. *AD*, *DE*, and *DF*

10.

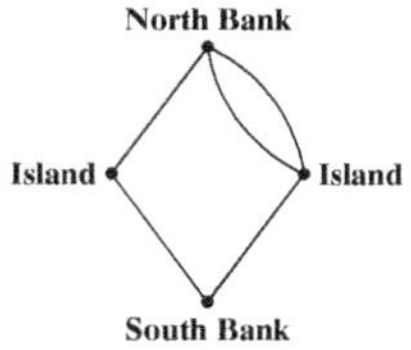

11.

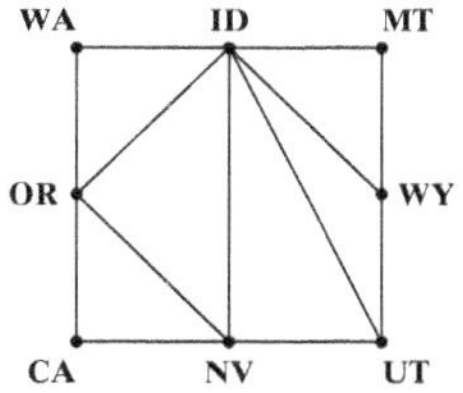

12.

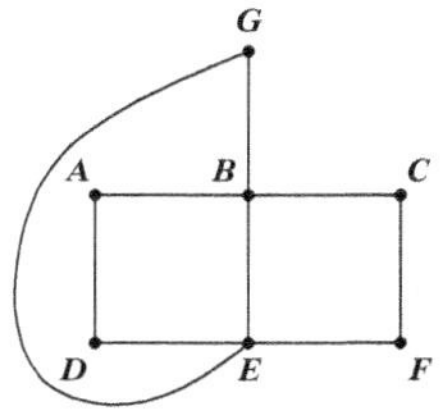

13. a. neither

14. a. Euler circuit

b. Possible answer:

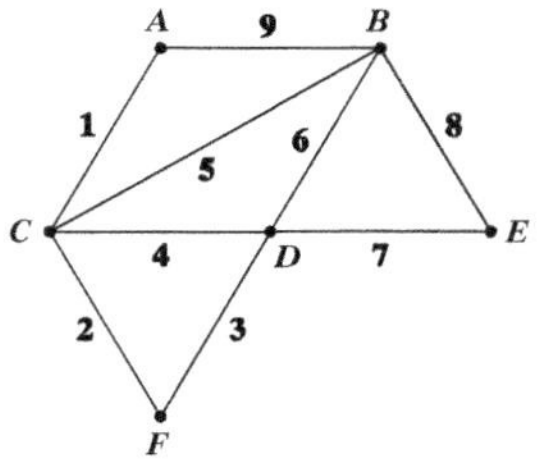

15. a. Euler path

b. Possible answer:

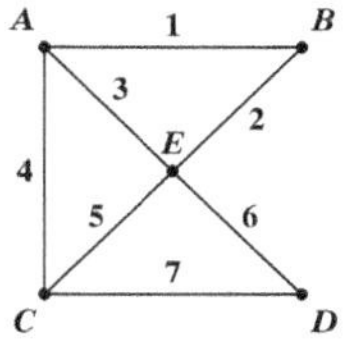

16.

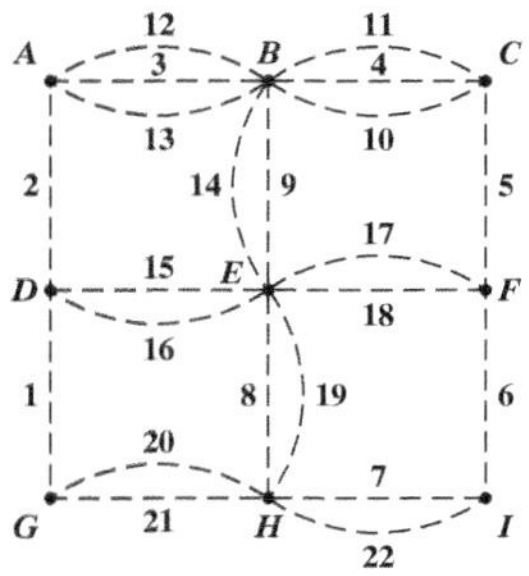

17.

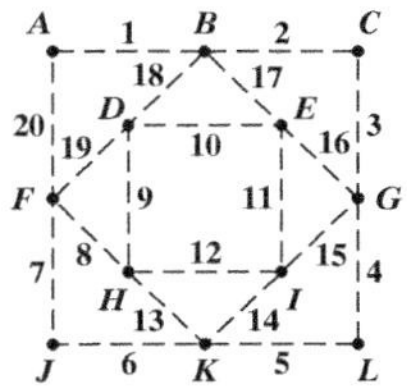

18. a. yes **b.**

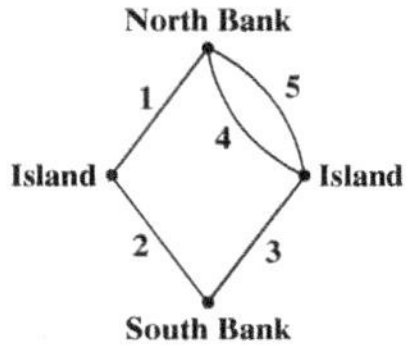

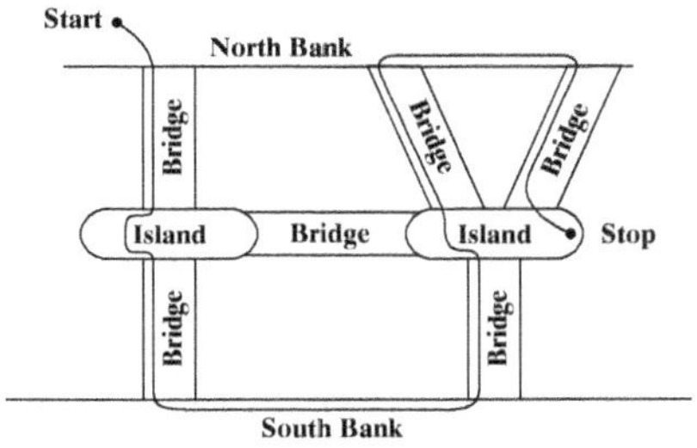

c. no

19. It is possible.

20. a. yes **b.** Possible answer:

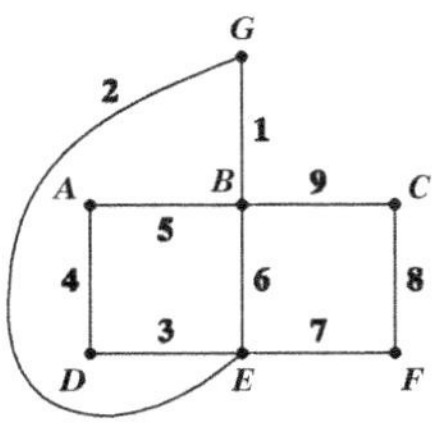

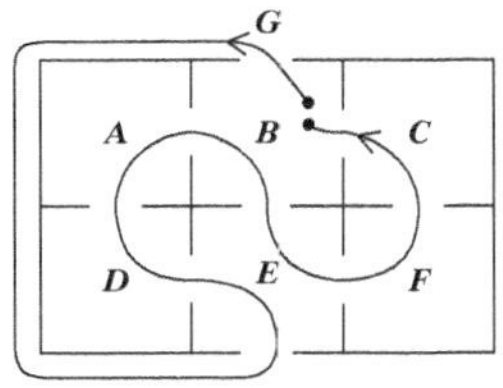

21. a.

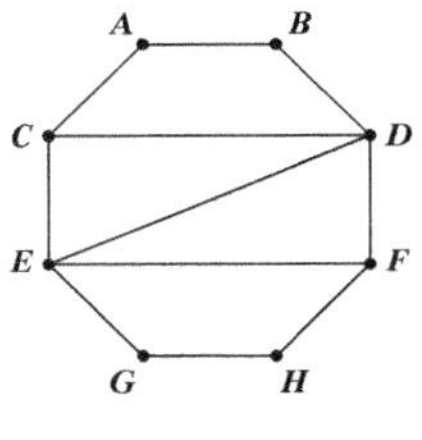

b. yes

c. The guard should begin at *C* and end at *F*, or vice versa.

22. Possible answer: *A*, *E*, *C*, *B*, *D*, *A*

23. Possible answer: *D*, *B*, *A*, *E*, *C*, *D*

24. a. no

25. a. yes **b.** 6

26. a. no

27. a. yes **b.** 24

28. *A*, *B*, *C*, *D*, *A*: 19; *A*, *B*, *D*, *C*, *A*: 18; *A*, *C*, *B*, *D*, *A*: 19; *A*, *C*, *D*, *B*, *A*: 18; *A*, *D*, *B*, *C*, *A*: 19; *A*, *D*, *C*, *B*, *A*: 19

29. *A*, *B*, *D*, *C*, *A* and *A*, *C*, *D*, *B*, *A*

30. *A*, *C*, *D*, *B*, *A*; 18

31. *A*, *B*, *E*, *D*, *C*, *A*; 24

32.

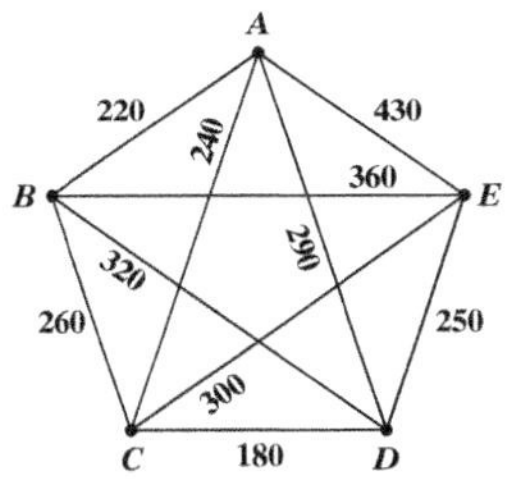

33. *A*, *B*, *C*, *D*, *E*; $1340

34. yes

35. No. It has a circuit.

36. No. It is disconnected.

37. Possible answer:

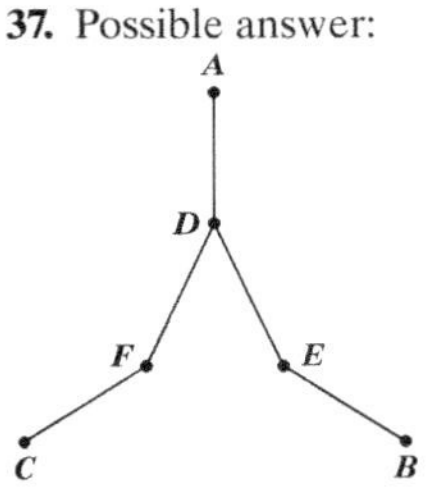

38. Possible answer:

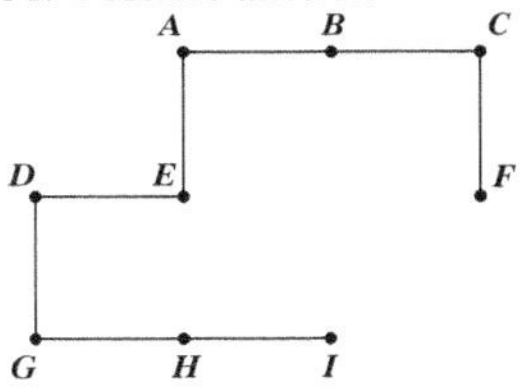

39. 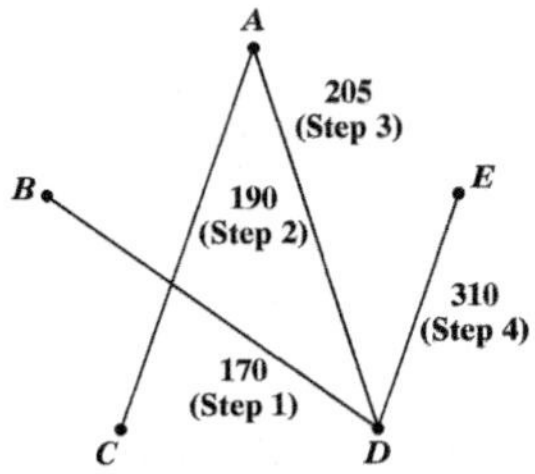

; 875

40. 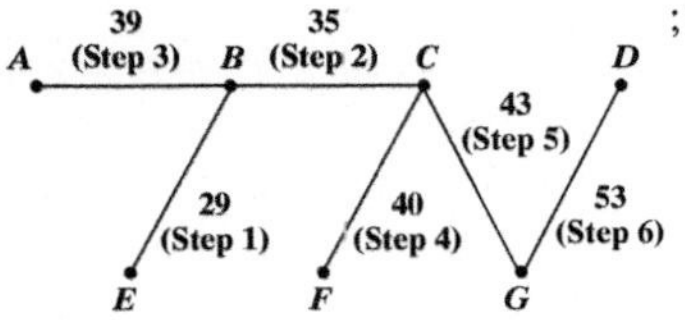

; 239

41. 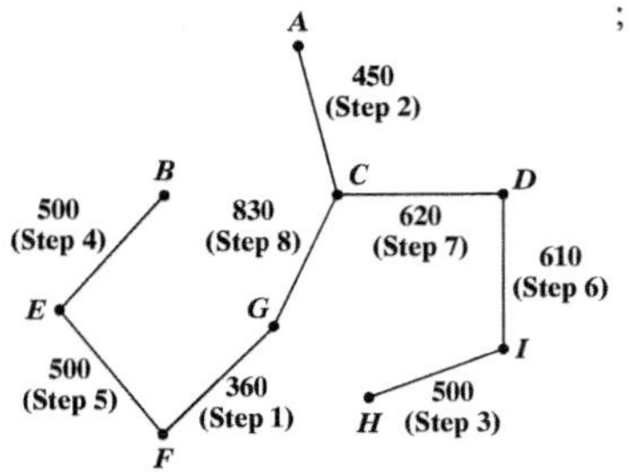

; 4370 mi

Chapter 14 Test

1. A: 2; B: 2; C: 4; D: 3; E: 2; F: 1 **2.** Possible answer: A, D, E and A, B, C, E **3.** B, A, D, E, C, B **4.** CF

5. 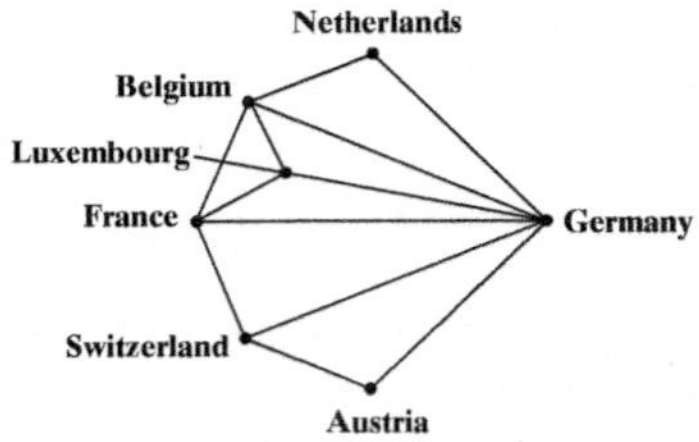

6. a. Euler path **b.** Possible answer: 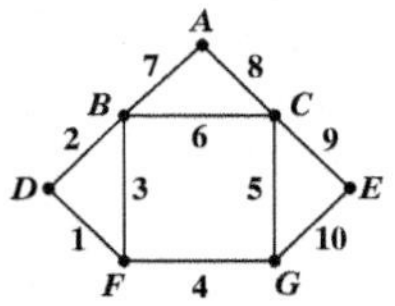

7. a. neither

8. a. Euler circuit **b.** Possible answer: 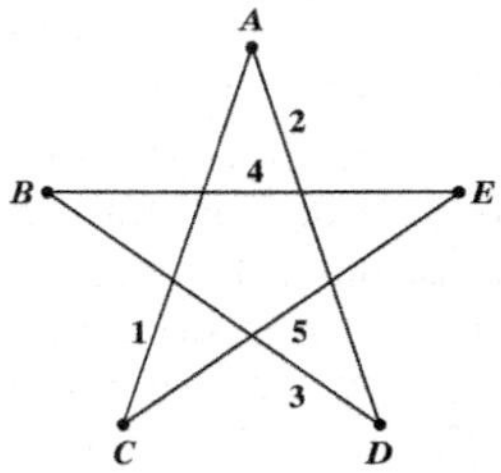

9. 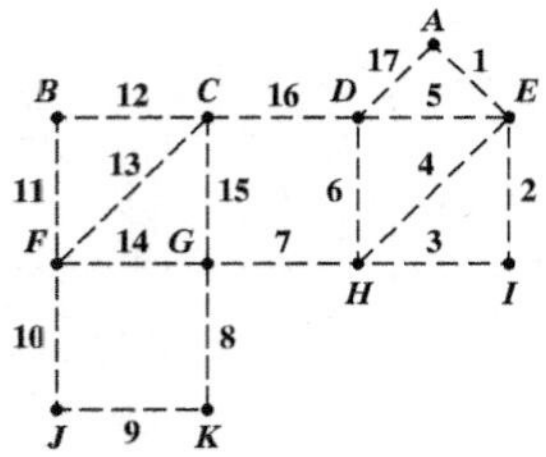

10. a. 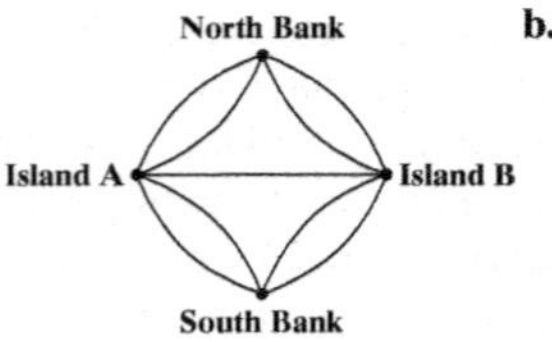

b. yes **c.** on one of the islands

11. a. 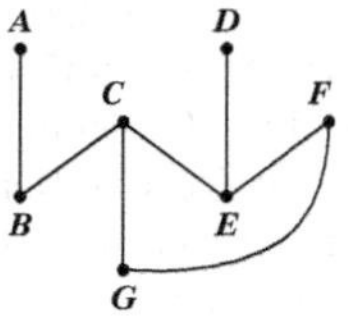

b. no

12. a. 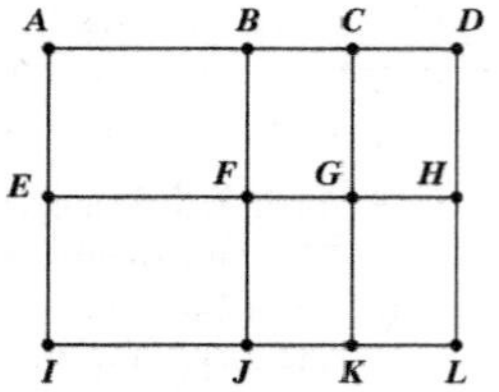

b. no

13. A, B, C, D, G, F, E, A and A, F, G, D, C, B, E, A

14. 24 **15.** 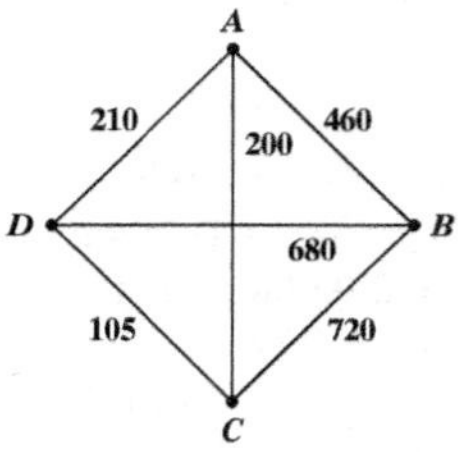

16. A, B, D, C, A or A, C, D, B, A; $1445 **17.** A, E, D, C, B, A; 33 **18.** no

19. Possible answer: 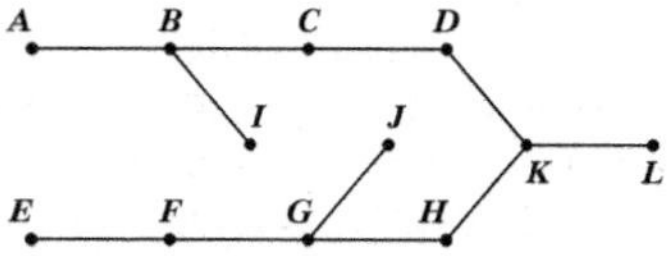

20. 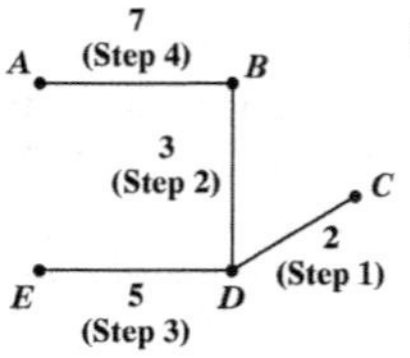

; 17

Credits

Photos

FM **p. viii** Robert F. Blitzer

CHAPTER 1 **p. 1 (l)** Cherezoff/Fotolia **p. 1 (r)** Reed Kaestner/Corbis **p. 2** Richard F. Voss **p. 7** Fred Murphy/BSIP SA/Alamy **p. 8** AF Archive/Alamy **p. 14** Joebelanger/iStockphoto **p. 17** Ken Fisher/Gettyimages **p. 23** Brian Snyder/Gettyimages **p. 29** Colin Anderson/Blend Images/Corbis **p. 30** Ohiophoto/iStockphoto **p. 35** Mike Hewitt/Getty Images **p. 43** Andreas Nilsson/Fotolia

CHAPTER 2 **p. 47** Michael Svoboda/iStockphoto **p. 48** Mood Board/SuperStock **p. 49** Aroon Vater/Alamy **p. 50** *Illustration 15 of 41 for Lewis Carroll's Through the Looking Glass and What Alice Found There* (1872), John Tenniel. Wood engraving by Brothers Dalziel/Pearson Education, Inc. **p. 55** Scientific Illustrators/Pearson Education, Inc. **p. 61** Nicole Ackerman **p. 64** Dave Stroup **p. 67** *Time and Time Again* (1981), P.J. Crook. Acrylic on canvas and wood (91.4 × 122). Private collection/Bridgeman Art Library **p. 70** Morandi/Milestone Media/Alamy **p. 75** Scientific Illustrators/Pearson Education, Inc. **p. 84** Shutterstock **p. 96** Aerialarchives.com/Alamy **p. 102** Ermolaev Alexandr/Fotolia

CHAPTER 3 **p. 113** Alistair Scott/Alamy **p. 114 (bl frame)** Trekandshoot/Alamy **p. 114 (bl)** Carsey-Warner/Everett Collection **p. 114 (r frame)** Trekandshoot/Alamy **p. 114 (r)** AF Archive/Alamy **p. 114 (tl frame)** Mikhail Olykaynen/Alamy **p. 114 (tl)** AF archive/Alamy **p. 119** AF Archive/Alamy **p. 122** Glowimages RM/Alamy **p. 135 (bl)** Everett Collection/Alamy **p. 135 (bm)** Everett Collection/Alamy **p. 135 (br)** Everett Collection/Alamy **p. 135 (tl)** Everett Collection/Alamy **p. 135 (tm)** Everett Collection/Alamy **p. 135 (tr)** Everett Collection/Alamy **p. 136 (b)** Martine Mouchy/Gettyimages **p. 136 (t)** Sborisov/Fotolia **p. 146** Photos12/Alamy **p. 150** Fotolia **p. 155** Scientific Illustrators/Pearson Education, Inc. **p. 162 (bl)** Allstar Picture Library/Alamy **p. 162 (blm)** AF Archive/Alamy **p. 162 (br)** Pictorial Press, Ltd./Alamy **p. 162 (brm)** Moviestore Collection, Ltd./Alamy **p. 162 (tr)** Douglas Pizac/AP Images **p. 172** Sdominick/iStockphoto **p. 174** Bettmann/Corbis **p. 177** *Deux Mystères* (The Two Mysteries) (1966), René Magritte. Oil on canvas, 65 × 80 cm. Private Collection. Photothèque R. Magritte/ADAGP/Art Resource, New York/© 2013 C. Herscovici, Brussels/Artists Rights Society (ARS), New York. **p. 180** Nick Ut/AP Images **p. 182** Paramount/Everett Collection **p. 188** Bettmann/Corbis **p. 192 (b)** *Betty* © 1992 Gerry Rasmussen and Gary Delainey. Reprinted with permission of Univeral Uclick for UFS. All rights reserved. **p. 192 (t)** *Peanuts* © 1977 Peanuts Worldwide LLC. Distributed by Universal Uclick. Reprinted with permission. All rights reserved. **p. 195 (l)** Interfoto/Alamy **p. 195 (r)** North Wind Picture Archives/Alamy **p. 202** *School of Athens* (Detail) (1510), Raphael. Stanza della Segnatura, Stanze di Raffaello, Vatican Palace. Scala/Art Resource, New York

CHAPTER 4 **p. 211** Shizuo Kambayashi/AP Images **p. 212** Dinosmichail/Fotolia **p. 214** RMN-Grand Palais/Art Resource, New York **p. 215** Franck Boston/Fotolia **p. 216** Rick Friedman/Corbis **p. 220** Fotolia **p. 222** The Print Collector/Heritage/The Image Works **p. 227** Compuinfoto/Fotolia **p. 230** Phil Degginger/Gettyimages **p. 232** IBM Corporate Archives **p. 233** Science Photo Library/Alamy **p. 236** Craig Fujii/AP Images **p. 238** Lee Snider/The Image Works

CHAPTER 5 **p. 247** Domen Colja/Alamy **p. 248** Chase Jarvis/Getty Images **p. 258** *Mirror II* (1963), George Tooker. Egg tempera on gesso panel, 20 × 20 in. Addison Gallery of American Art, Phillips Academy, Andover, Massachusetts. Gift of R. H. Donnelley Erdman [PA 1956]. © George Tooker. All Rights Reserved. **p. 262 (l)** Djarvik/Dreamstime **p. 262 (r)** Shutterstock **p. 272** Burwell Photography/iStockphoto **p. 284** CBS Television/Robert Voets/Album/Newscom **p. 287** *School of Athens* (Detail) (1510), Raphael. Stanza della Segnatura, Stanze di Raffaello, Vatican Palace. Scala/Art Resource, New York **p. 289 (b)** Joshua Haviv/Fotolia **p. 289 (t)** Chitose Suzuki/AP Images **p. 294** John G. Ross/Science Source/Photo Researchers, Inc. **p. 295** *The Persistence of Memory* (1931), Salvador Dali. Oil on canvas, 9 1/2 × 13 in. (24.1 × 33 cm). (162.1934). Given anonymously to the Museum of Modern Art/Digital image copyright © 2011 the Museum of Modern Art/Licensed by SCALA/Art Resource, New York/Artists Rights Society, New York **p. 298** *Foxtrot* © 2000 Bill Amend. Reprinted with permission of Universal Uclick. All rights reserved. **p. 299 (b)** *Hagar* ©1995 by King Features Syndicate, Inc. World rights reserved. **p. 299 (t)** *Foxtrot* © 2000 Bill Amend. Reprinted with permission of Universal Uclick. All rights reserved. **p. 300** Mary Evans/HBO/Your Face Goes Here Entertainment/Ronald Grant/Everett Collection **p. 301** Robert F. Blitzer **p. 304** Copyright © 1985 by Roz Chast/The New Yorker Collection/The Cartoon Bank **p. 305 (m)** A.B.Joyce/Science Source/Photo Researchers, Inc. **p. 305 (t)** Kenneth Libbrecht/Science Source/Photo Researchers, Inc. **p. 308** Rick Golt/Science Source/Photo Researchers, Inc. **p. 310** © 2007 Gordon Caulkins **p. 313 (b)** David Weintraub/Science Source/Photo Researchers, Inc. **p. 319** Mike Kiev/Fotolia **p. 322** iStockphoto/Thinkstock **p. 328** Shutterstock **p. 331** Harry Blair

CHAPTER 6 **p. 339** Newscom **p. 340** Flying Colours Ltd/Thinkstock/Getty Images **p. 342 (tl)** Megan Maloy/Image Source/AGE Fotostock **p. 342 (tr)** Erik Isakson/Tetra Images/Alamy **p. 346** Ikon Images/SuperStock **p. 350** Paramount/Everett Collection/Glow Images **p. 360** iStockphoto/Thinkstock/Getty Images **p. 365 (bl)** Alamy **p. 365 (blm)** Michael N. Paras/AGE Fotostock **p. 365 (bmr)** Image Source/AGE Fotostock **p. 365 (br)** Kurhan/Shutterstock **p. 365 (tl)** APphoto/Stuart Ramson **p. 365 (tlm)** S. Bukley/Newscom **p. 365 (tmr)** Newscom **p. 365 (tr)** Icon Sports Media181/Icon Sports Media/Newscom **p. 368** Robert F. Blitzer

p. 376 Steve Hamblin/Alamy **p. 382** Scientific Illustrators, Inc./ Pearson Education, Inc. **p. 386** Brian Goodman/Shutterstock **p. 394** Brian Goodman/Shutterstock **p. 398 (b)** Christian Delbert/Fotolia **p. 398 (t)** *Bathers at Asnières* (1884), Georges Seurat. Oil on canvas, 201 cm × 300 cm (79 in × 118 in). National Gallery, London/Art Resource, New York **p. 404 (l)** Everett Collection **p. 404 (r)** Phots12/Alamy

CHAPTER 7 **p. 407 (bl)** AF Archive/Alamy **p. 407 (br)** ZUMA Press/Alamy **p. 407 (tl)** ZUMA Press/Alamy **p. 407 (tr)** Neal Preston/Corbis **p. 408** Iofoto/Dreamstime **p. 413** © The New Yorker Collection 1995 Warren Miller from cartoonbank.com. All Rights Reserved. **p. 420** Matthew Ward/Dorling Kindersley, Ltd. **p. 433** Inspirestock/Brand X Royalty Free/Jupiter Images **p. 442** Sun/Newscom **p. 447** Terry/iStockphoto **p. 456** AP Images **p. 457** David Crawford/AP Images **p. 462** LWA/Sharie Kennedy/Blend Images/Alamy **p. 464 (l)** Seth Wenig/AP Images **p. 464 (r)** Rolff Images/Dreamstime **p. 465** Ladi Kirn/Alamy **p. 470 (l)** Pascal Rondeau/Getty Images **p. 470 (r)** Scott Kim **p. 476** Science Photo Library/SuperStock **p. 478** Scott Kim

CHAPTER 8 **p. 487** Exactostock/SuperStock **p. 488** MariusFM77/iStockphoto **p. 497 (l)** 20th Century Fox/Everett Collection **p. 497 (r)** Alex Slobodkin/iStockphoto **p. 498** David Breslauer/AP Images **p. 500 (b)** Thomas Peterson/ Alamy **p. 500 (b)** Hurst Photo/Shutterstock **p. 500 (tr)** TinCup/Alamy **p. 504** Jgroup/iStockphoto **p. 508** North Wind Picture Archives/Alamy **p. 513** Dononeg/iStockphoto **p. 523** Jeff Christensen/WireImage/Getty Images **p. 539** Moviestore Collections Limited/Alamy **p. 548** Grafissimo/ iStockphoto **p. 552** Mel Yates/Getty Images **p. 557** Piotr Marcinski/Shutterstock **p. 561 (l)** Inbj/Dreamstine **p. 561 (r)** Martin Harvey/Alamy

CHAPTER 9 **p. 575** Ecopic/iStockphoto **p. 576** Dave King/ Dorling Kindersley, Ltd. **p. 578 (dekadollar)** US Department of Treasury **p. 578 (dime)** Pearson Education, Inc. **p. 578 (dollar)** US Department of Treasury **p. 578 (hectodollar)** US Department of Treasury **p. 578 (kilodollar)** US Department of Treasury **p. 578 (penny)** Pearson Education, Inc. **p. 585** Radius Images/Corbis **p. 587** Mtilghma/Shutterstock **p. 596 (b)** Living Images/iStockphoto **p. 596 (t)** Kwest19/Dreamstime **p. 601** Katvic/Shutterstock

CHAPTER 10 **p. 609** Eric Isselée/Shutterstock **p. 610** Hemis/Alamy **p. 619** Everett Collection/Alamy **p. 630 (b)** WDG Photo/Shutterstock **p. 630 (t)** Pictorial Press, Ltd./Alamy **p. 631** Irina Tischenko/Shutterstock **p. 635** *Symmetry Drawing* E85 (1952), M.C. Escher. © 2013 The M.C. Escher Company, The Netherlands. All rights reserved. **p. 639** Red Cover/ Alamy **p. 643** Benjamin Shearn/Getty Images **p. 650** Chas Metiveir/Newscom **p. 652** Uyen Le/Photodisc/Gettyimages **p. 659** Imagebroker/Alamy **p. 664** Bruce Barrett/Getty Images **p. 668 (br)** Pecold/Shutterstock **p. 668 (tr)** Haider Azim/ Fotolia **p. 669 (t)** *The Family* (1963), Marisol Escobar. Wood, metal, graphite, textiles, paint, plaster and other accessories, 79 1/2 in. × 63 in. × 73 in. (201.93 cm × 160.02 cm × 185.42 cm). © 2011 Marisol Escobar/VAGA, New York/Art Resource, New York **p. 672** SSPL/The Image Works **p. 674** *Circle Limits III* (1959), M.C. Escher. © 2013 The M.C. Escher Company, The Netherlands. All rights reserved. **p. 675 (bl)** Daryl H. Hepting and Allan N. Snider "Desktop Tetrahedron" 1990. Computer generated at the University of Regina in Canada **p. 675 (bm)** SPL/Science Source/Photo Researchers, Inc. **p. 675 (br)** Keila Neokow/Eli Vokounova/Shutterstock **p. 677 (ml)** Iren Udvarhazi/Fotolia

CHAPTER 11 **p. 687 (l)** Corbis **p. 687 (m)** Corbis **p. 687 (ml)** Corbis **p. 687 (mr)** Corbis **p. 687 (r)** Corbis **p. 688** Nicolas McComber/Getty Images **p. 689** Christopher S. Howeth/Shutterstock **p. 691** Newscom **p. 694 (b)** Craftvision/Getty Images **p. 694 (t)** *Peanuts* © 1968 Peanuts Worldwide LLC. Distributed by Universal Uclick. Reprinted with permission. All rights reserved. **p. 702 (l)** AF Archive/ Alamy **p. 702 (r)** MARKA/Alamy **p. 709** Jupiter Images/ Comstock/Thinkstock **p. 712** Scientific Illustrators, Inc./ Pearson Education, Inc. **p. 715** Ronstik/iStockphoto **p. 718 (bl)** W. Breeze/Getty Images **p. 718 (t)** *Mirror II* (1963), George Tooker. Egg tempera on gesso panel, 20 × 20 in. Addison Gallery of American Art, Phillips Academy, Andover, Massachusetts. Gift of R. H. Donnelley Erdman [PA 1956]. © George Tooker. All Rights Reserved. **p. 719 (b)** Hulton Archive/Getty Images **p. 719 (bm)** Bryan Bedder/Getty Images **p. 719 (m)** CBS Photo Archive/Getty Images **p. 719 (t)** Robert F. Blitzer **p. 719 (tm)** Corbis **p. 720** Jack Kurtz/UPI Photo Service/Newscom **p. 725** F. Bettex/Lookandprint.com/Alamy **p. 729** Manoj Shah/ Getty Images **p. 732** Scott Serio/Icon SMI/Newscom **p. 737** Tom Salyer/Alamy **p. 738** Bettmann/Corbis **p. 749** Hill Street Studios/Blend Images/Getty images **p. 752** Car Culture/ Getty Images **p. 753** Monkey Business/Thinkstock

CHAPTER 12 **p. 765** Mstay/Getty Images **p. 766** Everett Collection **p. 767** The Protected Art Archive/Alamy **p. 780** Alex Bramwell/Getty Images **p. 794** Epic Stock Media/ Shutterstock **p. 779** Scientific Illustrators, Inc./Pearson Education, Inc. **p. 802** Marmaduke St. John / Alamy **p. 803** Peter Barritt/Alamy **p. 806** Janice Richard/iStockphoto **p. 809 (r)** Kemalbas/iStockphoto **p. 809(l)** Webeye/ iStockphoto **p. 813** Image Source/Getty Images **p. 816** Newscom **p. 818** Blend Images/Thinkstock **p. 821 (bl)** Newscom **p. 821 (ml)** AP Images **p. 821 (mr)** Library of Congress Prints and Photographs Division [LC-USZ62-133118] **p. 821 (r)** Mark Wilson/AP Images **p. 821 (t)** Newscom **p. 824** Stuart Westmorland/Getty Images **p. 829** Shirley Buchan/Pacific Coast News/Newscom

CHAPTER 13 **p. 839** Jgroup/iStockphoto **p. 840** Everett Collection **p. 845** Variety/Getty Images **p. 847** Will Schneekloth/ZUMA Press/Newscom **p. 852** Carr/MCT/ Newscom **p. 855** The Art Gallery Collection/Alamy **p. 859** Imaginechina/AP Images **p. 863** ClassicStock/Alamy **p. 866** Stocktrek Images, Inc./Alamy **p. 868** GL Archive/Alamy **p. 870** Image Asset Management, Ltd./Alamy **p. 871** Everett Collection/Alamy **p. 877 (l)** Buyenlarge/Archive Photos/Getty Images **p. 877 (r)** Buyenlarge/Archive Photos/Getty Images **p. 880** AFP/Getty Images **p. 881** Everett Collection/Alamy

CHAPTER 14 **p. 891** ART on FILE/Corbis **p. 892 (l)** Stephane Cardinale/People Avenue/Corbis **p. 892 (lm)** Imagebroker/Alamy **p. 892 (r)** James Leynse/Corbis **p. 892 (rm)** Chris Wattie/Corbis **p. 902** Peter Foley/EPA/Newscom **p. 914** Kevork Djansezian/Getty Images **p. 924** David Epperson/Getty Images

Text

CHAPTER 1 **p. 3** Bureau of Statistics **p. 8** Reprinted by permission of ICM Partners Copyright (2012) by William Poundstone

CHAPTER 2 **p. 66** Reprinted by permission of Dover Publications

CHAPTER 3 **p. 114** HarperCollins **p. 114** United States Patent and Trademark Office **p. 119** *Star Trek*, CBS, Inc. **p. 123** Ghandi quote Navajivan Trust **p. 140** Lewis Carroll **p. 169** Lewis Carroll **p. 174** Reprinted with heirs from the estate of Martin Luther King Jr., c/o Writers House as agent for the proprietor New York, NY. © 1963 Dr. Martin Luther King Jr © reserved 1991 Coretta Scott King. **p. 193** Reprinted with heirs from the estate of Martin Luther King Jr., c/o Writers House as agent for the proprietor New York, NY. © 1963 Dr. Martin Luther King Jr © reserved 1991 Coretta Scott King. **p. 194** *See, I Told You So*, Penguin Books, Ltd. (UK). **p. 194** *Anatomy of an Illness*, W.W. Norton & Co. **p. 204** C. Powell, Adam Clayton Powelll **p. 204** Rene Descartes

CHAPTER 5 **p. 257** *Contact*, Penguin Books, Ltd. (UK) **p. 284** *NUMB3Rs*, CBS Inc.

CHAPTER 8 **p. 487** Simon & Schuster, Inc.

Subject Index

Q

R

S